Fluid Mechanics and Its Applications

Volume 119

The purpose of this series is to focus on subjects in which fluid mechanics plays a fundamental role. As well as the more traditional applications of aeronautics, hydraulics, heat and mass transfer etc., books will be published dealing with topics which are currently in a state of rapid development, such as turbulence, suspensions and multiphase fluids, super and hypersonic flows and numerical modelling techniques. It is a widely held view that it is the interdisciplinary subjects that will receive intense scientific attention, bringing them to the forefront of technological advancement. Fluids have the ability to transport matter and its properties as well as transmit force, therefore fluid mechanics is a subject that is particulary open to cross fertilisation with other sciences and disciplines of engineering. The subject of fluid mechanics will be highly relevant in such domains as chemical, metallurgical, biological and ecological engineering. This series is particularly open to such new multidisciplinary domains. The median level of presentation is the first year graduate student. Some texts are monographs defining the current state of a field; others are accessible to final year undergraduates; but essentially the emphasis is on readability and clarity.

Springer and Professor Thess welcome book ideas from authors. Potential authors who wish to submit a book proposal should contact Nathalie Jacobs, Publishing Editor, Springer (Dordrecht), e-mail:

Nathalie.Jacobs@springer.com

Indexed by SCOPUS, EBSCO Discovery Service, OCLC, ProQuest Summon, Google Scholar and SpringerLink

More information about this series at http://www.springer.com/series/5980

Vadim L. Belenky · Kostas J. Spyrou
Frans van Walree · Marcelo Almeida Santos Neves
Naoya Umeda
Editors

Contemporary Ideas on Ship Stability

Risk of Capsizing

Springer

Editors
Vadim L. Belenky
David Taylor Model Basin/Naval Surface Warfare Center Carderock Division
West Bethesda, MD, USA

Kostas J. Spyrou
School of Naval Architecture and Marine Engineering
National Technical University of Athens
Athens, Greece

Frans van Walree
MARIN-Maritime Research Institute Netherlands
Wageningen, The Netherlands

Marcelo Almeida Santos Neves
Department of Naval Architecture and Ocean Engineering
Federal University of Rio de Janeiro
Rio de Janeiro, Brazil

Naoya Umeda
Department of Naval Architecture and Ocean Engineering
Osaka University
Suita, Osaka, Japan

ISSN 0926-5112 ISSN 2215-0056 (electronic)
Fluid Mechanics and Its Applications
ISBN 978-3-030-00514-6 ISBN 978-3-030-00516-0 (eBook)
https://doi.org/10.1007/978-3-030-00516-0

Library of Congress Control Number: 2018954616

This Springer imprint is published by the registered company Springer Nature Switzerland AG
The registered company address is: Gewerbestrasse 11, 6330 Cham, Switzerland

Preface

Ship stability is a topic combining scientific rigor with practicality. With floatability and strength, they are traditionally regarded as the most fundamental safety requirements in ship design. In the last decades, very significant progress was achieved toward understanding ship dynamic behavior under extreme conditions of operation. This progress is reflected in the current efforts at IMO for developing new stability criteria which are intended to provide fuller and more effective protection for the ships. Nonetheless, dynamic stability continues to be a challenging topic of research: The mathematical modeling of extreme ship motions still relies heavily on empiricism and efforts to approach it from basic scientific principles have some way to go until such models can stand alone. Capsize phenomena are nonlinear and the assessment of a ship's susceptibility for capsize calls for a probabilistic approach going clearly beyond the current state of the art. Experimental methods for testing dynamic stability need to be able to deal with the rare nature of capsize. Last but not least, new knowledge should be continually ingrained in the applied design rules and ship operation guidelines.

For these reasons, the current book serves an important purpose: to update the wider naval architecture community about the frontiers of current ship stability research. It is actually the third in a series started in the year 2000, with the second volume appearing in 2011. The initiative belongs collectively to the International Standing Committee for the Stability of Ships and Ocean Vehicles who overlook the organization of stability conferences and workshops. In recent years, the International Conference on the Stability of Ships and Ocean Vehicles (STAB) is held triennially, with international ship stability workshops (ISSW) taking place every year in-between. These are, generally, well-attended events, despite the fact that ship stability is a rather specialist topic. The current book covers the three events held from 2010 to 2012. Specifically, it is a collection of representative papers originally presented during the 11th Stability Workshop (Wageningen, 2010), the 12th Stability Workshop (Washington DC, 2011), and the 11th STAB Conference (Athens, 2012). All papers selected by the editorial committee went through an additional review process, with at least two reviewers allocated for each. Actually, many of the papers were significantly enhanced compared to their original version, In order to become up to date and reflect better the state of knowledge about stability that exists in the year of the book's publication.

The book is organized in four major divisions with each individual paper appearing as a different chapter. In these, divisions are covered in depth the mathematical modeling of ship motions, the study of extreme dynamic behavior, experimental research and, last but not least, the regulatory and ship operation sides. The structure of the book is summarized in detail in an introductory note written by the editor-in-chief Dr. Vadim Belenky who is especially thanked for leading the effort as well as for bearing most of the burden. The thanks extend of course to all the editorial committee members. Many thanks and congratulations are appropriate also for the authors, who contributed with their valuable material. We express our gratitude to the organizing committees and to the sponsors of the three stability events covered in this book. We acknowledge the important impact made by the expert reviewers whose names appear in the editor's introductory text as indication of our appreciation. Last but not least, we thank the participants whose consistent presence justifies our efforts.

Athens, Greece

Prof. Kostas J. Spyrou
On Behalf of the International
Standing Committee for the
Stability of Ships and Ocean Vehicles

Editor's Introductory Note

The purpose of this introductory note is twofold to attempt to summarize the contents of the book and to put those contents into a context that emphasizes the relationship between different chapters, which is always a challenge for such a collection of works.

As the reader was properly warned in the Preface, the assessment of a ship's susceptibility to capsizing goes beyond the current state of the art. Model tests and time-domain numerical simulations based on first principles are the main tools for the evaluation of large-amplitude ship motions because of significant nonlinearity of ship dynamics in severe seas. The appearance of novel, unconventional designs may call into question any stability assessment based solely on previous experience. As a result, the focus of recent efforts has been on developing numerical simulation tools capable of evaluating complex stability failures.

This book consists of four major divisions; each division is further subdivided into parts corresponding to detailed subject areas, while each part contains chapters with individual contributions. The four major divisions are:

1. Mathematical Model of Ship Motions in Waves (parts I through V, 15 chapters)
2. Dynamics of Large Motions (parts VI through VIII, 12 chapters)
3. Experimental Research (parts IX and X, 11 chapters)
4. Requirements, Regulations and Operations (parts XI through XIV, 17 chapters)

Mathematical Model of Ship Motions in Waves

A mathematical model is the centerpiece of any numerical tool. The first division of the book reports on progress in the development of mathematical models for large ship motions in waves.

The development of a new simulation code is an expensive, time-consuming, and risky endeavor. Supporting such tools is also not trivial. That is why there are few simulation codes capable of handling extreme ship motions, such as FREDYN (de Kat and Paulling 1989) and LAMP (Lin and Yue 1990; Shin et al. 2003); other relevant codes are reviewed in Peters et al. (2011). That is why the appearance of a new tool is a report-worthy event and also why there is only one chapter in Part I: "Mathematical Model of Ship Motions in Waves: New Simulation Tools", Chapter 1 describing Tempest—the newest addition in the family of numerical codes. A more thorough discussion of the types and capabilities of simulation tools can be found in Beck and Reed (2001) and Reed and Beck (2016).

Correct modeling of the wave environment is an instrumental part of any valid simulation of large ship motion and capsizing. The results of time-domain simulation can only be as good as the model of encountered waves, as well stated by Krylov's citation of Huxley that "*mathematics may be compared to a mill of exquisite workmanship, which grinds your stuff of any degree of fineness; but, nevertheless, what you get out depends upon what you put in…*". Practically all time-domain simulations use the classic model originally proposed in St. Denis and Pierson (1953), in which the instantaneous wave elevation at a point is represented with a Fourier series whose amplitudes are determined from a spectrum and whose phase-shift angles are uniformly distributed random variables. This model, frequently referred to as a Longuet-Higgins model, was originally intended for the linear seakeeping and wave loads problem. How much of a stretch is it to apply it to nonlinear simulations of stability in waves and what alternatives are available? Attempts to address this question are included in Part II, titled "Mathematical Model of Ship Motions in Waves: Environment."

Chapter 2 considers the autoregressive/moving average (ARMA) model for representing waves for the stability problem. ARMA model has a long history of application for modeling stochastic processes (Box et al. 2008) and has been proposed for ocean waves many years ago (Spanos 1983). One of the issues considered in Chap. 2 is how well ARMA model represents the physics of surface waves. Another issue is the computation of the hydrodynamic pressure field corresponding to ARMA-modeled waves. A method to calculate these pressures is described in Chap. 3.

Once the wave environment is defined, the next step is the consideration of the forces acting on a ship in waves. Part III reports on progress in the study of wave-body hydrodynamic forces, with the exception of roll damping, which deserves special attention and is considered in Part IV.

"Potential flow" remains the main approach for numerical models of wave-body hydrodynamics in practical simulations of stability in realistic waves, with forces of non-potential flow nature often included through lower-order models such as polynomials with empirical coefficients derived from model testing or CFD computation. This approach, frequently referred to as a "hybrid" method, allows computational performance sufficient for generating a representative volume of a response sample in irregular waves. The main objective of hydrodynamic computations is to obtain the values of the potential field. The gradient of the potential is velocity, from which pressure can be computed using Bernoulli's equation. The integration of pressure over the ship surface yields the forces acting on the ship. However, it is not the only possible way to get the forces from the potential.

Chapter 4 describes an alternative approach for computing forces directly from the potential without using Bernoulli's equation, which avoids a number of computational difficulties.

Models for vortex-related hull forces are frequently based on model tests or viscous flow calculations (e.g., RANS) in calm water, so the submerged portion of the ship hull remains constant. However, the submerged portion changes rather drastically when the ship is moving in large waves, and these changes may have a strong effect of such viscous forces. Chapter 5 proposes a model of how to account for the changing in the submerged geometry when calculating the hull lift and cross-flow drag forces, using just the calm water maneuvering data.

Another way to obtain forces from a model test or CFD calculation is described in Chap. 6—System Identification Method. The idea is to extract the forces from a path of the ship and the associated kinematic data, i.e., from the solution of the equations of motion. The problem is formulated as the inverse to the solution of a system of differential equations. Here, the solution and the structure of the equations are given—while the values of the coefficients are to be found. The approach has previously been used for calm water maneuvering using trials' data (Abkowitz 1980; Crider et al. 2008); Chapter 6 considers its use for maneuvering in waves.

Roll damping forces have large viscous and vortical contribution and cannot be modeled within the assumptions of potential flow; so they also have to be treated within the "hybrid" approach, where information of the non-potential components comes from a model test or viscous flow calculations. Vortical contribution, however, can be predicted by a lifting surface code that assumes inviscid flow. As roll damping has been a focus of many recent research efforts, it "deserved" its own section of this book, Part IV. The damping created by bilge keels is a matter of particular interest because, firstly—it is a significant portion of damping for many ships and secondly—it can be controlled by a designer to a certain extent. Chapter 7 describes a mathematical model for bilge keels based on a nonlinear low-aspect-ratio lifting surface theory. Chapter 8 considers the effect of shallow draft, large-amplitude roll motion, non-periodic roll and transitional motions. The results are compared with Ikeda's method (the *de facto* standard empirical method for roll damping assessment; see Ikeda et al. 1976). Chapter 9 provides comparisons between Ikeda's method and CFD calculations, looking into a number of effects, including maneuvering.

Empirical methods are simple to use, but they are generally developed using a limited set of experimental data for a narrow range of ships. Using empirical methods for a novel hull that may be outside of the range of the experimental data requires caution. The comparison with higher fidelity data (experiment or CFD) for such cases provides necessary verification of the applicability and robustness. Chapter 10 presents the results of a model test and full-scale measurement of a Panamax pure car and truck carrier at speed. Chapter 11 describes a comprehensive comparison between CFD results and empirical methods (Ikeda's method and neural network results based on Blume's roll damping measurements) for a twin-screw vessel.

When a ship is damaged, the mathematical model has to include additional forces related to water flowing into and out of the ship as well as flow inside the flooded compartments, which can be very complex. Part V reports recent progress in modeling those forces.

Chapter 12 focuses on the shipping of green water on deck and how it may trigger flooding. The proposed method is based on potential flow—the fundamental idea is to get an accurate wave profile along the ship hull, which will indicate the amount of water that can get shipped onto the deck and then flood an internal compartment through its openings. Chapter 13 looks into quasi-static and quasi-dynamic models of the forces created by flooded compartments of a cruise ship, and describes a numerical study supported by a model test.

Chapter 14 describes a CFD-based solution for the forces generated by fluid motions in a flooded compartment, including sloshing. The volume-of-fluid model was used within a Naiver–Stokes solver, while the external hydrodynamic forces were computed using potential flow. This hybrid approach is aimed at describing the dominant phenomena in the flooded compartment, mainly the nonlinear interaction between floodwater and ship motion (water/wave sloshing in the extended compartment). Further development of these ideas can be found in Gao et al. (2013).

Sloshing in flooded compartments is a formidable hydrodynamic problem due to the very complex nature of the flow. An alternative to traditional RANS solvers, which require a large mesh, is the method of smoothed particle hydrodynamics (SPH) that does not require any mesh. Chapter 15 describes how the SPH method can be implemented using graphic processing units (GPUs) that could be a much less expensive alternative to high-performance computers normally required for CFD. Note that sloshing may also be encountered in intact ships transporting liquid cargo in partially filled tanks.

In summary, the first division describes contemporary mathematical modeling of large-amplitude ship motions in waves. The models are typically of a hybrid nature, in which the wave and hydrodynamic forces are computed using potential flow methods, while CFD is more used in the modeling of forces of viscous and vortex nature, including maneuvering forces, forces in flooded compartments and roll damping forces. At the same time, Ikeda's method continues to provide a practical model for many cases and its applicability is being analyzed for more cases.

Dynamics of Large Motions

Having a valid mathematical model is a necessary but not a sufficient condition to obtain a practical assessment of dynamic stability. The results of numerical modeling must be interpreted to be properly understood. Correct interpretation identifies which ship parameters can be adjusted to achieve a desired effect. Without a correct interpretation, a naval architect must resort to trial and error. Dynamics provides the means to correctly interpret the results of numerical simulation.

The study of dynamics relies on suitable mathematical models. These models are sometimes schematic and generally simpler than the mathematical models of ship motions in waves, although the trend is for these models to become more complex as computational power is increased. In most cases, such "dynamical" mathematical models alone are not enough for obtaining an engineering solution, but the numerical values produced by such models reveal important qualitative trends—they provide guidance as to key design parameters and toward how the practical problem should be approached.

Parametric roll, which is a focus of Part VI, may also serve as an illustration of this approach. Parametric roll is a result of the parametric resonance that a ship experiences because of periodic stability change in waves. Any ship motion model with a body-nonlinear formulation for the hydrostatic and Froude–Krylov forces is capable of reproducing parametric roll. The "dynamic" model for parametric resonance is the Mathieu equation, which is a single degree of freedom (DOF) linear ordinary homogenous differential equation with a periodic coefficient. While the Mathieu equation is a very simplistic model of the phenomenon (its solution tends to infinity or stays at zero and does not even include a steady-state amplitude), it is an invaluable tool for interpretation. Analysis with Mathieu equations guided the successful reproduction of parametric roll in numerical simulations and model tests, as described by France et al. (2003).

Chapter 16 describes a mathematical model specifically developed for the numerical simulation of parametric roll. The objective was to produce quantitatively correct results, which were confirmed by favorable comparison with an experiment. The mathematical model has three degrees of freedom and is based on nonlinear strip theory. Froude–Krylov forces are computed by pressure integration over the instantaneous submerged body. Diffraction and radiation forces are found from a solution of the boundary integral equation for velocity potential. The hydrodynamic lift effect on the roll moment is also included. As it is specialized for parametric roll, the model is simpler than more universal ship motion codes.

Parametric resonance and Mathieu instability occur not only on single-hull ships. Spar platforms also can experience parametric resonance caused by periodic changes of the hydrostatic restoring forces. Chapter 17 describes a mathematical model of a spar platform and shows how Mathieu instability can develop following large-amplitude heave motions. Chapter 18 continues the discussion of Mathieu instability observed in offshore installations. It was found that a tension leg platform may exhibit parametric resonance in close proximity to an FPSO. The chapter describes a mathematical model of two moored bodies and summarizes a related model experiment. Energy exchange between sway and yaw was among the interesting phenomena found while solving this challenging two-body problem.

Part VII is focused on current developments related to surf-riding. Surf-riding is not a dangerous phenomenon by itself, but most ships are not directionally stable after being caught by a wave. This directional instability may result in broaching-to—an uncontrolled turn that may be sharp enough to create a dangerous heel angle or to capsize the ship. Mechanisms of surf-riding and broaching-to in regular waves are now well understood (Spyrou 1996), and the application of this understanding to a hybrid ship motion code has motivated a study of how conventional dynamical analysis techniques will work with a code. The continuation method is a technique of particular interest, as it allows the "big picture" to be seen. Chapter 19 describes a way in which the hydrodynamic memory associated with numerical hydrodynamics codes can be treated within the continuation method.

The consideration of surf-riding in irregular waves has proved to be a significant challenge. Chapter 20 describes a simple mathematical model for surf-riding in irregular waves. While considering how to identify surf-riding in irregular waves, a peculiar mechanism of catch-and-release was found. The transition to surf-riding in regular waves means an attraction to an equilibrium created by wave force, thrust, and resistance at the wave celerity. But what is the proper definition of wave celerity in irregular waves? What patterns of motion should one expect? Chapter 21 tries to answer such questions. The most practical way for calculating celerity uses the concept of instantaneous frequency in time and in space. Other possibilities include computing the velocity of characteristic points on the wave profile, say zero-crossing or maximum slope points. More details on that study are also available in Spyrou et al. (2014), while an abridged version in Chap. 21 is deemed necessary for completeness in reporting a contemporary state of the art.

Random variability, which is an essence of a realistic wave environment, brings many challenges. Part VIII tracks contemporary development in applied probabilistic methods suitable for ship stability applications. The practical implementation of randomness in seakeeping started with the seminal paper of St. Denis and Pierson (1953) that was based on linear dynamical systems under Gaussian excitation. As the linear system is a model for small motions, it cannot be applied to stability problems. Including nonlinearity of motion with the stochasticity of excitation is the central problem of the probabilistic approach to ship stability in realistic waves taking into account the rarity of stability failures.

One way to address this problem is to detect wave events that are likely to lead to large responses. The evaluation of the probability of such waves (or wave groups) separately from ship dynamic considerations significantly simplifies the problem. Chapter 22 considers the most basic statistical aspects of the problem: assessing the uncertainty of direct counting of events in time and the evaluating wave conditions that may lead to dynamic stability failure.

The wave that results in an extreme event may be difficult to find, as it may require a very long Monte Carlo simulation. Chapter 23 describes a technique for "manufacturing" such a wave by manipulating random phases; the technique is known as Design Load Generator (DLG), as it was originally developed for extreme loads. Chapter 23 gives a quick description of the DLG's application for parametric roll in near head seas, and more details can be found in Kim and Troesch (2013).

Sea waves have a group structure. That means that a large wave comes (most probably) with two other large waves that are slightly smaller; the largest wave is typically in the middle. Moreover, wave hydrodynamics make the nearest environment of a large wave almost deterministic. Using the statistical properties of wave groups, Themelis and Spyrou (2007) formulated a practical method for finding the extreme response, considering only critical wave groups, i.e., wave groups capable of invoking "interesting dynamics." Chapter 24 describes the practical application of the approach for a container ship. Other aspects of the wave group method development can be found in Anastopoulos et al. (2015) and Chap. 30.

By the most general definition, a stochastic process is a set of mutually dependent random variables. In the physical world, however, the dependence does not last forever. This provides a chance to simplify mathematical models of stochastic processes. White noise or Wiener process does not have any dependency at all, while for Markov process the current state depends only on the previous state, but not on the pre-previous state. If an excitation of a dynamical system can be described by white noise, its repose will be a Markov process. Best of all, the PDF of the response is a solution of the Fokker–Planck–Kolmogorov (FPK) equation. These models and solutions make a very powerful apparatus known as Itô calculus.

Chapter 25 describes two applications of these ideas for the dynamic stability of ship in irregular waves. Ocean waves cannot, of course, be modeled as white noise, but adding a forming filter in the form of a linear differential equation to the considered dynamical system obtains a response with Markovian properties. The first application is based on the moment equation technique and provides a moment of roll response higher than variance. The second application delivers a distribution of roll motion by solving the FPK equation with the path integral method.

The application of capsizing probability is not limited to intact ship stability assessment. The physics of the motion of a damaged ship is more complicated than for an intact ship, but on the other hand capsizing is no longer a rare event—this makes an approximate solution, from a probability perspective, viable. Chapter 26 describes a method for obtaining an analytical approximation of the capsize band, which is the critical significant wave height that leads to capsizing in 30 min with a probability greater than 0.5.

Chapter 27 shows an example of how unexpected the statistical properties of roll motion in waves can be. It was found that roll angles and rates actually are dependent, but not correlated, in simulations in stern quartering seas. This effect has not been encountered in beam seas.

In summary, the second division of the book describes recent progress in studying the dynamics of large ship motion in waves. The study of parametric roll in irregular waves has continued and been extended to offshore installations. Continuation methods for surf-riding and surging have been adopted to account for wave radiation effects, while the study of surf-riding in irregular waves has led to a formulation for celerity in irregular waves. In terms of stochastic dynamics, significant attention has been focused on a critical wave/wave group approach. Other topics include the application of Markov processes as well as an approximate model for damaged ships and some unexpected statistical properties of large-amplitude roll motions in stern quartering seas.

Experimental Research

An early record of contemporary science was when Galileo used an experiment to validate logic-based considerations. Since then, the role of experiments in research has been twofold: discovering new facts and confirming or rejecting the results of logical inference. In a more modern language, this can be called experimental evaluation and experimental validation. For both types of experimentation, testing technique is essential as it defines what can be seen in an experiment. Part IX describes the state of the art in experimental techniques and some notable results in experimental evaluation.

Capsizing in irregular waves is an especially challenging subject for experimental study. Its dependence on the phase of the roll motions makes test repeatability a very difficult objective. So, it makes sense first to study capsizing in regular waves in order to map the parametric domain. Chapter 28 describes a technology and results of the evaluation of dynamic stability in regular waves for systematically changing wave lengths, wave steepness, speed, and heading. The model test outcome was compared with numerical simulation using FREDYN. As the test was carried out in regular waves, the results are very consistent and help to define the ranges of parameters of concern for dynamic stability in waves.

Parametric roll in irregular waves is another nonlinear phenomenon for experimental study that is subject to repeatability problems. Chapter 29 describes a regular wave experiment on parametric roll in head seas. The model has a wide breadth and shallow draft, which leads to strong nonlinearity in restoring. As a result, the frequency range of parametric resonance is extended. Irregular wave tests were performed to complement the regular wave study, and parametric roll was observed when a wave group with the "correct" wave frequency was encountered.

The wave group approach for computation was introduced in Chap. 24, but it has very serious potential for experimental research as well. "Dangerous" wave groups are relatively short, so if initial conditions for the model test can be controlled, the test may turn out to be repeatable. Chapter 30 addresses technological issues for implementing the wave group method in a model basin. Once fully implemented, this approach has the potential to significantly cut test time as only "dangerous" wave groups need to be tested. A review of wave group applications is available in Anastopoulos et al. (2015).

Planing craft represents a difficult subject for experimental study due to their propensity for unstable behavior even in calm water. Chapter 31 describes a towing test that reveals the relationship between the hydrostatic, hydrodynamic, and total restoring moments of a craft at planing speeds. The results are presented for various model speeds at two model displacements.

A damaged ship is also a more complex subject for experimental research, as water in the flooded compartment as well as the water flowing in and out of the flooded compartment may considerably change the dynamics. Chapter 32 describes a fundamental experimental study of this problem. The model is a cylindrical body with open compartment, forced to roll in calm water. Not surprisingly, the effects of the inflow and outflow can be largely characterized as damping.

A not-so-surprising experimental result is good news, especially when it comes to validation, which is the main focus of Part X. Validation has become one of the central topics over the last decade as more engineering decisions are made based on numerical simulation. That takes the validation of numerical codes well outside the pure research domain, so one must expect such validations to become more procedural and formal.

Chapter 33 gives a "big picture" of the validation of a numerical code for assessment of dynamic stability. It comes with three main challenges: (1) validation metric and criteria, (2) uncertainty characterization, and (3) validation scope. The review of available answers to these challenges is the main contents of the chapter.

Extrapolation methods also require validation in terms of self-consistency: One must prove that the extrapolation technique produces correct results. Chapter 34 considers a procedure for validating extrapolations. The idea is to create a very large data set where rare events become observable, then take small subsets of this data and apply the extrapolation to these subsets. If the extrapolation repeatedly and correctly predicts the value observed from a large set, the extrapolation method is valid.

Dead-ship condition is a classical assumed situation for stability assessment. Chapter 35 describes a model test carried out to validate a 4-DOF (sway, heave, pitch, and roll) mathematical model for estimating the probability of capsizing in this condition. The model test was run with irregular waves and gusty wind generated by wind blowers. The judgment was made by an overlap of capsizing probability estimates obtained from model test and numerical simulation. The comparison was favorable for the 4-DOF mathematical model.

A probability estimate is generally a good metric for comparison, because it averages out small differences, although it requires large amounts of data. What if one wants to compare specific time histories between a model test and a code? The problem is to obtain reliable information on the wave acting on the model. Chapter 36 offers an approach to this problem by reconstructing the waves. This approach was used to validate a 6-DOF panel code called PANSHIP for the case of parametric roll in head seas.

Chapter 37 continues the discussion of parametric roll, describing a procedure and results of a benchmark study, carried out by the Stability in Waves Committee of the 26th International Towing Tank Conference (ITTC). Results of numerical simulation with six different numerical codes were compared with a single model test time history for each condition. The benchmark outcome was inconclusive: Due to practical non-ergodicity, a single record does not contain sufficient information for robust comparison. This latter point is an important observation, in and of itself, because this same deficiency has affected previous benchmarking efforts, which were inconclusive while providing no explanation as to why.

Chapter 38 describes the validation of a ship flooding simulation tool. The approach is to initially validate the flooding model and the vessel model separately and then couple the two models together for the final step in the validation process.

A series of model tests have been undertaken and the data obtained has been utilized as part of the validation process.

In summary, the third division of the book describes recent progress in experimental techniques and research, validation, and benchmarking. Five model test/experimental techniques are described: capsizing in waves, parametric roll, wave group approach, stability of high-speed craft, and inflow/outflow of water for a damaged ship. General validation procedures for numerical codes with model test data and a specific validation procedure for extrapolation were considered in details. Two validation experiments were described: dead-ship condition and for parametric roll. In addition, results of the 26th ITTC parametric roll numerical simulation benchmark are reported as well as validation of flooding simulations.

Requirements, Regulations, and Operations

Engineering practice plays a dual role for the development of applied science. It is a consumer and a motivator. The fourth division reports progress on the "practical side of the house." It is placed at the end of the book in the hope of showing how knowledge that was developed through theoretical considerations, experiments, and numerical observations will result in better regulations and safer practices.

Part XI covers the development of intact stability regulations, but it is not the only one to address this development. Many of the chapters, including those in the other parts of the book, have been motivated by regulatory development because regulations play a central role in providing marine vehicles with sufficient stability.

Goal-based standards (GBS) are high-level standards and procedures currently under development by the International Maritime Organization (IMO)—a specialized agency of the United Nations responsible for regulating international shipping and related matters. The central idea of GBS is to formulate a goal of safety, rather than prescribe how to achieve it (MSC Circ.1/1394). Chapter 39 describes research toward achieving the goal of safer container shipping. Large accelerations acting on a containership in heavy weather represent a serious hazard to human life, property, and the environment.

Modern design does not always allow hazards to be fully excluded through the design process, so safety provisions must be brought into operations in the form of operational guidance. Parametric roll of a containership provides a good example of this. The variation of stability in longitudinal waves is the main driver of this phenomenon. A buttock-flow type stern and flared bow are the hull form features that are "responsible" for significant stability variation in such waves. Levadou and van't Veer (2006) have shown that it is not possible to fully eliminate parametric roll through reasonable changes in the hull form, as the latter is dictated by the intended service of the ship and fuel efficiency requirements. For this reason, the responsibility for ensuring safety is increasingly shifting from design toward operation.

Another major development at IMO is the second-generation intact stability criteria. The idea is to apply modern computation assessments for stability failure types that are not covered or not completely covered by the current intact stability regulations. A discussion of the main ideas and overview can be found in Peters et al. (2011). Chapter 40 considers existing experience of numerical simulation that can be brought into a regulatory framework for these new criteria. In fact, many other chapters of this book were motivated by this IMO development, in particular Chaps. 16, 33 and 35.

Chapter 41 looks into recent developments of stability regulations for European sea–river ships. These vessels are mostly operated in inland waters, but are capable of sailing in coastal routes as well. The combination of these two operational modes makes regulatory development complicated, so a probabilistic approach is considered. The chapter also provides a review of current national and international stability regulations.

Part XII describes regulatory development in damage stability. It starts from an in-depth review of the development in Chap. 42. This chapter reflects the changes in assumptions used to revise the regulations since the 1960s. Special attention in this context is paid to ROPAX ships, whose subdivision has always been an issue.

Chapter 43 looks at the very same problem from a different perspective: prevention of collision. The idea is to put prevention into the center of regulations rather than simply mitigating its consequences. Crashworthiness is introduced as a prerequisite to survivability. Discussion of the influence of structural factors continues in Chap. 44. Here, the focus is on how structural degradation may affect safe return to port. Calculation results are presented for a ROPAX with side damage and for an Aframax tanker with asymmetric bottom damage.

Watertight doors are, without a doubt, a critical survival element of a ship. Chapter 45 presents a detail discussion on whether watertight doors should be allowed to be kept open during navigation. The watertight doors cannot stay open during navigation, but can be used for passenger and crew traffic in a controlled manner. The chapter provides a unique "inside" view on those deliberations.

Continuing with the subject of passenger vessels, Chap. 46 tracks the evolution of damage stability regulations, noting how progress in computational capabilities and mathematical modeling has increased the number of flooding scenarios that are considered: from one in mid-1960s to tens of thousands nowadays. The chapter presents a view of a way forward and makes the case to abandon "grandfather clauses" in regulations.

While going through the book, the reader probably noted more and more discussions on risk assessment/management approaches being considered for regulatory use. Taking a rational approach toward stability regulations, the question on how safe is safe enough will be inevitable. Chapter 47 looks into that question, considering the distribution of time before capsizing and the probabilistic characterization of passenger evacuation.

While regulations do play a central role in providing sufficient stability for a ship, the ultimate responsibility is shared between a designer and operator. Part XIII reports on new ideas and approaches to stability in design and operation. Chapter 48 discusses a new hull form concept that is meant to provide good stability at large roll angles, while keeping the metacentric height very small, resulting in very small motions in large waves. The chapter also looks into the role of water on deck and considers the influence of the deckhouse architecture on stability in severe weather.

As noted above, a modern design does not always allow a hazard to be completely excluded at the design stage. Operational guidance, developed with ship-specific data has become an important tool to ensure stability and safety during operation. How efficient this tool is going to be depends on how well a crew is trained to use it. Chapter 49 considers the crew training issues. It is noted that the traditional focus of formal ship handing training has been on maneuvering in calm water. Handing of a ship in heavy weather conditions is learned from experience. However, this experience is limited even for seasoned ship handlers, as those conditions usually are avoided. Use of a simulator for training is one possible solution.

The third element of providing operational stability in heavy weather—together with operational guidance and crew training—is onboard measurement and analysis. Onboard systems increase the crew's situational awareness in heavy weather, but to be useful those systems must run real time or faster. One idea discussed in Chap. 50 is using an autoregressive modeling procedure. These procedures are fast, as they use very few wave components—see also Chap. 2 of this book. In addition, Chap. 50 describes testing of the proposed scheme with a parametric roll model experiments.

Providing sufficient operational stability is not limited to heavy weather issues. Some operations require special engineering preparation and carefully monitored execution. One such operation was the transporting the damaged *USS Cole* with a semi-submersible heavy lift ship, which is described in Chap. 51. A semi-submersible heavy lift ship is an ocean-going vessel capable of submerging its open deck below the water's surface in order to allow another vessel to be floated over it. The operation is also known as "float-on, float-off" or FLO/FLO.

Part XIV discusses the stability of naval vessels. The physics of stability failure does not, of course, depend on whether a ship belongs to a navy or not, but stability requirements are managed differently for naval and commercial ships. Naval vessels are not covered by IMO instruments since they are either ships of war or troopships. Nevertheless, international cooperation plays an important role in naval stability assessment. The central place in this cooperation belongs to Cooperative Research Navies (CRNav) and its Naval Stability Standards Working Group (NSSWG).

Chapter 52 gives a review of naval stability standards and describes the motivations and objectives of international cooperation, including the forming of CRNav and the NSSWG. The main simulation tool for this cooperation is FREDYN. The chapter describes the work of the NSSWG toward a set of rational criteria for the stability of naval frigates and the limitations of these criteria. The relationship between the parameters of a GZ curve and the probability of capsizing has been one of the cooperative research objectives.

Chapter 53 continues the consideration of the NSSWG work, focusing on how naval stability requirements influence the design and acquisition process. Different methods for dynamics stability assessment are described, including relative probabilistic assessment (comparing to a ship with known stability safety) and direct (or absolute) assessment of capsizing probability. The probabilistic approach to dynamic stability naturally rolls into risk-based procedures.

Consideration of capsizing risk is continued in Chap. 54, which addresses the question of acceptable level of capsizing risk for a naval vessel. A concept of the tolerability of risk is introduced; boundaries of tolerability are compared for different types of human activities. Finally, a value of 1×10^{-4} annual capsize risk was proposed as a suitable level for the tolerable risk boundary for the loss of a naval frigate at sea.

The discussion of naval developments in the assessment of dynamic stability, probability of capsizing, and including stability hazard into overall risk assessment would have been incomplete without relating it to similar developments in the commercial fleet, namely GBS. This task is performed by Chap. 55, which concludes with an overarching scheme of risk assessment.

In summary, the fourth division of the book describes recent development in stability regulations, including goal-based standards. Notable developments include direct stability assessment within the framework of the second-generation IMO intact stability criteria, as well as crashworthiness, safe return to port, and role of watertight doors in damage stability regulations. Operational issues include the development of operational guidance, severe weather ship handing training, and onboard measurements. Discussion of specific aspects of naval stability requirements has led to risk-based approaches, interrelated with similar development for commercial fleet.

Acknowledgements

This book is a collection of works; the credit (or blame) for the selection of the texts of archive quality that provide a logical and representative story of the recent development in the field belongs to the editors: M. Neves, K. Spyrou, N. Umeda, F. van Walree, and V. Belenky.

As was already mentioned in the preface, all the chapters went through a thorough review by at least two reviewers. This was not a small work, it was a collective effort, so the editor would like to recognize the reviewers: M. Abdel-Maksoud, G. Bulian, B. Campbell, J. de Kat, A. Degtyarev, A. Francescutto, D. Greeley, D. Hayden, M. Hughes, A. Ibrahim, C. Judge, T. Katayama, M. Levine, A. Maki, M. Neves, J. Park, L. Pérez Rojas, W. Peters, V. Pipiras, A. Reed, I. Rychlick, A. Scott, T. Smith, K. Spyrou, A. Troesch, N. Umeda, F. van Walree, D. Vassalos, and K. Weems.

Finally, the editor would like to thank his "inner circle" colleagues and collaborators: A. Reed, T. Smith, K. Weems, W. Peters, V. Pipiras; the editor is especially grateful to K. Spyrou, who reviewed 20 chapters. Their support, reviews, and discussions were instrumental in the process of preparing this book for publication, which took longer than anticipated but is hoped that the outcome will be found worthwhile and useful to the readers.

References

Abkowitz M. A., 1980, "Measurement of Hydrodynamic Characteristics from Ship Maneuvering Trials by System Identification," *Transactions SNAME*, **88**:283–318.

Anastopoulos, P. A., K. J. Spyrou, C. C. Bassler & V. Belenky, 2015, "Towards an Improved Critical Wave Groups Method for the Probabilistic Assessment of Large Ship Motions in Irregular Seas," *Probabilistic Engineering Mechanics*, **44**:18–27.

Beck, R. F. & A. M. Reed, 2001, "Modern Computational Methods for Ships in a Seaway," *Transactions SNAME*, **109**:1–51.

Box, G. E. P., G. M. Jenkins & G. C. Reinsel, 2008, *Time series analysis: Forecasting and control*, 4th Edition. Wiley, xx+746 p.

Crider, D. A., J. M. Falzarano, A. M. Reed & J. S. Spencer (2008) Development of a Marine Kinematics Extraction Technique with Application to the Crown Princess Heeling Accident. *Transactions. SNAME*, **116**:75–90

de Kat, J. O. & J. R. Paulling, 1989, "The Simulation of Ship Motions and Capsizing in Severe Seas," *Transactions SNAME*, **97**:139–168.

France, W. N., M. Levadou, T. W. Treakle, J. R. Paulling, K. Michel & C. Moore, 2003, "An Investigation of Head-Sea Para-metric Roll and Its Influence on Container Lashing System," *Marine Technology*, **40**(1):1–19.

Gao, Z., Q. Gao & D. Vassalos, 2013, "Numerical study of damaged ship flooding in beam seas," *Ocean Engineering*, **61**:77–87.

Ikeda, Y., Y. Himeno & N. Tanaka, 1976, "On Roll Damping Force of ship: Effects of Friction of Hull and Normal Force of Bilge Keels," *Journal of Kansai Society of Naval Architects, Japan*, **161**:41–49.

IMO MCS Circ.1/1394, 2011, Generic guidelines for developing IMO goal-based standards, London.

Kim, D. H. & A. W. Troesch, 2013, "Statistical Estimation of Extreme Roll Responses in Short Crested Irregular Head Seas," *Transactions SNAME*, **121**:123–160.

Levadou, M. & R. van't Veer, 2006, "Parametric Roll and Ship Design," *Proceedings of the 9th International Conference on Stability of Ships and Ocean Vehicles (STAB 2006)*, Rio-de-Janeiro, Brazil, Vol. 1, pp. 191–206.

Lin, W.-M. & D. K. P. Yue, 1990, "Numerical solution for large-amplitude ship motions in the time domain," *Proceedings of the 18th Symposium on Naval Hydrodynamics*, Ann Arbor, Michigan, pp. 41–66.

Peters, W., V. Belenky, C. Bassler, K. Spyrou, N. Umeda, G. Bulian & B. Altmayer, 2011 "The Second Generation of Intact Stability Criteria An Overview of Development," *Transactions SNAME*, **119**:225–264.

Reed, A. M. & R. F. Beck, 2016, "Advances in the Predictive Capability for Ship Dynamics in Extreme Waves," *Transactions SNAME*, **124**:2–39.

Shin, Y. S, V. L. Belenky, W.-M. Lin, K. M. Weems & A. H. Engle, 2003, "Nonlinear time domain simulation technology for seakeeping and wave-load analysis for modern ship design," *Transactions SNAME*, **111**:557–578.

Spanos, P. D., 1983, "ARMA Algorithms for Ocean Wave Modeling," *Journal of Energy Resources Technology, Transactions of ASME*, **105**:300–309.

Spyrou, K. J., 1996, "Dynamic Instability in Quartering Seas: The Behavior of a Ship During Broaching," *Journal of Ship Research*, **40**(1):46–59.

Spyrou, K. J., V. Belenky, N. Themelis & K. M. Weems, 2014, "Detection of surf-riding behavior of ships in irregular seas," *Nonlinear Dynamics*, **78**:649–667.

St. Denis, M. & W. J. Pierson, Jr., 1953, "On the motions of ships in confused seas," *Transactions SNAME*, **61**:280–357.

Themelis, N. & K. J. Spyrou, 2007, "Probabilistic assessment of ship stability," *Transactions SNAME*, **115**:181–204.

Disclaimer

Contents

Part I
Mathematical Model of Ship Motions in Waves: New Simulation Tools

Chapter 1
TEMPEST—A New Computationally Efficient Dynamic Stability Prediction Tool

William F. Belknap and Arthur M. Reed

Abstract The US Navy has embarked upon the development of a new computational tool for simulating the responses of a ship operating in severe sea states. This new tool, TEMPEST, is designed to be computationally efficient to support real-time training simulators as well as high-resolution evaluation of surface-ship, dynamic-stability performance across a wide range of possible environmental conditions. TEMPEST aims to improve the state-of-the-art for real-time computations through the inclusion of nonlinear (body-exact) hydrodynamic perturbation forces and physics-based, viscosity-influenced lift and cross-flow drag forces. Slender-ship and low-aspect-ratio lifting-surface theories provide the ability to maintain computational efficiency while including the dominant nonlinearities within the dynamic stability problem. This paper argues for the efficacy of TEMPEST's theory in reconciling the need for accurate predictions with computational efficiency.

1.1 Introduction

Ship operability and safety are often linked to its motions in waves and eventually to its dynamic-stability risk. Evaluation of dynamic-stability risk is primarily achieved through the gathering of performance data in the wave environment and speed-heading condition of interest. The performance data can be obtained from model tests or simulations. Model tests are expensive, limited in flexibility (wave conditions, run length), and can have scale effects. If the design changes, or even the loading condition changes, an entirely new model test needs to be executed. Simulations offer the opportunity to include scale effects, provide nearly any environmental input desired, and are generally easier to re-run when geometry or loading conditions change. However, there is a significantly higher burden on simulations to validate

W. F. Belknap
Naval Sea Systems Command (NAVSEA), Washington, DC, USA

A. M. Reed (✉)
David Taylor Model Basin (NSWCCD), West Bethesda, MD, USA
e-mail: arthur.reed@navy.mil

V. L. Belenky et al. (eds.), *Contemporary Ideas on Ship Stability*, Fluid Mechanics and Its Applications 119, https://doi.org/10.1007/978-3-030-00516-0_1

the theory for full-scale ship performance. Regardless, there remains a need for the designers and/or regulatory authorities who need to evaluate dynamic-stability risk to have several tools at their disposal. Model tests, high-fidelity computational tools (like CFD), and fast simulations all have their roles.

The number of conditions that must be simulated depends upon the resolution to which dynamic stability needs to be characterized. If the failure modes are not known a priori, it may be necessary to obtain motion statistics over a complete range of environmental and ship-operating conditions. If the matrix of conditions includes multi-directional seas with two or more wave systems (swell is more than likely not correlated to the wind-driven system), the total number of simulations quickly grows. For a nominal speed-heading resolution of every 5 knots and every 15°, each environmental condition could have approximately 150 conditions for which extreme value statistics need to be generated. Because of this, there is a need for computational efficiency. However, computational speed does not provide the designer or regulatory authority any benefit if the answer is wrong. The goal then is to generate sufficiently accurate results as computationally efficiently as possible. The evolving understanding of the relevant physics allows for theory to be only as complex as needed. It is with this objective that the U.S. Navy has embarked upon the development of a new dynamic-stability simulation tool—TEMPEST.

1.2 Physical Problem

A simulation tool needs to be able to include the physical phenomena that are relevant to the full-scale problem. As such, the first step in developing a computational tool is to identify what the physical problem is and decompose it in a manner that can be modeled. At the highest level, the physical problem can be described by the ambient environment, the ship-control condition, and the forces acting on the ship. Figure 1.1 illustrates the physical problem to be modeled.

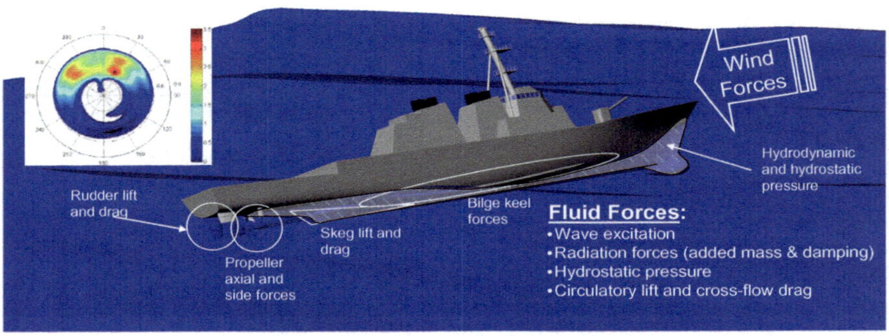

Fig. 1.1 Illustration of the physical problem to be simulated

1.2.1 Environment

The definition of the ambient environment for the dynamic stability problem must include both the wind and wave environments. In realistic sea conditions, the wave environment is generally considered multi-directional. An example polar spectrum showing two distinct wave systems is shown within Fig. 1.1. It is important to be able to include multiple wave directions in a computational model because of the unique physics that occur in such a situation. For example, one wave system may degrade transverse hydrostatic stability while another may provide a rolling moment.

Another aspect of the wave environment that is strongly correlated to dynamic-stability risk is the steepness of the seas. Steep seas have a more significant impact on the change in wetted geometry, which has a large effect on the forces acting on the hull. Within steep seas, nonlinear effects become stronger, such as the asymmetry of the wave profile and the nonlinear pressure and particle kinematics.

The wind environment may or may not be aligned with the wave systems, which produces another variable in the dynamic-stability-assessment matrix. Therefore, in addition to a reference mean speed, the wind environment includes a mean direction.

In order to determine the force on the ship due to wind in high sea states, the wind profile must be understood at the "local" scale, meaning that the effect of the nearby wave shadowing is included. This results in an apparent gustiness from the effect of being in the trough versus being on the crest. It is unclear whether or not capturing these effects has a significant effect on the final ship-motion results, but it is has been decided that the effects should be included until otherwise deemed unnecessary.

1.2.2 Ship Control

In a traditional "seakeeping" framework, the ship's speed and heading is considered known or prescribed. The solution of the seakeeping problem is the characterization of the motions about this nominally constant speed and heading. This framework is adequate and appropriate for determining the non-rare motion statistics, such as the RMS or significant values.

In the characterization of the large-amplitude, or rare-motion problem, it is necessary to consider the forces and responses that arise from large deviations from the constant speed-heading condition. These may include, but are not limited to, surf-riding and broaching. To allow for these, the ship must be *self-propelled* and *self-steered*. As such, the physical problem is best characterized as a maneuvering-in-waves problem.

To be self-propelled means that a propulsor model of some sort provides a thrust to balance the resistance forces present due to the air and water. Rather than prescribing a speed, the thrust and resistance, both of which can be time-dependent, determine the speed.

Self-steered means that a rudder, azimuthing propulsor, or other steering device is used to provide a yaw moment that counters a yaw moment induced by the aerodynamic and hydrodynamic forces on the hull. The time-changing balance of these forces and moments leads to the time-changing heading of the ship.

1.2.3 Forces

The forces acting on the vessel in the defined ambient environment for the ship under self-propulsion and self-steering control largely follow from the typically understood seakeeping and maneuvering problems. The unique aspects of the dynamic stability problem are the coupling of the forces and the effect of large-amplitude motion and/or large-amplitude waves.

The fluid forces on the hull consist of hydrostatic pressure, wave excitation (Froude-Krylov and diffraction), radiation forces (i.e., the added-mass and wave-making damping effects), resistance forces, and circulatory lift and cross-flow drag that arise from viscosity. In a high sea state, these forces can act on a hull with large changes in wetted geometry.

In the special case that the deck is submerged, the fluid flow must be treated as a "green water" problem. The green water problem describes the time delay in the force due to the time it takes for the fluid to cover the deck, as well as the shipping of water as the deck reemerges.

In addition to the bare hull, the bilge keels provide a lifting force and a cross-flow drag, as well as contribute to the added mass. As with other parts of the hull, the bilge keels can exit and re-enter the free-surface.

Propeller forces depend on the advance coefficient, J, which in turn is affected by the ambient environment (via wave-orbital velocities) and ship motions. In large waves the propellers can exit and re-enter the water, which will affect thrust and consequently speed of the ship. Furthermore, in the extreme motion and wave conditions present, large inflow angles of attack can result that lead to side forces that can be up to 40% or more of the axial force.

The rudder forces are coupled both with the propeller thrust and the ambient wave environment. As with other appendages, the rudders are subject to exit and re-entry through the free surface.

Finally, the wind environment imparts forces and moments on the exposed parts of the hull. The wind loads are dependent upon the time-changing, wind-speed profile acting on the ship.

1.3 Importance of Nonlinearity

There are a number of nonlinearities that manifest themselves in the prediction of motions of ships in extreme seas and dynamic stability. These range from: the equations of motion, to the geometry of the vessel, to the hydrodynamics as exemplified by the nonlinear free-surface boundary condition applied to the ambient wave field and the hydrodynamic disturbance (radiated and diffracted waves), and to Bernoulli's equation for pressure. The use of the fully nonlinear equations of motion is endemic among dynamic-stability codes, but otherwise there are as many differences as there are choices as to which nonlinearities are important and need to be included.

1.3.1 Hydrostatics and Froude-Krylov Forces

That nonlinearities are important for large-amplitude motion predictions has been recognized for many years, and is illustrated by the extensive use of "blended" methods that combine linear and nonlinear forces to predict large-amplitude vessel motions (Beck and Reed 2001). Blended methods typically incorporate nonlinear hydrostatic-restoring forces and nonlinear Froude-Krylov exciting forces due to the incident waves, with linear radiation and diffraction forces. Both the nonlinear hydrostatic-restoring forces and Froude-Krylov exciting forces account for body nonlinearities, particularly in the presence of large-amplitude waves and extreme motion responses.

The nonlinear hydrostatic-restoring forces arise from integrating the gZ term in Bernoulli's equation over the instantaneous wetted surface of the vessel in the incident waves, so there is little ambiguity as to what is to be computed (cf, de Kat and Paulling 1989). The issue here is how is the "incident wave" defined—is it purely linear, or does it include nonlinear (second-order or higher) terms? Since the mid 1800s, it has been known that steep second-order waves have higher crests and shallower troughs than linear waves (Stokes 1847), which will clearly affect the instantaneous wetted surface of the vessel and thus the hydrostatic-restoring force on the vessel. (More on the ambient wave description later in this section.)

The Froude-Krylov contribution to the exciting forces results from integrating the hydrodynamic terms of Bernoulli's equation ($\phi_t + 1/2\nabla\phi \cdot \nabla\phi$), which result from the incident waves over the immersed surface of the ship's hull. In this case, it is not as clear what terms should be integrated as it was for the hydrostatic term. Many codes linearize Bernoulli's equation to either ϕ_t or $\phi_t + U\phi$, where U is the forward speed of the vessel, either instantaneously or on the average. This leaves the possibility of significant variation in results for the Froude-Krylov component of the force without even considering the representation of the incident wave. Telste and Belknap (2008) and Belknap and Telste (2008) present and discuss some examples of this type of variation. The representation of the wave which will be presented later adds even more variation.

1.3.2 Hydrodynamic Forces

To develop an understanding of the hydrodynamic forces and moments on a vessel undergoing large-amplitude motions, a numerical experiment was performed using a variety of computational tools. These computational tools ranged from linear, to blended, to fully nonlinear. The complete experiment is documented in a massive report (15240 p.), Telste and Belknap (2008). Belknap and Telste (2008) and Reed (2009) contain summaries of the results.

In the numerical experiment, thousands of the force and moment calculations were made and compared for two hulls: oscillating in various modes of motion in calm water (Task 1), fixed in waves (Task 2), and simulating large-amplitude motions by contouring waves (Task 3). The results are presented in the form of time-history plots showing simulated forces and moments at two speeds, for a variety of headings and wave/motion amplitudes. It was not the purpose of the study to evaluate any one code relative to another, but rather to evaluate the differences between various complexities of theory; and in general, codes with a consistent level of theory produced quite consistent results.

Figure 1.2 shows a time history of ship-fixed vertical force from predictions for a hull undergoing forced heave in calm water at $F_N = 0.3$ and $\omega = 1.1$ rad/s, with heave amplitude/draft of 0.8. Many of these Task 1 force and moment predictions demonstrate the importance of nonlinearity in the radiation forces. An obvious indicator of nonlinearity is the departure of the components of force and moment from a simple sinusoidal form. This is seen in the predictions by the three nonlinear codes shown in Fig. 1.2. A surprising finding was that the body-exact strip theory is capable of capturing these important nonlinearities—comparable to the two fully nonlinear, 3-dimensional codes. This result provides hope for the development of fast codes to predict dynamic-stability failures on the order of real time.

Figure 1.3 provides a time-history of ship-fixed vertical forces [hydrodynamic (*i.e.*, radiation and diffraction); Froude-Krylov; hydrostatic] on a hull which is contouring waves in following seas at $F_N = 0$, $\lambda/L = 2$, and $H/\lambda = 1/20$. From these Task 3 computations, it was found that the hydrostatic and Froude-Krylov forces are an order of magnitude greater than the hydrodynamic forces. The hydrostatic and Froude-Krylov forces calculated by all of the codes are in remarkable agreement—there is no difference in the hydrostatic force, and the differences in the Froude-Krylov force predictions are small. The hydrodynamic forces show significant variation between the codes. As it was impossible to distinguish between the radiation and diffraction components of the hydrodynamic force, one cannot identify the sources of the difference. However, the hydrostatic and Froude-Krylov forces are 180° out of phase with each other, so they largely cancel each other. Thus the difference between the hydrostatic and Froude-Krylov forces is the same order of magnitude as the hydrodynamic force, which means an accurate calculation of the hydrodynamic force is very important.

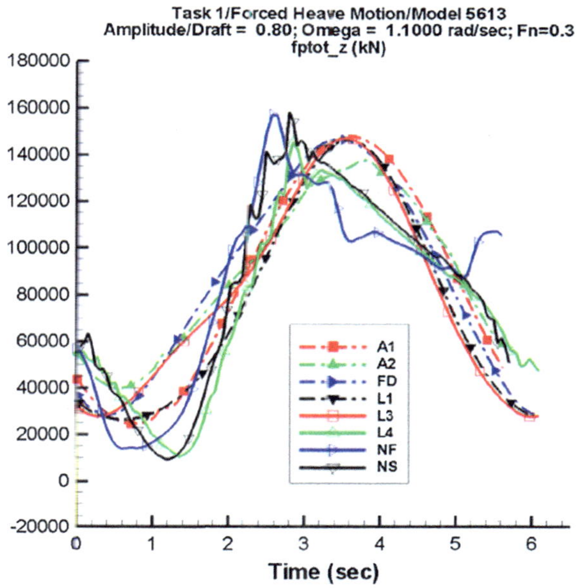

Fig. 1.2 Time-history of ship-fixed vertical force from Task 1 predictions for ONRTH hull undergoing forced heave at $F_N = 0.3$ and $\omega = 1.1$ rad/s, with heave amplitude/draft of 0.8 (Belknap and Telste 2008)

1.3.3 Second-Order Waves

As discussed earlier, nonlinear ambient-wave models have the potential to significantly influence predictions of dynamic stability. Two aspects of this are important: the shape of the wave profile; and the pressure within the wave. Stokes (1847) showed that the second order waves had steeper crests and shallower troughs than linear waves. According to linear theory, the pressure in wave crests (that portion of the wave above the calm free surface) is not zero at the free surface, which leads to significant errors in the predicted forces and moments on the ship's hull, particularly when the ship is in the wave crest in steep waves.

Figure 1.4 illustrates this for a wave of steepness (H/λ) of 1/10. It shows the pressure contribution from the zeroth- [$p_0/(\rho g) = -z$], first- [$p_1/(\rho g) = 2Ae^{\nu z}\cos\theta$], and second-order [$p_2/(\rho g) = -2\nu A2e^{2\nu z}$] terms in the pressure. As can be seen, the sum of the zeroth- and first-order pressure terms ($p_0 + p_1$) differs significantly from zero—providing an over prediction of the actual pressure at the free surface.

One method of dealing with this discrepancy with linear waves is the so called Wheeler stretching (Wheeler 1970), where the origin of the vertical coordinate is essentially shifted to the wave surface from the calm-water equilibrium surface, resulting in zero pressure at the free surface. The Wheeler-stretching approximation leads to much more realistic pressure distributions, and thus forces, than those forces which result from no stretching.

In the case where one is employing second-order wave theory to obtain realistic wave profiles in extreme seas, the use of second-order theory for the wave pressures

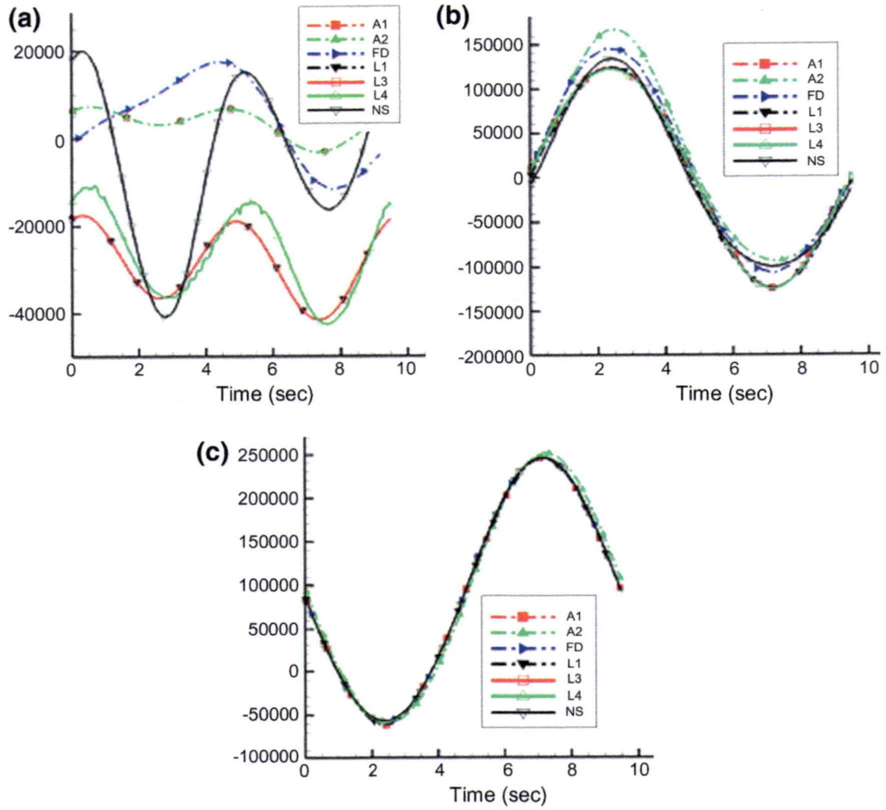

Fig. 1.3 Time history of ship-fixed vertical force from Task 3 predictions for Model 5514 hull, while contouring following seas at $F_N = 0$, $\lambda/L = 2$, $H/\lambda = 1/20$, **a** hydrodynamic force, **b** Froude-Krylov force, **c** hydrostatic force (Belknap and Telste 2008)

leads to accurate predictions of the pressure within the wave profile for regular waves.[1] The sum of the zeroth-, first-, and second-order pressure terms ($p_0 + p_1 + p_2$) in Fig. 1.4 provides an example of the second order pressure distribution, which comes quite close to zero at the free surface, much closer than the first-order approximation ($p_0 + p_1$).

A consistent implementation of second-order wave theory for irregular seas leads to sums containing exponentials of sum- and difference-frequency terms. The exponential-sum terms can become quite large near the wave crests, resulting in extremely unrealistic pressures near the free surface of wave crests. This is illustrated in Fig. 1.5 for two waves of differing frequencies such that the ratio of their wavelengths is 10.

[1]The second-order pressure equation does not require second-order wave theory; it can be used with linear wave theory.

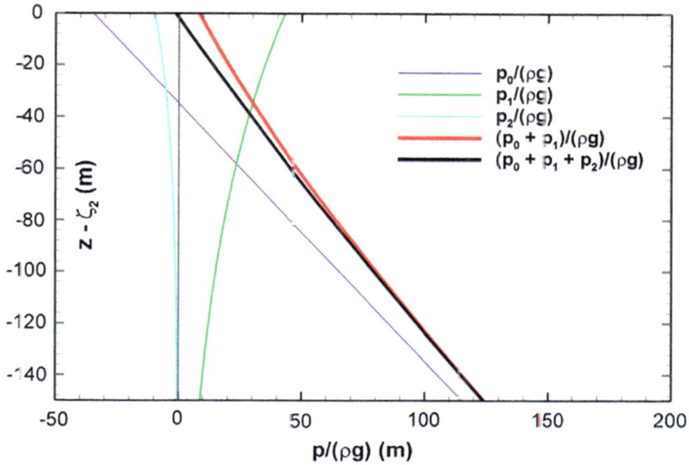

Fig. 1.4 Pressure under a wave crest through second order divided by ρg, as a function of the distance below the crest: $H/\lambda = 1/10$, $\zeta_1 = 2A\cos\theta$, $\zeta_2 = 2A^2\nu\cos(2\theta)$, $p_0/(\rho g) = -z$, $p_1/(\rho g) = 2Ae^{\nu z}\cos\theta$, $p_2/(\rho g) = -2\nu A2e^{2\nu z}$ (Courtesy of J. Telste)

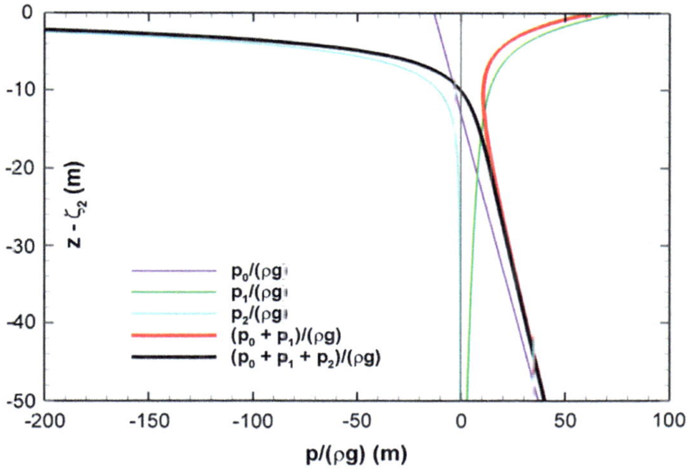

Fig. 1.5 Pressure under a wave crest through second order, $(p_0 + p_1 + p_2)/\rho g$, for the sum of two waves versus the distance below the crest, $z - (\zeta_1 + \zeta_2)$, for two frequencies: $\lambda_1/\lambda_2 = 10$, $H/\lambda = 1/10$, $(\theta_1, \theta_2) = (0, 0)$ (Courtesy of J. Telste)

There are several possible approaches that can be used to resolve the sum-frequency issue for irregular seas. One suggestion is to use a 2- or 3-term Taylor series expansion of the exponential rather than an exact-function evaluation. Stansberg et al. (2008) propose the use a low-pass filter applied to the linear horizontal velocity. The reason for such a filter is given by Gudmestad (1993), who states that the exponential term becomes very large near wave crests if the low-pass filter is not used.

1.3.4 Second-Order Forces

As a ship maneuvers in steeper and steeper waves, there are greater and greater interactions between seakeeping and maneuvering, to the point that one cannot predict maneuvering in steep waves by simply superimposing seakeeping and maneuvering in a linear fashion (cf, Reed 2009). One of the reasons for this is the fact that in steep waves the second-order hydrodynamic forces and moments (second-order drift forces and moments, and added resistance in waves) begin to play a significant part in the maneuvering behavior of the ship—slowing it down and speeding it up as it executes a turn in waves (Skejic and Faltinsen 2008). For this reason, it is important to have a comprehensive model of the physics that includes these forces. The Froude-Krylov forces and moments capture a portion of these forces and moments, but only the components due to ambient waves. There is a significant hydrodynamic component that must be captured accurately.

1.3.5 Nonlinear Dynamical System

Finally, it needs to be recognized that a ship undergoing large-amplitude motions in extreme seas represents a nonlinear dynamical system. As a consequence, the vessel response can change drastically with small increases in excitation—this is particularly true near and beyond the peak in the righting-arm curve, where the restoring moment remains essentially constant or even decreases as the heel angle (roll angle) increases. Conceptually this is easy to understand in calm water, but in a seaway, there is even more variability due to the ship being posed on a wave—as the wave passes along the hull the magnitude of the righting arm will fluctuate relative to the calm-water righting arm and the angle corresponding to the peak of the righting arm will vary. Whether the peak of the righting-arm curve increases or decreases in magnitude and the angle at which the peak occurs is a function of the shape of the hull above and below the calm-water waterline and the phase of the wave along the hull. Statistically, this says that there will be significant uncertainty as to the response of the ship under these circumstances. This has significant implications for the validation of computational tools and it is important for one to understand these concepts when validating the tools.

1.4 Code Approach Options

Having identified the components of the maneuvering-in-waves physical problem and understanding the importance of nonlinearity within the dynamic-stability problem, several modeling approaches were evaluated for implementation in TEMPEST. Vassalos et al. (1998) provide an overview of the numerical tools and approaches available for predicting dynamic-stability events. Further evaluation of options relied on experience with existing ship-motion computational tools, though physical considerations played a large role as well. One reason for this is that existing tools are fallible; e.g., some of the tools may not have been adequately verified, meaning that seemingly poor validation results can not be separated from potential bugs in the code. A key argument for developing a tool from scratch is that it allows for best verification practices (thorough documentation, unit tests, etc.) to be built in from the beginning.

Perhaps the first high-level-approach question to consider is whether to follow a complete flow solver (such as RANS or Euler VoF) or a potential flow-based track. While the option to compute a total solution of the fluid flow is attractive because it would include nearly the entire physical problem in a single computation, the computational cost is prohibitive given the number of conditions that need to be simulated. For that reason, a framework that follows the traditional seakeeping decomposition of a radiation and diffraction potential-flow solution added to a circulatory-lift solution is the only practical path. The argument for such an approach is that there is weak and/or one-way coupling between the hull radiation and diffraction (or "hydrodynamic disturbance") force and the lift and cross-flow drag on the appendages and the hull itself. While this assertion requires validation, there is no apparent alternative that meets computational speed requirements.

There are two basic paths that can be followed within the framework described above. One approach is to combine a maneuvering theory with a seakeeping theory, such as the two-time scale model employed by Skejic and Faltinsen (2008) that attempts to break the problem into its low-frequency part (maneuvering) and high-frequency part (seakeeping). The difficulty with this approach is avoiding any double-counting of forces. The attractiveness of this option is that trusted maneuvering models can be used. The second approach is to attempt to model the circulatory lift problem by itself, thereby avoiding double-counting issues. The challenge then is providing a robust model for this **force**.

Within the community of potential-flow approaches, a code can be described in simple terms by how 3-dimensional it is and how much nonlinearity is captured. In general, the more 3-dimensional and the more nonlinear a code, the less computationally efficient it will be. Table 1.1 provides a high-level view of the computational expense within the matrix of nonlinearity-assumption and slenderness-assumption ranges. "Linear" denotes potential-flow codes that are completely linear, whereas "Blended" includes nonlinear (body-exact) hydrostatic and Froude-Krylov forces. The term "Nonlinear" refers to codes with nonlinear hydrodynamic-disturbance forces as well as nonlinear hydrostatic and Froude-Krylov forces. The item "2D"

Table 1.1 Computational efficiency (computational seconds/simulated seconds)

	Linear	Blended	Nonlinear
2D	O(10^{-3})	O(10^{-1})	O(10^{0})
Slender ship			*est. O(10^{1})*
3D*	O(10^{1})	O(10^{1})	O(10^{3})

*Time-domain solution of hydrodynamic disturbance for linear and blended methods

Table 1.2 Capturing physics and nonlinearity

	Linear	Blended	Nonlinear
2D			
Slender ship			
3D			

is strip theory; "Slender ship" means strip theory with some 3D effects or corrections; "3D" refers to a fully 3D code. The cells of the table are colored green if the computational speed is considered acceptable for providing a sufficient level of data resolution for dynamic-stability risk characterization while red is considered unacceptable.

Table 1.2 is organized identically to Table 1.1, but rather than color-coding according to computational speed, the cells are color-coded based on an intuitive assessment of the code's ability to capture the relevant physical phenomena. This assessment largely follows the arguments laid out on the importance of nonlinearity to the dynamic stability problem.

These tables may provide simplistic views of the code-approach options for the solution of the hydrodynamic forces, but they help the theory developer navigate the solution space.

1.5 TEMPEST Approach

The philosophy driving the development of TEMPEST's theory has been to include all aspects of the maneuvering-in-waves physical problem as described earlier and model these components such that they capture the important nonlinearities. The review of code-approach options has given the development team confidence that a computationally-efficient approach is feasible as long as the simplifying assumption of ship slenderness is adopted. This is supported by Table 1.3, which provides an

Table 1.3 Estimated composite ranking of computational efficiency and ability to capture the relevant physics

	Linear	Blended	Nonlinear
2D			
Slender ship			
3D			

estimated composite ranking of the hydrodynamic-solution approaches within the criteria of accuracy and speed. As noted earlier, accuracy is weighted more heavily than speed, because quick but incorrect data is of no value to the user. The result is that the TEMPEST approach is based on a fully body-nonlinear hydrodynamic solution with advanced models: for the environment; for circulatory lift and for cross-flow drag on the hull and appendages; and for other superimposed forces.

1.5.1 Environment

As input to the force models, the modeling of the environment becomes just as important as the force models themselves. While the user generally describes the wave spectrum and wind speed, it is the environmental models that interpret these higher level inputs to provide ambient pressures and velocities at many places on the hull at every time step.

Waves In TEMPEST, the seaway is modeled by second-order waves with arbitrary directionality. Though the modelling of second-order waves adds significant computational cost relative to linear waves, it was determined that the steep waves that lead to dynamic-stability events are best captured by a second-order model. It is believed that the pressure and particle-velocity profiles obtained from the second-order model, while requiring additional validation, are more accurate than linear waves with Wheeler stretching in the "surf zone" above $z = 0$.

To alleviate some of the computational cost, FFT techniques are used to accelerate the computations. An additional feature of the TEMPEST wave model is the availability of an integral-equation iterative solution in the special case of unidirectional seas to find the linear input spectrum, when given the target second-order spectrum.

Long-term solutions may include a higher-order wave model that solves for the evolving wave field. This may significantly increase computational time, but may be necessary if the pressures and velocities are found to be not accurate enough in the steepest waves using lower-order wave models.

Wind The TEMPEST ambient wind environment model defines the vertical wind speed profile above the free surface at any point in space and time. The notable

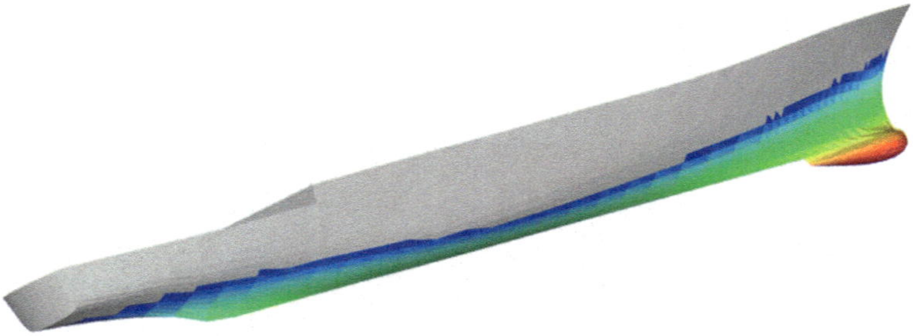

Fig. 1.6 Sample ambient-wave pressure on a 3D meshed hull

attribute of the TEMPEST wind model is that it attempts to account for the effects of shadowing near large steep waves. This model is currently in development using environmental data obtained from a North Sea oil rig.

1.5.2 Hydrodynamic Forces

The hydrodynamic forces acting on the ship are composed of:

- Hydrostatic and Froude-Krylov
- Hydrodynamic disturbance (radiation and diffraction)
- Green water on deck
- Resistance
- Bilge-keel
- Hull circulatory lift and cross-flow drag
- Propeller
- Rudder
- Wind

In all the force components, the effect of geometric nonlinearity is included by accounting for the position of the hull and appendages relative to the incident waves.

Froude-Krylov and Hydrostatic Forces The Froude-Krylov and hydrostatic forces are obtained by integrating the ambient-wave dynamic and static pressures, respectively, over the instantaneously wetted hull. The wetted hull is determined by the position of the ship and the undisturbed incident wave. To best capture the longitudinal force, the pressures are evaluated on 3D panels. An illustration of the body-exact Froude-Krylov plus hydrostatic pressure on a 3D mesh is given in Fig. 1.6.

Hydrodynamic-Disturbance Forces The force that captures the traditional sea-keeping radiation and diffraction forces is the hydrodynamic-disturbance force. TEMPEST obtains this disturbance force by solving the time-domain potential-flow boundary-value problem on the time-changing wetted surface of the hull. The conclusion of the theory development team was that applying a slender-ship approximation

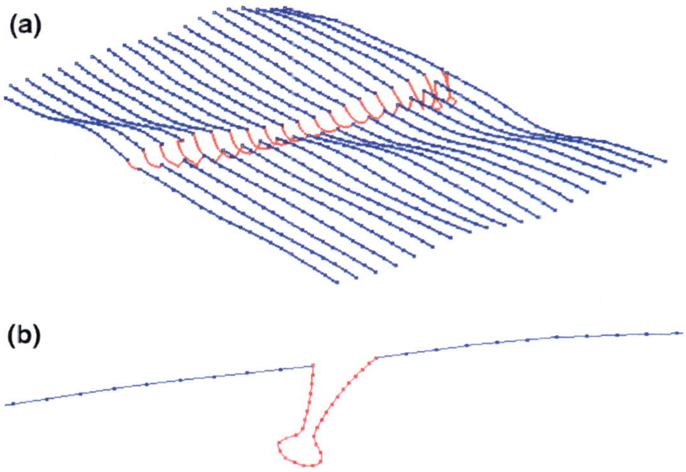

Fig. 1.7 Illustration of the body-exact strip theory problem for **a** the entire ship, and **b** a single 2D section

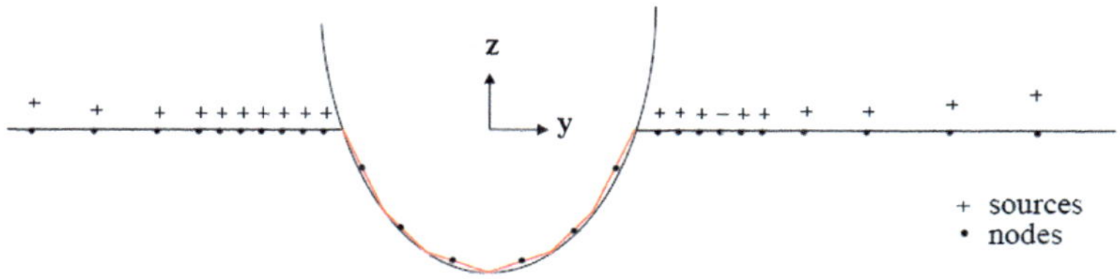

Fig. 1.8 Numerical solution of the time-domain boundary value problem for an example section (from Bandyk 2009)

would still capture the dominant physics while allowing the computations to occur at or near "real-time" speed. The theory behind this approach is given in a report to be published by Sclavounos et al. (2010).

The body-exact hydrodynamic disturbance solution in TEMPEST is being implemented in a two-phase process. In Phase 1, a strictly 2D approach is taken via a body-exact strip theory. Phase 2 implements a slender-ship theory, built upon body-exact strip theory that incorporates 3D effects.

The body-exact strip theory in Phase 1 follows the theoretical and numerical approach presented by Bandyk (2009). In this approach, impulsive and wave-memory problems are solved on 2D strips at each time step, an example of which is shown in Fig. 1.7. The boundary value problem is numerically solved by a 2D Rankine-panel method where the body section has sources distributed on 2D panels and the free surface uses desingularized panels. An example of this is shown in Fig. 1.8. Memory effects are automatically captured in the solution of the free-surface panels' source strengths.

In the Phase 2 hydrodynamic-disturbance potential solution, 3D effects are added through the use of a 3D time-domain Green function that operates on the impulsive source strengths determined on 2D sections. While this approach is presumably more computationally intensive than the body-exact strip theory, it may include 3D effects that are significant to the dynamic-stability problem. In this approach, as opposed to the body-exact strip theory in Phase 1, the wave-memory effects are obtained through evaluations of convolution integrals within the Green function. To address the computational burden, efficiency may be gained by simplifying the convolution integral functions and/or determining equivalent impulsive source-dipole line distributions within the interior of the wetted hull.

In both implementations of the body-exact hydrodynamic-disturbance problem, the force can be calculated from the velocity potentials through pressure integration or a momentum formulation. Calculating pressure for the 2D problem involves the difficult task of determining x-derivatives. Bandyk (2009) describes the use of radial-basis functions to overcome this difficulty. The momentum formulation (see Sclavounos et al. 2010) simplifies the force evaluation by requiring only a time-derivative on the integrated potentials.

Finally, hydrodynamic-drift forces that arise from the disturbed free-surface elevation are included. This is done in a simplified manner by evaluating a waterline integral that provides a hydrostatic correction due to the disturbance-wave elevation around the hull.

Green Water on Deck To account for the physics of deck submergence and re-emergence, a semi-empirical green-water model is included. This model has been implemented and successfully tested in LAMP (Liut et al. 2002). This model uses empirical relationships to get water height on deck given the deck-edge exceedence following Zhou et al. (1999). A notable deficiency of this green-water-on-deck model is that it does not capture the lag in elevation across the deck due to the flow of water on and off the deck. However, until it can be shown that the lag effect is important to the dynamic-stability problem, computational efficiency requirements dictate the use of this semi-empirical model.

Ship Resistance The TEMPEST resistance model uses a user-supplied resistance curve with the wave drag removed via a series of speed-calibration runs. The calibration runs remove any double-counting with the hydrodynamic-disturbance force. To account for body nonlinearity, the resistance curve is modified to account for the instantaneous wetted surface. The quasi-steady resistance is then obtained based on the instantaneous velocity through the water which includes the influence of wave orbital velocities.

Bilge-Keel Forces Low-aspect-ratio lifting-surface theory is the foundation of the TEMPEST bilge-keel force model (cf, Greeley and Peterson 2010). The work of Bollay (1936) inspired the model by showing that the trailing vortex sheet comes off the edge of the surface at an angle equal to half the angle of attack. By prescribing this trajectory of a trailing vortex sheet, a vortex-lattice method can be used to solve for the circulation strength and determine the (quasi) steady and unsteady forces due to lift. This method breaks down at angles of attack greater than about 50 (generally low-ship-speed conditions) where there is no true lift, so a Morison equation-based model is used. An "instantaneous" Keulegan-Carpenter (KC) number is estimated through the use of a short-time spectral analysis of normal velocity using a discrete Fourier transform. In large-amplitude roll cases, the effect of the bilge keels piercing the free surface is captured by means of a piece-wise damping model that accounts for various pieces of the hull entering and leaving the water (Bassler et al. 2010).

Hull Lift and Cross-Flow Drag Similar to the hydrodynamic-disturbance force, the hull lift and cross-flow drag-force model is being implemented in a two-phase manner. The initial model uses low-aspect-ratio lifting-surface theory to estimate time-changing (due to waves and motion) side-force and yaw-moment coeffi-cients. These coefficients are calibrated based upon user-supplied coefficients. This lift force is phased out over increasing drift angle, β, through a $\cos^2\beta$ multiplier that approximates stall. A cross-flow drag force is also calculated at each section for the time-changing geometry. This force follows a $\sin^2\beta$ behavior due to the fact that the only influence is the square of the cross-flow velocity. The cross-flow drag coefficients can be user-supplied or estimated based on shape coefficients. Reynolds-number dependence of cross-flow drag coefficients is included.

The second phase of the hull lift and cross-flow drag-force model implementation will apply the vortex-lattice techniques developed for the bilge-keel force model.

Propeller Forces The propeller forces are included as external forces to the hull. The key attribute of the TEMPEST propeller-force model is that it includes not just the axial force but also side forces when the inflow velocity provides an angle of attack to the propeller. The inflow velocity includes the effects of body velocity (including rotations), wave-orbital velocities, and an estimate of the viscous wake due to the presence of the hull.

The forces developed by the propellers due to the time-varying inflow are determined by a blade-element model. The blade-element model will properly account for partial or full emergence of the propeller. Pending more study, scale effects may be included to account for lost of thrust due to cavitation.

Rudder Forces The TEMPEST rudder-force model provides the forces due to lift and drag only. The contribution to the radiation and diffraction problem is not considered. Scale effects are accounted for by adjusting the angle of attack at which stall, with its fall off of lift, occurs as a function of Reynolds number. To account for body-nonlinearity, the rudder force is scaled by the immersed area of the rudder.

Wind Forces Wind forces are determined on the hull following a horizontal strip-theory approach similar to that given by Gould (1982). The benefit to a strip-theory approach is that it allows the use of an arbitrary wind-speed profile while still taking advantage of calibrated wind-drag and moment coefficients. Given the need to include non-traditional wind profiles due to the local presence of large, steep waves, such an approach is necessary.

1.6 Conclusions

TEMPEST is a new dynamic-stability simulation tool currently in development by the US Navy. The requirements of the tool are accuracy and computational speed.

After careful study of the physical problem, a comprehensive set of environment and force models has been described that is expected to provide a viable solution to the dynamic-stability prediction problem that advances the state-of-the-art. The fundamental argument behind the TEMPEST approach is the requirement for body-nonlinearity in all force models, including the hydrodynamic-disturbance force (radiation and diffraction).

The TEMPEST development will be followed by extensive validation at the component level and as a system.

Acknowledgements The authors would like to acknowledge the contributions of John Telste and Dr. Mike Hughes (NSWCCD), Dr. Dave Greeley (Applied Physical Sciences); and Prof. Paul Sclavounos (MIT), Prof. Bob Beck (U. Michigan), and the other members of the TEMPEST Theory Advisory Panel (TAP), who have contributed to the development of the TEMPEST theory. Dr. Pat Purtell (ONR) and Jim Webster (NAVSEA) have supported the TEMPEST effort.

References

Bandyk, P. (2009) A Body-Exact Strip Theory Approach to Ship-Motion Computations. Ph.D. Thesis, Univ. Michigan, Dept. Naval Architecture and Marine Engineering.

Bassler, C. C., A. M. Reed & A. J. Brown (2010) A Method to Determine Large-Amplitude Ship Roll Damping. *Proc. 11th Int'l Ship Stability Workshop*, 10 p.

Beck, R. F. & A. M. Reed (2001) Modern Computational Methods for Ships in a Seaway. *Trans. SNAME*, 109:1–51.

Belknap W. & J. Telste (2008) Identification of Leading Order Nonlinearities from Numerical Forced Motion Experiment Results. *Proc. 27th Symp. Naval Hydro.*, Seoul, Korea, 18 p.

Bollay, W. (1936) A New Theory for Wings with Small-Aspect Ratio. Ph.D. Thesis, Caltech, ii+ 86 p.

de Kat, J. O. & J. R. Paulling (1989) The simulation of ship motions and capsizing in severe seas. *Trans. SNAME*, 97:139–68.

Gould, R. W. F. (1982) *The Estimation of Wind Loads on Ship Superstructures*, Maritime Technology Monograph No. 8, The Royal Institution of Naval Architects.

Greeley, D. S. & B. J. Petersen (2010) Efficient Time-Domain Computation of Bilge Keel Forces. *Proc. 28th Symp. Naval Hydro.*, Pasadena, CA, 17 p.

Gudmestad, O. T. (1993) Measured and Predicted Deep Water Wave Kinematics in Regular and Irregular Seas. *Marine Structures*, **6**:1–73.

Liut, D. A., K. M. Weems, and W. M. Lin (2002) Nonlinear Green Water Effects on Ship Motions and Structural Loads. *Proc. 24th Symp. Naval Hydro.*, Fukuoka, Japan.

Reed, A. M. (2009) A Naval Perspective on Ship Stability, *Proc. 10th Intl. Conf. Stability of Ships and Ocean Vehicles*, St. Petersburg, Russia.

Sclavounos, P. D., J. G. Telste & A. M. Reed (2010) Nonlinear Response Modeling of a Vessel in Steep Random Waves. Carderock Division, Naval Surface Warfare Center Report NSWCCD-50-TR–2010/039.

Skejic, R. & O. M. Faltinsen (2008) A unified seakeeping and maneuvering analysis of ships in regular waves. *J. Mar. Sci. Tech.*, 13:371–394., DOI https://doi.org/10.1007/s00773-008-0025-2.

Stansberg, C. T., O. T. Gudmestad & S. K. Haver (2008) Kinematics under Extreme Waves. *J. Offshore Mechanics and Arctic Engineering*, 130(2) 020201.1–021013.8.
Stokes, G. G. (1847) On the theory of oscillatory waves, *Trans. Cambridge Philosophical Society*, Vol. 8, Part IV, No. XXXIII, pp. 441–455.
Telste, J. G. & W. F. Belknap (2008) Potential Flow Forces and Moments from Selected Ship Flow Codes in a Set of Numerical Experiments. Carderock Division, Naval Surface Warfare Center Report NSWCCD-50-TR–2008/040, 15,240 p.
Vassalos, D., M. Hamamoto, J. O. de Kat, D. Molyneux & A. Papanikolaou, (1998) The State of the Art in Modelling Ship Stability in Waves, Proc. 25th ATTC, Iowa City, Iowa, 8 p.
Wheeler, J. D. (1970) Method for calculating forces produced by irregular waves. *J. Petroleum Technol.*, **249**:359–367.
Zhou, Z. Q., J. O. de Kat & B. Buchner (1999) A Nonlinear 3-D Approach to Simulate Green Water Dynamics on Deck, *Proc. 7th Int'l Conf. Numerical Ship Hydro.*, Nantes, France.

Part II
Mathematical Model of Ship Motions in Waves: Environment

Chapter 2
Modeling of Incident Waves Near the Ship's Hull (Application of Autoregressive Approach in Problems of Simulation of Rough Seas)

Alexander B. Degtyarev, Arthur M. Reed and Vladimir Mareev

Abstract This chapter introduces the basics of the ARMA (Autoregressive Moving Average) model of short-crested wind waves. The model consists of an autoregressive component for temporal dependence and evolution and a two-dimensional moving average component for spatial dependence and propagation. A brief description of the validation of the model is given with special emphasis on the analysis of the dispersion relationship.

Keywords Autoregressive/moving average model (ARMA) · Short-crested waves

2.1 Introduction

Mathematical modeling of the behavior of a ship at sea requires a description of the wind driven wave field. The most popular models for describing wind waves are models based on the linear expansion of a stochastic moving surface as a system of independent random variables, such as the models by St. Denis and Pearson (1953), Rosenblatt (1957), Sveshnikov (1959) and Longuet-Higgins (1962). The most popular model is that of Longuet-Higgins, which is based on a stochastic approximation of the moving wave front as a superposition of elementary harmonic waves with random phases and random amplitudes, ε_n, c_n:

A. B. Degtyarev (✉) · V. Mareev
St. Petersburg State University, Saint Petersburg, Russia
e-mail: deg@csa.ru

A. M. Reed
David Taylor Model Basin (NSWCCD), West Bethesda, MD, USA

V. L. Belenky et al. (eds.), *Contemporary Ideas on Ship Stability*, Fluid Mechanics and Its Applications 119, https://doi.org/10.1007/978-3-030-00516-0_2

$$\varsigma(x, y, t) = \sum_n c_n \cos(u_n x + v_n y - \omega_n t + \varepsilon_n) \tag{2.1}$$

where the u_n and v_n are wave numbers; the frequency ω_n associated with wave numbers through a dispersion relation

$$\omega_n = \omega(u_n, v_n)$$

The phases ε_n are jointly independent random variables uniformly distributed in the interval $[0, 2\pi]$.

Longuet-Higgins' model is simple, easily computed, and incorporates the physical fundamentals of the wind waves in a consistent manner; see a brief argument in Degtyarev and Reed (2013). As noted in this reference, the model also has several deficiencies, mostly related to computational performance. In particular, its application becomes very computationally expensive in long records due to self-repeating effect (Belenky 2011). The effect is especially strong if a spectrum is broad and has several peaks.

The recent development of computationally efficient ship motion simulation tools (see Chap. 1 of this book) may greatly benefit from the computationally efficient way of modeling the wave environment with autoregressive model, the fundamentals of which are examined in this text. This chapter is focused on a short-term wave modeling; the long-term and synoptic scale modeling is considered in the previous book (Degtyarev 2011).

2.2 Basic Autoregressive Model of Ocean Waves

The basic idea of the autoregressive model (AR) is based on a Markov process, but the dependence is extended further in the past. AR presents a stochastic moving surface as a linear transformation of white noise with memory. ARs are commonly used in other areas of probabilistic mechanics and dynamics to model stationary ergodic Gaussian random processes with given correlation characteristics (Box et al. 2008), but they have not been applied extensively to wind waves. The formal mathematical framework of the ARM for waves was developed by Spanos (1983), Gurgenidze and Trapeznikov (1988) and Rozhkov and Trapeznikov (1990).

In the ARM, the time history of the wave elevation ζ_t at a fixed point can be presented as:

$$\zeta_t = \sum_{i=1}^{N} \Phi_i \zeta_{t-i} + \varepsilon_t \tag{2.2}$$

where t is time, N is the order of the model, Φ_i are autoregressive coefficients, ζ_{t-i} are the values of the elevation at the last N last time instants, and ε_t is a Gaussian white noise.

The autoregressive coefficients Φ_i can be estimated from the autocovariance function (ACVF) using the Yule-Walker equations, which are derived as follows. Start by multiplying both sides of (2.2) by ζ_t:

$$\zeta_t^2 = \zeta_t \sum_{i=1}^{N} \Phi_i \zeta_{t-i} + \zeta_t \varepsilon_t$$

Integrate both sides from 0 to T and multiply by $1/T$:

$$\frac{1}{T}\int_0^T dt \zeta_t^2 = \frac{1}{T}\int_0^T dt \zeta_t \sum_{i=0}^{N} \Phi_i \xi_{t-i} + \frac{1}{T}\int_0^T dt \zeta_t \varepsilon_t$$

In the limit as $T \rightarrow \infty$, the integral on the left-hand-side becomes the variance of the elevation, V_ζ, and the second integral on the right-hand side goes to zero, leaving:

$$V_\zeta = \frac{1}{T}\int_0^T dt \zeta_t \sum_{i=1}^{N} \Phi_i \zeta_{t-i}$$

Rearranging the order of integration and summation gives:

$$V_\zeta = \sum_{i=1}^{N} \Phi_i \frac{1}{T}\int_0^T dt \zeta_t \zeta_{t-i}$$

In the limit as $\mathrm{T} \rightarrow \infty$, $\frac{1}{T}\int_0^T dt \xi_t \xi_{t-i}$ becomes the autocovariance function K_ζ with lag i.

$$V_\zeta = \sum_{i=1}^{N} \Phi_i K_\zeta(i) \tag{2.3}$$

Equation (2.3) is the first of a series of linear equations for the unknown coefficients of the autoregression model for ζ_t. To develop the second equation, multiply both sides of (2.2) by ζ_{t-1}, resulting in:

$$\zeta_{t-1}\zeta_t = \zeta_{t-1} \sum_{i=1}^{N} \Phi_i \zeta_{t-i} + \zeta_{t-1} \varepsilon_t$$

As with the first equation, integrate both sides from 0 to T and multiply by $1/T$:

$$\frac{1}{T}\int_{0}^{T} dt\zeta_{t-1}\zeta_t = \frac{1}{T}\int_{0}^{T} dt\zeta_{t-1}\sum_{i=1}^{N}\Phi_i\zeta_{t-i} + \frac{1}{T}\int_{0}^{T} dt\zeta_{t-1}\varepsilon_t.$$

In the limit as $T\to\infty$, the integral on the left-hand-side becomes the autocovariance function of ζ_t with lag 1, $K_\zeta(1)$; and the second integral on the right-hand side goes to zero, leaving:

$$K_\zeta(1) = \frac{1}{T}\int_{0}^{T} dt\zeta_{t-1}\sum_{i=1}^{N}\Phi_i\zeta_{t-i}$$

Rearranging the order of integration and summation gives:

$$K_\zeta(1) = \sum_{i=1}^{N}\Phi_i\frac{1}{T}\int_{0}^{T} dt\zeta_{t-1}\zeta_{t-i}$$

In the limit, $\frac{1}{T}\int_{0}^{T} dt\zeta_{t-1}\zeta_{t-i}$ becomes the autocovariance function K_ζ with lag $i-$ 1,

$$K_\zeta(1) = \sum_{i=1}^{N}\Phi_i K_\zeta(i-1) \tag{2.4}$$

Equation (2.5) is the second in a series of linear equations for the unknown coefficients of the autoregression model for ζ_t. Generalizing this, multiply (2.2) by ζ_{t-n} and follow the same procedure used in deriving the first two linear equations, resulting in:

$$K_\zeta(n) = \sum_{i=1}^{N}\Phi_i K_\zeta(i-n) \tag{2.5}$$

Recognizing that the variance is the autocovariance function with a lag of zero, $V_\zeta = K_\zeta(0)$, Eq. (2.7) is valid for $n = 0, 1, 2, \ldots$. Thus, we have a general formula for defining the system of linear equations for the unknowns Φ_k, $k=1,\ldots, N$ needed to define the autoregressive model for ζ_t of order N, Eq. (2.2).

These equations for the coefficients of the autoregression equation are known as the Yule-Walker equations (Box et al. 2008). They are named after G. Udny Yule and George Walker who developed the first autoregressive models in the late 1920s and early 1930s (Yule 1927; Walker 1931).

The order of AR model is chosen to satisfy the condition that the difference between the variance of modeled process and the given variance is less than a certain threshold—usually 3–8%. The number of coefficients usually does not exceed 15, however it could be as low as 3 for the for a narrow-band process. Typical order for real waves consisted of wind waves and swell is about 8.

After the coefficients Φ_k have been computed, the variance of white noise is calculated as:

$$\sigma_{\varepsilon}^2 = c_{\zeta}^2 - \sum_{i=1}^{N} \Phi_i K_{\zeta}(i)$$

σ_{ζ} is the standard deviation of the wave elevation.

Autoregressive models for wind waves require a remarkably small number of components (N), to recreate the wave surface. The periodicity of such a model is determined solely by the period of random number generator used in the simulation, which fully resolves the self-repeating effect. The computational algorithm is relatively simple and converges fairly quickly. Furthermore, the AR is not limited to Gaussian processes. To get a non-Gaussian distribution, a nonlinear transformation of white noise with memory can be used; also one can use non-Gaussian white noise.

2.3 Autoregressive Moving Average Model of Ocean Waves

The autoregressive model (2.2) represents wave elevation at a point. The propagation of the wave is modeled by adding a moving average model to Eq. (2.2). The moving average model presents the current values of a stochastic process as a sum of current and previous random variables–values of Gaussian white noise ε_j

$$\zeta_t = \sum_{j=0}^{N} \Theta_j \varepsilon_{t-j} \tag{2.6}$$

where Θ_j are the coefficients expressing the dependence. These coefficients are can be found from the following equation (Box et al. 2008):

$$K_{\zeta}(i) = \sigma_{\varepsilon}^2 \sum_{j=i}^{N} \Theta_j \Theta_{j-i} \tag{2.7}$$

where σ_{ε} is the standard deviation of white noise. The standard deviation of white noise is found as:

$$\sigma_{\varepsilon}^2 = \frac{\sigma_{\zeta}^2}{1 + \sum_{j=0}^{N} \Theta_j^2} \tag{2.8}$$

Equations (2.7) and (2.8) make a system of nonlinear algebraic equations, which can be solved numerically with the Newton-Raphson method (Box et al. 2008).

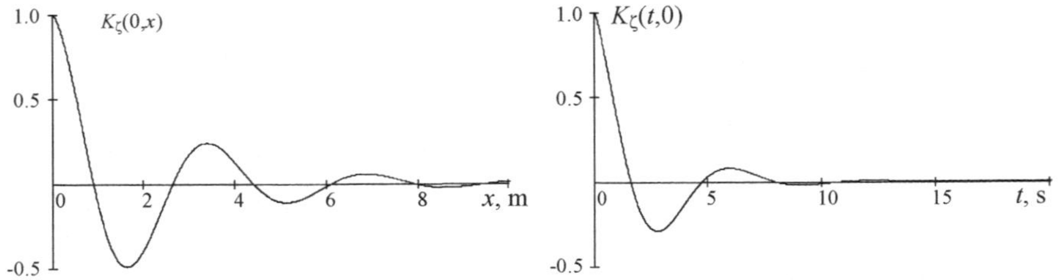

Fig. 2.1 Autocorrelation function versus x for $t=0$ and versus t for $x=0$

The moving average model has a similar function as the autoregressive model. The choice of one over another depends on the shape of spectrum. AR works better for a "peaky" spectrum—like a temporal spectrum of ocean waves. The moving average model was found to perform well numerically for spatial spectrum of ocean waves. A combination of autoregressive and moving average model is known in statistics as ARMA. The autoregressive part is used to reflect the temporal dependence while the moving average describes the spatial dependence:

$$\zeta_t = \sum_{i=1}^{N} \Phi_i \zeta_{t-i} + \sum_{j=0}^{M} \Theta_j \varepsilon_{t-j} \tag{2.9}$$

As an example, an ARMA model for long-crested progressive waves is developed for a nominal autocorrelation function of the following form:

$$K_\varsigma(x, t) = \exp(-0.42(t + x)) \cos(t - 1.8x)$$

Figure 2.1 plots the autocorrelation function versus x for $t=0$ and versus t for $x=0$.

An ARMA model was constructed with order $N=20$ for the auto-regressive term and $M=5$ for the moving average term, with a 1 s increment for time and a 1 m increment in space. Figure 2.2 shows the resulting waves simulated presented as a series of spatial "snapshots". The "base" points computed at the AMRA space/time increments are interpolated with a cubic spline to reveal the waves. Attempting to construct the ARMA using smaller increments may result in an ill-conditioned set of linear equations for the ARMA coefficients. This is probably caused by the accumulation of numerical error.

2.4 Short-Crested Ocean Waves

To describe short-crested ocean waves, the approach has been extended to any random scalar field (Degtyarev and Boukhanovsky 1995; Boukhanovsky et al. 1998, 2001):

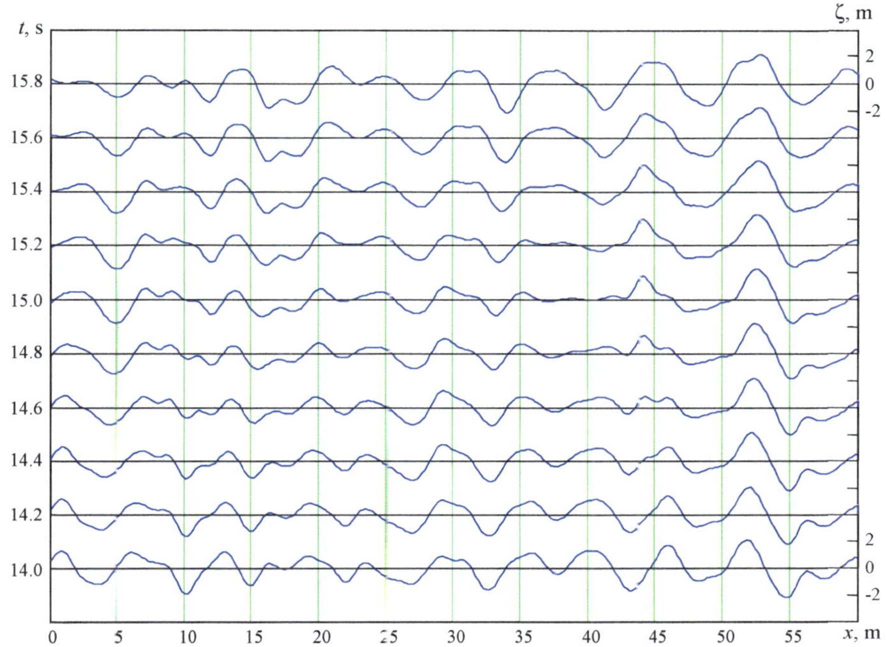

Fig. 2.2 Plane progressive wave—collection of spatial snapshots

$$\zeta_{(x,y,t)} = \sum_{i=1}^{N} \Phi_i \zeta_{(x,y,t-i)} + \sum_{j=0}^{M_x} \sum_{k=0}^{M_y} \Theta_{j,k} \varepsilon_{t-j,t-k} \tag{2.10}$$

where M_x is the order of the moving average model on coordinate x and M_y is the order of the moving average on coordinate y. The white noise values ε and coefficients Θ have double indexes.

The calculation of the autoregression coefficients Φ_i is no different than for the single-dimension case (2.5). However, it is difficult to apply the Newton-Raphson method for 2D spatial moving average model. Instead, a fixed point iteration can be used with the following formula:

$$\Theta_{i,m} = -\frac{K_\zeta(i,m)}{\sigma_\varepsilon^2} + \sum_{j=i}^{M_x} \sum_{k=m}^{M_y} \Theta_{j,k} \Theta_{j-i,k-m}$$

where $K_\zeta(i, m)$ is a spatial autocovariance function for indexes i and m, corresponding to coordinates x and y.

Details of the computational aspect of this model can be found in Gankevich and Degtyarev (2018).

2.5 Validation of Autoregressive Model

The Longuet-Higgins model (1) has a placeholder for the dispersion relationship, as both wave number and frequency have to be specified for each wave component. The ARMA model takes the spectra/ autocorrelation functions as inputs, so the dispersion relationship must be included implicitly through the modeled dependencies. Is that so?

This problem was addressed by Degtyarev and Reed (2013). Consider a point spectra of wave elevation and wave slope. The ratio of these spectra reveals the dispersion relationship:

$$S_\alpha(\omega) = (k(\omega))^2 S_\zeta(\omega) \quad \text{as} \quad \alpha(t) \approx \frac{d\zeta(x,t)}{dx}$$

where α is the angle of wave slope, x is the direction of wave propagation, ω is the wave frequency, S_ζ is the spectral density of wave elevations, S_α is the spectral density of angles of wave slope, and $k(\omega)$ is the wave number (spatial frequency). The dependence between the wave number and wave frequency is the dispersion relationship. It is found as:

$$k(\omega) = a\frac{\omega^2}{g}$$

where g is the acceleration of gravity and a is an empirical coefficient; $a = 1$ corresponds to the case of Airy theory for small-amplitude waves. Linear regression is used to find a for the estimated spectra.

Figure 2.3 shows results for wave measurements corresponding to sea state 5 and 8. The measurements were taken at a single-point in a towing tank of Kaliningrad Institute of Technology (Degtyarev and Boukhanovsky 1995). The estimates of the empirical parameter are greater than 1. As expected, the parameter increases with sea state, as the waves become larger and steeper, and the small-amplitude wave theory becomes less applicable.

The dispersion relationship is "embedded" in the estimated spectra of wave elevation and wave slope and will be "inherited" by corresponding auto-correlation functions. Thus, the coefficients Φ_i of the autoregressive model (2) will carry the information of the dispersion relationship. In principle, one could restore the records of wave elevations and wave slope angle with (2) and then compute their spectra and repeat the regression. However, as the autoregressive model is known to keep the input autocorrelation function, the result is not expected to differ from the one shown in Fig. 2.3. Thus the answer is: the correct dispersion relationship is a matter of the consistency of the input.

This analysis of the dispersion relationship is the first step towards validation. Once one is able to demonstrate that the modeled moving stochastic field has the physical properties of gravity waves in the water surface, the next step is verification of statistical properties.

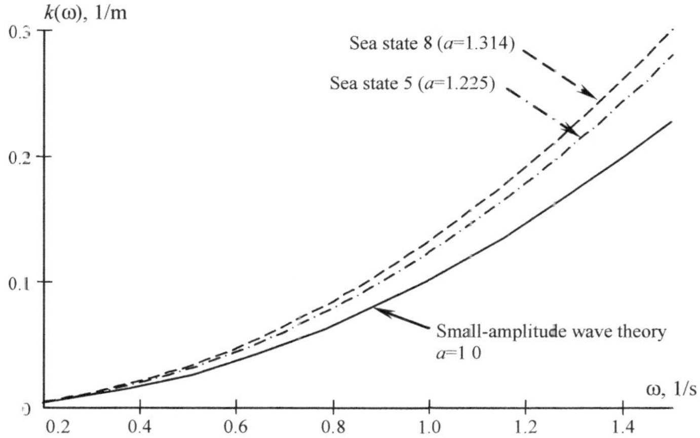

Fig. 2.3 Dispersion relationship estimates

Degtyarev and Reed (2013) describe this effort using simulated aerial photographs. Given a rectangle, ARMA is capable of generate wave elevations within this rectangle, corresponding to a particular instant of time. The resulting sample is similar to an aerial photograph of the sea surface, wherefrom the name has been taken.

The "true" values to compare with were taken from Davidan et al. (1978), which includes Welbull approximations for wave height, length, steepness and length of the crest developed from full scale measurements. The comparison was carried out using Q-Q plots and found ARMA to perform quite satisfactory.

Another validity check was done on the dependence between wave height and wave length. The comparison was carried out between sampled and "true" conditional mean values and between sampled and "true" conditional variance. The "true" values for conditional mean and variance were taken from Davidan et al. (1978). The results of the comparison were favorable for the ARMA model.

The importance of this successful comparison is that the model's input was limited by the spectra/ autocorrelation function and did not contain any explicit data on wave height, wave length and their dependence. Nevertheless, the model has managed to reproduce these "derived" characteristics correctly in a statistical sense; this is the best demonstration of the probabilistic validity of the model.

Other validation tests, described in Degtyarev and Reed (2013), include mixed ocean waves (wind waves plus swell).

2.6 Summary and Conclusions

This chapter has introduced the basics of Autoregressive Moving Average—ARMA—for short-crested ocean waves. As indicated by its name, ARMA consist of two parts: Autoregressive Model and Moving Average Model. Both models are intended to restore a record of a stochastic process from its autocorrelation function (or a spectrum) by filtering Gaussian white noise.

The ARMA model of short-crested ocean waves uses an autoregressive model to represent temporal dependence and evaluation and a two-dimensional moving average model to represent spatial dependence and propagation.

Compared to the conventional Longuet-Higgins model, the ARMA model requires considerably fewer coefficients for a record of the same length, which promises significant savings in computational cost. These computational cost savings are expected to be especially substantial for long records as the ARMA is not prone to the self-repeating effect. In addition ARMA is, in principle, capable of modelling non-Gaussian wave if a nonlinear transformation of white noise is used.

The ARMA is a statistical model; the correct reproduction of the physics of gravity surface waves depends on consistency of the input autocorrelation functions. The chapter gives a brief review of the statistical validation carried out for the ARMA model.

The ARMA provides only the wave elevation, so the practical application of the ARMA model for numerical simulation of ship motions in irregular waves requires a method to compute the hydrodynamic pressure and velocity field associated with the wave field (Degtyarev and Gankevich 2015). This is the topic of the next chapter.

Acknowledgements Dr. Degtyarev's work was supported by RFBR grants N 16-07-00886, 17-29-04288, project of St.Petersburg State University (id 28612502) and US Office of Naval Research Global Visiting Scientist Program under Dr. Woei-Min Lin. Dr. Paul Hess of ONR supported Dr. Reed's on this effort. This is much appreciated.

References

Belenky, V. (2011) "On Self-Repeating Effect in Reconstruction of Irregular Waves" Chapter 33 of Contemporary Ideas on Ship Stability, Neves, M.A.S., *et al.* (eds), Springer, ISBN 978-94-007-1481-6 pp. 589–598.

Box, G. E. P., G. M. Jenkins & G. C. Reinsel (2008) Time series analysis: Forecasting and control, 4th Edition. Wiley, xx + 746 p.

Bukhanovsky, A.V., Degtyarev, A.B., Lopatukhin, L.I., Rozhkov, V.A. (1998) "Probabilistic modeling of sea wave climate". Izvestiya - Atmospheric and Ocean Physics. Vol 34, No 2, pp. 235–239.

Boukhanovsky A., V. Rozhkov & A. Degtyarev (2001) Peculiarities of Computer Simulation and Statistical Representation of Time-Spatial Metocean Fields. In: Computational Science - ICCS, LNCS 2073, Springer, part I, pp. 463–472.

Davidan, I. N., L. I. Lopatuhin & V. A. Rozhkov (1978) Wind sea as a probabilistic hydrodynamic process. Leningrad, Gidrometeoizdat (in Russian).

Degtyarev, A., Boukhanovsky, A. (1995) On the Estimation of the Motion Stability in Real Seas. Proc. Intl Symp. Ship Safety in a Seaway: Stability, Maneuverability, Nonlinear Approach, Kaliningrad, Vol. 2, Paper 8, 10 p.

Degtyarev, A., (2011) "New Approach to Wave Weather Scenarios Modeling" Chapter 34 of Contemporary Ideas on Ship Stability Neves, M.A.S., et al. (eds), Springer, ISBN 978-94-007-1481-6 pp. 599–617.

Degtyarev, A.B., Reed, A.M. (2013) "Synoptic and short-term modeling of ocean waves" Intl. Shipbuilding Progress Vol. 60 pp. 523–553.

Degtyarev, A., Gankevich, I. (2015) "Hydrodynamic pressure computation under real sea surface on basis of autoregressive model of irregular waves". Physics of Particles and Nuclei Letters, Vol. 12, No. 3, pp. 389–391.

Gankevich, I., Degtyarev, A. (2018) Simulation of standing and propagating sea waves with three-dimensional ARMA model. In: Velarde M., Tarakanov R., Marchenko A. (eds) The Ocean in Motion. Springer Oceanography. Springer, Cham. https://doi.org/10.1007/978-3-319-71934-4_18.

Gurgenidze, A. T. & Y. A. Trapeznikov (1988) Probabilistic model of wind waves. In: Theoretical foundations and methods of calculating wind waves. Leningrad, Gidrometeoizdat, pp. 8–23.

Longuet-Higgins, M. S. (1962) The statistical analysis of a random, moving surface. Phil. Trans. Royal Soc. London. Series A, Mathematical and Physical Sciences, Vol. 249, No. 966, pp. 321–387.

Maddala, G. S. (1971) Generalized least squares with an estimated variance covariance matrix. Econometrica, Vol. 39, No. 1, pp. 23–33.

Rosenblatt, M. A (1957) Random model of the sea surface generated by the hurricane. J. Math., No. 6, p. 235–246.

Rozhkov, V. A. & Y. A. Trapeznikov (1990) Probabilistic models of oceanographic processes. Leningrad, Gidrometeoizdat (in Russian).

Spanos, P. D. (1983) ARMA Algorithms for Ocean Wave Modeling. J. Energy Resources Technology, Trans. ASME, Vol. 105 pp. 300–309.

St. Denis, M. & W. J. Pierson (1953) On the motion of ships in confused seas. Trans. SNAME, Vol. 61, pp. 280–354.

Sveshnikov A. A. (1959) Determination of the probability characteristics of three-dimensional sea waves. Math. Akad. Mech. and Engin., No. 3, pp. 32–41.

Walker, G. (1931) On Periodicity in Series of Related Term. Proc. Royal Soc. London, Ser. A, Vol. 131, pp. 518–532.

Yule G. U. (1927) On a Method of Investigating Periodicities in Disturbed Series, with Special Reference to Wolfer's Sunspot Numbers. Phil. Trans. Royal Soc. London, Ser. A, Vol. 226, pp. 267–298.

Chapter 3
Evaluation of Hydrodynamic Pressures for Autoregressive Model of Irregular Waves

Alexander B. Degtyarev and Ivan Gankevich

Abstract This paper proposes a new way of simulating the pressure field of the incident wave near a ship's hull. The approach is based on an autoregressive moving average (ARMA) model of the incident wave surface. This model retains all of the hydrodynamic characteristics of sea waves and allows the accurate solution of the potential flow problem and calculation of the hydrodynamic pressures below the surface. This chapter describes the solution of two-dimensional and three-dimensional problems.

Keywords Autoregressive/moving average model (ARMA) · Short-crested waves

3.1 Introduction

Direct assessment of ship stability in irregular waves requires the use of advanced hydrodynamic codes for numerical simulation of ship motions (e.g. see Beck and Reed 2001; Chap. 1 of this book). A model of irregular waves is an important component of these numerical simulations. Most current ship motion applications use model based on Longuet-Higgins (1962). Autoregressive moving average (ARMA) models have become a standard for modeling random excitation in many areas of probabilistic mechanics (Box et al. 2008; Spanos and Zeldin 1996), but the development of ARMA models for ship motions is still in progress (Spanos 1983; Bukhanovsky et al. 1998; Degtyarev 2011; Degtyarev and Reed 2013). The latter reference offers an analysis of the computational advantages of an ARMA wave model and suggests that it would make it a good fit for a new generation of computationally efficient tools (Chap. 2 of this book), thus renewing the interest.

The problem of modeling ocean waves in a form suitable for numerical simulation of ship motions is a complex one. Not only are the wave elevations a random moving surface—the computation of forces acting on ship requires the knowledge of

A. B. Degtyarev (✉) · I. Gankevich
St. Petersburg State University, Saint Petersburg, Russia
e-mail: deg@csa.ru

V. L. Belenky et al. (eds.), *Contemporary Ideas on Ship Stability*, Fluid Mechanics and Its Applications 119, https://doi.org/10.1007/978-3-030-00516-0_3

wave pressures around the ship hull. Modeling wave pressure is straightforward for a Longuet-Higgins model as an explicit expression for the 3-D pressure field is available, while the ARMA model provides only the water surface with required statistical characteristics. Thus, the calculation of the pressures becomes a separate problem. The fundamentals of this problem are considered in Degtyarev and Gankevich (2012) and further development is presented in Gankevich and Degtyarev (2015), Weems et al. (2016). This chapter is focused mostly on the fundamentals of the wave pressure problem under moving random surface.

The ARMA model of a moving wavy surface in three dimensions (2-D space+ 1-D time) is expressed as (Chap. 2 of this book):

$$\zeta(x, y, t) = \sum_{i=0}^{N} \Phi_i \zeta_{x,y,t-i} + \sum_{j=0}^{M_x} \sum_{k=0}^{M_y} \Theta_{j,k} \varepsilon_{x-j,y-k,t} \tag{3.1}$$

(M_x, M_y) is the order of the moving average model on coordinate x and M_y is the order of the moving average on coordinate y, t is time, N is the order of the autoregressive model, Φ_i are autoregressive coefficients, ζ_{t-i} are the values of the elevation at the previous N time increments, $\Theta_{j,k}$ are the coefficients expressing the spatial dependence through the moving average, and ε is Gaussian white noise.

3.2 Hydrodynamic Pressure Under the Wave Surface

To determine the evolution of the hydrodynamic pressure under the wave surface, consider the two-dimensional problem from wave theory. The traditional formulation is reduced to finding the wave potential (Kochin et al. 1964). The solution to this problem provides a complete definition of the hydrodynamic pressure of the wave surface:

$$\begin{aligned}
&\nabla^2 \varphi = 0 \\
&\frac{\partial \varphi}{\partial t} + \frac{1}{2}|\nabla \varphi|^2 + g\zeta = \frac{p_0}{\rho} \quad on \quad z = \zeta(x, y, t) \\
&\frac{D\zeta}{Dt} = \nabla \varphi \cdot \vec{n} \quad on \quad z = \zeta(x, y, t)
\end{aligned} \tag{3.2}$$

where φ is a velocity potential, g is gravity acceleration, p_0 is an atmospheric pressure, ζ is the free surface elevation, ρ is water density, D/Dt is a total derivative; vectors are identified by an arrow above the symbol; $\vec{n}$ is a normal vector, ∇ is a gradient operator, and ∇^2 is a Laplacian.

The Laplace equation for the velocity potential $\varphi(x, y, z, t)$ in the coordinate system shown in Fig. 3.1 is supplemented by two boundary conditions on the wave surface. These are the conditions that the pressure at the surface is equal to atmospheric pressure p_0 (dynamic boundary condition) and the continuity of fluid motion (kinematic

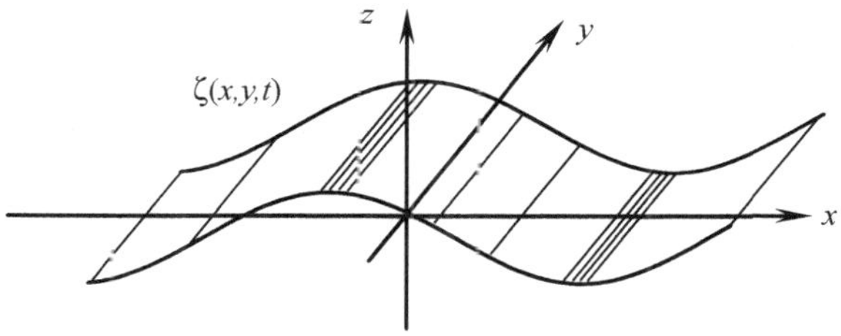

Fig. 3.1 The coordinate system

condition). The last condition states that a liquid particle belonging to the surface cannot go into (or out of) the fluid domain and must remain on the surface.

The complexity of (3.2) is that the boundary conditions are nonlinear and have to be satisfied at the unknown free surface. The system (3.2) can be reduced to Laplace's equation with one combined boundary condition by eliminating the unknown elevation of free surface (Kochin et al. 1964; Newman 1977). It is known that this formulation assumes the transfer of boundary conditions to the unperturbed surface $z = 0$.

The present case is different, however, as the free surface is known from the ARMA representation (3.1). This free surface is a result of statistical modeling that does not necessarily always describe real-world physics. The physicality of the ARMA model is dependent on the consistency of the auto covariance functions from which the ARMA model was developed; see the argument in Degtyarev and Reed (2013) or in Chap. 2 of this book. For the purpose of the numerical simulation of ship motions, any stationary realization of waves can be used, so the initial conditions for the system (3.2) may be taken random.

As the free surface is known, one of the boundary conditions in the system (3.2) can be dropped from further consideration. It is logical to exclude the first (dynamic) condition, as it contains a derivative of the potential over time. The Laplace equation itself and the second boundary condition do not contain derivatives of the unknown function of time. Note that the first boundary condition is usually linearized and used to find the free surface:

$$\zeta(x,\, y,\, t) = -\frac{1}{g}\frac{\partial \varphi}{\partial t} \tag{3.3}$$

Since the surface is already known, the first boundary condition for can be used to find the temporal derivative of the potential:

$$\frac{\partial \varphi}{\partial t} = \frac{p_0}{\rho} - \frac{1}{2}|\nabla \varphi|^2 - g\zeta \tag{3.4}$$

As a result, the system of Eq. (3.2) is reduced to solution of the Laplace equation with the kinematic boundary condition:

$$\frac{\partial^2\varphi}{\partial x^2}+\frac{\partial^2\varphi}{\partial y^2}+\frac{\partial^2\varphi}{\partial z^2}=0$$
$$\frac{\partial\zeta}{\partial t}+\frac{\partial\zeta}{\partial x}\frac{\partial\varphi}{\partial x}+\frac{\partial\zeta}{\partial y}\frac{\partial\varphi}{\partial y}=\frac{\partial\varphi}{\partial x}\cos(x,n)+\frac{\partial\varphi}{\partial y}\cos(y,n)+\frac{\partial\varphi}{\partial z}\cos(z,n)$$
$$on\, z=\zeta(x,y,t) \qquad (3.5)$$

Having in mind that

$$\cos(x,n)=\frac{\frac{\partial\zeta}{\partial x}}{\mp\sqrt{\left(\frac{\partial\zeta}{\partial x}\right)^2+\left(\frac{\partial\zeta}{\partial y}\right)^2+1}};\quad \cos(y,n)=\frac{\frac{\partial\zeta}{\partial y}}{\mp\sqrt{\left(\frac{\partial\zeta}{\partial x}\right)^2+\left(\frac{\partial\zeta}{\partial y}\right)^2+1}};$$
$$\cos(z,n)=\frac{1}{\pm\sqrt{\left(\frac{\partial\zeta}{\partial x}\right)^2+\left(\frac{\partial\zeta}{\partial y}\right)^2+1}}$$

The derivatives $\partial\zeta/\partial x$; $\partial\zeta/\partial y$; $\partial\zeta/\partial t$ and the angles between the normal vector to the surface and velocity of a liquid particle on the surface are known. These quantities can be evaluated from the ARMA model at each time instant at each point included in the computational grid.

Equation (3.5) is a mixed boundary value problem for the Laplace equation also known as the Robin's problem (Zachmanoglou and Thoe 1976).

3.3 Solution of 2D Problem

Consider a case of hydrodynamic pressures caused by a plane progressive wave, expressed as:

$$\zeta(x,t)=\sum_{i=0}^{N}\Phi_i\zeta_{x,y,t-i}+\sum_{j=0}^{M_x}\Theta_j\varepsilon_{x-j,t} \qquad (3.6)$$

The system of equations expressing the Robin's problem is reduced to 2D. Also taking into account that the angle of wave steepness changes in the interval [−0.142; 0.142] rad (for the steepest wave possible):

$$\varphi_{xx}+\varphi_{zz}=0$$
$$\zeta_t+\zeta_x\varphi_x=\frac{\zeta_x}{\sqrt{1+\zeta_x^2}}\varphi_x-\frac{1}{\sqrt{1+\zeta_z^2}}\varphi_z \quad on \quad z=\zeta(x,t) \qquad (3.7)$$

where indexes are used to identify derivatives:

$$\varphi_{xx} = \frac{\partial^2 \varphi}{\partial x^2}; \quad \varphi_{zz} = \frac{\partial^2 \varphi}{\partial z^2}; \quad \varphi_x = \frac{\partial \varphi}{\partial x}; \quad \varphi_z = \frac{\partial \varphi}{\partial z};$$
$$\zeta_t = \frac{\partial \zeta}{\partial t}; \quad \zeta_x = \frac{\partial \zeta}{\partial x}$$

Introduce direct and inverse Fourier transform that maps x to a new variable u:

$$\mathcal{F}(\varphi) = \Phi(u, z) = \int_{-\infty}^{\infty} \varphi(x, z) e^{-2\pi i x u} dx$$

$$\mathcal{F}^{-1}(\Phi) = \varphi(x, z) = \int_{-\infty}^{\infty} \phi(u, z)\, e^{2\pi i x u} du$$

The application of Fourier series to both sides of Laplace equation turns it into ordinary differential equation (using derivative properties of Fourier transform and swapping integration and differentiation):

$$\phi_{zz}(u, z) - 4\pi u^2 \phi(u, z) = 0 \tag{3.8}$$

The second-order linear ordinary differential Eq. (3.8) has a closed-form solution:

$$\phi(u, z) = A e^{2\pi u z} + B e^{-2\pi u z} \tag{3.9}$$

where A and B are arbitrary constants that can be found from boundary conditions. As the potential has to go to zero at the infinite depth, $B = 0$. To find the arbitrary constant A, apply the kinematic boundary condition on the free surface. A is a constant relative to z, but may depend on u. The derivatives of the potential can be expressed as:

$$\varphi_x = \frac{\partial}{\partial x} \mathcal{F}^{-1}(A(u) e^{2\pi u z}) = \int_{-\infty}^{\infty} \frac{\partial}{\partial x} \left(A(u) e^{2\pi u (z + i x)}\right) du = i \mathcal{F}^{-1}(2\pi u A(u) e^{2\pi u z})$$
$$\varphi_z = \frac{\partial}{\partial z} \mathcal{F}^{-1}(A(u) e^{2\pi u z}) = \int_{-\infty}^{\infty} \frac{\partial}{\partial z} \left(A(u) e^{2\pi u (z + i x)}\right) du = \mathcal{F}^{-1}(2\pi u A(u) e^{2\pi u z})$$
$$\tag{3.10}$$

Note that the difference between these derivatives is only multiplication by i, which is expected due to the circular trajectories of particles in wave within the potential theory. Also, it can easily be verified that the derivatives (3.9) are part of the solution of the Laplace equation by taking one more derivative by x and z respectively. This will produce identical expressions, but with the opposite sign, which will turn the Laplace equation into a true equality.

To complete the solution, the function $A(u)$ needs to be found from the kinematic boundary condition:

$$\mathcal{F}^{-1}(2\pi u A(u) e^{2\pi u z}) = \frac{\zeta_t \sqrt{1+\zeta_x^2}}{(1-\sqrt{1+\zeta_x^2})\zeta_x i - 1} \tag{3.11}$$

In order to preserve Fourier transform, the function being transformed must depend on u and not on x, but substitution $z=\zeta(x, t)$ makes it depend on x. To solve this problem we rewrite left hand side as a convolution:

$$\mathcal{F}^{-1}(2\pi u A(u)) * \mathcal{F}^{-1}(e^{2\pi u z})$$

and introduce a function $D(x, z)$ as:

$$D(x, z) = \mathcal{F}^{-1}(e^{2\pi u z}) = \delta(x + iz)$$

where δ is the Dirac delta function of complex argument. It is computed using its representation as a Lorentzian, noting that since the argument of Lorentzian is squared, the imaginary part vanishes. Introducing function D ensures that after substitution there will be no function which depends both on u and x and to which Fourier transform is applied. Applying Fourier transform to both sides of Eq. (3.11) with the new left hand side, one can express the function $A(u)$:

$$A(u, v) = \frac{1}{2\pi u \mathcal{F}(D(x, \zeta(x, t)))} \mathcal{F}\left(\frac{\zeta_t \sqrt{1+\zeta_x^2}}{(1-\sqrt{1+\zeta_x^2})\zeta_x i - 1} \right) \tag{3.12}$$

Substitution of Eq. (3.12) into (3.9) and application of inverse Fourier transform leads to the potential:

$$\begin{aligned} \varphi(x, y, z) &= \mathcal{F}^{-1}(\Phi) \\ &= \mathcal{F}^{-1}\left(\frac{e^{2\pi u z}}{2\pi u \mathcal{F}(D(x, \zeta(x, t)))} \mathcal{F}\left(\frac{\zeta_t \sqrt{1+\zeta_x^2}}{(1-\sqrt{1+\zeta_x^2})\zeta_x i - 1} \right) \right) \end{aligned} \tag{3.13}$$

In order to get the final answer, take the real part of the resulting complex-valued potential: $2Re(\varphi)$.

The solution cannot involve the imaginary part of the complex-valued potential for computational reasons. The $\exp(2\pi u z)$ term causes the integral in Eq. (3.13) to diverge for large wave numbers, which is a consequence of neglecting fluid viscosity in the original system of equations. To circumvent this, a range of wave numbers is computed numerically from the known wavy surface and used for the integration of the inverse Fourier transform, which can then be computed via FFT. This technique defines the velocity potential to be the real part of φ in the same manner as the one produced by linear wave theory formulae (when second-order elevation derivatives are omitted).

3.4 Approximate Solution of 3D Problem

To solve the Laplace equation for the 3D case, a 2D Fourier transform is used:

$$\mathcal{F}(\varphi) = \phi(u, v, z) = \int\limits_{-\infty}^{\infty}\int\limits_{-\infty}^{\infty} \varphi(x, y, z)e^{-2\pi i(xu+yv)}dxdy$$

$$\mathcal{F}^{-1}(\phi) = \varphi(x, y, z) = \int\limits_{-\infty}^{\infty}\int\limits_{-\infty}^{\infty} \phi(u, v, z)e^{2\pi i(xu+yv)}dudv$$

Similarly to the previous case, the application of the Fourier transform to each term of Laplace equation leads to the second-order ordinary differential equation relative to Φ:

$$\Phi_{zz} - 4\pi^2(u^2 + v^2)\Phi = 0 \quad (3.14)$$

The solution of Eq. (3.14) is expressed as

$$\Phi(u, v, z) = Ae^{2\pi z\sqrt{u^2+v^2}} + Be^{-2\pi z\sqrt{u^2+v^2}}$$

where A and B play the role of arbitrary constants that can be found from boundary conditions. As before, the potential has to go to zero at the infinite depth, so $B=0$:

$$\Phi(u, v, z) = A(u, v)e^{2\pi z\sqrt{u^2+v^2}} \quad (3.15)$$

The fluid velocities are then expressed as:

$$\begin{aligned}
\varphi_x &= i\mathcal{F}^{-1}\left(2\pi u A(u, v)e^{2\pi z\sqrt{u^2+v^2}}\right)\\
\varphi_y &= i\mathcal{F}^{-1}\left(2\pi v A(u, v)e^{2\pi z\sqrt{u^2+v^2}}\right)\\
\varphi_z &= \mathcal{F}^{-1}\left(2\pi\sqrt{u^2 + v^2}A(u, v)e^{2\pi z\sqrt{u^2+v^2}}\right)
\end{aligned} \quad (3.16)$$

One more differentiation of the velocities (3.16) and further substitution to the Laplace Eq. (3.5) turns the latter into the true equality.

To find the function $A(u, v)$, the solution (3.15) is substituted to the boundary condition in Eq. (3.5). Using a technique similar to the 2D case, the function $D(x, y, z)$ is defined as:

$$\mathcal{F}^{-1}\left(e^{2\pi z\sqrt{u^2+v^2}}\right) = D(x, y, z)$$

Replace u and v with $\sqrt{u^2+v^2}$ in inverse Fourier transforms of (3.16) to collect all transforms into one and apply forward Fourier transform to them. It can be done because:

- First, integration is done over positive wave numbers, so the sign of u and v is the same as the sign of $\sqrt{u^2+v^2}$.
- Second, the growth rate of exponent term of the integral kernel is much higher than that of u or $\sqrt{u^2+v^2}$ i.e.: $\sqrt{u^2+v^2}e^{2\pi(z\sqrt{u^2+v^2}+i(xu+yv))} \approx ue^{2\pi\pi(\sqrt{u^2+v^2}+i(xu+yv))}$ so the substitution has small effect on the magnitude of the solution.

Then,

$$A(u,v) = \frac{1}{2\pi\sqrt{u^2+v^2}\mathcal{F}(D(x,y,\zeta(x,y,t)))}\mathcal{F}\left(\frac{\zeta_t\sqrt{1+\zeta_x^2+\zeta_y^2}}{i(\zeta_x+\zeta_y)(1-\sqrt{1+\zeta_x^2+\zeta_y^2})-1}\right)$$

Finally, the potential is expressed as:

$$\begin{aligned}\varphi(x,y,z) &= \mathcal{F}^{-1}(\Phi)\\ &= \mathcal{F}^{-1}\left(\frac{e^{2\pi z\sqrt{u^2+v^2}}}{2\pi\sqrt{u^2+v^2}\mathcal{F}(D(x,y,\zeta(x,y,t)))}\mathcal{F}\left(\frac{\zeta_t\sqrt{1+\zeta_x^2+\zeta_y^2}}{i(\zeta_x+\zeta_y)(1-\sqrt{1+\zeta_x^2+\zeta_y^2})-1}\right)\right)\end{aligned} \tag{3.17}$$

Formulae (3.16) can be used to express the velocities. However, it is more computationally efficient to calculate the potential first and then get the derivatives using finite differences. Other computation aspects are considered in Gankevich and Degtyarev (2018). Once the potential field is available, the computation of hydrodynamic pressure is trivial.

3.5 Evaluation

The formula for the three-dimensional case was verified on the basis of the ARMA model against formula from linear wave theory. The ARMA model was used to generate short-crested waves using notional auto-covariance functions. The velocity potential field was computed using linear wave theory and formula (3.17).

The comparison showed that formula (3.17) gives the same field as the linear formula when angles of wave slope are assumed small in the kinematic boundary condition (3.5), i.e.

$$\zeta_t = -\varphi_z \quad on \quad z = \zeta(x,t)$$

When all the terms in the boundary condition (3.5) are retained, formula (3.17) gives a field with the same shape but slightly higher magnitude (19% in the considered case). The difference in amplitude depends on wave steepness, or, more precisely, the

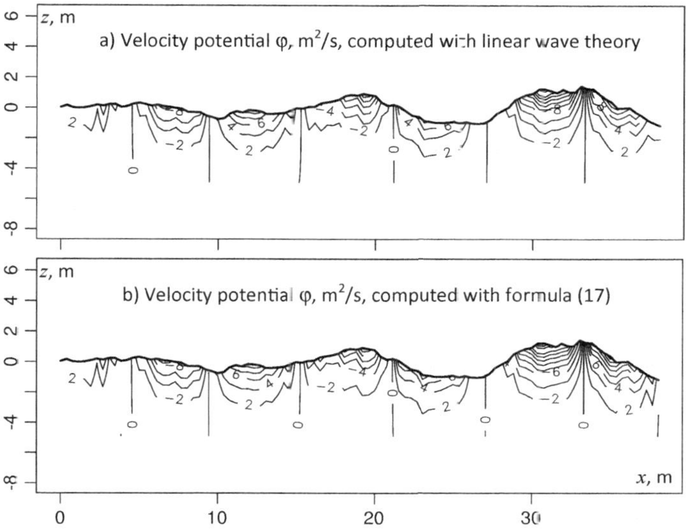

Fig. 3.2 Velocity potential field produced by linear wave theory (a) and formula (3.17) (b)

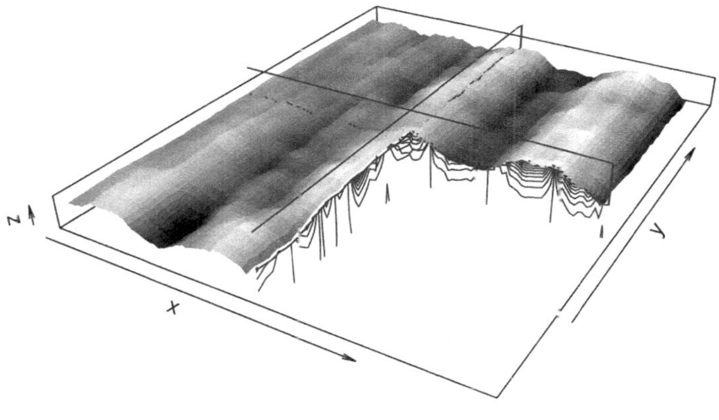

Fig. 3.3 Three-dimensional velocity potential field produced by formula (3.17)

values of spatial derivatives of the wave surface. Figure 3.2 shows the potential at an x-axis slice of the three-dimensional surface. Figure 3.3 shows the three-dimensional view of the wave surface along with velocity potential contours on both x- and y-axis slices.

3.6 Summary and Conclusions

Autoregressive/moving average model (ARMA) of sea waves may be seen as an attractive alternative to the traditional Longuett-Higgins model; it is computationally efficient and does not have limitations in terms of length of the record. However, the use of a wave model for numerical simulation of ship motions requires the hydrodynamic pressure field beneath the wave surface. The Longuett-Higgins model has this capability inherently as it is a solution of linear wave problem. To consider ARMA as a serious candidate for ship motion simulation, one needs to be able to compute those pressures efficiently.

ARMA provides a model of a moving random field. All of the physical properties of the waves are derived from temporal and spatial autocovariance functions. As was shown in Chap. 2, this is sufficient to create hydrodynamically valid wave surface. Thus, the ARMA surface can be considered a boundary condition for a potential flow problem. It is simpler that the wave problem in a hydrodynamic sense as the free surface in the kinematic boundary condition is given. It is known as Robin's problem in mathematical physics. The dynamic boundary condition is no longer necessary for the correct formulation of the problem.

One of the advantages of ARMA is that the model seamlessly propagates nonlinear properties reflected in the temporal and spatial autocovariance functions. Thus, the kinematic boundary condition has to be formulated for the normal velocity of liquid particle—without small angle assumption.

Fourier method is a convenient tool for the solution of the Laplace equation with numerically defined kinematic boundary condition. It provides a formula for potential and the velocities as quadratures, containing the direct and inverse Fourier transform. The formulae only require FFT for numerical evaluation. The chapter contains derivations of these formulae for both 2D and 3D cases.

Acknowledgements This work is partly supported by the project of St. Petersburg State University (id 28612502) and RFBR grants 16-07-00886, 17-29-04288.

References

Beck, R.F. and Reed, A.M. (2001). "Modern Computational Methods for Ships in Seaway", Trans. SNAME, Vol. 109 pp. 1–48.

Box, G. E. P., G. M. Jenkins & G. C. Reinsel (2008) Time series analysis: Forecasting and control, 4th Edition. Wiley, xx + 746 p.

Bukhanovsky, A.V., Degtyarev, A.B., Lopatukhin, L.I., Rozhkov, V.A. (1998) "Probabilistic modeling of sea wave climate". Izvestiya - Atmospheric and Ocean Physics. Vol 34, No 2, pp. 235–239.

Degtyarev, A., (2011) "New Approach to Wave Weather Scenarios Modeling" Chapter 34 of Contemporary Ideas on Ship Stability Neves, M.A.S., *et al.* (eds), Springer, ISBN 978-94-007-1481-6 pp. 599–617.

Degtyarev, A.B., Gankevich, I. (2012) "Evaluation of Hydrodynamic Pressures for Autoregression Model of Irregular Waves", Proc. 11th Int. Conf. Stability of Ships and Ocean Vehicles, Athens, Greece, 11 p.

Degtyarev, A.B., Reed, A.M. (2013) "Synoptic and short-term modeling of ocean waves" Intl. Shipbuilding Progress Vol. 60 pp. 523–553.

Gankevich, I., Degtyarev, A.B. (2015) "Computation of Pressures in Inverse Problem in Hydrodynamics of Potential Flow". Proc. 12th Int. Conf. Stability of Ships and Ocean Vehicles, Glasgow, UK, 5 p.

Gankevich, I., Degtyarev, A. (2018) Simulation of standing and propagating sea waves with three-dimensional ARMA model. In: Velarde M., Tarakanov R., Marchenko A. (eds) The Ocean in Motion. Springer Oceanography. Springer, Cham, https://doi.org/10.1007/978-3-319-71934-4_18.

Kochin, N. E., Kibel, I. A., Roze N. V., (1964) Theoretical Hydromechanics. Wiley 577 p.

Longuet-Higgins, M. S. (1962) The statistical analysis of a random, moving surface. Phil. Trans. Royal Soc. London. Series A, Mathematical and Physical Sciences, Vol. 249, No. 966, pp. 321–387.

Newman J.N. (1977) Marine Hydrodynamics, The MIT Press.

Spanos, P. D. (1983) ARMA Algorithms for Ocean Wave Modeling. J. Energy Resources Technology, Trans. ASME, Vol. 105 pp. 300–309.

Spanos P. D., Zeldin B. A. (1996) "Efficient iterative ARMA approximation of multivariate random processes for structural dynamics applications" Earthquake engineering and structural dynamics. Vol. 25, no. 5. pp. 497–507.

Weems, K., Reed, A.M., Degtyarev, A.B., Gankevich, I. (2016) "Implementation of an Autoregressive Wave Model in a Numerical Simulation Code," Proc. 31st Symp. on Naval Hydrodynamics, Monterey, California, 12p.

Zachmanoglou E.C., Thoe D.W. (1976) Introduction to Partial Differential Equations with Applications, Williams & Wilkins, Baltimore.

Part III
Mathematical Model of Ship Motions in Waves: Consideration of Forces

Chapter 4
Application of Computing Hydrodynamic Forces and Moments on a Vessel Without Bernoulli's Equation

Arthur M. Reed and John G. Telste

Abstract Traditionally the hydrodynamic force on a ship's hull is obtained by integrating the pressure over the hull, using Bernoulli's equation to compute the pressures. Due the need to evaluate Φ_t, Φ_x, Φ_y, Φ_z at every instant in time, this becomes a computational challenge when one wishes to know the hydrodynamic forces (and moments) on the instantaneous wetted surface of a vessel in extreme seas. A methodology that converts the integration of the pressure over the hull surface into an impulse, the time derivative of several integrals of the velocity potential over the surface of the vessel and possibly the free surface near the vessel is introduced. Some examples of applying the impulsive theory to 2- and 3-dimensional bodies are presented.

4.1 Introduction

Prof. Paul Sclavounos has developed a nonlinear slender-body model for the treatment of the potential flow problem governing the responses of a vessel in steep random waves (Sclavounos 2012; Sclavounos and Lee 2012; Sclavounos et al. 2019). Boundary value problems have been derived for the disturbance radiation and diffraction velocity potentials relative to the ship-fixed coordinate system. The evaluation of the sectional force (and moment) distributions based on the solution of these potential flow sectional boundary value problems is the subject of the present chapter. A sectional force method treats as unknown the sectional force distribution along the ship length as opposed to the local pressure, which is natural within a slender-body

This chapter is the union of two papers presented at a Stability Workshop (Reed & Telste 2011) and a STAB Conference (Reed 2012).

A. M. Reed (✉) · J. G. Telste
David Taylor Model Basin (NSWCCD), West Bethesda, MD, USA
e-mail: arthur.reed@navy.mil

V. L. Belenky et al. (eds.), *Contemporary Ideas on Ship Stability*, Fluid Mechanics and Its Applications 119, https://doi.org/10.1007/978-3-030-00516-0_4

framework. Combined with additive viscous models, these sectional force models lead to the evaluation of the integrated forces and moments which are input to the vessel nonlinear equations of motion.

4.2 Background

Traditionally, the derivation of correct sectional force distributions has played a central role in slender-body theory of aerodynamics and hydrodynamics. The direct application of Bernoulli's equation is complicated by a number of facts. The first is the need to evaluate gradients of the velocity potential, which may be a delicate computational task within a panel method. The second is the proper treatment of the longitudinal gradients of the ambient and disturbance potentials, which may not be possible to ignore in light of the slenderness approximations. A third fact which arises in connection with the present nonlinear time-domain slender-body theory is the proper interpretation of time derivatives with respect to the ship-fixed coordinate system and their careful treatment in the vicinity of the free-surface ship-hull intersection.

These complications with the direct application of Bernoulli's equation within a slender-body theory are mitigated if the integrated sectional forces are instead evaluated by the proper application of the momentum conservation theorem. This approach has several merits that have been taken into account in the development of strip theory and subsequent linear and nonlinear slender-body theories. Drawing upon the work of Lighthill (1960) and Newman and Wu (1973) on the swimming of slender fish, expressions can be derived for the sectional force distributions which are simple functions of the sectional integrals of the velocity potential. This important result circumvents in an elegant and robust manner the need to interpret the longitudinal convective terms in Bernoulli's equation. Moreover, the presence of a sectional integral of the velocity potential in the force expression suggest that this is the fundamental quantity needed for the evaluation of the sectional and total forces, as opposed to the local values of the pressure or velocity potential. This in turn may lead to simple—or even analytical—expressions for the sectional force distributions within a slender-body framework in a number of settings. Finally, this sectional force formulation allows for a simple and robust interpretation of time derivatives when the sectional wetted surface is time dependent as the vessel sections move in and out of the free surface.

4.3 The Boundary Value Problem and Its Decomposition

Let us assume as an earth-fixed reference a right-handed coordinate system (X, Y, Z) and a ship-fixed right-handed coordinate system (x, y, z) centered at an arbitrary point B with the xy-plane parallel to the calm water surface $Z = 0$ when the ship is

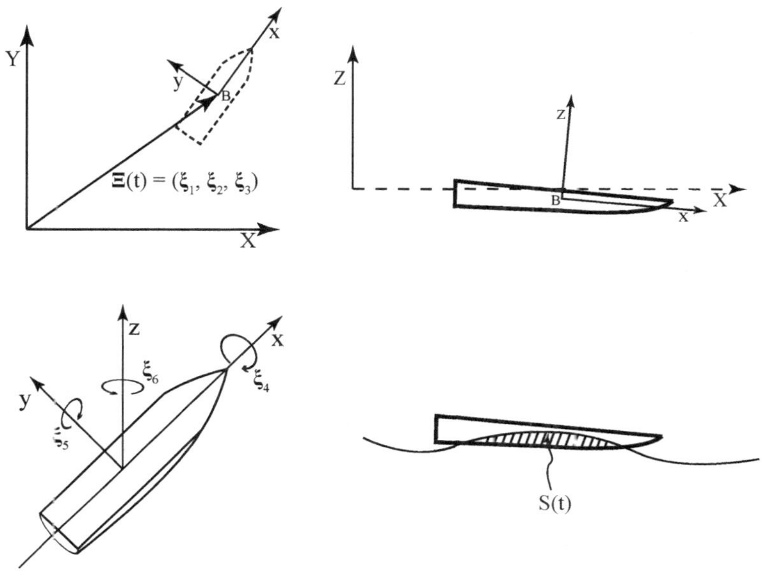

Fig. 4.1 Coordinate systems for the nonlinear ship response problem

at rest (Fig. 4.1). The ship position in space is completely defined by the rectilinear displacement vector $\Xi_B(t) = \xi_1(t)\boldsymbol{i} + \xi_2(t)\boldsymbol{j} + \xi_3(t)\boldsymbol{k}$ from the origin of the earth-fixed coordinate system to the origin of the ship-fixed coordinate system and the Euler angles defined in the order $[\xi_6(t), \xi_5(t), \xi_4(t)]$.

The free surface is assumed to be a single-valued function of the horizontal coordinates X and Y. Surface tension is negligible. The fluid is assumed to be homogeneous, incompressible, and frictionless. The fluid flow is assumed to be irrotational. These conditions are sufficient to guarantee the existence of a velocity potential.

4.3.1 The Total Velocity Potential

In the fluid surrounding the ship, the total velocity potential is Φ. It satisfies the Laplace equation

$$\nabla^2\Phi = 0$$

within the fluid domain bounded by the free-surface $Z = \zeta(X, Y, t)$ and the hull of the ship. The total potential satisfies at least a linear free-surface boundary condition on $Z = \zeta(X, Y, t)$.[1]

[1] As the free-surface boundary condition is not used in the development of the momentum theory for the force, the specific free-surface boundary condition chosen is not important.

4.3.1.1 Hull Boundary Condition for the Total Potential

Since the hull boundary condition is derived with vectors defined to be independent of frames of reference or coordinate systems, the appropriate boundary condition expressed in terms of either the earth-fixed or ship-fixed frame of reference is obtained from the components of vector equations.

4.3.1.2 Velocity of Points on the Hull

To obtain the hull boundary condition satisfied by Φ, we first consider a point fixed on the hull (fixed in the ship-fixed frame of reference). It has ship-fixed coordinates x, y, z and earth-fixed coordinates X, Y, Z. The vectors $\boldsymbol{x}$ and $\boldsymbol{X}$ from the origins of the ship-fixed and earth-fixed coordinate systems to the point, respectively, satisfy the equations

$$\begin{aligned}\boldsymbol{X} &= X\boldsymbol{i} + Y\boldsymbol{j} + Z\boldsymbol{k} \\ \boldsymbol{x} &= x\bar{\boldsymbol{i}} + y\bar{\boldsymbol{j}} + z\bar{\boldsymbol{k}} \\ \boldsymbol{x} &= \boldsymbol{X} - \Xi_B(t)\end{aligned}$$

where $\Xi_B(t) = \xi_1\boldsymbol{i} + \xi_2\boldsymbol{j} + \xi_3\boldsymbol{k}$ is the vector from the origin of the earth-fixed coordinate system to the origin of the ship-fixed coordinate system. Since the point is fixed on the hull, x, y, and z are independent of time. Consequently, the ship-fixed time derivative of $\boldsymbol{x}$ vanishes. We have

$$\boldsymbol{0} = \frac{d^*\boldsymbol{x}}{dt} = \frac{d\boldsymbol{X}}{dt} - \frac{d\,\Xi_B}{dt} - \boldsymbol{\Omega} \times \boldsymbol{x}$$

where d^*/dt and d/dt operating on a vector obtain the ship-fixed and earth-fixed time derivatives of the vector, respectively. The velocity $\boldsymbol{v}_{SHIP}$ of the point on the hull is then

$$\boldsymbol{v}_{SHIP} = \frac{d\boldsymbol{X}}{dt} = \frac{d\,\Xi_B}{dt} + \boldsymbol{\Omega} \times \boldsymbol{x}. \tag{4.1}$$

If the point slides along the hull surface, it is not fixed in the ship-fixed frame of reference and $d^*\boldsymbol{x}/dt \neq \boldsymbol{0}$. However, it is true that $\boldsymbol{n} \cdot d^*\boldsymbol{x}/dt = 0$ so that

$$\boldsymbol{n} \cdot \frac{d\boldsymbol{X}}{dt} = \boldsymbol{n} \cdot \left(\frac{d\,\Xi_B}{dt} + \boldsymbol{\Omega} \times \boldsymbol{x}\right) \boldsymbol{n} \cdot \boldsymbol{v}_{SHIP}$$

where $\boldsymbol{v}_{SHIP}$ is the velocity of a fixed point on the hull coinciding with the position of the sliding point at time t.

4.3.1.3 Equation for the Hull Surface

The hull surface $S_B(t)$ is rigid and therefore independent of time in the ship-fixed frame of reference. Points on the surface are those points whose ship-fixed coordinates x, y, z satisfy a mathematical equation of the form

$$h(x, y, z) = 0.$$

The unit normal $\boldsymbol{n}$ on the hull surface is defined by the equation

$$\boldsymbol{n} = \frac{\nabla h}{|\nabla h|}.$$

The gradient points in the direction of maximum increase of h. For that reason it is stipulated that $h(x, y, z) > 0$ for points inside the hull with ship-fixed coordinates x, y, z and $h(x, y, z) < 0$ for points outside the hull. Then $\boldsymbol{n}$ is guaranteed to point into the hull.

4.3.1.4 Hull Boundary Condition

To obtain the hull boundary condition, we now consider an arbitrary point with ship-fixed coordinates x, y, z and earth-fixed coordinates X, Y, Z. The point moves and traces out a smooth trajectory so that both the ship-fixed and earth-fixed coordinates are functions of time. The derivative of h following the point is

$$\begin{aligned} \frac{Dh}{Dt} &= \frac{\partial h}{\partial x}\frac{dx}{dt} + \frac{\partial h}{\partial y}\frac{dy}{dt} + \frac{\partial h}{\partial z}\frac{dz}{dt} \\ &= \nabla h \cdot \frac{d^* \boldsymbol{x}}{dt} \\ &= \nabla h \cdot \left(\frac{d\boldsymbol{X}}{dt} - \frac{d\Xi_B}{dt} - \boldsymbol{\Omega} \times \boldsymbol{x} \right). \end{aligned}$$

If the point is a fluid particle sliding along the surface of the hull, then $Dh/Dt = 0$, $d\boldsymbol{X}/dt = \nabla \Phi$, and

$$0 = \boldsymbol{n} \cdot \left(\nabla \Phi - \frac{d\Xi_B}{dt} - \boldsymbol{\Omega} \times \boldsymbol{x} \right).$$

The hull surface boundary condition requires that the normal velocity of a fluid particle on the hull surface match the normal velocity of the hull:

$$\boldsymbol{n} \cdot \nabla \Phi = \boldsymbol{n} \cdot \left(\frac{d\Xi_B}{dt} + \boldsymbol{\Omega} \times \boldsymbol{x} \right) \boldsymbol{n} \cdot \boldsymbol{v}_{SHIP}$$

where $\boldsymbol{v}_{SHIP}$ is given by (4.1).

4.3.2 Incident Wave Potential

In the absence of a body, the velocity potential would have been the ambient velocity potentialϕ_I, which satisfies the Laplace equation

$$\nabla^2\phi_I = 0$$

in the fluid below the free-surface elevation $Z = \zeta_I(X, Y, t)$.

The ambient wave velocity potential ϕ_I is assumed to satisfy the same free-surface boundary condition as the total velocity potential, but on $Z = \zeta_I(X, Y, t)$. It is also assumed that the incident wave elevation differs little from the total wave elevation except possibly near the vessel.

4.3.3 The Disturbance Velocity Potential

When $\zeta_I < \zeta$, it is assumed that ϕ_I can be analytically continued above $Z = \zeta_I$ to define a continuation everywhere outside the hull in the fluid below the free-surface elevation $Z = \zeta$. Then a disturbance potential ϕ_D is defined everywhere in this domain according to the equation

$$\phi_D = \Phi - \phi_I.$$

The difference in wave elevation between the total wave elevation around the hull and the ambient wave potential that would have existed in the absence of the ship is ζ_D. It obviously satisfies the equation

$$\zeta_D = \zeta - \zeta_I.$$

The free-surface boundary condition for the disturbance potential potential is derived from that of the total velocity potential, substituting $\phi_I + \phi_D$ and $\zeta_I + \zeta_D$ for Φ and ζ in the total velocity potential free-surface boundary condition and linearizing in ϕ_D and ζ_D.

4.3.3.1 Hull Boundary Condition for the Disturbance Potential

Using the assumed decomposition of the total potential as the sum of the incident wave potential and a disturbance potential, we obtain the equation

$$\boldsymbol{n} \cdot \nabla\phi_D = \boldsymbol{n} \cdot \boldsymbol{v}_{SHIP} - \boldsymbol{n} \cdot \nabla\phi_I.$$

where $\boldsymbol{v}_{SHIP}$ is given by (4.1).

4.3.3.2 Boundary Condition at Infinity for the Disturbance Potential

At infinity, the velocity due to the disturbance velocity potential approaches zero, and the free-surface waves generated by the interaction of the ship with the ambient waves radiate outward. This is the radiation boundary condition.

4.4 The Fluid Force on the Vessel

The purely three-dimensional case of a vessel oscillating in six degrees of freedom in steep ambient waves is considered. The most general fully nonlinear problem is formulated first leading to the treatment of special cases. The sectional force is evaluated first relative to the inertial frame and next relative to the ship-fixed coordinate system.

Figure 4.2 illustrates a 3D vessel undergoing rectilinear and rotational displacements in steep ambient waves. The fully nonlinear free-surface elevation $\zeta(t)$ is the sum of the ambient wave elevation and the disturbance caused by the vessel displacement and the corresponding total free surface is denoted by $S_F^T(t)$. The nonlinear wave elevation of the ambient wave alone is $\zeta_I(t)$ and the corresponding free-surface elevation is denoted by $S_F^I(t)$. The difference between the two free-surface elevations is assumed to be finite. Yet, this difference is expected to be small, except perhaps near the waterline. This assumption is essential for the derivation of an approximate form of the three-dimensional force acting on the vessel using the momentum theorem developed below. The fluid in the volume V bounded by the wetted surface S_B^T of the hull, the free surface S_F^T, and a control surface S_∞^T is considered:

$$S^T = S_B^T + S_F^T + S_\infty^T.$$

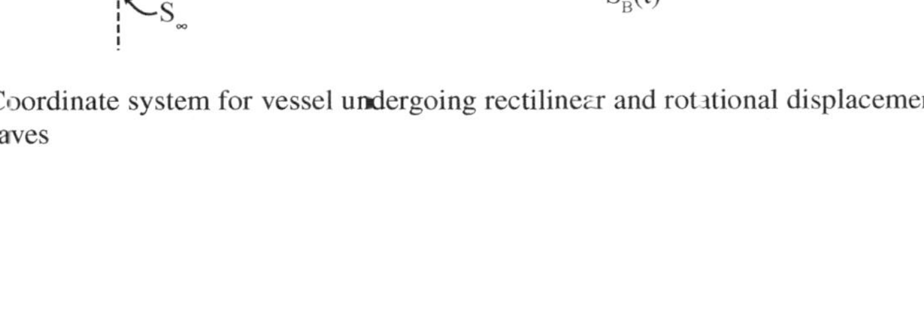

Fig. 4.2 Coordinate system for vessel undergoing rectilinear and rotational displacement in steep ambient waves

The control surface is fixed with respect to the earth-fixed coordinate system. It is also bounded until the end when the surface is moved to infinity in all directions. The rate of change of the fluid momentum in the volume is

$$\boldsymbol{F}^{\text{fluid}} = \rho \frac{d}{dt} \iiint\limits_{V} dV\, \nabla \Phi = \rho \frac{d}{dt} \oiint\limits_{S^T} dS\, \Phi \boldsymbol{n} \tag{4.2}$$

where Gauss' theorem has been used to convert the volume integral to a surface integral. (Here, partial derivatives with respect to time are earth-fixed where the earth-fixed coordinates X, Y, Z of a point in space are fixed.) According to the transport theorem (Serrin 1959; Smirnov 1964, §120), the rate of change of momentum is also given by the equation

$$\begin{aligned} \boldsymbol{F}^{\text{fluid}} &= \rho \iiint\limits_{V} dV\, \nabla \frac{\partial \Phi}{\partial t} + \rho \oiint\limits_{S^T} dS\, U_n \nabla \Phi \\ &= \rho \oiint\limits_{S^T} dS\, \frac{\partial \Phi}{\partial t} \boldsymbol{n} + \rho \oiint\limits_{S^T} dS U_n \nabla \Phi \end{aligned}$$

where $U_n = \boldsymbol{n} \cdot \boldsymbol{U}$ is the outward normal component of the velocity $\boldsymbol{U}$ of the surface S^T. Putting these results together, we obtain the equation

$$\begin{aligned} \rho \frac{d}{dt} \oiint\limits_{S^T} dS\, \Phi \boldsymbol{n} &= \rho \oiint\limits_{S^T} dS\, \frac{\partial \Phi}{\partial t} \boldsymbol{n} + \rho \oiint\limits_{S^T} dS U_n \nabla \Phi \\ &= \rho \oiint\limits_{S^T} dS \left(\frac{\partial \Phi}{\partial t} + \frac{1}{2} \nabla \Phi \cdot \nabla \Phi \right) \boldsymbol{n} + \rho \oiint\limits_{S^T} dS \left(U_n \nabla \Phi - \frac{1}{2} \nabla \Phi \cdot \nabla \Phi \boldsymbol{n} \right) \\ &= \rho \oiint\limits_{S^T} dS \left(\frac{\partial \Phi}{\partial t} + \frac{1}{2} \nabla \Phi \cdot \nabla \Phi \right) \boldsymbol{n} + \rho \oiint\limits_{S^T} dS \left(U_n - \frac{\partial \Phi}{\partial n} \right) \nabla \Phi \\ &= \rho \oiint\limits_{S^T} dS \left(\frac{\partial \Phi}{\partial t} + \frac{1}{2} \nabla \Phi \cdot \nabla \Phi \right) \boldsymbol{n} - \rho \iint\limits_{S^T_\infty} dS \frac{\partial \Phi}{\partial n} \nabla \Phi. \end{aligned} \tag{4.3}$$

The third equality is obtained by using Newman's identity

$$\oiint\limits_{S} dS \left[\nabla \varphi \frac{\partial \varphi}{\partial n} - \frac{1}{2} \nabla \varphi \cdot \nabla \varphi \boldsymbol{n} \right] = 0$$

which holds for any velocity potential φ within a volume enclosed by a surface S (Newman 1977, p. 134, Eq. 89). The last equality in (4.3) is obtained from the equations $U_n = 0$ on S^T_∞ and $U_n = \partial \Phi / \partial n$ on S^T_F and S^T_B. The total fluid force $\boldsymbol{F}_{\text{TOT}}$

acting on the body is the integral of $p\boldsymbol{n}$ over the wetted surface of the hull. Thus we obtain the equation

$$\begin{aligned}\boldsymbol{F}_{\mathrm{TOT}} &= -\rho \iint\limits_{S_B^T} dS \left(\frac{\partial \Phi}{\partial t} + \frac{1}{2}\nabla\Phi \cdot \nabla\Phi + gZ \right) \boldsymbol{n} \\ &= -\rho \frac{d}{dt} \oiint\limits_{S^T} dS\, \Phi \boldsymbol{n} - \rho \iint\limits_{S_\infty^T} dS \frac{\partial \Phi}{\partial n} \nabla \Phi \\ &\quad - \rho g \iint\limits_{S_B^T + S_F^T} dS\, Z \boldsymbol{n} + \rho \iint\limits_{S_\infty^T} dS \left(\frac{\partial \Phi}{\partial t} + \frac{1}{2}\nabla\Phi \cdot \nabla\Phi \right) \boldsymbol{n}\end{aligned} \tag{4.4}$$

for the total fluid force acting on the body. The fact that the pressure vanishes on S_F^T has been used to obtain this equation.

Equation (4.4) is an important intermediate result which was derived without invoking any approximations. It accomplishes one of the objectives of the momentum formulation, namely to reduce the definition of the force by pressure integration into integrals that are much easier to evaluate or further reduce as indicated below. The superscript T has been used to indicate surfaces for the total nonlinear problem.

The fluid that would have existed inside the volume bounded by the ambient wave free surface S_F^I and the control surface S_∞^I, if the ship had not disturbed the water, is now considered. The surface S_∞^I is slightly different from S_∞^T only due to the difference between the ambient wave elevation $\zeta_I(t)$ and the total nonlinear wave elevation $\zeta(t)$. The total bounding surface is S^I where

$$S^I = S_F^I + S_\infty^I.$$

Just as was done for the fluid in the volume V outside the hull below the surface S_F^T, one can consider the rate of change of the fluid momentum inside the volume bounded by S^I. It can be obtained from Eqs. (4.2) to (4.3) by letting the hull shrink to infinitesimal size. The integrals over S_B^T then vanish, S_F^T becomes S_F^I and S_∞^T becomes S_∞^I. The force acting on the vanishingly small ship is zero and is given by either side of the equation

$$\begin{aligned}0 &= -\rho \frac{d}{dt} \oiint\limits_{S^I} dS\, \phi_I \boldsymbol{n} - \rho \iint\limits_{S_\infty^I} dS \frac{\partial \phi_I}{\partial n} \nabla \phi_I - \rho g \iint\limits_{S_F^I} dS\, Z \boldsymbol{n} \\ &\quad + \rho \iint\limits_{S_\infty^I} dS \left(\frac{\partial \phi_I}{\partial t} + \frac{1}{2}\nabla\phi_I \cdot \nabla\phi_I \right) \boldsymbol{n}.\end{aligned} \tag{4.5}$$

Equation (4.5) is subtracted from (4.4) to obtain the equation

$$\begin{aligned}
\boldsymbol{F}_{\mathrm{TOT}} &= -\rho \iint\limits_{S_B^T} dS \left(\frac{\partial \Phi}{\partial t} + \frac{1}{2} \nabla \Phi \cdot \nabla \Phi + gZ \right) \boldsymbol{n} \\
&= -\rho \frac{d}{dt} \iint\limits_{S_B^T} dS\, \Phi \boldsymbol{n} - \rho g \iint\limits_{S_B^T} dS\, Z \boldsymbol{n} - \rho \frac{d}{dt} \left[\iint\limits_{S_F^T} dS\, \Phi \boldsymbol{n} - \iint\limits_{S_F^I} dS\, \phi_I \boldsymbol{n} \right] \\
&\quad - \rho g \left[\iint\limits_{S_F^T} dS\, Z \boldsymbol{n} - \iint\limits_{S_F^I} dS\, Z \boldsymbol{n} \right] - \rho \frac{d}{dt} \left[\iint\limits_{S_\infty^T} dS\, \Phi \boldsymbol{n} - \iint\limits_{S_\infty^I} dS\, \phi_I \boldsymbol{n} \right] \\
&\quad - \rho \left[\iint\limits_{S_\infty^T} dS\, \frac{\partial \Phi}{\partial n} \nabla \Phi - \iint\limits_{S_\infty^I} dS\, \frac{\partial \phi_I}{\partial n} \nabla \phi_I \right] \\
&\quad + \rho \left[\iint\limits_{S_\infty^T} dS \left(\frac{\partial \Phi}{\partial t} + \frac{1}{2} \nabla \Phi \cdot \nabla \Phi \right) \boldsymbol{n} \right. \\
&\quad \left. - \iint\limits_{S_\infty^I} dS \left(\frac{\partial \phi_I}{\partial t} - + \frac{1}{2} \nabla \phi_I \cdot \nabla \phi_I \right) \boldsymbol{n} \right].
\end{aligned}$$

It is argued that the sums of the terms within the last three pairs of square brackets are negligibly small when the control surfaces are moved infinitely far away from the ship. The force $\boldsymbol{F}_{\mathrm{DYN}}$ acting on the body due to the dynamic pressure is

$$\begin{aligned}
\boldsymbol{F}_{\mathrm{DYN}} &= -\rho \iint\limits_{S_B^T} dS \left(\frac{\partial \Phi}{\partial t} + \frac{1}{2} \nabla \Phi \cdot \nabla \Phi \right) \boldsymbol{n} \\
&\simeq -\rho \frac{d}{dt} \iint\limits_{S_B^T} dS\, \Phi \boldsymbol{n} - \rho \frac{d}{dt} \left[\iint\limits_{S_F^T} dS\, \Phi \boldsymbol{n} - \iint\limits_{S_E^I} dS\, \phi_I \boldsymbol{n} \right] \\
&\quad - \rho g \left[\iint\limits_{S_F^T} dS\, Z \boldsymbol{n} - \iint\limits_{S_E^I} dS\, Z \boldsymbol{n} \right] + \rho \frac{d}{dt} \iint\limits_{S_W^I} dS\, \phi_I \boldsymbol{n} + \rho g \iint\limits_{S_W^I} dS\, Z \boldsymbol{n}.
\end{aligned}$$

Here S_E^I is the portion of S_F^I that is outside the hull and S_W^I is the portion of S_F^I that is inside the hull:

$$S_F^I = S_E^I + S_W^I.$$

The functions ϕ_I and Φ are continued analytically about $Z = \zeta_I$ and $Z = \zeta$, respectively, so that the function ϕ_I is defined for $Z \leq \zeta$ and Φ is defined for $Z \leq \zeta_I$. Then Φ may be expanded about the ambient free-surface elevation $Z\zeta_I$. The dynamic force satisfies the approximation

$$\begin{aligned} \boldsymbol{F}_{\text{DYN}} &= -\rho \iint\limits_{S_B^T} dS \left(\frac{\partial \Phi}{\partial t} + \frac{1}{2} \nabla \Phi \cdot \nabla \Phi \right) \boldsymbol{n} \\ &\simeq -\rho \frac{d}{dt} \iint\limits_{S_B^T} dS\, \Phi \boldsymbol{n} - \rho \frac{d}{dt} \iint\limits_{S_E^I} dS\, \phi_D \boldsymbol{n} - \rho g \iint\limits_{S_E^I} dS\, \zeta_D \boldsymbol{n} \\ &\quad + \rho \frac{d}{dt} \iint\limits_{S_W^I} dS\, \phi_I \boldsymbol{n} + \rho g \iint\limits_{S_W^I} dS\, Z \boldsymbol{n} \end{aligned} \tag{4.6}$$

where $\boldsymbol{n}$ points into the body on S_B^T and upward ($\boldsymbol{n} \cdot \boldsymbol{k} > 0$) on S_E^I and S_W^I.

We now follow the steps taken in considering the rate of change of the momentum in the fluid outside the hull in Eqs. (4.2) and (4.3). However, this time we consider the rate of change of the momentum of the fluid inside the volume bounded by the surface S_B^I and the ambient free surface S_W^I that would have been the case if the ship had not disturbed the fluid. The surface S_B^I is the part of the hull surface that lies below the ambient free surface $Z = \zeta_F^I(t)$. The bounding surface S^{INT} is now the disjoint sum of the hull surface S_B^I and the nonlinear waterline S_W^I:

$$S^{\text{INT}} = S_B^I + S_W^I.$$

In this case, the velocity potential is ϕ_I. Since $U_n = \boldsymbol{n} \cdot \boldsymbol{U}$ and $\partial/\partial n = \boldsymbol{n} \cdot \nabla$, the final result given by (4.3) is unchanged if $\boldsymbol{n}$ is replaced by $-\boldsymbol{n}$. The normal is chosen to point into the volume enclosed by S^{INT} so that it matches the normal on S_B^T in previous equations. In (4.3), U_n is the same as $\partial \phi_I / \partial n$ on S_W^I. After rearranging terms, the equation corresponding to (4.3) is therefore

$$\begin{aligned} &\rho \oiint\limits_{S^{\text{INT}}} dS \left(\frac{\partial \phi_I}{\partial t} + \frac{1}{2} \nabla \phi_I \cdot \nabla \phi_I \right) \boldsymbol{n}' \\ &= \rho \iint\limits_{S_B^I} dS \left(\frac{\partial \phi_I}{\partial t} + \frac{1}{2} \nabla \phi_I \cdot \nabla \phi_I \right) \boldsymbol{n}' - \rho g \iint\limits_{S_W^I} dS\, Z \boldsymbol{n}' \\ &= \rho \frac{d}{dt} \oiint\limits_{S^{\text{INT}}} dS\, \phi_I \boldsymbol{n}' + \rho \iint\limits_{S^{\text{INT}}} dS \left(\nabla \phi_I \cdot \boldsymbol{n}' - \boldsymbol{U} \cdot \boldsymbol{n}' \right) \nabla \phi_I \\ &= \rho \frac{d}{dt} \oiint\limits_{S^{\text{INT}}} dS\, \phi_I \boldsymbol{n}' + \rho \iint\limits_{S_B^I} dS \left(\nabla \phi_I \cdot \boldsymbol{n}' - \boldsymbol{U} \cdot \boldsymbol{n}' \right) \nabla \phi_I \end{aligned} \tag{4.7}$$

where $\boldsymbol{n}'$ is an inward normal that points into the body on S_B^I and downward on S_W^I.

We now add (4.7) to (4.6) while accounting for the different meaning of $\boldsymbol{n}$ and $\boldsymbol{n}'$ on S_W^I in the two equations. The result is the disturbance force $\boldsymbol{F}_D$ given by the equation

$$\begin{aligned}\boldsymbol{F}_D &\simeq -\rho \iint\limits_{S_B^I} dS \left(\frac{\partial \phi_D}{\partial t} + \frac{1}{2}\nabla\phi_D \cdot \nabla\phi_D + \nabla\phi_D \cdot \nabla\phi_I\right)\boldsymbol{n} \\ &\simeq -\rho \frac{d}{dt} \iint\limits_{S_B^I} dS\, \phi_D \boldsymbol{n} - \rho \frac{d}{dt} \iint\limits_{S_E^I} dS\, \phi_D \boldsymbol{n} - \rho g \iint\limits_{S_E^I} dS\, \zeta_D \boldsymbol{n} \\ &\quad + \rho \iint\limits_{S_B^I} dS \left(\frac{\partial \phi_I}{\partial n} - U_n\right)\nabla\phi_I,\end{aligned}$$

which assumes that an integral over S_B^T is approximated well by an integral over S_B^I. This is the part of the dynamic force acting on the body that depends on ϕ_D. The part that depends on ϕ_I but not on ϕ_D is obtained from (4.7):

$$\begin{aligned}\boldsymbol{F}_{F\text{-}K} = &-\rho \iint\limits_{S_B^I} dS \left(\frac{\partial \phi_I}{\partial t} + \frac{1}{2}\nabla\phi_I \cdot \nabla\phi_I\right)\boldsymbol{n} - \rho \frac{d}{dt} \iint\limits_{S_B^I} dS\, \phi_I \boldsymbol{n} \\ &+ \rho \frac{d}{dt} \iint\limits_{S_W^I} dS\, \phi_I \boldsymbol{n} - \rho \iint\limits_{S_B^I} dS \left(\frac{\partial \phi_I}{\partial n} - U_n\right)\nabla\phi_I \\ &+ \rho g \iint\limits_{S_W^I} dS\, Z \boldsymbol{n}\end{aligned} \tag{4.8}$$

where $\boldsymbol{n}$ points into the body on S_B^I and upward on S_W^I. The sum of the nonlinear Froude-Krylov and disturbance forces is

$$\begin{aligned}\boldsymbol{F} &= -\rho \iint\limits_{S_B} dS \left(\frac{\partial \Phi}{\partial t} + \frac{1}{2}\nabla\Phi \cdot \nabla\Phi\right)\boldsymbol{n} \\ &\simeq -\rho \frac{d}{dt} \iint\limits_{S_B^I} dS\, \phi_D \boldsymbol{n} - \rho \frac{d}{dt} \oiint\limits_{S_B^I + S_W^I} dS\, \phi_I \boldsymbol{n}' - \rho g \iint\limits_{S_W^I} dS\, Z \boldsymbol{n}' \\ &\quad - \rho \frac{d}{dt} \iint\limits_{S_E^I} dS\, \phi_D \boldsymbol{n} - \rho g \iint\limits_{S_E^I} dS\, \zeta_D \boldsymbol{n}\end{aligned} \tag{4.9}$$

where the unit normal $\boldsymbol{n}'$ points into the body on S_B^I and downward on S_W^I.

The total force acting on the vessel may be obtained by adding the force due to the hydrostatic pressure in (4.9) as shown in the equation

$$
\begin{aligned}
\boldsymbol{F}_{\mathrm{TOT}} &= -\rho \iint_{S_B} dS \left(\frac{\partial \Phi}{\partial t} + \frac{1}{2} \nabla \Phi \cdot \nabla \Phi + gZ \right) \boldsymbol{n} \\
&\simeq -\rho \frac{d}{dt} \iint_{S_B^I} dS\, \phi_D \boldsymbol{n} - \rho \frac{d}{dt} \oiint_{S_B^I + S_W^I} dS\, \phi_I \boldsymbol{n}' - \rho g \oiint_{S_B^I + S_W^I} dS\, Z \boldsymbol{n}' \\
&\quad - \rho \frac{d}{dt} \iint_{S_E^I} dS\, \phi_D \boldsymbol{n} - \rho g \iint_{S_E^I} dS\, \zeta_D \boldsymbol{n},
\end{aligned}
\tag{4.10}
$$

where $\boldsymbol{n}'$ points downward on S_W^I and into the body on S_B^I. Three force components may be identified in (4.10).

4.4.1 Nonlinear Buoyancy Force

Applying the Gauss divergence theorem to the third term on the right side of (4.10), we obtain

$$
\widetilde{\boldsymbol{F}}_H = -\rho g \oiint_{S_B^I + S_W^I} dS\, Z \boldsymbol{n}' = \rho g \nabla(t) \boldsymbol{k} \tag{4.11}
$$

where $\boldsymbol{n}'$ points into the enclosed volume. The nonlinear hydrostatic force given by (4.11) acts in the vertical direction and on the volume of fluid enclosed by the ship wetted surface and the ambient wave surface interior to the vessel. This buoyancy force which results from the application of the momentum theorem differs from the conventional hydrostatic force that acts on the open wetted surface of the body and which in the nonlinear problem may not point in the vertical direction.

4.4.2 Momentum Froude-Krylov Force

The momentum Froude-Krylov force is the time derivative of the impulse integral involving just the ambient wave potential over the instantaneous ship surface:

$$
\widetilde{\boldsymbol{F}}_{\text{F-K}} = -\rho \frac{d}{dt} \oiint_{S_B^I(t) + S_W^I(t)} dS\, \phi_I \boldsymbol{n}'. \tag{4.12}
$$

Again $\boldsymbol{n}'$ points into the enclosed volume. The momentum Froude-Krylov force given by (4.12) differs from the conventional Froude-Krylov force $\boldsymbol{F}_{\text{F-K}}$ which involves the integral of the hydrodynamic pressure due to the ambient wave over the instantaneous ship wetted surface. Although a different force, expression (4.12) is simpler to evaluate numerically since it does not involve the time derivative and spatial gradients of the ambient velocity potential under the integral sign.

4.4.3 Momentum Radiation and Diffraction Force

The momentum disturbance force has a similar form to its Froude-Krylov counterpart and involves the disturbance radiation and diffraction velocity potentials under the integral sign in the definition of the corresponding impulse

$$\widetilde{\boldsymbol{F}}_D = -\rho \frac{d}{dt} \iint\limits_{S_B^I(t)} dS\, (\phi_{\text{RAD}} + \phi_{\text{DIF}})\boldsymbol{n}. \tag{4.13}$$

An advantage of (4.13) relative to the conventional definition of the nonlinear radiation and diffraction forces is that no time derivative and spatial gradients of the disturbance potentials are present under the integral sign in the definition of the disturbance impulse. This is a significant advantage of (4.13) which may be readily evaluated robustly assuming knowledge of just the values of the disturbance velocity potentials over the instantaneous ship wetted surface.

4.4.4 Interpretation of Momentum Hydrostatic and Froude-Krylov Forces

The momentum formulation derived above decomposes the total ideal fluid force into three components which are interpreted as the Momentum Hydrostatic, Froude-Krylov (F-K) and Disturbance Forces.

There exists an interdependence between the hydrostatic and Froude-Krylov forces, the understanding of which in the nonlinear ship response problem is essential for the study of the vessel stability problem in steep waves. As pointed out by Telste and Belknap (2008) and Belknap and Telste (2008), the nonlinear hydrostatic and Froude-Krylov force may cancel each other out in certain wave conditions, underscoring the significance of the accurate evaluation of these forces and the remaining disturbance forces. The discussion below explains how such a cancellation occurs.

4.4.4.1 The Momentum F-K Force

The derivation of the momentum hydrostatic and Froude-Krylov forces entailed no approximations in the use of Bernoulli's equation so they are considered exact, given an accurate representation of the kinematics of the ambient wave. The hydrostatic force always points upwards and its magnitude depends on the time dependent displaced volume of the vessel and is given by expression (4.11).

The nonlinear Froude-Krylov force given by (4.8) may be reduced further by adding and subtracting an integral over the nonlinear waterplane area of the vessel over the ambient wave free surface internal to the vessel:

$$\begin{aligned} \boldsymbol{F}_{\text{F-K}} &= -\rho \frac{d}{dt} \oiint\limits_{S_B^I(t)+S_W^I(t)} dS\, \phi_I \boldsymbol{n}' + \rho \frac{d}{dt} \iint\limits_{S_W^I(t)} dS\, \phi_I \boldsymbol{n}' \\ &= \widetilde{\boldsymbol{F}}_{\text{F-K}} + \rho \frac{d}{dt} \iint\limits_{S_W^I(t)} dS\, \phi_I \boldsymbol{n}'. \end{aligned} \tag{4.14}$$

Here $\boldsymbol{n}'$ points downward on S_W^I and into the body on S_B. In (4.14) the first integral is over a surface enclosing the time dependent volume of the vessel. The second integral is taken over the nonlinear waterplane area and will be seen to be the nonlinear extension of the Froude-Krylov hydrostatic-like restoring force acting on a floating vessel. For a submerged body this term vanishes. For a surface piercing body and in the limit of small amplitude waves which are long relative to the dimension of the vessel this term is proportional to the heave hydrostatic restoring coefficient $C_{33}\rho g A_W$, where A_W is the static waterplane area, times the ambient wave amplitude.

By applying Gauss's theorem, the first term may be reduced to a volume integral:

$$\boldsymbol{F}_{\text{F-K}} = \rho \frac{d}{dt} \iiint\limits_{\nabla(t)} dV\, \nabla\phi_I + \rho \frac{d}{dt} \iint\limits_{S_W^I(t)} dS\, \phi_I \boldsymbol{n}'. \tag{4.15}$$

The first term in (4.15) is the time rate of change of the linear momentum of all the fluid particles of an ambient wave enclosed by the time dependent volume of the vessel. In long waves the volume integral in (4.15) may to leading order be approximated by evaluating the ambient wave velocity vector at the centroid of the time dependent volume of the vessel. It is noted that the location of this centroid is time dependent.

The second integral in (4.15) has a familiar interpretation within linear theory. Recall that the linear dynamic free surface condition takes the form

$$\zeta_I = -\frac{1}{g}\left(\frac{\partial \phi_I}{\partial t}\right)_{Z=0}.$$

Substituting in (4.15), exchanging the time differentiations with the surface and volume integrations and taking into account that the unit vector points inside the volume we obtain the linearized version of the momentum Froude-Krylov force

$$\boldsymbol{F}_{\text{F-K, LINEAR}} = \rho \iiint_{\nabla} dV \, \frac{\partial}{\partial t} \nabla \phi_I + \rho g \boldsymbol{k} \iint_{S_W^I(t)} dS \, \zeta_I . \tag{4.16}$$

The first term in (4.16) is the inertia component of the momentum Froude-Krylov force which is equal to the integral of the acceleration of the ambient wave fluid particles within the linearized volume of the vessel below the calm water surface, multiplied by their density. The second term is the hydrostatic contribution which is proportional to the integral of the ambient wave elevation over the static waterplane area of the vessel. For long waves this integral may be approximated to leading order by the product of the waterplane area and the ambient wave elevation at the origin of the coordinate system. In this limiting case the hydrostatic component of the momentum Froude-Krylov force, per unit ambient wave elevation, reduces to the heave restoring coefficient which appears in the left hand side of the linearized vessel equations of motion. As expected, for submerged bodies the hydrostatic component of the momentum Froude-Krylov force vanishes.

In large amplitude waves the hydrostatic component of the Froude-Krylov force (4.15) may be comparable to the time dependent buoyancy force (4.11). Moreover, while the buoyancy force always points vertically upwards, the hydrostatic component of the Froude-Krylov force component has an oblique orientation which is a function of the inclination of the ambient wave surface contained in the unit normal vector. In the limit of linear theory this Froude-Krylov hydrostatic force points vertically upwards.

4.5 Application of Momentum Theory

In order to investigate the validity of the momentum theory based impulsive forces derived in Reed and Telste (2011), Sclavounos (2012) and Sclavounos and Lee (2012); and summarized above, the results of two studies will be presented. First there will be response-amplitude operators (RAO's) for a containership, and second there will be the results from a nonlinearity study for a two-dimensional circular cylinder that is compared with linear experimental results.

For these computations, a linearized version of the momentum theory is employed, where it is assumed the $-\rho \frac{d}{dt} \iint_{S_E^I} dS \, \phi_D \boldsymbol{n}$ in (4.10) is negligible, resulting in

$$
\begin{aligned}
\boldsymbol{F}_{\mathrm{LIN}} \simeq & -\rho\frac{d}{dt}\iint\limits_{S_E^I} dS\,\phi_D\boldsymbol{n} - \rho\frac{d}{dt}\oiint\limits_{S_B^I+S_W^I} dS\,\phi_I\boldsymbol{n}' \\
& - \rho g\oiint\limits_{S_B^I+S_W^I} dS\,Z\boldsymbol{n}' - \rho g\iint\limits_{S_E^I} dS\,\zeta_D\boldsymbol{n} \\
\approx & \iint\limits_{z=0} dS\left(\frac{\partial\phi}{\partial t} + g\zeta\right)\boldsymbol{k},
\end{aligned}
\tag{4.17}
$$

where the latter form of the equation represents the linear dynamic free-surface condition.

4.5.1 RAO's for S-175 Containership

To understand the general applicability of the impulse theory for the forces and moments on a ship, the linear motions of the S-175 containership are predicted in head and stern-quartering seas at two Froude numbers, $F_n = 0.0$ and 0.2. The S-175 containership (Watanabe et al. 1989) is a 175.0 m vessel, with a beam of 25.4 m and a draft of 9.5 m; it displaces 24,742 t. The body plan and bow profile of the vessel are shown in Fig. 4.3. This vessel has been used as a standard geometry for many seakeeping studies by the International Towing Tank Conference (ITTC), (*cf.*, O'Dea et al. 1992).

The motions of the S-175 are predicted using three or four different methods, three using pressure integration over the hull surface with Bernoulli's equation; the fourth being momentum-theory based impulse method. The first pressure integration

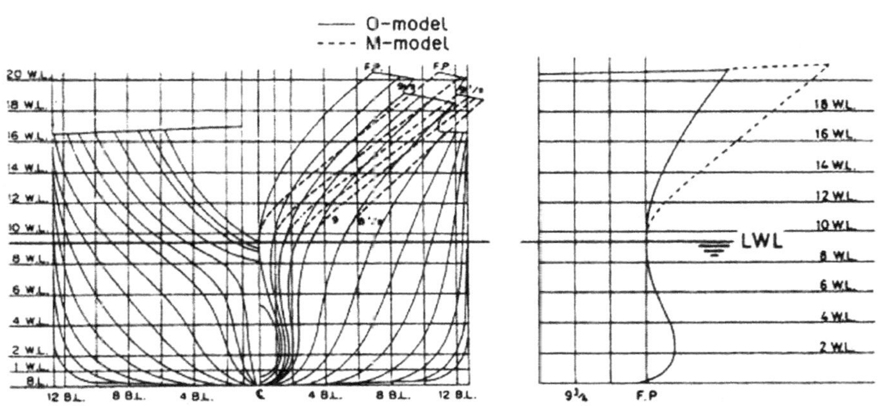

Fig. 4.3 Body plan and bow profile for S-175 containership (O-model). (Watanabe et al. 1989)

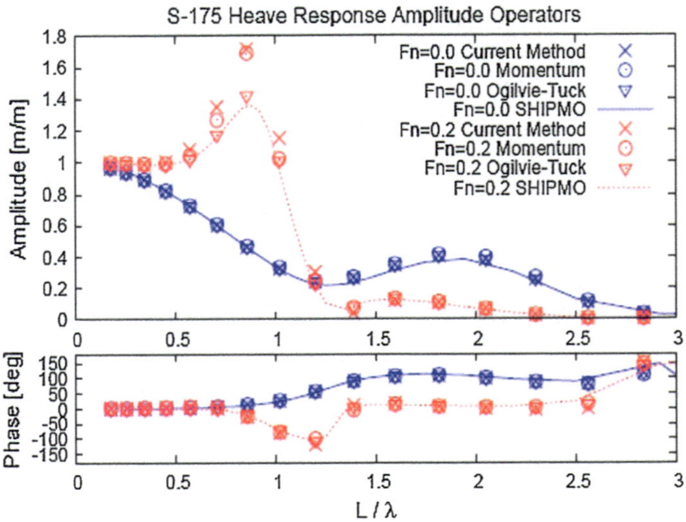

Fig. 4.4 Heave response-amplitude operator and phase angle as a function of L/λ for S-175 at $F_n = 0.0$ & 0.2 in head seas ($\beta = 180°$) [*Note* In the legend "Current method" means UMBest] (Courtesy of R. F. Beck, et al.; U. Mich.)

method is a conventional strip theory, similar to that of Salvesen et al. (1970), implemented in the University of Michigan code SHIPMO; the second is the conventional strip theory with the Ogilvie-Tuck corrections (Ogilvie and Tuck 1969); and the third is the University of Michigan nonlinear strip theory (UMBest)[2] (*cf.*, Bandyk 2009).

Starting with the traditional head-seas case, the heave and pitch response amplitude operators (RAO's) and phase angles of the responses of S-175 are predicted at $F_n = 0.0$ and 0.2. The heave and pitch RAO's and phase angles of the containership are presented in Figs. 4.4 and 4.5, respectively, as a function of wave length over ship length, L/λ.

At $F_n = 0.0$ the heave and pitch RAO's and phase angles from predictions by all four methods are in very close agreement over the entire range of L/λ's. However, at $F_n = 0.2$, there is a distinct increase in the RAO's at the peaks of the responses for both UMBest and momentum-theory results relative to the conventional strip theory and Ogilvie-Tuck predictions—the conventional strip theory and Ogilvie-Tuck results agree with each other, as do UMBest and momentum theory results. The phase angles for all four methods agree.

As a consequence of this simplest linear ship-motion prediction case, it can be concluded that the momentum formulation is correct. We shall now examine its performance in a full six-degrees-of-motion case.

The next comparison of the theories is for S-175 in stern-quartering seas at two Froude numbers, where motion responses in all six-degrees-of-freedom are expected.

[2]In the plots that follow, the UMBest results are labeled "Current Method".

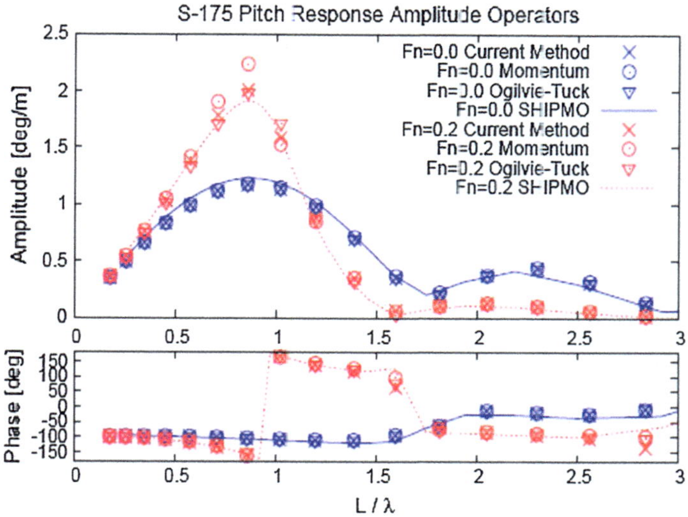

Fig. 4.5 Pitch response-amplitude operator and phase angle as a function of L/λ for S-175 at $F_n = 0.0$ & 0.2 in head seas ($\beta = 180°$) [*Note* In the legend "Current method" means UMBest] (Courtesy of R. F. Beck, et al.; U. Mich.)

In this case, we only have results for conventional strip theory and the nonlinear strip theory—no Ogilvie-Tuck predictions to compare with the momentum theory results. These results are presented as RAO's and phase angles versus L/λ in Figs. 4.6, 4.7, 4.8, 4.9, 4.10 and 4.11 for each mode of motion—surge, sway, heave, roll, pitch and yaw, respectively.

As was the case for head seas, at $F_n = 0.0$ the predictions in all modes of motion agree substantially for all three methods. Where there are slight deviations, the results from the nonlinear strip theory and the momentum formulation agree quite well.

For $F_n = 0.2$, there are significant differences in the results from the three prediction methods, particularly for sway and yaw—this is likely due to the effect of the method in which the "controller" is implemented and operates to keep the ship at speed and on heading, the results for each mode of motion will be discussed in order.

The $F_n = 0.2$ surge results (Fig. 4.6) agree well across the entire range of L/λ, except for a slight spread between all three methods of prediction for a short interval around $L/\lambda = 2.0$–2.25, where the surge response is relatively small anyway. The surge phases agree between all three prediction methods.

For smaller L/λ's, the sway results (Fig. 4.7) agree reasonably well, with the momentum method, the RAO being slightly higher than those of the conventional and nonlinear strip theories. However, for $L/\lambda > 1.5$, the RAO from the nonlinear strip theory increases relative to the other two methods, to exceed the conventional strip theory by a factor of roughly 2.5 times, but by $L/\lambda \sim 3$, all three methods are back in agreement. Except at the highest L/λ's, the phases agree. However, the L/λ

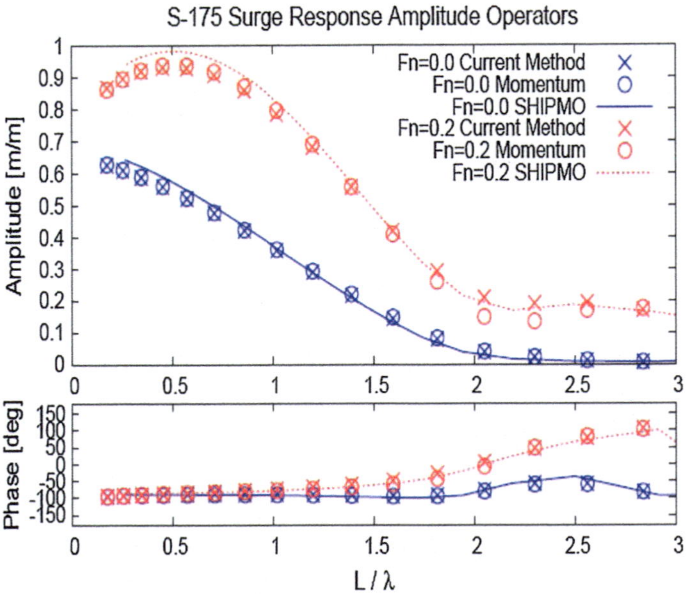

Fig. 4.6 Surge response-amplitude operator and phase angle as a function of L/λ for S-175 at $F_n = 0.0$ & 0.2 in Stern-quartering Seas ($\beta = 45°$) [*Note* In the legend "Current method" means UMBest] (Courtesy of R. F. Beck, et al.; U. Mich.)

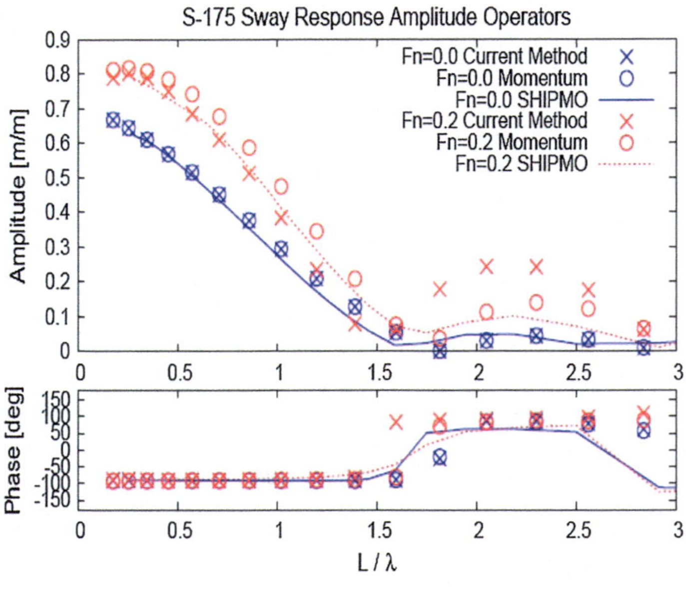

Fig. 4.7 Sway response-amplitude operator and phase angle as a function of L/λ for S-175 at $F_n = 0.0$ & 0.2 in Stern-quartering Seas ($\beta = 45°$) [*Note* In the legend "Current method" means UMBest] (Courtesy of R. F. Beck, et al.; U. Mich.)

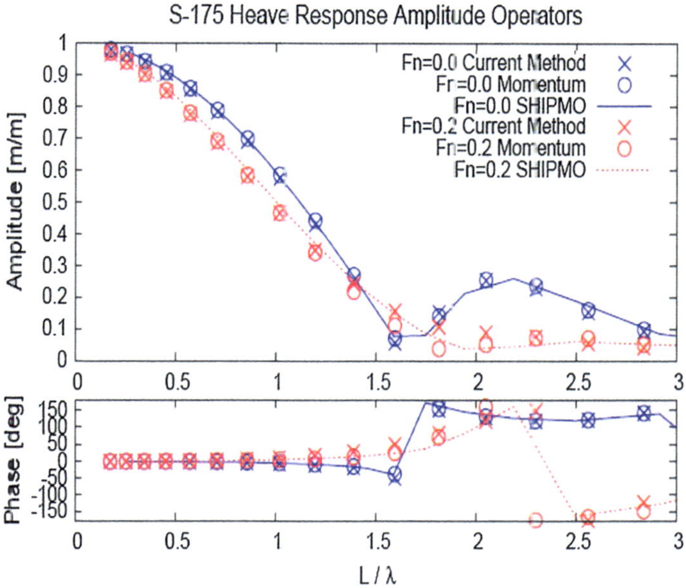

Fig. 4.8 Heave response-amplitude operator and phase angle as a function of L/λ for S-175 at $F_n = 0.0$ & 0.2 in Stern-quartering Seas ($\beta = 45°$) [*Note* In the legend "Current method" means UMBest] (Courtesy of R. F. Beck, et al.; U. Mich.)

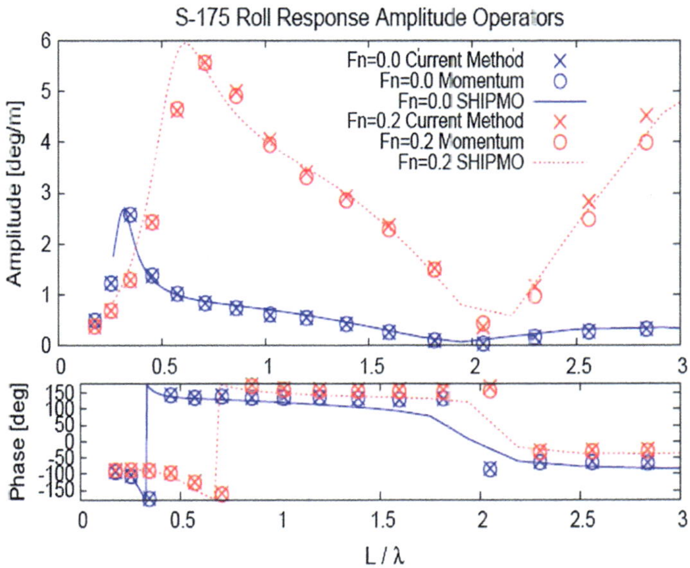

Fig. 4.9 Roll response-amplitude operator and phase angle as a function of L/λ for S-175 at $F_n = 0.0$ & 0.2 in Stern-quartering Seas ($\beta = 45°$) [*Note* In the legend "Current method" means UMBest] (Courtesy of R. F. Beck, et al.; U. Mich.)

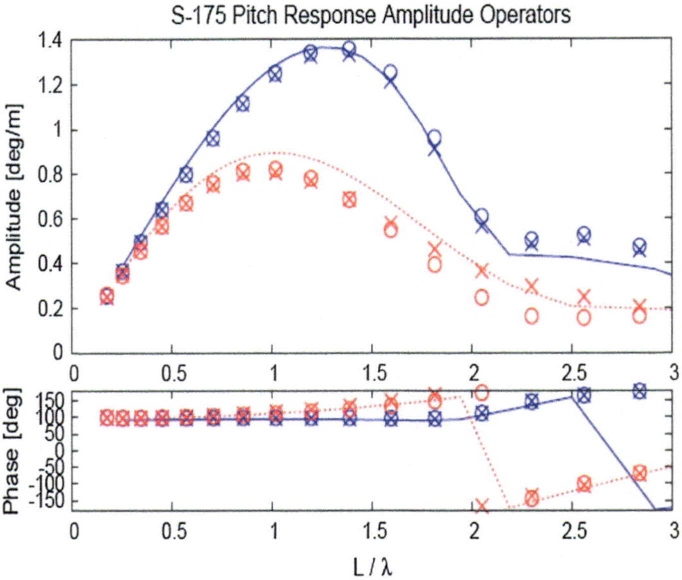

Fig. 4.10 Pitch response-amplitude operator and phase angle as a function of L/λ for S-175 at $F_n = 0.0$ & 0.2 in Stern-quartering Seas ($\beta = 45°$) — Legend is as given in Fig. 4.9 (Courtesy of R. F. Beck, et al.; U. Mich.)

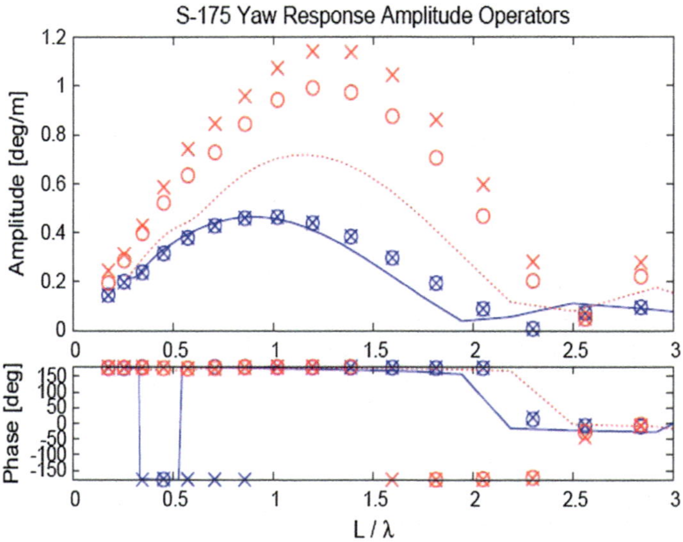

Fig. 4.11 Yaw response-amplitude operator and phase angle as a function of L/λ for S-175 at $F_n = 0.0$ & 0.2 in Stern-quartering Seas ($\beta = 45°$) — Legend is as given in Fig. 4.9 (Courtesy of R. F. Beck, et al.; U. Mich.)

at which the phase of the nonlinear strip theory jumps is lower than that of both conventional strip theory and momentum theory.

Examination of the predicted heave RAO's (Fig. 4.8) shows results that are very similar to those for surge, but with even smaller differences. Accounting for the wraparound in the phase angle, at $F_n = 0.2$ the phase for heave is very similar to that of surge, which is not the case for $F_n = 0.0$.

The comparison of the roll results (Fig. 4.9) is also similar to those of surge and heave. However, there is a slight shift to the right (higher L/λ) in the location of the peak of the RAO for the nonlinear strip theory and the momentum theory results relative to the location of the peak for conventional strip theory. It is noted that after achieving a minimum at around $L/\lambda = 2.0$, the RAO starts to increase again. This increase in RAO is probably a consequence of the roll moment created by the rudder's activation as a consequence of the controller that keeps the vessel on heading.

Through the resonant peak of the pitch RAO (Fig. 4.10), the momentum theory and the nonlinear strip theory agree quite well, and are slightly below the conventional strip theory results. However, for $L/\lambda > 1.75$, the nonlinear strip theory and momentum theory results start to separate, with the momentum theory results staying below the conventional strip theory results, while the nonlinear strip theory RAO increases to slightly above that of the conventional strip theory. By $L/\lambda = 2.75$, all three sets of predictions have converged to the same result.

At $F_n = 0.2$ the yaw results (Fig. 4.11) show the greatest differences in RAO responses of all the modes of motion across the three computational models. At the lowest and highest L/λ values, the three methods agree reasonably well. But in between the three sets of predictions show a resonant-like natural frequency peak, with the conventional strip theory having the smallest response and the nonlinear strip theory having the highest response by almost a factor of 2. The momentum theory results are in between, but closer to the nonlinear strip theory than to the conventional strip theory. Consistent with the sway observations, the peaks of the RAO for nonlinear strip theory and momentum theory are slightly to the right of the peak for the conventional strip theory. Again, this resonant like peak is probably a consequence of the heading controller. The reason that the deviations in the yaw responses are largest of all the modes of motion is because the rudder is most efficient at generating yaw responses, much more so than for sway or roll.

The momentum theory has now been tested in a much more challenging case than the head-seas case first tested—stern-quartering seas. Although the lateral-plane motions show some differences between the three prediction methods, it seems likely that these deviations are a result of the effects of the controller used to keep the vessel on course at this heading.

4.5.2 Nonlinearity Effects for 2-D Circular Cylinder

The previous RAO predictions have all been made for small-amplitude waves that result in small-amplitude motions. To test the large-amplitude aspects, and thus non-

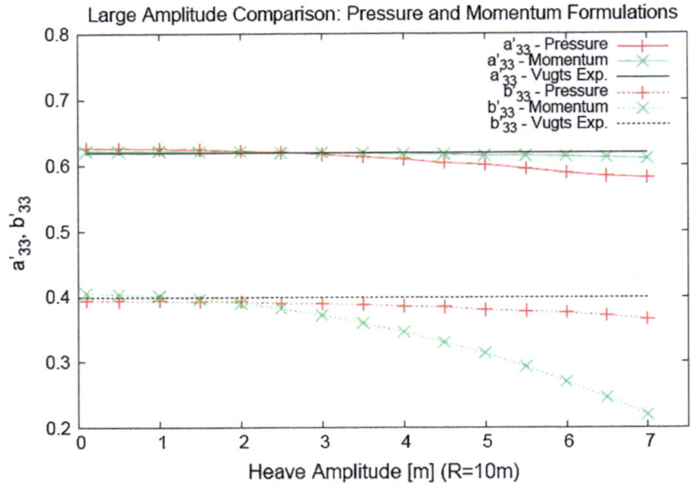

Fig. 4.12 Added mass & damping of Heaving 2-dimensional circular cylinder as a function of Heave amplitude, $\omega\sqrt{r/g} = 1.0$ (Courtesy of R. F. Beck, et al.; U. Mich.)

linear aspects, the momentum theory was used to predict the added mass and damping of a two-dimensional semicircular cylinder that was forced in heave of varying amplitudes. The results of this exercise are shown in Fig. 4.12 for a single non-dimensional frequency, $\omega\sqrt{r/g} = 1.0$.

Figure 4.12 presents the non-dimensional added mass ($a'_{33} = a_{33}/\rho A$, where ρ is the mass density of the fluid and A is the immersed area of the section) and damping ($b'_{33} = b_{33}/\rho A\sqrt{r/g}$, where r is the radius and g is the acceleration due to gravity) as a function of the motion amplitude which varies from 0.1 to 7 m for a 20 m diameter cylinder ($\xi_3/r = 0.01$–0.7). These results are predicted by both momentum theory and nonlinear strip theory with Bernoulli's equation for pressure integration. For reference, the added mass and damping measured by Vugts for small amplitude motions is also provided (Vugts 1968). For small heave amplitude, the added masses by the two computational methods are in full agreement and agree with Vugts' experiments. As the heave amplitude increases, the added mass by nonlinear strip theory decreases slightly relative to the experimental result, while the added mass by momentum theory agrees substantially with Vugts' experimental results. For small amplitude, the damping from the two prediction methods agree and are in agreement with Vugts' experimental results. However, as the heave amplitude increases, the damping by the nonlinear strip theory decreases slightly (by an amount comparable to the decrease in the added mass), while the damping by momentum theory decreases substantially.

It is hypothesized that the significant decrease in damping by the momentum theory is a result of using the linearized form (4.17) which neglects the integral of the potential over the free surface. Although this is only an untested hypothesis, it is consistent with the damping being a function of the waves being radiated away from the cylinder.

4.6 Summary and Conclusions

A new nonlinear momentum formulation for predicting the forces and moments on a body in waves has been developed (cf. Sclavounos 2012; Sclavounos and Lee 2012; Sclavounos et al. 2019). This formulation leads to the explicit decomposition of the total hydrodynamic force in nonlinear hydrostatics, Froude-Krylov and wave disturbance forces in steep random waves which are easily amenable to computation. All force components appear as time derivatives of the respective hydrodynamic impulses, defined as spatial integrals of the respective velocity potentials over the vessel instantaneous wetted surface, which do not requite the numerical evaluation of time derivatives of the velocity potential over the vessel wetted surface.

The new nonlinear momentum formulation has been used to predict response-amplitude operators for the S-175 containership in head and stern -quartering seas. The momentum theory predictions are found to be in substantial agreement with those by a nonlinear strip theory. The nonlinear aspects of the theory are tested against large-amplitude forced heave of a semicircular cylinder. The added mass computations are found to be in agreement with the results by Bernoulli's equation in conjunction with the nonlinear strip theory and experiments by Vugts. However, it is found that there is an unexpected decrease in the damping for increasing heave amplitude. It is hypothesized that this is a consequence of having employed a linearized form of the momentum theory for the calculations. This bears further investigation.

Acknowledgements As stated in the Introduction, this work is a summary of *some* of the significant work contained in Sclavounos (2012), Sclavounos & Lee (2012) and Sclavounos, et al. (2019). The significant contribution of Paul Sclavounos is very much appreciated. The many fruitful discussions with the Theory Advisory Panel (TAP) are also appreciated; as are the efforts of Prof. Robert F. Beck of the University of Michigan and his graduate students Xinshu Zhang, Jim Bretl, Piotr Bandyk and Rahul Subramanian who provided the computational results reported herein. This work was supported by Drs. L. Patrick Purtell and Paul Hess of the Office of Naval Research (ONR).

References

Bandyk, P. J. (2009) A Body-Exact Strip Theory Approach to Ship Motion Computations. Ph.D. Dissertation, Univ. Mich., Ann Arbor, MI, xii+122 p.

Belknap W. and J. Telste (2008) Identification o fLeading Order Nonlinearities from Numerical Forced Motion Experiment Results. *Proc. 27thSymp. NavalHydro.*, Seoul, Korea, 18 p.

Lighthill, M.J. (1960) A Note on the Swimming of Slender Fish.*J. Fluid Mech.*, 9:305–317.

Newman, J.N. (1977) *MarineHydrodynamics*. MIT Press, xiii + 402 p., Cambridge, MA.
Newman, J. N. & T. Y. Wu (1973) A Generalized Slender-Body Theory for Fish-Like Forms. *J. Fluid Mech.*, 57(4):673–693.
O'Dea, J. F., E. J. Powers and J. Zselecsky (1992) Experimental Determination of Nonlinearities in Vertical Plane Ship Motions. *Proc. 19th Symp. Naval Hydro.*, Seoul, Korea, pp. 73–91.
Ogilvie, T. F. and E. O. Tuck (1969) A rational strip theory of ship motions: Part I. Dept. of Nav. Arch. and Marine Eng., College of Eng., Univ. Michigan, Report No. 013, ix+92 p.
Reed, A.M. and J.G.Telste (2011) Computing Hydrodynamic Forces and Moments on a Vessel without Bernoulli's Equation. *Proc. 12th Int'l Ship Stability Workshop*, Washington, D.C., pp. 341–51.
Reed, A.M. (2012) Application of Computing Hydrodynamic Forces and Moments on a Vessel without Bernoulli's Equation.*Proc. 11th International Conference on the Stability of Ships and Ocean Vehicles*, Athens, Greece pp. 853–63.
Salvesen, N., E. O. Tuck & O. Faltinsen (1970) Ship motions and sea loads. *Trans. SNAME*, 78:250–87.
Sclavounos, P. D. (2012) Nonlinear Impulse of Ocean Waves on Floating Bodies. *J. Fluid Mech.*, 697:316–35.
Sclavounos, P. D. and S. Lee (2012) A Fluid Impulse Nonlinear Theory of Ship Motions and Sea Loads. *Proc. 29th Symp. Naval Hydro.*, Gothenburg, Sweden, 14 p.
Sclavounos, P.D., J.G.Telste and A.M.Reed (2019) Modeling of Nonlinear Vessel Responses in Steep Random Waves. Carderock Division, Naval Surface Warfare Center Report (to be published).
Serrin, J. (1959) Mathematical principles of classical fluid mechanics. *Encyclopedia of Physics,* Vol. VIII/Fluid Dynamics I, pp. 125–263, Springer-Verlag, Berlin.
Smirnov, V. I. (1964) *A Course of Higher Mathematics, Vol. II: Advanced Calculus.* Pergamon Press, 630 p., Oxford.
Telste, J.G. and W.F. Belknap (2008) Potential Flow Forces and Moments from Selected Ship Flow Codes in a Set of Numerical Experiments. Carderock Division, Naval Surface Warfare Center Report NSWCCD-50-TR-2008/040,15,240p.
Vugts, J. H. (1968) The hydrodynamic coefficients for swaying, heaving and rolling cylinders in a free surface, *Int'l Shipbuilding Prog.*, 15(167):251–76.
Watanabe, I., M. Ueno, H. Sawada (1989) Effects of Bow Flare Shape to the Wave Loads of a container ship. *J. Society Naval Architects of Japan*, 166:259–66.

Chapter 5
Modelling of Hull Lift and Cross Flow Drag Forces in Large Waves in a Computationally Efficient Dynamic Stability Prediction Tool

Michael J. Hughes, Paul J. Kopp and Ronald W. Miller

Abstract The US Navy is developing a new computationally efficient simulation tool to predict the responses of a ship operating in severe sea states. The tool computes the total force on the ship as the summation of component forces. An important component to the total force on the ship is the force from hull lift and cross-flow separation. These forces are predicted in calm water by maneuvering simulation tools but are often ignored by traditional seakeeping simulation tools. As viscous effects are important in the prediction of these forces, most maneuvering simulations are based on empirical data from calm water maneuvering tests. While these methods are valid in calm water the wetted shape of the hull changes significantly in large waves having significant influences on the hull lift and cross flow drag forces. In the present method a hull lift and cross flow drag force model is presented that accounts for the varying wetted geometry of the hull in waves. The method uses calm water maneuvering data from model tests and RANS calculations to calibrate the model. Proper modeling of the hull lift and cross flow drag force in large waves is very important for the prediction of some dynamic stability events such as broaching and broaching leading to capsize.

Keywords Tempest · Maneuvering in waves · CFD · Dynamic stability

5.1 Introduction

The US Navy is in the process of developing a new computational tool, called Tempest, for simulating the responses of a ship operating in severe sea states. An overview of this tool was provided in (Belknap and Reed 2010, see also Chap. 1 of this book). Tempest computes the total force on the ship at each time step as a summation of component forces. An important component force is that from circulatory lift and cross-flow separation on the ship's hull. This force component is sometimes referred

M. J. Hughes (✉) · P. J. Kopp · R. W. Miller
David Taylor Model Basin (NSWCCD), West Bethesda, MD, USA
e-mail: michael.j.hughes@navy.mil

V. L. Belenky et al. (eds.), *Contemporary Ideas on Ship Stability*, Fluid Mechanics and Its Applications 119, https://doi.org/10.1007/978-3-030-00516-0_5

to as the hull maneuvering force, as it is predicted in calm water by maneuvering simulation tools but is often ignored by traditional seakeeping simulation tools. As Tempest will be applied to a ship maneuvering in large waves, the maneuvering force model should not rely purely on empirical coefficients obtained from calm water model tests. To the greatest extent possible, the method should model the physics of the problem and be geometry based, but with the ability to be "calibrated" for a specific ship based on known full-scale or model-scale maneuvering data. An additional requirement is that the model be well-behaved within the context of a maneuvering-in-waves framework where the "seakeeping" forces are determined from potential flow assumptions (ideal fluid, no vorticity). Within this framework, there can potentially exist large heel and drift angles, as well as large sway and yaw rates. Furthermore, the maneuvering force/moment model needs to include the influence of the incident waves on the angle of attack of the ship hull relative to the local fluid velocity, though it must be consistent with an overall hydrodynamic model that does not provide the fluid velocity due to radiation and diffraction. The focus of this work is bare hull forces and moments since the rudder and propulsion forces are already included separately in Tempest, along with the wave induced seakeeping forces and moments, including, for example, added mass and slowly varying drift forces. As discussed in (Belknap and Reed 2010; Chap. 1) the theory in Tempest is being implemented in two phases. This paper describes the model that has been developed for the first phase. The second phase of the hull lift and cross flow drag force implementation will apply the vortex-lattice techniques developed for the bilge-keel force model (Greeley 2011, see also Chap. 7). The goal of the first phase model is to account for the time varying wetted geometry of the hull, while having the ability to tune the model to match available calm water maneuvering data. This method is expected to be faster and simpler than the model that will be implemented during the second phase. The first phase model is based partially on those of Ross (2008), Hooft (1994), and Hoerner (1965) and includes several modifications that are necessary to meet the requirements. Some example calculations are shown for the pre-contract DDG-51 hull form in calm water. This paper will present an overview of the model.

5.2 Approach Overview

The approach utilizes a blended model that is composed of two separate parts. The first treats the hull as a low aspect ratio airfoil in order to determine the normal force and moment on the ship due to circulatory lift. The other part of the overall approach is a cross-flow drag model that is based on an integrated 2D sectional analysis. Both portions account for the instantaneous submerged hull geometry in the presence of, potentially, large amplitude waves, orientation angles, and rotational/directional rates, while partially accounting for wave orbital velocities.

The normal force on the hull from circulatory lift and flow separation consists of a linear and quadratic term, expressed in the form of Eq. (5.1), where β is the drift

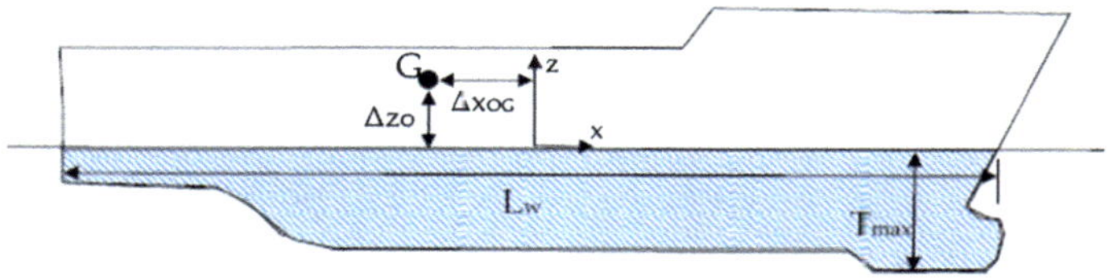

Fig. 5.1 Yawed earth-fixed frame for an upright ship in calm water

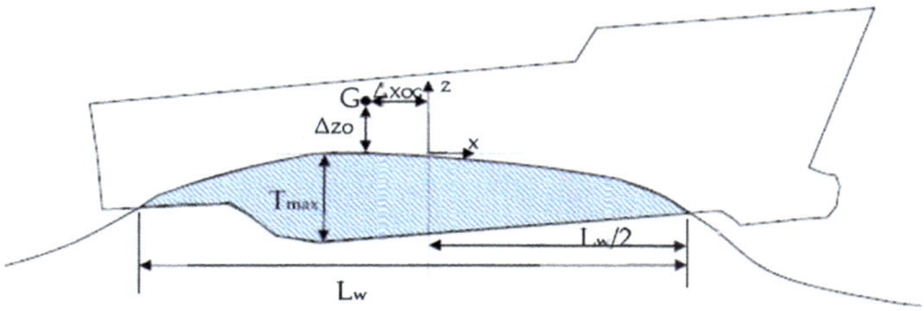

Fig. 5.2 Yawed earth-fixed frame for a ship in waves

angle. It is assumed that the linear term represents the circulatory lift on the hull while the quadratic term represents cross-flow drag.

$$C_N = C_{NL} \sin \beta \cos^2 \beta + C_{NN} |\sin \beta| \sin \beta \tag{5.1}$$

5.2.1 Coordinate Systems

The calculations for the circulatory lift on the hull are performed in a yawed earth-fixed frame with the x-axis aligned with the ship-fixed x-axis positive forward, the y-axis parallel to the calm water surface positive to port, and the z-axis aligned with the earth-fixed Z-axis positive up. The longitudinal origin is placed at the center of the instantaneous wetted length, the vertical origin is located at the waterline height of the section where the maximum draft is located and the transverse position is aligned with CG of the ship. This is illustrated in Fig. 5.1 for a ship in calm water and in Fig. 5.2 for a ship operating in waves.

For the 2D sectional calculations, it is useful to define the geometry of a hull section having arbitrary motion in the presence of waves, as illustrated in Fig. 5.3. In the case shown, the hull has a roll angle, ϕ, and local relative velocity vector, $\vec{V}_{rel}$, that is in the plane of the section. This vector includes the effects of the rigid body motions of the hull and makes an angle, α_v, relative to the horizontal plane. The instantaneous wave slope at the section is defined by the intersection of the water and the hull surface. The two intersection points, p_{i-1} and p_{i-2}, define the submerged shape of the section. The wave slope also defines an auxiliary wave slope coordinate

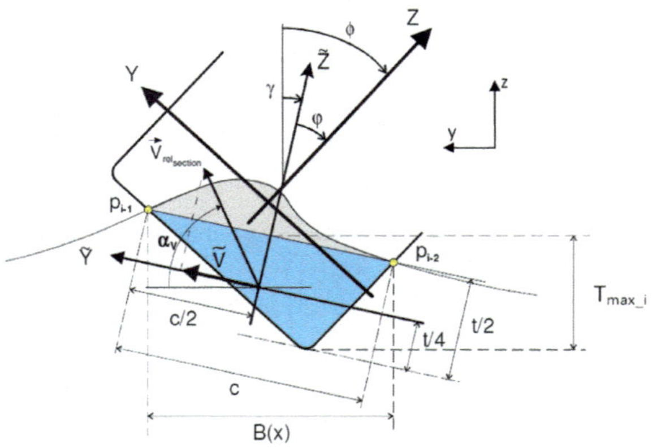

Fig. 5.3 Wave slope frame for sectional calculations

system $(\widetilde{X}, \widetilde{Y}, \widetilde{Z})$ in the plane of the section, centered at the midpoint of the chord of the submerged section. The angle between the wave slope system and the ship-fixed system is represented by the angle φ in Fig. 5.3. A separate auxiliary wave slope coordinate system is derived for each section at every time step.

5.3 Low Aspect Ratio Wing Theory

The circulation around the hull at a drift angle is analogous to the circulation of the flow about a low aspect ratio wing. The lift on the hull resulting from this circulation can be modeled by treating the submerged portion of the ship as a low aspect ratio foil, where the wetted length of the hull is the chord length and the maximum instantaneous draft is the span of the airfoil. Jones (1946) developed the expressions shown in Eq. (5.2) for the linear lift and drag on a low aspect ratio delta wing at a small angle of attack, α, where a_e is the effective aspect ratio that for a ship includes the wall effect of the free surface and is defined in Eq. (5.3).

$$C_L = \frac{\pi}{2} a_e \alpha, \quad C_D = \frac{C_L^2}{\pi a_e} \tag{5.2}$$

$$a_e = \frac{2T\text{max}}{L_W} \tag{5.3}$$

The lift and drag coefficients in Eq. (5.2) are "tuned" to match the linear terms corresponding to drift and yaw rate (Y_v, K_v, N_v, Y_r, N_r, and K_r) from a traditional calm water maneuvering model of the form developed by Abkowitz (1969). The tuning can be performed at a range of Froude numbers to include the influence of Froude number on the bare hull forces. The effect of the Munk moment will be accounted for in the tuned linear lift coefficients, so it is important to remove terms corresponding to the Munk moment from the hydrodynamic disturbance force model in Tempest. The dimensional linear normal force on the hull in the yawed earth-fixed frame from the drift angle and yaw rate is computed using Eq. (5.4). The speed dependent tuning coefficients defined in Eq. (5.5), C^L_{Nv} and C^L_{Nr}, are added to match the empirical data for the upright ship in calm water.

$$
\begin{aligned}
F^L_N &= F^L_{Nv} + F^L_{Nr} \quad \text{where,} \\
F^L_{Nv} &= C^L_{Nv}(Fn)\frac{\pi}{2}a_e \sin\beta\left(1/2\,\rho\,U^2LT\right)\cos^2\beta \\
F^L_{Nr} &= C^L_{Nr}(Fn)\frac{\pi}{4}a_e\frac{\tilde{r}L}{U}\left(1/2\,\rho\,U^2LT\right)\cos^2\beta
\end{aligned}
\tag{5.4}
$$

where,

$$
\begin{aligned}
C^L_{Nv}(Fn) &= -\frac{Y_v(Fn)L_{W\_calm}}{\pi/2\,a_e T_{max\_calm}} \\
C^L_{Nr}(Fn) &= \frac{Y_r(Fn)L_{W\_calm}}{\pi/4\,a_e T_{max\_calm}}
\end{aligned}
\tag{5.5}
$$

The calculations are performed in the yawed earth-fixed frame, so the drift angle and yaw rate should be defined in this frame. $\tilde{r}$ is the rotational velocity about the yawed earth-fixed z-axis. The local drift angle and local lateral plane velocity will first be computed at each section specified for the cross-flow drag calculations. The overall drift angle, β, and lateral velocity, U, used in Eq. (5.4) will then be obtained by averaging over the sections. The velocity U includes the velocity from the ship body motion and the longitudinal component of the wave orbital velocity. Only the (ship-fixed) longitudinal component of the ambient wave orbital velocity is included because the contribution to the total fluid velocity due to radiation and diffraction cannot be obtained within the current hydrodynamic framework in Tempest. Including the ambient wave velocity in the absence of the disturbance (radiation and diffraction) velocity would provide a higher velocity than reality in the transverse plane, which has strong effects on the hull forces due to the quadratic effect of the velocity on the force. However, because the disturbance velocity is assumed to be small in the longitudinal direction for slender ships, the orbital velocity is included in this direction. The $\cos^2\beta$ factor in Eq. (5.4) follows the method from Hooft (1994) where the force from the drift angle is normalized using the square of the axial component of the velocity. The $\cos^2\beta$ factor also has the advantage that it results in a maximum normal force at about 40°, which is a reasonable estimate for the stall angle, and causes the normal force to go to zero at a drift angle of 90°. With this behavior there is no need for an empirical stall model to be included. The values for L, T and a_e in Eq. (5.4) are the instantaneous wetted length and maximum draft, which are shown as L_W and T_{max} in Fig. 5.1, and the effective aspect ratio computed from those values for L and T using Eq. (5.3). The values for L, T and a_e used to compute the tuning coefficients in Eq. 5.5. are based on the initial calm water geometry as shown in Fig. 5.1.

The value of the maximum draft in the yawed earth-fixed frame is not obvious for sections with a roll angle in waves. The maximum draft at each section is computed as the vertical distance in the earth-fixed frame between the average waterline height and the deepest point on the section as shown in Fig. 5.3. The overall maximum draft (T_{max}) is defined as the largest value of T_{max_i} over all the sections defining the hull.

In addition to computing the side force acting on the hull the yaw and roll moment from the lift on the hull must be computed. The moments about the yawed earth-fixed z-axis and x-axis are computed by multiplying the computed normal force by a moment arm. The appropriate moment arm is determined from the linear derivatives for the yaw and roll moment in calm water, with the assumption that the distance from the center of pressure to the origin as a percentage of the wetted length and maximum draft remains constant as the ship maneuvers in waves. The formula for computing the moment about the yawed earth-fixed z-axis is given in Eq. (5.6).

$$M_Z^L = F_{Nv}^L C_{Mv}^L(Fn)L_W + F_{Nr}^L C_{Mr}^L(Fn)L_W \tag{5.6}$$

where,

$$C_{Mv}^L(Fn) = \frac{N_v(Fn)}{Y_v(Fn)}; C_{Mr}^L(Fn)\frac{N_r(Fn)}{Y_r(Fn)} \tag{5.7}$$

The formula for computing the moment about the yawed earth-fixed x-axis is:

$$M_X^L = F_{Nv}^{LK} C_{Kv}^L(Fn)T_{\max} + F_{Nr}^{LK} C_{Kr}^L(Fn)T_{\max} \tag{5.8}$$

where,

$$\begin{aligned} C_{Kv}^L(Fn) &= \frac{K_v(Fn)L_{W\_calm}}{Y_v(Fn)T_{\mathrm{max\_calm}}}; \\ C_{Kr}^L(Fn) &= \frac{K_r(Fn)L_{W\_calm}}{Y_r(Fn)T_{\mathrm{max\_calm}}} \end{aligned} \tag{5.9}$$

The linear normal forces used to compute the roll moment, F_{Nv}^{LK} and F_{Nr}^{LK}, are adjusted to remove the influence of the roll velocity on the normal force in order to avoid double counting of the roll moment computed from a separate roll damping model in Tempest, which computes a roll moment proportional to roll velocity. C_{Mv}^L and C_{Mr}^L represent the longitudinal position of the center of pressure of the normal force due to drift angle and turning as a percentage of the instantaneous wetted length. C_{Kv}^L and C_{Kr}^L represent the vertical position of the center of pressure of the normal force due to drift angle and turning as a percentage of the instantaneous maximum draft. Equations (5.6) and (5.8) compute the moment about the yawed earth-fixed frame referenced to the origin of that frame. To reference these moments to the center of gravity of the ship the moment arms are adjusted to account for the distance from the origin of the yawed earth-fixed frame to the center of gravity, which are shown as Δz_{OG} and Δx_{OG} in Fig. 5.2, representing the shift in the yawed earth-fixed z and x directions. Adding these values to Eqs. (5.6) and (5.8), the moments in the yawed earth-fixed frame referenced to the center of gravity are defined as:

$$M^L_{Z\_G} = F^L_{Nv}\left(C^L_{Mv}(Fn)L_W + \Delta z_{OG}\right) + F^L_{Nr}\left(C^L_{Mr}(Fn)L_W + \Delta z_{OG}\right) \quad (5.10)$$

$$M^L_{X\_G} = F^{LK}_{Nv}\left(C^L_{Kv}(Fn)T_{\max} - \Delta x_{OG}\right) + F^{LK}_{Nr}\left(C^L_{Kr}(Fn)T_{\max} - \Delta x_{OG}\right) \quad (5.11)$$

where Δz_{OG} is positive if the center of gravity is above the origin and Δx_{OG} is positive if center of gravity is forward of the origin.

A force in the yawed earth-fixed x direction from the induced drag is also included. The induced drag on a low aspect ratio foil is related to the circulatory lift by Jones (1946) using the second formula listed in Eq. (5.2). In this case the normal force will be used instead of the lift force and tuning coefficients will be included to calibrate with empirical data for the additional drag measured during steady drift and turning. It is assumed that this force can be applied at the ship center of gravity, and therefore does not contribute to the moments about the center of gravity. The induced drag force in the yawed earth-fixed x-direction is computed as:

$$\begin{aligned} F^L_X &= F^L_{Xv} + F^L_{Xr} \qquad \text{where,} \\ F^L_{Xv} &= C^L_{Xv}(Fn)\frac{\pi}{4}a_e \sin^2\beta\left(1/2\,\rho U^2 LT\right)\cos^2\beta \\ F^L_{Xr} &= C^L_{Xr}(Fn)\frac{\pi}{16}a_e\left(\frac{\tilde{r}L}{U}\right)^2\left(1/2\,\rho U^2 LT\right)\cos^2\beta \end{aligned} \quad (5.12)$$

The tuning coefficients are determined in this case from the quadratic coefficients X_{vv} and X_{rr}:

$$\begin{aligned} C^L_{Xv}(Fn) &= \frac{X_{vv}(Fn)L_{W\_calm}}{\pi/4a_e T\max_{\_calm}}; \\ C^L_{Xr}(Fn) &= \frac{X_{rr}(Fn)L_{W\_calm}}{\pi/16a_e T\max_{\_calm}} \end{aligned} \quad (5.13)$$

Input for X_{vv} and X_{rr} should both have negative values, since they are defined with a coordinate system with x forwards. All the forces and moments computed from the hull circulatory lift and drag model are computed first in the yawed earth-fixed frame, but are transformed to the ship-fixed frame before they are added to the other component forces and moments within the Tempest framework.

5.4 Cross-Flow Drag Model

At higher drift angles, cross-flow drag becomes the dominant viscous effect on the hull. The cross-flow drag force is obtained by first dividing the hull into a series of 2D sections. The local cross flow drag at each section is computed, and then

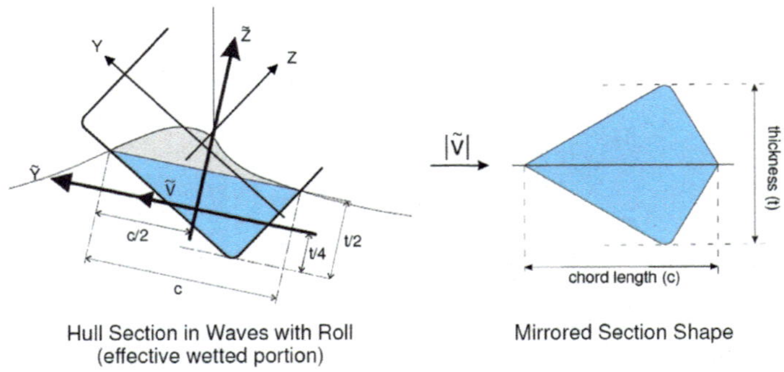

Fig. 5.4 Mirrored section shape used for 2D cross-flow drag calculations

the total cross-flow drag force is obtained by integrating the sectional forces along the length of the hull. The 2D sectional cross-flow drag calculations are performed in the local wave-slope frame for each section that was described earlier and is depicted in Fig. 5.3. The cross-flow drag force on each section is computed from the local cross flow velocity at the section and a drag coefficient, $C_d(x, Re)$, which is dependent on the shape of the section as well as the Reynolds number at the section. The local cross-flow velocity is taken as the velocity component parallel to $\widetilde{Y}$ axis of the wave-slope frame, and is depicted as $\tilde{v}$ in Figs. 5.3 and 5.4. The relative velocity used for the maneuvering calculations on each section will be computed at the center of the bounding box surrounding the section. This velocity includes the coupled effects of rigid body motions in 6 degrees of freedom. Since the transverse and vertical components of ambient wave orbital velocity are ignored in the present implementation of this cross-flow drag model, the sectional velocity in the y-z plane will only be due to rigid body motions. The section shapes used for the cross-flow drag calculations are obtained by reflecting the instantaneous wetted geometry about the approximated wave-slope line as shown in Fig. 5.4.

Two methods are available for obtaining the cross-flow drag coefficient $C_d(x, Re)$ for each section. In the first option, the instantaneous beam to draft ratio of the section is used to estimate $C_d(x, Re)$ based on wind tunnel data for elliptical or rectangular shaped cylinders. In the second option the user specifies the values for $C_d(x, Re)$ for each section at a range of heel angles and drafts, and the value at each time step for each section is then obtained through interpolation based on the instantaneous draft and roll angle relative to the wave slope frame. It is possible to use the second option for some sections and the first option for the other sections. Such an approach may be useful in the case of a ship with a large skeg, where the user could specify the coefficients directly for sections with the skeg, but use the first option on the sections forward of the skeg.

Hoerner (1965) provides a general discussion of sectional drag and shows the behavior of the drag coefficient for various sectional 2D shapes in steady flow. These shapes include parametrically varied rectangular, elliptic, and u/v shapes that are

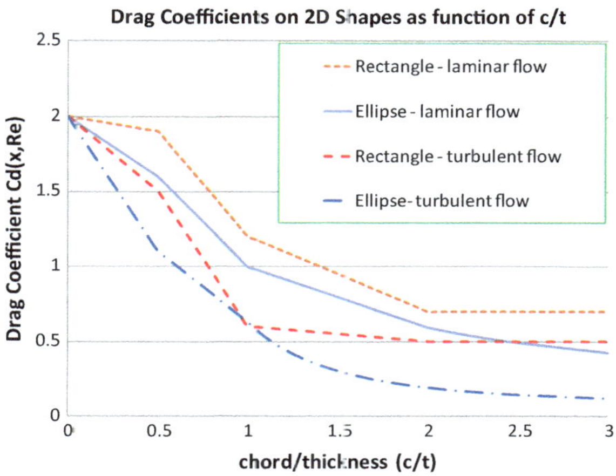

Fig. 5.5 Cd(x, Re) for elliptical and rectangular sections as a function of c/t, in laminar and turbulent flow

somewhat like ship hull section shapes. The data in Hoerner show that the dependence of the 2D cross-flow drag coefficient on Reynolds number is primarily a step function at the critical Reynolds number where the flow transitions from laminar to turbulent flow. For full-scale simulations the flow will be turbulent, but for model-scale simulations laminar flow may be present on some sections. Based on the data in Hoerner (1965), approximate curves for $C_d(x, Re)$ were generated for elliptical sections and rectangular sections with a moderate bilge radius. These curves are shown in Fig. 5.5 and are used to compute $C_d(x, Re)$ in the current method. The cross-flow drag coefficient is dependent on the instantaneous thickness and chord of the section, which are indicated by t and c respectively in Fig. 5.4 as well as whether the section shape more closely resembles an ellipse or a rectangle with rounded corners.

The general equations for the ship fixed reference frame sway and heave forces and yaw and pitch moments from the cross-flow drag model, adjusted for the presence of waves and ship motions are given by Eqs. (5.14)–(5.17). There is no surge force component from the cross-flow drag model.

$$F^{cf}_{Y\_G} = -\frac{\rho}{4}\int_0^L t(x)C_d(x, Re)\tilde{v}(x)|\tilde{v}(x)|\cos\varphi(x)dx \tag{5.14}$$

$$F^{cf}_{Z\_G} = -\frac{\rho}{4}\int_0^L t(x)C_d(x, Re)\tilde{v}(x)|\tilde{v}(x)|\sin\varphi(x)dx \tag{5.15}$$

$$M^{cf}_{Z\_G} = -\frac{\rho}{4}\int_0^L t(x)C_d(x, Re)\tilde{v}(x)|\tilde{v}(x)|\cos\varphi(x)x\,dx \tag{5.16}$$

$$\mathrm{M}_{Y\_G}^{cf} = \frac{\rho}{4} \int_{0}^{L} \mathrm{t(x)} C_d(\mathrm{x, Re}) \tilde{v}(\mathrm{x}) |\tilde{v}(\mathrm{x})| \sin \varphi(\mathrm{x}) \mathrm{x} d\mathrm{x} \tag{5.17}$$

The cos φ(x) and sin φ(x) terms appearing in the above equations provides the transformation of the sectional drag force back to the ship fixed coordinate system from the local $\widetilde{Y}\widetilde{Z}$ coordinate system. When applying Eqs. (5.14)–(5.17), the variable x must be defined in the ship-fixed frame.

There is also a roll moment generated by the ship-fixed sway force calculated above ($F_{Y\_G}^{cf}$). Initially it will be assumed that the force is evenly distributed along the $\widetilde{Z}$ axis. With this assumption the moment arm will act at 50% of the distance from the water surface to the half thickness (t/2) of the section, so the default value of zcp(x, Re) is 0.5. It is suggested that RANS computations be performed on a variety of sections to obtain better estimates of the moment arm. With the current assumptions, the roll moment is:

$$\mathrm{M}_{X\_G}^{cf} = -\frac{\rho}{4} \int_{0}^{L} \mathrm{t(x)} [z_{cp}(x, \mathrm{Re}) \frac{t(x)}{2} + \tilde{z}_{cg}] \\ C_d(x, \mathrm{Re}) \tilde{v}(\mathrm{x}) |\tilde{v}(\mathrm{x})| \cos \varphi(\mathrm{x}) d\mathrm{x} \tag{5.18}$$

where $\tilde{z}_{cg}$ is the distance in the direction of the $\widetilde{Z}$ axis from the projection of the center of gravity onto the section plane to the instantaneous waterline defined in the $\widetilde{Y}\widetilde{Z}$ system (the line connecting p_{i-1} and p_{i-2} in Fig. 5.3). $\tilde{z}_{cg}$ is positive when the CG is above the waterline.

5.4.1 *Longitudinal Attenuation of Cross-Flow Drag*

Equations (5.4)–(5.12) define the forces and moments from circulatory lift in the yawed earth-fixed frame. These forces are transformed to the ship-fixed frame and added to the forces and moments due to cross-flow drag, defined in Eqs. (5.14)–(5.18), to obtain the total hull lift and cross-flow drag forces and moments in the ship-fixed frame at each time step. Built into the models is a natural implied blending of the low aspect ratio airfoil forces and those from cross-flow drag. As the cross-flow drag force is roughly proportional to $\sin^2\beta$, the cross-flow drag force will be very small at low drift angles. It has been found through experiments, that at low and moderate drift angles, the distribution of cross flow drag force is such that this force acts mainly on the aft portion of the ship (see Hooft 1994). An attenuation factor is applied to the cross flow drag coefficients at each section to reduce their value on the forward portion of the ship at low drift angles. As the drift angle increases the attenuation is diminished. Figure 5.6, taken from Hooft (1994) shows a schematic of the attenuation factor for a range of drift angles. The schematic is based on segmented

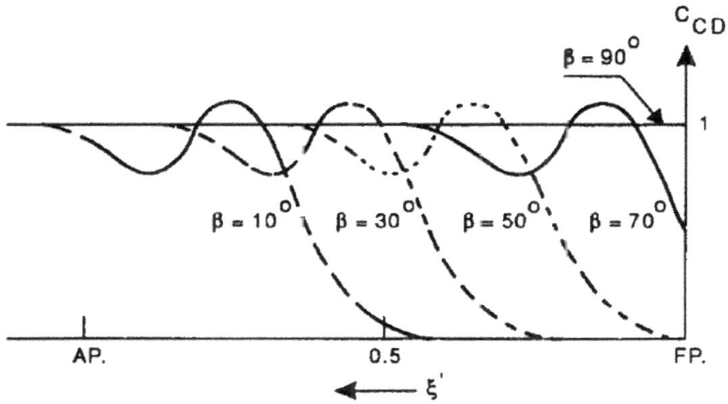

Fig. 5.6 Schematic indication of the longitudinal attenuation of Cd(x, Re) as a function of drift angle, from Hooft (1994)

model test data for a Series 60 hull; however, the model tests were performed only up to a drift angle of 20°. The curves shown for drift angles higher are based on extrapolation. At 90° drift there will be no attenuation and at drift angles higher than 90° the attenuation will be such that the cross-flow drag force will be shifted towards the forward portion of the ship. CFD analysis can be performed to produce curves of the attenuation factor along the length of the hull at various drift angles, and then interpolation can be used to obtain the attenuation factor at each section at the current drift angle during the simulation. The attenuation factor, $C_{T-Cd}(x, \beta)$, is used to determine a corrected cross-flow drag coefficient, C_{d_corr}.

$$C_{d\_corr}(\mathrm{x}, \mathrm{Re}, \beta) = C_{T-Cd}(x, \beta)\, C_d(\mathrm{x}, \mathrm{Re}) \tag{5.19}$$

The corrected coefficient, $C_{d\_corr}(x, Re, \beta)$, is used in place of $C_d(x, Re)$ in Eqs. (5.14)–(5.18).

5.5 Sample Calculation

The hull lift and cross-flow drag model has been implemented into the Tempest code and some preliminary calculations have been performed for the pre-contract DDG-51 hull form as represented by NSWCCD model 5415. The preliminary simulations have been performed using an approximate attenuation factor, $C_{T\text{-}Cd}(x, \beta)$, based on the limited Series 60 model test results discussed in Hooft (1994). For future simulations, RANS calculations will be performed to compute the attenuation factor. The RANS simulations will be performed over a large range of drift angles, with the results analyzed to provide the longitudinal distribution of the forces along the length of the hull in addition to the total forces. The simulations were not completed in time to

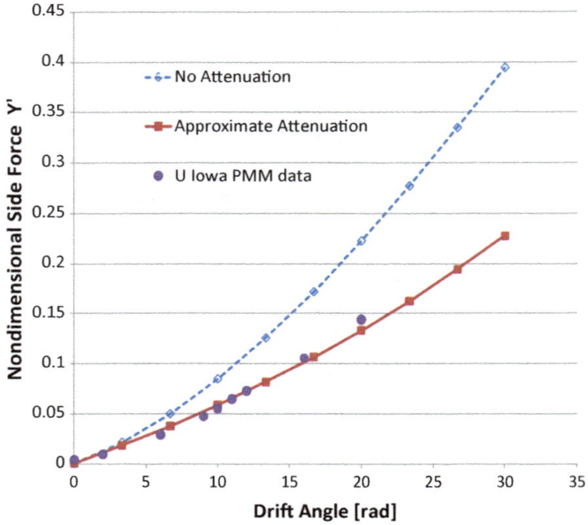

Fig. 5.7 Comparison of non-dimensional side force on the pre-contract DDG-51 at Fn = 0.28

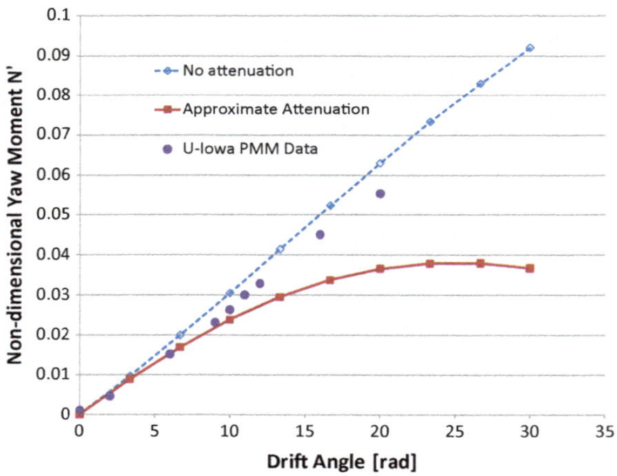

Fig. 5.8 Comparison of non-dimensional yaw moment on the pre-contract DDG-51 at Fn = 0.28

compute attenuation factors for the computations show in this paper. Results from some of the RANS simulations that have been completed have led to some new insights. For instance it is apparent that to predict the roll moment for some hull forms the model will need to include vertical component of forces (heave and pitch) in addition to lateral plane forces.

Results from the preliminary simulations performed in calm water have been compared with PMM model test results performed at the University of Iowa (Yoon 2009) and free running model tests performed at NSWCCD (Hayden 2006). The comparisons with the PMM tests were performed with the model fixed at the calm water draft and trim in both the Tempest simulations and the model tests. The Tempest simula-

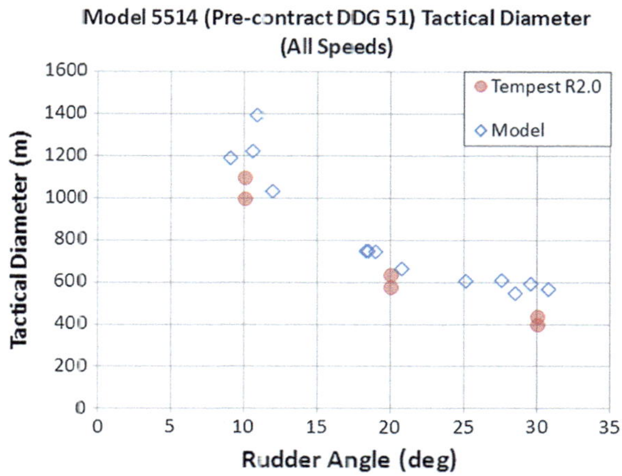

Fig. 5.9 Comparison of tactical turning diameter on the pre-contract DDG-51 versus rudder angle

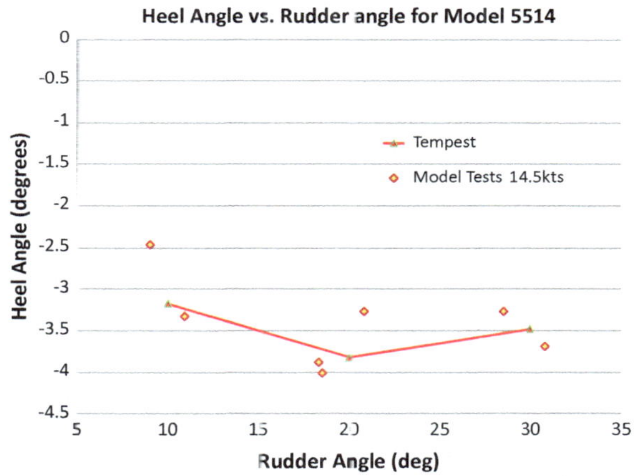

Fig. 5.10 Comparison of steady heel angle in a turn for the pre-contract DDG-51 versus rudder angle, 14.5 knots full-scale

tions were performed with and without the approximate attenuation factor switched on. Comparisons between the predictions and the model test results are shown in Figs. 5.7 and 5.8 for the non-dimensional side force and yaw moment respectively for the model traveling at a steady drift angle. The approximate attenuation factor improves the comparison for the side force and the correlation for the side force is very good for this case. However, the inclusion of the attenuation factor degrades the correlation for the yaw moment. It is anticipated that this correlation will improve once ship specific attenuation factors based on the RANS simulations are incorporated. Figure 5.9 shows the comparison between the predictions and the data from free running model tests for the tactical turning diameter. Figure 5.10 compares the steady heel angle in a turn for the same model tests. In both cases the correlation is good and falls within the scatter of the model test data.

5.6 Conclusions

A model has been developed which computes the hull lift and cross-flow separation forces on a ship maneuvering in waves. The model is computationally efficient and accounts for the changing wetted geometry of the ship hull, while retaining the ability to be calibrated to match available calm water maneuvering data. The model remains well-behaved even at extreme drift angles and yaw rates as a ship may encounter during broaching. Preliminary calculations show the model predictions compared reasonably well with calm water model test results.

Acknowledgements Dr. Pat Purtell (ONR) and Jim Webster (NAVSEA) have supported the work discussed in this paper.

References

Abkowitz, M. (1969), Stability and Motion Control of Ocean Vehicles, The MIT Press.

Belknap, W.F. and Reed, A.M. (2010), "TEMPEST – New Computationally Efficient Dynamic Stability Prediction Tool", Proc. of 11th Intl Ship Stability Workshop, Wageningen, the Netherlands.

Greeley, D.S. (2011), "Some Results from a New Time-Domain Bilge keel Force Model", Proc. of 12th Intl Ship Stability Workshop, Washington DC.

Hayden, D.D., Bishop, R.C., Park, J.T. and Laverty, S.M. (2006), Model 5514 Capsize Experiments Representing the Pre-Contract DDG51 Hull Form at End of Service Life Conditions, NSWCCD-50-TR-2006/020.

Hoerner, S.F. (1965), Fluid Dynamic Drag: Theoretical, Experimental and Statistical Information, Hoerner Fluid Dynamics.

Hooft, J. P. and Nienhuis (1994), "The Prediction of the Ship's Maneuverability in the Design Stage," SNAME Trans Vol 102, pp 419–445.

Jones, R.T. (1946) Properties of Low-Aspect-Ratio Pointed Wings at Speeds Below and Above the Speed of Sound, NACA Report No. 835.

Ross, A. (2008), "Nonlinear Manoeuvering Models for Ships: a Lagrangian Approach," PhD Thesis 2008:114.

Yoon, H. (2009), "Phase-averaged Stereo-PIV Flow Field and Force/Moment/Motion Measurements for Surface Combatant in PMM Maneuvers," Ph.D. thesis, The University of Iowa, Iowa City.

Chapter 6
Improved Maneuvering-Based Mathematical Model for Free-Running Ship Motions in Following Waves Using High-Fidelity CFD Results and System-Identification Technique

Motoki Araki, Hamid Sadat-Hosseini, Yugo Sanada, Naoya Umeda and Frederick Stern

Abstract Predicting maneuverability and stability of a free running ship in following and quartering waves are one of the most important topics to prevent broaching; however current mathematical models show quantitative errors with the experimental data while high-fidelity CFD simulations show quantitative agreement, which provides the opportunity to improve the mathematical models for free running ship dynamics in waves. In this study, both maneuvering coefficients and wave model in the mathematical model are improved utilizing system identification technique and CFD free running outputs. From turning circle and zigzag calm water CFD free running data, the maneuvering coefficients are estimated. The wave correction parameters are introduced to improve the wave model, which are found from a few forced and free running CFD simulations in waves. The mathematical model with the improved parameters shows much better agreement with experiments in both calm water and waves than the original mathematical model. The original mathematical model was based on the maneuvering coefficients estimated from several captive tests and wave forces calculated from linear Froude-Krylov forces and diffraction forces based on a slender ship theory.

M. Araki · Y. Sanada · N. Umeda
Department of Naval Architecture & Ocean Engineering, Osaka University, Osaka, Japan
e-mail: araki@nmri.go.jp

H. Sadat-Hosseini · Y. Sanada · F. Stern (✉)
IIHR—Hydroscience & Engineering, The University of Iowa, Iowa City, IA, USA
e-mail: frederick-stern@uiowa.edu

H. Sadat-Hosseini
e-mail: hamid.sadat@unt.edu

M. Araki
Ocean Engineering Department, National Maritime Research Institutes, Tokyo, Japan

H. Sadat-Hosseini
Department of Mechanical and Energy Engineering, University of North Texas, Denton, TX, USA

V. L. Belenky et al. (eds.), *Contemporary Ideas on Ship Stability*, Fluid Mechanics and Its Applications 119, https://doi.org/10.1007/978-3-030-00516-0_6

Keywords System identification · Constrained least square method
Maneuvering coefficients · Wave model · Free-running ship

List of Symbols

$a_{1,2,3,4}$	Tuning parameter for wave forces amplitude in surge
a_H	Rudder and hull hydrodynamic interaction coefficient in sway
B	Ship breadth
$b_{1,2,3,4}$	Tuning parameter for wave forces amplitude in sway
$c_{1,2,3,4}$	Tuning parameter for wave forces amplitude in roll
C_X	Nondimensionalized drift wave force in surge
C_Y	Nondimensionalized drift wave force in sway
C_N	Nondimensionalized drift wave moment in yaw
d	Ship draft
$d_{1,2,3,4}$	Tuning parameter for wave forces amplitude in yaw
Fr	Froude number
g	Gravitational acceleration
GZ	Restoring arm in roll
I_{xx}	Moment of inertia in roll
I_{zz}	Moment of inertia in yaw
J_{xx}	Added moment of inertia in roll
J_{zz}	Added moment of inertia in yaw
k	Wave number
K_p	Derivative of roll moment with roll rate
K_r	Derivative of roll moment with yaw rate
K_R	Rudder force in roll
K_s	Rotational index in nonlinear first order Nomoto's model
K_v	Derivative of roll moment with sway velocity
K_w	Wave moment in roll
K_w^{Dif}	Diffraction wave moment in roll
K_w^{FK}	Froude-Krylov wave moment in roll
K_{rrr}	Derivative of roll moment with cubed yaw rate
K_{rrv}	Derivative of roll moment with squared yaw rate and sway velocity
K_{rvv}	Derivative of roll moment with squared sway velocity and yaw rate
K_{vvv}	Derivative of roll moment with cubed sway velocity
K_ϕ	Derivative of roll moment with roll angle
L	Ship length
l_R	Longitudinal position of rudder center from center of ship gravity
m	Ship mass
m_{WD}	Shape parameter of Weibull distribution for wave drift forces/moment
m_x	Added mass in surge
m_y	Added mass in sway
N_r	Derivative of yaw moment with yaw rate

N_R	Rudder force in yaw
N_s	Nonlinear index in nonlinear first order Nomoto's model
N_v	Derivative of yaw moment with sway velocity
N_w	Wave moment in yaw
N_w^{Dif}	Diffraction wave moment in yaw
N_w^{FK}	Froude-Krylov wave moment in yaw
N_{rrr}	Derivative of yaw moment with cubed yaw rate
N_{rrv}	Derivative of yaw moment with squared yaw rate and sway velocity
N_{rvv}	Derivative of yaw moment with squared sway velocity and yaw rate
N_{vvv}	Derivative of yaw moment with cubed sway velocity
N_ϕ	Derivative of yaw moment with roll angle
p	Roll rate
r	Yaw rate
R	Ship resistance
T	Propeller thrust in surge
t_R	Rudder and hull hydrodynamic interaction coefficient in surge
T_s	Time constant index in nonlinear first order Nomoto's model
T_w	Wave period
u	Surge velocity
u_w	Wave particle velocity in surge
v	Sway velocity
v_w	Wave particle velocity in sway
W_{CN}	Drift wave moment in yaw
W_{CX}	Drift wave force in surge
W_{CY}	Drift wave force in sway
x_H	Rudder and hull hydrodynamic interaction coefficient in yaw
X_R	Rudder force in surge
X_{rr}	Derivative of surge force with squared yaw rate
X_{rv}	Derivative of surge force with yaw rate and sway velocity
X_{rr}	Derivative of surge force with squared sway velocity
X_w	Wave force in surge
X_w^{Dif}	Diffraction wave force in surge
X_w^{FK}	Froude-Krylov wave force in surge
Y_r	Derivative of sway force with yaw rate
Y_R	Rudder force in sway
Y_v	Derivative of sway force with sway velocity
Y_w	Wave force in sway
Y_w^{FK}	Froude-Krylov wave force in sway
Y_w^{Dif}	Diffraction wave force in sway
Y_{rrr}	Derivative of sway force with cubed yaw rate
Y_{rrv}	Derivative of sway force with squared yaw rate and sway velocity
Y_{rvv}	Derivative of sway force with squared sway velocity and yaw rate
Y_{vvv}	Derivative of sway force with cubed sway velocity
Y_ϕ	Derivative of sway force with roll angle

z_H	Height of hydrodynamic sway force application point from center of ship gravity
z_{HR}	Height of rudder force application point in roll
α_X	Tuning parameter for drift wave force in surge
α_Y	Tuning parameter for drift wave force in sway
α_N	Tuning parameter for drift wave moment in yaw
β_1	Tuning parameter for wave particle velocity in surge
β_2	Tuning parameter for wave particle velocity in sway
γ_R	Flow straitening coefficient
δ	Rudder angle
ε	Rudder effectiveness coefficient
$\varepsilon_{a1,2,3,4}$	Tuning parameter for wave forces phase lag in surge
$\varepsilon_{b1,2,3,4}$	Tuning parameter for wave forces phase lag in sway
$\varepsilon_{c1,2,3,4}$	Tuning parameter for wave forces phase lag in roll
$\varepsilon_{d1,2,3,4}$	Tuning parameter for wave forces phase lag in yaw
η_{WD}	Scale parameter of Weibull distribution for wave drift forces/moment
ζ_w	Wave amplitudes
λ	Wave length
ξ_G	Longitudinal position of center of ship gravity from a wave trough
ρ	Water density
ϕ	Roll angle
ψ	Yaw angle
ψ_0	Tuning parameter for wave drift wave force phase lag in surge
ω	Wave frequency

6.1 Introduction

Maneuverability and stability of a free running ship in waves are one of the most important topics considered in the ship design. Especially, in severe following and quartering waves, the ship is very likely to broach and capsize.

Mathematical models (MM) and recently computational fluid dynamic (CFD), numerically solving Navier-Stokes equation including viscous effects, are used to predict ship stability and maneuverability in calm water and waves. The MM approach in this paper means an approach consisting of two layered sub systems. In the lower layer, hydrodynamic forces mainly due to potential flow are calculated by solving partial differential equations of potential flow and hydrodynamic forces mainly due to viscosity flow are estimated with captive model experiments or empirical formulas. In the upper layer, ship motions are calculated by solving ordinary differential equations with initial conditions. Since short computational time is required to sweep out dangerous maneuvering and wave conditions from a huge number of suspect conditions, the MM method shows superior ability to the CFD; MM needs less than a minute for one free running simulation using a personal computer while CFD needs a few weeks or a month using a very expensive supercomputer.

Meanwhile, it is also very expensive and time-consuming to predict maneuvering and rudder coefficients from captive model tests which are necessary for MM method while CFD just needs ship geometry and propeller characteristics.

The MM free running simulations in calm water showed that MM was very sensitive to the accuracy of the maneuvering coefficients such that the scatter in the MM predictions were substantial for MM with coefficients estimated from different captive tests (Stern et al. 2011). Also, MM showed only qualitative agreement with model experiment (ME) free running results in following and quartering waves while CFD shows quantitative agreement (Sadat-Hosseini et al. 2011). Since CFD free running simulation can provide not only ship motion but also total forces/moments acting on the ship which are unknown during ME free running, it could give a chance to modify and tune MM to reduce the disagreement with ME free running results in calm water and waves.

Several mathematical models have been developed. Abkowitz (1964) developed a mathematical model to describe the hydrodynamic forces/moments acting on the ship with polynomial expressions using Taylor expansion on state variables. Christensen and Blanke (1986) developed 2nd order modulus expansions, which represent the hydrodynamic forces at angle of incidence: cross-flow drag Recently a new maneuvering model was developed from first principles of low aspect-ratio aerodynamic theory and Lagrangian mechanics (Ross et al. 2007). Meanwhile, the Maneuvering Mathematical Modeling Group (MMG) (Ogawa and Kasai 1978; MMG 1980) developed a mathematical model, which explicitly includes the individual open water characteristics of the hull/propeller/rudder and their interactions. Issues for improvement include both wave terms in the mathematical model and methods for obtaining the wave maneuvering coefficients. Usually, the maneuvering coefficients are assumed constant, which could be realistic for high encounter frequency wave conditions. In contrast, Son and Nomoto (1982) and Araki et al. (2010) showed a large variation of maneuvering coefficients in following waves in which the encounter frequency is very low.

System identification (SI) techniques are developed in control engineering to build mathematical models for dynamical systems by estimating maneuvering coefficients. The least square (LS) is the one of the simplest and the extended Kalman filtering (EKF) (Lewis 1986) is one of the most widely used methods in engineering. Nonaka et al. (1972) employed LS to estimate maneuvering coefficients from experimental free-running data with random rudder motions and the Abkowitz mathematical model. However, the estimated maneuvering coefficients were not accurate, which was due to some of the derivatives drifting to the wrong values known as the simultaneous drift problem (Kang et al. 1984). The constrained least square (CLS) method using the generalized reduced gradient algorithm developed by Lasdon et al. (1978) can help avoiding the simultaneous drift problem. EKF using full-scale trial data and the Abkowitz mathematical model was employed by Abkowitz (1980). The zigzag and turning simulation using estimated maneuvering coefficients showed reasonable agreement with the data. Rhee and Kim (1999) employed EKF for free-running trial data (zigzag, turning circle, large angle zigzag tests, etc.) and the MMG mathematical model to find the best trial type for system identification. The maneuvering coefficients reconstructed from the large angle zigzag test showed the smallest error with the original coefficients. Zhang and Zou (2011) employed support vector machine, one of the artificial intelligence methods, for zigzag test and the Abkowitz mathematical model for which the reconstructed coefficients showed close agreement with the original maneuvering coefficients. Several other researchers (for instance Shi et al. 2009) have employed EKF to estimate ship maneuvering coefficients.

Most of the studies were conducted for calm water and experimental data were used to improve the mathematical model by utilizing a system identification technique. The authors used CFD outputs to improve the mathematical model predictions in calm water (Araki et al. 2012). Hydrodynamic and rudder maneuvering coefficients included in MM were estimated from turning circle and zigzag CFD free running simulations trial data. The MM simulations using the predicted coefficients showed much better agreement with ME free running than those using coefficients estimated from captive model experiments and empirical prediction.

For maneuvering prediction in waves, the mathematical models often use the hydrodynamic maneuvering coefficients estimated from the experimental captive test in calm water. Also, the wave forces are considered as the summation of Froude-Krylov and diffraction forces. These result in differences between MM predictions and experimental free running data since the maneuvering coefficients variations due to waves and wave drift forces are important for MM prediction in waves (Son and Hamamoto 1982).

The objective of the present work is to employ the system identification technique with CFD outputs to improve MM predictions in following and quartering waves by tuning the maneuvering coefficients and wave forces. Since the ultimate goal of this study is to predict broaching with MM modified with SI, the 4DOF MM (Umeda et al. 2008) which is popular for broaching prediction model is applied as a basic MM for this study. The CLS system identification technique was used to estimate the manoeuvring coefficients and the wave forces using CFD outputs. Herein, only moderate wave conditions are considered which do not cause broaching. The wave forces/effects are found from CFD simulations. First CFD free running simulations in waves are executed. Second, CFD forced motion simulations in calm water are performed with imposing exactly same motions as the free running simulation. The wave forces/effects are estimated as the difference between the total force of the first and second simulations. The CFD wave forces/effects are compared with the conventional MM wave model based on slender body theory and used to tune MM wave forces/effects by the system identification technique. The improved MM and CFD free running simulation results are compared with that of ME. Here it should be pointed out that the MM and CFD simulations are done before the ME data are available.

Table 6.1 Principal particulars of the ONRTH

	Model scale
Length (L)	3.147 m
Breadth (B)	0.384 m
Depth (D)	0.266 m
Draft (d)	0.112 m
Displacement (W)	72.6 kg
Metacentric height (GM)	0.0424 m
Natural roll period (T_{ϕ})	1.644 s
Rudder area (A_R)	$0.012\ m^2 \times 2$
Block coefficient (C_b)	0.535
Vertical position of CoG from waterline (downward positive) (OG)	$-0.392 \times d$
Radius of gyration in pitch (κ_{yy})	$0.25 \times L$
Maximum rudder angle (δ_{max})	±35°

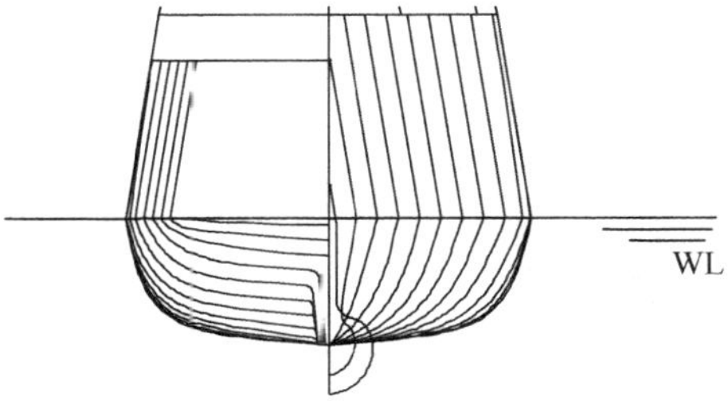

Fig. 6.1 Body plan of the ONRTH model

6.2 ME, CFD, and MM Methods

6.2.1 Subject Ship

The 1/49 scaled model of ONR tumblehome (ONRTH), was developed at Naval Surface Warfare Center Carderock Division (Bishop et al. 2005), appended with skeg, bilge keels, rudders, shafts with propeller shaft brackets and twin propellers was used for the free running experiments. The main particulars of the ONRTH ship are listed in Table 6.1. The details of the body plan and the model are shown in Figs. 6.1 and 6.2.

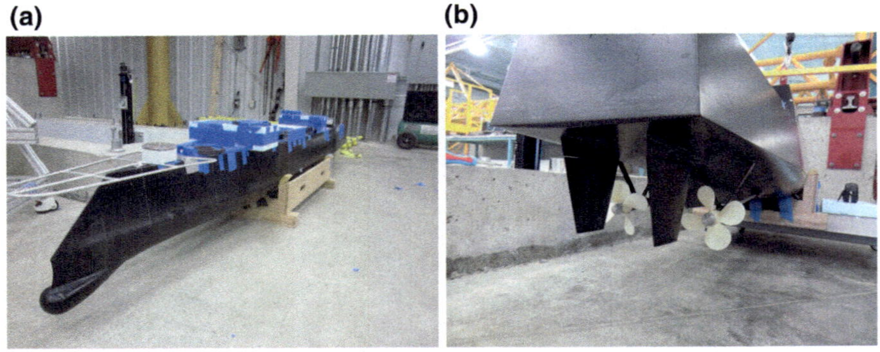

Fig. 6.2 Bow and stern of the ONRTH model: **a** bow; **b** stern

6.2.2 ME Method

All ME free running data was acquired in IIHR wave basin. The wave basin has dimensions of $40 \times 20\ \text{m}^2$ with 3 m water depth and is designed to test captive or radio-controlled model scale ships.

The model launch system enables specification and replication of the free running trial initial conditions. Roll, pitch, and yaw angles of the model ship were measured by a fiber optical gyroscope. Meanwhile, the plane trajectory of the model was recorded by the tracking system, which uses two-camera vision. The tracking cameras capture two LED lights placed on the deck of the model. In order to increase the reliability and accuracy of the 5DOF (Degree of Freedom) measurement and to enable measurement for all 6DOF of the free running model, i.e. the heave motion, a 6DOF visual motion capture system was added to the tracking system. A detailed description of the wave basin and wavemakers, carriage model tracking, 6DOF visual motion capture and free running 6DOF systems, model geometry and ballasting, and free running trials tests in calm water and waves is provided by Sanada et al. (2012).

The experimental procedure was as follows. First, the model ship was fixed on the launch system by electromagnetics while heave, roll, and pitch are free. After the propeller starts to rotate, the model was accelerated by the launch system to reach the target speed. Since the towing system acts as the hard spring there would be small oscillations for the surge motion of the towed ship. After the ship was at the target speed the model was towed for more distance until the bow was located on the wave crest. Then the model was released and the rudder controller was activated after few seconds to start maneuvering. The propeller rate was kept constant during free running. The ME and CFD trial conditions are shown in Table 6.2.

Table 6.2 ME and CFD free running test matrice

		Test	Nominal Fr	δ (deg)	ψ_c^* (deg)	H/λ	λ/L_{pp}
ME	Calm water	Turning circle	0.1, 0.2	25, 35	NA		
		Zigzag	0.1, 0.2	10, 20, 35	10, 20, 35		
		Large angle zigzag	0.2	35	90		
	Wave	Zigzag	0.1, 0.2	20	20	0.02, 0.03	1.0
		Straight running	0.1, 0.2	NA	NA	0.02, 0.03	1.0
		Course keeping	0.1, 0.2	Proportional control, P = 1	20	0.02, 0.03	1.0
CFD	Calm water	Turning circle	0.2	25	NA		
		Zigzag	0.2	20	20		
		Large angle zigzag	0.2	35	90		
	Wave	Zigzag	0.2	20	20	0.02, 0.03	1.0
		Straight running	0.2	NA	NA	0.02, 0.03	1.0
		Course keeping	0.2	Proportional control, P = 1	20	0.02, 0.03	1.0

ψ_c^*Target yaw angle

6.2.3 CFD Method

The code CFDShip-Iowa v4 (Carrica et al. 2010) is used for the CFD computations. The CFDShip-Iowa is an overset, block structured CFD solver designed for ship applications using either absolute or relative inertial non-orthogonal curvilinear coordinate system for arbitrary moving but non-deforming control volumes. Turbulence models include blended k-ε/k-ω based isotropic and anisotropic Raynolds Averaged Navier Stokes (RANS), and (detached eddy simulations) DES approaches with near-wall or wall functions. The discretized geometries of the twin propellers were not included in the simulations. Instead, a simplified body force model is used for the propeller which prescribes axisymmetric body force with axial and tangential components.

The propeller model requires the experimental open water curves and advance coefficients as input and provides the torque and thrust forces. The open water curves

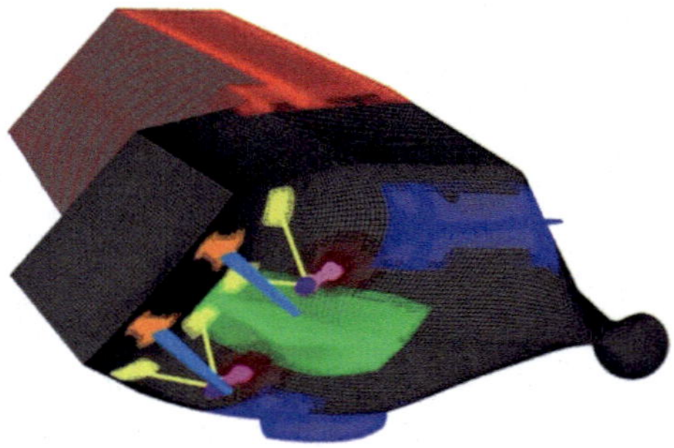

Fig. 6.3 CFD overset grids for ONRTH hull and appendages

are defined as a second order polynomial fit of the experimental $K_T(J)$ and $K_Q(J)$ curves. The advance coefficient is computed using ship speed with neglecting the wake effects. Herein, two PID controllers are used. The heading controller acting on the rudders are responsible to turn the rudders to keep the ship in the desired direction. The speed controller acting on the body force propeller model is responsible to rotate the propellers at appropriate propeller rate to keep the ship at the desired speed. The heading controller uses P = 1 for the proportional gain and zero for both the integral and derivative gains mimicking the experimental setup which uses a proportional heading control.

The CFD initial condition is different with ME in several ways. The CFD model was accelerated with infinite rate to the target speed unlike ME. Then the model was towed at target speed which was constant while the model was only free to heave and pitch and not roll until the wave trough was located at midship. After that, the model was released and rudder controller was activated immediately to start maneuvering. The differences between ME and CFD setup might cause some discrepancies between ME and CFD results.

The free model is appended with skeg, bilge keels, superstructure, rudders, rudder roots, shafts, and propeller brackets same as the ME model but not appended with actual propellers. The computational grids are overset with independent grids for the hull, superstructure, appendages, refinement, and background, and then assembled together to generate the total grid. The total number of grid points is 12.1 M for free model simulations. Details of the grids are shown in Table 6.3 and Fig. 6.3. The free running in waves and calm water verification studies have been done (Sadat-Hosseini et al. 2011; Araki et al. 2012) which showed quantitative agreement with ME results.

6.2.4 MM Method

4DOF maneuvering mathematical model was used for the MM simulations as shown in Eqs. (6.1)–(6.5). The low encounter frequency model (Umeda et al. 2008) is 4DOF surge-sway-roll-yaw model and is modified especially for surge equation and rudder

Table 6.3 Grids for free model simulations

Name	Size (grid points)	# of procs	Type
Hull S/P[a]	199 × 61 × 104 (1.26 M ×2)	12 (×2)	Double O
Skeg S/P	61 × 49 × 40 (0.12 M ×2)	1 (×2)	O
Bilge keel S/P	99 × 45 × 50 (0.23 M ×2)	2 (×2)	H
Rudder root collar S/P	121 × 35 × 28 (0.12 M ×2)	1 (×2)	O
Rudder root gap S/P	121 × 51 × 19 (0.12 M ×2)	2 (×2)	Conformal to collar
Rudder outer S/P	61 × 36 × 55 (0.12 M ×2)	1 (×2)	Double O
Rudder inner S/P	61 × 36 × 55 (0.12 M ×2)	1 (×2)	Double O
Rudder gap S/P	121 × 51 × 19 (0.12 M ×2)	2 (×2)	Conformal to inner and outer
Shaft collar S/P	39 × 50 × 57 (0.11 M ×2)	1 (×2)	O
Shaft proper S/P	74 × 41 × 37 (0.11 M ×2)	1 (×2)	O
Shaft tip S/P	110 × 117 × 100 (1.29 M ×2)	12 (×2)	O with end pole
Strut outer S/P	69 × 34 × 50 (0.12 M ×2)	1 (×2)	O
Strut inner S/P	69 × 34 × 50 (0.12 M ×2)	1 (×2)	O
Superstructure	165 × 61 × 85 (0.86 M)	8	Wrap
Refinement	145 × 81 × 113 (1.33 M)	12	Cartesian
Background	213 × 84 × 113 (2.02 M)	20	O
Total	(12.1 M)	116	

[a]S/P Starboard/Port

model (Araki et al. 2012). The model is developed in horizontal body axes which are shown in Fig. 6.4.

$$(m + m_x)\dot{u} - (m + m_y)vr = T(u;n) - R(u;n) + X_{vv}(u)v^2 + X_{vr}(u)vr + X_{rr}(u)r^2 + X_R(\delta, u, v, r) + X_W \quad (6.1)$$

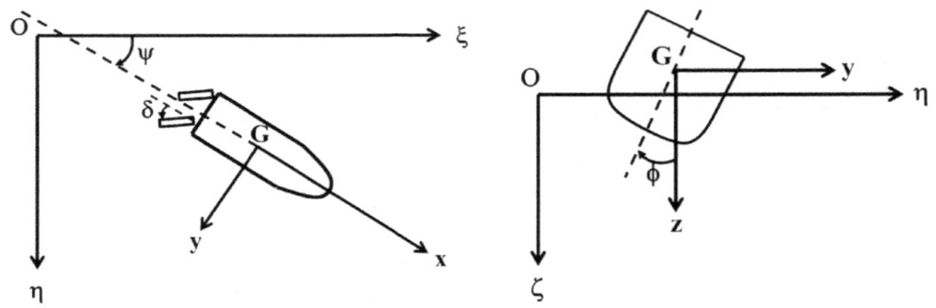

Fig. 6.4 Coordinate system for 4DOF MM

$$
\begin{aligned}
(m+m_y)\dot{v}+(m+m_x)ur &= Y_v(u)v+Y_r(u)r+Y_\phi(u)\phi \\
&\quad + Y_{vvv}(u)v^3+Y_{vvr}(u)v^2r+Y_{vrr}(u)r^2v \\
&\quad + Y_{rrr}(u)r^3+Y_R(\delta,u,v,r)+Y_W
\end{aligned} \tag{6.2}
$$

$$
\begin{aligned}
(I_{xx}+J_{xx})\dot{p} &= m_x z_H ur+K_v(u)v+K_r(u)r+K_p(u)p+K_\phi(u)\phi \\
&\quad - mgGZ(\phi)+K_{vvv}(u)v^3+K_{vvr}(u)v^2r+K_{vrr}(u)r^2v \\
&\quad + K_{rrr}(u)r^3+K_R(\delta,u,v,r)+K_W
\end{aligned} \tag{6.3}
$$

$$
\begin{aligned}
(I_{zz}+J_{zz})\dot{r} &= N_v(u)v+N_r(u)r+N_\phi(u)\phi+N_{vvv}(u)v^3+N_{vvr}(u)v^2r \\
&\quad + N_{vrr}(u)r^2v+N_{rrr}(u)r^3+N_R(\delta,\phi,u,v,r)+N_W
\end{aligned} \tag{6.4}
$$

Here

$$
[K_v\ K_r\ K_{vvv}\ K_{vvr}\ K_{vrr}\ K_{rrr}]^T = z_H[Y_v\ Y_r\ Y_{vvv}\ Y_{vvr}\ Y_{vrr}\ Y_{rrr}]^T \tag{6.5}
$$

In the mathematical model, resistance is estimated from a captive model experiment and the thrust is estimated from propeller open water tests in calm water as described in Umeda et al. (2008). Roll restoring moment ($mgGZ$) is estimated from hydrostatic calculations in calm water. Maneuvering coefficients including heel-induced hydrodynamic derivatives are estimated from calm water captive model experiments (Hashimoto et al. 2008). Roll damping is estimated from roll decay model tests (Umeda et al. 2008). For the ONRTH, the values of correction coefficients for rudder are empirically developed from other model experiments (Kose et al. 1981). The empirical values are also used for the interaction force coefficients induced on the hull by rudder nominal force. All maneuvering and rudder coefficients are listed in Table 6.4. In wave cases, the wave forces X_w, Y_w, K_w, and N_w are calculated from linear Froude-Krylov forces and diffraction forces based on slender ship theory for zero encounter frequency. It is known that the wave particle velocity affects the rudder and propeller inflow speed which is important for the maneuverability in waves. Therefore the effects of wave particle velocity to propeller thrust and rudder force are taken into account.

Table 6.4 Values of original and SI-calm maneuvering and rudder coefficients used in 4-DOF nonlinear MM

Coef.	Original	SI-calm	Coef.	Original	SI-calm
ε	1.0	0.75	Y_{vrr}	−0.80	0.32
γ_R	0.70	0.55	Y_{rrr}	0.174	0.080
l_R/L	−1.00	−0.95	Y_ϕ	−5.1E−04	−6.5E−04
t_R	0.30	0.10	J_{xx}	4.1E−05	0.0001
a_H	0.25	0.23	z_H	0.852	1.08
z_{HR}/d	0.854	0.802	K_p	−0.243	−0.203
x_H/L	−0.45	−0.52	K_ϕ	6.3E−04	1.0E−03
m_x	0.0131	0.0	J_{zz}	0.0079	0.0059
X_{vv}	−0.0858	−0.070	N_v	−0.0932	−0.0851
X_{vr}	0.0522	0.065	N_r	−0.0549	−0.0395
X_{rr}	−0.0213	−0.025	N_{vvv}	−0.532	−0.492
m_y	0.109	−0.070	N_{vvr}	−0.629	−0.805
Y_v	−0.30	−0.20	N_{vrr}	−0.139	−0.121
Y_r	−0.0832	0.07	N_{rrr}	−0.00446	−0.0065
Y_{vvv}	−1.77	−2.0	N_ϕ	−0.00511	−0.00989
Y_{vvr}	0.262	0.32			

However, the MM simulations using these coefficients and wave model show some discrepancy with the ME free running in calm water and waves. The authors tuned the calm water maneuvering and rudder coefficients values using the SI with CFD free running results in calm water. The results showed better agreement with the ME calm water data (Araki et al. 2012). The values of tuned coefficients are shown in Table 6.4. Herein, the wave model is tuned using the SI technique with CFD free running and captive results in waves.

6.3 SI Method and the Results

A constrained least square (CLS) method using generalized reduced gradient algorithm (Lasdon et al. 1978) is used for SI. The CLS method provided reasonable maneuvering coefficients from CFD calm water maneuvers (Araki et al. 2012). To predict wave forces/effects, it is necessary to extract the wave forces/effects from total hydrodynamic forces. To achieve this purpose, first 6DOF CFD free running simulations in waves are executed. Second, CFD forced motion simulations in calm water are performed with imposing exactly same motions as the free running simulation. Thus the wave forces/effects are estimated as the difference between the total force of the first and second simulations.

Fig. 6.5 Wave forces during $\psi_C = 20°$ course keeping maneuver in following waves with nominal Fr = 0.20, wave steepness 1/50 and wave length to ship length ratio 1.0

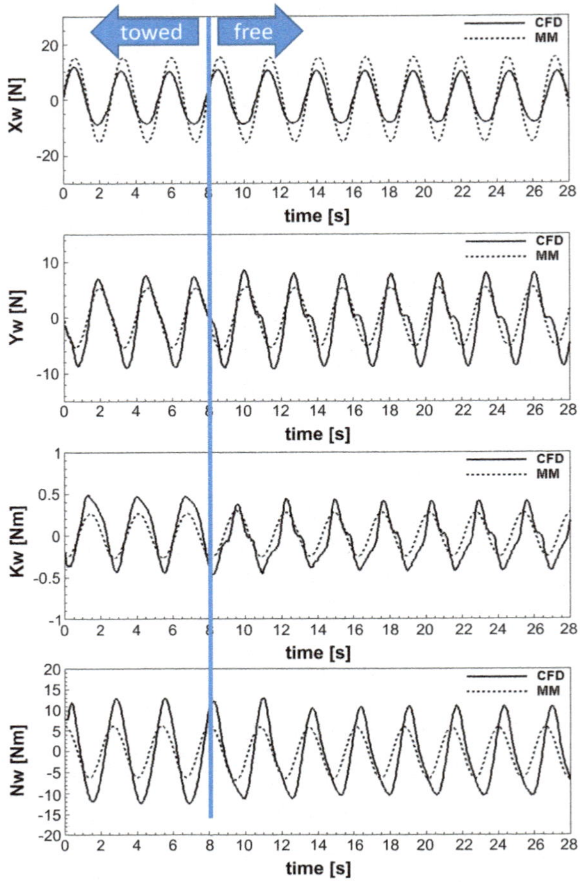

Figure 6.5 shows the extracted CFD wave forces/effects during $\psi_C = 20°$ course keeping maneuver in quartering waves with nominal Fr = 0.20, wave steepness 1/50 and wave length to ship length ratio 1.0. During the free running, the model is imposed with constant forward speed Fr = 0.20 and constant yaw angle 20° until 8.02 s then released to start course keeping maneuvers. Here the "MM" is the wave forces computed by MM wave model.

It is shown that MM overestimates surge wave force and underestimates sway, roll, and yaw wave forces. Although definitions of wave forces are different, these tendencies are found in previous research by Hashimoto et al. (2011) as well where they compared MM and captive ME wave exciting forces for ONRTH. Moreover the MM waveforms seem to be different from CFD waveforms especially after the model is released. This could because that MM wave model merely includes the Froude-Krylov and diffraction forces while CFD wave force includes all wave effects not just Froude-Krylov and diffraction forces but also wave drift forces and maneuvering coefficients variations due to waves. According to CFD extracted wave forces/effects, the wave correction parameters are estimated by SI methods to tune MM wave model based on Eqs. (6.6)–(6.8).

$$
\begin{aligned}
X_W &= a_1 X_W^{FK} + k\zeta_w \big(a_2 \sin(k\xi_G + \varepsilon_{a2}) \cdot X_{vv} \cdot v^2 \\
&\quad + a_3 \sin(k\xi_G + \varepsilon_{a3}) \cdot X_{vr} \cdot vr + a_4 \sin(k\xi_G + \varepsilon_{a4}) \cdot X_{rr} \cdot r^2\big) + W_{CX} \\
Y_W &= b_1 Y_W^{FK} + b_2 Y_W^{Dif} + k\zeta_w (b_3 \sin(k\xi_G + \varepsilon_{b3}) \cdot Y_v \cdot v \\
&\quad + b_4 \sin(k\xi_G + \varepsilon_{b4}) \cdot Y_r \cdot r) + W_{CY} \\
K_W &= c_1 K_W^{FK} + c_2 K_W^{Dif} + k\zeta_w (c_3 \sin(k\xi_G + \varepsilon_{c3}) \cdot K_v \cdot v + c_4 \sin(k\xi_G + \varepsilon_{c4}) \cdot K_r \cdot r) \\
N_W &= d_1 N_W^{FK} + d_2 N_W^{Dif} + k\zeta_w (d_3 \sin(k\xi_G + \varepsilon_{d3}) \cdot N_v \cdot v \\
&\quad + d_4 \sin(k\xi_G + \varepsilon_{d4}) \cdot N_r \cdot r) + W_{CN}
\end{aligned} \tag{6.6}
$$

where

$$
\begin{aligned}
W_{CX} &= \rho g \zeta_w^2 B^2 / L \cdot \sin(\psi/2 + \psi_0) \cdot \left(1 - e^{-10Fr}\right) \cdot C_X(T_w) \\
W_{CY} &= \rho g \zeta_w^2 B^2 / L \cdot \sin\psi \cdot C_Y(T_w) \\
W_{CN} &= \rho g \zeta_w^2 B^2 \cdot \sin\psi \cdot C_N(T_w)
\end{aligned} \tag{6.7}
$$

$$
C_{X,Y,N}(T_w) = \alpha_{X,Y,N} \cdot \left(\frac{m_{WD}}{\eta_{WD}}\right)\left(\frac{T_w}{\eta_{WD}}\right)^{m-1} \exp\left\{-\left(\frac{T_w}{\eta_{WD}}\right)^m\right\} \tag{6.8}
$$

$$
\begin{aligned}
u_w &= \beta_1 \cdot \zeta_w \omega \cos\psi e^{-kz} \cos(k\xi_G + kx\cos\psi) \\
v_w &= -\beta_2 \cdot \zeta_w \omega \sin\psi e^{-kz} \cos(k\xi_G + kx\cos\psi)
\end{aligned} \tag{6.9}
$$

The new wave model includes the major maneuvering coefficients variations and wave drift forces. For the simplification the wave drift coefficients shown in Eq. (6.8) are expressed as the Weibull distribution respect to wave periods. The shape and scale parameter m_{WD}, η_{WD} of Eq. (6.8) are determined from the Yasukawa's research (2006). Also the normal force of CFD rudder is used to tune the wave particle velocity effects to propeller and rudder as shown in Eq. (6.9).

Here a, b, c, d, ε in Eq. (6.6), ψ_0 in Eq. (6.7), α in Eq. (6.8), and β in Eq. (6.9) are tuned by the SI. These SI procedures are repeated to the other cases: straight running and 20/20 zigzag with nominal Fr = 0.20, wave steepness 1/50 and wave length to ship length ratio 1.0. The original wave correction coefficients and the tuned coefficients are shown in Table 6.5.

Table 6.5 Original wave correction coefficients and the coefficients estimated by SI using extracted CFD wave forces/effects data

Coef.	Orig	SI-wave	Coef.	Orig	SI-wave
a_1	1.0	0.758	c_1	1.0	1.26
a_2	0.0	16.33	c_2	1.0	0.355
a_3	0.0	0.855	c_3	0.0	0.510
a_4	0.0	0.132	c_4	0.0	0.195
ε_{a2}	0.0	0.391	ε_{c3}	0.0	−0.99
ε_{a3}	0.0	3.21	ε_{c4}	0.0	1.03
ε_{a4}	0.0	0.0	d_1	1.0	1.43
α_X	0.0	−46.3	d_2	1.0	0.403
b_1	1.0	1.35	d_3	0.0	1.01
b_2	1.0	0.423	d_4	0.0	0.213
b_3	0.0	2.18	ε_{d3}	0.0	0.982
b_4	0.0	0.496	ε_{d4}	0.0	−0.99
ε_{b3}	0.0	−0.552	α_N	0.0	−2.50
ε_{b4}	0.0	0.810	ψ_0	0.0	0.301
α_Y	0.0	−106.2	β_1	1.0	0.643
			β_2	1.0	0.425

6.4 Comparison Between EFD, CFD, and MM Free Runnings

6.4.1 Maneuvering in Calm Water

Comparison between ME, CFD, “MM-Orig.”, and “MM-SI-calm” trajectories are shown in Fig. 6.6. Here “MM-Orig.” indicates the MM simulation using the maneuvering coefficients estimated from captive model test and rudder coefficients predicted from empirical charts (Kose et al. 1981). “MM-SI-calm” indicates MM simulation using maneuvering and rudder coefficients estimated by SI using CFD calm water maneuvering data (Araki et al. 2012). Figure 6.6a shows the trajectories of turning circle tests ($\delta = 25°$, $Fr = 0.20$), Fig. 6.6b shows zigzag tests ($\psi_c/\delta = 20/20$, $Fr = 0.20$) trajectories, and Fig. 6.6c shows large angle zigzag tests ($\psi_c/\delta = 90/35$, $Fr = 0.20$).

The steady state variables and turning parameters for turning circle and 1st, 2nd overshoot angles and K_s, T_s, N_s steering quality indices for zigzag cases (ABS 2006) for both CFD and MM predictions are compared against ME ones and the overall errors are plotted in Fig. 6.7. The steering quality indices are computed from nonlinear first order Nomoto’s model (Norrbin 1963) shown in Eq. (6.10). As shown in Figs. 6.6 and 6.7, the errors of MM-SI-calm are much smaller than that of MM-Orig.

$$T_s\dot{r} + N_s r^3 + r = K_s\delta \tag{6.10}$$

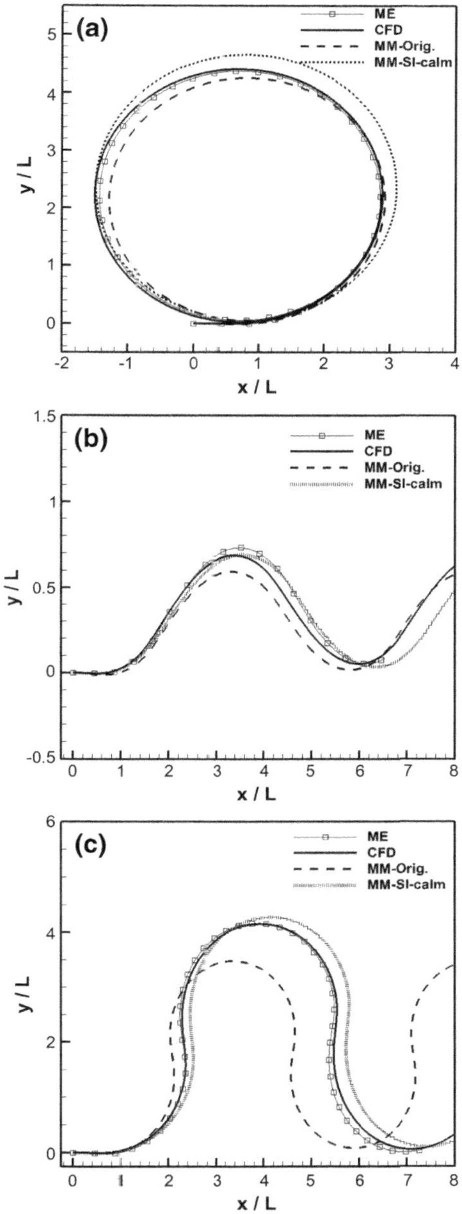

Fig. 6.6 Trajectories of ME, CFD, MM-Orig., and MM-SI free running in calm water: **a** $\delta = 25°$ turning circle; **b** 20/20 zigzag; **c** 90/35 large angle zigzag

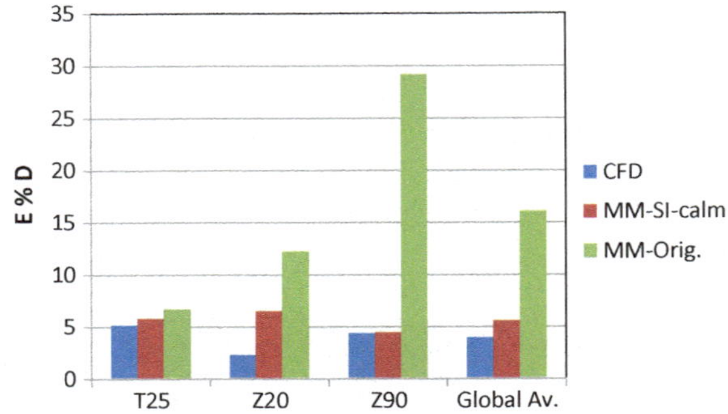

Fig. 6.7 Errors from ME free running results: "T25" $\delta = 25°$ turning circle; "Z20" 20/20 zigzag; "Z90" 90/35 zigzag; "Global Av." average of T25, Z20, and Z90 errors

6.4.2 *Straight Running in Following Waves*

Figure 6.8 shows the comparison between CFD, MM-SI-calm and MM-SI-wave straight running in following waves with nominal Fr = 0.20, wave steepness 1/50 and wave length to ship length ratio 1.0. Here "MM-SI-calm" indicates MM simulation using maneuvering and rudder coefficients estimated by SI from CFD calm water maneuvering data (Araki et al. 2012) with original wave model. "MM-SI-wave" indicates MM simulations using same maneuvering and rudder coefficients with "MM-SI-calm" but with new wave model shown in Eq. (6.6) which is the improved wave forces using CFD wave forces/effects data.

In Fig. 6.8, CFD shows remarkable agreement with ME especially for the surge and pitch motions. It is showing possibility to replace ME free running test with CFD simulations even in wave conditions. Here heave and pitch motions for the 4DOF (surge-sway-yaw-roll) MM are assumed to be the same as the static equilibrium positions of the ship in waves. MM shows larger heave and pitch motion than those of CFD and ME which indicates that 6DOF model could be desirable. In surge motion, CFD successfully reproduce the nominal speed loss due to waves. The MM-SI-calm fails to express the nominal speed loss and the surging amplitudes are larger than that of CFD and ME. The MM-SI-calm cannot represent nominal speed loss because the wave drift terms are not included in the MM-SI-calm model. Meanwhile MM-SI-wave successfully predicts nominal speed loss and surging amplitude within high degree of accuracy.

6.4.3 *Course Keeping and Zigzag in Following and Quartering Waves*

Course keeping and zigzag simulations in wave conditions are shown in this section. The simulation procedure is as follows. First the model is accelerated to the target ship speed with 2DOF (heave and pitch). After the model reaches to the speed, the

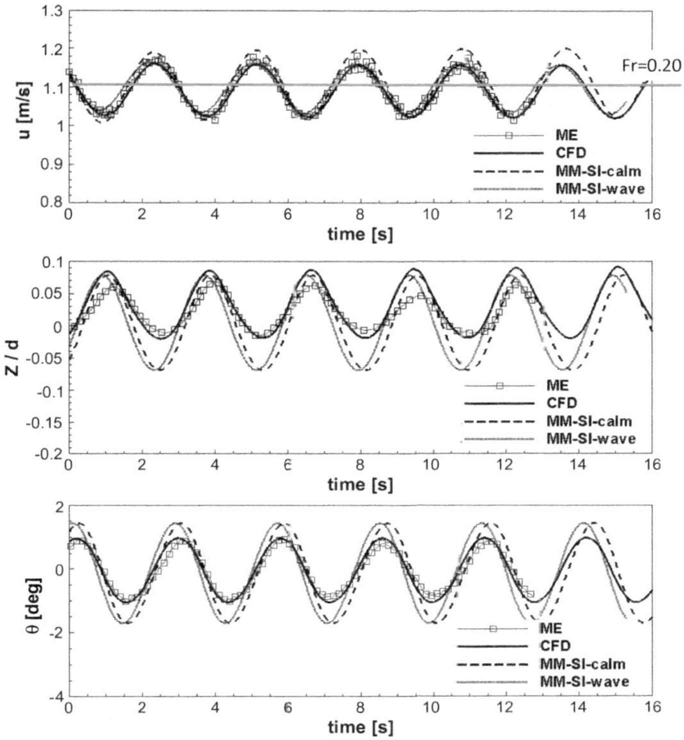

Fig. 6.8 Straight running in following waves with nominal Fr = 0.20, wave steepness 1/50, and wave length to ship length ratio 1.0

model is towed with constant speed for a while and released when the bow is located on the wave crest. The rudder control starts just after the model is released. In the ME, it should be noted that the towing time was very short because of the limitation of the facility's size. Moreover it was 3DOF (heave, pitch and roll) during towing in ME.

Figure 6.9 shows the comparison between CFD and MM-SI-calm $\psi_C = 20°$ course keeping in quartering waves with nominal Fr = 0.20, wave steepness 1/50 and wave length to ship length ratio 1.0. Here the ME and CFD rudder control start just after the model is released at a wave trough.

In the trajectory, CFD course deviation shows good agreement with ME which indicates that CFD well predicts the wave drift force. Although the ME shows wobbly trajectory compared to CFD due to large oscillations for sway motions. Due to the sway motion error, the roll motions show some difference between ME and CFD while the error is much smaller than that of sway motion. However CFD successfully predicts the surge and yaw motions in quartering waves. Paying attention to CFD and MM results, MM-SI-calm shows small course deviation compared to that of CFD.

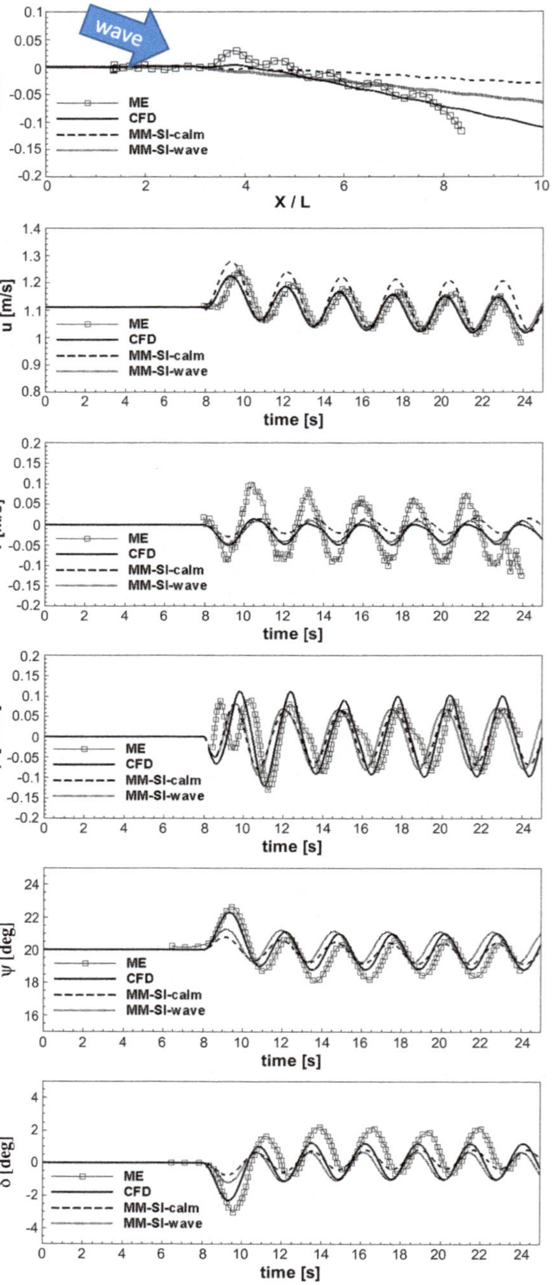

Fig. 6.9 $\psi_C = 20°$ course keeping maneuver in quartering waves with nominal Fr = 0.20, wave steepness 1/50, and wave length to ship length ratio 1.0

From the state variables comparisons, it is clear that MM-SI-calm has some discrepancy on the wave forces and wave drift effects compared to CFD and ME. MM-SI-calm's wave model overestimates the surge wave force and underestimates the sway, roll, and yaw wave forces. The MM-SI-wave shows better agreement with CFD than MM-SI-calm for state variables and the trajectory. The wave drift effects can be seen in sway motion prediction which improves the prediction of the course deviation.

Figure 6.10 shows the comparison between ME, CFD and MM 20/20 zigzag in following and quartering waves with nominal Fr = 0.20, wave steepness 1/50 and wave length to ship length ratio 1.0. The CFD results show good agreement with shifted ME for trajectory, surge, and yaw motions. In sway motion, CFD seems underestimating the wave force compared to ME which could explain the discrepancy of the roll motions. MM-SI-calm shows qualitative agreement with ME maneuver but not quantitative. MM-SI-calm overestimates surge wave force and underestimates sway, and yaw wave forces. The MM-SI-calm prediction of the zigzag trajectory is very close to the one predicted in calm water shown in Fig. 6.6b. This is due to the fact that the maneuvering coefficients oscillations and drift forces induced by waves are neglected in MM-SI-calm. The MM-SI-wave improves the prediction as it shows the oscillations on the state variables induced by the waves. Also, the speed loss is predicted well in MM-SI-wave such that the trajectory shows good agreement with CFD and ME ones.

6.5 Conclusions

System identification method using CFD free running data is shown to be an efficient approach for estimating maneuvering, rudder, and wave correction coefficients in the MM. Araki et al. (2012) show the reasonable maneuvering and rudder coefficients can be obtained from a few CFD free running data in calm water. However, the MM still shows some error predicting the ship motion in waves. The original MM includes the Froude-Krylov and diffraction forces as the wave forces and the wave particle velocity as the wave effect on the propeller and rudder which clearly fails to predict the oscillation amplitudes and the wave drift effects. Therefore the MM wave model is improved by adding correction parameters for Froude-Krylov, diffraction forces, and wave particle velocity. Moreover, effects of wave drift forces and maneuvering coefficient variations due to waves are taken into account and these correction parameters are predicted by CLS using the extracted CFD wave forces/effects data. The extracted CFD wave forces/effects data are generated from the CFD free running data in waves and CFD forced motion data in calm water. The MM simulations using the new wave model and estimated wave correction coefficients show much better agreement with CFD than the MM simulations using the original wave model. The CFD simulations are validated with ME free running results; CFD mostly shows quantitative agreement with ME which shows the possibility of replacing ME free running trials with CFD simulations.

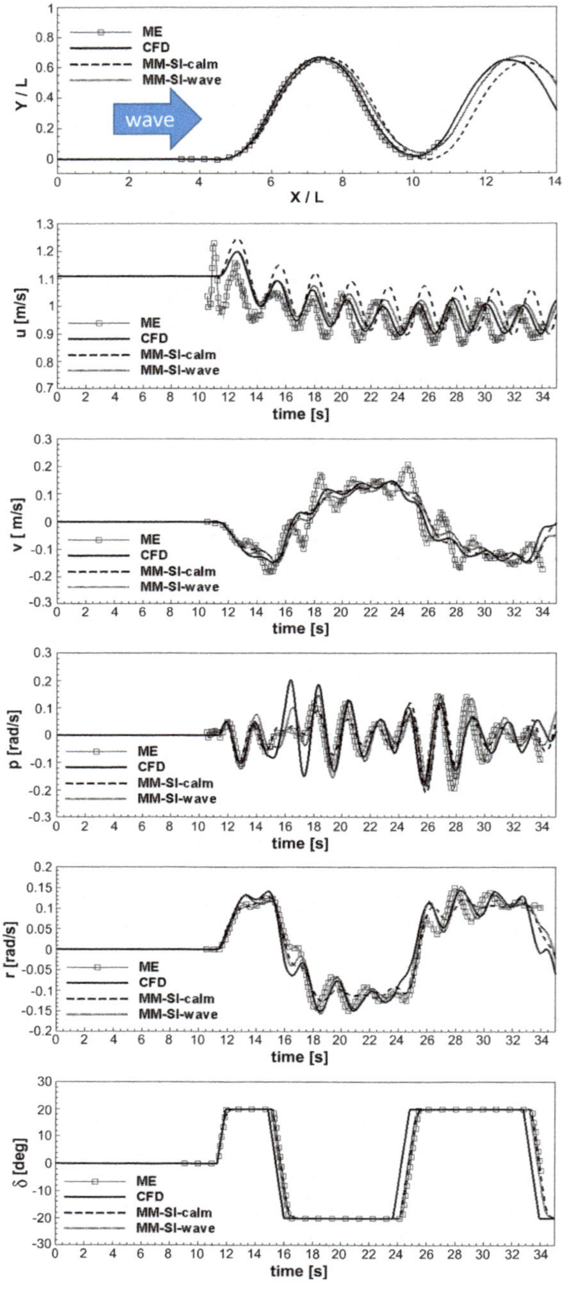

Fig. 6.10 20/20 zigzag in quartering waves with nominal Fr = 0.20, wave steepness 1/50, and wave length to ship length ratio 1.0

Several follow-up studies were conducted subsequent to this research to improve the mathematical model. Araki et al. (2013) improved the mathematical models for the wave forces, maneuvering coefficients variations, and rudder forces. However, the wave model provided too stable motions in severe waves and is unable to predict the instabilities such as broaching. Yoneda et al. (2017) and Mizumoto et al. (2018) conducted more studies to improve the wave model for severe condition. In future, more experiments and simulations will be conducted in IIHR wave basin and National Research Institute of Fisheries Engineering (Japan) to facilitate the mathematical model improvement for conditions with higher ship speed and tougher waves including irregular waves near to broaching conditions. The CFD and experimental studies will be used to include not only the wave drift force but also the nonlinear wave effects including the memory effect functions (Mikami and Kashiwagi 2009) and the rudder and propeller exposing effects in the mathematical model.

Acknowledgements This research was sponsored by the Office of Naval Research Grant N000141-21-05-6-8 and NICOP Grant N00014-09-1-1089 under the administration Dr. Patrick Purtell. The CFD simulations were conducted utilizing DoD HPC. The authors are grateful to Mr. K. Tanimoto and Ms. K. Takagi of Osaka University and Mr. A. Hanaoka of The University of Iowa, IIHR for assistance conducting the experiments.

References

Abkowitz, M.A., 1964, Lecture of ship hydrodynamics—steering and maneuverability, Hydro-and Aerodynamic Laboratory, Report No. Hy-5.

Abkowitz M. A., 1980, Measurement of Hydrodynamic Characteristics from Ship Maneuvering Trials by System Identification, *SNAME Transactions*, 88, 283–318.

American Bureau of Shipping, 2006, Guide for Vessel Maneuverability, Houston, TX.

Araki, M., Umeda, N., Hashimoto, H., Matsuda, A., 2010. "Broaching prediction using an improved system-based approach". *Proceedings of 28th Symposium on Naval Hydrodynamics*, pp. 56–68.

Araki, M., Sadat-Hosseini, H., Sanada, Y. Tanimoto, K., Umeda, N., and Stern, F., 2012, Estimating Maneuvering Coefficients Using System identification Methods with Experimental, System-based, and CFD Free-running Trial Data, *Ocean Engineering*, 51:63–84.

Araki, M., Sadat-Hosseini H., Sanada, Y., Umeda, N., and Stern, F., 2013, "System Identification using CFD Captive and Free Running Tests in Severe Stern Waves," *Proceedings 13th International Ship Stability Workshop*, Brest, France, September 23 – 26.

Bishop, R., Belknap, W., Turner, C., Simon, B., Kim, J., 2005, Parametric investigation on the influence of GM, roll damping, and above-water form on the roll response of model 5613, Report NSWCCD-50-TR-2005/027.

Carrica, P. M., Huang, J., Noack, R., Kaushik, D., Smith, B., Stern, F., 2010, "Large-scale DES computations of the forward speed diffraction and pitch and heave problems for a surface combatant", *Computer & Fluids*, 39(7):1095–1111.

Christensen, A., Blanke, M., 1986. A Linearized state-space model in steering and roll of high-speed container ship. Technical Report 86-D-574, Servolaboratoriet, Technical University of Denmark.

Hashimoto, H., Umeda, N., and Matsuda, A., 2011, "Broaching prediction of a wave-piercing tumblehome vessel with twin screw and twin rudder", *Journal of Marine Science and Technology*, 16(4):448–461.

Hashimoto, H., Stern, F., and Sadat-Hosseini, H., 2008, An application of CFD for advanced broaching prediction (2ndReport), *Conference Proceedings of the Japan Society of Naval Architects and Ocean Engineers*, Vol. 6, pp. 237–240.
Kang, C.G., Seo, S.H., Kim, J.S., 1984. "Maneuverability analysis of ship by system identification technique". *SNAK*, 21, No.4.
Kose, K., Yumuro, A., Yoshimura, Y., 1981, "Concrete of mathematical model for ship maneuverability", *Proceedings of 3rd Symposium on Ship Maneuverability*, Society of Naval Architects of Japan, pp. 27–80, (in Japanese).
Lasdon, L. S., Waren, A. D., Jain, A., Ratner, M., 1978, Design and testing of a generalized reduced gradient code for nonlinear programming, *Journal of ACM Transactions on Mathematical Software,* Volume 4 Issue 1.
Lewis, F. L., 1986, Optimal Estimation –with an introduction to stochastic control theory, John Wiley & Sons, Inc.
Mikami, T., Kashiwagi, M., 2009, "Time-domain strip method with memory-effect function considering body nonlinear wave-body interactions (2nd report)", *Journal of Marin Science and Technology*, 14(2) 185–199.
Mizumoto, K., Stern, F., Araki, M., Umeda, N., 2018, "CFD-based system identification for improving a system-based simulation model of broaching in stern quartering waves (tentative title)", *Proceeding of 13th International Conference on Stability of Ships and Ocean Vehicles STAB 2018,* Kobe, Japan, 92–98.
MMG, 1980. MMG report V. *Bulletin of Society of Naval Architects of Japan*. 616, 565–576.
Nonaka, K., Mori, M., Matsumoto, N., 1972. "Estimating Maneuvering Coefficients from Free-Running Trials". *Proceedings of 20th Meetings of Ship Research Institute*.
Norrbin, N. H., 1963, "On the design and analysis of zig-zag test on base of quasi linear frequency response", Technical Report B104-3, The Swedish State Shipbuilding Experimental Tank, Gothenburg, Sweden.
Ogawa, A., Kasai, H., 1978. "On the mathematical model of maneuvering motion of ship". *International Shipbuilding Progress* 25 (292), 306–319.
Rhee, K. P., Kim, K., 1999, A new sea trial method for estimating hydrodynamic derivatives, *Journal of Ship & Ocean Technology* 3, (3), 25–44.
Ross, A.,.T., Perez, and T., Fossen, 2007. A novel maneuvering model based on lowaspect-ratio lift theory and Lagrangian mechanics. *Proceedings of the IFAC Conference on Control Applications in Marine System (CAMS)*.
Sadat-Hosseini, H., Carrica, M. P., Stern, F., Umeda, N., Hashimoto, H., Yamamura, S., Mastuda, A., 2011, "CFD, system-based and ME study of ship dynamic instability events: surf-riding, periodic motion, and broaching", *Ocean Engineering*, Vol. 38, Issue 1, pp. 88–110.
Sanada, Y., Tanimoto, K., Takagi, K., Sano, M., Yeo, D.J., Toda, Y., Stern, F., 2012, "Trajectories of local flow field measurement around ONR tumblehome in maneuvering motion", *Proceedings of 29th Symposium on Naval Hydrodynamics*, Gothenburg, Sweden.
Shi, C., Zhao, D., Peng, J., Shen, C., 2009. "Identification of ship maneuvering model using extended kalman filtering". *International Journal Marine Navigation and Safety of Sea Transportation* 3 (1), 105–110.
Son K. H., Nomoto K., 1982, Combined behavior of manoeuvring and roll motion in following wave, Journal of the Society of Naval Architects of Japan, Vol.152, pp. 207–218.
Stern, F., Agdrup, K., Kim, S. Y., Hochbaum, A. C., Rhee, K. P., Quadvlieg, F., Perdon, P., Hino, T., Broglia, R., and Gorski, J., 2011, "Experience from 2008 – the first workshop on verification and validation of ship maneuvering simulation methods," *Journal of Ship Research*, vol. 55, No. 2, pp. 135–147.
Umeda, N., Matsuda, A., Hashimoto, H., Yamamura, S., and Maki, A., 2008, "Model experiments on extreme motions of a wave-piercing Tumblehome vessel in following and quartering waves", *Journal of the Japan Society of Naval Architects and Ocean Engineers*, Vol. 8, pp. 123–129.
Yasukawa, H., 2006, Simulation of Ship Maneuvering in Waves (1st report: turning motion), *Journal of the Japan Society of Naval Architects and Ocean Engineers*, Vol. 4, pp. 127–136, (in Japanese).

Yoneda, S., Hashimoto, H., Matsuda, A., Tahara, Y., Terada, D., Stern, F., 2017, "Investigation on the improvement of estimation accuracy of wave-exciting forces acting on ships in stern quartering waves", *Conference Proceedings of the Japan Society of Naval Architects and Ocean Engineers*, Vol. 24. (in Japanese).

Zhang, X.G., Zou, Z.J., 2011. "Identification of Abkowitz model for ship maneuvering motion using epsilon-support vector regression", *Journal of Hydrodynamics*, Ser. B 23 (3), 353–360.

Part IV
Mathematical Model of Ship Motions in Waves: Roll Damping

Chapter 7
Some Results from a New Time-Domain Bilge Keel Force Model

David S. Greeley

Abstract A new non-linear, time domain bilge keel force model was recently developed for inclusion in the new time-domain seakeeping/maneuvering in waves code TEMPEST, being developed by NSWCCD. This bilge keel force model combines a full unsteady extension of Bollay's non-linear low aspect ratio lifting surface theory for cases with adequate forward speed with a more conventional approach for cases with zero or low forward speed, using Morison's equation. This paper presents some representative results from the new bilge keel force model for a surface combatant for various roll amplitudes, roll periods, and forward speeds.

Keywords Bilge-keel forces · Time-domain seakeeping
Unsteady lifting surface theory

7.1 Introduction

The calculation of forces on ship bilge keels continues to be a challenging but important aspect of ship hydrodynamics, because of the large effect of these forces on rolling, and the consequent possibility of capsizing and loss of the ship. In the past, the estimation of these forces for ship motion simulations has traditionally relied on experimental data, and semi-empirical techniques derived largely from experimental data (Himeno 1981). More recently, an assault on this problem using modern computational techniques has begun to bear fruit. Before getting to the results which are central to this paper, it is worthwhile to review some "big-picture" aspects of the forces on bilge keels:

(1) Except for some exceptional cases such as yaw or rudder-induced rolling, ships roll because of the excitation from incident waves in the ocean. It is obvious from first principles that the forces on the bilge keels are not dependent strictly on the roll and roll rate (or even complicated functions of these variables alone!),

D. S. Greeley (✉)
Applied Physical Sciences Corporation, Concord, MA, USA
e-mail: dgreeley@aphysci.com

V. L. Belenky et al. (eds.), *Contemporary Ideas on Ship Stability*, Fluid Mechanics and Its Applications 119, https://doi.org/10.1007/978-3-030-00516-0_7

but rather on the relative motion between the bilge keels and the water. This relative motion includes *all* motions of the ship, the forward speed of the ship, the incident wave orbital velocities, and the velocities associated with the wave diffraction, the body radiation, and the steady waves generated by the ship. It makes little sense to get completely caught up with an elaborate analysis and fancy curve fits of free roll decay data from a still water model test (either stationary or with forward speed) in an attempt to discern new "physics" of rolling in a seaway, when this relative motion situation is ignored. A thorough analysis may someday indicate exactly under what conditions the relative motion problem may be ignored and roll decay data obtained from still water roll decay data constitutes the "right" input to a seakeeping code, but to the author's knowledge no such analysis has yet been done. The relative motion is largely ignored today because the tools to handle it are just being developed.

(2) Ship motion researchers commonly refer to possible hydrodynamic "memory" in a seakeeping problem as being associated only with free surface waves. However, unsteady shed vorticity (associated with time-varying circulation) also constitutes a powerful mechanism for hydrodynamic "memory". In the case of bilge keels, if it is assumed that only "linear" lift is generated (where the trailing vorticity is shed only from the trailing edge), this shed vorticity will have minimal impact. However, numerous URANS calculations and PIV measurements (Irvine 2006) have shown that for low aspect ratio bilge keels, trailing vorticity is shed from the entire *side* edge of the bilge keel, and this shed vorticity has a *large* impact on the forces developed on the middle and aft parts of the bilge keel. Bilge keel force models for zero or low forward speed that use the Morison equation implicitly capture this memory effect, in that the data used in the Morison equation depends on the history of the flow through the Keulegan-Carpenter parameter. For the situation with forward speed, either URANS calculations or an unsteady lifting surface theory that explicitly deals with side edge vortex shedding and the correct accounting for unsteady shed vorticity is necessary to capture this memory effect. The memory is dependent on all of the past motions of the ship and the incident waves, not just roll motion.

(3) It is common to recognize that bilge keels contribute to the added roll moment of inertia of the ship. It is also common to assume that this contribution is constant and may be easily estimated from the added mass of a 2D flat plate. However, if one uses a zero speed bilge keel force model based on the Morison equation, then the effect of shed vorticity on the bilge keel added mass (ship roll moment of inertia) is implicitly included because of the dependence of the inertia coefficient in the Morison equation on the Keulegan-Carpenter parameter. In the case with forward speed, the situation is more complicated and the added moment of inertia of the bilge keels, like the lifting forces which contribute to roll damping, are in fact dependent on the history of the motion because of the unsteady shed vorticity.

(4) It is typical within the ship motion community to refer to any force that cannot be modeled in a potential flow sense with a single-valued velocity potential as a "viscous" force. This implies, especially to newcomers to the field, that these

forces can only be addressed in the context of RANS or URANS calculations. Obvious examples are the side force and yaw moment on a yawed ship hull, and the forces on bilge keels. This is tantamount to ignoring all of the excellent work for calculating lifting potential flows about arbitary bodies done by John Hess and others, starting in the 1970s (Hess 1972), which became the main aerodynamic design tool for airplanes until Euler and RANS techniques supplanted them because these new techniques could deal with Mach numbers approaching or exceeding 1. In fact, many (but certainly not all) situations in ship hydrodynamics are perfect for the application of lifting potential flow techniques because of: (a) high Reynolds numbers, (b) low Mach numbers, and (c) sharp edges which fix flow separation locations. It is the author's opinion that the community would be better served if "lifting forces" were clearly differentiated from true "viscous" forces. There may certainly be challenges in prescribing or calculating the geometry of the shed vorticity, but this does not negate the usefulness of lifting potential flow techniques.

(5) For large roll angles, which are the most important to consider when considering ship safety, bilge keel interactions with the free surface are a distinct possibility, even to extent of partial or complete bilge keel emergence. It is almost certainly more important to try to account for this effect, at least approximately, than to get into protracted arguments about the correct functional form of the bilge keel damping for high roll angles so that low roll angle data may be extrapolated to high roll angles.

(6) Many of the considerations above argue for a time-domain treatment of the bilge keel force problem, in order to handle the inherent non-linearities and memory effects. Much current effort, including the author's, is in this direction. However, recognizing the usefulness of frequency-domain seakeeping calculations, it would be worthwhile to see if current and future advancements in time-domain calculations could be transferred to the frequency domain, perhaps using harmonic balance techniques.

7.2 New Bilge Keel Force Model

In Greeley (2010) a new time-domain bilge keel force model was presented, which was developed to be incorporated into the US Navy's new seakeeping/maneuvering in waves code TEMPEST (Belknap 2010). The main constraint on this new force model, besides the obvious requirement for maximum accuracy, was that it execute quickly so that the final TEMPEST code could run in near real-time. This obviously meant that URANS approaches could not be considered. In the end, a hybrid time-domain force model was developed that is based on the relative motion between the bilge keel and the water (as described above), and consists of two components:

(1) For zero or low forward speeds, where the unsteady angle of attack of the bilge keels exceeds 45°, the force model uses the Morison equation with the empirical

database for flat plates in unsteady flow presented by Sarpkaya (1996). Two different techniques were developed and evaluated for determining the Keulegan-Carpenter parameter from the relative motion history, which is a necessary input to the Morisson equation database.

(2) For those situations with forward speed where the angle of attack of the bilge keel is 45° or less, Bollay's theory (Bollay 1936) for low aspect ratio wings was extended for unsteady flow with an arbitrary distribution (in time and space) of normal and tangential onset velocities to the bilge keel. The vorticity, which is assumed to be continuously shed from the *side* edge (as well as the trailing edge) of the bilge keel according to Bollay's theory, is rigorously accounted for in an unsteady lifting surface fashion, and the full non-linear Bernoulli equation is used to compute the forces on the bilge keel. This model has been demonstrated to closely replicate URANS force results, even at an unsteady angle of attack of 24°.

The final bilge keel model contains the above two modules, all appropriate logic to switch between the force modules as appropriate, and a model for the unsteady pressure forces (due to bilge keel action) acting on the hull surface (which uses the exact shape of the hull adjacent to the bilge keels). In addition, the final model includes an approximate but physically-based model for the change in bilge keel forces as a bilge keel emerges from the water, and a model for the slam forces on the bilge keel as it re-enters the water.

This new bilge keel force model is currently being implemented into TEMPEST by the TEMPEST team, so no results are yet available from the new bilge keel model actually used with TEMPEST. We present below some computations using the new bilge keel force model for a ship rolling in calm water.

7.3 Example Calculations

The case chosen for the example calculations presented here is the U.S. Navy pre-contract DDG-51 hull form, as represented by NSWCCD model 5415. This hull form has been widely studied around the world. In particular, Miller (2008) has presented calm water roll decay measurements (at Fr = 0.0 and Fr = 0.280) for Model 5415 and corresponding CFD results, done using CFDShip-Iowa. These results were presented at model scale so that the experiments and CFD computations could be compared directly; we have done our example computations at ship scale, but we present the results in terms of non-dimensional coefficients so that model and full size results may be compared directly. The major results that we will use for comparison are roll decay coefficient (logarithmic decrement) versus average roll angle and roll period versus cycle number, as shown in their Fig. 9 (reproduced as Figs. 7.1a, b).

The excellent match between CFD and experiment (at the higher roll angles) allows one to have confidence in the contributions of the bilge keels to the motions, as determined by the difference in CFD computed motions with and without bilge

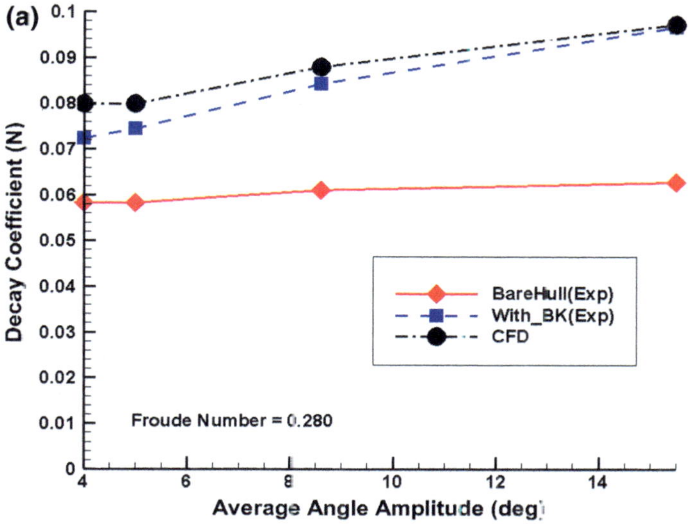

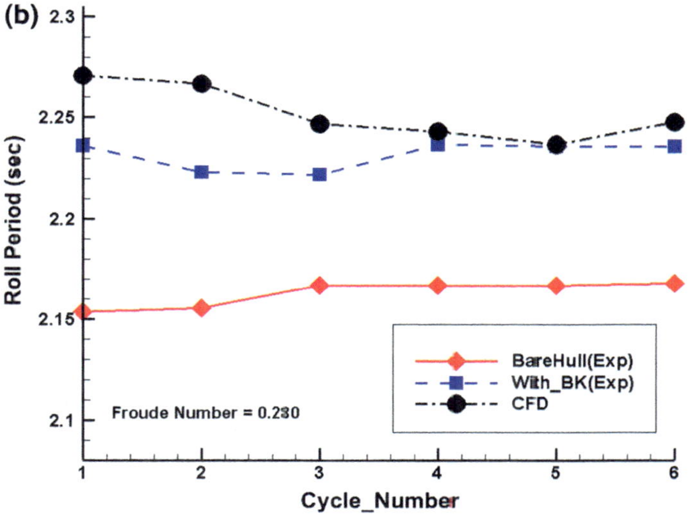

Fig. 7.1 **a** Decay coefficient for DTMB Model 5415 (from Miller 2008) (scale ratio = 24.84), **b** Roll period for DTMB model 5415 (from Miller 2008) (scale ratio = 24.84)

keels. From the data presented, we can pick off the difference in roll decay coefficient due to the bilge keels, and the difference in roll period due to the bilge keels. For convenience, these values have been converted to equivalent linear damping coefficients b_{44} and added moment of inertia coefficients a_{44}, and we present here their non-dimensional forms as recommended by Himeno (1981):

$$a'_{44} = \frac{a_{44}}{\rho \nabla B^2} \tag{7.1}$$

$$b'_{44} = \frac{b_{44}}{\rho \nabla B^2 \sqrt{\frac{2g}{B}}} \tag{7.2}$$

where B is the beam of the ship, and g is the acceleration due to gravity. These coefficients are terms in the single degree of freedom equation for the roll of the ship:

$$(I_{44} + a_{44})\ddot{\phi} + b_{44}\dot{\phi} + c_{44}\phi = M(t) \tag{7.3}$$

The inflow to the bilge keel for our example calculations is determined by using a lifting potential flow model, following the work of Hess (1972). The hull surface is paneled using source panels to enforce the zero normal velocity boundary condition on the ship hull, and an interior vortex lattice system is used to represent the lifting action of the sonar dome and hull. A non-linear, equal pressure Kutta condition is applied at the trailing edge of the sonar dome and skeg in order to determine the strengths of the interior vortex lattice system. Because the rolling motion considered is of relatively high frequency, the best approximation of the free surface available in this potential flow calculation is that the calm water surface be considered a plane of zero perturbation potential (Newman 1977): this may be realized by using a negative image above the waterplane for all of the singularities representing the hull. See Greeley and Willemann (2012, 2013).

The bilge keel calculations for each operating condition were done by computing the starboard bilge keel inflow (both tangential and normal velocity distributions along the bilge keel) for a series of roll angles and roll velocities corresponding to one sinusoidal roll cycle, using these velocities to "drive" the new bilge keel force model for several cycles to ensure that the hydrodynamic memory effects associated with the shed vorticity had reached a steady sinusoidal pattern, and then analyzing the force results for the last complete roll cycle in the calculations. Both the normal pressures on the bilge keel and the computed pressures acting on the nearby hull were used to compute the bilge keel roll moments, to be consistent with the previously mentioned CFD computations. A typical plot of the computed roll moments (for the starboard bilge keel only) is shown in Fig. 7.2.

Figure 7.2 shows the roll angle, the bilge keel roll moment from the unsteady lifting surface subroutine (**mxuls3**) which is the correct roll moment to use in this case, as well as the computed roll moment due to bilge-keel induced pressures on the hull (**hmx**). As is well known, the bilge keel forces decrease with forward speed, so the unsteady lifting surface computations (**mxuls3**) shows a smaller roll moment at this medium Froude number than the zero-forward-speed Morison equation (**mxme**) would indicate. Note that for both bilge keel force models, the computed time variation of the roll moment is *not* sinusoidal, despite the fact the imposed rolling motion is purely sinsoidal. For the unsteady lifting surface theory results (curve **mxuls3**), the

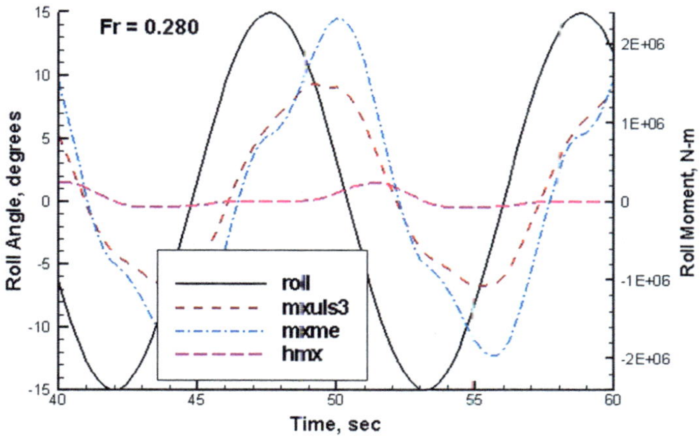

Fig. 7.2 Computed starboard Bilge Keel roll moments at Froude No. = 0.280 (full scale)

non-sinusoidal behavior is due to the complicated interactions between the vorticity shed off of the side edge of the bilge keel near the front of the bilge keel interacting with the rear portions of the bilge keel, as well as the fact that a significant non-linear lift component results from this side edge vortex shedding. [A comparison of this computed time variation of the forces with unsteady RANS results is shown in Greeley (2010)]. For the Morison equation results (curve **mxme**), the non-sinusoidal behavior of the forces is due to the quadratic drag force component inherent in this model.

The computed roll moments versus time were then least squares fitted with a model with 4 terms: one for the added moment of inertia, and 3 for the roll damping moment assuming that the roll damping can be expressed in the following form (Himeno 1981):

$$b_{44}(\dot{\phi}) = b_1\dot{\phi} + b_2\dot{\phi}\left|\dot{\phi}\right| + b_3\dot{\phi}^3 \tag{7.4}$$

The equivalent linear damping for this Froude number, roll frequency, and roll angle amplitude is then computed from these three components and used to compute the final b_{44}:

$$b_e(Fr, \omega, \phi_A) = b_1 + \frac{8}{3\pi}\omega\phi_A b_2 + \frac{3}{4}\omega^2\phi_A^2 b_3 \tag{7.5}$$

$$b_{44} = b_e(Fr, \omega, \phi_A) \tag{7.6}$$

Figure 7.3 shows the computed bilge keel contribution to the added moment of inertia and equivalent linear damping coefficients for the DDG-51 hull form at various Froude numbers, for a roll amplitude of 15° and a full scale roll period of 11.2 s. The added moments of inertia due to the bilge keels inferred from the data

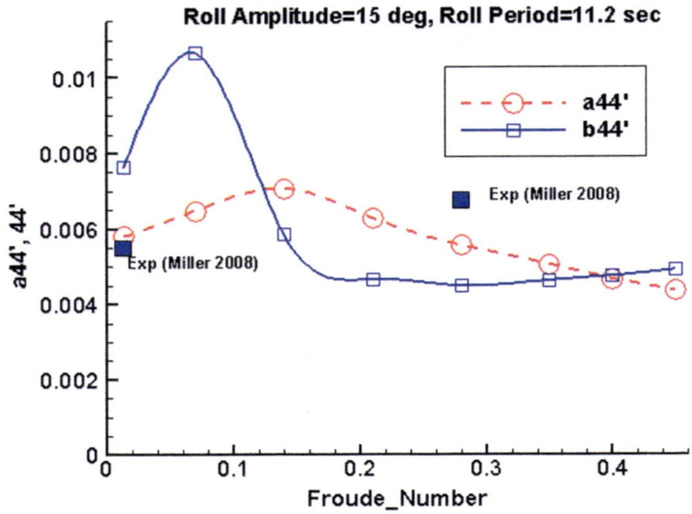

Fig. 7.3 Computed Bilge Keel contribution to added moment of inertia and equivalent linear damping

of Miller (2008) are two to three times the values shown in Fig. 7.3. The reason for this large discrepancy is unknown at present—it may be related to fact that during the experimental work and CFD computations, the model was forced to oscillate about its center of gravity. With the coupling of roll and sway due to added mass terms, perhaps a two degree of freedom model (roll and sway) is required to model the forced roll experiment and the influence of the bilge keels on the added moment of inertia in a meaningful way. Also, the Miller(2008) CFD calculations included a proper representation of the free surface waves, whereas the current calculations have made use of an approximate "negative image" representation of the free surface. In any event, this large discrepancy requires more study in the future.

Two experimental/CFD data points for the non-dimensional bilge keel damping are also shown in Fig. 7.3—these were deduced from the experimental/CFD results for the model shown in Fig. 7.1a. A perfect match between the current bilge keel model calculations and these points should not be expected, given the approximate nature of the bilge keel inflow calculations (not having a proper representation of the free surface in the hull potential flow computations). Still, the reasonable agreement for the bilge keel damping values is encouraging. It should also be noted that the current calculations are essentially for very high Reynolds numbers, while the CFD calculations are for the much smaller Reynolds numbers occuring at model scale, which may be responsible for some "scale effects".

The most unusual feature of the predicted damping due to the bilge keels shown in Fig. 7.3 is the large increase in damping expected at $Fr = 0.07$. This comes right out of the unsteady lifting surface model. This lifting surface model, since it only responds to the normal and tangential velocities supplied to it, does not care about the actual

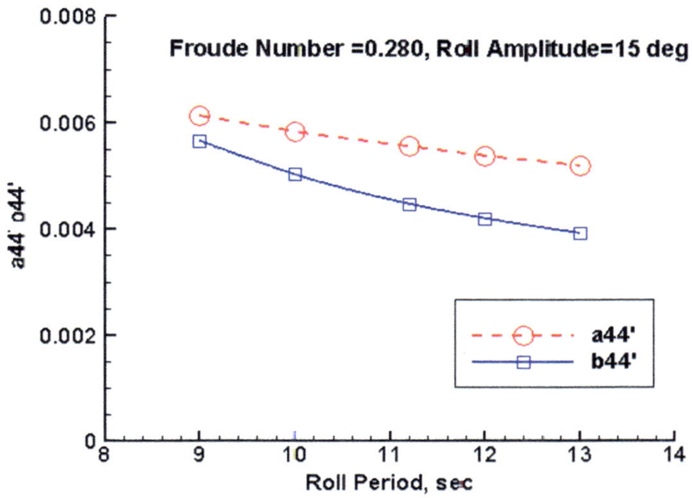

Fig. 7.4 Variation in Bilge Keel parameters with roll period (full scale)

Froude number. Rather, this Froude number of 0.07 with a roll amplitude of 15° and roll period of 11.2 s happens to correspond to an unsteady angle of attack and reduced frequency combination at which the vorticity shed from the side edges of the bilge keel on the forward third of the bilge keel has a *huge* impact on the load experienced on the after half of the bilge keel—the biggest computed unsteady pressures on the bilge keel are in fact near 75% chord! This kind of hump in the bilge keel damping is not expected according to generally used bilge keel force models or seen in most model testing. In fact, there is sometimes a dip seen in the bilge keel damping in this Froude number region. However, Irvine (2006) shows large scale effects on bilge keel forces in this Froude number range for geosims of Model 5415 tested at various facilities, with the smaller models showing significantly less damping than the larger models, so there are a number of issues that need examining here: (1) re-examination and possible additional testing and CFD runs for this Froude number (reduced frequency) range, over a range of Reynolds numbers, and (2) comparison (if possible) between model scale predicted roll damping and full scale ship roll damping in this Froude number range to make sure we understand the nature of possible scale effects, and possible shortcomings in the unsteady lifting surface model as currently implemented.

Additional computions with the new bilge keel model were done to look at variations in roll amplitude and roll period (for the full scale ship). Figure 7.4 shows the computed variation in a_{44} and b_{44} with roll period for a Froude number of 0.280 and a roll amplitude of 15°, while Fig. 7.5 shows a similar variation with roll amplitude for Froude number = 0.280 and a roll period of 11.2 s (full scale). As expected, since the roll damping actually has a large quadratic and cubic component, the equivalent linear damping is a strong function of roll angle.

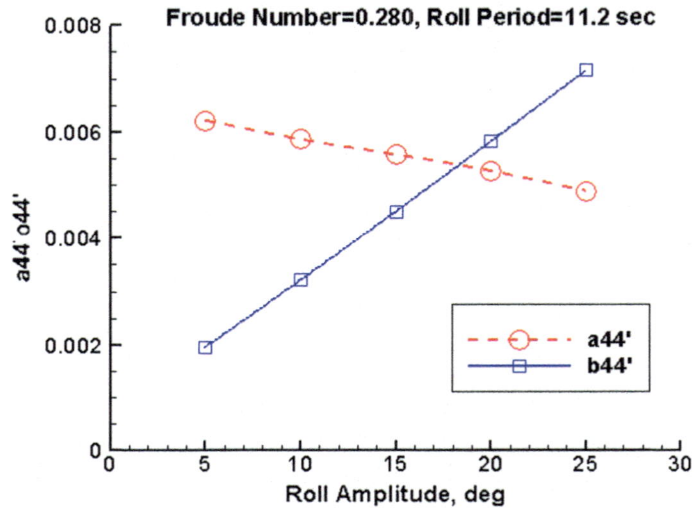

Fig. 7.5 Variation of Bilge Keel parameters with maximum roll angle (full scale)

7.4 Conclusions

Some example calculations from a new time-domain bilge keel force model have been presented and an attempt has been made to compare the results from this model with existing experimental and CFD results for DTMB Model 5415. Some encouraging agreement with existing experimental and CFD data has been obtained even with approximate calculations of the bilge keel inflow, but questions about the bilge keel force behavior, especially possible scale effects near a Froude number of 0.07 (for Model 5415), require further investigation. The interpretation of bilge keel added mass effects in the context of roll experiments where the roll axis passes through the center of gravity (regardless of added mass effects) also requires a re-examination.

Acknowledgements The development of the bilge keel model referred to in this paper was done initially with support from NAVSEA SE/TA funds (James Webster, Program Manager) and later support from ONR (Dr. Pat Purtell, Program Manager). Dr. William Belknap and Dr. Arthur Reed at NSWCCD were the technical monitors for the development of this bilge keel model. Simmy Willemann and Brian Petersen of APS assisted with the calculations shown in this paper. The opinions expressed here are those of the author.

References

Belknap, W.F. and Reed, A.M. (2010), "TEMPEST – New Computationally Efficient Dynamic Stability Prediction Tool", 11th International Ship Stability Workshop, Wageningen, June 2010.

Bollay, W. (1936), "A New Theory for Wings of Small Aspect Ratio", Caltech Ph.D. thesis, Pasadena, CA, 1936.

Greeley, D.S. and Petersen, B.J (2010), "Efficient Time-Domain Computation of Bilge Keel Forces", 28th ONR Symposium on Naval Hydrodynamics, Pasadena, CA, September 2010.

Greeley,D.S., and Willemann, S.D.(2012),"Surface Ship Maneuvering Forces in Calm Water and Waves", 29th ONR Symposium on Naval Hydrodynamics, Gothenburg, Sweden, 26–31 August 2012.

Greeley, D.S. and Willemann, S.D.(2013),'Surface ship maneuvering forces in calm water and waves", *International Shipbuilding Progress* 60, pp. 613–631.

Hess, J.L. (1972), "Calculation of Potential Flow About Arbitrary Three-Dimensional Lifting Bodies", McDonnell-Douglas Report No. MDC J5679-01, October 1972.

Himeno, Y. (1981), "Prediction of Ship Roll Damping – State of the Art", University of Michigan Department of Naval Architecture and Marine Engineering Report No. 239, September 1981.

Irvine, M., Atsavapranee, P., Carneal, J. *et al* (2006), "Comparisons of Free Roll Decay Tests for Model DTMB 5415/2340/5512, and Investigation of Lateral Hydrodynamic Loads on Bilge Keels", 26th ONR Symposium on Naval Hydrodynamics, Rome, Italy, September 17–22, 2006.

Miller, R.W., Bassler, C.C., Atsavapranee, P. and Gorski, J.J., (2008), "Viscous Roll Predictions for Naval Surface Ships Appended with Bilge Keels using RANS", 27th Symposium on Naval Hydrodynamics, Seoul, Korea, 5–10 October 2008.

Newman, J.N. (1977), Marine Hydrodynamics, MIT Press.

Sarpkaya, T. and O'Keefe, J.L. (1996), "Oscillating Flow About Two and Three Dimensional Bilge Keels", Journal of Offshore Mechanics and Arctic Engineering, Vol. 118, February 1996.

Chapter 8
Some Topics for Estimation of Bilge Keel Component of Roll Damping

Toru Katayama, Yuuki Yoshioka, Takahiro Kakinoki, Shugo Miyamoto and Yoshiho Ikeda

Abstract In this paper, two topics of roll damping estimation are introduced. In these topics, bilge-keel component of roll damping is focused, because this component is generally most part of viscous roll damping. First topic is the bilge-keel component of roll damping under shallow draft and large amplitude roll motion, and an estimation method of the draft effects based on Ikeda's method is proposed. Second topic is the bilge-keel component of roll damping under transitional and non-periodic rolling, an estimation method for time-domain simulation based on Ikeda's method is introduced and its estimated results are compared with measured results of roll damping in irregular forced rolling.

Keywords Bilge-keel component · Relative draft · Keulegan-Carpenter number
Drag coefficient · Transitional and non-periodic motion
Time-domain simulation · Parametric rolling

8.1 Introduction

In order to guarantee the safety of vessels, it is very important to understand the characteristics of roll motion and to estimate roll motion adequately. However, it is very complicated to calculate it because of difficulty of roll damping prediction due to significant viscous effects depending on vortex shedding.

It is well known that there is an estimation method of roll damping proposed by Ikeda et al. (1976, 1977a, b, 1978). However, some estimation problems are indicated in the previous studies (Tanaka et al. 1981, 1982; Ikeda et al. 1994; Hashimoto et al. 2008, 2009).

In this paper, two topics of estimation problems of the bilge-keel component of roll damping which is generally the largest part of total roll damping are introduced and discussed. The first topic is the effects of shallow draft. A forced rolling test

T. Katayama (✉) · Y. Yoshioka · T. Kakinoki · S. Miyamoto · Y. Ikeda
Graduate School of Engineering, Osaka Prefecture University, Sakai, Japan
e-mail: katayama@marine.osakafu-u.ac.jp

V. L. Belenky et al. (eds.), *Contemporary Ideas on Ship Stability*, Fluid Mechanics and Its Applications 119, https://doi.org/10.1007/978-3-030-00516-0_8

Table 8.1 Principle particulars of two-dimensional model

Length: L	0.80 m
Breadth: B	0.237 m
Depth: d	0.14465 m
Block coefficient: C_B	0.8 m
Bilge radius	0.035 m
Length × Breadth	0.01 m × 0.80 m

is introduced. And a simplified estimation method of this effects is indicated. The second topics is the effects of transitional and non-periodic motion. Under transitional motion, drag coefficient of a flat plate in the region of low Kc number is shown. Using the forced oscillation device, drag coefficient of the flat plate under transitional condition in periodic motion is indicated. Moreover, according to the basic idea of Ikeda's bilge-keel roll damping estimation method in frequency domain, using the proposed empirical formula of drag coefficient for flat plate, an estimation method of bilge-keel component of roll damping in time domain is introduced and its estimated results are compared with measured results of roll damping in irregular forced rolling.

8.2 Effects of Shallow Draft

8.2.1 Forced Rolling Test

In the previous study by Tanaka et al. (1981), it is pointed out that bilge-keel component of roll damping decreases when the draft is shallow. However, no formulation is proposed. Then, in this study, a forced rolling test is carried out by using two-dimensional model, and the characteristics of the effects of shallow draft on bilge-keel component is investigated to propose an empirical formula.

Table 8.1 shows the principal particulars of the model with bilge keel. Figure 8.1 shows some parameters for explaining experimental conditions. The measurements at systematically changed roll amplitudes, roll periods, drafts and height of roll axis (the center of rolling) are carried out. Bilge-keel component is obtained from subtraction measured data of hull without bilge keel from measured data of hull with bilge keel at the same condition.

8.2.2 Empirical Formula

Figures 8.2, 8.3 and 8.4 show the ratio of the measured results by the predicted ones. Horizontal axis shows $d_{\mathrm{bk}}/B_{\mathrm{bk}}$. Each figure shows the results of different $H_{\mathrm{bk}}/B_{\mathrm{bk}}$. Where d_{bk}, B_{bk} and H_{bk} are shown in Fig. 8.1. The ratios are indicated by different

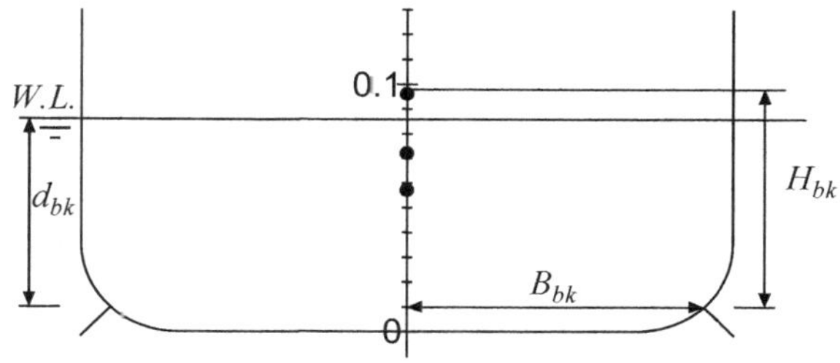

Fig. 8.1 Cross section of two-dimensional model

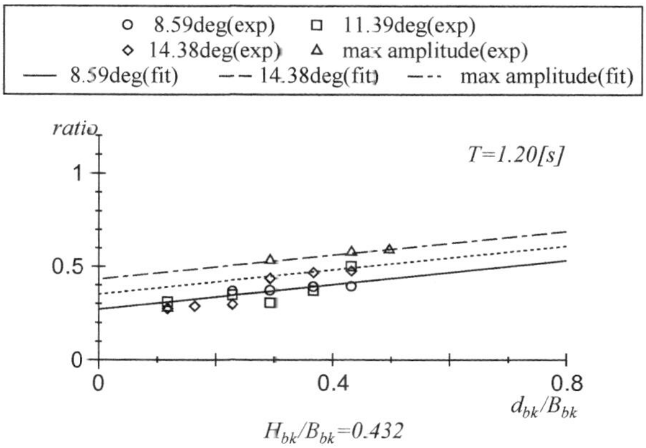

Fig. 8.2 Ratio of measured bilge-keel damping component to predicted one at height of roll axis $KG =$ 57 mm

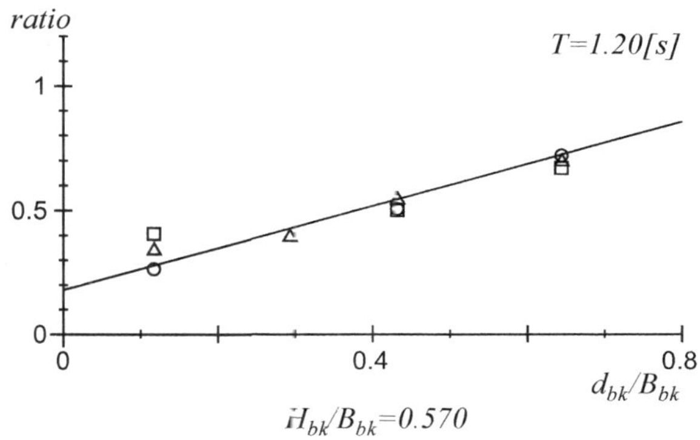

Fig. 8.3 Ratio of measured bilge-keel damping component to predicted one at height of roll axis $KG =$ 72 mm

marks for different roll amplitude. The maximum roll amplitudes are different for each height of roll axis and they are 17.71, 18.57, 21.0° respectively.

Figure 8.2 shows that the ratio increases linearly with increase of draft. For different roll amplitudes, the tendency is almost same quantitatively. Figure 8.3 shows that the ratio is almost same for different roll amplitudes. And the ratio increases linearly and its inclination is higher than Fig. 8.2. Figure 8.4 shows that the similar tendency as Fig. 8.2. And its inclination is the highest of all. If draft is deep enough,

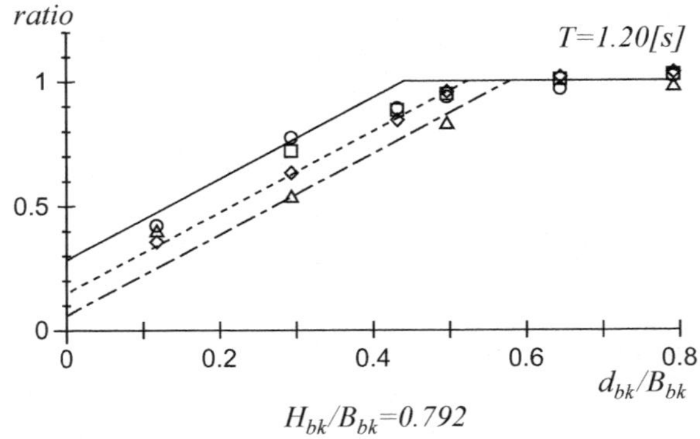

Fig. 8.4 Ratio of measured bilge-keel damping component to predicted one at height of roll axis $KG =$ 96 mm

the predicted results can agree with the measured results. It means that the ratio does not exceed 1.0 with increase of draft. Moreover, above-mentioned characteristics are not almost affected by roll period.

A fitting curve is obtained from the measured data. A correction factor is expressed as following equation.

$$
\begin{aligned}
C_{bk} = & \left(3.615\frac{H_{bk}}{B_{bk}} - 1.227\right)\frac{d_{bk}}{B_{bk}} \\
& + \left\{3.29\left(\frac{H_{bk}}{B_{bk}}\right)^2 - 5.35\frac{H_{bk}}{B_{bk}} + 1.98\right\}\phi_a^2 \\
& + \left\{2.48\left(\frac{H_{bk}}{B_{bk}}\right)^2 + 1.90\frac{H_{bk}}{B_{bk}} - 11.6\right\}\phi_a \\
& + \left\{2.77\left(\frac{H_{bk}}{B_{bk}}\right)^2 - 3.27\frac{H_{bk}}{B_{bk}} + 1.14\right\} \\
& \leq 1.0
\end{aligned}
\tag{8.1}
$$

where ϕ_a is in radian. Bilge-keel component is obtained by multiplying correction factor by bilge-keel component of Ikeda's method.

8.2.3 Calculated Results

For a post panamax container ship (Hashimoto et al. 2008, 2009), roll damping is calculated by Ikeda's method with the correction factor. When parametric rolling occurs at high wave height in head waves, large relative draft change is caused. In the roll damping calculation, the relative draft of each cross section at the moment, where roll is upright, is used. Figure 8.5 shows the calculated results. Total roll

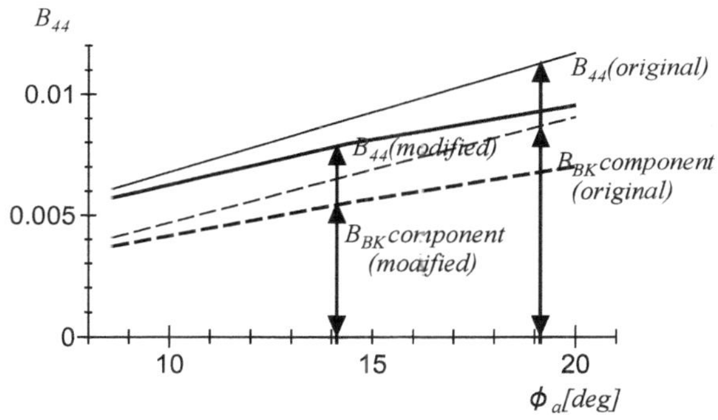

Fig. 8.5 Estimated results of roll damping including relative draft effects

damping decreases 6% at roll amplitude $\phi_a = 8.59°$, 11% at $\phi_a = 14.38°$, and 19% at $\phi_a = 20°$, for the results without considering the relative draft change. Shallow draft due to draft change in waves affects on bilge-keel component significantly.

8.3 Effects of Transitional and Non-periodic Rolling

8.3.1 Drag Coefficient of Flat Plate in Uniform Flow

Drag coefficient of a flat plate, which is assumed as bilge keels, in uniform flow is measured by Katayama et al. (2011). A strut and a flat plate are fixed by a load cell (shown in Fig. 8.6), and it is towed at constant forward speed. Towing speeds are from $U = 0.1$ to 1.0 m/s at 0.1 m/s space. Drag force acting on a flat plate D is obtained from deducting measured drag without the flat plate. Drag coefficient is calculated with the following equation.

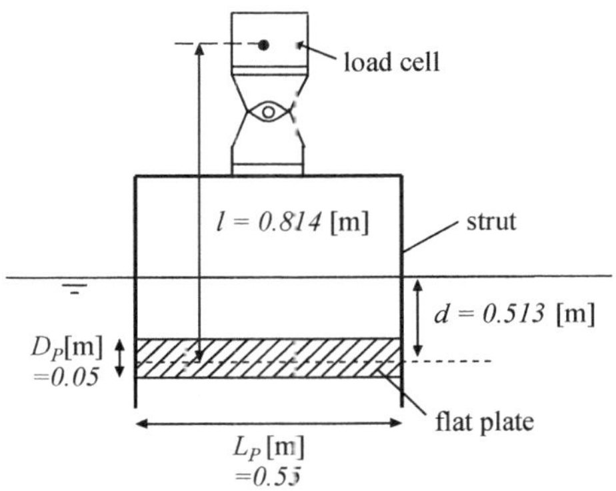

Fig. 8.6 Schematic view of the experimental device from the front. The moving direction of flat plate is directed from the back to the surface of the paper. Thickness of flat plate is 3 mm (Katayama et al. 2011)

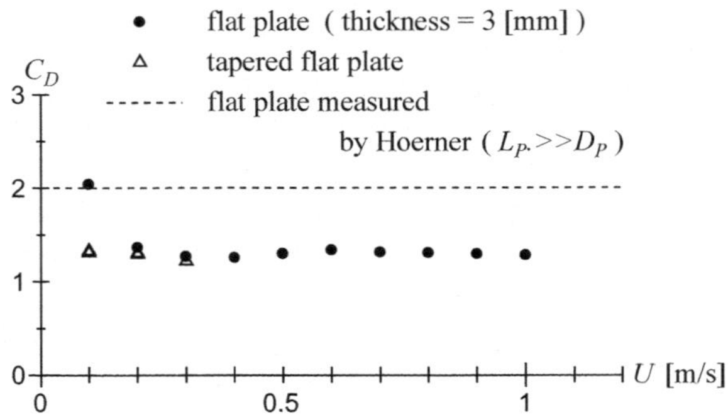

Fig. 8.7 Drag coefficients of flat plates in uniform flow (Katayama et al. 2011)

$$C_D = \frac{D}{0.5\rho S U^2} \tag{8.2}$$

where D, ρ, S and U denote drag force, density of fluid, area of flat plate and towing speed. In order to avoid low Kc number effects, measured data in the region $Kc > 100$ are used in the analysis of drag force. Kc number is expressed as follows,

$$Kc = Kc_d = \frac{2\pi\, y}{D_P} \tag{8.3}$$

where y and D_P denote forward moving distance and height of a flat plate shown in Fig. 8.6 ($L_P/D_P = 11$). It is noted that this Kc number is used as a non-dimensional coefficient of moving distance of flat plate. In Appendix 8.1, Kc number as the ratio between the drag and the inertia force is indicated.

Figure 8.7 shows the results. Drag coefficient of a flat plate ($L_P \gg D_P$) measured by Hoerner (1965) is also shown in Fig. 8.7. In order to remove low Reynolds number effects on drag force, drag force of a tapered flat plate is also measured. From this figure, it is confirmed that drag coefficient of a tapered flat plate is constant for change in forward speed, even if it is lower than Hoerner's results. In this study, a tapered flat plate is used.

8.3.2 Empirical Formula of Drag Coefficient of Flat Plate in Steady Oscillation

It is known that drag coefficients on oscillating flat plate at low Kc number ($Kc < 10$) is significantly changed by a slight change of Kc number (Tanaka et al. 1980; Kudo et al. 1980). Kc number of oscillating flat plate is expressed as follows,

$$Kc = Kc_a = \frac{2\pi\, y_a}{D_P} \tag{8.4}$$

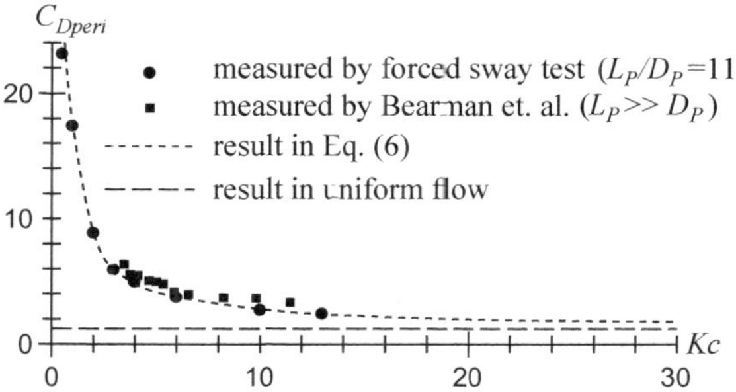

Fig. 8.8 Drag coefficients of flat plates in oscillatory flow (Katayama et al. 2011)

where y_a is amplitude of oscillation. However, the experimental results at low Kc number ($Kc<3$) is not found because of difficulty of measurement. Then, drag force of a flat plate at low Kc number is carefully measured.

The experimental device shown in Fig. 8.6 is oscillated and hydrodynamic force and forced motion are measured. And the same measurement is carried out for the strut without the flat plate. Drag force, which is proportional to motion velocity, is obtained from these data. Drag coefficient is calculated with the following equation.

$$C_{Dperi} = \frac{D_{Peri}}{0.5\rho S(y_a\,\omega)^2} \tag{8.5}$$

where, ω is circular frequency of forced oscillation, and D_{Peri} is drag force acting on a flat plate. D_{Peri} is obtained from deducting drag force without flat plate.

Figure 8.8 shows the results. Drag coefficient is about 20 at $Kc = 0.5$, and decreases with increase of Kc number, and becomes the value in uniform flow at about $Kc =$ 250. As the results, a fitting curve of drag coefficient Eq. (8.6) is determined by Katayama et al. (2011), and it is shown in Fig. 8.8 as a dotted line.

$$\begin{aligned}\frac{C_{Dperi}}{C_{D0}} &= (20.0e^{-..23Kc_a} + 2.86e^{-0.174Kc_a} + 1)\\ &\times\left(0.903 + \frac{1.2}{1 + 1.01^{Kc_a}}\right)\\ &(0 < Kc_a \leq 250)\end{aligned} \tag{8.6}$$

8.3.3 *Drag Coefficient Under One Direction Accelerating*

Experimental device shown in Fig. 8.5 is towed horizontally by a method of free fall of a weight shown in Fig. 8.9. In order to obtain drag force acting on a flat plate, two measurements with and without flat plate are carried out, and these data are analyzed after removing theoretical value of inertia and added inertia forces from both measured forces, respectively.

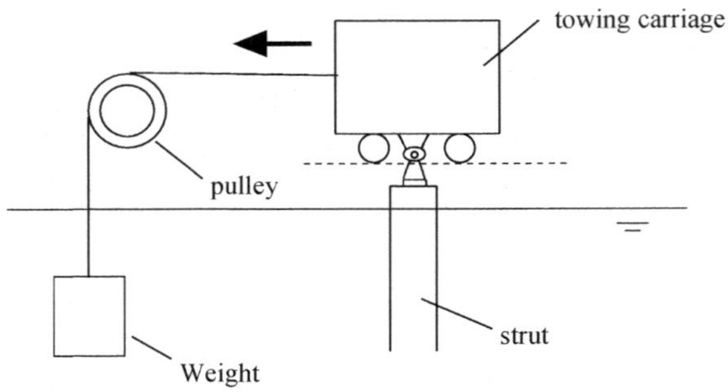

Fig. 8.9 Schematic view of experiment towed by free fall of a weight

$$
\begin{aligned}
D_{acc(S+P)}(t) &= F_{(S+P)}(t) - \left(m_{(S+P)} + \frac{\pi}{4} C_m\, \rho\, D_P^2\, L_P\right) \ddot{y}_{(S+P)}(t) \\
D_{accS}(t) &= F_S(t) - m_S \ddot{y}_S(t)
\end{aligned}
\tag{8.7}
$$

where $D_{acc(S+P)}(t)$ and $D_{accS}(t)$ are drag force and added inertia force acting on strut with or without flat plate. $F_{(S+P)}(t)$ and F_S(t) are measured forces acting on strut with or without flat plate, $m_{(S+P)}$ and m_S are mass of strut with or without flat plate, C_m is coefficient of added mass for flat plate (in this case C_m= 1.0), ρ is a density of fluid and $\ddot{y}_{(S+P)}(t)$ and $\ddot{y}_S(t)$ are acceleration determined by a time history of motion.

Coefficient of $D_{acc(S+P)}(t)$ or $D_{accS}(t)$ is calculated with the following equation.

$$
C_{Dacc} = \frac{D_{acc}(t)}{0.5 \rho S \dot{y}(t)^2} \tag{8.8}
$$

The coefficients in the both cases are obtained and fitting curve (8.9) and (8.10) are determined respectively by Katayama et al. (2011).

$$
\begin{aligned}
\frac{C_{Dacc(S+P)}}{C_{D0}} &= (14.3 e^{-1.80 Kc_d} + 4.41 e^{-0.37 Kc_d} \\
&\quad + 1.168) \times \left(0.908 + \frac{1.2}{1 + 1.01^{Kc_d}}\right) \\
&\quad (0 < Kc_d \leq 250)
\end{aligned}
\tag{8.9}
$$

$$
\begin{aligned}
\frac{C_{DaccS}}{C_{D0}} &= (10.4 e^{-1.03 Kc_d} + 0.30 e^{-0.17 Kc_d} \\
&\quad + 0.168) \times \left(0.908 + \frac{1.2}{1 + 1.01^{Kc_d}}\right) \\
&\quad (0 < Kc_d \leq 250)
\end{aligned}
\tag{8.10}
$$

where *Kc* number obtained from Eq. (8.3). It is noted that these coefficients are not function of time but K_C as a non-dimensional coefficient of moving distance.

From Eqs. (8.9) and (8.10), drag coefficient of a flat plate at one direction accelerating is calculated with the following equation and the results are shown in Fig. 8.10.

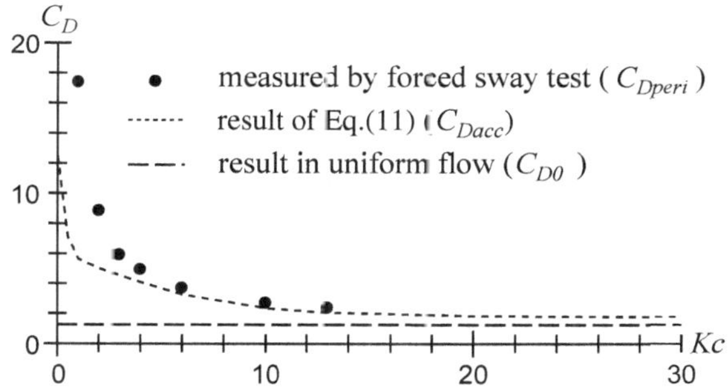

Fig. 8.10 Comparison of drag coefficients of a flat plate obtained by force measurement test in uniform flow, forced sway test and by direction accelerating test (Katayama et al. 2011)

$$\frac{C_{Dacc}}{C_{D0}} = \frac{C_{Dacc(S+P)}}{C_{D0}} - \frac{C_{LaccS}}{C_{D0}}$$
$$(0 < Kc_d \leq 250) \tag{8.11}$$

8.3.4 *Drag Coefficient Under Transitional Condition in Oscillatory Flow*

In this section, using the forced oscillating device, measurements of forces acting on a flat plate in each swing from rest is carried out.

Figures 8.11 and 8.12 show the results. Drag coefficient is gradually increasing from the first swing to the fourth swing. After the fourth swing, drag coefficient becomes constant. The phenomena may be caused by changing flow around hull (e.g. from laminar flow to turbulent flow), and it should be investigated in near future by using PIV or CFD. From the results, the formula of drag coefficient including the number of swing from rest is decided as the following equation by Katayama et al. (2011) .

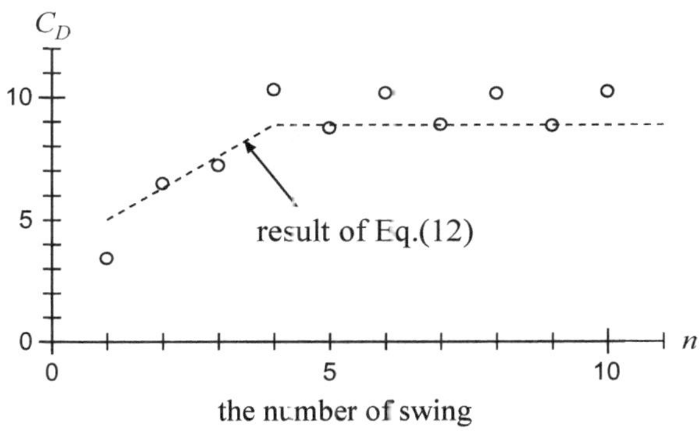

Fig. 8.11 Drag coefficient of flat plate versus the number of swing at Kc_a = 2.0 (Katayama et al. 2011)

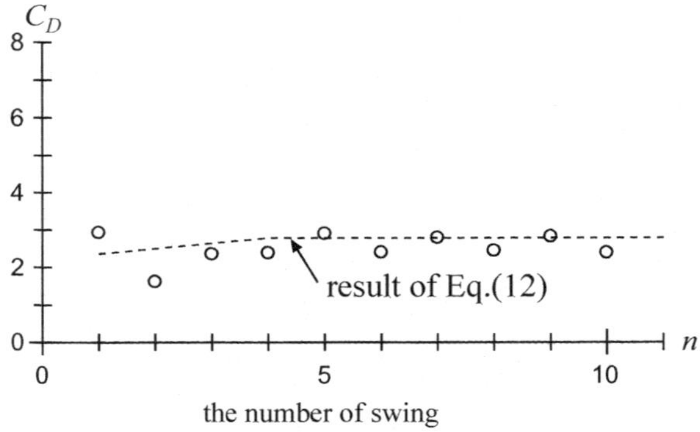

Fig. 8.12 Drag coefficient of flat plate versus the number of swing at $Kc_a = 10.0$

$$C_{Dn} = C_{Dacc} + \left(C_{Dperi} - C_{Dacc}\right)\frac{n-1}{3} \tag{8.12}$$

where n is the number of swing ($n = 1$, 2, 3 and 4). And Kc number in Figs. 8.11 and 8.12 is obtained from Eq. (8.4).

8.3.5 *Empirical Formula of Drag Coefficient of Flat Plate for 1st Swing in Steady Oscillation*

Using Eq. (8.11), the drag coefficient of flat plate for 1st swing in steady oscillation can be calculated according to the following equation

$$\int_0^{\frac{T}{4}} C_{D1} \sin^3 \omega t\, dt = \int_0^{\frac{T}{4}} C_{Dacc}(Kc_d) \sin^3 \omega t\, dt. \tag{8.13}$$

Equation (8.13) indicates energy integration for 1st swing in steady oscillation. Drag coefficient of flat plate for 1st swing in steady oscillation is obtained by following equation by Katayama et al. (2011).

$$\begin{aligned}\frac{C_{D1}}{C_{D0}} &= \left\{\begin{matrix} 5.42e^{-0.23Kc_a} + 13.2e^{-1.25Kc_a} - 1.96e^{-0.21Kc_a} \\ -8.72e^{-0.78Kc_a} + 1.0 \end{matrix}\right\} \\ &\quad \times \left(0.908 + \frac{1.2}{1 + 1.01^{Kc_a}}\right) \\ &\quad (0 < Kc_a \leq 250)\end{aligned} \tag{8.14}$$

Comparison among drag coefficients in steady oscillation, under one-way acceleration and in first swing of steady oscillation is shown in Fig. 8.13. In this calculation,

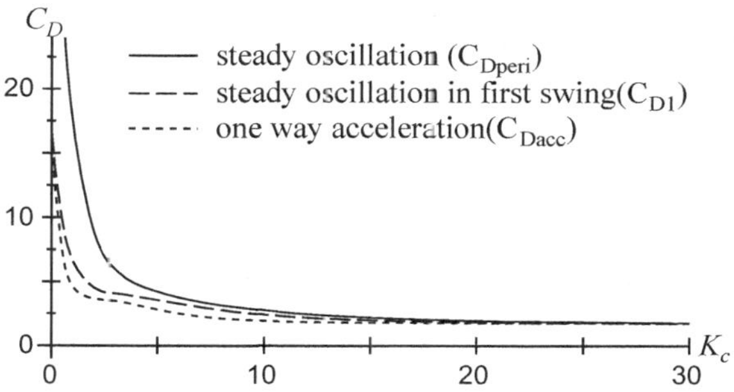

Fig. 8.13 Comparison among drag coefficients in steady oscillation, in first swing of steady oscillation and under one-way acceleration versus *Kc* number (Katayama et al. 2011)

drag coefficient in uniform flow $C_{D0} = 1.26$, which is measured value for the same flat plate shown in Fig. 8.7, is used.

8.3.6 Drag Coefficient Under Transitional Condition in Periodic Oscillation

In the previous section, it is confirmed that the drag coefficient in each swing of steady oscillation from rest is gradually increasing, and after the 4th swing the drag coefficient becomes constant. This characteristics are expressed as Eq. (8.12). In order to apply it to time domain estimation of drag coefficient, the following equation is proposed by Katayama et al. (2011),

$$\frac{C_{Dacc(n)}}{C_{D0}} = \frac{C_{Dacc}}{C_{D0}} \cdot \left[1 + \left(\frac{C_{Dperi}}{C_{D1}} - 1\right) \cdot \frac{n-1}{3}\right], \tag{8.15}$$

where n is the number of swing ($n = 1, 2, 3$ and 4). In Eq. (8.15), it is assumed that the drag coefficient in first swing is C_{Dacc} and the drag coefficient $C_{Dacc(n)}$ is increased from the 1st swing to the 4th swing according to the ratio of C_{Dperi} and C_{D1}.

Figure 8.14 shows comparison among the measured and the two estimated drag forces in time domain under steady oscillation in transitional condition. The first estimation method uses the drag coefficient, which changes in every time step, depending on *Kc* number expressed by Eq. (8.3), and the second estimation method uses a constant drag coefficient depending on *Kc* number expressed by Eq. (8.4). In the both estimation method drag forces are calculated by Eq. (8.7).

The result by the first estimation method is better agreement with measured results. In the lower figure of Fig. 8.14, the results of the first estimation method shows that drag coefficient changes in time step and its value is maximum at the start position of a swing. As the results, the estimated result of the first estimation method becomes larger than one of the second estimation method.

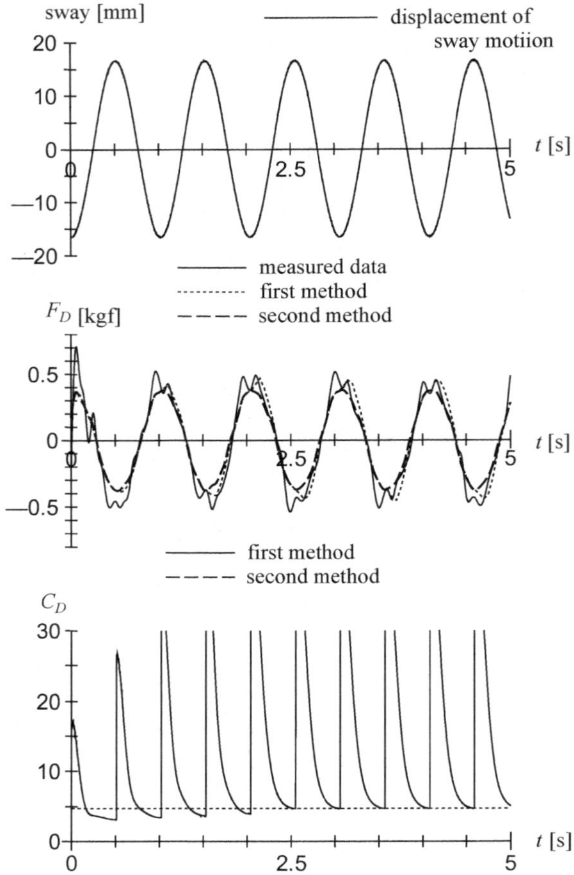

Fig. 8.14 Comparison among drag forces under steady oscillation in transitional condition. [First estimation method uses the drag coefficient, which changes in every time step, depending on *Kc* number expressed by Eq. (8.3). Second estimation method uses a constant drag coefficient depending on *Kc* number expressed by Eq. (8.4)]

Figure 8.15 shows comparison of drag forces under un-sinusoidal oscillation. In the figure, there are two estimated results as same as shown in Fig. 8.14. It is noted that drag coefficient of the second estimation method does not consider the memory effects, which is expressed by Eq. (8.15), and the drag coefficient is estimated at $n = 4$ in Eq. (8.15). The result of the first method shows better agreement with measured result, and the amplitude of drag force of the first estimation method is almost same as measured one because of the change of drag coefficient. However, drag coefficient is estimated at $n = 4$ in this case, then memory effects of drag coefficients are not clear, and more detailed measurement is desired.

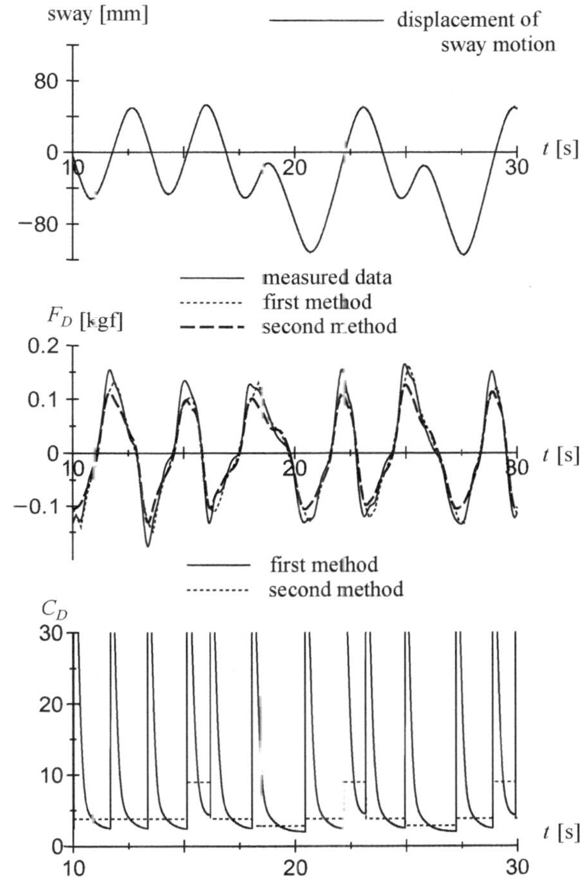

Fig. 8.15 Comparison among drag forces under un-sinusoidal oscillation. [First estimation method uses the drag coefficient, which changes in every time step, depending on *Kc* number expressed by Eq. (8.3). Second estimation method uses a constant drag coefficient depending on *Kc* number expressed by Eq. (8.4)]

8.4 Bilge Keel Roll Damping for Time Domain Simulation

Ikeda's method is one of discrete type estimation method. It is composed of wave, lift, frictional, eddy and appendages contributions (bilge keel, skeg, rudder etc.). In this section, an estimation method of the bilge-keel component in time domain is proposed to refer the basic concept of bilge-keel component of Ikeda's method (refer to Appendix 8.2). Moreover, the comparison between estimated and measured results is shown.

In periodic rolling, a drag coefficient acting on bilge-keels is expressed as Eq. (8.30) in Ikeda's method. This formula is obtained by fitting the measured drag coefficients of a flat plate under steady oscillation, which are shown in Fig. 8.16. In this figure, horizontal axis is *Kc* number and vertical axis is drag coefficient of flat plate under steady oscillation. Therefore Eq. (8.30) can be replaced with Eq. (8.6).

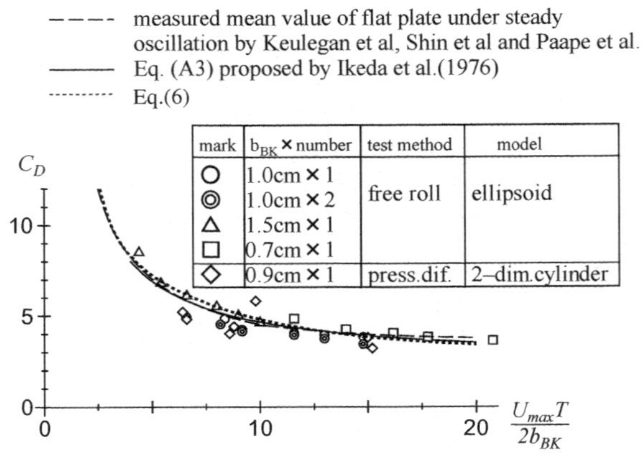

Fig. 8.16 Comparisons of drag coefficients of a bilge-keel and the flat plate, those areas are same

Moreover, if Eqs. (8.3) and (8.15) are applied, the normal force component of bilge-keels can be obtained by following equation

$$M_{BKN} = \frac{1}{2}\rho(2\, l_{BK}\, b_{BK})\, C_{Dacc(n)}\, l^2\, \dot{\phi}|\dot{\phi}|\, r\, f. \tag{8.16}$$

Using the same manner as the normal force component of bilge-keels, the hull surface pressure component of bilge-keel in time domain is obtained from Eq. (8.17) with replacing C_{D} in Eq. (8.35) to Eq. (8.15) and K_C number in Eq. (8.33) to Eq. (8.18).

$$M_{BKH} = \frac{1}{2}\rho\, l^2 f^2\, \dot{\phi}|\dot{\phi}| \int_G C_p \cdot l_p\, dG \tag{8.17}$$

$$Kc = \frac{\pi\, l\, \phi}{2b_{BK}} \tag{8.18}$$

Figure 8.17 shows the comparison between measured and estimated results. In the case, the results show that estimated result by the second method is better agreement with measured one than one by the first method.

8.5 Conclusion

In this paper, two topics of roll damping estimation problems are introduced.

In the first topics, the effects of shallow draft are investigated. Bilge-keel component of roll damping by Ikeda's prediction method is overestimated for lower roll axis and shallow draft. Based on the measured results, an empirical formula to the bilge-keel component is proposed. However, physical mechanism of the effects of shallow draft on the bilge keel component of roll damping has not been discussed. If the physical mechanism will become clear, it may be possible to correct Ikeda's method

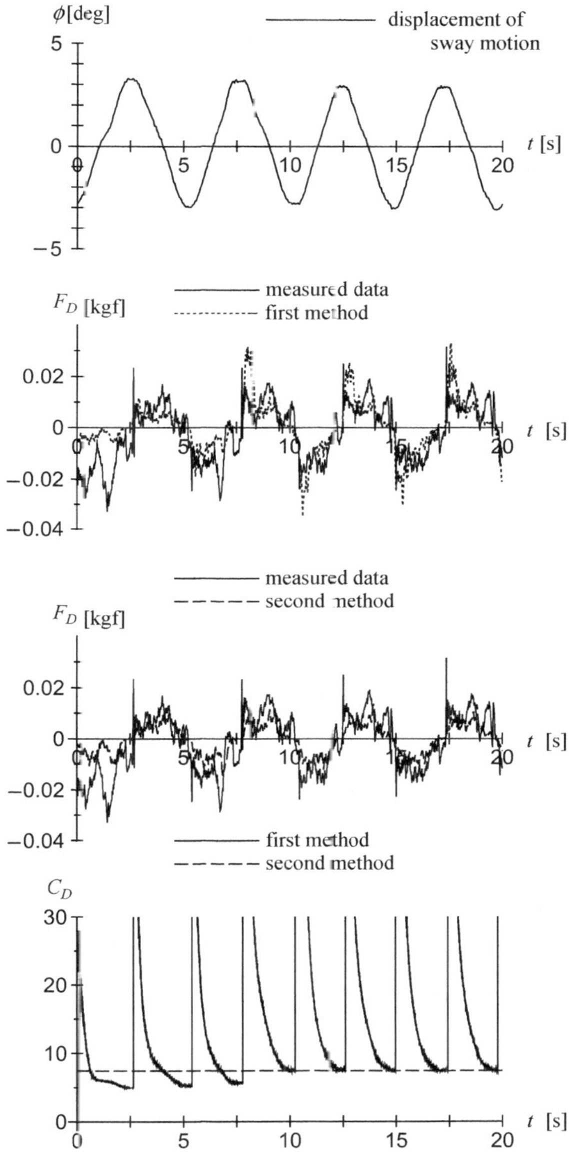

Fig. 8.17 Comparison among drag forces under un-sinusoidal rolling. [First estimation method uses the drag coefficient, which changes in every time step, depending on *Kc* number expressed by Eq. (8.3). Second estimation method uses a constant drag coefficient depending on *Kc* number expressed by Eq. (8.4)]

reasonably according to relevant physical reasons. The more detailed research work should be carried out in near future.

In the second topics, the effects of transitional motion are investigated. In the region at $Kc < 250$, drag coefficient of a flat plate under one direction accelerating is larger than that in uniform flow and smaller than that in steady oscillatory flow.

Moreover, in transitional condition under forced oscillation, the drag coefficients from 1st swing to 3rd swing are smaller than that in steady oscillatory flow. These facts may indicate that the characteristics of drag coefficient affect transitional and non-periodic rolling. Finally, based on the results, an empirical formula to the bilge-keel component by Ikeda's prediction method is presented. And, an estimated result is compared with a measured result and it shows good agreements depending on extremely large drag coefficient at low *Kc* number. However, until now, it is not necessary clear how to consider the memory effects, which is the effects of the vortexes created by previous swings. In the future work, more detailed measurement is required.

Appendix 1

Kc Number as the Ratio Between the Drag and the Inertia Force

Drag and inertia force acting on a flat plate ($L \times D$) are expressed by the following equations with coefficients C_D and C_M,

$$f_D = \frac{1}{2} C_D \rho D L U^2, \tag{8.19}$$

$$f_I = \frac{1}{4} C_M \rho \pi D^2 L \frac{d}{dt} U, \tag{8.20}$$

where ρ is density of fluid, U is relative fluid velocity to flat plate, time derivative of U is fluid acceleration.

K_C number for fluid force on an immersed body at rest a moving liquid is obtained from the ratio between the drag and the inertia force,

$$\frac{f_D}{f_I} = \frac{C_D}{\pi^2 C_M}\left(\frac{U_m T}{D}\right) = \frac{C_D}{\pi^2 C_M} K_C, \tag{8.21}$$

where

$$U = U_m \cos\left(\frac{2\pi}{T} t\right). \tag{8.22}$$

On the other hand, Keulegan and Carpenter, (1958) suggests the following interpretation as to the physical meaning of $U_m T/D$ for the eddy appearances. "If one define a length, l, as the distance that a fluid particle would move in one direction in the absence of the cylinder, $l = U_m T/\pi$. Thus,

$$\frac{U_m T}{D} = \frac{\pi \, l}{D}, \tag{8.23}$$

and accordingly the period parameter is proportional to the ratio of the distance traversed by a particle during a half cycle to the diameter of the cylinder. When the period parameter (which is K_C number) equals 15, l/D is 4.8. Perhaps when $U_m T/D$ is smaller than 15, the distance traveled by a particle is not large enough to form complete eddies. When it equals 15, the distance suffices to form a single eddy, and when much larger than 15 the greater distances allow the formation of numerous vortices of the Karman vortex street." As the results, it can be said that K_C number related to condition of flow separation caused by relative moving distance of a fluid particle to an object.

Kc number of oscillating flat plate in this paper is expressed as follows,

$$Kc = Kc_a = \frac{2\pi \, y_a}{D_P}, \tag{8.24}$$

where $2y_a$ is both amplitude and be equal to the above mentioned l.

For one direction accelerating test in this paper, *Kc* number can be expressed the following formula according to Eq. (8.23),

$$Kc = Kc_d = \frac{\pi \, y}{D_P}. \tag{8.25}$$

Because y is relative moving distance of fluid to flat plate. On the other hand, *Kc* number can be expressed the following formula according to Eq. (8.21), which is the ratio between the drag and the inertia force, when acceleration is constant.

$$Kc = Kc_d = \frac{4\pi \, y}{D_P} \tag{8.26}$$

Equations (8.25) and (8.26) is different formula from Eq. (8.24). In this paper, Eq. (8.27), which is the same formula as Eq. (8.24) is adapted as a non-dimensional coefficient of moving distance of flat plate.

$$Kc = Kc_d = \frac{2\pi \, y}{D_P} \tag{8.27}$$

Appendix 2

Bilge Keel Component of Original Ikeda's Method

The bilge keel component B_{44BK} is composed of two components:

$$B_{44\text{BK}} = B_{44\text{BKN0}} + B_{44\text{BKH0}} \tag{8.28}$$

The normal force component $B_{44\text{BKN0}}$ can be deduced from the experimental results of oscillating flat plates (Ikeda et al. 1976). The drag coefficient C_D of an oscillating flat plate depends on the K_C number.

$$Kc = \frac{\pi\, l\, \phi_A}{b_{BK}} \tag{8.29}$$

From the measurement of the drag coefficient, C_D, from free roll tests of an ellipsoid with and without bilge keels, the prediction formula for the drag coefficient of the normal force of a pair of the bilge keels can be expressed as follows:

$$C_D = 22.5\left(\frac{b_{\text{BK}}}{\pi\, l\phi_{\text{a}}}\right)\frac{1}{f} + 2.4. \tag{8.30}$$

where b_{BK} is the breadth of the bilge keel and l is the distance from the roll axis to the tip of the bilge keel. The equivalent linear damping coefficient $B'_{44\text{BKN0}}$ is:

$$B'_{44\text{BKN0}} = \frac{8}{3\pi}\rho\, l^3 \omega_e \phi_{\text{a}} b_{\text{BK}} f\, C_D \tag{8.31}$$

where f is a correction factor to take account of the increment of flow velocity at the bilge, determined from the experiments:

$$f = 1 + 0.3e^{\{-160(1-\sigma)\}} \tag{8.32}$$

From the measurement of the pressure on the hull surface created by the bilge keels, it was found that the coefficient C_{P}^{+} of pressure on the front face of the bilge keels does not depend on the K_C number. However, the coefficient C_{P}^{-} of the pressure on the back face of a bilge keel and the length of negative-pressure region do depend on the K_C number. From these results, the length of the negative-pressure region can be obtained as follows:

$$S_0/b_{\text{BK}} = 0.3\left(\frac{\pi\, l\phi_{\text{a}}}{b_{\text{BK}}}\right)f + 1.95 \tag{8.33}$$

assuming a pressure distribution on the hull as shown in Fig. 8.18.

The roll damping coefficient B'_{BKH0} can be expressed as follows (Ikeda et al. (1977a):

$$B'_{44\text{BKH0}} = \frac{4}{3\pi}\rho\, l^2 f^2 \omega_e \phi_{\text{a}} \int\limits_G C_{\text{p}} \cdot l_{\text{p}}\, dG \tag{8.34}$$

where G is length along the girth and l_{p} is the moment lever.

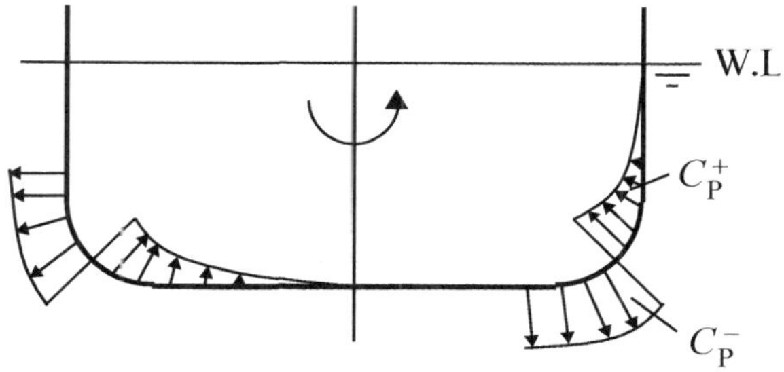

Fig. 8.18 Assumed pressure distribution on the hull surface created by bilge keels (Ikeda et al. 1977a)

The coefficient C_{p}^{+} can be taken approximately as 1.2 empirically. From the relation of $C_D = C_{\text{p}}^{+} - C_{\text{p}}^{-}$, the coefficient C_{p}^{-} can be obtained as follows:

$$C_{\text{p}}^{-} = 1.2 - C_D = -22.5\left(\frac{b_{\text{BK}}}{\pi l \phi_{\text{a}}}\right)\frac{1}{f} - 1.2 \tag{8.35}$$

The value of $\int_G C_{\text{P}} \cdot l_{\text{p}}\, dG$ in Eq. (8.34) can be obtained as follows:

$$\int\limits_G C_{\text{p}} \cdot l_{\text{p}} dG = d^2\left(-A_0 C_{\text{p}}^{-} + B_0 C_{\text{p}}^{+}\right) \tag{8.36}$$

where:

$$A_0 = (m_3 + m_4)m_8 - m_7^2$$
$$B_0 = \frac{m_2^2}{3(H_0 - 0.215m_1)} + \frac{(1 - m_1)^2(2m_3 - m_2)}{6(1 - 0.215m_1)} + m_1(m_3m_5 + m_4m_6)$$
$$m_1 = R/d$$
$$m_2 = \overline{OG}/d$$
$$m_3 = 1 - m_1 - m_2$$
$$m_4 = H_0 - m_1$$
$$m_5 = \frac{\{0.414H_0 + 0.0651m_1^2 - (0.382H_0 + 0.0106)m_1\}}{(H_0 - 0.215m_1)(1 - 0.215m_1)}$$
$$m_6 = \frac{\{0.414H_0 + 0.0651m_1^2 - (0.382 + 0.0106H_0)m_1\}}{(H_0 - 0.215m_1)(1 - 0.215m_1)}$$
$$m_7 = \begin{cases} S_0/d - 0.25\pi m_1, & S_0 > 0.25\pi R \\ 0, & S_0 \le 0.25\pi R \end{cases}$$
$$m_8 = \begin{cases} m_7 + 0.414m_1, & S_0 > 0.25\pi R \\ m_7 + 1.414m_1(1 - \cos(\frac{S_0}{R})), & S_0 \le 0.25\pi R \end{cases}$$

where l is a distance from roll axis to the tip of bilge keels and R is the bilge radius. These are calculated as follows:

$$l = d\sqrt{\left\{H_0 - \left(1 - \frac{\sqrt{2}}{2}\right)\frac{R}{d}\right\}^2 + \left\{1 - \frac{\overline{OG}}{d} - \left(1 - \frac{\sqrt{2}}{2}\right)\frac{R}{d}\right\}^2} \tag{8.37}$$

$$R = \begin{cases} 2d\sqrt{\frac{H_0(\sigma-1)}{\pi-4}}, & R < d \,\&\, R < \frac{B}{2} \\ d, & H_0 \geq 1 \,\&\, \frac{R}{d} > 1 \\ \frac{B}{2}, & H_0 \leq l \,\&\, \frac{R}{d} > H_0 \end{cases} \cdot \tag{8.38}$$

References

Ikeda Y, Himeno Y, Tanaka N (1976) On Roll Damping Force of ship: Effects of Friction of Hull and Normal Force of Bilge Keels. Journal of Kansai Society of Naval Architects, Japan, 161: 41–49

Ikeda Y, Komatsu K, Himeno Y, Tanaka N (1977a) On Roll Damping Force of Ship: Effects of Hull Surface Pressure Created by Bilge Keels. Journal of Kansai Society of Naval Architects, Japan, 165: 31–40

Ikeda Y, Himeno Y, Tanaka N (1977b) On Eddy Making Component of Roll Damping Force on Naked Hull. Journal of The Society of Naval Architects, Japan, 142: 54–64

Ikeda Y, Himeno Y, Tanaka N (1978) Components of Roll Damping of Ship at Forward Speed. Journal of The Society of Naval Architects, Japan, 143: 113–125

Tanaka N, Himeno Y, Ikeda Y, Isomura K (1981) Experimental study on Bilge-Keel Effect for Shallow-Draft Ship. Journal of the Kansai Society of Naval Architects, Japan, 180: 69–75

Tanaka N, Ikeda Y, Okada H (1982) Study on Roll Characteristics of Small Fishing Vessel Part I Measurement of Roll Damping, Journal of the Kansai Society of Naval Architects, Japan, 187: 15–23

Ikeda Y, Katayama T, Hasegawa Y, Segawa M (1994) Roll Damping of High Speed Slender Vessels. Journal of the Kansai Society of Naval Architects, Japan, 222: 73–81

Hashimoto H, Umeda N (2008) Preventing Parametric Roll with Use of Anti-Rolling Tank for a Large Containership in Head and Following Waves. Proc. Of the 4th Asia Pacific Workshop on Marine Hydrodynamics, 73–78

Hashimoto H, Sanya Y (2009) Research on Quantitative Prediction of Parametric Roll in Regular Waves. Journal of the Japan Society of Naval Architects and Ocean Engineers, 8: 361–364

Hoerner S F (1965) Fluid-Dynamic Drag

Katayama T, Yoshioka Y, Kakinoki T, Ikeda Y (2011) An Experimental Study on the Characteristics of Drag Force acting on a Flat Plate under Transitional and Irregular Oscillations. Journal of the Japan Society of Naval Architects and Ocean Engineers, 14: 55–62

Tanaka N, Ikeda Y, Himeno Y (1980) Experimental Study on Hydrodynamic Viscous Force Acting on Oscillating Bluff Body. Journal of the Kansai Society of Naval Architects, Japan, 179: 35–43

Kudo K, Kinoshita A, Nakawatari M (1980) Experimental Study on Hydrodynamic Viscous Force Acting on Oscillating Bluff Body. Journal of the Kansai Society of Naval Architects, Japan, 177: 83–90

Keulegan G H, Carpenter L H (1958) Forces on Cylinders and Plates in an Oscillating Fluid, Journal of Research of the National Bureau of Standards, 60 (5): 423–440

Chapter 9
Considerations for Bilge Keel Force Models in Potential Flow Simulations of Ship Maneuvering in Waves

Christopher C. Bassler, Ronald W. Miller, Arthur M. Reed and Alan J. Brown

Abstract Requirements for ship operations, both naval and commercial, may result in increased exposure to heavy weather and the occurrence of large amplitude motions. In order to enable evaluation of hull form designs, or to develop detailed ship specific operator guidance for these critical conditions, potential flow sectional, or strip-theory based, approaches remain the most practical method for fast ship motions simulations. However, some essential physical effects regarding the bilge keels are not captured by potential flow sectional formulations. To examine the relative importance of these effects, a series of unsteady RANS (URANS) computations were performed for the ONR Tumblehome model experiencing large amplitude roll motion at both zero and forward speed conditions, in calm water and in waves.

Keywords Bilge keels · Potential flow · Large amplitude motions

9.1 Introduction

Since their introduction in the mid-19th century, bilge keels remain an important hull form feature to increase damping and reduce the severity of roll motions experienced by a ship in waves (e.g. Froude 1865; Bryan 1900; Martin 1958; Kato 1965). This passive means of mitigating roll motion has become common for ships. Despite their importance, bilge keel models, particularly in fast numerical simulations using potential flow methods (cf. Beck and Reed 2001), are often simplified. Because of this, they are not necessarily accurate in heavy weather sea conditions, where the bilge keels are important for reducing the likelihood of large roll motions.

Potential flow methods for ship motions assessment depend on accurate modeling of roll damping to determine ship roll motion. However, the traditional semi-

C. C. Bassler (✉) · R. W. Miller · A. M. Reed
David Taylor Model Basin (NWSCCD), West Bethesda, MD, USA
e-mail: cbassler@gmail.com

A. J. Brown
Virginia Tech, Blacksburg, VA, USA

V. L. Belenky et al. (eds.), *Contemporary Ideas on Ship Stability*, Fluid Mechanics and Its Applications 119, https://doi.org/10.1007/978-3-030-00516-0_9

empirical roll damping models (Ikeda et al. 1978; Himeno 1981) do not explicitly account for the physical phenomena which occur during large amplitude ship motions, including the reduced effectiveness of the bilge keels (Bassler and Reed 2009; Reed 2009; Bassler et al. 2010a, b, 2011). Unless the bilge-keel model adequately captures the primary physical forces, ship motions may not be predicted accurately enough for design assessments or for the development of ship-specific operator guidance.

In order to obtain more accurate predictions of ship roll motion, high-fidelity codes may also be used. Improvements have been made in roll damping predictions using viscous flow codes (Yeung et al. 1998, 2000; Roddier et al. 2000; Seah 2007; Seah and Yeung 2008) and URANS codes (Korpus and Falzarano 1997; Miller et al. 2002, 2008; Wilson et al. 2006). However, currently, their computational requirements prevent their expansive use for early-stage design assessments, or assessments for a large number of conditions.

The roll motion of a ship is influenced by both its shape and appendages. Viscous related phenomena, such as flow separation from the bilge, skeg and bilge keels with the subsequent vortex formation, account for a large portion of roll damping. Bilge keels also generate a lift force with forward motion of the ship, which further tends to dampen the roll motion. For new ship designs, the effectiveness of the bilge keels for damping roll motion needs to be determined for a large range of roll angles and ship speeds.

Several important considerations are often neglected in the numerical tools that are needed for early-stage design evaluation, or to compute large numbers of realizations to develop ship-specific operator guidance. These include reduced effectiveness of bilge keels during large amplitude roll motion, the effectiveness of bilge keels while maneuvering in waves, and considerations for energy dissipation through shed vorticity from the bilge keels during these conditions.

Recent advances consider the effects of large amplitude motions for numerical ship motion performance assessments (e.g. Belknap and Reed 2010; Belknap et al. 2010). These advances have focused on the development and expansion of models for potential flow simulation tools with sectional formulations. However, additional effects of bilge keels due to vortex shedding, flow convection downstream, waves, and bilge keel emergence and submergence during large roll motion may be important, but not necessarily accounted for in the sectional formulations.

A series of unsteady Reynolds Averaged Navier Stokes (URANS) computations were performed for both 2-D and 3-D conditions of large amplitude ship roll motion, with and without forward speed, and in calm water and in waves. Comparisons were made to available experimental data for the 2-D calm water conditions at zero-speed. These results were then compared to the 3-D conditions to develop improved understanding of additional physical effects, including forward speed and waves, which should be considered for future developments of strip-theory approaches for ship motions prediction.

The purpose of this study was to examine high-fidelity numerical bilge keel force results and assess the relative importance of the additional physical effects which may need to be considered in future developments of sectional force component models for bilge keels. Some potential issues for modeling these effects in bilge keel force models are also discussed.

9.2 Bilge Keel Force Models

Existing bilge keel force models are typically based on a Morison-equation type formulation, or more advanced bilge keel models which consider unsteady vortex shedding. Both types of models are briefly discussed. However, there is still a need to improve these models, with consideration for large amplitude roll motions. Currently, bilge-keel force models using either the zero-speed (Morison-equation based) or forward speed formulations do not consider the physical phenomena which occur during large amplitude roll motions.

9.2.1 Morison-Equation Based Models

The Morison equation (Morison et al. 1950, 1953) is a robust engineering formulation for drag/inertia dominated problems of a body in an oscillatory fluid. However, it has limited application due to its semi-empirical form and its basis of using harmonically oscillating planar motion. The semi-empiricism requires precision in selecting values for the inertia and drag coefficients. For applications in non-harmonically oscillating fluids, such as ship motions in irregular waves, application of this method may be limited. Morison's equation appears to be most appropriate for conditions where the Keulegan-Carpenter number is less than 8, or greater than 25 (Sarpkaya and Isaacson 1981). Keulegan and Carpenter (1958) were the first to attempt to improve on the Morison equation, by specifying a remainder value, while still neglecting diffraction effects. Additional modifications have included consideration of higher-order harmonics (Sarpkaya 1981).

A Morison-equation-type formulation was used to study oscillating flow around 2-D and 3-D bilge keels, as represented by flat plates (Sarpkaya and O'Keefe 1996). However, wall effects that may be significant are not accounted for in typical flat plate formulations. Additional studies were carried out to examine the hydrodynamic forces on flat plates in forced oscillation (Klaka et al. 2007).

The unit normal force on the bilge keel can be modeled, using a Morison-type equation, as the sum of viscous drag and added mass due to roll motion. Initial models for the bilge keel force only considered the drag induced force on the bilge keel for zero speed (Lloyd 1998; Themelis 2008). However, even for zero speed conditions, the bilge keel force still retains some component due to added mass effects, particularly for thick span bilge keels.

A similar bilge keel force model formulation to describe the force on the bilge keel during roll decays was detailed in Irvine et al. (2006). This formulation was expanded with a suggested possible forward speed correction in Atsavapranee et al. (2007) and with considerations for multiple degree-of-freedom (roll and heave) motions and the effects of wave orbital velocities in Grant et al. (2007).

Potential flow ship motions codes typically use some variation of Ikeda's bilge keel component (Ikeda et al. 1978), which considers the bilge keel force as a combination of the normal force component and the hull interaction component and neglects wave-making due to the bilge keel, which may be significant for large amplitude roll motion (Bassler and Reed 2009).

9.2.2 Unsteady Lifting Surface Models

Additional models for potential flow codes have been developed using unsteady lifting surface theory, based on a vortex-lattice method. Liut (1999) and Liut and Lin (2006) used a vortex-lattice method to model arbitrary lifting shapes, such as rudders and fins. This is the approach applied in the *Large Amplitude Motions Program* (LAMP), and considers the lift force on the bilge keels (Lin and Yue 1990; Lin et al. 2006). Additionally, if the lifting surface stalls, then an eddy-making force is computed. This is equivalent to the force on a flat plate, with the flow normal to the surface of the plate. The typical angle of stall considered for the bilge keels is 12°.

Greeley and Petersen (2010) developed a bilge keel force model using an unsteady lifting surface (ULS) approach and showed favorable comparisons to the experimental and RANS results given in Miller et al. (2002) for a 3-D circular cylinder with instrumented bilge keels. In order to expand the application of this model to low speed conditions, the ULS model was coupled with a Morison-equation based approach for very low speeds and the bilge keel force is obtained by switching between the two models, depending on the speed conditions. This formulation is currently utilized in the potential flow ship motions code TEMPEST (Belknap and Reed 2010).

9.2.3 Large Amplitude Roll Models

For large amplitude ship roll motion, the bilge keels may become less effective, due to their interaction with the free surface and, for more severe motions, due to possible emergence. To further investigate the physical phenomena which occur during large amplitude roll motion, and to improve the modeling of these conditions in strip-theory based approaches, a series of experiments were performed (Bassler et al. 2010b) for a 2-D midship section model (DTMB Model 5699) derived from the ONR Topside Series hull forms (Bishop et al. 2005). These experiments used instrumented bilge keels to measure the force on the bilge keels through the forced roll oscillations, including conditions where large amplitude roll resulted in bilge keel emergence.

From the experiments, abrupt changes in the geometry of the body relative to the free surface were observed, which must be considered to accurately determine the properties of the dynamical system modeling ship roll motion (Bassler et al.

2010a). Because existing theoretical models were developed for small to moderate roll motions, the amount of energy dissipation for large amplitude roll motion may be over-estimated, resulting in under-predicted roll motion (Bassler and Reed 2009; Reed 2009).

In order to consider these physical phenomena in models for ship roll damping, and increase their applicability and robustness, a piecewise formulation for total ship roll damping was proposed (Bassler et al. 2010a, 2011). However, the piecewise model represented a simplified approach to the consideration of large amplitude effects for ship roll motion. For more accurate modelling, especially for more advanced potential flow numerical tools with discrete force models (e.g. Belknap and Reed 2010), an explicit model for the bilge keel force, with the consideration of large amplitude roll motion effects may be used.

9.3 Urans Simulations

To examine the relative importance of the physical effects identified earlier, a series of Unsteady Reynolds Averaged Navier Stokes (URANS) computations were performed.

9.3.1 Solver

Calculations were performed using the URANS solver CFDShip-Iowa, Version 4. CFDShip-Iowa is a general-purpose research RANS computational fluid dynamics code developed at the University of Iowa.

Basic solver numerical modeling details include 2nd-order upwind convective terms and 2nd-order central differenced viscous terms. For time discretization a 2nd-order backward difference scheme is used. CFDShip-Iowa uses predictor/corrector algorithms to couple the velocity and pressure and to enforce continuity. Turbulence modelling uses the blended k-ε/k-ω model of Menter (1994). The solver uses Message Passing Interface (MPI)-based domain decomposition for parallel processing. Details of the solution algorithm and numerical methods can be found in numerous references, including Carrica et al. (2006, 2007a, b). Details on previous applications of the code to problems of roll motion and bilge keel forces can be found in Miller et al. (2002, 2008).

9.3.2 Dynamic Overset Grids

CFDShip-Iowa uses dynamic overset grids to allow for the relative motion between blocks of meshes. Overset grids are used in a wide variety of applications, including calculations of a static hull embedded in a background grid and dynamically moving, fully-appended ships. Additionally, overset grids remove the necessity of point-to-point matching in structured grid systems. This capability alleviates some of the difficulties involved in creating structured grids for complex configurations, such as hulls appended with shafts and struts. The grid assembling tool SUGGAR (Structured Unstructured Generalized Grid Assembler), Version 2.73 (Noack 2005) was used.

SUGGAR may be run as a preprocessor for static calculations, or concurrently with CFDShip-Iowa, using calls to subroutines. The software, USURP, Version 2.39, (Boger and Dreyer 2006), was used to properly compute area and forces on overlapping surface regions.

For the calculations presented in this paper fixed yaw angles (±10°) and single degree-of-freedom (1-DoF) roll motion were prescribed, and a constant forward speed was imposed using uniform inflow into the computational domain. The axis of rotation was along the ship's center plane, through the vertical center of gravity. The body sway, heave, pitch, and yaw modes of motion were fixed for the 1-DoF roll calculations. At each time-step, the solver obtained the motion and sent the information to SUGGAR to create new grid assembling information. CFDShip-Iowa's implicit motion solver was utilized, resulting in updated grid motion at every inner-iteration of the time-step solution. Typically 3–5 inner iterations were performed at each time step. At each time step, CFDShip-Iowa uses the new value of roll displacement to re-orient the ship. SUGGAR was then run to update the grid assembly for the new orientation.

Overlapping boundary fitted grids were created for the port and starboard sides of the hull and the port and starboard bilge keel grids. A cross-section of the grid assembly shows the relative grid point densities of the hull, refinement, and background grids (Fig. 9.1), and also illustrates the use of dynamic overset grids. The moving hull grid assembly was embedded in a stationary intermediate refinement grid, which was embedded in a larger stationary background grid, representing a towing tank. The intermediate refinement grid is used to blend the grid sizes of the very fine boundary layer grid with the coarser background grid. As the hull grid rotates relative to the fixed background grids, new grid connectivity information is calculated by SUGGAR.

9.3.3 Computational Domain

The computational region is shown in Fig. 9.2. The surface discretization of the hull appended with bilge keels is shown in Fig. 9.3. No-slip boundary conditions were applied to the hull and appendages, far-field boundary conditions were applied to the side and bottom walls. A constant inlet velocity, $U/U_{ref} = U_S$, $V/U_{ref} = 0$ and $W/U_{ref} = 0$, represents the constant forward speed (Fig. 9.4). The velocities are extrapolated at the exit.

Details of the computational grid size are given in Table 9.1. The total number of grid points used for the URANS simulations was about 4.8 million. The skeg was

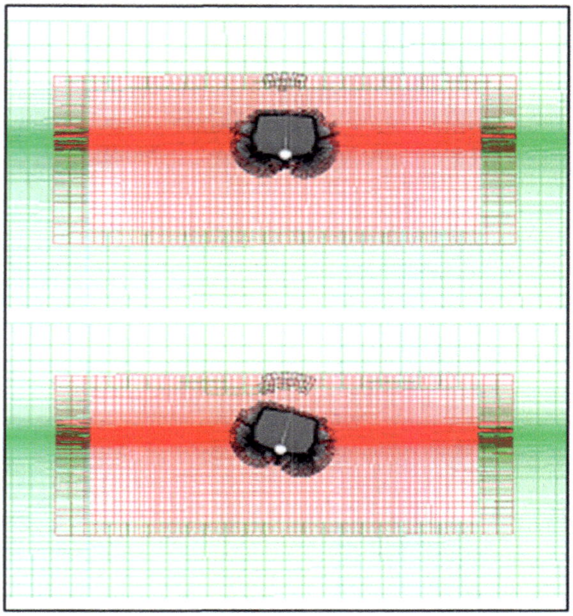

Fig. 9.1 SUGGAR grid assembly for DTMB Model 5613-1 at two roll positions, with relative grid densities between refinement and background grids

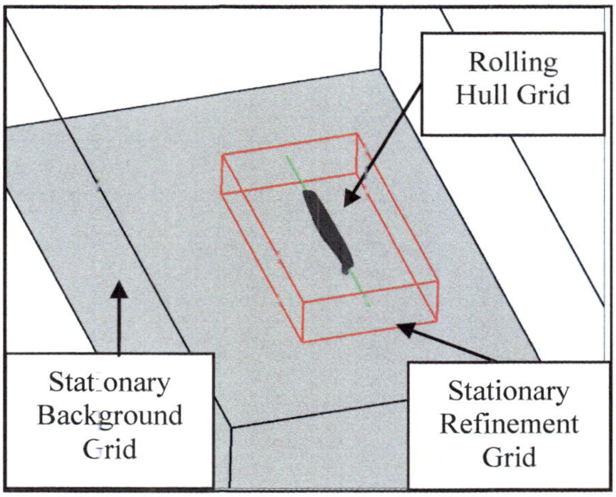

Fig. 9.2 Computational region: moving hull, grid, stationary refinement, and background grids

integrated into the hull grid. The domain was decomposed into 45 smaller blocks for parallel processing. One additional processor was used for SUGGAR.

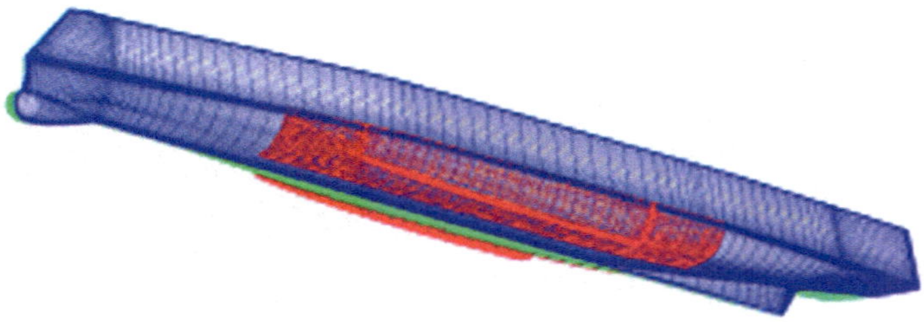

Fig. 9.3 Hull and bilge keels surface discretizations, shown for DTMB Model 5613-1

Fig. 9.4 CFDShip-Iowa boundary conditions

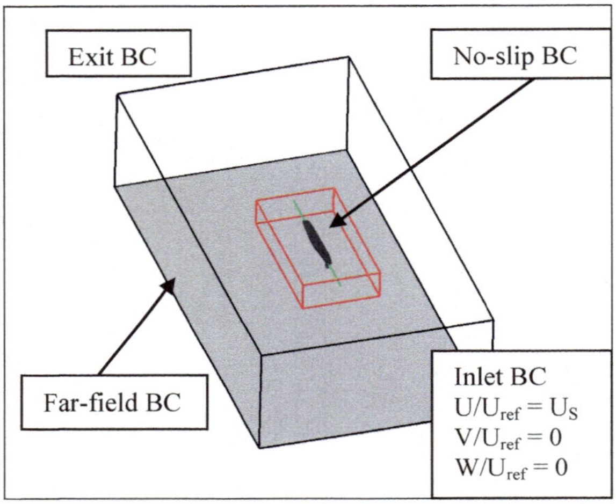

Table 9.1 DTMB Model 5613-1 grid sizes and decomposition

Block	# pts.	# proc.	# pts/proc.
Hull Stb	667,116	6	111,186
Hull Prt	667,116	6	111,186
BK Stb	363,750	3	121,250
BK Prt	363,750	3	121,250
Refinement	1,519,035	15	101,269
Background	1,241,240	12	103,436
Total	4,822,007	45	Avg = 107,155

9.3.4 Obtaining Segmented Forces on the Bilge Keel

CFDShip-Iowa sums the force/moment contributions from all of the wetted wall surface elements at each time step. In order to obtain segmented forces on the bilge keel, a post-processing routine was developed to create longitudinal segments distributions, Δx, over the hull and appendages. By decomposing the total force on all

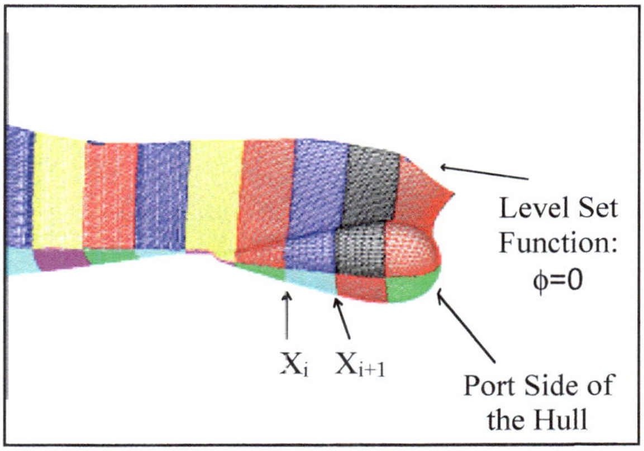

Fig. 9.5 Post-processing example of longitudinal segments for the hull, with the bow section shown, up to the instantaneous waterline

the wetted surface elements at each time step into these longitudinal segments, a force/Δx is obtained at each specified x location segment (Fig. 9.5).

The development of this post-processing routine[1] enabled changes to be made to the constraints, which do not necessarily conform to the surface meshes, or the regions of interest on the hull form, after the simulations have been performed without having to re-run them. Local regions, such as pressures on the bow dome or forces on the appendages (bilge keels, rudders, shafts, struts) may then be determined.

9.3.5 *Hull Form and Simulated Conditions*

Simulations were performed for both a 2-D and 3-D hull form. The ONR Topside Series, tumblehome topside configuration hull (Bishop et al. 2005) was used (Table 9.2). A 32nd scale model, DTMB Model 5613-1, had been used for previous experiments (Bassler et al. 2007) and simulations (Miller et al. 2008) of forced roll motions. A 2-D midship section model of this hull form (Fig. 9.6), DTMB Model 5699-1, was also constructed at the same scale ratio to carry out sectional experiments of large amplitude roll motion (Bassler et al. 2010b) and enable comparisons to the 3-D model at the same scale ratio. For simulations of DTMB Model 5613-1, 20 segments were specified (Fig. 9.7) to interrogate the bilge keel force during the prescribed roll motions.

Simulations were performed for forced 1-DoF roll motions, moderate to large amplitude, in calm water at zero speed and with forward speed. Additionally, simulations were performed in beam sea regular waves, with wavelength equal to ship length and a steepness of 1/100, for zero speed, and for steady drift in calm water at forward speed (Table 9.3).

[1]This routine was developed by Miller at NSWCCD in 2010.

Table 9.2 Barehull hydrostatic parameters for DTMB Model 5613-1 (Bassler et al. 2007)

Displacement	260.14 kg
Draft	0.172 m
Lpp	4.8125 m
Beam	0.5875 m
KG	0.172 m
Roll Gyradius/Beam	0.361
Scale ratio	32

Fig. 9.6 Midship section of the ONR Topside Series, Tumblehome configuration hull form with bilge keels (DTMB Model 5699-1)

Fig. 9.7 DTMB Model 5613-1 with bilge keel divided into 20 segments for sectional force analysis

Table 9.3 Simulated conditions for DTMB Model 5613-1

Froude number	Roll frequency (rad/s)	Motion amplitude (deg)	Sea conditions
0.0, 0.30	2.85	1-DoF Roll: 15, 25, 30, 35	Calm water
0.0	2.85	1-DoF Roll: 25, 35	Beam waves ($\lambda/L = 1.57$, $H/\lambda = 1/100$)
0.3	0	1-DoF Yaw: ± 10	Calm water

9.4 Effects to Consider for a Ship Maneuvering in Waves

Several physical effects were examined which are relevant to modeling a ship maneuvering in waves. These included 3D, forward speed, waves, maneuvering, and large amplitude roll effects. Each of these effects were examined individually, to assess their relative significant to the forces observed on the bilge keel.

9.4.1 3D Effects

URANS simulations were performed for the 3D ONR Tumblehome hull undergoing 1DoF forced roll oscillations at both zero and forward speed conditions, in calm water.

URANS simulations were performed for the 2-D midship section undergoing 1-DoF forced roll oscillations. Comparisons are shown between the 2-D URANS and 2-D experiment, as well as the midship section of the bilge keel force URANS results for the 3-D hull form at zero speed in calm water (Fig. 9.8). Comparisons were made to the experimental results of Bassler et al. (2010b). The distinct physical processes occurring during roll motion, as observed from the experimental measurements, were given in (Bassler et al. 2010b).

As shown, the bilge keel force results from the simulations and the experiments agreed well for roll amplitudes of 15 and 25°. However, for larger roll angles (30 and 35°), where the bilge keel interacts with and emerges from the free surface, some discrepancies are observed between the simulations and the experiments. The comparisons were made between filtered experimental results and unfiltered numerical results. Due to mechanical noise in the experimental results, filtered results for the measured bilge keel force are presented (Bassler et al. 2010b). For the large roll amplitudes, the general shape and peak values agree well for the 2-D midship section. However, the URANS simulations have more noise for the large amplitude cases, likely due to the finite difference method used for the simulations, where emergence and re-entry of the bilge keel results in nonlinearities in the instantaneous wetted portion of the bilge keel, including water run-off, entrained air collapse, and impulse loading from their emerging and immerging from the free-surface.

For roll amplitudes up to 30°, at zero speed, the differences between the 2-D and 3-D midship section cuts of the bilge keel force are small. For the 35° case, distinct differences occur between the 2-D and 3-D results for the portion of the roll cycle where the bilge keel is emerging and then re-entering the free surface. This is likely due to 3-D effects along the bilge keel due to the flow along chord, while the 2-D case does not have any longitudinal variation along the bilge keel during large amplitude roll motions.

9.4.2 Forward Speed Effects

URANS results are also presented to examine the influence of forward speed (Fn = 0.3 vice Fn = 0.0) on the midship section bilge keel force (Fig. 9.9). As observed, even for smaller amplitude roll motion, the differences in the unit bilge keel force are greater due to forward speed effects than due to 3-D effects. Overall, the greatest difference due to forward speed occurs during the peak of the roll cycle just after the bilge keel is fully submerged. For the portion of the roll cycle where the bilge keel is near the free surface, the zero and forward speed results agree quite well. This

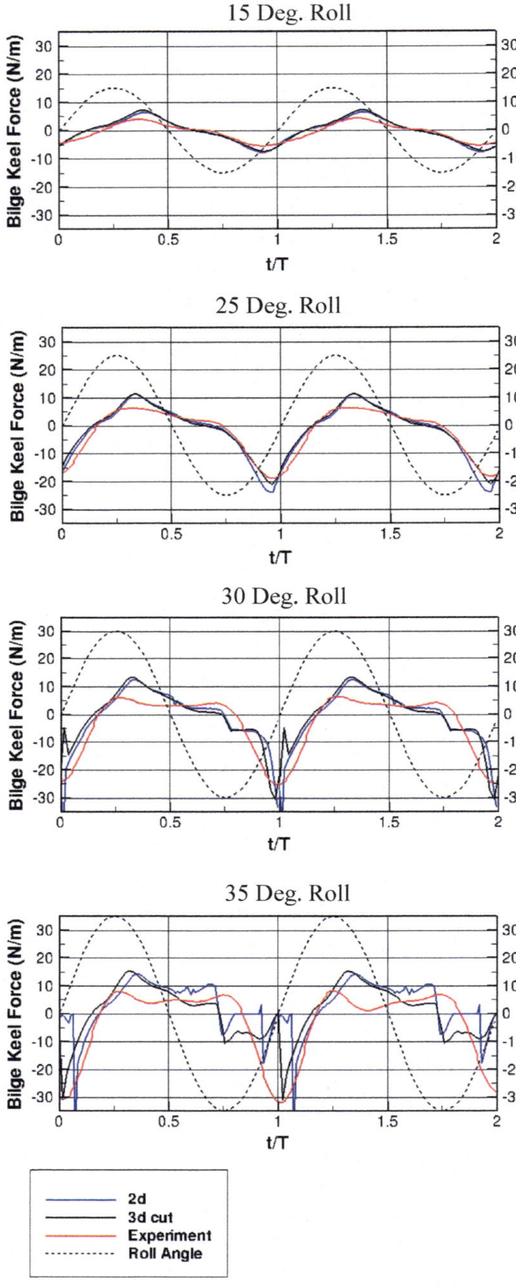

Fig. 9.8 Bilge keel force comparisons between 2-D (solid blue) and 3-D midship section cut (solid black) URANS simulations for DTMB Model 5613-1 and 2-D model experiments (solid red) at $Fn = 0.0$, $\omega = 2.85$ rad/s, for $\phi = 15$, 25, 30 and 35° roll

is likely due to the shedding of vorticity from the bilge keel at the fully submerged position in the roll cycle, and the reduction in force on the bilge keel due to the vortex when it is convected downstream in the forward speed condition. For the 25 and 30° roll cases, the difference in the bilge keel force at the peak of the roll cycle for zero and forward speed becomes larger. This is due to the interaction of the bilge keel with the free surface, and again the bilge keel force is less for the forward speed condition than for the zero speed condition, due to the downstream convection of vorticity. For the 35° roll case, the portion of the roll cycle corresponding to the emergence of the bilge keel agrees well between the zero speed and forward speed conditions. For this case, the impulse loading on the bilge keel during re-entry is significant during at both speeds. However, the influence of "water shipping" (or lingering) on the topside of the bilge keel is reduced for the forward speed case, compared to the zero speed case.

Results are also presented for both axial vorticity and bilge keel force at various longitudinal locations along the hull, for both speeds. As shown in Fig. 9.10, during roll motion the vortices shed from the bilge keels and bow dome remain clustered near their shed position for zero speed.

As expected, at forward speed, the vortices are convected downstream. Near the midship section, the magnitude of the vorticity is reduced with forward speed, compared to zero speed, but also the presence of the bow-dome vortices are observed (Fig. 9.11).

Vorticity and bilge keel forces were examined at four specific locations, to determine the longitudinal variation for each. These locations included the leading edge (LE) of the bilge keel, and forward ($x/L = 0.4$), midship ($x/L = 0.5$) and aft ($x/L = 0.6$) positions on the bilge keel (Fig. 9.12).

For 25° roll at zero speed (Fig. 9.13), all of the positions experience similar forces, and the force variation over a roll cycle is fairly small. However, for forward speed, the leading edge force differs significantly from that at the other three positions, especially for the portion of the roll cycle where the bilge keel is near to the free surface. The maximum force appears just before the maximum upward angular velocity ($t/T = 0.5$). For the maximum downward angular velocity, for large angles, free-surface re-entry effects occur due to the re-wetting of the leading edge of the bilge keel with the combination of the bow wave generated by the ship at forward speed.

This significant variation for the leading edge bilge keel section is because of the pressure differentials which exist at the leading edge at forward speed, corresponding to greater lift generated at the leading edge section. This is even further exacerbated for a low aspect ratio airfoil (such as a bilge keel). The variation between the forward, midship, and aft sections is also more significant than for the zero speed condition. The effect of re-entry after emergence of the bilge keel is more pronounced for the forward speed condition.

As seen from the bilge keel forces for 35° roll (Fig. 9.14) at zero speed, the forward, midship, and aft bilge keel sections do not have significant variations. Again, even for the zero speed condition, the leading edge section of the bilge keel has a more significant impact load on re-entry then for the other sections. This may be due to the pressure differential on the bilge keel at an even higher angle of attack of the leading

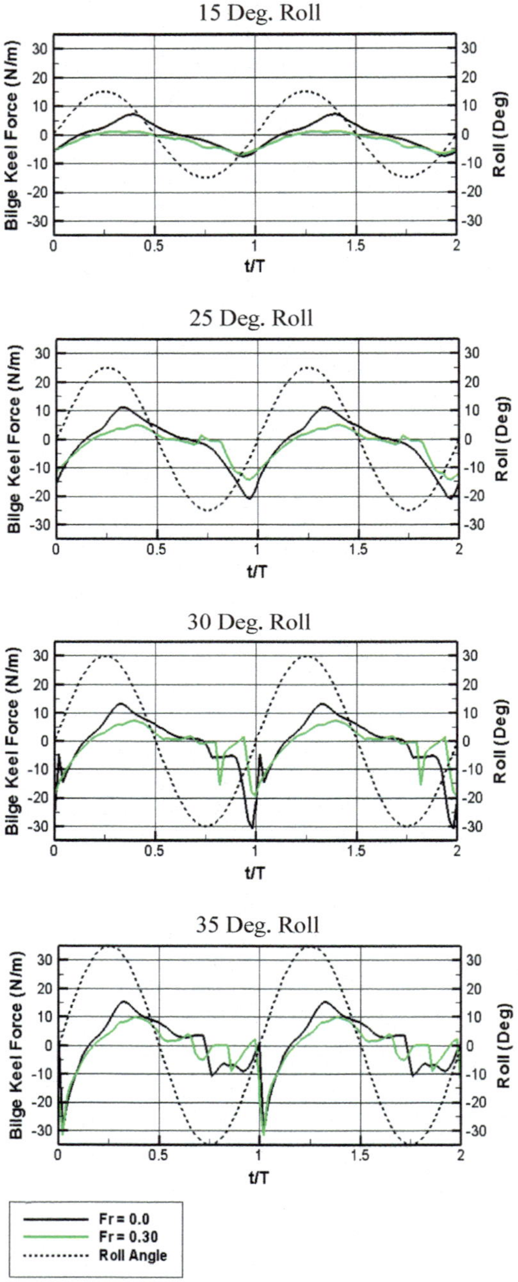

Fig. 9.9 Bilge keel force comparisons between 3-D URANS simulations for zero speed (solid black) and forward speed (solid green) for the midship section of DTMB Model 5613-1 at $\omega = 2.85$ rad/s, for $\phi = 15$, 25, 30 and 35° roll

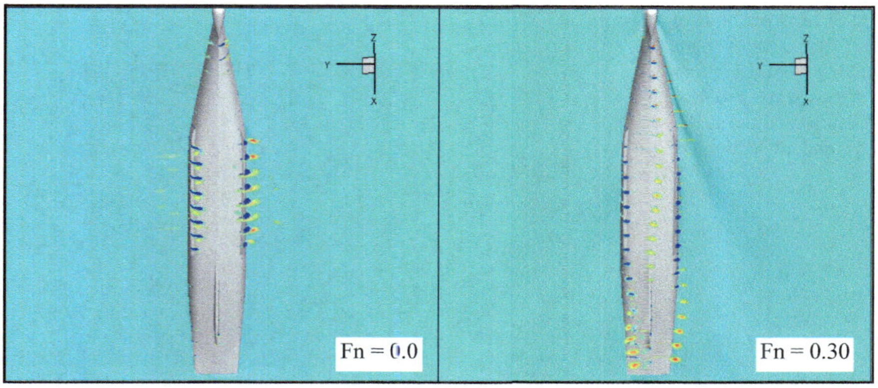

Fig. 9.10 Axial vorticity for DTMB Model 5613-1 at various longitudinal locations, for zero speed (left) and forward speed (right) conditions, $\phi = 25°$, $\omega = 2.85$ rad/s

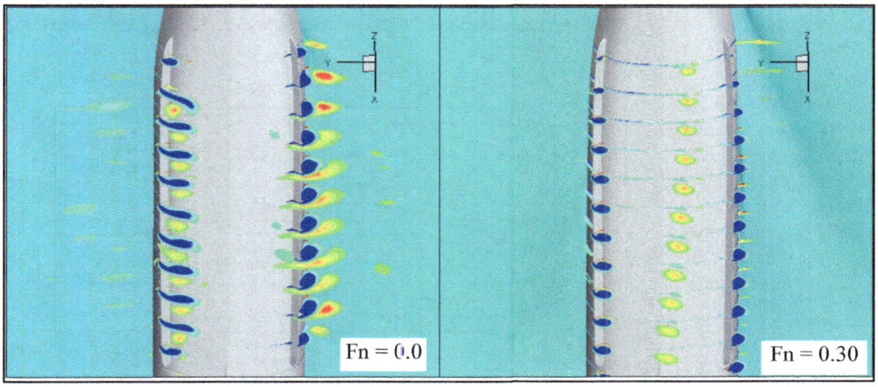

Fig. 9.11 Axial vorticity along the bilge keels of DTMB Model 5613-1, for zero speed (left) and forward speed (right) conditions, $\phi = 25°$, $\omega = 2.85$ rad/s

edge, as it follows the streamline along the hull. For the forward speed condition, significant variation is observed for the bilge keel force between the leading edge section and the other three sections. Similar to the 25° roll case, the bilge keel force on the leading edge section is larger for the portion of the roll cycle where the bilge keel is near to, and emerging from, the free surface. For the forward speed condition, the forward section, instead of the leading edge, of the bilge keel experienced the most significant impact loading on re-entry. This is due to the bow wave generated along the hull at forward speed, which reduced the disturbance of the bilge keel force on the leading edge section, even for the occurrence of large amplitude roll.

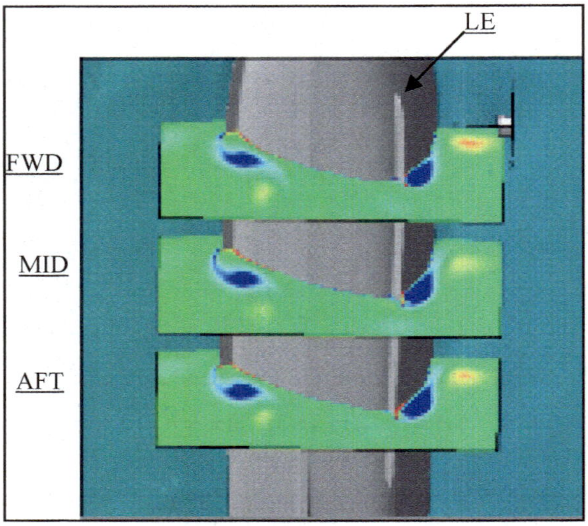

Fig. 9.12 Locations of bilge keel force sectional analysis: leading edge (LE), forward section (at $x/L = 0.4$), midship section (at $x/L = 0.5$), and aft section ($x/L = 0.6$)

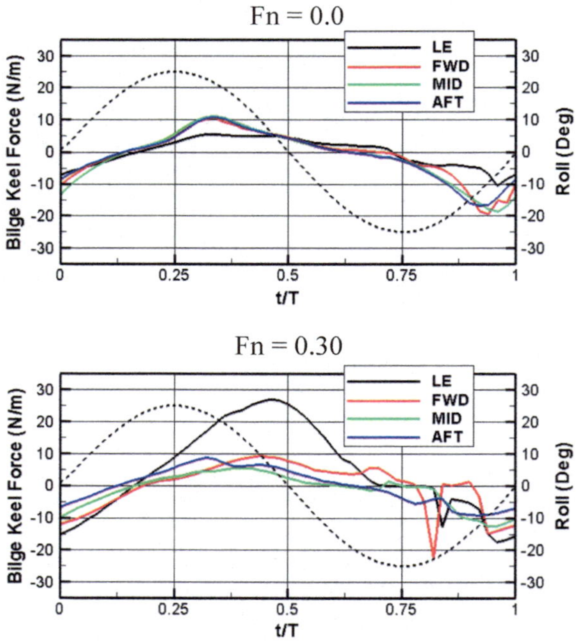

Fig. 9.13 Unit bilge keel force for DTMB Model 5613-1, at leading edge (black), forward (red), midship (green), and aft (blue) locations, for zero speed (top) and forward speed (bottom) conditions, $\phi = 25°$, $\omega = 2.85$ rad/s

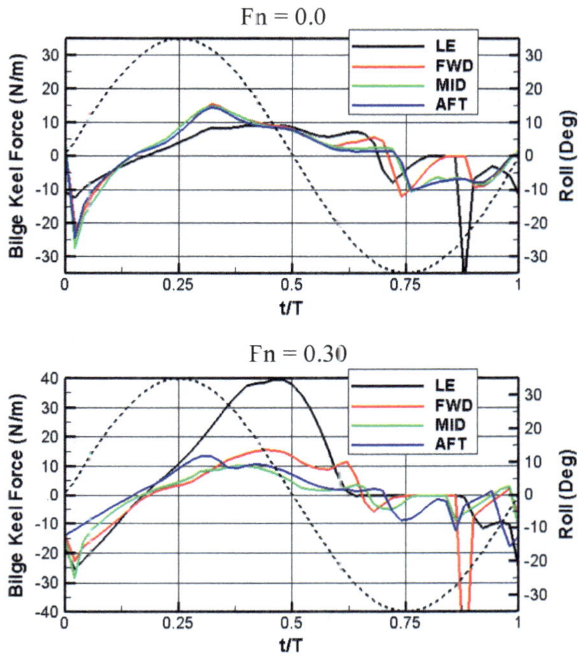

Fig. 9.14 Unit bilge keel force for DTMB Model 5613-1, at leading edge (black), forward (red), midship (green), and aft (blue) locations, for zero speed (top) and forward speed (bottom) conditions, $\phi = 35°$, $\omega = 2.85$ rad/s

9.4.3 Wave Effects

URANS simulations were also performed for the 3-D ONR Tumblehome hull undergoing 1-DoF forced roll oscillations, at zero speed, in beam seas regular waves, with $H/\lambda = 1/100$ and $\lambda/L = 1.0$. The ship was fixed in sinkage and trim to enable more direct comparisons with the calm water forced roll oscillation conditions. The waves impacted the hull from the port side and the variation in the force on the midship section of the bilge keel, was examined, as a function of wave phase

Four comparisons, based on the wave phase relative to the peak of the roll cycle, are shown for both the 25 and 35° forced roll oscillation conditions. These included when the wave crest (max), wave trough (min), front slope of the wave (front) and back slope of the wave (back), coincided with the port side bilge keel at the peak (maximum) of the forced roll cycle, Figs. 9.15, 9.16, 9.17 and 9.18, respectively. As shown in the plots, the synchronization between the wave and roll frequencies occurred at $t/T = 0.75$, when the port side bilge keel is closest to the free surface (the maximum of the roll cycle).

For the maximum condition, where the peak of the roll cycle corresponds to the wave crest, the force on the bilge keel varies least, since it remains submerged the longest. In this case, due to re-entry effects, the 35° (larger) roll amplitude results in greater force, than the 25° (smaller) roll amplitude. where the bilge keel remains submerged.

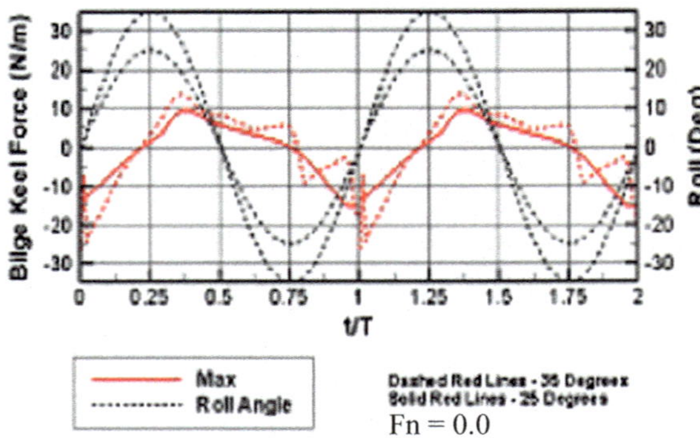

Fig. 9.15 Force on the midship section of the port side bilge keel for 25 (solid) and 35 (dotted) deg roll. Wave phase with crest corresponding to peak of the roll cycle (at t/T = 3/4)

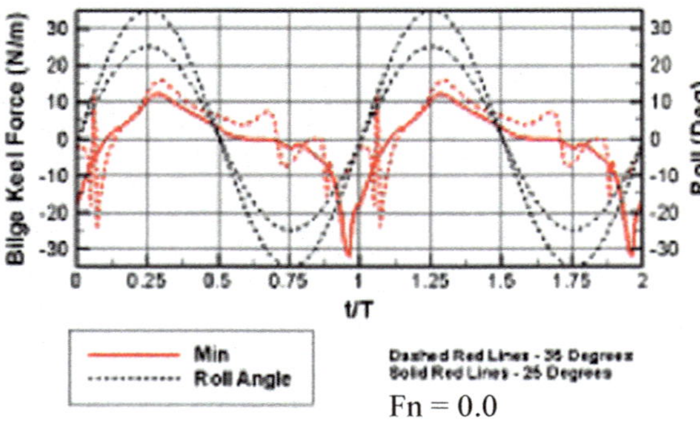

Fig. 9.16 Force on the midship section of the port side bilge keel for 25 (solid) and 35 (dotted) deg roll. Wave phase with trough corresponding to peak of the roll cycle (at t/T = 3/4)

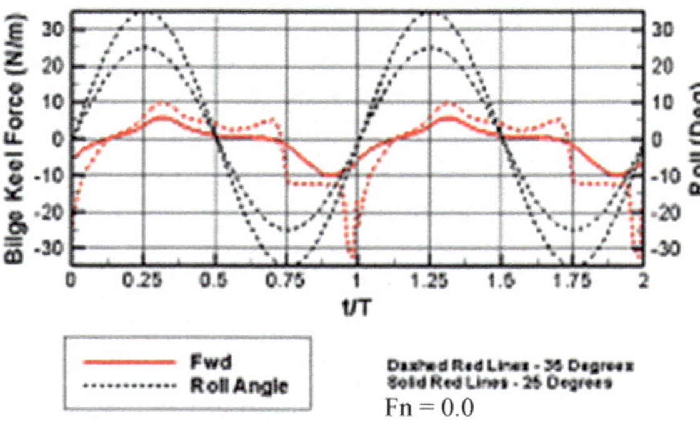

Fig. 9.17 Force on the midship section of the port side bilge keel for 25 (solid) and 35 (dotted) deg roll. Wave phase with front slope corresponding to peak of the roll cycle (at t/T = 3/4)

For the minimum condition, where the peak of the roll cycle corresponds to the trough of the wave, significant force on the bilge keel is observed due to impact loading on re-entry. In this condition, the peak force is largest, due to the higher relative velocity between the bilge keel and the free-surface—the bilge keel has greater velocity at re-entry. However, the peak force for the 35° case is slightly less because of the phase lag (and thus reduced time in the roll cycle) before re-entry.

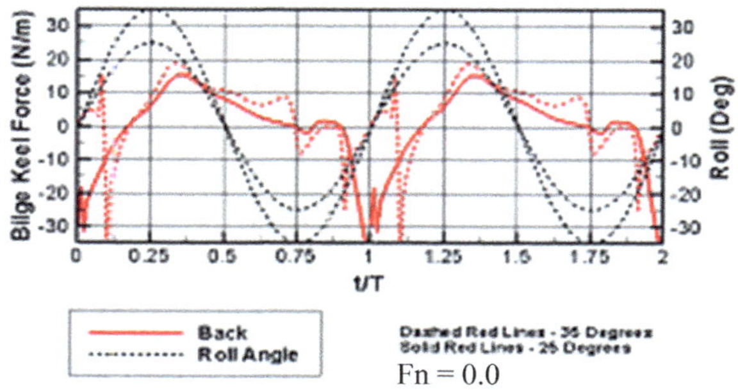

Fig. 9.18 Force on the midship section of the port side bilge keel for 25 (solid) and 35 (dotted) deg roll. Wave phase with back slope corresponding to peak of the roll cycle (at t/T = 3/4)

For the condition with the peak of the roll cycle corresponding to the front slope of the wave (midway between crest and trough), the 25° case has a reduced peak force because the bilge keel does not experience emergence. However, the 35° case shows impact loading due to emergence of the bilge keel. This is reduced compared to the back slope of the wave, due to the direction of the orbital velocities within the wave relative to the motion of the bilge keel.

For the condition with the peak of the roll cycle corresponding to the back slope of the wave (midway between trough and crest), both the 25° and 35° cases show large peak forces on the bilge keel, due to the relative velocity on the bilge keel at re-entry.

9.4.4 Maneuvering Effects

The effects of coupled motions in waves, including large roll motions, while also maneuvering, are important. Previous work has shown that the vortex shedding and cross-flow drag effects on the bilge keels during steady turning, even at only small heel angles, is already significant (Dai et al. 2009).

To examine the significance of cross-flow drag forces on the bilge keel, a series of URANS simulations were performed for the case of ±10° steady drift in calm water at forward speed. As seen from the pressure contours observed on the hull for windward (+10°) and leeward (−10°) drift conditions (Fig. 9.19), the bow dome experiences the largest pressure gradient, while the bilge keels also experience large pressure gradients at the leading edge of the windward side bilge keel (going into the flow). Meanwhile, the leeward side bilge keel is effectively sheltered by the hull. This is manifested in the longitudinal force distribution along the port side bilge keel for windward and leeward side conditions (Fig. 9.20). When the bilge keel is in the windward condition, a large force results on the leading edge, but is relatively constant along the remainder of the bilge keel, before tapering off at the trailing edge. When the bilge keel is on the leeward side, the sheltering by the hull results in almost no cross-flow drag force on the bilge keel.

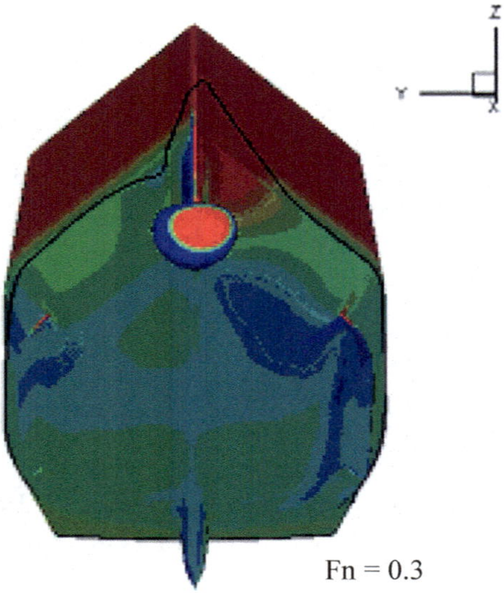

Fig. 9.19 Pressure contours on the submerged portion of ONRTH, for drift from the windward side (+10°), looking aft from the bow

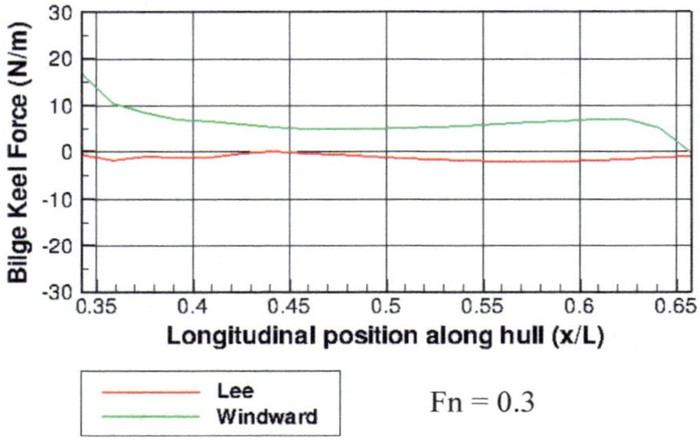

Fig. 9.20 Longitudinal distribution of force along the port side bilge keel for drift from the windward and leeward sides (±10°)

9.4.5 Large Amplitude Roll Motion Effects

For large amplitude roll conditions, the bilge keel may emerge from the water, resulting in lingering forces due to water-shipping effects and severe impact loading-type behavior due to re-entry during the roll motion. These abrupt changes may create difficulty for time-domain simulation of these types of motions in potential flow codes. Additional consideration must be given to how vorticity calculations will be re-started, how the bilge keel will be de/re-wetted upon emergence and re-entry, how hysteresis effects due to "water shipping" will be considered.

In previous studies of possible models for the transition behavior of the bilge keel during emergence and re-entry by the authors, analytical formulations using methods such as a step-function, Gompertz function, or generalized logistic function (Richard's curve) were considered. However, the semi-empirical nature of specifying coefficients for these models led them to be less than ideal for practical implementation in a robust potential flow simulation tool. Potential Flow Simulations.

From the decomposition of the physical effects for consideration in modeling the bilge keel forces during ship maneuvering in waves, several aspects were examined. These included 3-D, forward speed, wave, maneuvering (cross-flow drag due to steady drift), and large-amplitude roll motion effects. From the URANS simulations, the relative importance (greatest to least) of these physical effects was assessed and is summarized as follows: forward speed, large amplitude roll, wave effects, maneuvering, 3-D effects/longitudinal variation along the bilge keel.

Forward speed convects vorticity downstream, but zero (or low) speed allows the vorticity to linger near the bilge keels, which may be difficult for potential flow simulations to capture, due to interaction effects. Other forward speed issues such as the influence of vortex shedding from the sonar dome on the bilge keels must also be considered. Bilge keel vortices will convect downstream, possibly into the shafts, skeg, and finally propulsion and rudder sections fo the hull, depending on the present ship orientation (roll, drift angle, rates, etc.). An example of this for the bare hull with bilge keels, at Fn = 0.3, is given in Fig. 9.10. Accurate prediction of the vortices created by the bilge keels and the prediction of their interactions with other sections of the ship geometry are important.

9.4.6 Relative Importance and Implications for Potential Flow Simulations

Large amplitude roll motion may result in the biggest reduction in bilge keel performance while the ship is maneuvering in waves. Jumps in forces caused by bilge keel interaction with the free surface may also cause computational difficulties due to discontinuities across time-steps in potential flow simulations.

Wave effects were examined and wave phase was shown to have a significant effect on the bilge keel force, particularly when coupled with large amplitude roll motion, where the bilge keel may interact with the free surface during emergence and re-entry. Because of the variation in relative velocity between the bilge keel and the free surface, the peak loading on the bilge keel varies significantly with the phase of the motion relative to that of the wave.

Maneuvering imparts a significant force on the leading edge of the windward side bilge keel, and is important for consideration due to the impact of the inflow conditions on the bilge keel for computation of the vortex shedding in ULS methods for the bilge keel force.

From the cases examined, the 3-D effects on the bilge keel, aside from the leading edge, did not appear to be significant. Sectional formulations can be used for the bilge keel force, but care must be taken to account for forward speed, large amplitude roll, and maneuvering effects as they impact the forces on the leading edge of the bilge keel.

Morison-equation based approaches do not consider these 3-D effects, aside from the possibility of altering the coefficients at each time-step based on some predetermined specification related to forward speed, location along the bilge keel, angle of attack to the local flow, etc.. A blended approach (coupling Morison equation to a ULS method for forward speed), as proposed by Greeley and Petersen (2010), may help to address this issue. However, a blended approach will likely still have difficulties with re-wetting the sectional bilge keel geometries for large amplitude roll motions and re-starting the vorticity calculation after bilge keel re-entry has occurred. Particularly at low speed, where the trade-off between the ULS model and the Morison-equation based model occurs, and where vorticity is not being convected downstream as quickly. Under these conditions, the determination of the bilge keel force in a potential flow simulation will be difficult.

9.5 Conclusions

The purpose of this study was to assess effects which may need to be considered in future developments of force-component models for bilge keels. The comparisons performed for this study enabled the examination of effects due to vortex shedding, flow convection downstream, waves, maneuvering and bilge keel emergence and immergence during large roll motion. This investigation led to improved understanding of the effects which should be considered for sectional, or strip-theory based, approaches for ship motion predictions.

For the single hull form, bilge keel configuration, wave condition, and drift condition that were examined, for two speeds, the following observations were made:

- 2-D results from URANS and experiments generally agreed, even for large amplitude roll motion, although differences were likely due to the limitations of finite difference methods used in the URANS simulations
- 3-D effects were more significant for zero speed conditions, but were also apparent for the forward speed condition, where vorticity is convected downstream
- The effects of vorticity on the bilge keel force were reduced with forward speed; however, at forward speed an additional component from the bow dome was present and may have some influence on the bilge keel force for lower frequency oscillations
- At zero speed, there is not much longitudinal variation in the bilge keel force, except for the re-entry of the leading edge section for large amplitude roll

- At forward speed, lift and free surface effects (near to and emerging from) are greatest for the leading edge section
- At forward speed, the bow wave influences the location and magnitude of the impact loading on the bilge keel
- The presence of waves influences the force on the bilge keel, and depends on the phase of the wave relative to the roll cycle; while the bilge keel is deeply submerged, the influence of the waves is least.
- For large amplitude roll conditions in waves, when the bilge keel emerges from the free surface, the magnitude of the impact loading on the bilge keel during re-entry is reduced.
- The effect of different wave phases on the bilge keel force, such as bilge keel re-entry in a wave trough, which may lead to an increase in the bilge keel force compared to calm water and compared to other wave phases.
- In calm water, the effect of steady drift on the bilge keel forces was examined. The more complex conditions of unsteady turns in waves, while experiencing large amplitude coupled motions should be considered for future study.

Interactions of the vortices shed from the bilge keels may be significant, particularly for stability failure conditions. For these conditions, steep waves may result in large amplitude roll motions, while the ship is typically moving at slower speeds, or is in a dead ship condition, and the vortices will have a lingering presence. At zero or slow speeds, the influence of the vorticity may be more difficult to consider in sectional or strip-theory based approaches for ship motion predictions. Additionally, for more advanced bilge keel designs, including tip geometry variations, consideration must be given to the effects of bilge keel geometry on damping.

Future development of bilge-keel force-component models for potential flow ship motion simulations should consider forward speed effects.

Additional comparisons between the URANS results obtained in this study and the computationally efficient bilge keel force model proposed by Greeley and Petersen (2010) are planned.

The development of more physically robust bilge keel force models may enable more unconventional bilge keel designs to be evaluated. This will allow future numerical tools to evaluate ship designs with variations in bilge keel geometries to improve roll damping performance in sea conditions, including heavy weather.

Acknowledgements The authors would like to thank Dr. Pat Purtell (Office of Naval Research) for support of the work presented in this paper and acknowledge support for the experiments and additional analysis from the NSWCCD Independent Applied Research (IAR) Program, under the direction of Dr. John Barkyoumb. They appreciate Dr. Pablo Carrica (University of Iowa) for his continued guidance and support in using CFDShip-Iowa. They are grateful to the U.S. Department of Defense's High Performance Computing Modernization Program (HPCMP) office, which provided the computer resources at NAVO on the IBM P6.

References

Atsavapranee, P., J. B. Carneal, D. Grant, & A. S. Percival (2007), "Experimental Investigation of Viscous Roll Damping on the DTMB Model 5617 Hull Form," Proc. 26th Intl. Conf. on Offshore Mechanics and Arctic Eng, San Diego, CA.
Bassler, C., J. Carneal & P. Atsavapranee (2007) "Experimental Investigation of Hydrodynamic Coefficients of a Wave-Piercing Tumblehome Hull Form," Proc. 26th Intl. Conf. Offshore Mechanics and Arctic Engineering, San Diego, CA.
Bassler, C. C. & A. M. Reed (2009) "An Analysis of the Bilge Keel Roll Damping Component Model," Proc. 10th Intl. Conf. Stability of Ships and Ocean Vehicles, St. Petersburg, Russia.
Bassler, C. C., A. M. Reed, & A. J. Brown (2010a), "A Method to Model Large Amplitude Ship Roll Damping," Proc. 11th Intl. Ship Stability Workshop, Wageningen, The Netherlands.
Bassler, C. C., A. M. Reed, & A. J. Brown (2010b), "Characterization of Physical Phenomena for Large Amplitude Ship Roll Motion," Proc. 29th American Towing Tank Conf. (ATTC), Annapolis, MD, August.
Bassler, C. C., A. M. Reed, & A. J. Brown (2011) "A Piecewise Model for Prediction of Large Amplitude Ship Roll Damping," Proc. 30th Intl. Conf. Ocean, Offshore and Arctic Eng, Rotterdam, The Netherlands, June.
Beck, R. F. & A. M. Reed (2001) "Modern Computational Methods for Ships in a Seaway," Trans. SNAME, 109, pp. 1–51.
Belknap, W. & A. M. Reed (2010), "TEMPEST: A New Computationally Efficient Dynamic Stability Prediction Tool," Proc. 11th Intl. Ship Stability Workshop, Wageningen, The Netherlands.
Belknap, W., C. Bassler, M. Hughes, P. Bandyk, K. Maki, D. H. Kim, R. Beck, & A. Troesch (2010), "Comparisons of Body-Exact Force Computations in Large Amplitude Motion," Proc. 28th Symp. on Naval Hydro., Pasadena, CA,.
Bishop, R. C., W. Belknap, C. Turner, B. Simon, & J. H. Kim (2005), "Parametric Investigation on the Influence of GM, Roll Damping, and Above-Water Form on the Roll Response of Model 5613," Hydromechanics Dept. Technical Report, NSWCCD-50-TR-2005/027.
Boger D.A & J. J. Dreyer J.J. (2006), "Prediction of Hydrodynamic Forces and Moments for Underwater Vehicles Using Overset Grids," Proc 44th AIAA Aerospace Sciences Meeting, Reno, Nevada.
Bryan, G. H. (1900), "The Action of Bilge Keels," Trans. RINA 4.
Carrica P. M., R. V. Wilson, R. Noack, T. Xing, M. Kandasamy, J. Shao1, N. Sakamoto, & F. Stern (2006), "A Dynamic Overset, Single-Phase Level Set Approach for Viscous Ship Flows and Large Amplitude Motions and Maneuvering," Proc 26th Symp. on Naval Hydro., Rome, Italy.
Carrica, P.M., R. V. Wilson, R. W. Noack, & F. Stern, (2007a), "Ship Motions Using Single-Phase Level Set with Dynamic Overset Grids," Computers and Fluids, 36, pp. 1415–1433.
Carrica, P.M., R. V. Wilson, & F. Stern (2007b), "An Unsteady Single-Phase Level Set Method for Viscous Free Surface Flows," Intl. J. Numerical Methods in Fluids, 53, pp. 229–256.
Dai, C.M., Miller, R.W, & Percival A.S. (2009), "Hydrodynamic Effects of Bilge Keels on the Hull Flow During Steady Turns," Proc. 28th Intl. Conf. Ocean, Offshore and Arctic Engineering, Honolulu, Hawaii.
Froude, W. (1865), "On the Practical Limits of the Rolling of a Ship in a Seaway," *Trans. Institution of Naval Architects*, 6.
Grant, D. J., A. Etebari, & P. Atsavapranee (2007), "Experimental Investigation of Roll and Heave Excitation and Damping in Beam Wave Fields," *Proc. 26th Intl. Conf. on Offshore Mechanics and Arctic Engineering*, San Diego, CA.
Greeley, D. S. & B. J. Petersen (2010), "Efficient Time-Domain Computation of Bilge Keel Forces,"*Proc. 28th Symp. on Naval Hydro.*, Pasadena, CA, September.
Ikeda, Y., Y. Himeno, & N. Tanaka (1978), "A Prediction Method for Ship Roll Damping," Report of the Department of Naval Architecture, University of Osaka Prefecture, No. 00405.
Irvine, M., P. Atsavapranee, J. Carneal, A. Engle, S. Percival, R. Bishop, D. Grant, C. Lugni, F. Di Felice, J. Longo, & F. Stern (2006), "Comparisons of Free Roll Decay Tests for Model DTMB

5415/2340/5512, and Investigation of Lateral Hydrodynamic Loads on Bilge Keels," Proc. 26th Symp. on Naval Hydro., Rome, Italy,.
Himeno, Y. (1981), "Prediction of Ship Roll Damping-State of the Art," Dept. of Naval Architecture and Marine Engineering, Univ. of Michigan, Report 239.
Kato, H. (1965) "Effect of Bilge Keels on the Rolling of Ships." J. Soc. Naval Arch., Japan, 117, pp. 93–114.
Keulegan, G. M. & L. H. Carpenter (1958), "Forces on Cylinders and Plates in an Oscillating Fluid," J. Research of the National Bureau of Standards, 60.
Klaka, K., J. D. Penrose, R. R. Horsley, & M. R. Renilson (2007), "Hydrodynamic Tests on a Plate in Forced Oscillation," Ocean Engin., 34, pp. 1225–1234.
Korpus, R. A. & J. M. Falzarano (1997), "Prediction of Viscous Ship Roll Damping by Unsteady Navier-Stokes Techniques," J. Offshore Mech. & Arctic Engin, 119, pp. 108–113.
Lin, W. M., & D.K.P. Yue (1990), "Numerical Solutions for Large-Amplitude Ship Motions in the Time-Domain," Proc. 18th Symp. Naval Hydro., Ann Arbor, MI.
Lin, W. M., S. Zhang, K. Weems, & D. Liut (2006), "Numerical Simulations of Ship Maneuvering in Waves," Proc. 26th Symp. Naval Hydro., Rome, Italy.
Liut, D. A. (1999), Neural-Network and Fuzzy-Logic Learning and Control of Linear and Nonlinear Dynamic Systems, Ph.D. Dissertation, Virginia Tech.
Liut, D. A. & W.M. Lin (2006), "A Lagrangian Vortex-Lattice Method for Arbitrary Bodies Interacting with a Linearized Semi-Lagrangian Free Surface," Intl. Shipbuilding Progress, 53, pp. 1–32.
Lloyd, A. R. J. M. (1998), Seakeeping: Ship Behaviour in Rough Weather. London: Intl. Book Distributors, Ltd.
Martin, M. (1958) "Roll Damping Due to Bilge Keels." U. Iowa Institute of Hydraulic Research, Report.
Menter, F. R. (1994), "Two-Equation Eddy Viscosity Turbulence Models for Engineering Applications," *AIAA J.*, 32, pp. 1598–1605.
Miller, R. W., J. J. Gorski, & D. Fry (2002), "Viscous Roll Predictions of a Circular Cylinder with Bilge Keels," Proc. 24th Symp. Naval Hydro., Fukuoka, Japan.
Miller, R. W., C. C. Bassler, P. Atsavapranee, & J. J. Gorski (2008), "Viscous Roll Predictions for Naval Surface Ships Appended with Bilge Keels Using URANS," Proc. 27th Symp. Naval Hydro., Seoul, South Korea.
Morison, J. R., M. P. O'Brien, J. W. Johnson, & S. A. Schaaf (1950), "The Forces Exerted by Surface Waves on Piles," Petroleum Trans., AIME, 189, pp. 149–157.
Morison, J. R., J. W. Johnson, & M. P. O'Brien (1953), "Experimental Studies of Forces on Piles," Proc. 4th Conf. Coastal Engin.
Noack R. (2005), "SUGGAR: A General Capability for Moving Body Overset Grid Assembly," Proc 17th AIAA Computational Fluid Dynamics Conf., Toronto, Ontario, Canada.
Reed, A. M. (2009) "A Naval Perspective on Ship Stability," Proc. 10th Intl. Conf. Stability of Ships and Ocean Vehicles, St. Petersburg, Russia.
Roddier, D., S. W. Liao, & R. W. Yeung (2000), "On Freely-Floating Cylinders Fitted with Bilge Keels," Proc. 10th Intl. Offshore and Polar Engin. Conf.
Sarpkaya, T. (1981), "A Critical Assessment of Morison's Equation and Its Applications," Proc. Intl. Conf. on Hydro. Ocean Engin., Trondheim, Norway, pp. 447–467.
Sarpkaya, T. & M. Isaacson (1981), Mechanics of Wave Forces on Offshore Structures. New York: Van Nostrand Reinhold Co.
Sarpkaya, T. & J. L. O'Keefe (1996), "Oscillating Flow Around Two and Three-Dimensional Bilge Keels," J. Offshore Mechanics and Arctic Eng, 118, pp 1–6.
Seah, R. K. M. (2007), "The SSFSRVM Computational Model for Three-Dimensional Ship Flows With Viscosity," Ph.D. Dissertation, University of California Berkeley.
Seah, R. K. M. & R. W. Yeung (2008), "Vortical-Flow Modeling for Ship Hulls in Forward and Lateral Motion," Proc. 27th Symp. on Naval Hydrodynamics, Seoul, South Korea.
Themelis, N. I. (2008), "Probabilistic Assessment of Ship Dynamic Stability in Waves," Ph.D. Dissertation, National Technical University of Athens.

Wilson, R.V., P. M. Carrica, & F. Stern, (2006), "Unsteady RANS Method for Ship Motions with Application to Roll for a Surface Combatant," Computers and Fluids, 35, pp. 501–524.
Yeung, R. W, S. W. Liao, & D. Roddier (1998), "Hydrodynamic Coefficients of Rolling Rectangular Cylinders," Intl. J. Offshore and Polar Eng, 8(4).
Yeung, R. W., D. Roddier, B. Alessandrini, L. Gentaz & S. W. Liao (2000), "On the Roll Hydrodynamics of Cylinders Fitted with Bilge Keels," Proc. 23rd Symp. on Naval Hydro.

Chapter 10
Assessment of Ship Roll Damping Through Full Scale and Model Scale Experiments and Semi-empirical Methods

Carl-Johan Söder, Anders Rosén, Sofia Werner, Mikael Huss and Jakob Kuttenkeuler

Abstract This paper presents unique experimental set-ups in model scale and full scale for evaluating roll damping properties of a Panamax Pure Car and Truck Carrier at speed. The purpose of this study is to develop a method for the assessment of roll damping based on full scale trials and to validate the use of roll damping derived from model tests for full scale vessels. Experimental data are also used to assess a semi-empirical method that today provides input for the prediction of critical rolling events such as parametric rolling and severe rolling motions in general.

Keywords Roll damping · Parametric roll · Roll decay · Ikeda's method
Full-scale · Model-scale

C.-J. Söder
Wallenius Marine AB/KTH Royal Institute of Technology, Stockholm, Sweden
e-mail: cjsoder@kth.se

A. Rosén (✉) · J. Kuttenkeuler
KTH Royal Institute of Technology, Stockholm, Sweden
e-mail: aro@kth.se

J. Kuttenkeuler
e-mail: jakob@kth.se

S. Werner
SSPA Sweden AB, Stockholm, Sweden
e-mail: sofia.werner@sspa.se

M. Huss
Wallenius Marine AB, Stockholm, Sweden
e-mail: mikael.huss@walleniusmarine.com

V. L. Belenky et al. (eds.), *Contemporary Ideas on Ship Stability*, Fluid Mechanics and Its Applications 119, https://doi.org/10.1007/978-3-030-00516-0_10

10.1 Introduction

With ever increasing focus on maximizing payload and minimizing fuel consumption, modern Panamax Pure Car and Truck Carriers (PCTC) have evolved into highly optimized designs, incorporating very high centre of gravity for maximum cargo volume, combined with slender hulls for minimum resistance. High centre of gravity requires hull forms that provide high form stability, which is partially achieved by largely flared aft body sections with large water plane area. With slender and very form stable hulls PCTC's are very efficient volume carriers but also more vulnerable for events such as parametric rolling caused by stability variations in waves. In recent years several severe cases have been reported with modern PCTC designs, e.g. Palmquist and Nygren (2004) and Rosén et al. (2012), with extreme roll amplitudes causing significant risks to the crew and cargo on board.

The onset of parametric rolling requires that the stability varies in resonance with the natural roll period of the vessel as the waves travel along the hull and the vessel is pitching. In case of parametric resonance the roll damping is decisive for the roll amplitude. If the damping is sufficiently high parametric rolling will never develop, but if the damping is low relative to the stability variation large roll angles can develop rapidly.

To avoid incidents as those referred to in the previous, parametric rolling needs to be properly accounted for in design as well as in operation. In the second generation intact stability criteria, which are under development within the IMO, parametric rolling and other dynamic stability failure modes are addressed with a multi-level approach including simple vulnerability criteria, direct stability assessment, and formulation of ship specific operational guidance (e.g. Peters et al. 2011). In particular for the direct assessment level and for formulation of ship specific operational guidance, proper assessment of a ship's vulnerability to parametrical rolling is depending on accurate description of the roll damping.

Roll damping properties can either be estimated using semi-empirical formulations or by experiments. Direct calculations using RANS-CFD can, as shown by van't Veer and Fathi (2011), be used for investigating the flow around the hull and appendixes due to rolling motions, but is not yet widely used for quantitative estimates of the damping.

The most established semi-empirical method for roll damping prediction is Ikeda's method (1978) which is recommended by ITTC (2011). This method is derived based on a mixture of theory and systematic model testing using different kind of hull shapes and 2D sections. As to experiments roll damping properties can be derived using forced roll motions or free roll decay model tests, for the actual hull with appendixes as described in IMO (2006). According to Himeno (1981) scale effects of roll damping are mainly associated with skin friction related damping which however often constitute a negligible part of the total damping. Therefore, damping derived from model tests are normally transferred directly to full scale using a non-dimensional formulation expressed as the ratio between the actual damping and the critical damping of the rolling motion.

Fig. 10.1 The Panamax pure car and truck carrier that serves as case vessel in this paper

Little work has however been done for assessing the validity of model scale test results in full scale for large ships. In fact, Valle and Pérez Rojas (1997) and Valle et al. (2000) claimed considerable scale effects in roll damping for a fishing vessel when comparing full scale results with results from model tests. Furthermore; in Kawahara et al. (2009) it was found that the accuracy of Ikeda's method decreased for vessels with high centre of gravity and large flared stern sections—characteristics that are typical for modern PCTC's. Consequently, there appear to be a need for addressing the validity of current approaches, and possibly also a room for improvements.

This paper presents new approaches for assessing ship's roll damping at speed. Unique experimental set-ups in model scale in towing tank as well as in full scale with a Panamax PCTC, as seen in Fig. 10.1, are presented. Data generated is evaluated and the results are also used for benchmarking Ikeda's method for roll damping prediction.

10.2 Roll Decay Tests

Full-scale and model-scale roll decay tests have been performed on a Panamax PCTC, the Wallenius vessel m/v Faust with vessel particulars according to Table 10.1.

10.2.1 Model Scale Roll Decay Tests

The model tests were performed in the towing tank at SSPA Sweden AB in Gothenburg. The tank measures 260 m × 10 m and is 5 m deep. The model hull was produced in plastic foam and equipped with rudder and bilge keels but no propeller. The model

Table 10.1 Main particulars of m/v Faust

Length	[m]	220
Beam	[m]	32.3
Draft	[m]	9.5
Displacement	[m^3]	41,000
Bilge keel dimension	[m × m]	0.4 × 66
Block coefficient	[–]	0.66
Midship coefficient	[–]	0.93
Scale factor, model	[–]	29.6

was ballasted to the design draft condition as described in Table 10.1, with a metacentric height, GM, corresponding to 1.1 m for the full scale ship and a roll period corresponding to about 28 s. Draughts at the forward and aft perpendiculars were measured with rulers and GM was verified with inclination tests.

The roll decay tests were carried out by inclining the model to a heel angle of around 5°–10°, keeping it heeled for a moment and then releasing it. This was done at no speed as well as in speed, by towing the model in Froude numbers corresponding to full scale speeds of 12, 14 and 16 knots. The roll angle was measured with a gyro at a sampling frequency of 100 Hz. The accuracy of the gyro is 0.1° and the resolution 0.0023°.

During the tests at speed, the model was towed by the carriage through a solid rod attached amidships. The model was kept on course by attaching a connection fore and aft leaving the model free in heave and pitch but restrained in sway and yaw. The towing rod and the connection were located at a height corresponding to the assumed roll axis, and attached using joints that are free in rotation. Under the condition that the joints are placed exactly along the model's roll centre axis, the model can be considered as free in roll.

There is of course a risk that the roll motion is affected by the yaw restraint if the assumed roll centre deviates from the true roll centre. In order to investigate the effect of this yaw restraint, roll decay tests at zero speed were carried out with the model totally free. By comparing these tests with the corresponding tests of the restrained model, it was found that the influence of the yaw restraint on the roll damping is negligible at zero speed.

The connection joint locations were kept constant for all speeds. However, it is likely that the roll centre moves as the model sinks and trims with increasing speed. If the true roll centre deviates from the connections, forces will be transferred between the model and the connection and this may have some effect on the roll damping. Therefore, force transducers were mounted on the connections and the side forces were recorded throughout all tests. At 12 knots the side forces were of the same magnitude as at 0 knots. At the higher speeds, the side forces were larger than at 0 knots, which may indicate an effect on the decays. Even though the sizes of these forces were measured it is difficult to quantify their effect on the roll damping. A

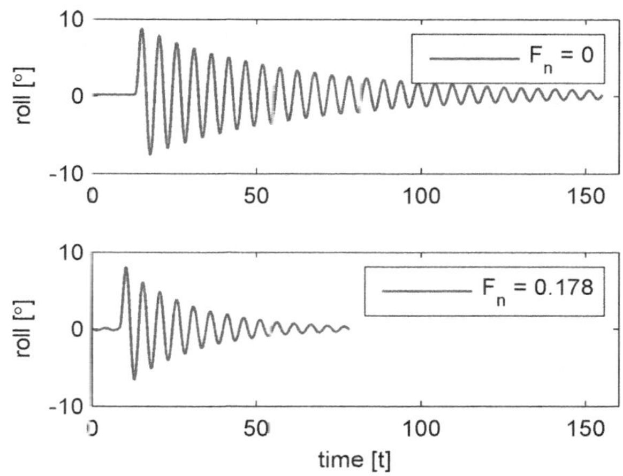

Fig. 10.2 Time series from model scale roll decay tests at speed corresponding to full scale speeds of 0 and 16 kn

consequence of this is that the data for tests carried out at 0 and 12 knots should be considered more reliable than the data from the 14 and 16 knots runs.

In Fig. 10.2 time series of roll motions can be seen from a selection of the model scale roll decay tests. The figure shows two different tests performed in Froude number corresponding to full scale speeds 0 and 16 knots. As can be seen the damping is very low at zero speed but increases significantly at speed.

10.2.2 Full Scale Roll Decay Tests

Full scale roll-decay tests were performed by inducing roll motion using controlled rudder impulses. For normal load cases, the vertical centre of gravity of PCTC's is high—as well as for most vessels that are sensitive for parametric rolling—giving a good lever for the rudder and distinct rudder-roll coupling as a consequence. Since the inertia of yaw is much larger than the inertia of roll, short rudder impulses can produce a roll motion without generating large yaw motion. In Fig. 10.3 a roll decay test conducted on board m/v Faust can be seen. The figure shows the rudder angle, the roll motion as measured by the gyro compass and the heading.

Figure 10.3 shows how two rudder impulses in opposite direction causes the vessel to roll with an initial amplitude of about 3° but only initiates a very limited change of heading. The duration of each rudder impulse was around half the natural roll period so the total duration of two rudder impulses was to be equal to the ships natural period of roll. The roll motion can be considered as freely oscillating as soon as the rudder is stationary in or near midships position. Tests were performed at three different speeds, around 14, 16 and 18 knots. During the tests the vessel was loaded to design draft. According to the on board loading computer GM was 1.7 m which was higher than the model test condition. The natural period of roll was measured to

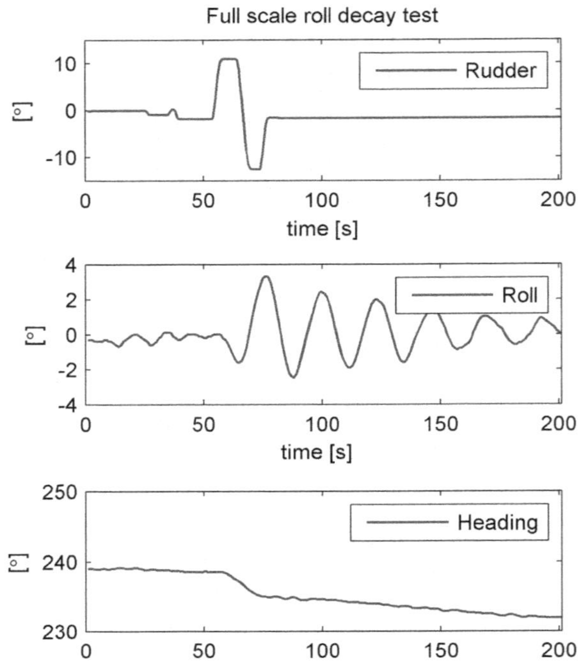

Fig. 10.3 Time series of rudder angle, roll angle and the heading during a full scale roll decay test onboard m/v Faust

about 23 s. Roll motions, water speed, rudder angle and heading were measured with a rate of 1 Hz with a resolution of 0.1°. All tests were performed in smooth weather conditions.

10.3 Roll Damping Evaluation

From roll decay tests roll damping coefficients are normally derived based on the logarithmic decrement of the roll peaks. However, this approach is sensitive to low-frequency disturbances and noise, which does not have to be a problem in controlled model test environment, but is difficult to avoid during full scale tests. An alternative and more robust approach, which utilizes full time series of roll decay tests and not only the peaks, is the numerical Parameter Identification Technique (PIT) as described in IMO (2006) and also used in Bulian (2004). In this approach a numerical solution to a one degree of freedom roll equation is fitted to the roll decay time series by tuning the parameters in the roll equation.

To evaluate the performed tests, a modified version of the PIT approach is developed. This modified approach adds a time-dependent second-degree polynomial to the fitting, that later can be separated from the solution to account for low frequency disturbances. The roll equation that is used for the evaluation has a linear-quadratic damping dependence and a linear restoring term. It should be noted that even though the approach could well handle roll equations with higher order of non-linearities in the damping term as well as a non-linear restoring term, the limited amplitudes at which the roll decay tests was conducted cannot motivate advantages of higher order models.

The equation of a freely oscillating roll motion with linear-quadratic damping can in non-dimensional form be expressed as

$$\ddot{\varphi} + 2\zeta\omega_0\dot{\varphi} + d|\dot{\varphi}|\dot{\varphi} + \omega_0^2\varphi = 0 \tag{10.1}$$

where φ is the roll angle, ω_0 is the non-damped natural frequency of roll, ζ is the linear damping coefficient and d is the quadratic damping coefficient. For convenience the equation is rewritten as a system of coupled first-order differential equations by introducing $y_1 = \varphi$ and $y_2 = \dot{\varphi}$ according to

$$\begin{cases} \dot{y}_1 = y_2 \\ \dot{y}_2 = -2\zeta\omega_0 y_2 - d|y_2|y_2 - \omega_0^2 y_1 \end{cases} \tag{10.2}$$

which can be solved in the time domain t using numerical integration if the initial values $\varphi(0)$ and $\dot{\varphi}(0)$ are given.

Roll decay tests are evaluated by tuning $\bar{k} = [\zeta, \omega_0, d, \varphi(0), \dot{\varphi}(0)]$ for best fit between the numerical solution and sampled roll decay data. This can be expressed as an unconstrained nonlinear optimization problem, by minimizing the sum square residual between the time samples from the decays φ_{m_i} and the numerical solution $\varphi_{n_i}(\bar{k}, t_i)$, according to

$$\min_{\bar{k}, p(t)} f(\bar{k}, p(t)) \tag{10.3}$$

where $p(t)$ is the previously mentioned second degree polynomial that is added in order to account for low frequency disturbances and

$$f = \sum_{i=0}^{t} \left(\varphi_{m_i} - \varphi_{n_i}(\bar{k}, t_i) - p(t_i)\right)^2 \tag{10.4}$$

For comparative purpose it is convenient to express the linear-quadratic damping as equivalent linear damping ζ_e. At a certain roll amplitude φ_a this linear equivalent damping constitutes the same damping over one period as the non-linear damping over the same period. In line with Himeno (1981) this equivalent linear damping is calculated as

$$\zeta_e = \zeta + \frac{4}{3\pi} d\varphi_a. \tag{10.5}$$

Fig. 10.4 Full scale and model scale roll decay tests evaluated with the linear-quadratic damping model. For comparative purpose the linear equivalent damping is displayed, calculated according to Eq. (10.5)

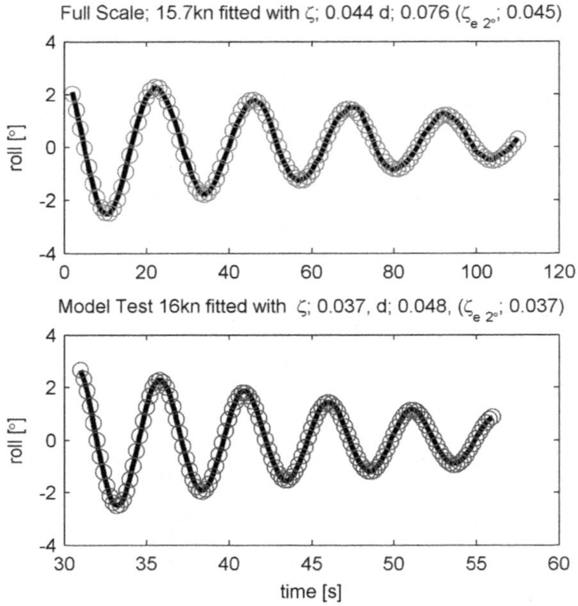

In Fig. 10.4 evaluations of full scale as well as model scale decays are exemplified. The derived linear-quadratic as well as linear equivalent damping for $\varphi_a = 2°$ are given in the diagram titles.

When comparing the linear damping term ζ with the linear equivalent damping term ζ_e in Fig. 10.4 it can be seen that the non-linearities are weak at amplitudes of 2°. Damping derived from all model tests and full scale tests are compared in Fig. 10.5 in terms of linear equivalent damping at 2°. For equally premised comparison the data in Fig. 10.5 comes from evaluations of test data with amplitudes in the range 4°–1° (as in Fig. 10.4).

As seen the evaluated damping from model tests and full scale tests show quite good agreement. This indicates a good potential for the full scale approach as such and also builds confidence in the procedure of applying roll damping evaluated based on model tests on full scale vessels. There is a trend though that the full scale evaluations give slightly higher damping results than the model tests.

10.4 Discussion

The model was fitted with continuous bilge keels while the real bilge keels on the ship are divided into five successive parts. The later configuration may have a positive effect on the damping at speed as discussed by Segal and Segal (2011). Additionally

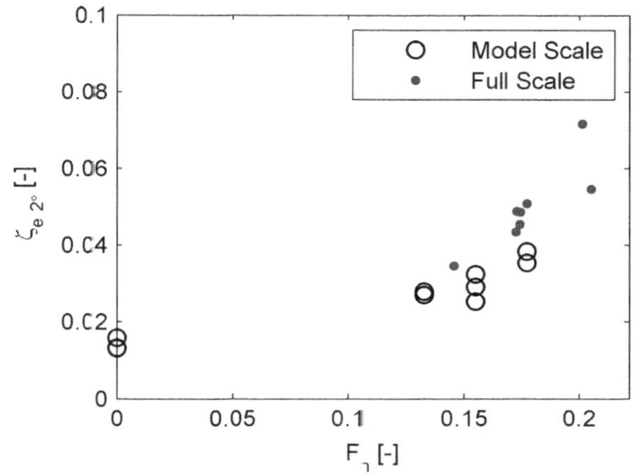

Fig. 10.5 Linear equivalent damping for $\varphi_a = 2°$ from model tests and full scale tests at different Froude numbers

the profiles of the full scale bilge keels are L-profiles (400 × 100 mm) which are likely to be slightly more efficient than the flat plate profiles (400 mm) of the model.

As previously mentioned, it is difficult to quantify the influence of the yaw restraint of the model in the towing tank for the higher speeds. In a similar manner slight yawing motions may have affected the full scale tests even though the tests were performed in such a manner that large yaw motions could be avoided.

There is of course scatter in the data associated to full scale measurements which have to be considered. As an effect of that the damping is very low small differences in input can have a large relative effect on the evaluations. However, in absolute figures the repeatability of the full scale tests appears to be quite good with a standard deviation of ~0.003 (0.3% of the critical damping) which actually is in the same magnitude as the standard deviation of the model tests.

According to IMO (2006) scale models with appendices shall have a minimum length of 2 m, the bilge keel height shall exceed 7 mm and the scale factor shall not be less than 1:75 to avoid viscous scale effects. These requirements are well fulfilled here. The requirement on bilge keel size is likely there to avoid that the boundary layer around the hull reduces the efficiency of the bilge keels in model scale. However, when considering the solution to Blasius equation it appears that the boundary layer thickness due to forward speed could be considerable during the model tests performed at speed. The boundary layer thickness t_b can be estimated based on the local Reynolds number Re_x at a longitudinal position x according to

$$t_b \approx 4.91x/\sqrt{Re_x}. \tag{10.6}$$

For the model tests performed at speed this gives a thickness up to, and for some cases more than, half the bilge keel height (depending on speed and position). This could have a negative impact on the bilge keel efficiency. In full scale the boundary layer is in the magnitude of some percentages of the bilge keel height.

Even though the full-scale vessel and model had the same draughts during the tests, the vertical centre of gravity was different, and as a consequence GM and the natural frequency was different which could have an influence on the damping.

In heavy weather it is common practice that captains on Wallenius PCTC's prefers bow heading wind to expose much lateral wind area while at the same time make use of the stabilizing forces of the strong apparent wind towards the hull to reduce the rolling motions. Considering that the lateral wind area of PCTC's are very high, aerodynamical damping may have a slight impact on the damping during the full scale trials, even though they were performed in calm weather.

As discussed by van't Veer and Fathi (2011), memory effects may play an important role when evaluating roll damping from roll decays. Roll damping may be overestimated during the initial phase of a roll decay test in the absence of the memory of previous rolling motions. In an attempt to investigate this effect the roll damping was evaluated for two different roll decay tests, the first with an initial amplitude of 9° and the second with an initial amplitude of 6°. Both tests were performed at the same speed (corresponding to 12 kn in full scale) and the roll damping was evaluated from time series when the amplitude was between 6° and 3°. No significant difference in evaluated damping could be observed between the two cases.

Water temperature differences affecting the viscous forces, increased hull surface roughness due to fouling and external forces disturbing the tests are other factors that may have an effect on the results.

10.5 Ikedas Method Versus Experimental Data

Even though first published in the late 1970s, Ikeda's method is still commonly used for roll damping prediction. In this method, roll damping in dimensional form B is calculated for a given amplitude, corresponding to linear equivalent damping, due to five different components; friction, wave generation, eddy making, linear-lift and bilge keel. For comparative purpose these components will here be presented non-dimensionalized according to

$$\zeta_e = \frac{1}{2\omega_0}\frac{B}{A} \tag{10.7}$$

where A is the inertia term that is estimated as $\rho g \nabla GM \omega_0^{-2}$, ρ is the water density, g is the constant of gravity and ∇ is the displacement of the vessel.

The method as described in Ikeda (1978) and Journée and Adegeest (2003) has been implemented and applied on the here studied ship. In Fig. 10.6 all components can be seen exemplified for $\varphi_a = 10°$, $T = 27.6$ s and $GM = 1.1$ m.

As seen in the figure the bilge keel damping is the dominating component at low speeds for the given condition. The contribution from bilge keels is caused by three different components: a pure drag normal force component, a bilge keel generated hull pressure component and bilge keel generated lift. At higher speeds the hull lift

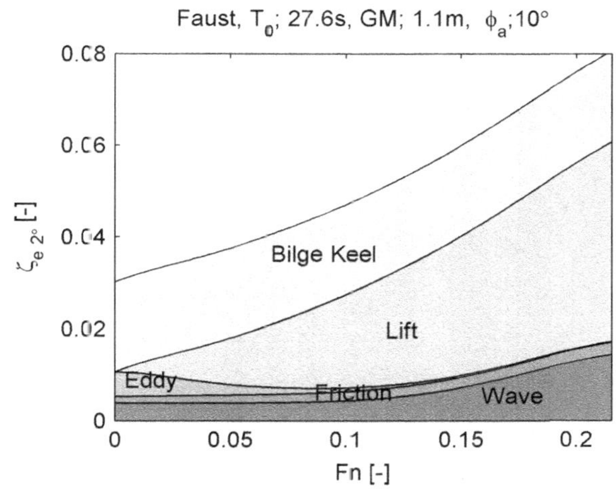

Fig. 10.6 Results from Ikeda's method for m/v Faust with the five different damping components; friction, wave, eddy, linear lift and bilge keel

component becomes dominant. This component is created when the submerged hull, periodically, has an angle of attack due to sway motions caused by rolling. With the magnitude of these sway motions being dependent on the distance of the roll centre above the water line, this component becomes particularly large for vessels with high centre of gravity. The lift component in Ikeda's method has been derived based on maneuvering experiments. The frictional component, which constitutes a small contribution, was derived from experiments with rotating cylinders where the viscous damping is dependent on the Reynolds number. This component also incorporates a small semi-theoretical correction for forward speed. The wave damping is the linear potential damping which typically is calculated using strip theory for zero forward speed and is modified with an empirical formula to incorporate forward speed effects. The eddy-making component represents damping due to vortex separation, and is almost negligible at forward speed.

In Kawahara et al. (2009) the validity of Ikeda's method was found questionable for vessels with high centre of gravity combined with buttock flow stern (largely flared stern sections), especially for small roll angles. With Faust being a buttock flow stern type PCTC with high centre of gravity, and with roll damping tests being conducted at small roll amplitudes, Ikeda's method should not be expected to have very good agreement with the tests.

In Fig. 10.7 the roll damping derived from the model tests are compared with Ikeda's method for the corresponding loading condition and scale. As previously stated it is important to keep in mind that the accuracy of the model tests performed at 14 and 16 knots may have been affected by the yaw constraint and should be considered with care. However the general observation is that Ikeda's method seems to overestimate the damping except for the smallest roll amplitudes at zero speed. As seen, the speed dependence according to Ikeda's method is stronger than what is shown by the model tests, indicating that the lift damping component, which constitutes the dominant part of the speed dependence, probably is overestimated

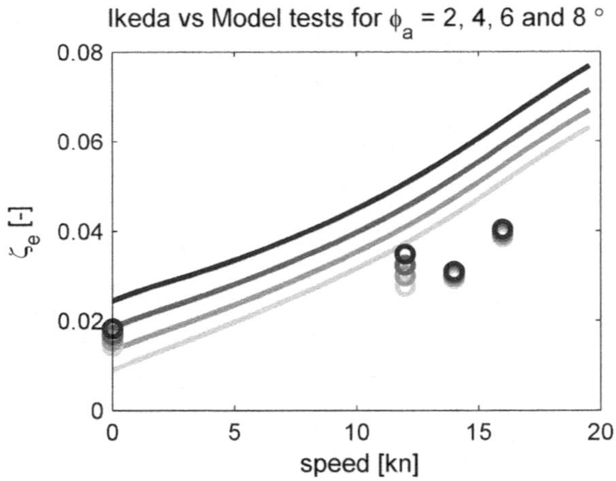

Fig. 10.7 Predicted damping for m/v Faust in model scale using Ikeda's method and model tests for roll amplitudes of 2, 4, 6 and 8°

for this ship. Furthermore, Ikeda's method seems to overestimate the non-linearities (amplitude dependence) that essentially are caused by the bilge keel component.

Both the lift damping and the bilge keel damping are strongly dependent on the location of the roll centre. In accordance with the prediction formula this is assumed to be located in the vertical centre of gravity. Recorded sway and yaw motions however indicate that the roll centre normally is located lower than the centre of gravity. If the height of the assumed centre of roll in the prediction formula is reduced, better agreement is achieved, thus indicating a potential for improvement of the method.

10.6 Conclusions

A method for assessing roll damping properties based on full scale roll decay tests has been developed. Evaluated roll damping from full scale tests shows good agreement with model tests for a Panamax Pure Car and Truck Carrier. This indicates a good potential for the approach as such and also builds confidence in the procedure of applying roll damping evaluated from model tests on full scale vessels. The tests presented here though indicate that the full scale damping is slightly higher than the model test damping. This should be further investigated.

The presented full scale tests have been performed in-service with the vessel loaded to the design condition. Alternatively, tests could be performed during the sea trial from the ship yard. However, as roll damping is dependent on the loading condition and speed, results will always be limited to specific cases. And as required test matrices to cover all possible operational conditions is practically impossible to perform, semi-empirical methods appear as an attractive option for inter- and extrapolation of test results to various conditions. Ikeda's semi-empirical method implemented for the studied vessel give physically relevant predictions of the roll damping. However, speed dependence and amplitude dependence was significantly overestimated by the method, likely as a consequence of that the method not has been adopted for modern volume carriers with very high centre of gravity such as PCTC's. Future work could therefore preferably aim at assessing the centre of gravity dependence of the lift damping and the bilge keel damping which are the two damping components in Ikeda's method that constitutes the major part of the non-linearities and the speed dependance.

An accurate description of the roll damping is crucial when predicting the risk of severe rolling events, such as parametric rolling. The here presented full-scale approach is well suited for providing roll damping input to methods for operational guidance (e.g. as presented in Ovegård et al. 2012) for accurate on-board decision support with respect to parametric rolling.

Acknowledgments This research has been financially supported by the Swedish Mercantile Marine Foundation (Stiftelsen Sveriges Sjömanshus) and the Swedish Maritime Administration (Sjöfartsverket) which are both gratefully acknowledged.

References

Bulian G., 2004, "Estimation Of Nonlinear Roll Decay Parameters Using An Analytical Approximate Solution Of The Decay Time History", International Shipbuilding Progress 51, no. 1, pp. 5–32.

Himeno Y., 1981, "Prediction of Ship Roll Damping- State of the Art", The University of Michigan College of Engineering, No. 239.

Ikeda, Y., Himeno, Y. & Tanaka, N., 1978, 'Components of roll damping of ship at forward speed", Journal of the Society of Naval Architects of Japan, Vol. 143.

IMO 2006, Interim Guidelines For Alternative Assessment Of The Weather Criterion, MSC.1/Circ.1200.

ITTC, 2011, "ITTC Recommended Procedures Numerical Estimation of Roll Damping", International Towing Tank Conference, Report 7.5-02-07-04.5.

Journée J.M.J. and Adegeest L.J.M., 2003, Theoretical Manual of "SEAWAY for Windows, TUD Report No. 1370 Revision: 14-12-2003.

Kawahara et al., 2009, "A Simple Prediction Formula of Roll Damping of Conventional Cargo Ships on the Basis of Ikeda's Method and Its Limitation", Proceedings of the 10th International Conference on Stability of Ships and Ocean Vehicles, Saint Petersburg, Russia.

Ovegård E., Rosén A., Palmquist M., Huss M., 2012, " Operational Guidance with Respect to Pure Loss of Stability and Parametric Rolling", 11th Intl Conf on the Stability of Ships and Ocean Vehicles (STAB 2012), Greece.

Palmquist M., Nygren C., 2004, "Recordings of Head-Sea Parametric Rolling on a PCTC", Annex in IMO SLF 47/INF.5.

Peters W, Belenky V, Bassler C, Spyrou K, Umeda N, Bulian G, Altmayer B, 2011, "The second generation intact stability criteria: an overview of development", SNAME Transactions, Vol. 121.

Rosén A., Huss M., Palmquist M., 2012, "Experience from Parametric Rolling of Ships", chapter in the book Fossen T.I., Nijmeijer H., Parametric Resonance in Dynamical Systems, Springer.

Segal Z. and Segal A., 2011, "Roll Stabilization with Short Wings', American Society of Naval Engineers, Naval Engineers Journal, pp. 45–54.

Valle J., Pérez Rojas L., 1997, "Study of ship roll decrement tests in calm water", Proceedings of the 6th International Conference on Stability of Ships and Ocean Vehicles international Conference, Varna, Bulgaria, p 121–128.
Valle J., Pérez Rojas L., Carrillo E.J., 2000, "Influences of Test Parameters on Roll Damping Coefficients", Proceedings of the 7th International Conference on Stability of Ships and Ocean Vehicles, Launceston, Tasmania, Australia, vol B, p 751–766.
van 't Veer R., Fathi F., 2011, "On the Roll Damping of an FPSO with Riser Balcony and Bilge Keels", 30th International Conference on Ocean, Offshore and Arctic Engineering, Rotterdam, The Netherlands.

Chapter 11
Roll Damping of a Twin-Screw Vessel: Comparison of RANSE-CFD with Established Methods

Sven Wassermann, Nikolai Köllisch and Moustafa Abdel-Maksoud

Abstract A RANSE-CFD method is applied to estimate the roll damping of a modern twin-screw RoPax vessel. The simulations are carried out in full scale and with an undisturbed water surface. The harmonic forced roll motion technique is implemented. The influence of ship speeds, the vertical position of the roll axis and roll amplitudes up to 35° are investigated. The interaction between the bilge keels and the ship hull is analyzed. The damping effects of further appendages are discussed. All simulation results are compared with the established method developed by Ikeda and a neural network method based on Blume's roll damping measurements. The established methods were developed based on studying results of single-screw ships. It can be concluded that both established methods provide acceptable results in certain ranges. For large roll amplitudes, the established methods are out of range and cannot deliver reliable results.

Nomenculture

b_{BK}	Bilge keel breadth
d	Ship draft
k	Velocity increment factor at bilge
l_{BK}	Bilge keel length
r_{BK}	Distance from roll axis to bilge keel
v	Transverse velocity component at bilge keel

S. Wassermann · M. Abdel-Maksoud (✉)
Institute for Fluid Dynamics and Ship Theory, Hamburg University of Technology, Am Schwarzenberg-Campus 4 (C), 21073 Hamburg, Germany
e-mail: m.abdel-maksoud@tuhh.de

S. Wassermann
e-mail: sven.wassermann@tuhh.de

N. Köllisch
Dynamics Group, Hamburg University of Technology, Schloßmühlendamm 30, 21073 Hamburg, Germany
e-mail: dynamics.group@tuhh.de

V. L. Belenky et al. (eds.), *Contemporary Ideas on Ship Stability*, Fluid Mechanics and Its Applications 119, https://doi.org/10.1007/978-3-030-00516-0_11

x	Relative motion of water in crosswise direction to bilge keel
A_{BK}	Bilge keel area
$\hat{B}$	Dimensionless roll damping coefficient
B	Equivalent roll damping coefficient
B_{wl}	Waterline breadth of the ship
B_{NBK}	Bilge keel damping coefficient, normal drag force part
B_{SBK}	Coefficient of hull-pressure damping due to bilge keels
B_W	Wave damping coefficient
$C_{D,BK}$	Drag coefficient for bilge keel
$C_{P,BK}$	Hull-bilge-keel pressure coefficient due to bilge keels
C_B	Block coefficient
C_W	Waterplane coefficient
Fr	Froude number of forward ship speed
F_{NBK}	Normal drag force of the bilge keel
KC_{BK}	Local Keulagan-Carpenter-Number for bilge keel
L_{OA}	Ship length over all
L_{WL}	Waterline length of the ship
RA	Distance to roll axis over undisturbed water surface
S	Wetted surface area of the ship
T	Roll period
α	Angle between an orthogonal line to the normal force and line of the lever
σ	Section area coefficient
ϕ	Roll angle
ϕ_H	Heel angle amplitude
ρ	Density
ω	Roll frequency
$\{\}_A$	Amplitude

11.1 Introduction

Calculating ship motions in natural waves with potential theory-based simulation methods has been a common practice for several decades. However, the accuracy of the calculated results for roll motion is strongly influenced by the chosen viscous damping. While potential theory-based methods are able to predict main energy dissipation in all other degrees of freedom, estimating damping for the roll motion is still a weak point in ship motion simulations. The reason for this drawback is that the magnitude of the wave damping compared to the dissipation caused by viscous effects is of similar orders of magnitude.

Table 11.1 Main dimensions of the RoPax ship hull

Waterline length	L_{WL}	(m)	176.0
Waterline breadth	B_{WL}	(m)	26.0
Draft	d	(m)	6.0
Waterplane coefficient	C_W	(−)	0.75
Block coefficient	C_B	(−)	0.57
Wetted surface area	S	(m^2)	4678.0

Extensive investigations on roll damping (Ikeda et al. 1976, 1977a, b, c; Himeno 1981; Schmitke 1978) were carried out in the mid-70s in Japan and North America,[1] which established the division of the phenomena into a wave, friction, eddy and lift component as well as a correction for appendages. A simple theoretical method based on experimental measurement results, also known as Ikeda's method, was developed based on these assumptions and is recommended by the ITTC International Towing Tank Conference (2011). It has become one of the most popular methods to correct the calculated energy dissipation using potential theory. Nevertheless, the empirical coefficients included in the method are based on former typical ship geometries.

Based upon harmonic excited roll motion (HERM) measurements,[2] Blume (1979) has published roll damping coefficients that continue to be used. The applicability of Ikeda's method and Blume's measurement results to a contemporary RoPax ferry is investigated in this paper. RANSE-based simulations using finite-volume method with various roll parameters have been carried out. The obtained results are compared with those of Ikeda's method and a neural network method based on Blume's results.

11.2 Numerical Simulations

11.2.1 Geometry

The investigated RoPax ferry with an overall length of $L_{OA} = 185.0\,\text{m}$ was designed by Potsdam Model Basin (SVA). The ferry is a twin-screw vessel, arranged with two shafts, two rudders and a large skeg (see Fig. 11.1). The main dimensions are summarized in Table 11.1, and the section plan is shown in Fig. 11.1.

[1] An overview of the differences between the work in Japan and North America is given in the discussion section of Schmitke's paper (1978) with Cox, Himeno and Schmitke (pp.41–46).

[2] Details of the HERM technique can be found in Handschel and Abdel-Maksoud (2014). A comparison to the widely-used decay technique is given by Wassermann et al. (2016).

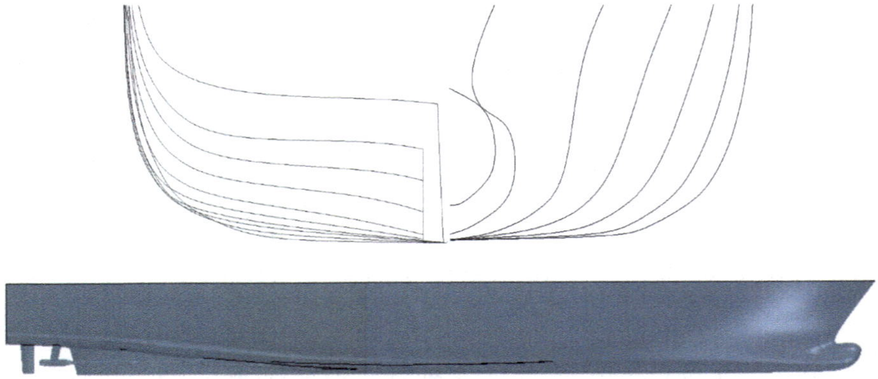

Fig. 11.1 Section plan and side view of the RoPax ferry

11.2.2 RANSE-CFD Method and Analysis of Roll Damping

The RANSE-solver STAR-CCM+ is used to simulate the incompressible flow around the ship with and without forward speed at different roll amplitudes. The volume of fluid method (VOF) is applied to calculate the position of the free water surface flow, see Fig. 11.2. The $k - \omega - SST$ turbulence model is used in all computations. The conservation equations for mass and momentum as well as the equations of the turbulence model and the volume fraction equation (the mixture of water and air) are discretized in integral form using the finite-volume method and solved in an iterative manner using the SIMPLE-algorithm (see Enger 2010). An Euler implicit scheme is used for time integration.

The flow field is divided into a rotating and a stationary region. The rotating region has a cylindrical shape. The roll axis of the ship is fixed and coincident with the cylinder axis. The outer boundaries of the stationary region have a rectangular shape. A sliding interface boundary condition is applied on the common surface between both regions. At the inflow boundary, the velocity is specified. At the outlet boundary, the hydrostatic pressure is applied. Except the geometry which is a regular wall, all other boundaries are slip walls.

An unstructured grid with mostly trimmed hexahedral cells is applied. At walls that have no slip condition, a prism layer is used due to the expected velocity profile. The grid is refined near the hull, the appendages as well as the free surface (see Fig. 11.3). The amount of grid cells used depends on the roll amplitude. For an amplitude of 25°, usually 4M cells are needed. The ship motion is realized by a harmonic forced roll motion. A detailed description of the numerical method with respect to the estimation of roll damping can be found in Handschel et al. (2012). A successful validation with experiments for the benchmarking container ship Duisburg Test Case is presented by Handschel et al. (2014).

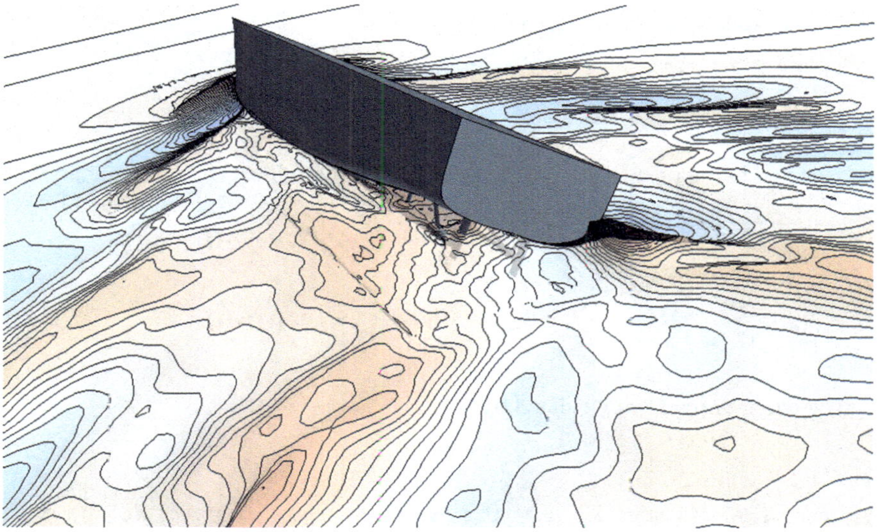

Fig. 11.2 Calculated wave system of the RoPax ferry with forward velocity of Froude number $Fr = 0.26$ and roll amplitude $\phi_A = 35°$

Fig. 11.3 RoPax ferry hull enclosed by cylindrical mesh domain. Mesh refinements are applied at the free water surface and near bilge keels. The sliding interface is shown as a circle; investigated roll amplitude $\Phi_A = 35°$

The equivalent roll damping moment coefficient,

$$B = \frac{E}{\pi \omega \phi_A^2}, \tag{11.1}$$

is estimated based on a Fourier series approach of the roll energy E. The analysis is based on the fact that only the damping part of the moment is phase-shifted by 90° to the roll angle. A Fourier series approximates the roll moment:

$$M(t) = \sum_{j=1}^{\infty} \left(C_{A,j} \sin(j\omega t) + C_{B,j} \cos(j\omega t) \right). \tag{11.2}$$

The energy E can be calculated by:

$$E = \int_{-\pi/\omega}^{\pi/\omega} M \, d\phi = \int_{-\pi/\omega}^{\pi/\omega} \sum_{j=1}^{\infty} \left(C_{A,j} \sin(j\omega t) + C_{B,j} \cos(j\omega t) \right) d\phi. \tag{11.3}$$

Solving the integral and inserting Eq. (11.3) into (11.1) results in:

$$B = \frac{C_{B,1}}{\omega \phi_A}. \tag{11.4}$$

The Fourier coefficient $C_{B,1}$ can be easily determined with a Fast-Fourier-Transformation algorithm.

11.3 Established Methods to Estimate Roll Damping

Ikeda divided the roll damping phenomena into the following parts: wave, friction, eddy and lift damping of the bare hull. The bilge keel effect was investigated separately and divided into three components. The first one is the normal-force acting on the keels; the second component is caused by the interaction between the bare hull and the bilge keel, and is acting on the hull; the interaction with the free surface as the third component is assumed to be small and is neglected.[3] Due to the semi-theoretical character of Ikeda's method, the application of this method should be considered with care. Especially in the ship design stage more straight forward solutions are often favored (Kawahara et al. 2012).

In addition to Ikeda's method, the results of a systematic experimental investigation carried out by Blume (1979) for three different bare hull forms are often used

[3] A detailed description of Ikeda's method, including the source code in *FORTAN*, can be downloaded at Ikeda's Laboratory, Osaka Prefecture University, Japan: http://www.marine.osakafu-u.ac.jp/~lab15/roll_damping.html (accessed: 2016-06-01).

to estimate the roll damping part in numerical ship motion simulations. Based on HERM measurements, Blume determined damping coefficient over a wide range of ship speeds and roll angles. The experimental results are summarized in a parametric form to make it useable for numerical simulations (see Salas Inzunza et al. 2001 and the Appendix) and will be referred to as SIMB^3-method. To estimate the roll moment of the bilge keel, Blume recommended the approach presented by Gadd (1964), which corresponds to the normal-force bilge keel part of Ikeda's method. The differences will be discussed in Sect. 11.3.2.

It needs to be mentioned that both established methods were developed for single-screw ships. Influences of two rudders, a skeg, shaft brackets and shafting lines were not taken into account.

The results of the RANSE-CFD simulations are compared with these two methods and the results are presented as dimensionless roll damping coefficients:

$$\hat{B} = \frac{B}{\rho \nabla B_{WL}^2} \sqrt{\frac{B_{WL}}{2g}}. \tag{11.5}$$

11.3.1 Bare Hull Roll Damping

SIMB^3-method: Blume measured the roll damping moment for three single-screw ship models (two Series 60 ships and a typical container ship of the 1970s) up to a roll angle of 20°. Compared to the RoPax vessel, the ships used in Blume's investigation did not have a flat stern or a large transom. It should be noted that these components can increase damping. Blume also investigated naval ships, which were considered in the development of the SIMB^3-method, see Salas Inzunza et al. (2001). All investigated models were equipped with rudders to hold the model on course straight ahead. Bilge keels were not considered. Blume assumed that a variation of the metacentric height $\overline{GM}$ has only a small influence on roll damping coefficient.[4] The influence was also neglected by Salas Inzunza et al. (2001). The range of validity of the SIMB^3-method is given in Table 11.2 and compared with the RoPax vessel.

In Fig. 11.4, simulations of the investigated hull geometry are compared with results of the SIMB^3-method. Deviations decrease for large ship forward velocities. This effect may be caused by the dominating lift damping effect at high Froude numbers. Deviations for zero speed are significant for roll amplitudes larger than 5°. Both the SIMB^3-method and the simulations indicate a weak nonlinear character over ship speed and a strong nonlinear character over the roll amplitude.

Ikeda's method: In a first step, CFD simulation results were compared with Ikeda's method for bare hull roll damping. For the wave damping part B_W for zero

[4]see discussion in Blume's paper (1979) on p. 23

Table 11.2 Range of validity of SIMB^3-method

	SIMB^3-method	RoPax
B_{WL}/d	2.55–4.88	4.33
C_B	0.48–0.80	0.53
$\omega^2 B_{WL}/2g$	0.10–0.25	0.25
ϕ_A °	5–20	5–35

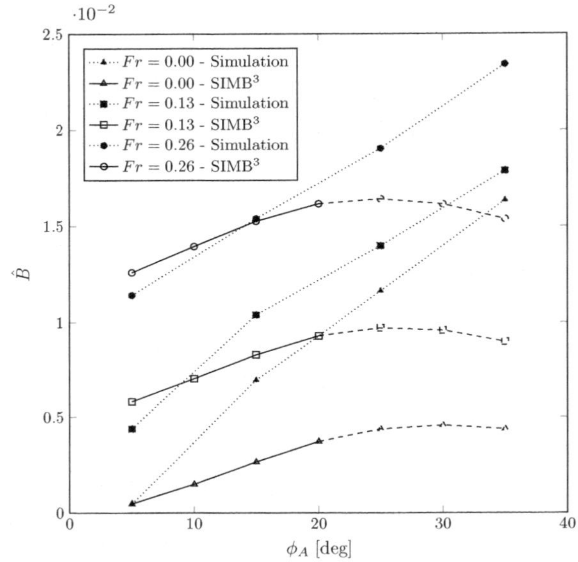

Fig. 11.4 Roll damping coefficients estimated by the SIMB^3-method and RANSE-CFD simulations for RoPax without bilge keels ($\omega = 0.435$ [rad/s]) — SIMB^3-method has a recommended range of validity for amplitudes not larger than 20°

velocity, an approach from Zhou (1987) based on the strip method *pdstrip*[5] is applied. The approach considers damping of several heel angles ϕ_H with

$$B_W(\phi) \approx B_{W,\phi_{H=0}} + \frac{1}{2} B^{''}_{W,\phi_H} \phi^2. \tag{11.6}$$

In addition to the zero heel wave roll damping $B_{W,\phi_{H=0}}$, $B^{''}_{W,\phi_H}$ considers the energy dissipation due to the submerged hull form with larger amplitudes.

Figure 11.5 shows the damping components over the ship speed. For full scale ships, the frictional part is vanishing. Because of large beam-draft-ratio and the position of the roll axis at waterline, the lift damping part is small and the wave damping is relatively large. The method was developed for smaller beam-draft-ratios with a maximum of approximately 2.5. Ikeda et al. (1977c) mentioned that his method is not sufficient for larger ratios. The characteristic wave damping component over the ship

[5]The source code of *pdstrip*, a *p*ublic *d*omain *strip* method, can be downloaded: https://sourceforge.net/p/pdstrip (accessed: 2016-06-01).

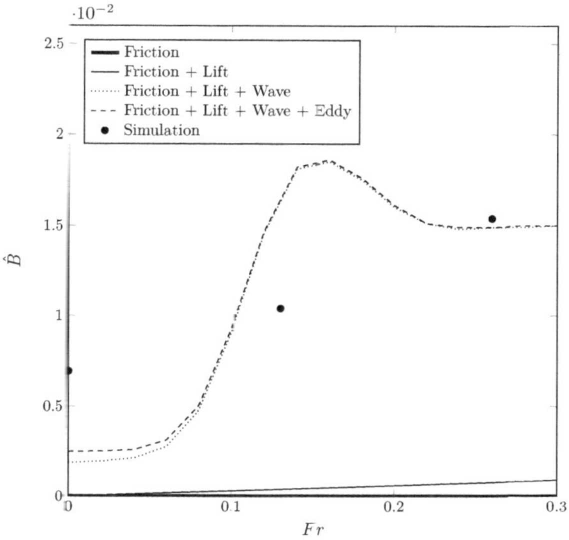

Fig. 11.5 Roll damping components of the Ikeda method compared with simulation results of $\phi_A = 15°$ and $\omega = 0.435\,[\text{rad/s}]$)

forward speed, the noticeable hump, can neither be confirmed by the SIMB^3-method nor by the finite-volume method simulations, see also Fig. 11.12. The noticeable hump also could not be observed in the boundary-element method simulation results from Falzarano et al. (2015).

11.3.2 Damping of Bilge Keels

Normal-force damping: Both Ikeda et al. (1976) and Blume (1979) recommend a force coefficient approach

$$F_{NBK} = \frac{1}{2}\rho A_{BK} C_{D,BK} v^2(t), \tag{11.7}$$

to estimate the normal drag force of the bilge keel. The velocity v is the relative transversal velocity component between hull and the surrounding water when the influence of the bilge keel on the flow is neglected:

$$v(t) = r_{BK} k \dot{\phi}(t). \tag{11.8}$$

The factor k is a flow velocity increment factor to take the local effect of the hull shape into account. Ikeda has suggested applying a sectional area coefficient σ formulation for sections with a small bilge radii ($\sigma > 0.975$). Blume formulates k as a function of ship breadth and bilge radius. Neither consider the roll center height. Söding (1991)

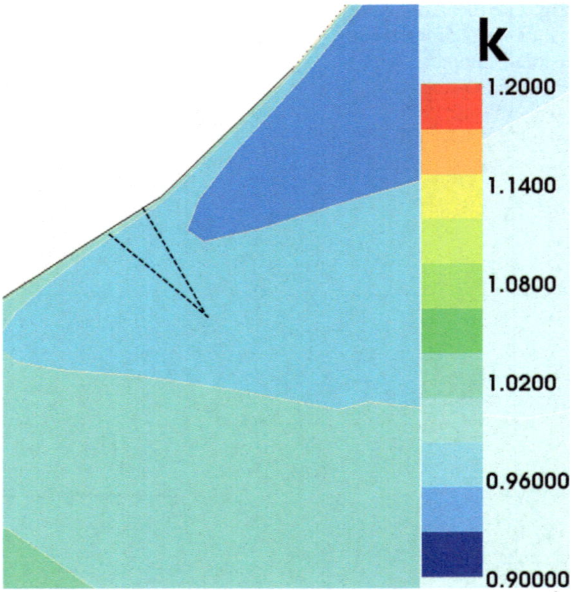

Fig. 11.6 Velocity increment factor k without the bilge keel

identified the influence of the height as significant. For the presented case, it will be shown that k is nearly one.

The factor k can be estimated by CFD simulations, see Fig. 11.6:

$$k = \frac{\nu_{A,sim}}{r_{BK}\omega\phi_A}. \tag{11.9}$$

The relative velocity amplitude $\nu_{A,sim}$ between the fluid and the bilge keel is normalized by the velocity amplitude of the bilge keel. This is done at 12 section planes along the keel for a simulation without bilge keel. In the position $\phi = 0°$, k is different for portside and starboard. Because of a harmonic rotating motion, a mean value $\bar{k}$ can be calculated (see Fig. 11.7).

The mean velocity increment factor in the simulation $\bar{k} \approx 1$ coincides well with Ikeda's suggestion regarding cases with section area coefficients of $\sigma < 0.975$. The midship section of the RoPax vessel has an area coefficient of $\sigma_m = 0.847$.

Using Eqs. (11.7) and (11.8), the roll moment for one bilge keel follows to:

$$M_{NBK}(t) = F_{NBK} \cdot r_{BK}\cos(\alpha) = \frac{1}{2}\rho\, A_{BK}\, C_{D,BK}\, r_{BK}^3\, k^2\dot{\phi}^2(t)\cos(\alpha). \tag{11.10}$$

The angle α is described in Fig. 11.8. Ikeda and Blume neglect this angle. By applying the energy equivalence principle, an equivalent damping moment for both bilge keels can be written as:

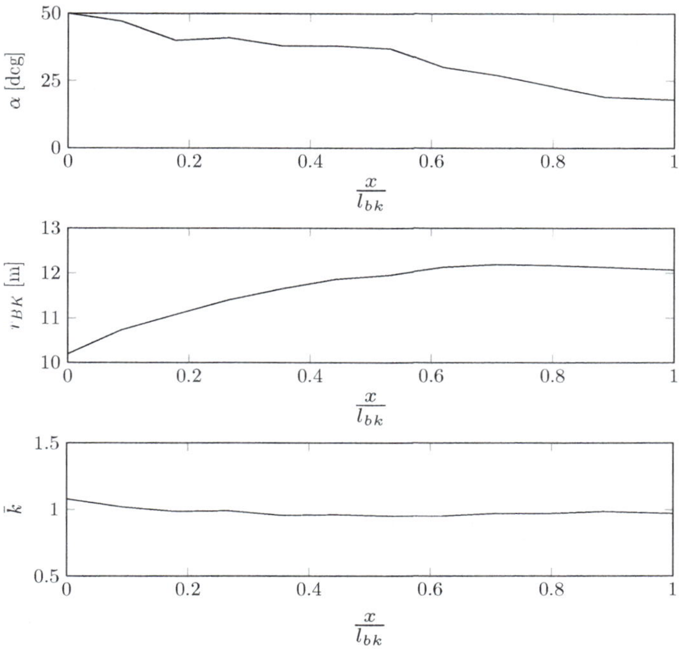

Fig. 11.7 Variation of α, radius r_{BK} and velocity increment factor $\bar{k}$ over non-dimensional bilge keel length $\frac{x}{l_{BK}}$ for the RoPax vessel

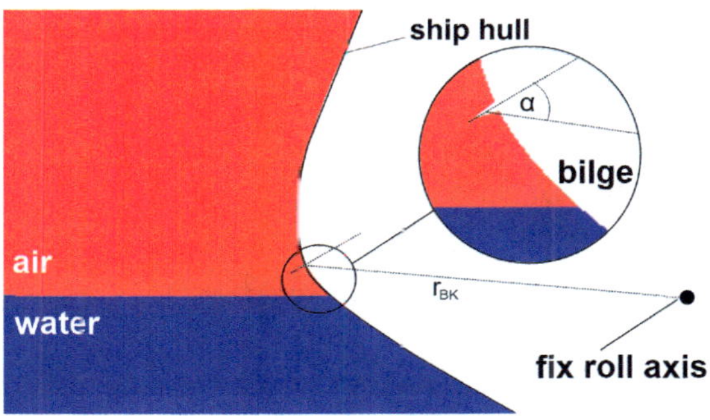

Fig. 11.8 Emerging bilge keel of RoPax vessel - main section with a roll amplitude of $\phi = 25°$

$$B_{NBK} = \frac{8}{3\pi} \rho \, A_{BK} \, C_{D,BK} \, r_{BK}^3 \, k^2 \, \omega \, \phi_A \, \cos(\alpha). \tag{11.11}$$

Compared to Blume, Ikeda recommends the calculation of the bilge keel damping coefficient in integral form over the bilge keel length l_{BK}. The distribution of α, r_{BK} and $\overline{k}$ over the bilge keel length l_{BK} is given in Fig. 11.7.

Both authors give formulations for the drag coefficient $C_{D,BK}$ that depend on the velocity increment factor k, the relative motion of the water orthogonal to the

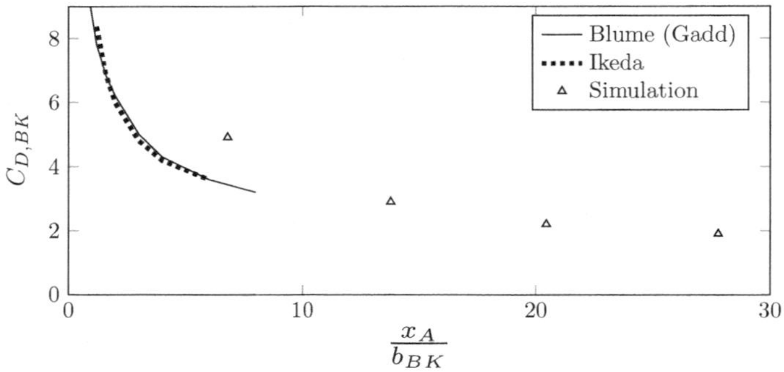

Fig. 11.9 Comparison drag coefficient $C_{D,BK}$ of bilge keel for RoPax vessel ($F_n = 0.13$)

bilge keel $x(t) = v(t)/\omega$ and the bilge keel breadth b_{BK}. On closer consideration, a connection to the Keulegan-Carpenter number (here for both bilge keels) becomes obvious, see Fig. 11.9:

$$KC_{BK} = \frac{v_A T}{2b_{BK}} = \frac{r_{BK} k \omega \phi_A 2\pi}{2b_{BK}\omega} = \pi \frac{x_A}{b_{BK}}. \tag{11.12}$$

The average drag coefficient $C_{D,BK}$ over one period for the normal force bilge keel damping (Fig. 11.9) can be calculated with CFD and estimated by the following formulation of the equivalent damping coefficient B_{NBK}:

$$C_{D,BK} = \frac{3\pi B_{NBK}}{8\rho b_{BK}\omega\phi_A \int_{lbk} \overline{k}^2 r_{BK}^3 \cos(\alpha)\, dl_{BK}}. \tag{11.13}$$

Even so, normal-force coefficients $C_{D,BK}$ for large KC-numbers are not well investigated, see Fig. 11.9.

Blume assumed that the forward speed is a negligible factor because the additional lifting effect due to an increase of forward speed is compensated by the decreasing normal-force damping. Ikeda (2004) gives a similar explanation if the ship forward speed is moderate.

Whereas Blume only uses $C_{D,BK}$ to consider the bilge keel effect, Ikeda divides the bilge keel damping into three parts, as mentioned above: the normal-force, the hull-pressure as well as the wave-interacting bilge keel damping. The wave interacting part is neglected in the original method.

Hull-pressure damping due to bilge keels: Ikeda analyzed the effect of eddies, produced by the bilge keels, which interact with the flow of the ship hull. An approach based on surface pressure measurement on a two-dimensional ship section is presented by Ikeda et al. (1977b). The induced damping by the surface pressure on the hull due to the bilge keels,

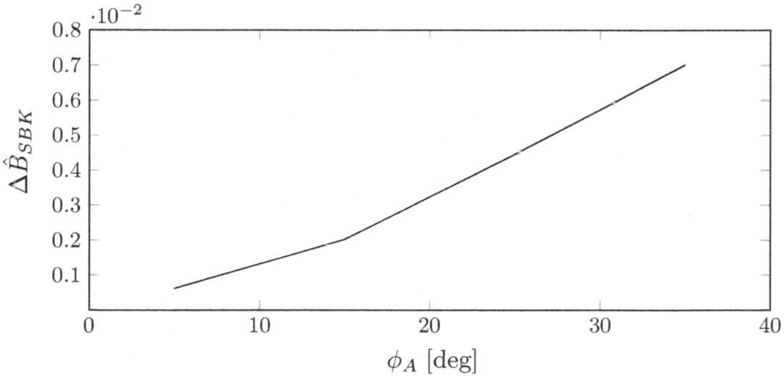

Fig. 11.10 Differences between results of CFD and Ikeda's method regarding hull interaction damping with bilge keels for the RoPax vessel ($Fr = 0.13$)

$$B_{SBK} = \frac{4}{3\pi}\rho\omega\phi_A \int_S r_{BK}^2 k^2 C_{P,BK} r \; ds, \qquad (11.14)$$

can be determined based on pressure coefficients $C_{P,BK}$. Here, B_{SBK} is estimated on two simulations, one with (w) and one without (wo) bilge keels. The normal force component of the bilge keels is subtracted from the estimated difference between the equivalent roll damping moment of the two simulations, which results in B_{SBK}.

The damping due to the bilge-keel-hull-pressure distribution in the simulation shows a similar tendency in comparison with the results obtained by Ikeda's method. Even so, Ikeda's method gives higher values of the bilge-keel-hull interaction component. Deviations between simulations and Ikeda's method are presented in Fig. 11.10 for the RoPax vessel.

Bilge keel damping due to water surface interaction: The systematic investigations carried out by Ikeda for bilge keel damping did not include roll angles higher than $17.2°(\phi_A = 0.3[\text{rad}])$. No explicit limitations for applying the method were given. Nevertheless, there is an angle for each ship where the bilge keel emerges. For the RoPax, this angle is $\phi \approx 20°$. An emerging or re-entering of bilge keels seems to be dominated by nonlinear effects, see Bassler and Reed (2009); Bassler et al. (2010). As an example, in Fig. 11.11 entrapped air due to a re-entering bilge keel can be seen for a simulation with large roll amplitude.

Himeno (1981) has presented results of increasing wave radiation due to the interaction of the bilge keel with the free surface. Comparisons of the measured and calculated results obtained by the linear theory showed that the additional damping underlie nonlinear effects of wave damping for larger roll angles. He noted that for ordinary ship hulls of that time (the 1970s), a wave-bilge keel effect can be safely neglected for bilge keel breadths of $b_{BK} = B_{WL}/60$ to $b_{BK} = B_{WL}/80$. The ratio of ship breadth and breadth of the bilge keel for the present RoPax is $B_{WL}/b_{BK} > 140$.

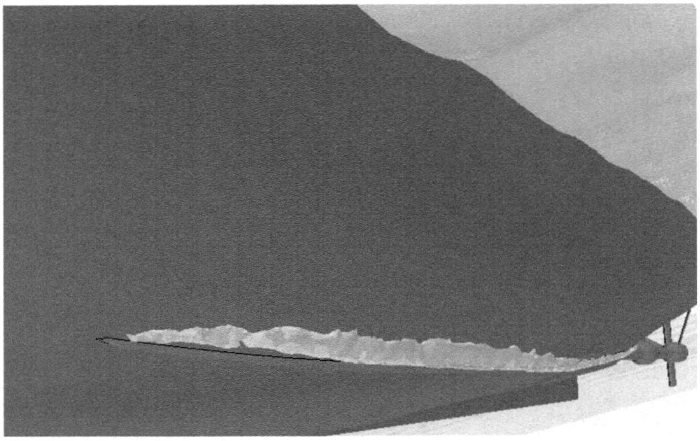

Fig. 11.11 Entrapped air in the simulation due to emerging and re-entering bilge keel - T = 14.6[s], $\phi_A = 35°$, Fn = 0.13

11.3.3 Rudders

Bertram (2011) assumed that the rudder component is considered to be independent from roll frequency and only depends linearly on the roll angle.

Blume (1979) measured the damping coefficients of hull forms equipped with rudders. Ikeda also included the rudder influence in his method, but the dimensions of the rudders were not specified in the existing published results; thus it is not clear in which component Ikeda considered the rudder influence.

Two CFD simulations, one with and one without rudders, were carried out. With a roll amplitude of $\phi_A = 35°$, a roll frequency of $\omega = 0.43$ [rad/s] and a ship velocity of $Fr = 0.26$, the contribution of the rudders was 0.81% of the total roll damping.

11.3.4 Shaft Brackets, Shafting Lines and Propellers

Because only single-screw hull forms were investigated by Ikeda and Blume, shaft brackets and shafting lines were not taken into account. The results of the CFD simulations on the estimated roll damping coefficients do not show distinctive differences between the simulation with and without shaft brackets and shafting lines. The flow on the house of the shafting line does not show any large separation areas. Investigations by Schmitke (1978) and a later investigation by Ikeda et al. (1994) were able to observe the same behavior.

Blume (1979) measured the damping coefficients with a rotating propeller, but influences due to a propeller will not be considered in Ikeda's method. The CFD simulations were carried out without propellers. It is assumed that propellers have no significant contribution to the roll damping moment.

11.4 Comparison of SIMB^3- and Ikeda's Methods with CFD

The simulations are carried out for different ship speeds and various roll amplitudes, see Fig. 11.12 for a roll period of $T = 14.6\,\mathrm{s}$. The roll axis is situated in the undisturbed free water surface plane.

The roll damping coefficients of the SIMB^3-method show similar characteristics to RANSE-CFD results. The gradient over the Froude number is nearly the same. For an amplitude of 5°, the results are in good agreement. Up to an amplitude of 25° and especially for higher Froude numbers, the SIMB^3-method gives damping coefficients in the same range of the CFD results. An extrapolation with Blume's method for roll amplitudes over 25° should be avoided. For zero forward velocity, the SIMB^3- and Ikeda's method give comparable results if bilge keels are neglected.

Ikeda's method estimates total roll damping coefficients for zero speed close to simulation results. Nevertheless, the ratio between bilge keel and total roll damping is different. The bilge keel damping is relatively large compared to CFD. One reason may be the large bilge-keel-hull interaction component in Ikeda's method and the neglected angle α for normal-force bilge keel damping. For large roll angles, Ikeda's method in the present configuration overestimates the roll damping compared to the simulations.

As expected, the roll damping increases with increasing height of roll axis to 5.3 m above the water surface, see Fig. 11.13. Compared with Ikeda's method, the calculated influence of the height of the roll axis on the roll damping coefficient is smaller.

11.5 Conclusion

RANSE-CFD simulations allow efficient calculations of roll damping coefficients. The presented numerical approach facilitates a detailed analysis of not only the total roll damping but also the contribution of each damping component.

A comparison between CFD calculations and the established SIMB^3- and Ikeda's method has been carried out. Although the established methods were developed for single-screw ships, they are able to give an acceptable approximation of the roll damping for twin-screw vessels in certain ranges. The results of Ikeda's method are more reliable at zero speed, the SIMB^3-method for setups with forward speed. For

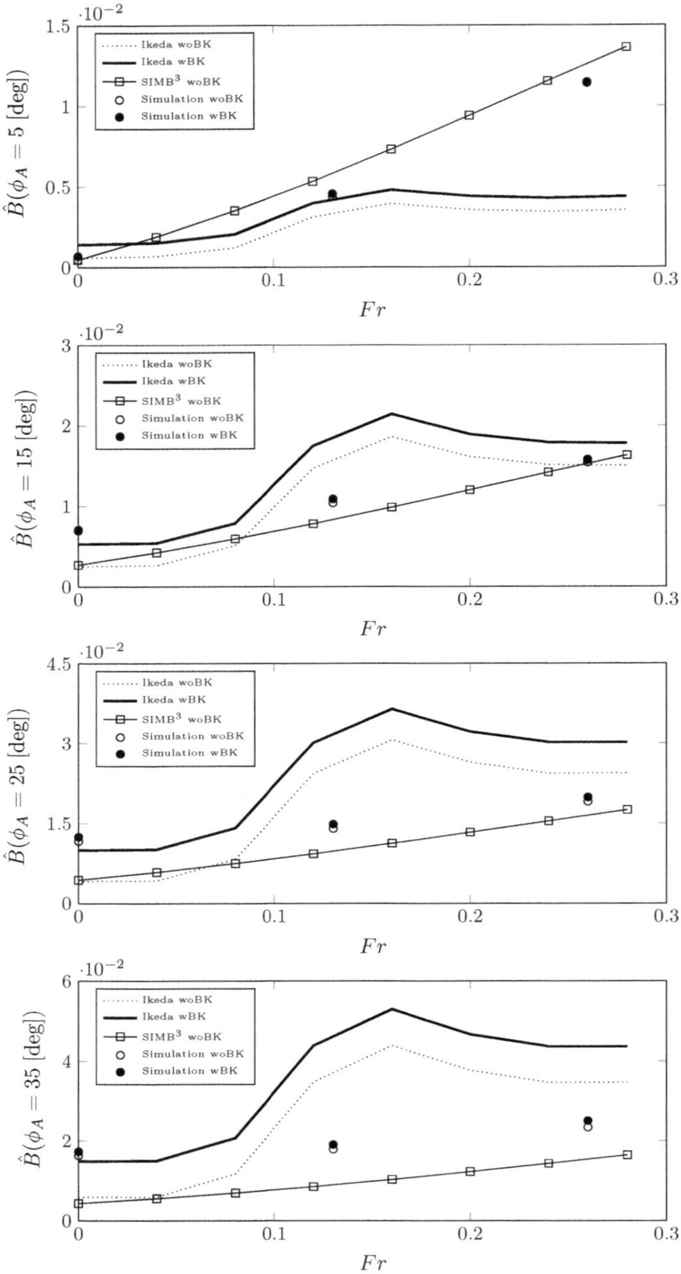

Fig. 11.12 Comparison of SIMB^3- and Ikeda's methods with simulations for $\phi_A = 5,\ 15,\ 25$ and $35°$ for RoPax vessel with a roll period $T = 14.6\,[\mathrm{s}]$

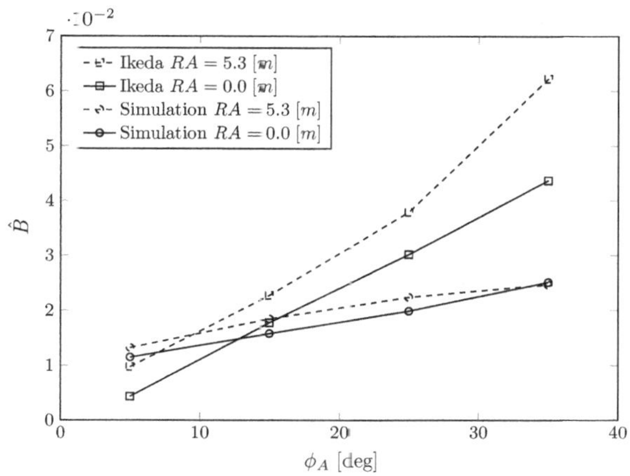

Fig. 11.13 Variation of roll axis height over waterline (*RA*) for RoPax vessel with $T = 14.6$ [s]

large roll amplitude simulations with forward speed, it should be mentioned that there is an urgent need to improve these methods. Modern CFD simulation methods seem to be a suitable instrument to face this challenge.

Acknowledgements This project was funded by the German Federal Ministry of Economics and Technology under the aegis of the BMWi-project *Best Rolldämpfung* within the framework program *Schifffahrt und Meerestechnik für das 21. Jahrhundert*. The authors would like to thank the project partners: Prof. Dr. B. el Moctar, H. Piehl and R. Kaiser (University Duisburg-Essen), Dr. M. Fröhlich (Potsdam Model Basin) and Dr. V. Shigunov (DNV-GL).

Appendix

SIMB³-method: Blume's roll damping measurements carried out at the Hamburg Ship Model Basin (HSVA) of various ship hulls of the 1970s and before were summarized as artificial neural network. This was developed by **S**alas **I**nzunza, **M**esbahi, **B**rink and **B**ertram Salas Inzunza et al. (2001) based on the measurement results of **B**lume: we call it here **SIMB³**-method.

For the artificial neural network, a sigmoid function is used:

$$sig(x) = \frac{1}{1 + e^{-x}}. \tag{11.15}$$

The non dimensional roll damping coefficient

$$\hat{B} = \sqrt{\zeta^2 \cdot \frac{2g\overline{GM}^2}{B_{WL}^3 \omega_0^2}} \tag{11.16}$$

depends on the gravity constant g, metacentric height $\overline{GM}$, ship breadth B_{WL}, roll resonance frequency ω_0 and the damping ratio ζ:

$$\zeta = 0.26525 \cdot sig[xx_1 + xx_2 + xx_3 + xx_4 + xx_5 + 1.121] - 0.071725 \quad (11.17)$$

with

$$xx_1 = -0.15923 \cdot sig(-0.54784 - 0.35004x_1 - 0.32394x_2 - 0.49683x_3 - 0.7495x_4)$$
$$xx_2 = 1.8997 \cdot sig(0.48293 - 2.71914x_1 - 5.87083x_2 + 5.55228x_3 - 0.99526x_4)$$
$$xx_3 = -0.45902 \cdot sig(-0.35086 - 0.3666x_1 - 0.21579x_2 - 0.78014x_3 - 1.1742x_4)$$
$$xx_4 = -2.0167 \cdot sig(4.2884 - 4.5154x_1 + 1.4302x_2 - 0.30797x_3 - 3.9884x_4)$$
$$xx_5 = -2.1800 \cdot sig(-0.09468 + 3.1056x_1 - 5.4142x_2 - 3.1332x_3 + 1.7851x_4)$$

and
$x_1 = 0.16807 \cdot (B_{WL}/d) - 0.12017$
$x_2 = 1.23456 \cdot C_B - 0.28765$
$x_3 = 1.33333 \cdot Fr + 0.3$
$x_4 = 0.026667 \cdot \varphi_a + 0.166667.$
To achieve the roll resonance frequency $\omega_0 = 0.435$ [rad/s] of the RoPax vessel, a metacentric height of $\overline{GM} = 2.09$ [m] is selected. The roll radius of gyration of the virtual and ship mass is assumed to be $i_\Phi = 0.4 \cdot B_{WL}$ [m].

References

Bassler C.C., Reed A.M. (2009) An Analysis of the Bilge Keel Roll Damping Component Model. In: Proc. 10th International Conference on Stability of Ships and Ocean Vehicles, St Petersburg, Russia

Bassler C.C., Reed A.M., Brown A.J. (2010) Characterization of Physical Phenomena for Large Amplitude Ship Roll Motion. In: The 29th American Towing Tank Conference, Annapolis, USA

Bertram V. (2011) Practical Ship Hydrodynamics. In: Butterworth Heinemann, 2nd revised edn.

Blume P. (1979) Experimentelle Bestimmung von Koeffizienten der wirksamen Rolldaempfung und ihre Anwendung zur Abschaetzung extremer Rollwinkel. In: Ship Technology Research / Schiffstechnik, Vol. 26 (in German)

Enger S., Peric M., Peric R. (2010) Simulation of flow around KCS-hull. In: A Workshop on Numerical Ship Hydrodynamics, Proceedings, Vol. II, Gothenburg, Sweden

Falzarano J., Somayajula A., Seah R. (2015) An overview of the prediction methods for roll damping of ships. In: Ocean Systems Engineering, Vol. 5.2

Gadd G.-E. (1964) Bilge Keels and Bilge Vanes. In: Technical Report of the National Physical Laboratory - Ship Division

Handschel S., Köllisch N., Soproni J.P., Abdel-Maksoud M. (2012) A numerical method for estimation of ship roll damping for large amplitudes. In: 29th Symposium on Naval Hydrodynamics, Gothenburg, Sweden

Handschel S., Fröhlich M., Abdel-Maksoud M. (2014) Experimental and Numerical Investigation of Ship Roll Damping by Applying the Harmonic Forced Roll Motion Technique. In: 30th Symposium on Naval Hydrodynamics, Tasmania, Australia

Handschel S., Abdel-Maksoud M. (2014) Improvement of the Harmonic Excited Roll Motion Technique for Estimating Roll Damping. In: Ship Technology Research / Schiffstechnik, Vol. 61

Himeno Y. (1981) Prediction of Ship Roll Damping - A State of the Art. In: Report 239, Dept. Of Naval Architecture and Marine Engineering University of Michigan, Ann Arbor, Michigan, USA

Ikeda Y., Himeno Y., Tanaka N. (1976) Ship Roll Damping - Frictional Component and Normal Pressure on Bilge Keel. In: J Kansai SNA, Vol. 161 (in Japanese, English translation available)

Ikeda Y., Himeno Y., Tanaka N. (1977a) On eddy-making component of roll damping force on naked hull. In: J Japan SNA, Vol. 142 (in Japanese)

Ikeda Y., Himeno Y., Tanaka N. (1977b) On Roll Damping Force of Ship Effect of Hull Surface Pressure Created by Bilge Keels. In: J Kansai SNA, Vol. 165 (in Japanese, English translation available)

Ikeda Y., Himeno Y., Tanaka N. (1977c) Components of roll damping of ship at forward speed. In: J Japan SNA, Vol. 143 (in Japanese, English translation available)

Ikeda Y., Katayama T., Hasegawa Y., Segawa M. (1994) Roll damping of high speed slender vessels. J Kansai Soc Naval Arch, Vol. 222

Ikeda Y. (2004) Prediction Methods of Roll Damping of Ships and Their Application to Determine Optimum Stabilization Devices. Marine Technology, Vol. 41

International Towing Tank Conference (2011) Recommended Procedures - Numerical Estimation of Roll Damping. (http://ittc.info/ Accessed: 2016-05-31)

Kawahara Y., Maekawa K., Ikeda Y. (2012) A Simple Prediction Formula of Roll Damping of Conventional Cargo Ships on the Basis of Ikeda's Method and Its Limitations. In: Journal of Shipping and Ocean Engineering, Vol. 2

Salas Inzunza M., Mesbahi E., Brink K.-E., Bertram V. (2001) Empirical Roll Damping Formula derived by Artificial Neural Network Applications. Jahrbuch Schiffbautechnische Gesellschaft, Vol. 95

Schmitke R.T. (1978) Ship sway, roll, and yaw motions in oblique seas. In: SNAME Transaction of the ASME, Vol. 86

Söding H. (1991) Die Wirkung von Rolldämpfungsflossen abhängig von ihrem Einbauort. In: Technical Report, Hamburg University, Germany (in German)

Wassermann S., Feder D.-F., Abdel-Maksoud M. (2016) Estimation of ship roll damping - A comparison of the decay and the harmonic excited roll motion technique for a post panamax container ship. In: Ocean Engineering, Vol. 120

Zhou Y.-Z. (1987) Die Berechnung der Rolldaempfung von Schiffen. In: Technical Report No. 474, Institut für Schiffbau der Universität Hamburg, Germany (in German)

Part V
Mathematical Model of Ship Motions in Waves: Damaged Ship

Chapter 12
Calculation Method to Include Water on Deck Effects

Nicolas F. A. J. Carette and Frans van Walree

Abstract Green water is an important issue regarding ships stability as it may dramatically change the loading of the ship compared to its dry deck condition. Until now, computational methods capturing this event are very time consuming as they often try to capture the complete dynamics of the flow over the vessel's structure and deck using CFD. Such methods are not practical when dealing with numerous lengthy time domain simulations for long term stability assessments MARIN has developed a fast method to be implemented in its 6 DOF time domain program FREDYN. This method has as objectives to be as fast as possible, even real time if achievable, but at the same time take into account correctly the mass of water flooding on the deck during green water events. The method is based on pre-computing the steady forward speed wave pattern and diffracted and radiated waves. The steady wave is computed for a series of sailing conditions using the in-house 3D linear panel code DAWSON. The diffracted and radiated waves are pre-computed using in-house 2D strip theory potential code SHIPMO for a series of frequencies and sailing conditions. A ship generated wave is then computed at each time step during the simulation using the current position and motions of the ship. This improves the computation of a realistic wave elevation consisting of the incident, steady, diffracted and radiated waves along the hull of the ship. This wave profile is then used to feed our flooding module which computes flows in tanks, compartments and through openings. This flooding model is based on a quasi-static Bernoulli formulation and empirical discharge coefficients. It is used to compute the flow over the bulwarks and through the freeing ports to the deck.

Keywords Time domain · Green water · Capsize · Calculations · FREDYN

N. F. A. J. Carette · F. van Walree (✉)
MARIN-Maritime Research Institute Netherlands, Wageningen, The Netherlands
e-mail: F.v.Walree@marin.nl

V. L. Belenky et al. (eds.), *Contemporary Ideas on Ship Stability*, Fluid Mechanics and Its Applications 119, https://doi.org/10.1007/978-3-030-00516-0_12

Notations

z	Incoming wave [m]
$\tilde{\zeta}$	Diffracted wave [m]
$\tilde{\zeta}$	Radiated wave [m]
ω	Wave frequency [s^{-1}]
κ	Wave number [–]
n	Index for frequency [–]
i	Index for section [–]
ε	Incoming wave phase [–]
$\tilde{\varepsilon}$	Diffraction wave phase [–]
ϕ_{rad}	Radiation potential [kg/ms^2]

12.1 Introduction

The capsize envelope obtained using time domain calculations appeared to be rather conservative during several risk analysis studies. This appeared to be strongly related to green water events happening too easily, too extremely and too often.

Until now, the Froude-Krylov forces were computed in FREDYN (de Kat and Paulling 2001) using the instantaneous waterline taking into account the ship motions and undisturbed incoming wave, and by this way these forces are taking care of the green water events. This is most of the time a conservative approach as it neglects diffraction, radiation and the forward speed wave which reduce the critical relative wave heights, this mostly for positions aft of the bow area.

The present new implementation proposes as first step to take into account the vessel and its motions on the water. The objective is to have a better estimation of the waterline to improve the calculation of the hydrostatic forces, including water on the deck.

12.2 Method

The effect of the ship on the water surface is divided in three components:

- Static forward speed wave
- Diffracted wave
- Radiated wave

Each component is computed separately at the beginning of the time step at several positions along the ship. By summing the three waves we obtain the perturbation wave profile that can be summed with the incoming wave. Points between calculation locations are obtained by spatial linear interpolation. If the point lies outside the waterline contour, for instance in case of bulb, closest approximation is used. By points we mean any location where the water height is needed such as, for instance, a panel on the hull for the Froude-Krylov forces or an opening into a flooded compartment.

12.2.1 Static Forward Speed Wave

The static wave is obtained by linear interpolation between series of wave profiles computed at different speeds, drafts and heel angles. The actual position and speed of the ship is then used to pick up the right databases. Draft and heel values must be extracted from low frequency motions. Wave patterns are computed once before the calculations using a 3D potential solver. From the patterns, only the values along the vessel are extracted to obtain the waterline.

12.2.2 Diffraction Wave

The diffracted wave profile is obtained for each section of the ship using databases of linear potential diffraction.

Using MARIN's 2D strip theory code SHIPMO, the diffraction potential is extracted at each section, at the waterline, for a series of wave frequencies, headings and speeds. The potential is saved as a complex number to allow for linear interpolation between the databases without losing the phase information. It is converted to a wave amplitude response operator in m/m. At each time step of the calculation a database of diffraction potentials is made, depending on the actual speed and heading of the ship. Then, for each incoming wave component n and at each section i, the instantaneous diffracted wave profile at each section is computed using (12.1).

$$\tilde{\zeta}_i = \sum_n \tilde{\zeta}_{in} z_r \sin(\omega_n t - \kappa_n + \varepsilon_n + \tilde{\varepsilon}_{in}) \qquad (12.1)$$

The diffracted waterline is then used further during the time step using spatial linear interpolation to every panel of the ship. The error in this case by the spatial interpolation is rather limited as the triggering factor for water on deck is the waterline itself which is as precise as there were sections in the calculations; the diffracted wave is not needed outside the ship where the spatial interpolation would introduce large errors.

Such pre-calculation followed by some spatial interpolation is used to save computation time as the sum of wave components is done only twice per section, one for port and one for starboard side, instead of doing it for every panel, relative location and flooding opening. Diffracted wave is actually the only wave that could be really computed at any point but the calculation time would be excessive using fine meshes and wave spectra.

12.2.3 Radiation Wave

The radiation wave is basically obtained in the same way than the diffraction wave except that there is here the need for retardation functions to go to the time domain.

For each section and wave encounter frequency, the radiation potential is extracted from potential solutions, for instance a SHIPMO calculation. The potential is converted to a wave amplitude response. Then a method similar to what is done with the added mass and damping is applied:

- The real part of the amplitude is divided by ω^2
- The imaginary part is divided by ω

We have thus similarly as for added mass and damping terms the following formula for the radiation wave components:

$$a_i(\omega) = \frac{Re\left(\phi_{rad_i}(\omega)\right)}{\rho g \omega^2} \tag{12.2}$$

$$b_i(\omega) = \frac{Im\left(\phi_{rad_i}(\omega)\right)}{\rho g \omega} \tag{12.3}$$

Converted to time domain functions using (12.4) and (12.5), they give "added mass" and "retardation function" of radiation wave amplitude.

$$A_i = a_i(\omega_\infty) + \frac{1}{\omega_\infty} \int_0^\infty B_i(\tau) \sin \omega_\infty \tau d\tau \tag{12.4}$$

$$B_i(\tau) = \frac{2}{\pi} \int_0^\infty \left(b_i(\omega) - b_{i_\infty}\right) \cos \omega \tau d\omega \tag{12.5}$$

The retardation functions are saved for each section and side for the whole calculation. Using correlation with the time history of motions we can thus compute the radiated wave at each section using (12.6).

$$\breve{\zeta}_i(t) = A_i \ddot{x} + B_i(\infty)\dot{x} + \int_0^\infty B_i(\tau)\dot{x}(t-\tau)d\tau \tag{12.6}$$

As for the diffracted wave, the radiated wave profile is saved for each section at the waterline for both sides during a complete time step and spatially interpolated to any point on the ship.

12.2.4 Calculation

When using only the static wave correction, the calculation can still be done in real time on a PC with a quad core CPU at 2.66 GHz.

The diffraction calculation strongly depends on the number of wave components. On a dual core PC, the calculation time doubles with 80 wave components compared to the calculation without correction. The difference tends to reduce as the interpolation between the databases becomes less and less the bottleneck.

The radiation correction has not been fully tested but non constant time step is the most expensive factor as the retardation functions have to be recomputed for each section every time it changes. Otherwise it costs at every time step two correlations per section.

12.3 Tank Testing

The validation of the present method is based on a series of tests carried out at MARIN using a model of the DDG51 (European version), see van Walree (2010). Tests were carried out with a captive and free sailing model. The loading condition was such that stability was low giving a high capsize risk.

12.3.1 Captive Tests

The tests with a captive model were done to look at steady forward speed wave and diffracted wave. The tests were done at different speeds and heel angles in regular waves of various frequencies and amplitudes. Relative wave elevation were recorded at several locations along the model (Table 12.1).

Table 12.1 Regular wave captive tests

Speeds	18, 24	Knots
Heel angles	10, 20	deg
Amplitudes	1.25, 1.875	m
Frequencies	0.546, 0.598, 0.661, 0.739	rad/s

Fig. 12.1 High roll motion without capsize and very low amount of green water in stern quartering seas

12.3.2 Free Sailing Tests

Free sailing tests were done, in high stern quartering seas to look at green water events. Conditions were such that capsize risk was high during the standard time domain simulations but rather low during the tests. Tests were done at two headings (300 and 330°) and three speeds (12, 18 and 24 knots) in irregular waves (Fig. 12.1).

12.4 Validation

12.4.1 Steady Wave

The steady wave implementation was validated by comparing the wave profile computed to the average wave elevation during the tests. At the speeds of interest one can observe a large trough at amidships increasing the margin against green water events. This was until now absolutely not taken into account. One can also notice that the heel angle does not have a strong effect on the wave profile in these conditions. The method clearly improves the estimation of the waterline to the original undisturbed wave compared with the experimental measurements (Fig. 12.2).

12.4.2 Diffracted Wave

The maximum wave measurements along the hull have been compared to the maximum amplitude of the potential diffracted wave summed to the incoming and steady waves. The following figures give the profiles of maximum wave elevation along the ship for different conditions during experiments and calculations compared to the

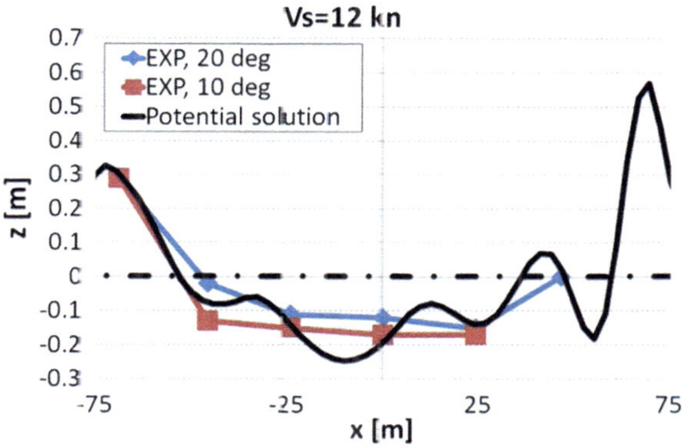

Fig. 12.2 Computed waterline compared to experimental steady wave profile during captive tests for different heel angles at 12 knots

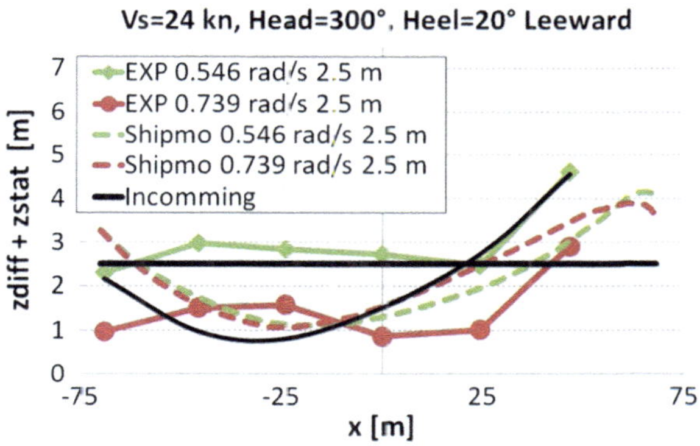

Fig. 12.3 Maximum wave elevation along captive vessel in regular waves: experimental and computed (leeward)

deck line and incoming wave for both leeward and windward sides (Figs. 12.3 and 12.4).

In case of large roll angles, taking diffraction and steady wave into account improves the estimation of the green water events (see Fig. 12.3). The diffraction most of the time reduces the water elevation along the vessel, and combined with the steady wave very often avoids the water to flood on the deck. However, the effect of the frequency on the diffraction seems often underestimated by strip theory. The diffracted wave is also overestimated at the aft of the ship, but this is a typical drawback from linear theory with forward speed.

Finally, the disturbed wave amplitude on windward seems underestimated for some configurations, this appeared using both strip theory or 3D diffraction (PRECAL), but this is not critical when looking at capsize risk due to green water as most capsize over predictions are on the leeward side.

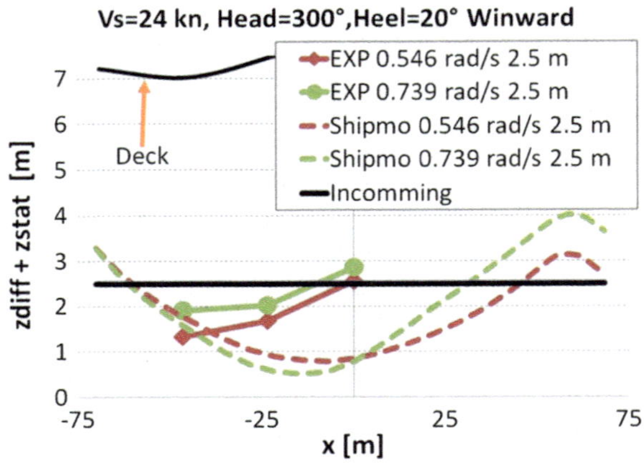

Fig. 12.4 Maximum wave elevation along captive vessel in regular waves: experimental and computed (windward)

12.4.3 Radiation Wave

The radiation was not used during these calculations as first attempts gave unrealistically high waves. This probably comes from a lack of a forward speed correction. The radiation potential is solved for a series of encounter frequencies but is valid at zero speed, the effect of radiated waves being washed backwards when sailing is not taken into account. Depending of the velocity, the retardation function at one section should become more and more dependent of the ones in front. Another solution would be to compute the potential radiation wave databases at forward speed using an exact solution and have a set of retardation functions for different speeds as it is done for the damping.

12.5 Results

A series of free sailing time domain calculations were done with and without steady and diffracted wave correction. For each condition a series of five runs of half an hour was done.

Without correction, almost in all conditions very high capsize risk is observed. Most of the capsizes happen very soon and fast. They are always due to excessive amounts of water on deck. For most simulations the deck is almost constantly wet on the leeward side. As the encounter frequencies were quite low, if a wave crest exceeds the freeboard at amidships, it will stay there and induce large and increasing roll angle until capsize occurs. This process appears as a static loss of stability triggered by the first freeboard exceedance event (Fig. 12.5).

When the correction is applied, the threshold of the capsize event is definitely increased. One can observe much less capsizes, most of the time those capsizes are now due to broaching. If water on deck occurs, the volume of trapped water is maybe

Fig. 12.5 Capsize risk with and without wave correction for 30 min sailing at 300° heading compared to experiments

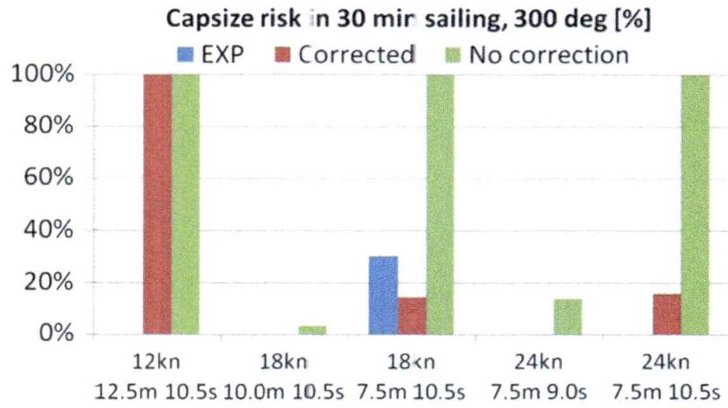

Fig. 12.6 Capsize risk with and without wave correction for 30 min sailing at 330° heading compared to experiments

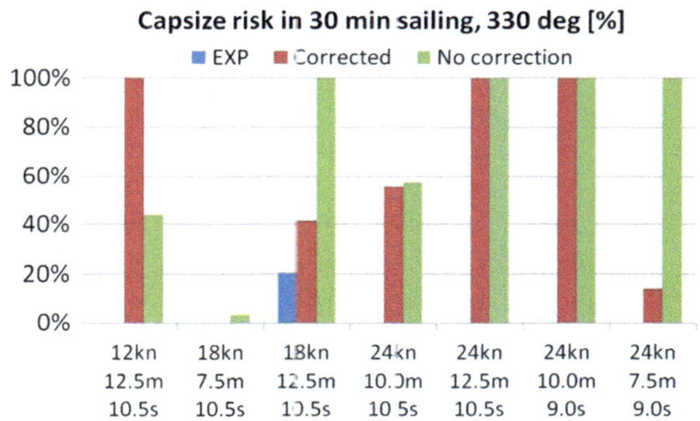

still overestimated due to the absence of a model computing the well known dam break motion of the green water which retards the progression of the water at the beginning of the green water event. This may explain why the correction seems still not sufficient in very large waves. However in those cases, they were also very steep and breaking, which anyhow cannot be captured with linear waves (Fig. 12.6).

The reduction of capsize risk is of course accompanied by a reduction of the roll. We can see that this reduction results in a better matching of the experiments most of the time (Fig. 12.7).

12.6 Conclusions

The correction of the waterline for forward speed gives, for a very reasonable computation time, a much better threshold for freeboard exceedance. This helps improving capsize risk analysis at high speeds. The effect of the heel angle on the wave profile is limited in a normal rolling range. This should be checked up to very high heel angles to know if the database really needs to depend on the heel angle. A dependence on trim could be easily included but raises the question of how to extract its value from the pitch and wave slope.

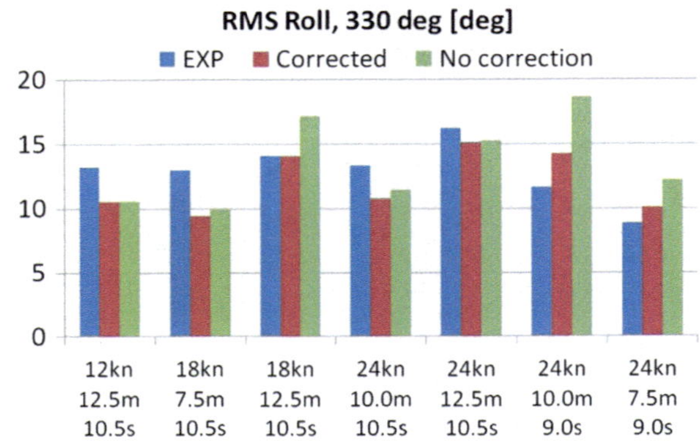

Fig. 12.7 Standard deviation of roll with and without wave correction compared to experiments

The correction for the diffraction slightly improves the asymmetry of the waterline between wind- and leeward sides. This would be even more important for headings closer to beam seas and slightly higher wave frequencies. The correction improves the capsize risk prediction by lowering the waterline in the conditions tested.

As already mentioned, the radiation was not used during these calculations as first attempts gave unrealistically high waves, probably due to a wrong forward speed effect when using strip theory. Two ways are seen, either a correction on the retardation functions or a corrected potential solution.

The case of very high or breaking waves seems still to be an issue. This could maybe be corrected by applying the radiation correction as large amplitude motions generally push the water away from the deck, retarding the flooding. Another correction could also come from a larger database of radiation and steady waves for very large heel angles. Finally, representing the deck by a floodable compartment might introduce some delay in the flooding of this one by using appropriate discharge coefficients and openings representing the flow over the bulwarks. On the other hand, breaking waves cannot be computed using linear wave spectra whatever method is used to correct them.

As this new method relies on steady, diffraction and radiation databases, any improved method to estimates these components would immediately improve the calculation of the instantaneous waterline without need of a reimplementation.

Acknowledgements This research is performed for the Cooperative Research Navies. The permission of the CRN to publish the results is gratefully acknowledged.

References

de Kat J.O. and Paulling J.R., 'Prediction of extreme motions and capsizing of ships and offshore vehicles', Proc. of the 20th OMAE Conference, Rio de Janeiro, June 2001.

Van Walree F., 'Seakeeping deck Edge Immersion Model Tests', MARIN Report 22810-1-CPS, 2010.

Chapter 13
Study on the Motions and Flooding Process of a Damaged Ship in Waves

Seokkyu Cho, Honggun Sung, Sayoung Hong, Bowoo Nam, Sungchul Hwang and Youngsik Kim

Abstract To study the motions and flooding process of a damaged cruiser, a series of experiments and numerical calculations have been performed in calm water and in waves. Two damaged scenarios are selected to investigate effect on the motions and flooding process; midship section and fore section. The results of the experiment, relating to the quasi-static numerical model and the quasi-dynamic numerical model are compared . Numerical simulations are then conducted using quasi-static and quasi-dynamic models. The quasi-dynamic model adopts the mass-spring system for internal water motion description and the model considers the dynamics of free surface as ship motion. The flooding water with free surface in midship section changed the heave and roll of the cruiser and roll RAOs in waves because the center of flood water amidships locates on the starboard side and the flooded water with large free surface area in the upper compartment generates sloshing flow. The developed quasi-dynamic model reproduces these flooding water motion in calm water and waves.

Keywords Cruiser · Damaged · Flooding · Experiment · Sloshing · Damping

---

S. Cho (✉) · H. Sung · S. Hong · B. Nam · S. Hwang · Y. Kim
KRISO (Korea Research Institute of Ships and Ocean Engineering), 32 Yuseong-daero 1312 beon-gil, Yuseong-gu, Daejeon 34103, Korea
e-mail: skcho33@kriso.re.kr

H. Sung
e-mail: hgsung@kriso.re.kr

S. Hong
e-mail: sayhong@kriso.re.kr

B. Nam
e-mail: bwnam@kriso.re.kr

S. Hwang
e-mail: schwang@kriso.re.kr

V. L. Belenky et al. (eds.), *Contemporary Ideas on Ship Stability*, Fluid Mechanics and Its Applications 119, https://doi.org/10.1007/978-3-030-00516-0_13

13.1 Introduction

Ship accidents may occur due to various reasons; collision, running a ground, malfunctioning of an engine, attack, etc. When a ship is damaged for certain reason, she loses her function and safety. So, the evaluation of the motions and assessment of stability is very important. Many efforts have been also made for the development of numerical methods for the behaviour of damaged ship. These numerical methods have been validated and improved by the international benchmark studies such as those done by ITTC and HARDER project. Up to now it is believed that the numerical methods are able to predict the overall tendency of the damaged ship motions and flooding process to an extent compared with experiments. But reliable prediction is difficult because the underlying phenomena are very complicated and highly nonlinear due to the various factors such as geometry of damaged compartment, flooding process and waves etc. To improve the accuracy of the numerical methods and the understanding of the mechanism of flooding process, data of various damaged scenarios need more thorough numerical simulations and experiments. Also it is generally believed that the physics of damaged ship can be analyzed by experiments more realistically.

In this study a series of experiments and numerical analyses have been carried out for the behaviour of a damaged cruiser in waves. Two damaged configurations are selected to study the damage effects. The one is the mid-section part which has 6 compartments. The second is the fore-section part which has 4 compartments. The starboards of hull are damaged for two damage conditions. The flooding tests were performed for the transient process and the flood water height was measured by 19 water height sensors. To study the effect of flood water and damage compartment, model tests were carried out in various wave conditions. The motion tests in waves were carried out after the compartments are completely flooded. The experiments indicate that the internal compartment influences the transient flooding process and roll motion. When there is water with free surface in compartments and the ship moves at the natural frequency of internal water in the compartment, coupling of internal water and ship motion occurs. The numerical simulations were conducted using quasi-static model, quasi-dynamic model and CFD. The quasi-dynamic model adopts the mass-spring for internal water motion description. The model considers the dynamics of free surface as ship motion. This mass-spring equation is explicitly coupled with ship motion equation. The quasi-dynamic model shows the intermediate results of CFD and quasi-static model.

13.2 Model Experiment

The model tests were performed in KRISO ocean engineering basin ($L \times B \times D$: $56 \times 30 \times 4.5$ m). The model ship is a cruiser and the hull data of cruiser is provided by SSRC. The contents of model test are as follows.

Table 13.1 Particulars of cruiser

Items	
Length, Lpp	247.2 m
Beam, B	35.5 m
Draft, T	8.3 m
Displaced weight	56,541.5 ton
KG	16.393 m
GM	2.388 m
Natural roll period	21.07 m
Gyration of roll	14.814 m
Gyration of pitch	61.925 m

- Motion in regular and irregular waves
  Intact, damaged conditions
- Flooding process in calm water
  Intact, damaged conditions
- Free decay in calm water
  Intact, damaged (opened, closed)

13.2.1 Ship Model

The object ship is a cruiser. The main particulars are summarized in Table 13.1 and Figs. 13.1 and 13.2 show lines and model of the cruiser. The model was fitted with bilge keels. Its length is 75 m and height is 0.50 m in prototype. They are symmetrically located about the mid ship at half the bilge girth. The inclination with the vertical is 45°. The model was around 5 m long corresponding to a scale of 50.

13.2.2 Damage Compartment

Two damaged scenarios were chosen. The one (DAM1) is that mid-section part is damaged, which has 6 compartments. The second (DAM2) is that fore section part is damaged, which has 4 compartments. These damaged parts are little different with the original inner compartment of the cruiser. The compartments were simplified for model tests. The inlet of damaged compartment is located at the starboard side, the length is 6 m and the height is 5 m. The general arrangements of the damaged compartment are shown in Fig. 13.3.

The damage models are shown in Fig. 13.4. The material of the damaged model is acryl and thickness is 5 mm. The coordinates of compartments and inner connections can be found in Cho et al. (2009). The origin is amidships (10 St.) in x, center in y

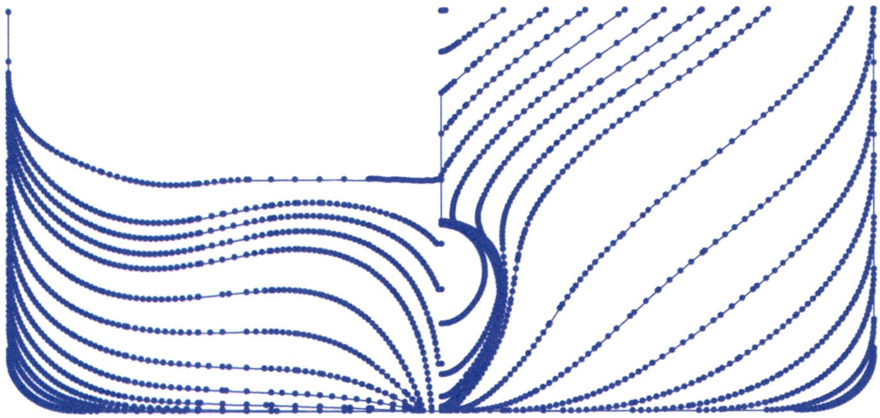

Fig. 13.1 Lines of cruiser

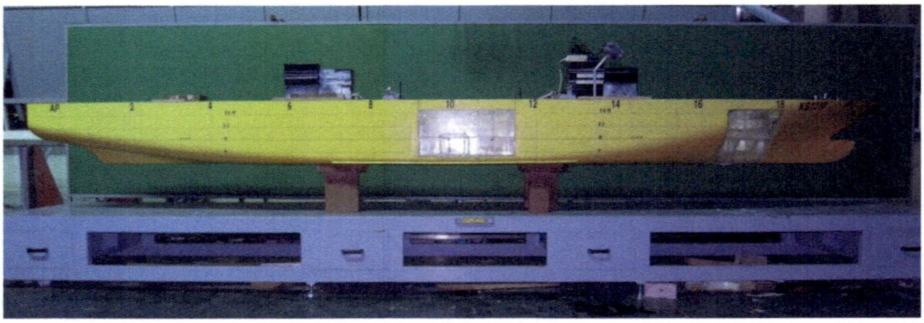

Fig. 13.2 Cruiser model

and baseline in z direction. The inlet of DAM1 is from 3.4 to 8.4 m from baseline and the top of inlet is above the water free surface. The top of DAM2 inlet is 8.05 m from the keel. The inlet is opened by pulling the door suddenly during model test.

13.2.3 Environmental Conditions

The characteristics of damaged cruiser in waves are investigated. To study the effects of flood water and in/out flow through damage inlet, motions of cruiser and flooding heights in compartments are measured. In order to study the effects of wave height on the roll RAO, 4 regular wave heights (1, 3, 5, 7 m) are used. The wave conditions are as follows.

- Regular waves
  Frequency: 0.2–1.1 rad/s

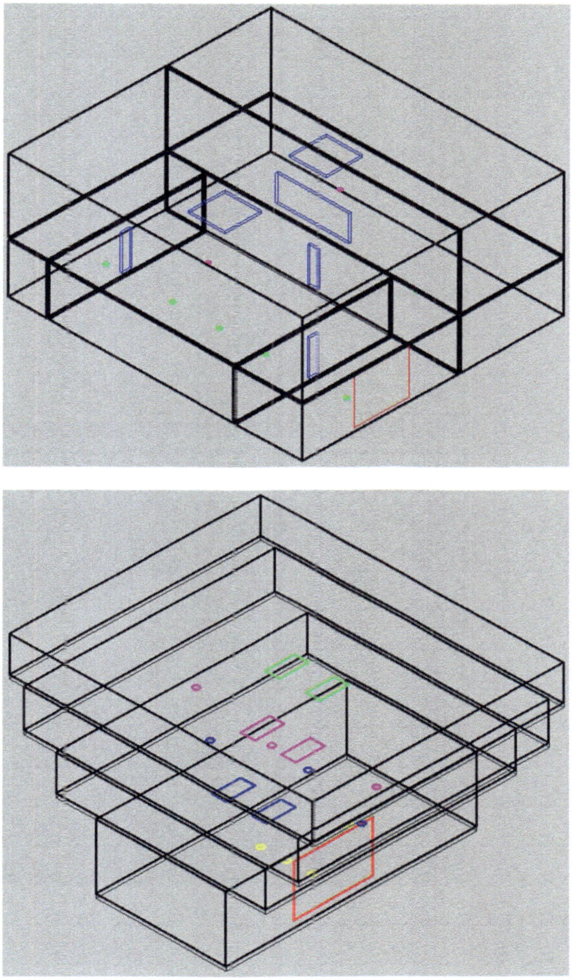

Fig. 13.3 Arrangement of damage compartments (CP10/11, CP17)

Height: 1, 3, 5, 7 m

- Irregular waves: JONSWAP (gamma = 3.3)
  Irregular wave 1: H1/3 = 1 m, Tp = 5sqrt(H1/3)
  Irregular wave 1: H1/3 = 3 m, Tp = 5sqrt(H1/3)

13.2.4 Measurement System

To analyze the behaviour of damaged ship, the motions of the ship and water in compartment must be measured. The 6 dof motion of ship are measured by non-

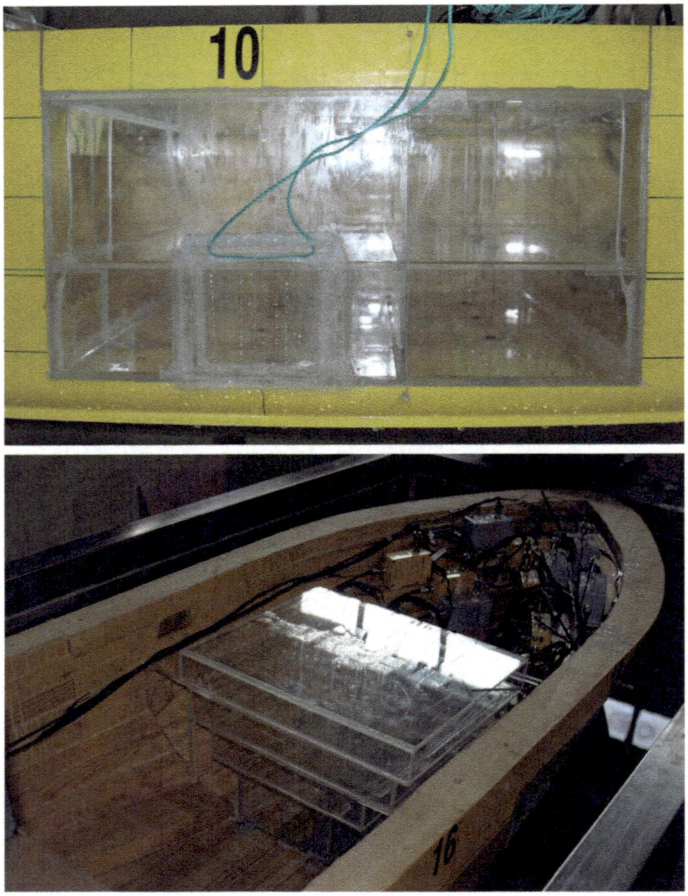

Fig. 13.4 Damage compartment model

contact optical system (RODYM6D). The flooding flows in each compartment are measured by capacity type wave probes. The number of wave probe is 10 in CP10/11 and 6 in CP17. Video cameras are also used to record the flooding process. The RBM1 is in CP10-R1S next to damage inlet (Fig. 13.5). The locations of wave probes can be found in Cho et al. (2009) .

13.3 Numerical Method

In order to analyze the flooding numerically, the quasi-static model (Cho et al. 2009) and quasi-dynamic model are used. The quasi-dynamic model is lumped mass-spring

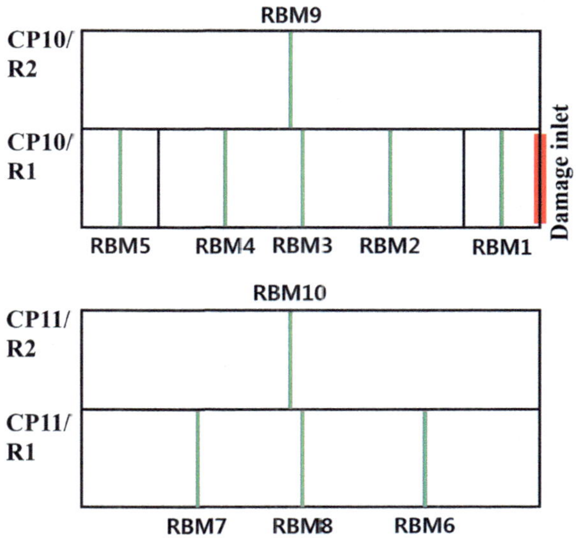

Fig. 13.5 Locations of flooding height sensor

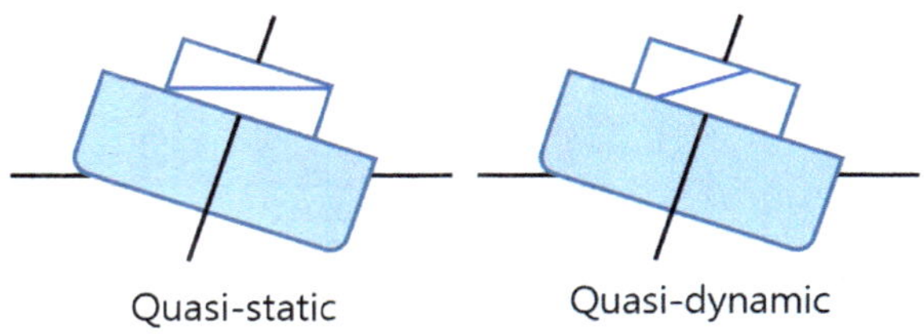

Fig. 13.6 Free surface description for quasi model

system. This model calculates the free surface angle with respect to the horizontal plane while the ship moves. Figure 13.6 shows the concept of quasi-static and quasi-dynamic model. The quasi-dynamic model equation coupled with ship motion is solved. The 4th order Runge-Kutta method is used for time integration.

$$a_1 \dddot{y} + a_2 \dot{y} + a_3 y = -b_1 \ddot{x} - b_3 x \tag{13.1}$$

where y is free surface angle, x ship roll, a and b equation coefficients.

$$v = \dot{y}$$

$$\dot{v} = \frac{-b_1 \ddot{x} - b_3 x - a_2 v - a_3 y}{a_1} \tag{13.2}$$

13.4 Test Results and Discussion

13.4.1 Experimental Results

13.4.1.1 Free Decay Test

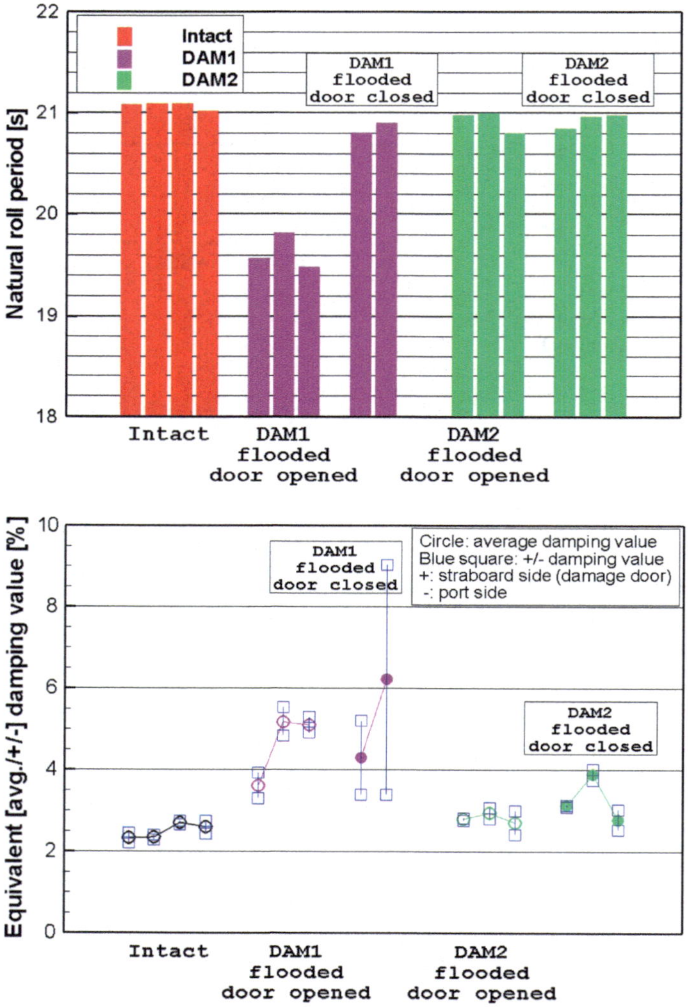

Fig. 13.7 Results of free decay

Figure 13.7 shows the results of roll free decay test. The natural roll period of intact ship is 21 s. The period of DAM1 decreases to 19.8 s because of flooding, heeling and free surface, etc. When the inlet is closed after flooding, there is no in/out flows though the inlet and the period of DAM1 for closed inlet is 20.8 s. The period of DAM2 is almost the same as intact ship. The equivalent damping values are plotted in Fig. 13.7. The damping values of intact and DAM2 range from 2 to 4%. The damping values of DAM1 increase from 4 to 6% (Table 13.2). The damping value of DAM1 is proportional to the magnitude of initial angle. When inlet is opened, flooding starts at the starboard side and water accumulates much more at the starboard side then the ship heels to the right. After flooding, roll is not symmetric and the roll and damping is not symmetric anymore. There is flow in/out through the CP10-R1S during free decay of DAM1. This indicates that the estimation of damping is difficult when a damaged part is severe. In case of the closed condition, there is no flow in/out through opening. The roll is affected by only internal water motion. The natural periods are very similar but the damping of DAM1 increases due to the flooding water motion.

13.4.1.2 Flooding Test in Calm Water

The flooding test was performed in calm water for DAM1 and DAM2. Figures 13.8 and 13.9 shows the height of flooding water compartments and motions. The flooding through the inlet starts at CP10-R1S (RBM1) and continues to CP10-R1C (RBM2, 3, 4) and CP10-R1P(RBM5). The water instantly fills up CP10-R1S. After filling of CP10-R1S, water propagates to next compartments. The RBM6, 7 and 8 show the flow from CP10 to CP11. The required time for flooding of second floor, CP10/11-R2 is about 240 s. Figure 13.9 shows the motions with flooding. Roll motion begins at the same time with flooding and reaches the steady state (~400 s) after filling of CP10-R1S/C/P. The flooding process of DAM2 is quite simple due to simple geometry and configuration. The flooding starts at CP17-R1 and flooding water reaches to the bottom of CP17-R2.

13.4.1.3 Tests in Waves

The motion tests in waves were carried out in the condition that the compartments were flooded. This gives the same initial condition in different waves. The results of motions in waves are shown in Figs. 13.10, 13.11 and 13.12. The wave amplitudes of regular wave are 1, 3, 5, 7 m to investigate the effect of nonlinearity of the incident waves on the roll motion. The roll motions are significantly influenced by wave amplitude and damage conditions. Interestingly enough, the effect of wave amplitude on roll motion also appears in intact condition. The peak value of roll RAO decreases at resonance frequency when wave amplitude increases. In case of DAM1, roll RAOs are changed due to internal water motion and inflow/outflow. The resonance frequency moved from 0.3 to 0.33 rad/s due to sloshing. The effect of internal water motion appears for wave amplitude 3, 5, 7 m and sloshing occurred in

Table 13.2 Initial roll angles and equivalent damping values of free decay test

	Intact		DAM1 open		DAM1 closed		DAM2 open		DAM2 closed	
	θ initial	δ	θ initial	δ	θ initial	δ	θ initial	δ	θ initial	δ
1st	4.80	2.33	3.20	3.62	3.10	4.30	6.90	2.80	7.17	3.12
2nd	6.10	2.35	6.90	5.19	4.90	6.23	6.67	2.95	6.05	3.88
3rd	7.50	2.71	6.70	5.11			7.00	2.71	6.20	2.79
4th	7.95	2.60								

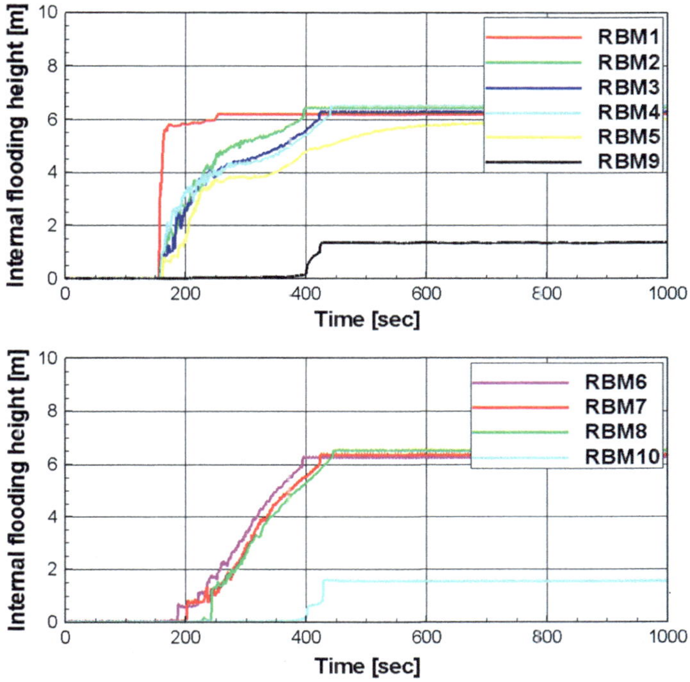

Fig. 13.8 Height of flooding water in DAM1 compartments

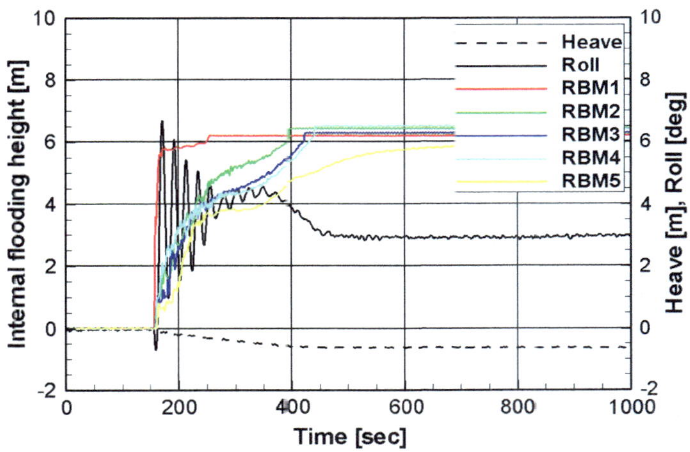

Fig. 13.9 Motions of ship with flooding of DAM1

CP10/11-R2. This is sloshing in low filling ratio. When wave amplitude is 1 m, the internal water motion is small and sloshing doesn't occur. In order to excite sloshing

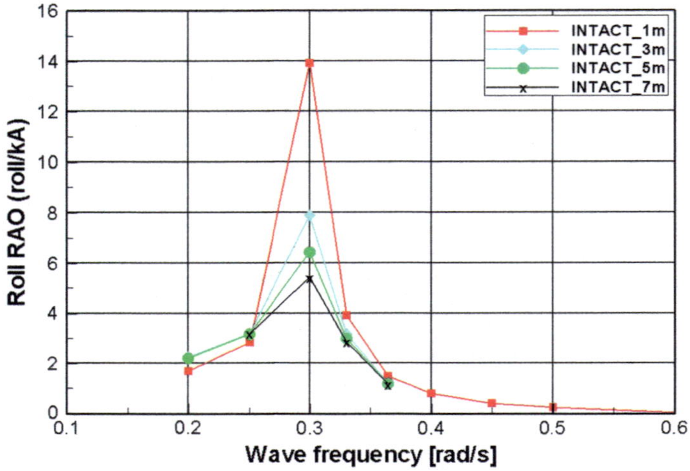

Fig. 13.10 Roll RAO of intact

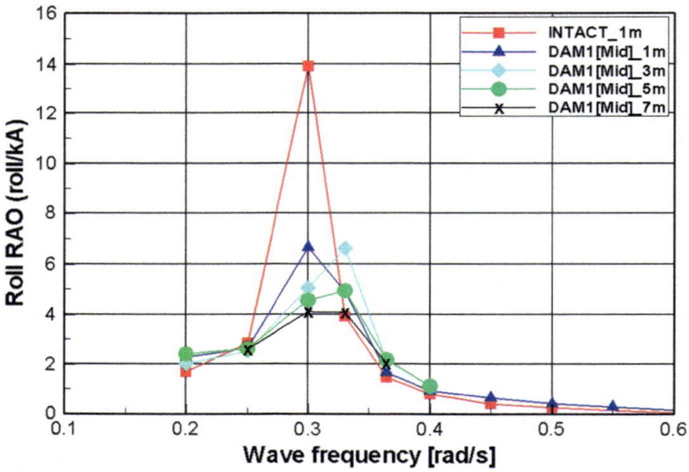

Fig. 13.11 Roll RAO of DAM1

in a considerable level, waves more than 3 m should be incident because the ship heels 4° to starboard. Figure 13.13 shows the effect of opening and in/outflow. The difference between open and close may be explained by the magnitude of sloshing effect. In case of close condition, the effect of sloshing is strong. When inlet is opened, sloshing is weakened because the damping increase due to in/out flow. In case of DAM2, roll RAOs is similar with intact RAOs. Although sloshing in CP17-R2 occurs, there is no significant influence of flooding because of small amount of water.

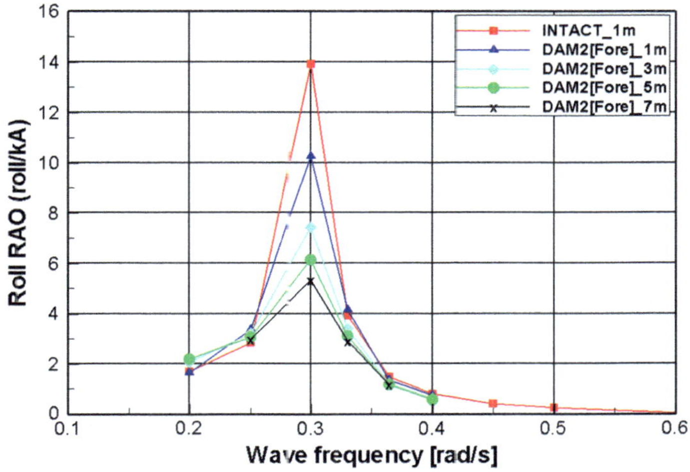

Fig. 13.12 Roll RAO of DAM2

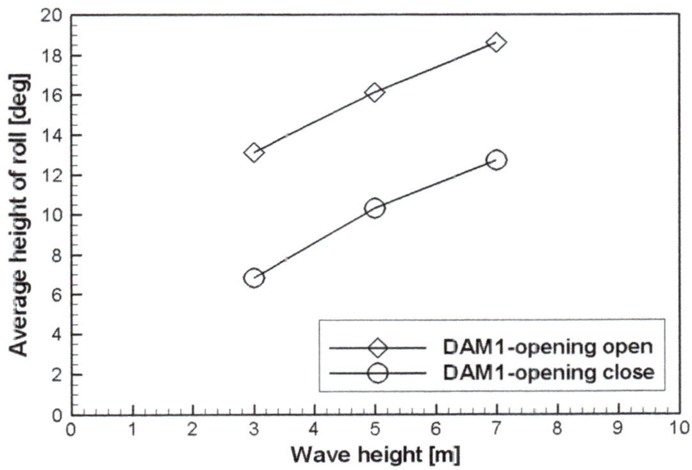

Fig. 13.13 Effect of opening on the roll motion (DAM1, $\omega = 0.33$ rad/s)

Figures 13.14, 13.15, and 13.16 show the roll motion and internal water motion in CP10/11-R2 for the condition of wave frequency 0.30 and 0.33 rad/s. The position of water height measurement (RBM9/10) is at the center in compartments. The initial value of water height is zero in flooded situation. The positive value stands for increasing and negative value decreasing. When wave height is 1 m, flooding water doesn't reach to port side wall and sloshing doesn't occur. But in case of wave height 5 m, flooding water reached port side wall. When wave frequency is 0.33 rad/s, the

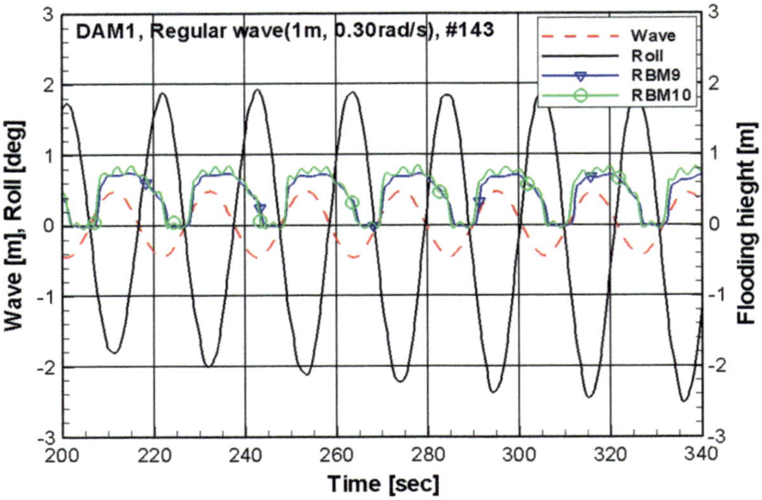

Fig. 13.14 Motion and flooding of DAM1 (Regular wave test, A 1 m, ω 0.3 rad/s)

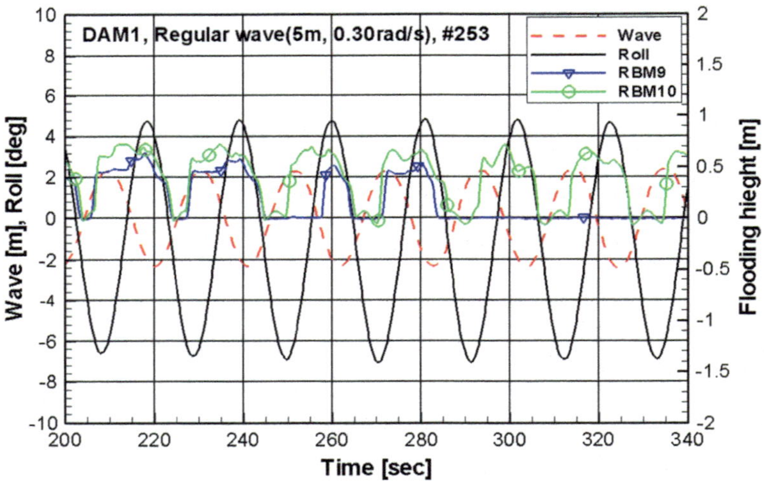

Fig. 13.15 Motion and flooding of DAM1 (Regular wave test, A 5 m, ω 0.3 rad/s)

coupling of sloshing and roll is more strong. Table 13.3 shows the phase of roll and incoming wave due to sloshing (Figs. 13.17 and 13.18).

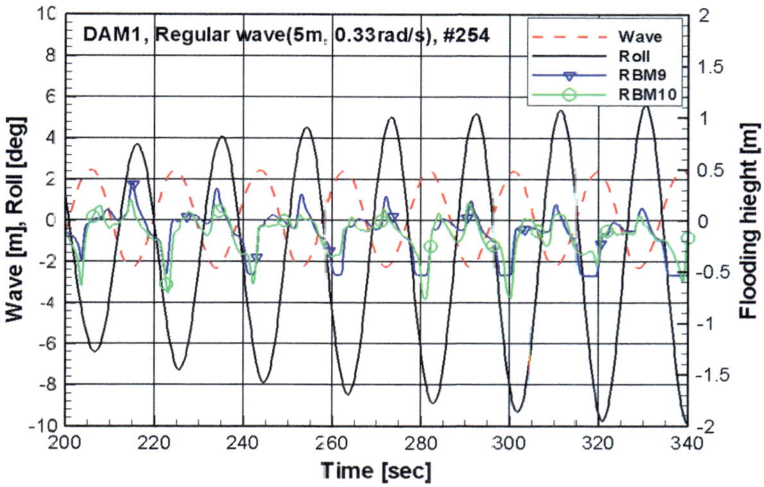

Fig. 13.16 Motion and flooding of DAM1 (Regular wave test, A 5 m, ω 0.33 rad/s)

Table 13.3 Phase of wave and roll

Wave frequency	0.2–0.25 rad/s	**0.3–0.36** rad/s	0.4 rad/s
Intact	90°	0°	−90°
DAM1	90°	**180°**	−90°

Bold indicates that the value is in special range to represent the phase relation of ship motion and flooding

13.4.2 Numerical Simulation Results

Figures 13.19, 13.20, 13.21, 13.22 and 13.23 show the simulation results for ITTC tanker model. The model test results were from MARIN, Surship3 is KRISO results using quasi-dynamic model, and Coupled cal. stands for viscous CFD results. The free decay results of quasi-dynamic model are similar with experiments and CFD. This indicates that the quasi-dynamic model can calculate the dynamics of free surface. Also regular wave test shows the reasonable results. The merit of quasi-dynamic model is very fast calculation in comparison to viscous CFD. The required time is almost same as quasi-static model.

The damaged problem is calculated by the quasi-dynamic model. The results are shown in Figs. 13.24, 13.25, 13.26, 13.27 and 13.28. The transient flooding process is well represented by the model. The flooding heights are compared and the numerical results agree with experiments. But the calculated roll is different. This may be due to the different amount of flooding water in CP10R1S. The increase of numerical result in CP10R1S is almost step jump to saturation limit and CP10R1S is full. But experiment shows CP10R1S is not filled once and is full after 150 s. This lag may be occurred due to the air compression and numerical model limit.

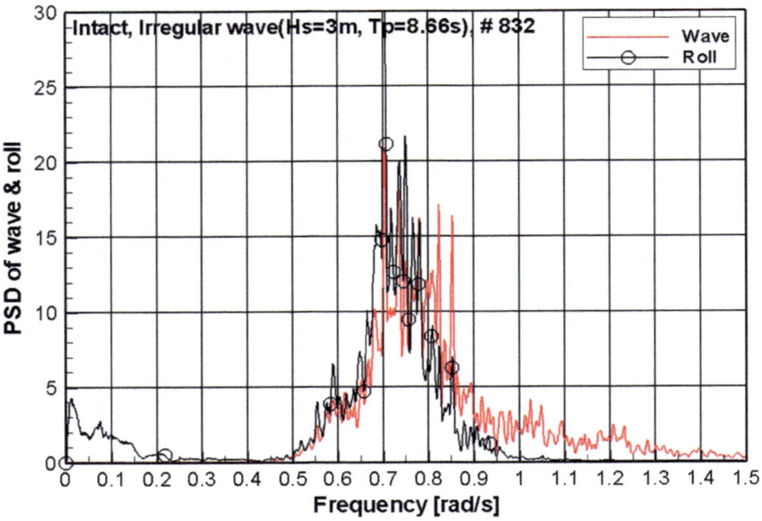

Fig. 13.17 PSD of intact in irregular wave (Hs 3 m)

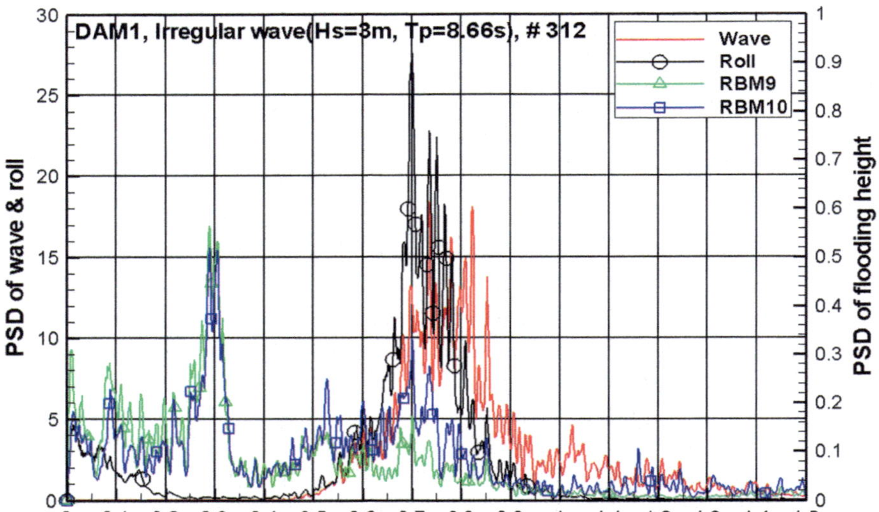

Fig. 13.18 PSD of DAM1 in irregular wave (Hs 3 m)

Figures 13.27 and 13.28 show the regular wave results. Roll RAOs show similar tendency of experiments. The viscous CFD calculation (Figs. 13.29 and 13.30) shows very similar results also.

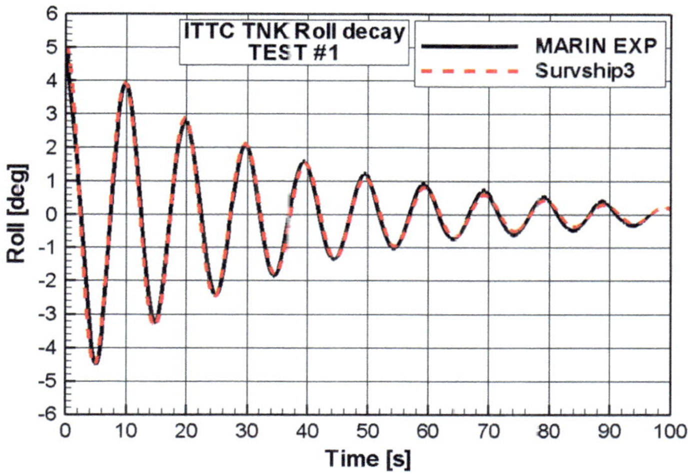

Fig. 13.19 Free decay (h = 0 m)

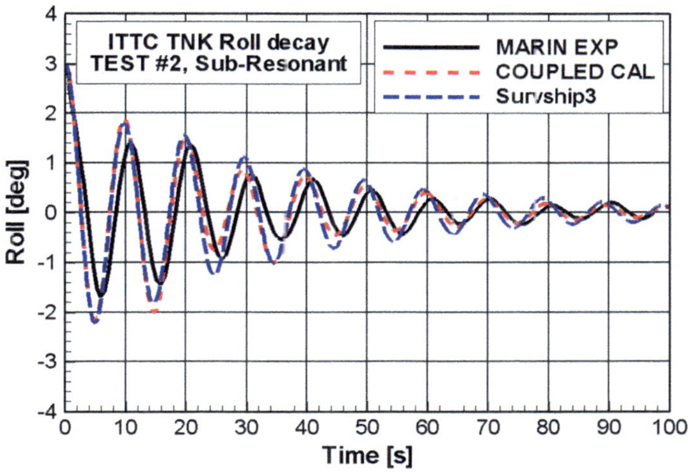

Fig. 13.20 Free decay-sub resonance (h = 3 m)

13.5 Conclusions

The experiments and numerical analysis have been performed for investigating the behaviour of damaged cruiser in waves. The influences of damage configuration, internal water motion, wave height and flow in/out are considered. The transient process and motion behaviour in waves are analyzed. The transient flooding process is measured in each compartment. The effect of flooding on the ship motion appeared in roll motion. Although the amount of water in the upper compartment is small,

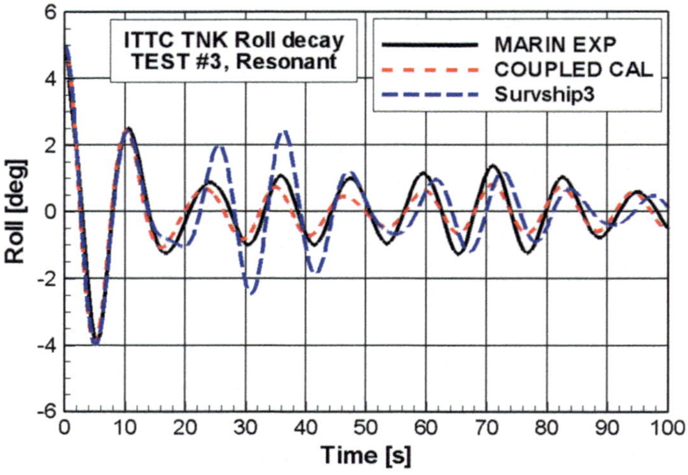

Fig. 13.21 Free decay-resonance (h = 4 m)

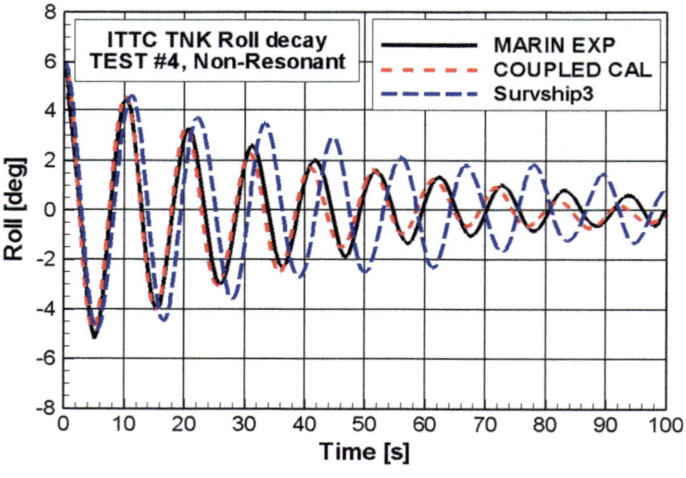

Fig. 13.22 Free decay-non-resonance (h = 16 m)

sloshing is occurred and the effect is significant. Quasi-dynamic model and viscous CFD shows quite good results. The physics and phenomena are more explained and understood by both the experiment and numerical analysis.

Acknowledgments This study is a part of research program, "Development of simulation technologies for dynamics stability of ships" supported by Ministry of Knowledge Economy of Korea and "Development of Technology of Design and Performance Evaluation of the mooring System for Deepsea Offshore Plant (ES2140)" supported by Korea Research Institute of Ships and Ocean Engineering.

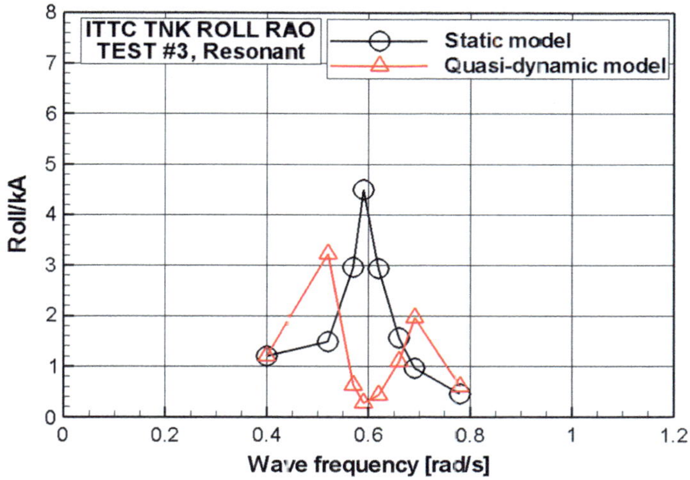

Fig. 13.23 Roll RAO-non-resonance (h = 4 m)

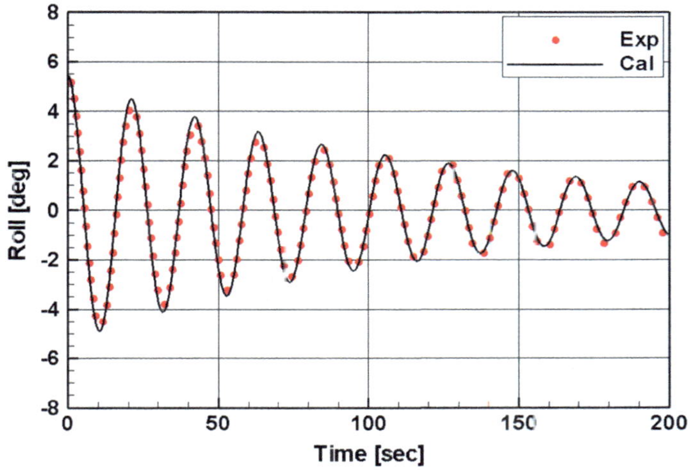

Fig. 13.24 Roll free decay of cruiser

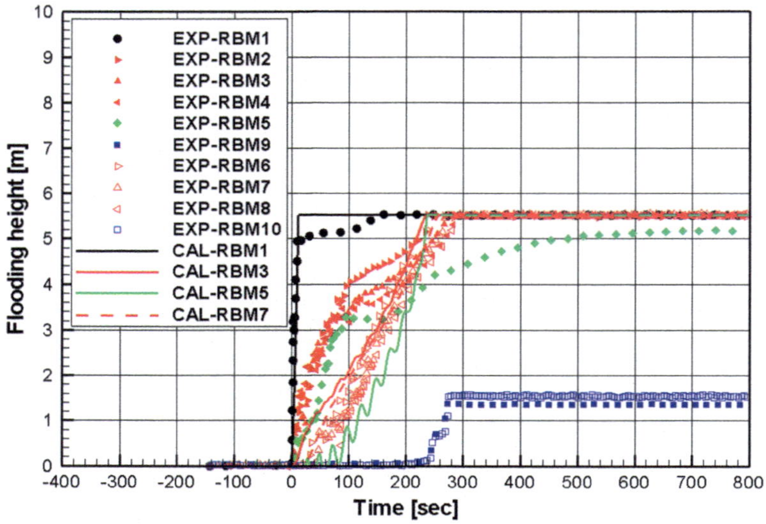

Fig. 13.25 Comparison of flooding heights for DAM1

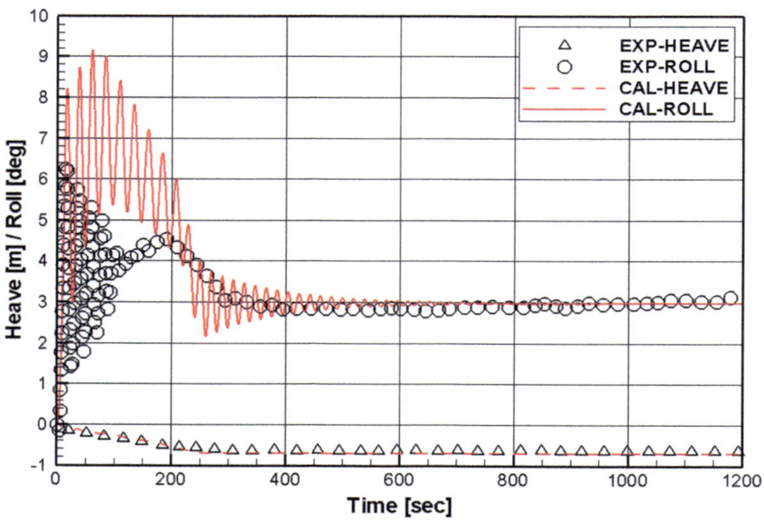

Fig. 13.26 Comparison of motions for DAM1

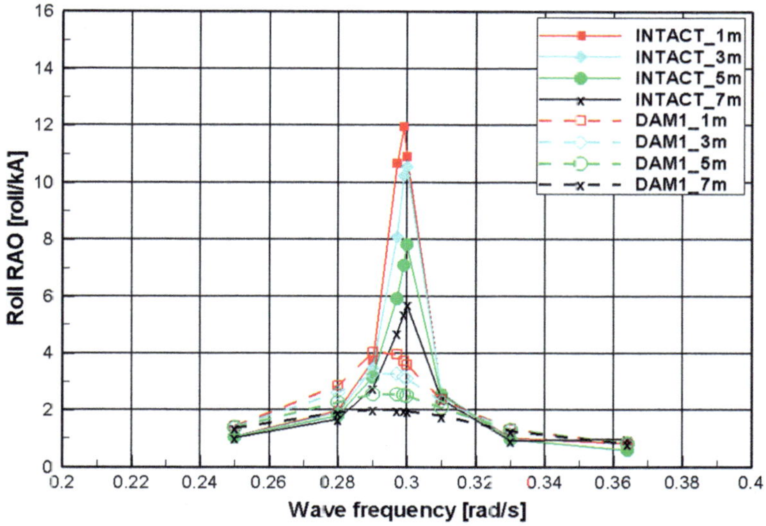

Fig. 13.27 Roll RAO for intact and DAM1—Quasi-dynamic model

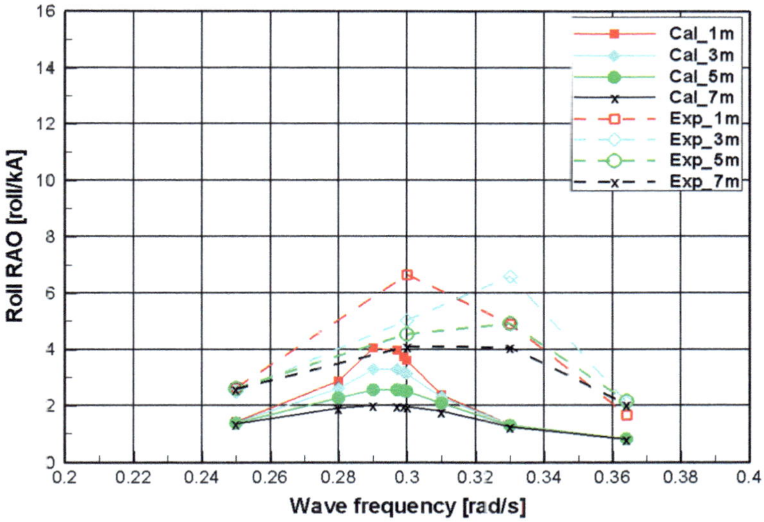

Fig. 13.28 Roll RAO comparison for DAM1—Quasi-dynamic model

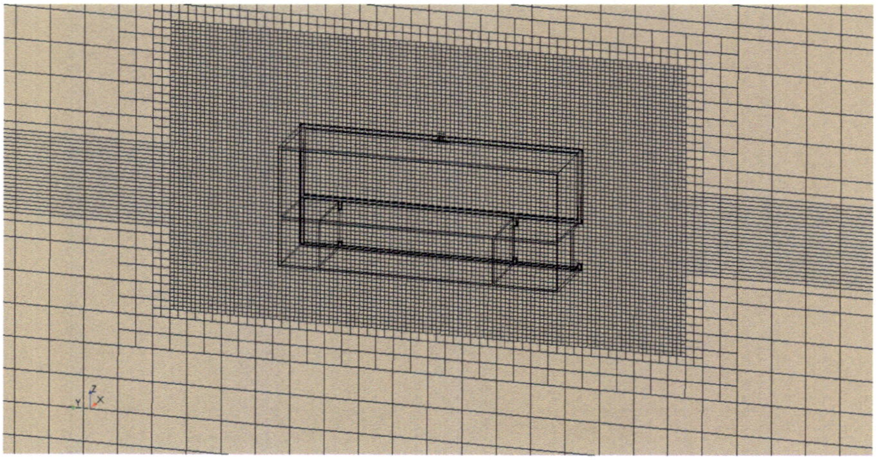

Fig. 13.29 CFD Mesh for viscous simulation

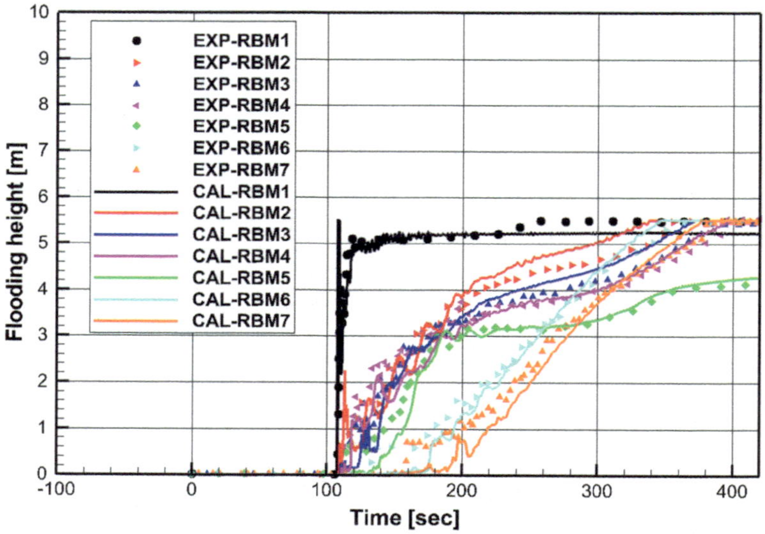

Fig. 13.30 CFD calculation of flooding height for DAM1

References

Cho, S., Sung, H., Nam, B., Hong, S. and Kim, K., 2009, "Experimental Study on Flooding of a Cruiser in Waves", Proc. 10th International Conference on Stability of Ships and Ocean Vehicles, St. Petersburg, pp. 747–753.

Palazzi L. and de Kat J., 2004, "Model Experiments and Simulations of a Damaged Ship with Air Flow Taken Into Account", Marine Technology, Vol. 41, No. 1, pp. 38–44.

Papanikolaou, A. and Spanos, D., 2002, "On the Modelling of Floodwater Dynamics and its effects on Ship Motions", Proc. 6th International Ship Stability Workshop, Webb Institute, New York.

Papanikolaou, A. and Spanos, D., 2005, "24th ITTC Benchmark Study on Numerical Prediction of Damage Ship Stability in Waves Analysis of Final Results", Proc. 8th International Ship Stability Workshop, Turkey.

Ruponen, P. 2007, "Progressive flooding of a damaged passenger ship", Doctor thesis, Helsinki university of technology.

van Walree F. and N. Carette, 2008, "Benchmark study of numerical codes for the prediction of time to flood of ship-phase 2", Proc. 10th International Ship Stability Workshop, Daejeon, Korea.

Chapter 14
Numerical Study of Damaged Ship Motion in Waves

Zhiliang Gao, Qiuxin Gao and Dracos Vassalos

Abstract An integrated numerical method, which couples a seakeeping solver and a Navier-Stokes (NS) solver with the volume of fluid (VOF) model, has been developed to study the behavior of a damage ship in waves. The dynamics of water flooding and sloshing in the compartments were calculated by the NS solver, while the hydrodynamic forces induced by the sea wave on the external hull surface were calculated using the seakeeping solver. To validate its performance, the solver was applied to the flooding problem of a damaged Ro-Ro ferry in regular beam seas. The computed results are satisfactory in comparison with the experimental data.

Keywords Flooding · Damaged ship motion · Floodwater motion
Interactive dynamics

14.1 Introduction

When a ship is damaged in waves, the ship behaviour is not only influenced by the excitation of sea wave but also influenced by the internal liquid loads due to water flooding and sloshing. Simultaneously, the hydrodynamics of flooding and sloshing is also affected by the ship motion. Limited understanding of these interactive dynamics impedes the study on damaged ship stability.

Numerical studies on this intricate dynamic problem have been conducted since the 1990s. Mathematical models presented in the earlier works are normally based on the potential flow theory with a simple model for floodwater motion (Vassalos and Turan 1994; Papanikolaou et al. 2000; Jasionowski 2001; Palazzi and de Kat 2004). The ship hydrodynamic forces due to external wave excitation are calculated using the potential flow method. The viscous effects are treated by semi-empirical approaches. The inflow and outflow of water through the openings is determined by

Z. Gao · Q. Gao · D. Vassalos (✉)
The Ship Stability Research Centre, Department of Naval Architecture, Ocean and Marine Engineering, University of Strathclyde, Glasgow, Scotland, UK
e-mail: d.vassalos@strath.ac.uk

V. L. Belenky et al. (eds.), *Contemporary Ideas on Ship Stability*, Fluid Mechanics and Its Applications 119, https://doi.org/10.1007/978-3-030-00516-0_14

the modified empirical Bernoulli's equation. The non-linear sloshing effect inside the compartment is neglected, and the internal water surface is assumed to be either horizontal or a freely movable plane. Numerical tools based on the above assumptions of floodwater motion can not precisely predict the behaviour of a damaged ship upon flooding, as reported in the ITTC benchmarking study for damaged ship stability (ITTC report 2002). To model the floodwater motion more physically, Santos and Guedes Soares (2008) employed the shallow water equation to calculate the internal water dynamics. The improved model has the ability to address the motion of internal water displaying non-linear behaviour. However, this method can not fully account for the influence of compartment's internal layout on the floodwater motion and is ineffective if the depth of internal water is larger compared to the width of compartment.

Over the past few years, with improvements in the capabilities of high-performance computers, the computational fluid dynamics (CFD) method based on solving the Navier-Stokes (NS) equation with the volume of fluid (VOF) model (Hirt and Nichols 1981) has been increasingly applied to the flooding problem of a damaged ship. By employing this sophisticated method, all of the flow characteristics and parametric effects can be considered in the numerical simulation. Cho et al. (2005) , Nabavi et al. (2006) and Strasser (2010) used the CFD method to investigate the effect of damaged opening geometry, compartment internal layout, turbulent flow or air compression on the flooding process. Gao et al. (2004, 2010a) employed the CFD method to analyse the hydrodynamics of a damaged ship section under forced heave or roll motions.

Numerical simulation of damaged ship flooding in waves solely based on CFD method is time-consuming. On the other hand, the potential flow method is practical and efficient to solve general seakeeping problems of ship. To ensure high fidelity in flooding simulations while reducing the computational cost, it is rational to conceive the idea of coupling the CFD and potential flow methods, i.e., the floodwater dynamics is calculated using the CFD method while the ship hydrodynamics induced by sea wave is predicted with the potential flow method. Woodburn et al. (2002) developed a coupled model based on this idea to assess the survivability of a damaged ship in waves.

In this study, an integrated numerical method, which couples a seakeeping solver based on potential flow method and an NS solver with the VOF model, was developed to study the behavior of a damage ship in waves. To assess its performance, the method was applied to solve the flooding problem of a damaged Ro-Ro ferry in regular beam seas. The computed results were validated against the experimental data.

14.2 Mathematical Model

The ship is considered as a rigid body with six degrees of freedom (6-DOF), and its motion is governed by the following linear and angular momentum equations described in the body-fixed coordinate system:

$$m(\dot{\boldsymbol{u}}_c + \omega \times \boldsymbol{u}_c) = \boldsymbol{F} \tag{14.1}$$

$$\boldsymbol{J}_c \cdot \dot{\omega} + \omega \times \boldsymbol{J}_c \cdot \omega = \boldsymbol{M}_c \tag{14.2}$$

where m is the mass of the ship; C refers to the centre of m; $\boldsymbol{u}C$ is the velocity vector of C; $\boldsymbol{F}$ is the resultant vector of external forces acting on the ship; $\boldsymbol{J}C$ is the tensor of inertia moments of the ship with respect to C; $\boldsymbol{\omega}$ is the angular velocity vector of the ship; and $\boldsymbol{M}C$ is the resultant vector of external moments acting on the ship with respect to C.

Within the framework of potential flow theory, the components of external forces and moments can be generalized as follows:

$$\begin{aligned} F_i = {} & (F_{\mathrm{FK}})_i + (F_{\mathrm{D}})_i + (F_{\mathrm{R}})_i + (F_{\mathrm{B}})_i \\ & + (F_{\mathrm{G}})_i + (F_{\mathrm{W}})_i, \quad i = 1, 2, \ldots, 6. \end{aligned} \tag{14.3}$$

where i denotes the components of the external forces or moments (moment understood for i=4, 5, 6); FFK is the Froude-Krylov force; FD is the diffraction force; FR is the radiation force; FB is the buoyancy force; FG is the gravitational force; and FW is liquid load due to the motion of floodwater inside compartments.

14.3 Numerical Method

14.3.1 Seakeeping Solver

An in-house seakeeping solver PRCTEUS3 (Jasionowski 2001) is used to calculate the hydrostatic and hydrodynamic forces induced by the sea wave on the external hull surface. The hydrostatic and Froude-Krylov forces are evaluated by integrating the pressure over the instantaneous wetted surface of the ship. The diffraction and radiation forces are first derived from the linear potential flow theory in frequency domain, applying strip theory, and then transformed into time domain applying convolution and spectral techniques, respectively. Viscous effects on the roll motion are treated through a semi-empirical approach, known in roll damping research as the "Ikeda" method Ikeda et al. (1978), Components of roll damping of ship at forward speed, Trans. JSNA 143.

14.3.2 NS Solver

A finite-volume-discretisation based NS solver (Gao et al. 2010b) , which is developed to solve the problems of incompressible two-phase flow, is employed to calculate the dynamics of water flooding and sloshing. Figure 14.1 shows the computational domain, which includes the ship's floodable compartments and an external flow

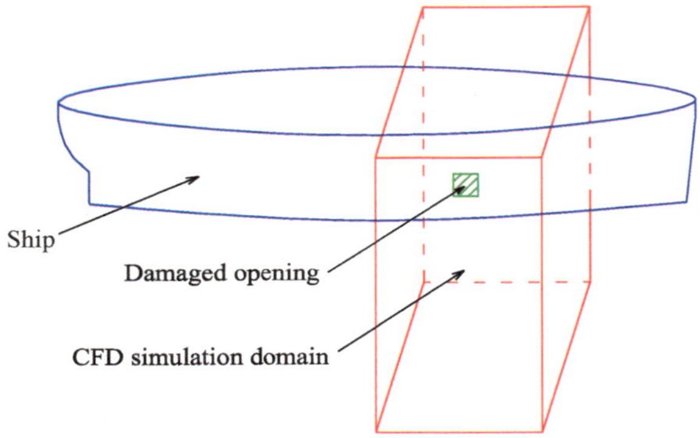

Fig. 14.1 Sketch of the CFD simulation domain

region around the damaged section of the ship. A VOF family algorithm, CICSAM, is used to capture the free surface. The well-known SIMPLE algorithm is employed for pressure-velocity coupling. The ship motion is tackled with the dynamic mesh technique. Turbulence modelling is omitted.

14.3.3 Solution Procedure

The entire flooding problem is solved by a newly developed integrated method that couples the aforementioned seakeeping and NS solvers. The overall solution procedure is shown in Fig. 14.2.

14.4 Validation Test

14.4.1 The Test Ship

A Ro-Ro ferry, known as PRR1 in the literature (ITTC report 2002), was adopted herein for the validation test. The main particulars of the ferry are given in Table 14.1. Figure 14.3 shows its general arrangement, in which the parts depicting shadow denote the floodable compartments.

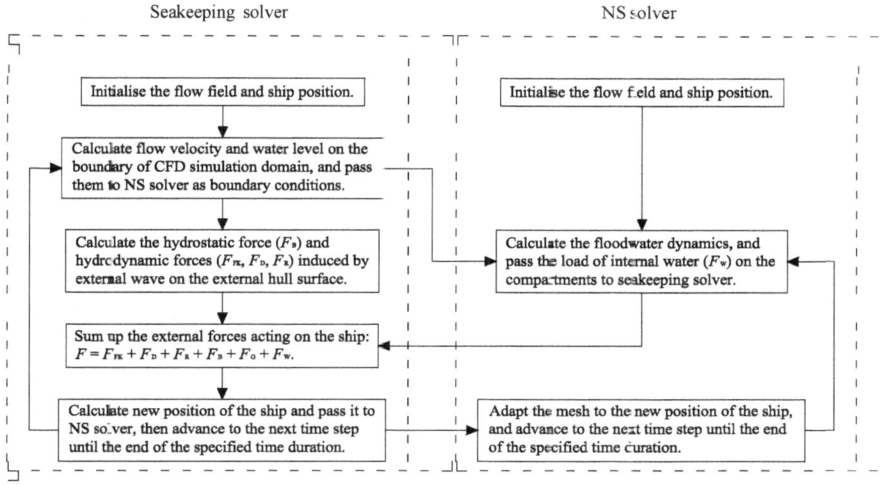

Fig. 14.2 Outline of solution procedure of the integrated method

Table 14.1 Main particulars of PRR1

Length between perpendiculars (*L*pp)	170.00 m
Breath (*B*)	27.80 m
Draft (*T*)	6.25 m
Depth to car deck (*D*cd)	9.00 m
Damaged length (*L*dam)	8.10 m
Centre of gravity above base (KG)	12.892 m
Metacentric height (GM)	2.63 m
Displacement (Δ)	17,301.7 t

14.4.2 Motion of Intact Ship

Before applying the integrated method to the damaged ship flooding, we first tested the ability of PROTEUS3 to predict the motion of an intact ship in waves. The cases of PRR1 in regular beam seas with wave heights (Hw) of 1.2 and 2.4 m were tested, respectively. Figure 14.4 shows the comparisons of roll response amplitude operators (RAO) obtained by PROTEUS3 and model test (ITTC report 2002). The computed RAOs are over-predicted at wave frequencies close to the natural roll frequency (approximately 0.49 rad/s) of the ship. In the range of other frequencies, good agreement between the numerical and experimental results is observed.

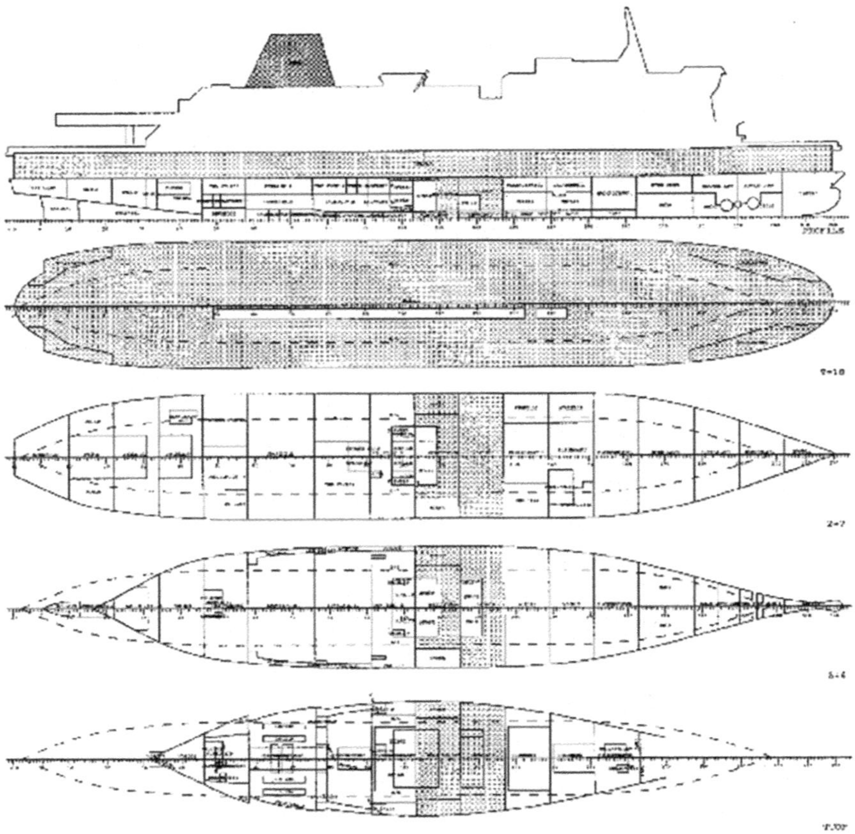

Fig. 14.3 General arrangement of PRR1 [from ITTC report (2002)]

14.4.3 Motion of Damaged Ship

As reported in the literature (ITTC report 2002), the behaviour of PRR1 in damaged condition has been extensively studied by various research groups. However, because all numerical tools developed by the participants cannot properly address the motion of floodwater inside the compartments, neither the peak response frequency nor its magnitude in the computations agrees with the experimental ones, as shown in Fig. 14.5. Thus, the same case, i.e., PRR1 in regular beam seas in damaged condition, was used here for the validation of our integrated method. A wave height of 1.2 m was selected in the test. Figure 14.6 shows the computational domain for the CFD simulation. The total number of mesh elements was 245,048. Only 4-DOF of the ship (sway, heave, roll and pitch) was considered in the numerical simulation. For each wave frequency, the simulation ran up till the ship motion became stable. It roughly

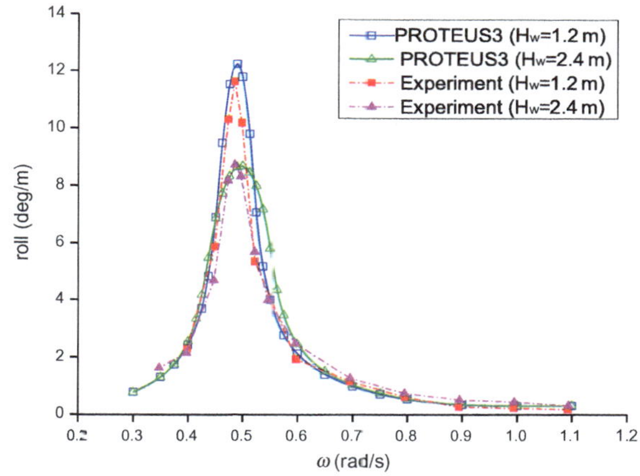

Fig. 14.4 Comparison of roll RAO of PRR1 in intact condition

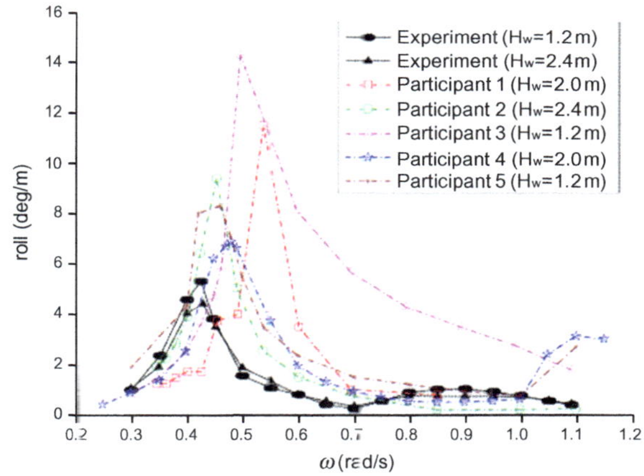

Fig. 14.5 Results of the ITTC benchmarking study for PRR1 in damaged condition

took 69 CPU hours to complete a 200-s simulation on a dual-cores (Intel Core2 @ 3.0 GHz) personal computer.

A snapshot of internal water motion is shown in Fig. 14.7, at the instant corresponding to the maximum heeling of the ship. Figure 14.8 shows the comparison of roll RAOs obtained by the integrated method and model test. The frequencies of peak response in the computation and experiment are approximately 0.4 and 0.42 rad/s, respectively, both of which shift moderately from 1.49 rad/s in the case of intact ship. On the other hand, the peak response is weakened significantly due to the presence of internal water which increases the roll damping, and its magnitude is over-predicted by the present method. For the wave frequencies which are less than 0.7 rad/s, the change trend of computed RAOs is consistent with its experimental counterpart. As the wave frequency increases further, a second peak of RAO is observed in the experiment, while it did not appear in the computation. Such discrepancy may be attributed

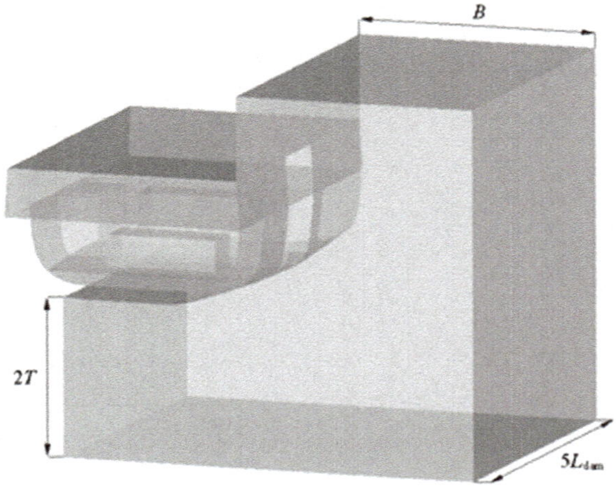

Fig. 14.6 CFD simulation domain for the case of PRR1 in damaged condition

Fig. 14.7 Snapshot of floodwater motion inside the compartments (Hw = 1.2 m, ω = 0.4 rad/s)

to the grid resolution which is not fine enough to accurately capture the non- linear free surface motion inside the compartments. Further analysis on the disagreement between the numerical and experimental results is on-going.

14.5 Conclusion

An integrated numerical method that couples a seakeeping solver and an NS solver was developed to study the behavior of a damage ship in waves. Preliminary results shown in the benchmarking study are encouraging. The shift of peak response frequency and increase of damping due to the presence of internal water is predicted reasonably well in comparison with the experimental data. More validation tests will be carried out in the next step.

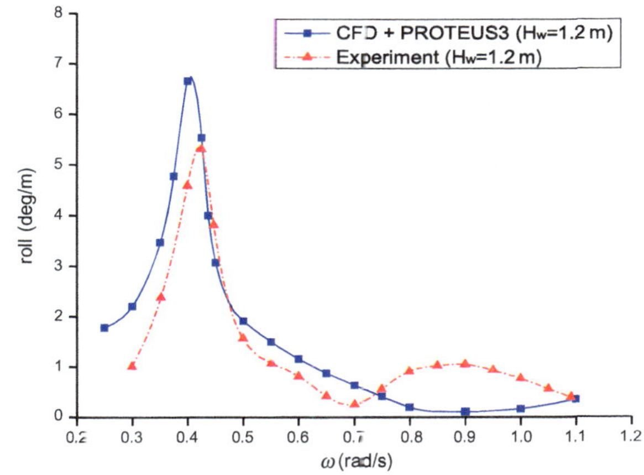

Fig. 14.8 Comparison of roll RAO of PRR1 in damaged condition

References

Cho, S. K., Hong, S. Y., Kim, Y. H., Lee, K. J., (2005)."Investigation of dynamic characteristics of the flooding water of the damaged compartment of an ITTC RORO passenger ship", Proc of the 8th Intl Ship Stability Workshop, Istanbul.

Gao, Q., Gao, Z., Vassalos, D., (2010a). "Numerical study of damage ship flooding", Proc. of the 4th Intl Maritime Conf. on Design for Safety and 3rd Workshop on Risk-Based Approaches in the Marine Industries, Trieste.

Gao, Q., Kara, F., Shigunov, V., Vassalos, D., (2004). "Numerical simulation of damage ship flooding", Proc of the 7th Numerical Towing Tank Symp, Hamburg.

Gao, Z., Vassalos, D., Gao, Q., (2010b). "Numerical simulation of water flooding into a damaged vessel's compartment by the volume of fluid method", Ocean Engineering, 37 (16), 1428–1442.

Hirt, C. W., Nichols, B. D., (1981). "Volume of fluid method for the dynamics of free boundaries", J. of Computational Physics, 39 (1), 201–225.

Jasionowski, A., (2001). An integrated approach to damage ship survivability assessment, Ph.D. Thesis, University of Strathclyde.

Nabavi, Y., Calisal, S. M., Akinturk, A., Klaptocz, V., (2006). "A computational investigation of the three dimensional geometric parameters' effects on the discharge rate of a ship opening", Proc of the 9th Intl Conf. on Stability of Ships and Ocean Vehicles (STAB2006), Rio de Janeiro, pp. 617–624.

Palazzi, L., de Kat, J. O., (2004). "Model experiments and simulations of a damaged ship with air flow taken into account", Marine Technology, 41 (1), 38–44.

Papanikolaou, A., Zaraphonitis, G., Spanos, D., Boulougouris, E., Eliopoulou, E., (2000). "Investigation into the capsizing of damaged Ro- Ro passenger ships in waves", Proc. of the 7th Intl Conf on Stability of Ships and Ocean Vehicles, STAB2000, Launceston, Tasmania, pp. 351–362.

Santos, T. A., Guedes Soares, C., (2008). "Study of damaged ship motions taking into account floodwater dynamics", J of Marine Science and Technology, 13 (3), 291–307.

Strasser, C., (2010). Simulation of progressive flooding of damaged ship by CFD, Ph.D. Thesis, Universities of Glasgow and Strathclyde.

The Specialist Committee on Prediction of Extreme Ship Motions and Capsizing, (2002). "Final report and recommendations to the 23rd ITTC", Proc. of the 23rd Intl Towing Tank Conf, Venice, pp. 633–649.

Vassalos, D., Turan, O., (1994). "A realistic approach to assessing the damage survivability of passenger ships", Trans SNAME, 102, 367–394.
Woodburn, P., Gallagher, P., Letizia, L., (2002). "Fundamentals of damage ship survivability", Trans RINA 144, 143–163.

Chapter 15
3D GPU SPH Analysis of Coupled Sloshing and Roll Motion

Luis Pérez Rojas and Jose L. Cercos Pita

Abstract The coupled roll motion response of a single degree of freedom system to which a passive anti-roll tank has been attached is considered and its performance studied numerically with a 3D GPU SPH code, aimed at simulating the sloshing flows occurring inside the tank. Results are compared with experiments from Bulian et al. (2010), in which 2D simulations were also presented. Progress achieved thereafter is documented, mainly consisting in the implementation of a parallelized solver that runs on a GPU card, which allows the simulation of low resolution 3D and high resolution 2D computations.

Keywords SPH · Anti-roll tanks · Single degree of freedom systems · SDOF GPU · Sloshing

15.1 Introduction

Ship motions are affected by sloshing flows occurring inside her liquid cargo tanks, as has already been documented both numerically and experimentally in the literature. Nam et al. (2009) for instance, use a finite difference method to model the flow inside the tanks and a linear sea-keeping code for the vessel motions simulation. However, a multi-modal approach to sloshing presents large problems in resonance condition due to the lack of an intrinsic dissipation mechanism. Mesh based simulation techniques often struggle with sloshing flows due to the tremendous fragmentation that takes place at the free surface in such condition and the large numerical diffusion occurring at the free surface for long simulations. Due to this, meshless methods like SPH (Smoothed Particle Hydrodynamics) become an attractive option to simulate these flows, albeit to tackle 6 degrees of freedom, a full 3D SPH solver is necessary.

L. Pérez Rojas (✉) · J. L. Cercos Pita
CEHINAV, ETSIN, UPM, Madrid, Spain
e-mail: luis.perezrojas@upm.es

J. L. Cercos Pita
e-mail: jl.cercos@upm.es

V. L. Belenky et al. (eds.), *Contemporary Ideas on Ship Stability*, Fluid Mechanics and Its Applications 119, https://doi.org/10.1007/978-3-030-00516-0_15

The onset of sloshing flows inside tanks has been used in order to dampen the roll motion. This anti-roll tank concept is equivalent to the tuned sloshing damper or tuned liquid damper (TLD) concept in Civil Engineering. In the existing literature, two approaches can be found to characterise the behaviour of a TLD exposed to external excitations. The first one consists in imposing a periodic motion on the TLD by using a shaking table or a forced roll motion device and measuring the response in terms of lateral force or moment Souto-Iglesias et al. (2006), Tait et al. (2005). The other approach, more complex, and the one the present paper deals with, is to consider the motion response of the coupled system and tank-structure, subjected to external excitation in terms of force, moment or even induced motion to the tank interfaced with an elastic structure. With this second approach the damping characteristics, inertia and restoring terms are also relevant in the dynamic analysis. Realistic motions of the structure that are the outcome of this process can be compared with design limit states Attari and Rofooei (2008), Bulian et al. (2010).

In this paper, the SDOF system presented by Bulian et al. (2010), to which a partially filled tank has been attached, is considered. The roll motion is modelled by means of an "exact" (from the dynamics point of view) 1-DOF approach. The moment created by the fluid with respect to the rolling axis is simulated and results for different roll angles are compared. It is important to underline that the present experimental/numerical approach removes the difficulties usually encountered in a correct modelling of the actual ship roll motion. Indeed, when numerical simulations are compared with experimental tests carried out on ship models excited by waves, it is almost never completely clear where the real source of discrepancy between experimental results and numerical prediction comes from, i.e. whether the reason is to be sought in the modelling of sea-ship interaction or in the modelling of sea-tank interaction. In the present tests, being the dynamics of the mechanical system practically known "exactly" (at least at a reasonable level of accuracy, with some question mark on damping at small rolling angles), any significant discrepancy is likely to be sought in the simulation method.

The numerical simulations have been performed using the SPH particle method. SPH has been successfully applied to shallow depth sloshing problems with periodic oscillation in sway Landrini et al. (2003) and roll Souto-Iglesias et al. (2006) motions. It had also been applied to the coupled motion problem in Bulian et al. (2010), showing promising results. Nevertheless, in Bulian et al. (2010), the computations were carried out in 2D whilst 3D computations are presented herein. A significant progress in this regard has been recently achieved through the use of graphical cards (GPUs) that can perform massive parallel SPH computations Herault et al. (2010) Rey-Villaverde et al. (2011). The main advantage of these cards is that they are extremely cheap and incorporate of the order of 500 processors each, substantially speeding up SPH computations. It would be extremely interesting to obtain an accurate description of the effect of the flow inside the tank to the system using a GPU based SPH solver, since SPH is, in principle, able to deal with highly distorted free surface flows. Checking whether this is feasible is the main objective of the present work.

In this work, the experiments performed are first introduced. Secondly, the GPU based SPH implementation is described. Thirdly, the simulation results are presented and compared with the experimental ones. Finally, some conclusions are drawn and future work threads hinted.

15.2 Experiments and Mechanical Model

15.2.1 Experiments

The experiments were conducted with the tank testing device of the CEHINAV group (see Souto-Iglesias et al. (2011) for a detailed description). The standard forced motion configuration of the device, used regularly in the design of anti-roll tanks, was modified by disconnecting the driving electrical engine from the tank holding structure, in order to allow a free motion of the tank. The tank is rectangular, 900 × 508 × 62 mm. A horizontal linear guide consisting of a controllable electrical engine that laterally moves a weight with a specified motion has been attached. This weight is intended to generate a heeling moment in order to reproduce the wave action on the roll motion. The water depth (H) whose first sloshing frequency matches the first own frequency of the structural system ω_0 (3.26 rad/s) has been chosen for the experiments and the same frequency has been chosen for the weight movement. This resonance condition is the hardest to tackle since in this condition the system accumulates energy in every cycle which then has to be dissipated by the fluid through internal dissipation and breaking in order to reach a steady state condition. The amplitude of the weight motion has been chosen as A = 100 mm. This combination of frequency and amplitude were the ones analyzed in greater detail in Bulian et al. (2010). Three different liquids have been used, namely: water, sunflower oil and glycerin, covering 3 different orders of Reynolds numbers, as it can be seen in Table 15.1. In the case of water, the dissipation comes from breaking and from internal dissipation while the larger viscosity of oil and glycerin does not allow such breaking to take place thus inducing much smaller dampening effects than water. The experiments have been considered relevant by the SPHERIC ERCOFTAC Interest group on SPH as a benchmark for validation and further information about the experimental data can be found in the SPHERIC site and in Bulian et al. (2010).

Table 15.1 Reynolds number for the three test cases

	Re
Water	97,546
Oil	1748
Glycerine	118

15.2.2 Analytical Model of the System

An analytical model of the SDOF structural system used in the experiments is needed in order to incorporate it into the structural part of the SPH code. This model was obtained by deducing the coefficients after carefully analysing a set of tests with the empty tank and deriving a data-consistent damping term model. The analytical model used to describe the behaviour of the system is, in general, as follows:

$$\left[I_0 + m\,\xi_m^2(t)\right]\ddot{\phi} + 2\,m\,\dot{\xi_m}(t)\,\dot{\phi} - g\,S_G\,\sin(\phi) + m\,g\,\xi_m(t)\,\cos(\phi) = Q_{\text{damp}}(t) + Q_{\text{fluid}}(t), \quad (15.1)$$

$$Q_{\text{damp}}(t) = -K_{df}\,sign(\dot{\phi}) - B_\phi\,\dot{\phi} \quad (15.2)$$

where:

- ϕ [rad] is the roll angle.
- g [m/s^2] is the gravitational acceleration.
- I_0 [kg m^2] is the polar moment of inertia of the rigid system.
- m [kg] is the mass of the moving weight.
- $\xi_m(t)$ [m] is the instantaneous (imposed) position of the excitation weight along the linear guide (tank-fixed reference system).
- $\dot{\xi_m}(t)$[m/s] time derivative of $\xi_m(t)$ [m].
- $S_G = M_R\,\eta_G$ [kg m] is the static moment of the rigid system with respect to the rotation axis.
- M_R [kg] is the total mass of the rigid system.
- η_G [m] is the (signed) distance of the centre of gravity of the rigid system with respect to the rotation axis (tank-fixed reference system).
- $Q_{\text{damp}}(t) = -K_{df}\,sign(\dot{\phi}) - B_\phi\,\dot{\phi}$ [N m] is the assumed form of roll damping moment with a:
- A dry friction term $-K_{df}\,sign(\dot{\phi})$ with K_{df} [N m] being the dry friction coefficient.
- A linear damping term $-B_\phi\,\dot{\phi}$ with:
- B_ϕ [N m/(rad/s)] being the linear damping coefficient.
- $Q_{\text{fluid}}(t)$ [N m] is the fluid moment.

By using a set of inclining as well as decay tests, the unknown parameters have been experimentally determined, including the natural frequency of the rigid system ω_0. The values of these parameters can be found in Table 15.2.

15.2.3 Dissipation Indicator

The rotating mechanical system stores energy in kinetic and potential forms. There is a transfer between these forms of energy during each rotation cycle. This accumulation is significantly reduced with the fluid action, which dissipates part of that

Table 15.2 Mechanical parameters of the rigid system

Quantity	Units	Value
S_G	kg m	−29.2
I_0	kg m^2	26.9
K_{df}	N m	0.540
B_ϕ	N m/(rad/s)	0.326
ω_0	rad/s	3.26

energy in every cycle. Since the potential energy is proportional to the square of the rotation angle amplitude, a reasonable indicator of the TLD dampening performance is defined as the ratio of the empty tank motion amplitude and the partially filled tank motion amplitude, as in Bulian et al. (2010).

15.3 GPU-SPH Formulation

15.3.1 General

The recent implantation of graphic process units (GPUs) in scientific computation has drastically increased the processing speed of several applications. Not many years ago, parallel computing was restricted to super-computing centres or large and expensive clusters. Nowadays, thanks to the arrival of GPU multicore processors originally designed for graphic processing, massively parallel processing is becoming increasingly more accessible and cheaper for the developer.

Increasing the efficiency of the algorithms involved not only depends on the specific hardware improvements, but also on the new approaches aimed at maximizing available resources and minimizing costs. In the case of GPU processors, it is necessary to note that the computational power lies in its specialization. The GPU multicore architecture is designed for highly efficient graphic processing. To explore the degree of adaptability of the GPU technology to certain algorithms which simulate large particle systems, first we analyse which steps of the SPH code are more suitable to be parallelized, as well as different strategies for the parallelization of the main subroutines. This requires the evaluation of any problematic aspect and the consequent speed-up and scalability obtained.

Since the SPH methodology generally uses an explicit resolution scheme, their algorithms are easily parallelized to its minimum unit (particles, cells). However, there are certain subroutines for which GPU parallelization is not immediate; in those cases, different strategies can be implemented focusing on obtaining the maximum increase of the CPU versatility. Although it is always possible to use the CPU in those subroutines whose parallelization is problematic, this should be avoided due to the relatively high latencies associated with data transfers between CPU and GPU and the consequent reductions in computational performance.

For that reason, in order to maximize the computational performance, recent advances include the hybridization of both technologies - so called APU (Accelerated Processor Units) - is becoming a very promising tendency due to the significant reduction of the overhead due to the memory transfers.

15.3.2 GP-GPU: OPEN CL Implementation

Traditionally, the GP-GPU has been developed using special languages (shaders) as GLSL, CG, or HLSL and incorporated as extensions of the OpenGL and Direct3D APIs. Learning GP-GPU programming not only required graphic programming as a prerequisite, but also required a considerable expertise in APIs (Application Program Interface) and graphic languages. Currently, developing GP-GPU applications is done with the help of programming environments specifically designed to develop this type of codes such as the CUDA or OpenCL platforms. Moreover, the latter can be used to completely shortcut the limitations introduced by CUDA codes due mainly to the fact of being a proprietary framework that only works with NVidia, because of its specific design to unify the acceleration of codes when heterogeneous platforms are considered.

Regarding the GPU architecture, a scheme of the processors distribution in the graphic cards used, is shown in Fig. 15.1. Basically, the GPU is distributed in a set of multiprocessors. Each multiprocessor typically hosts 8 scalar processors with NVidia architectures. From the viewpoint of parallel codes, the first important concept to consider is related to the kernel functions (analogous to the shaders in the graphic computing context). When a kernel is called, it launches a vector of N threads where each thread is executed in a different processor. In turn, every thread executes the instructions found in the kernel function sequentially. Once the kernel function is called, the N threads perform the instructions in parallel. Threads are grouped into blocks of threads. The threads associated in a specific block are executed in a common multiprocessor (8 single processors) where they can share variables and make use of the shared memory space associated to each multiprocessor.

When a kernel is called, blocks of threads are listed and distributed in the available multiprocessors. The threads of a block are executed simultaneously on a single multiprocessor, while multiple blocks could be executed concurrently in one multiprocessor. Once all the threads of a block have been processed, new blocks are launched in the vacant multiprocessors. One multiprocessor can concurrently execute hundreds of threads. To efficiently manage the large number of threads, it uses a special architecture called SIMT (Single Instruction, Multiple Thread). The total amount of threads is divided into 32 unit packs called warps. In the SIMT architecture, the threads of a warp execute the same kernel instruction at the same time. In

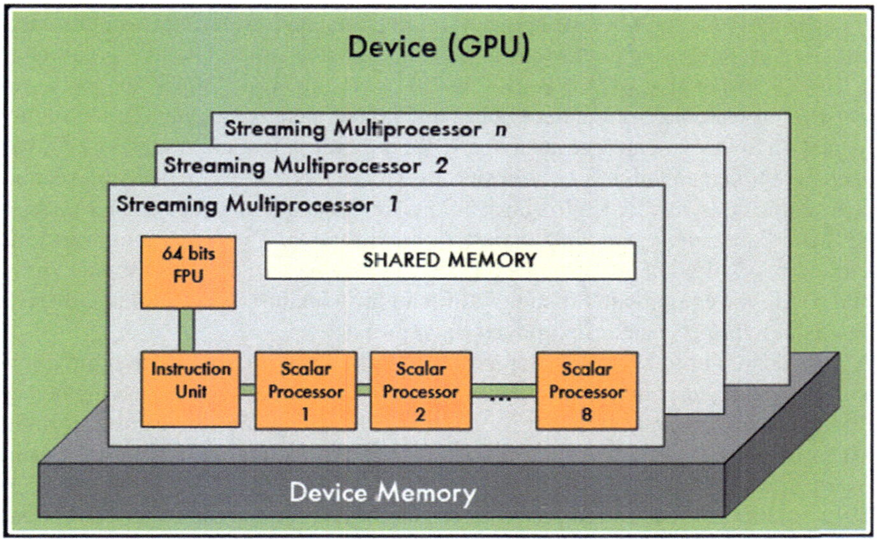

Fig. 15.1 Basic architecture of a GPU card

order to make a parallel implementation, achieve maximum transfer rates and avoid bottlenecks, it is crucial to understand the GPU memory hierarchy in order to manage different memory spaces.

15.3.3 SPH Parallel Code

Smoothed Particle Hydrodynamics (SPH) is a Lagrangian method, with no computational mesh that has been widely employed to study free-surface flows Monaghan (1994). Recent comprehensive reviews can be found in Monaghan (2012) and Violeau and Rogers (2016).

Due to the large number of interactions for each particle at each time step, when SPH codes are computed on a single CPU they usually require a large computational time. When millions of particles are required to accurately compute a physical process, only a parallel computing setup can guarantee efficient computational times. Due to the inexorable development of the market of video games and multimedia, the GPU power and streaming multi-processor technology has increased much faster than CPUs. Thus, GPUs now appear as an accessible alternative to accelerate SPH models using a powerful parallel programming model where the graphics cards are used as the execution device. Their performance can be compared with large cluster machines. A huge advantage is the price and the easy maintenance GPUs require in comparison with large multi-core systems. The capability of GPUs to handle SPH was shown by the pioneer work of Harada et al. (2007).

Particularly for the SPH method, several implementations have proven the capabilities of graphic oriented devices to perform massive computations Herault et al. (2010), Crespo et al. (2015), Cercos Pita (2015). In this work, AQUAgpusph, a free 3D SPH software licensed under GPLv3 and accelerated using OpenCL, developed by the CEHINAV research group at the UPM in Madrid, is used Cercos Pita (2015). OpenCL implementation is chosen due to its great flexibility. All major vendors have adopted this standard, allowing a unique implementation to be used for massive parallel computations in a wide variety of architectures. Therefore, more powerful devices from different vendors can be selected, having a remarkable impact on the final cost of the equipment. For more information about the software, we kindly refer the reader to http://canal.etsin.upm.es/aquagpusph.

A weakly compressible SPH implementation has been chosen for the simulations, using a Wendland kernel, a second order Leap-Frog time integration scheme and Monaghan, Cleary and Gingold's Monaghan and Gingold (1983) viscosity formula (MCG formula from now on). The no-slip boundary condition is simulated using several rows of fixed fluid particles attached to the solid boundary Macia et al. (2011). Although we kindly refer the reader to Rey-Villaverde et al. (2011) and Souto-Iglesias et al. (2006) for further details on the GPU-SPH implementation used in the present work, lets point out some details about its dissipation mechanisms.

In Colagrossi et al. (2011) it was demonstrated that in the continuum, MCG viscosity formula provides the correct viscous dissipation for free surface flows. This viscosity formulation was originally devised as an artificial viscosity but was later shown to be a consistent Newtonian viscous term for incompressible flows Hu and Adams (2006). When comparing the kinematic viscosity with the artificial viscosity, the following relation is obtained in 2D and an equivalent one with a denominator of 15 in 3D:

$$\nu = \frac{1}{8}\alpha h c_s \tag{15.3}$$

The α factor should be no less than 0.01 if time integration is expected to remain stable Monaghan (1994), with c_s being 10 times the maximum expected velocity and h the smoothing length. This link sets the minimum value of the kinematic viscosity for a certain resolution (keep in mind that in practical terms the smoothing length h is proportional to the typical particle distance Δx). Since the computational effort was limited to 2 days per case (around 3×10^5 particles), it was not feasible to perform a full resolution 3D computation for the water and oil cases.

15.4 Results

15.4.1 General

In order to limit the computational effort, the simulations have been run up to 30 seconds ($\sim$15 oscillation cycles). Simulations have been performed on the GPU

AMD Radeon HD 7970 and GeForce GTX Titan. The reasons to use GPU instead of CPU architecture are discussed in 15.3.3. Speed-up achieved when using GPU instead of CPU can be up to 20x. Results are presented in terms of the angle across the time history and comparing experimental with simulation results for each fluid. Free surface shape results are also presented.

15.4.2 Glycerine

For the glycerine case, the fluid adheres to the walls of the tank due to its large dynamic viscosity (0.934 Pa/s, around 1000 times larger than water). Therefore, the effect of the front and aft walls may have an effect in the simulations when comparing the 3D with the 2D case. The stability criterion of Eq. 15.1 is fulfilled for the 3D simulation with 300,000 particles. The maximum experimental angle after 30 s is 17°, while the 2D simulation angle only reaches 7° (Fig. 15.2). This is a consequence of the dissipation being much larger in the 2D simulation compared to the experiments due to the absence of the front and aft wall boundary layers. Therefore in this case the 2D hypothesis is not acceptable.

The 3D simulation result is much closer to the experimental one (13° compared to 17 in the experiments) although identifying the origin of the discrepancies requires further research work. In both the 2D and 3D simulations there is a slight lag compared to the experimental result.

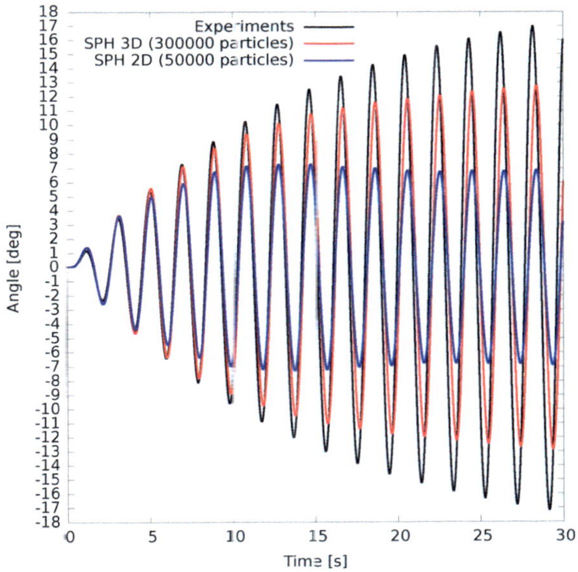

Fig. 15.2 Glycerin case roll angle

15.4.3 Oil

The 2D case can be run at full resolution in regards to correctly modelling the dynamic viscosity of the fluid (0.045 Pa/s) by using 200,000 particles. Around 5 million particles would be necessary to run a 3D case and this was not possible at this stage due to the time necessary to obtain a reasonable estimation of the damping effect (30 s). The maximum experimental angle was 12°, while the maximum angle in the 2D simulation was 10°. The accuracy is therefore reasonable. The time history of the experiments and simulations can be appreciated in Fig. 15.3. In the simulation there is a slight lag compared to the experimental result.

15.4.4 Water

The effective viscosity that can be reached for this case with 100,000 particles in 2D was 0.068114 Pa/s, more than one order of magnitude larger than real water viscosity and quite similar to oil. It was therefore not possible to run a 3D simulation. In the water case, the influence of both the front and back walls is negligible and the 2D approximation should therefore provide reasonable results.

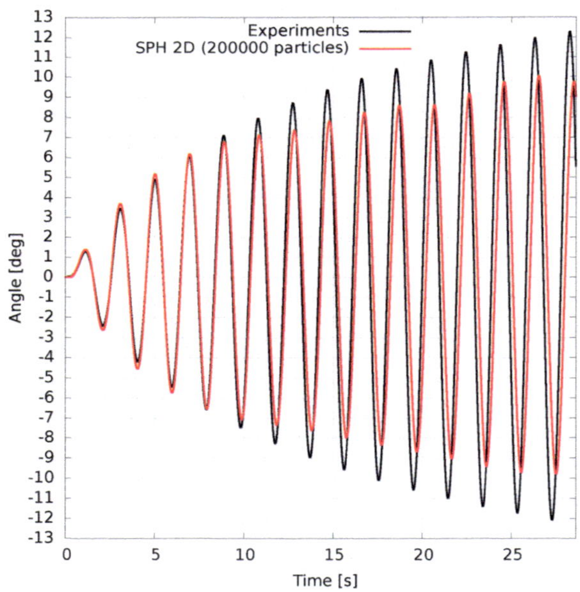

Fig. 15.3 Oil case roll angle

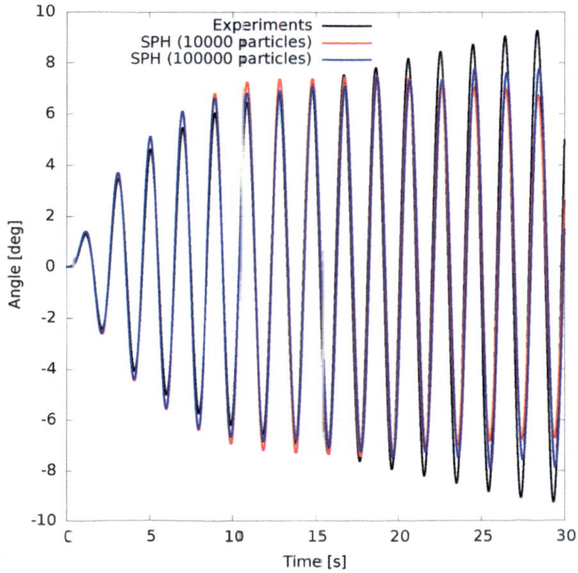

Fig. 15.4 Water results

The maximum experimental angle was 9.5°, while the maximum angle in the 2D simulation was 7.9°. The accuracy is therefore reasonable. The time history of the experiments and simulations can be appreciated in Fig. 15.4, where two resolutions results are presented. In the simulation, no lag is present compared to the experimental result.

15.4.5 Free Surface Shape

Together with the roll angle time history it is relevant to analyse the free surface evolution for each case. In Fig. 15.5, the free surface is presented for the 2D simulations of the three liquids when t = 9.43 s ($t/T_0 = 4.9$, T_0 being the first sloshing period). In this instant the tank is in its horizontal position for the water case while there is a slight deviation from this position for the oil and a more significant one for the glycerine case. The free surface pattern is quite different for the glycerine when compared to the water and oil cases, for which, as already discussed, the effective numerical viscosity was finally similar for this resolution. The glycerine shape is single valued while a breaking wave is observed in the oil case and the building up of a breaking wave can be appreciated for the water case.

As the simulation evolves to a point of extreme roll angle (t = 10 s, $t/T_0 = 5.2$, Fig. 15.6), a strong wave run-up with overturning waves and breaking takes place for the oil and the water simulations while a mild run-up occurs for the glycerine. The matching of these cases with the experiments can be observed by comparing Fig. 15.6 with Figs. 15.7 and 15.9.

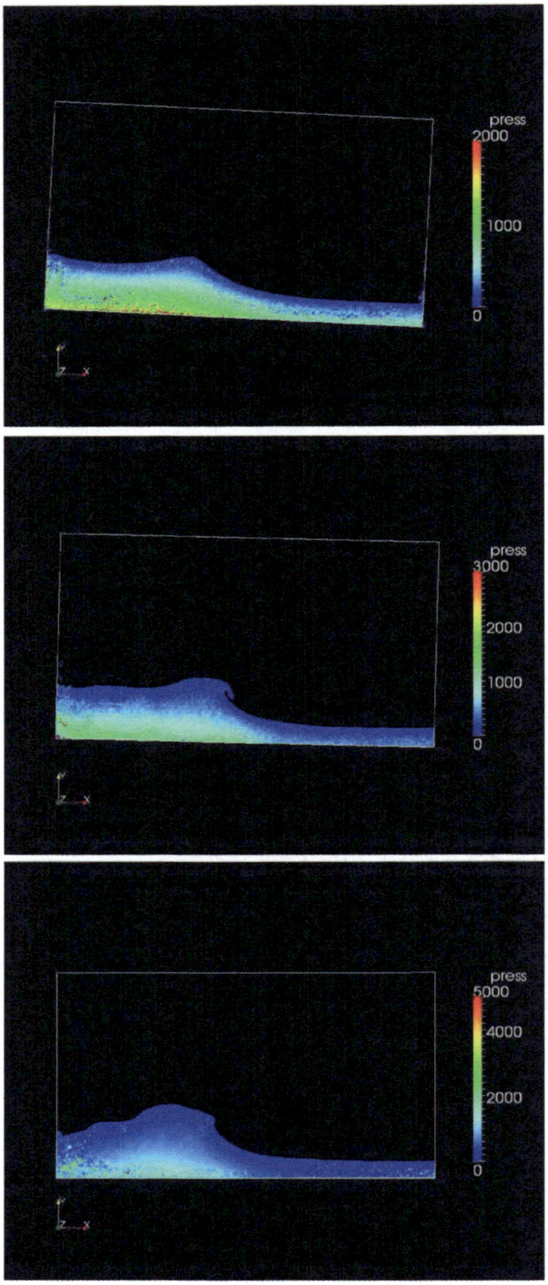

Fig. 15.5 Glycerin (top), oil (mid), water (bottom), t = 9.43 s

Fig. 15.6 Glycerin (top), oil (mid), water (bottom), t = 10 s

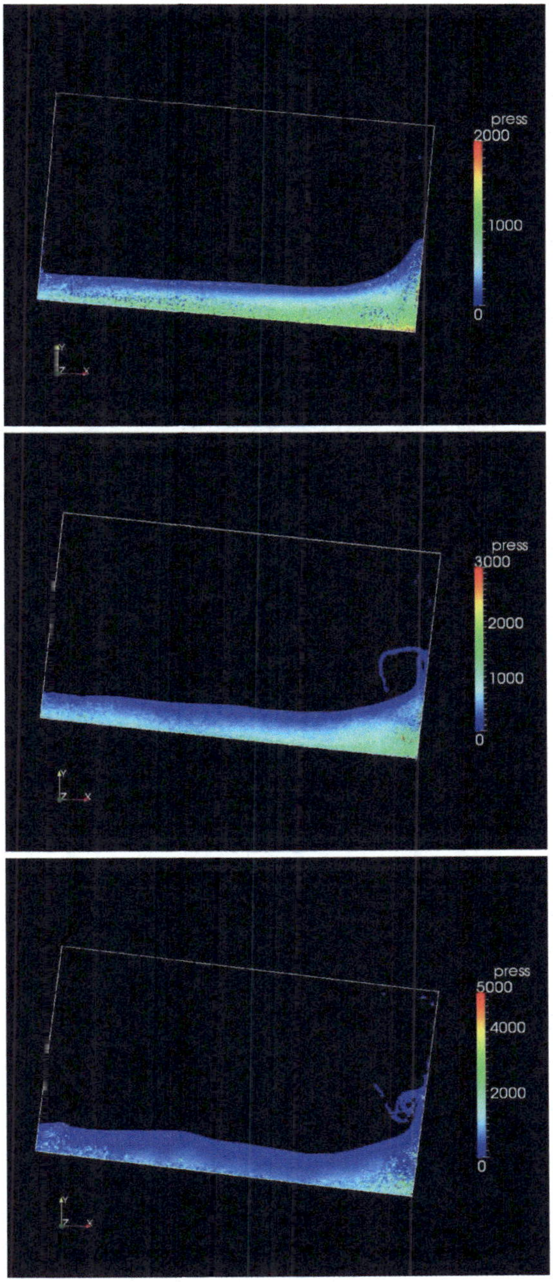

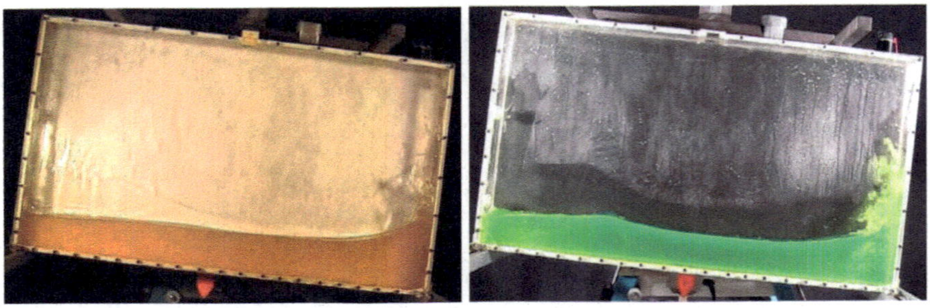

Fig. 15.7 Experiments, oil (left) and water (right) at t = 10 s

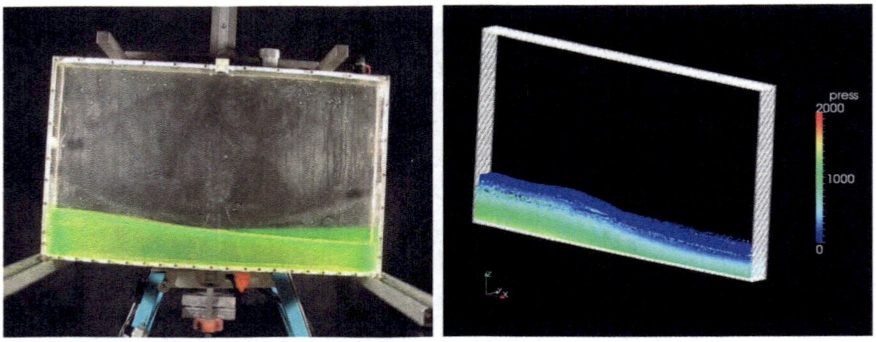

Fig. 15.8 Glycerin, t = 9.43 s, experiment (left) and 3D simulation (right)

The objective of the present paper is to assess the capabilities of SPH to deal with full 3D flows coupling problems in resonance conditions, where attenuating the vessel motion by dissipating energy through sloshing is the main target. 3D results of the computations with glycerine are presented in Figs. 15.8 and 15.9. As discussed in Sect. 15.4.2, the 2D simulation presents much larger dissipation than the experimental one due to the absence of the effect of the back and front walls of the container. If the free surface shape of the glycerine in 2D for t = 9.43 s in Fig. 15.5 is compared with the 3D one in Fig. 15.8, the matching with the experiments is much better for the 3D case, which is coherent with the roll angle found for both cases, as presented in Fig. 15.2. This tendency is confirmed by looking at the t = 10 s frames. In the 2D simulation shown in Fig. 15.6, a mild but significant wave run-up is apparent. This is not the case in 3D as can be seen in both the experiments and 3D simulations (Figs. 15.8 and 15.9).

Another interesting feature that can be observed is how the fluid sticks to the wall in Figs. 15.8 and 15.9, thus confirming the importance of the front and back walls of the container in flow dynamics.

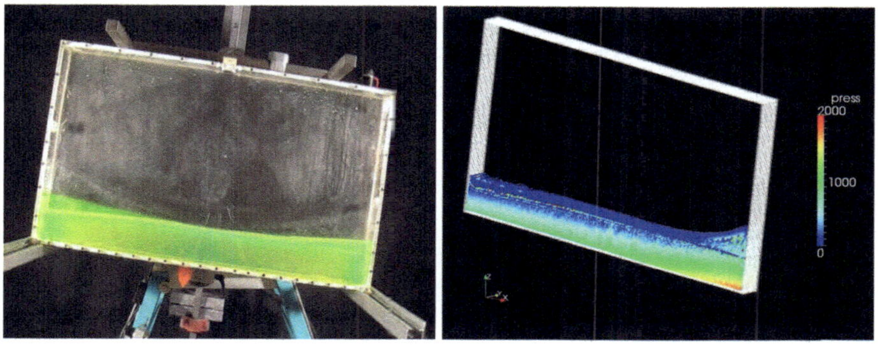

Fig. 15.9 Glycerin, t = 10 s, experiment (top) 3D simulation (bottom)

15.5 Conclusions

The roll motion response of a single degree of freedom (SDOF) structural system for which an "exact" analytic mechanical model is available and to which a rigid rectangular partially filled liquid tank was attached has been considered. The coupled sloshing and SDOF system motion in resonance conditions has been numerically studied with a 3D GPU based SPH model and compared with the experimental results.

In order to characterise the type flow dynamics effects on the response curves, simulations have been performed with liquids of different viscosity, concluding that increasing the viscosity prevents the onset of breaking waves. The capabilities of SPH to treat this coupling problem have been assessed. From the comparisons with the experiments, it seems that SPH is able to capture part of the dissipation effects due to wave breaking which is reflected in reasonably accurate damping reduction ratios. Nevertheless, there are intrinsic limitations in the stability of the method that limit the effective Reynolds number which can be reached for a certain resolution. Further work has to be done in this regard.

The next step along this path is to incorporate the SPH model of the tanks internal flow into a 6DOF ship motions model.

Acknowledgements The research leading to these results has received funding from the Spanish Ministry for Science and Innovation under grant TRA2010-16988, "Caracterización Numérica y Experimental de las Cargas Fluido-Dinámicas en el transporte de Gas Licuado".

References

Antuono A, Souto-Iglesias A, Le Touzé D (2011) Theoretical analysis and numerical verification of the consistency of viscous smoothed-particle-hydrodynamics formulations in simulating free-surface flows. Phys. Rev. E 84:26705+

Attari N K A, Rofooei F R (2008) On lateral response of structures containing a cylindrical liquid tank under the effect of fluid/structure resonances. J. Sound Vibration 318: 4–5, 1154–1179

Bulian G, Souto-Iglesias A, Delorme L, Botia-Vera E (2010) SPH simulation of a tuned liquid damper with angular motion. J. Hydraul. Res. 48: 28–39

Cercos Pita J L (2015) AQUAgpusph, a new free 3D SPH solver accelerated with OpenCL. Comput. Phys. Commun. 192: 295–312

Crespo A J C, Dominguez J M, Rogers B D, Gómez-Gesteira S, Longshaw R, Canelas R, Vacondio R, Barreiro A, García-Feal O (2015) DualSPHysics: Open-source parallel CFD solver based on Smoothed Particle Hydrodynamics (SPH) Comput. Phys. Commun. 187: 204–216

Harada T, Koshizuka S, Kawaguchi Y (2007) Smoothed particle hydrodynamics on GPUs. Proceedings of the Computer Graphics International Conference 63–70

Herault A, Bilotta G, Dalrymple R A (2010) SPH on GPU with CUDA. Journal of Hydraulic Research 48:74–79

Hu X Y, Adams N A (2006) Angular-momentum conservative Smoothed Particle Hydrodynamics for incompressible viscous flows. Phys. Fluids 18:702–706

Landrini M, Colagrossi A, Faltinsen O M (2003) Sloshing in 2D flows by the SPH method. Proceedings of the 8th International Conference on Numerical Ship Hydrodynamics

Macia F, Antuono M, Gonzalez L M, Colagrossi A (2011) Theoretical analysis of the no-slip boundary condition enforcement in SPH methods. Progr. Theoret. Phys. 125: 1091–1121

Monaghan J J (1994) Simulating free surface flows with SPH. J. Comput. Phys 110: 399–406

Monaghan J J (2012) Smoothed particle hydrodynamics and its diverse applications. Annu. Rev. Fluid Mech. 44:323–46

Monaghan J J, Gingold R A (1983) Shock simulation by the particle method SPH. J. Comput. Phys 52: 374–389

Nam B-W, Kim Y, Kim D-W, Kim Y-S (2009) Experimental and numerical studies on ship motion responses coupled with sloshing in waves. J. Ship. Res. 53:68–82

Rey-Villaverde A, Cercos Pita J L, Souto-Iglesias A, Gonzalez L M (2011) Particle methods parallel implementations by GP-GPU strategies. Proceedings of the II International Conference on Particle-based Methods - Fundamentals and Applications, PARTICLES 2011

Souto-Iglesias A, Botia-Vera E, Martin A, Perez-Arribas F (2011) A set of canonical problems in sloshing part 0: Experimental setup and data processing. Ocean Eng. 38: 1823–1830

Souto-Iglesias A, Delorme L, Rojas P L, Abril S (2006) Liquid moment amplitude assessment in sloshing type problems with SPH. Ocean Eng. 33: 11–12

Tait M J, El Damatty A A, Isyumov N (2005) An investigation of Tuned Liquid Dampers Equipped with Damping Screens under 2D Excitation. Earthq. Eng. Struct. D. 34(7):719–735

Violeau D, Rogers B D (2016) Smoothed Particle Hydrodynamics (SPH) for free-surface flows: past, present and future. J. Hydraul. Res. 54:1 1–26

Part VI
Dynamics of Large Ship Motions: Parametric Roll

Chapter 16
Prediction of Parametric Rolling in Irregular Head Waves

Hirotada Hashimoto and Naoya Umeda

Abstract For providing a benchmark data for numerical codes for parametric roll prediction, a model experiment of a post-Panamax C11 class containership whose hull form is slightly modified from its original but opened for public was conducted and significant parametric rolling in irregular head waves was recorded. A 3DOF (degrees of freedom) numerical model based on the nonlinear strip theory is developed. The numerical code for time domain simulations is developed for the prediction of large amplitude of parametric roll both in regular and irregular head waves. Comparisons between the model experiment and the numerical simulations show good agreement under the consideration of dispersion due to practical non-ergodicity. By utilizing the developed numerical model, it was demonstrated that small height of the bilge keels is a major reason why significant parametric roll could happen for a C11 class post-Panamax containership in 1998.

Nomenclature

a_{ij}	Added mass/added moment of inertia
b_1, b_2	Linear and quadratic coefficients of roll damping
b_{ij}	Wave-making damping coefficient
F_i^{DF}	Diffraction force
F_i^{FK+B}	Froude-Krylov force and buoyancy
F_n	Froude number
g	Gravity
H	Wave height
I_{xx}	Roll moment of inertia
I_{yy}	Pitch moment of inertia
m	Ship mass

H. Hashimoto (✉) · N. Umeda
Department of Naval Architecture and Ocean Engineering, Graduate School of Engineering, Osaka University, 2-1, Yamadaoka, Suita, Osaka 565-0871, Japan
e-mail: hashimoto@port.kobe-u.ac.jp

V. L. Belenky et al. (eds.), *Contemporary Ideas on Ship Stability*, Fluid Mechanics and Its Applications 119, https://doi.org/10.1007/978-3-030-00516-0_16

t	Time
ξ_G	Longitudinal position of centre of gravity
ϕ	Roll angle
λ	Wave length
θ	Pitch angle
ω	Angular frequency
ζ	Heave displacement

16.1 Introduction

Since parametric roll is a well-known phenomenon as one of dangerous modes which could lead to stability failures. It has been recognized that parametric roll is a great threat to containerships and pure car and truck carriers with serious accident reports (France et al. 2003; Hua et al. 2006). There are also investigations of parametric roll of fishing vessels and passenger ships (Neves et al. 2011; Munif et al. 2011). The second generation Intact Stability criteria, which is under development at the International Maritime Organization (IMO) as physics-based and goal-based criteria for intact ships, is required to cover this phenomenon as well as other stability failure modes, i.e. pure loss of stability, surf-riding/broaching, dead ship and excessive acceleration. There has been a strong demand for developing numerical models which can directly assess stability of ships to those failure modes. Therefore a lot of numerical models have been proposed so far for the parametric roll prediction. However numerical results show significant scatters among numerical codes in benchmark studies on parametric roll prediction in irregular waves as reported by SAFEDOR (Spanos and Papanikolaou 2009) and ITTC (Reed 2011), see also Chap. 37 of this book. Organizations and researchers developed their numerical codes and often claimed that their codes are validated with their own model experiments. It is difficult to make a systematic comparison of different organizations/researchers because hull forms used for the validation of each code are generally not available for different organizations mainly because of commercial reasons.

In order to respond to this situation, a model experiment using a post-Panamax C11 containership is conducted to provide a benchmark data for parametric roll. The hull form is slightly modified by MARIN (Levadou and van't Veer 2006) from its original which experienced head-sea parametric roll with its maximum roll angle of 40° at the North Pacific in 1998 (France et al. 2003). Model experiments are conducted and roll motions are recorded in regular and long-crested irregular head waves at a towing tank of Osaka University. In the case of irregular waves, model experiments are executed for several significant wave heights, Froude numbers and realisations. As a result, large amplitude parametric roll was successfully observed in both regular and irregular head waves.

A coupled heave-roll-pitch model is developed for the prediction of parametric roll in longitudinal waves. A nonlinear Froude-Krylov force, which is a major component of parametric roll excitation in waves, is calculated for the instantaneous submerged hull in each time step. Radiation and diffraction forces are considered as the function of roll angle based on a nonlinear strip theory. This numerical model can be applied to the prediction of parametric roll in irregular longitudinal waves, with the superposition of elementary waves for representing irregular wave surfaces, on the basis of the small amplitude wave theory. Then, numerical results are compared with the model experiment to validate the presented numerical code. The numerical results show fairly good agreement with the measured results of parametric roll in both regular and irregular head waves for the modified C11 class post-Panamax containership.

We have examined the reason why serious parametric roll could happen for the C11 class post-Panamax containership based on the developed numerical model. Roll restoring variation and roll damping are compared with those of other post-Panamax containership. As a result, a major reason can be presumed that the height of bilge keels of the C11 post-Panamax containership is much smaller than that of other post-Panamax containership while the amplitude of roll restoring variation is not so.

16.2 Mathematical Model

A coupled 3DOF (degrees of freedom) mathematical model based on a nonlinear strip theory is developed for the prediction of large-amplitude parametric roll in head and following waves. In this model, a nonlinear Froude-Krylov force is calculated by integrating wave pressure up to the wave surface. Dynamic components, i.e. radiation and diffraction forces, are calculated for the under-water hull changing with instantaneous roll angle. Two-dimensional hydrodynamic forces are calculated by solving a boundary integral equation for the velocity potential. Diffraction forces are calculated by the STF (Salvesen–Tuck–Faltinsen) method (Salvesen et al. 1970). An end term effect is considered in the calculation of the radiation force because a hydrodynamic lift effect on the roll moment cannot be neglected when a ship has advance speed.

In numerical simulations, a ship motion is obtained by time integration of differential equations of ship motion with the fourth order Runge-Kutta method. Linear and quadratic roll damping coefficients are used, which are determined from experimental data of roll decay tests.

16.2.1 Regular Seas

The Froude-Krylov force is calculated by integrating the hydrostatic and wave pressures for the wetted hull surface. The radiation force is calculated at the encounter frequency of the ship to incident waves for the heave and pitch motions, whilst that for the roll motion is done at half the encounter frequency by assuming a condition of the principal parametric roll. The diffraction force is calculated at the encounter frequency for all the motions. The two-dimensional hydrodynamic forces are calculated for each section taking account of instantaneous roll angle, to take account of the effects of largely changeable submerged hull geometry on the linear cross-coupling forces and the diffraction force. Therefore the coupling effects with respect to roll are implicitly included in this sense. The coupled 3DOF equations for parametric roll in regular longitudinal waves are expressed as Eqs. 16.1–16.3.

$$\begin{aligned}&(m + a_{33}(\phi))\ddot{\zeta} + b_{33}(\phi)\dot{\zeta} + a_{34}(\phi)\ddot{\phi} + b_{34}(\phi)\dot{\phi} + a_{35}(\phi)\ddot{\theta} + b_{35}(\phi)\dot{\theta}\\&= F_3^{FK+B}(\xi_G/\lambda, \zeta, \phi, \theta) + F_3^{DF}(\phi) - mg\end{aligned} \tag{16.1}$$

$$\begin{aligned}&(I_{xx} + a_{44}(\phi))\ddot{\phi} + b_1\dot{\phi} + b_2\dot{\phi}\left|\dot{\phi}\right| + a_{43}(\phi)\ddot{\zeta} + b_{43}(\phi)\dot{\zeta} + a_{45}(\phi)\ddot{\theta} + b_{45}(\phi)\dot{\theta}\\&= F_4^{FK+B}(\xi_G/\lambda, \zeta, \phi, \theta) + F_4^{DF}(\phi)\end{aligned} \tag{16.2}$$

$$\begin{aligned}&\left(I_{yy} + a_{55}(\phi)\right)\ddot{\theta} + b_{55}(\phi)\dot{\theta} + a_{53}(\phi)\ddot{\zeta} + b_{53}(\phi)\dot{\zeta} + a_{54}(\phi)\ddot{\phi} + b_{54}(\phi)\dot{\phi}\\&= F_5^{FK+B}(\xi_G/\lambda, \zeta, \phi, \theta) + F_5^{DF}(\phi)\end{aligned} \tag{16.3}$$

16.2.2 Irregular Seas

The nonlinear Froude-Krylov force is calculated by integrating hydrostatic and wave pressures up to the irregular wave surface. The wave pressure and the diffraction force are calculated with the linear superposition principle. In general, a number of elementary waves is 200 and their frequencies are determined from the ITTC wave spectrum (Eq. 16.4). Here unequally divided frequencies, having the same area in each division, are used to avoid the so-called self-repeating effect. For the diffraction forces, transfer functions with respect to a combination of the frequencies and heel angles (up to the vanishing angle) are pre-calculated. The radiation forces of roll are calculated at the natural roll frequency and those of heave and pitch are done at a representative encounter wave frequency. The most frequent encounter frequency is generally used which can be obtained from the peak frequency of the wave spectra. The coupled 3DOF equations for parametric roll in irregular longitudinal waves are expressed as Eqs. 16.5–16.7.

$$\mathrm{S}(\omega) = 173\frac{H_{1/3}^2}{T_{01}^4}\omega^{-5}\exp\left(-\frac{691}{T_{01}^4}\omega^{-4}\right) \tag{16.4}$$

$$\begin{aligned}&(m + a_{33}(\phi))\ddot{\zeta} + b_{33}(\phi)\dot{\zeta} + a_{34}(\phi)\ddot{\phi} + b_{34}(\phi)\dot{\phi} + a_{35}(\phi)\ddot{\theta} + b_{35}(\phi)\dot{\theta}\\&= F_3^{FK+B}(t, \zeta, \phi, \theta) + F_3^{DF}(t, \phi) - mg\end{aligned} \tag{16.5}$$

$$\begin{aligned}&(I_{xx} + a_{44}(\phi))\ddot{\phi} + b_1\dot{\phi} + b_2\dot{\phi}\left|\dot{\phi}\right| + a_{43}(\phi)\ddot{\zeta} + b_{43}(\phi)\dot{\zeta} + a_{45}(\phi)\ddot{\theta} + b_{45}(\phi)\dot{\theta}\\&= F_4^{FK+B}(t, \zeta, \phi, \theta) + F_4^{DF}(t, \phi)\end{aligned} \tag{16.6}$$

$$\begin{aligned}&\left(I_{yy} + a_{55}(\phi)\right)\ddot{\theta} + b_{55}(\phi)\dot{\theta} + a_{53}(\phi)\ddot{\zeta} + b_{53}(\phi)\dot{\zeta} + a_{54}(\phi)\ddot{\phi} + b_{54}(\phi)\dot{\phi}\\&= F_5^{FK+B}(t, \zeta, \phi, \theta) + F_5^{DF}(t, \phi)\end{aligned} \tag{16.7}$$

16.3 Model Experiment

It is desired to provide a benchmark data for the validation of numerical codes developed for parametric roll prediction. Therefore a model experiment is conducted using a 1/100 scaled model of a Post-Panamax C11 class container ship, whose hull form is slightly modified by MARIN from its original but opened for public, to measure time histories of parametric roll in regular and irregular head waves at a towing tank of Osaka University. The ship model is towed by a towing carriage by means of sufficiently soft elastic ropes connected to the bow. Principal particulars and body plan of the subject ship are shown in Table 16.1 and Fig. 16.1 respectively. Before the measurements of parametric roll, roll decay tests are performed with and without forward velocity to estimate roll damping coefficients.

Firstly the model experiment is conducted in regular head waves with $\lambda/L = 1.0$, 1.3, 1.6 and with wave steepness of 0.01, 0.02 and 0.03. The Froude numbers tested are 0.0, 0.05, 0.10 and 0.15. Time histories of parametric roll are measured by an onboard gyro scope and the steady amplitude of parametric roll is obtained.

Secondly the model experiment is conducted in irregular head waves with a wave mean period, T_{01}, of 9.99 s, and three significant wave heights, $H_{1/3}$, of 5.22, 7.82, 10.43 m. Parametric roll of the subject ship is recorded at zero forward velocity for all the significant wave heights. The model tests with forward velocity ($Fn = 0.05$, 0.1) are also conducted for the case of $H_{1/3}$ of 7.82 m. The time durations of the measurement are 4200, 2400, 1200 s in full scale for the cases of $Fn = 0.0$, 0.05, 0.1, respectively. Here, the time duration for the case without forward velocity is determined by an experimental guidance using a running standard deviation (Umeda et al. 2011). In cases of irregular wave, direct evaluation of the numerical model

Table 16.1 Principal particulars of a C11 post-Panamax containership

Length between perpendiculars: L_{PP}	262.0 m
Breadth: B	40.0 m
Depth: D	24.45 m
Mean draught: T	11.5 m
Block coefficient: C_b	0.56
Metacentric height: GM	1.965 m
Natural roll period: T_ϕ	25.1 s

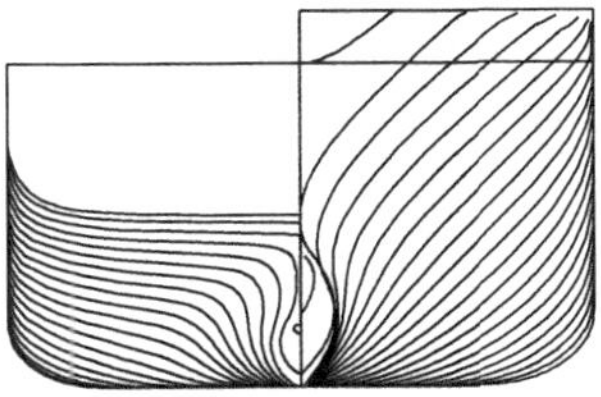

Fig. 16.1 Body plan of a C11 post-Panamax containership

using one long record is difficult because of practical non-ergodicity of parametric roll (Belenky 2004; Bulian et al. 2008). Therefore model runs in irregular waves are repeated with different phase sets of elementary waves.

16.4 Results and Discussion

16.4.1 Regular Head Seas

Numerical results of parametric roll in regular head waves are compared with the model experiment as shown in Fig. 16.2. The initial roll angle and the roll rate are set as 5 deg and 0 deg/sec in the simulations. In the experiment, the steady roll amplitude increases with wave steepness nonlinearly and changes strongly with the Froude number, i.e. encounter frequency. The numerical simulation generally shows good agreement with the model experiment in the steady amplitude and the occurrence region of parametric roll. The nonlinear relationship between parametric roll and the wave steepness is well reproduced. Some discrepancies near the onset of parametric roll could be explained as a local bifurcation, in other words initial condition dependence. These comparisons demonstrate that the 3DOF coupled model of heave-roll-pitch based on a nonlinear strip theory can predict parametric roll in regular head waves with sufficient accuracy.

16.4.2 Irregular Head Seas

Numerical simulations were executed for the same conditions of mean wave period, significant wave height, Froude number and number of realisation as the model experiment. The comparisons of maximum roll angle between the model experiment

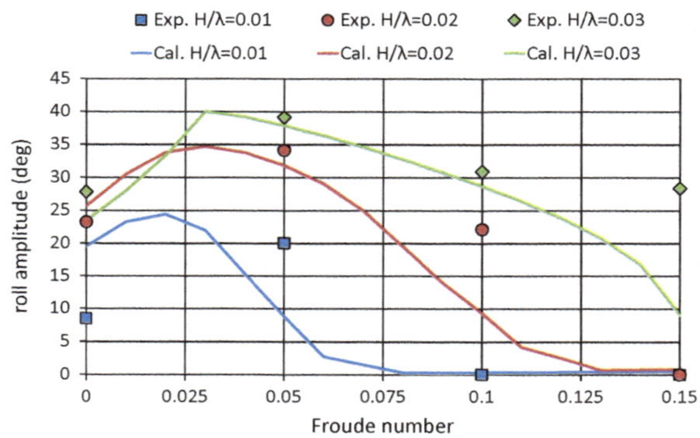

Fig. 16.2 Comparison of steady amplitude of parametric roll in regular head waves with $\lambda/L = 1.0$ between experiment and simulation

and the numerical simulation for $H_{1/3}$ = 7.82 m and Fn = 0.0, 0.05, 0.1 are shown in Figs. 16.3, 16.4 and 16.5, respectively. The scattering of the maximum roll angle among realizations is found in both the experiment and the simulation despite the long-time duration due to practical non-ergodicity of nonlinear rolling (Belenky et al. 1998). In order to check the existence of the self-repeating effect, the auto-correlation function is calculated for a time history of incident irregular waves. In Fig. 16.6, the self-repeating effect is not found for the entire duration thanks to the non-uniform frequency distribution, in which the ratio of frequencies is not a ratio of integer (Thompson and Stewart 2002). The ensemble mean and the confidence interval of maximum roll angles are compared between the experiment and the simulation with respect to Froude number and significant wave height as presented in Figs. 16.7 and 16.8. Although the instantaneous roll angles for the entire duration does not follow the Gaussian distribution, the ensemble of maximum roll angles of parametric roll generally follows the Gaussian distribution because the maximum roll angles mostly appear in certain stationary states. Therefore the confidence interval can be obtained using the estimated standard deviations. Numerical results generally show reasonably good agreement and the ensemble means are in the 95% confidence interval of the experiment for the conditions where relatively large parametric roll happens while it is not so for the cases of $H_{1/3}$ = 5.22 m and Fn = 0.0, and $H_{1/3}$ = 7.82 m and Fn = 0.1. From the discussion above, it is presumed that these conditions are close to the threshold of parametric roll. It is known that maximum roll angles of parametric roll scatter near the threshold. Therefore, the agreement would be improved if a number of realisation increases for these conditions. The standard deviations of roll and pitch angles are calculated with $H_{1/3}$ = 7.82 m and Fn = 0.0, and plotted in Fig. 16.9. The standard deviation of roll angle scatters significantly due to practical non-ergodicity while that of pitch angle does not. This important trend of parametric roll in irregular waves can be well reproduced by the numerical simulation.

In the experiment, wave elevations were recorded by a wave probe attached to a towing carriage. The incident irregular wave profile can be reproduced by the Fourier transformation with the measured data. The numerical simulations are executed with the reproduced wave surface train, and calculated time histories of roll and pitch are compared with the measured ones. An example of the comparisons is shown in Fig. 16.10. The development of parametric roll can be predicted accurately at the beginning and the end of the time history where largest and 2nd largest parametric roll happen. It can be concluded that the developed numerical model has sufficient accuracy for parametric roll prediction through the comparison results with the experiment. Validation of the mathematical model in terms of statistical properties of non-ergodic parametric roll can be found in the literature (Hashimoto and Umeda 2012).

Next a maximum roll angle of parametric roll per 1 h at the North Atlantic is calculated for the C11 containership without forward speed with wave statistics. Time domain simulations are repeated with 10 realisations for each wave condition, and the maximum roll angle is shown in the table of Fig. 16.11. If a maximum roll angle exceeds 90°, it is regarded as capsize and written as “cap”. Large amplitude parametric roll is confirmed for a wide range of wave condition. Probabilities of

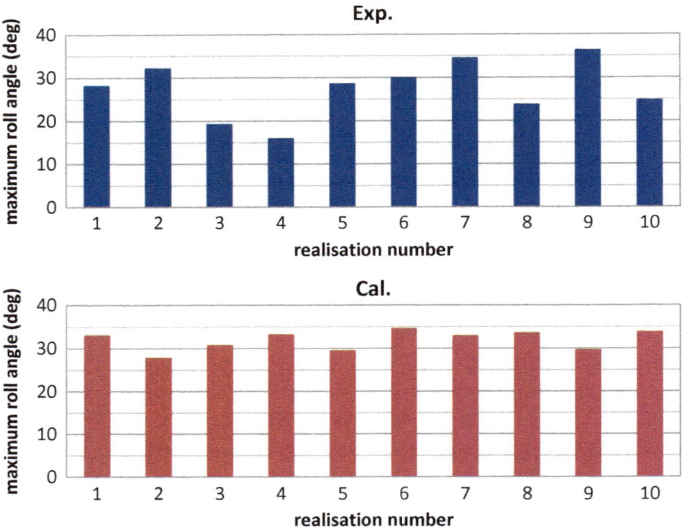

Fig. 16.3 Maximum roll angle of parametric roll with $T_{01} = 9.99$ s, $H_{1/3} = 7.82$ m and $Fn = 0.0$

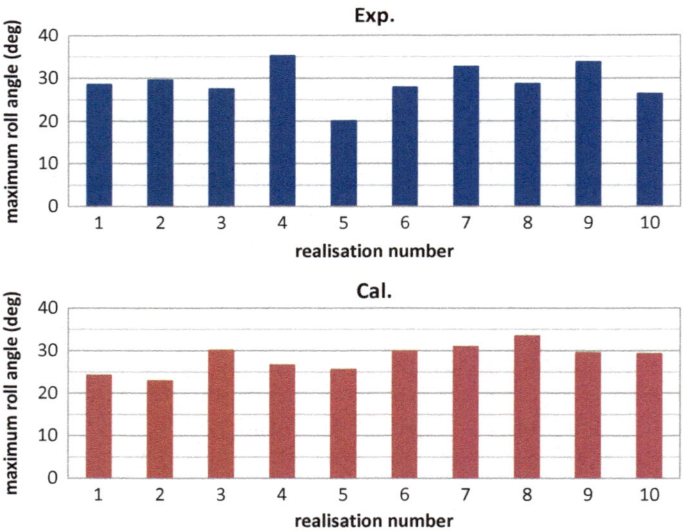

Fig. 16.4 Maximum roll angle of parametric roll with $T_{01} = 9.99$ s, $H_{1/3} = 7.82$ m and $Fn = 0.05$

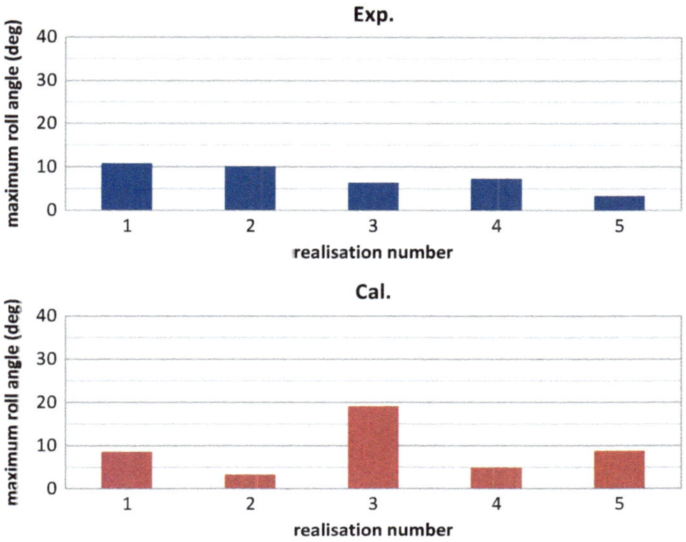

Fig. 16.5 Maximum roll angle of parametric roll with $T_{01} = 9.99$ s, $H_{1/3} = 7.82$ m and $Fn = 0.1$

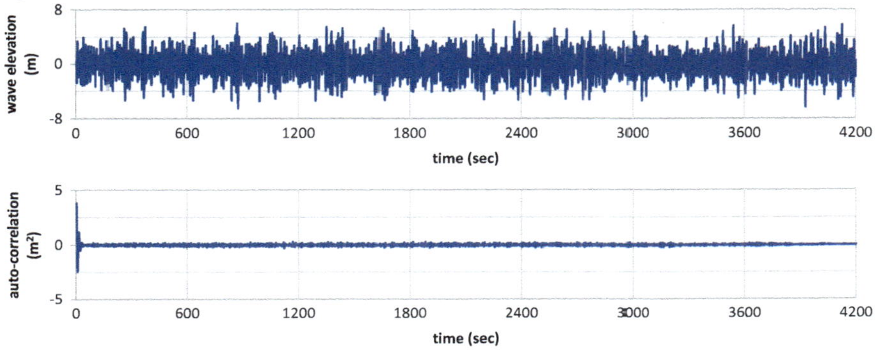

Fig. 16.6 An example of incident irregular waves and auto-correlation function

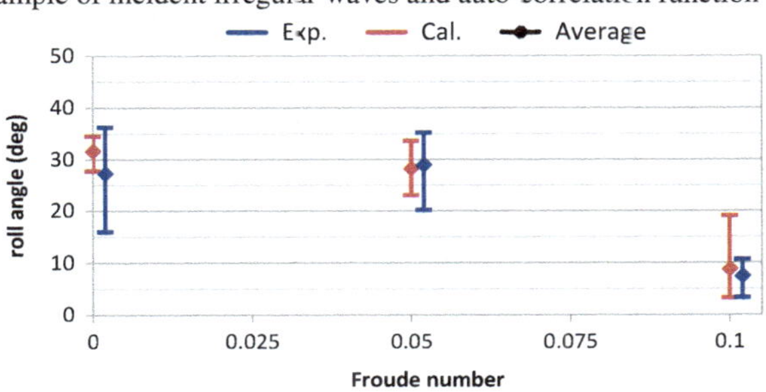

Fig. 16.7 Average of maximum roll angle of parametric rolling with respect to Froude number with $T_{01} = 9.99$ s, $H_{1/3} = 7.82$ m

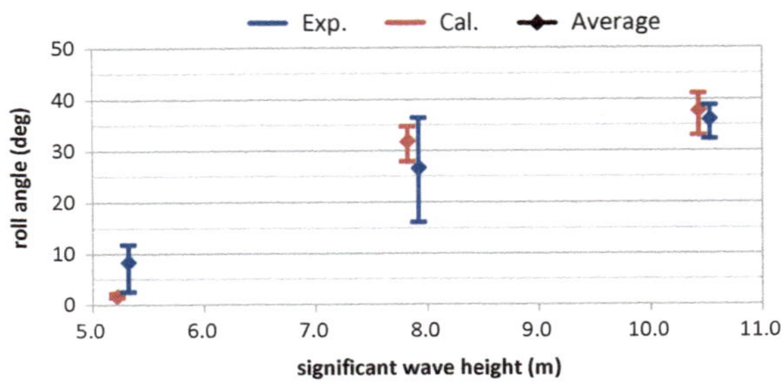

Fig. 16.8 Average of maximum roll angle of parametric rolling with respect to significant wave height with $T_{01} = 9.99$ s and $Fn = 0.0$

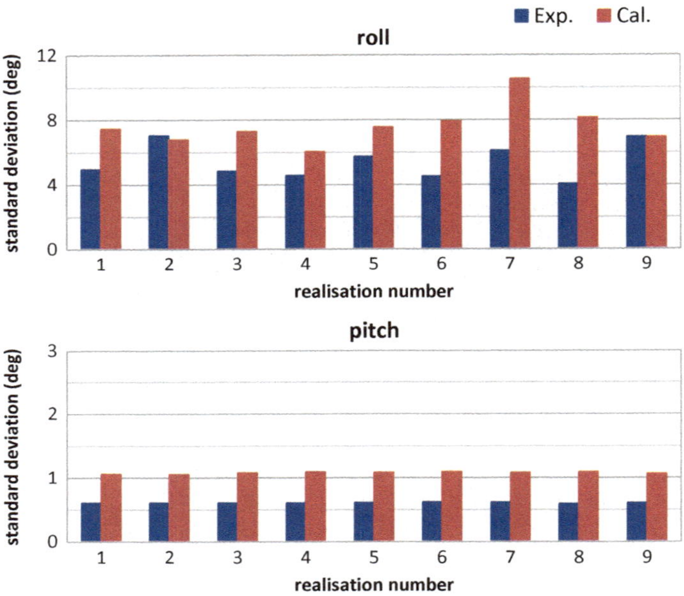

Fig. 16.9 Standard deviation of roll and pitch angles with $T_{01} = 9.99$ s, $H_{1/3} = 7.82$ m and $Fn = 0.0$

parametric roll exceeding critical angles are presented in Fig. 16.12. The exceeding probability is very high even though a large roll angle is set as the critical value. This is because the numerical simulations are performed at zero forward speed, e.g. heave-to condition, and it would be the most dangerous situation to suffer serious parametric roll. The similar calculations are expected using actual ship speed in waves to examine effects of advanced speed on the exceeding probability for a practical design aspect.

Since the maximum roll angle of parametric roll of the C11 class post-Panamax containership is sometimes beyond the range of roll decay tests, rationality of curve-

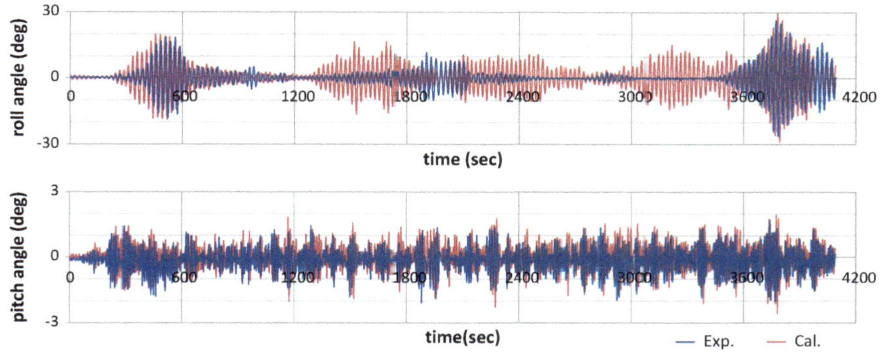

Fig. 16.10 Time histories of roll and pitch angles with $T_{01} = 9.99$ s, $H_{1/3} = 7.82$ m and Fn = 0.0

H_{13} (m) \ T_{01} (s)	1.5	2.5	3.5	4.5	5.5	6.5	7.5	8.5	9.5	10.5	11.5	12.5	13.5	14.5	15.5	16.5	17.5	18.5
0.5			0.0	0.0	0.0	0.0	0.0	0.0	0.0	0.0	0.0	0.0						
1.5				0.0	0.0	0.0	0.0	0.0	0.0	0.0	0.0	0.0	0.0	0.0	0.0			
2.5				0.0	0.3	0.5	4.0	3.0	3.9	4.1	3.2	2.6	2.1	2.2	1.6	1.5		
3.5				1.0	2.1	1.0	8.2	9.0	12.2	10.9	6.8	6.7	5.0	4.5	3.4	3.1	4.3	
4.5					4.4	2.7	13.0	14.9	16.9	16.0	13.1	11.4	9.3	7.1	5.8	6.0	8.2	
5.5					6.5	6.6	17.1	20.5	22.9	21.9	17.9	15.9	15.0	9.9	9.7	11.5	10.6	9.4
6.5					15.5	7.0	19.7	24.9	27.8	27.1	24.0	20.8	19.3	15.2	12.1	14.0	17.4	12.1
7.5						13.9	25.1	28.7	31.5	30.7	28.5	25.5	23.3	20.2	17.7	17.6	20.3	16.8
8.5						cap	24.2	31.9	35.4	33.3	31.1	23.8	28.6	25.0	22.1	22.7	22.6	21.0
9.5						cap	cap	35.6	37.3	35.4	34.4	31.7	30.0	28.2	25.4	25.4	27.7	24.1
10.5							cap	37.6	38.8	37.0	36.0	34.3	32.2	28.7	28.2	27.7	29.2	25.5
11.5							cap	cap	40.6	38.0	36.6	35.8	33.8	30.8	32.0	30.0	29.4	28.2
12.5							cap	cap	42.3	39.5	38.9	35.7	34.9	32.2	33.5	30.4	31.5	
13.5								cap	41.8	40.9	40.6	38.1	35.9	33.2	34.3	29.7	32.1	
14.5								cap	44.1	41.9	39.8	37.7	36.9	34.1	33.6	31.9		
15.5									cap	43.5	40.7	40.0	38.5	35.8	35.7	33.7		
16.5										48.0	43.3	39.1	39.4	36.4	34.7			

Fig. 16.11 Maximum roll angle of parametric roll per 1 h in irregular head waves appeared at the North Atlantic

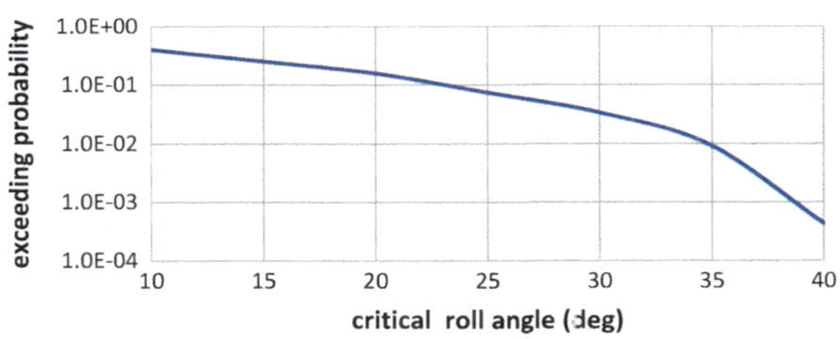

Fig. 16.12 Probability of parametric roll exceeding a critical roll angle at the North Atlantic

fittings of extinction curve using linear and quadratic terms is not guaranteed for large amplitude parametric roll. Therefore numerical simulations are additionally attempted using a curve-fitting with 3rd order polynomial. The calculated results are shown in Fig. 16.13. The difference of the maximum roll angle is not negligibly small between the numerical simulations using 2nd and 3rd order polynomials. Further

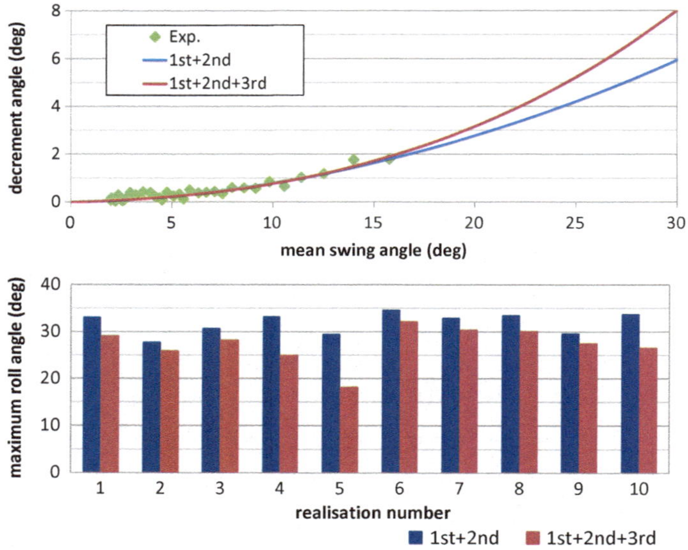

Fig. 16.13 Roll extinction curve and maximum roll angle of parametric roll using 2nd and 3rd order polynomial curve-fittings

discussions on this matter, that is extrapolation of a roll extinction curve up to large roll angle, are desirable in future.

16.5 Reasons for Large Amplitude Parametric Roll

Since serious parametric roll is observed both in the model experiment and the numerical simulation, we attempt to investigate a reason why the C11 class post-Panamax containership could suffer such large amplitude parametric roll from viewpoints of roll restoring variation and roll damping, which are the elements determining the occurrence of parametric roll, by comparing with other post-Panamax containership. The body plan and the principal particulars of this containership are shown in Fig. 16.14 and Table 16.2. Experimental investigation of parametric roll of this containership can be found in the literature (Hashimoto et al. 2011).

16.5.1 Roll Restoring Variation

Since roll restoring variation is a source of parametric roll excitation, the amplitude of roll restoring variation in regular head waves is calculated for the C11 class—and

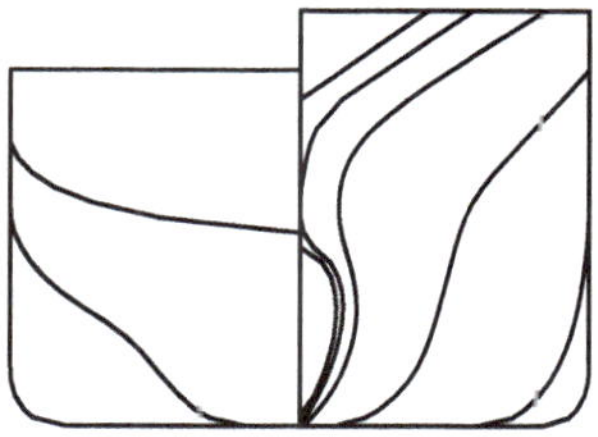

Fig. 16.14 Body plan of the other post-Panamax containership

Table 16.2 Principal particulars of the other post-Panamax containership

Length between perpendiculars: L_{PP}	283.8 m
Breadth: B	42.8 m
Depth: D	24.0 m
Mean draught: T	14.0 m
Block coefficient: C_b	0.63
Metacentric height: GM	1.06 m
Natural roll period: T_ϕ	30.3 s

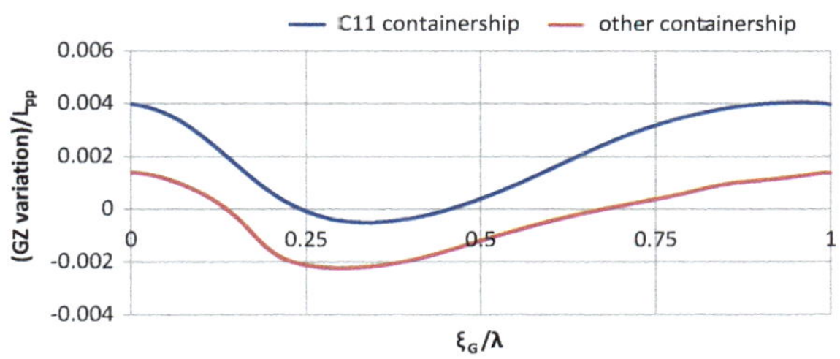

Fig. 16.15 Roll restoring variation at heel angle of 20° in regular head waves with H/λ = 0.03, λ/L = 1.0 and Fn = 0.0

the other post-Panamax containerships. The calculated amplitudes of roll restoring variation of the C11 class containership is larger than that of other containership by 26% as shown in Fig. 16.15. This comparison result indicates that the C11 class post-Panamax containership has a worse hull form from a view point of parametric roll occurrence.

16.5.2 Roll Damping

Since the C11 class post-Panamax containership has small bilge keels, time domain simulation is conducted with roll damping replaced by the other post-Panamax containership to examine the effect of bilge keel. The bilge keel size of two containerships is shown in Table 16.3. The area of bilge keel of the C11 class containership is smaller than half of that of the other containership. Figure 16.16 shows the comparison of extinction coefficients determined from roll decay tests for each ship. Figure 16.17 shows an interesting result that the maximum roll angle of parametric roll drastically decreases if the roll damping of the other containership is used instead of the original one. This numerical result clearly demonstrates that the smallness of bilge keel is a major cause leading to serious parametric roll of the C11 class post-Panamax containership.

Table 16.3 Bilge keel size

	C11 containership	Other containership
Bilge keel area: A_{BK}	30.6 m^2	84.3 m^2
$\frac{A_{BK}}{L_{PP} \times B} \times 100$	0.58%	1.38%

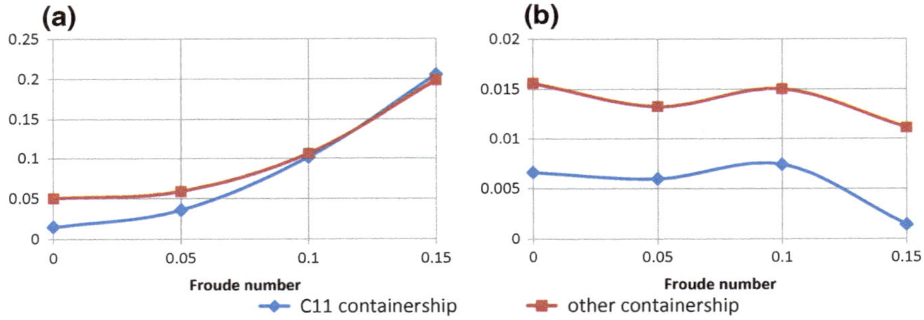

Fig. 16.16 Linear (**a**) and quadratic (**b**) extinction coefficients determined from roll decay tests

Fig. 16.17 Maximum roll angle of parametric roll with roll damping of the C11 and other post-Panamax containerships with T_{01} = 9.99 s, $H_{1/3}$ = 7.82 m and Fn = 0.0

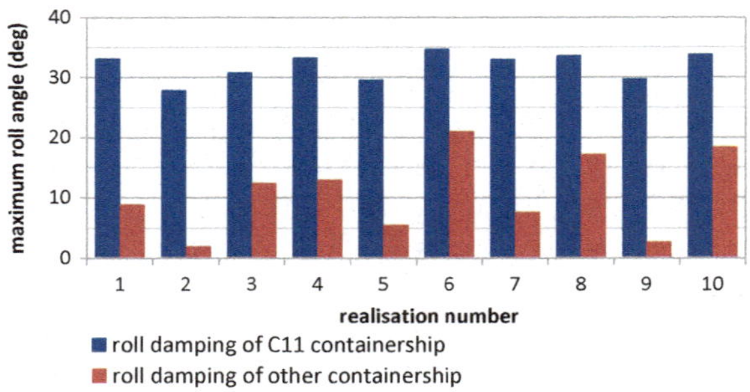

16.6 Conclusions

A model experiment of a C11 class post-Panamax containership was conducted for providing a benchmark data to develop numerical simulation tools for the direct stability assessment of parametric roll discussed in the second generation intact stability criteria at IMO, and large amplitude parametric roll is clearly observed in both regular and irregular head waves.

A coupled heave-roll-pitch numerical model was developed based on a nonlinear strip theory, and numerical results are compared with the model experiment. As a result, the developed numerical model shows good agreement with the experiment in the steady amplitude of parametric roll in regular head waves and in the ensemble mean of maximum roll angles of parametric roll in irregular head waves.

Comparison studies with other post-Panamax containership demonstrate a major cause of serious parametric roll of the C11 class post-Panamax containership that the height of bilge keel is much smaller than usual containerships.

Acknowledgements This work was supported by a Grant-in Aid for Scientific Research of the Japan Society for Promotion of Science (No. 20760558 & 21360427). The authors express their sincere gratitude to the ITTC (International Towing Tank), as well as MARIN, for providing geometric data of the modified C11 class containership for works on the development of the second generation intact stability criteria. The authors thank also Messrs. Yu Sanya and Yasuhiro Sogawa for their great contribution to this work.

References

Belenky, V. L., (2004), On Risk Evaluation at Extreme Seas. Proceedings of the 7th International Ship Stability Workshop, pp. 188–202, Shanghai.

Belenky, V. L., Degtyarev, A. B., Boukhanovsky, A. V., (1998), Probabilistic Qualities of Nonlinear Stochastic Rolling, Ocean Engineering, Vol. 25, pp. 1–25.

Bulian, G., Francescutto, A., Umeda, N. and Hashimoto, H. (2008), Qualitative and Quantitative Characteristics of Parametric Ship Rolling in Random Waves in the Light of Physical Model Experiments, Ocean Engineering, Vol. 35, pp. 1661–1675.

France, W. N., Levadou, M., Treakle, T. W. et al., (2003), An Investigation of Head-Sea Parametric Roll and Its Influence on Container Lashing System, Marine Technology, 40(1), pp. 1–19.

Hashimoto, H. and Umeda, N., (2012), Validation of a Numerical Simulation Model for Parametric Rolling Prediction Using a PCTC, Proceedings of the 11th International conference on Stability of Ships and Ocean Vehicles, Athens, Vol. 1, pp. 109–118.

Hashimoto, H., Umeda, N. and Matsuda A., (2011), Experimental Study on Parametric Roll of a Post-Panamax Containership in Short-Crested Irregular Waves, Contemporary Ideas on Ship Stability and Capsizing in Waves (Belenky, V.L., de Kat, J.O., Spyrou, K. and Umeda, N. eds.), Springer, pp. 267–276.

Hua, J., Palmquist, M. and Lindgren, G., (2006), An Analysis of the Parametric Roll Events Measured Onboard the PCTC AIDA, Proceedings of the 9th International conference on Stability of Ships and Ocean Vehicles, Rio de Janeiro, Vol. 1, pp. 109–118.

Levadou, M. and van't Veer, R., (2006), Parametric Roll and Ship Design, Proceedings of the 9th International Conference on Stability of Ships and Ocean Vehicles Rio-de-Janeiro, Vol. 1, pp. 191–206.

Munif, A., Ikeda, Y., Fujiwara, T. and Katayama, T., (2011), Parametric Roll Resonance of a Large Passenger Ship in Dead Ship Condition in All Heading Angles, Contemporary Ideas on Ship Stability and Capsizing in Waves (Belenky, V.L., de Kat, J.O., Spyrou, K. and Umeda, N. eds.), Springer, pp. 331–345.

Neves, M. A. S., Pérez, N. A., Osvaldo, M. L. and Rodríguez, C. A., (2011), An Investigation of Head-Sea Parametric Rolling for Two Fishing Vessels, Contemporary Ideas on Ship Stability and Capsizing in Waves (Belenky, V.L., de Kat, J.O., Spyrou, K. and Umeda, N. eds.), Springer, pp. 231–252.

Reed, A. M., (2011), 26th ITTC Parametric Roll Benchmark Study, Proceedings of the 12th International Ship Stability Workshop, Washington D.C., pp. 195–204.

Salvesen, N., Tuck E. O. and Faltinsen, O., (1970), Ship Motions and Sea Load, TSNAME, Vol. 78.

Spanos, D.A. and Papanikolaou, A., (2009), Benchmark Study on Numerical Simulation Methods for the Prediction of Parametric Roll of Ships in Waves, Proceedings of the 10th International Conference on Stability of Ships and Ocean Vehicles, St. Petersburg, pp. 1–9.

Thompson, J. M. T. and Stewart, H. B., (2002), Nonlinear Dynamics and Chaos, second edition, John Wiley & Sons. Led (Chichester), pp. 191–192.

Umeda, N., Hashimoto, H., Tsukamoto, I. and Sogawa, Y., (2011), Estimation of Parametric Resonance in Random Waves, Parametric Resonance in Dynamical Systems (Fossen, T. and Nijmeijer, H. eds.), Springer, pp. 45–62.

Chapter 17
Investigation on Parametrically Excited Motions of Spar Platforms in Waves

Claudio A. Rodríguez and Marcelo A. S. Neves

Abstract As offshore oil exploration goes into deeper waters, spar platforms appear as a good alternative for oil field developments due to their inherent hydrodynamic behaviour regarding its vertical motions response in waves. Spar's long natural periods in heave and pitch guarantee the good linear responses of the associated motions; however, nonlinear unstable motions can be triggered. Many numerical and experimental investigations have put forward the reasoning that this kind of floating structure is prone to parametrically induced motions due to changes in their pure hydrostatic restoring. Based on an analytical model, the present paper demonstrates that the main contribution to parametric excitation comes from variations in the pressure field associated to the incident wave and not from purely nonlinear hydrostatic actions. A nonlinear mathematical model based on Taylor series expansions, Neves and Rodríguez (Ocean Eng 33(14):1853–1883, 2006), is employed to explain the underlying mechanism that leads to the phenomenon of Mathieu instability in vertical deep drafted cylinders. Analytical expressions are derived for the nonlinear hydrostatic and Froude-Krilov actions. A set of coupled time-dependent equations is obtained. Based on that, general conditions for the appearance of principal resonances are derived. These analytical results are verified by means of numerical simulations. Numerical analyses are carried out for regular wave conditions and a domain of parametric amplifications is obtained.

Keywords Spar platforms · Stability · Parametric roll · Nonlinear dynamics

17.1 Introduction

As offshore oil exploration goes into deeper waters, spar platforms appear as a good alternative for oil field developments due to their inherent hydrodynamic behavior regarding its vertical motion responses in waves. Generally speaking, spar's long

C. A. Rodríguez · M. A. S. Neves (✉)
LabOceano, COPPE/UFRJ, Rio de Janeiro, Brazil
e-mail: masn@peno.coppe.ufrj.br

V. L. Belenky et al. (eds.), *Contemporary Ideas on Ship Stability*, Fluid Mechanics and Its Applications 119, https://doi.org/10.1007/978-3-030-00516-0_17

natural periods in heave and pitch guarantee good linear responses of these motions. However, nonlinear unstable motions can be triggered when heave resonance produces large heave motions even with relatively small wave excitation. Under these circumstances and due to nonlinear restoring coupling among the vertical modes, roll motions can also be amplified. This phenomenon in spar platforms has been studied in earlier works (Haslum and Faltinsen 1999; Rho et al. 2002; Koo et al. 2004; Hong et al. 2005) as Mathieu type instabilities associated with time dependent characteristic of roll/pitch hydrostatic restoring.

In the pertinent literature a common assumption is that Froude-Krilov forces and moments are not relevant for the stability problem of spar platforms, Bin-Bin et al. (2011), Koo et al. (2004). One of the aims of the present paper is to examine the validity of such simplifying hypothesis. One of the consequences of disregarding wave passage effect is that roll and pitch restoring actions become the same and some confusion may then be established between pitch (directly excited) and the internally excited roll motion.

In a more general perspective, a literature review reveals that there are many different mathematical models for spar dynamic analysis. Liao and Yeung (2001), Koo et al. (2004), Liaw et al. (1993), Liaw (1994) models consider the roll (pitch) affected by heave but heave not affected by roll (pitch). In this context the authors of the present paper felt it would be appropriate to revisit the problem of parametric resonance of spar platforms employing a purely analytical mathematical model previously developed for parametric rolling of ships and recently expanded to a proper modelling of deep drafted bodies, Rodríguez (2010), Rodríguez and Neves (2012a) . Hopefully, as an analytical model, it may contribute to a better understanding of the problem, to be achieved in a rational way.

Employing a panel method Neves et al. (2008) advocated the need for heave-roll-pitch complete coupling and the consideration of nonlinear Froude-Krilov in all modes for a robust modelling of parametric roll excitation of spar platforms. The relevance of non-linear Froude-Krilov actions in inducing unstable roll motions had already been introduced by Neves and Rodríguez (2006) in the context of parametric rolling of ships. Kleiman and Gotlieb (2008) applied the same reasoning for ships with vertical walls. Liu et al. (2010) numerically investigated heave/pitch nonlinear coupling including nonlinear wave passage effects in both regular and irregular waves.

Two-fold objectives for the paper have been defined: (a) consolidate an analytically derived coupled model for parametric resonance of deep drafted spar platforms in waves and (b) demonstrate that the analytical model is capable of revealing complex nonlinearities under resonant conditions.

Haslum and Faltinsen (1999) reported on some few test results with a very small 1:300 model scale of a spar, in which large angles have been reached somewhat away from the Mathieu tuning corresponding to the first region of instability. The paper gave an indication that parametric excitation may take place at large drafted vertical circular cylinders, but their experimental results were scarce, not providing a consistent set of data for comparisons. In the present study, a simplified form of a spar platform tested by Hong et al. (2005) is used for the numerical investigations, but only qualitative reference is made to these results.

The present analytical description of the restoring actions takes into account non-linear coupling terms in the heave, pitch and roll modes, wave elevation profile and wave pressure field variation along the hull, Rodríguez and Neves (2012a). With this new modelling for restoring actions, spar motions are numerically investigated. Results are presented for typical unstable conditions reported in other authors' previous experimental works. Parametric amplification domain (PAD), a plot of instability regions with response amplitudes due to parametric resonance for different excitation frequencies and wave amplitudes, is computed for the platform. Some interesting dynamical characteristics are revealed and discussed, in particular the quantitative contribution of pure hydrostatics and Froude-Krilov to the inception of unstable motions.

17.2 Equations of Motion

The floater's behaviour in waves can be described by a set of equations based on the second Newton's law. As main interest is on vertical motions (heave, roll and pitch) and the resonant phenomena associated to them in longitudinal waves, only 3-DOF need be considered and the restoring actions (including wave effects) should be described nonlinearly. Forces/moments of other nature such as hydrodynamic reactions (added mass and potential damping), or diffraction forces/moments can be described using the linear approach. Viscous effects in the damping force and moments are considered in an approximated way, based on empirical evidence.

Body and wave motions are described by making use of coordinate axes positioned at the mean water-surface. The equations of motion for the spar platform become:

$$(m + Z_{\ddot{z}})\ddot{z} + Z_{\ddot{\theta}}\ddot{\theta} + Z_{\dot{z}}\dot{z} + Z_{\dot{\theta}}\dot{\theta} + Z(z, \phi, \theta) = Z_{w0}\cos(\omega t + \alpha_z) \quad (17.1)$$

$$(J_x + K_{\ddot{\phi}})\ddot{\phi} + K_{\dot{\phi}}\dot{\phi} + K(z, \phi, \theta) = 0 \quad (17.2)$$

$$(J_y + M_{\ddot{\theta}})\ddot{\theta} + M_{\ddot{z}}\ddot{z} + M_{\dot{z}}\dot{z} + M_{\dot{\theta}}\dot{\theta} + M(z, \phi, \theta) = M_{w0}\cos(\omega t + \alpha_\theta) \quad (17.3)$$

where the double dot terms denote accelerations, single dot terms denote velocities and coefficients with subscripts associated to these terms are added mass and damping coefficients, respectively. Hull mass is m and J_x and J_y are mass moments of inertia in roll and pitch, respectively. For damping terms, instead of adopting a quadratic damping model to represent viscous effects, it was considered an equivalent linear damping approach in percentage of the critical damping (2.0% for heave, and 1.9% for roll and pitch). The adopted damping levels are similar to those obtained by Hong et al. (2005). Z_{w0} and M_{w0} represent the amplitudes of the linear wave exciting force and moment in heave and pitch, respectively, while α_z and α_θ denote the phase angles of the heave and pitch wave exciting actions with respect to the wave elevation.

17.3 Restoring Actions

Restoring actions in the case of a SPAR platform have been derived in Rodríguez and Neves (2012a). These may be expressed as:

$$\begin{aligned} Z(z,\phi,\theta) &= Z_H + Z_W \\ &= Z_{CW}(z) + \frac{1}{2}Z_{\phi\phi z}\phi^2 z + \frac{1}{2}Z_{\theta\theta z}\theta^2 z \\ &\quad + Z_{\zeta z}(t)z + Z_{\zeta\theta}(t)\theta + Z_{\zeta\phi\phi}(t)\phi^2 \end{aligned} \tag{17.4}$$

$$\begin{aligned} K(z,\phi,\theta) &= K_H + K_W \\ &= K_{CW}(\phi) + \frac{1}{2}\left[K_{zz\phi}z^2 + K_{\theta\theta\phi}\theta^2\right]\phi \\ &\quad + \left[K_{\zeta\phi}(t) + K_{\zeta z\phi}(t)z + K_{\zeta\phi\theta}(t)\theta + K_{\zeta\zeta\phi}(t)\right]\phi \end{aligned} \tag{17.5}$$

$$\begin{aligned} M(z,\phi,\theta) &= M_H + M_W \\ &= M_{CW}(\theta) + \frac{1}{2}M_{zz\theta}z^2\theta + \frac{1}{2}M_{\phi\phi\theta}\phi^2\theta \\ &\quad + M_{\zeta z}(t)z + M_{\zeta\theta}(t)\theta + M_{\zeta\phi\phi}(t)\phi^2 \end{aligned} \tag{17.6}$$

where $Z_{CW}(z) = \rho g A_0 z$ is the linear restoring coefficient in heave, $K_{CW}(\phi)$ and $M_{CW}(\theta)$ are the pure calm water restoring moments (here adjusted up to third order) in roll and pitch, respectively. Clearly, its linear term is $\rho g \nabla_0 \overline{GM_0}$, where ρ is water density, g is gravity acceleration, A_0 is the waterline area, ∇_0 is the displaced volume and $\overline{GM_0}$ is the initial metacentric height. In the above equations subscripts H and W indicate restoring actions due to hydrostatic and wave pressure fields, respectively. Wave term coefficients are denoted by their first subscript ζ, which corresponds to:

$$\zeta = A_w \cos(kX - \omega t) \tag{17.7}$$

where A_w and ω are wave amplitude and frequency, respectively, and k is wave number. Equation (17.7) is a sinusoidal long-crested wave train system described in a *CXYZ* inertial reference frame; the well known corresponding incident velocity potential (assuming deepwater) is given by:

$$\Phi_I = A_w \left(\frac{g}{k}\right)^{1/2} e^{kZ} \sin(kX - \omega t) \tag{17.8}$$

Analytical expressions (valid for general hull forms) of the non-zero nonlinear Froude-Krilov coefficients appearing in Eqs. (17.4–17.6) are given in Tables 17.1, 17.2 and 17.3 for the heave, roll and pitch modes, respectively, see also Rodríguez and Neves (2012a).

The expressions presented in those tables are functions of the geometric parameters of the transverse sections of the hull, so that $\bar{x}$, $\bar{y}$ are coordinates of waterline, S_b is the transversal submersed area, and $\bar{z}_b$ is the vertical coordinate of its centroid.

Table 17.1 Heave F-K restoring coefficients

$Z_{\zeta z}(t) = 2\rho g \int\limits_L \left(\frac{\partial \bar{y}}{\partial \bar{z}} - \bar{y} k e^{k\bar{z}_b}\right)\zeta dx$
$Z_{\zeta\theta}(t) = -2\rho g \int\limits_L \left(\frac{\partial \bar{y}}{\partial \bar{z}} - \bar{y} k e^{k\bar{z}_b}\right)\bar{x}\zeta dx$
$Z_{\zeta\zeta z}(t) = -2\rho g \int\limits_L \left(\frac{\partial \bar{y}}{\partial \bar{z}}\right) k e^{k\bar{z}_b}\zeta^2 dx$
$Z_{\zeta zz}(t) = \rho g \int\limits_L \left(\frac{\partial \bar{y}}{\partial \bar{z}}\right) k e^{k\bar{z}_b}\zeta dx$
$Z_{\zeta z\theta}(t) = -2\rho g \int\limits_L \left(\frac{\partial \bar{y}}{\partial \bar{z}}\right) k e^{k\bar{z}_b}\bar{x}\zeta dx$
$Z_{\zeta\phi\phi}(t) = -\rho g \int\limits_L \left[2\bar{y}\left(\frac{\partial \bar{y}}{\partial \bar{z}}\right)^2 + \bar{y} - \bar{y}^2\left(\frac{\partial \bar{y}}{\partial \bar{z}}\right) k e^{k\bar{z}_b}\right]\zeta dx$
$Z_{\zeta\zeta\theta}(t) = 2\rho g \int\limits_L \left(\frac{\partial \bar{y}}{\partial \bar{z}}\right) k e^{k\bar{z}_b}\bar{x}\zeta^2 dx$
$Z_{\zeta\theta\theta}(t) = \rho g \int\limits_L \left(\frac{\partial \bar{y}}{\partial \bar{z}}\right) k e^{k\bar{z}_b}\bar{x}^2\zeta dx$

Table 17.2 Roll F-K restoring coefficients

$K_{\zeta\phi}(t) = 2\rho g \int\limits_L \left(\bar{y}^2\frac{\partial \bar{y}}{\partial \bar{z}} - \frac{1}{3}\bar{y}^3 k e^{k\bar{z}_b} - S_b\bar{z}_b k e^{k\bar{z}_b}\right)\zeta dx$
$K_{\zeta\zeta\phi}(t) = \rho g \int\limits_L \left[2\bar{y}\left(\frac{\partial \bar{y}}{\partial \bar{z}}\right)^2 + \bar{y} - 2\bar{y}^2\left(\frac{\partial \bar{y}}{\partial \bar{z}}\right) k e^{k\bar{z}_b}\right]\zeta^2 dx$
$K_{\zeta z\phi}(t) = -\rho g \int\limits_L \left[4\bar{y}\left(\frac{\partial \bar{y}}{\partial \bar{z}}\right)^2 + 2\bar{y} - 2\bar{y}^2\left(\frac{\partial \bar{y}}{\partial \bar{z}}\right) k e^{k\bar{z}_b}\right]\zeta dx$
$K_{\zeta\phi\theta}(t) = \rho g \int\limits_L \left[4\overline{xy}\left(\frac{\partial \bar{y}}{\partial \bar{z}}\right)^2 + 2\overline{xy} - 2\overline{xy}^2\left(\frac{\partial \bar{y}}{\partial \bar{z}}\right) k e^{k\bar{z}_b}\right]\zeta dx$

It is important to notice that the above coefficients have, for general hull forms, two contributions:

(a) one is purely geometrical due to form variations (with longitudinal integrals of flare at mean water-line $\partial\bar{y}/\partial\bar{z}$ and other geometrical characteristics) and,
(b) the other one is due to wave elevation and pressure attenuation (so-called Smith effect). The Smith effect terms have been derived following the reasoning of Paulling (1961).

In the case of a spar with vertical walls, the geometrical terms containing $\partial\bar{y}/\partial\bar{z}$ in Tables 17.1, 17.2 and 17.3 are zero. Therefore, the only non-zero coefficients are those defined in Eqs. 17.4–17.6, which are summarized as follows:

$$Z_{\phi\phi z} = Z_{\theta\theta z} = K_{zz\phi} = M_{zz\theta} = \rho g A_0 \tag{17.9}$$

$$K_{\theta\theta\phi} = M_{\phi\phi\theta} = \rho g I_0 \tag{17.10}$$

Table 17.3 Pitch F-K restoring coefficients

$M_{\zeta z}(t) = -2\rho g \int_L \left(\frac{\partial \bar{y}}{\partial \bar{z}} - \bar{y}ke^{k\bar{z}_b}\right)\bar{x}\zeta dx$
$M_{\zeta\theta}(t) = 2\rho g \int_L \left(\frac{\partial \bar{y}}{\partial \bar{z}} - \bar{y}ke^{k\bar{z}_b}\right)\bar{x}^2\zeta dx$
$M_{\zeta\zeta z}(t) = 2\rho g \int_L \left(\frac{\partial \bar{y}}{\partial \bar{z}}\right)ke^{k\bar{z}_b}\bar{x}\zeta^2 dx$
$M_{\zeta zz}(t) = -\rho g \int_L \left(\frac{\partial \bar{y}}{\partial \bar{z}}\right)ke^{k\bar{z}_b}\bar{x}\zeta dx$
$M_{\zeta z\theta}(t) = 2\rho g \int_L \left(\frac{\partial \bar{y}}{\partial \bar{z}}\right)ke^{k\bar{z}_b}\bar{x}^2\zeta dx$
$M_{\zeta\phi\phi}(t) = \rho g \int_L \left[2\bar{y}\left(\frac{\partial \bar{y}}{\partial \bar{z}}\right)^2 + \bar{y} - \bar{y}^2\left(\frac{\partial \bar{y}}{\partial \bar{z}}\right)ke^{k\bar{z}_b}\right]\bar{x}\zeta dx$
$M_{\zeta\zeta\theta}(t) = -2\rho g \int_L \left(\frac{\partial \bar{y}}{\partial \bar{z}}\right)ke^{k\bar{z}_b}\bar{x}^2\zeta^2 dx$
$M_{\zeta\theta\theta}(t) = -\rho g \int_L \left(\frac{\partial \bar{y}}{\partial \bar{z}}\right)ke^{k\bar{z}_b}\bar{x}^3\zeta dx$

$$Z_{\zeta z}(t) = 2Z_{\zeta\phi\phi}(t) = -2\rho g \int_L \left(\bar{y}ke^{k\bar{z}_b}\right)\zeta dx \tag{17.11}$$

$$Z_{\zeta\theta}(t) = M_{\zeta z}(t) = 2\rho g \int_L \left(\bar{y}ke^{k\bar{z}_b}\right)\bar{x}\zeta dx \tag{17.12}$$

$$K_{\zeta\phi}(t) = -2\rho g \int_L \left(\frac{1}{3}\bar{y}^3 + S_b\bar{z}_b\right)ke^{k\bar{z}_b}\zeta dx \tag{17.13}$$

$$K_{\zeta z\phi}(t) = -2\rho g \int_L \bar{y}\zeta dx \tag{17.14}$$

$$K_{\zeta\phi\theta}(t) = 2M_{\zeta\phi\phi}(t) = \rho g \int_L 2\overline{xy}\zeta dx \tag{17.15}$$

$$K_{\zeta\zeta\phi}(t) = \rho g \int_L \bar{y}\zeta^2 dx \tag{17.16}$$

$$M_{\zeta\theta}(t) = -2\rho g \int_L \left(\bar{y}ke^{k\bar{z}_b}\right)\bar{x}^2\zeta dx \tag{17.17}$$

where I_0 is the second area moment of the waterplane. The coefficients defined in Eqs. (17.9–17.17) may be separated into three main groups:

(a) $Z_{\phi\phi z}$, $Z_{\theta\theta z}$, $K_{zz\phi}$, $K_{\theta\theta\phi}$, $M_{zz\theta}$ and $M_{\phi\phi\theta}$ are hydrostatic coefficients of third order terms;

(b) $K_{\zeta z\phi}$, $K_{\zeta\phi\theta}$, $K_{\zeta\zeta\phi}$ and $M_{\zeta\phi\phi}$ are moment coefficients associated to third order terms containing contributions from wave elevation ζ; and,
(c) $Z_{\zeta z}$, $Z_{\zeta\theta}$, $K_{\zeta\phi}$, $M_{\zeta z}$ and $M_{\zeta\theta}$ are second order coefficients representing influence of pressure variations due to pressure attenuation (Smith effect).

As discussed in Neves et al. (2008), parametric excitation in conditions close to the exact Mathieu tuning ($T_{n4} = 2T_w$) is governed by second order terms in Eqs. (17.4–17.6). Therefore it may be concluded that parametric excitations in spar platforms are strictly governed by the influence of the Smith effect. A complementary tentative conclusion derived from Eq. (17.13), to be verified by means of future specific experiments, is that the amplitude of roll parametric excitation shall be larger for spar platforms with larger drafts.

17.4 Numerical Simulations

A vertical cylindrical shape is used in the simulations as a simplified spar platform. The main characteristics are described in Table 17.4. An illustration of the cylinder is given in Fig. 17.1. The cylinder was taken with the same main dimensions and inertial characteristics of the spar investigated in Hong et al. (2005), but without consideration of the moon**pool**.

The set of differential Eqs. (17.1–17.3) with restoring actions as defined in Eqs. (17.4–17.17) can be efficiently solved for different combinations of wave periods and wave amplitudes. Steady-state roll amplitudes are captured and plotted in the form of a numerical mapping in which a colour-scale is introduced to reflect the intensity of roll responses, see Rodríguez and Neves (2012b). Figure 17.2 shows the obtained numerical mapping. Similar plots may in principle be obtained reflecting the heave and pitch tendency to display unstable responses.

Table 17.4 Spar main characteristics

Parameter	Value
Diameter, D (m)	37.20
Depth, H (m)	213.20
Draught, T (m)	198.10
Displacement, Δ (t)	220,691
Metacentricheight, $\overline{GM_o}$ (m)	10.08
Roll gyration radius, r^*_{xx} (m)	59.20
Pitch gyration radius, r^*_{yy} (m)	59.20
Heave natural period, T_{n3} (s)	29.91
Roll/pitch natural period, T_{n4}, T_{n5} (s)	49.00

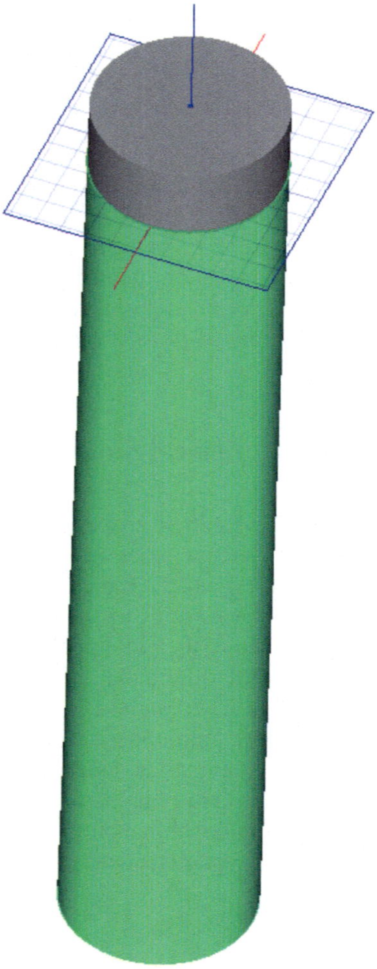

Fig. 17.1 Spar hull sketch

The domain in Fig. 17.2 does not display upper frontiers as is usual with many numerical assessments of ship hulls's limits of stability; but it has a back-bone clearly inclined to the left. This is compatible with effects associated with nonlinear detuning dependent on wave amplitude, as discussed in Rodríguez and Neves (2012b) in the case of a fishing vessel and a container vessel. The present results point-out to a spar with more peculiar results than the previously investigated spar reported in Rodríguez and Neves (2012a).

With the aim of getting a deeper insight of the coupling taking place, in the following few figures a closer look will be devoted to the heave, roll and pitch time-series at two selected points of the domain of parametric excitation.

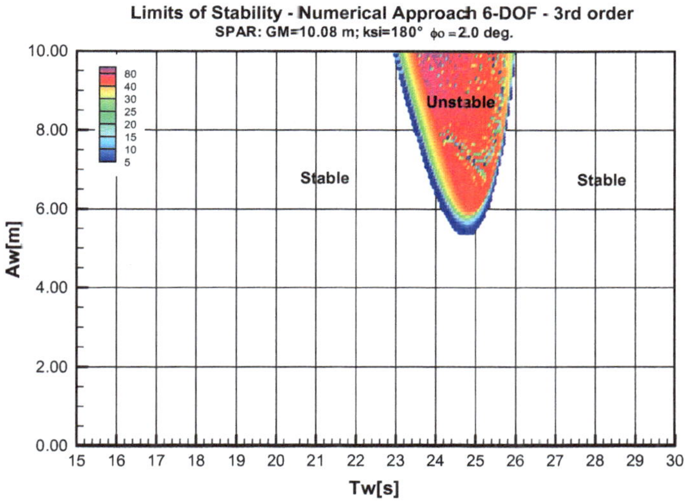

Fig. 17.2 Domain of roll parametric amplification in the region close to $T_{n4} = 2T_w$

Figure 17.3 shows the heave, roll and pitch responses for wave period $T_w = 24$ s, wave amplitude $A_w = 7.5$ m. The period in this case is slightly below half the roll natural period and slightly below the heave natural period. All modes have reached steady state responses after a slow growing process. The roll motion response period is clearly double the heave and pitch response periods. Roll amplitudes reach 40°, whereas pitch motions remain limited in the range of 2.2°. The graphs display results for both with Smith effect consideration and without Smith effect. In case Smith effect is not considered, roll mode is not excited at all. Pitch motion is not affected by the consideration of Smith effect. The heave mode is significantly affected when this effect is considered, mainly due to the coupling with roll motion which exists in this case.

Figure 17.4 shows time series for (a) instantaneous metacentric height ($\overline{GM}$) contributions: purely hydrostatic and wave induced, (b) restoring moment and (c) restoring force in heave, all of them computed for the same conditions of Fig. 17.2. Both, hydrostatic and Froude-Krilov terms take into account the Smith effect. In the upper graph, it is also possible to notice that the instantaneous hydrostatic metacentric height has only very small oscillations around the large initial metacentric height, $\overline{GM_0} = 10.08$ m. On the other hand, instantaneous metacentric height due to wave field pressures has larger amplitudes. Both functions respond mainly at wave period. It is evident that these oscillatory changes of the metacentric height due to wave field pressures will act as an internal parametric excitation. The middle graph gives the roll restoring moment, that is, the instantaneous $\overline{GM}$ multiplied by $\rho g \nabla_0$ times the instantaneous roll angle. The interesting results are that the hydrostatic contribution to the roll restoring moment takes place at a period slightly lower than the roll natural period; and the wave contribution has two periods. Lower graph shows heave restoring force which is realized at wave period.

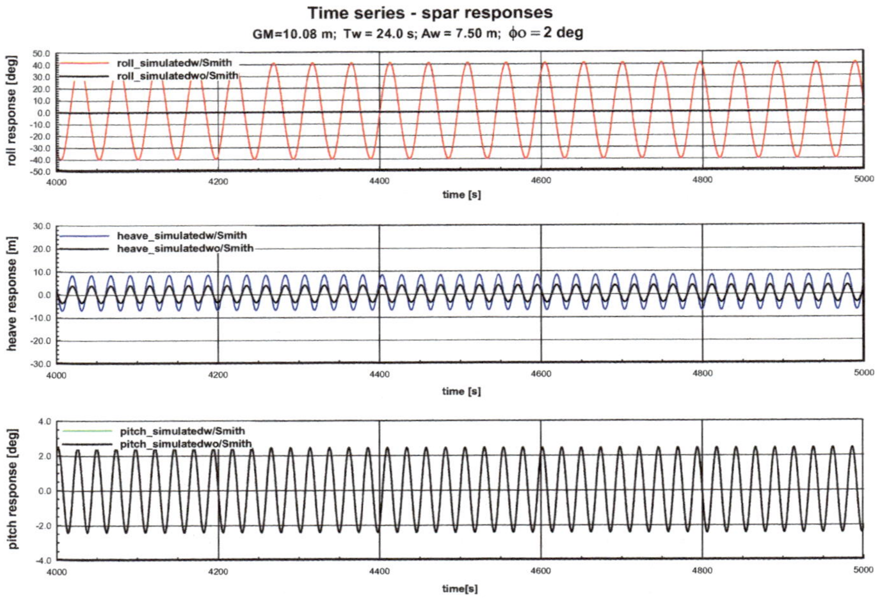

Fig. 17.3 Heave, roll and pitch responses, $T_w = 24$ s, $A_w = 7.5$ m

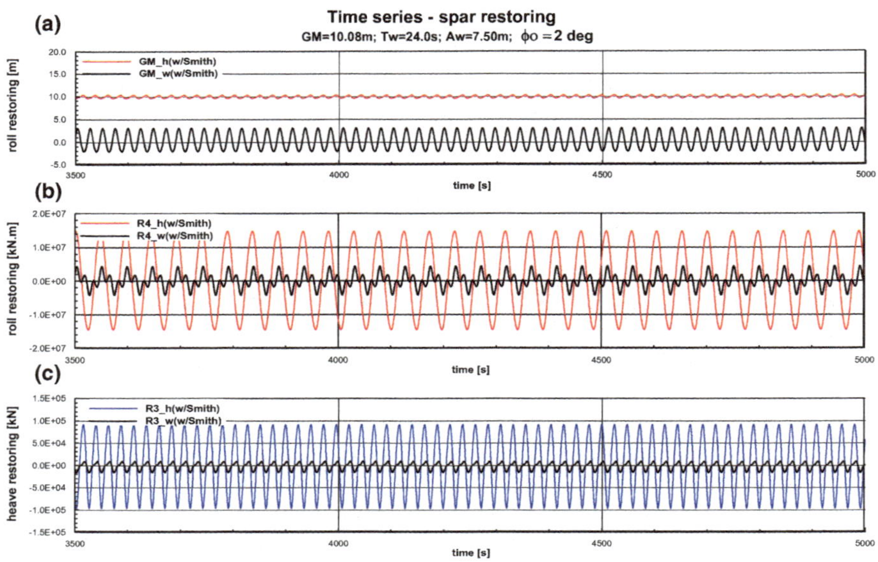

Fig. 17.4 **a** Instantaneous *GM* contributions: due to ship motions and due to wave pressure; **b** Roll restoring moment; **c** Heave restoring force, $T_w = 24$ s, $A_w = 7.5$ m

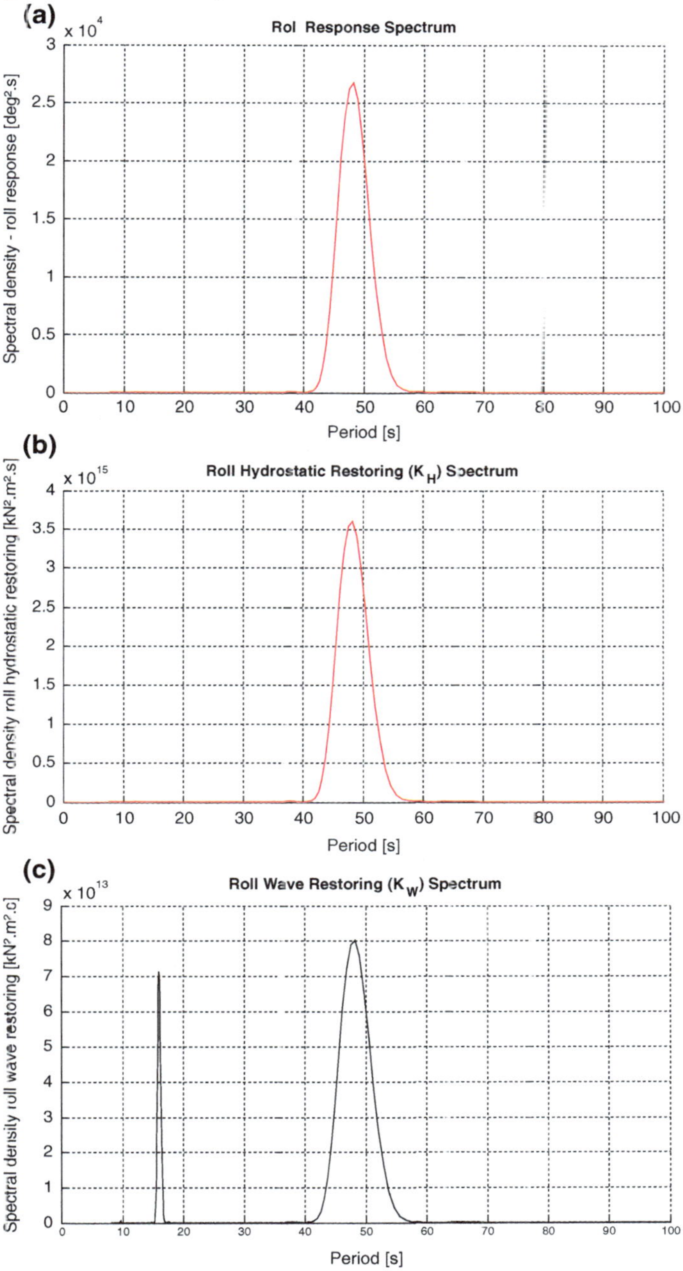

Fig. 17.5 Roll spectra for $T_w = 24$ s, $A_w = 7.5$ m: **a** roll response (from Fig. 17.3); **b** hydrostatic restoring moment; **c** wave restoring moment

In order to clarify the dynamical characteristics of the different responses, frequency decomposition of three roll-related signals are given in Fig. 17.5. In the two upper graphs it is observed that both roll motion and hydrostatic restoring moment spectra are centered close to roll natural period (in fact, double the wave period).

On the other hand frequency decomposition of roll wave restoring moment shows two contributions, a main one close to double the exciting period and a second one at about $T \approx 16$ s, a super-harmonic contribution at a third of the roll response period.

Now, let the exciting period be slightly above half the roll natural period, $T_{\mathrm{w}} = 25.5$ s, and the wave amplitude be the same as before. The results are given in Fig. 17.6. Now, a much more intense exchange of energy between the heave and roll modes takes place, whereas the pitch motion remains practically unaffected.

The coupling between heave and roll is of such nature that as roll amplification starts to develop (at roll natural period) heave motion increases (at wave period) correspondingly. Subsequently, as the heave motion reaches much larger values it induces a period detuning to the roll motion, which then progressively loses energy, leading, through coupling, to energy losses in heave.

Figure 17.7 shows the corresponding roll spectra for the conditions of Fig. 17.6, $T_{\mathrm{w}} = 25.5$ s, $A_{\mathrm{w}} = 7.5$ m. Roll spectrum and roll hydrostatic restoring moment spectrum are centered around a period which is double the wave period. Roll wave restoring moment spectrum has two contributions, a main one close to double the exciting period and a second one at about $T \approx 17$ s, a super-harmonic contribution at a third of the roll response period.

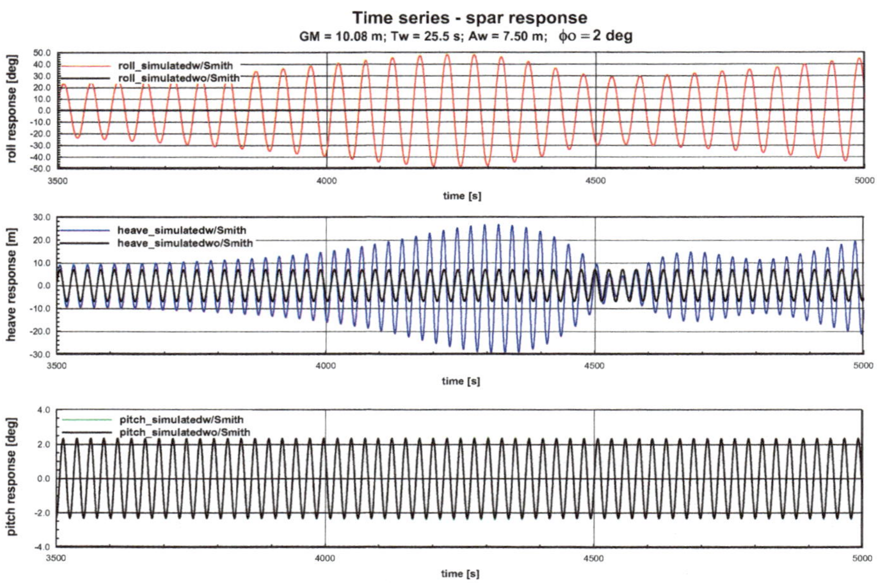

Fig. 17.6 Heave, roll and pitch motions, $T_{\mathrm{w}} = 25.5$ s, $A_{\mathrm{w}} = 7.5$ m

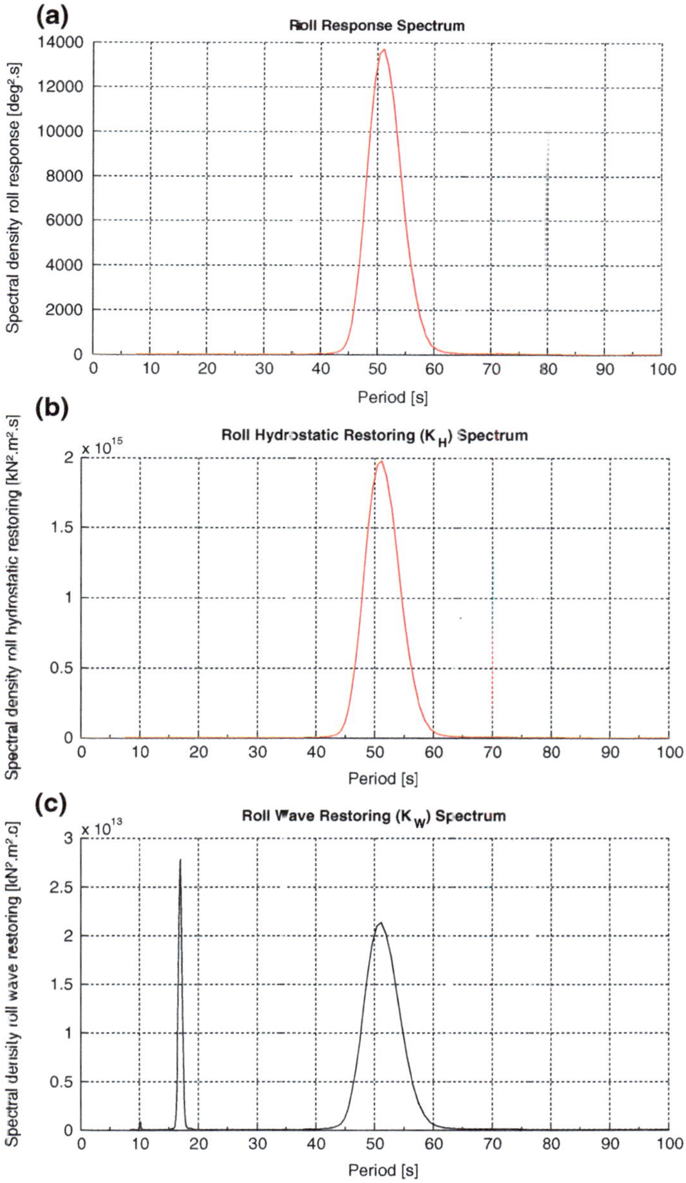

Fig. 17.7 Roll spectra for $T_w = 25.5$ s, $\mathcal{A}_w = 7.5$ m: **a** response; **b** hydrostatic restoring moment; **c** wave restoring moment

17.5 Conclusions

Unstable motions of a deep drafted spar platform have been investigated employing an analytical mathematical model previously developed for typical hull forms. It has been shown that in the case of deep drafted bodies additional terms associated with Smith effect are required in order to model the appearance of parametric roll.

Numerical limits of stability have been obtained; they display a back-bone clearly inclined to the range of lower periods. A dynamical analysis has been performed showing unexpected (at least for this type of vertically walled floater) levels of coupling between heave and pitch motions. Strong and complex interactions are obtained for periods close to the exact Mathieu tuning, $T_{n4} = 2T_w$.

It has also been shown that at this Mathieu tuning, hydrostatic pressure field does not induce any parametric amplification. Wave pressure field with its characteristic attenuation is the key element in the internal excitation of spar platforms.

It is concluded from the numerical/analytical analysis that a full assessment of spar responses in waves should involve heave, roll and pitch modes in a coupled frame. Finally, it has been found that externally and internally excited pitch motions do not undergo significant unstable responses. The situation is not the same in the case of internally excited roll motion, which may reach undesirable amplifications.

Future research shall contemplate:

(a) a dedicated experimental programme for validating the present research work;
(b) determination of a more precise damping model from roll decrement tests;
(c) check the analytically derived coefficients against their counterparts obtained from a more time consuming panel method algorithm;
(d) extension of the mathematical model to irregular seas, to be accomplished by means of frequency decomposition of wave-terms.

Acknowledgements The present investigation is supported by CNPq within the STAB project (Nonlinear Stability of Ships). The Authors also acknowledge financial support from CAPES, FAPERJ and LabOceano.

References

Bin–Bin Li, Jin-Ping Ou and Bin Teng, 2011. "Numerical Investigation of Damping Effects on Coupled Heave and Pitch Motion of an Innovative Deep Draft Multi-SPAR." Journal of Marine Science and Technology, Vol. 19, No 2, pp-231–244.

Haslum, H.A., and Faltinsen, O.M., 1999. "Alternative Shape of Spar Platform for Use in Hostile Areas", Proceedings of Offshore Technology Conference, Paper No. OTC10953, Houston.

Hong, Y., Lee, D., Choi, Y., Hong, S. and Kim, S., 2005. "An Experimental Study on the Extreme Motion Responses of a SPAR Platform in the Heave Resonant Waves." In: Proceedings of the 15th International Offshore and Polar Engineering Conference (ISOPE'2005), Seoul, Korea.

Kleiman, A. and Gotlieb, O., 2008. "Nonlinear Dynamics and Internal Resonances of a Ship with a Rectangular Cross-Section in Head Seas." In: Proceedings of the 27th ASME International

Conference on Offshore Mechanics and Arctic Engineering, OMAE 2008, Paper OMAE2008-57691, Lisbon, June, 15–20.
Koo, B.J., Kim, M.H. and Randall R.E., 2004. "Mathieu Instability of a Spar Platform with Mooring and Risers". Ocean Engineering, (31), pp 2175–2208.
Liao, S.-W. and Yeung, R. W., 2001. "Investigation of the Mathieu Instability of Roll Motion by a Time-Domain Viscous-Fluid Method". The 16th International Workshop on Water Waves and Floating Bodies. Hiroshima, Japan, April
Liaw, C. Y., Bishop, S. R. and Thompson, J. M. T., 1993, "Heave-Excited Rolling Motion of a Rectangular Vessel in Head Seas". International Journal of Offsore and Polar Engineering, vol. 3, no. 1 (Mar), pp. 26–31.
Liaw, C. Y., 1994. "Dynamic Instability of a Parametrically Excited Ship Rolling Model", International Journal of Offsore and Polar Engineering, vol. 4, no. 2 (Jun), pp. 106–111.
Neves, M.A.S. and Rodríguez, C.A., 2006. "On Unstable Ship Motions Resulting from Strong Non-Linear Coupling", Ocean Engineering, Vol. 33, no. 14 (Oct.), pp. 1853–1883.
Neves, M.A.S., Sphaier S.H., Mattoso, B.M., Rodríguez, C., Santos, A., Vileti, V. and Torres, F.G.S., 2008. "Parametric Resonance of Mono-column Structures". In: Proceedings of the 6th Osaka Colloquium on Seakeeping and Stability of Ships, 26–28th March, Osaka, Japan.
Paulling, J.R., 1961. "The Transverse Stability of a Ship in a Longitudinal Seaway". Journal of Ship Research, vol. 4, no. 4 (Mar.), pp. 37–49
Rho, J.B., Choi, H.S., Lee, W.C.Shin, H.S. and Park, I.K., 2002. "Heave and Pitch Motions of a SPAR Platform with Damping Plate." In: Proceedings of the 12th International Offshore and Polar Engineering Conference (ISOPE'2002), Kitakyushu, Japan.
Rodríguez, C.A., 2010. "On the Nonlinear Dynamics of Parametric Rolling". D.Sc. Thesis, COPPE - Universidade Federal do Rio de Janeiro, Brazil. (in Portuguese).
Rodríguez, C.A. and Neves, M.A.S., 2012a. "Nonlinear Instabilities of SPAR Platforms in Waves". Paper OMAE2012-83577, Rio de Janeiro, July.
Rodríguez, C.A. and Neves, M.A.S., 2012b. "Domains of Parametric Roll Amplification for Different Hull Forms". In: Fossen, T.I., Nijmeijer, H. (eds.), "Parametric Resonance in Dynamical Systems", https://doi.org/10.1007/978-1-4614-1043-0, Chapter 6, pp 107–127.
Yumin Liu, Hongmei Yan and Tin-Woo Yung, 2010. "Nonlinear Resonant Response of Deep Draft Platforms in Surface Waves", Paper OMAE2010-20823, Shanghai, China.

Chapter 18
A Study on Unstable Motions of a Tension Leg Platform in Close Proximity to a Large FPSO

Luis Alberto Rivera, Marcelo A. S. Neves, Roberto E. Cruz and Paulo de Tarso T. Esperança

Abstract The present paper elaborates on model experimental results obtained from tests conducted with a TLP connected to a nearby positioned FPSO. The tests revealed, at a given range of wave periods, the onset of unexpected large oscillatory yaw motions of the platform whereas the FPSO remained rather stable when the TLP was directly excited in sway. The paper summarizes the model experiments emphasizing the types of coupled motions taking place. It is observed that as the yaw motion develops increasing amplitudes the sway motion is reduced, pointing out to an interesting exchange of energy between the sway and yaw modes. A mathematical model is proposed to describe the main aspects of the two-body moored system. In principle a 12 DOF model is contemplated. Numerical simulations are compared to the time series obtained from the experiments showing adequate agreement.

Keywords Parametric resonance · Non-linear coupling
Tension leg wellhead platform

18.1 Introduction

Model experimental results obtained from tests with a TLP (in fact a Tension Leg Wellhead Platform) connected to a nearby positioned FPSO surprisingly revealed, at a given range of wave exciting periods, the onset of quite large oscillatory yaw motions whereas the FPSO remained rather stable when the TLWP was directly excited in sway. In such wave incidence the TLWP's yaw mode was not directly excited. Details of the experimental setup and results may be found in the NMRI Report, see Maeda et al. (2008) and in the reference Cruz et al. (2012).

The aim of the present paper is to contribute to a better understanding of these large yaw oscillations through numerical and analytical modeling. The line of investigation adopted here is that these large yaw motions are the result of internal excitations of the

L. A. Rivera · M. A. S. Neves (✉) · R. E. Cruz · P. de T. T. Esperança
COPPE/UFRJ, Rio de Janeiro, Brazil
e-mail: masn@peno.coppe.ufrj.br

V. L. Belenky et al. (eds.), *Contemporary Ideas on Ship Stability*, Fluid Mechanics and Its Applications 119, https://doi.org/10.1007/978-3-030-00516-0_18

two-body system induced by the directly excited sway motions, thus characterizing the occurrence of parametric resonance.

The paper summarizes the model experiments emphasizing the types of coupled motions taking place. An interesting exchange of energy between the sway and yaw motions is observed: as yaw motion develops increasingly unstable motions at twice the period of the waves, the directly excited sway motion (taking place at wave period) is reduced. This could be recognized as a revealing aspect of strongly non-linearly coupled parametric resonant motions. After the description of the experimental results of the two-body moored system (in principle a 12 DOF problem) the proposed mathematical model is introduced. Physical arguments are invoked for numerically solving the problem in 7 DOF. Using this reduced model numerical simulations are compared to experimental results, see Cruz et al. (2012). In general, these comparisons show good agreement, indicating that the main nonlinear couplings are well captured by the numerical model. In order to provide a better understanding of the complex dynamic structure responsible for the observed TLWP's yaw amplifications, an analytical mathematical modeling of the surge, sway and yaw coupling is introduced. The final aim here is to develop an analytical tool that would allow for a rationally exercised dissection of the system, thus having an anatomical description of the system for a critical analysis. For this purpose the restoring actions introduced by the mooring lines, tendons and connecting lines will be decomposed by means of multivariable Taylor series. It is expected that this methodology will help to clarify what are the mooring lines inducing internal excitation to the system, therefore contributing to improved design of the two-body system free from unacceptable unstable motions.

18.2 The FPSO and TLWP

Figure 18.1 shows the upper view of the layout set up during the tests: on the left the 50×50 m Tension Leg Wellhead Platform (TLWP), the large FPSO (length 297 m) on the right. The FPSO model is moored by four oblique horizontal mooring lines displayed at the bow and stern of the hull. The FPSO is connected to the TLWP by two lines, indicated as M7 and M8 in Fig. 18.2. The TLWP has two oblique horizontal mooring lines displayed in the opposite side of its connections to the FPSO. The main mooring system of the TLWP is provided by four vertical tendons. See Cruz et al. (2012) for complete details on the bodies and mooring systems. In all cases considered in the present report the two-body system was directly excited by regular waves coming from the left to the right. The model scale adopted in the experiments was 1:100. The two-body system is described in principle by 12 degrees of freedom (nomenclature defined in Table 18.2).

Figure 18.2 illustrates the FPSO (M1 to M4) horizontal moorings, 4 TLWP tendons (T1 to T4), 2 TLWP mooring lines (M5 and M6) and two lines connecting the two units (M7 and M8). TLWP tendons length and diameter (real scale) are 975 and 1.1 m, respectively.

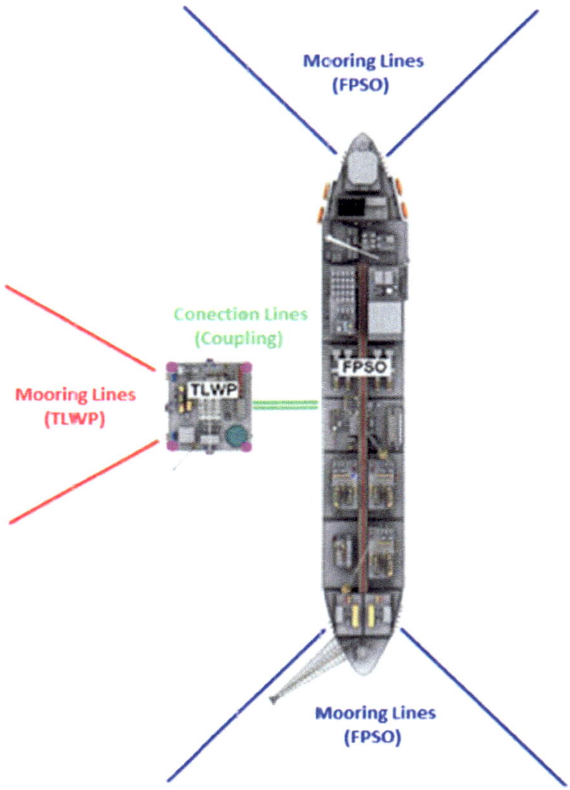

Fig. 18.1 Two-body arrangement showing the mooring set-up

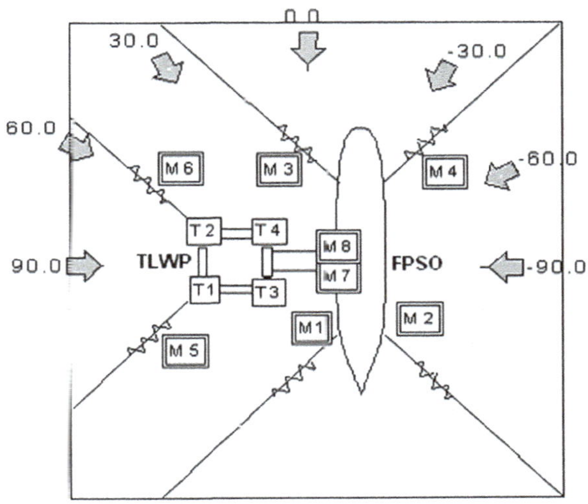

Fig. 18.2 Mooring setup at the tests. Connecting lines M7 and M8 have both 50 m in length

18.3 Analysis of Test Results

Steady state motion amplitudes (divided by wave amplitude) observed in the whole range of tested periods for the FPSO in surge, sway, heave and roll modes. showed limited responses in the whole range of periods. The surprising characteristic that appeared in the tests was the appearance of large yaw motions for the TLWP in the range of periods 12–15 s (yaw natural period was 33.6 s). These are typically resonant motions, which take place despite the fact that the yaw motion was not directly excited in the series of experiments under discussion. The TLWP surge mode was also observed to be larger at this range of periods. It is important to notice that the surge mode, as the yaw mode, is not externally excited. Taking into account these unexpected features it may be relevant to investigate the sway characteristics of the TLWP, which is an externally excited mode. In particular, a peculiar two-peaked pattern of responses in the range of periods from 10 to 17 s, well above the sway natural period (Tny = 7.1 s). Additionally, it was observed that at the sway natural period there was no large response. Such pattern of responses indicates some level of strong nonlinear coupling between the sway and yaw modes in which an increase in yaw motions induces a corresponding reduction of the sway motions, possibly an exchange of energy between the two modes. This suggests that nonlinear parametric resonance may be governing the dynamics of the platform. A better perception of the type of prevailing resonant motion is gained when the experimental time series of sway and yaw motions—obtained from a resonant condition—are plotted together, as shown in Fig. 18.3. In this case the wave period is $T = 14$ s. The incident wave is originated at about $t = 300$ s. The figure shows that asymmetric sway motion starting at about 450 s develops at this wave period. After some cycles, symmetric yaw motions develop increasing amplitudes, reaching some stable steady state motion at around 800 s. It is observed that the yaw motion is responding at a period corresponding to nearly double the wave period, suggesting that parametric resonance in yaw is being developed fed by the directly excited sway motion. Another interesting aspect may be recognized: as the large yaw motions develop, there is a complementary reduction in the sway motion, suggesting an interesting strong nonlinear coupling between the two modes. This is in general the observed picture in the range of wave periods corresponding to yaw amplifications. In order to contribute to a better understanding of these complicated couplings and instabilities a mathematical model has been developed. It is expected that through analytical and numerical modeling an engineering diagnosis of the observed instabilization process may be reached and consequently modifications in the project may be proposed through rational analysis.

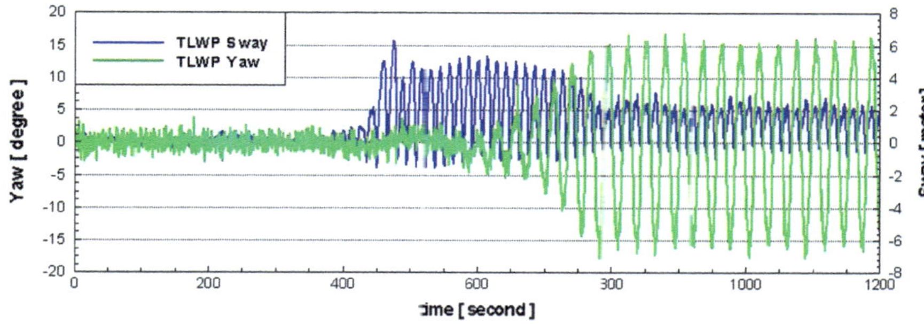

Fig. 18.3 TLWP sway and yaw time series from tests, period T = 14 s, wave height H = 2.8 m

18.4 Mathematical Model

The two-body system is defined by a 12-DOF nonlinear model expressed as:

$$(\mathbf{M} + \mathbf{A})\ddot{\mathbf{s}} + \mathbf{B_L}\dot{\mathbf{s}} + \mathbf{B_{NL}}(\dot{\mathbf{s}}) + \mathbf{c_r}(\mathbf{s}, \zeta) = \mathbf{c_{ext}}\left(\zeta, \dot{\zeta}, \ddot{\zeta}\right) \quad (18.1)$$

where the displacement vector is:

$$\mathbf{s}(t) = \left[x_1 \; y_1 \; z_1 \; \phi_1 \; \theta_1 \; \psi_1 \; x \; y \; z \; \phi \; \theta \; \psi \right]^T$$

In Eq. (18.1) **M** is a 12 × 12 matrix which describes hull inertia characteristics. **A** is also a 12 × 12 matrix, whose elements represent hydrodynamic generalized added masses. $\mathbf{B_L}$ and $\mathbf{B_{NL}}(\dot{\mathbf{s}})$ describe the hydrodynamic reactions dependent on the FPSO and platform velocities (damping). The latter incorporates non-linear terms due to the great influence of viscosity on tendons, pontoons and columns of the TLWP and roll motion of the FPSO. $\mathbf{c_r}(\mathbf{s}, \zeta)$ is a 12 × 1 vector which describes non-linear restoring forces and moments dependent on the relative motions between the platform and the mooring lines. It also considers the influence of the wave passage on the tendons tension. On the right hand side of Eq. (18.1), the generalized vector $\mathbf{c_{ext}}(\zeta, \dot{\zeta}, \ddot{\zeta})$ represents linear wave external excitation, composed of the Froude-Krilov and diffraction wave forcing terms, dependent on wave heading, excitation frequency ω_w, wave amplitude A_w and time t.

Taking into account the prevailing physics observed in the tests and the particular transversal wave excitation considered, without loss of generality the 12-DOF problem will be restricted to the following 7-DOF coupled problem: x_1 (surge), y_1 (sway), z_1 (heave) and ϕ_1 (roll) for the FPSO and x (surge), y (sway) and ψ (yaw) for the TLWP.

18.5 Hydrodynamic Coefficients and Non-linear Damping

Matrices **A** and **B** are computed by means of WAMIT computer code. The same applies to vector $\mathbf{c_{ext}}(\zeta, \dot{\zeta}, \ddot{\zeta})$. In addition to potential damping, additional damping has been considered for the following motions: (a) horizontal motions of TLWP; (b) horizontal motions of FPSO; roll motion of FPSO. Drag quadratic terms in Morison formula have been used as an approximation for calculation of non-linear damping of the TLWP, being applied to columns, tendons and pontoons. These actions are computed by taking into account the instantaneous velocities of the body, and disregarding the influence of fluid velocities. For some specific wave periods the TLWP responded with large angular displacements in yaw. For this reason a quadratic damping model was considered for this platform motion mode.

Assuming an angular displacement ψ with corresponding yaw velocity $\dot{\psi}$ and linear velocities $\dot{x}$ and $\dot{y}$, the drag forces and moment for each TLWP column are expressed as:

$$X(\dot{x}, \dot{y}, \dot{\psi}) = -\frac{1}{2}\rho S_c C_{dc}\left[\dot{x} - \dot{\psi}.R.\sin(\psi + \beta)\right].$$
$$\left\{\dot{x}^2 + \dot{y}^2 + \dot{\psi}^2.R^2 - 2.\dot{\psi}.R[\dot{y}.\cos(\psi + \beta) - \dot{x}.\sin(\psi + \beta)]\right\}^{1/2} \tag{18.2}$$

$$Y(\dot{x}, \dot{y}, \dot{\psi}) = -\frac{1}{2}\rho S_c C_{dc}\left[\dot{y} + \dot{\psi}.R.\cos(\psi + \beta)\right].$$
$$\left\{\dot{x}^2 + \dot{y}^2 + \dot{\psi}^2.R^2 - 2.\dot{\psi}.R[\dot{y}\ \cos(\psi + \beta) - \dot{x}\ \sin(\psi + \beta)]\right\}^{1/2} \tag{18.3}$$

$$N(\dot{x}, \dot{y}, \dot{\psi}) =$$
$$-\frac{1}{2}\rho S_c C_{dc}\left[\dot{\psi}.R^2 + \dot{y}.R\ \cos(\psi + \beta) - \dot{x}.R\ \sin(\psi + \beta)\right].$$
$$\left\{\dot{x}^2 + \dot{y}^2 + \dot{\psi}^2.R^2 - 2.\dot{\psi}.R[\dot{y}\ \cos(\psi + \beta) - \dot{x}\ \sin(\psi + \beta)]\right\}^{1/2} \tag{18.4}$$

where $\beta = n.\pi/4;\ n = 1, 3, 5, 7$. Finally, these are summed up to obtain the total drag forces and moment. In the above expressions ρ is density, S_c and C_{dc} are the representative area and drag coefficients of column, respectively, R is the distance from the centre of each column to the unit origin and β is the local angle of the element in relation to TLWP reference.

The procedure for calculating the drag force and moment in a tendon is similar to the preceding one. But in this case we need to integrate the force along the tendon length because the local velocity changes from zero at the ground up to a maximum at the TLWP. As a result of integration the total force on each tendon is one third of the force on the column:

$$X\left(\dot{x}, \dot{y}, \dot{\psi}\right) = -\frac{1}{6}\rho S_t C_{dt}\left[\dot{x} - \dot{\psi}.R.\sin(\psi + \beta)\right].$$
$$\left\{\dot{x}^2 + \dot{y}^2 + \dot{\psi}^2.R^2 - 2.\dot{\psi}.R\left[\dot{y}^2.\cos(\psi + \beta) - \dot{x}.\sin(\psi + \beta)\right]\right\}^{1/2} \tag{18.5}$$

$$Y(\dot{x}, \dot{y}, \dot{\psi}) = -\frac{1}{6}\rho S_t C_{dt}\left[\dot{y} + \dot{\psi}.R.\cos(\psi + \beta)\right]$$
$$\left\{\dot{x}^2 + \dot{y}^2 + \dot{\psi}^2.R^2 - 2.\dot{\psi}.R\left[\dot{y}^2\cos(\psi + \beta) - \dot{x}\sin(\psi + \beta)\right]\right\}^{1/2} \tag{18.6}$$
$$N(\dot{x}, \dot{y}, \dot{\psi}) = -\frac{1}{6}\rho S_t C_{dt}\left[\dot{\psi}.R^2 + \dot{y}.R\cos(\psi - \beta) - \dot{x}.R\sin(\psi + \beta)\right]$$
$$\left\{\dot{x}^2 + \dot{y}^2 + \dot{\psi}^2.R^2 - 2.\dot{\psi}.R\left[\dot{y}^2\cos(\psi + \beta) - \dot{x}\sin(\psi + \beta)\right]\right\}^{1/2} \tag{18.7}$$

finally, they are summed up to obtain the total drag forces and moment on tendons.

Regarding the TLWP pontoons: in order to simplify the calculation the coupled term between angular and linear velocities is not considered. A transformation matrix is used to transport translational velocities and drag forces from the local system to the global system:

$$N(\dot{\psi}) = \frac{-1/2\rho.S_p.L^3C_{dp}}{32}\dot{\psi}|\dot{\psi}| \tag{18.8}$$
$$X(\dot{x})_{local} = -\rho h L C_d \dot{x}_{local}|\dot{x}_{local}| \tag{18.9}$$
$$Y(\dot{y})_{local} = -\rho h L C_d \dot{y}_{local}|\dot{y}_{local}| \tag{18.10}$$
$$\begin{bmatrix}\dot{x}_{local}\\ \dot{y}_{local}\end{bmatrix} = \begin{bmatrix}\cos(\psi) & \sin(\psi)\\ -\sin(\psi) & \cos(\psi)\end{bmatrix}.\begin{bmatrix}\dot{x}\\ \dot{y}\end{bmatrix} \tag{18.11}$$
$$\begin{bmatrix}X(\dot{x})\\ Y(\dot{y})\end{bmatrix} = \begin{bmatrix}\cos(\psi) & -\sin(\psi)\\ \sin(\psi) & \cos(\psi)\end{bmatrix}.\begin{bmatrix}X(\dot{x})_{local}\\ Y(\dot{y})_{local}\end{bmatrix} \tag{18.12}$$

For the quadratic damping of the FPSO the drag coefficients (Cd) were experimentally obtained by means of captive tests of the hull under current action. The drag is therefore given as:

$$X_1(\dot{x}_1) = -0.5\,\rho\,S\,Cd\,\dot{x}_1|\dot{x}_1| \tag{18.13}$$
$$Y_1(\dot{y}_1) = -0.5\,\rho\,S\,Cd\,\dot{y}_1|\dot{y}_1| \tag{18.14}$$

In the case of the FPSO roll motion, instead of adopting a quadratic damping, it was considered an additional linear damping of 1.5% of the critical damping to represent viscous effects. This damping level was obtained by adjusting the theoretical roll response curve to the experimental response curve of the FPSO under beam regular wave excitation.

18.6 Restoring Actions and Numerical Analysis

These are composed of hydrostatic and individual mooring forces and moments (horizontal for the FPSO and TLWP, vertical for the tendons of TLWP). As TLWP vertical motions were neglected, the unit will not present hydrostatic restoring forces or moment. The platform will only present horizontal restoring forces and moment due to mooring lines, connecting lines and tendons. Hydrostatic restoring forces and moments are considered to be linear. Individual horizontal and vertical mooring actions are considered to be non-linear restoring actions. In the numerical analysis restoring forces and moments due to FPSO and TLWP mooring lines, TLWP tendons and connecting lines between the two bodies have been modeled taking into account their elastic characteristics and modes of displacement.

The numerical model has been described in detail in Cruz et al. (2012). In this section of the present paper some results from the numerical model are compared to experimental results. The interest here is to verify whether the numerical model is capable of capturing the complex coupling between the TLWP sway and yaw modes observed at the range of wave periods in which yaw amplification takes place, as shown in Fig. 18.3. For this purpose Fig. 18.4 shows comparisons of experimental and numerical time series of these motions for T=14 s, H=2.8 m. It is observed that the essential aspects of the coupling are captured by the numerical model: (a) sway motions responses are observed at wave period, yaw motions at double the wave period; therefore, the essential nonlinear sway-yaw exchange of energy is captured; (b) as the yaw amplitudes start to amplify the sway amplitudes move from an attractor of large amplitudes to another one of markedly lower sway amplitudes; (c) final sway and yaw amplitudes are in good agreement with the experimental results; (d) there are some differences between the transient phase of numerical and experimental amplification and decaying processes; these are clearly due to lack of information about the initial conditions prevailing in the experiments. These transient phase discrepancies may be mitigated when different initial conditions are considered in the numerical simulations, as discussed in Rodriguez and Neves (2012).

Figure 18.5 shows a comparison of TLWP's surge motion time series for $T = 12$ s, $H = 2.4$ m. It is observed that there is acceptable agreement both in terms of steady state response amplitude and period. These motions are slightly over-estimated, with responses taking place at double the exciting period, following the same pattern of the TLWP yaw motions.

TLWP sway motions mainly respond at wave period, whereas the surge and yaw respond at twice the excitation period. TLWP yaw motions are considerably large; numerical results compare very well with those from experiments. It is verified that surge responses take place at the same period as yaw. It has been pointed out that in the TLWP case neither surge nor yaw are directly excited by the incoming waves. As mentioned, the yaw natural period is $T_n\psi = 33.6$ s whereas the surge natural period is much larger, $T_{nx} = 64.5$ s. Therefore, for the exciting range of wave periods in which yaw amplifications are observed the first Mathieu tuning is attained, indicating that yaw motion is undergoing parametric resonance. At the same time yaw motion is leading surge motion at the same period of oscillation. This observed tendency may be appreciated in Fig. 18.6 which shows the surge, sway and yaw motions for $T =$ 13 s, $H = 2.6$ m: large sway motions induce yaw amplification which forces surge motion to respond at the yaw oscillating period.

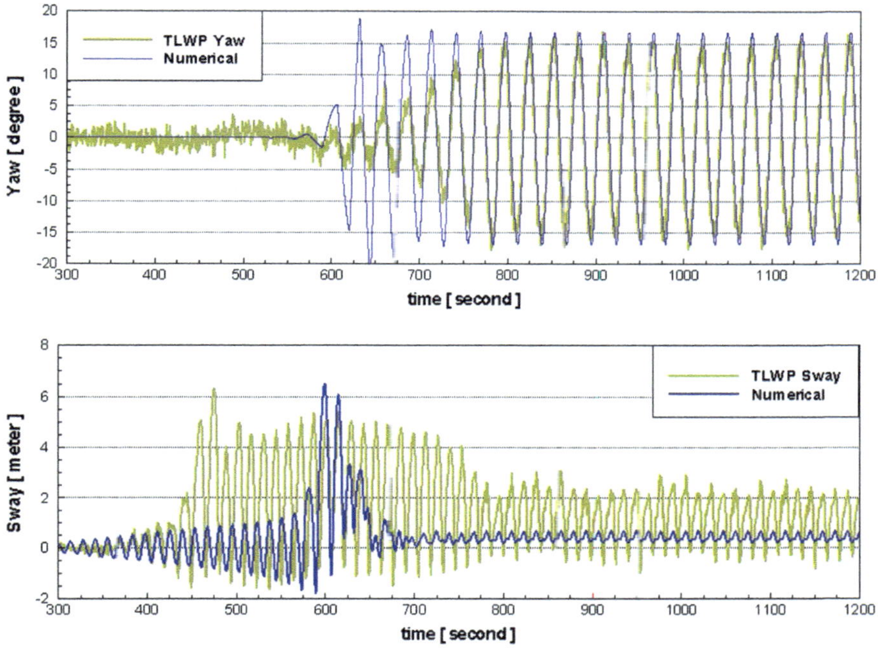

Fig. 18.4 Comparison of numerical and experimental sway and yaw motions

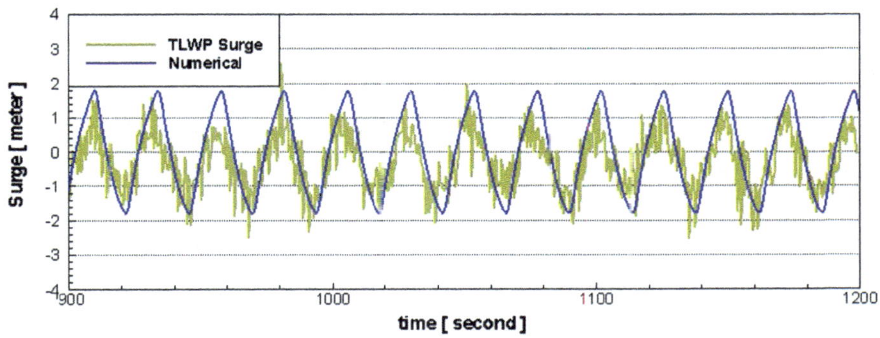

Fig. 18.5 Comparison of numerical and experimental TLWP surge motion, T = 12 s, H = 2.4 m

18.7 Analytical Model

In the present problem restoring terms are due to mooring lines, tendons and connecting lines. These expressions may be summed-up and expressed in terms of multivariable Taylor series expansions taken with respect to the mean equilibrium position. Expansions are here defined with terms up to third order of the expansions. As the

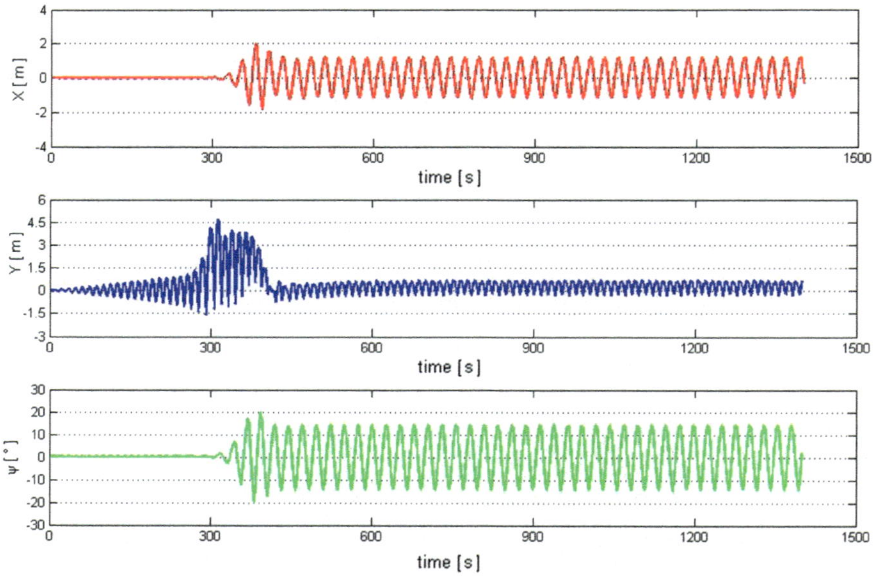

Fig. 18.6 Surge, sway and yaw motions, T = 13 s, H = 2.6 m

focus here lies in the surge, sway and yaw motions of the TLWP, it is assumed that the FPSO is fixed at its mean position for the assessment of the connecting actions.

Adopting a nomenclature similar to that of Neves and Rodríguez (2007), non-linear restoring actions may be expressed as:

$$\mathbf{c_r} = \mathbf{c}^{(1)}_{r(\mathbf{s})} + \left(\mathbf{c}^{(2)}_{r(\mathbf{s})} + \mathbf{c}^{(2)}_{r(\zeta)}\right) + \mathbf{c}^{(3)}_{r(\mathbf{s})}$$
$$= \begin{bmatrix} X^{(1)}_{r(\mathbf{s})} \\ Y^{(1)}_{r(\mathbf{s})} \\ N^{(1)}_{r(\mathbf{s})} \end{bmatrix} + \begin{bmatrix} X^{(2)}_{r(\mathbf{s})} + X^{(2)}_{r(\zeta)} \\ Y^{(2)}_{r(\mathbf{s})} + Y^{(2)}_{r(\zeta)} \\ N^{(2)}_{r(\mathbf{s})} + N^{(2)}_{r(\zeta)} \end{bmatrix} + \begin{bmatrix} X^{(3)}_{r(\mathbf{s})} \\ Y^{(3)}_{r(\mathbf{s})} \\ N^{(3)}_{r(\mathbf{s})} \end{bmatrix} \quad (18.15)$$

where superscripts (18.1)–(18.3) refer to first, second and third order restoring terms, respectively:

$$X^{(1)}_{r(\mathbf{s})} = X_x x + X_\psi \psi$$
$$Y^{(1)}_{r(\mathbf{s})} = Y_y y$$
$$N^{(1)}_{r(\mathbf{s})} = N_\psi \psi + N_x x \quad (18.16)$$

Second order actions are composed of two terms. Subscripts (s) refer to body motions, whereas subscripts (ζ) refer to wave passage effects on the tendons described above. According to the derivations, second order restoring actions (motions) may be defined as:

$$\begin{aligned} X_{r(\mathbf{s})}^{(2)} &= X_{xy}xy + X_{y\psi}y\psi \\ Y_{r(\mathbf{s})}^{(2)} &= \frac{1}{2}\left[Y_{xx}x^2 + 2Y_{x\psi}x\psi + Y_{yy}y^2 + Y_{\psi\psi}\psi^2\right] \\ N_{r(\mathbf{s})}^{(2)} &= N_{xy}xy + N_{y\psi}y\psi \end{aligned} \tag{18.17}$$

and third order restoring actions as:

$$\begin{aligned} X_{r(\mathbf{s})}^{(3)} &= \frac{1}{6}\begin{bmatrix} X_{xxx}x^3 + 3X_{xx\psi}x^2\psi + 3X_{yyx}y^2x \\ +3X_{yy\psi}y^2\psi + 3X_{\psi\psi x}\psi^2x + X_{\psi\psi\psi}\psi^3 \end{bmatrix} \\ Y_{r(\mathbf{s})}^{(3)} &= \frac{1}{6}\begin{bmatrix} Y_{yyy}y^3 + 3Y_{xxy}x^2y + 3Y_{\psi\psi y}\psi^2y \\ +6Y_{xy\psi}xy\psi \end{bmatrix} \\ N_{r(\mathbf{s})}^{(3)} &= \frac{1}{6}\begin{bmatrix} N_{xxx}x^3 + 3N_{xx\psi}x^2\psi + 3N_{yyx}y^2x \\ +3N_{yy\psi}y^2\psi + 3N_{\psi\psi x}\psi^2x + N_{\psi\psi\psi}\psi^3 \end{bmatrix} \end{aligned} \tag{18.18}$$

Since platform vertical motions are not considered, the length of the tendons may be assumed to remain constant for small horizontal displacements; however, it does not imply that there is no variation of the tension on the vertical tendons. The tension may be varying as a result of the cyclic wave vertical force on the platform. Thus, a restoring force depending on the tension will be varying in time. The expressions for the tendons restoring forces and moment due to a linear variation are given as:

$$X_T = \left[\frac{4T}{L_0}\right]x;\quad Y_T = \left[\frac{4T}{L_0}\right]y;\quad N_T = \left[\frac{4T.R^2}{L_0}\right]\psi \tag{18.19}$$

where R is the distance of the tendon to the centre of reference at the platform. Considering the wave vertical force on the platform for each tendon:

$$T(t) = T_0 + \frac{Z_w(t)}{4} \tag{18.20}$$

Each of the above contributions will be composed of a pre-tension T_0 plus a harmonic contribution due to wave action. In regular waves, vertical wave exciting force on the platform may be assumed as a simple harmonic:

$$Z_w(t) = Z_{w0}\cos(\omega_w t + \gamma_{zw}) \tag{18.21}$$

Following the same nomenclature, the wave passage contributions are represented as:

$$X^{(2)}_{r(\zeta)} = X_{\zeta x}(t)x$$
$$Y^{(2)}_{r(\zeta)} = Y_{\zeta y}(t)y$$
$$N^{(2)}_{r(\zeta)} = N_{\zeta\psi}(t)\psi \tag{18.22}$$

These are second order terms in the sense that they represent first order vertical wave force $Z_w(t)$ multiplied by first order displacements x, y and ψ. The implication here is that, as indicated above, in principle the tendons may induce small time-dependent internal excitations to the surge, sway and yaw modes of the platform.

Each coefficient in Eqs. (18.16–18.19) corresponds to constant derivatives, each one being the sum of three contributions: mooring lines, tendons and connecting lines. The derivatives were obtained by means of MAPLE computer code. In general the obtained expressions of these coefficients are too long to be presented here. For this reason, only the corresponding numerical values will be shown here. Therefore, considering the above expressions for Eqs. (18.16, 18.17, 18.18 and 18.22), complete restoring coupling is provided between surge, sway and yaw modes for the TLWP. Numerical values of all constant restoring coefficients are given in Tables 18.1, 18.2 and 18.3 (already including the Taylor series multipliers) for the separated influence of tendons, mooring lines and connecting lines, respectively.

Sub-harmonic resonance may be induced mainly by second order terms, as discussed by Neves and Rodríguez (2007) , corresponding to responses at twice the exciting period, the first zone of instability of the Ince-Strutt diagram. Obviously

Table 18.1 Restoring coefficients—tendons

		X	Y	N
1st order	x	−130.3		
	y		−130.3	
	ψ			−94,045.1
3rd order	x³	−7.0E−02		
	x².y		−7.0E−02	
	x².ψ			−1.0E+02
	y³		−7.0E−02	
	y².x	−7.0E−02		
	y².ψ			−1.0E+02
	ψ³			−2.1E+4
	ψ².x	−1.0E+02		
	ψ².y		−1.0E+02	
	x.y.ψ			

Table 18.2 Restoring coefficients—mooring lines

		X	Y	N
1st order	x	−51.4		283.4
	y		−226	
	ψ	283.4		−36,763
2nd order	x^2		−8.7E−02	
	xy	−1.7E−01		-8.6
	xψ		−8.6E+00	
	y^2		−6.3E−02	
	yψ	−8.6E+00		−8257.7
	ψ^2		−4.1E+03	
3rd order	x^3	−5.5E−05		−5.0E−03
	$x^2.y$		4.1E−05	
	$x^2.\psi$			−6.2E+00
	y^3		9.9E−05	
	$y^2.x$	4.1E−05		7.0E−03
	$y^2.\psi$	7.0E−03		6.6E−01
	ψ^3	−2.2E+02		−1.3E+05
	$\psi^2.x$	−6.2E+00		−6.7E+02
	$\psi^2.y$		6.6E−01	
	x.y.ψ		1.4E−02	

first order terms do not induce parametric resonance and third order terms induce resonance at the wave period and a complementary period shift, as demonstrated in Neves and Rodríguez (2007). It is realized that Eqs. (18.17 and 18.22) are therefore the ones with direct interest in the present analysis. They reveal that, at least in principle, all three modes may undergo coupled parametric resonance. A more specific argument may be given to substantiate the central role of parametric resonance in the observed instabilities if only the sway-yaw coupling is discussed: considering the second order yaw restoring term $N_{y\psi}y\psi$ defined in the third equation of the set of Eq. (18.17) coupling the sway and yaw modes, it may be seen that its linearized version may be cast as a time dependent linear term $N_{y\psi}(y_o\cos(\omega_W t))\psi$, which describes an amplitude of parametric excitation in the canonical form of the well-known Mathieu equation, proportional to both the sway amplitude y_o and to the connecting line coefficient $N_{y\psi}$.

Tendons alone cannot induce parametric amplification, as all their second order terms are zero, as shown in Table 18.1. On the other hand, wave passage effects on their tensions tend to be very small when compared with the effect of mooring lines and connecting lines. The main reason is that the vertical wave force on the TLWP is very small in the range of interest here, i.e., periods between 12 and 15 s. This is

Table 18.3 Restoring coefficients—connecting lines

		X	Y	N
1st order	x	−240		6600
	y		−31,626	
	ψ	6600		−796,134
2nd order	x^2		313.86	
	xy	672.3		−17,262
	$x\psi$		−1.7E+04	
	y^2			
	$y\psi$	−17,262		1,344,428.3
	ψ^2		672,214.1	
3rd order	x^3	−6.3E+00		172.6
	$x^2.y$		1.3E+01	
	$x^2.\psi$	517.9		−22,760
	y^3			
	$y^2.x$	1.3E+01		−345.3
	$y^2.\psi$	−345.3		9494.3
	ψ^3	360,871		−27,756,920
	$\psi^2.x$	−22,760		1,062,612
	$\psi^2.y$		9494.3	
	$x.y.\psi$		−6.9E+02	

due to a typical platform design configuration usually adopted for minimizing heave responses, as discussed in De Conti et al. (2009).

Tables 18.1, 18.2 and 18.3 demonstrate that parametric amplifications of the TLWP are largely governed by the mooring lines and connecting lines. In particular, it is observed that coefficient $N_{y\psi} = 1{,}344{,}428.25$ kN for the connecting lines, given in Table 18.3, is much larger than $N_{y\psi} = -8257.7$ kN given in Table 18.2, which is the corresponding value for the mooring lines. This is the main coefficient in the yaw equation responsible for the transfer of externally excited sway motion into internally excited yaw motion. It is then concluded that the connecting lines are the main agent of parametric amplification. On the other hand it may be concluded that the term $Y_{\psi\psi}\psi^2$ in Eq. (18.17) is the one responsible for the reduction in the sway motion that takes place as yaw motions amplify. Again, consulting Tables 18.2 and 18.3, it is possible to conclude that the connecting lines are mainly responsible for the strong coupling observed in the experiments.

18.8 Conclusions

A mathematical model for the analysis of parametric resonance in yaw of a TLWP platform has been described. It has been shown that the encountered amplifications are consistent with instabilities of the Mathieu type. Numerical simulations are well compared to experimental results. A strong coupling between sway and yaw is evident from the experiments. This important coupling is captured in the numerical simulations. It has been shown that the restoring system provided by the mooring lines, tendons and connecting lines leads to sub-harmonic motions not only in the yaw mode, but also in the surge mode. It has also been shown that parametric excitation resulting from oscillatory loading of the tendons due to the vertical wave force on the platform is negligible. It is concluded that the connecting lines are the main agent for yaw parametric amplification of the TLWP. Hydrodynamic loading, due to the close proximity of the two bodies, is obviously asymmetric. The relevance of these asymmetries is a topic for future research.

References

Cruz, R.E., Neves, M.A.S., Rivera, L.A. and Esperança, P.T.T., 2012, "An Investigation on the Excitation of Yaw Parametric Resonance of a Tension Leg Platform". Paper OMAE2012-83574, Rio de Janeiro, June 2012.

De Conti, M., Andrade, B. and Birk, L., 2009, "Differentiation between the Lower and Upper Parts of Columns of a Semi-Submersible for Heave Response Improvement". Marine Systems & Ocean Technology, Vol 4, No 2, pp. 63–72.

Maeda, K., Uto, S., Kuroda, T. 2008," TLWP-FPSO coupled model tests in NMRI", National Maritime Research Institute, Japan, Report OCN-07-03.

Neves, M.A.S., Rodríguez, C.A. (2007) "Influence of Nonlinearities on the Limits of Stability of Ships Rolling in Head Seas". *Ocean Engineering*, Vol 34, pp. 1618–1630.

Rodríguez, C.A. and Neves, M.A.S., 2012, "Domains of Parametric Roll Amplification for Different Hull Forms". In: Parametric Resonance in Dynamical Systems, Chapter 6, Ed.: T.I. Fossen and H. Nijmeijer, Springer Science, New York.

Part VII
Dynamics of Large Ship Motions: Surf-riding

Chapter 19
Bifurcation Analysis of Ship Motions in Steep Quartering Seas, Including Hydrodynamic "Memory"

Ioannis Tigkas and Kostas J. Spyrou

Abstract Steady-state ship dynamics in steep harmonic waves encountering the ship from stern quartering direction is under investigation. Bifurcation analysis is performed by applying a numerical continuation method. Stationary as well as periodic states are traced, as selected control parameters are varied. Regions with coexistence of different ship responses are identified. The main novelty of the paper lies in the extension of the continuation analysis to a 6-DOF model, for a quartering sea environment, with inclusion of memory effects within a potential flow framework. Complete, vessel-specific stability diagrams, for horizontal plane motions, are produced in an automated and time-efficient manner. These could provide useful guidance to ship masters for avoiding the occurrence of surf-riding and broaching-to.

Keywords Manoeuvring · Surf-riding · Broaching-to · Bifurcation · Homoclinic Continuation · Nonlinear dynamics

Nomenclature

A	Wave amplitude
$A_{ij}(\omega)$	Added mass coefficient
A_R	Rudder area
a_{ψ}, a_r	Proportional, differential gain
$B_{ij}(\omega)$	Damping coefficient
c	Wave celerity
F_N	Rudder normal force

I. Tigkas
Streamlined Naval Architects (Previously at, National Technical University of Athens), Piraeus, Greece

K. J. Spyrou (✉)
School of Naval Architecture and Marine Engineering, National Technical University of Athens, Athens, Greece
e-mail: k.spyrou@central.ntua.gr

V. L. Belenky et al. (eds.), *Contemporary Ideas on Ship Stability*, Fluid Mechanics and Its Applications 119, https://doi.org/10.1007/978-3-030-00516-0_19

Fn	Froude number
H	Wave height
H/λ	Wave steepness
$I_x,\ I_y,\ I_z$	Roll, pitch and yaw ship mass moment of inertia
$K,\ M,\ N$	Moments in roll, pitch and yaw respectively
$K_{ij}(\tau)$	Impulse response function
K_T	Propeller thrust coefficient
k	Wave number ($k = 2\pi / \lambda$)
L	Ship length
m	Ship mass
q, p, r	Pitch, roll and yaw angular velocity in a body-fixed system, respectively
$S(x_s, T_s)$	Vertical hull sectional area below instantaneous waterline
t	Time
t_p	Thrust deduction coefficient
t_r	Rudder's time constant
$T_s(x_s x_O, t, z, \theta)$	Draught of ship at vertical section S
u, v, w	Surge, sway and heave velocity in a body-fixed system, respectively
U_R	Inflow velocity at rudder
$X,\ Y,\ Z$	Forces in surge, sway and heave respectively
x	Longitudinal distance travelled by the ship, with respect to a system fixed at a wave trough
x_O	Longitudinal distance travelled by the ship in an earth-fixed system
x_S	Longitudinal distance of a vertical ship section S in the body-fixed system
$x_G,\ z_G$	Longitudinal distance from amidships and vertical distance from keel of ship's centre of gravity, respectively

Greek Letters

δ	Rudder angle
Λ	Rudder aspect ratio
θ	Pitch angle
λ	Wave length
ρ	Water density
φ	Roll angle
ψ	Heading angle
ψ_r	Desired heading angle
ω_e	Encounter frequency
ω	Wave frequency

19.1 Introduction

Instability phenomena in severe astern seas are feared by seafarers from the times when wind was the prime mover (Spyrou 2010). But it is only since the '40s that they started receiving deeper attention within the scientific community (Davidson 1948; Rydill 1959). Shortly it was realised that, key phenomena which are responsible for the occurrence of instability in following/quartering seas have a strongly nonlinear nature (e.g. Weinblum and St. Denis 1950; Grim 1951, 1963; Wahab and Swaan 1964). Simulation, i.e. direct numerical integration of the equations of motion, is a very straight-forward technique for predicting ship responses from some assumed initial conditions. As a matter of fact, it has been used in almost all investigations having practical orientation (Motora et al. 1981; Fuwa et al. 1981; Renilson 1982; Hamamoto 1988, 1989; de Kat and Paulling 1989; Hamamoto et al. 1994). However, simulation alone is generally ineffective for understanding, in a deeper sense, a dynamical system's behaviour and for identifying stability limitations. Novel techniques that can target more globally and more directly a system's potential for exhibiting rich and unconventional dynamic behaviour need to be implemented. One technique that can help to unravel behavioural changes as some system parameters are varied is numerical "continuation" (e.g. Krauskopf et al. 2007).

In earlier studies this technique was applied for studying surf-riding and broaching-to on the basis of a 4-degree-of-freedom (DOF) model with no "memory effect" (Spyrou 1995, 1996a, b; Belenky and Sevastianov 2007). In a more recent paper, the authors have performed continuation analysis of periodic motions in exact following seas (surging, heaving and pitching), including potential flow memory effects, for a ship that is on the verge of capture to surf-riding (Spyrou and Tigkas 2011). This work is further expanded here by performing combined continuation of stationary and periodic responses for the 6-DOF model with memory. It is demonstrated that, by using the so-called *codimenion-2* continuation method, practically useful stability diagrams of the system can be directly produced.

The dynamical system of ship behaviour in stern-quartering waves plays host to nonlinear phenomena that cannot be fully captured via simulation alone. Continuation is a numerical method extracting automatically the limit-states (stable or unstable) and also the bifurcations of the system, hence producing the "big picture" of system's global response. Conventionally, continuation is applied for dynamical systems described by ordinary differential equations (ODEs) and having no "memory". No memory means that, current loads on the body can be determined completely by the current (instantaneous) values of the system's variables and there is no need one to invoke their past values. As well-known, wave radiation installs in the system hydrodynamic memory which needs to be dealt with during the investigation of ship dynamics. Hence, the integration of the memory effect within a continuation study of ship motions in stern-quartering waves can be considered as an important development for the future studies in this field.

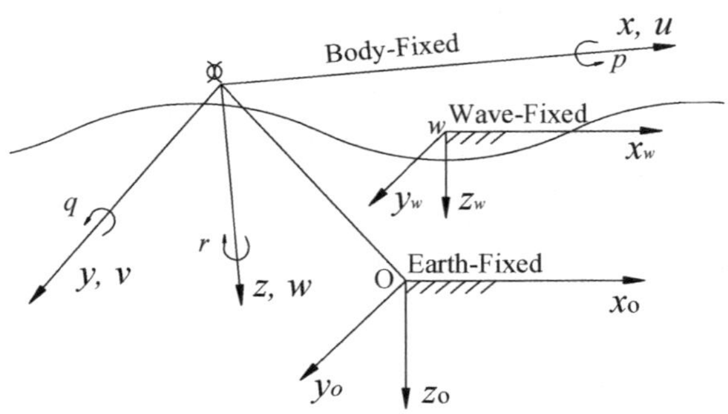

Fig. 19.1 Body-fixed, wave-fixed and earth-fixed coordinate systems

19.2 Mathematical Model

19.2.1 Equations of Motions

Long harmonic waves are considered to propagate from following/quartering direction relatively to the ship, which is moving with forward speed while performing also parasitic motions in all 6 degrees of freedom, due to waves' effect. A non-inertial system is placed at the intersection of the initial waterline, the centerplane and the middle of the ship, with its longitudinal axis running along the ship, the transverse directed to the left and the vertical axis looking downwards. This system follows the translations and rotations of the moving ship (body-fixed system) and it is used for monitoring ship velocities and accelerations (SNAME 1952). Furthermore, two inertial systems are used in secondary role: one fixed at a wave trough, and thus moving with the wave celerity; and also an earth-fixed system (see Fig. 19.1). Assuming that the ship behaves as a rigid-body, it is well-known that the equations of motions can be expressed with respect to the body-fixed system as follows[1]:

[1]All symbols are explained in the Nomenclature at the end of the paper.

Surge:

$$m\left[\dot{u} + qw - rv - x_G(q^2 + r^2) + z_G(pr + \dot{q})\right] = X - mg\sin\theta$$

Sway:

$$m[\dot{v} + ru - pw + z_G(qr - \dot{p}) + x_G(qp + \dot{r})] = Y + mg\sin\varphi\cos\theta$$

Heave:

$$m\left[\dot{w} + pv - qu - z_G(p^2 + q^2) + x_G(rp - \dot{q})\right] = Z + mg\cos\varphi\cos\theta$$

Roll:

$$I_x\dot{p} - mz_G(\dot{v} + ru - pw) - mx_Gz_G(\dot{r} + pq) = K$$

Pitch:

$$I_y\dot{q} + (I_x - I_z)rp + mz_G(\dot{u} + qw - rv) - mx_G(\dot{w} + pv - qu) + mx_Gz_G(p^2 - r^2) = M - mgx_G\cos\varphi\cos\theta$$

Yaw:

$$I_z\dot{r} + (I_y - I_x)pq + mx_G(\dot{v} + ru - pw) + mz_Gx_G(rq - \dot{p}) = N \quad (19.1)$$

19.2.2 Hull Forces and Moments

The external forces and moments acting on the ship are expressed in a modular form as a summation of hull reaction, rudder, propeller and wave excitations. For low frequencies of encounter, added mass and damping coefficients are calculated by the "strip theory" method of Clarke (1972). Whilst "old fashioned", this method is easily integrated in a nonlinear dynamical system continuation analysis. Firstly are computed the potential sway added mass coefficients for a number of (time-varying) hull sections up to the wave surface by multi-parameter conformal mapping. Then by integrating along the length of the hull the forces and moments can be found and expressed as accelerations and velocity derivatives by partial differentiation (Tigkas 2009). Using pre-processing we obtain the added-mass coefficient of a section as a polynomial function of the instantaneous sectional draught. In this method it is assumed that only the wave profile corresponding to the ship's specific longitudinal position on the wave and the ship's heave and pitch responses, influence the draught of each section and consequently the zero-frequency added mass properties. An example of the obtained results is shown in Fig. 19.2. A universal problem is the lack of accurate calculation of the viscous part of the hydrodynamic derivatives that is usually considerable at the sections near the stern. For this reason, in some occasions we applied an empirical hybrid approach, extracting the viscous part from the semi-empirical zero-frequency derivatives and adding it to the "potential" coefficients obtained by Clarke's method.

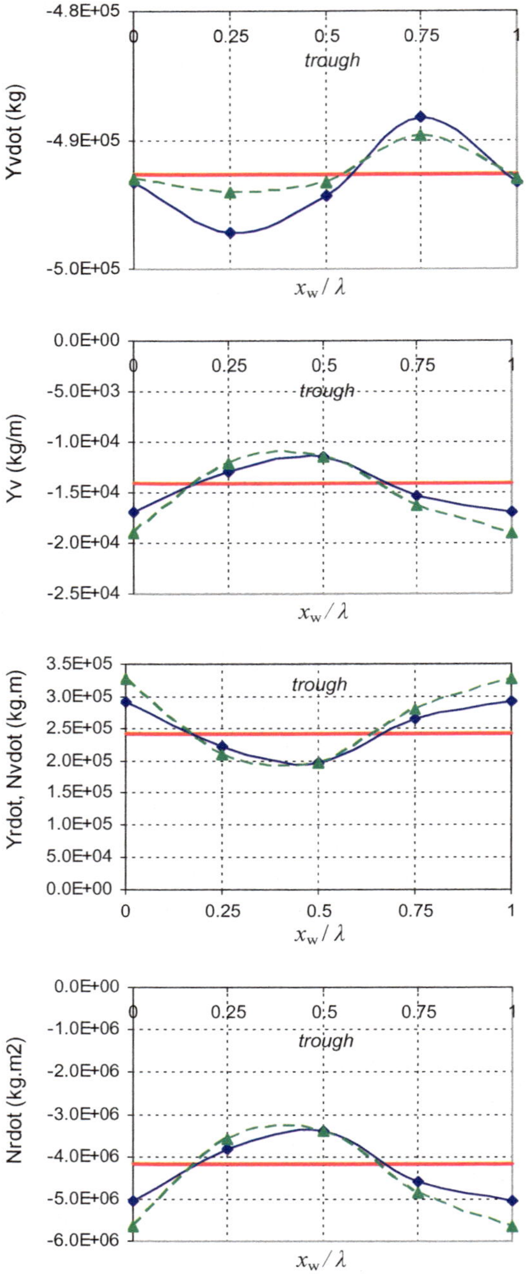

Fig. 19.2 Linear hydrodynamic derivatives for the fishing vessel investigated in Sect. 19.4, at different longitudinal positions on a wave starting from a wave crest ($\lambda/L = 2$)

Fig. 19.2 (continued)

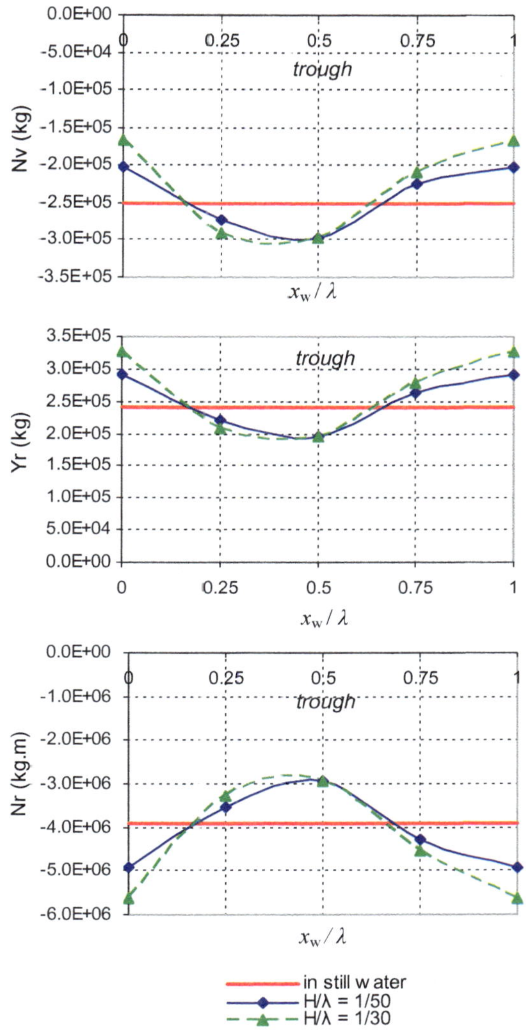

19.2.3 Memory Effects

To investigate ship motions not only for the stationary condition of surf-riding (zero-frequency of encounter) but for a rather wide range of wave encounter frequencies, "memory effects" should be considered, accounting for fluid's hydrodynamic radiation loads that are caused in mode i due to ship motion in mode j. According to Cummins (1962) the radiation force can be expressed as:

$$F_{Mij}(t) = -A_{ij}(\infty)\dot{v}_j(t) - \underbrace{\int_0^\infty K_{ij}(\tau)v_j(t-\tau)d\tau}_{s_{ij}(t)} \tag{19.2}$$

The impulse response (or retardation) function K_{ij} in the above equation can be written as follows (Ogilvie 1964; see however also detailed presentation in Taghipour 2008):

$$K_{ij}(\tau) = \frac{2}{\pi}\int_0^\infty [B_{ij}(\omega_e) - B_{ij}(\infty)]\cos(\omega_e\tau)d\omega_e \tag{19.3}$$

The integro-differential equations (IDEs) appearing in Eq. (19.2) can be approximated by a system of ODEs by means of a standard state-space approximation technique (Tick 1959; Schmiechen 1975; Jefferys 1984). As has been shown by Schmiechen (1975), the following finite set of recursive first-order linear ODEs can replace Eq. (19.2):

$$\begin{aligned}
\dot{s}_{ij}(t) &= s_{ij(1)}(t) - a_{ij(k)}s_{ij(k)}(t) - b_{ij(k)}v_j(t) \\
\dot{s}_{ij(1)}(t) &= s_{ij(2)}(t) - a_{ij(k-1)}s_{ij}(t) - b_{ij(k-1)}v_j(t) \\
\dot{s}_{ij(2)}(t) &= s_{ij(3)}(t) - a_{ij(k-2)}s_{ij}(t) - b_{ij(k-2)}v_j(t) \\
&\ldots \quad \ldots \quad \ldots \quad \ldots \\
&\ldots \quad \ldots \quad \ldots \quad \ldots \\
\dot{s}_{ij(k)}(t) &= -a_{ij(0)}s_{ij}(t) - b_{ij(0)}v_j(t)
\end{aligned} \tag{19.4}$$

The coefficients $a_{ij(m)}$ and $b_{ij(m)}$ are calculated from curve fitting in frequency domain, on the basis of the values of the added mass and damping terms. However, some attention is required in this identification scheme, for ensuring that the recursive system of Eq. (19.4) is stable (Tigkas 2009).

It should be noted however that the developed procedure calculates the hydrodynamic memory accounting also for the zero frequency effect that has already been calculated by the applied strip - theory method. In order to sort out this overlap issue, from each associated linear hydrodynamic derivative we subtract the corresponding potential part at zero-frequency that was calculated by the potential seakeeping code, that has been used in order to extract the added mass and damping coefficients.

19.2.4 Wave and Hydrostatic Loads

The wave excitation loads of a rigid body in regular harmonic waves, assuming inviscid and irrotational flow, can be linked with the effects of the incident and the diffracted wave potentials. The Froude-Krylov excitation loads (including hydrostatic) are determined by the integration of the fluid pressure of the submerged portion of the hull, up to the free surface of the incident wave. However, the contribution of the diffracted wave excitation loads are not taken into account in this study. Their calculation in a very low to medium encounter frequency range and the efficient integration of such a scheme with continuation analysis is a demanding task on its own that will be resolved in a future research study. Nevertheless, the inclusion of diffraction loads is only expected to alter quantitatively but not qualitatively the results.

19.2.5 Rudder and Propeller

Kose's (1982) model is the basis for calculating rudder forces. As for propulsor's force, a polynomial fit of available propeller performance data is used for approximating the thrust coefficient K_T as function of propeller's rate of rotation. Wave effects on the rudder and on the propeller are produced by the variation of draught at those locations and by the change in the inflow velocity. Rudder's area and aspect ratio "seen" by the water are obtained from its instantaneous draught. Sometimes the propeller might ventilate or even emerge out of the water with significant efficiency loss (Koushan 2006; Paik et al. 2008). However such losses can vary considerably depending on stern's layout, propeller characteristics and other design characteristics. A simple model calculating the loss of thrust could not be deduced.

When the rudder angle is not taken as fixed, a standard proportional-differential (PD) controller is used, whose equation is expressed in an O.D.E. form as follows:

$$\dot{\delta} = t_r \left[-\delta - a_\psi (\psi - \psi_r) - a_\psi b_r r \right] \tag{19.5}$$

19.3 Adaptation of Mathematical Model to Enable Continuation

To conduct continuation analysis a number of transformations were obliged on the described mathematical model:

(a) The investigated dynamical system needed to come into the generic ODE form $\dot{\mathbf{x}} = \mathbf{F}(\mathbf{x}, \mathbf{a}; t)$ where $\mathbf{x}$ is the vector of state variables and $\mathbf{a}$ is the vector of system's parameters. Furthermore, all state variables of the system should be bounded as time progresses. As a matter of fact, continuation analysis could not be attempted before transforming the equations of the system in such a way, so that all state variables take values within bounded limits and, a steady state recognised by the algorithm as such, irrespectively of whether this is stationary or periodic, can be reached.

(b) The last criterion states that the ODEs that constitute the state-space representation should have no explicit dependence upon the time variable. This may include special transformations on the system's form of equations, if the investigated dynamical system is excited by a time-varying force, as for example the wave load.

The first problem is realised when a state variable of the system increases monotonically to infinity as time progresses. The variable that renders impossible for the system to reach a recurring state and is hence imposing non-conformity to the first requirement, is the distance travelled by the ship in the longitudinal direction. In our mathematical model, this is a state variable used for defining the relative position of the ship on the wave and it appears in the module where the Froude-Krylov forces and moments are calculated. To demonstrate the transformations involved, let's consider the surge wave force. By using the relationship $x = x_0 - ct$, the position-dependent part of this force can be written as:

$$X_{FK} = -\rho g A k \cos \psi \int_{-L/2}^{L/2} S(x_s, T_s) e^{-kT_s(x_s, x_0, t, z, \theta)/2} \sin k(x_s \cos \psi + x) dx_s \tag{19.6}$$

After trigonometric expansion it is written as follows:

$$X_{FK} = -\rho g A k \cos \psi \int_{-L/2}^{L/2} S(x_s, T_s) e^{-kT_s(x_s, x_0, t, z, \theta)/2}$$
$$[(\cos kx \cdot \sin kx_s \cos \psi) + (\sin kx \cdot \cos kx_s \cos \psi)] dx_s \tag{19.7}$$

Two dummy variables $a = \sin kx$ and $b = \cos kx$ are introduced, replacing the cyclic functions of x in Eq. (19.7). The periodic nature of a and b means that these variables are inherently bounded, unlike x which is monotonically increasing. For consistency an extra pair of ODEs needs however to be added (see also Doedel et al. 1997):

$$\begin{aligned} \dot{a} &= a - \omega_e b - a(a^2 + b^2) \\ \dot{b} &= \omega_e a + b - b(a^2 + b^2) \end{aligned} \tag{19.8}$$

The introduced pair stands basically for a harmonic oscillator that, despite of increasing by two the number of variables of the system, at steady-state it bears no effect on the behaviour relative to the original system (Spyrou and Tigkas 2011). This method is applied for all position-dependent forces and moments in all directions of ship motion.

With the above transformations the model can be interfaced with the continuation algorithm where in our case such is MATCONT (Dhooge et al. 2003). The mathematical principles and the main capabilities of this algorithm for investigating nonlinear dynamical systems have been discussed in our earlier works and need not to be repeated (see for example, Spyrou et al. 2007; Spyrou and Tigkas 2007, 2011).

19.4 Investigation Results

The ship investigated is the well-known 34.5 m long Japanese fishing vessel ('purse-seiner') that has also been examined several times in the past in free running and captive model tests, for numerical analysis of its dynamics and for benchmarking evaluations (see for example Umeda et al. 1995; ITTC 2005). Her added mass and potential damping coefficients at each encounter frequency were identified by the commercial code Trident Waveload (2006). Integrations concerning hull geometry were carried out on the basis of 20 transverse stations along the hull length. For the memory fits discussed earlier, the number of the required linear first-order O.D.E.s of the filter has been investigated and a few values of the order k (see Sect. 2.3) were tested, in each case evaluating the quality of the fit produced (Fig. 19.3). The order $k = 3$ seems to be a popular choice in relevant work, e.g. see Holappa and Falzarano (1999). Here it was judged that it provides satisfactory accuracy [see also Spyrou and Tigkas (2011) for more detailed explanation].

The standard state-space form of the mathematical model after implementing the above described actions is consisted of 77 ODEs. This is considered as a very high number of equations for a continuation study.

19.4.1 Stationary Responses

In Fig. 19.4 are shown the obtained "equilibrium headings" for the entire range of rudder angles and for several nominal Froude numbers. These equilibrium headings (actually these constitute surf-riding states) are unstable unless rudder control exists, in which case the parts of the curve nearer to the trough are stabilised. But for the basic case of a ship without active control, *saddle-type* instabilities are formed between LP1 and LP1$'$ as also between LP2 and LP2$'$. Equilibrium diagrams for the other modes of motion are also shown in Fig. 19.5.

Since LP1 points determine the range of potential stabilisation by rudder control state laws, their locus defines in fact the domain where surf-riding can be experienced in practice. Such a curve can be obtained by *codimension-2* continuation of the *fold* LP1, varying simultaneously Fn and δ. The result is shown in Fig. 19.6. The branch of LP2 has also been included in this diagram. Even though these latter points do not receive an immediate practical interpretation, their "behaviour" is interesting: at the zero rudder angle and for $Fn = 0.42$, a *cusp* point is formed by the tangential contact of the *saddle-node* branches corresponding to LP2's evolution.

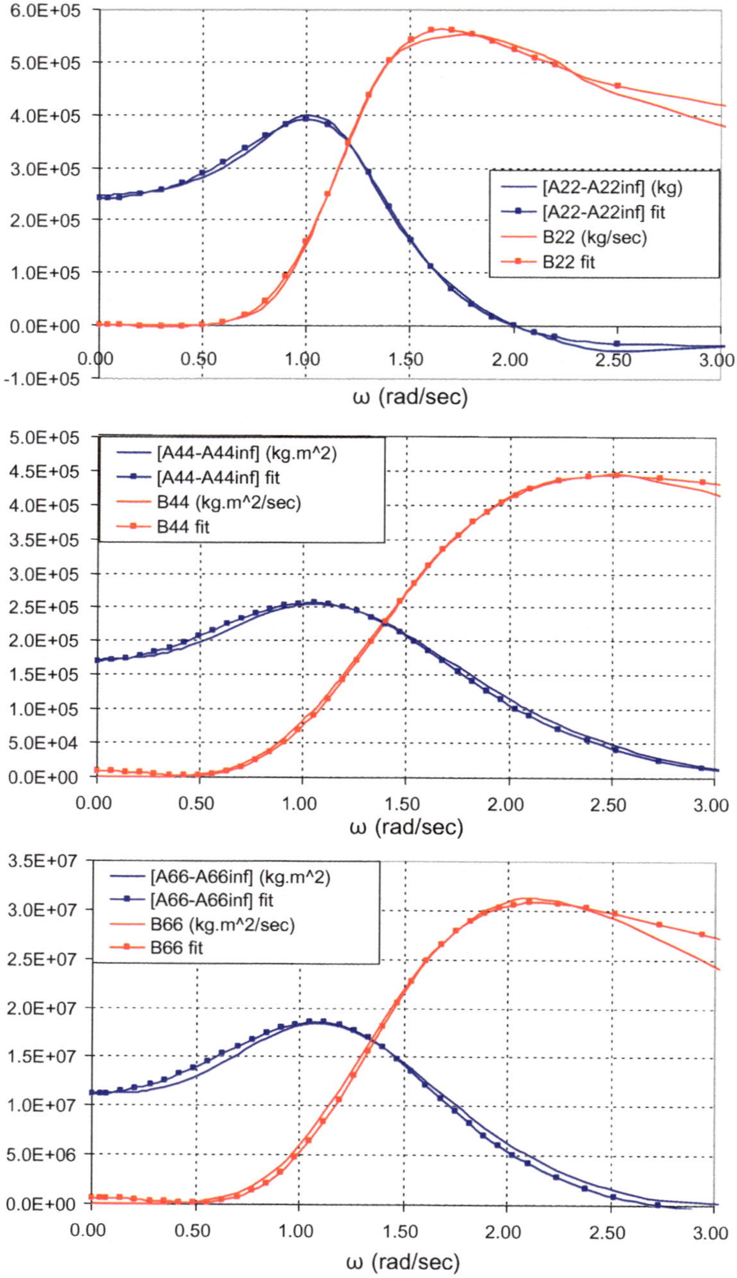

Fig. 19.3 Examples of fits for added mass and damping coefficients for the purse seiner. (Fits for heave, pitch and cross terms of pitch with surge can be found in Spyrou and Tigkas 2011)

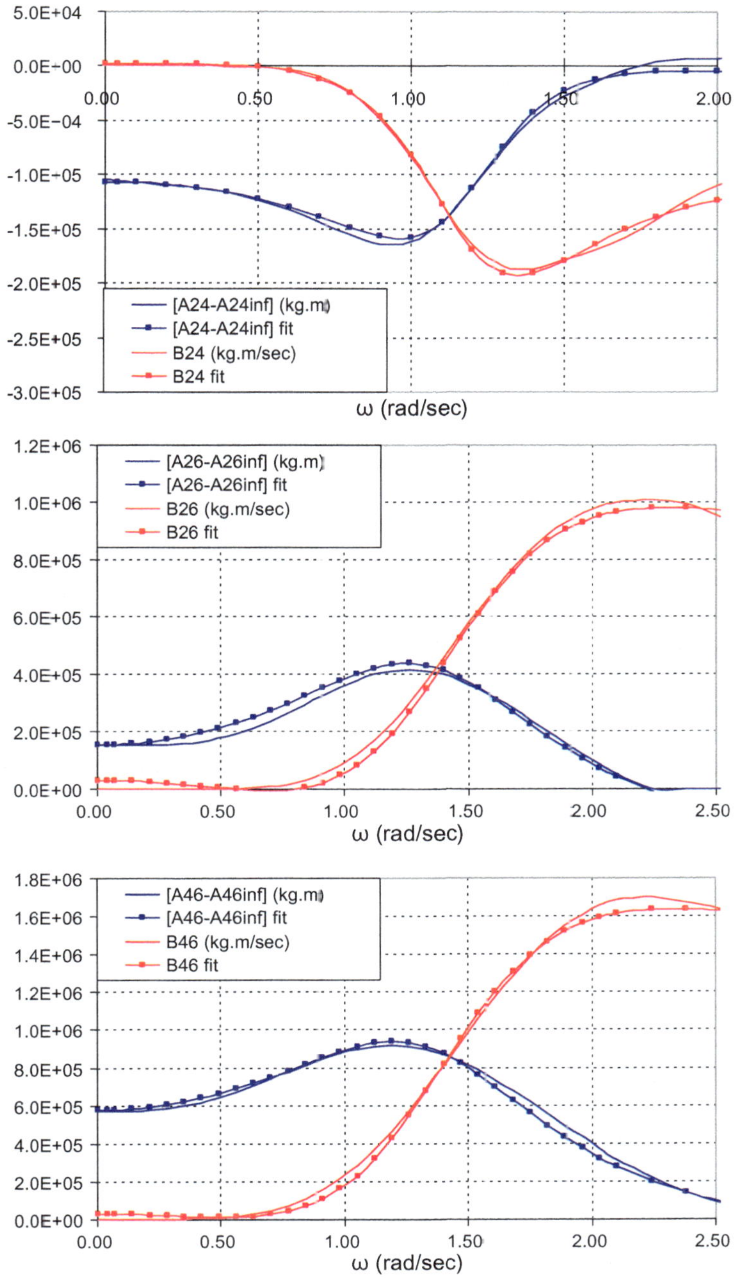

Fig. 19.3 (continued)

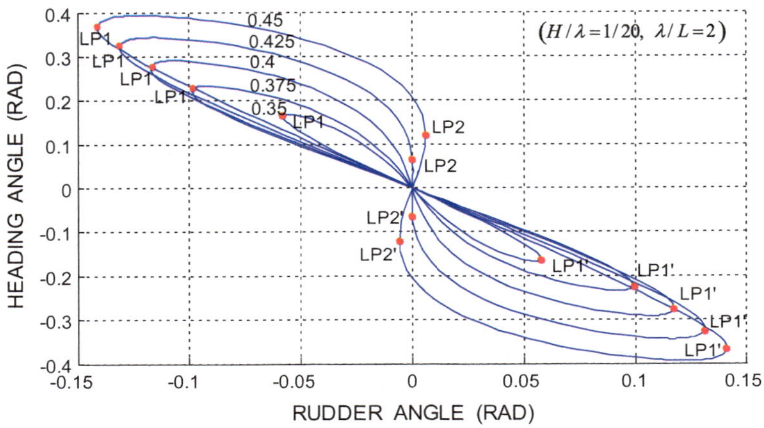

Fig. 19.4 Equilibrium headings versus rudder angles, based on the 6-DOF model

A diagram such as the one of Fig. 19.6, relating speed, wave characteristics and rudder angles resulting in surf-riding behaviour, could be useful for a ship master who would be aware of the operational parameters' values that are conducive to surf-riding. Of course, given the deterministic nature of the method, the accuracy of the input that could realistically become available to him is critical, especially regarding wave height and period.

The influence of proportional gain on the realized heading in the condition of surf -riding in quartering seas is illustrated in Fig. 19.7. Accordingly, the upper parts of the curves in Fig. 19.7 from 0 to LP are always stable, whereas the lower parts are occupied by saddles. Again, LP points indicate the transition of stability and are occurring this time at the maximum possible commanded heading angles for each selected proportional gain. As one expects, larger commanded angles of heading will not correspond to a reduced heading error (commanded heading minus actual heading), but rather practically mean the ship to start turning. It is also evident that at low values of proportional gain, the heading error especially in relatively large commanded heading angles is also large and thus the controller's function is problematic. The locus of limit points (LP) is finally obtained by *codimension-2* continuation when varying simultaneously the commanded heading and the proportional gain.

Another useful investigation is to determine the locus of lowest nominal Froude numbers for which surf-riding can be realised in quartering waves. This can also be obtained by *codimension-2* continuation, varying Fn and ψ_r. In the scenario of Fig. 19.8, the control settings were: $t_r = 3$, $a_\psi = 3$, and $b_r = 1$ Several diagrams of this kind for different heights and lengths can also assist on-board decision-making.

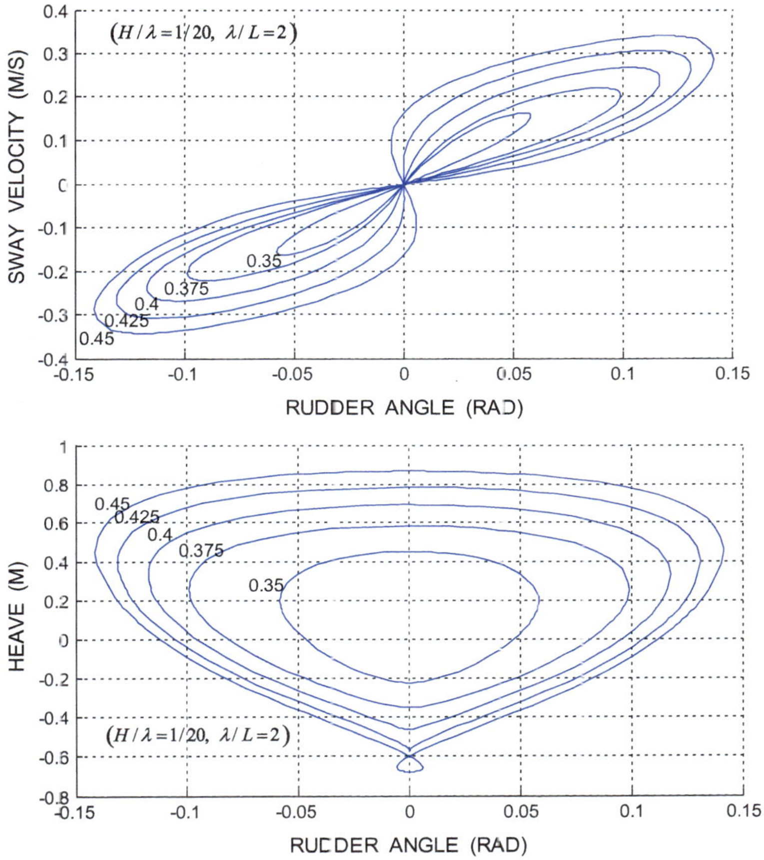

Fig. 19.5 Equilibrium sway velocity (upper) and heave displacement (lower) corresponding to rudder angles, for a range of nominal Froude numbers

19.4.2 Periodic Responses with Active Steering Control

The key element of the current effort targets the periodic responses in quartering waves and furthermore, the conditions that are eligible for broaching-to behaviour not only as a natural consequence of surf-riding but also by a more direct escape mechanism (Spyrou 1996b, 1997). However controller's settings influence the amplitude of yaw oscillation as well as the maximum commanded heading angle up to which the ship remains controllable.

To run effectively the continuation of these periodic orbits for the complete model in 6-DOF we needed to reduce the system of equations to 42 ordinary differential equations (ODEs) by neglecting several memory terms that have a lesser effect on the outcome. Thus, the most influential memory loads kept in this simplified system

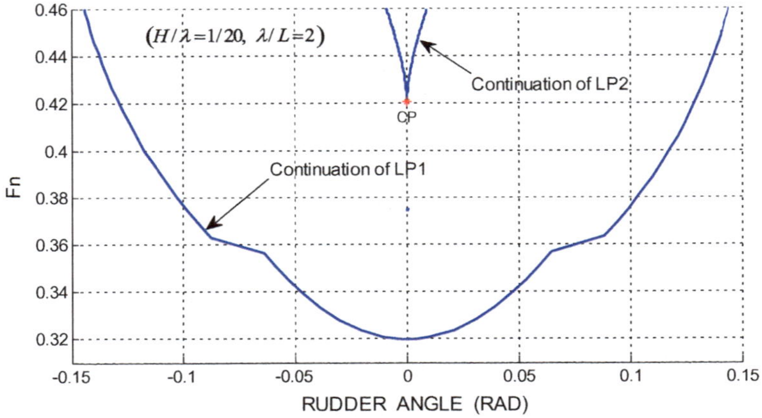

Fig. 19.6 Stability diagram in following/quartering seas, showing the range of rudder angles where surf-riding is sustainable

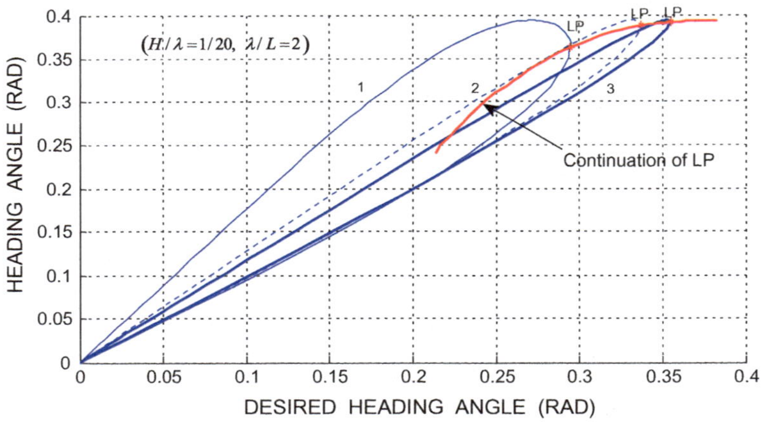

Fig. 19.7 Difference between commanded and actual headings for three different proportional gains (1, 2 and 3), $Fn = 0.45$ and differential gain equal to 1

comprise; s_{22}, s_{26}, s_{33}, s_{35}, s_{44}, s_{55}, s_{53}, s_{66} and s_{62}. Below only two characteristic examples of the several obtained results will be discussed.

Consider firstly the evolution of periodic motions for a fixed commanded heading angle of 15°. For low nominal Froude numbers, ship motion is basically linear but as the speed is increased, it becomes increasingly asymmetrical. As well known, at some stage the periodic behaviour abruptly stops due a *homoclinic connection*. In Fig. 19.9 is shown the contact between the stationary (surf-riding) and the periodic motions that is responsible for this phenomenon. Continuation produces a unique picture of this spectacular encounter between qualitatively different ship states.

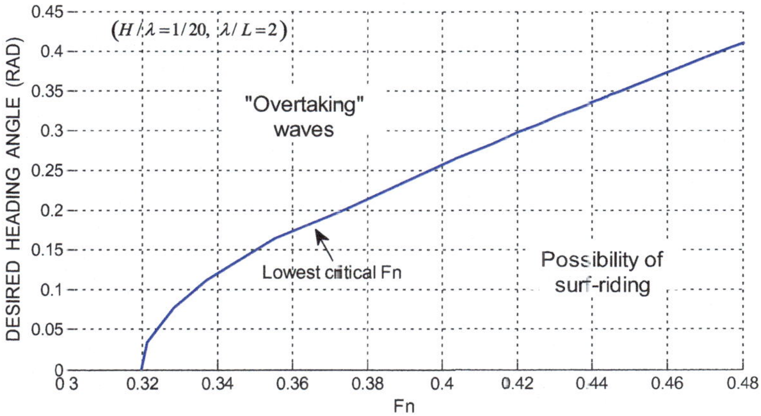

Fig. 19.8 *Codimension-2* bifurcation diagram produced by continuation and showing the lowest nominal Froude number values for which surf-riding can be realised, for a wide range of commanded heading angles

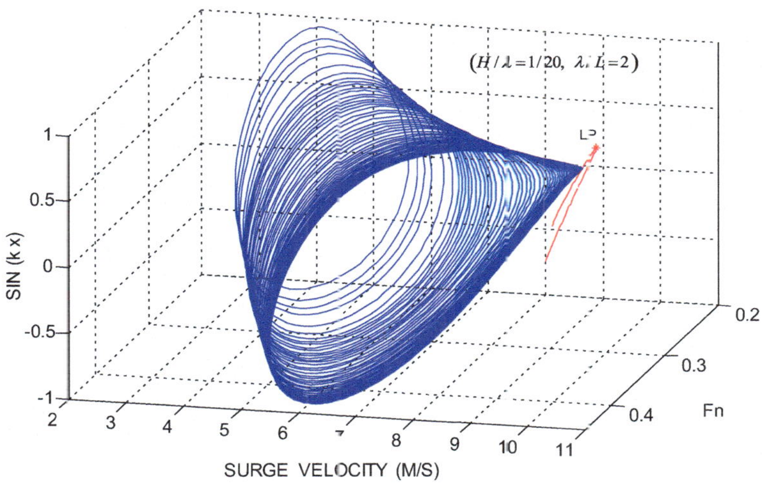

Fig. 19.9 3-D view of the evolution of limit cycles collapsing onto the branch of saddles, for a fixed 15° desired heading angle

The sequence of phenomena is qualitatively similar to what happens in an exactly following sea environment (see for comparison Spyrou and Tigkas 2011). However, the critical Froude numbers are moved to slightly higher values since the experienced surge wave force in quartering waves is reduced. Control settings were selected like in Sect. 4.1, sufficient for keeping the ship on a mean heading very close to the commanded one.

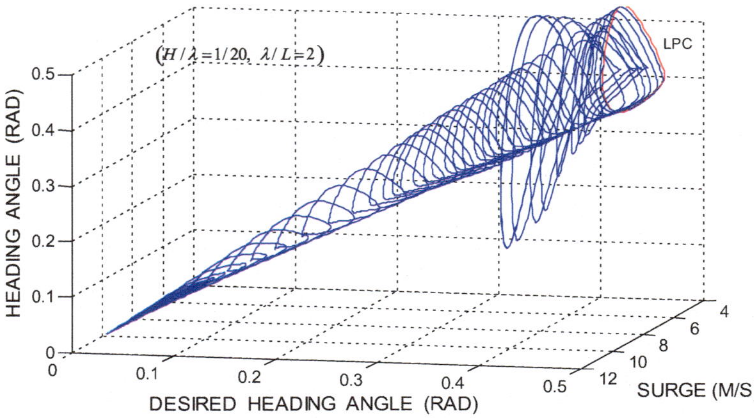

Fig. 19.10 3-D view of the evolution of limit cycles by increasing the desired heading angle at nominal $Fn = 0.4$

In the second scenario presented the interest is on the evolution of the periodic motion amplitudes for a fixed nominal Froude number, as the commanded heading angle is increased to values beyond the surf-riding range. As observed in Figs. 19.10 and 19.11, the amplitudes of both yaw and surge oscillations increase when the commanded heading angle is also increased. Whilst this behaviour continues for the lower range of commanded heading angles, at a critical value of the commanded heading angle a *limit point of cycles* (*fold of cycles*) is encountered and the stable limit cycles are turned unstable. Thus a sufficient condition for a discontinuous jump to a distant state is established. The created unstable limit cycles continue their evolution "backwards" with increasing amplitude. This phenomenon indicates a different broaching-to scenario at *LPC*, which occurs not due to surf-riding, but directly from periodic oscillations by an over-increased heading angle. Such behaviour is not identified for the first time. The direct type of broaching (found also in the literature as "cumulative broaching", due to the oscillatory growth of yaw preceding the final turn) has been discussed a few times in the past and the phenomena explaining its inception were described in Spyrou (1997).

19.5 Concluding Remarks

A 6-DOF mathematical model containing modules for the effect of waves, hull reaction including hydrodynamic memory, propeller and rudder has been described. The mathematical model is the offspring of an earlier manoeuvring-type model with 4-DOF and it was brought into a form making it amenable to bifurcation analysis. A continuation analysis algorithm was interfaced with this mathematical model in order to fully capture the stationary and periodic ship motions, with special emphasis

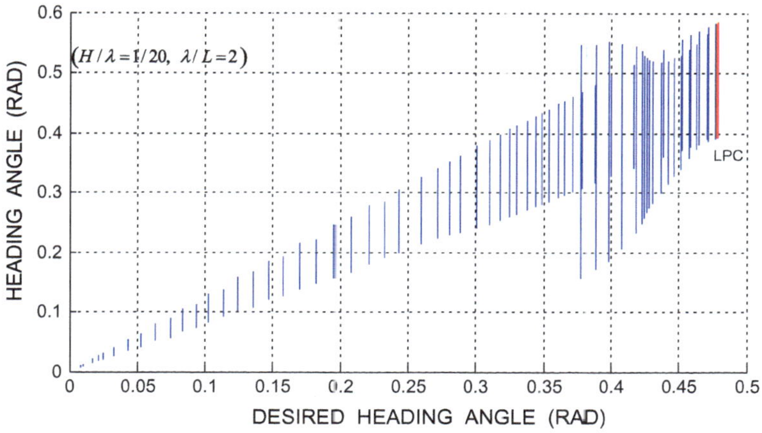

Fig. 19.11 Bifurcation diagram showing the increase of yaw amplitude as the commanded heading angle is increased, at nominal $Fn = 0.4$

given to their interactions in phase-space. Loci, in parameters' space, of bifurcation points have also been traced. These loci represent stability boundaries and the presented approach essentially shows how one can identify such boundaries through an automated procedure. Design diagrams relating to controller's tuning; and ultimately a complete booklet offering operational guidance for averting broaching-to phenomena in steep quartering seas, can also be produced.

Despite the progress achieved so far, several areas requiring further attention have emerged. For example, the nonlinear mathematical model can be definitely improved; especially the part related to the calculation of the manoeuvring derivatives and the diffraction wave loads within a user-friendly framework for applying continuation analysis and other nonlinear dynamics techniques. The combined effects of wind and waves can also be studied. Despite that this appears to be straightforward given the current development and previous work studies by the authors (e.g. Spyrou et al. 2007), the wealth of dynamical phenomena that may arise from a combination of excitations, merits in our view a dedicated study. Another, more ambitious, direction of research is the extension of the presented methods of analysis for a probabilistic environment.

Acknowledgements Ioannis Tigkas would like to thank "Alexander S. Onassis" Public Benefit Foundation for a scholarship that supported his Ph.D. studies at NTUA.

References

Belenky, V.L. and Sevastianov, N.B., 2007, Stability and Safety of Ships. Risk of Capsizing. Published by The Society of Naval Architects and Marine Engineers, ISBN 10: 0939773619.

Clarke, D., 1972, "A Two-Dimensional Strip Method for Surface Ship Hull Derivatives: Comparison of Theory with Experiments on a Segmented Tanker Model", Journal of Mechanical Engineering Science, Vol. 14, No. 7, pp. 53–61.

Cummins, W.E., 1962, "The Impulse-Response Function and Ship Motions", Schiffstechnik, Vol. 9, No. 47, pp. 101–109.

Davidson, K.S.M., 1948, "A Note on the Steering Of Ships in Following Seas", Proceedings of 7th International Congress of Applied Mechanics, London.

de Kat J.O. and Paulling J.R., 1989, "The Simulation of Ship Motions and Capsizing in Severe Seas", SNAME Annual Meeting, Presentation, No. 5.

Dhooge A., Govaerts W., Kuznetsov Yu. A., Mestrom W., Riet A.M. and Sautois B., 2003, "MATCONT and CL_MATCONT: Continuation Toolboxes for MATLAB". Report of Gent and Utrecht Universities.

Doedel E.J., Champneys A.R., Fairgrieve T.F., Kuznetsov Y.A., Sandstede B. and Wang X.J., 1997, "AUTO97-00: Continuation and Bifurcation Software for Ordinary Differential Equations (with HomCont)", User's Guide, Concordia University, Montreal, Canada.

Fuwa T., Sugai K., Yoshino T. and Yamamoto T., 1981, "An Experimental Study on broaching of Small High Speed Boat", Journal of the SNA of Japan, Vol. 150.

Grim O., 1951, "Das Schiff in von Achtern Auflaufender See", Jahrbuch S.T.G., Bd. 45.

Grim O., 1963, "Surging Motion and Broaching Tendencies in a Severe Irregular Sea", Deutsche Hydrographische Zeitschrift, Bd. 16.

Hamamoto M., 1988, "Study on Ship Motions and Capsizing in Following Seas: 1st Report - Equations of Motion for Numerical Simulation", Journal of the SNA of Japan, Vol. 163, pp. 173–180.

Hamamoto M., 1989, "Study on Ship Motions and Capsizing in Following Seas: 2nd Report - Simulation of Capsizing", Journal of the SNA of Japan, Vol. 163, pp. 123–130.

Hamamoto M., Fujino M. and Kim Y.S., 1994, "Dynamic Stability of a Ship in Quartering Seas", Proceedings of STAB 94, Melbourne, Florida.

Holappa K.W. and Falzarano, J.M., 1999, "Application of Extended State-space to Non-linear Ship Rolling", Journal of Ocean Engineering, Vol. 26, p. 227–240.

ITTC, 2005, "Report of Specialist Committee on Ship Stability in Waves", Proceedings of ITTC Conference, Edinburgh.

Jefferys, E.R., 1984, "Simulation of Wave Power Devices", Journal of Applied Ocean Research, No. 6, Vol. 1, pp. 31–39.

Kose K., 1982, "On a New Mathematical Model of Manoeuvring Motions of A Ship and its Applications", Journal of International Shipbuilding Progress, Vol. 29, No. 336.

Koushan, K., 2006, "Dynamics of Ventilated Propeller Blade Loading on Thrusters due to Forced Sinusoidal Heave Motion", Proceedings of 26th Symposium on Naval Hydrodynamics, Rome, Italy.

Krauskopf B., Osinga H. G. and Galan-Vioque J. (editors), 2007, "Numerical Continuation Methods for Dynamical Systems", Springer Publications.

Motora S., Fujino M., Koyanagi M., Ishida S., Shimada K. and Maki T., 1981, "A Consideration on the Mechanism of Occurrence of Broaching-to Phenomena", Journal of the SNA of Japan, Vol. 150, pp. 84–97.

Ogilvie T.F., 1964, "Recent Progress towards the Understanding and Prediction of Ship Motions", Proceedings of 5th Symposium on Naval Hydrodynamics, Bergen, Norway, pp. 3–79.

Paik B.G., Lee J.Y. and Lee S.J., 2008, "Effect of Propeller Immersion Depth on the Flow around a Marine Propeller", Journal of Ship Research, Vol. 52, No. 2, pp. 102–113.

Renilson M., 1982, "An Investigation Into the Factors Affecting the Likelihood of Broaching-to in Following Seas", Proceedings of STAB 82, Tokyo, pp. 17–28.

Rydill L.J., 1959, "A Linear Theory for the Steered Motion of Ships in Waves", Transactions of RINA, pp. 81–112.

Schmiechen M., 1975, "Equations for Non-Quasi-Steady Ship Motions", Seakeeping Committee Report, Proceedings of 14th ITTC Conference, Ottawa, Canada.

SNAME, 1952, "Nomenclature for Treating the Motion of a Submerged Body Through Fluid", Technical and Research Bulletin, No. 1–5.

Spyrou K.J., 1995, "Surf-Riding Yaw Instability and Large Heeling of Ships in Following/ Quartering Waves", Journal of Ship Technology Research, Vol. 42, pp. 103–112.

Spyrou K.J., 1996a, "Dynamic Instability in Quartering Seas: The Behaviour of a Ship During Broaching", Journal of Ship Research, Vol. 40, No. 1, pp. 46–59.

Spyrou K.J., 1996b, "Homoclinic Connections and Period Doublings of a Ship Advancing in Quartering Waves", Journal of Chaos, Vol. 6, pp. 209–218.

Spyrou K.J., 1997, "Dynamic Instability in Quartering Seas – Part III: Nonlinear Effects on Periodic Motions", Journal of Ship Research, Vol. 41, No. 3, pp. 210–223.

Spyrou, K.J., 2010, "Historical Trails of Ship Broaching-to", Transactions of RINA, Vol. 152, Part A4, pp. 163–173.

Spyrou K.J., Tigkas I. and Chatzis A., 2007, "Dynamics of a Ship Steering in Wind Revisited", Journal of Ship Research, Vol. 51, No. 2.

Spyrou K. and Tigkas I.G., 2007, "Principles and Application of Continuation Methods for Ship Design and Operability Analysis", Proceedings of 10th PRADS Symposium, Houston, Texas, USA.

Spyrou K.J. and Tigkas I.G., 2011, "Nonlinear Surge Dynamics of a Ship in Astern Seas: "Continuation Analysis" of Periodic States with Hydrodynamic Memory", Journal of Ship Research, Vol. 55, No. 1, pp. 19–28.

Taghipour R., Perez T. and Moan T., 2008, "Hybrid Frequency–Time Domain Models for Dynamic Response Analysis of Marine Structures", Journal of Ocean Engineering, 35, pp. 685–705.

Tick L.J., 1959, "Differential Equations with Frequency-Dependent Coefficients", Journal of Ship Research, Vol. 3, No. 2, pp. 45–46.

Tigkas I.G., 2009, "Nonlinear Dynamics Analysis of the Directional Instabilities of Ships in Wind and Waves", PhD Thesis, National Technical University of Athens, Greece.

Trident F-D Waveload, 2006, "User's Guide and Theory Manual", Martec Ltd, Halifax, Nova Scotia, Canada.

Umeda N., Hamamoto M., Takaishi Y., Chiba Y., Matsuda A., Sera W., Suzuki S., Spyrou K.J. and Watanabe K., 1995, "Model Experiments of Ship Capsize in Astern Seas", Journal of the SNA of Japan, Vol. 177, pp. 207–218.

Wahab R. and Swaan W.A., 1964, "Course-keeping and Broaching of Ships in Following Seas", Journal of International Shipbuilding Progress, Vol. 7, No. 4, pp. 293–301.

Weinblum G. and St. Denis M., 1950, "On the Motions of Ships at Sea", Transactions of SNAME, Vol. 58, pp. 184–248.

Chapter 20
Modeling of Surf-Riding in Irregular Waves

Vadim L. Belenky, Kostas J. Spyrou and Kenneth M. Weems

Abstract Surf-riding is an important phenomenon for the evaluation of ship dynamic stability, as it is related to one of the principal mechanisms of broaching-to. The evaluation of the probability of surf-riding in irregular waves is a necessary step toward determining the probability of broaching-to following surf-riding. To facilitate the probabilistic study of surf-riding, a simple model of surging and surf-riding in irregular waves of variable bandwidth is introduced. This model can be used to identify patterns of surf-riding in irregular waves.

Keywords Surf-riding · Irregular waves

20.1 Introduction

The dynamics of broaching-to originating from surf-riding in regular waves is now well understood from a global dynamical systems' viewpoint (Spyrou 1996). A cornerstone of this phenomenon is the appearance of a pair of surf-riding equilibria of which the one that attracts in surge may be a repeller in yaw, depending on the effectiveness of rudder control. Another key issue is the possible dominance of this equilibrium in state space due to a "homoclinic connection" bifurcation that renders surf-riding inevitable.

The physical mechanism of surf-riding includes the appearance of dynamical equilibria and a ship's attraction to the stable equilibrium (Kan 1990). The equilibria appear when the wave surging force becomes large enough to offset the difference between the ship's thrust and resistance at speed equal to wave celerity. The equi-

V. L. Belenky (✉) · K. M. Weems
David Taylor Model Basin (NSWCCD), West Bethesda, MD, USA
e-mail: vadim.belenky@navy.mil

K. J. Spyrou
School of Naval Architecture and Marine Engineering, National Technical University of Athens, Athens, Greece

V. L. Belenky et al. (eds.), *Contemporary Ideas on Ship Stability*, Fluid Mechanics and Its Applications 119, https://doi.org/10.1007/978-3-030-00516-0_20

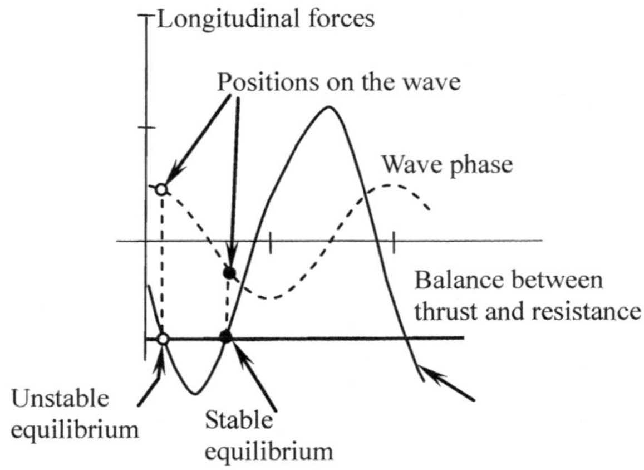

Fig. 20.1 On the appearance of dynamic equilibria

librium points are the positions of the ship on the wave where the forces balance exactly.

To illustrate this, consider surf-riding in regular waves and plot the wave-induced surging force as a function of the ship's position on the waves, taken with the opposite sign; see Fig. 20.1. In this plot, the horizontal axis is the position of the ship's center of gravity ahead of the wave crest, the dashed line is the wave profile, and the solid line is the wave surging force, with a negative value indicating a forward (accelerating) force.

Since the commanded speed is insufficient to propel the ship at wave celerity in calm water, additional wave force is necessary to drive the ship at wave celerity. If the amplitude of the wave surging force exceeds the absolute value of the balance between thrust and resistance, two intersection points appear, as shown in Fig. 20.1. Those will be called "surf-riding equilibria" (knowing that this is not an exact condition of equilibrium); one shows stability features (black point, located around the wave trough) and the other behaves as unstable (empty point, located around wave crest).

While these considerations are well-established in the field, they are repeated here in order to highlight the difference between regular and irregular waves. If Fig. 20.1 is considered as a snapshot of forces in an irregular wave, all of the elements of the surf-riding problem can be transferred from regular waves to irregular waves. However, as these elements may vary in time and space, the problem should be defined simultaneously in space and time—creating a spatial-temporal framework for surf-riding in irregular seas, see Fig. 20.2.

As the wave surging force and celerity change with the randomness of the irregular wave, these dynamic equilibria appear and disappear and move with unsteady speed. As a result, these equilibria points cannot be solutions of the equations of motion and, strictly speaking, do not have all the properties of equilibria in a physical sense. Nevertheless, they still define the topology of phase space. This brief argument indicates a significant difference between surf-riding and broaching in regular and irregular waves.

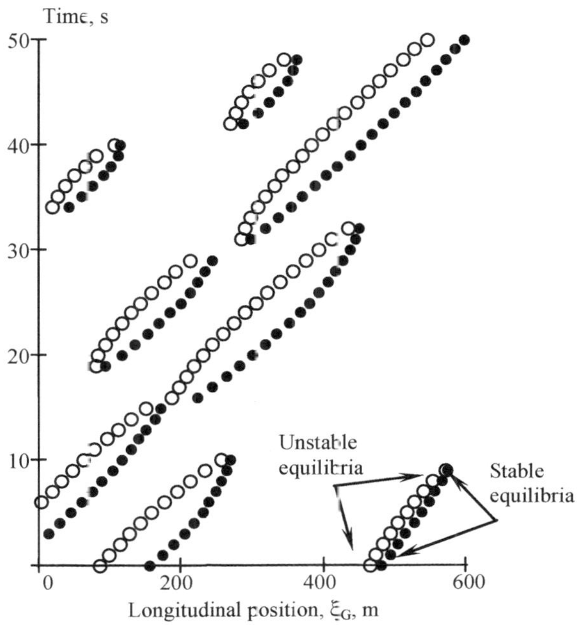

Fig. 20.2 Random surf-riding in spatial-temporal framework

While the overview of this problem is available from Belenky et al. (2016), the present text is focused on the simplest mathematical model for surf-riding in irregular waves, originally described in Belenky et al. (2011, 2012).

20.2 Surging in Irregular Waves

Consider a simple model for one-degree-of-freedom nonlinear surging:

$$(M + A_{11})\ddot{\xi}_G + R(\dot{\xi}_G) - T(\dot{\xi}_G, n) + F_X(t, \xi_G) = 0 \qquad (20.1)$$

Here M is the mass of the ship, A_{11} is the longitudinal added mass, R is the resistance in calm water, T is the thrust in calm water, n is the number of propeller revolutions, F_X is the surging component of the Froude-Krylov wave force, and ξ_G is the longitudinal position of the center of gravity in the Earth-fixed coordinate system. The dot above the symbol stands for temporal derivative. The following polynomial presentations are used for the resistance and thrust in calm water (for compatibility with Spyrou 2006):

$$R(U) = r_1 U + r_2 U^2 + r_3 U^3$$
$$T(U, n) = \tau_1 n^2 + \tau_2 n U + \tau_3 U^2 \quad (20.2)$$

The irregular waves are presented as a spatial–temporal stochastic process using the standard Longuet-Higgins model, based on the linear dispersion relation:

$$\zeta_W(t, \xi) = \sum_{i=1}^{N} a_i \cos(k_i \xi - \omega_i t + \varphi_i) \quad (20.3)$$

where a_i is the amplitude, k_i is the wave number, ω_i is the frequency, ξ is a coordinate in the direction of wave propagation, and φ_i is a random, uniformly distributed phase shift of the wave component i. As the model is meant at this stage to be qualitative, a linear wave-body formulation is appropriate. Therefore

$$F_X(t, \xi_G) = \sum_{i=1}^{N} A_{Xi} \cos(k_i \xi - \omega_i t + \varphi_i + \gamma_i) \quad (20.4)$$

Since a body-linear formulation is adopted, the amplitude A_{Xi} and phase shift γ_i are available from response amplitude and phase operators:

$$A_{Xi} = a_i RAO(k_i) \quad (20.5)$$

$$RAO(k_i) = \rho g k_i \sqrt{\left(\int_{-0.5L}^{0.5L} C(x, k_i) \cos(k_i x) dx \right)^2 + \left(\int_{-0.5L}^{0.5L} C(x, k_i) \sin(k_i x) dx \right)^2} \quad (20.6)$$

$$C(x, k_i) = 2 \int_{-d}^{0} \exp(k_i z) b(x, z + d) dz \quad (20.7)$$

where x and z are measured in the ship fixed coordinate system (positive forwards of amidships and upward from the base line), $b(x, z)$ is the molded local half-breadth, ρ is a mass density of water, L is a length of the ship, and d is the amidships section draft. Figure 20.3 shows the RAO of the surging wave force for the tumblehome ship from the ONR topside series (Bishop et al. 2005). The phase shift γ_i is presented as Fig. 20.4.

$$\gamma_i = \arctan\left(\frac{\int_{-0.5L}^{0.5L} C(x, k_i) \sin(k_i x) dx}{\int_{-0.5L}^{0.5L} C(x, k_i) \cos(k_i x) dx} \right) \quad (20.8)$$

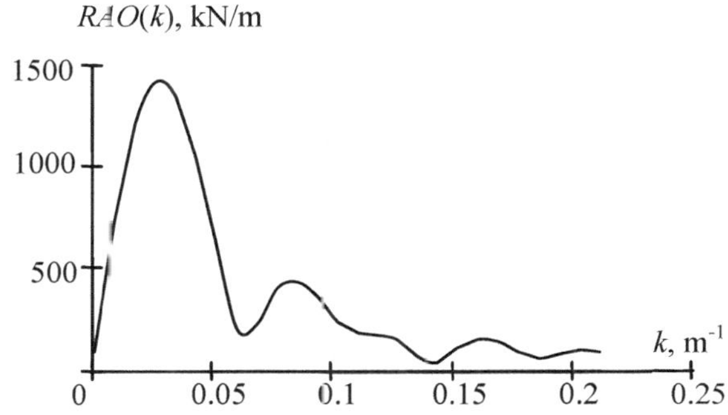

Fig. 20.3 RAO of surging force

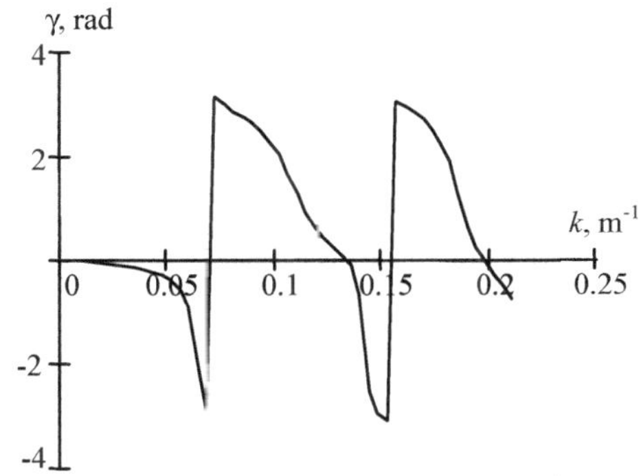

Fig. 20.4 Phase shift of surging force

20.3 Spectral Bandwidth: Modeling and Influence

To observe the appearance and development of the qualitative difference between surf-riding in regular and irregular seas, it may be useful to start from a spectrum with extremely narrow bandwidth. This idea has been implemented in the form of a "bilinear filter."

$$F(\omega_i) = \begin{cases} 0 & \omega < b_{low}\Delta\omega \\ \frac{\omega-\omega_m-b_{low}\Delta\omega}{b_{low}\Delta\omega} & b_{low}\Delta\omega \le \omega < \omega_m \\ -\frac{\omega-\omega_m+b_{up}\Delta\omega}{b_{low}\Delta\omega} & \omega_m \le \omega \le b_{up}\Delta\omega \\ 0 & \omega > b_{up}\Delta\omega \end{cases} \qquad (20.9)$$

where ω_m is the modal frequency of the spectrum while $\Delta\omega$ is the frequency step. The filter consists of two lines: the low frequency corresponds to the index b_{low} and the high frequency index is b_{up}. These two indices are parameters for controlling the spectrum bandwidth. To keep the variance of the wave elevation constant, a normalization coefficient is used:

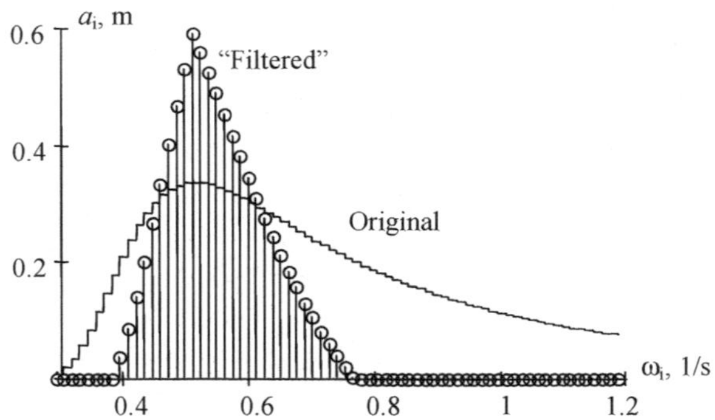

Fig. 20.5 Example of changing wave spectrum bandwidth with bilinear filter

$$K_N = \frac{\sum_{i=1}^{N} a_i^2}{\sum_{i=1}^{N} a_i^2 F(\omega_i)} \tag{20.10}$$

A sample result of the bilinear filter is shown in Fig. 20.5. After discretizing a Bretschneider spectrum with 174 frequencies, a total filtered spectrum is created by selecting the lower boundary 10 frequencies below the modal frequency and the upper boundary 20 frequencies above the modal frequency. This corresponds to a decrease of the spectrum bandwidth parameter from 0.703 to 0.21.

The filtering allows regular wave simulations to be carried out in an irregular wave framework; it is enough to set the indices of the high and low frequency boundaries to the modal frequency and only one component will remain in Formula (20.3). The random phase in Formula (20.3) allows the effect of initial conditions to be observed. It can be clearly seen in Fig. 20.6, which plots the time history of the surge velocity for wave spectrum and speed setting but with different wave phases. The speed setting refers to the ship's calm water speed for the specified constant propeller rate, which is also the initial speed in the simulation. For this regular wave case with a speed setting of 27.5 kn (14.1 m/s), there is a co-existence of surging (Fig. 20.6a) and surf-riding (Fig. 20.6b) which can be realized through different initial phases.

Adding just one additional frequency significantly changes the picture. As pointed out in Spyrou et al. (2016, 2018), the dynamic behavior becomes very rich—it includes oscillatory surf-riding, period doubling and chaotic motions. The type of observed dynamical behavior is dependent on how far the frequencies are apart and the ratio of the amplitudes.

Figure 20.7 shows the wave elevation with three frequencies. As the frequencies are very close to each other (see insert in Fig. 20.7), the waves are almost harmonic with slowly changing amplitude. Figure 20.8 shows the ship response. It starts with acceleration to 19 m/s and apparent surf-riding until about 300 s when it turns to periodic surging. Then there is another acceleration around 600 s, followed by apparent surf-riding. Thus, the considered tri-chromatic case seems to model the "catch-and-release" type of surf-riding, expected for the irregular waves, see Fig. 20.2.

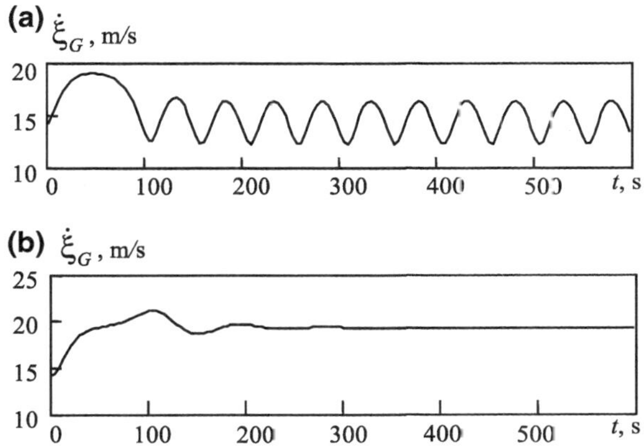

Fig. 20.6 Simulation in regular wave—co-existence of surging (**a**) and surf-riding (**b**)

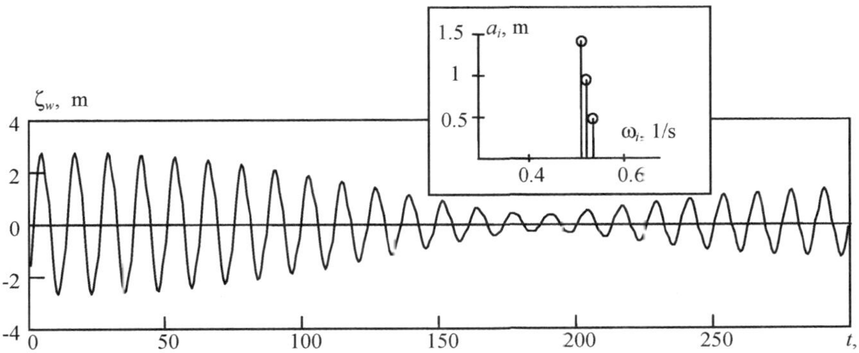

Fig. 20.7 Wave elevation at the origin composed of 3 frequencies

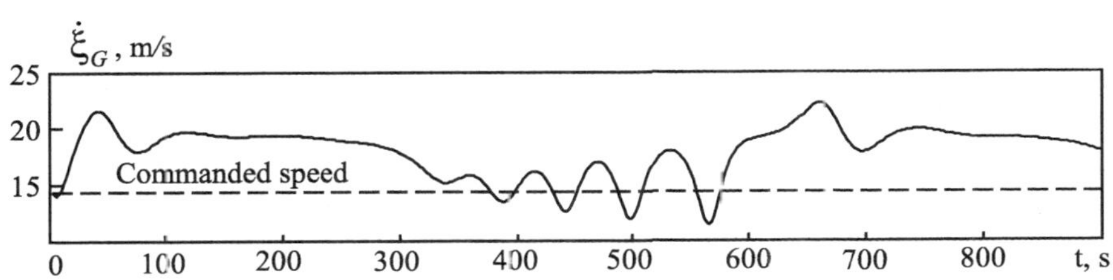

Fig. 20.8 Simulation with 3 frequencies

Figure 20.9 shows the response for the original Bretschneider spectrum. Both records contain large amplitude asymmetric oscillation, sometimes with flattened peaks. This asymmetric behavior may be a manifestation of surf-riding in irregular seas. To prove this, however, one needs to find the positions and evolution of the surf-riding equilibria.

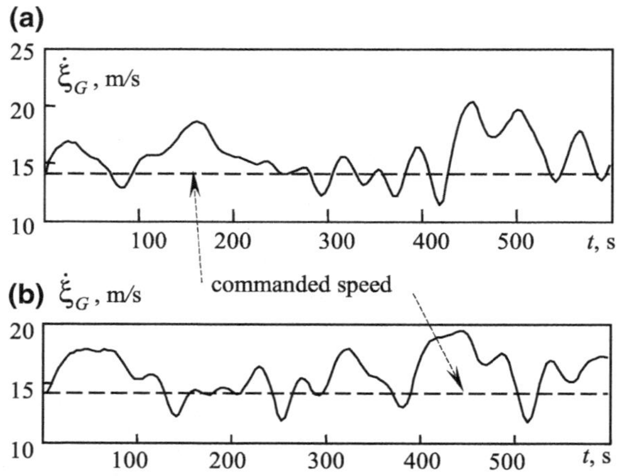

Fig. 20.9 Simulation with original Bretschneider spectrum

20.4 Surf-Riding in Tri-Chromatic Seas

The very definition of celerity in irregular waves is actually a very deep problem and is given full consideration in the next chapter of the book, as well as in Spyrou et al. (2012, 2014), which discusses the formulation of a practical definition of wave celerity and the implementation of schemes for evaluating it for theoretical and numerical analysis. An extremely simplified version of the approach is used here, in which the local celerity is defined by identifying the three profile zero-crossing points that are closest to the ship and tracking their velocities from time step to time step.

Once this approximate concept of wave celerity is accepted, the next objective is to see the relation between the visible surf-riding behavior and appearance of equilibrium. Tri-chromatic case, shown in Figs. 20.7 and 20.8, is chosen for further study. While still very simplistic, simulations using this wave and the same simplified ship surging model show three transitions—two captures and one release—over the passing of six waves. The time histories of the celerities of these six waves are shown in Fig. 20.10.

Along with a larger (about 1 m/s magnitude) changes in wave celerity, there are quite dramatic peaks with three secondary peaks on the top. These secondary peaks may be artifacts of the simplified wave tracking scheme and/or results of waves overtaking one another.

Figure 20.11a shows the time history of the instantaneous ship speed and the wave celerity of the "current" wave, the wave closest to the ship, at any given time. One can see that the celerity curve at Fig. 20.11a is a combination of all six time histories in Fig. 20.10, and can be discontinuous as the current wave changes.

Figure 20.11a also shows the temporal boundaries for waves. During periodic surge motion, they almost always coincide with a down-crossing if the commanded speed line. Figures 20.11b through 20.11q are the spatial snapshot corresponding to

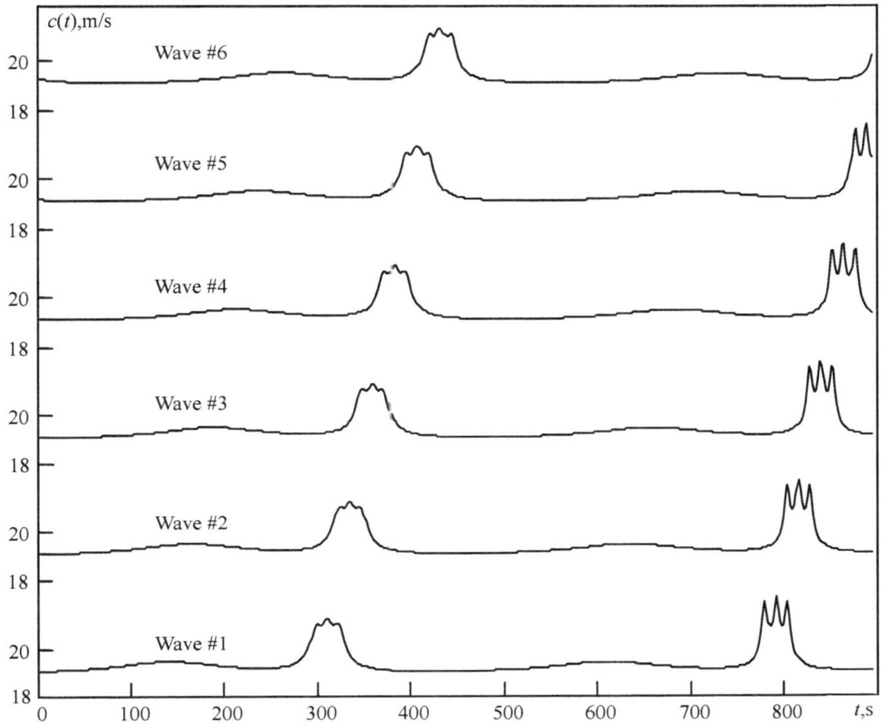

Fig. 20.10 Time histories of wave celerities

specific instant of time as noted in the captions. These instances of time are identified in Fig. 20.11a with lettered arrows referring to the respective spatial snapshot.

On each spatial snapshot plot, the blue line shows the wave profile around the ship, with the horizontal position of the diamond marking the ship's position relative to the wave. The circles on the wave profile mark three zero-crossing points that are tracked to estimate wave celerity, of which the outermost circles define the spatial boundary of the "current" wave. The direction of the ship (and wave) motion is to the right. The vertical position of the diamond indicates the ship speed, while the middle line (zero line of the wave profile) marks the commanded speed in calm water. The increase of speed is downwards.

Each spatial snapshot also contains a plot for surging wave force (red curve) and the balance between the available thrust and resistance at the current wave celerity in the same scale (lower brown line).

The first spatial snapshot, Fig. 20.11b, corresponds to the initial conditions, with the instantaneous speed equal to the commanded speed. The ship has just encountered wave #1 and is located just within its boundary. The surf-riding equilibria exist, since

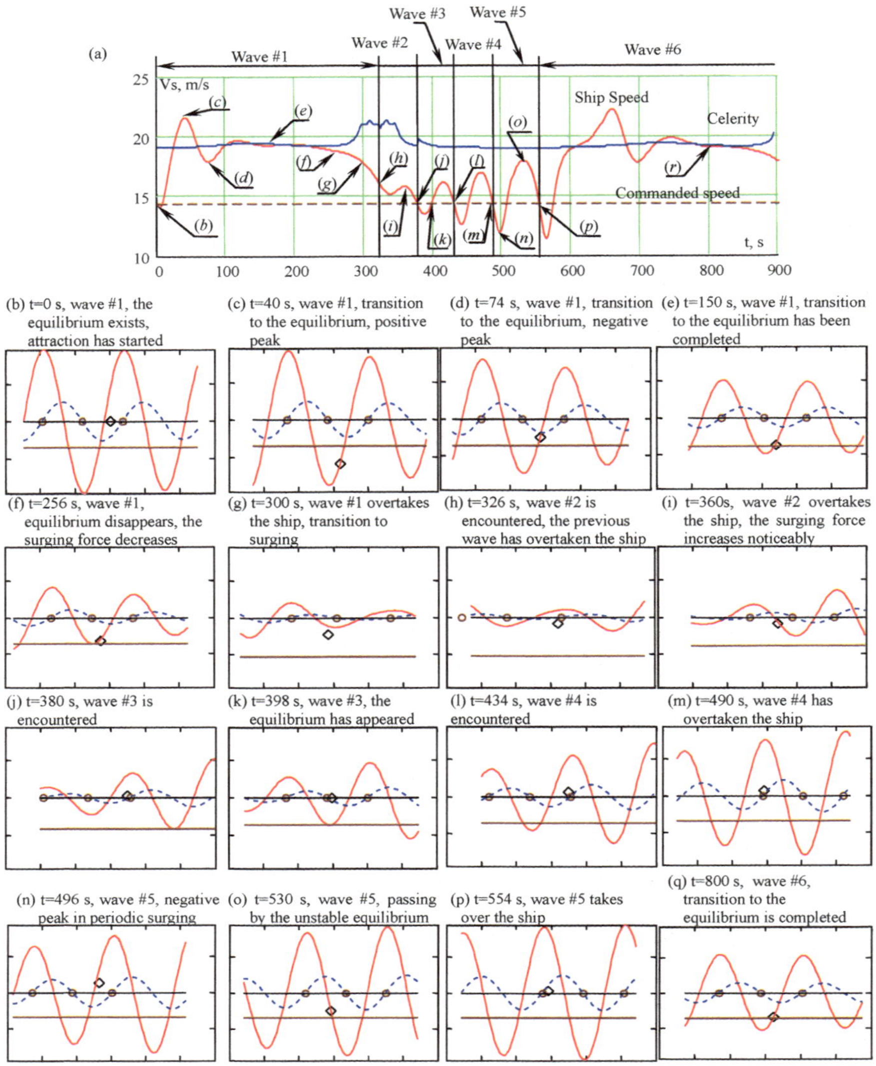

Fig. 20.11 Two captures and one release from surf-riding in a three-component irregular wave; time history (**a**) and spatial snapshots (**b**–**q**)

the surging force crosses the line corresponding to the balance between the thrust and resistance.

The stable surf-riding equilibrium attracts the dynamical system and one oscillation period is seen in Fig. 20.11a until approximately t = 100 s. The next two spatial snapshots, Fig. 20.11c, d, correspond to the positive and negative peaks during this transition, respectively. The transition is completed and the dynamical system reaches the stable equilibrium at around t = 150 s, in Fig. 20.11e.

Looking at Fig. 20.11b through 20.11e, one can see that the amplitude of the surging force is decreasing due to a reduction of the wave amplitude. This tendency leads to the disappearance of the surf-riding equilibria around t = 256 s and to the release of the ship from surf-riding (Fig. 20.11f).

As the ship slows down (Fig. 20.11g), the wave #1 overtakes her, and wave #2 is encountered at around t = 326 s (Fig. 20.11h). The ship experiences the first almost periodic surge with the positive peak corresponding to the spatial snapshot in Fig. 20.11i. As expected, wave #2 overtakes the ship quite quickly and wave #3 is encountered around t = 380 s (Fig. 20.11j).

The modulation of wave amplitude and surging force then reverses and they begin to increase. This is may be already seen in Fig. 20.11h, but becomes quite apparent in Fig. 20.11i, j. New surf-riding equilibria appear around t = 398 s (Fig. 20.11k).

The existence of the surf-riding equilibria has an immediate influence on the surge motions, which become asymmetric with wider positive peaks and sharper negative ones (Spyrou 2006). Symmetry is observed during the passing of waves #4 and #5, during which the surf-riding equilibria exists continuously (Fig. 20.11l through 20.11o). Figure 20.11o shows how the dynamical system passed near the unstable surf-riding equilibrium, but the ship is not yet "caught" and wave #5 takes over (Fig. 20.11p). The ship is finally "caught" by wave #6 and at around t = 800 again reaches the stable surf-riding equilibrium (Fig. 20.11q).

20.5 Summary and Conclusions

A simple model of surging and surf-riding in irregular waves has been formulated. A bilinear filter is introduced to control the bandwidth of the waves, so the connection to surging/surf-riding response in regular waves can be made. This filter was used for a visual analysis of the influence of the spectrum bandwidth on the surging/surf-riding response. An asymmetric response was observed that may incorporate the surf-riding behavior. The next logical step is to find the locations of the surf-riding equilibria in time and space. That step has led to the necessity of defining wave celerity in irregular waves.

The formulation of a viable definition of wave celerity in irregular waves and the development of robust methods for evaluating it in numerical theoretic analysis is a quite substantial task and represents its own area of research. For the present study, a simple definition was adopted, in which the wave celerity was defined as a speed of zero-crossing points of a wave and evaluated using a point tracking scheme. This approximate definition of wave celerity in irregular waves nevertheless allowed the characterization of the surf-riding equilibria (or the lack thereof) for some numerical case studies involving irregular waves with three components.

Acknowledgments The work described in this paper has been funded by the Office of Naval Research (ONR) under Dr. Patrick Purtell and Dr. Woei-Min Lin, NSWCCD Independent Apply Research (IAR) program under Dr. Jack Price and by ONR Global under Dr. Richard Vogelsong and Dr. Salahuddin Ahmed.

References

Belenky, V., Spyrou, K. J., Weems, K. M., 2011, "Split-Time Method for Surf-Riding and Broaching-To," Proceedings of 12th International Ship Stability Workshop, (ISSW 2011), Washington DC, pp. 163–168.

Belenky, V., Spyrou, K.J., Weems, K.M., 2012 "Evaluation of the Probability of Surf-Riding in Irregular Waves with the Time-Split Method," Proceedings of 11th International Conference on Stability of Ships and Ocean Vehicles (STAB 2012), Athens, Greece, pp. 29–37.

Belenky, V., Weems, K. M. Spyrou, K. J., 2016 "Towards a Split-Time Method for Estimation of Probability of Surf-Riding in Irregular Seas," Ocean Engineering Vol. 120 pp. 264–273.

Bishop, R. C., W. Belknap, C. Turner, B. Simon, and J. H. Kim, 2005, "Parametric Investigation on the Influence of GM, Roll damping, and Above-Water Form on the Roll Response of Model 5613". Report NSWCCD-50-TR-2005/027, Bethesda, Washington DC.

Kan, M., 1990, "Surging of Large Amplitude and Surf-Riding of Ships in Following Seas," Selected papers in Naval Architecture and Ocean Engineering, The Society of Naval Architects of Japan, 28, Ship and Ocean Foundation, Tokyo, 14 p.

Spyrou, K.J., 1996, "Dynamic Instability in Quartering Seas: The Behavior of a Ship During Broaching," Journal of Ship Research, Vol. 40, No. 1, pp. 46–59.

Spyrou, K.J., 2006, "Asymmetric Surging of Ships in Following Seas and its Repercussions for Safety," Nonlinear Dynamics, Vol. 43, pp. 149–172.

Spyrou, K.J., Belenky, V., Themelis, N., Weems, K.M., 2012, "Conditions of Surf-riding in an Irregular Seaway," Proceedings of 11th International Conference on Stability of Ships and Ocean Vehicles (STAB 2012), Athens, Greece, pp. 323–336.

Spyrou, K.J., Belenky, V., Themelis, N., Weems, K.M., 2014 "Detection of Surf-riding Behavior of Ships in Irregular Seas," Nonlinear Dynamics, Vol. 78, No. 1, pp 649–667.

Spyrou, K.J., Kontolefas, I., Themelis, N., 2016, "Dynamics of the Surf-Riding Behavior of A Ship in A Multi-Chromatic Sea Environment" Proc 31st Symp. Naval Hydrodynamics, Monterey, California, USA, 12 p.

Spyrou, K.J., Themelis, N., Kontolefas, I., 2018, "Nonlinear surge motions of a ship in bi-chromatic following waves", Communications in Nonlinear Science and Numerical Simulation, Vol. 56, pp. 296–313.

Chapter 21
Definitions of Celerity for Investigating Surf-Riding in an Irregular Seaway

Kostas J. Spyrou, Vadim L. Belenky, Nikos Themelis and Kenneth M. Weems

Abstract As is well-known, if the speed of a ship operating in high, fairly regular, following waves exceeds wave celerity, then surf-riding is realized. This motivates one to approach the calculation of the probability of surf-riding in irregular seas as a threshold exceedance problem. However, it is unknown whether such a simple phenomenological rule, using the celerity as threshold, could also be applicable for the ship dynamics associated with a stochastic wave environment. To clarify this, a suitable definition of wave celerity for an irregular seaway needs first to become available. In this chapter, we define celerity as the velocity of propagation of a fixed slope value of the wave profile. This leads to the concept of instantaneous celerity, opening up a window to the literature of instantaneous frequency in signal processing. As it turns out, instantaneous celerity is not always a consistently smooth and bounded curve. Other definitions of local celerity are also conceivable. We tested a few different selections, obtaining time-dependent celerity curves for various types of waves. Relaxing the requirement for a narrow-band spectrum, we offer some clues about the effect of spectrum's bandwidth on celerity. In a further step, simultaneous treatment of the "wave" and "ship" processes is implemented, in order to investigate the potential of applying a local celerity condition for surf-riding's prediction. Various patterns of ship motion, before and into surf-riding, are observed.

Keywords Surf-riding · Wave celerity · Irregular waves

21.1 Introduction

For developing a probabilistic evaluation of a ship's tendency for surf-riding, a significant challenge lies in the definition of wave celerity for an irregular sea environment. Specifically, it is not known whether some wave velocity notion pertaining to

K. J. Spyrou (✉) · N. Themelis
School of Naval Architecture and Marine Engineering,
National Technical University of Athens, Athens, Greece
e-mail: k.spyrou@central.ntua.gr

V. L. Belenky · K. M. Weems
David Taylor Model Basin (NSWCCD), West Bethesda, MD, USA

V. L. Belenky et al. (eds.), *Contemporary Ideas on Ship Stability*, Fluid Mechanics and Its Applications 119, https://doi.org/10.1007/978-3-030-00516-0_21

a multi-frequency wave profile could be used as an unambiguous speed threshold for predicting surf-riding, in the same manner that celerity is used for regular seas. A simple concept of mean celerity is supplied by the so-called "drift velocity" which is calculated by averaging wave lengths and wave periods and then taking their ratio. Moreover, various velocity distributions related to the waves have been discussed in the literature: Longuet-Higgins (1957) proposed approximate closed-form expressions of the velocity distribution for the maxima and minima of the profile of a short-crested sea, for a Gaussian sea. Given a spectrum, the velocity distributions of the so-called "specular points" of the profile, where the gradients in the two directions of wave propagation match desired constant values, can also be obtained. More recent elaborations on the velocity distributions of various points of random wave surfaces are described in Baxevani et al. (2003), Aberg and Rychlik (2006, 2007), and elsewhere. In a study that aimed to calculate the probability of encounter of waves that could cause broaching-to, Aberg and Rychlik (2007) selected the "wave centre" (the point of downward zero-crossing) for defining wave velocity in a Gaussian sea. Its distribution was derived for both fixed and moving (with constant speed) observers by using Rice's formula. They also discussed a celerity distribution based on the half wave length for an approximate, deterministic, dispersion relationship.

Unfortunately, statistical models of wave celerity are of little use for deriving the probability of surf-riding. This is because the process of the difference between celerity and ship speed becomes strongly nonlinear when the surf-riding tendency appears (Belenky et al. 2012, see also Chap. 20 of this book). In the current chapter certain recent results are reviewed, focusing first on possible general methods of calculation of wave celerity for irregular seas. Then, a specific celerity condition for surf-riding's prediction is evaluated (Spyrou et al. 2014).

21.2 Equation of Ship Motion

A wave elevation process evolving in time and unidirectional space is considered:

$$\zeta(x;t) = \int_0^{\infty} \sin\left(kx - \omega t + \varepsilon^{(r)}(\omega)\right)\sqrt{2S(\omega)d\omega} \approx \sum_{i=1}^{n} A_i \sin\left[k_i(x - c_i t) + \varepsilon_i^{(r)}\right] \tag{21.1}$$

Symbols are defined as follows: ζ is the elevation; x is the distance of some considered point of the profile from the axes origin; t is the time instant; ω, k are, respectively, the wave frequency and wave number; $\varepsilon^{(r)}$ is the random phase of the ω-harmonic wave component ($\varepsilon_i^{(r)}$ being its value at the i discrete frequency ω_i); A_i, k_i, c_i are the wave amplitude, number and celerity, respectively, of the discrete wave component with frequency ω_i; and $S(\omega)$ is the wave spectrum. The usual deep-water dispersion relation $\omega = \sqrt{gk}$ is applied for each harmonic wave component. The superscript (r) indicates that the phase $\varepsilon_i^{(r)}$ at $t = x = 0$ is a random number, uniformly distributed in $[0, 2\pi]$. Note that the "neat" integral representation of the elevation in Eq. (21.1) is not a Riemann-type integral (Pierson 1955; see also discussion in Kinsman 1984).

For a non-moving observer, ship's surge motion could be expressed through the following equation (Spyrou 2006):

$$\underbrace{(m - X_{\dot{u}})\ddot{\xi}}_{inertia\ M(\ddot{\xi})} - \underbrace{\left(\tau_2\dot{\xi}^2 + \tau_1 n\dot{\xi} + \tau_0 n^2\right)}_{thrust\ T(\dot{\xi};n)} + \underbrace{\left(r_1\dot{\xi} + r_2\dot{\xi}^2 + r_3\dot{\xi}^3\right)}_{resistance\ R(\dot{\xi})}$$
$$+ \int_0^\infty \underbrace{\cos\left[\frac{\omega^2}{g}\xi - \omega t + \varepsilon^{(r)}(\omega) + \varepsilon_f(\omega)\right] X_w(\omega)\sqrt{2S(\omega)d\omega}}_{wave\ force\ F(\xi;t)} = 0 \tag{21.2}$$

The following new symbols appear in (21.2): ξ is the distance of the ship from the observer; $m, -X_{\dot{u}}$ are, respectively, the ship's mass and her surge added mass; n is propeller's rate of rotation; $X_w(\omega)$ and $\varepsilon_f(\omega)$ are, the RAO value and the phase of the linear wave surge force corresponding to the ω-harmonic wave. The wave dispersion relationship for deep water has been used: $k = \omega^2/g$. Since the surge wave force is approximated by a sum of discrete harmonic force components, this equation can be expressed, after some rearrangement, as:

$$(m - X_{\dot{u}})\ddot{\xi} + r_3\dot{\xi}^3 - (r_2 - \tau_2)\dot{\xi}^2 + (r_1 - \tau_1\eta)\dot{\xi}$$
$$+ \sum_{i=1}^{n} f_i \cos\left[k_i\xi - \omega_i t + \varepsilon_i^{(r)} + \varepsilon_{f_i}\right] = \tau_0 n^2 \tag{21.3}$$

The above equation depends explicitly on time t, as for a parametrically excited system. For a regular sea, the time-dependence can be removed by setting the observer moving with the wave celerity. But this does not seem practical for an irregular sea, since the new system would be now non-inertial (for an investigation on this see Spyrou et al. 2014).

21.3 Instantaneous Celerity for Gaussian Formulation

When the wave profile is irregular, the frequency and wave number change continually. One might consider an expression of celerity based on the ratio of wave frequency over wave number. However, this requires meaningful definitions of the localized wave frequency (i.e. one associated with a short segment of the signal) in space and in time. One could extract such quantities by applying a short-time Fourier transform or by processing the wave profile simultaneously in the time and frequency domains by a continuous wavelet transform. However, due to the renowned problem of resolution when the time duration of the considered signal's segment is very short [there can be either temporal or spectral localization, but not both (Gabor limit); see Gabor (1946)] these approaches may have an applicability issue. In signal processing, the idea of "instantaneous frequency" was introduced many years ago (for a review see Boashash 1992). The matter acquires greater importance for non-stationary signals. Mandel (1974) proposed that the average frequency of a spectrum derived by the Fourier transform is equal to the time average of the instantaneous frequency. A calculation of celerity by using the concept of instantaneous frequency is presented in Spyrou et al. (2016). Here a more direct approach has been implemented, described next.

A first observation is that, the mapping of wave phase to the local gradient of the profile is not as apparent as it is in the periodic case. A quite generic idea is that celerity could be the rate at which some local wave profile quantity is propagated. The selection of the propagated quantity may result in different values of celerity. Longuet-Higgins (1957) presented definitions of "velocities of zeros" based on a fixed value of the elevation ζ, or of its gradient $\frac{\partial \zeta}{\partial x}$, as follows: consider a function h determined from the wave profile, obtaining the value h_0 at $(x_0; t_0)$. On the locus of $h = h_0$, the differential dh is naught by definition, leading to an expression of velocity:

$$dh = h_0 - h_0 = 0 = \frac{\partial h}{\partial x}dx + \frac{\partial h}{\partial t}dt \quad \rightarrow \quad c_{h_0} = -\frac{\partial h}{\partial t} \Big/ \frac{\partial h}{\partial x} \tag{21.4}$$

If the spatial gradient $h = \frac{\partial \zeta(x;t)}{\partial x}$ was used in the above, the expression of celerity would become:

$$c(x, t) = -\frac{\frac{\partial^2 \zeta}{\partial x \partial t}}{\frac{\partial^2 \zeta}{\partial x^2}} \tag{21.5}$$

Let us consider now a deep water wave elevation process represented by a large sum of harmonics, as described by Eq. (21.1). Consider also a point on the wave profile corresponding to a time instant t_0 and located at a longitudinal distance x_0 from the origin. The instantaneous spatial wave gradient at x_0 is:

$$\left.\frac{\partial \zeta(x; t_0)}{\partial x}\right|_{x_0} \equiv a(x_0; t_0) = \sum_{\iota=1}^{n} A_i k_i \cos\left[k_i(x_0 - c_i t_0) + \varepsilon_i^{(r)}\right] \tag{21.6}$$

It is possible to trace the locus $x_a(t)$ of the slope value a on the $x - t$ plane, by simple time-marching of Eq. (21.6). In the first time step, for instance, Eq. (21.7) below would be solved to get δx, given a δt (the reverse is in fact computationally more efficient—such cases are encountered later):

$$\sum_{\iota=1}^{n} A_i k_i \cos\left[k_i[x_0 + \delta x - c_i(t_0 + \delta t)] + \varepsilon_i^{(r)}\right] = a(x_0; t_0) \tag{21.7}$$

For subsequent steps, the solution $x_0+\delta x$ will replace x_0, while $t_0+\delta t$ will replace t_0 and the next nearby solution will be identified. Eventually, the solution traces a curve $f(t;\ a)$ started from $(x_0,\ t_0)$:

$$x_a(t) = f(t;a)|_{x_0,t_0} \tag{21.8}$$

where the subscript on the right-hand side indicates the starting point of the iterative solution process. The time derivative of $x_a(t)$ should produce a celerity function uniquely associated with each pair (x_a, t):

$$c_a(t) = \dot{x}_a(t) = \frac{df(t;a)}{dt} \tag{21.9}$$

For irregular waves this velocity is obviously time-varying. Furthermore, a different celerity value should, in principle, be expected for different points of the profile located between successive up-crossings.

Approximate calculations of celerity in an irregular seaway might also be contemplated. For example, consider a pair (x, t) that defines a point of the irregular profile in space-time. Fix t and measure, in space, the distance between successive up-crossings (or down-crossings) lying immediately before and after x, calling this distance $\tilde{\lambda}$. Subsequently, fix x and measure the corresponding distance in the time domain (before and after the considered instant t), noting it as $\tilde{T}$. Then an alternative definition of celerity would be:

$$\tilde{c} = \frac{\tilde{\lambda}}{\tilde{T}} \tag{21.10}$$

One should recall here that, counting distances between successive crests or troughs might lead to a substantially different result if the wave spectrum is not very narrow-banded. A feature of irregular waves is that, individual crests (or troughs) can emerge or vanish. The generation of troughs with positive elevation as well as of crests with negative elevation, are phenomena that are expected to make celerity behave in a strongly nonlinear fashion. It is reminded that, for "Gaussian" elevations, the ratio of the number of zero crossings N_0 to the number of crest and troughs N_1 is defined as follows (in this case, ε is the spectrum's width, with no relation with the phase symbol ε used earlier):

$$\frac{N_0}{N_1} = \sqrt{1-\varepsilon^2} \tag{21.11}$$

Some of the presented definitions of celerity are implemented and evaluated next.

Example 1 As a first case, consider a wave comprised of three frequencies defined in Table 21.1. The waves are selected to be reasonably steep, yet linear. In Fig. 21.1 appears the wave profile in space–time; while in Fig. 21.2 are plotted several corresponding loci of solutions $x_a(t)$. Curves of a similar nature, drawn however for the zero crossing points (which obviously do not maintain the slope) have been presented earlier by Sjo (2001). Recalling that the celerity corresponding to any one point of these curves is given by the gradient dx_a/dt, x_a has been traced with respect to t for four different local wave slope values: 0 (i.e. a crest or a trough), 0.01, 0.03 and 0.05. The solutions appear in Fig. 21.2 as several bunches of four contour lines (one contour line per value of slope). Note that, at any time instant, each value of slope can be realised at several locations, so are observed families of solutions.

In contrast to a periodic wave case, these constant slope lines are neither straight nor parallel to one another. While the gradient, and thus celerity, is mostly not changing much, there are "strange" regions that basically correspond to the merging of crests. Note in particular the backward turn of some lines—corresponding to the more extreme value of local wave slope 0.05—that leads to their unification with the corresponding curve of the previous group. It reflects the fact that in the considered wave cycle this value of wave slope was not realized—i.e. the wave was not steep enough. The annihilation (or creation) of such points of constant slope is referred-to in the literature as "twinkle" (Kratz and León 2005).

Example 2 Now consider a wave elevation process defined by the JONSWAP spectrum. This spectrum was discretised using 15 frequencies distributed equidistantly

Table 21.1 Selected linear wave components

k (m^{-1})	0.04	0.025	0.05
$\omega = \sqrt{gk}$ (s^{-1})	0.626	0.495	0.70
A (m)	1.0	1.5	0.7
Ak	0.04	0.038	0.035
$\varphi = kx - \omega t + \varepsilon$	0.1	0.8	0.5

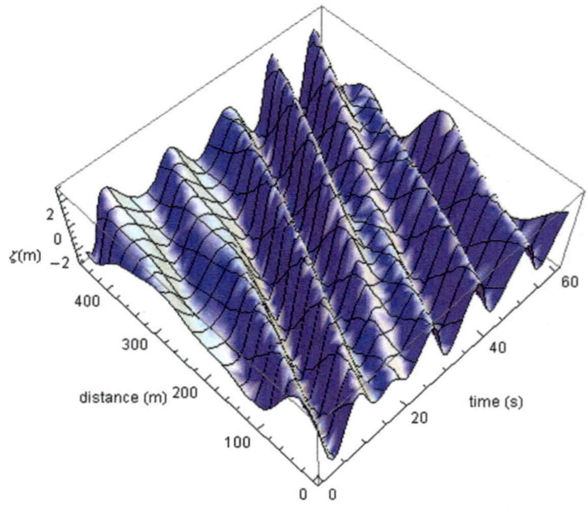

Fig. 21.1 Example 1 wave in space-time

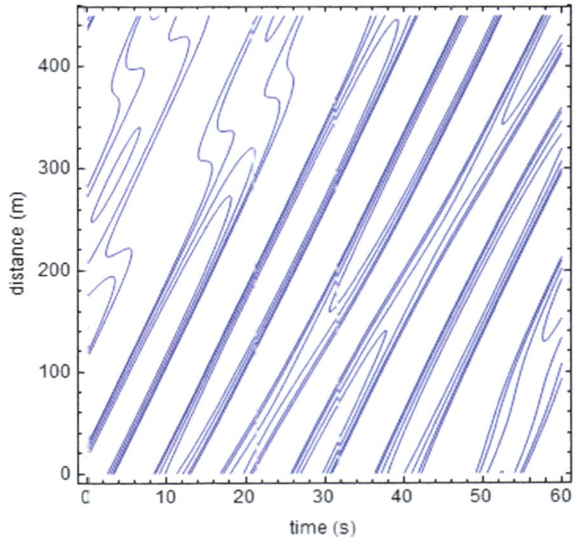

Fig. 21.2 Bunches of constant slope contours, obtained from repetitive solution of Eq. (21.7), for the three-frequency wave considered in Example 1. The slopes are 0, 0.01, 0.03 and 0.05

Table 21.2 Location of propagated wave profile points at $t = 0$ s

Slope	$x_0(m)$
	Around first crest
0	13.23
1/100	11.47

between 0.63 and 1.43 of the peak value. For the prescribed range, the bandwidth parameter ε receives the rather moderate value of 0.315. In Fig. 21.3 $x_a(t)$ contours for two slope values (0 and 0.02) are shown, using the solution $x_{a=0}(t) = f(t;0)|_{x_0,0}$.

Celerity curves produced by propagating the first crest lying right in front of the axes origin, and also a profile point with 1/100 slope, are shown in Fig. 21.4. Table 21.2 specifies the initial ($t = 0$) ordinates of these points. The variability of celerity is prevalent and spikes are noted actually indicating singular behavior. Such wave velocity jumps have been observed also earlier (see for example Baxevani et al. 2003). While the curve could be smoothed or clipped, evaluation of its importance for the ship motion phenomena of interest is required.

Moreover, the calculated local celerity values are compared to those obtained by applying the approximate expression $\tilde{c} = \tilde{\lambda}/\tilde{T}$, applied between peaks. Exact and approximate values of celerity are compared, for a slope value 1/100, in Fig. 21.5. The ratio of apparent half-length to apparent half-period gave satisfactory predictions. Such a conclusion is deduced if one compares the position of unfilled squares against the continuous line; and similarly, that of the unfilled circles against the dashed line. It is noted that a similar comparison for the filled symbols (that are based on whole apparent lengths and periods) revealed larger discrepancies. The examination of other points of the profile did not change this conclusion.

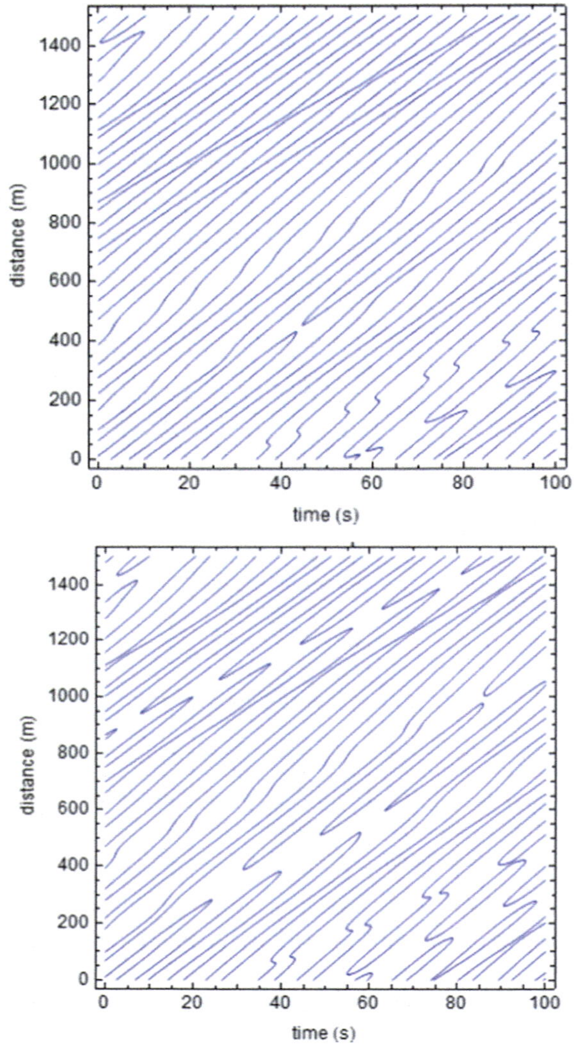

Fig. 21.3 Contour plots referring to Example 2, for slope values $a = 0$ (upper) and $a = 0.02$ (lower)

Example 3 In the third example, a portion of the JONSWAP spectrum, located around its peak, was assumed. The considered spectral area was successively widened, thus using bandwidth as the free parameter (Fig. 21.6).

The celerity curve corresponding to the first crest after the origin was derived, for three realisations of each one of the 5 tested bandwidths (see Fig. 21.7). The three realisations differed only in their (random) phases. As expected, the variability of celerity in time is intensified as the bandwidth is enlarged. This trend is consistent for all considered samples.

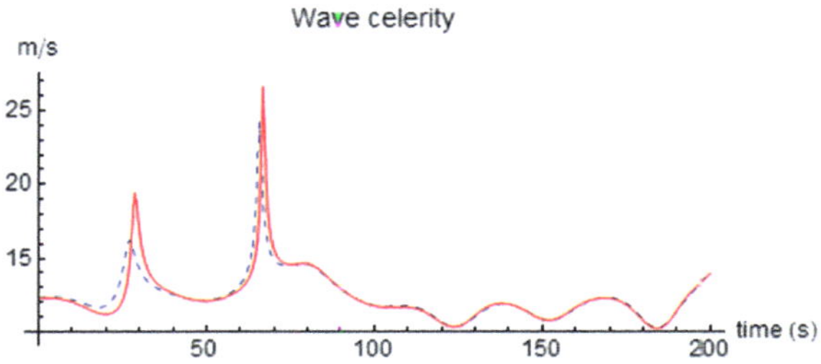

Fig. 21.4 Celerity curves corresponding to Example 2, for wave slope $a = 0$ (dashed line) and $a = 0.01$ (continuous line), started from the vicinity of the first crest

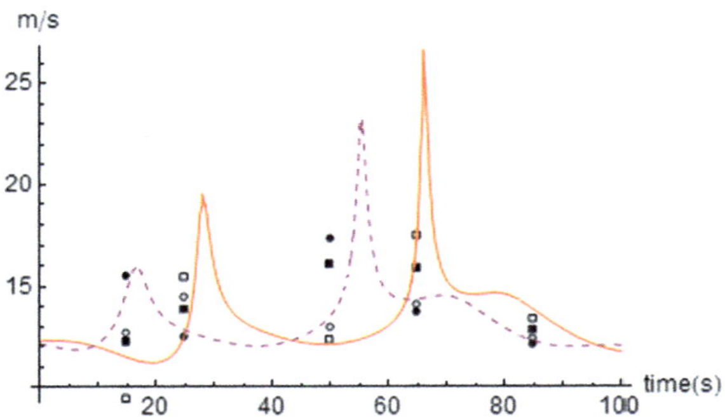

Fig. 21.5 Comparison of celerities for the JONSWAP wave of Example 2 (see also Table 21.2). The two lines represent the celerities of the points of the downslope with $a=0.01$ (dashed line for the point near to crest and continuous line for the point near to trough). Each unfilled symbol refers to a celerity calculation based on the distance between successive peaks (crest to trough, or trough to crest). Filled symbols refer to distance between successive crests. Circles refer to the higher point (after the crest) and rectangles to the lower one (before the trough) (Spyrou et al. 2014)

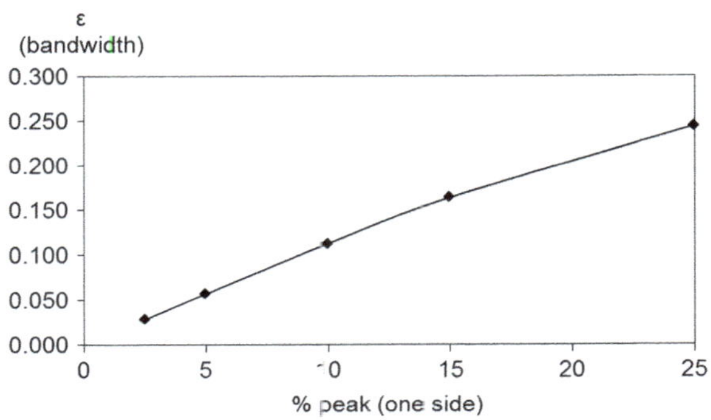

Fig. 21.6 The 5 examined cases of spectrum bandwidths examined in Example 3

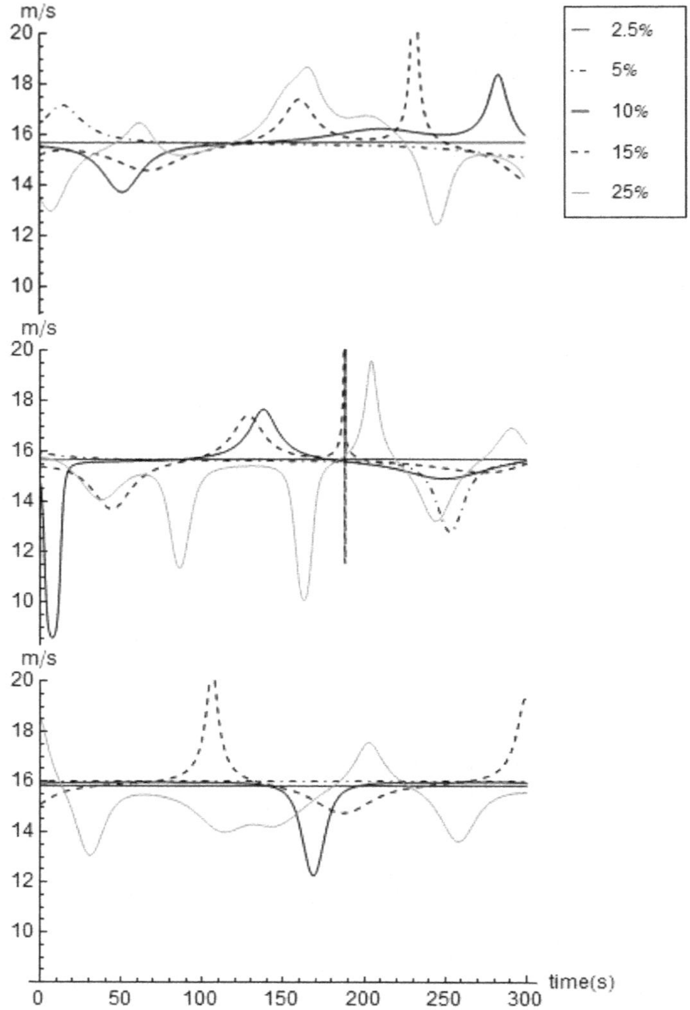

Fig. 21.7 Celerity curves for Example 3, obtained by propagating the first crest for five different bandwidths. The three diagrams differ only in the random phases of the component waves

21.4 Celerity Definition for Surf-Riding Assessment

The definition of instantaneous celerity suffers from a number of drawbacks. First, no matter what slope value is selected, there will be apparent wave cycles where any given value will not be realized. However, these waves will not be particularly steep and thus are unlikely to create surf-riding. The spikes, which correspond to points where celerity jumps to infinity because of the annihilation of the points with the considered slope value, represent a bigger concern.

To avoid a large number of these spikes, one could employ a celerity corresponding to the steepest point of wave down-slope located nearest to the ship. Such a choice has an important advantage: it is likely to be very close to the point of the profile where the maximum Froude-Krylov surge force is realized. As is well-known, the magnitude of the surge wave force is a critical factor for the occurrence of surf-riding (Kan 1990). Often, maximum surge wave force and maximum slope are realized with a small phase difference. Whilst the computational burden is not substantially different between the point of maximum wave slope and the point of maximum wave force, the first is a local wave characteristic that is, in a physical context, observable.

At least one point of maximum slope can be found on the down-slope of every apparent cycle, yet it can degenerate to zero slope if its vicinity (e.g. its distance from the neighbouring crest) shrinks to a zero length. However, it is believed that such encounters represent relatively mild conditions for the ship and such singular points should therefore be unimportant for the surf-riding probability calculation. Moreover, for a relatively narrow band sea, such events become rare. It should be noted that there can be more than one point of locally maximum slope on a down-slope.

As a ship advances with respect to a wave profile, or as the wave profile overtakes the ship, the targeted nearest point of maximum slope will change in a discrete stepped manner at least once every apparent encounter period The obtained celerity curve will therefore present points of discontinuity featuring stepped changes.

In order to calculate the celerity based on the propagation of the point of maximum slope in the vicinity of the ship, let $\xi(t)$ be the ship's position at some arbitrary time instant t, and let's search for points of maximum wave slope that lie near to $\xi(t)$ and on a down-slope. The following equation is solved for $x_{a_{max}} = f(t; a_{max})$, using Newton iterations started from the current position $\xi(t)$ of the ship:

$$\frac{\partial^2 \zeta\left(x_{a_{max}}, t\right)}{\partial x^2} = 0 \tag{21.12}$$

Simultaneously, the following inequality should be satisfied to ensure that the point is on the down-slope:

$$\frac{\partial^3 \zeta\left(x_{a_{max}}, t\right)}{\partial x^3} < 0 \tag{21.13}$$

For a 4th order finite difference approximation of the derivative, k more points are determined, separated by a time interval δt:

$$\frac{\partial^2 \zeta\left(x_{a_{max}}, t + \delta t\right)}{\partial x^2} = 0 \quad \rightarrow \quad x_{a_{max}}^{(k)} = f(t + k \cdot \delta t; a_{max}) \tag{21.14}$$

Since the time step δt is selected to be very small, the same initial guess $\xi(t)$ should practically suffice unless the initial point is very near to one of the special

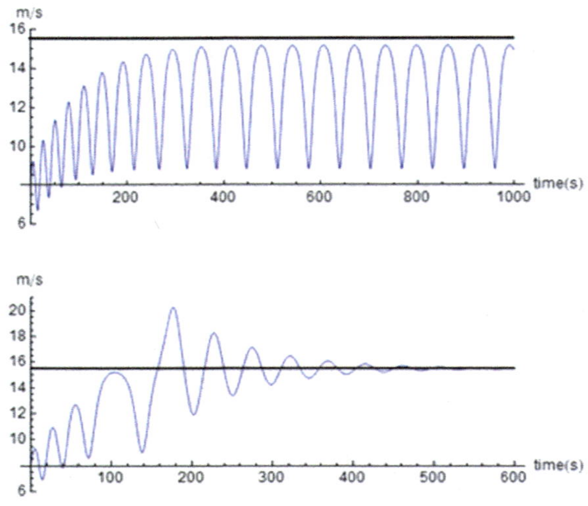

Fig. 21.8 Surge velocity (fluctuating lines) and wave celerity (straight horizontal lines) for: $\zeta = 3.8$ m, $\lambda = 154$ m. In the upper graph the Froude number is $Fn = 0.295$ while in the lower $Fn = 0.324$. The expected and the calculated celerity lines coincide

points mentioned earlier, where a stepped change takes place. Thereafter the celerity can be approximated by the formula:

$$c \approx \frac{-x_{a_{max}}^{(2)} + 8x_{a_{max}}^{(1)} - 8x_{a_{max}}^{(-1)} + x_{a_{max}}^{(-2)}}{12\delta t} \tag{21.15}$$

In order to ensure that the located points truly lie near to the ship, the following inequality condition can also be imposed on the solution $x_{a\max}^{(k)}$:

$$\left| x_{a_{max}}^{(k)} - \xi(t) \right| < d \tag{21.16}$$

where d can be a suitable fraction of the instantaneous wave length.

In the analysis of ship motion data, the celerity is calculated for each time step of the simulation time history by tracking the point of maximum wave slope nearest this ship at that time. The examples below demonstrate the tracking scheme for simulations of the tumble-home ship from the ONR Topside series (L = 154 m) in regular, bi-chromatic and irregular waves. In these examples, the simulation time step was $\Delta t = 1$ while the tracking time interval was $\delta t = 10^{-3}$.

Example 4 Here the scheme is applied for simulating regular waves, in order to verify that the expected regular wave celerity is successfully captured. The calculated celerity curve appears in Fig. 21.8 as a horizontal thick line. In this initial implementation, the procedure is to find a maximum slope point by solving Eq. (21.12) with the inequality condition (21.13).

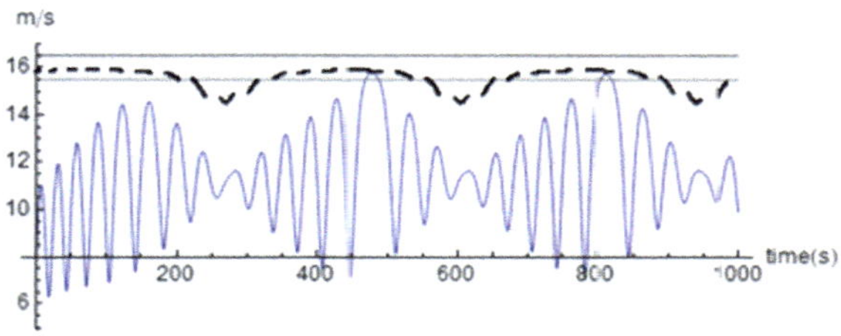

Fig. 21.9 Surging in a bi-chromatic sea, very near to surf-riding occurrence. The two present harmonic wave components had: $\zeta_1 = 2.5$ m, $\lambda_1 = 175$ m and $\zeta_2 = 3.4$ m, $\lambda_2 = 154$ m ($Fn = 0.28$). The two horizontal lines indicate their celerities. What appears as broken line is the sequence of calculated local celerity values (Spyrou et al. 2014)

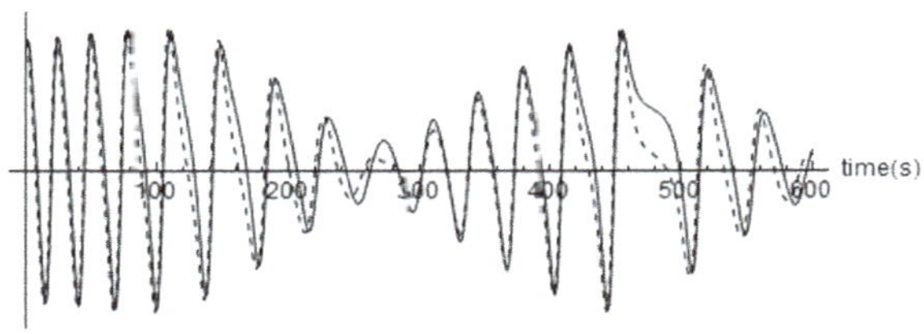

Fig. 21.10 Example of phasing of wave slope (straight line) versus Froude-Krylov force (dashed line) (Spyrou et al. 2014)

Example 5 In this example, bi-chromatic waves consisted of two wave components are considered. The first scenario has one component with a length equal to the ship length and a second component that is longer. A limiting case is presented in Fig. 21.9 where ship's nominal speed is slightly below the critical one required for surf-riding. As seen, surge velocity has come very close, yet did not cross the calculated celerity curve.

In Fig. 21.10 have been plotted the time histories of Froude-Krylov surge force and wave slope. Both are measured at the instantaneous position of the middle point of the ship. Figure 21.11 shows contours of the maxima of these quantities on a space-time plot. Both figures suggest that the maximum wave slope is a good indicator of the location where the maximum of the wave surge force arises.

Moreover, the bi-chromatic wave simulation was repeated using the same wave but with a slightly higher nominal speed (Fig. 21.12). This speed plot clearly reveals crossings of the celerity curve. In this case, the mean speed is considerably higher than the nominal speed and also, it remains quite close to the celerity curve. This is prima facie realization of surf-riding like behavior; with the remark that the observed motion has a persistently strong oscillatory character.

Another combination of wave components, with moderate steepness, led to similar patterns of ship behavior; hence the observed behavior can be considered as quite generic (Fig. 21.13).

Example 6 The final example of the celerity calculation is an irregular sea case again based on the JONSWAP spectrum ($H_S = 3.5\,m,\ T_P = 10\,s$). A region of the

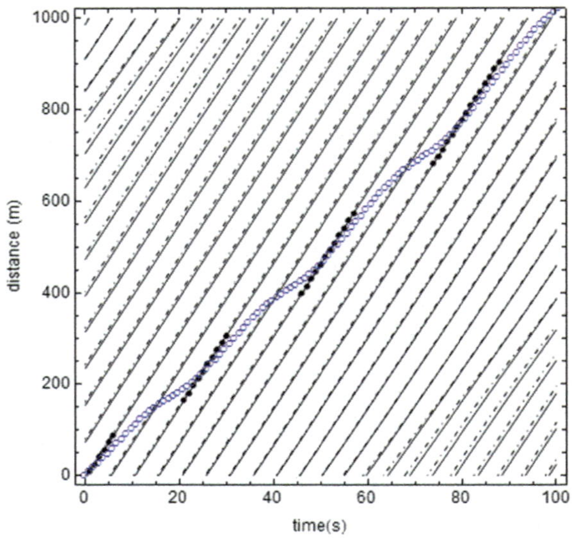

Fig. 21.11 Contour plots in space-time for maxima of wave slope and Froude-Krylov force. Straight lines indicate the maximum slopes while the dashed lines refer to the maximum force. Empty dots indicate model's instantaneous position. Black dots show the nearest identified point of maximum slope for successive positions of the model (Spyrou et al. 2014)

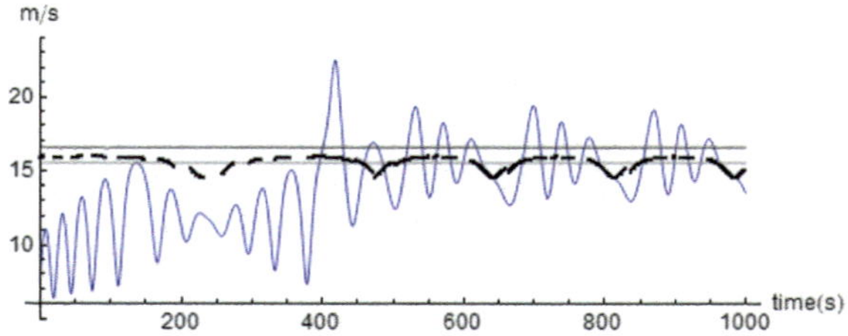

Fig. 21.12 Crossing of the time-varying celerity threshold by the surge velocity curve, for nominal Froude number 0.288

spectrum around the peak frequency has been used, corresponding to wave lengths that are very close to the ship length (between 148 and 164 m). Wave realisations were based on six frequencies, with random phases. Ship behavior for nominal speeds just below and well into the surf-riding regime are shown in Fig. 21.14. For reference, the horizontal line indicates the wave celerity corresponding to the mean wave frequency.

The initial implementation of the celerity calculation was producing results only for points of maximum slope on the down-slope adjacently to the ship. This explains the broken celerity curve in Fig. 21.14 (upper). In Fig. 21.14 (lower), in the celerity calculation were considered points on both the up- and down-slopes, producing a complete, though not necessarily continuous, curve.

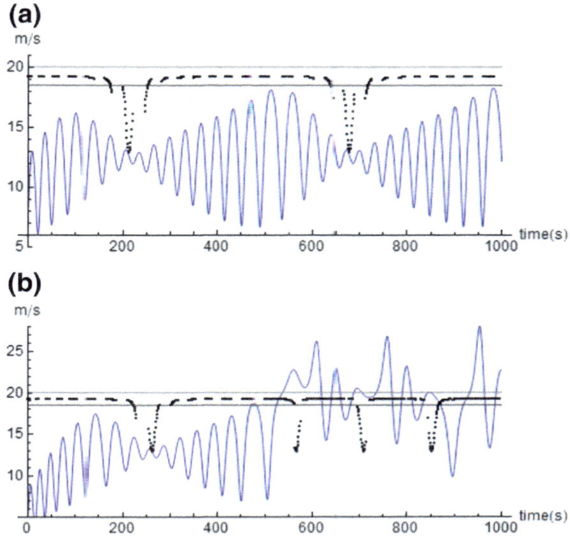

Fig. 21.13 Change of surge motion in a bi-chromatic following sea: **a** just before surf-riding occurrence with $Fn = 0.311$ (upper); **b** crossing of celerity threshold leading to surf-riding, for $Fn = 0.321$ ($\zeta_1 = 3.8$ m, $\lambda_1 = 258$ m and $\zeta_2 = 3$ m, $\lambda_2 = 220$ m)

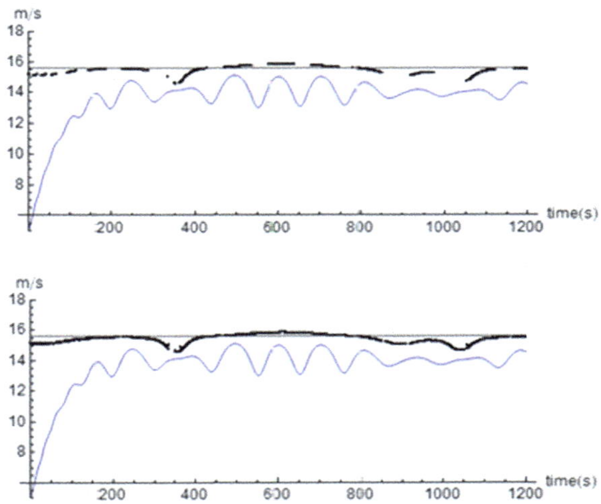

Fig. 21.14 Celerity (dotted curve) and surge motion based on crude discretisation of the central part of a JONSWAP spectrum, by 6 frequencies around the peak. Froude number was 0.283. In the upper diagram the celerity is calculated for the points of maximum slope found in the down-slope nearest to the ship. In the lower diagram the celerity calculation is based on maximum slopes of both down and up-slopes (Spyrou et al. 2014)

21.5 LAMP Implementation

Following its initial verification using the 1-DOF model of surging, the irregular wave celerity calculation is being implemented in the Large Amplitude Motions Program (LAMP) where it will be used as part of a probabilistic model of surf-riding (Belenky et al. 2012, see also Chap. 20 of this book). LAMP is a nonlinear time-domain computer simulation code for ship wave motions and loads that is built around a body-nonlinear 3-D potential flow solution of the wave-body hydrodynamic interaction

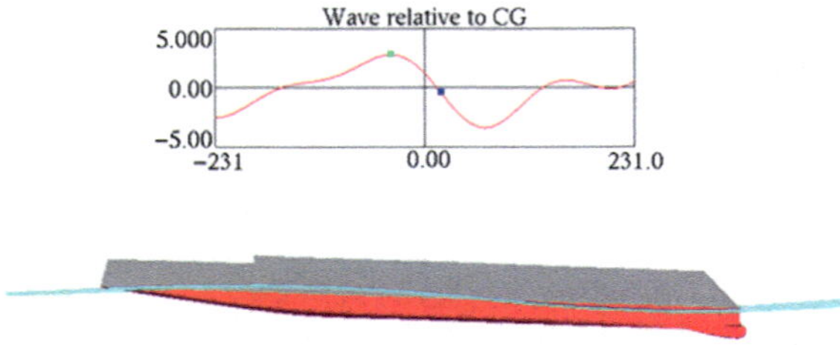

Fig. 21.15 LAMP simulation of ONR Topsides tumblehome hull in irregular following seas

problem and incorporates models for viscous, lift and propulsor effects. Previous studies have demonstrated that LAMP can simulate the principal phenomena of surf-riding and broaching-to (Spyrou et al. 2009).

The LAMP implementation of the wave celerity calculation generally follows the scheme described above, but has been adapted for LAMP's more general irregular wave models including oblique, short-crested (multi-direction) and nonlinear incident waves. The wave point that is tracked is the maximum wave slope in the direction of ship travel; it is tracked only in that direction, and the resulting wave celerity is calculated in this direction. In this manner, the tracked point is related to the force in the surging direction and the "celerity" characterizes the speed that the ship must reach, at its current heading, to stay at this point of maximum surging force. Note that this "directional" definition of celerity can be quite different from a "physical" wave celerity obtained by observing the motion of a wave feature from a global view point. In fact, for a ship travelling at an angle to long-crested seas, LAMP's "directional" celerity will be larger than the wave "physical" celerity, going to infinity in beam seas. However, this is correct for the consideration of surf-riding, as a ship would need to go infinitely fast to maintain its relative position to the crest or other feature of a beam wave.

Another difference in the LAMP implementation of the wave celerity calculation is that it always searches for the nearest maximum slope point on the down-slope even if an up-slope maximum is closer. This is done by pre-computing the elevation and its derivatives on a regular spatial interval δx_s in the travel direction and identifying intervals where a down-slope maximum can be found. A careful selection of the interval size can also significantly reduce the number of wave evaluations required to find and track the maximum slope points, which can be very important for wave models that include many components and/or nonlinear terms.

Figure 21.15 shows a snapshot of a LAMP simulation for the tumblehome hull form from the ONR Topsides series running in long-crested irregular waves. The plot shows the wave profile at that time instance along the ship's travel direction with marks for the points of maximum down-slope and elevation (crest). The wave in this case is derived from a Bretschneider spectrum with $H_S = 7\,m,\ T_m = 12\,s$.

At the lower propeller speed shown in Fig. 21.16, the ship speed oscillates near its calm water speed and below the wave celerity, and no surf-riding is observed. At the higher propeller speed, there are three episodes where the ship's speed upcrosses the wave celerity and is captured into surf-riding for a period of time, matching the

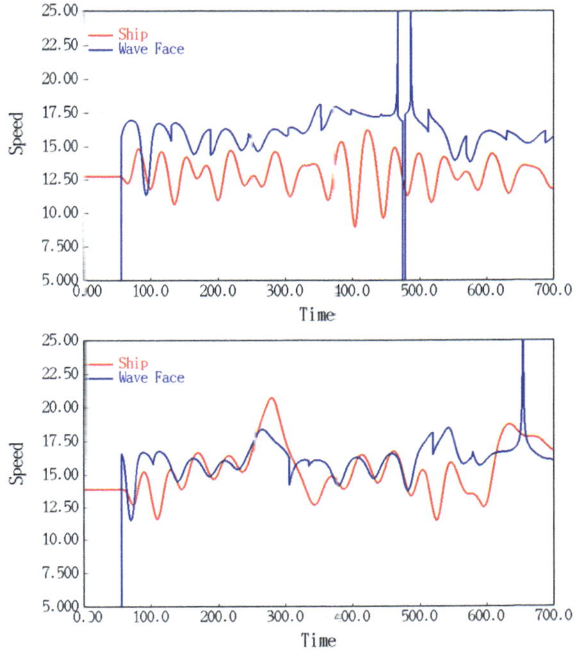

Fig. 21.16 Ship speed and wave celerity for propeller speed of 2.2 rps (upper diagram) and 2.4 rps (lower diagram)

wave speed while surf-riding. While the initial wave celerity jump at 60 s is a plotting artifact, the other "spikes" in wave speed are cases where the phasing of the wave component produces a very rapid local translation of the point of maximum slope. At ~480 s of the first record (Fig. 21.16), an overtaking of one wave by a following one actual causes the maximum point to briefly slide back up the wave face and results in a negative celerity. The identification and handling of such situations is one objective of the current development. Despite these problem points, the present scheme seems to provide a practical calculation of wave celerity that is suitable for characterizing surf-riding in irregular waves.

21.6 Conclusions

Various definitions of wave celerity for irregular seas have been discussed. Then, a scheme has been developed for the calculation of instantaneous wave celerity that could be applied towards the calculation of the probability of vessel surf-riding in a steep following seaway. The scheme is based on determining the point of maximum wave slope on the downslope of the wave nearest the ship and tracking it for a series of short time increments. The choice of the point of maximum slope for calculating celerity has been motivated by the fact that, it is generally found close to the point of maximum wave surge force that increases the propensity for surf-riding. The tracking scheme has been implemented and demonstrated in both a 1-DOF model of ship surging and in the LAMP 6-DOF time domain seakeeping code. Initial results indicated that this scheme can reliably produce wave celerity values that can indeed be used to identify surf-riding in irregular waves.

Acknowledgements The research reported in this chapter was funded by the Office of Naval Research (ONR) under Dr. Patrick Purtell and by ONR Global under Dr. Richard Vogelsong. This support is gratefully acknowledged by the authors.

References

Aberg, S., and Rychlik, I., 2006, "Role of wave velocity for encountered wave statistics," Proceedings of the Sixteenth (2006) International Offshore and Polar Engineering Conference, San Francisco, California, May 28–June 2.

Aberg, S., and Rychlik, I., 2007, "Doppler shift approximations of encountered wave statistics," Ocean Engineering, Vol. 34, 2300–2310.

Baxevani, A., Podgórski, K., and Rychlik, I., 2003, "Velocities for moving random surfaces", Probabilistic Engineering Mechanics, Vol. 18, 251–271.

Belenky, V.M., Spyrou, K.J., and Weems, K.M., 2012, "Evaluation of the Probability of Surf-Riding in Irregular Waves with the Time-Split Method," Proceedings of the 11th International Conference on the Stability of Ships and Ocean Vehicles (STAB 2012), Athens, Greece.

Boashash, B., 1992, "Estimating and Interpreting The Instantaneous Frequency of a Signal – Part 1: Fundamentals," Proceedings of the IEEE, Vol. 80, 520–538.

Gabor, D., 1946, "Theory of Communication," Journal of the IEE, Vol. 93, 429–457.

Kan, M., 1990, "Surging of large amplitude and surf-riding of ships in following seas," Selected Papers in Naval Architecture and Ocean Engineering. The Society of Naval Architects of Japan, 28.

Kinsman, B., 1984, Wind Waves: Their Generation and Propagation on the Ocean Surface. Dover Publications Inc., New York, ISBN 0-486-64652-1.

Kratz, M.F., and León, J.R., 2005, "Curve crossings and specular points, d'apres Longuet-Higgins," www.mendeley.com/research/curve-crossings-specular-points-dapres-longuethiggins/#page-1.

Longuet-Higgins, M.S., 1957, "The Statistical Analysis of a Random, Moving Surface," Philosophical Transactions of the Royal Society of London. Series A, Mathematical and Physical Sciences, Vol. 249, 966, 321–387.

Mandel, L., 1974, "Interpretation of instantaneous frequency," American Journal of Physics, Vol. 42, 840–846.

Pierson, W.J., Jr., 1955, "Wind-generated gravity waves", in Advances in Geophysics, Vol. 2, Academic Press Inc. (New York), 93–178.

Sjo, E., 2001, "Simultaneous distributions of space-time wave characteristics in a Gaussian sea", Extremes, Vol. 4, 263–288.

Spyrou, K.J., 2006, "Asymmetric surging of ships in following seas and its repercussions for safety", Nonlinear Dynamics (ed. A.H. Nayfeh), Kluwer, Vol. 43, 149–272.

Spyrou, K.J, Weems, K.M., and Belenky, V., 2009, "Patterns of Surf-Riding and Broaching-to Captured by Advanced Hydrodynamic Modelling," Proceedings of the 10th International Conference on Stability of Ships and Ocean Vehicles (STAB2009), St. Petersburg, Russia.

Spyrou, K.J., Belenky, V., Themelis, N. & Weems, K.M., 2014, "Detection of surf-riding behavior of ships in irregular seas", Nonlinear Dynamics, Vol. 78, 649–667.
Spyrou, K.J., Themelis, N. & Kontolefas, I., 2016, "Numerical statistical estimates of ship broaching-to", Journal of Ship Research, Vol. 60, Issue 4, 219–238.

Part VIII
Dynamics of Large Ship Motions: Stochastic Treatments

Chapter 22
Estimating Dynamic Stability Event Probabilities from Simulation and Wave Modeling Methods

M. Ross Leadbetter, Igor Rychlik and Karl Stambaugh

Abstract Predicting the dynamic stability of ships in severe wave environments is challenging, not only due to the complex non-linear hydrodynamics, but also the need to characterize the rarity of events. The latter involves conducting enough simulations to calculate associated small probabilities or alternate approaches for estimating the rarity of events. This paper presents techniques for calculating probabilities of occurrence of rare dynamic stability events using direct counting, Poisson distribution fitting techniques and estimating the dynamic event probabilities. The latter probability estimate is obtained by defining dangerous wave conditions that produce rare events through hydrodynamic simulations and estimating their probabilities of occurrence through joint probability distributions or simulations of the wave environment. The accuracy of these calculations is discussed. An example application is presented using a U.S. Coast Guard Cutter along with information useful for operator guidance in heavy weather. A recommendation is presented for further work on defining the limiting probabilities one might use for design or operational criteria.

Keywords Ship stability · Dynamic stability · Simulation probabilities

22.1 Introduction

Intact dynamic stability is an important issue in design and operation particularly of smaller Naval ships and U.S. Coast Guard Cutters operating in demanding weather and wave conditions. Considerable longstanding research effort has been devoted to the estimation of dynamic event probabilities and associated risks for specific vessel

M. R. Leadbetter
University of North Carolina, Chapel Hill, NC, USA

I. Rychlik
Chalmers University, Gothenburg, Sweden

K. Stambaugh (✉)
U.S. Coast Guard, SFLC NAME, Baltimore, MD, USA
e-mail: karl.a.stambaugh@uscg.mil

V. L. Belenky et al. (eds.), *Contemporary Ideas on Ship Stability*, Fluid Mechanics and Its Applications 119, https://doi.org/10.1007/978-3-030-00516-0_22

types in given operational conditions. This paper discusses techniques for calculating probabilities of dynamic stability events, with emphasis on loss of intact stability and capsize in a dynamic wave environment.

There are varied motivations for risk evaluation (deployment in operations requiring high speeds and maneuverability in heavy seas, operator guidance in escaping storms, longer-term missions in less hostile environments, and so on), as well as the relative importance of determining precise numerical estimates, vis-à-vis reliance on pure operator experience. The direct counting approach for estimation of capsize probability of a vessel in a given operational period is simply to simulate a number of "runs" of the specified duration under conditions of interest and to obtain the proportion of these for which capsize occurs, yielding a natural (binomial) estimate whose statistical properties are known. Moreover, given reliable modeling of vessel motion, such simulation is applicable to capsize from any mode resulting from existing sea conditions in that simulation run, i.e. one does not have to model specific capsize mechanisms but allow the simulation program to determine what if any type of capsize occurs in the given sea. Simulation and direct counting statistics work especially well for severe seas where many runs can be expected to yield capsizes—the relevant estimation statistics is summarized in Part 1. However, for calmer less severe seas it may require excessive numbers of runs to produce even one capsize, resulting in potentially prohibitive simulation, amenable to the use of alternative methods emanating from the seminal work of de Kat et al. (1994) as will be described in Part 2, along with illustrations of their use.

22.2 Part 1 Simulation Estimates and Their Precision

22.2.1 The Probability Estimates

One obvious and time-honored method for estimation of capsize probability of a vessel is to simply simulate a number of "runs" of given duration under conditions of interest and to obtain the proportion of these for which capsize occurs. Given good ship motion simulation, this yields a natural binomial estimate whose statistical properties are known. If n runs each of duration T are involved the capsize probability p_T in the time duration T is estimated by:

$$\hat{p}_T = r/n, \tag{22.1}$$

where r is the number of the n runs for which capsize occurs. In severe sea states this can be accomplished by reasonably few simulations since capsizes will tend to occur relatively frequently, but for calmer less severe seas excessively many simulations may be required to achieve reasonable values of r and hence a meaningful capsize probability estimate.

A natural assumption which we make is that disjoint time intervals are independent as far as capsize is concerned—in the specific sense that the probability that capsize will not occur in either of two disjoint intervals is the product of the probabilities of no capsize for each interval. Based on this, the capsize probabilities for intervals of different lengths are simply related to each other. For example it is immediate that if $p = p_1$ is the capsize probability in unit time (e.g. 1 h) then p_T (the capsize probability for T hours) is related to p by the formula:

$$p_T = 1 - (1 - p)^T \approx Tp \tag{22.2}$$

$$(\text{i.e } p \approx p_T/T)$$

to a close approximation when p is small. When considering lower sea states, we will use the approximation to provide a simple means of calculation of the capsize probability in any period from that in any other period. The same relationships may be used for the *estimates* of the probabilities involved.

22.2.2 Estimation Precision for P_T

Since the number of capsize runs r is a binomial random variable with parameters n, p_T, $\hat{p}_T$ has mean p_T i.e. is unbiased, and binomial law variance:

$$\operatorname{var} \hat{p}_T = p_T(1 - p_T)/n \approx p_T/n \tag{22.3}$$

if p is small (low sea state conditions). Hence the precision of the estimator $\hat{p}_T$ of p_T may be summarized by its standard deviation $\sigma_T = \sqrt{p_T/n}$ estimated as:

$$\hat{\sigma}_T = \sqrt{\hat{p}_T(1 - \hat{p}_T)/n} \approx \sqrt{\hat{p}_T/n} \tag{22.4}$$

This of course can be used to gauge how satisfactory $\hat{p}_T$ is as an estimator of p_T. Further, confidence limits and intervals can be simply constructed for p_T by assuming that the binomial distribution for r has reached its limiting normal approximation and hence that $\hat{p}_T$ is approximately normal. These however, may be less accurate approximations in view of the small binomial probability and it may be desirable to use exact intervals obtained, e.g. from Swogstat.org.

22.2.3 Capsize Probabilities in Standard (e.g. Unit Time) Periods

The above formulae thus enable an appraisal of the accuracy of $\hat{p}_T$ as an estimator of the capsize probability p_T in a run of any length T. The feature of this is that the capsize probability in time T is estimated by simulation of (n) runs of that same length T. For comparison and mission risk evaluation, it is convenient to have tables for capsize probabilities in standard (e.g. unit—1, 12 h etc.) periods of time, as a function of sea state. For example, for unit time the above procedure can be carried out exactly by simulating n runs of unit duration, but n may have to be very large to achieve good accuracy (perhaps prohibitively so for low sea states). A further potential disadvantage of short runs is the possibility that part of the time may be required for "startup" before the actual desired ocean conditions are reached and initial conditions have an influence on response and related statistics. However, as noted above, the value of p_T for any T determines its value for any other T such as $T = 1$, by (22.2). In particular for unit duration to estimate $p = p_1$ one may instead estimate p_T for a much larger value of T using an appropriate number n of runs as above and infer the value of p or its estimate via (22.2). It may then in fact be shown that all such choices lead to the same accuracy in estimation of p, provided p is not too large. For example, when p is small the accuracy depends only on the *total projected simulation time* $M = Tn$. That is if one chooses a total simulation time M say, the same estimate of p is obtained by choosing any run length T and taking $n = M/T$ runs of length T with the same estimation error. This is intuitively reasonable since we might think of a long run of length M formed from M/T consecutive runs of length T. This intuitive reasoning does not apply to large values of p—when numbers of runs have capsizes, when resulting in wasted computer "downtime" following capsize in a run.

22.2.4 Runs Required to Obtain a Given Accuracy in Estimating p

As in the preceding section for estimating a general p_T, the standard deviation σ_1 of $\hat{p}_1$ is an appropriate measure of the accuracy of determination of $p = p_1$. This may be estimated by (22.5) for a prescribed total run time M. On the other hand if one does not know the value of p one may wish to determine the total run time so that it may be estimated with a given (at least estimated) accuracy $\hat{\sigma}_1$. If $\hat{p}$ were already known this could be obtained simply from (22.5) with $M = nT$ to give:

$$M = \hat{p}/\sigma_1^2 \tag{22.5}$$

in which σ_1 is the accuracy (standard deviation) desired. Since the purpose at hand is to obtain a simulation estimate of p with prescribed accuracy for given sea conditions, a natural procedure will be to use (22.5) to determine simulation requirements with an estimate $\hat{p}$ obtained in some way other than simulation such as the Poisson Regression estimate of Part 2. It is not yet known how well such estimates will work for low sea states, but they are available and can be expected to give at least "ballpark" values, hence good guidance in the choice of total simulation time M. Of course once the resulting simulations are run, they can be used to validate the use of alternative and potentially less computer intensive estimation methods for low sea states. To illustrate this consider a series of simulations all with significant wave height 5 m, having mean wave period T_1, speeds and headings as in Table 22.1. These are sea conditions used in simulations run by U.S. Coast Guard. The fifth column contains "Poisson regression estimates" (Part 2) of the 1 h capsize probabilities computed using improvements of the methods in Åberg et al. (2008), and the final column uses these as preliminary estimates to give the total simulation time to give a 50% relative standard deviation (rsd) for the 1 h simulation estimate.

22.3 PART 2 Wave Modeling—Poisson Regression Methods

22.3.1 Generalities

We now turn attention to calmer sea conditions for which an excessive simulation effort may be required and describe alternative primarily theoretical involving mathematical (stochastic) modeling of sea states and the frequency of occurrence of waves deemed to be threatening to the stability of a vessel. This more theoretical approach requires more detailed understanding of the *particular* capsize mode involved and of wave characteristics which may thus threaten vessel stability. This may be obtained from a combination of theoretical and empirical studies—e.g. simulations for a moderate number of cases. Here we focus primarily on capsize of vessels caused by instability from large waves in following and stern quartering seas—of particular concern for smaller naval vessels. A U.S. Coast Guard cutter is used as an example. Our basic approach emanates from that of de Kat et al. (1994) in which ranges of wave parameters (wavelength, wave-height) prone to cause capsize were suggested and the probabilities of waves with such characteristics evaluated for specific standard sea surface spectra (but without direct consideration of motion of the vessel or its interaction with waves). Our aims here to obtain closer, much less conservative estimates by development of such ideas in important directions including: (i) characterization of relevant geometrical properties of waves which actually cause capsize when they encounter the vessel, based on U.S. Coast Guard studies of simulation histories of vessel operation in which capsize occurs, and their relationship to the vessel at capsize, and (ii) development of Poisson regression methods for prediction of intensities of critical waves from sea state and vessel parameters. Here intensities to refers to the frequency of the critical wave(s) and related event for the time frame and average encounter period of the specific seaway.

Table 22.1 Per hr capsize probabilities p for different sea states and total simulation time needed for estimating p with rsd = 0.5 computed using an improved poisson regression model from Leadbetter and Rychlik (2007)

Hs (m)	T_1 (s)	Heading (°)	Speed (knots)	Capsize prob p (per hour)	Sim time needed for rsd = 0.5 (h)
5	8	15	10	1.35×10^{-3}	3.0×10^{3}
5	8	15	15	8.57×10^{-3}	4.7×10^{2}
5	8	45	10	4.52×10^{-3}	8.8×10^{2}
5	8	45	15	3.51×10^{-3}	1.1×10^{3}
5	8	60	10	2.49×10^{-2}	1.6×10^{2}
5	8	60	15	2.14×10^{-2}	1.9×10^{2}
5	10	15	10	1.69×10^{-5}	2.4×10^{5}
5	10	15	15	1.17×10^{-5}	3.4×10^{5}
5	10	45	10	6.65×10^{-5}	6.0×10^{4}
5	10	45	15	5.37×10^{-5}	7.4×10^{4}
5	10	60	10	4.87×10^{-4}	8.2×10^{3}
5	10	60	15	4.28×10^{-4}	9.3×10^{3}
5	12	15	10	5.57×10^{-8}	7.2×10^{7}
5	12	15	15	4.10×10^{-8}	9.8×10^{7}
5	12	45	10	2.51×10^{-7}	1.7×10^{7}
5	12	45	15	2.09×10^{-7}	1.9×10^{7}
5	12	60	10	2.30×10^{-6}	1.7×10^{6}
5	12	60	15	2.06×10^{-6}	$1.9 \cdot 10^{6}$
5	14	15	10	3.56×10^{-11}	1.1×10^{11}
5	14	15	15	2.73×10^{-11}	1.5×10^{11}
5	14	45	10	1.80×10^{-10}	2.2×10^{10}
5	14	45	15	1.53×10^{-10}	2.6×10^{10}
5	14	60	10	2.01×10^{-9}	2.0×10^{9}
5	14	60	15	1.81×10^{-9}	2.5×10^{9}

22.3.2 Dangerous Waves in Following Seas

There are a number of physical phenomena leading to capsizing or capsize modes; like pure loss of stability, broaching-to, parametric resonance in roll, etc., but there number is not limited. Simulation or these events with an appropriate simulation program (e.g. FREDYN de Kat et al. 1994) is a method for estimating probability of capsize and related events under moderately high sea conditions because it captures the non-linear aspects of hull shape and empirical maneuvering model in the time

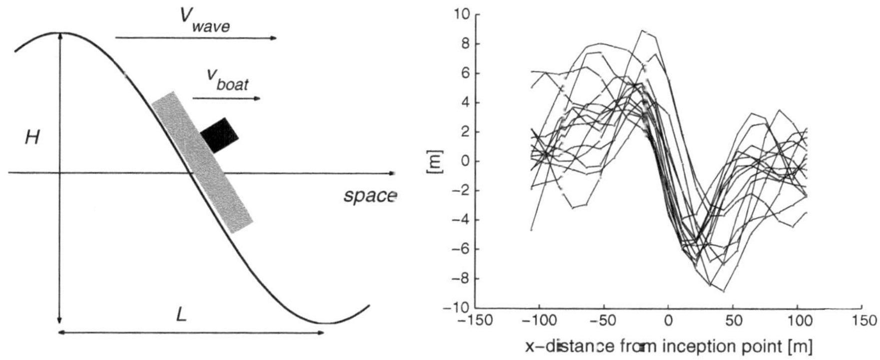

Fig. 22.1 Illustration of an overtaking wave (left). Observed waves triggering capsize (right)

domain to model the physical response to large waves. This approach counts capsizes caused by whatever mode occurs in a given sea and vessel operational conditions, and the event then becomes only statistically dependent to causes of particular modes. For less severe seas capsizes may be rare and one may wish to focus on particular capsize modes of interest, model mathematically the sea—ship motion conditions that are likely to lead to capsize from that mode, and estimate theoretically the probability of such combination for specific sea spectra and vessel operation. In particular we consider vessel capsize in following seas, where larger waves can lift the vessel and cause directional instability such as surf-riding and broaching-to, leading to capsize.

Specifically in de Kat (1994) it is observed from simulations that certain specific ranges of wave amplitude and wavelength tended to engender capsize and the probability that a wave would have parameters in this range is evaluated for given standard sea spectra from known theory by Longuet Higgins (1957). The assumption that such a wave would lead to capsize, thus provided capsize probability estimates. However, the wave calculations were done as would be observed from a ship at rest and wave-ship encounter interaction was not considered yielding a very conservative estimate. It is clear that better estimates of capsize probability should be obtainable from a more detailed understanding of what constitutes a “dangerous wave”, i.e. what geometrical properties make it more likely to cause capsize when it encounters a vessel from behind. Further, it seems likely that the probability of such a wave causing a capsize will depend on factors such as the position and motion of the vessel relative to the overtaking wave when encounter is initiated. Therefore, U.S. Coast Guard simulations of vessel tracks and parameters were studied and times of capsize recorded along with the shape of the last wave preceding the capsize event, which we refer to as “triggering waves”. Typical examples of these are shown in Fig. 22.1 from which it can be seen that a common of the triggering waves is the similar (steep) slope between peak and trough.

Influenced by this, it seems reasonable to define a wave to be “dangerous” if its downward slope lies within some small range (typified by those of Fig. 22.1 as the wave passes the centre of the vessel, in a sense to be made precise below. By studying the example shown in Fig. 22.1, we define a wave to be "dangerous" if its downward slope lies within some small range, viz. between −0.4 and −0.2 as the wave passes the center of the vessel. Our proposed approach is to calculate the rate $\mu_D = \mu_D(\theta)$ at which dangerous waves are expected to overtake the vessel and further adjust this by the estimated probability that a dangerous wave will cause capsize to obtain the probability of capsize in a given time period. For this we take the mathematically equivalent view point of regarding a vessel as "righted" after capsize and evaluate probability as estimated by the expected number of capsizes, which will depend on the type of ship, and operating conditions: sea state, heading, speed, duration etc.

We summarize the operating conditions in a vector of parameters θ, such as vessel type, heading α; speed v, and sea surface, modeled as a stationary Gaussian field having Bretschneider (1959) power density spectrum. This spectrum is parametrized by the so called significant wave-height H_s and mean wave period T_1, viz. $\theta = (H_s, T_1, \alpha, \text{v})$. Let $\lambda = \lambda(\theta)$ denote the capsize intensity (rate), measured in years^{-1} (hours^{-1}) as convenient, under the operational conditions θ. The probability p_T of capsize in a given time T can be immediately obtained from the capsize rate $\lambda(\theta)$, as $p_T = 1 - e^{-\lambda T}$.

22.3.3 Estimation of Event Intensity $\lambda(\theta)$

The proposed approach is to use the estimated event rate intensities $\lambda(\theta)$ in the range about 10^{-2}, which can be accurately estimated using 12 h mission simulation, to model $\lambda(\theta)$. Then the method is validated by means of extrapolation to operational conditions θ with smaller risks for capsizing, about 10^{-3}. (Such intensities can be estimated using the simulation program.) Then, if the model satisfactorily predicts such risks, one could expect to use it to predict the annual probabilities and risk. The first step is to calculate $\mu_D = \mu_D(\theta)$ mathematically from the assumed sea state, spectral form, and vessel motion parameters. The detailed calculations are too lengthy to re-produce here; but, may be found in references Leadbetter and Rychlik (2007), Åberg et al. (2008), Rychlik et al. (2006) and Rychlik and Leadbetter (2009).

Two components are involved. First step is to calculate the rate of occurrence of waves overtaking the vessel, and then adjust this by the probability that the specified steep wave slope will be attained, to give the expected dangerous wave occurrence rate $\mu_D = \mu_D(\theta)$. The second and final step is to adjust this by a factor representing capsize probability from a dangerous wave, requiring estimation from a number of capsize runs. This involves a "Poisson Regression" (or log-linear model, with Poisson distributions which is more natural for counting data than the customary linear regression with normal distributions).

Specifically we propose the model:

$$\lambda(H_s, T_1, \alpha) = b_1 (H_S/T_1)^{b_2} \mu_D(\theta), \qquad (22.6)$$

EXAMPLE

Probability of Capsize Calculation
USCG Cutter
1 hr Exposure, Stern Quartering Seas, 15 kts
Tp (sec)

		7.5	8.5	9.7	10.9	12.4	13.9	15	16.4	18	20
	0.5	0.000E+00	0.000E+00	0.000E+00	0.000E+00	0.000E+00	0.000E+00	0.000E+00	0.000E+00	0.000E+00	0.000E+00
	1.5	3.669E-08	0.000E+00	0.000E+00	0.000E+00	0.000E+00	0.000E+00	0.000E+00	0.000E+00	0.000E+00	0.000E+00
	2.5	1.560E-08	0.000E+00	0.000E+00	0.000E+00	0.000E+00	0.000E+00	0.000E+00	0.000E+00	0.000E+00	0.000E+00
	3.5		1.670E-05	5.350E-07	0.000E+00	0.000E+00	0.000E+00	0.000E+00	0.000E+00	0.000E+00	0.000E+00
			R-16	R-12		Weather Routing Limit					
Hs (m)	4.5			3.100E-03	3.170E-04	9.080E-06	1.060E-07	0.000E+00	0.000E+00	0.000E+00	0.000E+00
				R-18	R-15	R-12					
	5.5			5.680E-02	1.070E-02	8.250E-04	3.550E-05	2.300E-06	3.860E-08	0.000E+00	0.000E+00
					R-19	R-15		R-12			
	6.5				9.500E-02	1.340E-02	1.230E-03	1.570E-04	7.500E-06	1.190E-07	0.000E+00
							R-20	R-15	R-12		
	7.5					8.710E-02	1.320E-02	2.590E-03	2.390E-04	9.510E-06	7.160E-08
								R-20	R-15		R-12
	8.5										

Fig. 22.2 Example of probability of capsize calculations in stern quartering seas using the "Dangerous Wave" approach. R—is significant roll amplitude

This model is then fitted to actual observed capsize rates from a series of simulation runs, fitting the parameters b_1, b_2 by standard Poisson regression methods. The procedure has been programmed in Excel for these applications, and examples are given in Fig. 22.2 for a 1 h exposure timeframe. Probabilities less than 10E-8 are negligible and set equal to zero.

22.3.4 Validation of Poisson Regression or Other Estimation Methods

For more severe seas the Poisson Regression estimation (or any proposed alternative estimation method) may be validated by comparison with simulation, which is known to provide reliable results. It is no longer possible to make comprehensive comparisons for low sea states as noted relative to the required simulation effort evident from Table 22.1 at least in view of the excessive simulation times which may be needed for a few cases requiring longer simulations that can be done routinely. Such "spot checks" do not all have to be done on the same computer of course, since as shown it is only the total simulation time that affects the value of the (simulation) estimate and its accuracy. Hence, the time needed can be reduced if many computers are available—even if for short periods such as weekends or other normally idle time. Figure 22.3 presents a comparison of the Direct Counting (Part 1) and the "Dangerous Wave" (Part 2) approaches described in this paper for an example U.S. Coast Guard Cutter.

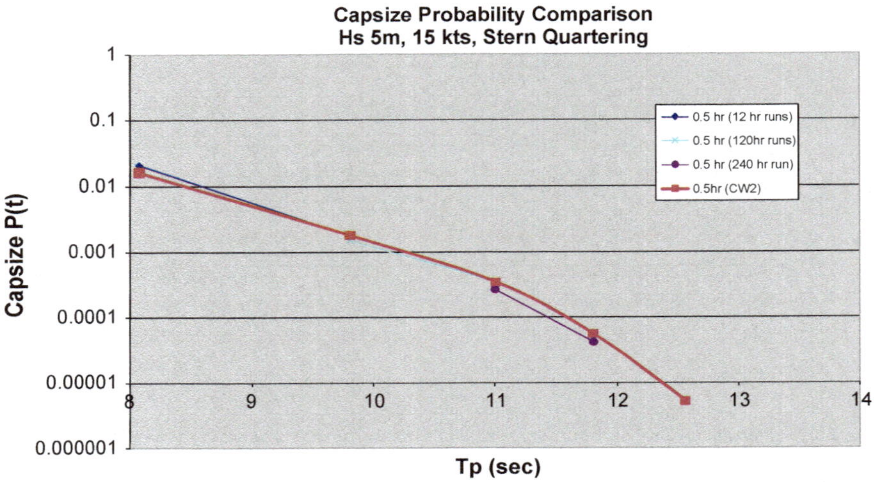

Fig. 22.3 Comparison between "direct counting" and "dangerous wave" approaches for predicting probability of capsize

22.4 Conclusions and Recommendations

This paper discusses two approaches for estimating the statistics of dynamic events associated with ships operating in diverse wave environments and their specific application depending on the desired accuracy and simulation time available and practical. In particular these include:

(1) Direct counting approaches provide accurate statistical information for estimating the probabilities of dynamic events in severe wave conditions, and enable confidence bounds to be determined for the estimates.
(2) Characterization of the wave conditions, wave–vessel interactions, and related probability of occurrence in a given seaway associated with dynamic events is found to be a useful approach to avoid excessive simulation time required to predict rare events, especially for less severe seas.
(3) Further work is recommended to compare predicted probabilities to actual experience in severe steep wave conditions in order to benchmark the predictions for operator guidance.

Disclaimer The opinions expressed herein are those of the authors and do not represent official policy of the U.S. Coast Guard.

References

Åberg S., Leadbetter M.R., Rychlik I. (2008) Palm Distributions of Wave Characteristics in Encountering Seas. *Ann. Appl. Probab.,* 18, pp. 1059–1084.

Bretschneider, C. L., 1959. Wave variability and wave spectra for wind-generated gravity waves. U.S. Army Corps of Engineers, Beach Erosion Board, Tech. Mem., No. 113, 192 pp. (Available from Maury Oceanographic Library, Stennis Space Center, MS 39522-5001).

de Kat, J.O., Brower, R., McTaggart, K, Thomas, W.L 1994 "Intact ship survivability in extreme waves: criteria from a research and Navy perspective" *STAB 94, Fifth Internat. Conf. on Stability of Ships and Ocean Vehicles* Melbourne, FL.

Longuet Higgins M.S., (1957) The statistical analysis of a random, moving surface. *Phil Trans Roy SocA*, 29: 321–387.

Leadbetter, M.R., Lindgren, G., Rootzen, H., (1983) *Extremes and Related Properties of Random Sequences and Processes*, Springer-Verlag.

Leadbetter, M.R, Rychlik, I., "Poisson Regression models for estimation of capsize risk", Tech. Memo 2:4, USCG Task 2, June 2007.

Rychlik I., Aberg S., Leadbetter M.R. (2006) Note on the intensity of encountered waves, *Marine Structures*, 34 pp. 1561–1568.

Rychlik, I., Leadbetter, M.R. (2009) Estimating Capsize Risk for a Vessel in a Following Sea, Bulletin Int. Stat. Ins.

Chapter 23
Stochastic Wave Inputs for Extreme Roll in Near Head Seas

Dae-Hyun Kim and Armin W. Troesch

Abstract An approach to generate the extreme value distribution of parametric roll in near head sea conditions is presented using a Design Load Generator (DLG), a process to approximate the extreme value distribution of a Gaussian random variable. Statistics of the roll amplitudes of a Joint High Speed Sealift (JHSS) concept hull calculated from the DLG and the Large Amplitude Motion Program (LAMP) are compared in Weibull space with the results from limited Monte Carlo simulations. The interpretation of the exposure period of the DLG results is included.

23.1 Preface

The work presented in the following sections was originally presented at the 12th International Ship Stability Workshop. It represents an early attempt to correlate parametric roll with the extreme response of an artificial linear oscillator excited by stern relative motion. The process and results are a complement to a later, more in-depth, publication – *Statistical Estimation of Extreme Roll Responses in Short Crested Irregular Head Seas, 2013 SNAME Transactions, Society of Naval Architects and Marine Engineers, USA* (Kim and Troesch 2013) – by the same authors.

In both works, the idea that the extreme value distribution of extreme roll responses can be calculated by examining seaways where a surrogate process yields the responses of interest was used. But there were two major differences. Rather than the artificial linear oscillator, rare wave groups in short crested seaways, identified by a novel derived process based on the time-varying metacentric height GM(t), were shown to efficiently capture rare instances of parametric roll. The other important difference was that the DLG methodology had been completely redeveloped such that it could statistically accurately reproduce the extreme value distribution of the

D.-H. Kim · A. W. Troesch (✉)
Department of Naval Architecture and Marine Engineering,
The University of Michigan, Ann Arbor, MI, USA
e-mail: troesch@engin.umich.edu

D.-H. Kim
e-mail: daehyun@umich.edu

V. L. Belenky et al. (eds.), *Contemporary Ideas on Ship Stability*, Fluid Mechanics and Its Applications 119, https://doi.org/10.1007/978-3-030-00516-0_23

surrogate process even in short crested seaways. The value of the paper included here is its demonstration of the relationship between extreme relative motion in near-head seas and large roll and the concise explanation of an application of the DLG methodology. For more information about the DLG methodology and its application, refer to, for example, Kim et al. (2011), Kim (2012).

23.2 Introduction

The underlying physics of parametric roll has been studied extensively by researchers. As a result, many aspects of this nonlinear phenomenon can now be explained reasonably clearly (e.g., France et al. 2003; Shin et al. 2004). However, the prediction of parametric roll in a stochastic seaway is a question to be answered with further research. One recent effort uses the first order reliability method (FORM), a popular approach in the field of structural reliability, to calculate the most probable time evolution of parametric roll yielding a known roll angle at a known time (Jensen and Pedersen 2006). Although the nonlinearity associated with parametric roll is partially recovered with a closed-form roll response model used to approximately find the design point, the hydrodynamic model for the roll response is a simplified one and the wave model is still linear. Moreover, the wave episode leading to the most probable roll response time series is essentially a regular wave except the neighborhood of the maximum roll. Whether this near regular wave behavior is due to the small number of wave frequencies (i.e., 15–50) in the wave model or the intrinsic nature of the FORM solution, which is just a point on the failure surface with the shortest distance from the origin, is not clear. One of the limitations inherent with this method is that the closed-form roll response model will play a crucial role in the calculation of the reliability index, which may result in significant deviations in the exceedance probability of the most probable time evolution of the known response (see Vidic-Perunovic 2011).

A more desirable approach would be to use the time domain seakeeping tools that can capture the onset of parametric roll with higher fidelity. For example, a 3D time domain body nonlinear code, Large Amplitude Motion Program (LAMP) (Shin et al. 2003), has been successfully applied to the prediction of parametric roll (see, e.g., France et al. 2003; Shin et al. 2004). Then, Monte Carlo simulations in theory should be able to generate the statistics of parametric roll in irregular seaways. However, several questions still persist. For example, what kind of probability distribution would parametric roll follow? Is the roll response with parametric roll an ergodic process? (see, e.g., Shin et al. 2004) In addition, Monte Carlo simulations in general are very expensive. The purpose of the current paper is to present an approach to the extreme value distribution of parametric roll as practically as possible using a Design Load Generator (DLG). The DLG has been previously applied to estimate the statistics of not only Gaussian or slightly non-Gaussian processes but also of a highly non-Gaussian process (Kim and Troesch 2010; Kim et al. 2010; Alford et al. 2011).

23.3 Methods

23.3.1 Design Load Generator

The DLG is a process to approximate the extreme value distribution of a Gaussian random variable without expensive Monte Carlo simulations. For example, a random variable X sampled from Eq. (23.1) follows the Gaussian distribution as N goes to infinity due to the central limit theorem.

$$x(t) = \sum_{j=1}^{N} a_j \cos(\omega_j t + \epsilon_j) \tag{23.1}$$

Here ϵ_j is a uniformly distributed random variable between $-\pi$ and π. The amplitude a_j is obtained from a single-sided spectrum $S(\omega_j)$ as

$$a_j = \sqrt{2S(\omega_j)\Delta\omega_j} \tag{23.2}$$

The standard deviation of the random process, σ, may be calculated from the area under the spectrum $S(\omega_j)$ as

$$\sigma^2 = \sum_{j=1}^{N} S(\omega_j)\Delta\omega_j = \sum_{j=1}^{N} \frac{1}{2} a_j^2 \tag{23.3}$$

The probability density function (PDF) of the random variable X is then

$$f_X(x) = \frac{1}{\sigma\sqrt{2\pi}} \mathrm{e}^{-x^2/(2\sigma^2)} \tag{23.4}$$

The largest value out of m zero-mean Gaussian random samples is also a random variable and designated as X_m. The PDF of X_m can be theoretically derived as

$$f_{X_m}(x) = m\left(\frac{1}{\sigma\sqrt{2\pi}} \mathrm{e}^{-x^2/(2\sigma^2)}\right)\left(\Phi(\frac{x}{\sigma})\right)^{m-1} \tag{23.5}$$

By differentiating Eq. (23.5) with respect to x, the most probable (or the peak value) $\bar{x}$ can be obtained: the number of observations m is related to the peak extreme value of the random variable by

$$\frac{1}{m} \sim 1 - \Phi\left(\frac{\bar{x}}{\sigma}\right) \quad \text{as} \quad m \to \infty \tag{23.6}$$

where $\Phi(\cdot)$ is the cumulative distribution function of the standard normal distribution. Eq. (23.6) leads to the definition of the target extreme event (TEV), which is

$$TEV = \frac{\bar{x}}{\sigma} = \frac{\text{design response}}{\sigma} \tag{23.7}$$

It is evident that TEV is closely related to the exposure periods of the extreme events: a higher TEV represents a rarer event.

Let X'_m be a new random variable modeled in the DLG by

$$x'_m(0) = \sum_{j=1}^{N} a_j \cos(\omega_j \cdot 0 + \epsilon'_{m_j}) = \sum_{j=1}^{N} a_j \cos(\epsilon'_{m_j}) \tag{23.8}$$

where the extreme is assumed to occur at $t = 0$ without loss of generality. Once σ and m (or TEV) are fixed, the DLG calculates the PDF of ϵ'_{m_j} such that X'_m approximates X_m (Alford et al. 2011). Since the PDF of ϵ'_{m_j} is given, the shape of $x'_m(t)$ around $t = 0$ is readily available. Assuming X is the Gaussian response from a linear system, the corresponding Gaussian input wave time series is readily available, from which any nonlinear time domain simulation can further be conducted. When the correlation coefficient between $x'_m(t)$ and the corresponding nonlinear response is 1, the two responses have the same exposure period (i.e., m). As the correlation weakens, the DLG-generated nonlinear response becomes the lower bound of the real nonlinear response of the same exposure period. Then, the question becomes what linear response is closely correlated to the parametric roll response of a vessel. In this paper, an imaginary damped mass spring oscillator of which forcing is the relative motion of ship's stern with respect to the input wave is considered. To be specific, it is hypothesized that the response of the oscillator would be closely related to the parametric roll. Since the output response of the oscillator is still linear, the exposure period of the DLG-based extreme response of the oscillator can easily be calculated. Choosing the parameters of the oscillator will be discussed in the next section.

23.3.2 Requirements for the Onset of Parametric Roll

In order to devise the parameters of the oscillator, the onset of parametric roll is first considered. Parametric roll is said to occur when a few critical requirements are satisfied. The conditions summarized in France et al. (2003) are repeated in Table 23.1.

In Table 23.1, ω_{roll} is the roll natural frequency, ω_e is the wave encounter frequency, λ is the wave length, L is the ship length, η_o is the input wave height, and η_c is a threshold value. In addition, a ship is said to be more vulnerable to head sea parametric roll when its stern area is wide and flat and bow flare is pronounced. In this sense,

Table 23.1 Typical requirements for parametric roll

Requirements
$\omega_{roll} \sim 0.5\omega_e$
$\lambda \sim O(L)$ (or $0.8L \leq \lambda \leq 1.2L$)
$\eta_o \geq \eta_c$
Roll damping is low

the example vessel used in this study, the Joint High Speed Sealift (JHSS), has those characteristics that may contribute to parametric excitation. However, the simulation of extreme parametric roll for the JHSS through Monte Carlo simulations for practical purpose is usually not feasible. The numerical model and principal dimension of the JHSS is shown in Fig. 23.1 and Table 23.2, respectively. A technical report of experiments that includes the hydrostatic stability curves of this hull form (JHSS Model 5663) is available in Piro et al. (2012).

In the current research, the roll response of the JHSS is simulated using LAMP. Depending on the level of approximations, four solver options (LAMP1 ~ LAMP4) are available in LAMP. LAMP2 is an approximate non-linear model with body-exact

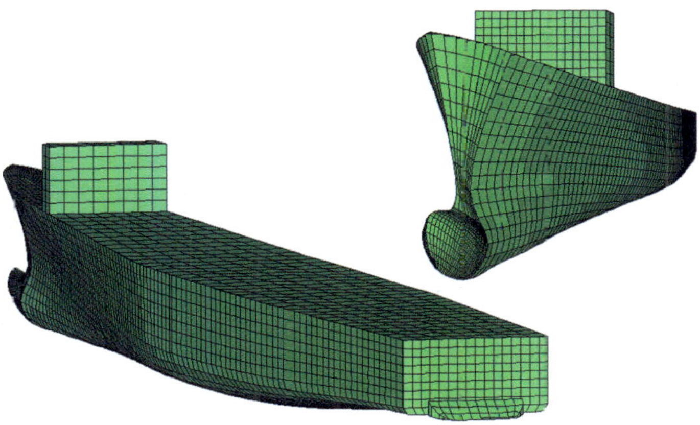

Fig. 23.1 Joint high speed sealift

Table 23.2 Principal dimensions of JHSS

Parameter	Value
LOA	303 m
Beam	32.0 m
Draft	8.65 m
Displacement	35122 tonnes
Model number	5663

Table 23.3 Example test case

Parameter	Value
H_{sig}	11.5 m
T_{modal}	14.0 s
$S(\omega)$	Bretschneider
Ship speed V	9.8517 m/s
Heading angle β	170°
TEV	4.5

Froude-Krylov and hydrostatic forces and the perturbation potential solved over the mean wetted hull surface. LAMP4 is the most realistic solver option because all three forces (i.e., hydrostatic force, Froude-Krylov force, and the forces due to the perturbation potential) are calculated over the instantaneous free surface. However, previous research has shown that LAMP2 can capture parametric response (e.g., France et al. 2003; Shin et al. 2004) and LAMP2 is significantly faster than LAMP4. Consequently, LAMP2 is used in this paper. It should be mentioned that the roll damping model in the LAMP simulations is a crucial system parameter to be studied carefully. The tuning of numerical roll damping model would ensure more accurate calculation of roll damping force. Since there is no roll experimental data available for this concept hull, however, attempts to find a better roll damping model have not been made. Therefore, the default Kato roll damping model (Kato 1965) implemented in LAMP is utilized. The comparison between the default Kato model and a tuned damping model for a container vessel is presented in France et al. (2003).

The test condition chosen for this paper is given in Table 23.3, which is based on Table 23.1 and a numerical roll decay test. The roll decay test using LAMP2 is shown in Fig. 23.2. The significant wave height and modal period correspond to a sea state 8; a condition that JHSS can reasonably be expected to encounter several times during its operational lifetime.

23.3.3 Linear Oscillator

The DLG methodology assumes that the wave environment leading to extreme nonlinear responses will be similar to the wave environment that leads to extreme values of a surrogate process. This approach has been shown to successfully produce a lower bound on the estimation of the probability of nonlinear extreme responses. The accuracy of this lower bound estimate depends on whether the surrogate process captures the physics of the onset of the nonlinear extreme responses. Refer to Kim and Troesch (2013) for a more thorough explanation of this idea for the prediction of parametric rolling.

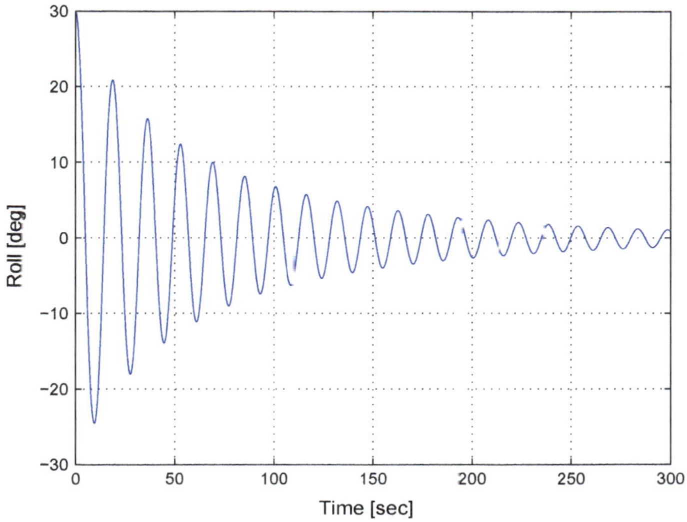

Fig. 23.2 Roll decay test with LAMP2 at $V = 9.8517$ m/s

Consider, as a surrogate process, an imaginary damped mass spring oscillator that can be expressed as

$$m\ddot{x} + c\dot{x} + kx = \alpha m z_o(t) \tag{23.9}$$

$z_o(t)$ is the relative motion of a point on ship's stern with respect to the incident wave and α is an arbitrary constant. The linear relative motion $z_o(t)$ may be expressed as

$$z_o(t) = \eta_3(t) + l_\xi \eta_5(t) - \eta_\xi(t) \tag{23.10}$$

where $\eta_3(t)$ is the heave displacement, $\eta_5(t)$ is the pitch angle in radians, and $\eta_\xi(t)$ is the incident wave height at a location ξ, and l_ξ is the longitudinal distance between the ship center of gravity and the point ξ on the ship stern. Assuming a harmonic input forcing with a frequency ω,

$$\ddot{x} + 2\zeta\omega_n\dot{x} + \omega_n^2 x = \alpha \tilde{z}_o e^{i\omega t} \tag{23.11}$$

where the natural frequency ω_n and the damping ratio ζ are simply

$$\omega_n^2 = \frac{k}{m} \quad \text{and} \quad \zeta = \frac{c}{2m\omega_n} \tag{23.12}$$

Introducing the complex frequency function $H_z(i\omega)$ defined as the response of the oscillator divided by the input relative motion,

$$H_1(i\omega) = \frac{\omega_n^2 \tilde{x}}{\alpha \tilde{z_o}} = \frac{1}{1 - (\omega/\omega_n)^2 + i(2\zeta\omega/\omega_n)} \tag{23.13}$$

The complex transfer function $H(i\omega)$ is then defined as the response divided by the input incident wave becomes

$$H(i\omega) = H_0(i\omega)H_1(i\omega) \tag{23.14}$$

where $H_0(i\omega)$ is an usual relative motion transfer function. $H(i\omega)$ is the transfer function to be applied in the DLG. Since TEV is the ratio of maximum response to the standard deviation σ of the process, α is set equal to ω_n^2 without loss of generality. The natural frequency ω_n is set as twice the roll natural frequency from the roll decay test and the damping ratio ζ used in this paper is 0.04.

23.4 Results and Discussion

To assess the likelihood of the JHSS experiencing large roll in near-head seas (i.e., $\beta = 170°$), completed in this research are LAMP simulations based on an ensemble of the DLG wave sequences and comparable Monte Carlo simulations based on 300 uniformly distributed random wave phase components. The DLG/LAMP simulations are based on an ensemble of short time series (200 s in length) identified by the DLG process. The number of wave frequencies N is 101. The Monte Carlo simulations consist of 500 fifteen minute time histories (−20 s at the beginning of each record to remove transients) spanning an effective exposure period of approximately 125 h. For the Monte Carlo simulations, N is 301. This strategy is adopted to avoid repetition in the incident wave profiles (see, e.g., Belenky 2011). By taking the average of 500 sequences of the Monte Carlo simulations, the approximate root mean square (RMS) values for the heave, pitch, and roll responses are calculated to be approximately 1.57 m, 1.70° , and 1.14°, respectively. The roll mean period, based on zero-upcrossings, is about 13.5 s, which is consistent with the roll natural period based on the roll decay test.

The DLG/LAMP simulations are conducted for three degrees of freedom - heave, roll, and pitch - vessel motions responding to 999 DLG identified incident wave profiles configured to produce a TEV of 4.5 for the linear oscillator given in Eq. (23.11). A 4.5 TEV maximum amplitude for the oscillator has an approximate Rayleigh exceedance probability of 1/25,000. The corresponding mean exposure period is approximately 85 h, when the mean frequency, $\omega_{mean} = 0.5145$ rad/sec, calculated from the response spectrum of the oscillator is converted to actual time. An example DLG/LAMP realization is shown in Fig. 23.3. Note that in Fig. 23.3, the maximum incident wave (η_0) elevation at midship was approximately 4.8 times the wave RMS, the maximum heave (η_3) elevation was approximately 6.3 times the heave RMS, and the maximum pitch (η_5) elevation was approximately 4.7 times the pitch RMS, while the maximum roll (η_4) elevation was approximately 34.4 times the roll RMS!

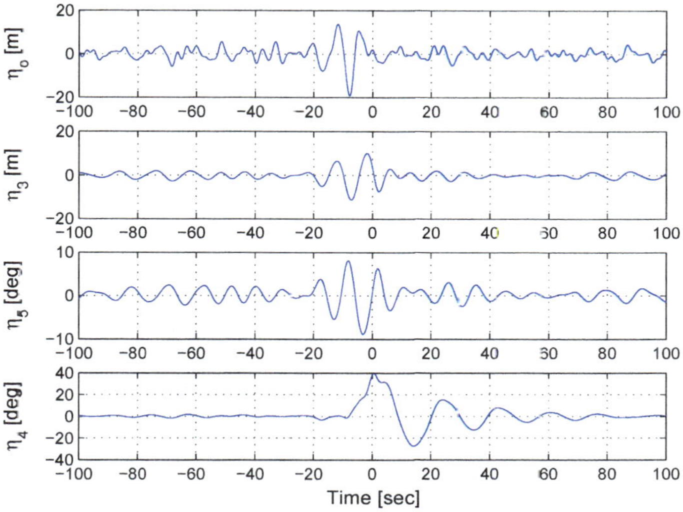

Fig. 23.3 Example DLG/LAMP realization ($TEV = 4.5$)

As mentioned, the DLG approximates the extreme value distribution of the response X of the oscillator. To show the level of approximation by the current DLG, the histogram of 999 realizations and the expected histogram from Eq. (23.5) are compared in Fig. 23.4a. The maximum roll amplitudes in the DLG ensemble (LAMP2) are shown in Fig. 23.4b. Note that the most probable roll maximum in Fig. 23.4b is approximately 9°. Also note that there are 4 of the 999 realizations that exceed 39°, including three capsize sequences. A snapshot of one of the LAMP2 capsize sequences is shown in Fig. 23.5.

Fig. 23.6 combines the result of the Monte Carlo simulations and the DLG in Weibull space. It should be noted that the extreme value PDF of the roll is exaggerated. This figure supports, to a degree, the hypothesis upon which the linear oscillator in Eq. (23.5) is based. That is, large, nonlinear, non-Gaussian roll in near head seas is correlated with the response of a linear oscillator with a natural frequency of twice the roll natural frequency and excited by large stern relative motion. In addition, while the most expected roll maximum shown in Fig. 23.4b lies on the very proximity of the curve traced by the Monte Carlo simulations in Weibull space, there is a finite probability that the JHSS will experience conditions during the same exposure period that lead to capsize. The probability can be summarized as follows:

For the operational condition given in Table 23.3, *the most probable roll amplitude is approximately* 9°, *which represents a lower bound for the exposure period of 85 h. In addition, considering all the roll motions in the 999 member ensemble, extreme values greater than* 39°, *including three capsize realizations, occur approximately 0.4% of the time.*

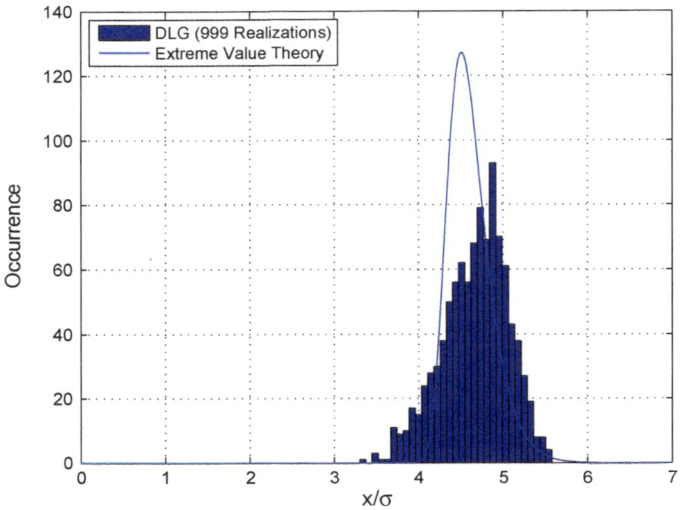

(a) X at $TEV = 4.5$

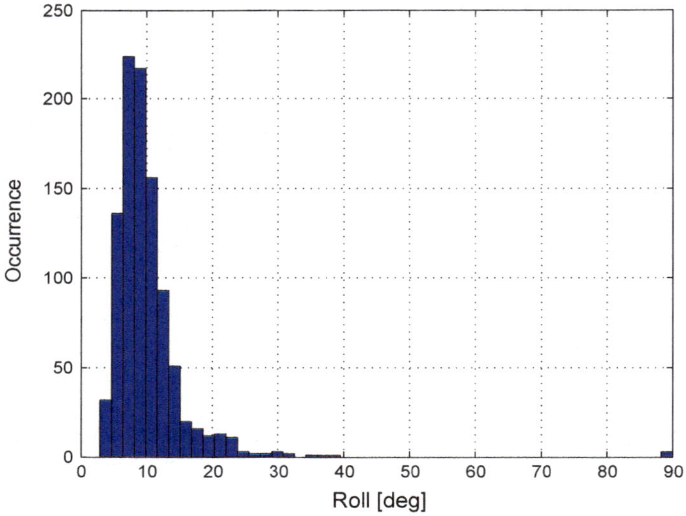

(b) Maximum Roll Response ζ_4 at $TEV = 4.5$

Fig. 23.4 Histograms of 999 DLG generated extreme response X of linear oscillator and corresponding roll response ζ_4

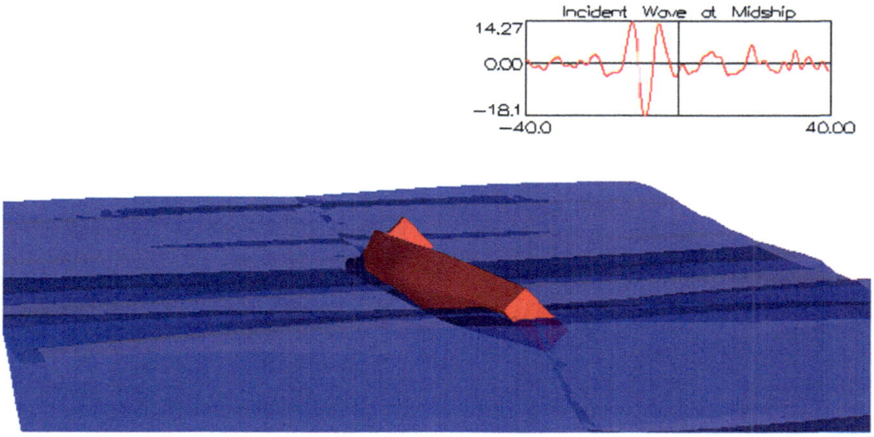

Fig. 23.5 Example LAMP simulation (One of three capsize cases)

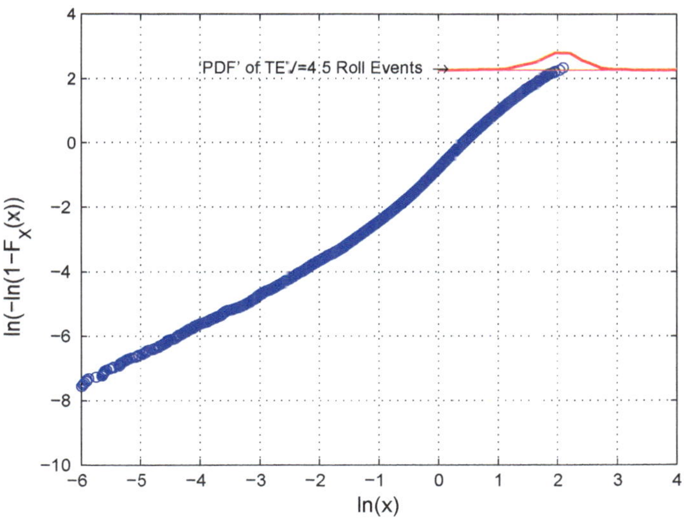

Fig. 23.6 Weibull plots for roll monte carlo simulations

23.5 Conclusion

This study calculates the extreme value distribution of parametric roll for a given exposure period. Since the DLG is based on a Gaussian process, the extreme value distribution of the extreme roll is a lower bound of the *exact* extreme value distribution. The DLG can generate an ensemble of short design wave time series, from which the extreme value distribution of the corresponding nonlinear responses can be calculated relatively efficiently. For example, about 85,000 h worth of Monte Carlo simulations are required to obtain a comparable histogram presented in this paper. As such, the DLG method can provide a valuable tool for designers in establishing useful operational guidance.

Acknowledgements The authors would like to gratefully acknowledge the support of the Office of Naval Research grants: "Design Loads Generator - Ship Design and Simulation Tools for Complex Marine Systems", Award Number N00014-08-1-0594, under the direction of Kelly Cooper and "Support for the Theory Advisory Panel (TAP)", Award Number N00014-09-1-0978, under the direction of Dr. Pat Purtell. Special thanks go to William Belknap and Allen Engle at NSWCCD for JHSS hull geometry and technical reports on statistical techniques; Woei-Min Lin and Ken Weems from SAIC for technical support on LAMP; and Matt Collette from the University of Michigan.

References

Alford, L.K., Kim, D.H., Troesch, A.W.: Estimation of extreme slamming pressures using the non-uniform fourier phase distributions of a design loads generator. Ocean Engineering **38**(5-6), 748–762 (2011)

Belenky, V.L.: On Self-Repeating Effect in Reconstruction of Irregular Waves. Contemporary Ideas on Ship Stability pp. 589–598 (2011)

France, W.N., Levadou, M., Treakle, T.W., Paulling, J.R., Michel, R.K., Moore, C.: An investigation of head-sea parametric rolling and its influence on container lashing systems. Marine Technology **40**(1), 1–19 (2003)

Jensen, J.J., Pedersen, P.T.: Critical wave episodes for assessment of parametric roll. Proceedings of the Ninth International Marine Design Conference (IMDC) pp. 399–411 (2006)

Kato, H.: Effect of Bilge Keels on the Rolling of Ships. J. Soc. Nav. Archit. Japan **117** (1965)

Kim, D.H.: Design Loads Generator: Estimation of Extreme Environmental Loadings for Ship and Offshore Applications. Ph.D. thesis, The University of Michigan (2012)

Kim, D.H., Alford, L.K., Troesch, A.W.: Probability Based Simulation of Stern Slamming Design Events. In: The 2010 Conference on Grand Challenges in Modeling and Simulation (2010)

Kim, D.H., Engle, A.H., Troesch, A.W.: Estimates of Long-Term Combined Wave Bending and Whipping for Two Alternative Hull Forms. Transactions of Society of Naval Architects and Marine Engineers **119** (2011)

Kim, D.H., Troesch, A.W.: An application of Design Load Generator to predict extreme dynamic bending moments. 29th International Conference on Ocean, Offshore and Arctic Engineering (OMAE2010) (49118), 635–643 (2010). https://doi.org/10.1115/OMAE2010-20378. URL http://link.aip.org/link/abstract/ASMECP/v2010/i49118/p635/s1

Kim, D.H., Troesch, A.W.: Statistical Estimation of Extreme Roll Responses in Short Crested Irregular Head Seas. Transactions of Society of Naval Architects and Marine Engineers **121** (2013)

Piro, D., Brucker, K.A., O'Shea, T.T., Wyatt, D., Dommermuth, D., Story, W.R., Devine, E.A., Powers, A.M., Fu, T.C., Fullerton, A.M.: Joint High Speed Sealift (JHSS) Segmented Model Test Data Analysis and Validation of Numerical Simulations. Tech. rep., Naval Surface Warfare Center Carderock Division (2012)

Shin, Y.S., Belenky, V.L., Lin, W.M., Weems, K.M., Engle, A.H.: Nonlinear Time Domain Simulation Technology for Seakeeping and Wave-Load Analysis for Modern Ship Design. Transactions of Society of Naval Architects and Marine Engineers **111**, 557–583 (2003)

Shin, Y.S., Belenky, V.L., Paulling, J.R., Weems, K.M., Lin, W.M.: Criteria for parametric roll of large containerships in longitudinal seas. Transactions-Society of Naval Architects and Marine Engineers **112**, 14–47 (2004)

Vidic-Perunovic, J.: Influence of the GZ calculation method on parametric roll prediction. Ocean Engineering **38**(2-3), 295–303 (2011). https://doi.org/10.1016/j.oceaneng.2010.11.002. URL http://dx.doi.org/10.1016/j.oceaneng.2010.11.002

Chapter 24
Critical Wave Groups Versus Direct Monte-Carlo Simulations for Typical Stability Failure Modes of a Container Ship

Vladimir Shigunov, Nikos Themelis and Kostas J. Spyrou

Abstract In the second Generation Intact Stability Criteria currently developed at IMO, the process of direct stability assessment (DSA) and providing operational guidance (OG) are interlaced with a requirement of performing a large number of numerical simulations. However, extreme roll motions that are generally behind stability failures are rare events as any extreme responses. An additional significant difficulty is that roll response as stochastic process is usually non-Gaussian, therefore, close-form expressions for the probability of extreme roll responses, based on spectral moments, are in general not applicable. A practical approach proposed recently exploits the idea that extreme events occur due to the encountering of extreme wave groups (critical wave episodes). This could alleviate the need for a large number of simulations by focusing on the systematic identification of those deterministic wave sequences that generate unacceptable roll responses. Taking a first step towards a systematic validation process of the wave groups method, the present study compares the exceedance probabilities of 40° roll angle and of $g/2$ lateral acceleration, computed by the critical wave groups method with Monte-Carlo simulations for a large containership. The nonlinear seakeeping code *rolls* is used as mathematical model of ship motion. Typical loading conditions where various stability failure modes can occur are examined.

Keywords Numerical simulations · Monte carlo method · Critical wave groups

---

V. Shigunov
DNVGL SE, Hamburg, Germany
e-mail: vladimir.shigunov@dnvgl.com

N. Themelis · K. J. Spyrou (✉)
School of Naval Architecture and Marine Engineering,
National Technical University of Athens, Athens, Greece
e-mail: k.spyrou@central.ntua.gr

N. Themelis
e-mail: nthemelis@naval.ntua.gr

V. L. Belenky et al. (eds.), *Contemporary Ideas on Ship Stability*, Fluid Mechanics and Its Applications 119, https://doi.org/10.1007/978-3-030-00516-0_24

24.1 Introduction

Application of numerical simulations for direct stability assessment is currently discussed at IMO as an important part of the second Generation Intact Stability Criteria. The use of numerical simulations for probabilistic assessment of dynamic stability requires sufficiently accurate mathematical models of nonlinear ship dynamics, as well as practical rational probabilistic procedures able of predicting probabilities of rare extreme motion events for non-Gaussian processes.

A practical solution for the latter problem proposed by Themelis and Spyrou (2007) exploits the idea that extreme events occur due to the encountering of extreme wave groups (critical wave episodes). Following this approach, the identification of the critical wave groups in terms of their height, period and duration is possible on the basis of nonlinear deterministic ship motion analysis, whereas the estimation of the probability of encountering specific wave groups is based on statistical Gaussian seaway models.

The method is by its nature approximative in its consideration of the ship-wave encounter, because, at its current stage of development, certain regular wave profiles are examined. On the other hand, the method is very versatile, because it combines arbitrarily sophisticated nonlinear analysis of ship dynamics (and thus takes into account any stability failure mode captured by the motion analysis method) with accurate probabilistic analysis of the seaway.

The present study is a step towards the comparison of this method with the direct stability assessment through numerical simulations with the nonlinear seakeeping code *RollS*, for typical stability failure modes of a modern post-panamax container ship in a range of practically relevant initial *GM* values. Short-term average exceedance rates of the maximum (over the ship) lateral acceleration value $g/2$, as well as of roll angle 40°, are determined from Monte-Carlo simulations. They are compared against the exceedance rate obtained by using the critical wave groups approach.

24.2 Critical Wave Groups

According to the principle of the critical wave groups method, the ship can be assumed as performing initial motions of small to moderate amplitude in the considered mode(s). Then extreme behavior is realised due to the encounter of a wave group. Thus the critical wave groups identification process supplies in fact the threshold excitation that generates un-desirable ship behavior. Such wave groups are identified for frequencies spanning the usual range of wave frequencies. Each group is characterised by its run length, period and height. However, the choice of fixed height does not imply that groups physically display such a property. Instead, it specifies the critical height above which ship behavior exhibits at least one unacceptable characteristic. As for the discrete period, it should be seen as a representative of the small range of periods around it. It is important to note that, in the implementation of this method, all wave groups that result in undesirable dynamic response are extracted beforehand. Then, probabilities can be determined of encountering conditions "worse" than the critical, for the seaway situations that exist in the considered area of ship operation.

The process of consecutive waves is modeled as a first order "autoregressive model" which is in fact equivalent to assuming the Markov chain property for the waves, a well-established characteristic of sea waves. However, it is essential that no similar assumption is necessary concerning roll response.

The targeted probability is calculated by the next product

$$p = p(T|H > H_{cr}) \cdot p(H > H_{cr}) \tag{24.1}$$

where $p(T|H > H_{cr})$ is the conditional probability of encountering n successive waves with periods T in a specific range and heights above a threshold level H_{cr}, and $p(H > H_{cr})$ is the probability that n successive waves have heights exceeding this critical threshold.[1] For the calculation of the first probability, the conditional multivariate normal probability density function (pdfc) is used (Wist et al. 2004), combined with Tayfun's (1993) work on the joint probability density function (pdf) of large wave height and its associated period. On the other hand, the calculation of $p(H > H_{cr})$ is based on the bivariate Rayleigh pdf of two successive wave heights, Battjes and Van Vledder (1984), combined with the Markov chain property. This modelling approach can handle efficiently both the period and the height of successive waves and represents a key step beyond the Kimura-type modelling of wave groups that is not sensitive to the period, Kimura (1980).

24.3 Ship and Loading Conditions

A modern 8000 TEU container ship was selected for the study. Such vessels are presently the work horses of east-bound container shipping routes, and many of them might be employed in west-bound routes with more harsh weather conditions after the modernisation of Panama channel. Many years of full-scale measurements on board several vessels of this size are available, thus their loading and speed profiles are well known. Moreover, roll damping parameters are known from model tests.

The selected vessel has a length between perpendiculars of about 320.0 m, waterline breadth of about 43.0 m and design speed of about 25.0 knots. Vessels of this size operate most frequently in partial loading conditions with *GM* from about 2.5 to about 4.5 m; such loading conditions are relatively safe with respect to both parametric and synchronous resonance. Therefore, a wider range of loading conditions was studied, Table 24.1, including (1) nearly full load with *GM* of 1.2 m, which might

[1]The slightly unorthodox inequality in the above means that each component of vector $\mathbf{H}$ obtains greater value than the corresponding value in vector $\mathbf{H}_{cr}$. However, in the current implementation, all entries of vector $\mathbf{H}_{cr}$ receive the same value H_{cr}.

Table 24.1 Loading conditions

Draught T_m [m]	14.44	12.84	11.36
Metacentric height GM [m]	1.2	3.8	7.5
Natural roll period T_φ [s]	30.1	18.5	12.6

Table 24.2 Selected scenarios

Case	GM [m]	Fr [−]	μ [°]	T_1 [s]
01	7.50	0.16	98.0	11.5
02	1.26	0.09	50.0	10.0
03	1.26	0.12	49.0	9.2
04	1.26	0.04	180.0	13.4
05	3.80	0.16	60.0	12.7

be vulnerable to parametric roll, (2) a "typical" loading condition with GM of 3.8 m and (3) ballast loading condition with a very high GM of 7.5 m, expected to lead to large lateral accelerations due to synchronous rolling.

24.4 Selection of Speeds and Seaway Parameters

Table 24.2 summarises the scenarios selected for comparison. In this table, GM means the initial metacentric height, Fr the Froude number, μ the wave direction (0° following waves, 90° waves from steering board, 180 degree head waves) and T_1 is the mean wave period.

Synchronous roll is most relevant in beam waves, where the added resistance is rather low and thus forward speed can be rather high, see case 01 in Table 24.2. The corresponding critical seaway parameters for this loading condition are shown with a black point in Fig. 24.1.

Principal parametric resonance occurs typically at low GM in following waves (at low forward speeds) to quartering waves (at higher forward speeds, which are more realistic in waves from the stern directions), Shigunov et al. (2009). Cases 02 and 03 in Table 24.2 represent this failure mode, and Fig. 24.2 shows these cases on polar plots; two cases with different ship forward speeds are selected to study the influence of the ship speed.

Principal parametric resonance also occurs at low GM in head waves at low forward speeds, which are relevant in rather high head waves. Case 04 in Table 24.2 and Fig. 24.3 illustrate the selected seaway conditions and the ship forward speed.

Finally, case 05 in Table 24.2 and Fig. 24.4 represent direct excitation case for a "typical" loading condition.

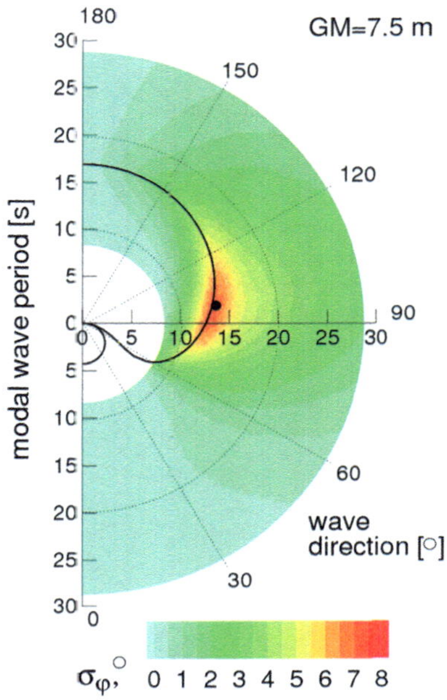

Fig. 24.1 Calculated root-mean-square (rms) of roll angle for loading condition with $GM = 7.5$ m at $Fr = 0.16$ in irregular waves versus modal wave period and wave direction (radial and circumferential co-ordinates, respectively). Black line indicates resonance condition

24.5 Monte-Carlo Simulations

To study the influence of the significant wave height on the accuracy of the critical wave groups method, significant wave height was systematically varied with a small step for each of the selected combinations 01–05 of wave direction, wave period and ship speed. Simulations were performed in long-crested seaways described by the JONSWAP spectrum with the peak parameter $\gamma = 3.3$. For each combination of ship speed and seaway parameters, 50–500 simulations were carried out with different initial phases of seaway components, until maximum over the ship lateral acceleration exceeded the value g/2. In the other series of simulations (denoted as cases 01a–05a, with the conditions corresponding to those in cases 01–05), exceedance of the roll angle of 40° was considered as the extreme event. The average exceedance period was calculated by averaging of all exceedance periods. Numerical code *RollS*, proposed by Söding (1982), was used for simulations (see Petey 1986, for details of the method and Shigunov et al. 2009, for validation examples).

24.6 Calculation of Hydrodynamic Database

For each of cases 01–05 and 01a–05a, respectively, a corresponding hydrodynamic database of roll responses in regular waves was computed. The period of regular

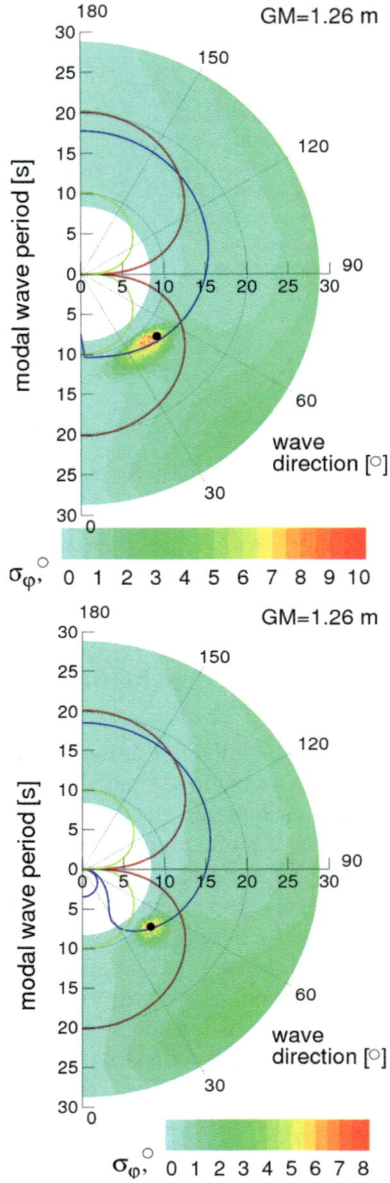

Fig. 24.2 Calculated rms of roll angle for $GM = 1.2$ m at $Fr = 0.09$ (top) and 0.12 (bottom). The black points indicate selected representative scenarios for parametric resonance in quartering waves; blue line indicates resonance condition while yellow and green lines limit the area of suitable wave lengths

waves was varied in a broad range with a step of 0.5 s; *GM*, forward speed and wave direction were kept constant. For consistency, the same simulation method *rolls* as used for the Monte-Carlo simulations was also used for the computation of

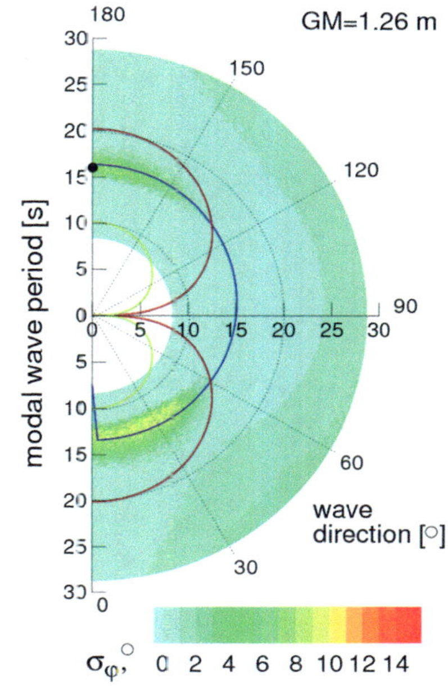

Fig. 24.3 Calculated rms of roll angle for $GM = 1.2$ m and $Fr = 0.04$ in irregular waves. Black point indicates selected representative case for parametric resonance in bow waves. Blue line indicates resonance condition, yellow and green lines limit the area of suitable wave lengths

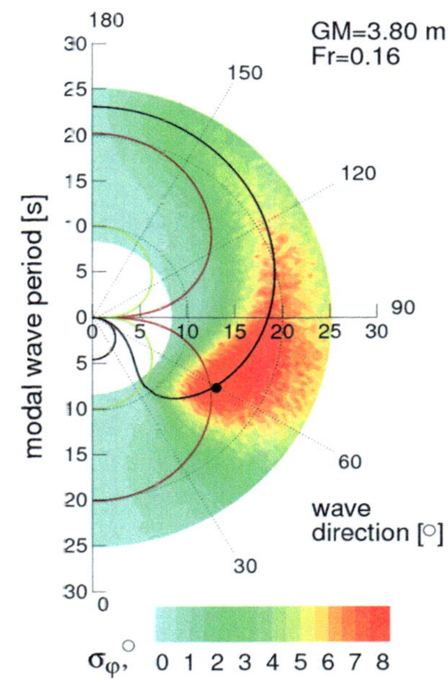

Fig. 24.4 Calculated rms of roll angle for $GM = 3.8$ m at $Fr = 0.16$ in irregular waves

the hydrodynamic database. The initial condition was an upright ship with zero roll velocity.

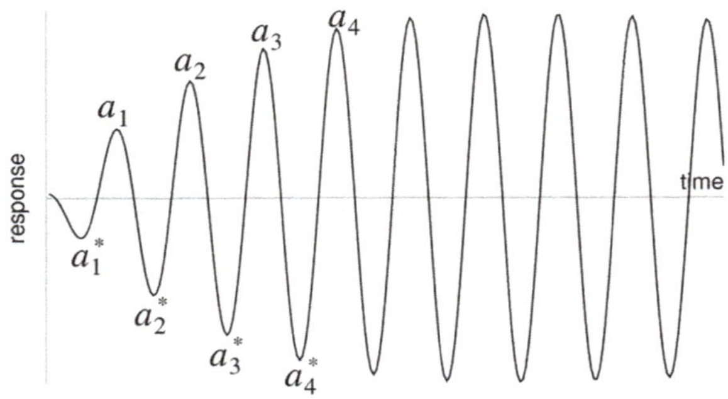

Fig. 24.5 Post-processing of responses in regular waves

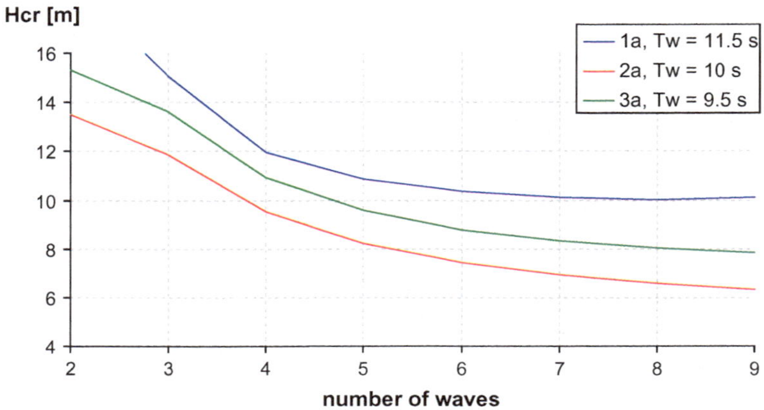

Fig. 24.6 Characteristics of critical wave groups for 40° roll angle exceedance

Time histories of responses (maximum over the ship lateral acceleration or roll angle) are processed to obtain response maxima, i.e. $\max(a_n, -a_{n*})$, Fig. 24.5, per cycle of oscillation. Figure 24.5 explains the definitions used: the reaction period n from the start of simulation and n^* for negative peaks. Post-processing considered 9 roll cycles. The wave height was varied until the response maximum, in each response period in turn, exceeded the threshold ($g/2$ for lateral acceleration or 40° for roll amplitude). In this way, nine "critical" wave height values $H_{cr}(T, n)$, $n = 1, \ldots, 9$ were identified, each of which led to a response amplitude equal to or greater than the threshold, during the corresponding wave encounter. Figures 24.6 and 24.7 present examples of the identified critical wave heights. The corresponding scenario and the threshold are also indicated. The curves shown were obtained for the wave group period equal to the mean period of the seaway, Table 24.2.

In general, exceedance of 40° roll angle requires higher waves than those required to exceed $g/2$ lateral acceleration. Figure 24.6 also indicates that higher waves are required for 40° roll angle exceedance in beam seas (case 01a) compared to quartering

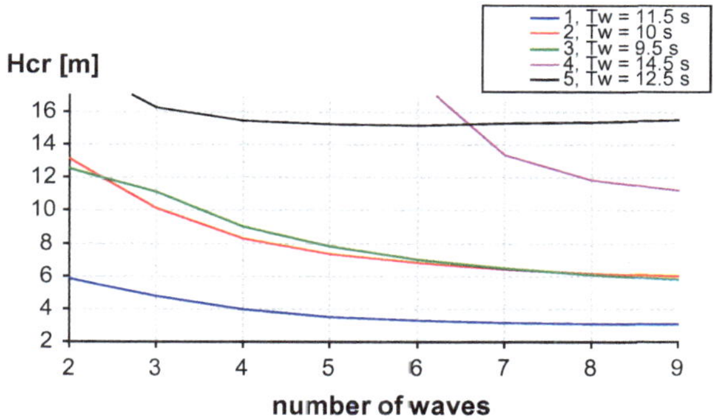

Fig. 24.7 Characteristics of critical wave groups for $g/2$ lateral acceleration exceedance

seas parametric roll (at a low forward speed with $Fr = 0.09$—case 02a, and at a higher forward speed with $Fr = 0.12$—case 03a). Figure 24 7 indicates that the exceedance of acceleration threshold in beam seas is realised in lower waves (case 01) compared to the two quartering sea parametric roll scenarios 02 and 03.

The critical wave heights for case 04 (parametric rolling in head seas at low speed) and case 05 (direct excitation in a typical loading condition) are much higher than those in cases 01, 02 and 03. Note that the identified critical wave heights for cases 04a and 05a (these correspond to the 40° roll angle threshold) were found to be extremely high even for high run length (they are not shown in Fig. 24.6), and thus their encounter probability should be very low.

24.7 Exceedance Rate and Hourly Probability of Exceedance

Figures 24.8 and 24.9 show the calculated exceedance rates per case and threshold (roll angle and lateral acceleration) obtained by the two methodologies, as a function of the inverse significant wave height squared. For the 40° roll angle threshold (Fig. 24.8), case 02a presents the best example of agreement between the two approaches. For case 03a, the results are in a satisfactory agreement for $1/h_s^2$ greater than 0.015 m^{-2}, but for the smaller values of $1/h_s^2$ (i.e. large h_s), the wave groups approach predicts higher rates. On the other hand, the exceedance rates obtained with Monte-Carlo simulations for case 01a are consistently higher than those of the wave groups approach.

Figure 24.9 shows the results related to the acceleration threshold (cases 01 to 05). Case 05 demonstrates the best agreement here, and cases 02 and 03 agree well on average. Moreover it is observed that, for the more severe sea states examined, the "wave groups" approach predicts more frequent exceedances for cases 02 and 03. The largest discrepancies arise for cases 01 and 04, where Monte Carlo simulations predict far more frequent exceedances in the whole range of wave heights.

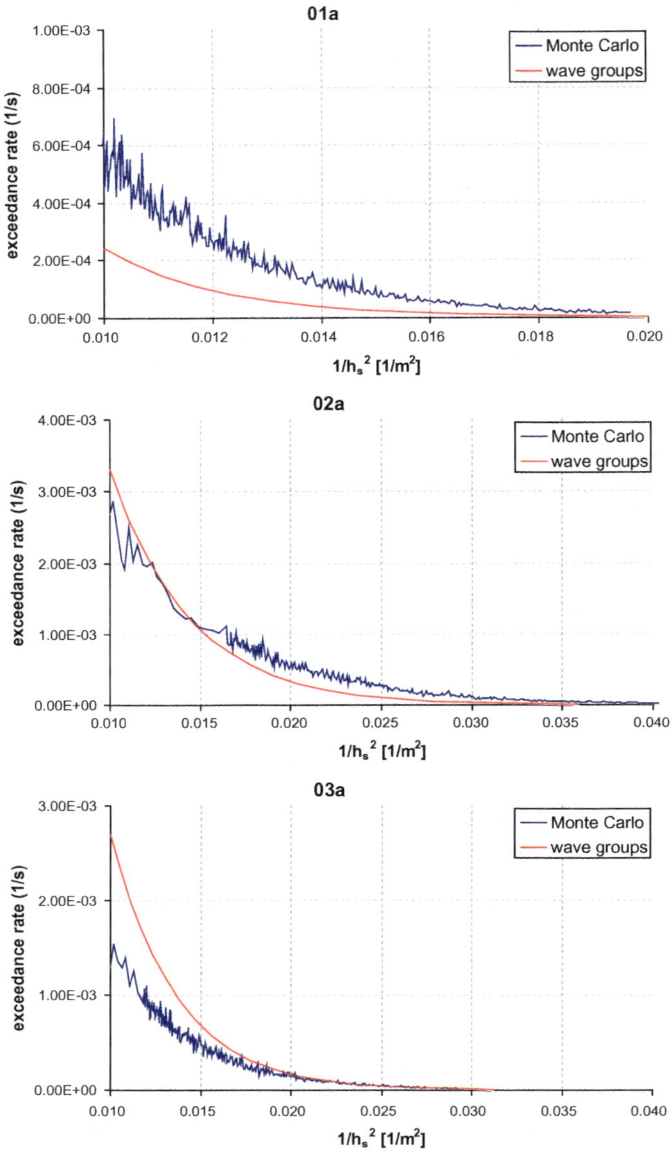

Fig. 24.8 Comparison of exceedance rate of 40° roll angle between Monte-Carlo simulations and wave groups approach

Knowing the exceedance rate r, the probability of at least one exceedance during a given exposure time T (e.g. one hour) can be calculated assuming Poisson law for the flow of the exceedance events. Both exceedance rate and probability of exceedance

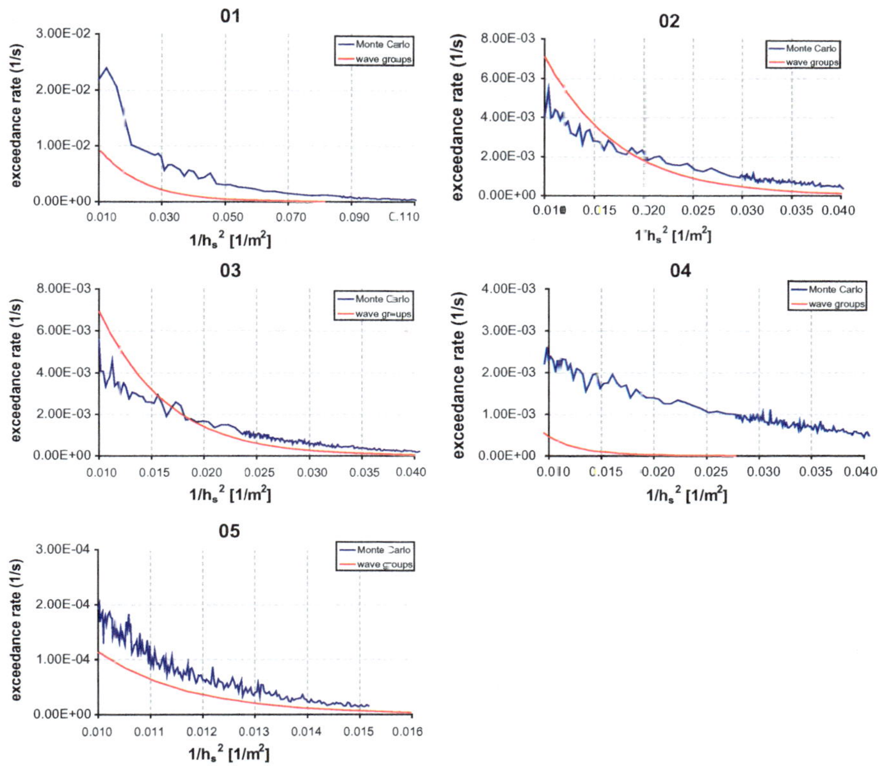

Fig. 24.9 Comparison of exceedance rate of $g/2$ lateral acceleration between Monte-Carlo simulations and critical wave groups approach

per given time have advantages, and offer different comparison viewpoints on the results. The next formula can be used:

$$p(T) = 1 - e^{-rT} \tag{24.2}$$

The applicability of the Poisson flow assumption depends upon the validity of certain conditions, namely that

- only one event can happen at a given time;
- the probability of event happening at a particular time instant is infinitely small; and
- events are independent of each other.

Whereas the first two conditions are satisfied for roll motion and related processes, the last condition is not, because exceedance events of a certain large roll angle tend to appear in groups. In order to eliminate the influence of this strong auto-correlation of roll motion, average estimates of the exceedance time period were derived from

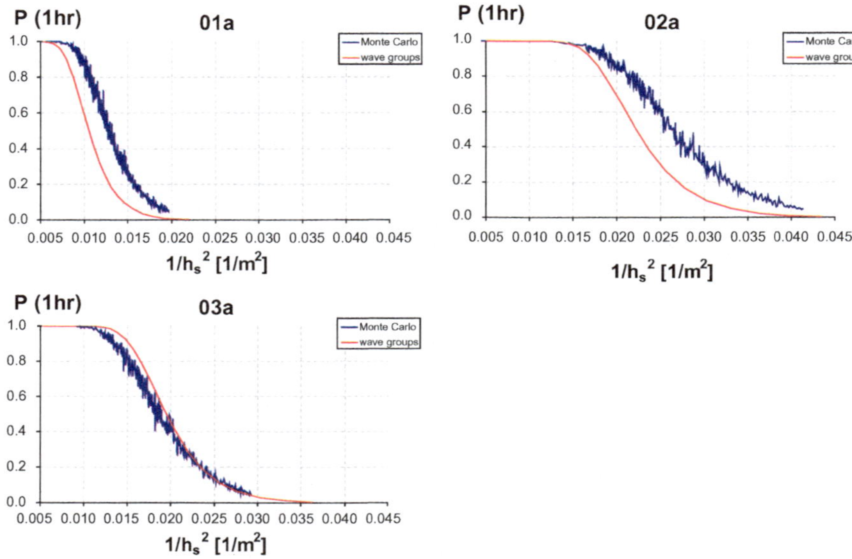

Fig. 24.10 Comparison of exceedance probability per hour of roll angle 40° between Monte-Carlo simulations and critical wave groups approach

multiple realisations of the same seaway, as proposed by Söding (1987), see Shigunov (2009) for application, in such a way that each simulation was continued only until the first exceedance event. Then the ship was returned to the upright position, and the simulation was repeated in the same seaway with the new set of random phases, frequencies and directions of seaway components until the next exceedance event. The estimation of the expected exceedance period is found as the average of the exceedance periods obtained in all simulations.

Figure 24.10 shows the respective results for the 40° threshold for each case considered. Case 03a seems to show the best agreement between the two approaches in terms of the hourly probability. Larger deviation of the rate occurs for values of $1/h_s^2$ less than 0.015 m^{-2} where the hourly probability is close to 1.0 anyway. Cases 01a and 02a could be also considered as adequately close. The average relative difference between the two methods in terms of h_s corresponding to the same exceedance probability is 9.4 and 7.8% for cases 01a and 02a, respectively.

Figure 24.11 shows results for the *g*/2 lateral acceleration threshold. Case 05 demonstrates the best agreement between the two methods, similarly to what was observed for the exceedance rate. Relative differences between the two methods in terms of h_s corresponding to the same hourly exceedance probability are 15.6 and 10.45% for cases 02 and 03, respectively.

On the contrary, there is a significant disagreement in the exceedance of acceleration threshold for the beam-sea scenario (case 01) and for the head-sea scenario (case 04). Specifically, for the beam-sea scenario, the 90% exceedance probability

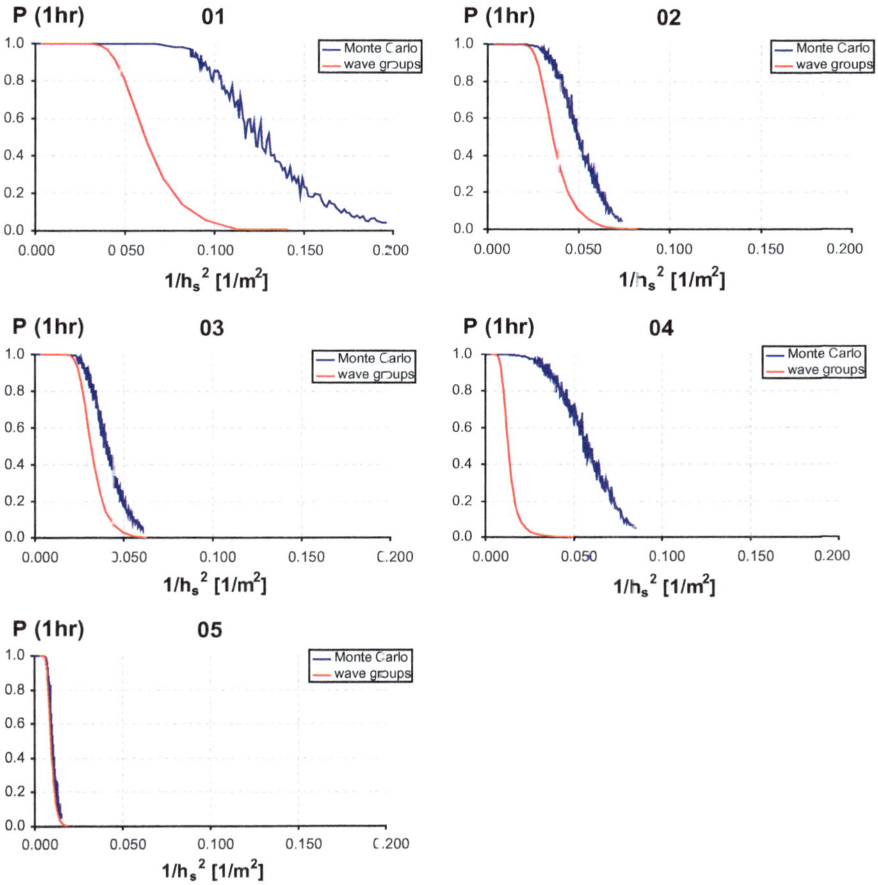

Fig. 24.11 Exceedance probability per hour of *g*/2 lateral acceleration from Monte-Carlo simulations and wave groups approach

within one hour exposure corresponds to significant wave heights 3.31 and 4.78 m for the Monte-Carlo simulations and wave groups method, respectively. For the head-sea scenario, the corresponding wave heights are 5.2 and 10.6 m, respectively. Table 24.3 shows the significant wave heights corresponding to 90% hourly probability of exceedance for all examined cases.

The order of the magnitude of probabilities seems to be the same between the two approaches, with the exception, however, of case 04 related to the acceleration threshold for head-sea parametric rolling. Further work will be required in order to identify the reason of the differences in this case.

Table 24.3 Significant wave heights required for 90% hourly probability of exceedance

Case	Monte-Carlo	Wave groups	Relative difference (%)
01a	10.11	11.25	10.13
02a	7.10	7.63	6.89
03a	8.50	8.13	−4.62
01	3.31	4.63	28.43
02	5.30	6.13	13.47
03	5.70	6.30	9.52
04	5.20	10.60	50.94
05	11.75	11.90	1.26

It is recalled that a Monte-Carlo method captures the statistic of a process without identifying the phenomenon, whilst the wave groups methods addresses each specific phenomenon individually.

For the 40° roll angle exceedance criterion, case 02a appears to produce the most frequent exceedances, while case 01a the least frequent. On the other hand, both approaches indicate that case 01 is the most critical for the acceleration threshold, followed by cases 02 and 03. These trends follow the trend of the required wave heights in Figs. 24.6 and 24.7.

24.8 Conclusions

A comparison of exceedance rates and exceedance probabilities obtained by Monte-Carlo simulations and the critical wave groups approach was carried out. Typical stability failure modes have been studied. Thresholds were set in terms of roll angle and lateral acceleration. In some cases satisfactory agreement was shown, e.g. for parametric rolling in quartering seas (cases 02 and 03) and direct excitation (case 05), but in some other cases, such as head seas parametric rolling, the difference was not negligible. A factor that can be governing certain discrepancies, especially when these are within an order of magnitude, is the different initial phasing, because in the wave groups method, the ship is assumed initially upright and with a fixed phase with respect to the first wave crest. An assessment of the effect of the initial phase on the results of the wave groups method can be found in Themelis and Spyrou (2008). The average exceedance rate in Monte-Carlo simulations was derived by averaging over a large number of realisations, in each of which the ship was assumed initially upright, but the phases of wave realisations changed randomly.

A source of large quantitative differences can be the fact that Monte-Carlo simulations do not discriminate between phenomena and record threshold exceedance events irrespectively of the underlying causes, which could, in principle, be more than one per realisation. On the other hand, the wave groups method is implemented for each scenario with one specific phenomenon in mind.

A further factor that could produce higher exceedance rate in Monte-Carlo simulations is the possible non-monotonic increase of roll amplitude with time, which can occur due to the passage of a wave group having an intermediate wave with height below the critical one. The critical wave group approach as used in this paper excludes such events, as it sets the same value for the heights of the waves in the group (vector **H** mentioned in Sect. 24.3). A further study taking into account intermediate variation of the heights of the groups and their associated probabilities of occurrence will consider this effect.

References

Battjes, J. A., and van Vledder, G. Ph., 1984, "Verification of Kimura's theory for wave group statistics", Proceedings of 10th Int. Conf. on Coastal Engineering, pp. 642–648.

Kimura, A., 1980, "Statistical properties of random wave groups", Proceedings of 17th Int. Conf. on Coastal Engineering, Sydney, Australia, pp. 2955–73.

Petey, F., 1986, "Forces and moments due to fluid motions in tanks and damaged com-partments", Proceedings of 3rd Int. Conf. on Stability of Ships and Ocean Vehicles STAB'86, Gdansk.

Shigunov, V., 2009, "Operational Guidance for Prevention of Container Loss", Proceedings of 10th Int. Conf. on Stability of Ships and Ocean Vehicles STAB 2009, St. Petersburg.

Shigunov, V., el Moctar, O., and Rathje, H., 2009, "Conditions of parametric roll", Proceedings of 10th Int. Conf. on Stability of Ships and Ocean Vehicles STAB 2009, St. Petersburg.

Söding, H., 1982, "Leckstabilität im Seegang", Report No. 429, Institut für Schiffbau, Hamburg.

Söding, H., 1987, "Ermittlung der Kentergefahr aus Bewegungssimulationen", Schiffstechnik, Vol. 34, pp. 28–39.

Tayfun, M. A., 1993, "Joint distributions of large wave heights and associated periods", Journal of Waterway, Port, Coastal and Ocean Engineering, Vol. 119, pp. 261–273.

Themelis, N., and Spyrou, K. J, 2007, "Probabilistic assessment of ship stability", SNAME Transactions, Vol. 115, pp. 181–204.

Themelis, N., and Spyrou, K. J., 2008, "Probabilistic assessment of ship stability based on the concept of critical wave groups." Proceedings of 10th International Ship Stability Workshop, Daejeon, Korea.

Wist, H., Myrhaug, D., and Rue, H., 2004, "Statistical properties of successive wave heights and successive wave periods", Applied Ocean Research, Vol. 26, pp. 114–136.

Chapter 25
Solving the Problem of Nonlinear Ship Roll Motion Using Stochastic Dynamics

Jeffrey M. Falzarano, Zhiyong Su, Arada Jamnongpipatkul and Abhilash Somayajula

Abstract Due to nonlinear viscous damping and the softening characteristic of the stiffness, the roll motion of a ship exhibits complex dynamics. Specifically predicting the probabilistic characteristics of roll response in an irregular seaway is still a challenging problem and continues to be of interest for both practitioners and researchers. In this work two techniques from the theory of stochastic dynamics are applied to study the probabilistic nature of roll motion in irregular seas. The first method is the "Moment Equation method" where the roll response moment equation is formulated from a six dimensional state space rolling model with a fourth order linear filter using the Itô differential rule. The resulting moment equations are solved using a cumulant neglect technique. Alternatively in the second approach, the probability density function of the rolling response is evaluated by solving the corresponding Fokker Planck Equation of the system using "Path Integral method".

25.1 Introduction

While the heave and pitch motions of a ship can be predicted accurately using linear theory, the same is not true for roll motion. Due to the softening roll stiffness and a nonlinear viscous damping, the roll motion of a ship exhibits complex dynamics in both regular and irregular seas.

J. M. Falzarano (✉) · A. Somayajula
Texas A&M University, College Station, TX, USA
e-mail: jfalzarano@civil.tamu.edu

A. Somayajula
e-mail: s.abhilash89@gmail.com

Z. Su
COTEC Offshore Engineering Services, Houston, TX, USA
e-mail: lincsu@neo.tamu.edu

A. Jamnongpipatkul
Houston Offshore Engineering, Houston, TX, USA
e-mail: plummy23@neo.tamu.edu

V. L. Belenky et al. (eds.), *Contemporary Ideas on Ship Stability*, Fluid Mechanics and Its Applications 119, https://doi.org/10.1007/978-3-030-00516-0_25

In the past, significant focus has been given to the problem of predicting roll response in regular head or beam seas (Falzarano et al. 1992; Thompson 1997; Spyrou and Thompson 2000). However with increasing number of incidents reporting large amplitude of roll motions at sea, the focus is shifting to predicting the roll motion in irregular seas (Su et al. 2011; Su and Falzarano 2011; Hsieh et al. 1994; Jiang et al. 2000; Somayajula and Falzarano 2015a, 2016). The International Maritime Organization (IMO) is revamping the current intact stability code which is based on prescriptive static stability to make way for a second generation of stability code based on dynamic stability. Two of the four focus areas in this new stability code are the directly excited roll motion when the ship has lost power (dead ship condition) and the indirectly excited roll motion when the ship is moving with forward speed into longitudinal seas (parametric roll).

Between these two focus areas on roll motion, the prediction of parametric roll in irregular seas is definitely more complicated in terms of modeling (Moideen et al. 2012, 2013, 2014; Somayajula and Falzarano 2014, 2015b; Somayajula et al. 2014), simulation (Somayajula and Falzarano 2015a; Guha et al. 2016) and analytical analysis. However, the dead ship condition is also a challenging problem and has been an active area of interest. The dead ship condition is usually modeled using a single degree of freedom roll equation of motion (Webster 1989) characterized by a cubic softening spiring, a linear and quadratic viscous damping (Falzarano et al. 2015) and a random external excitation. The large amplitude ship rolling motion under random beam sea has been analyzed by various researchers using a variety of methods. These methods can broadly be classified as Markov and non-Markov methods.

The non-Markov methods include statistical equivalent linearization (Roberts and Spanos 2003), perturbation methods, Monte Carlo methods, Melnikov methods (Falzarano et al. 1992; Hsieh et al. 1994; Jiang et al. 1996, 2000) andFalzarano the Vakakis method (Vishnubhotla et al. 2000; Falzarano et al. 2004; Vishnubhotla and Falzarano 2009; Falzarano et al. 2010).

The Markov methods particularly rely on approximating the response as a Markov process. While real processes rarely exhibit the Markov property, often it is possible to approximate a real process as one using stochastic averaging or expressing the excitation as a response to a linear filter. The Markov approximation allows for the application of various analysis techniques of stochastic dynamics to the real processes which are otherwise restricted to Markov processes. Some of the Markov methods include the stochastic averaging method (Roberts and Vasta 2000; Su and Falzarano 2013), moment closure (Francescutto and Naito 2004; Su and Falzarano 2011) and the direct solution of the Fokker-Planck-Kolmogorov (FPK) equation (Naess and Moe 2000; Chai et al. 2015).

While the stochastic averaging helps reduce the dimension of the original system, the other methods in general result in an increase in the total number of states in the system. Particularly for the application of moment closure or solving the FPK equation numerically, it is important to express the system such that it is excited by Gaussian white noise. Since a general wave excitation spectrum is dependent on the wave spectrum and has a limited bandwidth, it cannot be directly approximated as a Markov process. However, the wave excitation itself can be expressed as an output of a linear filter whose input is an ideal Gaussian white noise. Introducing the filter increases the dimension of the system. However, this higher dimensional system is excited by an ideal white noise and hence the transition probability density is governed by the Fokker Planck Equation.

A method of transformation between ideal white noise and colored noise has been developed using filter technology. Using linear filters, any type of excitation can be handled by the Fokker-Planck-Kolmogorov equation.

The cumulant neglect method which discards cumulants higher than a particular order N is adopted in this paper to close the moment equations. If $N = 2$, then the method is defined as Gaussian closure, otherwise the method is non-Gaussian closure. By setting the higher order cumulants to zero, the higher order moment can be expressed by the lower order moments. Much of the literature discussing the application of moment closure techniques to the problem of ship rolling are restricted to Gaussian closure due to the difficulty in tracking the higher order closure (Francescutto 1990; Francescutto and Naito 2004). Specifically for high dimensional systems with large number of states, the cumulant neglect method becomes tedious. This tedium is the motivation for the development of an automatic tool in this work that addresses the difficulty in handling the higher dimensional state space stochastic models and higher order closure.

In this paper we also investigate the alternate approach of numerically solving the FPK equation to obtain the probability density function of the response. Particularly, the path integral method is chosen to solve the FPK numerically. In order to apply path integral method to understand the roll motion, the wave excitation is treated as regular waves perturbed by Gaussian white noise (Lin and Yim 1995). Due to the nonlinearity present in the system, there is a possibility of stochastic chaotic motion. Lin and Yim (1995) and Yim and Lin (2001) studied the stochastic chaotic motion of ship under periodic excitation with the disturbance approximated by Gaussian white noise from a probability perspective. The joint probability density functions of roll angle and roll angular velocity was calculated by applying the path integral method to solve the stochastic differential equations governing ship rolling motion. Lin and Yim (1995) found that the steady-state joint probability density functions can reflect the existing chaotic attractor on the Poincaré section and also the roll response in the heteroclinic region can be related to the capsizing through the joint probability density functions. To examine the chaotic characteristic of nonlinear roll motion in an unpredictable sea state, one cannot avoid dealing with the probabilistic approaches.

Gaussian white noise has become an important factor in these studies. The response of a dynamical system, roll angle and roll angular velocity in ship rolling study, under periodic excitation and Gaussian white noise can be modeled as a Markov process whose transition probability density function is governed by the FPK equation. Solving the FPK equation would provide an alternative means to express the evolution of the probability density function.

25.2 Nonlinear Ship Rolling Analysis using Stochastic Dynamics

25.2.1 Application of Gaussian and Non-gaussian Closure of Moment Equation using Cumulant Neglect Method

Since the moment equations are derived from the Fokker Planck equation, which is based on the theory of diffusion process, the excitation force for the Itô system has to be pure white noise. However, the random excitation term of rolling motion cannot be modelled as white noise. One method to handle the non-white excitation is to use a shaping filter, which is driven by Gaussian white noise. Consider a Linear Constant-Coefficient Differential Equation (LCCDE system) as follows (Stark and Woods 2002):

$$\begin{aligned} a_N Y^N(t) \ &+ a_{N-1} Y^{N-1}(t) + \cdots + a_0 Y^0(t) = b_M X^M(t) \\ &+ b_{M-1} X^{M-1}(t) + \cdots + b_0 X^0(t) \qquad -\infty < t < \infty \end{aligned} \tag{25.1}$$

The frequency response function or transfer function for this linear system shown in Eq. (25.1) is given by Eq. (25.2).

$$H(\omega) = \frac{B(\omega)}{A(\omega)} \text{ where } a_0 \neq 0 \tag{25.2}$$

$$B(\omega) = \sum_{m=0}^{M} b_m (j\omega)^m \tag{25.3}$$

$$A(\omega) = \sum_{n=0}^{N} a_n (j\omega)^n \tag{25.4}$$

where j represents the imaginary unit defined as $j = \sqrt{-1}$. The relationship between the spectra of the input excitation $X(t)$ and the output response $Y(t)$ in the frequency domain is given by Eq. (25.5).

$$S_{YY}(\omega) = |H(\omega)|^2 S_{XX}(\omega) \tag{25.5}$$

The bandwidth, characteristic frequency, etc., can be easily adjusted by changing the coefficient of the linear filter. Three filters were introduced in this paper following Francescutto and Naito (2004):

$$\frac{d^2Y}{dt^2} + \alpha_1 \frac{dY}{dt} + \beta_1 Y = \gamma_1 W(t) \tag{25.6}$$

$$\frac{d^2Y}{dt^2} + \alpha_2 \frac{dY}{dt} + \beta_2 Y = \gamma_2 \dot{W}(t) \tag{25.7}$$

$$\frac{d^4Y}{dt^4} + \lambda_3 \frac{d^3Y}{dt^3} + \lambda_2 \frac{d^2Y}{dt^2} t + \lambda_1 \frac{dY}{dt} + \lambda_0 Y = \gamma_3 \ddot{W}(t) \tag{25.8}$$

Where W (t) is the Gaussian white noise input or excitation of the dynamical system and Y (t) is the response of this linear dynamical system. The response of this filter system is actually the input excitation to the ship roll system. The frequency response function is defined by Eq. (25.2). A fourth order linear differential equation could be designed as a higher order filter. This filter can also be viewed as a cascade of two linear filters (Spanos 1983; Francescutto and Naito 2004).

All coefficients of the filter are determined through a non linear curve fitting method after setting $S_{ff}(\omega) = S_{YY}(\omega)$,where the $S_{ff}(\omega)$ is the wave excitation force spectrum. Fitting results are shown in Fig. 25.1. The original curve represents the target spectrum $S_{ff}(\omega)$. *Filter 1, Filter 2, Filter 3* represent $S_{YY}(\omega)$ obtained using the filters shown in Eqs. (25.6), (25.7) and (25.8) respectively.

By combining the filter shown in Eq. (25.8) with a classical nonlinear single degree of roll model, the roll motion in random beam seas can be modeled as a set first order differential equations as shown in Eq. (25.9). In general, any generic excitation spectrum can be reformulated using an appropriately chosen filter. Note that in Eq. (25.9), $x_3(t) = f(t)$ and $W(t)$ represents the Gaussian white noise defined

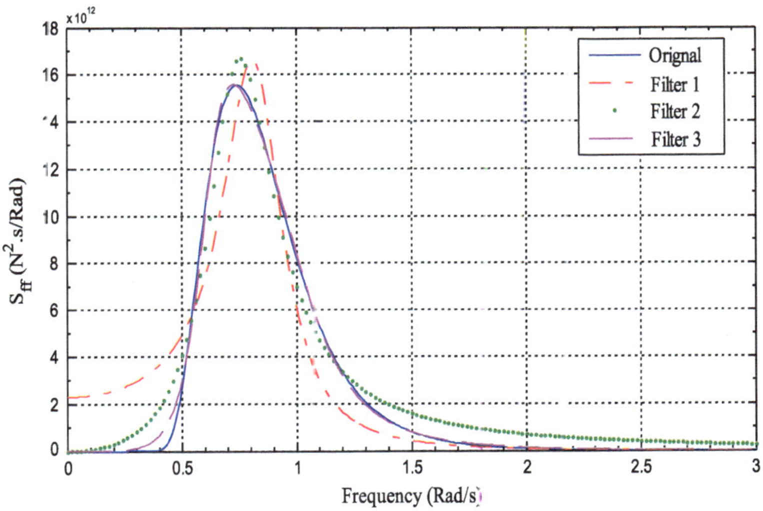

Fig. 25.1 Comparison of original force spectrum with filtered spectrum

as the generalized derivative of Brownian motion. With the Markov assumption, Eq. (25.9) forms a vector Itô's differential equations, e.g. Eq. (25.10).

$$\begin{cases} \dot{x}_1 = x_2 \\ \dot{x}_2 = -2\mu x_2 - \delta x_2^3 - \omega_0^2 x_1 - \alpha_3 x_1^3 + \varepsilon x_3(t) \\ \dot{x}_3 = x_4 - \lambda_3 x_3 \\ \dot{x}_4 = x_5 - \lambda_2 x_3 + \gamma_3 W(t) \\ \dot{x}_5 = x_6 - \lambda_1 x_3 \\ \dot{x}_6 = -\lambda_0 x_3 \end{cases} \tag{25.9}$$

$$dX = F(X,t)dt + G(X,t)dB \tag{25.10}$$

Following Itôs differential rule (Itô, 1951), the moment equation can be formed as below,

$$\frac{\partial}{\partial t}E[\phi] = \sum_{i=1}^{N} E\left[\frac{\partial}{\partial x_i}(F_i\phi)\right] + \sum_{i=1}^{N}\sum_{j=1}^{N} E\left[\frac{\partial^2}{\partial x_i \partial x_j}\{(GQG^T)_{ij}\phi\}\right] \tag{25.11}$$

The moment equations will form a closed system of equations after application of the closure of the cumulant. To apply the Gaussian Cumulant neglect method, 27 moment equations up to second order are derived by Eq. (25.11) for system, Eq. (25.9) and analyzed by neglecting cumulants higher than second order. Figures 25.2, 25.3 and 25.4 express the effect of nonlinear damping coefficient, nonlinear stiffness and linear damping on the rolling response with the Gaussian cumulant neglect.

In order to better understand the nonlinear effect, the authors have developed an automatic cumulant neglect tool using characteristic functions to investigate the non-Gaussian response up to the 4th order cumulant neglect (Su and Falzarano 2011; Su et al. 2011). The automatic cumulant neglect tool takes advantages of two different representations of the characteristic functions of random variables to automatically neglect higher order cumulants. The higher order moments could also be derived directly by applying higher order neglect.

25.2.2 Application of Path Integral Method

The perturbed waveforms may be modeled as regular waves with Gaussian white noise as the external disturbance as in Eq. (25.10). With some noise intensity, the response appears random as shown in the Poincaré map of Fig. 25.5.

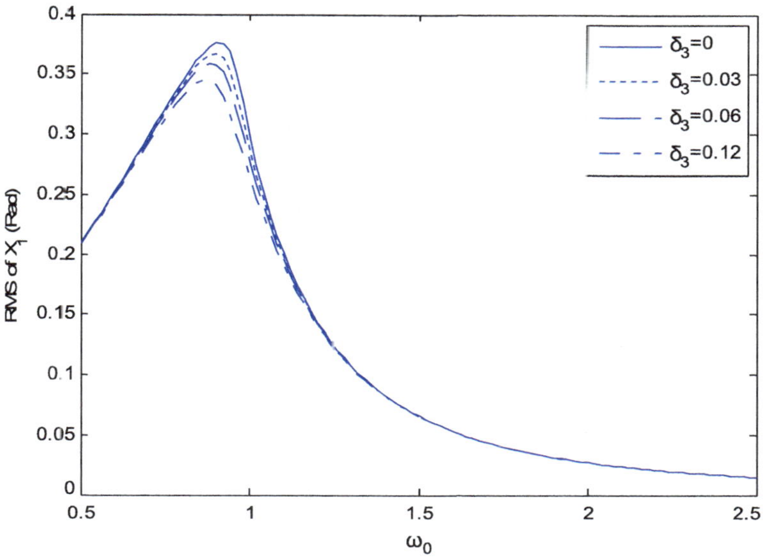

Fig. 25.2 Effect of nonlinear damping coefficient on the root mean square (RMS) of rolling displacement

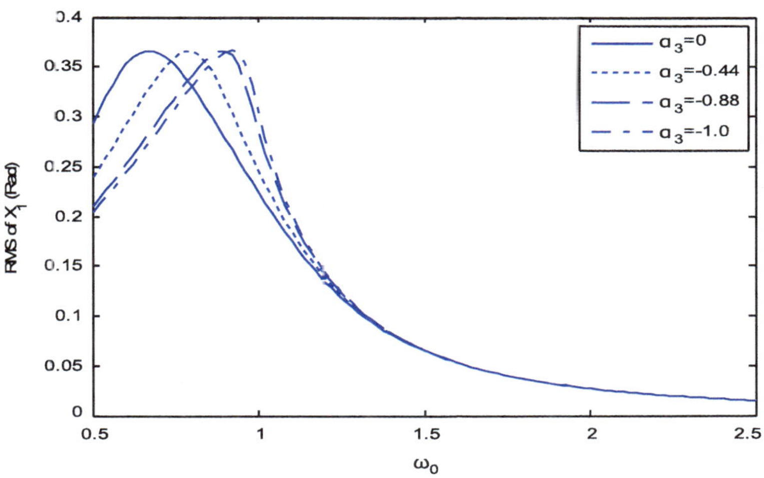

Fig. 25.3 Effect of nonlinear stiffness coefficient on the root mean square (RMS) of rolling displacement

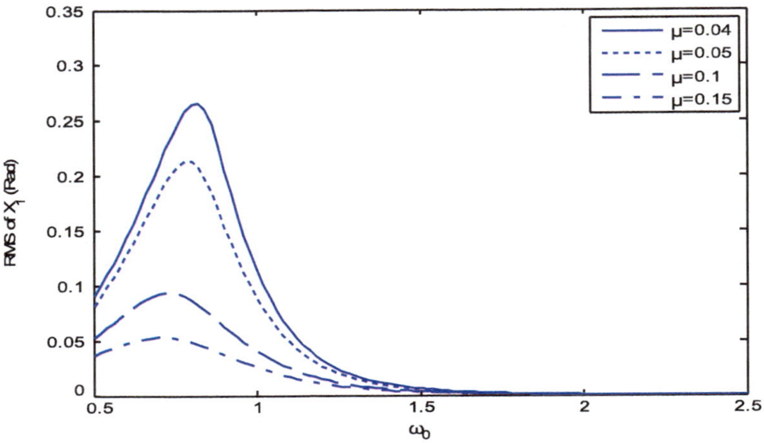

Fig. 25.4 Effect of linear damping coefficient on the root mean square (RMS) of rolling displacement

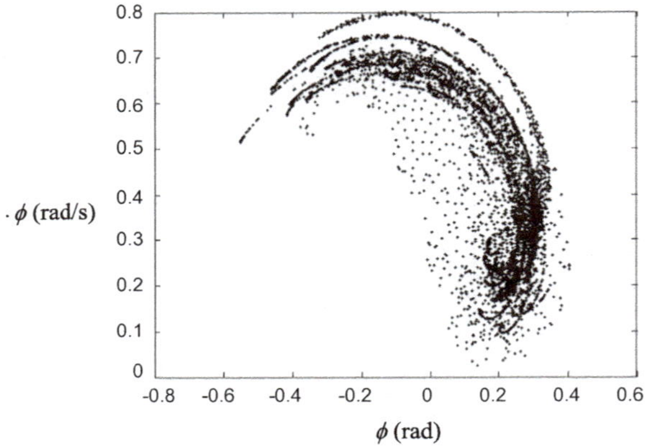

Fig. 25.5 The Poincaré map with noise intensity

$$\ddot{x} + \mu\dot{x} + \delta\dot{x}^3 + x - \alpha x^3 = \frac{H}{2}\cos(\omega t) + \sqrt{D}N(t) \tag{25.12}$$

Note that the coefficients of linear and nonlinear damping terms are set as constant by assuming a fixed frequency. The parameters values are set as constant to the following values: $\mu = 0.1321$, $\delta = 0.02656$ and $\alpha = -0.9018$.

The evolution of the probability density function is another way to describe the behavior of the nonlinear roll motion in random waves. The behavior of the noisy forced ship roll motion under periodic excitation with Gaussian white noise can be modeled as a Markov process. The probability density function of a Markov process satisfies the Fokker-Planck equation. The associated Fokker-Planck equation governing the evolution of the probability density function of the roll motion is derived and

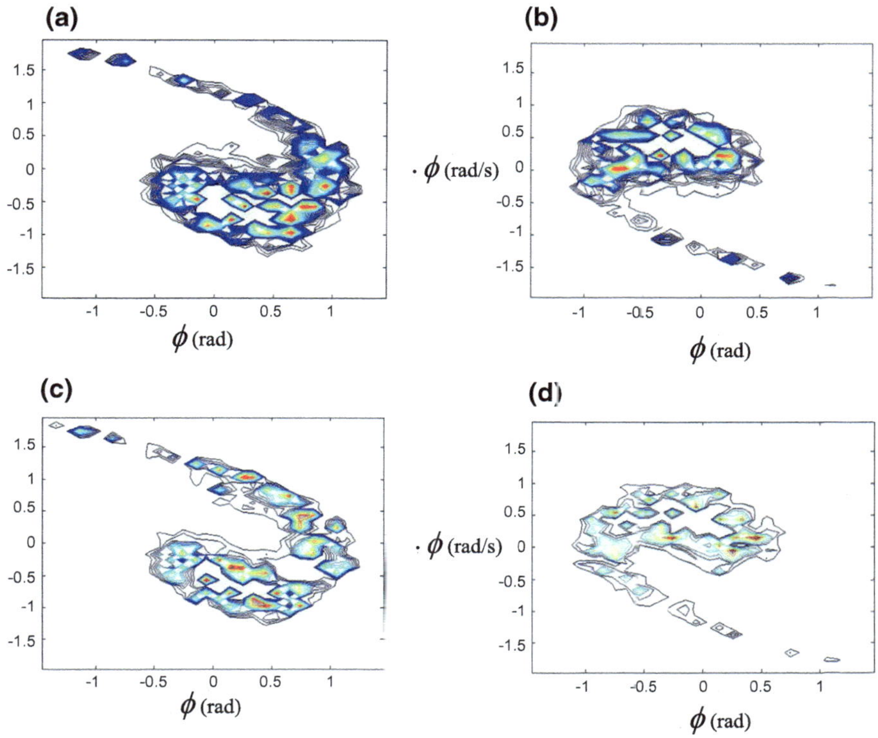

Fig. 25.6 Evolution of contour plot of the joint probability density function with $(H, D, \omega) = (0.3, 0.01, 0.97)$ at time **a** $t = 12.95$ s **b** $t = 29.13$ s **c** $t = 45.32$ s **d** $t = 74.45$ s

numerically solved by the path integral method based on Gauss-Legendre interpolation to obtain joint probability density functions in the state space (Jamnongpipatkul et al. 2011).

In the presence of a small random disturbance to the external period excitation, the imprint of the Poincaré map is preserved and can be identified via the joint probability density function on the Poincaré section as shown in Figs. 25.6, 25.7 and 25.8. The probability density function indicates the preferred locations of the trajectories in the average sense.

The time-dependent probability of ship rolling restricted within the safe domain is provided in Fig. 25.9. It is found that the probability decreases as time progresses and it decreases much more quickly for the high intensity of the white noise. At t = 45.32 s, the probability is about 45% when D = 0.01. It is about 20% when D = 0.05, and it is only about 2% when D = 0.1. The ship will finally leave the safe domain and capsize in the probabilitys view.

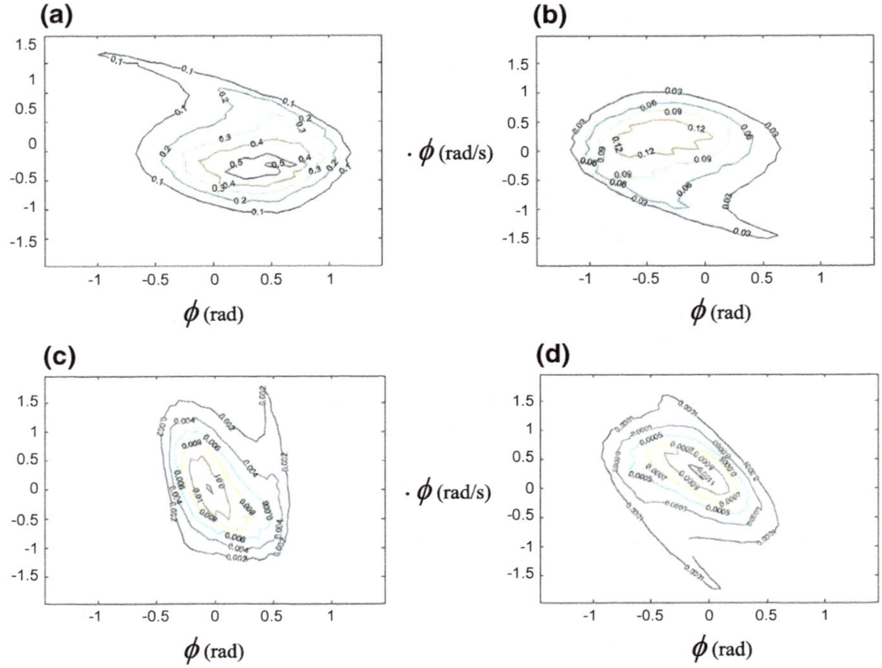

Fig. 25.7 Evolution of contour plot of the joint probability density function with $(H, D, \omega) = (0.3, 0.05, 0.97)$ at time **a** $t = 12.95$ s **b** $t = 29.13$ s **c** $t = 45.32$ s **d** $t = 74.45$ s

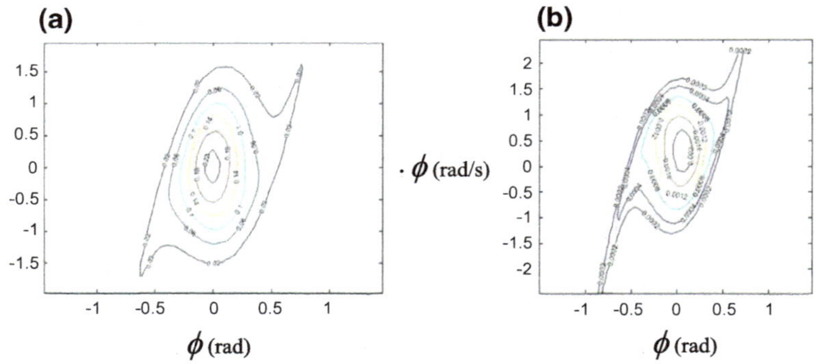

Fig. 25.8 Evolution of contour plot of the joint probability density function with $(H, D, \omega) = (0.3, 0.1, 0.97)$ at time **a** $t = 12.95$ s **b** $t = 29.13$ s

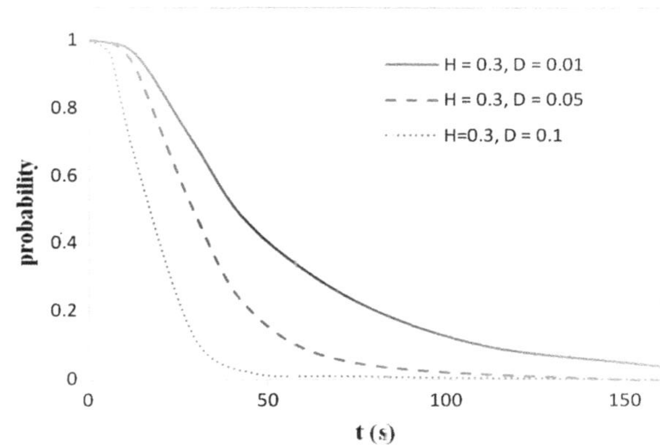

Fig. 25.9 Probability of ship rolling within the safe domain when $\omega = 0.97$ rad/s

25.3 Conclusions

This paper describes the study of probabilistic nature of ship roll motion in irregular beam seas. The stochastic nonlinear dynamic behavior and the probability density function of roll motion in the random beam waves are studied. Moment equations are calculated for a high dimensional dynamical system obtained by expressing wave excitation as the output of a linear filter excited by Gaussian white noise. The corresponding moment equations are solved using both Gaussian and Non Gaussian cumulant neglect methods. Alternatively, the probability density function for the response is also evaluated using the path integral method. Particularly various excitation types consisting of varying amounts of noise are investigated. The probability density function of rolling response is evaluated and the time-dependent probability of ship roll being restricted within the safe domain is provided. It is found that the value of joint probability density decreases gradually as time progresses. For higher intensity of white noise, it decreases much more quickly which means that probabilistically, given enough time the ship will leave the safe domain and result in a capsize.

Using the linear filter approach, the moment equation method can be applied to analyze ships being excited by waves modeled as a real colored noise. The probabilistic nature of the response is obtained by solving for the moments of the response using the cumulant neglect method. However, it is difficult to find any bifurcation phenomena and probability density function using this approach. On the contrary, the path integral method provides the transient response probability density function, when considering only harmonic wave with disturbance noise. Even though significant work has been done in the past, solving the high dimensional Fokker Planck Equation for ship roll motion continues to be a challenging task.

Acknowledgements The work has been funded by the Office of Naval Research (ONR) T-Craft Tools development program ONR Grant N00014-07- 1-1067 with program manager Kelly Cooper.

References

Chai W, Naess A, Leira BJ (2015) Stochastic Dynamic Analysis and Reliability of a Vessel Rolling in Random Beam Seas. Journal of Ship Research 59(2):113–131

Falzarano J, Vishnubhotla S, Cheng J (2004) Nonlinear Dynamic Analysis of Ship Capsizing in Random Waves. In: 14th International Offshore and Polar Engineering Conference, The International Society of Offshore and Polar Engineers, Toulon, France, vol 1, pp 479–484

Falzarano J, Somayajula A, Seah R (2015) An overview of the prediction methods for roll damping of ships. Ocean Systems Engineering 5(2):55–76

Falzarano JM, Shaw SW, Troesch AW (1992) Application of Global Methods for Analyzing Dynamical Systems To Ship Rolling Motion and Capsizing. International Journal of Bifurcation and Chaos 02(01):101–115, https://doi.org/10.1142/S0218127492000100

Falzarano JM, Vishnubhotla S, Juckett SE (2010) Combined Steady State and Transient Analysis of a Patrol Vessel as Affected by Varying Amounts of Damping and Periodic and Random Wave Excitation. Journal of Offshore Mechanics and Arctic Engineering 132(1):014,501, https://doi.org/10.1115/1.4000390

Francescutto A (1990) On the Non-linear Motions of Ships and Structures in Narrow Band Sea. In: IUTAM Symposium on Dynamics of Marine Vehicles and Structures in Waves, Elsevier, London, UK, pp 291–304

Francescutto A, Naito S (2004) Large amplitude rolling in a realistic sea. International shipbuilding progress 51(2):221–235

Guha A, Somayajula A, Falzarano J (2016) Time domain simulation of large amplitude motions in shallow water. In: 21st SNAME Offshore Symposium, Society of Naval Architects and Marine Engineers, Houston, February

Hsieh SR, Troesch aW, Shaw SW (1994) A Nonlinear Probabilistic Method for Predicting Vessel Capsizing in Random Beam Seas. Proceedings of the Royal Society A: Mathematical, Physical and Engineering Sciences 446(1926):195–211, https://doi.org/10.1098/rspa.1994.0099

Jamnongpipatkul A, Su Z, Falzarano JM (2011) Nonlinear ship rolling motion subjected to noise excitation. Ocean Systems Engineering 1(3):249–261, https://doi.org/10.12989/ose.2011.1.3.249

Jiang C, Troesch AWA, Shaw SWS (2000) Capsize criteria for ship models with memory-dependent hydrodynamics and random excitation. Philosophical Transactions of the Royal Society A: Mathematical, Physical and Engineering Sciences 358(1771):1761–1791, https://doi.org/10.1098/rsta.2000.0614

Jiang CB, Troesch AW, Shaw SW (1996) Highly nonlinear rolling motion of biased ships in random beam seas. Journal of Ship Research 40(2):125–135

Lin H, Yim SC (1995) Chaotic roll motion and capsize of ships under periodic excitation with random noise. Applied Ocean Research 17(3):185–204, https://doi.org/10.1016/0141-1187(95)00014-3

Moideen H, Falzarano JM, Sharma S (2012) Parametric roll of container ships in head waves. International Journal of Ocean Systems Engineering 2(4):239–255, https://doi.org/10.12989/ose.2012.2.4.239

Moideen H, Somayajula A, Falzarano JM (2013) Parametric Roll of High Speed Ships in Regular Waves. In: Proceedings of ASME 2013 32nd International Conferences on Ocean, Offshore and Arctic Engineering, ASME, vol 5, p V005T06A095, https://doi.org/10.1115/OMAE2013-11602

Moideen H, Somayajula A, Falzarano JMJ (2014) Application of Volterra Series Analysis for Parametric Rolling in Irregular Seas. Journal of Ship Research 58(2):97–105, https://doi.org/10.5957/JOSR.58.2.130047

Naess A, Moe V (2000) Efficient path integration methods for nonlinear dynamic systems. Probabilistic Engineering Mechanics 15(2):221–231, https://doi.org/10.1016/S0266-8920(99)00031-4

Roberts JB, Spanos PD (2003) Random Vibration and Statistical Linearization. Dover Publications, Mineola, New York

Roberts JB, Vasta M (2000) Markov modelling and stochastic identification for nonlinear ship rolling in random waves. Philosophical Transactions of the Royal Society A: Mathematical, Physical and Engineering Sciences 358(1771):1917–1941, https://doi.org/10.1098/rsta.2000.0621

Somayajula A, Falzarano J (2015a) Large-amplitude time-domain simulation tool for marine and offshore motion prediction. Marine Systems & Ocean Technology 10(1):1–17, https://doi.org/10.1007/s40868-015-0002-7

Somayajula A, Falzarano JM (2014) Non-linear Dynamics of Parametric Roll of Container Ship in Irregular Seas. In: Proceedings of ASME 2014 33rd International Conferences on Ocean, Offshore and Arctic Engineering, San Francisco, pp 1–10, https://doi.org/10.1115/OMAE2014-24186

Somayajula A, Falzarano JM (2015b) Validation of Volterra Series Approach for Modelling Parametric Rolling of Ships. In: Proceedings of ASME 2015 34th International Conferences on Ocean, Offshore and Arctic Engineering, American Society of Mechanical Engineers, St. John's, NL, Canada

Somayajula A, Guha A, Falzarano J, Chun HH, Jung KH (2014) Added resistance and parametric roll prediction as a design criteria for energy efficient ships. International Journal of Ocean Systems Engineering 4(2):117–136, https://doi.org/10.12989/ose.2014.4.2.117

Somayajula AS, Falzarano JM (2016) A comparative assessment of simplified models for simulating parametric roll. Journal of Offshore Mechanics and Arctic Engineering https://doi.org/10.1115/1.4034921

Spanos PTD (1983) ARMA Algorithms for Ocean Wave Modeling. Journal of Energy Resources Technology 105(3):300, https://doi.org/10.1115/1.3230919

Spyrou KJ, Thompson JMT (2000) The nonlinear dynamics of ship motions: a field overview and some recent developments. Philosophical Transactions of the Royal Society of London Series A: Mathematical, Physical and Engineering Sciences 358(1771):1735–1760

Stark H, Woods JW (2002) Probability and random processes with applications to signal processing. Prentice Hall, Upper Saddle River, NJ

Su Z, Falzarano JM (2011) Gaussian and non-Gaussian cumulant neglect application to large amplitude rolling in random waves. International Shipbuilding Progress 58:97–113, https://doi.org/10.3233/ISP-2011-0071

Su Z, Falzarano JM (2013) Markov and Melnikov based methods for vessel capsizing criteria. Ocean Engineering 64:146–152, https://doi.org/10.1016/j.oceaneng.2013.02.002

Su Z, Falzarano JM, Su Z (2011) Gaussian and Non Gaussian Response of Ship Rolling in Random Beam Waves. In: Proceedings of the12th International Ship Stability Workshop, Washington DC, USA, pp 189–193

Thompson JMT (1997) Designing Against Capsize in Beam Seas: Recent Advances and New Insights. Applied Mechanics Reviews 50(5):307, https://doi.org/10 1115/1.3101710

Vishnubhotla S, Falzarano J (2009) Effect of More Accurate Hydrodynamic Modeling on Calculating Critical Nonlinear Ship Rolling Response. In: Lecture Notes in Applied and Computational Mechanics, vol 44, pp 269–274, https://doi.org/10.1007/978-3-642-00629-6_27

Vishnubhotla S, Falzarano J, Vakakis A (2000) A new method to predict vessel/platform critical dynamics in a realistic seaway. Philosophical Transactions of the Royal Society A: Mathematical, Physical and Engineering Sciences 358(1771):1967–1981, https://doi.org/10.1098/rsta.2000.0623

Webster W (1989) Motion in Regular Waves - Transverse Motions. In: Lewis E (ed) Principles of Naval Architecture Vol III, SNAME, Jersey City, New Jersey

Yim SCS, Lin H (2001) Unified Analysis of Complex Nonlinear Motions via Densities. Nonlinear Dynamics 24(1):103–127, https://doi.org/10.1023/A:1026583521930

Chapter 26
The Capsize Band Concept Revisited

Nikolaos Tsakalakis, Jakub Cichowicz and Dracos Vassalos

Abstract A concept for analytical representation of the capsize rate, a measure directly related to damage ship survivability, has attracted attention ever since the first attempts were made to explain the behaviour of a damaged ship in waves. Attempts in the late 1990s helped to enhance understanding and facilitate characterisation of phenomena pertaining to capsize probability and time to capsize in given environments and loading conditions, but a consistent verifiable formulation is still lacking. In this respect, pursuing an analytical approach to express the capsize rate offers many advantages, time efficiency being amongst the most important. In an era when stability/survivability calculations are required to be carried out in real time, there is a need for a model accounting for the random nature of capsize whilst achieving accuracy close to that of time-domain simulations with simple hydrostatic calculations. This study is an attempt to establish a new methodology for survivability assessment by means of a multivariable analytical model based on numerical simulations, validated against the results of physical model tests.

Keywords Damage stability · Capsize band · Critical wave height · Ro-Pax

26.1 Introduction

The concepts of capsize boundary and capsize band lie at the core of damage survivability assessment of ships. The s-factor used to derive the Attained Index of subdivision corresponds to the 50% probability of survival in damaged condition at specific sea state for 30 min (Jasionowski et al. 2002). The significant wave height corresponding to the sea state at which the capsize rate (P_f) equals 0.5 in 30 min long trials is referred to as *critical significant wave height, Hs_{crit}*. The capsize band, in turn, reflects the marginal nature of the capsize phenomena and by analogy to statis-

---

N. Tsakalakis · J. Cichowicz · D. Vassalos (✉)
The Ship Stability Research Centre, Department of Naval Architecture, Ocean and Marine Engineering, University of Strathclyde, Glasgow, Scotland, UK
e-mail: d.vassalos@strath.ac.uk

V. L. Belenky et al. (eds.), *Contemporary Ideas on Ship Stability*, Fluid Mechanics and Its Applications 119, https://doi.org/10.1007/978-3-030-00516-0_26

tics it can be interpreted as a confidence interval about Hs_{crit}. In fact, the capsize band is not a confidence interval in a strict sense[1]—it is rather a measure of dispersion of capsizes, separating sea states in which the capsize rate (i.e. the conditional probability of capsize given H_S) is very low from those in which the rate is very high. In other words, the capsize band reflects a well-known fact that there is no distinct boundary separating safe from unsafe sea states; instead there is rather a transition zone within which capsize is possible. The presence of the band also implies that although there must be sea states at which the vessel will never capsize and that there must be sea states at which she would inevitably capsize, due to limited resolution of physical or numerical experiments the lower and upper boundaries can only be expressed by means of limits. Such asymptotic nature requires the use of some threshold values of P_f outside of which occurrence of capsize will either be virtually impossible or practically certain. Making use of analogy to statistics again, such limiting sea states corresponding to threshold values of P_f, can be interpreted as confidence limits.

Although the capsize rate, P_f, is a function of many variables, such as sea state, loading condition and damage characteristics, it has been observed that in all cases it follows a clear and recurring trend. This has triggered the pursuit for its analytical representation that could be used in parametric studies on capsize phenomena in order to derive universal formulae for probability of capsize and corresponding time to capsize.

Understandably, such studies require a vast number of experiments to be performed, which sets practical limits on the achievable resolution and accuracy of the results. In this paper, the authors present a brief account of the current state-of-the-art, discuss advantages and shortcomings and propose an alternative approach, which can offer significant reduction of effort (normally expended in numerical simulations and model experiments) whilst retaining comparable accuracy of the outcome.

26.2 Approach

26.2.1 Software Tools

Numerical experiments supporting this work have been carried out with PROTEUS3, the in-house developed software[2] that has been successfully employed over many years in a number of research and commercial projects. It has been referenced a number of times, benchmarked against experimental data and other numerical codes

[1]Given that the significant wave height at which the capsizes are observed is a random variable, the confidence interval would simply be a band of wave heights containing most of the area under the p(Hs|capsize) probability density function curve. Instead, the boundaries of the capsize band are expressed with the use of the following equalities: $(H_S)_{low} = H_S|_{P_f(H_S)=\alpha}$ and $(H_S)_{high} = H_S|_{P_f(H_S)=1-\alpha}$, where α is some (small) number.

[2]PROTEUS 3 is a software suite for time-domain seakeeping analysis of intact and damaged ships http://www.brookesbell.com/service/software/design-software/proteus-simulation-sofware.

successfully and has aided greatly in our understanding of capsize phenomena in damage conditions. OriginPro8—a powerful statistical package—was used for processing the results, parametric studies and development of the methodology.

26.2.2 Ship Models

Two models of Ro-Pax vessels have been studied for the purpose of this paper, of 89 and 170 m in length between perpendiculars. The first ship (EUGD01-R2) is being extensively tested numerically at the moment for the on-going[3] EU Project GOALDS (Project GOALDS, 2009–2012) that aims to re-engineer the probabilistic rules formulation for damage survivability of passenger ships. Physical model experiments were conducted for this ship in the course of the project. The larger Ro-Pax has been used in previous research projects, including HARDER (Project HARDER 2000–2003), which provided the foundation of the current probabilistic regulatory framework for damage stability. Results of physical model experiments carried out on this vessel are being used for validation of the numerical code (Figs. 26.1, 26.2; Table 26.1).

The two chosen ships cover different regions of the design space to ensure universal application of the results. The PRR01 was designed for the transport of, primarily, vehicles across short routes such as the English Channel; the vessel was re-designed to carry a larger number of passengers during the building stage. The second ship was designed for transport of a small number of both passengers and vehicles within an island archipelago in short-crested, choppy seas.

Table 26.1 Main particulars of models

Model	PRR01	EUGD01-R2	
Passengers	1420	622	
L_{OA}	194.3	97.9	m
L_{BP}	176	89	m
Breadth	25	16.4	m
Deepest subdivision load line	6.55	4	m
Depth to bulkhead deck	9.1	6.3	m
Displacement	16,558	3445	tn
Service speed	21.0	19.5	kn

[3]At the time of writing.

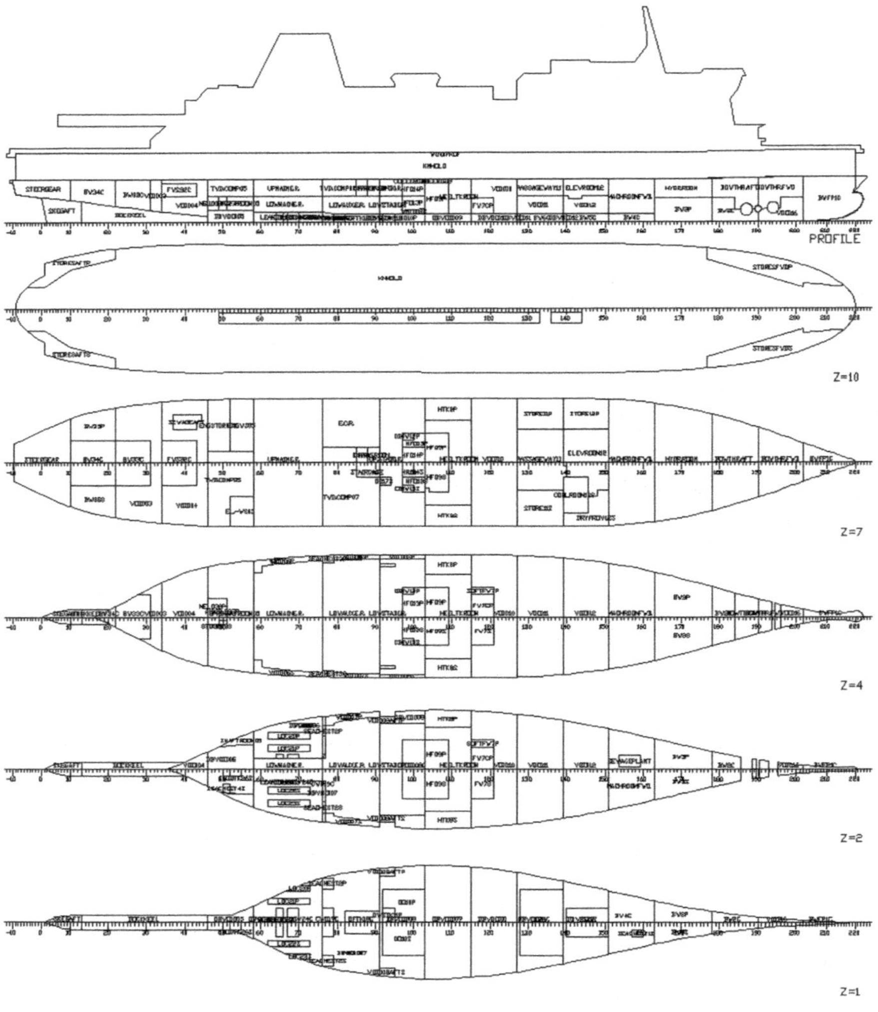

Fig. 26.1 NAPA (NAPA is a software package for ship design and operations https://www.napa.fi/) model of watertight subdivision of PRR01

26.2.3 Numerical Experiments

Accurate representation of the capsize rate characteristic across the entire capsize band, requires adequate resolution. Therefore, it was deemed necessary to use at least 10 measurements within the transition zone, performed by increasing Hs in small steps, varying from 0.1 to 0.25 m depending on the width of the capsize band. For each wave height, Pf was determined on a basis of at least 20 wave realisations to maintain at least 5% resolution. The larger ship was tested in seven and the smaller in five

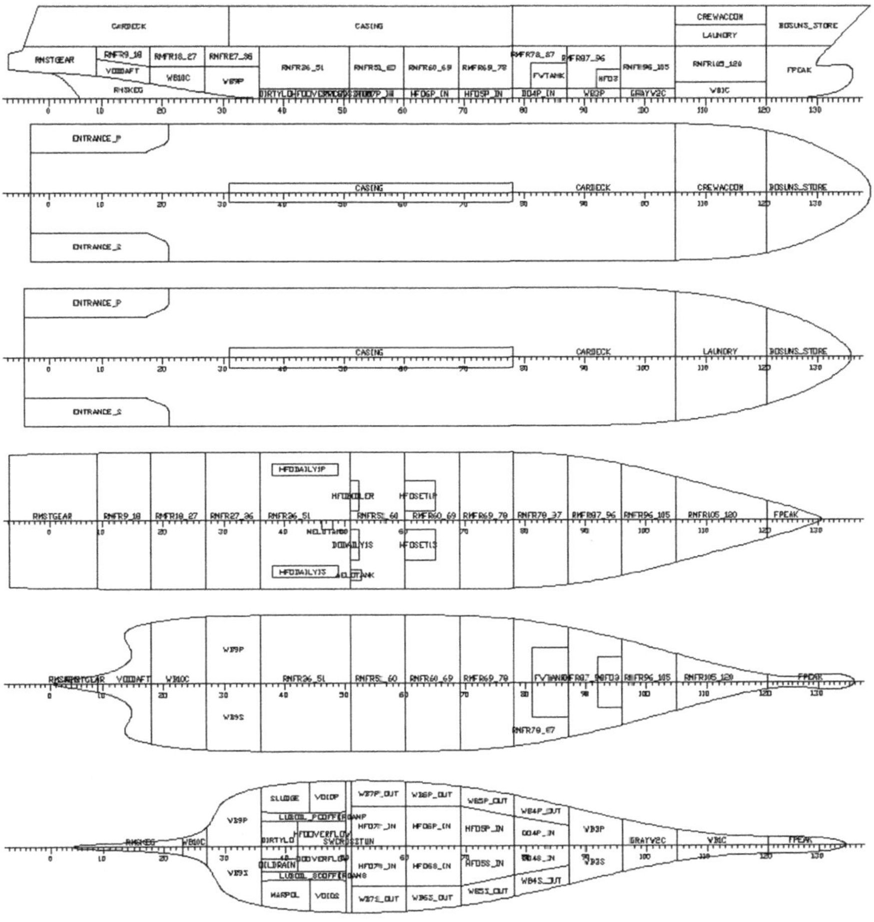

Fig. 26.2 NAPA model of watertight subdivision of EUGD01-R2

different loading conditions, including variations of draught and KG. Additionally, the survivability of the smaller vessel was studied in two distinct damages and various wave spectra. Waves were modelled using JONSWAP spectrum of slope (height to length ratio) equal to 1/20 and 1/25, respectively. Each realisation was limited to 1800 s, which is the maximum time currently required by regulations for evacuation of a vessel. Complete time history of the motions and water accumulation (including water on Ro-Ro deck) was measured and recorded (e.g. Fig. 26.3). No wind effect was included in the experiments. All simulations started with the ship in the damage equilibrium position.

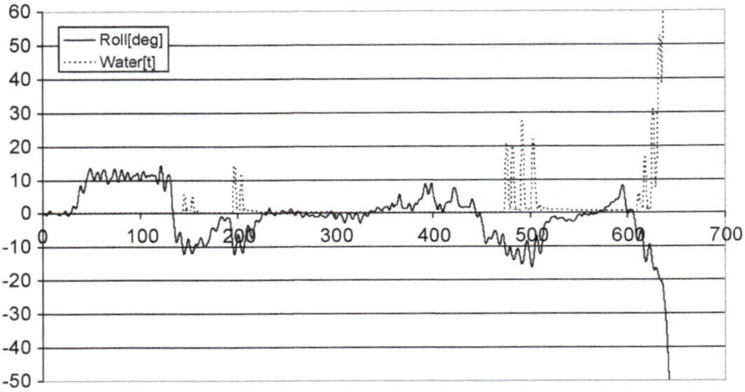

Fig. 26.3 Time history of Roll motion and water accumulation as recorded from PROTEUS3 for a capsize case

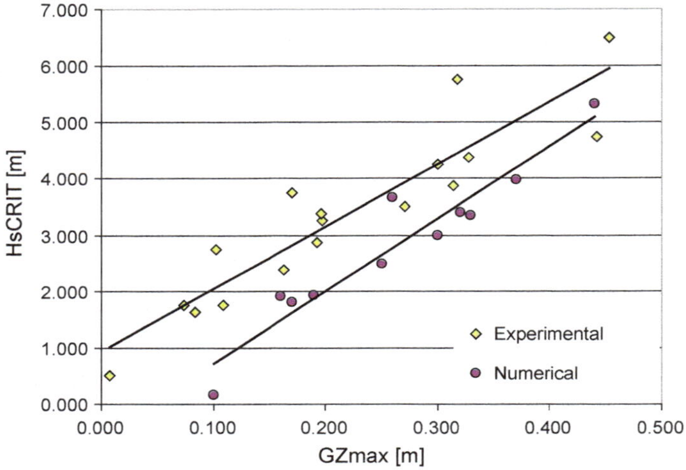

Fig. 26.4 Experimental versus numerical results for model PRR01. Solid lines represent trendlines

26.2.4 Numerical Code Validation

Given the relative ease of use of numerical tools it is possible to carry out hundreds of verifiable simulations in a short period of time. Within the present study, the outcome of numerical software was benchmarked against available experimental data from project HARDER (availability of data was one of the reasons for selecting PRR01 as sample ship). Comparison between numerical and experimental results shows satisfactory agreement (Fig. 26.4).

It should be noted here that the quantitative agreement between the results was considered of minor importance with emphasis being put on the observed trends. For the purpose of this work it was decided that as long as the differences are systematic an exact match is not required and no further numerical model calibration was performed (particularly as observations show that numerical results are conservative).

26.3 Probability of Capsize

26.3.1 Capsize Rate

The term capsize rate (P_f) is used to denote the approximation[4] of the probability of capsize of a damaged ship at specific sea state, given loading conditions and duration of trials. Predictably, for a given number of realisations,[5] capsize rate will vary from 0 for very small[6] to 1 for very large waves. Between minimum H_S for which $P_f = 0$ and maximum H_S for which $P_f = 1$, P_f can take any value ranging from 0 to 1. Following adopted convention (Vassalos et al. 1997), critical wave height corresponds to the significant wave height for which capsize rate is 0.5 during 30 min trials.

Disregarding the experimental errors, it is obvious from Fig. 26.5 that data follow a specific pattern throughout the range. The evident trend common to all observations made across the entire H_S range led previous attempts to approach this characteristic by making use of its similarity to the integral of a normal Gaussian distribution—Cumulative Density Function (CDF) (Jasionowski et al. 2007). A major advantage of such approach is that the normal distribution is a well known function and statistical tools can be readily applied to the recorded data in order to find an interval around critical H_S, which could be interpreted as capsize band by use of standard deviation of the derivative of capsize rate. The biggest downside of this method is that it requires numerical differentiation of recorded data, i.e. it involves computation of the derivative of the capsize rate, P_f. As differentiation of infrequent data unequally distributed along the H_S range may introduce large uncertainties, the approach is practically limited to large[7] data sets.

26.3.2 Non-linear Regression

A pursuit for a more convenient functional representation of the capsize rate resulted in a parametrically defined sigmoid function that turned out to be an attractive alter-

---

[4]This follows the classical definition of probability, expressed as the ratio of favourable experiment outcomes over the total number of trials. It would become a probability of capsize (conditional on loading condition and wave parameters) if the number of trials approached infinity.

[5]A time series of seakeeping either by means of numerical simulations or physical model tests.

[6]Relative to the critical significant wave height.

[7]Word *large* in this context refers rather to computational or experimental effort than actual, numerical size of the data.

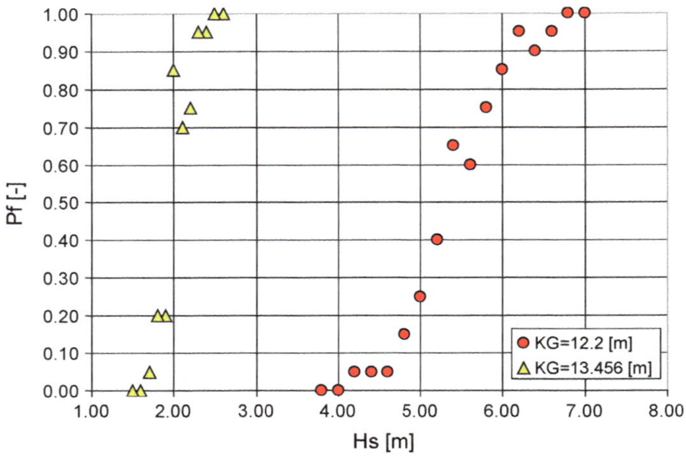

Fig. 26.5 Capsize rate values for different loading conditions (PRR01)

native to the Gaussian distribution. Boltzmann's[8] sigmoid allows direct regression of measured rates, without the need for prior numerical differentiation. The resulting function can be differentiated easily afterwards to derive the requisite information on the capsize band. The Boltzmann's sigmoid function $y(x)$ is given by means of four parameters: A_1, A_2, x_0 and d_x.

$$y(x) = \frac{A_2 + (A_1 - A_2)}{1 + e^{\frac{x - x_0}{dx}}} \tag{26.1}$$

where:

A_1 asymptotic lower limit
A_2 asymptotic upper limit
x_0 ordinate of centre of symmetry
d_x time constant.[9]

By nature of the capsize rate observations, the first two parameters can be constrained to 0 and 1, respectively, which leaves just two parameters requiring estimation and allows for, after some basic manipulation, the expression of P_f as a function of H_S, x_O and d_x (26.2). The derivative of P_f with respect to H_S is given as in (26.3)

[8]The reference follows the nomenclature of OriginPro software package http://www.originlab.com/doc/Origin-Help/Boltzmann-FitFunc.

[9]The parameter dx is referred to by analogy to dynamic system response to step input. In context of current application is a span parameter (related to slope at inclusion point).

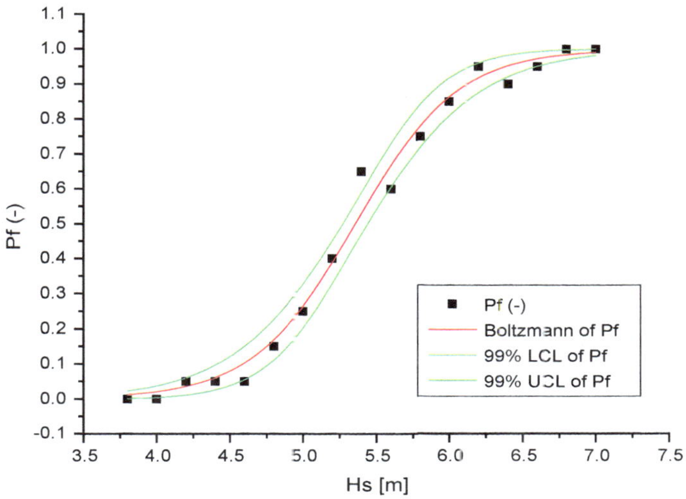

Fig. 26.6 Fitted sigmoid and 99% confidence boundaries (PRR01 with KG = 12.2 m)

Table 26.2 Parameters of sigmoid regression to Pf for T = 6.25 m, KG = 12.200 m, even keel

Parameters		
	Value	Standard error
A1	0	0
A2	1	0
x0	5.35778	0.02832
dx	0.34893	0.02503

$$P_f(H_S) = \frac{e^{\frac{H_S - x_0}{dx}}}{1 + e^{\frac{H_S - x_0}{dx}}} \tag{26.2}$$

$$\frac{dP_f}{dH_S} = \frac{e^{\frac{H_S - x_0}{dx}}}{dx\left(1 + e^{\frac{H_S - x_0}{dx}}\right)^2} \tag{26.3}$$

Figures 26.6 and 26.7 depict an example of Boltzmann's sigmoid fitted to the experimental data as well as residuals of fitting. Statistical data describing goodness of fit are presented in Tables 26.2 and 26.3.

Results of employing this technique to data deriving from numerical simulations performed at different KGs are presented in Fig. 26.8. It can be readily seen that increasing KG causes a shift of P_f characteristics towards lower sea states with a more rapid transition from low to high capsize rates (probability distribution becoming narrower as KG increases). This implies that as survivability decreases the transition from the region considered safe to that considered as unsafe is faster. The performance

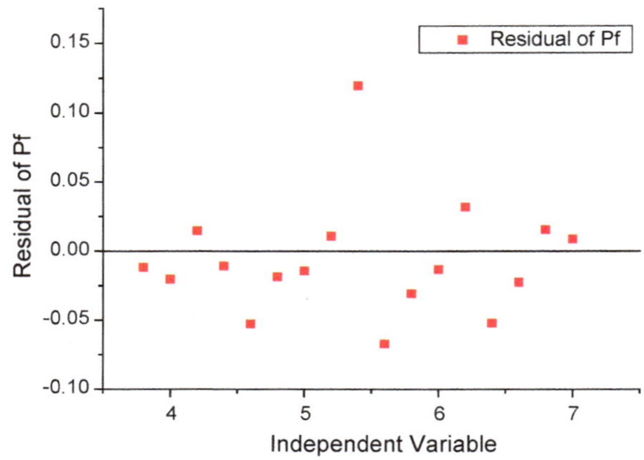

Fig. 26.7 Residuals of Pf sigmoid fitting (PRR01 with KG = 12.2 m)

Table 26.3 Statistics of sigmoid regression

Statistics	
Number of points	17
Degrees of freedom	15
Reduced chi-sqr	0.00192
Residual sum of squares	0.02873
Adj. R-square	0.98814

parameters of this particular probability distribution against other ship characteristics can be established in the same manner with the scope to detect any dependencies between survivability and specific design variables.

26.3.3 *Estimation of the Capsize Band*

The previous observation can be quantitatively confirmed by use of critical significant wave height and capsize band parameters. The first quantity is associated with x_0 parameter of the regression's sigmoid function whereas the latter can be easily calculated using Eq. (26.1). By analogy to statistics the capsize band can be interpreted as the range of the probability distribution, spreading either side of the capsize boundary ($P_f = 0.5$), symmetrically. In a more straightforward interpretation limits of the capsize band simply determine boundaries outside which capsize rate is either so high or so low that capsize in given H_S is either certain or unlikely, beyond upper and below lower limits, respectively. In order to determine such limits, it is convenient to take some small number α, and find those values of $H_{S,}$ which satisfy the following conditions:

$$(H_S)_{low} = H_S|_{P_f(H_S)=\alpha} \quad (26.4)$$

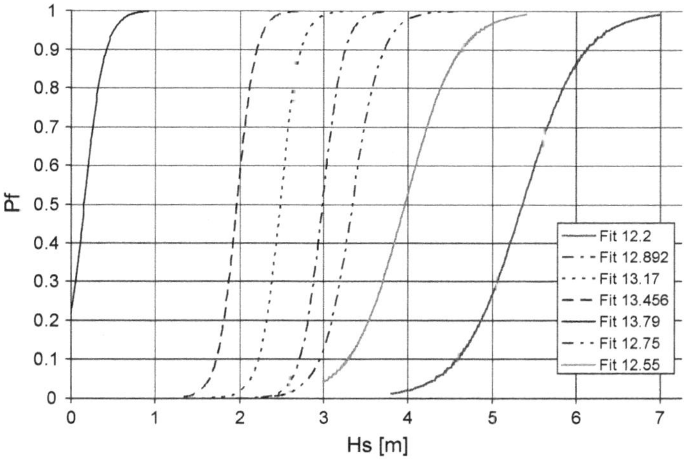

Fig. 26.8 Capsize rate for various critical significant wave heights of PRR01 (the legend entries indicate the loading conditions at which the experiments were performed, e.g. "Fit 12.2" denotes the fit to the capsize rate curve at KG = 12.2 m)

and

$$(H_S)_{high} = H_S|_{P_f(H_S)=1-\alpha} \tag{26.5}$$

The boundaries $(H_S)_{low}$ and $(H_S)_{high}$ can be calculated using the inverse P_f function, given as:

$$H_S(P_f) = x_0 + dx \cdot \ln\left(\frac{P_f}{1 - P_f}\right) \tag{26.6}$$

Lower and higher limits of the capsize band, given as $H_S(P_f = a)$ and $H_S(P_f = 1 - a)$ are equal to:

$$H_S(P_f = \alpha) = x_0 + dx \cdot \ln\left(\frac{\alpha}{1 - \alpha}\right) \tag{26.7}$$

and

$$H_S(P_f = 1 - \alpha) = x_0 + dx \cdot \ln\left(\frac{\alpha}{1 - \alpha}\right) \tag{26.8}$$

The following figure demonstrates these limits, calculated with the parameter. α = 0.05 (Fig. 26.9).

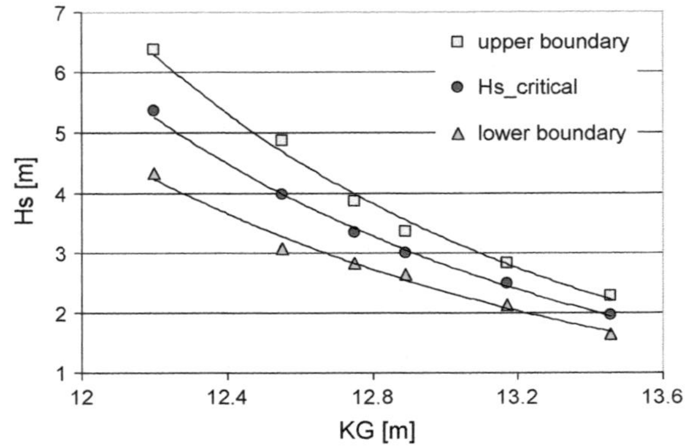

Fig. 26.9 Capsize band versus KG

26.3.4 Parameterisation

Attempts to derive a simple analytical function to represent capsize boundaries and capsize band revealed new possibilities for parameterisation of the formula to populate a family of functions, which could be used as a universal tool for survivability assessment in both design and operational stages. In case of the sigmoid, the two defining parameters, i.e. x_0 and d_x can be expressed by means of wave characteristics (other than H_S, which is explicitly present in the P_f formulae) or parameters related to loading condition, damage extent etc. Understandably, parametric studies require extensive and systematic simulation (testing) effort but some rough examples may be presented here. They may also shed some light on sensitivity problems associated with these studies. A single-variable parameterisation of the sigmoid's x_0 and d_x using KG as a parameter is presented in Figs. 26.10 and 26.11.

Obviously, the family of sigmoids describing the capsize rate should be populated with as many parameters as necessary, including also those specific to the damaged ship, e.g. residual freeboard, water head on a car deck etc. to enhance its function-

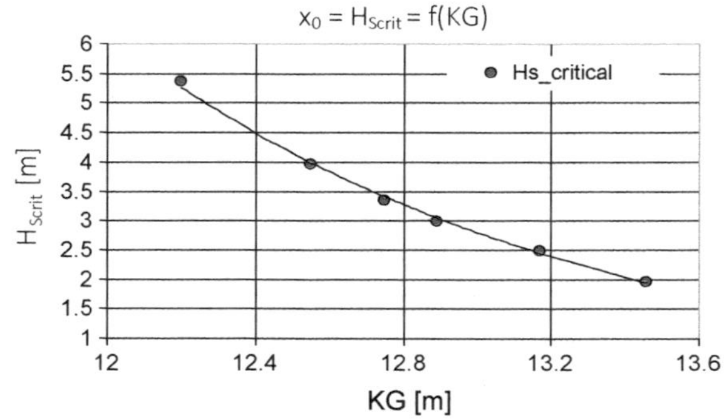

Fig. 26.10 Plot of critical significant wave height (corresponding to parameter x_0) versus KG (intact ship)

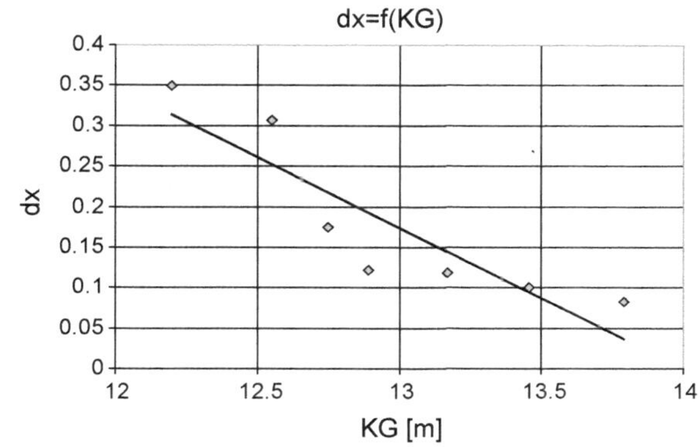

Fig. 26.11 Bandwidth parameter versus KG (intact ship)

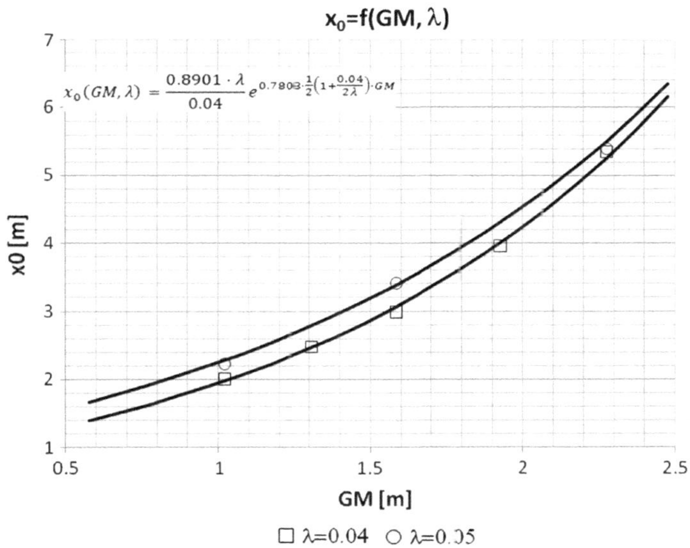

Fig. 26.12 Bi-variate parameterisation of critical significant wave height

ality. For the purpose of this work, the parameters investigated are associated with the intact ship characteristics, leaving aside damage-related quantities, until more research output is available. The following figure shows an example of decomposition of critical significant wave height with respect to (intact) *GM* and wave slope λ (Fig. 26.12).

26.4 Linear Approximation

However convenient the sigmoid regression is to use, it also comprises some significant drawbacks. To start with, something that is particularly evident in cases of very narrow capsize band is that the goodness of fit depends strongly on the quality of data in the proximity of tail asymptotes. Unfortunately, due to limited resolution of experimental data, these regions bear the highest uncertainty (Fig. 26.13).

Assuming that the data in proximity of the critical value, lying in the middle of the range of P_f should be the most reliable, an attempt has been made to simplify the approach and to use linear regression instead of non-linear, with encouraging results. It can be noticed that some cases demonstrate higher goodness of fit for linear regression than for a sigmoid. In order to achieve this, though, the tails of the series needed to be omitted as that is where the non-linear behaviour is dominant. However, it was observed that removing "tails" from the data set has no major impact on the result.

This, as demonstrated in Table 26.4 by comparison of the residual sum of squares for one sample dataset, makes this approach really attractive. A major concern whilst using linear regression is related to the capsize band and its analogy to the confidence interval. It is obvious that relying on statistical measures of goodness of fit may overshadow the fact that linear regression does not bring any information about the "tails" of the capsize rate distribution and therefore any prediction of capsize band based on this method should be approached carefully. However, closer examination of the linear regression and its affiliation with the sigmoid reveals some important virtues. Linear regression of the data close to x_0 will actually result to the tangent of the sigmoid at the inclusion point (x_0, 0.5). Therefore, for the linear regression parameters α (slope) and β (intercept) the following relation holds:

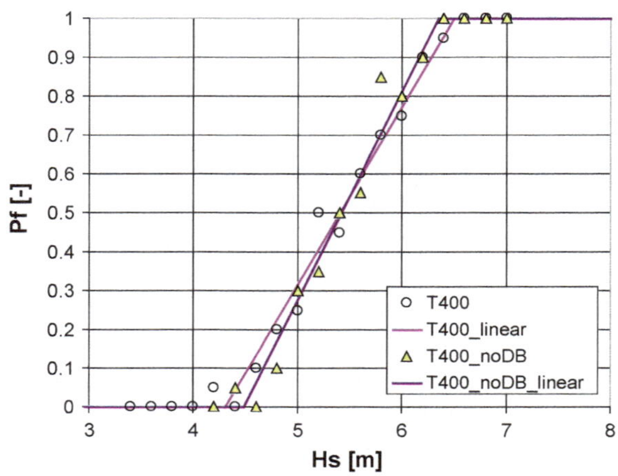

Fig. 26.13 Linear regression for different damage cases (EUGD01-R2)

Table 26.4 Sigmoidal versus linear regression

	Sigmoid	Linear
Number of points	19	13
Degrees of freedom	17	11
Reduced chi-squares	0.00211	
Residual sum of squares	0.03595	0.03247
Adj. R-square	0.98703	0.97626

Table 26.5 Impact of slope estimate on capsize band and Hs critical

Sigmoid value (Pf)	Hs [m]		
	Sigmoid fit	Sigmoid fit with x_0 and dx based on linear regression (estimate 1)	Sigmoid fit with x_0 and dx based on linear regression (estimate 2)
0.05	1.28691	0.99266	1.16301
0.5 (H_{Scrit})	1.68031	1.70087	1.68481
0.95	2.07372	2.40909	2.20661
Bandwidth	0.78681	1.41643	1.0436

$$\alpha = \left.\frac{dP_f}{dH_S}\right|_{H_S=x_0} = \frac{e^{\frac{H_S-x_0}{dx}}}{dx\left(1+e^{\frac{H_S-x_0}{dx}}\right)^2} = \frac{1}{4dx} \tag{26.9}$$

$$y(x_0) = \alpha x_0 + \beta = \frac{1}{2} \tag{26.10}$$

The parameters for the bandwidth and centre of symmetry of the sigmoid function can be derived directly from the linear regression formula:

$$dx = \frac{1}{4\alpha} \tag{26.11}$$

$$x_0 = \frac{0.5-\beta}{\alpha} \tag{26.12}$$

Finally, since all the parameters required for the sigmoid representation can be evaluated on the basis of a linear fit, it is sufficient to apply linear regression to the observations and once x_0 and d_x are estimated, the capsize band limits can be calculated with the use of Eqs. 26.7 and 26.8, respectively. The results presented in Table 26.5 and in Fig. 26.14 below demonstrate use of linear estimates of x_0 and dx to construct the capsize rate curve. The bandwidth parameter denotes the range of sea states containing 90% of capsizes, i.e.

$$(H_S)_{high} - (H_S)_{low} = H_S|_{P_f(H_S)=0.95} - H_S|_{P_f(H_S)=0.05}.$$

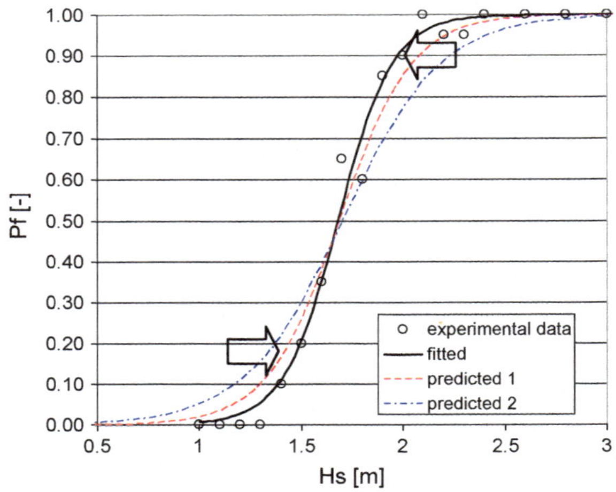

Fig. 26.14 Fit convergence—accurate estimate of slope at x_0 results in closer match

Such approach, based on linear regression, has some rather serious implications. First of all, it allows use of formulae derived for the sigmoid curve, well representing observed phenomena, but without the necessity of non-linear (least-squares) regression. Furthermore, as discussed earlier, experimental results in close proximity to 0 and 1 asymptotes are expected to suffer due to large uncertainties and in general, they require higher resolution. On the contrary, points corresponding to moderate capsize rates are usually following the trend better. An approach based on linear regression makes it possible to disregard those regions entirely or just the parts that might be ambiguous. In the latter case (partial reduction) it is important that the remaining data preserve basic characteristics of the distribution, such as symmetry around x_0. Given that sufficient resolution is available around the x_0 region, the resulting sigmoid function should be very accurate. The benefit of this approach is that one could derive an approximate capsize band, having nothing more than 2 measurements of the capsize rate, as long as they are different than 1 and 0—ideally—and the smaller measurement corresponds to lower H_S. Of course, this should only be treated as an indication and a more accurate calculation of the slope of the probability distribution at it's centre would have to be available for reliable results.

26.5 Conclusions

This paper presents an alternative approach to the representation of the behaviour of a damaged ship in waves. The approach adopted for analytical approximation of the capsize band has both benefits (speed) and drawbacks (uncertainty) but some compromise is not only inevitable but also necessary in most engineering applications—particularly those that are exceptionally labour intensive and costly.

The characteristics of the probability distribution that describes the behaviour of Ro-Ro ships in boundary conditions have been identified and an analytical model describing the capsize band has been developed.

Furthermore, the way to utilise the outcome to predict the critical wave height has been demonstrated. In addition, the capability to facilitate these characteristics in the design process as constrains and/or objectives have been discussed.

Lastly, the merits of having an analytical approach to describe such a complex phenomenon are indisputable. The amount of realisation performed numerically for this work is counted in thousands, so the amount of work saved by such an approach is massive.

References

HARDER (1999–2003). "Harmonization of Rules and Design Rational". Project funded by the European Commission, DG XII-BRITE, 2000–2003.

Jasionowski, A., Vassalos, D., Guarin, L.. "Time-Based Survival Criteria for Passenger Ro-Ro Vessels", 6th International Ship Stability Workshop, Webb Institute, 2002.

Jasionowski, A., Vassalos, D., Scott, A., Ship Vulnerability to flooding, 3rd international conference for maritime safety, Berkeley, California 2007.

Vassalos, D., Turan, O. and Pawlowski, M.: "Dynamic Stability Assessment of Damaged Ships and Proposal of Rational Survival Criteria", Journal of Marine Technology, Vol. 34, No. 4, pp 241–269, October 1997.

Chapter 27
Dependence of Roll and Roll Rate in Nonlinear Ship Motions in Following and Stern Quartering Seas

Vadim L. Belenky and Kenneth M. Weems

Abstract The changing stability of a ship in waves may have a significant influence on the probabilistic properties of roll in irregular following and quartering seas. In particular, nonlinear effects may lead to dependence between roll angles and rates, which will have significant repercussions on the application of the theory of upcrossings for evaluating the probability of a stability failure related to roll motion such as capsizing. The roll response of a ship in a stationary seaway is a stationary stochastic process. For such a process, the roll angle and its first derivative are, by definition, not correlated and are often assumed to be independent. However, this independence can only be assumed a priori for normal processes, and the nonlinearity of large-amplitude roll motions can lead to a deviation from normal distribution. In the present work, the independence of roll angles and rates is studied from the results of numerical simulations from the Large Amplitude Motion Program (LAMP), which includes a general body-nonlinear calculation of the Froude-Krylov and hydrostatic restoring forces. These simulations show that, for the considered case, roll and roll rate are independent in beam seas, even though the distribution of the roll response is not normal. However, roll angles and roll rates for stern quartering seas are not independent.

Keywords Independence · Correlation · Roll motions · Stern quartering seas

27.1 Introduction

The problem of dependence between roll and roll rate originally appeared during an attempt to extend a closed-form solution for capsizing probability developed from a 1-DOF (degree of freedom) roll model with an advanced numerical simulation code (Belenky et al. 2016). While further development of the numerical split-time method did not actually require any modeling of joint distribution, understanding the

V. L. Belenky (✉) · K. M. Weems
David Taylor Model Basin (NSWCCD), West Bethesda, MD, USA
e-mail: vadim.belenky@navy.mil

V. L. Belenky et al. (eds.), *Contemporary Ideas on Ship Stability*, Fluid Mechanics and Its Applications 119, https://doi.org/10.1007/978-3-030-00516-0_27

probabilistic relationship between roll angles and roll rates may be important in its own right. The independence between displacement and velocity of ship motion is a standard assumption in frequency domain seakeeping and may also be relevant for the time domain.

When the stochastic processes of roll and roll rate are independent, the following formula is correct:

$$f\left(\phi,\dot{\phi}\right)=f\left(\dot{\phi}\right)\cdot f(\phi) \tag{27.1}$$

where ϕ is the roll angle, a dot above a symbol means a temporal derivative, and $f()$ is a probability density function (PDF). Correlation is one measure of dependence. It is defined through the joint second moment, which is known as covariance (cov), or the correlation coefficient *r:*

$$\operatorname{cov}(\phi,\dot{\phi})=\int_{-\infty}^{\infty}\int_{\infty}^{\infty}\left(\phi-E_{\phi}\right)\dot{\phi}\,f\left(\phi,\dot{\phi}\right)d\phi\,d\dot{\phi}=E\left(\left(\phi(t)-E_{\phi}\right)\dot{\phi}(t)\right) \tag{27.2}$$

$$r_{\phi,\dot{\phi}}=\frac{\operatorname{cov}(\phi,\dot{\phi})}{\sqrt{V_{\phi}\cdot V_{\dot{\phi}}}}, \tag{27.3}$$

where V_{ϕ} and $V_{\dot{\phi}}$ are the variances of roll and roll rate, respectively, and E_{ϕ} is the mean value of the roll angle, while $E(..)$ is an averaging operator. After substitution of Eq. (27.1) into Eq. (27.2), both the covariance and correlation coefficient equal zero if the processes are independent.

In general, a stationary process is not correlated with its derivative. Consider the auto-covariance function, R_{ϕ} (covariance of the value of the process at a given instant of time t with the value of the same process after some time lag τ):

$$R_{\phi}(\tau)=E\left(\left(\phi(t)-E_{\phi}\right)\left(\phi(t+\tau)-E_{\phi}\right)\right) \tag{27.4}$$

The derivative of the auto-covariance function with respect to the time lag is the covariance between the process and its derivative value after time lag τ:

$$\frac{dR_{\phi}(\tau)}{d\tau}=E\left(\left(\phi(t)-E_{\phi}\right)\left(\dot{\phi}(t+\tau)\right)\right)=\operatorname{cov}(\phi(t),\dot{\phi}(t+\tau)) \tag{27.5}$$

The maximum of the auto-covariance is at the time lag $\tau=0$: $R_{\phi}(0)=V_{\phi}$. Thus, its derivative must be zero, making roll and roll rates uncorrelated processes:

$$\frac{dR_{\phi}(\tau=0)}{d\tau}=0\quad\Rightarrow\quad\operatorname{cov}(\phi,\dot{\phi})=0 \tag{27.6}$$

Absence of correlation does not necessarily mean independence. However, if two random variables or stochastic processes have normal distributions and are uncorrelated, they are independent. Consider the two-dimensional normal distribution:

$$f(\phi, \dot{\phi}) = \frac{1}{2\pi\sqrt{V_\phi V_{\dot{\phi}}}\sqrt{1 - r^2_{\phi,\dot{\phi}}}} \exp\left(-\frac{1}{2(1 - r^2_{\phi,\dot{\phi}})}\left(\frac{(\phi - E_\circ)^2}{V_\phi} + \frac{\dot{\phi}^2}{V_{\dot{\phi}}} - \frac{2r_{\phi,\dot{\phi}}(\phi - E_\phi)\cdot\dot{\phi})}{\sqrt{V_\phi V_{\dot{\phi}}}}\right)\right) \tag{27.7}$$

If the correlation coefficient, r, equals zero, the joint distribution in Eq. (27.7) becomes simply the product of two Gaussian distributions, which indicates independence according to Eq. (27.1).

27.2 Numerical Study

27.2.1 Numerical Simulations

The numerical simulations for the present study consist of three series of calculations performed with Office of Naval Research (ONR) Topside series Tumblehome configuration (Bishop et al. 2005): zero speed with a wave heading angle of 90° (beam seas), 15 knots with a wave heading of 45° (stern quartering seas), and 14 knots with a wave heading of 40°. Long-crested irregular seas were generated from a Bretschneider spectrum with a significant wave height 7.5 m and a modal period of 15 s. The waves were discretized with 790 frequency components, which provide a wave representation free of self-repeating effect for 40 min. Each set consisted of about 200 records corresponding to independent realizations of the seaway. The ship motion is calculated by LAMP's blended body-nonlinear formulation in which the hydrostatic and Froude-Krylov forces were computed on the instantaneous wetted hull surface, while the wave-body disturbance forces, which include forward speed, diffraction and radiation effects, were computed by a 3-D potential-flow panel method over the mean wetted surface (Lin and Yu 1990; Shin et al. 2003). The ship was free to move in heave, roll, and pitch but constrained in surge (constant speed), sway, and yaw. An empirical model was applied for roll damping.

In addition to the ship-motion data, this irregular wave representation was evaluated by the incident wave elevation and its derivatives at a fixed or constant speed point in order to provide a corresponding data set that is known to be Gaussian. These wave elevation data are employed as a test sample with known distribution.

27.2.2 Estimation of Correlation

While the result of the estimation of the correlation between the process and its derivative is expected to be zero, the actual objective here is to try a method of assessing the statistical uncertainty on a problem with a known answer.

As the formulation seems to be rather standard, a solution may not be found in the existing literature, e.g. Priestley (1981). Existing methods for uncertainty of the variance and covariance estimates employ an assumption of normality of the underlying process, which is not relevant for the problem at hand.

Calculation of the estimate of the correlation coefficient is straightforward:

$$\hat{r}_{\phi,\dot{\phi}} = \frac{c\hat{o}v(\phi,\dot{\phi})}{\sqrt{\hat{V}_{\phi}\cdot\hat{V}_{\dot{\phi}}}} \tag{27.8}$$

The "hat" symbol above a notation indicates an estimate of the quantity. The absence of the "hat" symbol implies a theoretical or "true" value, which is normally not known. The exception is the mean value of the derivative, which is known to be zero for a stationary process.

The estimate from Eq. (27.8) is a random number, so a confidence interval is required to assess the statistical uncertainty of this estimate resulting from the finite size of the available sample. The confidence interval contains the "true" value with the given confidence probability, β. Most engineering applications accept $\beta = 0.95$. The derivation of the formula for the confidence interval for the correlation coefficient of two generic stationary processes is given in the Appendix.

Calculation of the confidence interval of the correlation coefficient between roll and roll rates is simpler than in the case of two generic processes as the mean value of roll rate is known exactly: $E_{\dot{\phi}} = 0$. As a result, all the terms containing the variance or covariance of the mean value of roll rate are identically zero, and the covariance matrix for the estimate of the correlation coefficient is only 4×4. Numerical results are shown in Fig. 27.1. Assuming normal distribution for the estimates of the correlation coefficients, boundaries of the confidence interval are computed as:

$$\hat{r}_{up,low} = \hat{r} \pm Q_{\beta}\left(\frac{1+\beta}{2}\right)\sqrt{\mathrm{var}(\hat{r})}$$

where Q_{β} is a quantile function for normal distribution; $Q_{\beta} = 1.96$ for $\beta = 0.95$. All of the confidence intervals contain zero, so the statistical technique is not rejected by the theory

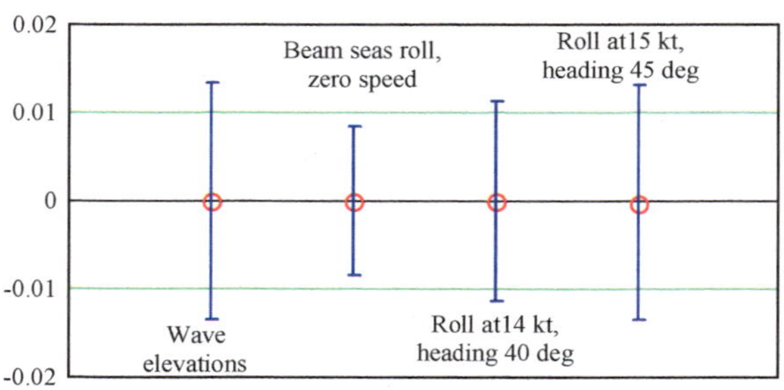

Fig. 27.1 Estimates of correlation coefficients between the processes and derivatives

27.2.3 Correlation of Squares

Correlation coefficient reflects the linear aspect of the dependence. For any signs of dependence for processes with zero correlation, an estimation of the correlation coefficient of squares of the processes and their respective derivatives is considered. Wave elevation is again considered as a control sample. If two processes (wave elevation and its temporal derivative) are independent, their squares must be independent as well, as squaring of a value cannot create a dependence with another value.

$$\hat{r}_{\phi^2,\dot{\phi}^2} = \frac{c\hat{o}v(\phi^2, \dot{\phi}^2)}{\sqrt{\hat{V}_{\phi^2} \cdot \hat{V}_{\dot{\phi}^2}}} \tag{27.9}$$

The technique described in the Appendix computes the confidence interval for Eq. (27.9). Figure 27.2 shows the numerical results on correlation of processes squares. The correlation coefficient of squares of wave elevations and their derivatives still show zero with very good accuracy. The correlation coefficient for roll in beam seas is essentially zero as its confidence interval contains zero pretty close to the estimate. However, cases at both 40 and 45 degrees heading do not have zero in their confidence interval, indicating dependence without correlation for roll and roll rate in stern quartering seas.

27.2.4 Modeling of the Joint Distribution

In order to further visualize and study the influence of dependence on the upcrossing rate, the moving average method is applied. The moving average allows smoothing of the histogram and possesses sufficient flexibility to represent the unusual features of the PDF of a significantly nonlinear response.

The calculation starts with estimates of the conditional variances of roll angle for a series of values of roll rate. A series of conditional histograms for roll are then computed by Scott's (1979) formula for the width of a bin of the histogram:

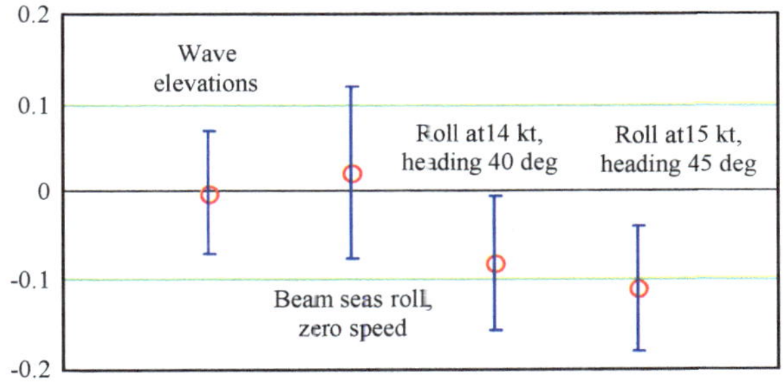

Fig. 27.2 Estimates of correlation coefficient between the squares of processes and squares of derivatives

$$W(\dot{\phi}) = \frac{3.5\sqrt{V_{\phi}(\dot{\phi})}}{\sqrt[3]{N(\dot{\phi})}} \tag{27.10}$$

where $V_{\phi}(\dot{\phi})$ is the conditional variance of roll estimated for a particular value of roll rate and $N(\dot{\phi})$ is the total number of points from the conditional histogram for this value of roll rate. The moving average is then applied to each of the conditional histograms:

$$H_m = \frac{1}{k}\sum_{n=1}^{k} H0_{n+m} \tag{27.11}$$

where $H0$ is an original histogram and k is a number of neighboring bins for averaging. The averaged points H are used with linear interpolation to create a continuous piecewise linear function, which is then corrected to meet the normalization condition as well as equality to the conditional variance.

Figure 27.3 shows examples of these conditional distributions for the incident wave data set ζ and illustrates how the method is capable of recovering the shape of the normal distribution.

Once the conditional distributions are complete, the procedure is applied once more for the marginal distributions of the derivatives. The joint distribution can then be approximated as a product of two piecewise linear functions (An averaged 2D histogram may be applied.):

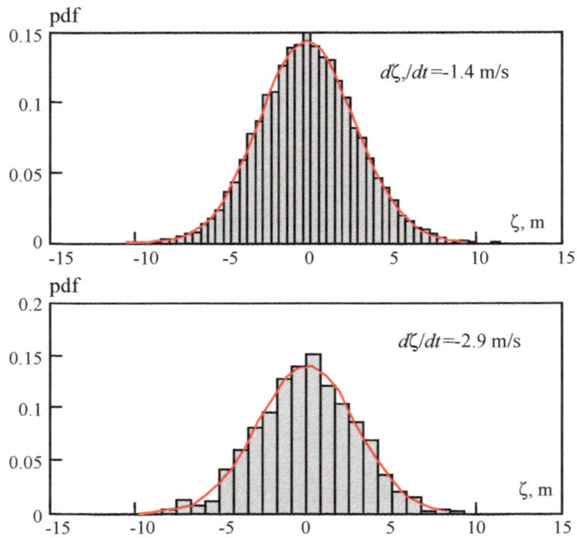

Fig. 27.3 Conditional distribution of wave elevations for two particular values of the time derivative

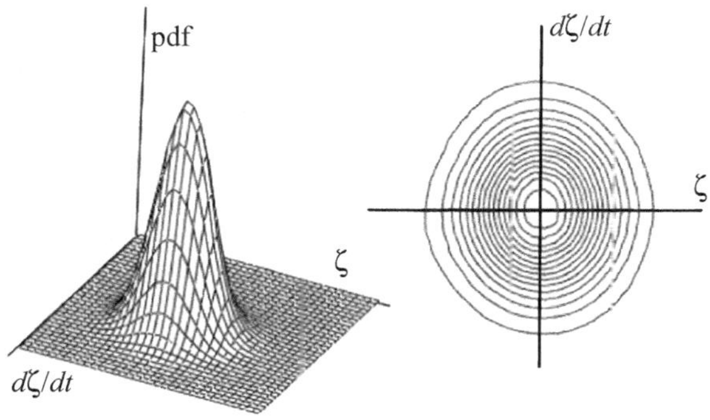

Fig. 27.4 Joint PDF of wave elevations and their temporal derivatives and horizontal section

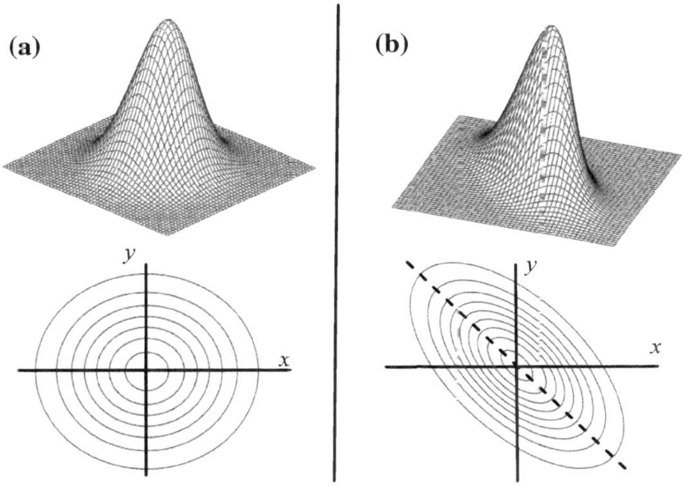

Fig. 27.5 Theoretical gaussian joint distributions of two random variables **a** without correlation and **b** with correlation

$$f(\phi, \dot{\phi}) = f(\phi|\dot{\phi}) \cdot f(\dot{\phi}) \quad (27.12)$$

The resultant joint distribution and its horizontal sections are shown in Fig. 27.4. Visually, both shapes appear to be smooth and look similar to the theoretical Gaussian distribution shown in Fig. 27.5a.

27.2.5 Visual Inspection

Correlation between two values can be seen in the joint distribution of the values as a "rotation of axis," as shown in Fig. 27.5, which plots theoretical Gaussian distributions with and without correlation.

This "rotation of the axes" can be most clearly seen in the lower plot of a series of horizontal cross-sections of the joint distributions. The correlation leads to asymmetry relative to the horizonal axis. Figures 27.6, 27.7, and 27.8 contain similar plots for the roll angles and rates from the LAMP simulations. These joint distributions were smoothed with the moving average method described earlier.

All three plots retain symmetry relative to the horizontal axis, taking into account that the plots are based on moving average method. Nevertheless, the horizontal sections for the stern quartering seas appear to be "squeezed," resulting in a somewhat rhomboidal pattern. This may be a visual sign of dependence without correlation.

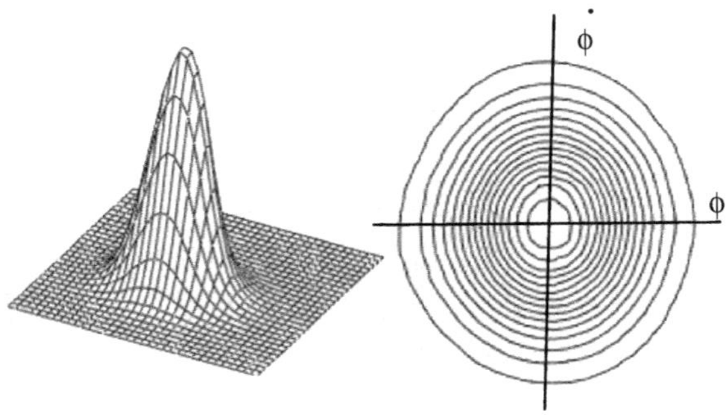

Fig. 27.6 Joint distribution for beam seas (heading 90°) at zero speed

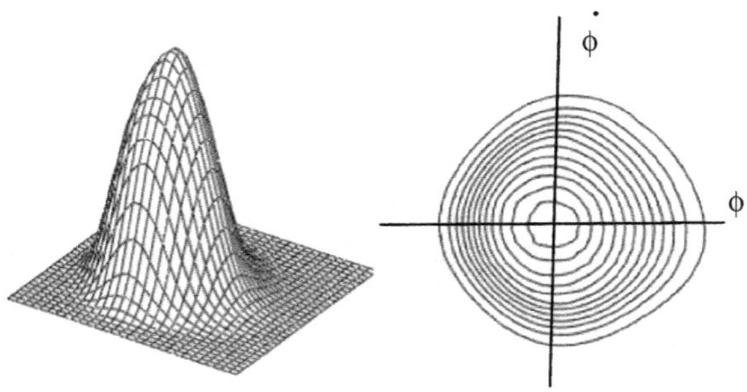

Fig. 27.7 Joint distribution of roll angles and rates for heading 45° at 15 Knots

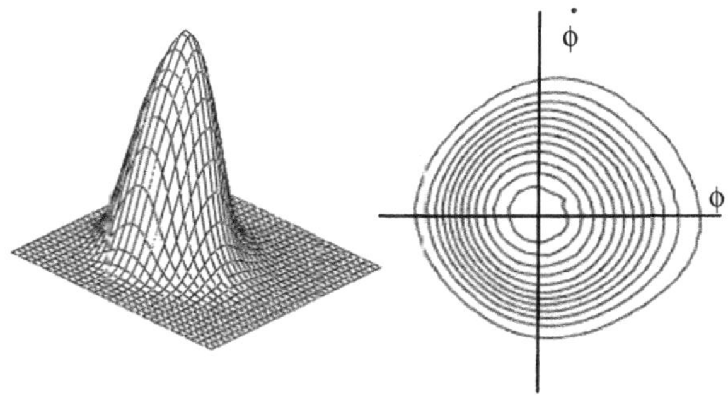

Fig. 27.8 Joint distribution of roll angles and rates for heading 40° at 14 Knots

27.2.6 Application to Upcrossing Rate

A rate of upcrossing may be sensitive to the dependence between the process and its derivative. The upcrossing through a level ϕ_{m0} is an event of reaching this level with positive derivative at this instant. The rate of upcrossing, ξ, expresses the probability of this crossing event per unit of time and can be computed as (e.g. Cramer and Leadbetter 1967):

$$\xi = \int_0^\infty \dot{\phi} \cdot f(\phi = \phi_{m0}, \dot{\phi})\, d\dot{\phi} \tag{27.13}$$

If the process and its derivative are assumed independent, Eq. (27.13) becomes:

$$\xi = f(\phi = \phi_{m0}) \int_0^\infty \dot{\phi} \cdot f(\dot{\phi})\, d\dot{\phi} \tag{27.14}$$

If the processes of roll and roll rates are assumed to be normal, the integral in Eq. (27.14) can be expressed in a closed-form:

$$\xi = \frac{1}{2\pi}\sqrt{\frac{V_{\dot{\phi}}}{V_{\phi}}} \exp\left(-\frac{(\phi_{m0} - E_{\phi})^2}{V_{\phi}}\right) \tag{27.15}$$

Approximate joint distributions (Figs. 27.4, 27.6, 27.7 and 27.8) modeled with smoothed histogram, Eqs. (27.11) and (27.12) and Fig. 27.3 can compute an approximate value of the upcrossing rate with Eq. (27.14) and without Eq. (27.13) assuming independence of the process and its derivative. The difference between the two quantities is an indicator of the influence of the dependence. Additionally, the upcrossing rate can be estimated from the time series by counting instances of crossing a level ϕ_{m0} with positive derivative. The estimate of the upcrossing rate is expressed as:

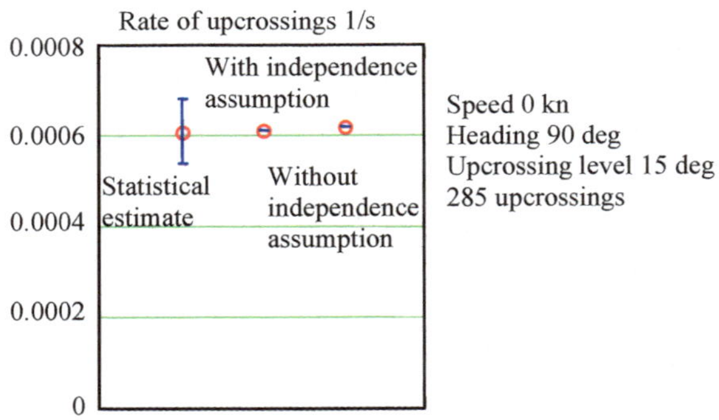

Fig. 27.9 Upcrossing statistics of roll angles for beam seas case

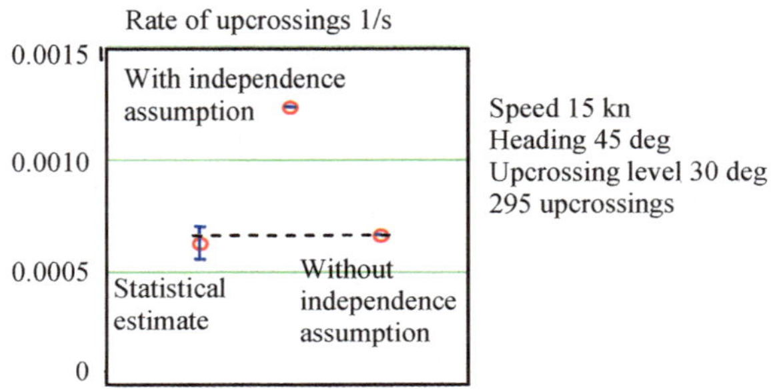

Fig. 27.10 Upcrossing statistics of roll angles for stern quartering seas case, heading 45°

$$\hat{\xi} = \frac{N_U}{N \cdot N_R \cdot \Delta t} \tag{27.16}$$

where N_U is number of observed upcrossing events, N_R is number of records in the ensemble, N is a number of data points in each record, and Δt is the time increment between data points.

The estimate of the upcrossing rate has a binomial distribution; each time instant either produces an upcrossing event or not. However, the direct application of binomial distribution for calculation of the confidence interval for large samples encounters numerical difficulties in computing the quantile of binomial distribution (Belenky et al. 2016) . Here, the binomial distribution of the upcrossing rate estimate is approximated with a normal distribution with the following formulae for the estimated mean and variance:

$$\hat{E}(\hat{\xi}) = \hat{\xi};\ v\hat{a}r\left(\hat{\xi}\right) = \frac{\hat{\xi} \cdot (1 - \hat{\xi})}{N \cdot N_R \cdot \Delta t} \tag{27.17}$$

Upcrossing statistics were evaluated for the three ship motion data sets: beam seas, zero speed (Fig. 27.9); 45° heading at 15 knots (Fig. 27.10); and 40° heading at 14 knots (Fig. 27.11). An upcrossing level was chosen in order to keep the observed number of upcrossings around 300.

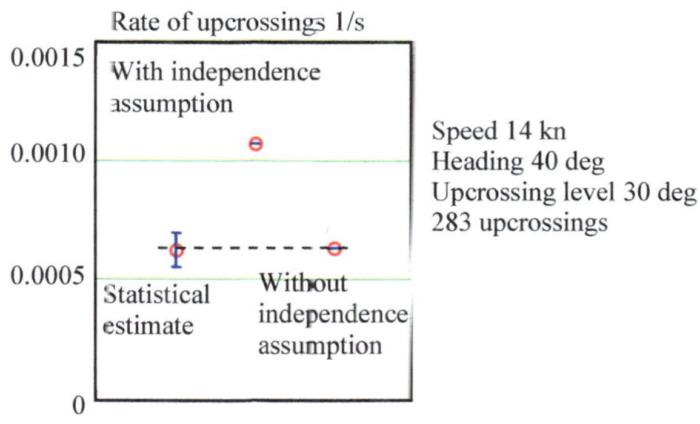

Fig. 27.11 Upcrossing statistics of roll angles for stern quartering seas case, heading 40°

Strictly speaking, both of the calculated upcrossing rates with Eqs. (27.13) and (27.14) are also random numbers and should be presented with their own confidence interval, as they are computed with an approximate distribution based on histogram, and are therefore also subject to statistical uncertainty. However, in the absence of an established technique for computing the confidence interval for these quantities, the confidence interval of the estimate from Eq. (27.16) can be used as an indicator of the influence of the statistical uncertainty.

Figures 27.9, 27.10 and 27.11 clearly show that the independence assumption may be applicable in beam seas and but not in the two stern quartering sea cases. The upcrossing rate calculated with the independence assumption in stern quartering seas overestimates the upcrossing rate significantly. This is a consequence of the dependence of roll and roll rate in stern quartering seas.

The manifestation of dependence between roll and roll rate seems to be more dramatic in Figs. 27.10 and 27.11 as compared to Fig. 27.2 The estimate of correlation between squares is computed over the entire sample, while the computation of the rate of upcrossing is computed further in the tail of the distribution, where non-Gaussian properties seem to be stronger in stern quartering seas.

27.3 Summary and Conclusuons

This chapter describes a study of the probabilistic properties of nonlinear roll motions in irregular ocean waves generated by a time-domain potential flow simulation code with a body-nonlinear formulation for Froude-Krylov and hydrostatic forces, while diffraction and radiation was computed up to a mean waterplane. Roll damping was modeled with externally computed coefficients. Numerical simulation was carried out with three degrees of freedom—heave, roll and pitch—at zero and constant forward speed in long-crested irregular beam and quartering seas. The focus of the study was on the dependence between roll and roll rates.

Four large data sets were employed for the study: beam seas at zero speed, two cases of stern quartering seas (heading of 40° and 45°) at moderate forward speed and wave elevations at a point, the latter of which provides a control data set that is known to be Gaussian. Each data set consisted of multiple records of the process itself and its temporal derivatives. The dependence was studied with three different approaches:

- Estimate of coefficient of correlation between roll and roll rate for all the cases were found to be zero (as theory suggests). Estimate of coefficient of correlation between squared values of roll and roll rate was found to be zero for Gaussian process (as theory suggests) and for roll in beam seas; but non-zero for both stern quartering cases. Thus, dependence between roll and roll rate exists for stern quartering case.
- Visual inspection of joint distribution models based on smoothed histograms. All cases were found to be symmetric relative to the horizontal axis (as theory suggests). Rotational invariance was observed for Gaussian distribution (as theory suggest) and for roll in the beam seas. A "squeezed" shape was observed for stern quartering cases. This "squeezed" (i.e. without rotational invariance) but symmetrical shape may be how the dependence without correlation manifests itself visually.
- Upcrossing rate was computed by joint distribution modeled with smoothed histograms complemented with statistical estimation of the upcrossing rate. Calculation was carried out with and without the assumption of independence between the process and its derivative. Both calculated rates were found to be statistically identical to the observed rate for the Gaussian case (as theory suggest) and for the beam seas case. For the stern quartering cases, only the calculation without the assumption of independence matched the statistical estimate. Thus, the dependence between roll and roll rate is important for stern quartering case.

As theory suggest, no correlation was found between roll angles and roll rates. However, the absence of correlation between roll angles and roll rates does not mean that they are independent for all cases. The independence of roll angles and rates is justified for a linear system where a normal distribution is applicable. It is a mere assumption for all other cases; in the case studied for beam seas this assumption seems to be applicable, while in stern quartering seas it seems to be not applicable.

At the time when this chapter was written, no clear explanation for the observed dependence was available. However, a hypothesis may be suggested that as the dependence between roll angles and roll rates was observed in stern quartering seas on the ONR tumblehome topside configuration, it may be related to the stability variation in waves.

Acknowledgements The work described in this paper has been funded by the Office of Naval Research (ONR) under Dr. Patrick Purtell in 2011. The authors are grateful and happy to recognize contributions from their colleagues: Prof. Matthew Collette (formerly of SAIC, currently University of Michigan) discovered problems predicting the upcrossing rate with independence assumption; Prof. Ross Leadbetter (University of North Carolina) pointed out the possibility of dependence without correlation and hinted towards use of squares; and Prof. Pol Spanos (Rice University) motivated the authors to pursue this issue and has contributed to many fruitful discussions. Re-analysis of the results of simulation was carried out in 2017 and funded by NSWCCD Independent Applied Research (IAR) program under Dr. Jack Price. The authors are grateful to Prof. Vladas Pipiras (University of North Carolina) for fruitful discussions and, in particular, for help with confidence interval of the estimate of correlation coefficient. Finally, the authors are grateful Dr. Joel Park (NSWCCD) who made many useful editorial comments.

Appendix: Confidence Interval of Correlation Coefficient

Distribution of the Estimate

A confidence interval is considered for the estimate of the correlation coefficient between two generic stationary ergodic processes X and Y without making any assumptions about their distribution. The only assumption made for the joint distribution of the processes is that their covariance and variances do exist. The data sample presented in form of two sets (ensembles) of N_R records, each of which contains N data points:

$$X = \{x_{n,k}\}; \quad Y = \{y_{n,k}\}; \quad n = 1, \ldots, N; \quad k = 1, \ldots, N_R \tag{27.18}$$

k is a counter for records, while n is a counter for points inside a record. The estimate of the correlation coefficient is expressed as:

$$\hat{r}_{XY} = \frac{\hat{cov}(X, Y)}{\sqrt{\hat{V}_X \cdot \hat{V}_Y}}, \tag{27.19}$$

where:

$$\hat{cov}(X, Y) = \frac{1}{N_R \cdot N} \sum_{k=1}^{N_R} \sum_{n=1}^{N} (x_{n,k} - \hat{E}_X)(y_{n,k} - \hat{E}_Y) \tag{27.20}$$

$$\hat{V}_X = \frac{1}{N_R \cdot N} \sum_{k=1}^{N_R} \sum_{n=1}^{N} (x_{n,k} - \hat{E}_X)^2, \; \hat{V}_Y = \frac{1}{N_R \cdot N} \sum_{k=1}^{N_R} \sum_{n=1}^{N} (y_{n,k} - \hat{E}_Y)^2 \tag{27.21}$$

$$\hat{E}_X = \frac{1}{N_R \cdot N} \sum_{k=1}^{N_R} \sum_{n=1}^{N} x_{n,k}; \; \hat{E}_Y = \frac{1}{N_R \cdot N} \sum_{k=1}^{N_R} \sum_{n=1}^{N} y_{n,k} \tag{27.22}$$

As the size of the sample is rather large, the bias in the covariance and variance estimate is expected to be small.

As with any random number, the estimate of Eq. (27.19) is characterized by a probability distribution. In order to justify the choice of the distribution of the estimate of correlation coefficient, consider it as a deterministic function of random arguments. Both covariance and variance estimates are dependent on the mean value estimates, so it is convenient to use known properties of covariance and variance to present the estimate of Eq. (27.19) as:

$$\hat{r}_{X,Y} = \frac{\hat{E}_{X\cdot Y} - \hat{E}_X \hat{E}_Y}{\sqrt{(\hat{E}_{X^2} - \hat{E}_X^2)(\hat{E}_{Y^2} - \hat{E}_Y^2)}}, \tag{27.23}$$

where:

$$\hat{E}_{X\cdot Y} = \frac{1}{N_R \cdot N} \sum_{k=1}^{N_R} \sum_{n=1}^{N} x_{n,k} y_{n,k} \tag{27.24}$$

$$\hat{E}_{X^2} = \frac{1}{N_R \cdot N} \sum_{k=1}^{N_R} \sum_{n=1}^{N} x_{n,k}^2; \ \hat{E}_{Y^2} = \frac{1}{N_R \cdot N} \sum_{k=1}^{N_R} \sum_{n=1}^{N} y_{n,k}^2 \tag{27.25}$$

The deterministic function, Eq. (27.23), has 5 random arguments. The distribution of the mean value estimates from Eqs. (27.22) and (27.24) can be taken normal based on the Central Limit Theorem. The distribution of the means of squares of Eq. (27.25) can be approximated by normal for sufficiently large sample (similar argument in Levine et al. 2017).

Equation (27.23) is a nonlinear function; deriving its exact distribution may be difficult even for normally distributed arguments. Fortunately, it is not necessary as its linear approximation with Taylor series has a correct asymptotic behavior. The first-order Taylor expansion works well for small deviation from the initial point. If Eq. (27.23) is expanded in the vicinity of the "true" values of its random arguments, it will tend to the "true" value of the function with increase of the sample size. Thus, Taylor expansion can be applied for the considered problem as the sample size here is not small. This technique is known in statistical literature as "delta-method," see e.g. Bickel and Doksum (2001).

To avoid bulky equations, the following vector notation is applied for the true values and their estimates:

$$\vec{Z} = \left(E_{X\cdot Y} \ E_{X^2} \ E_{Y^2} \ E_X \ E_Y \right)^T \tag{27.26}$$

$$\widehat{\vec{Z}} = \left(\hat{E}_{X\cdot Y} \ \hat{E}_{X^2} \ \hat{E}_{Y^2} \ \hat{E}_X \ \hat{E}_Y \right)^T \tag{27.27}$$

where T means transposition operation. The first-order Taylor series expansion is expressed as:

$$\hat{r}_{X,Y} = g(\widehat{\vec{Z}}) \approx g(\vec{Z}) + \vec{D} \cdot \left(\vec{Z} - \widehat{\vec{Z}} \right) \tag{27.28}$$

where $\vec{D}$ is the vector of derivatives:

$$\vec{D} = \frac{1}{\sqrt{(E_{X^2} - E_X^2)(E_{Y^2} - E_Y^2)}} \cdot \left(1, -\frac{E_{X\cdot Y} - E_X E_Y}{2(E_{X^2} - E_X^2)}, -\frac{E_{X\cdot Y} - E_X E_Y}{2(E_{Y^2} - E_Y^2)}, \frac{E_{X\cdot Y} E_X - E_{X^2} E_Y}{E_{X^2} - E_X^2}, \frac{E_{X\cdot Y} E_Y - E_Y^2 E_X}{E_{Y^2} - E_Y^2}\right)^T \quad (27.29)$$

As Eq. (27.28) is linear and normal distribution is assumed for all the arguments, the function g and the estimate $\hat{r}_{XY}$ have normal distribution. The boundaries of the confidence interval are computed as:

$$\hat{r}_{XY\,up,low} = E(\hat{r}_{XY}) \pm K_\beta \sqrt{\mathrm{var}(\hat{r}_{XY})} \quad (27.30)$$

where $E(\hat{r}_{X,Y})$ is a mean value of the estimate, $\mathrm{var}(\hat{r}_{X,Y})$ is its variance and K_β is non-dimensional half-length of the confidence interval, computed as $0.5(1+\beta)$ quantile of the standard normal distribution ($K_\beta = 1.96$ for $\beta = 0.95$). The calculation of mean and variance of the estimate is considered in the next section of the Appendix.

Mean and Variance of the Correlation Coefficient Estimate

Normal distribution has two parameters: mean value and variance. Both need to be determined to define the distribution. To find the mean value, apply the average operator $E(..)$ to both sides of Eq. (27.28). As the elements of the vector of true value are not known, substituting it with the vector of estimates is the usual practice:

$$E\left(\hat{r}_{X,Y}\right) \approx E\left(g(\vec{Z}) + \vec{D} \cdot \left(\vec{Z} - \widehat{\vec{Z}}\right)\right) = g(\vec{Z}) = r_{X,Y} \approx \hat{r}_{X,Y} \quad (27.31)$$

A similar approach is employed for variance, applying the variance operator var(..) to both sides of Eq. (27.28), substituting true values with their estimates, and using well-known formula for a variance of linear combination of random variable, taking into account that these random variables may be dependent:

$$\mathrm{var}\left(\hat{r}_{X,Y}\right) \approx \mathrm{var}\left(g(\vec{Z}) + \vec{D} \cdot \left(\vec{Z} - \widehat{\vec{Z}}\right)\right) = \sum_{m=1}^{5} \sum_{p=1}^{5} \hat{c}_{m,p} q_{m,p} \quad (27.32)$$

where $\hat{c}_{m,p}$ are the values of the matrix of estimates of coefficients and $q_{m,p}$ are the elements of covariance matrix of the vector $\widehat{\vec{Z}}$:

$$\hat{\mathbf{C}} = \left\{\hat{c}_{m,p}\right\} = \widehat{\vec{D}} \cdot \widehat{\vec{D}}^T; \quad m = 1, \ldots, 5; \quad p = 1, \ldots, 5 \quad (27.33)$$

$$\mathbf{Q} = \left\{q_{m,p}\right\} = \mathrm{cov}(\widehat{\vec{Z}}); \quad m = 1, \ldots, 5; \quad p = 1, \ldots, 5 \quad (27.34)$$

The elements of the covariance matrix $\mathbf{Q}$ are essentially covariance values between different mean estimates, i.e. covariance between two sums. The explicit expression for the covariance matrix $\mathbf{Q}$ is as follows:

$$\mathbf{Q} = \begin{pmatrix} \mathrm{var}\left(\hat{E}_{X\cdot Y}\right) & \mathrm{cov}\left(\hat{E}_{X\cdot Y}, \hat{E}_{X^2}\right) & \mathrm{cov}\left(\hat{E}_{X\cdot Y}, \hat{E}_{Y^2}\right) & \mathrm{cov}\left(\hat{E}_{X\cdot Y}, \hat{E}_{X}\right) & \mathrm{cov}\left(\hat{E}_{X\cdot Y}, \hat{E}_{Y}\right) \\ \mathrm{cov}\left(\hat{E}_{X\cdot Y}, \hat{E}_{X^2}\right) & \mathrm{var}\left(\hat{E}_{X^2}\right) & \mathrm{cov}\left(\hat{E}_{X^2}, \hat{E}_{Y^2}\right) & \mathrm{cov}\left(\hat{E}_{X^2}, \hat{E}_{X}\right) & \mathrm{cov}\left(\hat{E}_{X^2}, \hat{E}_{Y}\right) \\ \mathrm{cov}\left(\hat{E}_{X\cdot Y}, \hat{E}_{Y^2}\right) & \mathrm{cov}\left(\hat{E}_{X^2}, \hat{E}_{Y^2}\right) & \mathrm{var}\left(\hat{E}_{Y^2}\right) & \mathrm{cov}\left(\hat{E}_{Y^2}, \hat{E}_{X}\right) & \mathrm{cov}\left(\hat{E}_{Y^2}, \hat{E}_{Y}\right) \\ \mathrm{cov}\left(\hat{E}_{X\cdot Y}, \hat{E}_{X}\right) & \mathrm{cov}\left(\hat{E}_{X^2}, \hat{E}_{X}\right) & \mathrm{cov}\left(\hat{E}_{Y^2}, \hat{E}_{X}\right) & \mathrm{var}\left(\hat{E}_{X}\right) & \mathrm{cov}\left(\hat{E}_{X}, \hat{E}_{Y}\right) \\ \mathrm{cov}\left(\hat{E}_{X\cdot Y}, \hat{E}_{Y}\right) & \mathrm{cov}\left(\hat{E}_{X^2}, \hat{E}_{Y}\right) & \mathrm{cov}\left(\hat{E}_{Y^2}, \hat{E}_{Y}\right) & \mathrm{cov}\left(\hat{E}_{X}, \hat{E}_{Y}\right) & \mathrm{var}\left(\hat{E}_{Y}\right) \end{pmatrix} \quad (27.35)$$

As these covariance (and variance) values usually are not known, they are substituted by their respective estimates. The calculation of these estimates is considered in the next section of the Appendix with $\mathrm{cov}\left(\hat{E}_X, \hat{E}_Y\right)$ as an example.

A3. Covariance of Two Mean Value Estimates

Consider a covariance $\mathrm{cov}\left(\hat{E}_X, \hat{E}_Y\right)$ as an example. As covariance is essentially an integral Eq. (27.2), integration and summation can be swapped in order:

$$\mathrm{cov}\left(\hat{E}_X, \hat{E}_Y\right) = \frac{1}{N_R}\sum_{k=1}^{N_R}\frac{1}{N^2}\sum_{n=1}^{N}\sum_{m=1}^{N}\mathrm{cov}(x_{n,k}, y_{m,k}) \quad (27.36)$$

As both processes are ergodic, the covariance is the same for all the records, and Eq. (27.36) turns into a sum of elements of a matrix:

$$\mathrm{cov}\left(\hat{E}_X, \hat{E}_Y\right) = \sum \begin{pmatrix} R_{XY}(\tau_0) & R_{XY}(\tau_1) & \ldots & R_{XY}(\tau_{N-2}) & R_{XY}(\tau_{N-1}) \\ R_{XY}(\tau_{-1}) & R_{XY}(\tau_0) & \ldots & R_{XY}(\tau_{N-3}) & R_{XY}(\tau_{N-2}) \\ \ldots & \ldots & \ldots & \ldots & .. \\ R_{XY}\left(\tau_{-(N-2)}\right) & R_{XY}\left(\tau_{-(N-3)}\right) & \ldots & R_{XY}(\tau_0) & R_{XY}(\tau_1) \\ R_{XY}\left(\tau_{-(N-1)}\right) & R_{XY}\left(\tau_{-(N-2)}\right) & \ldots & R_{XY}(\tau_{-1}) & R_{XY}(\tau_0) \end{pmatrix} \quad (27.37)$$

$R_{XY}(\tau)$ is a cross-covariance function between the processes X and Y taken at the time lag τ, which is a difference between the n-th and m-th time increments:

$$R_{XY}(\tau_{n-m}) = \mathrm{cov}(x_n, y_m); \quad \tau_{n-m} = t_n - t_m \quad (27.38)$$

where t stands for time. The time lag τ can be negative, when $n < m$; negative indexes correspond to negative values of time lags. Unlike the auto-covariance function, the cross-covariance function is not symmetrical.

Equation (27.37) makes sense when both process X and Y are stationary – their covariance depends only on the time difference and does not depend on a particular instance of time. As a result, the elements of the matrix in Eq. (27.37) are identical along the diagonals. Then, the matrix can be summed by diagonals instead of by rows and columns:

$$\text{cov}\left(\hat{E}_X, \hat{E}_Y\right) = \frac{1}{N^2}\left(R_{XY}\left(\tau_{-(N-1)}\right) + 2R_{XY}\left(\tau_{-(N-2)}\right) + \ldots + (N-1)\cdot R_{XY}(\tau_{-1}) \right. \\ \left. + N \cdot R_{XY}(\tau_0) + (N-1)\cdot R_{XY}(\tau_1) + \ldots + 2R_{XY}(\tau_{N-2}) + R_{XY}(\tau_{N-1})\right) \tag{27.39}$$

Thus:

$$\text{cov}\left(\hat{E}_X, \hat{E}_Y\right) = \frac{1}{N}\sum_{n=-(N-1)}^{N-1}\left(1 - \frac{|n|}{N}\right)R_{XY}(\tau_n) \tag{27.40}$$

The cross-covariance function is known for the following property:

$$R_{XY}(-\tau) = R_{YX}(\tau) \tag{27.41}$$

This allows Eq. (27.37) to be rewritten for positive indexes only:

$$\text{cov}\left(\hat{E}_X, \hat{E}_Y\right) = \frac{R_{XY}(0)}{N} + \frac{1}{N}\sum_{n=1}^{N-1}\left(1 - \frac{n}{N}\right)\left(R_{XY}(\tau_n) + R_{YX}(\tau_n)\right) \tag{27.42}$$

Equations (27.40) and (27.42) can also express variance of the mean—see Priestley (1981)

$$\text{var}\left(\hat{E}_X\right) = \text{cov}\left(\hat{E}_X, \hat{E}_X\right) = \frac{R_{XY}(0)}{N} + \frac{2}{N}\sum_{n=1}^{N-1}\left(1 - \frac{n}{N}\right)R_{XX}(\tau_n) \tag{27.43}$$

Alternatively, taking account of the symmetry of the auto-covariance function $R_{XX}(\tau)$, the variance of the mean can be presented as (compare to relevant formula from Levine et al. 2017):

$$\text{var}\left(\hat{E}_X\right) = \frac{1}{N}\sum_{n=-(N-1)}^{N-1}\left(1 - \frac{|n|}{N}\right)\cdot R_{XX}(\tau_n) \tag{27.44}$$

The true cross-covariance function $R_{XY}(\tau)$ is not known and has to be substituted by its estimate:

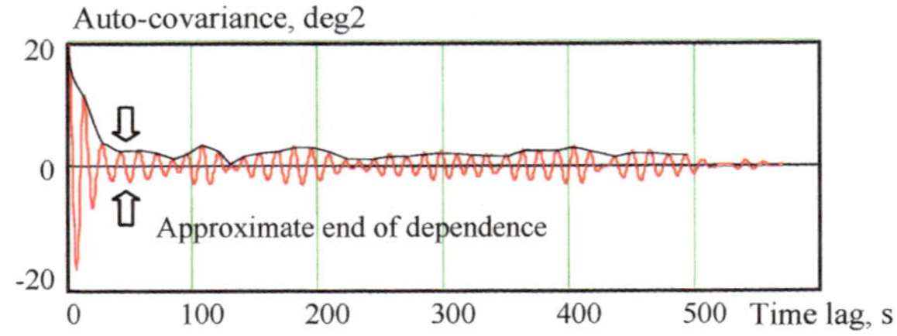

Fig. 27.12 Selection of the window width for auto-covariance function estimate for roll motion

$$\hat{R}_{XY}(\tau_m) = \frac{w(\tau_m)}{N_R \cdot N} \sum_{k=1}^{N_R} \sum_{n=1}^{N-m} \left(x_{n,k} - \hat{E}_X\right) \cdot \left(y_{n+m,k} - \hat{E}_y\right) \tag{27.45}$$

where w is a windowing function. Similar to the estimate of the auto-covariance function, the cross-covariance is subject to a deterioration of accuracy with the increase of time lag as the number of available points drops. Following Levine et al. (2017), a triangle (Bartlett) window is applied here:

$$w(\tau) = \begin{cases} 1 - \frac{|\tau|}{\tau_W} \; if \;\; |\tau| \le \tau_W \\ 0 \quad otherwise \end{cases} \tag{27.46}$$

Justification of the choice of the width τ_W of Bartlett windows is the subject of current interest. Levine et al. (2017) propose the first minimum of an envelope of the auto-covariance function as the first approximation for the width, see Fig. 27.12. This method was used for all auto- and cross-covariance estimates in this study.

The reasoning behind the window width choice is as follows. The auto-covariance function reflects how much dependence of the current state of the dynamical system propagates into the future. The dependence in ship motion is driven by the dependence in the excitation, the inertia of the ship and hydrodynamic memory. All of these phenomena fade with time, no physical reason exists for the dependence to increase. Thus, the increase of the dependence has to be treated as numerical artifact that should be suppressed.

References

Belenky, V., Weems, K., and Lin, W.-M. 2016 "Split-time method for estimation of probability of capsizing caused by pure loss of stability", Ocean Engineering, Vol. 122, pp. 333–343.

Bishop, B, Belknap, W., Turner, C., Simon, B., Kim, J., 2005, "Parametric Investigation on the Influence of GM, Roll Damping, and Above-Water Form on the Roll Response of Model 5613," Technical Report NSWCCD-50-TR-2005/027, Naval Surface Warfare Center Carderock Division, West Bethesda, Maryland USA.

Bickel, P. J. and Doksum, K. A., 2001 Mathematical Statistics. Basic Ideas and Selected Topics Volume 1, Prentice Hall, Upper Saddle River, NJ, 2nd ed., 556 p.

Cramer, H., and Leadbetter, M.R., 1967, Stationary and Related Stochastic Processes, John Wiley, New York.

Levine, M. D., Belenky, V., Weems, K.M. and Pipiras. V. 2017, "Statistical Uncertainty Techniques for Analysis of Simulation and Model Test Data", Proceedings of 30th American Towing Tank Conference, Carderock, Maryland, USA.

Lin, W.-M., Yue, D. K. P., 1990. "Numerical Solutions for Large Amplitude Ship Motions in the Time-Domain" Proceedings of the 18th Symposium on Naval Hydrodynamics, Ann Arbor, Michigan, USA, pp. 41–66.

Shin, Y.S, Belenky, V.L., Lin, W.M., Weems, K.M. and A. H. Engle, 2003 "Nonlinear time domain simulation technology for seakeeping and wave-load analysis for modern ship design" SNAME Trans. Vol. 111, pp. 557–578.

Priestley, M. B., 1981, Spectral Analysis and Time Series, Vol. 1, Academic Press, New York.

Scott, D.W., 1979, "On Optimal and Data-based Histograms," Biometrika, 66(3):605–610.

Part IX
Experimental Research: Techniques

Chapter 28
Regular Wave Testing as a Crucial First Step for Dynamic Stability Evaluation

David D. Hayden, Richard C. Bishop and Martin J. Dipper

Abstract The DDG51 pre-contract hull form as represented by Model 5514 was evaluated for capsize events at End of Service Life Load Limit for the Righting Arm Limiting and Intact 100 Knot Wind Limiting KG Conditions. The hull was evaluated using a matrix of various following sea headings and Froude numbers at various regular wave lengths and steepness. The results provide a good comparison point to computer simulations and a way to verify regimes of concern relating to dynamic stability. The domains of reduced dynamic stability can be used to determine future areas of study for random or deterministic irregular wave testing. The model test also provides data to compare measured roll decay and maneuvering characteristics against simulation predictions for validation. Experimental techniques and model layout are described. Suggestions for future improvements in regular wave testing are provided. Even though regular wave dynamic stability testing provides a basic evaluation of dynamic stability, the results can be used as a key starting point for additional testing or as a basis for operational guidance.

Keywords Regular waves · Parametric evaluation · Dynamic stability

28.1 Introduction

Laboratory testing for dynamic stability can assume many forms, with different type of wave environments. A free running or very lightly tethered model, will yield the purest results when testing for 3-D dynamic stability events. Instrumentation should be suitable to capture all motions, control surfaces, and the wave environment. With respect to waves the experiment can utilize steep regular waves, random irregular waves with a spectral shape known to produce steep wave fronts by the nature of the height and period, or deterministic irregular wave sequences. At its simplest level a free running model operating in dynamic stability type conditions in regular waves

D. D. Hayden (✉) · R. C. Bishop · M. J. Dipper
David Taylor Model Basin (NSWCCD), West Bethesda, MD, USA
e-mail: david.hayden@navy.mil

V. L. Belenky et al. (eds.), *Contemporary Ideas on Ship Stability*, Fluid Mechanics and Its Applications 119, https://doi.org/10.1007/978-3-030-00516-0_28

Fig. 28.1 View of model 5514 during capsize testing in MASK basin

can provide great insight to performance in other more realistic seaways. This type of testing will be described.

28.2 Experiment Design

28.2.1 Model Scale and Facility

Selection of model scale is typically based upon the need for waves with sufficient steepness at wavelengths likely to create a dynamic stability event. The ability to achieve speed, power and instrument the model, and provide sufficient run length should be considered in model scale selection. The ability to meet the data and control needs and still meet ballast conditions is often the most challenging part of preparation. Model 5514 is a ten foot model requiring vigilance with respect to ballasting. But when testing at end of service life (EOSL) the ballast condition is much more obtainable, since it is the ship condition at maximum displacement and highest vertical center of gravity. The model operating in the basin is shown in Fig. 28.1. The Maneuvering and Seakeeping Basin (MASK) provides a long run length suitable to initiate dynamic stability events in regular waves, and still generally have time for the dynamic event to occur prior to reaching the edge of the basin. Plastic net fences are installed along the beach edges of the basin to capture the model in the event that the model operator can not stop the model in time. The schematic of the MASK basin is shown in Fig. 28.2.

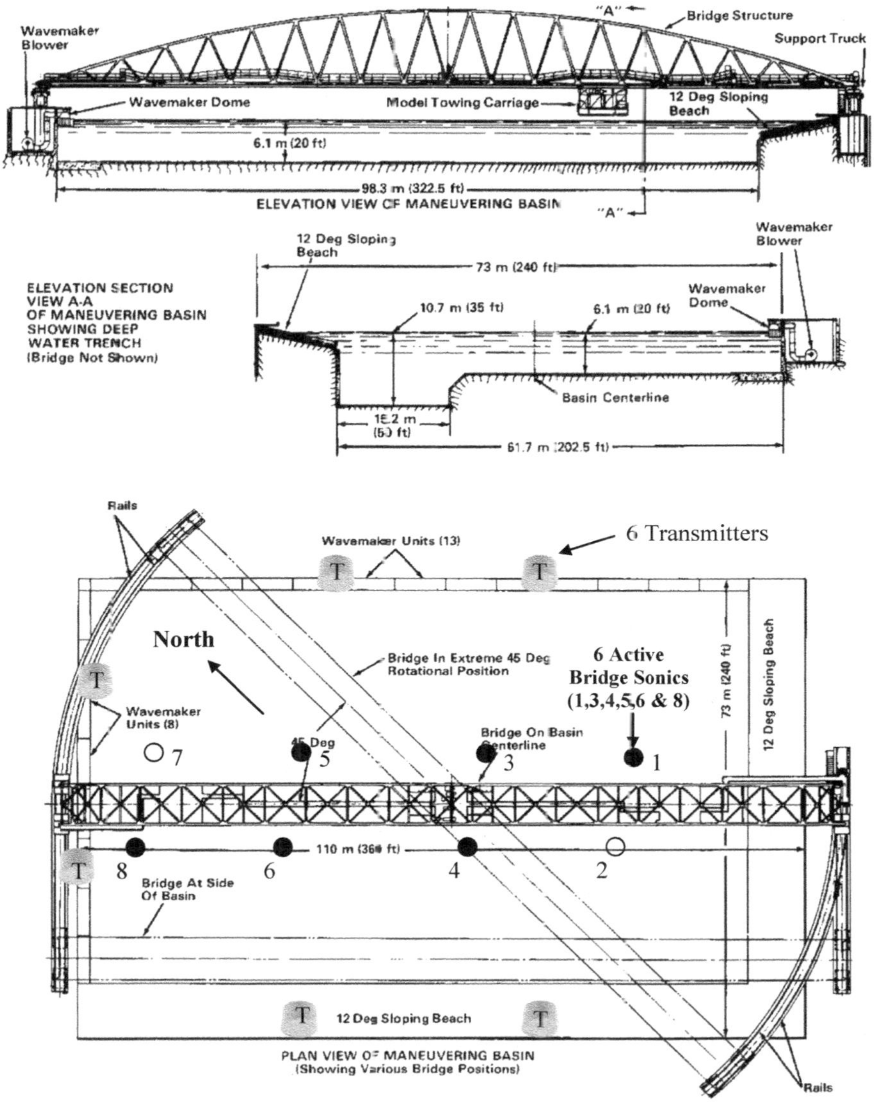

Fig. 28.2 Sketch of maneuvering and seakeeping (MASK) basin

28.2.2 *Test Matrix*

The test matrix can be designed based upon simulations or past experimental experiences with the conditions which cause dynamic stability events. The most likely wave conditions to cause dynamic stability events are λ/L values surrounding 1.0,

Table 28.1 Desired wave conditions as defined by test matrix and model geometry

Full scale test conditions				
Ship length: 466.00 ft				
Wave/hull length	Wave length (ft)	Wave length (m)	Wave period (s)	Wave freq (Hz)
0.75	349.50	6.71	8.26	0.1210
1.00	466.00	8.95	9.54	0.1048
1.25	582.50	11.18	10.67	0.0938
1.50	699.00	13.42	11.68	0.0856
Model scale conditions—lambda = 46.6				
Ship length: 10.00 ft				
Wave/hull length	Wave length (ft)	Wave length (m)	Wave period (s)	Wave freq (Hz)
0.75	7.50	2.29	1.210	0.826
1.00	10.00	3.05	1.398	0.716
1.25	12.50	3.81	1.563	0.640
1.50	15.00	4.57	1.712	0.584
Wave heights				
Wave H/L	Wave length/hull length			
	0.75	1.00	1.25	1.50
1/20	11.4 cm	15.2 cm	19.1 cm	22.9 cm
1/15	15.2 cm	20.3 cm	25.4 cm	30.5 cm
1/10	22.9 cm	30.5 cm	38.1 cm	45.7 cm
Wave H/L	Wave length/hull length			
	0.75	1.00	1.25	1.50
1/20	4.50 in	6.00 in	7.50 in	9.00 in
1/15	6.00 in	8.00 in	10.00 in	12.00 in
1/10	9.00 in	12.00 in	15.00 in	18.00 in

where L is the ship length at the waterline and λ is the wavelength. For this test λ/L values from 0.75 to 1.50 had been choosen for testing, with wave steepnesses ranging from H/λ = 1/20 to H/λ = 1/10, where H is the trough to crest wave height.

The corresponding target regular wave heights are provided in Table 28.1. Even though linear wave theory is assumed for λ based upon period, there is no appreciable change in λ for the steep non-linear wave since it is still in deep water when testing in the MASK. The model was tested at speeds of Fn = 0.0, 0.1, 0.2, 0.3, and 0.4. Headings aft of beam to pure following were tested. In this case the conditions tested were based upon simulation results from FREDYN 9.3.

Fig. 28.3 Midship OBC, GM pole, and instruments

28.2.3 Instrumentation

Instrumentation should be sufficient to capture the motions of interest required to define the event. Accelerations were measured at the center of gravity (CG), bow, and stern. Angular rates, roll, pitch, and heading via a gimbaled flux-gate compass were used to measure angular characteristics. The rudder angle was measured via a rotary potentiometer, and shaft RPM was measured via optical pulsar and a in-house circuit board with an Analog Devices frequency to voltage converter. The wave environment was recorded with six sonic wave surface probes suspended from the MASK bridge. The bridge was rotated to a 30° angle to provide better basin wave probe coverage. The model position and track were recorded with an ArcSecond Constellation 3DI indoor GPS system. A single ArcSecond receiver was placed on the forward deck of the model and is visible in Fig. 28.1. The approximate spacing and position of the wave sonics and ArcSecond transmitters are shown in Fig. 28.2. The instruments were filtered, sampled, and stored via an and on board computer (OBC) and associated filter cards and analog to digital converter (ADC). The OBC and some of the CG instruments are shown in Fig. 28.3.

28.3 Experimental Procedure

28.3.1 Pre Experiment Preparation

Prior to testing the model was outfitted and ballasted. The powering and steering systems were somewhat fixed based upon the ship arrangements. The instrument location was fixed for those channels to be measured at a specific location. The

Table 28.2 Model mass properties end of service life load limit, intact 100 kt wind limiting KG

	Full scale desired		Model scale achieved	
	Values english	Values metric	Values english	Values metric
Displacement	9400.00 LTSW	9549.21 m-ton SW	202.46 lbs FW	91.92 kg FW
LCG wrt Station 10	−5.90 ft	−1.80 m	−1.52 in	−3.86 cm
KG	28.06 ft	8.55 m	7.12 in	18.08 cm
Draft	21.37 ft	6.51 m	NA	NA
Trim	0.36 ft	0.11 m	NA	NA
$GM_{w/FS}$	3.00 ft	0.91 m	0.76 in	1.93 cm
k_{pitch}	116.50 ft	35.51 m	29.73	75.52 cm
k_{pitch}/LBP	0.250	0.250	0.248	0.248
k_{roll}	24.09 ft	7.34 m	6.67	16.94 cm
$k_{roll}/Beam_{WL}$	0.383	0.383	0.412	0.412
Roll $period_{zero\ speed}$	15.41	15.41	2.421 s	2.421 s

Table 28.3 Model mass properties end of service life load limit, righting arm limiting KG

	Full scale desired		Model scale achieved	
	Values english	Values metric	Values english	Values metric
Displacement	9400.00 LTSW	9550.85 m-ton SW	202.46 lbs FW	91.92 kg FW
LCG wrt station 10	−5.90 ft	−1.80 m	−1.52 in	−3.86 cm
KG	26.97 ft	8.22 m	6.81 in	17.29 cm
Draft	21.37 ft	6.51 m	NA	NA
Trim	0.36 ft	0.11 m	NA	NA
$GM_{w/FS}$	4.09 ft	1.25 m	1.04 in	2.63 cm
k_{pitch}	116.50 ft	35.51 m	29.65	75.31 cm
k_{pitch}/LBP	0.250	0.250	0.247	0.247
k_{roll}	24.09 ft	7.34 m	6.29	15.97 cm
k_{roll}/Beam WL	0.383	0.383	0.388	0.388
Roll $period_{Zero\ speed}$	13.19	13.19	1.971 s	1.971 s

OBC and batteries were then located to provide the approximate CG and gyradius characteristics. The remaining ballast weights were used to fine tune the ballast conditions. Additionally a pole is located at the longitudinal CG (LCG) location with weights which can be slid up and down the pole to adjust the metacentric height (GM). The GM pole was located to have minimal effect on other desired ballast variables. The GM was verified by performing an incline experiment. The results of the ballasting effort are shown in Tables 28.2 and 28.3. The measurement uncertainty of the data collection is provided in Table 28.4, and is calculated by methods defined in ISO Uncertainty Guide (1995).

28.3.2 Model Characterization Runs

The roll period and roll decay were measured at all speeds to be tested. This measurement procedure was performed by depressing the model on the gunwale while the model is lightly tethered under the MASK carriage, then releasing the model and measuring the roll decay after release This procedure was repeated for both port and starboard sides, and for varying angles of initial roll. The roll period and amount of roll decay can be used to compare against anticipated roll period and roll decay from simulation. The difference between measured roll period and FREDYN calculated roll period is also used as a way to adjust the coefficients of the autopilot algorithm at the operational speeds.

The speed of the free running model was set by performing a calm water speed calibration. The propulsion motor was controlled with a voltage command from the OBC. The model was run at set voltage commands parallel to the long bank of the MASK across a measured distance to create a speed measurement. This is done for 12–20 increments across the range of speed operations. Regression analysis was then performed relating measured speed, shaft RPM, and throttle command. The throttle voltage commands at each target speed can then be calculated and used to set the target ordered speed at each matrix condition of concern. Once the calm water roll periods and speed settings were defined, then other operations of the test matrix could be performed.

28.3.3 Maneuvering Runs

Though not required for dynamic stability, the calm water maneuvering characteristics as measured by turning circles and zig-zags can be quite useful for comparison against simulation efforts. The effects of the control surfaces, bilge keels, and ballast condition are all interrelated and present when performing calm water maneuvers. These results provide a way to look at scale effects, and how such influences correlate during simulations and other scaled tests. Both turning circles and zig-zags were performed as presented by Hayden et al. (2006).

28.3.4 Regular Wave Runs

Regular wave dynamic stability runs were performed at the conditions of interest. The model was held stationary from a punt at the edge of the basin while the desired

Table 28.4 Uncertainty estimates for instrumentation

				Type A uncertainty		Type B		Total	Total
Parameter	Units	Min	Max	N	Std Dev	U95	U95		(% Max)
Acceleration, bow transverse	g	−1	+1	321	0.00668	0.0007	0.0081	0.0082	0.82
Acceleration, bow vertical	g	−1	+1	321	0.00379	0.0004	0.0010	0.0011	0.11
Acceleration, cg longitudinal	g	−1	+1	321	0.00049	0.0001	0.0020	0.0020	0.20
Acceleration, cg transverse	g	−1	+1	321	0.00649	0.0007	0.0033	0.0033	0.33
Acceleration, cg vertical	g	−1	+1	321	0.00236	0.0003	0.0108	0.0108	1.08
Acceleration, stern transverse	g	−1	+1	321	0.00694	0.0008	0.0098	0.0098	0.98
Acceleration, stern vertical	g	−1	+1	321	0.00337	0.0004	0.0209	0.0209	2.09
Angle, Pitch	deg	−50	+50	321	0.177	0.0198	0.64	0.64	1.28
Angle, Roll	deg	−90	+90	321	0.370	0.0413	0.77	0.77	2.67
Angle, Rudder	deg	−30	+30	321	0.029	0.0033	0.80	0.80	2.67
Heading	deg	0	360	321	0.733	0.0818	1.00	1.07	0.28
Model Speed	m/s	0.55	2.19	321	0.00836	0.0009	0.12	0.12	5.66
Propeller Shaft Speed	rpm	264	1093	321	4.28	0.48	1.78	1.84	0.17
Rudder Angular Rate, Max.	deg/s	−67.5	+67.2				1.61	1.61	2.40
Wave Height #1	mm	−381	+381				2.29	2.29	0.60
Wave Height #3	mm	−381	+381				3.42	3.42	0.90
Wave Height #4	mm	−381	+381				2.61	2.61	0.69
Wave Height #5	mm	−381	+381				3.22	3.22	0.85
Wave Height #6	mm	−381	+381				3.40	3.40	0.89
Wave Height #8	mm	−381	+381				1.89	1.89	0.50

wave field was allowed to build. The model operator would set the autopilot heading, and bring the propellers up to speed using the pre-determined throttle command. Model tenders in the punt would hold the model at the desired heading relative to the waves, and then release the model on command from the model operator once there was sufficient wave coverage in the basin. The tracker data, model and wave data collection, and video would start prior to model release in order to record initial conditions. The model would accelerate away from the punt after release. The operator would make a verbal and computer collection note ("Model Mark") when the model was on speed and heading.

The model would be allowed to translate the basin using the autopilot and throttle commands until capsize occurred or it was necessary to stop the model because of basin limitations. Additional personnel would observe the model and make notes of surfing, broaching, wave captures, bow roots, or other roll events indicative of dynamic instabilities. The model operator would declare end of run to the video and data personnel. Example of a tracker image for a capsize event (Run 396) in following seas at $Fn = 0.4$, $\lambda/L = 1.25$, $H/\lambda = 1/10$ is shown in Fig. 28.4. The model was observed to surf on the crest of the wave, broach to starboard, and then capsize. Evidence of the broach and capsize event can be evidenced by the curved trajectory to starboard at the end of the run.

All pre-defined test conditions were performed a minimum of one time. If critical dynamic stability behavior was present for a condition, then that condition would be repeated. If a capsize or dangerous roll was observed, then the condition would be repeated three or more times. If the model consistently capsized, then that condition would be labeled as a capsize condition. If dangerous roll consistently occurred, but capsize did not always occur, then that condition would be classified as "In Danger of Capsize." "Dangerous Roll" was classified as any maximum roll greater than 75% of the positive area under the GZ curve.

28.4 Results

The results of the test were presented as performance matrices which summarized the run numbers, maximum rolls, wave conditions, and comparison to FREDYN results

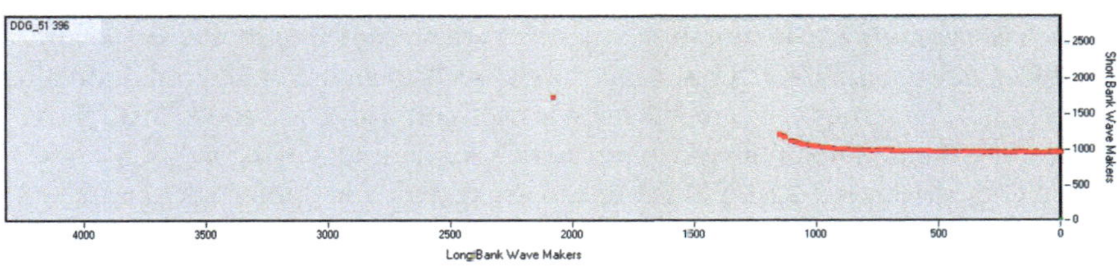

Fig. 28.4 Example tracker plot for run 396

Table 28.5 Roll angle matrix for intact 100 kt wind limiting KG = 8.55 m, eosl load condition

λ/L	λ/h	Fn	χ (deg)							
			0	15	30	45	60	75	90	105
0.75	20	0.0								
		0.1								
		0.2								
		0.3								
		0.4								
	15	0								
		0.1								
		0.2								
		0.3								
		0.4								
	10	0								
		0.1								
		0.2				26.2, 25.8	11.6, 12.2			
		0.3			32.0, 30.8	37.9, 34.2	19.2, 21.0	17.2, 18.7		
		0.4			76.1, **43.2**, 90, 68.9, 61.9, 90, 90	90, 90, 67.4, 59.8, 61.8	26.6, 27.0			
1	20	0.0								
		0.1								
		0.2								
		0.3								
		0.4								
	15	0.0								
		0.1								
		0.2								
		0.3				20.2, 21.9	21.0, 23.4			
		0.4				32.1, 31.9	32.0, 31.9			
	10	0.0	25.6						28.1	
		0.1			22.2		14.3, 15.7	15.3, 12.9		
		0.2			34.6, 32.8	20.4, 18.9	19.4, 11.3	20.1, 18.9		
		0.3		29.6, 29.0	32.8, 42.1	27.5, 25.9	26.4, 26.9	21.8, 25.4		
		0.4		90, 90, 90, 90	90, 90, 90, 90	37.2, 33.4	38.2, 31.1	25.9, 34.6		
1.25	20	0.0								
		0.1								
		0.2								
		0.3				24.4, 21.9				
		0.4								
	15	0.0	19.0							
		0.1								
		0.2				19				
		0.3			29.9	25.8, 27.5	26.8			
		0.4			42.1, 42.0	40.8, 42.2	33.3, 34.4			
	10	0.0	28.4						30.5	
		0.1		34.3		32.5	30.6	30.1	27.6	
		0.2		26.9	35.3	39.1	36.7	34.8		
		0.3		33.3	38.7, 40.7	58.3, 48.1	45.1, 45.0	39.8		
		0.4	64.5, 90, 90	90, 90, 90, 90	90, 90, 90, 90	74.4, 90, 90, 90	69.0, 67.4, 90, 90, 61.2, 54.8	42.6		
1.5	20	0.0								
		0.1								
		0.2								
		0.3				29.3				
		0.4				37.5	37			
	15	0.0								
		0.1								
		0.2		14.9		32				
		0.3		21.8	30	35.2	41.4			
		0.4			37.2	43.2	43.5			
	10	0.0								
		0.1								
		0.2								
		0.3								
		0.4								

Key:		
	36.6	Max Roll Angle < 49.4 deg (75% of GZ curve)
	55.2	Max Roll Angle > 49.4 deg: in danger of capsize (capsize, dangerous roll)
	90.0, 54.3, 48.1	Capsize condition (capsize, dangerous roll, acceptable roll)

for each test condition. Maximum roll matrices are presented for both load condition in Tables 28.5 and 28.6. Each roll value corresponds to an individual run. Individual capsizes and dangerous rolls are indicated at each cell with color coded text. The cell was classified based upon the predominate behaviour at that cell condition.

The comparison to FREDYN 9.3 results is presented in Tables 28.7 and 28.8. It can be seen that FREDYN generally overpredicts the capsize event for the typical surface combatant.

Table 28.6 Roll angle matrices for righting arm limiting KG = 8.22 m, EOSL load condition

λ/L	λ/h	Fn	χ (deg)							
			0	15	30	45	60	75	90	105
		0.0								
		0.1								
		0.2								
		0.3								
	20	0.4								
		0								
		0.1								
		0.2								
		0.3								
	15	0.4								
		0.0								
		0.1								
		0.2								
		0.3			32.5, 34.5	36.4	39.5			
0.75	10	0.4			48.1, 45.4	45.7, 44.7, 44.2, 41.7, 45.7	45.7			
		0.0								
		0.1								
		0.2								
		0.3								
	20	0.4								
		0.0								
		0.1								
		0.2								
		0.3								
	15	0.4								
		0.0								
		0.1								
		0.2								
		0.3		23.6	35.5	45.8				
1	10	0.4		43.3, 48.6	45.2	52.3, 46.8				
		0.0								
		0.1								
		0.2								
		0.3								
	20	0.4								
		0.0								
		0.1								
		0.2								
		0.3								
	15	0.4								
		0.0								
		0.1								
		0.2								
		0.3	11.5	24.2	32.5	43.6	47.4	42.2, 45.6		
1.25	10	0.4	52.0, 39.3	44.6, 41.9	46.1 42.6	44.3, 45.0, 43.9	46.32	41.8, 28.5, 43.3		

Key:		
	36.6	Max Roll Angle < 49.4 deg (75% of GZ curve)
	55.2	Max Roll Angle > 49.4 deg: in danger of capsize (capsize, dangerous roll)
	90.0, 54.3, 48.1	Capsize condition (capsize, dangerous roll, acceptable roll)

Table 28.7 Test versus FREDYN results for intact 100 kt wind limiting KG = 8.55 m, EOSL load condition

λ/L	λ/h	Fn	χ (deg) 0	15	30	45	60	75	90	105
		0.0								
		0.1								
		0.2								
		0.3								
	20	0.4								
		0.0								
		0.1								
		0.2								
		0.3								
	15	0.4								
		0.0								
		0.1								
		0.2				C_f	C_f			
		0.3			C_f	C_f	C_f	C_f		
0.75	10	0.4			C_f	C_f	C_f			
		0.0								
		0.1								
		0.2								
		0.3								
	20	0.4								
		0.0								
		0.1								
		0.2								
		0.3				C_f	C_f			
	15	0.4				C_f	C_f			
		0.0	C_f	C_f	C_f	C_f	C_f	C_f	C_f	
		0.1			C_f		C_f	C_f		
		0.2			C_f	C_f	C_f	C_f		
		0.3		C_f	C_f	C_f	C_f	C_f		
1	10	0.4		C_f	C_f	C_f	C_f	C_f		
		0.0								
		0.1								
		0.2								
		0.3				C_f				
	20	0.4								
		0.0	C_f							
		0.1								
		0.2				C_f				
		0.3			C_f	C_f	C_f			
	15	0.4			C_f	C_f	C_f			
		0.0	C_f	C_f	C_f	C_f	C_f	C_f	C_f	C_f
		0.1		N_f		C_f	C_f	C_f	C_f	
		0.2		C_f	C_f	C_f	C_f	C_f		
		0.3		C_f	C_f	C_f	C_f	C_f		
1.25	10	0.4	C_f	C_f	C_f	C_f	C_f	C_f		
		0.0								
		0.1								
		0.2								
		0.3				C_f				
	20	0.4				C_f	C_f			
		0.0								
		0.1								
		0.2		C_f		C_f				
		0.3		C_f	C_f	C_f	C_f			
	15	0.4			C_f	C_f	C_f			
		0.0								
		0.1								
		0.2								
		0.3								
1.5	10	0.4								

Key:

Model Test Results
- Max Roll Angle < 49.4 deg (75% of GZ curve)
- Max Roll Angle > 49.4 deg: in danger of capsize
- Consistent capsize condition.

FREDYN Prediction
- C_f FREDYN predicted capsize prior to commanded time of 900 seconds.
- N_f FREDYN predicted no capsize for commanded time of 900 seconds.
- T_f FREDYN prediction terminated due to operational limitations of software.

Table 28.8 Test versus FREDYN results for righting arm limiting KG = 8.22 m, EOSL load condition

λ/A	λ/η	Fn	χ (δεγ)							
			0	15	30	45	60	75	90	105
		0.0								
		0.1								
		0.2								
		0.3								
	20	0.4								
		0.0								
		0.1								
		0.2								
		0.3								
	15	0.4								
		0.0								
		0.1								
		0.2					C_f			
		0.3			N_f	C_f	C_f			
0.75	10	0.4			C_f	C_f	C_f			
		0.0								
		0.1								
		0.2								
		0.3								
	20	0.4								
		0.0								
		0.1								
		0.2								
		0.3								
	15	0.4					C_f			
		0.0	T_f	C_f	C_f	C_f	C_f	C_f	C_f	
		0.1					C_f	C_f		
		0.2				C_f	C_f	C_f		
		0.3		N_f	C_f	C_f	C_f	C_f		
1	10	0.4		C_f	C_f	C_f	C_f	C_f		
		0.0								
		0.1								
		0.2								
		0.3								
	20	0.4								
		0.0								
		0.1								
		0.2				C_f	C_f			
		0.3				C_f				
	15	0.4				C_f	C_f			
		0.0	C_f	C_f	C_f	C_f	C_f	C_f	C_f	C_f
		0.1				C_f	C_f	C_f	C_f	
		0.2		C_f	C_f	C_f	C_f	C_f		
		0.3	N_f	C_f	C_f	C_f	C_f	C_f		
1.25	10	0.4	C_f	C_f	C_f	C_f	C_f	C_f		

Key:		
Model Test Results	(light green)	Max Roll Angle < 49.4 deg (75% of GZ curve)
	(yellow)	Max Roll Angle > 49.4 deg: in danger of capsize
	(red)	Consistent capsize condition.
FREDYN Prediction	C_f	FREDYN predicted capsize prior to commanded time of 900 seconds.
	N_f	FREDYN predicted no capsize for commanded time of 900 seconds.
	T_f	FREDYN prediction terminated due to operational limitations of software.

28.5 Uncertainty Analysis

Uncertainty analysis as applied for this report is based upon the ISO Uncertainty Guide (1995). The analysis consists of two parts: (1) Type A evaluation and (2) Type B evaluation. For this report, all uncertainties are defined at the 95% confidence limit. Typically, these experiments were highly unsteady and random in character, and the standard deviations were not typically computed. Consequently, most of the uncertainty was determined by Type B evaluation method.

28.6 Conclusions

The regular wave, dynamic stability results yield a very consistent result with respect to performance at any particular operational condition. The operational areas of concern are well defined by the "Danger of Capsize" and "Capsize Condition" cells. There are no cells at separate locations from the concentrated concern areas at Fn = 0.4. There is one capsize cell at Fn = 0.3 for a stern quartering condition, for the 100 kt wind limiting KG. There are no "Capsize Condition" cells for the righting arm limiting KG condition.

These results could be used to create operational guidance for the ship being evaluated. The matrices indicate that if the ship speed is operationally limited to Fn = 0.2 and below for extreme following seas, then there is very limited possibility of a dynamic stability event.

Additionally if other tests in irregular seas are to be performed, then the irregular seas will most likely need to contain wave components which are near the same steepness as that observed for regular wave testing. The results indicate that wave steepnesses near $H/\lambda = 1/10$ will be required, and that these conditions will have to be run at the speed and heading indicated by regular wave results.

Acknowledgements The authors would like to acknowledge the many contributions of naval architects, electronic and mechanical engineers and technicians whom have improved NSWCCD dynamic stability testing capabilities since the early evolutions. The technical efforts from Code 51 and 55 personnel have gradually built upon the previous effort. The contributions of each and all is greatly appreciated in this effort.

References

Hayden, D.D., Bishop, R.C., Park, J.T., Laverty, S.M., 2006, Model 5514 Capsize Experiments Representing the Pre-Contract DDG51 Hull Form At End of Service Life, Report No. NSWCCD-50-TR-2006/20, Carderock Div, NSWC.

ISO 1995, "Guide to the Expression of Uncertainty in Measurement," 1995, International Organization for Standardization (ISO), Geneva, Switzerland.

Chapter 29
An Experimental Study on Characteristics of Rolling in Head Waves for a Vessel with Nonlinear GZ-curve

Toru Katayama, Shugo Miyamoto, Hirotada Hashimoto and Yoshifumi Tai

Abstract In this study, the characteristics of rolling in head waves for a vessel with strong nonlinear GZ-curve, which includes parametric rolling, are investigated. Rolling is measured for systematically changed wave length and height under the same forward speed, which is service speed in heavy weather. As a result, the range of T_e/T_ϕ (T_e and T_ϕ are encounter wave period and the roll natural period of the model) when oscillatory rolling occurs is wider than that of previous results by Taguchi et al. (Model Experiment on Parametric Rolling of a Post-Panamax Containership in Head Waves. Proceedings of the Ninth International Conference on Stability of Ships and Ocean Vehicles, 2006), and the range spreads out wide area of $T_e/T_\phi > 0.5$. Especially, in the range of $T_e/T_\phi > 0.5$, oscillatory rolling is caused by large wave height and roll amplitude becomes larger than that at $T_e/T = 0.5$. It is supposed that the result is caused by change of roll natural period caused by nonlinear GZ-curve. In order to confirm it, numerical simulations are carried out for several variations of GZ-curve. Additionally, roll measurements in irregular waves with Pierson-Moskowitz spectrum are also carried out.

Keywords Parametric rolling · Wide-breadth and shallow-draft
Strongly nonlinear-GZ-curve

T. Katayama (✉) · S. Miyamoto
Graduate School of Engineering, Osaka Prefecture University, Sakai, Japan
e-mail: katayama@marine.osakafu-u.ac.jp

S. Miyamoto
e-mail: ss103029@edu.osakafu-u.ac.jp

H. Hashimoto
Graduate School of Engineering, Osaka University, Sakai, Japan
e-mail: hashimoto@port.kobe-u.ac.jp

Y. Tai
Imabari Shipbuilding Co., Ltd, Imabari, Japan
e-mail: tai.yoshifumi@imazo.com

V. L. Belenky et al. (eds.), *Contemporary Ideas on Ship Stability*, Fluid Mechanics and Its Applications 119, https://doi.org/10.1007/978-3-030-00516-0_29

29.1 Introduction

Ballast water for vessels is significant to prevent their stability and propulsion performance losses, which are caused by their light-draught. However, the International Maritime Organization (IMO) adopted International Convention for the Control and Management of Ships' Ballast Water and Sediments in 2004, because of avoiding its environmental impact.

Then, in order to solve this problem, new hull forms are being designed, which use a small amount of ballast water or don't need ballast water. For example, one of new designed hull has shallow draft and wide breadth. Such hull may show different roll characteristics from the conventional vessels.

In the previous study (Tai et al. 2011), for a shallow draft and wide breadth vessel, its roll characteristics in beam waves have been investigated experimentally. The results show that its roll natural period is changed by its roll amplitude and its roll resonance occurs in different wave period depending on different wave height.

In this study, the roll characteristics of the same model (Tai et al. 2011) in head waves (including parametric rolling) are investigated. Roll measurements in regular head waves are carried out. From the results, it is found that large amplitude rolling occurs at a wider range of encounter frequencies than that of a conventional vessel, and it is supposed that it is caused by strong nonlinearity of its *GZ*-curves. Then, in order to be clear the reasons of the results, numerical simulations are carried out. Additionally, the roll characteristics in irregular waves with Pierson-Moskowitz spectrum are also investigated.

29.2 Model and Measuerment

Figure 29.1 and Table 29.1 show a body plan and principal particulars of a model. Figure 29.2 shows the calculated *GZ*-curve of the model in real scale. In the same figure, the linear *GZ*-value of the model is also shown. The calculated *GZ*-curve shows strong nonlinear characteristics in the range of small roll angles. This is because bilge of the model is exposed above the water surface from the range of small roll angles.

Figure 29.3 shows measured roll natural period of the model by a free decay test, which is carried out for various initial heel angles. This figure shows that the roll natural period becomes longer according to increase in roll amplitude. And this is caused by the strong nonlinearity of the *GZ*-curve shown in Fig. 29.2.

In order to investigate the occurrence of oscillatory rolling, towing tests in regular head waves are carried out at the towing tank of Osaka Prefecture University (length 70 m, breadth 3 m, depth 1.5 m). Figure 29.4 shows a schematic view of the experiment. A model is attached to the towing carriage with two elastic ropes (Hashimoto et al. 2007) . The model is towed at constant forward speed in head waves, and a small disturbance which is heel caused by hand is given. Roll motion is measured

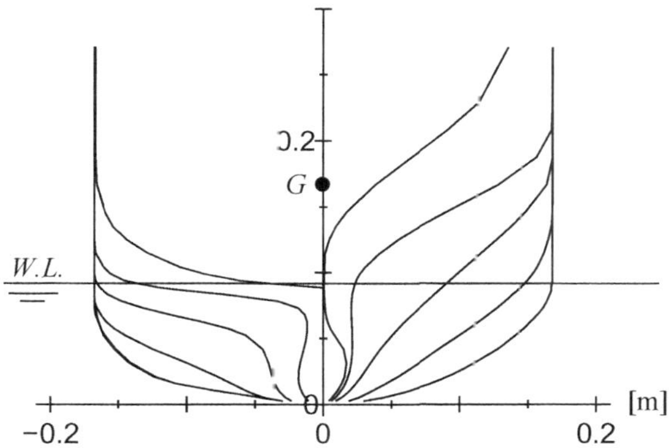

Fig. 29.1 Body plan of the model

Table 29.1 Principal particulars of the model

	Ship	Model
L_{pp} [m]	192	2.0
B[m]	32.26	0.336
draft: d[m]	9.0	0.0938
KG [m]	17.0	0.177
GM [m]	1.81	0.0189
Roll natural period: T_{ϕ} [sec]	18.42	1.88
Breadth of bilge-keels	0.7	0.0073
Position of bilge-keels	s.s. 3.34–s.s. 5.59	

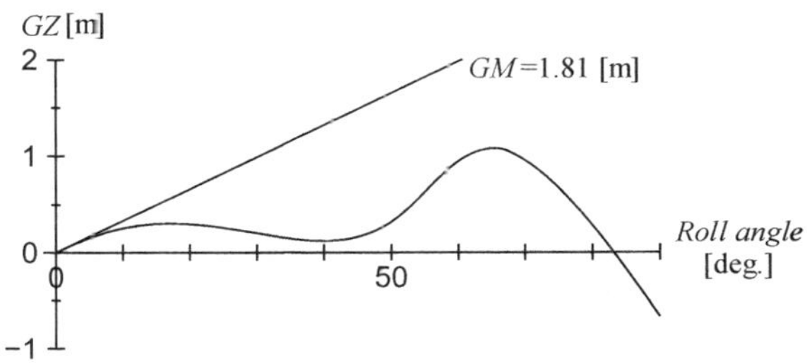

Fig. 29.2 Calculated *GZ*-curve of the model in real scale

by a gyroscope (CROSSBOW NAV440) and wave elevation is also measured by a servo type wave height meter attached to the towing carriage.

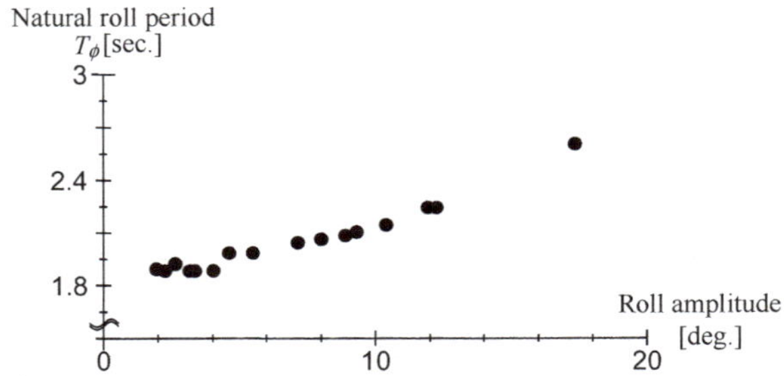

Fig. 29.3 Measured roll natural period of the model

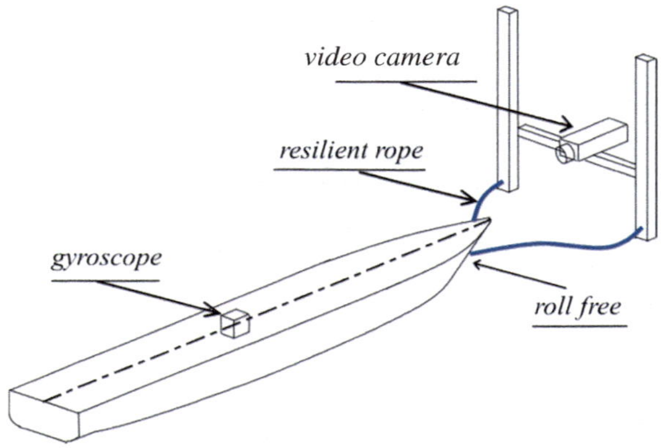

Fig. 29.4 Schematic view of the experiment

29.3 Roll Measurement in Regular Head Waves

Table 29.2 shows the experimental conditions. When the model is towed at $F_n = 0.083$ in head wave of $T_w = 1.05$ s, the encounter wave period is half of the roll natural period ($T_\phi = 1.88$ s) for small amplitude. This F_n is 7 kts in the real scale.

Figure 29.5 shows examples of typical time histories of measured rolling in regular head waves. The upper figure shows the result that oscillatory rolling does not occur. The middle figure shows the result that oscillatory rolling occurs. The bottom figure shows the result that oscillatory rolling occurs after a small disturbance which is

Table 29.2 Experimental condition

F_n	0.083
Wave period: T_w [sec]	0.99–1.55
T_e/T_ϕ	0.42–0.71
Wave height: H_w[m]	0.01–0.05

T_e:encounter wave period, T_ϕ:roll natural period

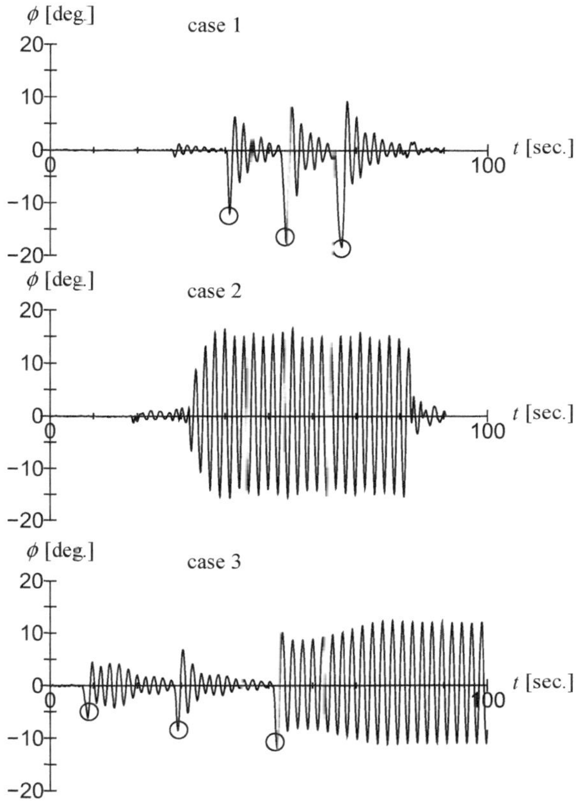

Fig. 29.5 Time histories of measured roll motions in regular head waves. (○:disturbance)

heel caused by hand, this phenomenon is investigated as a bi-stability mechanism by Spyrou et al. (2008).

Figure 29.6 shows measured roll amplitudes in regular head waves. In this figure, its horizontal axis is the ratio of encounter wave period to the roll natural period for small amplitude. At $T_e/T_\phi = 0.5$, oscillatory rolling occurs for any wave heights. For small wave height $H_w = 0.01$ m, oscillatory rolling does not occur except at $T_e/T_\phi = 0.5$.

Figure 29.7 shows the measured roll amplitude of a conventional post-panamax container vessel by Taguchi et al. (2006) in order to compare with Fig. 29.6. From the comparisons, it is noted that the range of T_e/T_ϕ where oscillatory rolling occurs in Fig. 29.6 is wider than that of Fig. 29.7 even if its amplitude is smaller, and the range of Fig. 29.6 spreads wider area of $T_e/T_\phi > 0.6$. Especially, at the range over $T_e/T_\phi = 0.5$. in Fig. 29.6, oscillatory rolling is caused by large wave height, and its amplitude becomes larger than that at $T_e/T_\phi = 0.5$. It is supposed that the spread of occurrence of large amplitude rolling to wider area of $T_e/T_\phi > 0.5$ in Fig. 29.6 is caused by the strong nonlinearity of GZ-curves.

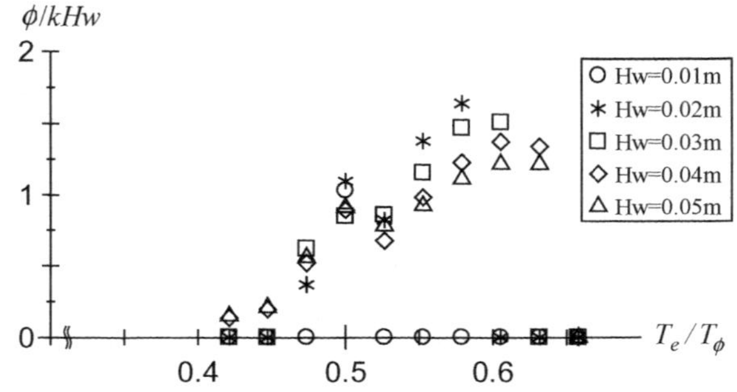

Fig. 29.6 Measured roll amplitude in regular head waves at $F_n = 0.083$. (Case 3: $T_e/T_\phi = 0.53, 0.55, 0.58$ at $H_w = 0.02$ m, $T_e/T_\phi = 0.61$ at $H_w = 0.03$ m, $T_e/T_\phi = 0.63$ at $H_w = 0.04$ m)

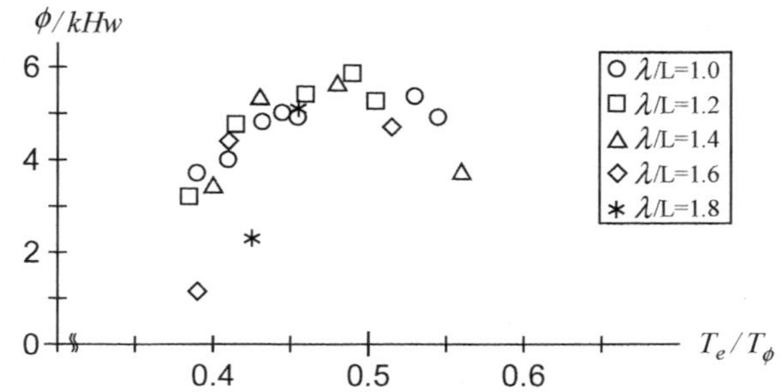

Fig. 29.7 Measured roll amplitude in regular head waves at $H_w = 0.11$ m. (citation from Fig. 6 in Taguchi et al. 2006)

Table 29.3 Experimental conditions

λ/L_{pp}	1.0		
F_n	0.000	0.032	0.077
T_e/T_ϕ	0.60	0.55	0.5

Table 29.3 shows the experimental condition for the roll measurement in regular head waves with constant wave length ($\lambda/L_{pp} = 1$). When model is towed at $F_n = 0.077$, its encounter wave period is half of the roll natural period for small amplitude ($T_\phi = 1.88$ s). According to decrease in forward speed, its encounter period becomes long.

Figure 29.8 shows the measured roll amplitude. In this figure, roll amplitude increases according to decrease in forward speed, even if T_e/T_ϕ becomes larger than 0.5.

In order to make its reason clear, the characteristics of roll damping is investigated by using Ikeda's method, and roll amplitude is estimated by using a simplified estimation for amplitude of parametric rolling (Katayama et al. 2009) .

Figure 29.9 shows the estimated roll damping. The roll damping decreases according to decrease in forward speed. It is caused by decrease in the lift component of roll damping which is caused by forward speed.

Figure 29.10 shows the estimated roll amplitude by Eq. (29.1) (Katayama et al. 2009).

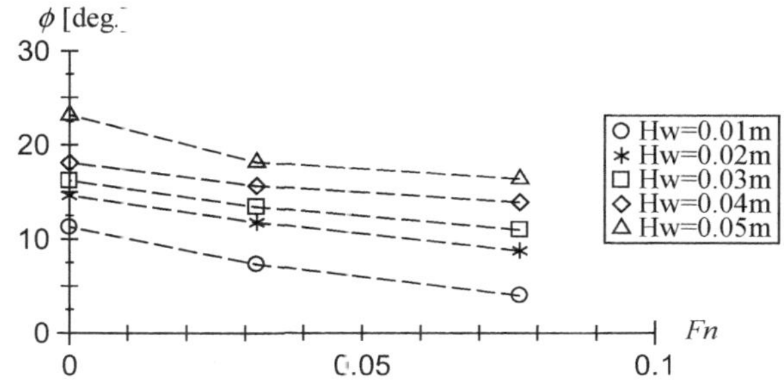

Fig. 29.8 Measured roll amplitude

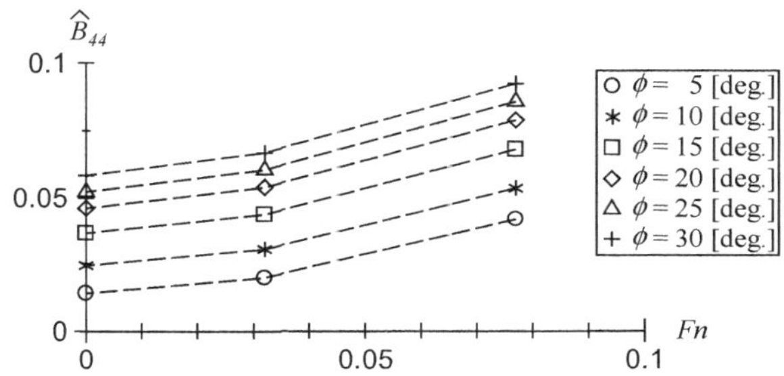

Fig. 29.9 Estimated roll damping coefficient by Ikeda's method

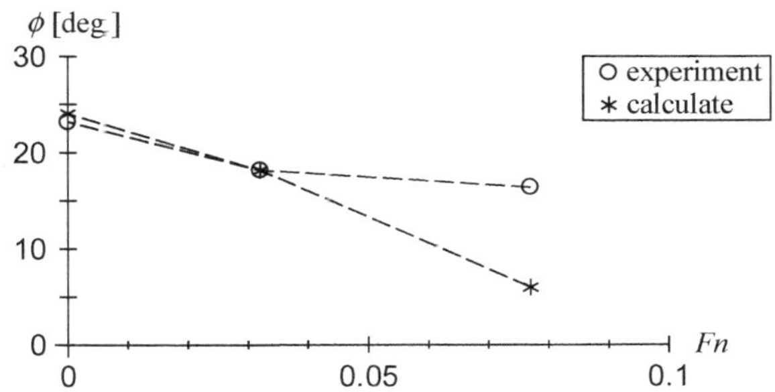

Fig. 29.10 Estimated roll amplitude by Eq. (29.1). (H_w = 0.05 m)

$$B_{44}(\phi_a) = \frac{W \Delta GM}{2\omega_e} \tag{29.1}$$

The measured results are also shown in the same figure. Both results show the same tendency for change of forward speed.

From these results, it is understood that decrease in forward speed causes decrease in roll damping, and roll amplitude becomes larger. On the other hand, roll natural period becomes long according to increase in roll amplitude as shown in Fig. 29.3. Then rolling resonates with increased wave encounter period by decrease in forward speed, and parametric rolling occurs.

29.4 Roll Measuring in Irregular Head Waves

Equation (29.2) is one of Pierson-Moskowitz spectrum.

$$S(\omega) = \frac{8.1 \times 10^{-3} \times g^2}{\omega^2} \exp\left(- \frac{3.11/H_{1/3}^2}{\omega^4} \right) \tag{29.2}$$

To make irregular waves, the spectrum is divided into 1000 equally in $\omega =$ 1–15 rad/sec and a sine wave of each frequency component is superposed. In addition, the phase difference of each frequency component is given as random numbers. The number of measurements is decided by the convergence of variance of rolling (Appendix).

In this study, roll measurements are carried out for three significant wave heights, and Fig. 29.11 shows the power spectral densities obtained from the measured wave elevations.

Figure 29.12 shows the probability distributions of the measured roll amplitudes. In this measurement, roll amplitude is not over 10 deg at significant wave height = 0.04 m. The results show that the roll amplitude increases with increase in significant wave height.

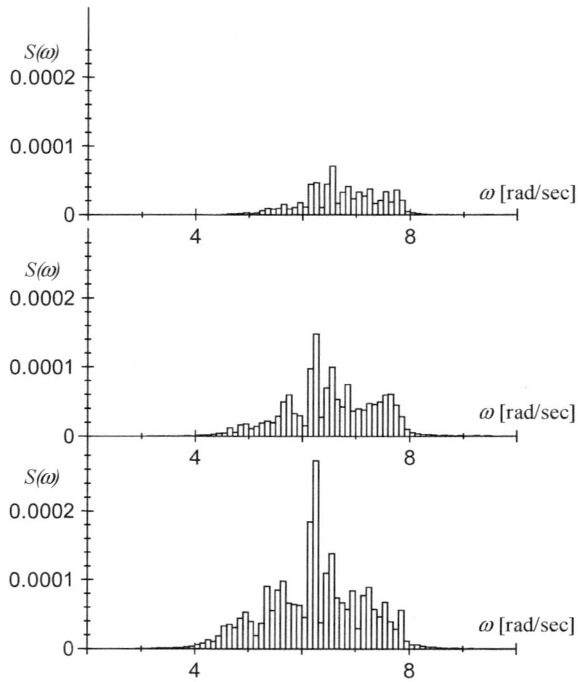

Fig. 29.11 Measured power spectral density of wave height. (1/3 significant wave height = 0.04, 0.05, 0.06 m)

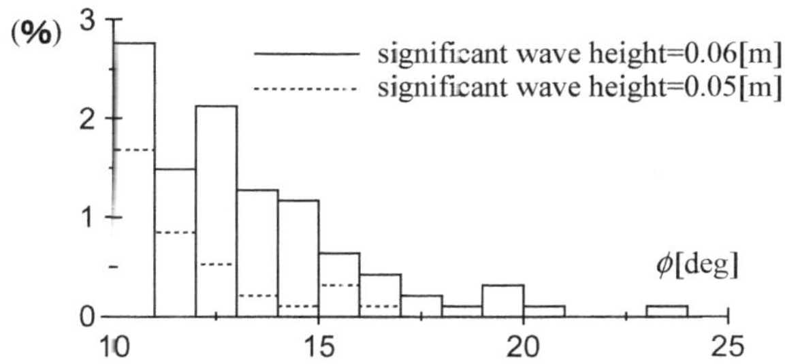

Fig. 29.12 Probability distribution of roll amplitude. (significant wave height = 0.05, 0.06 m)

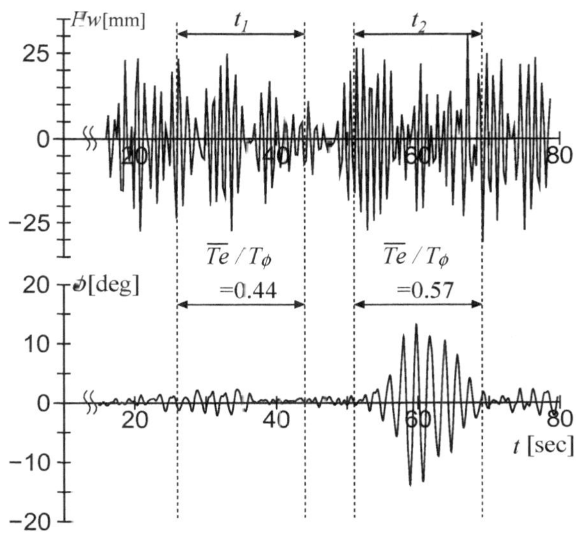

Fig. 29.13 Time histories of measured wave height and roll motion in irregular head waves at $H_w = 0.06$ m.)

Figure 29.13 shows time histories of measured wave height and rolling at $H_w = 0.06$ m. In this figure, the ratios of mean encounter wave period to the roll natural period for small amplitude (T_e/T_ϕ) at t_1 and t_2 are also shown. The mean wave heights at t_1 and t_2 are almost same. From this figure, it is found that oscillatory rolling does not occur at t_1 ($T_e/T_\phi < 0.5$), and oscillatory rolling occurs at t_2 ($T_e/T_\phi > 0.5$).

In the Pierson-Moskowitz spectrum expressed by Eq. (29.2), increase in significant wave height causes increase in mean wave period, therefore the occurrence probability of a wave group containing frequency component of $T_e/T_\phi > 0.5$ mainly also becomes higher. On the other hand, the roll natural period becomes longer according to increase in roll amplitude. Therefore, rolling may resonate the wave group containing frequency component of $T_e/T_\phi > 0.5$ mainly.

29.5 Roll Measuring in Head Waves of Two Different Wave Period

In the roll measurement in irregular head waves, when mean wave period of group waves is $T_e/T_\phi > 0.5$, oscillatory rolling occurs. In order to investigate its reasons, the roll measurements are carried out in the waves which are generated by the superposition of two sine waves having different periods. One of two different periods is half ($T_e/T_\phi = 0.5$) of the roll natural period for small amplitude. Moreover, the wave heights of two sine waves are different. (In the following, the component with higher wave height is called Main-wave.) Experimental conditions are shown in Table 29.4 with the results.

In Table 29.4, measured roll amplitude is shown. The results show that oscillatory rolling occurs when period of Main-wave is $T_e/T_\phi = 0.50$–0.61. As the results, it is confirmed that rolling can resonate with a wave with $T_e/T_\phi = 0.50$–0.61 and it is supposed that oscillatory rolling may occur in a frame of wide line.

29.6 Numerical Simulation

By utilizing the numerical simulation model (Hashimoto and Umeda 2010) , the effects of nonlinearity of *GZ*-curve on the occurrence of parametric rolling are investigated. In the simulation model, a 3DOF sway-heave-roll motion is solved in time-domain based on a nonlinear strip method in which dynamic components, i.e. radiation and diffraction forces, are calculated for an asymmetric submerged hull due to roll. Hydrodynamic forces for the heave and diffraction modes are determined

Table 29.4 Measred results

	T_e/T_ϕ ($H_w = 0.01$ m)							
T_e/T_ϕ ($H_w =$ 0.04 m)		0.39	0.45	0.50	0.55	0.61	0.66	0.71
	0.39			×				
	0.45		×	×				
	0.50	10.4	14.1		18.4	15.4	18.3	16.8
	0.55			8.3				
	0.61			16.0				
	0.66			×			×	
	0.71			×				×

ϕ [deg]: oscillatory rolling occurs
×: oscillatory rolling does not occur
Colored: oscillatory rolling occurs in regular waves

at the encounter frequency and those for the sway and roll modes are done at the half the encounter frequency.

In order to control the nonlinearity of *GZ*-curve, the breadth-draft ratio (*B/d*) of original model is changed under the constant displacement. Figures 29.14 and 29.15 show modified body plans and *GZ*-curves. *GM* for each modified model is adjusted to the same *GM* as the original model. Moreover, in order to make sure the characteristics of ΔGM of the original and modified models, the estimated ΔGM by the simplified method (Umeda et al. 2011) are shown in Fig. 29.16.

Figure 29.17 shows the numerical results for the original model. The calculated results show large amplitude parametric rolling occurs in the wide range over $T_e/T_\phi = 0.5$ and the numerical result shows a good agreement with the measured results.

Figures 29.18, 29.19 and 29.20 shows the calculated results for the modified models. These results show the roll amplitudes in the range over $T_e/T_\phi = 0.5$ become larger according to the increase in nonlinearity of *GZ*-curve.

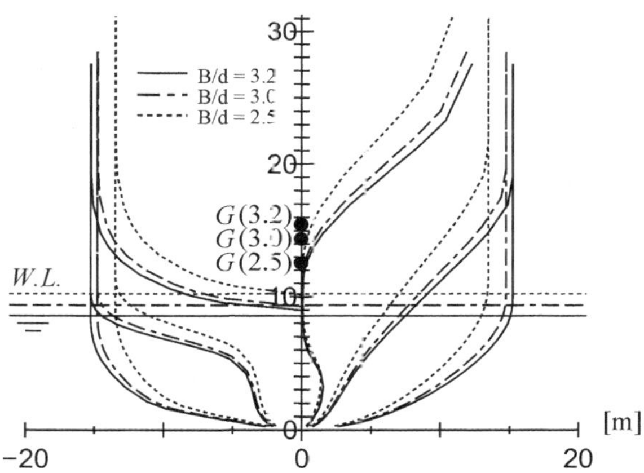

Fig. 29.14 Body plan in real scale. (*B/d* = 3.2, 3.0, 2.5)

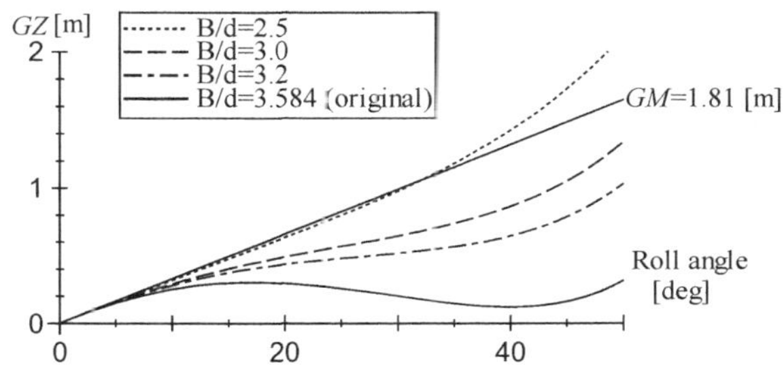

Fig. 29.15 *GZ*-curve of the model in real scale. (*B/d* = 3.584, 3.2, 3.0, 2.5)

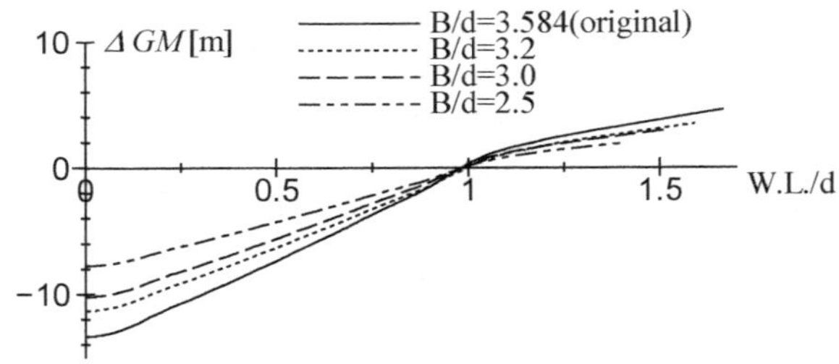

Fig. 29.16 Variation of GM caused by change of water surface of the models in real scale. (*B/d* = 3.584, 3.2, 3.0, 2.5)

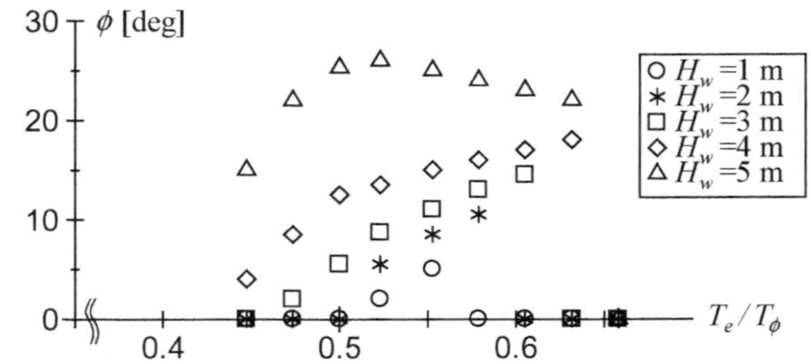

Fig. 29.17 Calculated roll amplitude in regular head waves at *Fn* = 0.083. (*B/d* = 3.584)

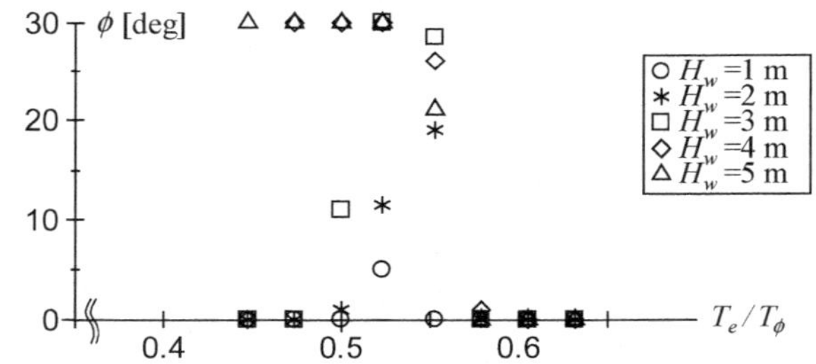

Fig. 29.18 Calculated roll amplitude in regular head waves at *Fn* = 0.083. (*B/d* = 3.2)

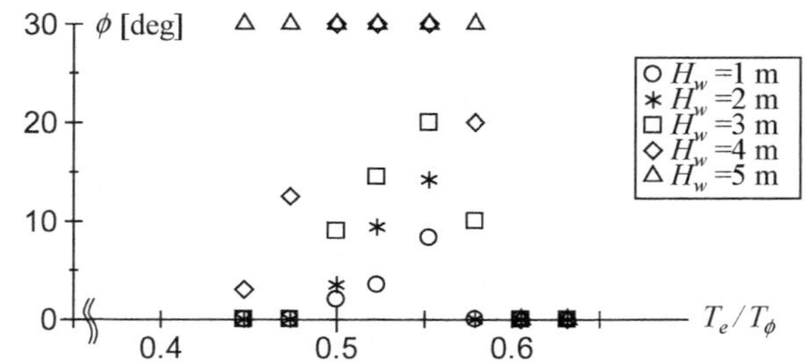

Fig. 29.19 Calculated roll amplitude in regular head waves at *Fn* = 0.083. (*B/d* = 3.0)

29.7 Conclusions

In this study, the roll characteristics of a wide breadth and shallow draft vessel in head waves are investigated. And the following conclusions are obtained.

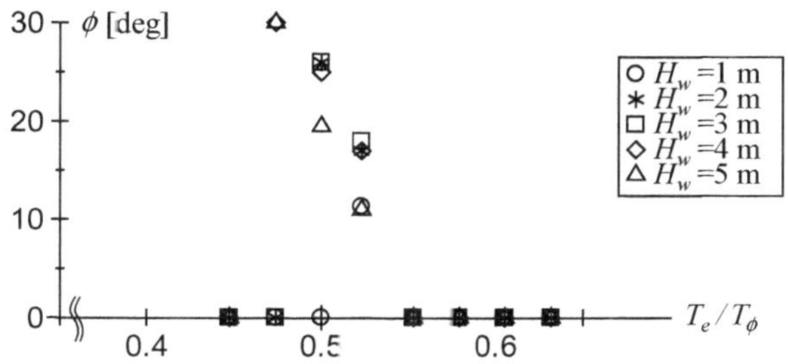

Fig. 29.20 Calculated roll amplitude in regular head waves at $Fn = 0.083$. ($B/d = 2.5$)

- In regular head waves, a oscillatory rolling occurs at wider range of encounter frequency than that of a conventional vessel. Especially, in the range of $T_e/T_\phi > 0.5$, a oscillatory rolling is caused by large wave height and roll amplitude becomes large.
- Numerical simulations demonstrate that the large amplitude parametric rolling occurs in the wide range over $T_e/T_\phi = 0.5$ because of the strong nonlinearity of GZ-curves.
- There is the possibility that slowdown increases roll amplitude, when parametric rolling occurs at $\lambda/L_{pp} = 1.0$.
- It is confirmed by simplified estimation that this phenomenon is caused by the following scenario. Slowdown decreases roll damping, and roll amplitude becomes large. On the other hand, roll natural period becomes longer with increase in roll amplitude. Therefore, rolling resonates with increased wave encounter period by slowdown.
- In irregular head waves with a Pierson-Moskowitz spectrum, a oscillatory rolling is caused by resonating with the group waves which contain the frequency component of $T_e/T_\phi > 0.5$ mainly.

Appendix

The Number of Times of Test in Irregular Head Waves

From wave height, pitch motion and roll motion, each variance and ensemble mean by Eqs. (29.3), (29.4). Figure 29.21 shows calculated result. In these figure, a vertical axis is a ratio ensemble mean to standard deviation. Upper and middle figure uses 20 samples and bottom figure uses 30 samples. (1 sample = 40 s) From upper and middle figure, it is found that wave height and pitch motion convergence with 5 samples. On the other hand, roll motion does not converge until 15 samples. So, this study uses 20 samples.

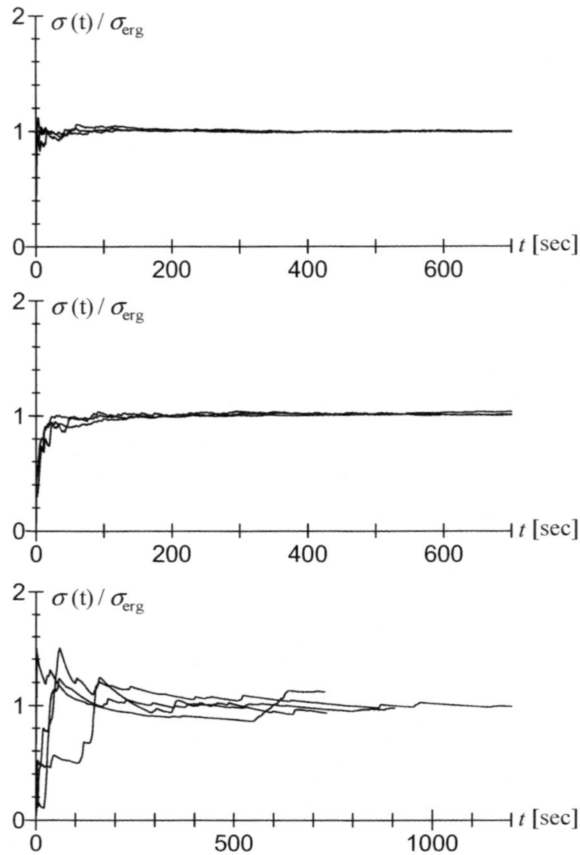

Fig. 29.21 Standard deviation of wave elevation, pitching and rolling. (1/3 significant wave height = 0.04, 0.05, 0.06 m)

$$\sigma^2(t) = \sum_{j=1}^{n_s} \frac{(X_j(t_j) - \bar{X}(t))^2}{n-1} \tag{29.3}$$

$$\sigma_{erg}^2 = \frac{\sum_{i=1}^{N} \sum_{j=1}^{n} \frac{(X_i(t_j) - \overline{X_i})^2}{n-1}}{N} \tag{29.4}$$

References

Taguchi H, Ishida S, Sawada H, Minami M (2006) Model Experiment on Parametric Rolling of a Post-Panamax Containership in Head Waves. Proceedings of the Ninth International Conference on Stability of Ships and Ocean Vehicles.

Hashimoto H, Umeda N (2010) A Study on Quantitative Prediction of Parametric Roll in Regular Waves. Proceedings of the 11th International Ship Stability Workshop: 295–301.

Tai Y, Ohashi T, Katayama T, Ikeda Y (2011) An Experimental Study on Stability of a Shallow Draft Vessel in Beam Seas. Journal of the Japan Society of Naval Architects and Ocean Engineers 14: 47–54.

Hashimoto H, Sueyoshi M, Minegaki S (2007) An Estimation of the Anti-Rolling Tank Performance for Parametric Rolling Prevention. Journal of the Japan Society of Naval Architects and Ocean Engineers 6: 305–311.

Ikeda Y (2004) Prediction Methods of Roll Damping of Ships and Their Application to Determine Optimum Stabilization Devices. Marine Technology 41:89–93.

Katayama T, Taniguchi T, Umeda N (2009) An Experimental Study on Parametric Rolling of a High Speed Trimaran in Head Seas. Journal of the Japan Society of Naval Architects and Ocean Engineers 10:57–63.

Spyrou KJ, Tigkas I, Scanferla G, Pallikaropoulos N, Themelis N (2008) Prediction potential of the parametric rolling behaviour of a post-panamax containership, Ocean Engineering, 35, 11, pp. 1235–1234.

Umeda N, Hashimoto H, Tsukamoto I, Sogawa Y (2011) Estimation of Parametric Roll in Random Seaways. In Fossen TI, Nijimeijer H (eds) Parametric Resonance in Dynamical Systems. Springer, 45–52

Chapter 30
Experimental Ship Dynamic Stability Assessment Using Wave Groups

Christopher C. Bassler, Martin J. Dipper, Jr. and Mark Melendez

Abstract The assessment of ship performance in heavy weather, particularly dynamic stability performance, is an important but difficult assessment to make. Traditional experimental assessment methods using regular and random waves provide insight into dynamic stability performance, but may not identify, or provide a means to mitigate, specific modes of dynamic stability failure. Assessment using deterministic wave groups may provide repeatability and systematic exposure important for the assessment of ship designs, as well as aid in development and validation of numerical simulation tools. The deterministic grouped wave approach, when used to define ship behavior in heavy weather, can also be useful in the development of ship-specific operator guidance.

Keywords Wave groups · Deterministic model testing · Ship design
Operator guidance

30.1 Introduction

Assessment of ship performance in heavy weather remains an important but difficult undertaking, due to the significant degree of nonlinearities associated with ship response to large steep waves. For heavy weather operations, it is important to assess ship performance related to crew performance and safety, mission effective-ness, and ultimately platform survivability.

Experimental assessments continue to be necessary to assess the performance of new designs, evaluate response in the most severe conditions, and enable the continued development and validation of numerical tools. An efficient, yet accurate, method is still needed for use in the early-design stage when many hull form concepts, or parametric variations in hull form, are still being evaluated. Additionally, later in the ship design process, an efficient and accurate assessment method is needed for

---

C. C. Bassler (✉) · M. J. Dipper, Jr. · M. Melendez
David Taylor Model Basin (NSWCCD), West Bethesda, MD, USA
e-mail: cbassler@gmail.com

V. L. Belenky et al. (eds.), *Contemporary Ideas on Ship Stability*, Fluid Mechanics and Its Applications 119, https://doi.org/10.1007/978-3-030-00516-0_30

the development of ship-specific operator guidance. In the foreseeable future, model experiments will remain a necessary component of gaining new knowledge of ship performance, evaluating ship designs, and the development ship-specific operator guidance for the most severe wave conditions.

In this paper, an experimental procedure is discussed to systematically evaluate dynamic stability performance of a ship in heavy weather conditions. The procedure relies on the ability to experimentally generate deterministic wave sequences, based on ship-specific information. Using deterministic wave sequences, determined from ship-specific wave properties, a systematic model experiment program may be developed to reproduce and repeat conditions which will result in an undesired motions response. These properties also enable the determination of the probability of occurrence of rare wave events, which have critical characteristics, resulting in undesirable ship response. These deterministic wave sequences will enable efficient realizations in the basin, allowing for the investigation of parametric design changes, as well as assisting with the development of ship-specific operator guidance.

30.2 Current Dynamic Stability Assessment Procedures

A short discussion of two primary methods for dynamic stability assessment is provided. These include testing in regular waves, and testing in random waves. For the purpose of this discussion, both wave environment approaches employ remotely controlled ship models. The advantages and disadvantages of each approach are briefly summarized. For most new ship designs, experimental testing in a basin will remain the method of choice.

30.2.1 Regular Wave Testing

Traditional dynamic stability assessment methods, experimental as well as numerical, often use regular waves, which are fundamental and easily represented mathematically. An example of a ship model test from the Naval Surface Warfare Center, Carderock Division (NSWCCD) Maneuvering and Seakeeping (MASK) Basin with DTMB Model #5514 (Hayden et al. 2006) is shown in Fig. 30.1.

For many dynamic stability related phenomena, an initial assessment made with regular waves of varying steepness ($H/\lambda \sim 1/10$) and varying length ($\lambda/L \sim 1.0$) can provide an indication of seakeeping and dynamic stability performance. These fundamental experiments enable a distinction to be made between adequate ship designs and infeasible ship designs. However, because regular waves are not directly representative of the natural wave environment in which a ship operates, subtle performance issues may not be able to be assessed or distinguished.

For hull forms which may have particular design or operational constraints, regular wave testing may not provide insight into performance in specific limiting seaway

Fig. 30.1 Experimental testing of ship performance in heavy seas using regular waves

conditions. For these hull forms, regular wave testing is only the first step in the assessment process and can be used to compare new hull forms to previous designs. The operational experience from previous designs can be used to determine the general scope of potential operational restrictions for the new hull form, and eliminate infeasible or unsatisfactory hull form designs from consideration.

30.2.2 Random Wave Testing

The next level of complexity in experimental assessment uses random waves, which are more representative of the natural wave environment. These more realistic conditions can provide additional insight into the behavior of a ship in sea conditions deemed critical to its hull design and load condition. For most dynamic stability related phenomena, this next level assessment made with random waves of varying significant height, modal period, and spectral shape, can provide a more realistic indication of seakeeping and dynamic stability performance.

Random wave testing relies on long exposure times in the model basin to observe critical, but statistically rare, events which can occur in a given seaway. To achieve long exposure times, multiple test realizations in the seakeeping basin are required. Random wave testing can be used to determine response to a limited range of severe seaway conditions and provide more details regarding potential operational restrictions for the hull form.

However, these long exposure times may be practically limited by the ability of the wave-maker in the experimental basin to generate specific, more complex wave conditions, and also by concerns over reflections and self-repeating of the generated waves. Additionally, the time (and cost) associated with experimental testing limits the number of seaway conditions to which the ship may be exposed.

Limits associated with run time, reflections, and statistical uncertainty, may be alleviated by the use of large-scale model testing in open-water conditions, to enable long exposure times. However, the lack of control over weather conditions, and the techniques used to measure environmental conditions in open water testing become technical challenges which must be addressed.

Once a rare event is observed in the basin, repeatability can become an issue in random wave testing. Because of the nonlinearity of ship motion response at large angles of pitch and roll, there is sensitivity to initial wave and model attitude conditions. In random wave testing, lack of control of initial conditions of both the environment and the ship motions prior to encountering a critical event, may result in significant difficulty in reproducing the event, such as a large roll angle, in a particular seaway condition. Some gross approximation of the initial conditions, such as relative phase with the wave, can be made, but the specific initial conditions for the ship at the initialization of an encounter with a wave sequence resulting in a critical event are generally not provided. This not only makes probabilistic assessment of the dynamic stability performance difficult, but also is a challenge for systematically assessing parametric design changes and validating numerical simulations of ship performance in the seaway.

30.3 Difficulties in Dynamic Stability Assessment

As discussed, critical ship motions are most often the result of either extremely high or extremely steep waves, or a particular sequence of waves. However, the rarity of occurrence of these wave events makes the assessment of ship response in these conditions difficult. The response of a ship to these critical wave events is expected to result in large amplitude motion, and to be significantly influenced by nonlinearities from wave forcing, damping, and hydrostatic restoring. As is well known, when a dynamical system has significant nonlinearities, its behavior becomes very sensitive to initial conditions (Poincaré 1890; Lorenz 1963). Depending on the initial conditions, the ship response to a large wave may range from small motions, to catastrophic motions including possibly capsizing. The difficulty with the assessment of dynamic stability "failures," is as result of both their rarity of occurrence and the significant nonlinearity of the ship response in these seaway conditions, and both must be addressed simultaneously.

30.3.1 Problem of Rarity and the Principle of Separation

Assessment of the dynamic stability behavior in random seaway conditions constitutes the general problem of rarity—when the time between "failure" events is long, compared to a specific time-scale of interest, such as roll period (Belenky et al. 2008, 2010). As discussed in Belenky, et al., the principle of separation considers distinguishing the nonlinear phenomena which results in an undesired ship response from the conditions which lead to its occurrence. This enables the possibility of modeling of the ship response as a combination of two sub-problems: non-rare and rare. The non-rare problem is used to determine the probability of occurrence of conditions which may lead to severe response, and determining the distribution of the appropriate initial conditions. The rare problem is used to determine whether large responses occur for a particular set of initial conditions.

As further discussed in Belenky (2008, 2010), the main assumption behind the principle of separation is that a mechanical system can be "restarted" at a particular moment of time, *if* the state variables at that instant are fully determined. This is an assumption for the ship response, because the hydrodynamic memory effect cannot be fully considered in this form.

30.3.2 Assessment of Parametric Design Changes

In addition to the difficulty of assessing dynamic stability, due to the nonlinearities and rarity, systematic assessment of parametric design changes are also difficult. Just as the ship performance is highly sensitive to initial conditions, subtle changes in hull form geometry or appendages (such as bilge keels, rudders, etc.) may have significant impact on the dynamic stability performance. In order to assess this impact in realistic seaway conditions, repeatability becomes very important.

An example of this was the study of the effect of topside geometry variation on dynamic stability performance, using the ONR Topside Series (Fig. 30.2), which were designed at NSWCCD by Dipper, Campbell, and Belknap (Bishop et al. 2005) and include flared, tumblehome, and wall-sided variants (DTMB Model #5613, 5613-1, and 5613-2). Similar to DTMB Model #5514, they have become an international standard for ship performance testing of naval combatant type hull forms.

Fig. 30.2 ONR Topside Series Hull Forms, including tumblehome (top left, bottom), wall-sided (top middle, middle), and flared (top right, top) topside variations

30.3.3 Validation of Numerical Simulations

Repeatability is important for assessing the effect of parametric design changes, and is also important for validation of numerical simulations. The use of numerical simulations can become a practical method for assessing a large number of ship designs, or for developing ship-specific operator guidance. However, numerical tools with the required fidelity to assess design changes or provide the level of accuracy necessary for detailed operator guidance are still in a developmental stage. Therefore, it is currently necessary to use experimental assessments for development and validation of these numerical simulation tools. A detailed discussion regarding precisely what constitutes validation is outside the scope of this paper, but is a subject of much recent development.

In order to provide useful data for this, initial conditions, wave conditions, and ship response must all be measured and recorded at a significant level of detail. As discussed, this is currently difficult in the random wave testing methods typically used for more physically realistic ship dynamic stability assessments.

30.4 Wave Group Approach for Dynamic Stability Assessment

Because of the practical limitations of providing long exposure times and a broad range of seaway conditions for ship models in basin experiments, several alternative methods have previously been proposed. A more detailed review is given in Bassler et al. (2008, 2009). At this time, the approach that appears most likely to address the aforementioned issues for experimental dynamic stability assessment is the wave group approach.

The concept for this method is to extract a sequence of waves which can result in large amplitude excitation of the ship model and evaluate the dynamic response to these particular sequences of waves, or "wave groups," with random initial conditions. A definition for this type of wave sequence, or wave group, from the ship response perspective is proposed in Bassler et al. (2010, 2010a). The first complete implementation of this type of approach with quantitative results was proposed during the SAFEDOR project (Spyrou and Themelis 2005; Themelis and Spyrou 2007, 2008). A similar approach has also been followed by Umeda et al. (2007).

30.4.1 Overview

The goal of using the wave group approach for experimental evaluation of ship dynamic stability performance is to:

(1) be able to evaluate and compare specific hull form design variants and loading conditions
(2) determine specific potential operational restrictions for a given hull form design,
(3) verify simulation tools and aid with further development
(4) provide an assessment process that can be used to eliminate infeasible or unsatisfactory hull form designs.

The wave group method relies on two primary components: precise generation of a deterministic wave-field, and control of initial conditions.

30.4.2 Wave-Field Generation

Within one seaway spectrum, an infinite number of seaway time series realizations can be generated. However, only a small subset may have wave sequences which result in critical ship motions. These wave sequences are then the conditions of interest for ship dynamic stability performance and extreme event assessment.

In order to address the previously discussed issues for systematic exposure and repeatability, the wave-field generation process must be deterministic, such that the time and location that wave groups will occur in the model basin is known. It has been shown that groups of large waves, as well as single large waves, can be reproduced deterministically in an experimental basin (e.g. Davis and Zarnick 1964; Takezawa and Takekawa 1976; Clauss 2000; Clauss et al. 2008; Bassler et al. 2008, 2009). However, modern paddle-type wave-makers can be used to generate more complex seaways in a deterministic manner (Fig. 30.3). A new wave-maker is currently under construction for NSWCCD (Hayden et al. 2010), as shown in Fig. 30.4, and will provide improved capability to deterministically generate complex wave-fields.

It is also important to characterize the local wave-field surrounding the ship up to and including the dynamic stability failure event. Typically an array of ultrasonic wave probes is used in the MASK basin. However, to provide a more dense set of high-resolution local wave-field measurements, several other techniques may be considered, such as LIDAR or the Global Laser Rangefinder Profilometry (GLRP) technique, which was developed at NSWCCD (Atsavapranee et al. 2005; Carneal et al. 2005, 2005a; Carneal and Atsavapranee 2006). GLRP was used to characterize

Fig. 30.3 A sample visualization of a complex seaway generated in the NSWCCD MASK basin, using the new wave-maker

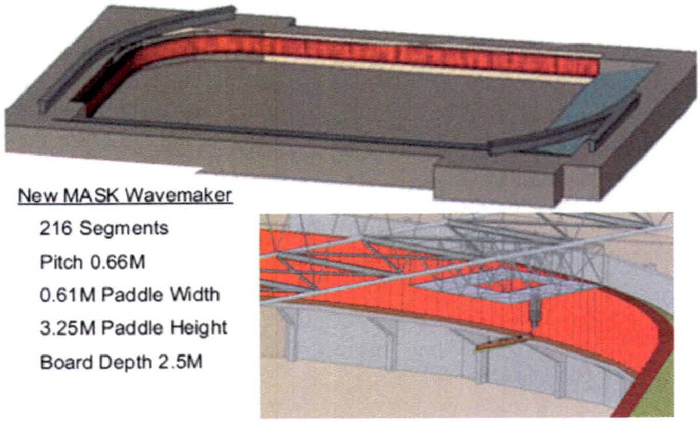

Fig. 30.4 NSWCCD MASK basin with new wave-maker (top), with model release mechanism shown attached to the basin carriage in the Northwest corner (bottom)

deterministic realizations of steep waves in the MASK basin by Bassler et al. (2008, 2009)—see Fig. 30.6.

Sub-surface measurement methods, such as particle image velocimetry (PIV), have also been used to characterize wave kinematics (Minnick et al. 2010, 2011, 2011a). In addition to the surface field measurement methods, this can provide a complete quantitative picture of the waves in the near-field to the ship model. Then, a complete data set of wave conditions (free-surface and kinematics) will be available to compare with ship model performance, and also provide a more complete characterization for the development and validation of numerical simulation tools. Future wave-tracking methods may even be developed to provide detailed measurement of the instantaneous waterline along the hull, which will enable additional insights into the primary mechanism for dynamic stability failures due to righting arm variations in waves.

Coupled with the precise local wave-field measurements, the model position can be tracked in the basin using a laser-tracker system, such as the *iGPS* system currently used at NSWCCD in the MASK basin. The *iGPS* system uses several spinning eye-safe laser transmitters with infrared LED flashes to provide stable constellation references along the perimeter of the MASK basin. In addition, an onboard detector and Position Calculation Engine (PCE) are used in the model to determine the XYZ coordinates of the detector's position in the basin, with respect to the known location of the transmitters. Using three or more detectors on a model provides six degree-of-freedom (6 DOF) motion data for any object in the MASK basin.

30.4.3 Initial Condition Control—A Model Release Mechanism

With the ability to repeatedly generate deterministic wave-fields, initial condition control is the next important step. For initial condition control, a model release mechanism can be used to enable repeatable, measurable initial conditions for a free-running remote controlled ship model in the basin. Previous attempts included mechanisms based on using electro-magnets or solenoids, but proved to be infeasible, due to concerns over instrumentation interference issues and reliability issues, so a new concept was developed.

For dynamic stability testing, a model release mechanism was designed (Figs. 30.5 and 30.6) to allow the ship model to be free to pitch, heave and roll, but fixed in surge, sway and yaw. Additionally, the mechanism could also allow for the option to fix roll in various heel conditions. For precise initial condition control, the mechanism should negate as much as practically possible, the inertial impact of the mechanism on the motions of the model. Wave measurements near the model would enable accurate recording of the wave conditions with respect to the model at the instant of release. The mechanism was also intended to provide the ability to release the model at a controlled point in time, with respect to at least four phase positions in waves (crest, trough, front slope, and back slope). Lastly, the mechanism was also designed with the consideration that it should allow for easy retrieval and rest of the model between realization conditions in the basin.

For each deterministic wave sequence condition of interest, a minimum of thirty-six variations are recommended for investigation. These include three speeds, three headings, and four wave phase realizations. The three speeds would consist of the target speed of interest and speed conditions slightly slower and slightly faster. The three heading conditions would consist of the heading of interest (e.g. stern-quartering

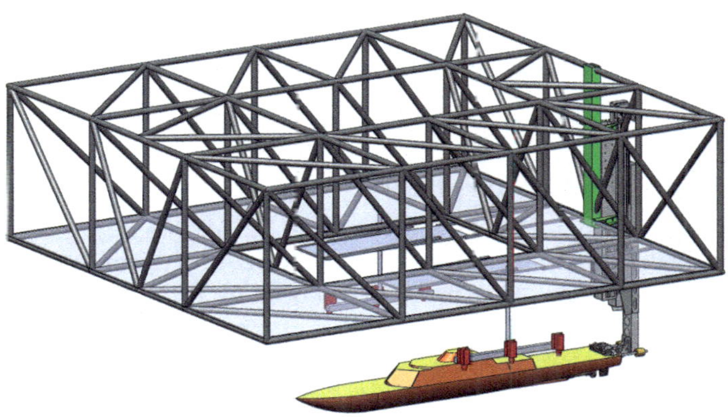

Fig. 30.5 Notional ship model attached the carriage with the model release mechanism and wave probes for local wave-field measurement shown (in red)

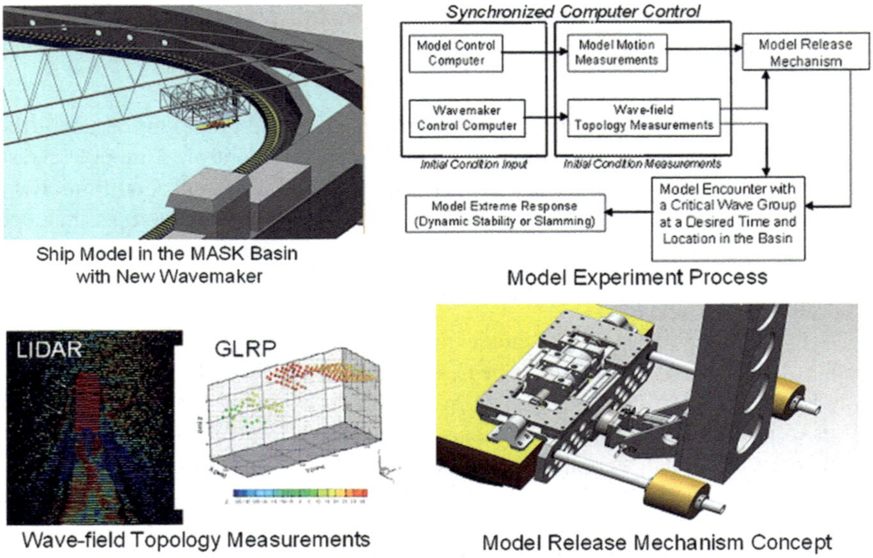

Fig. 30.6 Model release mechanism concept and process

seas) and headings slightly aft and slightly abeam. The four wave conditions would be the crest, trough, front slope, and back slope of the wave. A pre-determined acceleration profile would be programmed to enable the model to achieve the desired speed condition after release from the mechanism.

In addition to the typical instrumentation packages for ship motions, including gyroscopes and accelerometers, the remote control models at the appropriate scale can also be outfitted with instrumentation to record forces on appendages, while also recording instantaneous ship speed and propeller rpm. Pressure panels can also be installed on the hull and superstructure to measure wave-impact loads at specific regions of interest, or to measure deck wetness during green water occurrence.

30.4.4 *Experimental Technique*

After initial model tests are performed in regular waves to determine general behavior of the ship in a seaway, additional model experiments using the wave group technique can be performed to determine probabilistic risk assessment, to investigate parametric design changes, or to develop ship-specific operational guidance. The ship specific properties can also be used to identify wave sequence parameters that are more likely to result in undesirable dynamic stability performance. An example of this for ship roll motion is given in Bassler et al. (2010, 2010a).

These parameters can also be used to search for probability of occurrence of wave sequences of interest in particular geographic regions, enabling the coupling of the risk of dynamic stability failure events to occur, with likelihood of occurrence of wave sequences that may result in dynamic stability failure, assuming a particular set of initial conditions (Themelis and Spyrou 2007; Bassler et al. 2010a; Belenky et al. 2010).

The overall technique described is shown in Fig. 30.6. Synchronized computer control of the model and the wave-maker are necessary to enable the specified initial conditions to be achieved, in a given sea condition of interest. A feedback loop of the measured state conditions of the ship model and wave conditions near-field of the model can be used to determine the precise moment to release the model.

Because the wave-field in the basin is deterministic, for each set of initial conditions of the model, the wave phase can be systematically varied to examine the sensitivity of the end-state response for the wave sequence of interest.

30.4.5 Potential Benefits of the Wave Group Approach

The wave group approach provides a practical method to experimentally assess ship dynamic stability performance. Systematically exposing the ship model to wave conditions which yield a high probability of failure can provide a more accurate and detailed determination of dynamic stability performance. This process lends itself to parametric modeling, allowing assessment of the parametric change in design by comparing performance in the same wave sequence. Because of the measurement of initial conditions and a deterministic wave-field, the data from this method of conducting experiments may be more helpful for conducting probablistic risk assessment and for the validation of numerical tools used to assess dynamic stability performance.

The wave group approach may be used to provide guidance for operational environments. The method can provide a way to link the ship performance with the likelihood of encountering critical wave conditions for a specific ship. Then, dangerous combinations of loading conditions and operational parameters (ship speed and heading) can be identified.

The wave group approach to experimental dynamic stability assessment appears to have the potential for increased accuracy, as well as time (and cost) savings, in experimental testing over traditional methods using random waves. However, regular wave testing should remain a crucial first-step for experimental assessment of dynamic stability performance.

30.5 Conclusions

A method using wave groups for experimental dynamic stability assessment was discussed. The method can reduce testing time in the model basin when a dynamic stability problem has been identified or a design change has been implemented. This process can improve the representation of the particular critical seaway conditions of interest, increase confidence in the dynamic stability assessment of ship performance, and increase confidence in safety in severe seaway conditions.

Using this method, parametric design variations may be examined in the same deterministic wave sequences, which can be precisely repeated for each design variation. This deterministic method also provides for more precise comparisons with simulation tools, enabling validation and further tool development. Quantification of initial conditions and the position of the model relative to the wave field at an instant in time also aides in simulation tool development. Because of the short run times inherent in the deterministic evaluation process, unsteady Reynolds-Averaged Navier Stokes (URANS) methods can also become more practical for conducting comparative assessments.

Longer-term, this deterministic method may also be considered for the identification of specific wave sequences that are critical to the ship. Then, with accurate on-board wave-field sensor measurements, and with faster than real-time nonlinear wave propagation and ship motion prediction methods, the possibility for real-time ship-specific operator guidance may be realized.

Although not yet examined, the wave group method coupled with model instrumentation techniques may also have the possibly to be used for the assessment of other rare events, such as slamming. However, this is the subject of future work.

Acknowledgements The authors would like to express their appreciation for the support of the work in the paper from the Naval Innovative Science and Engineering Program (NISE) at NSWCCD, under the direction of Dr. John Barkyoumb. Additionally, the authors also appreciate support from Mr. James Webster (NAVSEA) and Dr. Pat Purtell (Office of Naval Research) for previous research and development, which provided the foundation and initialization of this work. The authors would also like to thank Mr. John Hoyt and Mr. Dan Hayden (NSWCCD) and Prof. Kostas Spyrou (NTUA) for their helpful discussions regarding the contents of this paper.

References

Atsavapranee, P., J. B. Carneal, C. W. Baumann, J. H. Hamilton, and J. W. Shan (2005), "Global Laser Rangefinder Profilome-try (GLRP): A Novel Optical Surface-Wave Measurement System," Hydrome-chanics Dept. Technical Report, NSWCCD-50-TR-2005/022.

Bassler, C. C. G. E. Lang, S. S. Lee, J. B. Carneal, J. T. Park, and M. J. Dipper (2008), "Formation of Large-Amplitude Wave Groups in an Experimental Basin," Hydrome-chanics Dept. Technical Report, NSWCCD-50-TR-2008/025.

Bassler, C. C., M. J. Dipper, and G. E. Lang (2009), "Formation of Large-Amplitude Wave Groups in an Experimental Basin," *Proc. 10th Intl. Conf. on Stability of Ships and Ocean Vehicles*, St. Petersburg, Russia.

Bassler, C. C., V. Belenky, and M. J. Dipper (2010), "Characteristics of Wave Groups for the Evaluation of Ship Response in Irregular Seas," *Proc. 29th Intl. Conf. on Ocean, Offshore, and Arctic Engineering*, Shanghai, China.

Bassler, C.C., V. Belenky, and M.J. Dipper (2010a), "Application of Wave Groups to Assess Ship Response in Irregular Seas," *Proc. 11th Int. Ship Stability Workshop*, Wageningen, The Netherlands.

Belenky, V., J. O. de Kat, and N. Umeda (2008), "Toward Performance-Based Criteria for Intact Stability," *Marine Technology*, 45(2):101–123,.

Belenky, V., C. Bassler, M. Dipper, B. Campbell, K. Weems, and K. Spyrou (2010), "Direct Assessment Methods for Nonlinear Ship Response in Severe Seas," *Proc. ITTC Intl. Workshop on Seakeeping*, Seoul, Korea, 19–21 October.

Bishop, R.C., W. Belknap, C. Turner, B. Simon and J.H. Kim (2005), "Parametric Inves-tigation on the Influence of GM, Roll Damping, and Above-Water Form on the Roll Response of Model 5613," Hydrome-chanics Dept. Technical Report, NSWCCD-50-TR-2005/027.

Carneal, J. B., P. Atsavapranee, C. W. Baumann, J. H. Hamilton, and J. Shan (2005), "A Global Laser Rangefinder Profilometry System for the Measurement of Three Dimensional Wave Surfaces," *Proc. ASME Fluids Engineering Division Summer Meeting*, Houston, TX, USA, The Netherlands 19–23 June.

Carneal, J. B., P. Atsavapranee, and J. T. Curight (2005a), "Global Laser Rangefinder Profilometry: Initial Test and Uncertainty Analysis," Hydromechanics Dept. Technical Report, NSWCCD-50-TR-2005/069.

Carneal, J. B. and P. Atsavapranee (2006), ' Global Laser Rangefinder Profilometry: Initial Test and Uncertainty Analysis," *Proc. ASME Joint European Fluids Summer Meeting*, Miami, FL, USA, 17–23 July.

Clauss, G. F. (2000), "Tailor-made Transient Wave Groups for Capsizing Tests," *Proc. 7th Intl. Conf. on Stability of Ships and Ocean Vehicles*, Launceston, Tasmania, Australia.

Clauss, G. F., C.E. Schmittner, and J. Hennig (2008), "Systematically Varied Rogue Wave Sequences for the Experimental Investigation of Extreme Structure Behavior," *J. Offshore Mechanics and Arctic Engineering*, 130, May.

Davis, M.C. and E.E. Zarnick (1964), "Testing Ship Models in Transient Waves," *Proc. 5th Symp. on Naval Hydro.*, Bergen, Norway, 10–12 September.

Hayden. D. D., R. C. Bishop, J. T. Park, and S. M. Laverty (2006), "Model 5514 Capsize Experiments Representing the Pre-Contract DDG51 Hull Form at End of Service Life Conditions," Hydromechanics Dept. Technical Report, NSWCCD-50-TR-2006/020, April

Hayden, D. D., J. G. Hoyt III, M. Melendez, H. J. Moeller, Y. Bargman, S. Carpenter, and S. R. Turner (2010), "Naval Surface Warfare Center's Wavemaker Modernization Program," *Proc. 29th American Towing Tank Conference (ATTC)*, Annapolis, MD, USA, August.

Lorenz, E.N., "Deterministic Non-Periodic Flow," *J. Atmosperic Sci.* 20, 130–141, 1963.

Minnick, L., C. Bassler, S. Percival, and L. Hanyok (2010), "Large-scale Wave Kinematics Measurements of Regular Waves and Large-Amplitude Wave Groups," *Proc. 29th Int. Conf. on Ocean, Offshore and Arctic Engineering*, Shanghai, China.

Minnick, L. M., C. C. Bassler, S. Percival, and L. W. Hanyok (2011), 'Characterization of Regular Wave, Irregular Waves, and Large-Amplitude Wave Group Kinematics in an Experimental Basin" Hydrome-chanics Dept. Technical Report, NSWCCD-50-TR-2011/012, February

Minnick, L, C. Kent, C. Bassler, S. Percival, and L. Hanyok (2011a), "Kinematics of Experimentally Generated Severe Wave Conditions and Implications for Numerical Models," *Proc. 30th Intl. Conf. on Offshore Mechanics and Arctic Engineering*, Rotterdam, The Netherlands, 19–24 June

Poincaré, H.J. (1890), "Sur le problème des trois corps et les équations de la dynamique," *Acta Mathemctica*, 13, 1–270.

Spyrou, K. J. and N. Themelis (2005), "Probabilistic Assessment of Intact Stability," *Proc. 8th Intl. Ship Stability Workshop,* Istanbul, Turkey.

Takezawa, S. and M. Takekawa (1976), "Advanced Experiment Technique for Testing Ship Models in Transient Water Waves, Part I: The Transient Test Technique on Ship Motions in Waves," *Proc. 11th Symp. on Naval Hydrodynamics*, University College, London, UK.

Themelis, N. and K. J. Spyrou (2007), "Probabilistic Assessment of Ship Stability," *Trans. Society of Naval Architects and Marine Engineers*, Vol. 115 pp. 181–206.

Themelis, N. and K. J. Spyrou (2008), "Probabilistic Assessment of Ship Stability Based on the Concept of Critical Wave Groups," *Proc. 10th Intl. Ship Stability Workshop*, Daejeon, Korea, 23–25 March.

Umeda, N., M. Shuto, and A. Maki (2007), "Theoretical Prediction of Broaching Probability for a Ship in Irregular Astern Seas," *Proc. 9th Intl. Ship Stability Workshop*, Hamburg, Germany.

Chapter 31
Dynamic Transverse Stability for High Speed Craft

Carolyn Q. Judge

Abstract Even in calm water, high-speed vessels can display unstable behaviors, such as chine-walking, sudden large heel, and porpoising. Large heel results from the loss of transverse stability due to high forward speed. When a planing craft begins to plane, the hydrodynamic lift forces raise the hull out of the water, reducing the underwater submergence. The available righting moment due to the hydrostatic buoyancy is, therefore, reduced. As the righting moment due to hydrostatic buoyancy is reduced, the righting moment due to hydrodynamic effects becomes important. These hydrodynamic righting effects are related to the hydrodynamic lift. This paper explores the relationship between the hydrostatic righting moment, the hydrodynamic righting moment, and the total roll restoring moment of a planing craft operating at planing speeds. A series of tow tests using a prismatic hull with a constant deadrise of 20° measured the righting moment at various angles of heel and at various model velocities. The model was completely constrained in heave, pitch, sway, roll, yaw, and surge. The underwater volume is determined from the known hull configuration and underwater photography of the keel and chine wetted lengths. The results presented include the total righting moment with the hydrostatic and hydrodynamic contributions for various model speeds at two model displacements.

31.1 Introduction

For planing hulls, the forces acting on the hull are dominated by the complex hydrodynamics of planing. The dynamic lift reduces the submergence of the hull, allowing small motions to result in large changes in wetted surface area. Linearity assumptions with respect to forces and moments acting on the hull are then less valid since hydodynamic forces dominate in the planing regime. The problem of dynamic stability of high speed planing craft has been known for many years. Codega and Lewis (1987) described a class of high-speed planing boats that exhibited dynamic insta-

C. Q. Judge (✉)
United States Naval Academy, Annapolis, MD, USA
e-mail: judge@usna.edu

V. L. Belenky et al. (eds.), *Contemporary Ideas on Ship Stability*, Fluid Mechanics and Its Applications 119, https://doi.org/10.1007/978-3-030-00516-0_31

bilities, such as the craft trimming by the bow, rolling to a large angle of heel to port, and broaching violently to starboard. Blount and Codega (1992) presented data on boats that exhibited non-oscillatory dynamic instabilities and suggested quantitative criteria for development of dynamically stable planing boats.

Recently there has been research into the stability of planing craft specifically in the transverse plane. Katayama et al. (2007) found that instability is strongly influenced by the running attitude of the hull at high speeds. Katayama et al. measured the rise, trim angle, pitch, and roll and measured the rise and trim angles for the same models when fixed at zero roll. The underwater surface was recorded and both models were found to experience instability when the wetted surface remained close to the keel (the waterplane became narrow). Katayama et al. developed a sectional roll restoring moment equation based on formulas proposed by Smiley (1952) and Payne (1994) that showed the roll restoring force decreased due to the point of action approaching the keel line.

Ranzenbach and Bowles (2010) performed dynamic inclining tests on three different hull shapes to establish transverse dynamic stability. Each model was ballasted to 5 degrees of static heel and then run down the tank at several different speeds in this asymmetric ballasted condition. The hull was defined as dynamically unstable when the heel angle increased significantly and the hull as dynamically stable if the heel angle reduced at planing speed. The RANS CFD (Reynolds Averaged Navier Stokes Computational Fluid Dynamics) program used to analyze the tested models was not able to correctly match the experimental results quantitatively, but did correctly predict the propensity of a particular hull form to exhibit dynamic instability via the dynamic inclining test.

Lewandowski (1996) developed the roll restoring moment for a planing hull as a static restoring moment combined with a dynamic restoring moment. He found that the roll stability of a hard-chine planing craft was generally increased by dynamic effects for vertical center of gravity positions below a critical value. This paper compares the experimental results for roll restoring moment with predictions based on the method given in Lewandowski (1996).

Faltinsen (2005) presents a 2.5D (2D + t) method for calculating the dynamic roll moment using the forces on symmetric wedges impacting a calm water surface as predicted by Wagner (1932). Previous work by the author (Judge 2000) developed a method for predicting forces and moments on wedges falling with geometric asymmetry or with horizontal impact velocity. This paper compares the dynamic moment predicted using a 2.5D low-order strip theory and the forces and moments on vertically impacting wedges. Both a method based on Wagner (1932) and one based on Judge (2000) solutions for a vertically impacting wedge are considered.

Testing was done in the United States Naval Academy Hydromechanics Laboratory's 380 foot tow tank. A prismatic wooden model with a deadrise of 20° was fixed in pitch, heave, heel, and sway, and towed through a range of constant speeds in calm water. The characteristics of the planing hull model and testing conditions are given in Table 31.1. The test matrix consisted of a series of static tests with the

Table 31.1 Testing conditions

Length overall	1.524 m (5 ft)					
Chine beam	0.448 m (1.47 ft)					
Deadrise	20°					
Displacement	13.5 kg (29.8 lb)			26.5 kg (58.4 lb)		
LCG (fwd transom)	0.305 m (1.0 ft)			0.594 m (1.95 ft)		
Model speeds, Fn_b	2.9	3.6	4.3	2.9	3.6	4.3
Trim, τ	3.8°	3.3°	2.9°	4.0°	3.3°	2.7°
Transom depth (relative to calm water)	0.09 m (0.30 ft)	0.08 m (0.25 ft)	0.07 m (0.22 ft)	0.06 m (0.19 ft)	0.05 m (0.16 ft)	0.04 m (0.15 ft)
Keel wetted Length/beam ($\phi = 0°$)	1.91	1.95	1.88	2.82	2.60	2.64
Chine wetted length/beam ($\phi = 0°$)	0.18	0.0	0.0	1.30	0.98	0.65

model fixed in heel. The heave (lift) force, sway force, and heeling moment were recorded and underwater photographs taken for each run. This paper compares the roll moment measured during the static heel tests with different predictions for total roll moment based on heel angle.

31.2 Predictions for Roll Restoring Moment

Lewandowski (1996) developed a formula for the roll restoring moment from an evaluation of the roll moment due to hydrostatic forces and the moment due to dynamic lift. The static restoring moment is based on the transverse waterplane area moment of inertia, the vertical distance from the assumed center of buoyancy to the center of gravity, and the buoyancy force due to the "underwater" volume while planing. Since these quantities vary with speed, the static restoring moment is speed dependent and generally decreases with increasing speed. The dynamic roll restoring moment is based on the empirically determined dynamic lift from Brown (1971). This dynamic lift is used to compute the contributions of the port and starboard sides using an "effective" deadrise angle. The "effective" deadrise angle is the hull deadrise angle minus the roll angle on the side rolled down and the hull deadrise angle plus the roll angle on the side rolled up. The center of pressure is determined from Smiley (1952).

Another way of considering the static roll restoring moment is to look at the volume of fluid displaced when the hull is at its planing trim and heave position.

Then the hydrostatic roll restoring moment can be calculated from this volume. This method is applied (called Underwater Volume based on Flat Surface Water Intersection or UVFWS) and compared with the hydrostatic roll restoring moment calculated from Lewandowski's equations.

Two predictions for dynamic roll restoring moment are compared with the Lewandowski prediction. Both methods use a 2D + t approach using the forces and moments acting on a wedge vertically impacting a flat water surface. One method is based upon Wagner's (1932) solution for the vertical impact of symmetric wedges (solved for asymmetric impact using the approach described in Faltinsen 2005). The other approach is based on the solution of an asymmetric wedge impacting a flat water surface using the solution described in Judge (2000).

31.2.1 Static Roll Restoring Moment

For a planing hull, the static restoring moment is difficult to predict or measure experimentally. For a displacement hull, the roll restoring moment is due to the hydrostatic pressure acting on the underwater portion of the hull. For a planing hull there is an "underwater" portion that can be defined relative to the undisturbed free surface. However, because of water pile-up (where the free surface encounters the forward portion of the hull) and separation of the water flow at the chines and transom, the wetted surface is different than predicted by the intersection of the hull and the undisturbed free surface. In addition, the pressure acting on the wetted surface cannot be considered simply hydrostatic. Therefore, it is difficult to determine how the forces acting on the wetted surface relate to a traditional roll restoring moment for a displacement hull.

31.2.1.1 Prediction of Static Restoring Moment Described in Lewandowski (1996)

Lewandowski (1996) developed a "static" roll restoring moment by considering the waterplane area created from the planing wetted surface. He assumes the roll restoring moment behaves linearly with roll angle, such that the static contribution to the roll restoring moment is

$$M^*_{static} = (-\rho g I_T + BG \cdot \Delta_s) \cdot \phi. \tag{31.1}$$

I_T is the transverse waterplane area moment of inertia, BG is the vertical distance from the assumed center of buoyancy to the center of gravity, Δ_s is the "static lift" due to the hydrostatic pressure on the hull, and ϕ is the heel angle. The hydrostatic pressure is assumed to act on the full wetted surface area of the hull. The transverse waterplane area moment of inertia for a prismatic hull is

$$I_T = \frac{b^3}{48}(L_K + 3L_c) \tag{31.2}$$

where L_K and L_c are the wetted keel and chine lengths, respectively, and b is the average wetted chine beam. The vertical distance from the "underwater" volume to the center of gravity is

$$BG = KG - \frac{b}{6}\tan\beta\left(1 + \frac{L_c}{L_K}\right) \tag{31.3}$$

where β refers to the actual deadrise of the hull and KG is the distance from the keel to the center of gravity. Lewandowski uses an expression for the "static lift" from Brown (1971) that is based on integrating the hydrostatic pressure acting on the equivalent wetted surface for a stationary hull,

$$\Delta_s \approx 0.25\rho g b^3 \lambda^2 \sin 2\tau \tag{31.4}$$

where τ is the trim angle and λ is the mean wetted length to beam ratio

$$\lambda \equiv \frac{L_K + L_c}{2b}. \tag{31.5}$$

The transom is not wet as the water breaks clear and, therefore, pressure acting on the transom is atmospheric. This is expected to reduce the static pressure forces on the defined underwater hull and Brown (1971) found a factor of 0.624 to be appropriate. Thus, the equation for the static roll restoring moment is

$$M_{static} = 0.624 \cdot M^*_{static}. \tag{31.6}$$

31.2.1.2 Underwater Volume based on Flat Water Surface Intersection with the Planing Hull (UVFWS)

As mentioned, the actual pressure acting on the wetted surface is not the same as the hydrostatic pressure acting on the hull of a displacement vessel. The expression developed above is an estimate of the equivalent hydrostatic lift determined from treating the wetted surface on the planing hull the same as the wetted surface on a displacement hull and applying an empirically determined reduction to account for the dry transom while the vessel is at speed. For comparison, consider the same trim and heave as the hull at planing speed, but the roll restoring moment is evaluated assuming the hull is stationary. In other words, the underwater volume is determined from the intersection of the flat water surface and the planing hull (referred to in the figures as UVFWS — Underwater Volume based on Flat Water Surface). The sinkage and trim of the model are held fixed at the values for zero roll, then the model is rolled about an axis through the center of gravity and parallel to the keel. If the hull is not moving there is no water pile-up and the transom is wet. The displacement of this imaginary stationary hull would not equal the weight of the planing craft, but a hydrostatic roll restoring moment at each angle of heel can be determined. This static roll restoring moment is comparable to the estimated "static" roll restoring moment determined from the dynamic wetted lengths as described above.

31.2.2 Dynamic Roll Restoring Moment

The dynamic roll restoring moment is due to the dynamic lift forces acting on the planing surface. The magnitude of the lift force differs between the port and starboard sides because of the roll angle. The dynamic nature of the pressure force complicates the calculation of the location of the action of this force.

31.2.2.1 Method for Prediction of Dynamic Restoring Moment based on Lewandowski (1996)

Lewandowski (1996) estimated these lift forces using empirical relationships developed by Brown (1971). To determine the difference in lift forces between the starboard and port sides when the hull begins to heel, the "effective" deadrise angles are used (see Fig. 31.1). For a hull heeled toward the starboard side,

$$\beta_{e,stbd} = \beta - \phi \tag{31.7}$$

$$\beta_{e,port} = \beta + \phi. \tag{31.8}$$

The mean wetted length for the port and starboard sides depends on the chine wetted length. When the hull heels, the chine wetted length changes, so the mean wetted length becomes different for the port and starboard sides. The method assumes the keel wetted length does not change with roll angle.

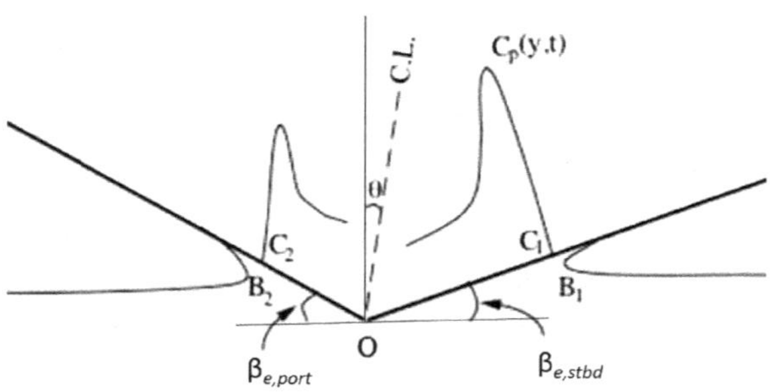

Fig. 31.1 Sketch showing the "effective" deadrise angles of a prismatic hull heeled at angle ϕ

Lewandowski gives the dynamic lift on the port side as

$$F_{d,port} = \frac{1}{2}\rho U^2 b^2 \sin 2\tau \left[\frac{\pi}{4}(1 - \sin \beta_{e,port}) \cos \tau \frac{\lambda_{port}}{1 + \lambda_{port}} + \frac{1.33}{4}\lambda_{port} \cos \tau \sin 2\tau \cos \beta_{e,port} \right] \quad (31.9)$$

and the starboard side dynamic lift can be found in the same way. To find the dynamic roll restoring moment, the location of the action of the lift force with respect to the center of gravity needs to be determined. Lewandowski uses an expression from Smiley (1952) for the location of the dynamic pressure on a prismatic hull. Lewandowski gives the dynamic roll restoring moment about the center of gravity as

$$M_{dynamic} = (F_{d,port} - F_{d,stbd}) \cdot (0.8\frac{\pi b}{8 \cos \beta} - KG \sin \beta). \quad (31.10)$$

31.2.2.2 Method for Prediction of Dynamic Restoring Moment based on Faltinsen (2005)

Another way to determine the dynamic roll restoring moment is to consider the dynamic lift force predicted using Wagner's two-dimensional water impact method (Wagner 1932) combined with the 2.5D (2D + t) theory presented in Faltinsen (2005). Wagner's method results in most of the lift force occurring in the wetted region where the chines are dry. In the aft portion of the hull, where the chines are wet, the dynamic lift force is small compared with the forward portion of the wetted surface. For this paper, only the lift generated in the "chines-dry" portion of the hull will be included. Wagner's method provides an equation for the dynamic pressure due to slamming of a two-dimensional wedge impacting the calm water surface as,

$$p - p_a = \rho V \frac{c}{\sqrt{c^2 - y^2}} \frac{dc}{dt} + \rho \frac{dV}{dt}\sqrt{c^2 - y^2} \quad (31.11)$$

where V is the vertical velocity of the wedge impacting the water surface, c is the instantaneous intersection between the free surface and the deadrise surface, and y is the transverse location where the pressure is acting. Figure 31.2 shows a physical representation for the variables c and V.

For a symmetric impact with a wedge of deadrise angle β, falling with constant vertical velocity, V, point c is determined as a function of time as

$$c(t) = \frac{\pi V t}{2 \tan \beta}. \quad (31.12)$$

A two-dimensional wedge impacting a calm surface can be used to solve for the dynamic lift on a prismatic planing surface using a method described as 2D + t by

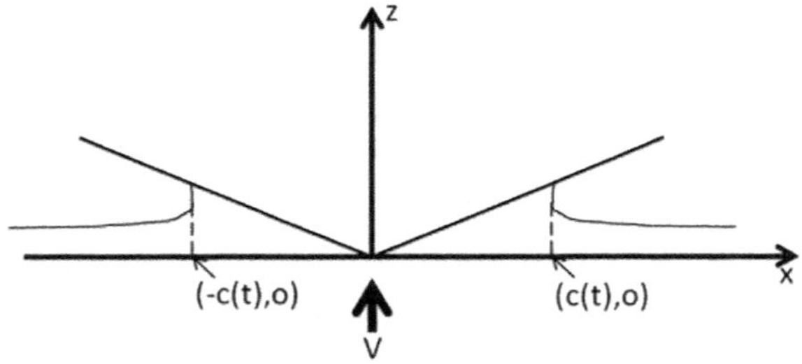

Fig. 31.2 Sketch showing showing variables c and V from Eq. (31.10)

Faltinsen (2005). The falling wedge in time represents the sections of the prismatic hull in space. The vertical fall velocity, V, and time for the impacting wedge relates to the forward velocity, U, and longitudinal position for the prismatic hull as

$$Vt = (U\tau)(\frac{x}{U}) = \tau x \tag{31.13}$$

where τ is the trim angle of the prismatic hull in radians and x is the longitudinal position along the hull. Following the procedure described in Faltinsen (2005) to determine the steady heel restoring moment due to dynamic forces, the roll restoring moment about the center of gravity due to the hydrodynamic pressure is given by

$$\begin{aligned} M_{dynamic} =& \rho U^2 b^3 \tau \left[\frac{1}{24} \left(\frac{1}{\cos^2 \beta_{e,stbd}} - \frac{1}{\cos^2 \beta_{e,port}} \right) \right. \\ & \left. - \frac{KG}{b} \left(\frac{\pi}{16} (\frac{\sin\beta}{\cos\beta_{e,stbd}} - \frac{\sin\beta}{\cos\beta_{e,port}} \right) \right] \end{aligned} \tag{31.14}$$

where β is the deadrise of the hull and $\beta_{e,port}$ and $\beta_{e,stbd}$ are defined as above.

31.2.2.3 Method for Prediction of Dynamic Restoring Moment based on Judge (2000)

Judge (2000) presents a method for solving for the forces and moments acting on an impacting wedge allowing for either geometric asymmetry (vertical impact velocity) or hydrodynamic asymmetry (includes both vertical and horizontal impact velocity). The mathematical model approximates the transverse flow characteristics over the bottom of an asymmetric impact. The method of two-dimensional vortex distributions is employed to model the boundary-value problem, which is solved through discretization of the surfaces and an iterative solution technique. The solution method directly solves the asymmetric impact problem (as opposed to solving two symmetric wedge impact problems and combining the solutions to determine the asymmetric result) and allows for calculation of the forces and moments even after the chines are wet. Once the chines are wet, the forces and moments decrease significantly, so this contribution would not be expected to change the dynamic roll restoring moment by very much. The vertical velocity of the wedge, V, is related to the forward velocity of the model, U, by

$$V = U\tau. \tag{31.15}$$

The theory for the asymmetric impact solution is limited to small roll angles, so results are limited to heel angles of 10° and smaller.

31.2.3 Total Roll Restoring Moment

The total roll restoring moment is a combination of the static and dynamic contributions. The total roll restoring moment predicted by Lewandowski's method is the sum of the static and dynamic roll moments calculated from Lewandowski (1996). For the static roll moment calculated from the UVFWS, the total roll restoring moment can be determined by adding a dynamic roll restoring moment predicted based on either Faltinsen (2005) or Judge (2000).

31.3 Analytical and Experimental Comparisons

Plot (a) in Figs. 31.3, 31.4, 31.5, 31.6, 31.7 and 31.8 compares the static roll restoring moment predicted by Lewandowski and the restoring moment for the planing model considered in calm water using the planing trim and heave (UVFWS). For the lighter displacement, the Lewandowski static moment, which uses the dynamic wetted lengths, gives very similar results to the UVFWS static prediction based on zero hull speed and calm water. For the heavier displacement the static roll moment predicted by Lewandowski's method is close to double the prediction on a stationary hull in calm water (UVFWS). This implies that the reduction factor of 0.624 for the dry transom reduces the hydrostatic lift less than the increase in the hydrostatic lift due to the water pile-up. In Fig. 31.4a, the static roll restoring moment predicted using Lewandowski's method accounts for almost the entire experimentally measured roll restoring moment.

Plot (b) in Figs. 31.3, 31.4, 31.5, 31.6, 31.7 and 31.8 compares the dynamic roll restoring moment predicted using Lewandowski's method with the dynamic roll restoring moment predicted by the method based on Faltinsen (2005) and Judge (2000). Lewandowski's method prediction of the dynamic moment contribution is consistently greater than the total measured moment at all roll angles for both displacements. The method is particularly inaccurate at higher roll angles, at the higher model speeds, and for the lighter displacement. In these cases, the Lewandowski method predicts a sharp change in slope around 20° (the model's deadrise angle). The dynamic roll restoring moment from the method described in Faltinsen (2005)

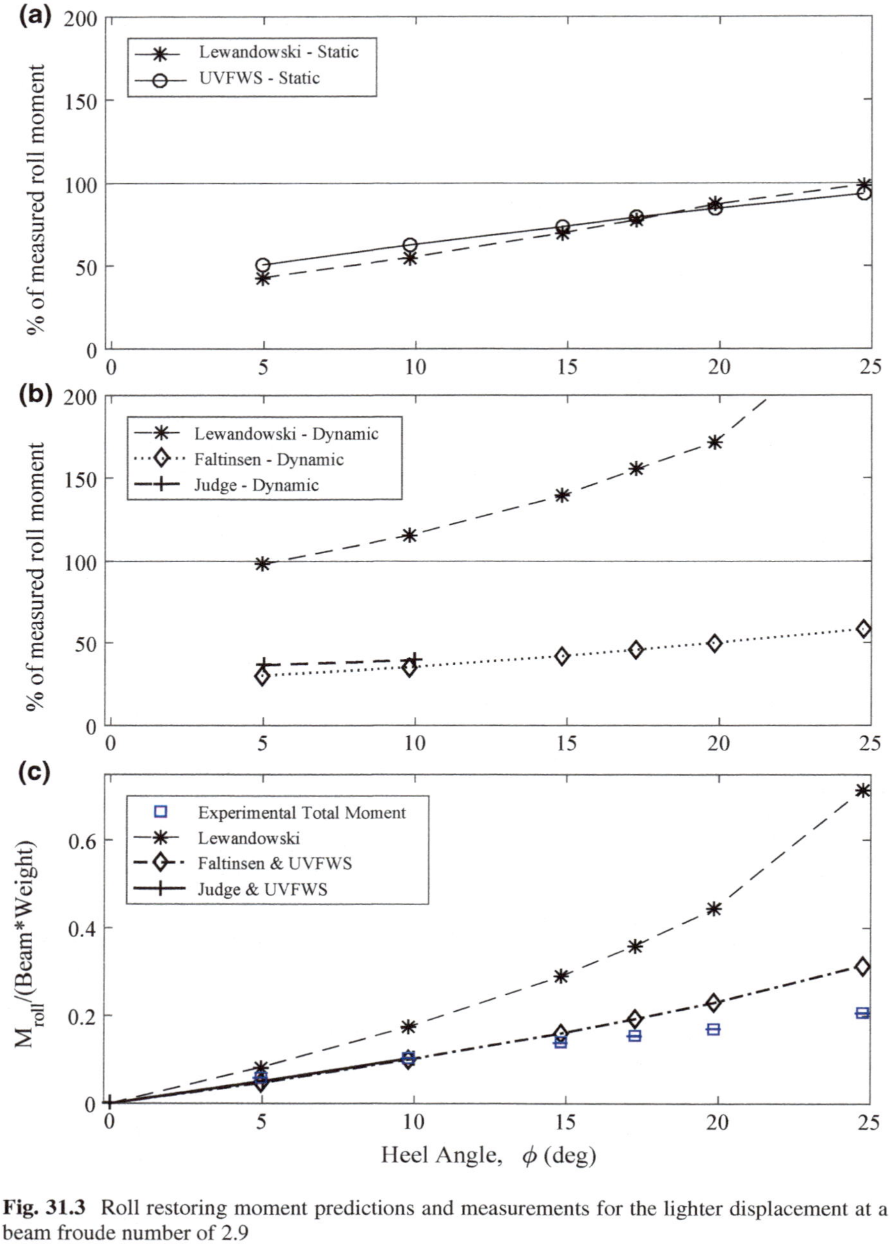

Fig. 31.3 Roll restoring moment predictions and measurements for the lighter displacement at a beam froude number of 2.9

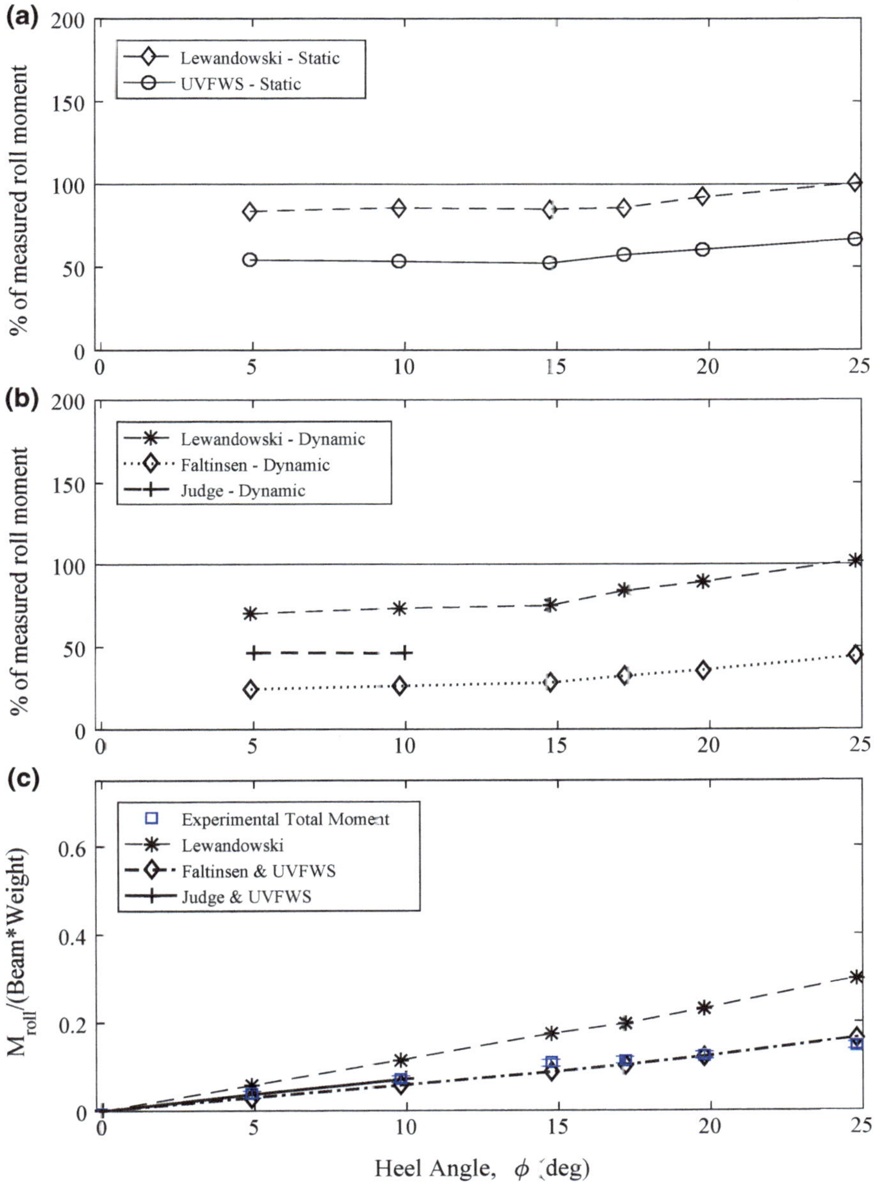

Fig. 31.4 Roll restoring moment Predictions and measurements for the heavier displacement at a beam froude number of 2.9

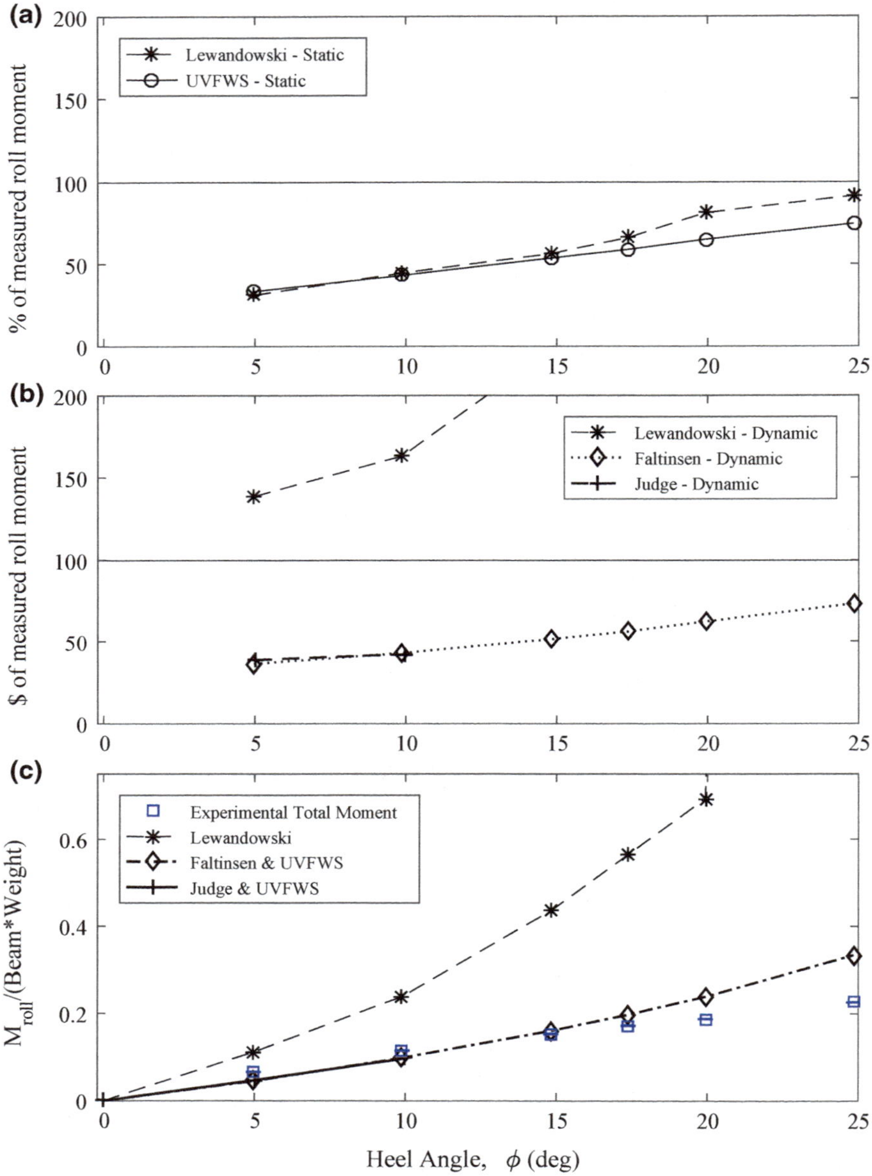

Fig. 31.5 Roll restoring moment predictions and measurements for the lighter displacement at a beam froude number of 3.6

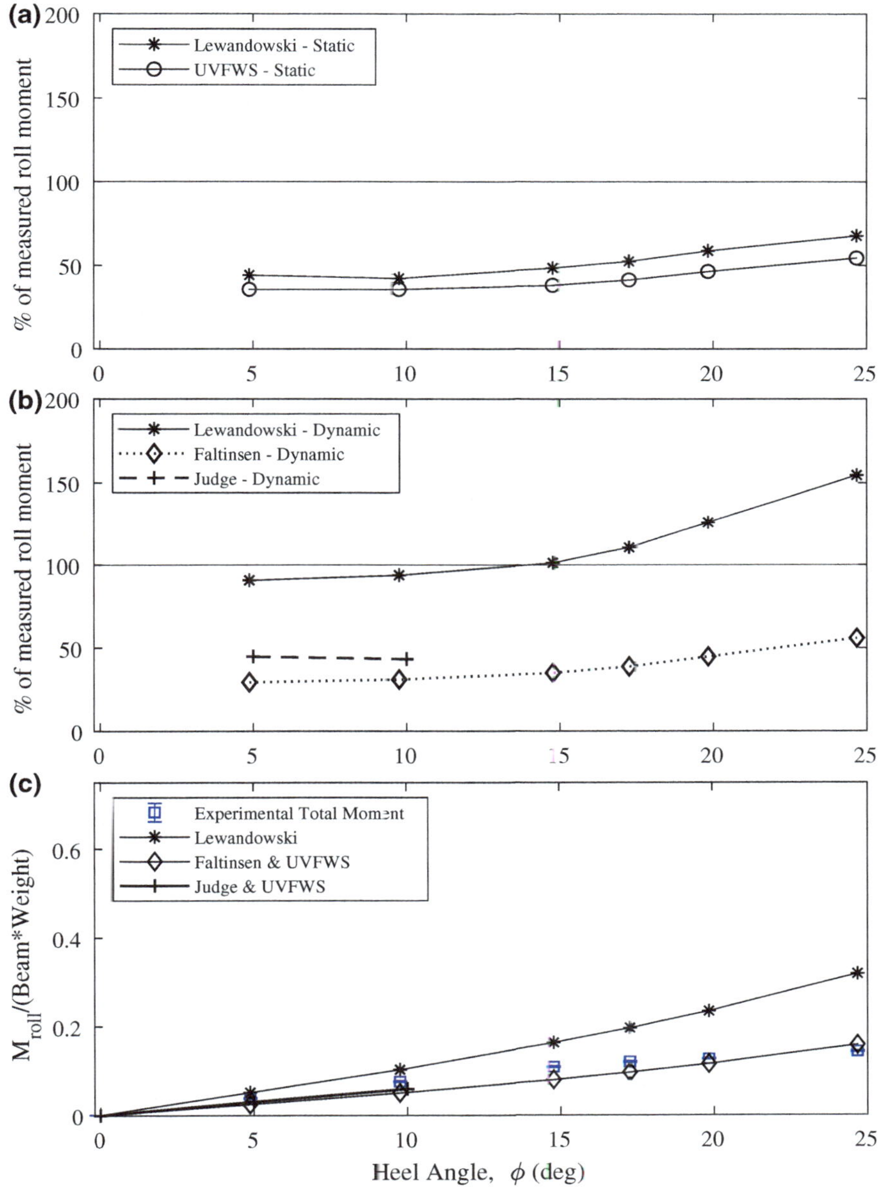

Fig. 31.6 Roll restoring moment predictions and measurements for the heavier displacement at a beam froude number of 3.6

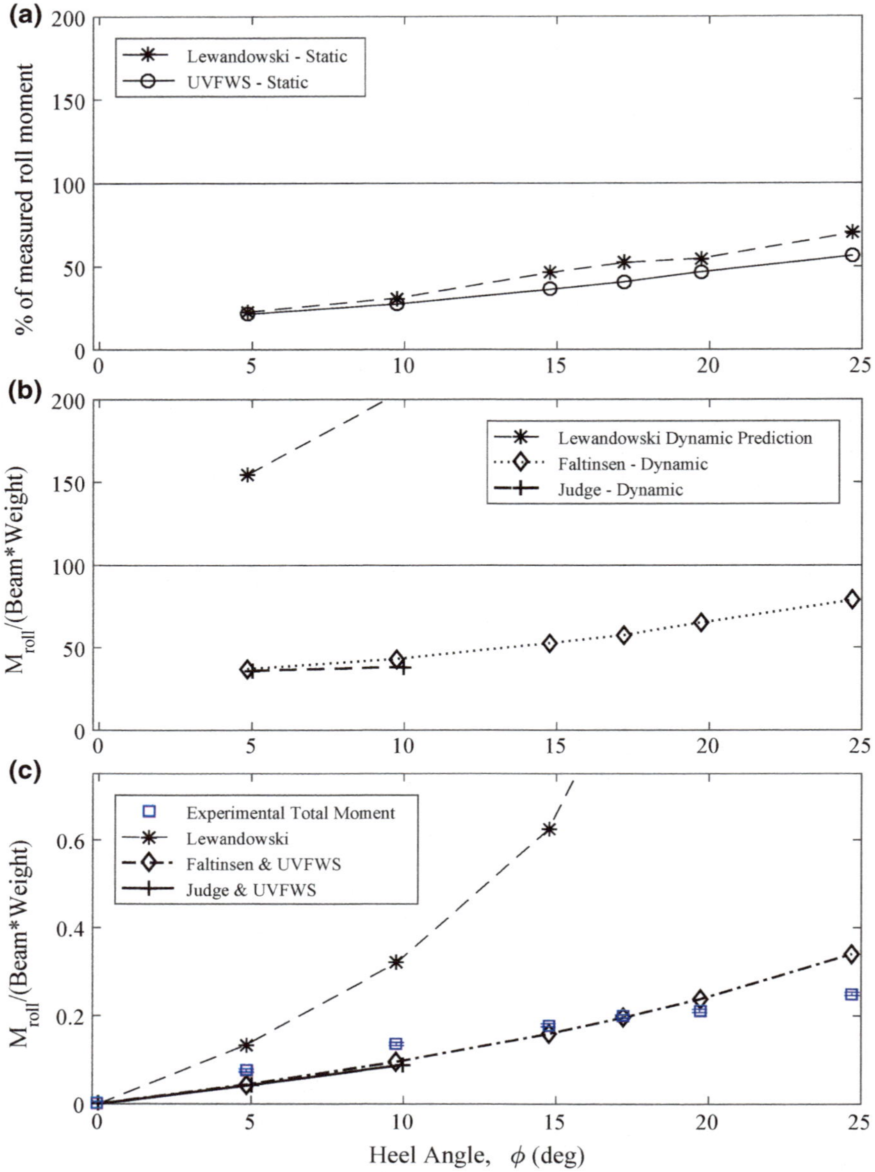

Fig. 31.7 Roll restoring moment predictions and measurements for the lighter displacement at a beam froude number of 4.3

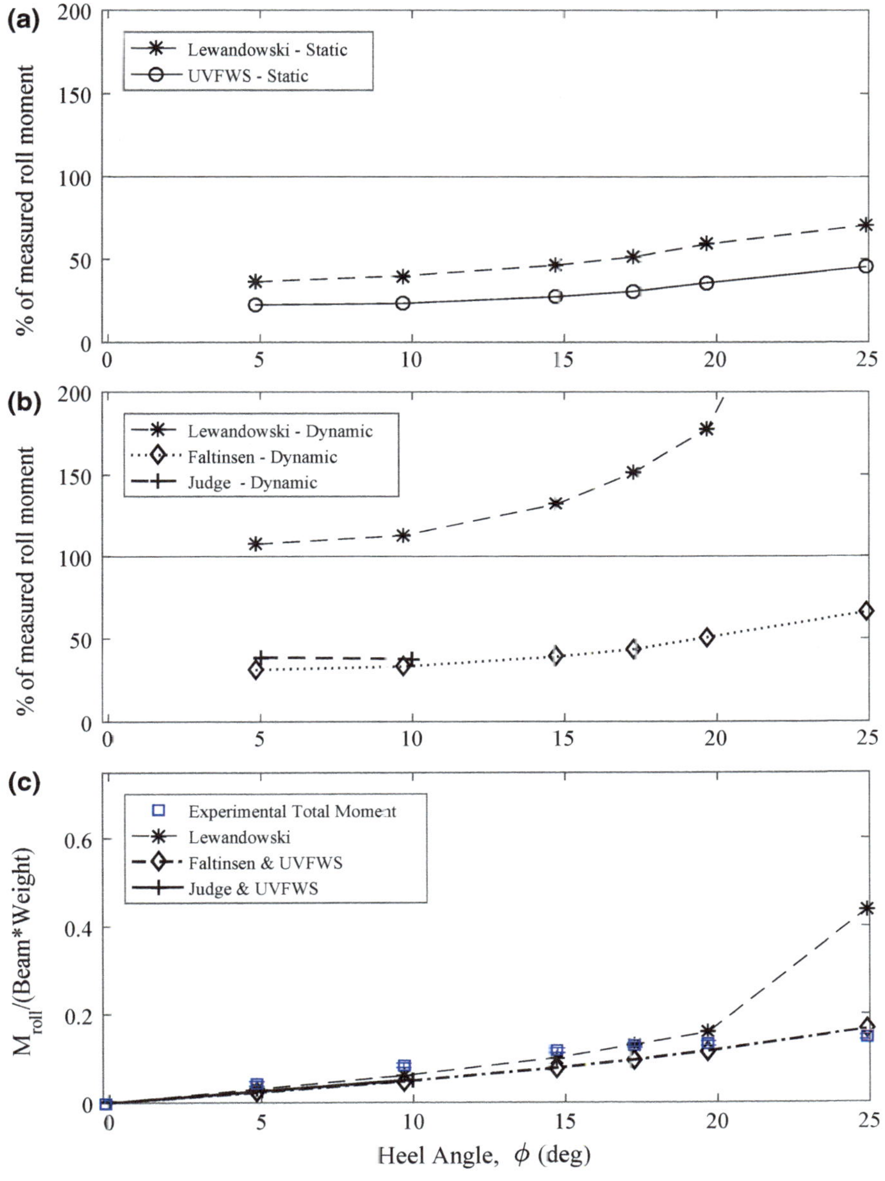

Fig. 31.8 Roll restoring moment predictions and measurements for the heavier displacement at a beam froude number of 4.3

shows a similar trend to the measured moment, and it is less than the total moment in all conditions. The dynamic roll restoring moment predicted based on Judge (2000) follows the same trends as the prediction based on Faltinsen (2005). The Judge predicted dynamic roll moment shows the same trends at the low heel angles as the Faltinsen predicted dynamic roll moment. For the heavier displacement at the highest speed and for the lighter displacement at the lower two speeds, there is not a noticeable difference between this method and the Faltinsen method. At the highest speed for the lighter displacement the dynamic roll restoring moment from the Judge prediction is slightly less than the Faltinsen prediction. One noticeable difference between the Faltinsen and Judge dynamic roll restoring moments is that the Judge prediction includes the contribution from the chines-wetted portion. The fact that the dynamic roll restoring moment for the lighter displacement at the highest speed is lower for the Judge prediction indicates that solving the problem as an asymmetric impact (instead of as a combination of two symmetric solutions) can be important.

The dynamic roll restoring moments calculated from the different methods are for constant deadrise hulls and once the chine is lower than the keel (i.e. when the roll angle is greater than 20°), these methods are no longer valid. They are included in the plots simply for reference. However, the results from Lewandowski's equations show difficulties at roll angles much lower than 20°. For all calculations, this dynamic roll moment prediction becomes a larger percentage of the total roll moment as the model speed increases, as would be expected.

Plot (c) in Figs. 31.3, 31.4, 31.5, 31.6, 31.7 and 31.8 shows the experimentally measured roll moment compared with the combination of static and dynamic restoring moments shown in plots (a) and (b). The error bars for the experimental measurements were determined by doubling the largest roll moment standard deviation at each model speed. The total roll restoring moment predicted using the method described in Lewandowski (1996) over-predicts the roll moment, although this is primarily due to the dynamic component. The dynamic component is much larger than the static contribution for all cases. For comparison, the dynamic roll restoring moment calculated from the method described in Faltinsen (2005) and the method based on Judge (2000) is combined with the roll restoring moment determined for the stationary hull (UVFWS). Compared with the Lewandowski results, these approximations give reasonable predictions for the total roll restoring moment. The Faltinsen and UVFWS predicted roll moment curve has slightly more curvature with roll angle than the measured roll moment. The approximation tends to slightly underpredict the total roll moment for the heavier displacement. For the lighter displacement, this combination of dynamic and static roll restoring moment tends to under-predict the moment at small angles and over-predict the total moment at larger angles. As the total roll moment determined from the dynamic roll restoring moments predicted from the methods based on Faltinsen (2005) and Judge (2000) are both combined with the static roll restoring moment predicted from the UVFWS method, the differences in total roll restoring moment for these methods match the trends for the differences for the dynamic roll restoring component. As the Faltinsen plus UVFWS total roll moment tends to slightly under-predict the measured roll moment, when the Judge dynamic roll moment is larger the Judge plus UVFWS total roll moment compares better. In the one case where the Faltinsen dynamic roll moment is larger, the Faltinsen plus UVFWS total roll restoring moment compares better.

31.4 Conclusions

The prediction method for roll restoring moment developed by Lewandowski (1996) significantly over-predicts the total roll restoring moment. The static moment predicted from Lewandowski's method is similar to the hydrostatic roll moment predicted by the planing hull's trim and heave position (assuming the vessel was not actually moving) for the lighter displacement model. However, for the heavier displacement model the Lewandowski equations predict a static roll restoring moment that is significantly greater than the equivalent restoring moment from hydrostatic analysis. The comparison with the UVFWS static analysis indicates that the effect of the water pile-up is greater than the reduction in pressure due to the dry transom. The Lewandowski static roll restoring moment is based on the transverse waterplane area moment of inertia, the vertical distance from the assumed center of buoyancy to the center of gravity, and the buoyancy force due to the "underwater" volume while planing. All of these quantities are evaluated assuming a symmetric underwater shape where the starboard chine wetted length is equal to the port chine wetted length. However, the chine wetted lengths change as the hull heels and the high side wetted length can go to zero while the low side wetted length continues to increase with roll angle. Therefore, it might be possible to improve on this formulation if the actual underwater volume (based on the dynamic wetted lengths) was used.

The prediction method described in Lewandowski (1996) for the dynamic roll moment does account for the different chine wetted lengths when determining the dynamic lift on each side of the hull. The method assumes the keel wetted length does not change with roll angle and measurements made of the keel ⊕ wetted length during the experiment showed this to be a reasonable assumption. Lewandowski's method assumes the center of pressure does not change with roll angle (the actual deadrise rather than the effective deadrise is used for calculated the lever arm). This is clearly not valid at the larger roll angles. Similarly, the wetted surface is assumed to reach the chines on both sides and this was not the case for the port side (when experiencing starboard heel) for the heavier displacement at large heel angles nor was it the case at any heel angles for the lighter displacement. There is a large difference in magnitude between the dynamic restoring moments predicted using the methods based on Faltinsen (2005) and Judge (2000) and the Lewandowski prediction. The Faltinsen and Judge dynamic roll restoring moment predictions are much more reasonable when compared with the experimental data.

It is interesting to note the combination of a purely hydrostatic analysis of the hull in the planing heave and trim condition combined with the prediction of dynamic restoring moment derived from Wagner's calculations or the method based upon Judge (2000) gives reasonable agreement with the experimental results. This indicates the contribution from hydrostatics is less than expected from the wetted surface area. It is clear the interaction of the hydrostatic and hydrodynamic pressure forces acting on a planing hull is not fully understood. This remains an important area for future research.

Acknowledgements The author wishes to acknowledge the work of Bill Beaver and Dan Rhodes, of the Hydromechanics Laboratory at the United States Naval Academy, who put considerable effort and talent into construction of the Forced Roll Mechanism and data collection for this ambitious project. The support of the Office of Naval Research is gratefully acknowledged through the support of Program Officer Dr. Robert Brizzolara.

References

D.L. Blount and L.T. Codega. Dynamic stability of planing boats. *Marine Technology*, 29 (1), 1992.

P.W. Brown. An experimental and theoretical study of planing surfaces with trim flaps. Technical Report SIT-DL-70-1463, Stevens Institute of Technology, Hoboken, N.J., 1971.

L. Codega and J. Lewis. A case study of dynamic instability in a planing hull. *Marine Technology*, 24 (2), 1987.

O. Faltinsen. *Hydrodynamics of high-speed marine vehicles*. Cambridge University Press, 2005.

C.Q. Judge. A theory for asymmetric vessel impact with horizontal impact velocity, 2000. Thesis submitted to University of Michigan, Ann Arbor, MI.

T. Katayama, M. Fujimoto, and Y. Ikeda. A study on transverse stability loss of planing craft at super high forward speed. *International Shipbuilding Progress*, 54, 2007.

E.M. Lewandowski. Prediction of the dynamic roll stability of hard-chine planing craft. *Journal of Ship Research*, 40 (2), 1996.

P.R. Payne. Recent developments in 'added-mass' planing theory. *Ocean Engineering*, 21 (3), 1994.

R. Ranzenbach and J. Bowles. Full forms with low cruise speed resistance and high speed dynamic stability. In *Proceedings, 29^{th}American Towing Tank Conference*, 2010.

R.F. Smiley. A theoretical and experimental investigation of the effects of yaw on pressures, forces, and moments during seaplane landing and planing. Technical Report Technical Note 2817, NACA, 1952.

H. Wagner. The landing of seaplanes. Technical Report Technical Note 622, 254, NACA, 1932.

Chapter 32
Experiments on a Floating Body Subjected to Forced Oscillation in Calm Water at the Presence of an Open-to-Sea Compartment

Jakub Cichowicz, Dracos Vassalos and Andrzej Jasionowski

Abstract This paper presents results of the physical experiments carried out at the SSRC, aiming at measurements of hydrodynamic reactions on a cylindrical body forced to roll in calm water with an open-to-sea compartment The research addresses the problem of ship-floodwater interaction—an issue of fundamental importance in predicting roll damping for ships in damaged condition.

Keywords Damaged ship hydrodynamics · Ship-floodwater interaction
Physical experiments

32.1 Introduction

The methodology of forced oscillations induced by an internal forcing apparatus as well as its validation for intact ship measurements have been already presented (Cichowicz et al. 2009, 2010)—the results depicted here illustrate applicability of the technique to more complex dynamical systems. Although the measurements on a floating body have certain disadvantages, e.g. limitation in imposed modes of oscillation, on the other hand they offer a unique possibility to study complex, multi-modal response of the vessel—a feature of major significance in the research of multi-mass dynamic systems, which generally cannot be fully explored with tests on a constrained model. Specifically, the authors discuss technical aspects of the measurements with particular emphasis on accuracy and uncertainty assessment. This is followed by broader-context considerations on mathematical modelling, post-processing of data and general remarks regarding dynamics of damaged ships and applicability of the demonstrated technique to systematic experimental research.

J. Cichowicz · D. Vassalos (✉) · A. Jasionowski
The Ship Stability Research Centre, Department of Naval Architecture, Ocean and Marine Engineering, University of Strathclyde, Glasgow, Scotland, UK
e-mail: d.vassalos@strath.ac.uk

V. L. Belenky et al. (eds.), *Contemporary Ideas on Ship Stability*, Fluid Mechanics and Its Applications 119, https://doi.org/10.1007/978-3-030-00516-0_32

Table 32.1 Particulars of the tested model

	Dim.	Intact	Dam.
L	m	1.5	1.5
B	m	0.695	0.695
T	m	0.158	0.157
KMT	m	0.343	0.344
KG (dry)	m	0.220	0.297
KG (flooded)	m	–	0.232
GM	m	0.123	0.047
Mass	kg	156.8	116.7
Kxx	Nm/rad	189.2	53.79
Roll inertia (air)	kg.m^2	10.4	4.8
Radius of in. (i_{xx})	m	0.258	0.251
i_{xx}/B	–	0.370	0.36
Scale	–	40	40

32.2 Experiments

32.2.1 Experiments Set-up

The experiments presented in this paper were conducted at the Kelvin Hydrodynamics Laboratory (KHM), the testing facilities of the University of Strathclyde.[1] The tested model was a 1.5 m (60 m full scale) cylindrical section of a RoPax ferry. Its particulars are shown in Table 32.1.

The freely-floating model has been forced to roll in calm water by an internal gyroscopic device, designed and manufactured at the KHM. As discussed in (Cichowicz et al. 2009), the periodic (pure) moment to sustain motion generated by the apparatus can be assumed harmonic and therefore, at least in the case of small motions of the intact ship, the measured hydrodynamic reaction can be expressed in terms of orthogonal components—added inertia and damping. The case of damaged ship will be discussed later.

Ship motions were recorded with the use of optical motion capture system (Qualisys^{TM}) but for reference measurements of the phase lag, a single axis accelerometer was also fitted to the model. The component of the (total) rolling moment was measured by a single axis 500 lb transducer (Fig. 32.1).

The draft of the model was kept constant in both conditions. However, due to some inaccuracy in positioning of the replacement masses (i.e. weights placed inside the floodable compartments to account for floodwater mass) the estimated position of the centre of gravity of the ship-floodwater system was 0.012 m (0.48 m full scale) higher than in the case of the intact ship.

[1] http://www.strath.ac.uk/na-me/facilities/cmh/.

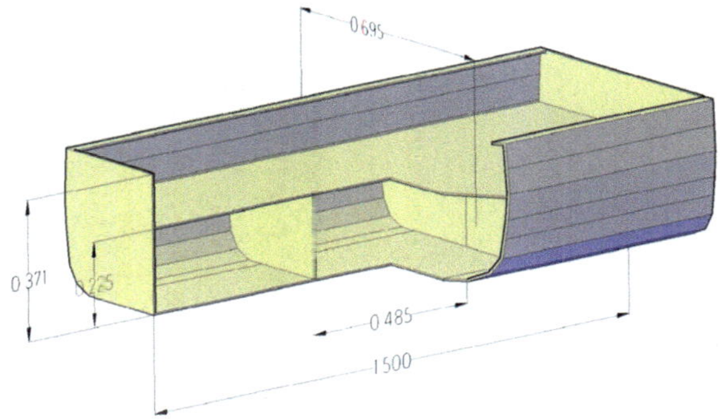

Fig. 32.1 Main dimensions and internal arrangement of the model

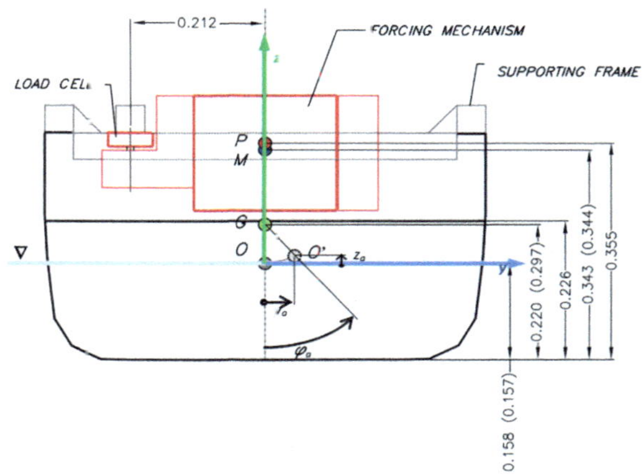

Fig. 32.2 The configuration of the system. Values in parentheses as in flooded conditions. Capital letters denote: O—origin of the body-fixed coordinate frame, G—centre of gravity, M—metacentre, P—pivoting point of the gyro

During the damaged ship measurements, the flooded compartment was (on one side) open to sea with 0.071 m (2.82 m) freeboard and a prismatic damage opening of 0.203 m (8.1 m) width (Fig. 32.2).

32.2.2 *Intact Ship Measurements*

The intact ship measurements were performed primarily to validate the experimental set-up and the components of the hydrodynamic reaction were derived with use of a simple, single DoF model (Fig. 32.3).

The analysis performed during initial tests in intact condition showed that a major source of uncertainty is related to the phase lag prediction and for that reason the spectral techniques proved to be of insufficient resolution, hence the steady-state parts of the time histories were approximated with sinusoidal fit. The errors in coefficients' predictions were incorporated into the derived hydrodynamic components by means of a standard differential model. In the following, only errors associated with roll

Fig. 32.3 The model during intact ship measurements

damping coefficients are presented as the remaining quantities exhibit much lesser sensitivity.

As can be readily seen from Fig. 32.4 both, amplitude and phase lag, characteristics compare very well across experimental and numerical (based on potential flow code) predictions but from the relative amplitude graph it can be noticed that numerical data is shifted slightly towards lower frequencies (i.e. numerical results suggest slightly lower than measured natural frequency). There is, however, no observable distinction between experimental characteristics obtained for 2° and 5° amplitude of roll (Fig. 32.5).

Regardless of the close match in amplitude-phase characteristics, experimentally derived damping and added inertia show significant dissimilarity with respect to the numerical prediction (Fig. 32.2). The added inertia coefficient is smaller over the entire frequency range and unlike the numerical data, it diminishes at low frequencies (the numerical prediction exhibits asymptotic behaviour at both extremities of the frequency range). The damping coefficient, on the other hand, is larger than the numerically predicted and its modal frequency is higher than the numerical counterpart. The exact nature of these discrepancies was not determined but a possible explanation is that this is caused by introducing additional constraints on position of axis of rotation. In analytical approach a cylinder section oscillates about axis lying in the calm water surface. However, as the floating cylinder of an arbitrary section oscillate about the *natural,* and in general case elevated, axis of rotation. In both cases a cylinder is a single DoF oscillator but in theoretical approach there is an additional constraint imposed on the system $v_A = 0$. Hence, from mechanical perspective theoretical and physically tested systems are different.

Finally, although it can be argued that there is indication of increased (nonlinear) damping at higher amplitude motion in the "peak region", the underlying large errors do not substantiate such conclusion. In any case, the experimental results demonstrate good repeatability and reasonable accuracy.

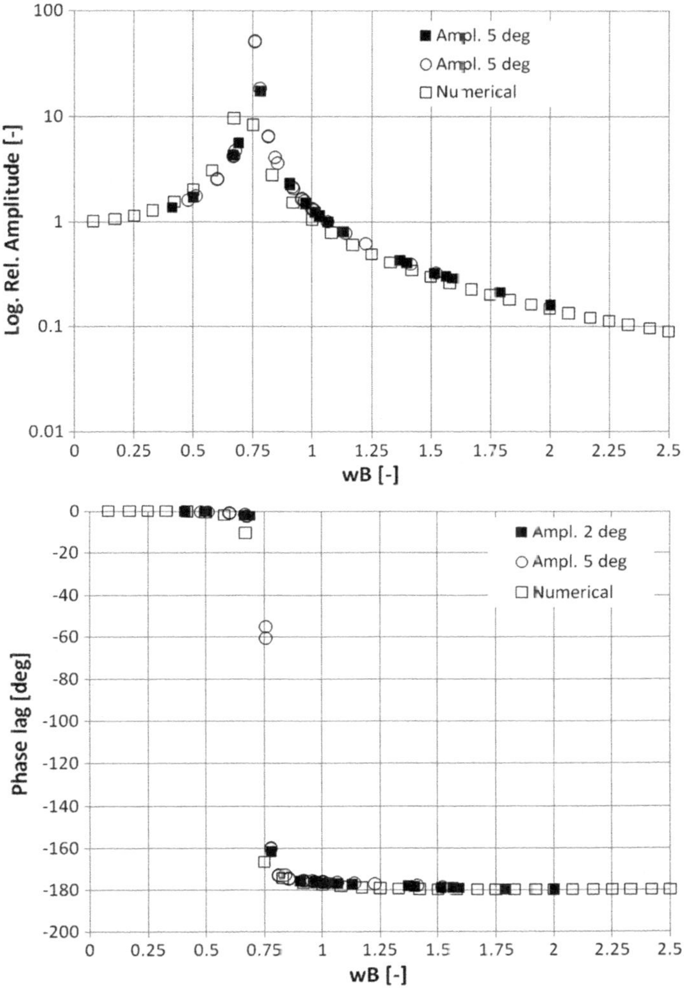

Fig. 32.4 Amplitude–phase characteristics of the intact ship

32.2.3 Damaged Ship Measurements

As mentioned earlier, although the draught in both conditions was kept virtually the same, the centre of gravity of the flooded hull was slightly (12 mm) higher in the damaged condition and therefore direct comparison of the results might be questionable. Nevertheless, as it will be reasoned in the following, the observed behaviour of flooded ship was so utterly different compared to intact vessel that the KG difference is overshadowed by other factors.

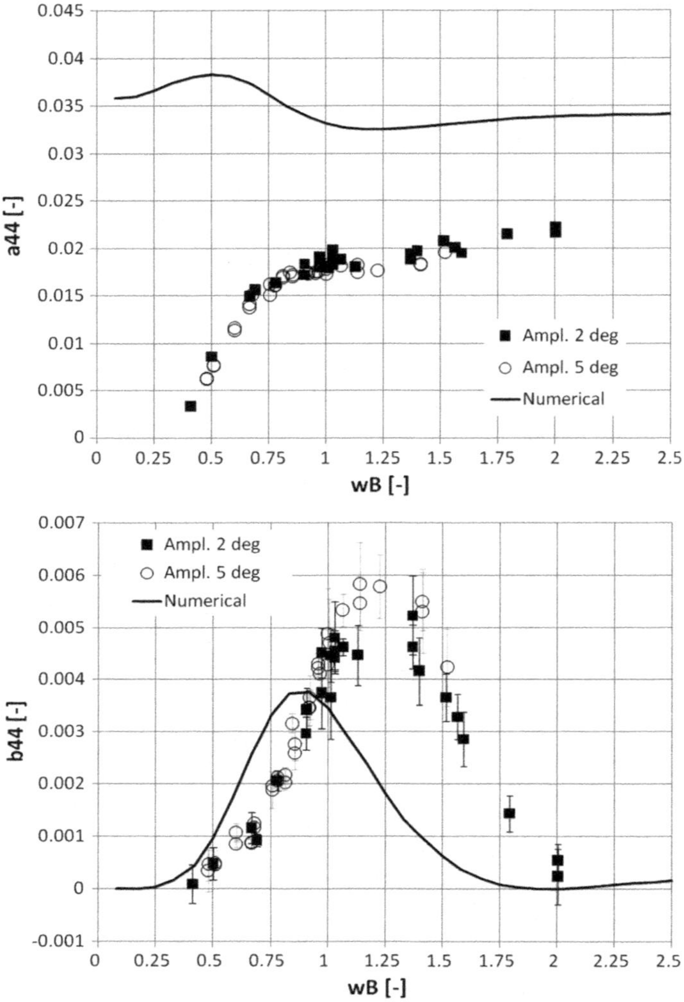

Fig. 32.5 Added inertia and damping (The symbols wB (ω_B), a44 (a_{44}) and b44 (b_{44}) denote non-dimensionalised circular frequency, added inertia and damping, respectively.) derived with single DoF linear model

The components of the hydrodynamic reaction of flooded ship were derived with use of the same single DoF as in the case of the intact ship. Obviously, the model is very simplified and cannot accurately represent the underlying physics and all the fine details of ship-floodwater interactions. However, it had been found that in the tested cases the interactions were governed mainly by low-order effects and the higher order effects, when present, were so small that the more complex models failed to capture them.

Furthermore, although it was observed that the measured hydrodynamic reaction was periodic but not harmonic, it had also been found that multi-harmonic models (e.g. spring-coupled masses) did not perform any better than the single DoF model. Obviously, use of the “wrong” model resulted in large epistemic uncertainties and prohibited fully-quantitative assessment, yet it allowed for assessment of the scale of the effects associated with the ship-floodwater interaction. It is also understood that use of terms *added inertia* and *damping* in the case of a damaged ship might not be appropriate. Nevertheless, these terms will be used in the following to denote approximate harmonic components in phase with roll angular acceleration and velocity, respectively.

32.3 Results

32.3.1 Amplitude-Phase Characteristics

The first feature that can be readily observed in the amplitude-phase characteristics shown in Fig. 32.6 is a significant shift of the natural response towards lower frequencies, compared to intact ship. On one hand, the frequency shift results from lower GM and hull mass but on the other hand, given that dry-hull inertia is halved, the natural period of the flooded ship is still longer than expected—this indicates higher (by a factor of two) added inertia.

Furthermore, the damaged ship RAO is bi-modal with the second peak corresponding to the sloshing natural frequency at about $\omega_B = 1.2$ [−] (6.14 rad/s)[2]. It is noteworthy that the impact of sloshing on the response amplitude is rather modest whereas the phase lag is considerably more affected with a maximum lag decrease of about 25°. These, i.e. small increase in RAO combined with a large decrease in phase lag accounts for a 15-fold increase in damping just by taking the ratio of the relevant sines (sin(−153)/sin(−178°)).

Finally, although it is not easy to notice, the phase lag characteristic is not a smooth function at $\omega_B = 1.5$ [−], where it “crosses” −180°. The implication of this (given the single DoF model) is negative estimate of damping coefficient at higher frequencies.

32.3.2 Added Inertia and Damping

The added inertia and damping, or more appropriately, harmonic components of the hydrodynamic force in phase with roll angular acceleration and velocity are presented in Fig. 32.7. It can be readily noticed that both components are much

[2]The first two sloshing natural frequencies derived on the basis of the linear model for a rectangular compartment are 5.2 and 10.8 rad/s, respectively.

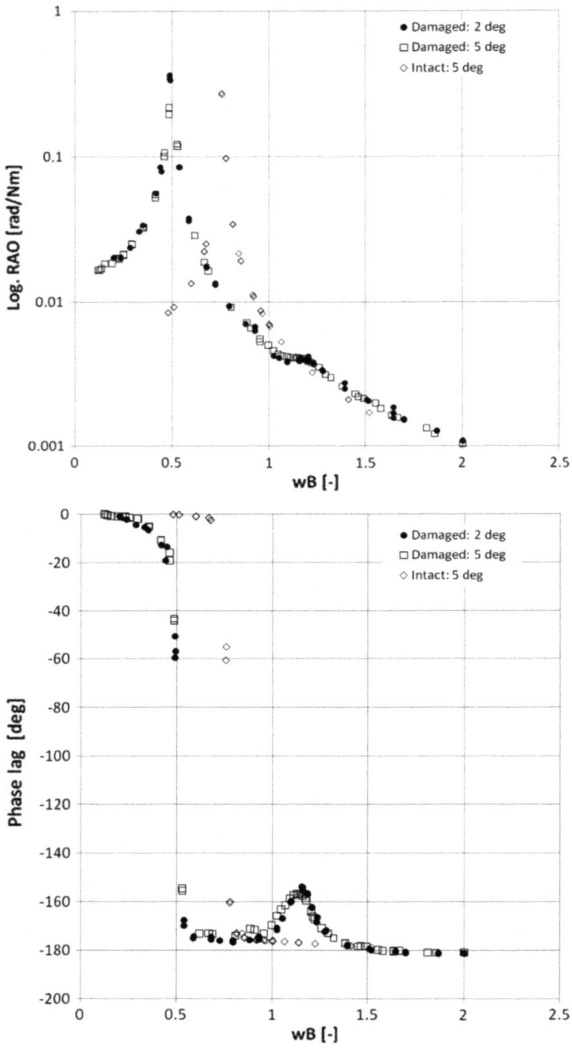

Fig. 32.6 Amplitude-phase characteristics of the flooded ship

larger than in the case of intact ship—damping up to tenfold higher than the intact-ship value. Furthermore, it can be speculated that both characteristics are dominated by hull-floodwater interaction (large dip/peak at sloshing natural frequency).

There are two frequency ranges that attract particular attention—relatively low ($\omega_B < 0.5$) and relatively high ($\omega_B > 1.5$)—at which either added inertia or damping become negative, respectively, based on the logic of the simple model being used here.

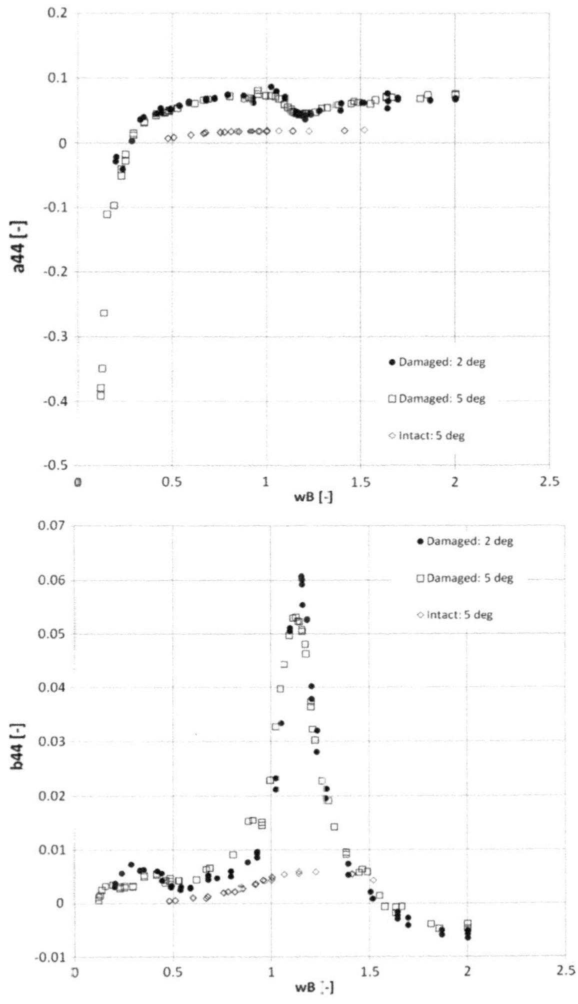

Fig. 32.7 Added inertia and damping coefficients—flooded ship

At the low-end of the frequency range the measurements indicate negative added inertia and there can be a few reasons for this—most likely it is caused by slightly underestimated restoring. At the high-end of the frequency range the damping becomes negative, which results from the aforementioned behaviour of the phase-lag characteristics.

In any case at ($\omega_B = 1.5$ [$-$]) there is a sharp change in the response characteristics caused by hull-floodwater interaction.

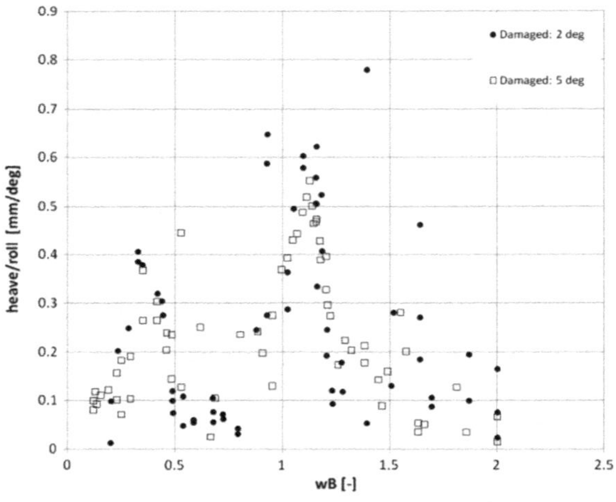

Fig. 32.8 Relative amplitude of heave

32.3.3 Hull-Floodwater Interaction

The amplitude-phase and "added inertia–damping" characteristics show clearly that the hull response can be divided into two modes—the first, at relatively low frequencies, dominated by hull dynamics and, the second, predominantly affected by floodwater motions. In the former, the floodwater surface remains calm and virtually horizontal, so its impact on the ship response can be assumed static. In the latter surface of the floodwater undergoes violent motions.

Interestingly, the impact of sloshing seems to be unrelated to the amplitude of roll at frequencies close to and higher than sloshing resonance. The behaviour, however, is not the same at lower frequencies where increased amplitude causes "deformation" of the hull-dominated part of the damping characteristics. In any case it is clear, that as relative motions of hull and floodwater increase, roll motion becomes heavily damped. Furthermore, near the sloshing resonance the flow through the opening becomes so violent that, among other effects, it generates a complex system of surface waves. The coupled motions, i.e. the roll-induced heave, Fig. 32.8, and sway, Fig. 32.9, also indicate bi-modal behaviour with two peaks corresponding to hull and floodwater (sloshing) natural frequencies. The amplitude of sway implies that the natural axis of rotation lies (depending on frequency) between the dry hull KG and the flooded system KG except at low frequencies where higher than expected sway amplitudes suggest sideway rectilinear motions ("sliding").

Deriving from the above, it has been shown that the hull-floodwater interaction has a major impact on roll damping and added inertia whereas other modes of motion are not affected significantly. There is, however, one additional mode of motion that is induced by floodwater dynamics—namely drift.

The drift, as expected, is not observed in the intact ship measurements as there is symmetry in both hull geometry and excitation. The situation changes dramatically

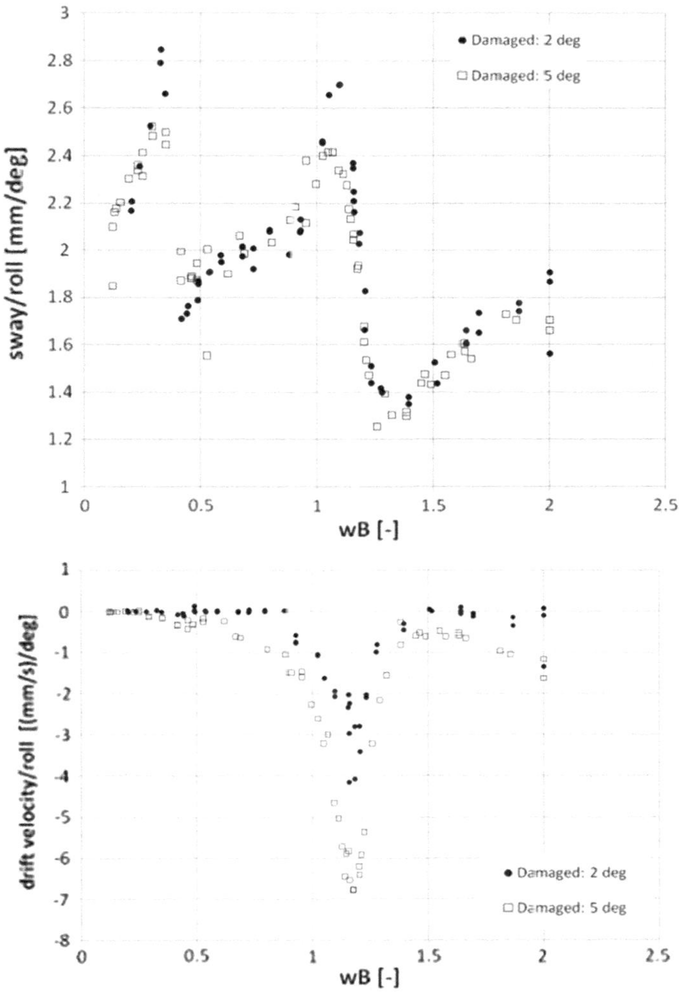

Fig. 32.9 Relative amplitude of sway (top) and drift velocity (bottom)

with the presence of a large opening at the side of the ship. Given the prismatic shape of the opening, its impact on the restoring (due to the asymmetry in pressure distribution on the hull surface) can be considered negligible at small amplitude motions but the same cannot be said with respect to the dynamic effects.

Beginning with 2° roll, it can be noticed that the frequency ranges at which the drift is being observed coincide with the natural frequencies of the hull and the floodwater (sloshing). In case of the former, the drift velocity is small, practically difficult to note, whereas at the higher frequency the drift velocity reaches almost 10 mm/s (0.12 knots full scale). Outside the relatively narrow frequency bands in the proximity

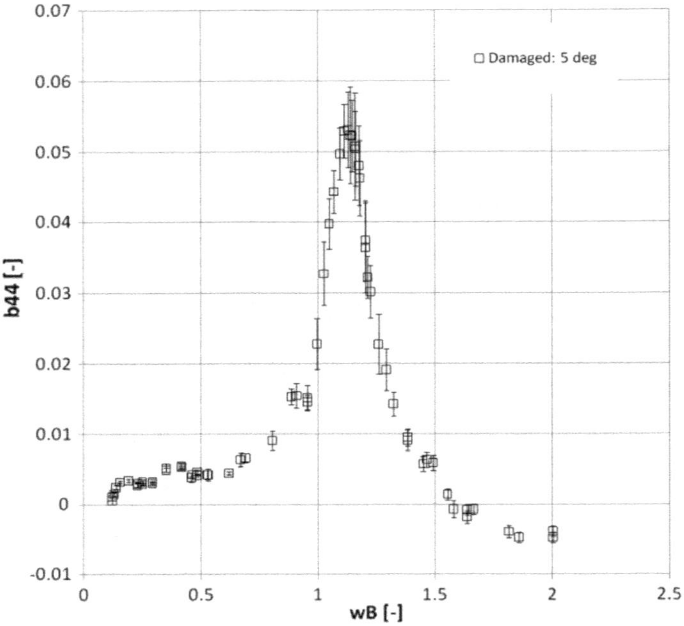

Fig. 32.10 Uncertainties in roll damping coefficient caused by propagation of least-squares errors in fit coefficients

of the resonant frequencies, there is no drift detected. When the amplitude of roll becomes larger, at 5°, drift velocity at the natural frequencies increases significantly (3.5-fold increase of the drift velocity with 2.5-fold increase of the roll amplitude) and the frequency bands of non-zero drift widen, spanning practically over the entire frequency range.

Furthermore, after significant decrease beyond the sloshing natural frequency, it seems to be increasing again for frequencies ($\omega_B > 1.5$ [−]). This, combined with the observed C^0 continuity of the phase lag and related characteristics, suggest a sharp change in the character of the flow through the opening and/or floodwater dynamics that take place at this frequency range. Moreover, with some caution, it can be stated that increasing motion amplitude has rather minor impact on roll hydrodynamic reaction—it causes, observed mainly at lower frequencies, a deformation of the damping characteristics of the hull (rigid-body) "component". Apart from this, energy dissipation through roll and roll-induced sway and heave is linearly dependent on roll amplitude.

Finally—although at this stage it can be speculative—one could expect that removing asymmetry from the system, for example by closing the opening, might lead to a completely distinct (e.g. in terms of the observed linearity of coefficients/induced motion amplitudes) impact of floodwater sloshing on roll and other relevant oscillatory motions, due to inability of energy dissipation through drift (see for example: Murashige and Aihara 1998).

32.3.4 Uncertainty in the Results

As mentioned earlier, it has been initially assumed that the single DoF model would present a major source of errors and in principle it would prohibit quantitative assessment. Although this assumption is generally valid, the analysis has shown that more complicated systems have failed to provide better quality assessment. Starting from the artificial decoupling of roll from the roll-induced sway and heave it can be stated that their impact on the final results is expected to be small (Cichowicz et al. 2010) compared to errors in phase lag derivation (and moment amplitude at roll/sloshing natural frequency).

Furthermore, the errors associated with the least-squares averaging (fitting) show clearly that even if the measured force and hull motions are not purely harmonic the resultant uncertainties are small to moderate. Obviously, it is understood that ignoring coupling of roll with other modes of motion does not account for errors in the coupled modes. Nevertheless, it can be reasoned[3] that even if a proper model were available, the actual errors might not have been considerably larger.

In any case, although there are a number of mathematical models of damaged ship available (see: de Kat 2000; Rakhmanin and Zhivitsa 2000; Faltinsen and Timokha 2002), usually the numerical or analytical solutions do not match the experimental data very well. This could be observed in the case of calm-water experiments (Jasionowski and Vassalos 2002) just as in the case of in-waves tests (Kong and Faltisen 2008). Therefore it is believed that the single DoF model should be sufficient to determine nature and scale of the dominant dynamic effects.

Additionally, an attempt made to decompose recorded signals into harmonic components following either coupled system solution or second order harmonic decomposition has not been successful. In the particular experimental set-up, more comprehensive models happened to be of little use—they did not improve accuracy of fitting but required much more computational effort and far too often failed to converge. On one hand, this can account for the common convergence problems of many of the least-squares algorithms. On the other hand, small-scale effects violating the assumption of harmonic force/response, are very difficult to capture due to the noise in the recorded signals and often very limited time of the steady-state measurements. The latter is particularly important in the case of low-frequency oscillations with prolonged transients where there is a risk of short-time wave reflections.

Moreover, as Fig. 32.11 indicates, the least-squares errors in the moment amplitude are generally well below 1% and it confirms the harmonic nature of the measured moment. On the other hand, recorded roll exhibits much higher inaccuracies with an average error of 10% and substantial standard deviation. Thus, the response needs to

[3]Based on the uncertainty assessment performed for coupled, intact ship, model.

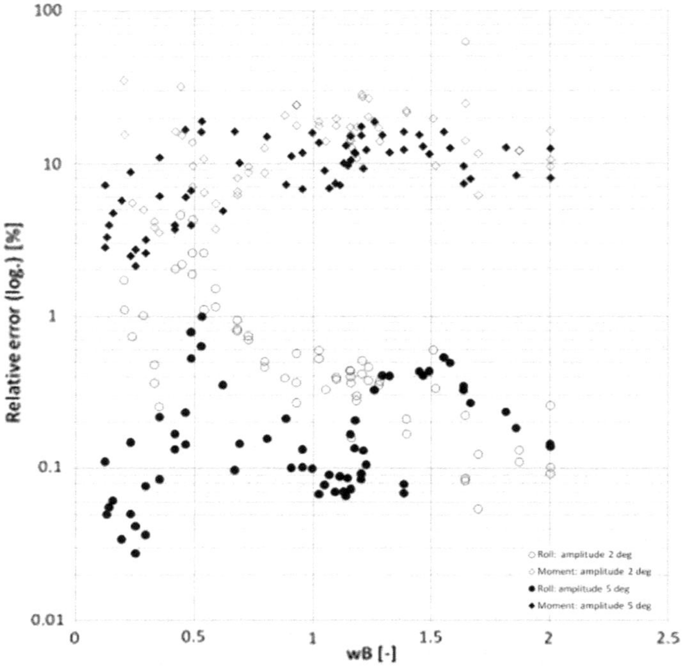

Fig. 32.11 Relative errors in least-squares amplitude predictions for roll motion and total measured moment for 2° and 5° roll motion, respectively

be carefully examined as there is strong evidence that it may depart significantly from the assumed harmonic pattern. It is noteworthy however, that even very large errors in roll amplitude prediction have negligible impact on the damping errors outside the sloshing resonance range Fig. 32.10).

Finally, although initially the authors intended to apply FFT-based techniques - these proved to be, given relatively low sampling frequency, of insufficient resolution (Fig. 32.12).

32.4 Concluding Remarks

The results of the exercise presented in this paper do not come as a surprise, as damping effects of free surface have been studied for a very long time (Moaleji and Greig 2007) but it is thought that the experiments reported here have shed more light on the complex phenomena of the hull—floodwater interaction. Use of an unconstrained model in calm-water allowed observing the scale of the effects associated with complex flow through an opening on the hull and floodwater dynamics and their impact

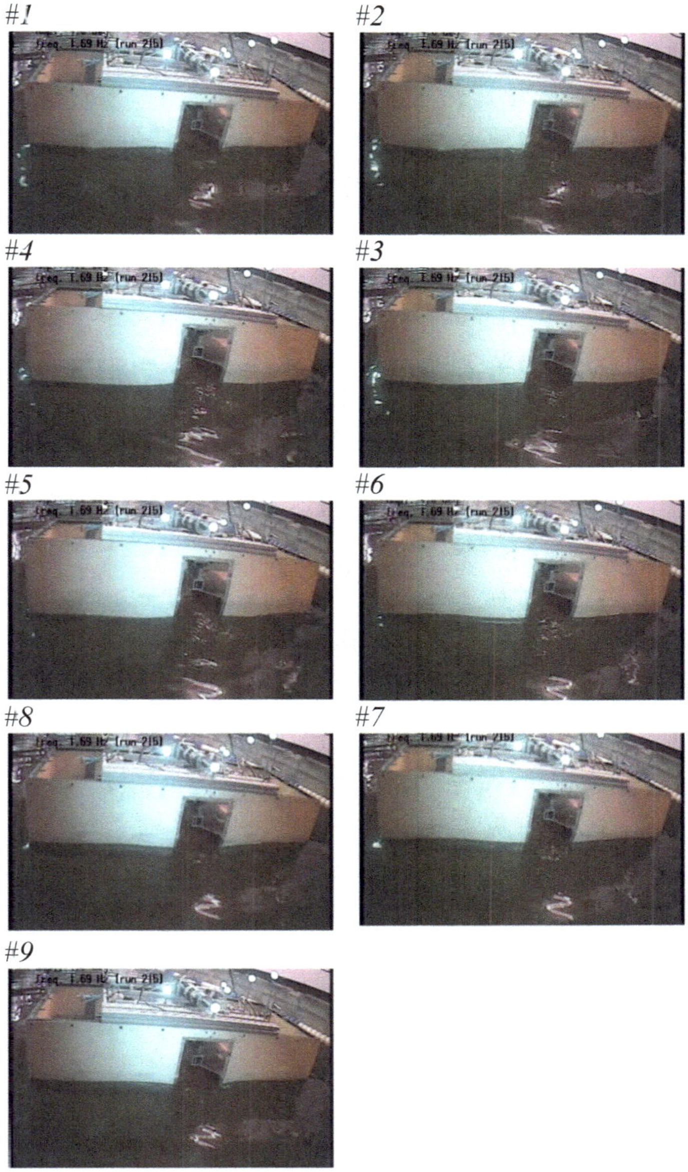

Fig. 32.12 Sequence of video frames (numbered above the pictures) showing pattern of the waves created by the flow through the opening at 1.69 Hz roll frequency (approximately 1 cycle)

on the behaviour of the ship. On the other hand, it is clear that unless the experiments presented here are followed by more systematic research, the results cannot be generalised or quantified, particularly due to the fact that a suitable mathematical model does not exist (to the authors knowledge).

Acknowledgements The authors would like to express their gratitude to the management and staff of the Kelvin Hydrodynamics Laboratory of UoS. They also acknowledge and appreciate the travel support offered by the AuxNavalia project (www.auxnavalia.org), which is funded by the Transnational Cooperation in the European Atlantic Area (www.coop-atlantico.com).

References

Cichowicz, J., Jasionowski, A., Vassalos, D.: *"Uncertainty Assessment in Experiments on a Floating Body in Forced Roll Motion in Calm Water"*, Proceedings of the 10th International Stability Workshop, 2010.

Cichowicz, J., Vassalos, D., Jasionowski, A.: *"Hydrodynamics Of Damaged Ship In Roll Mode Of Motion – An Experimental Approach"*, Proceedings of the 1st International Conference on Advanced Model Measurements Technology for the EU Maritime Industry, 2009.

de Kat, J. O.: *"Dynamics of a ship with partially flooded compartment"*, Contemporary ideas on ship stability, Elsevier Science Ltd., 2000.

Faltinsen, O. M., Timokha, A. N.: *"Asymptotic modal approximation of nonlinear resonant sloshing in a rectangular tank with small fluid depth"*, Journal of Fluid Mechanics, Vol. 470, pp. 319–357, 2002.

Kong, X. J., Faltinsen, O. M. *"Coupling between Ship Motions and Flooding of a Damaged Ship in Waves"*. Proceedings of the 8th International Conference on Hydrodynamics (ICHD'08), 2008.

Jasionowski, A., Vasslos, D.: *"Damaged ship hydrodynamics"*, Proceedings of the 6th International Ship Stability Workshop, Webb Institute, 2002.

Moaleji, R., Greig, A. R.: *"On the development of ship anti-rolling tanks"*, Ocean Engineering, Vol. 34, pp. 103–121, 2007.

Rakhmanin, N. N., Zhivitsa, S. G.: *"The influence of liquid cargo dynamics on ship stability"*, Contemporary ideas on ship stability, Elsevier Science Ltd., 2000.

Murashige, S., Aihara, K., "*Experimental study on chaotic motion of a flooded ship in waves*", Proceedings of the Royal Society A: Mathematical, Physical and Engineering Sciences 454, pp. 2537–2553, 1998.

Part X
Experimental Research: Techniques: Validation and Benchmarking

Chapter 33
Model Characteristics and Validation Approach for a Simulation Tool Supporting Direct Stability Assessment

Arthur M. Reed, William F. Belknap, Timothy C. Smith and Bradley L. Campbell

Abstract Significant challenges exist in the validation and formal acceptance of dynamic stability simulation tools, which can have a limiting effect on operations or concept design due to a need for conservatism. These challenges primarily consist of validation metrics and criteria, uncertainty characterization, and defining the scope of conditions (both environmental and operating) that must be examined in order to ensure that the simulation tool is valid for all conditions of interest. In discussing these challenges, this paper proposes approaches to the problem of validation and formal acceptance that can be applied in future efforts.

33.1 Introduction

The modernization of intact ship stability criteria includes the notion of "direct assessment," which is the use of simulation tools to quantitatively investigate the probabilities associated with a given vulnerability. The direct assessment approach is only brought to bear when a vulnerability is expected, which means the intent of the assessment is to provide a significantly higher level of insight into the dynamic stability risk than is available through empirical and heuristic means.

This chapter is the union of two papers presented at a Stability Workshop (Belknap et al. 2011) and a STAB Conference (Belknap et al. 2012)

A. M. Reed (✉) · T. C. Smith · B. L. Campbell
David Taylor Model Basin (NSWCCD), West Bethesda, MD, USA
e-mail: arthur.reed@navy.mil

W. F. Belknap
Naval Sea Systems Command (NAVSEA), Washington, DC, USA
e-mail: william.belknap@navy.mil

T. C. Smith
e-mail: timothy.c.smith1@navy.mil

B. L. Campbell
e-mail: bradley.campbell@navy.mil

V. L. Belenky et al. (eds.), *Contemporary Ideas on Ship Stability*, Fluid Mechanics and Its Applications 119, https://doi.org/10.1007/978-3-030-00516-0_33

Peters et al. (2012) provide a comprehensive look at how direct assessment applies simulation tools to the problem, including a summary of methods for modeling ship motions, see also Chap. 40. This includes not only 6-Degree-of-Freedom (6-DoF) models intended to simulate the fully coupled ship motions problem, but also reduced-DoF models that target specific dynamic stability failures. Additionally, the requirement for validation of the simulation tools is discussed.

But while the drive toward stability criteria based on direct assessment is built upon the notion that simulation tools will be validated and accredited for this purpose, the development of the simulation tools that are able to provide sufficiently accurate results in a timely manner is still an area of research. Similarly, the means of validating a given model is also an area of research and development, given the significant challenges involved with the nonlinearity of the dynamics and rarity of the extreme motions.

Perhaps the most difficult subset of the direct assessment simulation options is the full 6-DoF maneuvering-in-waves simulation. This is the simulation approach that is required when the precise failure mechanism is not known a priori, or when a reduced order model has not be validated for particular vulnerability. But while the task of simulating 6-DoF maneuvering-in-waves ship motions can be considered difficult, it is the requirement to do so in a computationally efficient manner that presently leaves these simulation tools in the realm of research and development. The requirement for computational efficiency arises from the need to directly assess ship motions for a wide variety of environmental conditions and for several ship speeds and relative wave headings. Depending on the number of conditions to be simulated and the quantity of simulated data required for each condition (extrapolation methods are needed to ultimately provide failure probabilities—see Peters et al. (2012) and Belenky and Campbell (2012) for further discussion), “computational efficiency” may mean faster than real time to on the order of 10–100 times slower than real time. Such a requirement rules out the use of 3D unsteady RANS and nonlinear potential flow approaches.

Given the breadth of possible environmental and operating conditions (wave height, period, and directionality; wind speed and relative direction), numerical simulation tools are an attractive, if not necessary, option for predicting the dynamic stability performance of a ship. However, in order for a simulation tool to be useful, it must be validated and its limitations understood.

The very challenges that lead to the difficulty in modeling the dynamic stability hydrodynamic problem (see Belknap and Reed 2010, Chap. 1) also contribute to the difficulty in validating the simulation tool. Not only are the stability failure events of interest exceedingly rare, but they are governed by a nonlinear dynamical system. Additionally, the conditions of interest are so vast that the validation domain is essentially limitless.

This paper is not the first to propose an approach to solving the dynamic stability validation problem. In fact, Grochowalski and Jankowski (2009) provide a validation vision that is perhaps the most complete published to date. The present paper addresses three key challenges that are encountered when such a state of the art process is applied. These challenges are:

1. Validation metrics and criteria
2. Uncertainty characterization
3. Validation scope

It is hoped that the following discussion of these three issues will spur debate within the ship stability community that will result in a higher quality validation of dynamic stability simulation tools.

33.2 Validation Metrics and Criteria

While seemingly straight forward, establishing the validation metrics and criteria for what constitutes acceptable agreement is a significant problem. It is generally not sufficient for a group of subject matter experts to simply agree that the simulation results and model data (for example) are "consistently close enough." Rather, given how the numerical tool is going to be used, definitive quantifiable criteria must be established that can be used to defend the conclusions of the decision makers.

33.2.1 Specific Intended Uses

Within the framework of direct assessment of stability risk, the most important step in ensuring that the validation activities are appropriate and properly defined in scope is defining the "Specific Intended Use(s)" (SIUs) of the simulation tool. The SIUs state exactly what problem is being directly assessed and what quantities of interest are being predicted. Therefore, the SIUs will govern what physics need to be modeled, though in general limitations in ship speed are the only clearly beneficial limitations from the perspective of eliminating physics. This is because modeling all 6 degrees of freedom is required to properly include coupling effects (forces and phase). Key aspects that should be clarified in the SIUs are:

- Type of vessel (and appendages)
- Environmental conditions
- Operating conditions (ordered ship speeds and relative wave headings)
- Loading conditions
- Motions and stability failure events of interest
- Statistical quantities of interest
- How the simulation motion data and statistics will be used

Having defined the SIUs, the succeeding question is one of acceptance criteria. These are the conditions that must be met for the simulation tool to be considered acceptable for use in performing a direct assessment of stability risk. Smith (2012) describes how the acceptance criteria are derived from the SIUs, but that ultimately it is up to the certification bodies to determine how accurate is accurate enough.

In order to support the certification bodies in making such a determination, the validation task must produce the evidence that not only can the quantities of interest be predicted to within a given level of quantifiable "error" (the "Quantitative Validation" task), but also that it has done so in a manner that is consistent with the physical problem. This latter point is required so that the certifying authority can have confidence that validation extends beyond the set of direct comparison points submitted for validation. It is accomplished through validation at the physical element level, which does not have quantifiable accuracy requirements. Therefore, this is considered "Qualitative Validation."

33.2.2 Qualitative Validation

As emphasized by Grochowalski and Jankowski (2009), validation of a dynamic stability code's underlying component physics is necessary because it is not only the large amplitude motion event physics that are important to validate but also the transient behavior that leads to the event in the first place. The component physics can be thought of as both the component force models (presuming the theoretical model employs force superposition) and/or elements of the dynamic stability hydrodynamic problem domain. This domain includes, but is not limited to:

- Wind and wave environment modeling
- Roll damping
- Calm water maneuvering
- Seakeeping
  - Radiation problem
  - Diffraction problem
- Nonlinear stiffness (buoyant restoring forces)
- Appendage forces (including propellers and rudders)
- Maneuvering in small to moderate waves
- Drift forces (including added resistance)
- Surf-riding incidence

33.2.2.1 Elemental Tests

Elemental tests (or comparisons to model data) provide insight into the code's ability to capture the overall physics of the ship motion problem. They also provide confidence that the quantitative comparison results obtained with available model data may be assumed characteristic of the code and applicable for similar conditions for which model data is not available. The results of the elemental tests provide evidence to inform final decision making. Subject matter experts provide general guidelines about the comparisons; this guidance will include both qualitative and quantitative characteristics of good correlation.

The following elemental tests are useful elements for qualitative of validation:

- Roll decays
- Calm water Zig-zag maneuvers
- Calm water turning circles
- Turning circles in regular waves
- Acceleration from rest tests
- Generation of response amplitude operators (RAOs) for comparison with model data (if available)
- Integrity values

Standard maneuvering and seakeeping analyses of the motion time histories will be performed on the code and model-data time histories to provide comparison quantities guidance.

Integrity values plotted on polar and surface plots provide insight into the code's ability to capture the ship's capsize boundary. An integrity value is the ratio of the number of runs which did not include a dynamic stability failure event divided by the total number of runs examined. This metric allows for comparisons between model test and simulation in which the ship response is highly sensitive to initial conditions. Since the initial conditions under which each model test was performed cannot be known precisely, a range of simulations is performed in an attempt to cover the range of possibilities.

The integrity value elemental test is included on the list above to specifically address the known dynamic stability concerns associated with a ship operating in stern quartering seas. Characterization of the ship's response in these conditions from irregular seas model data is challenging, so integrity value plots (using regular waves model-test results) provide the necessary additional insight into the code's ability to capture this aspect of the physics. Integrity value plots will be discussed further and an example provided in Sect. 33.3.2.

There are two primary objectives in validating at the elemental level:

1. Providing insight into what may be causing the simulation results to differ from the validation data in the SIU-defined conditions
2. Providing confidence that the physics are well-modeled in conditions for which direct validation data are not available

These objectives are satisfied by performing elemental validation tests that isolate the active elements of interest. For example, an elemental validation test to assess roll damping prediction is to run a roll-decay comparison, where the qualitative validation metrics are the roll decrement coefficient and roll period as a function of roll amplitude. More examples of elemental validation tests and their qualitative metrics are provided in Table 33.1. An example series of plots showing the qualitative comparison for the regular wave dynamic stability elemental validation runs is shown in Fig. 33.1. Figure 33.1a shows the comparison of roll motion through a phase plot.

Table 33.1 Example elemental tests and their qualitative metrics

Test	Element(s)	Metric(s)
Roll Decay	Roll damping, nonlinear stiffness	Decrement coefficients and periods
Calm water turning circle	"Maneuvering" forces, appendage forces	Diameters, steady heel, steady drift angle
Calm water zig-zag	"Maneuvering" forces, radiation force, appendage forces	Overshoots, maximum roll angle
Regular wave seakeeping (non-steep waves)	Radiation force, diffraction force, roll damping	Motion transfer functions: amplitude and phase
Regular wave dynamic stability (steep waves)	(Several)	Motion time histories, integrity values, maximum value scatter
Forced motion	(Several, but isolated)	Force time histories: amplitude and phase

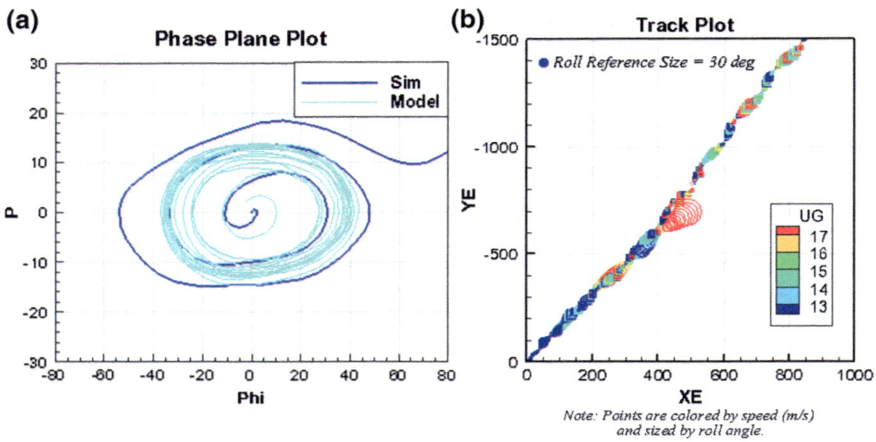

Fig. 33.1 Example visualization of regular wave dynamic stability run time history data: **a** roll phase plane plot; **b** track-speed-roll plot

Figure 33.1b shows a joint visualization of the predicted and measured track, speed (scatter symbol color), and roll amplitude (scatter symbol size).

What makes these elemental validation tests part of "qualitative validation" is that there are no obvious requirements for just how close the simulation data must match the validation data to affect the SIUs. If a calm water tactical diameter has 50% error, will that mean the simulation tool is unusable for predicting instances of parametric roll? The answer is likely "no." However, such an error indicates that hull lift and/or rudder and propulsor forces are not well modeled when the ship has a drift angle, which may lead to poor ability to predict broaching.

The elements that must be tested depend on the physical phenomena the SIUs call for capturing. Ideally, a systematic procedure is followed that lists the failure

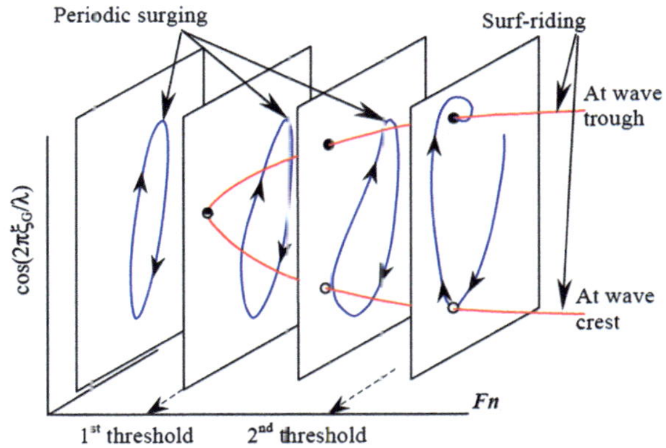

Fig. 33.2 Change from surging to surf-riding behavior with increasing speed (Spyrou 1996)

mechanisms expected to be captured (*e.g.*, pure loss of stability, parametric roll, surf-ride and broaching, synchronous roll, breaking wave impact) and then validates that component models are appropriate for capturing each failure mechanism. If parametric roll is intended to be captured, the simulation tool must adequately model nonlinear stiffness and roll damping, in addition to regular seakeeping forces.

If grouped waves are required to initiate the event, does the wave model produce grouped waves? If surf-riding is expected to be captured, does the simulation tool properly model resistance in a zero-encounter frequency steep wave? Is the loss of rudder lift captured in a steep following sea? These are just some examples of the questions that must be asked to ensure that the proper elemental tests are performed.

Finally, it should be noted that the nonlinear time series analysis techniques that have been proposed for use as validation tools (see McCue et al. 2008) fall into the realm of qualitative validation. An example of this is the incidence of surf-riding in following regular wave conditions as speed increases across a threshold. This phenomenon was presented in Spyrou (1996) and is depicted in Fig. 33.2.

While these techniques can be very illuminating in showing fundamental ways in which the time series are different, there is not a defined value to which these metrics need to agree in order for the simulation tool to be useful.

Qualitative validation is then the aggregate of these elemental/component validation tests which provide confidence in the simulation tool.

33.2.3 *Quantitative Validation*

The key measures that will be used by decision makers to determine whether or not a simulation tool can be accepted (*i.e.*, meets the requirements for a given SIU) are those for which quantitative criteria are established. The only defensible quantitative criteria are those directly tied to the SIU, which means that the metrics need to be the very same quantities that the tool is intended to predict. This ideally means multi-directional seas with wind, but due to difficulty in obtaining and working with that kind of data, long-crested irregular wave (no wind) runs may be all that is available. This requirement is similarly noted by Vassalos et al. (1998) and Grochowalski and Jankowski (2009).

To the degree that real-world conditions are not directly tested (referred to as "direct validation"), the less-than-ideal validation data set must be used for quantitative validation in conjunction with the elemental validation tests in an approach referred to as "indirect validation."

The problem of establishing acceptance criteria on these metrics is deceptively difficult. In all likelihood, there is an acceptable level of error that can be tolerated and accounted for, but establishing what is acceptable may be an iterative process. The more difficult questions to answer are whether or not all modes of motion must concurrently meet the acceptance criteria, or just certain key channels, and how to define an overall acceptable level of quantitative validation performance where the "passing" trends are inconsistent.

The last item to be discussed within the context of validation metrics and acceptance criteria is the issue of uncertainty characterization. Because this issue represents a significant challenge by itself, it will simply be stated that the validation metrics and acceptance criteria must acknowledge the uncertainty in the system and take steps to address it. The next section discusses this particular challenge.

33.3 Uncertainty Characterization

As stated earlier, the dynamic stability problem represents a stochastic nonlinear process where the events of interest can be extremely rare. Because of this, the quantities of interest are not known exactly, but rather have an associated uncertainty. The validation of dynamic stability simulation tools must therefore properly address this associated uncertainty.

33.3.1 Stochastic Process Uncertainty

The dynamic stability problem represents a particular challenge in defining the variance of mean or variance estimates, due to the fact that ship motion is a continuous random process. Care must be taken to ensure the independence of data sets, which means that the length of runs (both model data and simulation data) and the number of component frequencies used to create the exciting wave fields must be appropriately selected. The approach to addressing the challenges related to stochastic process uncertainty in the dynamic stability problem is discussed in detail by Smith (2011, 2012). Belenky et al. (2015) provide details on the calculation of the variance of the mean and the variance of the variance values needed to calculate the confidence bands on the measured or predicted motion quantities, see also Chap. 34 of this book.

33.3.2 Nonlinear Process Uncertainty

More widely discussed in dynamic stability research are the issues related to dealing with a nonlinear system. Particularly, it is well acknowledged that initial conditions can have a significant impact on the trajectory of the model (or simulated ship). It is the uncertainty associated with the variation in results that can occur that must be addressed. Free-running model-test data that are used for validation purposes more often than not do not provide precise information on the initial conditions relative to the wave field at some arbitrarily start time. If the initial conditions are not known precisely for a given model-test run (where nonlinear dynamics are believed to be in play), several model-test runs at random initial conditions are the minimum requirement for testing initial condition dependency and providing data that is useful for validation.

On the simulation side, the solution is to obtain simulation runs over a wide range of initial conditions to ensure the level of initial condition dependency is determined. It is recommended that, at a minimum, the varied initial conditions include:

- Wave phase
- Roll angle
- Roll rate

An example of this validation process is shown in Figs. 33.3, 33.4 and 33.5 for the elemental validation test of steep regular wave dynamic stability runs (free running condition at a constant *ordered* speed and heading).

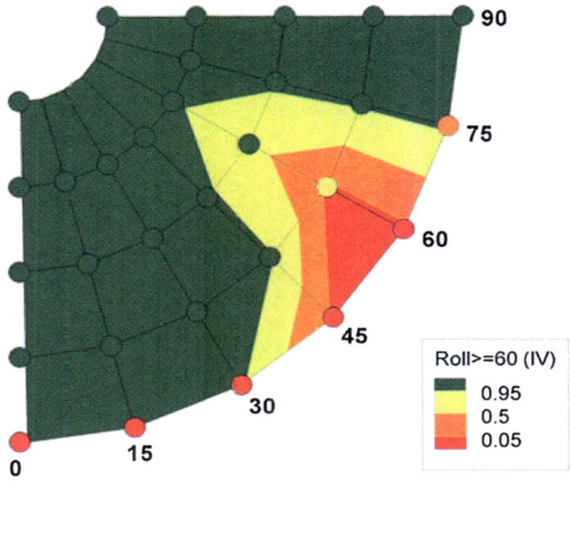

Fig. 33.3 Regular wave dynamic stability elemental validation data using contours of integrity values (IV) for 60° roll angle exceedance over a range of ship speeds (F_n = 0.0–0.4) and relative wave headings (μ = 0° to 90°) [sim.—surface; model data—circles)

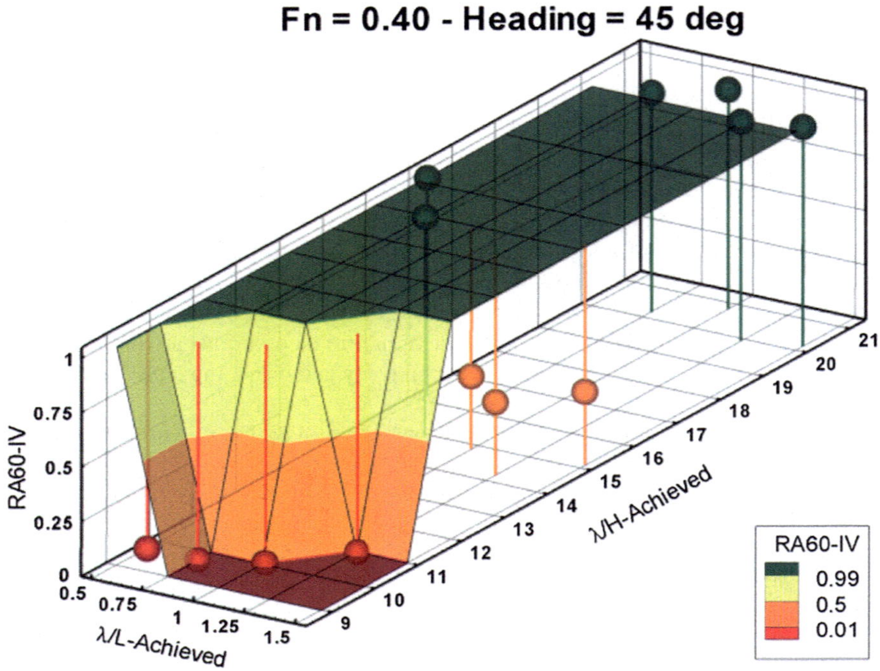

Fig. 33.4 Integrity surface for regular wave validation data (integrity values for 60° roll) as a function of wave steepness and length (simulation—surface; model data—spheres)

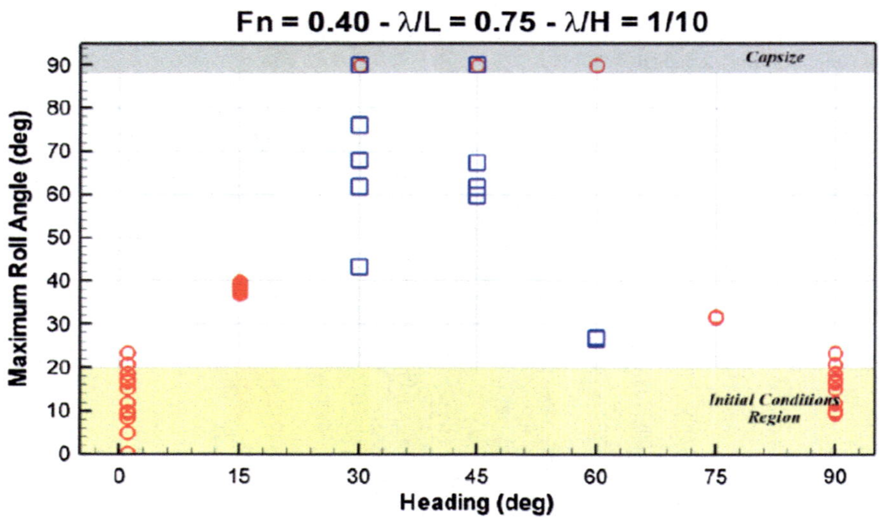

Fig. 33.5 Maximum roll values as a function of relative wave heading at a single speed for regular wave dynamic stability case (simulation—red circles; model data—blue squares)

Figure 33.3 presents a quarter-polar plot of the validation comparison for a single regular wave condition [wave length (λ) = 0.75 × ship length (L); wave height (H) = 0.1λ], where the surface mesh coloring represents the 60° roll event integrity value (IV = number of runs with events/total number of runs) for the simulation tool prediction. The scatter data points represent the model data, which are also colored by integrity value. Ideally, the simulation surface mesh will have the same color as the model-data scatter points, representing an agreement in integrity values.

Figure 33.4 presents this example validation condition (30° relative wave heading, $F_n = 0.4$, $\lambda/L = 0.75$, $H/\lambda = 1/10$) in a different dimension. The simulation data was run at much finer resolution in wave steepness than the model data in order to develop the integrity surface shown in the figure. This allows the model data to be plotted at the actual achieved wave conditions. In this example condition, it can be seen that the average achieved wave steepness in the model experiment was actually slightly less than $H/\lambda = 1/10$). This places the model result within the steep region of the simulation integrity surface, which would change the validation conclusion to "good agreement." Where there is a difference, further investigation is warranted.

Figure 33.5 shows the actual distribution of maximum roll across relative wave heading for a single speed ($F_n = 0.4$). At the 30° relative wave heading, Fig. 33.5 shows that the simulation predicts all runs at this condition to have a roll greater than 60° (as seen by the red surface at that point), whereas the model data point has some cases where roll exceeds 60° and others that do not, depending on initial conditions. Further examination of Fig. 33.5 shows that, in fact, all of the simulation runs go well beyond 60° to the point where they all capsize, whereas the model data are scattered between 42° and capsize. This would imply that the simulation is not providing a good prediction.

Clearly, there is a sharp transition across relative wave heading, such that the simulation tool predicts a "green" condition at 15° relative wave heading. Therefore, it is reasonable to wonder about uncertainty in the ordered heading. But what about uncertainty in the actual achieved wave conditions in the model basin? The results plotted in Figs. 33.3 and 33.4 overlay data such that it is assumed all represent behavior in wave conditions with the ideal wave steepness of 1/10. However, the wave makers in any model basin are an imperfect system, in that the actual realized wave steepness (and possibly wave length) may be off by some amount.

The narrative above represents the effect of "input parameter uncertainty," which is the third uncertainty-related challenge to be addressed within the dynamic stability validation problem.

33.3.3 Input Parameter Uncertainty

In describing the validation uncertainty related to the nonlinear dynamical process of the dynamic stability problem, it was shown in an example case that uncertainty in the knowledge of the input parameters (wave height, in the previous example) can have a dramatic impact on the validation conclusions drawn. Particularly in the quantitative validation stage where direct comparisons are made with real-world data, input uncertainty can be a significant challenge.

There are many parameters that are provided as input to the simulation validation cases that in many cases in the past have been taken as known quantities. In reality, these quantities represent best estimates that are based on direct measurement of an unchanging variable (such as mass properties) or statistical analysis of a random process (such as the seaway). As the state-of-the-art in validation of dynamic stability simulation tools is expanded, the issue of input uncertainty must be addressed. In other words, validation conclusions must be based on a comprehension of the propagation of uncertainty due to errors in the assumed input parameters.

The issue of addressing error propagation in the validation of engineering models has been studied by Hills and Trucano (1999). They specifically address the effect of error propagation in statistical validation of nonlinear systems, which means their work is particularly relevant to the validation of dynamic stability simulation tools. One approach offered by Hills and Trucano to assessing the impact of error propagation is to perform Monte Carlo simulations that examine at a system level the impact of perturbations to a matrix of input parameters.

To get a first look at the impact of input parameter uncertainty on the dynamic stability validation problem, a sample Monte Carlo based study was conducted that examined notional input errors in the validation of a stern quartering high sea-state condition. The study was not designed to be a comprehensive examination of the impact to be expected across typical validation cases, but rather to provide a snapshot on a single case as an exercise in the process.

The input parameters to be varied within the study were taken as representative sources of error expected in the validation of a real-world scenario. These were taken to be:

- Relative wave heading
- Significant wave height
- Modal period
- GM

The magnitudes of the errors are hypothetical, but are representative of the size of 95% confidence intervals. Using the "best estimate" input values as the basis, the Monte Carlo approach parametrically varied the four sources of error to include simulations at the upper and lower bounds of the input values. These input values are shown in Table 33.2. Even with this coarse resolution, the resulting matrix of conditions is $3 \times 3 \times 3 \times 3 = 81$ discrete simulation conditions to capture the potential error in validating a single validation condition.

The results were processed to examine the variation in the mean up-crossing rate for roll, as this is a typical quantity of interest for dynamic stability analysis. Figure 33.6 shows all simulation results for the roll mean up-crossing rates as determined from envelope peaks over threshold (EPOT) analysis (see Campbell and Belenky 2010). The solid blue line represents the simulation results based on the best estimates of the input parameters. The scatter points (at select roll levels of

Table 33.2 Input value uncertainties for Monte-Carlo simulation matrix

	Lower bound	Best estimate	Upper bound
Heading	35°	45°	55°
$H_{1/3}$	11.0 m	11.5 m	12.0 m
T_m	13.0 s	14.0 s	15.0 s
GM	$GM_0 - 0.05$ m	GM_0	$GM_0 + 0.05$ m

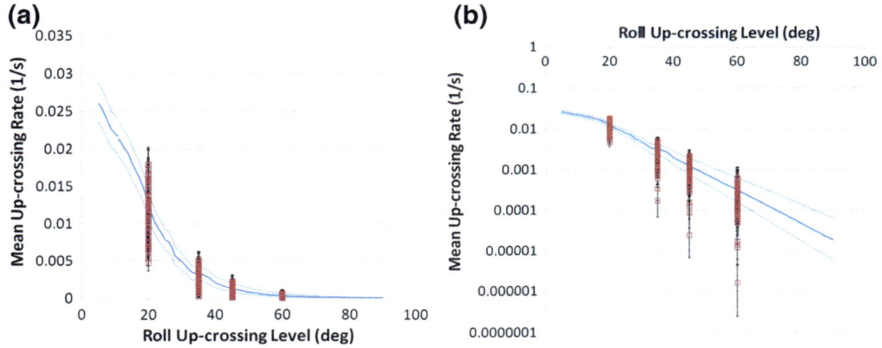

Fig. 33.6 Simulation results for the mean up-crossing rates for roll using the best estimate of input parameters (solid blue line with dashed blue lines for 95% confidence intervals) and the parametrically varied input parameters (red scatter points with 95% confidence intervals)—presented on: **a** linear scale and **b** log scale

20°, 35°, 45°; 60°) show the range of variation in the results that can be expected when the uncertainty in the input values are accounted for.

The results show that, for at least this hypothetical case, the variation in the mean up-crossing rates can vary on a scale considerably greater than the statistical process uncertainty (shown by the "best estimate" 95% confidence intervals).

Based on this single example result, it is reasonable to consider input value uncertainty when interpreting validation results. However, it must be emphasized that the present study is a single hypothetical example and the size of the effect of input error is likely to be a function of the physics. It is anticipated that certain parameters will have more of an effect than others depending on the dynamics of the problem (*e.g.*, resonant conditions versus pure excitation, etc.), which are condition dependent.

More work must be undertaken to relate the joint probabilities of the various input errors to the final total confidence interval size. The example calculations presented in this paper represent the first attempt by the authors to investigate other degrees of uncertainty in the validation problem.

33.4 Validation Scope

The final validation challenge addressed involves answering a question that must always be asked in a validation task: When is the validation complete?

The only defensible answer to the above question is that there is no good answer. Furthermore, the answer should *not* be "when the available validation data are exhausted."

It has been argued that validation tasking should include qualitative validation and quantitative validation. The scope of the qualitative validation is more easily defined in that it should include the full range of ship speeds and address all motion components of interest (as defined by the SIU). If the qualitative validation is incomplete, the ability to perform "indirect validation" will be limited or considered higher risk. The scope for quantitative validation is considerably more challenging to define. This is generally because the number of environmental conditions a ship will see is infinite. Not only are there combinations of significant wave height and modal period (as well as spectral shape) to examine, but if bi-directional seas and relative wind direction are part of the environment defined in the SIU statement, the matrix of conditions quickly grows in size.

It is not reasonable to expect that direct validation (with quantitative metrics) will be performed for all conditions for which the simulation tool will be accepted. First and foremost, a simulation tool would not be required if trusted data existed for every condition of interest. Also, the variability of the physics between directly validated conditions may not be significant. Even so, a sufficient range of directly validated cases must be obtained. For those conditions that are not directly validated, an argument must be made that the physics in play for the non-tested conditions are not unique. The elemental validation cases and the neighboring quantitative validation cases will be relied upon to provide confidence that the simulation tool is valid at these "interior" conditions.

In the end, the validation scope is likely to be a judgment call by the subject matter experts. The resolution of the validation domain may need to be adjusted as the validation process proceeds.

33.5 Conclusions

This paper has addressed the key challenges in the validation process for dynamic stability simulation codes. These challenges are:

- Developing the validation metrics and criteria
- Characterizing and incorporating uncertainty
- Defining the validation scope

The key point that is made relative to successful establishment of validation metrics and criteria is that the Specific Intended Use of the simulation tool must be completely and precisely defined. The metrics and the acceptance criteria should flow directly from the SIU. The metrics and criteria can be divided between qualitative and quantitative categories. While the quantitative metrics are the most valuable for the decision makers in determining the usefulness of the simulation tool, the qualitative validation tests are essential for diagnostics and in support of "indirect" validation. And while the quantitative metrics are the most useful, they must be well thought out and, if possible, incorporate an allowable error that is tied to the SIU.

The challenges associated with uncertainty characterization involve stochastic process uncertainty, nonlinear dynamics uncertainty, and input parameter uncertainty. All three can factor heavily into the validation process and interpretation of results. At the moment, the influence of input parameter uncertainty remains an area of research, though an example study using hypothetical errors in the input parameters has demonstrated that the magnitude of the scatter due to input error can plausibly be larger than stochastic process uncertainty alone.

Finally, difficulty arises in determining when the body of validation work is sufficiently complete. Particularly when the vast matrix of possible environmental conditions is considered, it becomes obvious that a direct validation cannot be performed for all operational (speed, heading, load condition) and environmental conditions. It is concluded that indirect validation will fill the void, but that direct validation must be done for conditions where the physics are sufficiently unique.

33.5.1 Future Challenges

This paper does not represent an exhaustive list of challenges remaining in the validation of dynamic stability simulation tools. Many difficulties will need to be addressed in future efforts, particularly as the "realism" of the validation data increases. For example, the non-stationarity of real-world environmental conditions will require special treatment when these data are used for validation. Another challenge to be addressed is validation for full-scale ships, when the vast majority of available validation data comes from model experiments. Also, there is the very real scenario of dealing with validation results where 100% of validation conditions do not pass the criteria, and there is not a clear trend as to where the simulation is not properly modeling the physics, is the simulation invalidated? Is there an acceptable passing rate across all conditions and what should it be?

In this chapter, no attempt has been made to deal with an important aspect of the validation of a simulation tool for regulatory purposes, that is the ability of a person or organization to use the tool correctly, which addresses several aspects of both the tool (documentation, self-consistency checks, example, etc.) and the organization (training or even certification).

These and other future challenges will be discussed in the coming years, but at present, the validation of dynamic stability simulation tools stands to benefit from an improved understanding of metrics and criteria, uncertainty characterization, and proper establishment of validation scope.

References

Belenky V. & B. L. Campbell (2012) "Statistical Extrapolation for Direct Stability Assessment." *Proc. 11th Int'l. Conf. Stability of Ships & Ocean Vehicles (STAB '12)*, 16 p., Athens, Greece.

Belenky, V., V. Pipiras & K. Weems (2015) "Statistical Uncertainty of Ship Motion Data." *Proc. 12th Int'l. Conf. Stability of Ships & Ocean Vehicles (STAB '15)*, pp. 891–902, Glasgow, UK.

Belknap, W. F. & A. M. Reed (2010) "TEMPEST — A New Computationally Efficient Dynamic Stability Prediction Tool." *Proc. 11th Int'l. Ship Stability Workshop*, pp. 185–197, Wageningen, The Netherlands.

Belknap, W. F., A. M. Reed & M. J. Hughes (2012) "Model Characteristics and Validation Approach for a Simulation Tool Supporting Direct Stability Assessment." *Proc. 11th Int'l. Conf. Stability of Ships & Ocean Vehicles (STAB '12)*, 16 p., Athens, Greece.

Belknap, W. F., T. C. Smith & B. Campbell (2011) "Addressing Challenges in the Validation of Dynamic Stability Simulation Tools." *Proc. 12th Int'l. Ship Stability Workshop*, 9 p., Washington, DC.

Campbell, B. L. & V. Belenky (2010) "Assessment of Short-Term Risk with Monte-Carlo Method." *Proc. 11th Int'l. Ship Stability Workshop*, pp. 85–92, Wageningen, The Netherlands.

Grochowalski, S. & J. Jankowski (2009) "Validation Methodology for Simulation Software of Ship Behaviour in Extreme Seas." *Proc. 10th Int'l. Conf. Stability of Ships & Ocean Vehicles (STAB '09)*, pp. 409–420, St. Petersburg, Russia.

Hills, R. G. & T. G. Trucano (1999) "Statistical Validation of Engineering and Scientific Models: Background." SAND99-1256, Sandia National Laboratories, Albuquerque, NM, 92 p.

McCue, L. S., W. R. Story & A. M. Reed (2008) "Nonlinear Dynamics Applied to the Validation of Computational Methods." *Proc. 27th Symp. Naval Hydro.*, 10 p., Seoul, South Korea.

Peters, W. S., V. Belenky & A. M. Reed (2012) "On Regulatory Framework of Direct Stability Assessment." *Proc. 11th Int'l. Conf. Stability of Ships & Ocean Vehicles (STAB '12)*, 14 p, Athens, Greece.

Smith, T. C. (2011) "Statistical Data Set Comparison for Continuous, Dependent Data." *Proc. 12th Int'l. Ship Stability Workshop*, 7 p., Washington, DC, USA.

Smith, T. C. (2012) "Approaches to Ship Motion Simulation Acceptance Criteria." *Proc. 11th Int'l. Conf. Stability of Ships & Ocean Vehicles (STAB '12)*, Athens, Greece.

Spyrou, K. J. (1996) "Dynamic instability in quartering seas: The Behavior of a Ship During Broaching." J. Ship Research, 40(1):46–59.

Vassalos, D., M. Hamamoto, J. O. de Kat, D. Molyneux & A. Papanikolaou (1998) "The State of the Art in Modelling Ship Stability in Waves." *Proc. 25th ATTC*, 8 p., Iowa City, Iowa.

Chapter 34
Validation Approach for Statistical Extrapolation

Timothy C. Smith

Abstract Statistical extrapolation is the estimation of rare, extreme responses from data sets that consist primarily, if not entirely, of lower (non-rare) values. The validation of statistical extrapolation involves elements common to all validation efforts with the additional difficulty of needing to determine the true value and its uncertainty. The determination of the true value requires an extensive amount of data in order to observe multiple rare events In some cases, the desired extreme events may be so rare that validation is forced to accept a less rare event such as a lower threshold value. This paper reviews the basics of simulation validation and focuses on the challenges of the validation of statistical extrapolation.

34.1 Introduction

For many years, the numerical simulation of ship motions has been used to determine relative performance of various design options (Bales 1980) with idealized wave spectra. The state of numerical simulation of ship motion, maneuvering and seakeeping, is approaching the point of predicting performance in an absolute sense for operational guidance. The use of simulation in many, if not all, areas of engineering is a given. Prediction tools for early design work can afford to trade accuracy for speed and can identify trends, if not absolute values. Later stages of design require simulations of greater accuracy. The required fidelity or accuracy of the simulation is related to the simulation's purpose. The degree of accuracy is a function of the assumptions made in the development of theory on which the simulation is based, the numerical implementation of the theory, and the expertise of the user. For the purpose of this discussion, user expertise will not be examined. In all cases, there is a desire to know how well the simulation represents the real world (i.e. the accuracy of the simulation). Simulation validation provides this knowledge.

T. C. Smith (✉)
David Taylor Model Basin (NSWCCD), West Bethesda, MD, USA
e-mail: timothy.c.smith1@navy.mil

V. L. Belenky et al. (eds.), *Contemporary Ideas on Ship Stability*, Fluid Mechanics and Its Applications 119, https://doi.org/10.1007/978-3-030-00516-0_34

Simulation validation is carried out as part of a verification, validation, and accreditation (VVA) or verification, validation, and uncertainty quantification (VVUQ) effort. Verification deals with the correctness of the software coding process and theory implementation. Validation is the comparison of simulation results to real world data in order to determine how close the simulation results represent the real world data or "true value." Accreditation is the official determination of a simulation tool's applicability to a particular problem as defined by specific intended uses (SIUs). Uncertainty quantification combines the numerical error of the simulation, input error (not matching test conditions), and random process error. The true value is the benchmark data set used for comparison to the simulation. The true value may be scale-model test data, higher fidelity simulations, full-scale trials data or other representative data sets appropriate to the problem.

The validation of numerical simulations is addressed by various professional societies and governmental bodies for many engineering disciplines. There are established verification and validation outlines, guides, and processes to follow when performing numerical simulation verification and validation (AIAA 1998; ASME 2009; ITTC 2011; IEEE 2012). The 26th International Towing Tank Conference (ITTC 2011) recognized the need for verification and validation of seakeeping computer codes and prepared guidelines to that purpose. The ITTC guidelines clearly lay out verification and validation tasks and activities going so far as listing what to consider at a component level. The US Navy describes a generic formal process with review panels and extensive supporting documentation, but leaves details of comparison metrics to the analysts (DoN 2004; DoD 2009; Sec. of Navy 1999). Griffith and Locke (2006) also discuss these VVA processes. These processes and guides are often generalized with details left to the engineers actually performing the verification and validation. Validation at its core consists of a comparison between the simulation and the "true value," and becomes the basis for a validation decision.

To be used for a particular application, such as direct assessment of parametric rolling, the validated simulation needs to be accredited for that use. Validation is necessary, but not sufficient, for accreditation. The first step is to select a simulation code with the desired simulation capability. The accreditation process then clearly defines the boundaries of where the simulation may be used and not used. These boundaries are defined in the SIUs (Belknap et al. 2012; DoN 2004; Reed and Zuzick 2015). The SIUs serve to limit the scope of the validation to a reasonable and useful domain. The SIUs define the operational and environmental conditions and type of response and statistics where the simulation is applicable. For instance, a typical linear seakeeping simulation would have a SIU such as "The simulation will be applied for all speeds and heading for Sea States 5 and less to produce significant single amplitude statistics." This limits simulation applicability to an expected linear response domain and precludes using the simulation for extreme events and nonlinear behavior.

The SIUs are codified into acceptance criteria, which provide a means to evaluate the simulation code with respect to the SIUs. Acceptance criteria specify the required validation data for comparison and parameters to compare. For example, consider a simulation to predict low speed maneuvering in shallow water with current and wind.

In this case, wind and current drag coefficients, thruster/propeller performance, full-scale trials tracks would all be pertinent data allowing for validation of specific force modules and overall simulation performance.

The acceptance criteria state the required level of accuracy for the simulation to be accurate enough for use; validation is the statement of that accuracy. Depending upon simulation use, "accurate enough" can range from similar motion properties (qualitative) of random behavior to matching a set of time histories to a specified accuracy (quantitative). If a simulation code passes the acceptance criteria, it can be accredited for use in the region defined by the SIUs.

When comparing simulation data to benchmark data, Belknap et al. (2011) make the distinction between qualitative and quantitative validation. Qualitative data comparisons are made by examining trends and expected behavior, often at a component or force model level, such as roll damping or Froude-Krylov force. It is often difficult to determine quantitatively how much a specific component affects the final or total answer. In these cases, benchmark data are often highly constrained to focus on one particular element, such as roll decay tests. Favorable qualitative comparisons give the user confidence in the simulation, but cannot replace quantitative acceptance criteria.

Quantitative validation is a direct comparison of simulation and benchmark data with some defined metric of accuracy. There is a difference between acceptance criteria for validation and for accreditation. In both cases, the simulation is compared against benchmark data to ascertain its accuracy. However, for validation, the purpose is a diagnostic identification of areas requiring improvement with the goal of improving the overall accuracy of the simulation. This can focus on specific aspects or elements of the simulation. For accreditation, the acceptance criteria are a quantification of acceptable accuracy for SIUs.

Uncertainty quantification affects the development of acceptance criteria in terms of appropriate or attainable levels of accuracy. The acceptance criteria specify which parameters will be compared and how they are to be compared. Development of acceptance criteria can be the subject of lengthy debate as they need to adequately reflect the SIUs and be complete, rational, and achievable. Specifics of acceptance criteria are discussed later. The acceptance criteria along with comparison data sets are used in the validation process to make an accreditation recommendation. A lack of data (or high uncertainty in the data) will limit the extent and certainty of the validation.

As ship motions in irregular waves are stochastic in nature, all of the elements of VVA have to be formulated in a probabilistic sense as the quantities are random fields, stochastic processes or random numbers. This presents a philosophical challenge: all of the statements in the validation process have probabilistic meaning and, in principle, nothing can be stated for certain, so even in the best cases there is only a significant probability that the method is valid. On top of the "usual" uncertainty related to data sources, approximation, etc., the statistical uncertainty, which is related to finite volume of data, has to be evaluated and propagated through the validation process. Belenky et al. (2013, 2015) considers methods to assess the statistical uncertainty for ship motions.

Focusing on the phenomena of large motion and capsizing, the simulation of these phenomena requires advanced, hydrodynamic blended method prediction tools due to the non-linearity involved (de Kat and Pauling 1989; Lin and Yue 1990; Shin et al. 2003). Furthermore, the ITTC parametric roll study (Reed 2011; ITTC Stability in Waves 2011, Chap. 37; Belenky and Weems 2012) showed that the uncertainty can be quite large due to practical non-ergodicity. These phenomena are also rare in realistic weather and loading conditions. Observation of a dangerously large motion or capsizing may require a very long and prohibitively expensive simulation. Statistical extrapolation methods (Campbell et al. 2016; Belenky et al. 2016; Anastopoulos et al. 2015) use simulations of practical length to estimate probability of rare event. This further increases the difficulty in understanding the result of the validation effort and achieving a definitive result.

A primary difficulty for validation of statistical extrapolation is determining the true value if the required numerical simulation is too long to be practical. A practical solution is to use a very fast simulation tool that would be only qualitatively valid for the SIU under consideration. An example of such tool is described by Weems and Wundrow (2013). The tool simulates heave, roll and pitch, using instantaneous volume instead of pressure integration for hydrostatic and Froude-Krylov forces. All other forces are modeled with polynomials. The approach yields sufficient calculation speed to observe capsizing and large motions in realistic conditions and reproduces essential physics of capsizing in waves i.e. is qualitatively valid (Weems and Belenky 2015). A limited validation was necessary to determine if the representative data set is truly representative. With the fast code, it is possible to produce a very large number of motion records, sufficient to observe the rare phenomenon in statistically significant numbers. An estimate of probability of the rare phenomenon from this large data set is the true value for the validation comparison. In this case, the true value has its own uncertainty because the data set while very large is still not infinite. For the sake of consistency, the small subset of the same records is to be used to produce the extrapolated value to be compared with the true value. This chapter focuses on the formulation of procedures and acceptance criteria for validation of extrapolation methods.

34.2 Three Tier Acceptance Criteria

Acceptance criteria are derived from the SIUs and define what is compared, the nature of the comparison, and what is an acceptable or passing comparison. However, acceptance criteria can quickly become very complex as an SIU becomes fully defined. For instance, validating linear seakeeping predictions quickly leads to questions as to which motions or subset of motions to compare. Then, should those comparisons be qualitative, quantitative, statistically-based, expert opinion-based, or some mix. Determining what constitutes an acceptable comparison can be a difficult task in itself.

A three tier hierarchy acceptance criteria that matches typical benchmark data structure provides a convenient means for formulating acceptance criteria (Smith et al. 2014). The three tiers are: parameter, condition, and set. A parameter is a specific response or motion that is being compared. This is the most elemental comparison to true value; the comparison is specified as an accuracy metric. The accuracy metric specifies how the comparison in made and what constitutes passing. A condition defines the metadata for the parameter, for example ship loading condition, wave height, wave period, relative wave direction and ship speed The set or overall tier specifies the number and type of conditions to accept the simulation. So the acceptance criteria specifies how parameters are compared and what constitutes passing as the parameter, condition, and set metrics are rolled up.

Parameter acceptance criteria deal with the comparison of a single parameter, for example, roll angle, deck wetness probability or turning circle trajectory. The selection of parameter depends on available data and SIUs. The condition and set acceptance criteria reflect a validation or accreditation philosophy rather than a strict comparison of data. It is at these levels that subjectivism is most apparent, and perhaps most needed as requiring 100% matching of all conditions in the entire set is likely impossible. Depending on the benchmark data structure and SIUs, the condition and set tiers may be duplicative and only two tiers should be used.

34.2.1 Parameter Criteria (Tier 1)

Parameter criteria focus on a single parameter and make use of comparison accuracy metrics. To represent the SIUs, a mix of accuracy metrics maybe appropriate depending on data and SIU requirements. This is an attempt to use to best elements of the other approaches, while balancing or eliminating the disadvantages. An obvious mixed acceptance criterion is using engineering judgment to set some minimum level of concern and overlapping uncertainty bars above that level of concern. This approach can become quite elaborate as more considerations are included. Care needs to be taken to ensure that the interaction between the different pieces does not result in an undesirable cumulative acceptance criterion, even though all individual pieces seem reasonable.

Many fluid simulation validation efforts have focused on computational fluid dynamics (CFD) looking at steady state phenomenon (ASME 2009; Oberkampf and Blottner 1997; Stern et al. 1999, 2005). Some typical accuracy metrics that account for the random nature of ship motions are confidence interval overlap, statistical hypothesis testing, subject matter expert judgment, sum of errors, and correlation metrics. This list should be considered representative and not exhaustive.

Campbell et al. (2016) and Glotzer et al. (2017) describe the calculation of confidence interval for extrapolated rate of exceedance. Special care is needed for non-ergodic processes where the uncertainty needs to be determined through repeated simulations (Reed 2011, Chap. 37), so extrapolation should be done on a number of records. Uncertainty bands for the true value can be calculated using the normal distribution approximation to the binomial distribution (Campbell et al. 2016). Confidence interval overlap is used as a parameter criterion in the example considered further in this chapter. Maximum conservative distance (MCD) is the difference between the extrapolation upper boundary of confidence interval of the extrapolated and the true value; MCD provides an alternative to confidence interval overlap.

The use of statistical hypothesis tests is to determine if two estimates could come from two samples belonging to the same general population (Bendat and Piersol 1966; Priestly 1981). Smith (2011) proposed a concept of equivalent number of degrees of freedom (data points or samples) to apply the standard statistical significant tests such as Student's t, or F-test to dependent data. Similarly, by specifying the limit on the difference and solving for the critical level of significance to pass the test, the probability associated with success of the comparison is produced (Smith and Zuzick 2015) .

For sum of error metrics, Stern et al. (1999) and Oberkampf and Blottner (1997) suggest a weighted vector sum of the uncertainties in a manner similar to standard uncertainty analysis (Coleman and Steele 1989). The simulation is considered acceptable when its uncertainty band overlaps the benchmark uncertainty band.

There are many correlation metrics available to compare trajectories or data sets. Ammeen (1994) surveys many of existing approaches and discusses their merits for trajectory comparison. The Distance-Weighted Average Angle Error measure was determined to most closely represent subject matter expert opinion. Lee et al. (2016) provides a validation example using Average Angle Error Measure (AAM) to compare maneuvering simulation trajectories.

34.2.2 Condition Criteria (Tier 2)

Acceptance criteria for a condition are generally based upon having a specified number of parameters/motions/components passing based on individual motion/component accuracy metrics. The SIUs specify which motions to include in the criteria. As mentioned in Belknap et al. (2011), when considering the total simulation, it is desired that all the motions be acceptable simultaneously. This precludes considering the simulation acceptable when matching vertical acceleration, but not matching speed or heading at which the vertical acceleration is calculated.

However, motions are not uniformly sensitive to variations in speed, heading, wave height, modal period, or wave spreading. Requiring all motions, including speed and heading, to pass simultaneously may penalize the simulation unduly. This can be accounted for by adjusting the parameter acceptance criteria to different accuracy levels depending upon sensitivity.

Alternatively, requiring a certain number or mix of motions, but not all motions, is a potential compromise provided there is not a consistently failing motion. The presence of a consistently failing motion indicates a theory or implementation problem with the simulation.

34.2.3 Set Criterion (Tier 3)

The set or overall criterion deals with acceptance of the total simulation or component rather than just a specific condition or case. This can be simply characterized as a passing percentage of all cases examined or even a weighted sum of the passing cases based on importance. For instance, if beam seas performance was most important, then those conditions could be weighted more than other cases. This would ensure the simulation performed adequately where it was most important. The benchmark data set also provides a de facto weighting, which should be evaluated against the desired weighting.

It should be noted that setting an overall acceptance criterion also sets a required pass rate for the individual parameter/motion/components. Intuitively, it is apparent the higher the required condition passing requirement (overall criterion), the higher the individual motion or component pass rate that is required to meet the overall criterion. What can be overlooked is that requiring the simultaneous passing of multiple individual parameters, motions, or components quickly leads to not passing the overall acceptance criterion despite relatively high individual motion or component pass rates.

34.2.4 Example Criteria Formulation

As an example of acceptance criteria formulation, consider the validation of an extrapolation method to predict extreme, rare motion values. The SIU is relatively simple and general—determine the mean crossing rate (the upper boundary of confidence interval is often used in practical cases) at a specified comparison level on specified exposure time for a given motion or acceleration, speed, heading and environmental conditions. Using a single comparison level is certainly simplest but allows for an extrapolation to fail at all other levels of interest and still be acceptable. However, using multiple comparison levels may give too much weight to levels that are unimportant.

Following the multi-tier approach from Smith (2012), a overlap check of confidence intervals at the comparison level would be parameter; multiple confidence interval overlap checks would be condition; and multiple conditions for different motions, speeds, heading, and seaway data sets would be the overall or set tier. This is not the only possible tier breakdown, for instance multiple confidence interval comparisons could be used as the parameter (Tier 1), a unique speed-heading-seaway a condition, and all the conditions would be the set tier. Acceptance also depends on the data set used to make the extrapolation. As a result, multiple extrapolations from multiple data sets, $N_{EX} = 100$ or more, should be used to evaluate acceptance.

The extrapolation would be checked against the true value at that comparison level. The check could be made based on confidence interval overlap, a more rigorous statistical test (Smith 2011), or even simply the distance between mean crossing rates. For the confidence interval overlap as Tier 1, one can consider a random variable x:

$$x_i\left(P_\beta, \vec{S}, \phi_j\right) = \begin{cases} 1\ if & \left[\lambda^E_{low}; \lambda^E_{up}\right] \cap \left[\lambda^T_{low}; \lambda^T_{up}\right] \neq \varnothing \\ 0\ if & \left[\lambda^E_{low}; \lambda^E_{up}\right] \cap \left[\lambda^T_{low}; \lambda^T_{up}\right] = \varnothing \end{cases} \tag{34.1}$$

where i is the index of extrapolation data set, ϕ_j is the comparison level and λ are the crossing rate confidence interval boundaries distinguished with superscripts for extrapolation (E) and true value (T); P_β is confidence probability, $\vec{S}$ is a vector of values describing environmental (significant wave height, modal period, etc.) and loading conditions (KG, draft, trim), speed and heading.

Specifying an allowable distance between mean crossing rates requires knowing the effect of the distance on the final use of the extrapolation. A sensitivity study on the final use of the extrapolation can help set these values. Also, the allowable distance should not be less than the true value uncertainty (where the true value uncertainty is that due to random process error). Using an allowable distance less than this could conceivably not accept another realization of the true value. This is indicative of overly strict acceptance criteria.

With multiple extrapolations, the percentage of passing extrapolations needs to be specified to pass at the comparison level. Even with multiple extrapolations, this is still a random process dependent on the data sets:

$$C2\left(P_\beta, \vec{S}, \phi_j\right) = \frac{1}{N_{EX}} \sum_{i=1}^{N_{EX}} x_i\left(P_\beta, \vec{S}, \phi_j\right) \tag{34.2}$$

Equation (34.2) represents the condition criterion (Tier 2), as it reflects how good the extrapolation is. If the extrapolation method is ideal, the criterion $C2$ will equal the given confidence probability P_β. Real-world extrapolation methods may miss the true value just because they use statistical estimates that are random variables. Thus the acceptance condition is formulated as:

$$C2\left(P_\beta, \vec{S}, \phi_j\right) \geq B2 \tag{34.3}$$

where $B2$ is a boundary (or standard) for the acceptance at the Tier 2.

The boundary, B2, needs to be set to a level that accounts for the uncertainty. For extrapolations carried out with the confidence probability P_β, it is not reasonable to set $B2$ to be more than P_β. Otherwise, the acceptance may not be reached purely because of natural statistical uncertainty of the random variable x, which has nothing to do with validity of the extrapolation method, so

$$B2 < P_\beta \tag{34.4}$$

However, statistical uncertainty is not the only imperfection of an extrapolation method intended for nonlinear dynamical system under random excitation. Inevitably, other assumptions are made. Some of them may be related to use of limit distributions, like Generalized Pareto Distribution (GPD) (Coles 2001) for extreme values or even normal for estimates. Then the result becomes dependent on how quickly the actual distribution converges to its limit, which is not always known. Other assumptions may involve dynamics; such as the response to a wave group made of sinusoidal waves represents the response to a real-world wave group. Thus, it makes sense to set the standard lower than the confidence probability, for example:

$$B2 = 0.9 \quad if \quad P_{\beta} = 0.95 \tag{34.5}$$

The averaging in Eq. (34.2) also brings additional statistical uncertainty that can be dealt with by the calculation of the confidence interval for the estimate of $C2$ which can be used to set $B2$. This can be done using the binomial distribution for the number of extrapolations, as Eq. (34.1) can be considered as a Bernoulli trial:

$$\begin{aligned} C2_{low} &= \frac{Q_B\left(0.5(1 - P_{\beta}), N_{EX}, p\right)}{N_{EX}} \\ C2_{up} &= \frac{Q_B\left(0.5(1 + P_{\beta}), N_{EX}, p\right)}{N_{EX}} \end{aligned} \tag{34.6}$$

where Q_B is the quantile function or inverse of binomial cumulative distribution function (CDF) for N_{EX} Bernoulli trials with the desired confidence probability, p. For example, for 100 data sets with 95% probability and taking the 95% confidence, the lower quantile is 90, or 0.90 as stated in Eq. (34.5). $B2$ needs to be adjusted in a similar fashion for other cases using Eq. (34.6). Finally the Tier 2 acceptance condition is written as:

$$C2(P_{\beta}, \vec{S}, \phi_j) \geq C2_{low} \text{ where } B2 = C2_{low} \tag{34.7}$$

The third tier, overall acceptance, deals with how many conditions need to pass for overall acceptance of an extrapolation method.

Interestingly, the actual environmental conditions used do not matter as many different environmental conditions will produce similar levels of rarity (order of magnitude for the rate of failures). The key is that the extrapolation method is valid for a wide range of rarities or low probability tail behaviors. Therefore, the acceptance criteria need to specify simulations that cover the different tail behaviors rather than operational conditions. There may be a correlation between tail behaviors and operational conditions, but that is not guaranteed. Thus, all conditions or cases would be required to pass for overall acceptance as the total number of cases can be relatively small.

$$C3(P_\beta) = \prod_{k=1}^{N_{CND}} \begin{cases} 1 \; if \quad C2\left(P_\beta, \vec{S}_k, \phi_j\right) \geq B2 \\ 0 \; if \quad C2\left(P_\beta, \vec{S}_k, \phi_j\right) < B2 \end{cases} \tag{34.8}$$

$$C3(P_\beta) \geq B3 \tag{34.9}$$

where $B3$ is the acceptable overall condition pass rate. $B3 = 1$ for an all conditions passing criterion.

Thus, the acceptance criteria formulation involves the determination of which parameters to compare, their manner of comparison, and appropriate accuracy levels at each tier. In this case, the parameter to compare is the confidence interval at the comparison level (Eq. (34.1) for Tier 1. Tier 2 and Tier 3 acceptance metrics are calculated with Eqs. (34.2) and (34.8), respectively. The acceptance criteria are set in Eqs. (34.1), (34.7), and (34.9) for Tiers 1, 2, and 3, respectively.

34.2.5 Statistical Extrapolation Example

As an example of acceptance criteria application, consider the validation of an extrapolation method to estimate mean crossing rates at high levels of ship motion (Smith and Campbell 2013; Smith 2014; Smith and Zuzick 2015). For this simple case, that objective statement is the SIU with some clarification on what constitutes a "ship motion," "high levels," and range of seaway-speed-relative wave headings.

Most motion criteria provide limits for roll, pitch, vertical and lateral accelerations in manned spaces. High levels of motion are generally at some stability limit, a machinery limit, or a rare level such as 95% confidence of not exceeding in 1000 encounters. The combinations of seaway, ship speeds, and relative wave heading are usually representative of the ship's operational profile. For high motion levels, more weight is put on higher sea states and speed-heading combinations at which a ship would operate at in those conditions. A reason to choose a wider range of speeds and headings is to validate a more general case simulation and find its applicability limitations.

The complete SIU is a general purpose ship motion extrapolation method to estimate mean crossing rates at high levels of ship motion. The ship motions of interest are roll, pitch, vertical and lateral acceleration at the bridge. A steep, representative Sea State 7 will provide high levels of motions (9.5 m significant wave height, 15 s modal period, using Bretschneider spectrum). The actual motion levels will be determined from the data sets. The headings ranged from near following (15°) to bow seas (135°). The headings were 15°, 30°, 45°, 60°, 90°, and 135°. The speed was 12 knots for all cases. This evaluation domain provides enough information to say the extrapolation method is promising but should not be considered an exhaustive validation as only one speed and seaway were examined.

The acceptance criteria reflect the structure of the benchmark data. Validation requires a comparison between the extrapolated and true value. The three tier acceptance criteria that codify the mean crossing rate comparison for the 4 motions and 6 seaways, speed, heading combinations (1 seaway × 1 speed × 6 headings) are as follows:

Tier 1: 95% confidence interval overlaps with direct counting value of motion estimate.
Tier 2: A condition defined as a unique motion-seaway-speed-heading combination passes if the percentage of Tier 1 passes is more than B2 = 0.90.
Tier 3: All conditions pass to accept extrapolation method (B3 = 1). Due to the limited number of conditions, all conditions must pass to have confidence in the method to be generally useful.

Hundreds of thousands of hours of ship motions were simulated using the 3DOF simplified simulation tool in Sea State 7 for the range of heading (Weems and Wundrow 2013; Weems and Belenky 2015). The total exposure time was accumulated by ensembling many half-hour simulations. Each simulation had a unique set of random phases to generate a unique and independent wave realization. Time histories of roll, pitch, vertical acceleration, and lateral acceleration were used to estimate an exceedance rate at several motion levels. These estimates are the true values for each motion-seaway-speed-heading combination The length of exposure time varies between headings as a matter of convenience. The difference in exposure time does not affect the validation results beyond potentially limiting the maximum level of comparison. Around 100 GPD extrapolations were made for each heading-motion combination; needed for Tier 2. The use of multiple statistically independent extrapolations allows for a direct check on confidence interval formulation. The data sub-set exposure time was either 50 or 100 h depending on the number of peaks extracted, with 50 h being used for cases with more peaks. This was due to a memory limitation on the analysis software.

As an added point of complexity, the confidence interval can be calculated in more than one way. Two methods were examined as part of this example: boundary method and logarithm method. The boundary method uses the boundaries of confidence interval of GPD shape and scale parameter estimates to compute respective boundaries of the extrapolated value. The "logarithm" method uses an assumption of lognormal distribution of the GPD shape and scale parameters to determine the confidence interval.

With a defined acceptance criteria and data sets to work with, the validation begins with Tier 1 comparisons of the extrapolations to the true value. The level of interest was selected as the highest level in the true data set that had more than 30 data samples. Thirty samples are enough to have meaningful uncertainty. With fewer than 30 samples, the uncertainty becomes very large and the true value is not stabilized.

The comparison was based on overlap of the confidence intervals of the extrapolated and true values. Confidence probability was 0.95. Figure 34.1 shows an example of the roll angle parameter comparisons for 30° heading at the level of 30° angle.

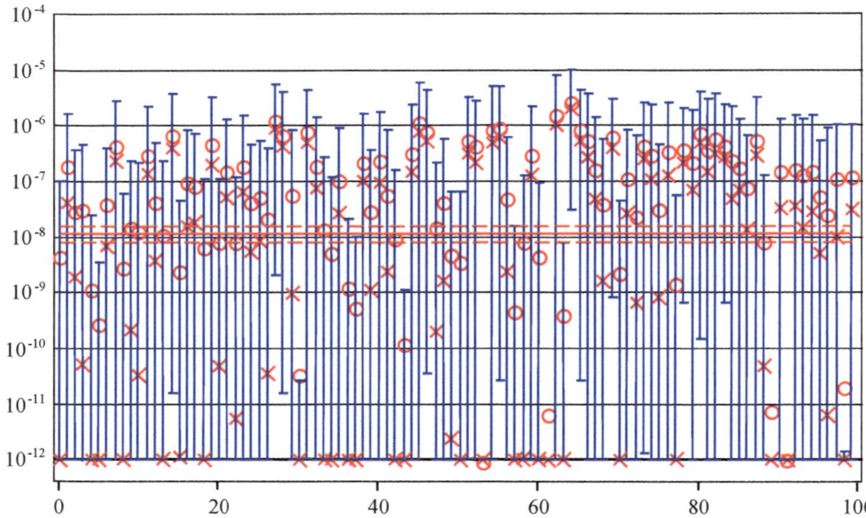

Fig. 34.1 Comparison between the set of extrapolated and true-value exceedance rates for roll angle by confidence interval overlap at the level of interest 30°, heading at 30°, using logarithm CI (91%) (true value 1.111E−08 1/s)

In this figure, the true value is represented by a solid horizontal line (1.11×10^{-8}). The confidence interval of each extrapolation is represented by a vertical line and represents a single Tier I comparison. The extrapolation captures the true value if the vertical line crosses the horizontal true value line. The estimate of mean value of crossing rate is denoted by a circle and the most probable crossing rate is denoted by a cross. The most probable crossing rate is the crossing rate associated with the peak of the probability density function. When both the mean value estimates and most probable values are greater than the true value a conservative bias is indicated. The confidence intervals are asymmetric relative to the mean or most probable crossing rate. The GPD can have zero probability, which results in asymmetric confidence intervals that have very small lower confidence limits. This can result in automatically capturing the true value if the extrapolation is at all conservative, that is, larger than the true value.

The entirety of Fig. 34.1 represents a Tier 2 comparison comprised of 100 Tier 1 comparisons (vertical lines). The expected passing rate percentage is the same as the confidence interval due to use of confidence interval overlap for the parameter comparison. Due to the finite number of data sets, the passing rate can vary from 90 to 100% and still be acceptable; though the mean rate across all the conditions should be close to the confidence level. The upper limit is actually 99% from Eq. (34.6); however, 100% was used because the upper limit was so close to the maximum possible value that the results became overly sensitive to which data sets were used due to using only 100 data samples (extrapolations).

Table 34.1 shows the passing rates for all the conditions for the two different confidence interval (CI) formulations. At 15° heading, there were no lateral acceleration data over the minimum level of interest (0.2 g), so no data are presented on that row. Conditions that pass are bold; failing conditions are italicized. Smith (2014) indicates both CI approaches were acceptable based on roll and pitch. The addition of lateral and vertical acceleration shows a difference between the two CI approaches. The logarithm CI has many instances where the true value capture rate is less than 90% and fails the Tier 1 comparison (parameter). Glotzer et al. (2017) found the boundary CI method overestimated the width of the confidence interval width for roll and pitch; acceleration was not examined.

In terms of acceptance criteria, the boundary CI approach had all the parameter comparisons pass. Therefore, all the Tier 2 conditions pass and Tier 3, overall acceptance, automatically passes if all condition comparisons, Tier 2, are acceptable. In this case, even an alternate Tier 2 definition requiring all the motions to pass for a condition to pass results in overall acceptance.

It is perhaps more instructive to look at the logarithm CI results. Here some of the acceleration parameter comparisons are not acceptable; true value capture rates less than 90%. As a result, some conditions do not pass (18/23 pass) and overall acceptance fails as well. However, this does show there is a heading range that is acceptable; aft of beam seas. There could be a limited acceptance with the restricted range of headings. This may also highlight a difference in performance due to behavior of the distribution tail; whether it is heavy or light.

As an alternative, the mean conservative distance is a metric which uses the upper confidence limit on an extrapolated sample value. This metric estimates how much conservatism (or over-prediction) is present in the simulation results. For validation of a simulation tool against model data for ship guidance, this quantity may be more important than overall total confidence interval overlap. The difficulty is agreeing to what is an acceptable value. In this example, only one case was over three orders of magnitude and almost all were over two orders of magnitude. At first glance, this appears to be completely unacceptable as a 100% difference is usually considered unacceptable. However, for exceedance rates of extreme values the uncertainty is inherently high and 1 in a billion is essentially the same as 10 in a billion. The acceptable MCD can be determined by the level at which the over conservatism produces an undesired operational restriction or life-time risk level.

The MCD is calculated from the upper confidence limit; the 95% upper confidence limit suggests that that 97.5% of the time, the true value is smaller than the upper confidence limit value (two sided confidence interval). Re-analysis of the existing data would show how successfully both methodologies estimate this quantity. Because this investigation would be focused on only the upper interval limit, the overall methodology validation conclusions may differ from those related to formulation of the entire confidence interval.

This example demonstrates the many factors influencing the comparison: CI method, comparison level, and comparison metric

Table 34.1 Confidence interval overlap results

Heading	Motion	Avg. GPD threshold	Comparison level	# Points at threshold	Exposure hours	Boundary CI		Logarithm CI	
						Pass (%)	MCD	Pass (%)	MCD
15	Roll	6.947	15	53	230,000	**96**	119.210	*84*	60.867
15	Pitch	7.359	12	69	230,000	**99**	278.512	**94**	166.585
15	Lat accel	No data > 0.2 g							
15	Vert accel	0.125	0.2	468	230,000	**100**	89.878	*86*	49.892
30	Roll	12.877	30	40	100,000	**96**	178.616	**91**	101.038
30	Pitch	7.296	11.5	46	100,000	**99**	420.751	**97**	244.296
30	Lat accel	0.091	0.2	16	100,000	**100**	1061.000	**98**	591.799
30	Vert accel	0.133	0.25	13	100,000	**100**	955.188	**96**	484.258
45	Roll	17.094	60	30	230,000	**99**	193.110	**94**	112.282
45	Pitch	7.012	11.5	28	230,000	**100**	503.342	**98**	269.962
45	Lat accel	0.135	0.3	37	230,000	**98**	237.340	**94**	118.744
45	Vert accel	0.158	0.25	518	230,000	**99**	74.643	**90**	41.558
60	Roll	18.754	50	49	100,000	**100**	276.284	**100**	169.168
60	Pitch	6.257	9.5	71	100,000	**100**	330.874	**91**	179.786
60	Lat accel	0.157	0.35	19	100,000	**98**	415.630	**97**	213.923
60	Vert accel	0.194	0.3	169	100,000	**100**	193.528	*86*	98.556
90	Roll	16.055	32.5	41	230,000	**99**	329.773	**99**	192.832
90	Pitch	1.517	2.5	170	230,000	**100**	136.716	**94**	68.382
90	Lat accel	0.154	0.25	43	230,000	**94**	136.448	*86*	74.551
90	Vert accel	0.270	0.4	287	230,000	**100**	98.103	**96**	52.179
135	Roll	12.350	17.5	186	230,000	**100**	92.851	**92**	60.694
135	Pitch	4.909	7	172	230,000	**100**	134.019	**94**	70.853
135	Lat accel	0.137	0.25	25	230,000	**96**	424.676	*88*	232.106
135	Vert accel	0.283	0.4	192	230,000	**98**	150.486	**96**	85.572
Average value						**99**		**93**	

34.3 Conclusion

Verification, validation, and accreditation (VVA) or verification, validation, and uncertainty quantification (VVUQ) of numerical simulation is an important topic addressed by many organizations, government agencies and class societies. The VVA/VVUQ process is a formalized comparison between simulation results and benchmark data. The process involves Specific Intended Uses (SIU) to define the scope of the effort. The acceptance criteria are the codification of the SIUs.

At this time there does not appear a universal answer to acceptance criteria approach. The concept of a three tiered acceptance criteria—parameter, condition, and set—were introduced to ensure a definitive statement could be made of the simulation's applicability. A number of different accuracy metrics were proposed as parameter acceptance criteria. The basic traits of acceptance criteria are: alignment with SIUs, definitive statement of acceptance, matching benchmark data structure (multi-tiered), completeness, and a capability to evaluate for performance.

One of the main challenges in validation of numerical simulation of ship motions is a diversity of physical phenomena that may be associated with different speeds, headings, loading and environmental condition (e.g. parametric roll appears at certain combination of these conditions only). To handle this diversity of phenomena, a three-tier structure of validation and acceptance criteria is proposed here. The Tier 1 "parameter" covers a specific value (e.g. exceedance rate of roll angle at specified level); Tier 2 "condition" covers all the levels for this condition, while Tier 3 "set" formulates requirements for performance at all the practical conditions.

When the SIU includes assessment of rare phenomena, such as large motion or capsizing, statistical extrapolation becomes a part of post-processing. The reason is that the sample of sufficient size is too expensive to produce with the regular simulation tool due to the rarity of the phenomena in realistic weather conditions. The extrapolation method needs to be validated as well. Validation of extrapolation method requires a very fast simulation tool, capable of qualitatively reproducing the phenomenon with the computation speed that makes this rare phenomenon observable. This large sample is used to estimate exceedance rate that is used as a true value. The small subset of this sample is used to run extrapolation and validate extrapolation method.

Comparison between the extrapolated and true value is done by recording the overlap of the corresponding confidence intervals. Average passing rate over a large number of comparison is expected to be close to confidence probability and can be used to validate the confidence interval computation method. Alternatively a distance between upper boundaries of extrapolated and true values can be used as an acceptance criterion. The example showed the interaction of acceptance criteria at the different tiers. Specifying the simultaneous passing of parameters for a condition forces the individual parameter pass rates to be high in order to pass the condition. Most of the focus of acceptance criteria is at the parameter level as that is more easily understood. However, to validate or accredit a simulation requires a condition and set acceptance criteria.

The methodology described here does not address required levels of accuracy, merely how to make the comparison. Determining the required levels of accuracy is left to certification bodies.

References

Ammeen, E., 1994, "Evaluation of Correlation Measures," CRDKNSWC-HD-0406, March.

American Society of Mechanical Engineers, 2009, "Standard for Verification and Validation in Computational Fluid Dynamics and Heat Transfer," American Society of Mechanical Engineers, New York.

American Institute for Aeronautics and Astronautics, 1998, Guide for the Verification and Validation of Computational Fluid Dynamics Simulations, AIAA-G-077-1998, Reston, VA.

Anastopoulos, P.A., Spyrou, K.J., Bassler, C.C. and V. Belenky, 2015, "Towards an Improved Critical Wave Groups Method for the Probabilistic Assessment of Large Ship Motions in Irregular Sea" Probabilistic Engineering Mechanics Vol. 44, pp 18–27.

Bales, N.K., 1980, "Optimizing the Seakeeping Performance of Destroyer-Type Hulls," Proc. 13th Symp. Naval Hydrodynamics, Tokyo, Japan.

Belenky, V. and K.M. Weems, 2012, "Probabilistic Properties of Parametric Roll", Chapter 6 of Parametric Resonance in Dynamical Systems, Fossen, T. I. and H. Nijmeijer, eds., Springer, ISBN 978-1-4614-10423-0, pp. 129–146.

Belenky, V., Pipiras, V., Kent, C., Hughes, M., Campbell, B. and Smith, T., 2013, "On the Statistical Uncertainties of Time-domain-based Assessment of Stability Failures: Confidence Interval for the Mean and Variance of a Time Series," Proc. 13th Intl. Ship Stability Workshop, Brest, France, pp. 251–258.

Belenky, V., Pipiras, V. and Weems, K., 2015, "Statistical Uncertainty of Ship Motion Data," Proc. 12th Intl. Conf. on Stability of Ships and Ocean Vehicles, Glasgow, UK

Belenky, V., Weems, K. and Lin, W.M. 2016, "Split-time Method for Estimation of Probability of Capsizing Caused by Pure Loss of Stability" Ocean Engineering, Vol. 122, pp. 333–343.

Belknap, W.F., Smith, T.C., Campbell, B.L., 2011, "Addressing Challenges in the Validation of Dynamic Stability Simulation Tools," Proc. 12th Intl. Ship Stability Workshop, Washington, DC., USA pp 81–90.

Belknap, W, Belenky, V., Campbell, B. and Smith, T., 2012, "A Guide for Validation of Simulation-Based Approach for Predicting Ship Stability Failure Probabilities" Proc. 29th Symp. Naval Hydrodynamics, Gothenburg, Sweden

Bendat, J. and Piersol, 1966, Measurement and Analysis of Random Data, John Wiley & Sons.

Campbell, B., Belenky, V. and Pipiras, V., 2016, "Application of the Envelope Peaks over Threshold (EPOT) Method for Probabilistic Assessment of Dynamic Stability," Ocean Engineering, Vol. 120, pp. 298–304.

Coleman, H. W. and Steele, W. G., 1989, Experimentation and Uncertainty Analysis for Engineers, John Wiley & Sons.

Coles, S., 2001, An Introduction to Statistical Modeling of Extreme Values. Springer-Verlag, London.

de Kat, J. O. and Paulling, J. R., 1989, "The Simulation of Ship Motions and Capsizing in Severe Seas," Trans. SNAME vol. 97.

DoD, 2009, DoD Modeling and Simulation, (M&S) Verification, Validation, and Accreditation, (VVA), Instruction 5000.61, 9 Dec 2009.

DoN, 2004, Department of the Navy Modeling and Simulation, (M&S) VVA Implementation Handbook, 30 March 2004.

IEEE, 2012, Standard for System and Software Verification and Validation, 1012–2012, Rev o.

International Towing Tank Conference, 2011, "Loads and Responses Seakeeping Verification and Validation of Linear and Weakly Nonlinear Seakeeping Computer Codes," 26th ITTC Seakeeping Committee Procedure 7.5 – 02 07-02.5 rev 1.

ITTC Specialist Committee on Stability in Waves, 2011, Final Report and Recommendations to the 26th ITTC, Proc. 26th International Towing Tank Conference, Vol II. Rio de Janeiro, Brazil.

Glotzer, D., Pipiras, V., Belenky, V., Campbell, B., Smith, T., 2017, "Confidence Interval for Exceedance Probabilities with Application to Extreme Ship Motions," REVSTAT Statistical J. Vol. 15, No 4, pp. 537–563.

Griffith, A. A. and Locke, W. M., 2006, An Analysis of the U.S. Navy Verification, Validation, and Accreditation, (VVA) Process for Modeling and Simulation, M(&S) Used for Operation Test (OT) of Surface Ships and Weapons, Thesis Naval Post Graduate School, June 2006.

Lee, E., Fullerton, A., Geiser, J., Schleicher, C., Merril, C., Weil, C., Jiang, M., Stern, F., and Mousaviraad, M., 2016, "Experimental and Computation Comparisons of the R/V Athena in Calm Water," Proc. 31st Symp. Naval Hydrodynamics, Monterey, California, USA.

Lin, W.M., Yue, D.K.P., 1990, "Numerical Solutions for Large Amplitude Ship Motions in the Time-Domain." Proc. 18th Symp. Naval Hydrodynamics, Ann Arbor, Michigan, USA. pp. 41–66.

Oberkampf, W., Blottner, F., 1997, Issues in Computational Fluid Dynamics Code Verification and Validation. Sandia National Laboratories Report SAND95-1352.

Priestley, M. B., 1981, Spectral Analysis and Time Series, Vol. 1, Academic Press, New York.

Reed, A. M., 2011, "26th ITTC Parametric Roll Benchmark Study," Proc. 12th Intl Ship Stability Workshop, Washington DC., USA

Reed, A. M. and Zuzick, A., 2015, "Direct Assessment Will Require Accreditation - What This Means," Proc. 12th Intl. Conf. on the Stability of Ships and Ocean Vehicles, Glasgow, UK, pp. 51–78

Sec. of Navy, 1999, "Verification, Validation, and Accreditation (VVA) of Models and Simulations," SECNAVINST 5200.40, 19 April 1999.

Shin, Y.S., Belenky, V., Lin, W.M. Weems, K.M. and Engle, A.H., 2003, "Nonlinear Time Domain Simulation Technology for Seakeeping and Wave-Load Analysis for Modern Ship Design," Trans. SNAME, Vol. 111.

Smith, T.C., 2011, "Statistical Data Set Comparison for Continuous, Dependent Data" Proc. 12th Intl. Ship Stability Workshop, Washington DC, USA, pp. 75–80.

Smith, T.C., 2012, "Approaches to Ship Motion Simulation Acceptance Criteria" Proc. of 11th Intl Conference of the Stability of Ships and Ocean Vehicles, Athens, Greece. pp. 101–114.

Smith, T. and Campbell, B. 2013, "On the Validation of Statistical Extrapolation for Stability Failure Rate," Proc. 13th Intl. Ship Stability Workshop, Brest, France.

Smith, T., Campbell, B., Zuzick, A., Belknap, W., and Reed, A., 2014, "Approaches to Validation of Ship Motion Prediction Tools and Extrapolation Procedures for Large Excursion of Ship Motions in Irregular Seas" Proc. 7th Intl. Conf. of Computational Stochastic Mechanics, Santorini, Greece. ISBN: 978-981-09-5347-8

Smith, T. C., 2014, "Example of Validation of Statistical Extrapolation Example of Validation of Statistical Extrapolation," Proc. 14th Intl. Ship Stability Workshop, Kuala Lumpur, Malaysia.

Smith, T. and Zuzick, A., 2015, "Validation of Statistical Extrapolation Methods for Large Motion Prediction," Proc. 12th Intl. Conf. on Stability of Ships and Ocean Vehicles, Glasgow, UK.

Stern, F., Wilson, R., Coleman, H., Paterson, E., 1999, Verification and Validation of CFD Simulations, Iowa Institute of Hydraulic Research Report No. 407.

Stern. F., Wilson, R., and Shao, J., 2005, "Quantitative V&V of CFD Simulations and Certification of CFD Codes," Intl. J. Numerical Methods in Fluid. Vol 50: pp 1335–1355.

Weems, K. and Wundrow, D., 2013, "Hybrid Models for Fast Time-Domain Simulation of Stability Failures in Irregular Waves with Volume-Based Calculations for Froude Krylov and Hydrostatic Force", Proc. 13th Intl. Ship Stability Workshop, Brest, France.

Weems, K. and Belenky, V., 2015, "Fast Time-Domain Simulation in Irregular Waves With Volume-Based Calculations for Froude Krylov and Hydrostatic Force" Proc. 12th Intl Conference on Stability of Ships and Ocean Vehicles, Glasgow, UK.

Chapter 35
Total Stability Failure Probability of a Ship in Beam Wind and Waves: Model Experiment and Numerical Simulation

Takumi Kubo, Naoya Umeda, Satoshi Izawa and Akihiko Matsuda

Abstract To establish second-generation intact stability criteria, the International Maritime Organization requires experimentally validated numerical simulation models for stability under the dead ship condition. Here, a beam wind and wave condition is selected as the worst-case scenario and the total-stability-failure probability is quantified. The authors developed a coupled sway–heave–roll–pitch numerical model and compared it with physical experiments of a ship model in artificial irregular beam waves and fluctuating beam wind. The results indicate that the probability of total stability failure estimated by the simulation exists within the confidence interval range of those estimated by the experiment.

35.1 Introduction

At the International Maritime Organization (IMO) (2012), second-generation intact stability criteria are now under development. These criteria aim to prevent total stability failure (capsizing) and partial stability failure (e.g. cargo shift) due to pure loss of stability, parametric rolling, broaching, excessive acceleration and harmonic resonance under the dead ship condition with physics-based approaches. They will comprise two-layered vulnerability criteria and a direct assessment for each failure mode, and ships will be requested to comply with at least one of these criteria because the use of expensive numerical simulations as direct assessments needs to

---

T. Kubo · N. Umeda (✉) · S. Izawa
Osaka University, Osaka, Japan
e-mail: umeda@naoe.eng.osaka-u.ac.jp

T. Kubo
e-mail: psktk@yahoo.co.jp

S. Izawa
e-mail: tg5012is@docomo.ne.jp

A. Matsuda
National Research Institute of Fisheries Engineering, Hasaki, Japan
e-mail: amatsuda@fra.affrc.go.jp

V. L. Belenky et al. (eds.), *Contemporary Ideas on Ship Stability*, Fluid Mechanics and Its Applications 119, https://doi.org/10.1007/978-3-030-00516-0_35

be minimised to realize a feasible application of the new scheme. It is also essential that numerical simulations used for the direct assessment shall be validated using physical model experiments.

Other than harmonic resonance under the dead ship condition, it was agreed to allow a failing ship to comply with all three levels to function with operational guidance based on the method used for higher-level criterion. This is because these failure modes can be avoided with appropriate operational means. The dead ship condition is their exception because it assumes that all propulsive power is lost so that no operational means, such as propeller revolution and rudder deflection, are available to avoid danger. In other words, it is crucial to guarantee stability safety under the dead ship condition because a ship shall have at least one safe option to escape from all other failure modes by operational means.

If a ship is nearly longitudinally symmetrical and the wind direction is the same as the wave direction, a ship without any propulsive power suffers beam wind and waves. This can be regarded as the worst scenario in case of dead ship condition. It was theoretically confirmed by one of the authors (Umeda et al. 2007). If a ship is allowed to yaw, it could escape the worst-case scenario because of the wind-induced yaw moment and/or the second-order wave-induced yaw moment (Umeda et al. 2007). For the vulnerability criterion, at the IMO, it was agreed that the current weather criterion but with the extended wave steepness table would be used for the level 1 criterion and that an analytical calculation of the stability failure probability under stochastic wind and wave excitation with uncoupled roll model would be used for the level 2 criterion (IMO 2012). In the case of the IMO weather criterion, random wind is modelled with a gusty wind with an assumed occurrence probability (IMO 2008). Therefore, it is necessary for the level 2 criterion to directly use fluctuating wind to overcome the drawback of the level 1 criterion. Multiple numerical simulation techniques have been reported for a direct stability assessment (e.g. Vassalos et al. 2004); however, no numerical models have been authorised by the IMO.

One reason for this is that model experiments of capsizing ship in artificial random beam wind and waves have not been available previously for this purpose, to the best knowledge of the authors. For example, Shaughnessy et al. (1994) and Ogawa et al. (2006) executed model experiments of ships in beam wind and irregular waves under moored and drifting conditions, respectively. They, however, used non-fluctuating wind for their experiments. In the field of offshore structures, model experiments with random wind were reported but without waves (Kajita and Tanaka 1986). Indeed, the generation of a fluctuating wind is a challenge in tank testing.

In response, some of the authors (Umeda et al. 2011) executed physical model experiments of ship capsizing in irregular beam waves but with a non-fluctuating wind. Therefore, the authors upgraded the wind blower to generate fluctuating wind and conducted experiments of ship total stability failure in an environment of both random waves and wind.

Furthermore, a coupled sway–heave–roll–pitch numerical model was newly developed to directly assess the stability under the dead ship condition. The numerical simulation model was then compared with the above-mentioned experimental data.

35.2 Numerical Simulation Models

35.2.1 Uncoupled Roll Model

The roll motion was modelled with the following nonlinear and uncoupled equation of absolute rolling angle of a ship under stochastic wave excitation and steady wind moment (Kubo et al. 2010). Usually ship motions in beam seas are modelled with equations of coupled motions in sway and roll with wave radiation forces and diffraction forces. Watanabe (1938), however, proposed a one-degree of freedom equation of the roll angle, as follows:

$$(I + J)\ddot{\phi} + D(\dot{\phi}) + WGZ(\phi) = M_{wind}(t) + M_{wave}(t), \tag{35.1}$$

where ϕ is the absolute roll angle, I the roll moment of inertia of a ship, J is the added roll moment of inertia, D is the roll damping moment, W is the ship weight, GZ is the righting arm and a dot indicates differentiation with respect to time. $M_{wind}(t)$ is the wind induced moment consisting of the steady and fluctuating wind moment and $M_{wave}(t)$ is the wave exciting moment based on the Froude-Krylov assumption. This is because the roll diffraction moment and roll radiation moment due to sway can cancel each other out when the wavelength is sufficiently longer than the ship breadth. More exactly speaking, this approximation is accurate enough when the sway velocity is almost equal to the horizontal component of wave particle velocity (Tasai 1965).

The uncoupled equation of the absolute roll motion can be rewritten by dividing by the virtual moment of inertia as follows:

$$\ddot{\phi} + d(\dot{\phi}) + \omega_0^2 \cdot kf(\phi) = \omega_0^2 \cdot (m_{wind}(t) + m_{wave}(t)) \tag{35.2}$$

where

$$d(\dot{\phi}) = \frac{D(\dot{\phi})}{I + J} \tag{35.3}$$

$$\omega_0 = \sqrt{\frac{W \cdot GM}{I + J}} \tag{35.4}$$

$$\mathrm{kf}(\phi) = \frac{\mathrm{GZ}(\phi)}{\mathrm{GM}} \tag{35.5}$$

$$m_{wind}(t) = \frac{M_{wind}(t)}{W \cdot GM} \tag{35.6}$$

$$m_{wave}(t) = \frac{M_{wave}(t)}{W \cdot GM} \tag{35.7}$$

and GM is the metacentric height.

The wind induced roll moment is modelled as follows:

$$M_{wind}(t) = 0.5\,\rho_{air} C_m(\varphi)\{U_{w0} + U_w(t)\}^2 A_L(\phi) H_c(\phi) \tag{35.8}$$

where ρ_{air} is the air density, U_{w0} is the mean wind velocity, U_w is the fluctuating wind velocity component, A_L is the lateral windage area, H_C is the height of the centre of the lateral windage area and C_m is the aerodynamic drag coefficient. The fluctuating wind velocity is determined from the following equation:

$$U(t) = \sum_{i=1}^{N_w} b_i \sin(\sigma_i t + \psi_i), \tag{35.9}$$

where

$$b_i = \sqrt{2 S_{wind}(\sigma_i) d\sigma}, \tag{35.10}$$

S_{wind} is the wind velocity spectrum, σ_i is the wind velocity circular frequency, ψ_i is a random phase and N_w is the number of components of the wind velocities.

The wind velocity in metres per second is modelled by the Davenport spectrum as follows (Davenport 1961)

$$S_{wind}(\sigma_i) = 4K \frac{U_{w0}^2}{\sigma_i} \frac{X_D^2}{(1 + X_D^2)^{4/3}}, \tag{35.11}$$

where

$$K = 0.003 \tag{35.12}$$

$$X_D = 600 \frac{\sigma_i}{\pi U_{w0}}. \tag{35.13}$$

The choice of the wind velocity spectrum could require further discussion but here we used most typical one following the discussion at the IMO (Bulian and Francescutto 2004).

The wave exciting moment is modelled as follows:

$$M_{wave}(t) = W GM \gamma \Theta(t), \tag{35.14}$$

where $\gamma\Theta(t)$ is the effective wave slope, which can be calculated using strip theory. Similar to the fluctuating wind velocity, the wave slope is calculated as follows:

$$\Theta(t) = \sum_{i=1}^{N_{wave}} \frac{\omega_i^2}{g} a_i \sin(\omega_i t + \varphi_i), \tag{35.15}$$

where

$$a_i = \sqrt{2 S_{wave}(\omega_i) d\omega}, \tag{35.16}$$

S_{wave} is the wind velocity spectrum, ω_i is the wave circular frequency, g is the gravitational acceleration, ϕ_i is a random phase and N_{wave} is the number of wave components.

The sea elevation in metres is modelled using the spectrum recommended by the 15th International Towing Tank Conference (ITTC):

$$S_{wave}(\omega_i) = \frac{A}{\omega_i^5}\exp\left(\frac{-B}{\omega_i^4}\right), \tag{35.17}$$

where

$$A = 172.75\frac{H_{1/3}^2}{T_{01}^4}, \tag{35.18}$$

$$B = \frac{691}{T_{01}^4}, \tag{35.19}$$

$H_{1/3}$ is the significant wave height (m) and T_{01} is the mean wave period (s).

35.2.2 Coupled Sway–Heave–Roll–Pitch Model

A more rigorous approach is to model the six degrees-of-freedom motion as a rigid body in a three-dimensional space. If surge and yaw motion is constrained to realize the beam sea condition in model experiments, sway-heave-roll and pitch motions should be modelled. The body-fixed coordinate system, G-x, y, z, where the origin is situated at the centre of the ship's gravity and the x-axis points towards the bow, the y-axis to starboard and the z-axis downwards, and the inertia coordinate system, O-ξ, η, ζ which moves with a constant lateral speed in the direction of the wind and waves, are used. The ξ-axis is in the direction of wave travel and the ζ-axis points downwards.

The equations of coupled motion are given by

$$\begin{aligned} m\ddot{x}_2 &= F_2(x_2, x_3, x_4, x_5, \dot{x}_2, \dot{x}_3, \dot{x}_4, \dot{x}_5, \ddot{x}_2, \ddot{x}_3, \ddot{x}_4, \ddot{x}_5, t) \\ m\ddot{x}_3 &= F_3(x_2, x_3, x_4, x_5, \dot{x}_2, \dot{x}_3, \dot{x}_4, \dot{x}_5, \ddot{x}_2, \ddot{x}_3, \ddot{x}_4, \ddot{x}_5, t) \\ I_{44}\ddot{x}_4 &= F_4(x_2, x_3, x_4, x_5, \dot{x}_2, \dot{x}_3, \dot{x}_4, \dot{x}_5, \ddot{x}_2, \ddot{x}_3, \ddot{x}_4, \ddot{x}_5, t) \\ I_{55}\ddot{x}_5 &= F_5(x_2, x_3, x_4, x_5, \dot{x}_2, \dot{x}_3, \dot{x}_4, \dot{x}_5, \ddot{x}_2, \ddot{x}_3, \ddot{x}_4, \ddot{x}_5, t) \end{aligned} \tag{35.20}$$

where the suffixes 2, 3, 4 and 5 denote sway, heave, roll and pitch motions measured from O-ξ, η, ζ. The ship mass is indicated by m and the moment of inertia due to the i-th ($i = 1, \ldots, 6$) direction is I_{ii}. The external forces or moments in the i-th direction, F_i, are assumed to consist of the buoyancy (B), Froude-Krylov (FK), radiation (R), diffraction (D), gravity (G), wind (WD) and hydrodynamic reaction (HR) components as follows:

$$F_i = F_i^B + F_i^{FK} + F_i^R + F_i^D + F_i^G + F_i^{WD} + F_i^{HR} \tag{35.21}$$

where

$$F_i^B = \rho g \int_{AE}^{FE} dx \int_{S_H(x_2,x_3,x_4,x_5)} (-zn_i)ds \tag{35.22}$$

$$F_i^{FK} = -\rho g \int_{AE}^{FE} dx \int_{S_H(x_2,x_3,x_4,x_5)} \sum_{k=1}^{N_w} a_k \exp\left(-\frac{\omega_k^2}{g}\zeta\right) \cos\left(-\frac{\omega_k^2}{g}\xi - \omega_k t + \varphi_k\right) n_i ds \tag{35.23}$$

$$F_i^R = \sum_{j=2}^{5} (-A_{ij}(x_4)\ddot{x}_j - B_{ij}(x_4)\dot{x}_j) \tag{35.24}$$

$$\begin{aligned} F_i^D &= \sum_{k=1}^{N_w} a_k f_{ki}^D(x_4) \cos(-\omega_k t - \varepsilon_{ki}^D(x_4) + \varphi_k) \\ F_3^G &= mg \\ F_2^G &= F_4^G = F_5^G = 0 \end{aligned} \tag{35.25}$$

$$\begin{aligned} F_2^{WD} &= \frac{1}{2}\rho\{U_{w0} + U(t)\}^2 A_L(\phi) C_m(\phi) \\ F_4^{WD} &= \frac{1}{2}\rho\{U_{w0} + U(t)\}^2 A_L(\phi) H_c(\phi) C_m(\phi) \\ F_3^{WD} &= F_5^{WD} = 0 \end{aligned} \tag{35.26}$$

AE is the ship aft end, FE is the ship fore end, n_j is the directional cosine of the hull surface, S_H is the wetted surface, A_{ij} is the added mass due to the j-th motion in the direction of the i-th motion, B_{ij} is the wave making damping coefficient due to the j-th motion in the direction of the i-th motion, f_{ki}^D is the amplitude of the wave diffraction force in the i-th direction with the k-th frequency and ε_{kj}^D is the phase difference of the wave diffraction force in the i-th direction with the k-th frequency. The hydrodynamic reaction force was added to maintain the mean position of the sway direction.

In this numerical model, the radiation force coefficients are estimated with strip theory (Lee and Kim 1982) using the natural roll frequency for the sway and roll and the peak wave frequency of the heave and pitch. The roll radiation coefficients were estimated from the roll decay model tests. The diffraction forces are calculated from strip theory (Salvesen et al. 1970). The roll restoring moment is obtained in a conventional manner. The aerodynamic coefficient is estimated using an empirical method (Fujiwara et al. 1998) and its heel effect is calculated from momentum theory.

35.3 Model Experiment

Model experiments of a ship model in beam wind and waves were conducted in a seakeeping and manoeuvring basin of the National Research Institute of Fisheries Engineering (NRIFE). The subject ship used was a hypothetical ship known as CEHIPAR2792. Its principal particulars and restoring arm curves are shown in Table 35.1 and Fig. 35.1, respectively. Its 1/70-scaled model has a flat plate on the upper deck to realise the windage area and the area centre height of the superstructure without additional buoyancy. The ship model was not equipped with bilge keels, propellers, shaft brackets or rudders. An optical fibre gyroscope inside the model was used to detect the roll, pitch and yaw angles. If bilge keels are added, probability of total stability failure for a ship complying with the current weather criterion could be extremely small so that experimental validation is not feasible.

Irregular water waves were generated using plunger-type wave makers with the ITTC spectrum. As shown in Fig. 35.2, the specified spectrum was satisfactorily realised. The fluctuating wind was generated by a wind blower in the wave direction. The wind blower, as shown in Fig. 35.3, consisted of 36 axial flow fans and was controlled by invertors with a v/f control law. The relationship between the frequency for this control and the wind velocity was adjusted by measuring the steady heel

Table 35.1 Principal particulars of the CEHIPAR2792 ship

Displacement	24,585.7	Ton
Length between perpendicular	205.7	m
Breadth	32	m
Draught	6.6	m
Metacentric height at upright condition: GM	2	m
Natural roll period at upright condition	18.36	s

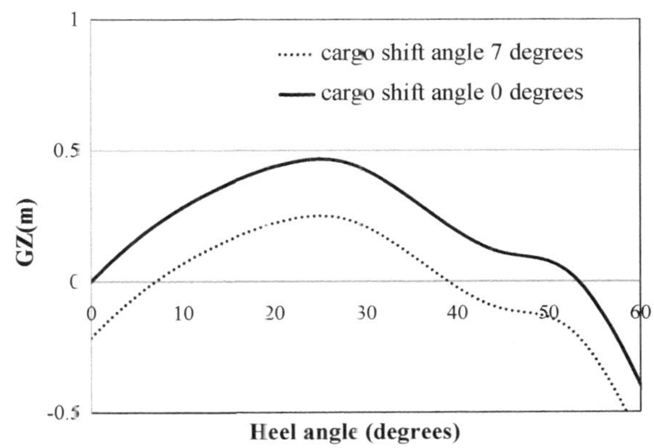

Fig. 35.1 GZ curves for the CEHIPAR2792 ship

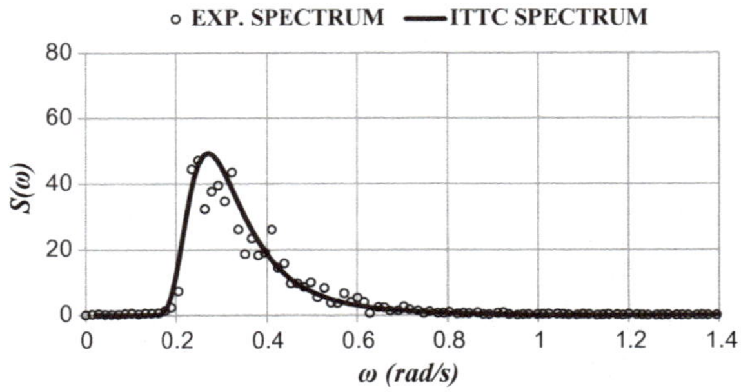

Fig. 35.2 The generated wave spectrum

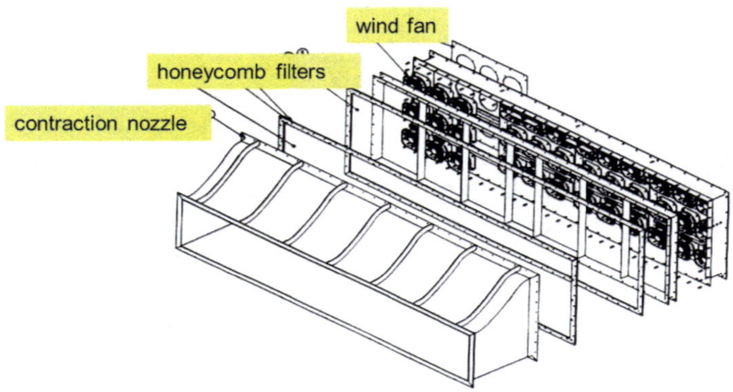

Fig. 35.3 The wind blower

Fig. 35.4 The wind velocity spectrum

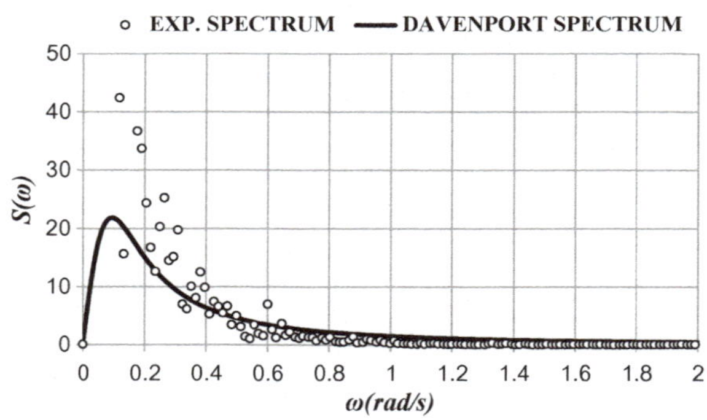

angle of the ship model under non-fluctuating wind with Eq. (35.8). The wind velocity spectrum was designed using the Davenport spectrum. The measured spectrum agrees with the specified one in the region near the ship's natural roll frequency, i.e. 0.34 rad/s, as shown in Fig. 35.4.

Fig. 35.5 Layout of the experimental setup

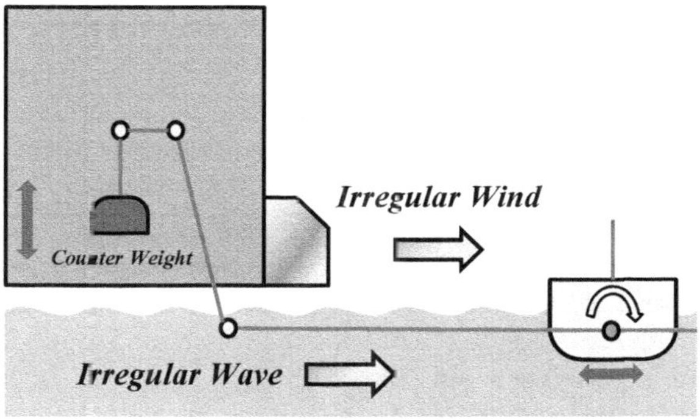

The model was kept orthogonal to the wind and wave direction by a wire system, which softly restrains the drift and yaw. Here the wire system was connected to the ship model at the bow and stern where the height was set equal to the calm water surface based on the measured hydrodynamic reaction force and the moment in a captive model test of the subject ship. The mean of the fluid dynamic force in the sway direction was cancelled by a counter weight, as shown in Fig. 35.5. If we allow the drift, we need a much longer model basin and accurate tracking capability of the wind blower, which were not available even in most major towing tanks.

The experimental condition in full scale was specified as a wind velocity of 24.5 m/s, a significant wave height of 11.04 m, a mean wave period of 16.48 s and a duration of 1 h. Furthermore, to simulate cargo shifts, three different constant heel angles, i.e. 6°, 8° and 10°, were added due to the lateral shifts of the weights on board. The wave elevation was measured using a servo needle-type wave probe, and the wind velocity was measured using a hot wire sensor. The instantaneous model position relative to the basin was measured by a total station system. This system consisted of a theodolite, an optical distance measuring device and a prism, which was on the model.

The experiment was repeated multiple times, and then the total stability failure probability was estimated as the ratio of the number of total stability failure events to the total number of experiments. The confidence interval was estimated assuming a binomial distribution.

For making this experiment feasible, the experimental condition should be severe enough because total stability failure probabilities of realistic ships are sufficiently small. Thus, the mean wind velocity and the significant wave height were set as a worst situation in the North Atlantic.

35.4 Comparison Between the Experiment and the Calculations

The results of a comparison between the experiment and the calculations with confidence intervals of 95% of confidence probability are shown in Fig. 35.6. The confidence interval was estimated assuming a binomial distribution (e.g. Tellkamp and Cramer 2002). Here, the number of experimental runs with the 1 degrees-of-freedom (DoF) model and the 4 DoF model for each heel angle are 1000 and 300, respectively. As a result, their confidence intervals are very small. Conversely, the number of experimental runs for each heel angle is 25 or less; therefore, the confidence intervals for these conditions are not as small. As a whole, when the heel angle due to the cargo shift increases, the total stability failure increases. The 1 DoF model roughly agrees with the experiment; however, in the case of a heel angle of 8°, the confidence interval of the 1 DoF model and that of the experiment do not overlap. By contrast, the confidence interval of the 4 DoF model and that of the experiment are completely overlapping for all three cases. Therefore, it can be concluded that the agreement between the experiment and 4 DoF model is satisfactory and that the 4 DoF model slightly overestimates the results of the experiment. This means that the 4 DoF simulation model used here provides a conservative prediction at least for this subject ship and the tested environmental conditions.

To investigate the reason for the discrepancy between the experiment and the simulation models, the roll spectra were calculated as shown in Fig. 35.7. The results indicate that the roll spectral densities of the simulation models are larger than that of the experiment in the region of the frequency ranging from 0.25 to 0.3 rad/s. The difference between the two simulation models is not as significant. Therefore, the roll damping moment could depend on the frequency even though both models use

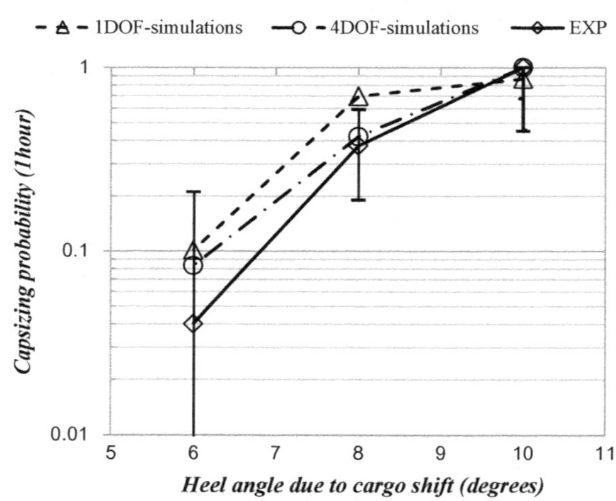

Fig. 35.6 Estimated probability of the total stability failure in one hour given the confidence intervals. Here, wind velocity is 24.5 m/s, the significant wave height is 11.04 m and the mean wave period is 16.48 s

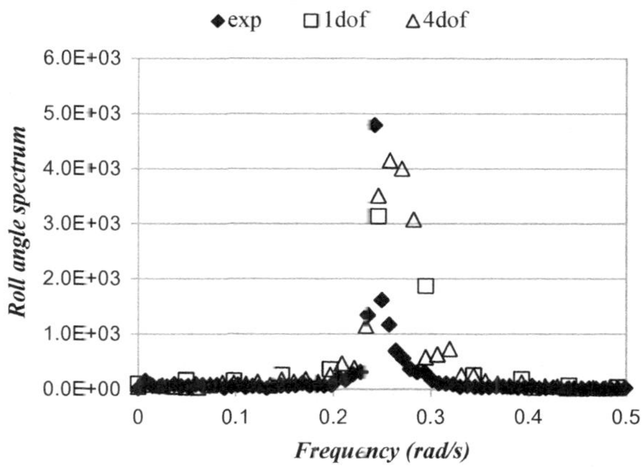

Fig. 35.7 The roll spectra. Here, the cargo shift angle is 8°, the wind velocity is 24.5 m/s, the significant wave height is 11.04 m and the mean wave period is 16.48 s

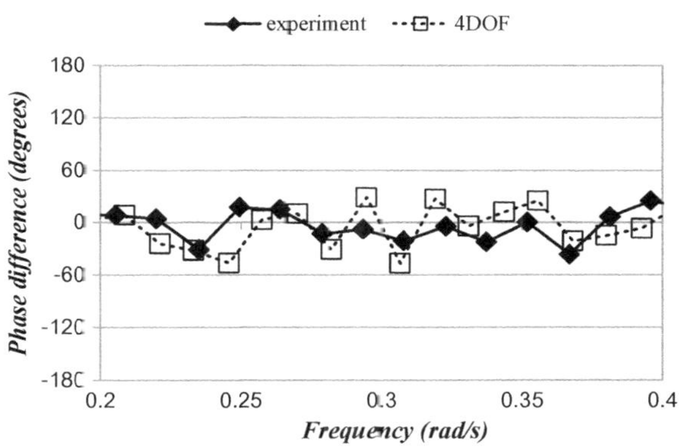

Fig. 35.8 Phase difference between the roll angle and sway velocity. Here, the cargo shift angle is 6°, the wind velocity is 24.5 m/s, the significant wave height is 11.04 m and the mean wave period is 16.48 s

the roll damping at the natural roll frequency. It may be desirable to model the effect of frequency on the roll damping in the future.

In addition, the phase difference between the roll angle and the sway velocity was estimated using the cross spectrum of these variables as shown in Fig. 35.8. The phase difference between the two is generally small; however, it is not exactly zero, as we assumed in the derivation of the 1 DoF model. This could be the reason why the 4 DoF model outcomes improved compared to the 1 DoF ones; however, this should be discussed in the context of the accuracy of spectral analysis in the future.

35.5 Concluding Remarks

A coupled sway–heave–roll–pitch model in irregular beam wind and waves, based on potential flow theory, was presented to quantify the total stability failure probability using a Monte Caro simulation. The physical model experiment was executed for stochastic wind and waves. The comparison between the existing and presented simulation models and the experiment indicates that the probability of total stability failure estimated by the presented simulation model exists within the confidence interval of those estimated by the experiment. The existing uncoupled model does not show similar agreements to the presented model.

Acknowledgements This study was supported by Grant-in Aids for Scientific Research from the Japan Society for the Promotion of Science (Nos. 21360427 and 15H02327) and was partly carried out as a research activity of the Stability Project of the Japan Ship Technology Research Association in the fiscal year of 2011, funded by the Nippon Foundation. The authors sincerely thank these organisations. The authors also thank Prof. Alberto Francescutto and Dr. Gabriele Bulian from the University of Trieste for kindly providing data of the subject ship. The authors are grateful to Dr. Daisuke Terada from NRIFE and Mr. Keisuke Yamane, Mses. Hisako Kubo, Ayumi Morimoto and Fuka Yoshiyama from Osaka University for their assistance during the experiments. The authors would like to thank Enago (www.enago.jp) for the professional English language review.

References

Bulian, G. and Francescutto, A. 2004. "A Simplified Modular Approach for the Prediction of the Roll Motion Due to the Combined Action of Wind and Waves." Journal of Engineering for the Maritime Environment, Vol. 218, pp. 189 – 212.

Davenport, A. G., 1961, "The spectrum of horizontal gustiness near the ground in strong winds", Journal of the Royal Meteorological Society, Vol. 87, pp. 194–211.

Fujiwara, T., Ueno, M. and Nimura, T., 1998, "Estimation of Wind Forces and Moments Acting on Ships", Journal of Society of Naval Architects of Japan, Vol. 183, pp. 77–90, (in Japanese).

IMO, 2008, "Explanatory Notes to the International Code on Intact Stability", MSC.1/Circ. 1281, London.

IMO, 2012, "Development of Second Generation Intact Stability Criteria - Report of the Working Group (part I)", SLF 54/WP.3, London.

Kajita, E., and Tanaka, K., 1986, "Experimental Techniques for Behaviors of Offshore Structures Under Extreme Environmental Conditions", Proceedings of the 3rd Marine Dynamics Symposium, The Society of Naval Architects of Japan, pp. 375–395, (in Japanese).

Kubo, T., Maeda, E., and Umeda, N., 2010, "Theoretical Methodology for Quantifying Probability of Stability Failure for a Ship in Beam Wind and Waves and its Numerical Validation", Proceedings of 4th International Maritime Conference on Design for Safety, Trieste, pp. 1–8.

Lee, C. M., and Kim, K. H., 1982, "Prediction of Ships in Damaged Condition in Waves", Proceedings of the 2nd International Conference on Stability of Ships and Ocean Vehicles, Tokyo, pp. 287–301.

Ogawa, Y., de Kat, J. O., and Ishida, S., 2006, "Analytical Study of the Effect of Drift Motion on the Capsizing Probability under Dead Ship Condition", Proceedings of the 9th International Conference on Stability of Ships and Ocean Vehicles, Rio de Janeiro, Vol. 1, pp. 29–36.

Salvesen, N., Tuck, E. O., and Faltinsen, O., 1970, "Ship Motions and Sea Load", Transaction of the Society of Naval Architects and Marine Engineers, Vol. 78, pp. 250–287.

Shaughnessy, J., Nehrling, B. C., and Compton, R. H., 1994, "Some Observations on Experimental Techniques for Modeling Ship Stability in Wind and Waves", Proceedings of the 5th International Conference on Stability of Ships and Ocean Vehicles, Melbourne, Vol. 3, pp. 21–37.

Tasai, F., 1965, "On the Equation of Rolling of a Ship", Bulletin of Research Institute for Applied Mechanics, Kyushu University, Vol. 26, pp. 51–57 (in Japanese).

Tellkamp, J. and Cramer, H. 2002, "A Methodology for Design Evaluation of Damage Stability", Proceedings of the 6th International Ship Stability Workshop, New York, pp. 1.5.1–1.5. 9.
Umeda, N., Koga, S., Ueda, J., Maeda, E., Tsukamoto, I., and Paroka, D., 2007, "Methodology for Calculating Capsizing Probability for a Ship under Dead Ship Condition", Proceedings of the 9th International Ship Stability Workshop, Hamburg, pp. 1.2.1–1.2.19.
Umeda, N., Izawa, S., Sano, H., Kubo, H., and Yamane, K., 2011, "Validation Attempts on Draft New Generation Intact Stability Criteria", Proceedings of the 12th International Ship Stability Workshop, Washington D.C., pp. 19–26.
Vassalos, D., Jasionowski, A., and Cichowicz, J., 2004, "Issues Related to the Weather Criterion", International Shipbuilding Progress, Vol. 51, No. 2/3, pp. 251–271.
Watanabe, Y., 1938, "Some Contributions of the Theory of Rolling." Transactions of Institution of Naval Architects, Vol. 80, pp. 408–432.

Chapter 36
Deterministic Validation of a Time Domain Panel Code for Parametric Roll

Frans van Walree and Pepijn de Jong

Abstract Validation of simulation methods for dynamic stability is hampered by the fact that dynamic stability phenomena can be quite rare. In order to obtain sufficient statistical confidence in both experimental data and simulation results long duration time histories are required for a range of operational conditions. This is at most times not feasible from a practical point of view. One way of validating time domain simulation methods for dynamic stability phenomena is by deterministic validation. This means that the simulation is run in the same wave sequence as used during the model experiments. Ideally, a one to one comparison between experiments and simulations is then possible. A difficulty in such an approach is, in case of irregular waves, the reconstruction of the experimental wave train in the simulation tool. Even if this were successful, the encountered wave train in the simulations will deviate from the experimental one because it is inevitable that the position in the horizontal plane will differ from the experimental one after some time. The paper describes the deterministic validation of a non-linear, 6-DoF time domain panel code for parametric roll. The paper explains the method for reconstructing the experimental wave train in the simulation method and how to circumvent the problem of the deviation in horizontal position. Finally, comparisons between experimental and simulated time traces are given for the motions in the vertical plane.

Keywords Parametric roll · Time domain simulation · 6-DoF panel method Determinsitic validation

F. van Walree (✉)
MARIN-Maritime Research Institute Netherlands, Wageningen, The Netherlands
e-mail: F.v.Walree@marin.nl

P. de Jong
Delft University of Technology, Delft, The Netherlands

V. L. Belenky et al. (eds.), *Contemporary Ideas on Ship Stability*, Fluid Mechanics and Its Applications 119, https://doi.org/10.1007/978-3-030-00516-0_36

36.1 Introduction

The operability and safety of a ship and its cargo depends, amongst others, on its behavior in waves. Under certain conditions in bow and bow quartering seas resonant (parametric) roll may occur which can lead to large roll angles. This can lead to loss of cargo and endangers the ship and its crew. The occurrence of parametric roll can be investigated by means of scale model tests. Provided the tests are properly executed, they offer the most reliable information on dynamic stability.

Issues in the use of model testing are the costs, the limited statistical reliability of the required tests in irregular waves, the limited flexibility and the fact that the test results are not always easy to understand. The limitations in the physical representation relate to viscous effects in the components of the hull resistance with an effect on the propeller loading and speed loss in waves, in some of the smaller components of the roll damping and in components of the manoeuvring reaction forces. The neglect of wind on the roll damping, the wind heel and on the propeller loading and related steerage has an effect in bow quartering seas. Issues that are modeled implicitly correctly are the natural crest-trough a-symmetry in steep waves, the presence of breaking waves, the wave induced forces on the propeller and rudder, rudder and propeller ventilation and down-stream effects of vortices from the bilges and bilge keels on the rudder.

In order to understand the physics of dynamic stability, numerical modeling has been pursued for quite some time. Although the latest CFD techniques have undoubtedly the largest potential, they have not met the expectations yet. This is partly due to the problems of modeling the generation, propagation and absorption of steep waves in a limited computational domain and partly due to the local physical character of issues like spilling wave crests on deck, roll damping from bilge keels and rudder stall and ventilation and the role of the propeller herein. In combination with the required domain size, this yields an extreme computational effort.

In between the above two techniques are hybrid time domain methods, which combine the efficiency of potential flow theory with empirical modules covering the non-linear aspects of manoeuvring and roll damping. After successful validation, these models are particularly useful to investigate dynamic stability.

The present paper deals with deterministic validation of such a hybrid simulation method for a container ship subject to parametric roll. A brief description of the simulation method is given first and the experimental arrangement is outlined. Next, the method to reconstruct and use the experimental wave train in the simulation method is described and finally a comparison between experimental and simulated motion responses is given.

36.2 Simulation Method

The time domain panel method PANSHIP, see De Jong and van Walree (2009) and van Walree and Carette (2010), is characterized by:

- A 3D transient Green function to account for linearized free surface effects with the exact forward speed effects for the mean wetted surface, yielding mean, radiated and diffracted wave components along the hull and a Kutta condition at the stern,
- A 3D panel method to account for Froude-Krylov forces on the instantaneous submerged body,
- A cross-flow drag method to include viscosity effects, resistance (in waves) is obtained from panel pressure integration each time step applying an empirical coefficient to the local flow velocity,
- Propulsion and steering using propeller open water characteristics, semi-empirical lifting surface characteristics and propeller, rudder and hull interaction coefficients,
- Viscous roll damping due to hull and bilge keels according to the ITTC Recommended Procedure (2011). These viscous roll damping components are added to the inherent wave making roll damping present in the time domain panel method,
- Autopilot steering with rudders.

PANSHIP is used at MARIN mainly for seakeeping predictions for fast and unconventional ships; however it can also deal with low speed ships.

36.3 Model Tests

The model tests were performed at MARIN on a scale 55 C11 class container ship, the same hull form and experimental data have been used for the ITTC benchmark on parametric roll. For a detailed description of the experimental arrangement and the results of this benchmark study one is referred to Reed (2011), see also Chap. 37 of this book. The model was tested with propeller, rudder and bilge keels. The model was free to move in six degrees of freedom and was self-propelled and steered by an autopilot. The main particulars of the ship are given in Table 36.1 while Fig. 36.1 shows the hull form of the model.

Table 36.2 shows the experimental test conditions.

The nominal speed was 5 knots for all three tests. The duration of each test was about 2500 s full scale. The experiments showed no parametric roll for the lowest wave height (Test 307002), appreciable parametric roll for the intermediate wave height (Test 307001) and heavy parametric roll for the highest sea state (Test 307004).

Table 36.1 Main particulars of the vessel

Designation	Symbol	Magnitude
Length between perpendiculars	L_{PP}	262.00 m
Breadth	B	40.00 m
Depth	D	24.45 m
Draft moulded on FP	T_F	11.72 m
Draft moulded on AP	T_A	12.86 m
Displacement weight	Δ	76,020 t
Transverse metacentric height	GM_T	2.08 m
Natural roll period	T_ϕ	25.20 s

Table 36.2 Tests in irregular head seas

Test no.	Wave conditions	
	Significant wave height (m)	Peak period (s)
307002	3.500	14.40
307001	4.125	14.40
307004	5.250	14.40

36.4 Deterministic Waves

A way to circumvent the need for lengthy model tests and simulations when validating simulation methods for predicting rare events is to run the simulations in the same wave train as the experiments, i.e. deterministic validation. This is very relevant to parametric roll, as shown by Reed (2011), see also Chap. 37 of this book.

The presently adopted procedure for deterministic validation starts with determining the wave spectrum components from the experimental wave train. During the model tests, the wave height was measured by two wave probes attached to the carriage which tracked and followed the model. One probe was mounted in front of the model while the second one was mounted at the side of the model at a distance sufficient to avoid interference from ship radiated waves.

The wave spectral densities S for each wave spectrum were determined by means of spectral analysis of the wave train signals measured while travelling at very low speed through the basin (without the model present). This yields an "average" wave spectrum valid anywhere in the basin. Next, the phase angles ε are determined by means of a non-linear minimization procedure (IMSL routine RNLIN). In this procedure the difference between the measured and reconstructed wave trains, ζ_m and ζ_e respectively, is minimized at each time step by varying the phase angles. The measured wave train ζ_m is that measured during the actual model tests. The object function F at time t, and the reconstructed wave train at wave probe #j are defined by:

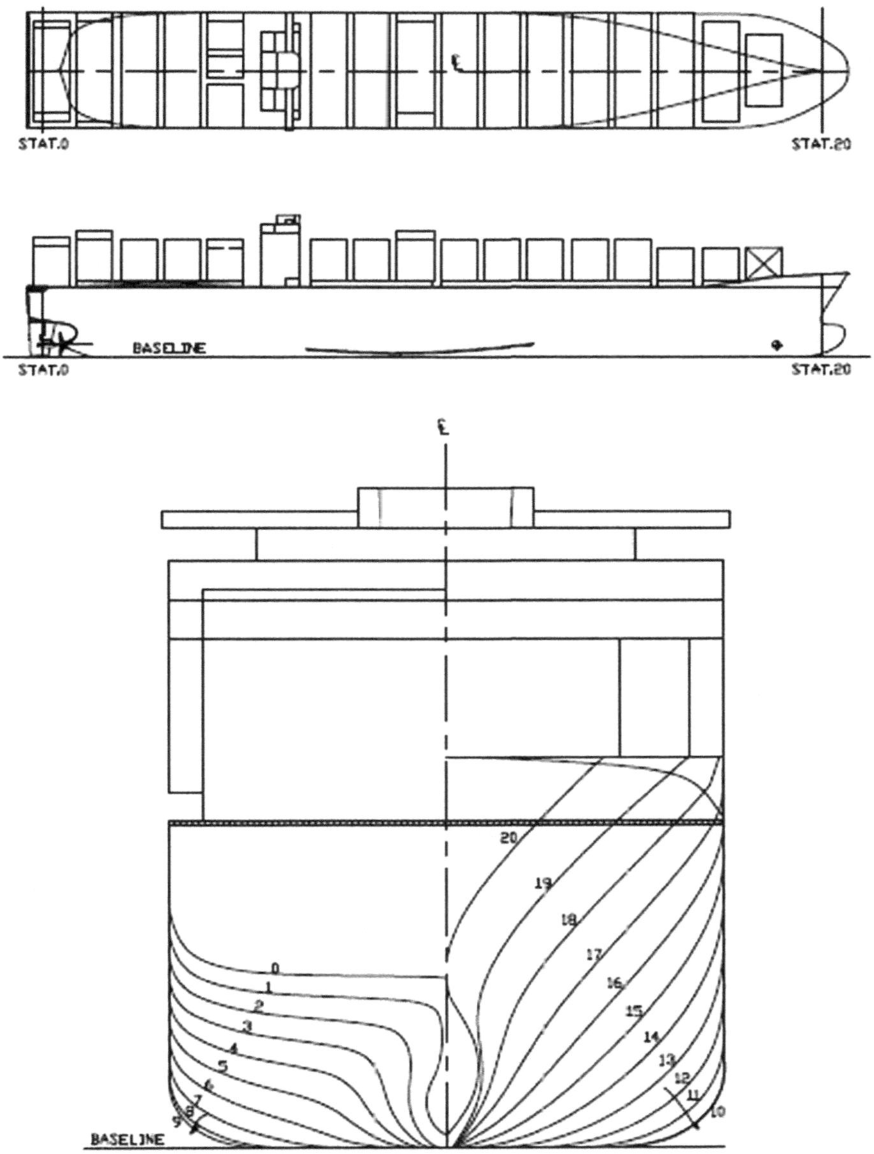

Fig. 36.1 General arrangement and small scale body plan

$$F_j(t) = \zeta_{mj}(t) - \zeta_{ej}(t) \tag{36.1}$$

$$\zeta_{ej}(t) = \sum_{i=1}^{n} A_i \cos(k_i x_j - \omega_i t + \varepsilon_i) \tag{36.2}$$

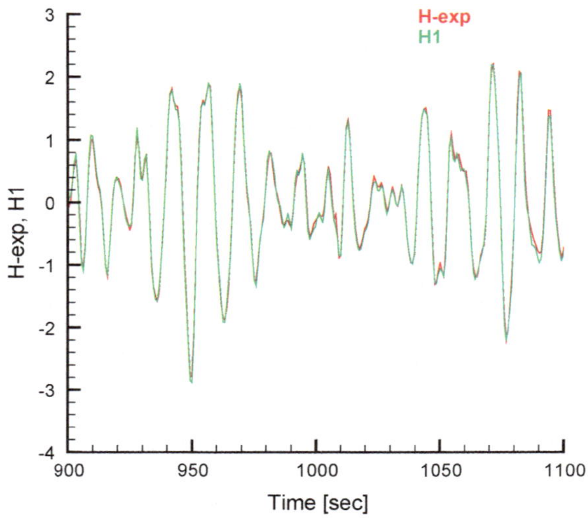

Fig. 36.2 Comparison between measured and reconstructed wave trains (red: measured; green: reconstructed)

where $A_i = \sqrt{2\Delta\omega_i S_i}$ is the wave amplitude of spectral component i, $k = \omega^2/g$ is the wave number and ω is the wave frequency. The position of wave probe j in the wave field, x_j, is given by

$$x_j = x_{0j}(t)\cos(\psi) + y_{oj}(t)\sin(\psi) \tag{36.3}$$

and (x_{0j}, y_{0j}) is the basin fixed position of the wave probe and ψ is the wave direction.

The object function F is minimized using the observations (measurements) at the two wave probe positions sequentially, yielding the phase angles ε. The length of a single model test run was 2500 s with a time step of 0.074 s (full scale values). The number of observations used per run was about 3300 per wave probe. The wave height time traces were cut in to 20 parts of 125 s length and for each part the phase angles were determined for 80 spectral components.

The assumption is now that Eq. (36.2) is valid for arbitrary positions (x, y) in the neighborhood of the ship. Figure 36.2 shows a comparison between the measured wave train and the reconstructed wave train at one of the wave probes for a typical segment of the time trace.

The reconstructed wave spectral properties (amplitudes and phase angles) were imported in the simulation method so that simulations in the experimental wave train could be performed. The propeller RPM was set to have the same mean velocity as during the model tests. The experimentally used autopilot coefficients were used during the simulations as well.

In principle one can now perform the deterministic validation study. However, simulation methods can not predict the motions of the ship perfectly and sooner or later the x-y track of the ship will deviate from that in the experiments. A different

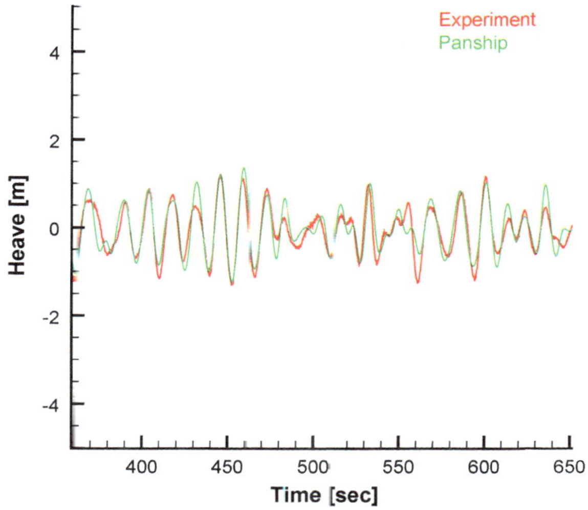

Fig. 36.3 Comparison heave Test 307004 (detail)

position in the wave field means that a different wave will be met and the deterministic comparison needs to stop.

This problem can be circumvented by forcing the position and speed in the horizontal plane in the simulation method to be equal to that of the experiments. This is a viable solution for the present head sea simulations, but for simulations in oblique seas the vertical plane motions will be affected due to this forcing, which is undesirable. A more elegant and general approach is to use an interpolation table for the model *x*-*y* position versus time when determining the wave kinematics in the simulation method. In this way the actual tracks may be different, but still the wave kinematics are evaluated at the position corresponding to that of the experiments. This latter method was used in the present study.

36.5 Deterministic Validation Results

With respect to the initial conditions for the simulations, the positions in six degrees of freedom and the forward speed were set to those measured in the model tests. The roll damping coefficients were tuned to match calm water roll decay tests, for the nominal forward speed of 5 knots. The duration of the simulations was 1800 s.

The first comparison concerns Test 307004, i.e. the highest sea state. Figures 36.3, 36.4, 36.5 and 36.6 show comparisons for heave, pitch and roll versus time. The signals in red are the experimental time traces, the green lines denote the simulation time traces obtained from PANSHIP.

It is seen that the comparison between the measured and simulated heave, pitch and roll motions is fairly good.

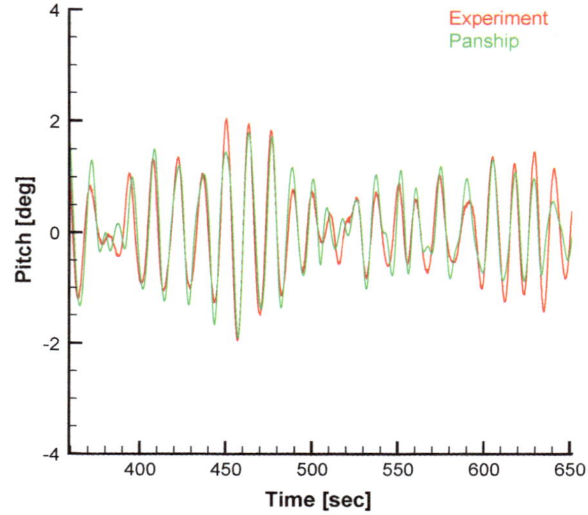

Fig. 36.4 Comparison pitch Test 307004 (detail)

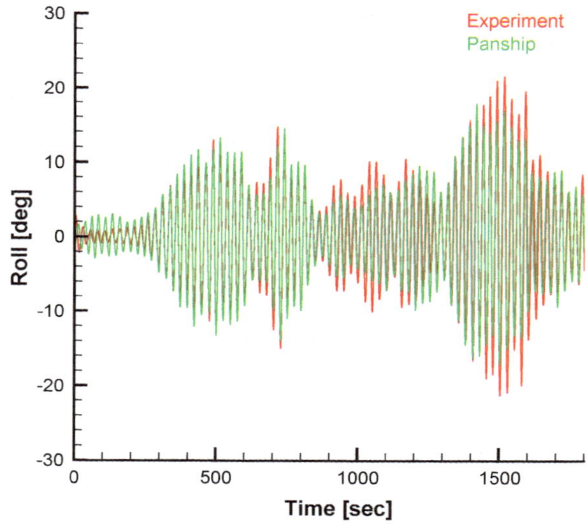

Fig. 36.5 Comparison roll Test 307004

Figures 36.7 and 36.8 show a comparison of the forward speed for Test 307004. Although the simulation method only partially accounts for second order wave forces (added resistance in waves) the global comparison in Fig. 36.7 looks reasonable, i.e. low frequent forward speed variations are similar. Figure 36.8 shows that the wave frequent variations in speed tend to be underestimated in the simulation method.

For the intermediate wave height (Test 307001) Figs. 36.9 and 36.10 show the comparison for roll. It is seen that the resemblance is again fairly good, although around 800 and 1500 s of simulation differences start to appear. Probably the less

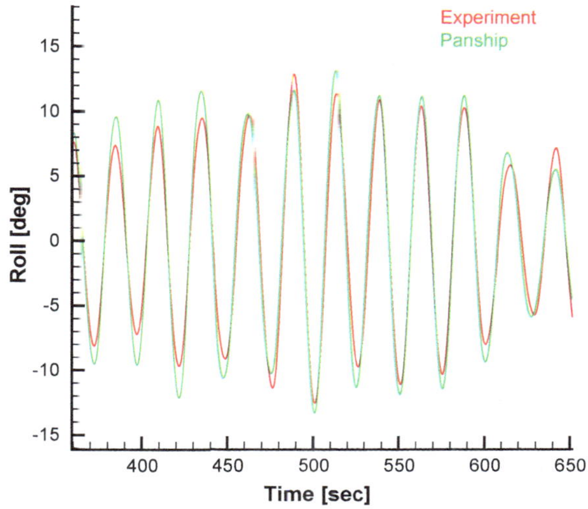

Fig. 36.6 Comparison roll Test 307004 (detail)

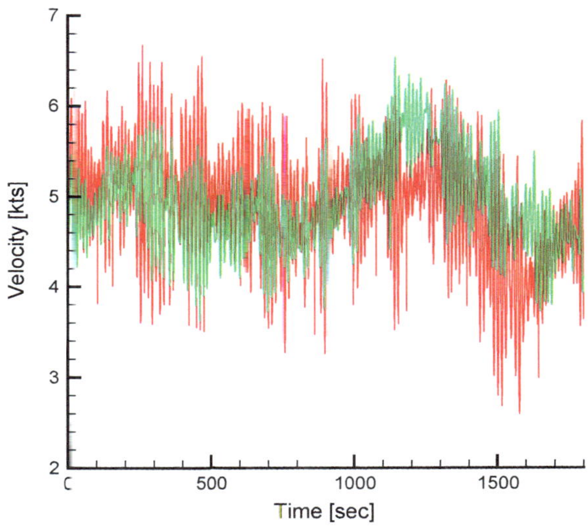

Fig. 36.7 Comparison velocity Test 307004

strong roll forcing for this case makes the results more sensitive to inevitable differences between the simulation method and reality.

The last comparison given is for the lowest sea state (Test 307002) in Figs. 36.11, 36.12 and 36.13. The heave and pitch motions are again in good agreement. While parametric rolling is indeed virtually absent in the simulation results, the method can not accurately reproduce the low amplitude roll motions. This may be due to differences in course keeping, a slight asymmetry in the model configuration, small differences in the wave reproduction, the speed independent roll damping and the

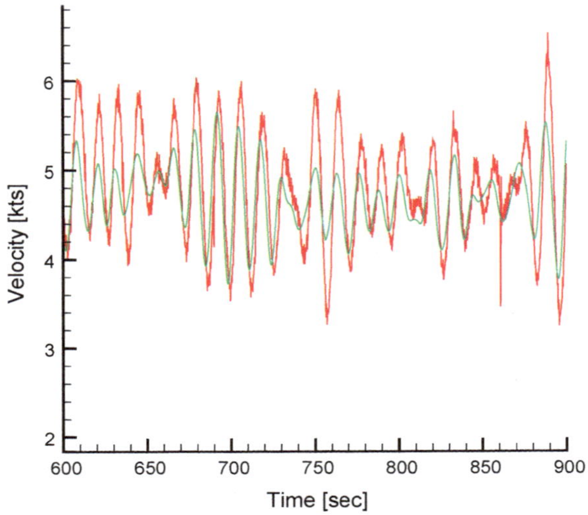

Fig. 36.8 Comparison velocity Test 307004 (detail)

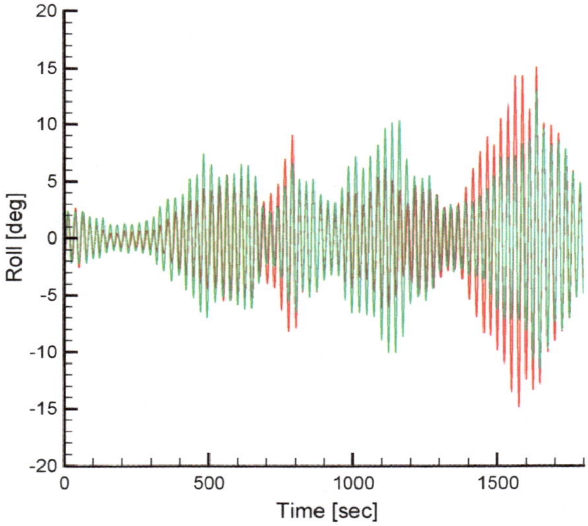

Fig. 36.9 Comparison roll Test 307001

fact that the memory effect at the start of the simulations was not initialised. The latter may have caused the decaying initial rolling motion in the first segment of the time trace during the build-up of the memory effect.

Fig. 36.10 Comparison roll Test 307001 (detail)

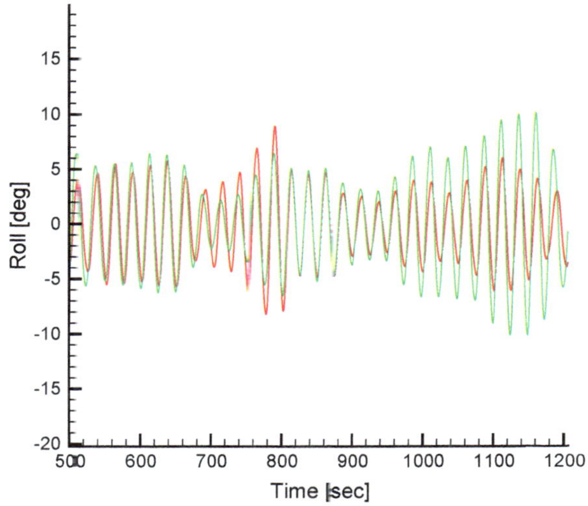

Fig. 36.11 Comparison heave Test 307002 (detail)

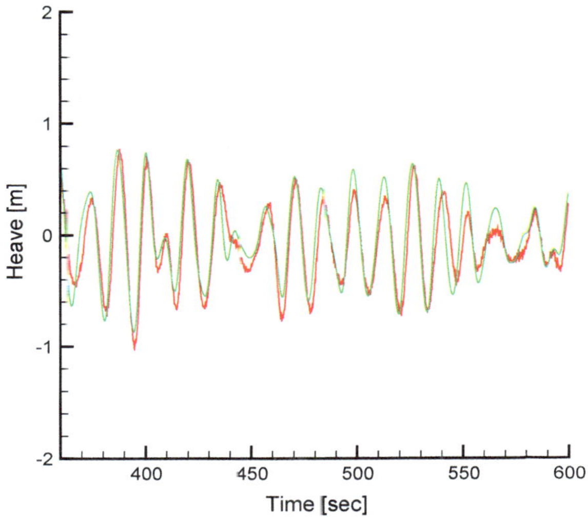

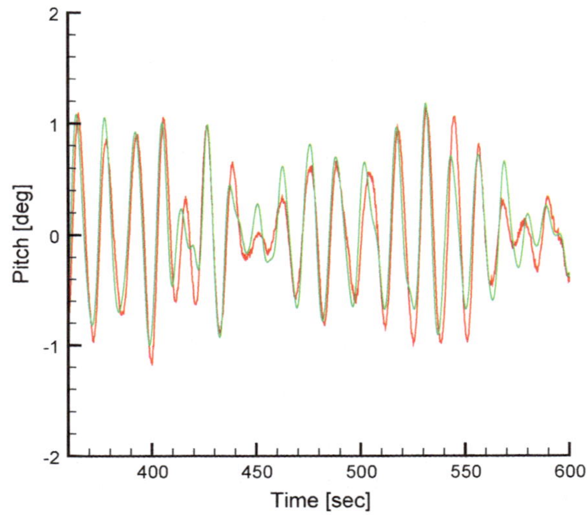

Fig. 36.12 Comparison pitch Test 307002 (detail)

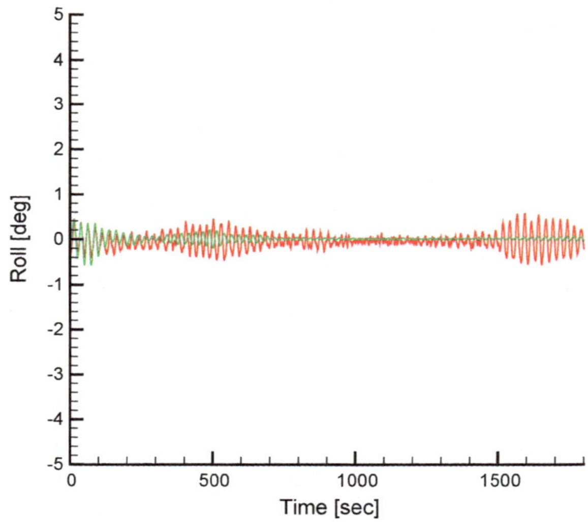

Fig. 36.13 Comparison roll Test 307002

36.6 Conclusions and Future Work

A method has been developed to reconstruct experimental wave trains in a time domain simulation method.

The problem of inevitable deviations between the position of the model and simulated ship in the wave field has been circumvented by using the experimental position in the wave field to determine the wave kinematics.

The experimental and simulated time traces show a fairly good resemblance for heave, pitch and roll motions. The best resemblance for roll is found for the highest sea state.

Future work will focus on deterministic validation for a high speed ship operating in steep stern quartering seas. In this work the focus will not only be on roll but also on the motions in the horizontal plane (sway and yaw).

References

Reed A.M. (2011), "26th ITTC Parametric Roll Benchmark Study", Proceedings of the 12th International Ship Stability Workshop, Washington DC, USA, June 2011.

De Jong P. and van Walree F. (2009), "The Development and Validation of a Time Domain Panel Method for the Seakeeping of High Speed Ships ", Proceedings of the 10th International Conference on Fast Sea Transportation, FAST2009, Athens, Greece, October 2009.

van Walree F. and Carette N.F.A.J. (2010), "Validation of Time Domain Seakeeping Codes for a Destroyer Hull Form Operating in Steep Stern-quartering Seas", ITTC Workshop on Seakeeping, Seoul, South Korea, October 2010.

International Towing Tank Conference (2011), "Numerical Estimation of Roll Damping", ITTC Procedure 7.5-02-07-04.5.

Chapter 37
26th ITTC Parametric Roll Benchmark Study

Arthur M. Reed

Abstract The 26th ITTC Specialist Committee on Stability in Waves was assigned the task of conducting a benchmark of numerical simulation methods for the prediction of the parametric rolling of ships in head seas. The vessel chosen for the benchmark and the organizations which participated in the benchmark and their simulation tools are described. The results of the benchmark are presented.

37.1 Introduction

For the 26th Session of the International Towing Tank Conference (ITTC),[1] running from 2009 to 2011, the Stability in Waves Committee was assigned the task of conducting a benchmark of numerical simulation methods for the prediction of the parametric rolling of ships in head seas. Participants in the study were to be qualified organizations from both inside and outside the ITTC.

This study aimed to evaluate numerical simulation methods currently employed for the prediction of the parametric rolling of ships in waves, and to assess the current level of accuracy of the relevant numerical prediction methods and computer codes by comparison with model-experimental data.

This study was designed to capture the capabilities of the benchmarked numerical methods for ship responses in realistic random sea conditions. The performance of the methods for the selected loading and wave conditions was assessed in comparison to relevant experimental data as well as with respect to the relative performance of each participating method.

The study comprised the simulation of the behaviour of a containership in three steepnesses of longitudinal head waves at one ship loading condition. For the selected

[1] Each Session of the ITTC lasts three years, with four meetings of each committee, culminating with a general meeting in September of the last year of the Session.

This chapter was originally presented at a Stability Workshop (Reed 2011).

A. M. Reed (✉)
David Taylor Model Basin (NSWCCD), West Bethesda, MD, USA
e-mail: arthur.reed@navy.mil

V. L. Belenky et al. (eds.), *Contemporary Ideas on Ship Stability*, Fluid Mechanics and Its Applications 119, https://doi.org/10.1007/978-3-030-00516-0_37

conditions, the excitation of roll motion is expected as a result of parametric resonance. The simulated motions in six degrees-of-freedom (6-DoF) were to be recorded and submitted for review to the study coordinator.

37.2 Background

In 1999, the Sub-Committee on Ship Stability in Waves of the 23rd ITTC was tasked to perform benchmark experiments for stability in extreme seas and to evaluate computational methods against these experiments. The study attempted to evaluate both a Japanese fishing vessel and a container ship. However, the only experiments that were successfully completed were for the fishing vessel. The experiments and numerical comparisons were for: calm water roll decays at two large initial heel angles; calm water turning circles and zig-zag maneuvers at various rudder angles; for rolling in extreme regular beam waves of two wave lengths at zero speed; and motions in near following and quartering waves at three headings, three Froude numbers, in regular waves of one wavelength and three steepnesses. Ultimately, seven different organizations provided motion predictions to the study.

The results of these benchmark experiments and numerical predictions were reported out at the meeting of the 24th ITTC in 2005 (Spyrou et al. 2004; Spyrou 2005; ITTC 2005). The conclusion of the benchmark comparisons were that none of the codes participating in the ITTC study were capable of consistently accurately predicting the measured motion parameters (roll damping, natural periods, responses, etc.).

In 2008, the Ship Design Laboratory of the National Technical University of Athens (NTUA-SDL) coordinated a benchmark study for parametric rolling under the European Union research project SAFEDOR (Design, Operation and Regulation for Safety) (SP.7.3.9, FP6). This project evaluated the performance of the computer codes of 13 participants from around the world. These Participants predicted roll decays, and the onset of parametric rolling and its magnitude for the ITTC-A1 container ship in 19 different conditions. These conditions included 2 GM values, 4 speeds, 3 headings (following seas, near head seas, and head seas), 4 wave amplitudes, and 3 regular wave frequencies and irregular seas. The regular waves were run as individual frequencies and as a sum of waves of the three frequencies.

Spanos and Papanikolaou (2009a, b) report the results of the SAFDOR parametric rolling benchmark study. They assessed the over all agreement of the predictions with the experiments as low, due to the wide spread of the results, However, they concluded that the state-of-the-art was satisfactory, because the best of the predictions was consistently significantly better than the average of all of the predictions.

Neither of the benchmark studies discussed above attempted to determine the uncertainty of the experiments or the numerical predictions. Thus it was not really possible to assess the quality of the predictions and ascertain their accuracy relative to the experiments—were the confidence bands so large that one could have "driven a truck through them" or were they so small that "one would have needed a magnifying glass to see them"? Thus the 26th ITTC parametric rolling benchmark was established with the objective of assessing predictions *with* confidence bands.

37.3 The Ship Model

The vessel chosen for the 26th ITTC study was a model of a C11 class container ship, MARIN Model 8004-2 (Levadou and Van't Veer 2006; Paulling 2007; MARIN 2005, 2009). MARIN provided the hull definition for the study, which was supplied to the benchmark participants in 3 formats. The fully appended hull was tested, with propeller and rudder. The ship was free to move in 6-DoF—surge, sway, heave, roll, pitch and yaw. Thus, the ability of the simulation to maintain course and heading was a part of the benchmark.

The model is a 55th-scale model of a notional 262 m container ship. Its full-scale principal dimensions and mass properties are provided in Table 37.1. Figure 37.1 provides the plan view, profile and body plan of the vessel. Figure 37.2 provides a photograph of the overall model, including the modelling of containers on deck.

The container ship is fitted with a single horn-type rudder and a single 5-bladed propeller. Figure 37.3 provides a photograph of the stern configuration, including propeller and rudder.

The actual propeller used on the model was a Wageningen B-Series 5.59 propeller. Its characteristics were provided to the participants both graphically and as a table.

The ship model was fitted with bilge keels that were aligned with the streamlines on the hull. Participants were provided with a table providing the gross parameters of the bilge keels, and a drawing of the bilge keels with an embedded table that provided

Table 37.1 Main particulars and mass properties of the vessel

Designation	Symbol	Magnitude
Length between perpendiculars	L_{PP}	262.0 m
Breadth	B	40.0 m
Depth	D	24.45 m
Draft molded at FP	T_F	11.719 m
Draft molded at AP	T_A	12.856 m
Displacement (weight)	Δ	76020 t
Centre of gravity fwd of station 0	L_{CG}	122.78 m
Centre of gravity above keel	KG	18.40 m
Transverse metacentric height	G_{MT}	2.075 m
Transverse radius of gyration in air	k_{XX}	16.73 m
Longitudinal radius of gyration in air	k_{YY}	62.55 m
Natural roll period	T_ϕ	25.2 s

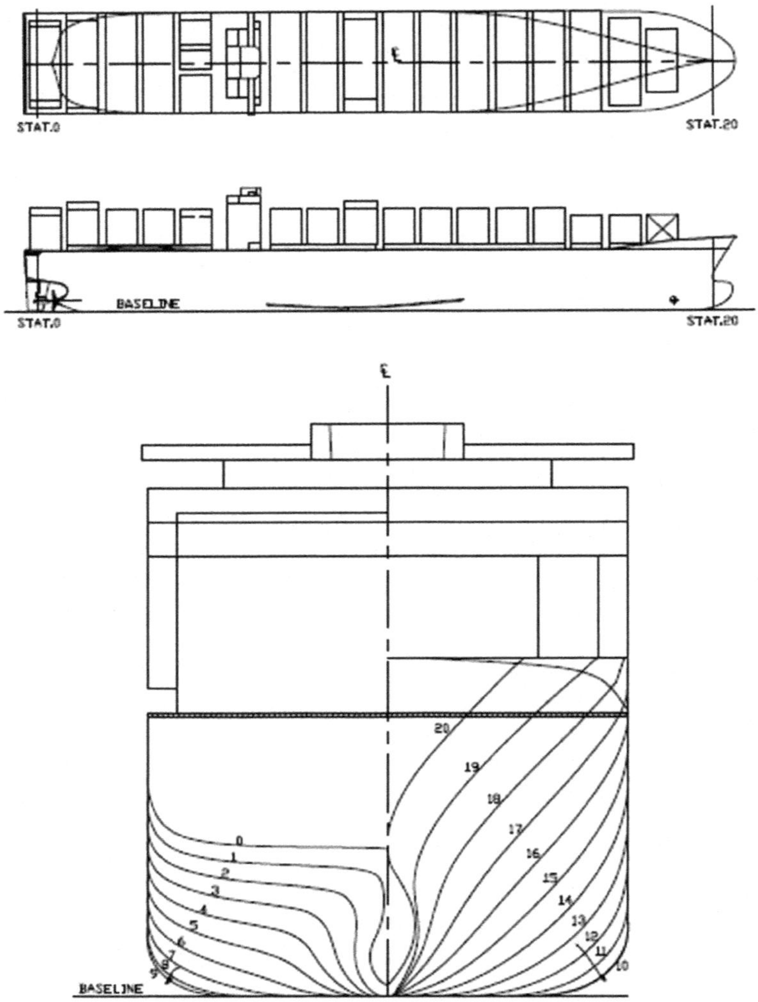

Fig. 37.1 General arrangement and small scale body plan for MARIN Model 8004-2 (MARIN 2009)

their placement along the girth of the hull. The bilge keels can be seen in Figs. 37.1 and 37.2.

The model was fitted with a simple PID autopilot that operated based on the model's yaw, yaw rate, and sway. Although it is not traditional to most autopilots, the sway component was used to keep the model located in the middle of the basin during the runs.

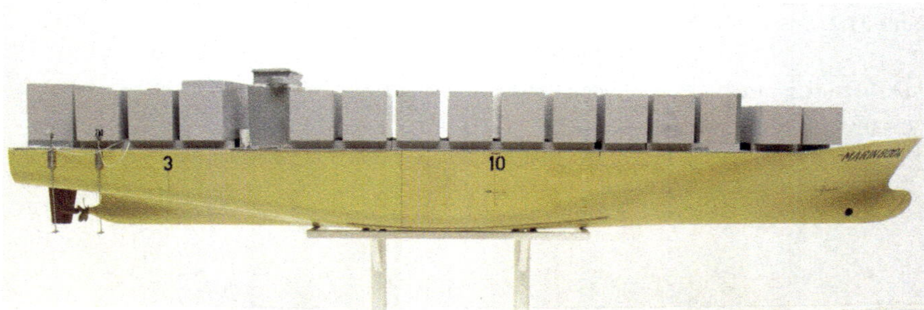

Fig. 37.2 Side view of MARIN Model 8004-2 (MARIN 2009)

Fig. 37.3 Stern-quartering view of MARIN Model 8004-2 showing rudder and propeller (MARIN 2009)

37.4 Parametric Roll Benchmark Cases

The parametric roll benchmark experiments comprised of two parts. The first was a set of roll decay tests in calm water at zero speed and at 5 kn, the speed of the parametric roll experiments. The second component was three parametric roll runs at 5 kn in random seas of differing significant wave height for the same wave modal period.

The conditions for the three parametric roll runs that are being benchmarked are provided in Table 37.2. They are all head seas runs at a nominal speed of 5 kn, in spectra with the same modal period but with different significant wave heights.

The wave spectra with phase information that could be used to reconstruct the wave train time histories were provided for the three runs were provided, as were

Table 37.2 Tests in irregular head seas

	Wave conditions	
MARIN test number	Significant wave height (m)	Peak period (s)
307001	4.125	14.4
307002	3.5	14.4
307004	5.25	14.4

time histories of the actual encountered heights for each run. These wave heights were measured nominally 349 m (full scale) forward of station 10 on the centerline (nominally in the sense that the distance is correct on the average, because the surge motion of the model is not taken into account).

For the parametric roll runs, the model was positioned approximately 8.25 km full-scale (150 m model scale) from the wave maker and the wave maker was started. When the waves reached the model, the model and carriage were brought up to speed, and the model proceeded along the tank under the control of the autopilot; tracked by the carriage. Data collection commenced once the model had reached speed.

During all tests, the model was self-propelled at a propeller RPM that was the equivalent of 5 kn full scale in calm water. (The resistance curve, required for the simulations of these cases, was provided to the participants.) Connections between model and carriage consisted only of free-hanging wires for relay of measurement signals and to supply power. These cables did not restrict the motions of the model significantly.

MARIN's Basic Measurement System (BMS) was used for the data acquisition, with a sample rate of 100Hz, model scale.

Figures 37.4, 37.5, 37.6, 37.7, 37.8 and 37.9 show the encountered wave and roll time histories for Runs 307001, 307002 and 307004. As can be seen from these figures, the wave elevation time histories increase moderately from Run 307002 to Run 307001 to Run 307004. The roll for Run 307002 is negligible; while Runs 307001 and 307004 both show the occurrence of significant parametric roll.

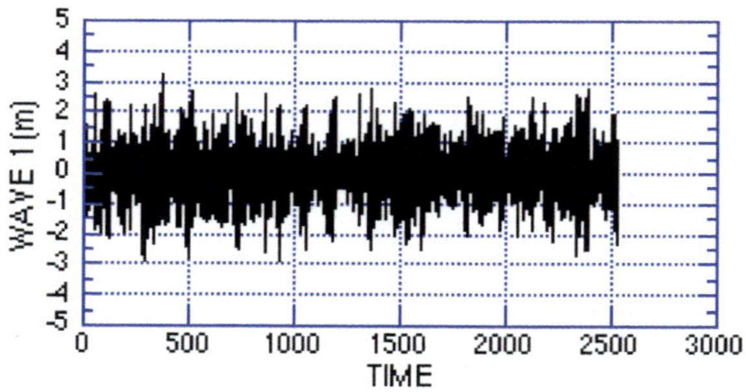

Fig. 37.4 Wave time history for MARIN Model 8004-2, Run 307001

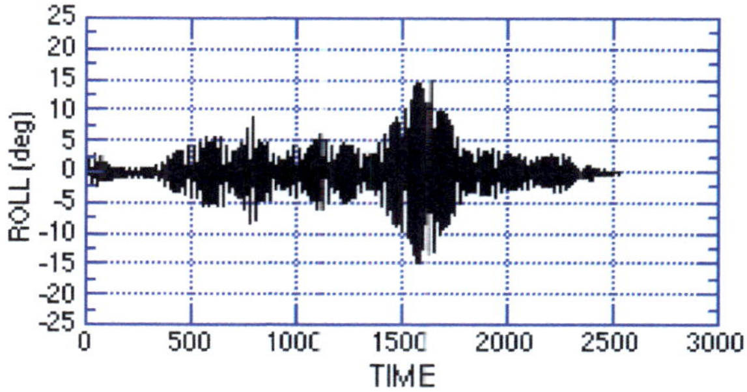

Fig. 37.5 Roll time history record for MARIN Model 8004-2, Run 307001

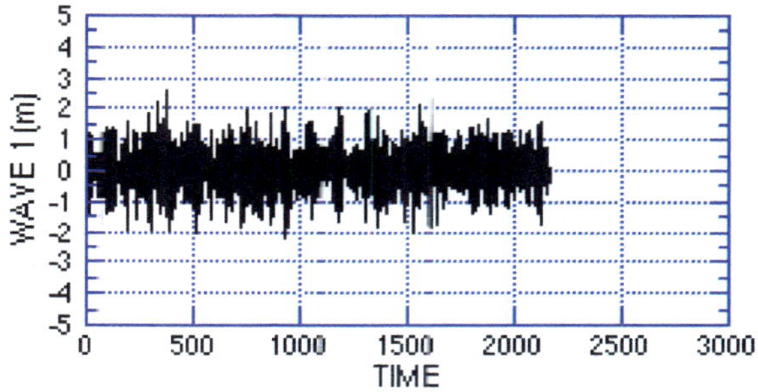

Fig. 37.6 Wave time history for MARIN Model 8004-2, Run 307002

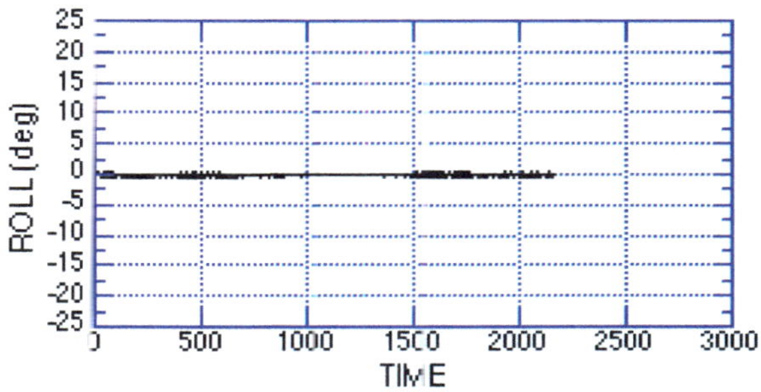

Fig. 37.7 Roll time history for MARIN Model 8004-2, Run 307002

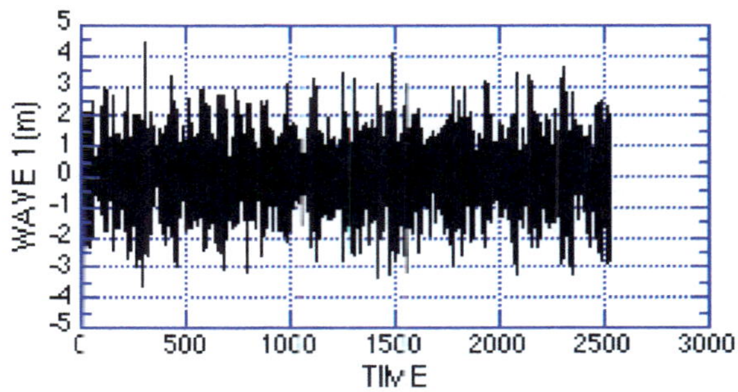

Fig. 37.8 Wave time history for MARIN Model 8004-2, Run 307004

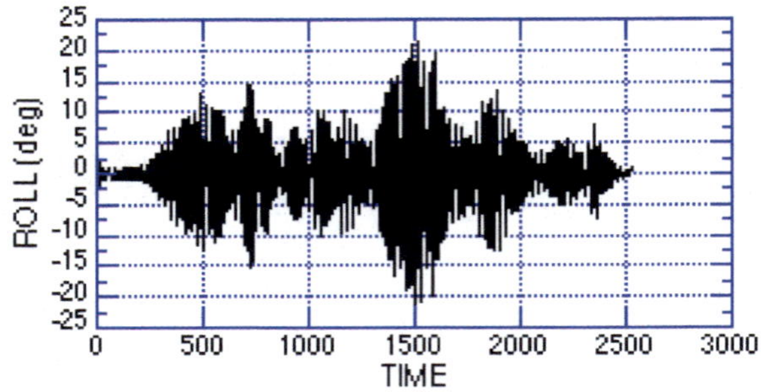

Fig. 37.9 Roll time history for MARIN Model 8004-2, Run 307004

37.5 Benchmark Simulations

Comparisons of the simulations of parametric roll with the experimental results were to be made based on the statistics of the simulated waves and of the predicted 6-DoF motions; and where possible, on comparisons of the actual 6-DoF motion predictions. For the three cases that were to be compared, participants were asked to perform one set of simulations using the "default" roll damping model from their simulation tool, and if they "tuned" their roll damping against the roll decay data, a second set of simulations using that tuned roll damping. Additionally, participants were asked to submit their roll decay predictions from their "default" roll decay model and their "tuned" roll decay model.

Initially, ten organizations indicated an interest in participating in the benchmark study, and ultimately, six organizations provided results predicted using seven different computer codes. The six organizations and their computer codes are listed in Table 37.3.

Brief descriptions of these computational tools are provided below. With the exception of ROLLSS, all of the codes are blended codes that compute the exact nonlinear Froude-Krylov exciting forces due to the incident wave and the exact hydrostatic restoring forces over the instantaneous wetted surface of the vessel, and employ linear computations of the radiation and diffraction forces for the vessel up

Table 37.3 Organizations participating in the benchmark study and their respective computer

Organization	Code(s)
David Taylor Model Basin (NSWCCD)	FREDYN, v9.8
HSVA	ROLLSS
MARIN	FREDYN, v10.1
Osaka University	OU-PR
Science Applications International Corporation	LAMP 3
Seoul National University	SNU-PARAROLL, WISH

to the mean waterline. ROLLSS employs a different blending where all forces except for those in roll are calculated linearly. The descriptions, with references, follow:

FREDYN v9.80 and 10.1 (de Kat and Paulling 2001; Hooft 1987) simulates the dynamic behaviour of a steered ship subjected to waves and wind. All six degrees of freedom are computed in the time domain, where the motions can be large up to the point of capsize. Nonlinearities arise from rigid-body dynamics with large angles and fluid flow effects. A linear strip theory approach is used to compute the hydrodynamic forces acting on the hull.

ROLLSS (Söding 1982; Kröger 1986; Petey 1988; Brunswig et al. 2006) is a code for the simulation of parametric rolling. While the pitch, heave, sway and yaw motions are computed by a linear strip method, and the surge motion by a simple nonlinear approach, the roll motion is computed nonlinearly in time domain using the righting arm curves for static stability in waves.

OU-PR (Osaka University simulation program for Parametric Rolling) (Hashimoto and Umeda 2010; Hashimoto et al. 2011, Chapt. 16) is a time domain simulation program for prediction parametric rolling in regular and long-crested irregular waves, which uses a 3-DoF coupled of heave-roll-pitch model. The 2-D radiation and diffraction hydrodynamic forces are calculated for the submerged hull with the instantaneous roll angle taken into account. The roll radiation force is calculated at the natural roll frequency and those in vertical modes (heave and pitch) are at the peak of the mean wave frequency. Linear and quadratic roll damping is determined from results of a roll decay test if available. Otherwise, they are determined by Ikeda's semi-empirical method.

LAMP 3 (Lin and Yue 1990; Shin et al. 2003; Lin et al. 2006; Yen et al. 2008; Yen 2010) predicts the motions and loads of a ship operating in a seaway. In LAMP's simulations, the wave-body hydrodynamic forces are calculated using a 3-D Rankine potential flow panel method with a linearized free-surface boundary condition to solve the wave-body interaction problem in the time domain, while forces due to viscous flow and other "external" effects such as hull lift, propulsors, rudders, etc. are modeled using other computation methods, or with empirical or semi-empirical formulas. LAMP's calculations include 2nd and higher order "drift" forces in the horizontal plane (Zhang et al. 2009). These drift forces play an important role in the horizontal-plane motions for the prediction of course keeping in waves.

SNU-PARAROLL (Kim and Kim 2010b, 2011) employs a linear impulse-response-function approach to compute the radiation and diffraction forces. The impulse response function approach is basically the conversion of the frequency-domain strip-theory solution into the time domain. In this method, the conversion is limited to the radiation force. The excitation force includes the nonlinear Froude-Krylov and restoring force and moment on the instantaneous wetted surface as well as the linear diffraction force. The wetted surface is defined as the hull surface wetted by the body motion and the incident wave.

WISH (computer program for nonlinear Wave-Induced loads and SHip motion) (Kim et al. 2009; Kim and Kim 2010a, b) is a three-dimensional Rankine panel method used to study nonlinear roll motions. In this method, the total velocity potential is decomposed into three components: the basis flow; the incident wave; and disturbance velocity potentials. In this weakly-nonlinear approach, the disturbed component of the wave and velocity potentials are assumed to be small. The kinematic, dynamic free surface and body boundary conditions are linearized. The basis flow-wave induced motion terms (m-terms) are hard to compute, since they require second-order differentials of the basis flow. In this code, the second-order differentials are converted to first-order differentials using Stoke's theorem.

37.6 Benchmark Comparisons

37.6.1 Statistical Methodology

The motions that have no restoring force (surge, sway, and yaw) can be significantly affected by the actions of the autopilot and the propulsion algorithm. Taking advantage of the fact that these effects will, in general, be at much lower frequency than the wave encounter frequency, the surge sway and yaw motions are decomposed into "low-frequency" and "wave-frequency" components, to separate the maneuvering and autopilot related contribution to the motions from the direct response to the wave excitation.

The analysis of the results consisted of statistical analysis of the wave elevation and the 6-degree-of-freedom motions.[2] The statistical quantities that are computed are given below:

1. Mean value: $\overline{u}$ (MEAN)

$$\overline{u} = \frac{1}{N}\sum_{n=1}^{N} u_n$$

   where u_n is the nth sample of the signal and N is number of samples.
2. Variance of the Mean: V_M (VAR MEAN),

$$V_M = V\frac{1}{N}\sum_{l=-N+1}^{N-1}\left(1 - \frac{|l|}{N}\overline{R}_{|l|}\right), \tag{37.1}$$

   where $\overline{R}_i$ is the ith value of the autocorrelation function of the signal and V is the variance of the signal.
3. Variance: $V\ (= \sigma^2)$ (VAR),

$$V = \frac{1}{N}\sum_{n=1}^{N}(u_n - \overline{u})^2$$

[2]The methodologies for performing the statistical analysis have been advanced since this study was performed, See Belenky et al. (2015) for more refined analysis techniques.

4. Variance of the Variance: V_V (VAR VAR),

$$V_V = V^2 \frac{2}{N} \sum_{l=-N+1}^{N-1} \left(1 - \frac{|l|}{N} \overline{R}_{|l|}^2\right), \tag{37.2}$$

used to compute a confidence interval for the value of the variance.

Based on the ideas and method of Belenky et al. (2007), the computation of the variance of the mean (V_M) and the variance of the variance (V_V) allows the computation of confidence bounds for the mean and variance of the experiments and computations. The computation of V_M and V_V requires an estimate of the autocorrelation function $\overline{R}$ of the measured or computed response.

The autocorrelation function can be computed directly from the signal or from the spectrum of the signal. If $\overline{R}(t)$ is computed directly from the signal, $\overline{R}(t)$ must be truncated due to the loss of statistical confidence in the function's values for large lags. The computation of $\overline{R}(t)$ from the spectrum requires that the spectrum be smoothed before $\overline{R}(t)$ is computed.

The recommended method for computing $\overline{R}(t)$ is from the spectrum. The auto-covariance function is the cosine transform of the spectrum, so it can be computed as follows:

$$C_{uu}(t) = \int_0^\infty d\omega\, S(\omega) \cos \omega t.$$

The autocorrelation function $\overline{R}(t)$ is produced by dividing the autocovariance function $C_{uu}(t)$ by the approximate variance V' ($C_{uu}(0)$), so that $\overline{R}(0)$ is identically 1.

$$\overline{R}(t) = \frac{1}{V'} C_{uu}(t) = \frac{1}{C_{uu}(0)} C_{uu}(t).$$

This normalization also accounts for the computational and/or truncation errors that are likely to accrue in the computation of the spectrum and the autocovariance function from the spectrum, which result in a difference between the approximate variance V' [$=C_{uu}(0)$] and the true variance of the signal, V.

The spectrum of the encountered wave train, complete motions (heave, roll, and pitch) and wave-frequency decomposed motions (surge, sway, and yaw) are computed in the usual manner using Fourier transforms. As seen in Fig. 37.10, this spectrum will be quite jagged; and the autocorrelation function computed from this jagged spectrum will not continuously decrease for large lag times, Fig. 37.11.

Smoothing the spectrum before computing the autocorrelation function improves the quality of $R(t)$ and reduces the growth of the envelope for large lag times. The smoothing can be performed by using a digital filter, or more easily by means of a "boxcar" filter, a simple 5-point running average filter that sets the new value of point n to be the average of the 5 points centered about point n. The spectrum smoothed with a 5-point boxcar filter is shown in Fig. 37.10 and the resulting autocorrelation function is given in Fig. 37.12. A boxcar filter with more points (7, 9 or 11) produces

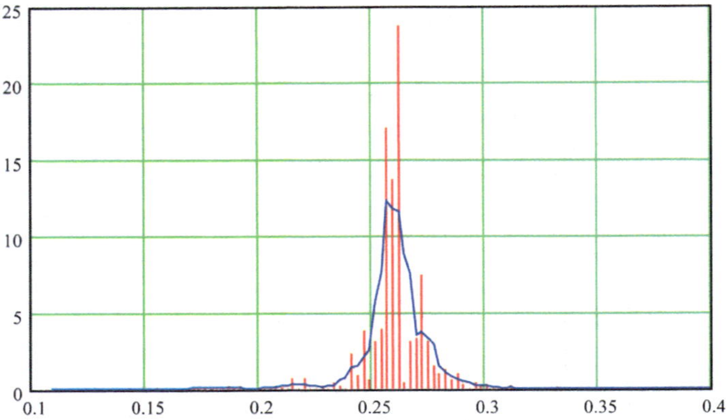

Fig. 37.10 Raw roll spectrum and roll spectrum smoothed with 5-point boxcar filter

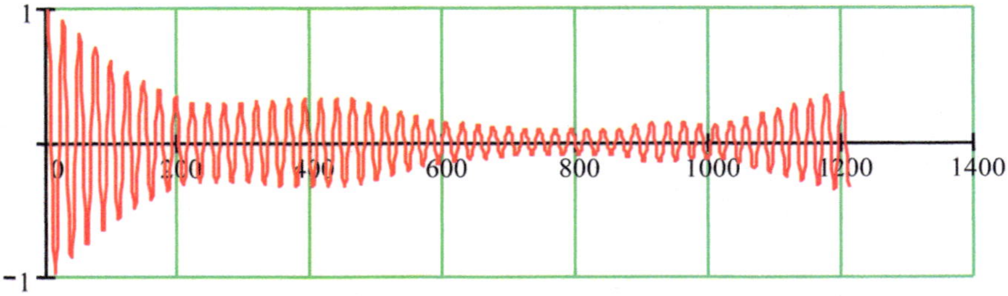

Fig. 37.11 Roll autocorrelation function from raw spectrum

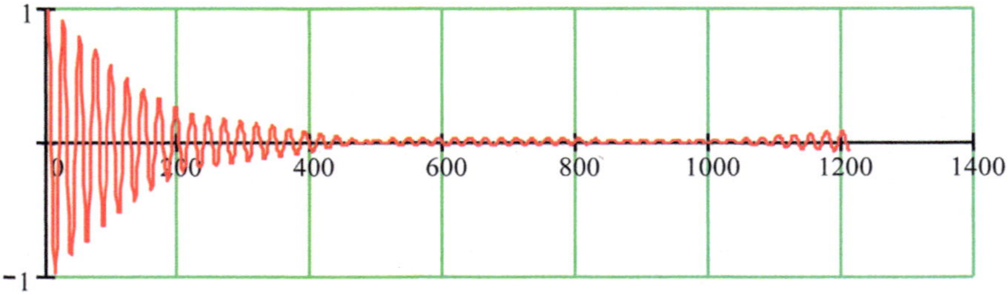

Fig. 37.12 Roll autocorrelation function from spectrum smoothed with 5 point boxcar filter

a smoother spectrum and better looking autocorrelation function, but has little effect on the computed values of V_M and V_V.

Once the autocorrelation function has been computed, it is straightforward to compute the variance of the mean, V_M and variance of the variance V_V of the measured or computed signal, using (37.1) and (37.2), respectively.

37.6.2 Comparisons

Although all participants provided the statistics for the waves and all 6-DoF of motion, the allotted space does not allow all of the results to be presented. Thus, only the wave and roll statistics are being provided. Figure 37.13 shows the variance of the wave time histories from the experiments and each of the predictions along with the corresponding 95% confidence bands. Figure 37.14 shows similar results for roll. On these plots, the experiments are denoted by "Ex" and the predictions have been randomly assigned the letters "A" to "G".

Examining the variance of the wave height shown in Fig. 37.13, it is seen that the variance of the wave heights are consistent with the experimental results for all three runs. Additionally, all of the predictions are well within the 95% confidence band of the experimental results.

The statistics for roll shown in Fig. 37.14 exhibit much less agreement between the predictions and the experimental results.

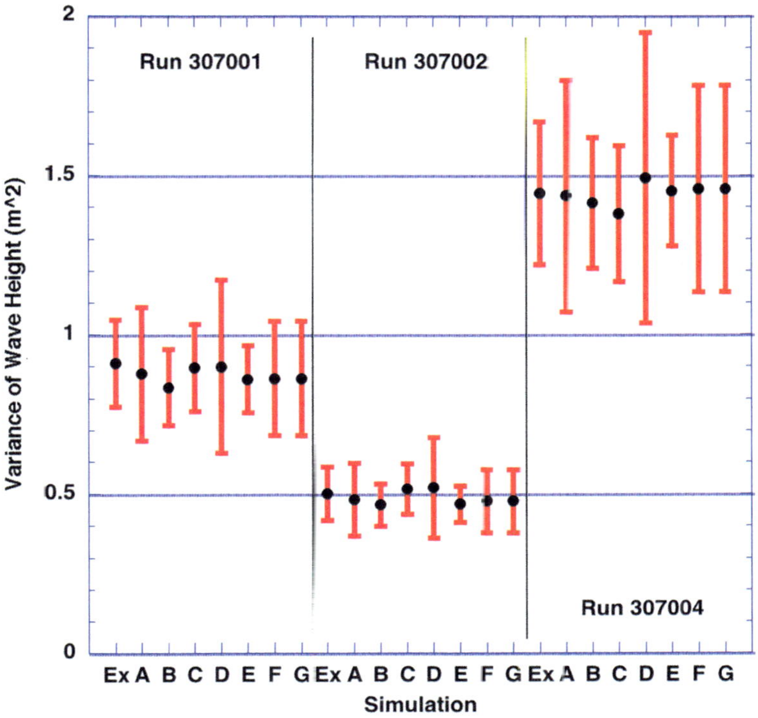

Fig. 37.13 Variance of wave height with 95% confidence bands for MARIN Model 8004-2, Runs 307001, 307002, and 307004, as predicted from the experiments (Ex) and computations (A–G)

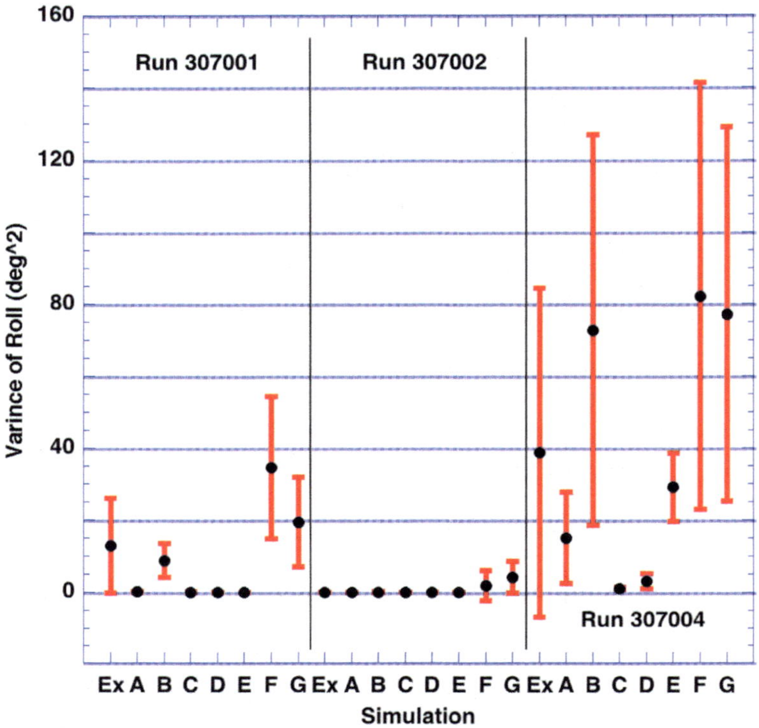

Fig. 37.14 Variance of roll angle with 95% confidence bands for MARIN Model 8004-2, Runs 307001, 307002, and 307004, as predicted from the experiments (Ex) and computations (A–G)

For Run 307002, only Methods F and G show any notable roll response. None of the experiments or computations except for Methods F and G have any significant variance of the variance.

In the case of Run 307001, the experiments show significant variance and 95% confidence bands, while Methods B, F, and G show significant variance and have significant 95% confidence bands. None of the other methods show either any variance or 95% confidence bands. In fact the upper 95% confidence limits for the predictions by Methods A, C, D and E are below the 95% confidence band for the experiments. Methods B, F and G have 95% confidence bands comparable to that of the experiments, which overlap the experimental 95% confidence band—statistically, all of these results are the same.

The results for Run 307004 are different from both of the previous runs. The experimental results show a lower 95% confidence limit, which is negative, an impossibility, which indicates that the usual normal distribution assumptions are not applicable. (The same comment applies to Method B in Run 307001 and Methods F and!G in Run 307002.) In fact the 95% confidence band limits must be computed assuming a truncated normal distribution, and this will lead to upper and lower limits of 89.5 deg^2 and 2.7 deg^2, respectively. With the exception of Method C the variance of all of the predictions are within the 95% confidence band of the experiments, and are statistically equivalent.

37.7 Benchmark Conclusions

The results of the SAFEDOR Parametric Roll Benchmark (Spanos and Papanikolaou 2009a, b) provide no confidence bands, but show levels of scatter in the range of roll predictions consistent with those seen in Fig. 37.14—they have more participating codes and more cases. Thus the scatter in the variance of the predictions should not have been unexpected.

Belenky and Weems (2012) have conducted a study where they used `LAMP 2` to predict the head-sea motions of a C11 class container vessel, looking for parametric roll. They produced 50 realizations of the same JONSWAP spectrum, where each realization consisted of 1500 s of data. They then computed the 95% confidence interval for the roll variance for each individual realization and for the ensemble of all 50 realizations. Figure 37.15 shows the variance of the roll along with the confidence interval for each individual run and for the ensemble of 50 runs.

Figure 37.15 shows the degree of variability that can occur from run to run. The variance of the first and second realizations differ by more than a factor of two (29 deg^2 vs. 72 deg^2), and the largest and smallest vary by a factor of five (15 deg^2 vs. 77 deg^2). As can be seen, the ensemble confidence band is significantly narrower than those of the individual records.

The reason for the dramatic differences in the variance of the records is the fact that parametric roll is a consequence of a group of waves of length close to the ship length and a speed-heading combination that results in an encounter frequency

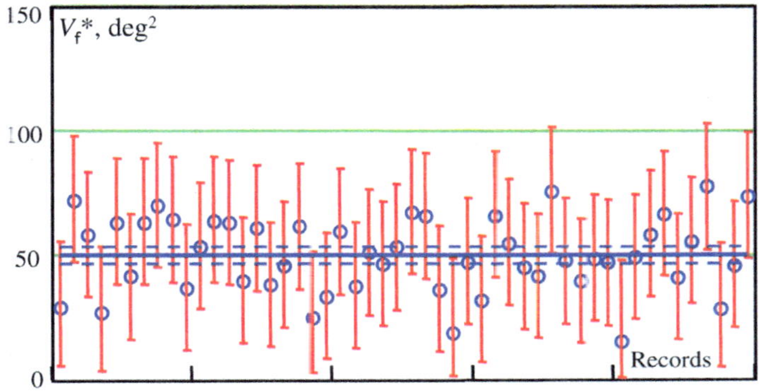

Fig. 37.15 Estimates of variance values of records and ensemble (from Belenky and Weems 2012)

that is twice the roll natural frequency. The practical implications of this are that an individual record contains little statistically independent data. Thus, these waves have a narrow spectral peak that results in an autocorrelation function that does not decay quickly with lag time.

To characterize the solution to this problem, Belenky & Weems go on to study the number of records that must be ensemble averaged in order to produce a "tight" statistical characterization of the variance of the roll. This is demonstrated in Fig. 37.16, where the convergence of the variance of the ensemble and its confidence interval is shown as a function of the number of records included in the ensemble average.

As can be seen, after approximately 20 records there is little change in either the variance or the confidence band for the ensemble and, there is significant convergence after as few as 4 or 5 records. Thus, from a practical perspective, as few as 7–10 realizations should produce usefully convergent results.

In principle, this problem can be overcome by producing significantly longer records, but this is not in general realistic. It is a manifestation of what Belenky and Weems call practical non-ergodicity, "meaning that several independent records must be used in order to devise any judgment on the statistical characteristics of the parametric roll response." To characterize the solution to this problem, Belenky & Weems go on to study the number of records that must be ensemble averaged in order to produce a "tight" statistical characterization of the variance of the roll—the convergence of the variance of the ensemble and its confidence interval as a function of the number of records included in the ensemble average. They find that after approximately 20 records there is little change in either the variance or the confidence band for the ensemble and there is significant convergence after as few as 4 or 5 records. Thus, from a practical perspective, as few as 7 to 10 realizations should produce usefully convergent results.

From the perspective of the benchmark study, this indicates that it is not possible to draw any conclusions regarding the performance of any simulation method from a single realization. One may conclude that those methods that have very narrow

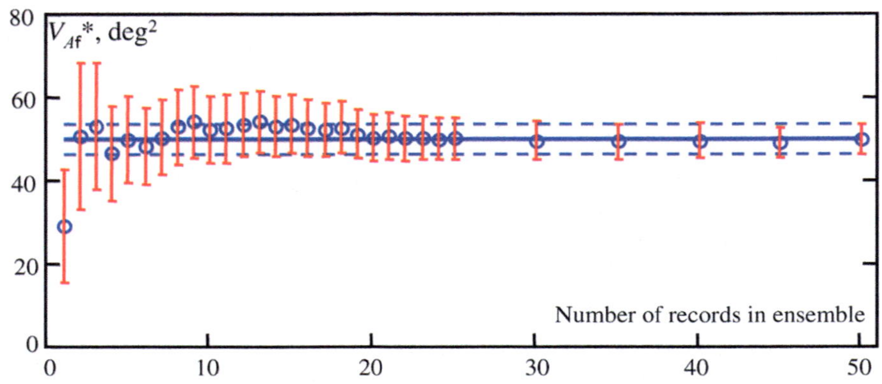

Fig. 37.16 Convergence of ensemble estimate of variance (from Belenky and Weems 2012)

confidence bands on their results are probably not fully capturing the physics of parametric roll. For either experiments or computations, it will take the results from 7 to 10 realizations at the same significant wave height to determine convergent results. Further, as converged statistical results are obtained, more sophisticated means of comparison than the variance of the roll amplitude, such a roll exceedance rates with the appropriate confidence bands should be used to compare methods.

References

Belenky, V., V. Pipiras, and K. Weems (2015) "Statistical Uncertainty of Ship Motion Data." *Proc. 12th Int'l. Conf. Stability of Ships and Ocean Vehicles (STAB'15)*, pp. 891–902, Glasgow, UK.

Belenky, V. & K. M. Weems (2012) "Probabilistic Properties of Parametric Roll." Chapter 6, In: *Parametric Resonance in Dynamical Systems*, Fossen, T. I. & H. Nijmeijer, eds., Springer, pp. 129–146.

Belenky, V., K. M. Weems & W.-M. Lin (2007) "A Probabilistic Procedure for Evaluating the Dynamic Stability and Capsizing of Naval Vessels, Phase 1: Technology Demonstration." SAIC Report ASTD 08-017, 211 p.

Brunswig, J., R. Pereira & D. Kim (2006) "Validation of Parametric Roll Motion Predictions for a Modern Containership Design." *Proc. 9th Int'l Conf. Stability of Ships & Ocean Vehicles*, Rio de Janeiro, Brazil, 14 p.

de Kat, J. O. & J. R. Paulling (2001) "Prediction of extreme motions and capsizing of ships and offshore vehicles." *Proc. 20th Int'l Conf. Offshore Mechanics & Arctic Engin.*, Rio de Janeiro, Brazil, 12 p.

Hashimoto H. & N. Umeda (2010) "A study on Quantitative Prediction of Parametric Roll in Regular Waves." *Proc. 11th Int'l Ship Stability Workshop*, Wageningen, The Netherlands, 7 p.

Hashimoto H., N. Umeda & Y. Sogawa (2011) "Prediction of Parametric Rolling in Irregular Head Waves." *Proc. 12th Int'l Ship Stability Workshop*, Washington, DC, USA.

Hooft, J. P. (1987) "Mathematical description of the manoeuvrability of high-speed surface ships." MARIN report No. 47583-1-MO.

ITTC (2005) "The Specialist Committee on Stability in Waves, Final Report and Recommendations to the 24th ITTC." *Proc. 24th ITTC*, Vol. II, pp. 369–407, Edinburgh, UK.

Kim, K. H. & Y. Kim (2010a) "Comparative Study on Ship Hydrodynamics Based on Neumann-Kelvin and Double-Body Linearizations in Time-Domain Analysis." *Int'l J. Offshore & Polar Engin.*, Vol. 20, No 4, pp. 265–274.

Kim, K. H., Y. Kim & M. S. Kim (2009) "Numerical Analysis on Motion Responses of Adjacent Multiple Floating Bodies by Using Rankine Panel Method." *Int'l J. Offshore & Polar Engin.*, Vol 19, No 2.

Kim, T. & Y. Kim (2010b) "Multi-level Approach of Parametric Roll Analysis." *Proc. ITTC Workshop on Seakeeping*, Seoul, Korea, 21 p.

Kim, T. & Y. Kim (2011) "Numerical Study on Difference-Frequency-Induced Parametric Roll Occurrence." *Int'l J. Offshore & Polar Engin.*, Vol. 21, No. 1, pp. 1–10.

Kröger, H. P. (1986) "Rollsimulation von Schiffen im Seegang." *Schiffstechnik*, Vol. 33

Levadou M. & R. Van't Veer (2006) "Parametric Roll and Ship Design." *Proc. 9th Int'l Conf. Stability of Ships & Ocean Vehicles*, Rio de Janeiro, Brazil.

Lin, W.-M. & D. K. P. Yue (1990) "Numerical Solutions for Large-Amplitude Ship Motions in the Time-Domain." *Proc. 18th Symp. Naval Hydro.*, Ann Arbor, MI, USA, pp. 41–66.

Lin, W.-M., S. Zhang, K. Weems & D. Liut (2006) "Numerical Simulations of Ship Maneuvering in Waves." *Proc. 26th Symp. Naval Hydro.*, Rome, Italy.

MARIN (2005) "Parametric Roll Tests On C11 Post PANAMAX Container Vessel." MARIN Report No. 17701-2-SMB, Vol. 1, 28 p.
MARIN (2009) "Parametric Roll Tests On C11 Post PANAMAX Container Vessel, Data Report." MARIN Report No. 17701-2-SMB, Vol 2, 280 p.
Paulling, J. R. (2007) "On Parametric Rolling of Ships." *Proc. 10th Int'l Symp. Practical Design of Ships & Other Floating Structures*, Houston, TX, USA, 9 p.
Petey, F. (1988) "Ermittlung der Kentersicherheit lecker Schiffe im Seegang aus Bewegungssimulationen." Report Nr. 487, Institut für Schiffbau der Universität Hamburg
Reed, A. M. (2011) "26th ITTC Parametric Roll Benchmark Study." *Proc. 12th Int'l. Ship Stability Workshop*, Washington, D.C, USA, pp. 195–204.
Shin, Y. S., V. L. Belenky, W.-M. Lin, K. M. Weems & A. H. Engle (2003) "Nonlinear Time Domain Simulation Technology for Seakeeping and Wave-Load Analysis for Modern Ship Design." *Trans. SNAME*, Vol. 111, pp. 557–78.
Söding, H. (1982) "Leckstabilität im Seegang." Report Nr. 429, Institut für Schiffbau der Universität Hamburg
Spanos, D. & A. Papanikolaou (2009a) "Benchmark Study on Numerical Simulation Methods for the Prediction of Parametric Roll of Ships in Waves." *Proc. 10th Int'l Conf. Stability of Ships & Ocean Vehicles*, St. Petersburg, Russia, pp. 627–36.
Spanos, D. & A. Papanikolaou (2009b) "SAFEDOR International Benchmark Study on Numerical Simulation Methods for the Prediction of Parametric Rolling of Ships in Wave." NTUA-SDL Report, Rev. 1.0, 60 p.
Spyrou, K. J. , N. Themelis & I. Tigkas (2004) "Benchmark Study of Numerical Codes for the Prediction of Intact Stability of Ships in Extreme Seas: Phase I." Report of the Sub-Committee on Ship Stability in Waves to the 24th ITTC, 59 p.
Spyrou, K. J. , N. Themelis & I. Tigkas (2005) "Benchmark Study of Numerical Codes for the Prediction of Intact Stability of Ships in Extreme Seas: Results of 'Phase II'." Report of the Sub-Committee on Ship Stability in Waves to the 24th ITTC, 42 p.
Yen, T. G., D. A. Liut, S. Zhang, W.-M. Lin, & K. M. Weems (2008) "LAMP Simulation of Calm Water Maneuvers for a U. S. Navy Surface Combatant." *Proc. Workshop on Verification & Validation of Ship Maneuvering Simulation Methods*, Copenhagen, Denmark.
Yen, T. G., S. Zhang, K. Weems & W.-M. Lin (2010) "Development and Validation of Numerical Simulations for Ship Maneuvering in Calm Water and in Waves." *Proc. 28th Symp. Naval Hydro.*, Pasadena, CA, USA.
Zhang, S., K. M. Weems & W.-M. Lin (2009) "Investigation of the Horizontal Drifting Effects on Ships with Forward Speed." *Proc. ASME 28th Int'l Conf. Ocean, Offshore & Arctic Engin.*, Honolulu, HI, USA.

Chapter 38
An Approach to the Validation of Ship Flooding Simulation Models

Egbert L. Ypma and Terry Turner

Abstract A methodology has been developed to validate a Ship Flooding simulation tool. The approach is to initially validate the flooding model and the vessel model separately and then couple the two models together for the final step in the validation process. A series of model tests have been undertaken and data obtained has been utilised as part of the validation process. Uncertainty in the model test measurements and the geometry of the physical model play a crucial role in the validation process. Therefore, an important element is an assessment of the uncertainties that play a role in this process together with how they propagate and eventually influence the end result. The aim was to develop a practical engineering approach trying to use the data that was available and making educated guesses where it could not be avoided. It is by no means intended to be a full-fledged theoretical elaboration on uncertainty propagation. This paper provides an overview of the methodology adopted for the validation of the ship flooding simulation tool and presents some of the preliminary results from this study.

Keywords Time domain · Flooding · Simulation · Damaged stability · Validation Uncertainty determination · Chaotic · Non-linear

38.1 Introduction

To accurately predict the progressive flooding of a damaged vessel and its effect on the ships motion two tightly numerical coupled methods are required to be developed: a vessel motion model and an internal fluid motion model. If the vessel changes its orientation, the internal floodwater distribution changes and vice versa. In addition,

E. L. Ypma (✉)
MARIN-Maritime Research Institute Netherlands, Wageningen, The Netherlands
e-mail: research@marin.nl; e.ypma@marin.nl

T. Turner
Defence Science & Technology (DST), Canberra, Australia

V. L. Belenky et al. (eds.), *Contemporary Ideas on Ship Stability*, Fluid Mechanics and Its Applications 119, https://doi.org/10.1007/978-3-030-00516-0_38

the changing distribution of the floodwater changes the dynamics of the vessel (centre of gravity, total mass and mass inertia).

An added complexity in trying to accurately simulate the flooding phenomenon is the highly non-linear, even chaotic nature of the flooding process. Small variations in this flooding process, e.g. how the water progresses through an opening, can influence the final result.

Due to the highly chaotic nature of the flooding process it is vital that the numerical model represents the experimental model as closely as possible. However, to obtain an exact numerical representation of the physical model is very difficult or even impossible. Differences may occur due to the limited accuracy of the production process of the physical model, modelling errors made in the translation from 'real' world to 'simulated' world, and the uncertainty (or limited accuracy) of the measurements. All these factors must be considered and could, in combination with the non-linear process, potentially lead to large differences between measurements and simulations. When there is a requirement to numerically model an actual full scale vessel other factors also must also be considered. The internal geometry of a full scale ship is extremely complicated and it would be very difficult to account for all the small details that may influence the flooding process. Issues such as leaking doors and collapsing air ducts are highly random events that can never be fully accounted for numerically. For this reason several "acceptable" assumptions are required to be made when numerically modelling full scale ships.

Due to the chaotic nature of the flooding process and the various areas of uncertainties in both model scale and full scale vessels, a method for progressive flooding tools must be developed to account for these uncertainties at an acceptable level. The following three questions need to be carefully considered when defining this method:

1. How can a validation process be defined such that it is possible to conclude whether a simulation tool is sufficiently accurate?
2. Is it possible to define general rules to model the internal ship-geometry in such a way that the simulation tool predicts extreme events sufficiently accurate (both statistically and in magnitude)?
3. What is the best way to deal with uncertainties in the validation process?

This paper will provide a brief overview of the numerical tool FREDYN (MARIN 2014), its progressive flooding modelling capability and will also outline the approach undertaken by both the Maritime Institute Netherlands (MARIN), and Defence Science and Technology (DST), Australia, for the validation of the progressive flooding module.

38.2 Ship Motion and Progressive Flooding Simulation Model

38.2.1 Background

The Cooperative Research Navies group (CRNav) , was established in 1989 to initiate a research program focussed on increasing the understanding of the dynamic stability of both intact and damaged naval vessels sailing in waves. The group has representatives from Australia, Canada, France, The Netherlands, the United Kingdom and the United States. The CRNav aims to increase the understanding of the stability of naval vessels from a more physics based approach rather than an empirically derived one. To manage this process the CRNav has formed the Naval Stability Standards Working Group (NSSWG). One of the objectives of the NSSWG is to investigate the applicability of the quasistatic, empirically based Sarchin and Goldberg stability criteria for modern navy vessels in waves and to develop a shared view on the future of naval stability assessment, see (de Kat et al. 1994) and Chap. 52 of this book.

The main focus of the NSSWG has so far been on intact stability and this has until now been a fairly comprehensive and complex task. With the development of the flooding module within FREDYN, future programs of work will be more and more focused on the damage stability of naval vessels in waves.

The flooding simulation model was developed and implemented by MARIN and funded by the CRNav group. In 2009 and 2010, Defence Science and Technology (DST), Australia, in collaboration with the University of Australia, Australian Maritime College (AMC), undertook a research program to support MARIN in the validation of the progressive flooding module.

In theory the flooding module can be interfaced to any time domain ship motion simulation program. Currently, however, it is interfaced to FREDYN, jointly developed by MARIN with the CRNav (McTaggart 1999). FREDYN was used for all the examples in this paper.

38.2.2 The Simulation Model

To enable an accurate simulation of the flooding of a damaged vessel operating in waves the simulation model describing the motions of the vessel and the model that determines the progressive flooding mechanism must be closely coupled to each other. Figure 38.1 shows an example of the typical information that is required to be interchanged between the simulation models.

In the scenario with significant flooding onboard a vessel, it is not unusual for sudden large changes in the vessels motions to occur. For an accurate simulation of this event both the flooding model and the vessel motion model must take this into account. The simulation must also be able to calculate a changing mass, centre of gravity and inertia over time. The accuracy of the roll damping model utilised is vital

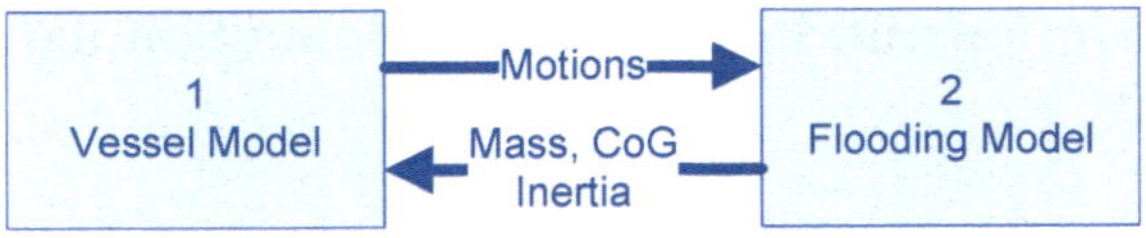

Fig. 38.1 Schematic showing example of information interchanged between vessel model and flooding model

in modelling this scenario. The relative wave height at the damage location is also required to be accurately predicted.

38.2.3 FREDYN

FREDYN is an integrated seakeeping and manoeuvring ship simulation tool capable of predicting large ship motions in extreme conditions. The development of FREDYN has been jointly funded since 1989 by MARIN and the CRNav. Over the years a substantial effort was made to validate and improve the code with model test experiments.

FREDYN uses a frequency domain tool as a preprocessor to calculate the frequency dependent added mass, damping and diffraction coefficients. Using a panellised hull form the program is capable of calculating the Froude-Krylov forces on the instantaneous wetted hull. An appropriate roll damping model can be selected from a number of different models, and it can be tuned to satisfaction when roll decay data is available. A lot of effort was spent over the years to validate the roll damping model. A wide variety of simulation components is available to model the vessel's propulsion, manoeuvring characteristics including: rudders, fins, skegs, bilge keels, various propeller types, waterjets and trim-flaps. The forces calculated by these sub-models are partly empirically based.

The environment can be modelled by wind and by various wave systems coming from different directions. The wind and wave systems can be specified by making a selection from one of the available spectra. Together with the user specified parameters this will generate a (random) wave sequence (or varying wind speed).

The equations of motions are based on Newton's second law: the forces and moments of all the sub-models are calculated, transferred to the space-fixed reference frame attached to the ship's centre of gravity, and summed. FREDYN allows for time-varying mass properties to be able to deal with the potentially large mass fluctuations caused by the flooding process.

FREDYN also has the capability to calculate the internal loads of the ship as a rigid body. This capability includes all the forces acting on the ship, including forces caused by the flooding of the internal geometry.

Over the last year the FREDYN was completely restructured resulting in a modular and highly configurable simulation program that can be easily extended by additional modules. The software framework underlying FREDYN is used for all the time domain simulation programs developed at MARIN, including the bridge simulators. This makes it possible to develop, reuse and share simulation components.

Verification and validation is an ongoing effort to improve the quality of the code and its results.

38.2.4 Flooding Module

The flooding module is used to calculate the flow of water and air through a user specified geometry. It assumes a horizontal fluid surface at all times. The effect of air-compressibility and its effect on the flow of water is fully taken into account.

The compartment geometry is represented by tank-tables that are generated prior to the simulation. A tank-table for a compartment tabularises the relationship between heel, and trim of the vessel, level and volume of the fluid in the compartment and its centre of gravity and inertia matrix. Linear interpolation on actual heel, trim and level values is used to find intermediate values.

Any number of openings can be specified connecting two compartments or a compartment to the sea. A single opening consists of four corner-points that specify the size and orientation of the opening. For each opening a constant discharge coefficient for water and a separate discharge coefficient for air has to be specified. If required, the user can specify a leaking pressure and area, a collapse pressure and a start-and/or stop time.

It is also possible to define a duct between two compartments, or between a compartment and the sea.

Pumps can be defined connecting two compartments. A pump is defined by a capacity diagram to describe the relation between the flow and the discharge head. A level switch can be specified to switch the pump on or off.

Bernoulli's equation for incompressible media is used to determine the flow velocity of fluid along a stream line from the centre of a compartment (A) to the opening (B).

$$p_B - p_A + \frac{1}{2}.\rho_w.v_B^2 + g.\rho_w.(h_B - h_A) = 0$$

In this application of Bernoulli's equation the ρ_w is constant and equal for all compartments. The velocity in point A, the centre of the compartment, is neglected. The variables p_A and p_B are the air pressures above the fluid in the compartment (A) and on the other side of the opening (B). After determining the velocity in the opening the mass flow through an entire opening is determined by integration over the height of the opening (along the local vertical):

$$\dot{m}_w = \rho_w \,.\, C_{d,w} \,.\, \int_0^H width(z)\,.\,v_B(z)\,.\,dz$$

where $C_{d,w}$ is the discharge coefficient for water specified by the user for this opening. It takes all the losses into account caused by contraction, pressure losses etc.

A similar procedure is used to determine the mass flow of air through an opening and the Bernoulli's equation for compressible flow is used:

$$\frac{p_0}{\rho_0}\,.\,(\ln(p_B) - \ln(p_A)) + \frac{1}{2}\,.\,v_B^2 = 0$$

The flooding process is considered isothermal, hence Boyle's law applies and thus the density of air is assumed to vary linearly with the pressure.

The pressure correction method developed for air and water flows by Ruponen (2007) is used to solve the coupled flow of fluid and air through a complex user defined geometry. The pressure correction method is using the equation of (mass) continuity and the linearised equations of Bernoulli to correct the water levels and air-pressures in an iterative method until the error in mass flow drops below a user specified minimum. Upon conversion both the equations of mass continuity and momentum are satisfied.

38.3 Model Tests

The model tests undertaken at AMC were performed in two phases. Prior to undertaking these experiments, it was expected that the tests were going to be complex and the lessons learned from first phase would have to be incorporated into the second phase of testing. The original plan was to focus in phase one on calm water and a simple compartment geometry, while the focus in phase two was planned to be on waves and a more complex compartment geometry.

Preliminary comparisons between the measured data and simulations showed a much more complex situation and much larger differences than were initially expected. Investigations were then undertaken to understand the cause of these differences and reduce the uncertainties found. The results shown in this paper are from Phase one.

The test programme was setup in such a way that the complexity of the scenario was gradually increased. A priori simulations were performed to determine the most interesting loading conditions. Prior to the flooding tests roll decays for various roll angles at zero speed were undertaken. The roll decay data was used to tune the roll damping of the simulation model. Fully constrained model tests were then performed with an initial heel and trim of zero to enable the assessment of the accuracy of the geometry definition of the compartment model and the location of the level probes.

Fig. 38.2 A photograph of the generic destroyer model

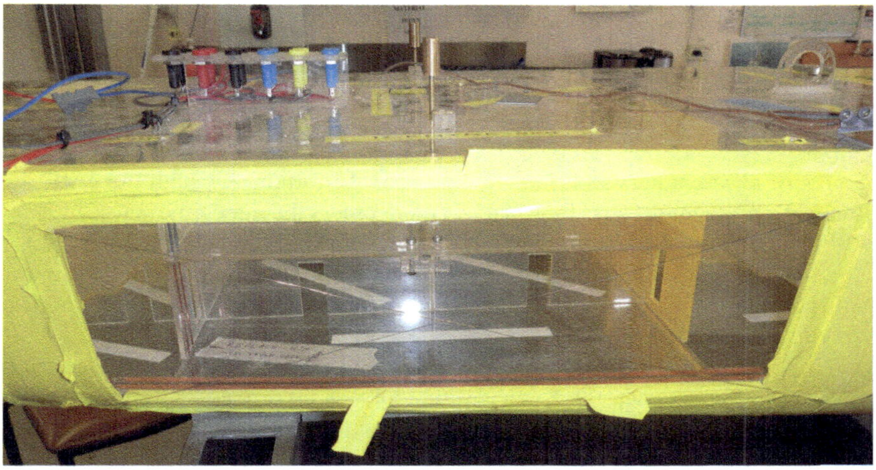

Fig. 38.3 A photograph showing the generic destroyer damage opening

The experimental model used, shown in Fig. 38.2, was a generic destroyer design (scale 1:45, length 3.27 m) with an internal compartment geometry. The model was built in such a way that the damaged compartment block could be replaced with a more complex compartment arrangement at a later stage (phase 2) to research the influence of the geometry complexity on the flooding process and vessel motions.

The compartment arrangement comprised of three decks, twenty five compartments of which six had level measurements and two had air pressure measurements, see Appendix for an schematic of the geometry configuration. In addition to these measurements, the six motions of the vessel were also recorded along with both internal and external video. The compartment block was located aft of amidships.

A large, three compartment damage opening was designed to cause very significant flooding. The flooding was initiated by puncturing a latex membrane that covered the large damage opening as seen in Fig. 38.3.

A similar geometry was used for model tests in the FLOODSTAND project (Ypma 2010). The tests defined for that project were done with a constrained model with a

Table 38.1 Vessel particulars (model-scale)

Property	Value	Uncertainty (2σ–95%)	Units
Length	3268.0	1.5	mm
Lpp	2962.0	1.5	mm
Beam (moulded)	412.0	1.0	mm
Taft	118.0	1.0	mm
Tfore	118.0	1.0	mm
Mass (intact)	68.63	0.10	kg
VCG	173.0	1.0	mm
LCG (fwd of APP)	1418.0	1.0	mm
Roll radius of gyration	128.0	10.0	mm
Pitch radius of gyration	715.0	2.0	mm

number of different heel and trim combinations. Unfortunately, both the AMC and FLOODSTAND model tests took place in more or less the same time period, making it difficult to apply the lessons learned.

A single loading condition was tested. The stability was tuned such that significant roll motions were expected to occur. See Table 38.1 for the vessel particulars (including uncertainties).

During the model tests many precautions were taken to reduce the uncertainty of the results. After each run a calculation tool was used to check the equilibrium levels in each (measured) tank with respect to each other and to the still water plane. After each run a check of the (level) calibration was done. A run was repeated when spurious results were suspected.

38.3.1 "As-Built" Compartment Geometry Assessment

After the tests the model was carefully remeasured to ascertain the 'as-built' condition, see Fig. 38.4, particularly the location and size of the openings and the vertical location of the decks. Both of these have a big impact on the flow of water (and air).

All compartment and level probe measurements were undertaken with the internal geometry fixed inside the model. The model itself was carefully fixed in a measurement rig and the datum reference was APP, Centreline and Keel. To be able to access the geometry holes were drilled where required, see Fig. 38.4. The measurements resulted in accurate data for the vertical locations of the decks, bulkheads, openings and hatches: the estimated uncertainty in all measurements was 0.25 mm (2σ–95%). From the measurements it could be concluded that the decks could not be considered as horizontal planes in the 'as-built' situations: the difference between maximum and minimum were as much as 6.0 mm for the vertical coordinate.

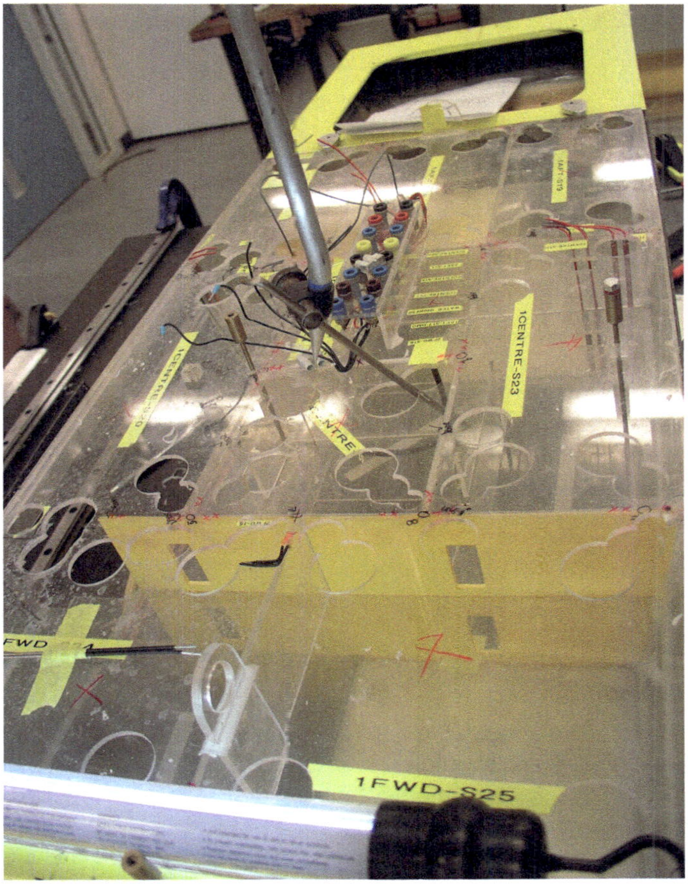

Fig. 38.4 "As-built" assessment

During the analysis of the fully constrained calm water model tests the water level measurements in starboard compartment S12 (the damaged compartment) and the port compartment S15, were added to the vertical coordinate of the decks within these compartments. In the constrained, horizontal condition the water level in these compartments, which are located on the same deck, see Appendix, should measure in the same plane: i.e. the still water plane. As the model is constrained with zero heel and trim, the still water plane will be 118.0 mm above the keel (the draught). This is shown in Fig. 38.5.

The probe measurements measure the water level above the lower deck of those compartments. By combining the vertical deck coordinate with the measured water level and comparing with the draught, a check on the consistency of the combined measurements could be obtained.

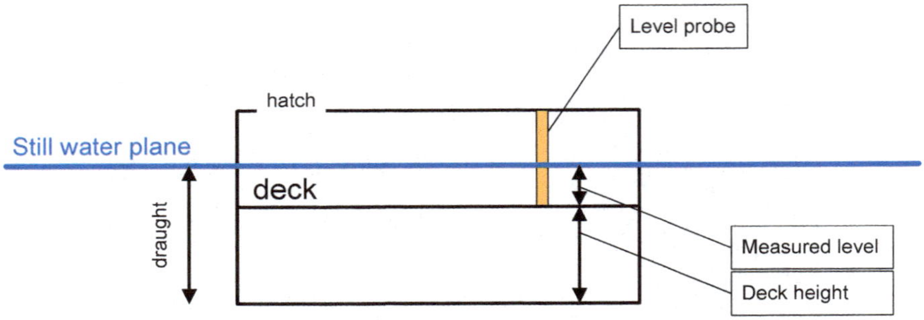

Fig. 38.5 Schematic showing measurement comparisons

Taking into account the measurement uncertainties of the deck location measurement (0.25 mm), the overall uncertainty in the water level measurements and the uncertainty in the draught measurement (1.0 mm) results in the best-estimate of the vertical deck location with an associated accuracy. The result is a vertical location of 91.2 ± 0.35 mm.

38.3.2 Measurement Uncertainty and Consistency

For the unconstrained condition a number of repetitions were done to check the repeatability of the results. In all these cases, the initial condition was meticulously checked to ensure that the same starting configurations were used. Figure 38.6 shows the roll versus time for these runs.

In the initial stages of the flooding process (until ~15 s) the roll angles are repeatable. After this stage, for some unknown reason, the roll motions of the vessel are not reproducible between repeat runs. The difference in the roll responses between runs is larger than the uncertainty values for roll measurements (0.1°, 95%).

The air-pressure measurement in compartment S12 was also checked for repeatability, see Fig. 38.7. A slight over pressure in S12 exists after ~40 s, indicating that an air bubble has formed above the water plane preventing the flood water level to rise to the still water plane. If the air-pressure is converted to a corresponding water column height by dividing the pressure by the density of the water and the gravity constant, then the calculated value of the water level should be 2 mm lower than the still water plane.

The last consistency check that was done was to convert the level measurements in the connected, ventilated compartments (S11, S15, S17) from the vessel coordinate system to the earth fixed coordinate system (correcting for roll, pitch and heave). All these compartments should have a water level that equals the still water plane. This is shown in Fig. 38.8 where the still water plane is marked by the line at a level of

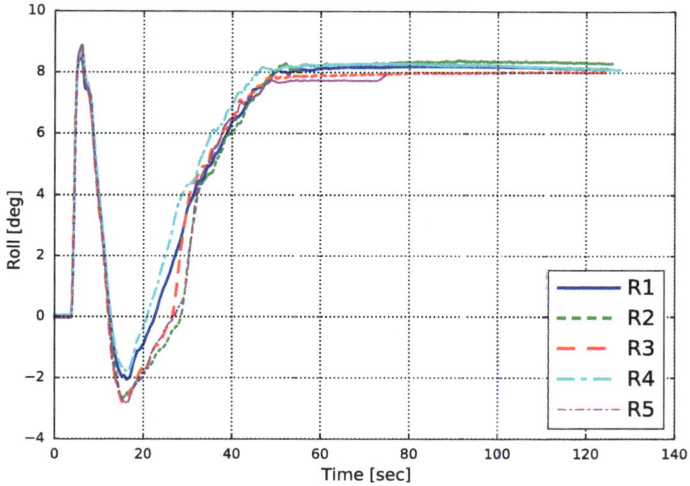

Fig. 38.6 Unconstrained runs—roll angle comparison

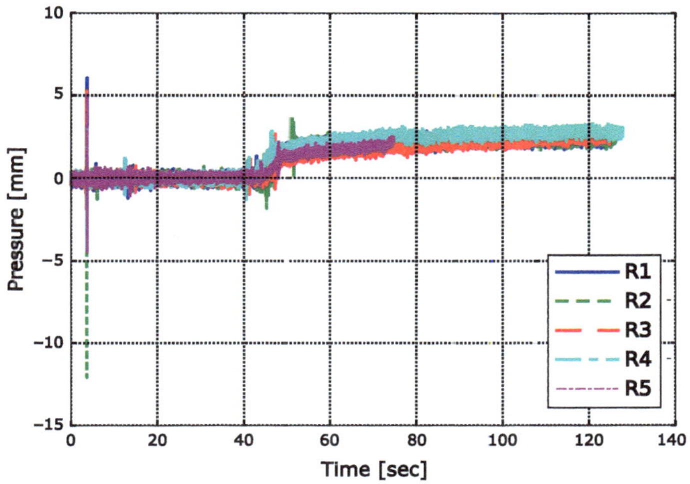

Fig. 38.7 Unconstrained runs—air pressure in S12

0 mm and the converted measurements for those compartments are given for five measurement runs in the unconstrained condition.

The level measurements that were used as input for the still water plane calculation and its uncertainty are the level measurements in the equilibrium condition of five different unconstrained runs.

The mean values (red dots) and the uncertainties are given (black lines). The uncertainties propagate through the whole calculation using a specifically designed

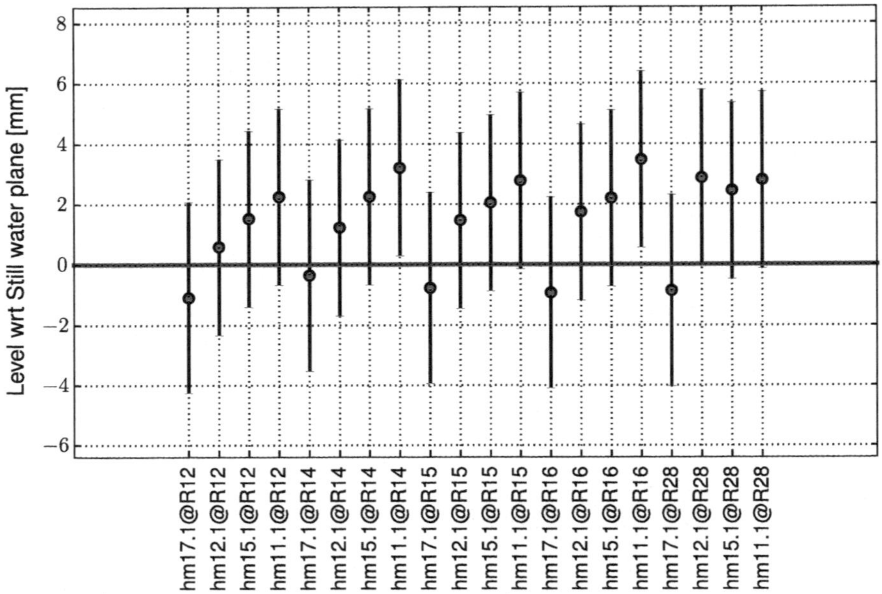

Fig. 38.8 Measured level comparison with still water plane

software module (Lebigot 2015). This means that the uncertainty in the calculated value, in this case the vertical distance to the still water plane, takes into account the uncertainties of all the measured values used in the calculation. The basis is a linear propagation uncertainties, based on:

$$\text{given}\, z = f(\underline{x}): \quad \sigma_z^2 = \sum_{i=0}^{n-1} \left(\frac{\partial f(\underline{x})}{\partial x_i} \right)^2 . \sigma_{x_i}^2$$

where σ_z^2 is the propagated standard deviation and $\sigma_{x_i}^2$ is the standard deviation of input parameter x_i. The vector $\underline{x}$ contains the measured parameters that are used in the calculation of $f(\underline{x})$ (probe location, position and attitude, level, draught). They are assumed to be uncorrelated.

Note that the propagated uncertainty, see Fig. 38.8, does not seem to depend upon the probe location as the variation in uncertainty between the probes is minimal. The 95% value is about 1.5 as high as the measurement uncertainty of the probe alone.

When the probe level values within each run are sorted with respect to the mean value in R12, a pattern seems to emerge. It cannot be contributed to a random measurement uncertainty: the differences seem to be consistent for each measurement run. A combination of one or more systematic uncertainties might have been the cause of this variation. Interestingly, the probes hm11.1 and hm15.1 are separated by a transverse bulkhead only (3 mm), yet there is a difference of ~1 mm in the mea-

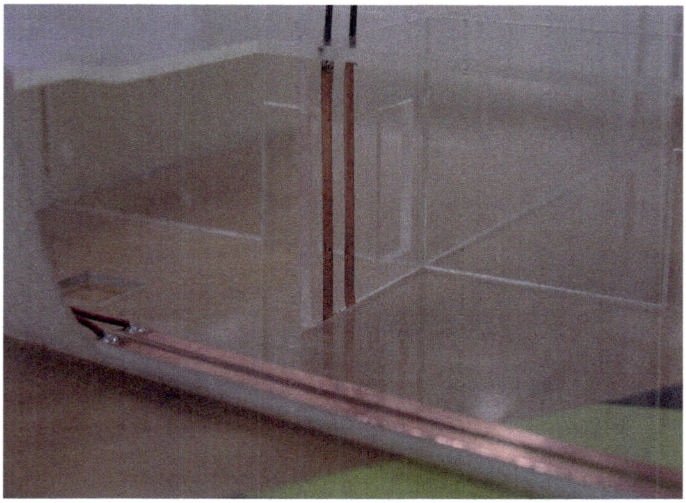

Fig. 38.9 Level probe hm12.1

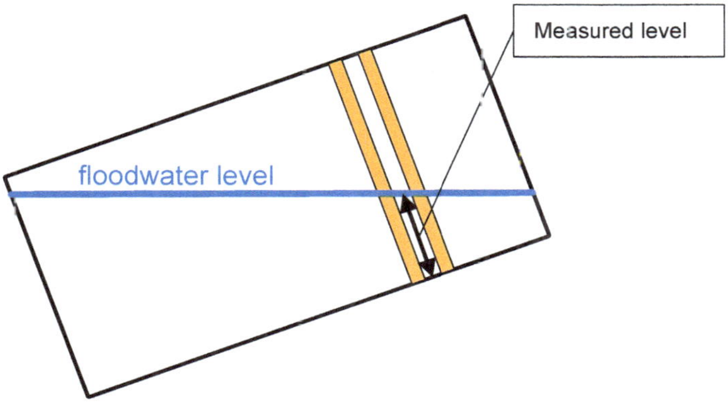

Fig. 38.10 Level measurement in inclined situation

surement of the still water plane. What might have contributed is that the calibration of the probes was done in a horizontal position whereas the equilibrium position is used to do the calculation. In the equilibrium position the geometry is substantially listed (~8°) and also has a moderate trim (~−0.7°). Note that all the level probes (except hm6.1 in the double bottom) were attached to a transverse bulkhead.

When inspecting the probe in hm12.1 in Fig. 38.9 it is clear that the width of the probe is considerable (~20 mm). The difference between the wetted length of the two copper strips in an inclined position might not have been negligible in the calibration, see Fig. 38.10. Unfortunately, this could not be checked anymore.

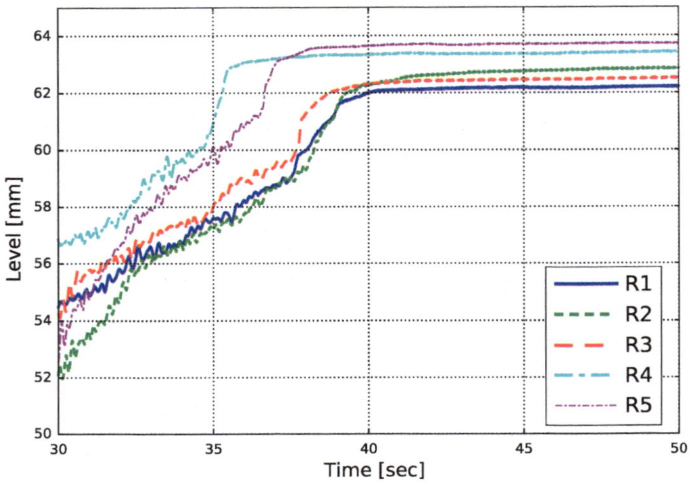

Fig. 38.11 Unconstrained runs—hm12.1 level comparison (zoomed in)

Interestingly enough the air pressure build-up in compartment S12 cannot be distinguished in Fig. 38.8: it was expected to have resulted in a consistently lower level measurement (~−2 mm) of hm12.1 because the compressed air above the floodwater prevents the level to rise to the still water plane. When inspecting the hm12.1 level measurement (see Fig. 38.11), focussing on the moment the water reaches the upper compartment deck, it seems that for all runs the maximum possible level measurement (the compartment height of 60.0 mm) is exceeded. Therefore, all hm12.1 level measurements seem to give a value that is too high. This might be caused by the construction of the probe wires through the upper deck of S12 (see Fig. 38.9): water might have leaked through the deck penetrations required to attach the electrical cable wires to the probe.

In general, although differences were found, the uncertainty bar of most of the converted deck locations includes the still water plane, see Fig. 38.8. Only hm11.1 seems to be a bit of an outlier.

All the causes and effects described here are difficult to model in the simulation. However, they should somehow be taken into account and assessed when comparing the simulation results and interpreting the differences between simulations and measurements. In addition, they will also play a role when modelling a real life scenario. In fact, even more uncertainties will play a role here, such as open/closed doors, collapsing structures and floating and moving debris, to mention some. All of these combined might have similar effects on the flooding process as has been described in (Ruponen 2017). In his paper he describes the effect of a random collapse pressure and open/close status of fire doors on the time to capsize of a passenger ship for various damage scenario's. There seems to be a wide variation in the time to capsize, especially for the grounding damage scenario.

38.4 The Validation Process

38.4.1 Introduction

As previously discussed it is extremely difficult to model and capture all the phenomena that occur during the flooding of a vessel. For this reason there will always be slight variations in the experimental results when compared to the numerical results. There are several factors that contribute to the differences observed. These include:

- Measurement accuracy (motions, levels)
- Determination of the hydrostatic and dynamic properties of the model
- Production accuracy of the test model (both the internal and external geometry)
- Choices made during the modelling of the internal geometry (deck and bulkhead positions and thickness, permeability of compartments, discharge coefficients)
- Imperfections caused by mathematical modelling (empirics!).

The uncertainties can be grouped in three main categories:

- Uncertainties caused by the physical model and the measurements
- Uncertainties caused by flooding model algorithm
- Uncertainties caused by the vessel model algorithm.

The first category plays a role during the model testing, the second two are tightly coupled and play a role during the simulations. The result of the validation process will be a comparison between the measurements and the simulation data. In general, when they are 'acceptably' close then the conclusion is justified that the simulation application performs well. The first key problem in view of the nature of the process and the uncertainties that play a role is how to define 'acceptable'. The second key problem is that if the result is not 'acceptable' then which sub-model has to be changed to improve the result? Changing a sub-model might involve changes to the modelled geometry (e.g. change the deck height, change discharge coefficient values) or the algorithm itself (e.g. model the momentum of the incoming water).

To disentangle the tightly coupled vessel and flooding simulation system as shown in Fig. 38.12.

To a configuration that splits the tight coupling between the vessel model and the flooding model in a serial approach, see Fig. 38.13.

The issue with the uncertainty covered by the physical model and measurements can be considered by firstly having a clear understanding of the uncertainties involved, secondly by designing a clever model test setup and programme, and thirdly, a meticulous preparation and execution of the model-tests. It is also important to undertake a series of simulations to determine the influence that these uncertainties have on the overall result. Worthwhile mentioning is also the approach taken by Khaddaj-Mallat et al. (2010). He applied DoE methodology (Design of Experiments) to efficiently assess the effect of parameter variations on the flooding process of a partially submerged geometry.

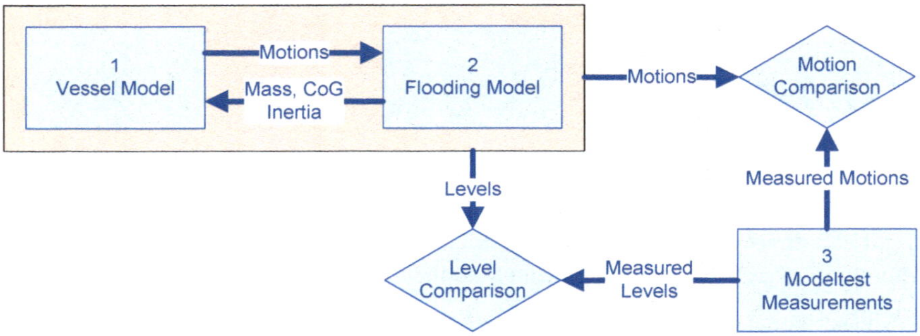

Fig. 38.12 Comparisons in the tightly coupled configuration

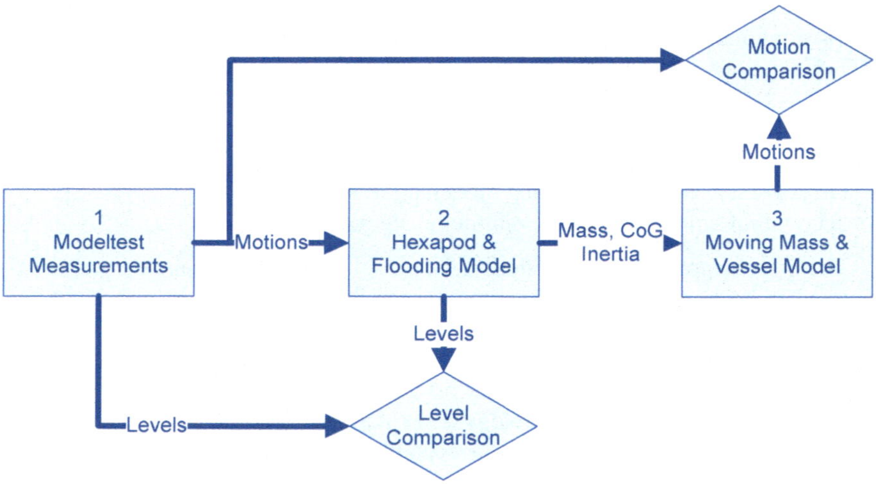

Fig. 38.13 Comparisons in the decoupled configuration

38.4.2 Component and Interface Verification

Prior to using the measurements in this approach, the validation method was verified by replacing the model test measurements with the data obtained from a simulation with the fully coupled system. The fully coupled system provides a set of consistent data (motions and floodwater levels). The process flow of this test is shown in Fig. 38.14. Here, the first step is to do a single simulation with FREDYN and the flooding module. This gives the ship motions, the varying floodwater mass, its centre of gravity, its inertia and the floodwater levels in each compartment.

These motions are input to the hexapod application driving the same flooding geometry as used for FREDYN. This will again produce the total floodwater mass, its centre of gravity and its inertia. Also the floodwater levels per compartment

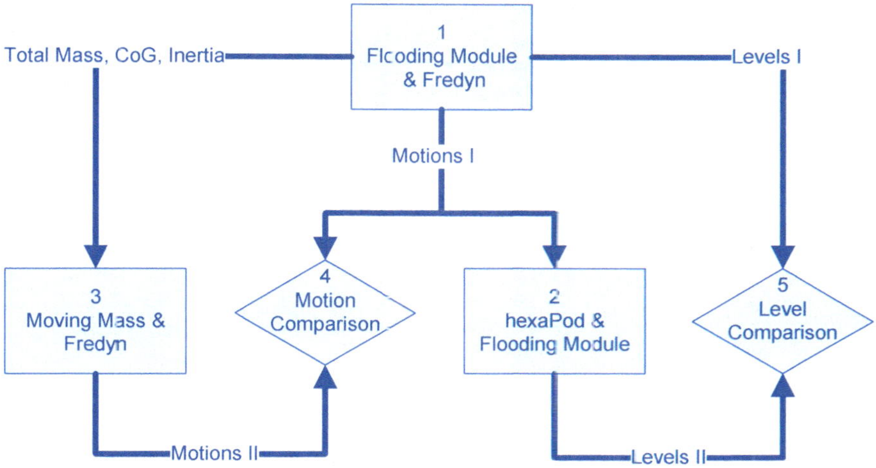

Fig. 38.14 Interface verification

are compared. This is to check whether the motions in the hexapod application are correctly used to drive the flooding module.

Finally, the total floodwater mass, its centre of gravity and inertia are used as input to drive a "Moving Mass" simulation where the same ship as used for the FREDYN simulation is exposed to the changing floodwater mass. The outputs are the motions of the ship. This is to check if a changing mass (and its CoG and inertia) is correctly applied to the vessel model in FREDYN.

The motions as calculated by steps one and three in Fig. 38.14 should compare very well as modelling errors play no role: the same geometry is used. It is a test of the interfaces and coordinate transforms involved.

The same applies to the comparison of the levels calculated in step one and the levels calculated in step two.

38.4.3 Steps in the Validation Process

The problems caused by the tight coupling between the flooding and vessel motion models can be alleviated by separating the validation of the two models. To be able to do that the validation process has to be split into several steps:

1. **Fully constrained model**

In calm water, fully constraining the model in a pre-described heel, trim and draught allows for a check of the flooding module without the dynamics of the vessel. Setting up the vessel model in a fully constrained condition and flooding it at a series of cleverly chosen draughts will allow for a crosscheck of the as-built deck height. The

Fig. 38.15 Flowchart showing the approach for the prescribed motion phase

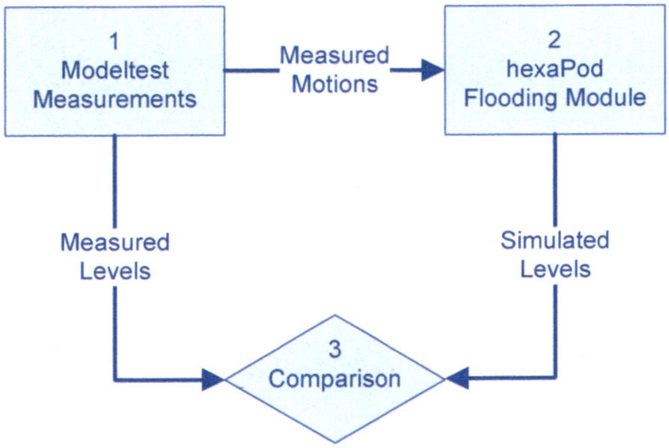

water levels inside connected, flooded compartments will rise to the still water plane in the equilibrium position provided the air is free to flow out of that compartment. This is described for a single draught in Sect. 38.3.

By flooding the constrained model at a heel and trim value different from zero, the performance of the flooding module for inclined openings can be validated.

2. **Force the flooding geometry with the measured motions from the model test**

In this step (shown in Fig. 38.15) the motions measured experimentally are prescribed onto the flooding module. The water levels and volumes in each compartment are calculated over time (HexaPod approach, see also see also Khaddaj-Mallat et al. (2010) for a Hexapod application in the model-test). These levels are then compared to those obtained experimentally. If the rise of the fluid-level in a compartment is different between the measured and the calculated results, the discharge coefficient(s) of the compartment openings can be tuned until the simulations better predict the measured values. However, for geometries with an increasing number of compartments and openings this will become an increasingly difficult, if not impossible task due to the many dependencies between the water levels, fluid and air velocities in the different compartments.

This process step is a validation that the flooding model is working satisfactory and a check if the modelling assumptions are plausible. Some examples of these assumptions are: the quasi static approach does not take sloshing into account, and the effect of inflow momentum of the floodwater in the initial stage is also neglected (Manderbacka and Ruponen 2016). Also, the initial violent inflow is modelled in the simulation by the Bernoulli equation. The assumptions that allow the application of this equation are violated to a certain extent (see Sect. 38.2).

3. **Use of a time varying floodwater mass and centre of gravity**

One of the outcomes of step 2 is a single, time varying, floodwater mass (having a centre of gravity, inertia and a weight) that moves in time through the vessel. This output can then be used as input to the vessel model. This approach is allowed as

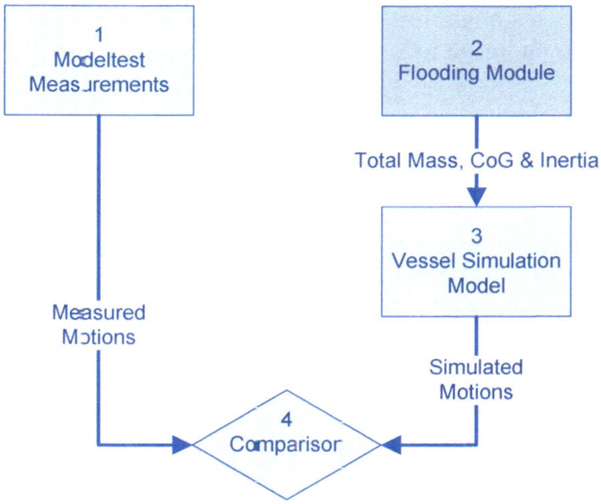

Fig. 38.16 Flowchart showing the approach for the moving mass phase

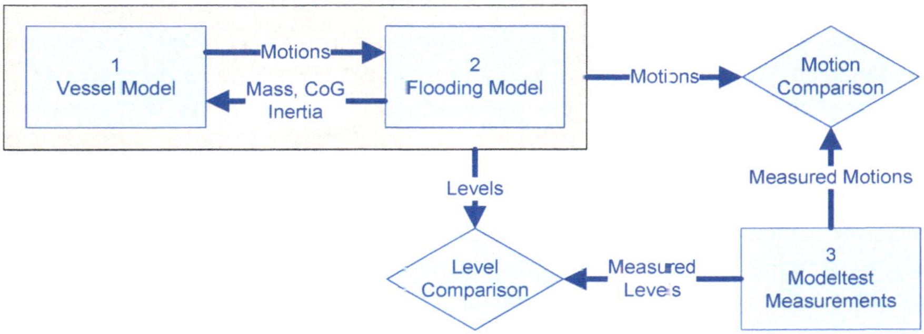

Fig. 38.17 Flowchart showing the approach for the full simulation phase

long as the vessel is only excited by this wandering and changing floodwater mass and will not be applicable when the vessel is also subject to other external forces such as waves.

The outcome of this step, the vessel motions, are then compared with the measured vessel motions obtained from the model test. A schematic diagram depicting this process is shown in Fig. 38.16. It is a test of the interface between the flooding module and the vessel simulation program.

4. **Close the loop and combine both models**

The final step in the validation process is to undertake a simulation with the complete model i.e. vessel model and flooding model, and compare the predicted levels and motions with the measured model test data. A schematic outlining this process is shown in Fig. 38.17. This is the same situation as depicted in Fig. 38.12.

Differences between the output of this step and the previous steps give an indication of the dynamic properties of the coupled system. Coupling two highly non-linear systems might result in an apparent chaotic system where initial, small differences in input result in large differences in output.

38.5 Validation Results

The validation methodology approach, as previously described, has been applied to the data obtained with the first phase of the model test done at AMC. The validation was performed in following steps:

1. The vessel roll damping.
2. Component and interface verification.
3. Fully Constrained.
4. Forced Motions.
5. Moving Mass.
6. Unconstrained Analysis.

All plots and other data are given in full scale. Where it has added value, the roll is plotted as a reference signal at the bottom of each plot. It makes it more easy to identify events since the roll is the most dominating motion. The uncertainty range for the measurements is given where this is applicable. Unfortunately, the uncertainty of the output for the simulation model, given the input and its uncertainties, is unknown at the moment: it will certainly not be zero but an actual value is difficult to ascertain. Not only the uncertainty of the input data play a role here (geometry of the hull and appendages, internal geometry, etc.) but also the choices and simplifications used to implement the vessel and flooding models. The last determine the accuracy of the simulation model. Finally, the implementation of the algorithms itself will have an influence. For complete validation this aspect is also very important but unfortunately it is outside the scope of this document.

38.5.1 Vessel Roll Damping Model

The roll decay data was used to tune the roll damping model. All flooding tests were done at zero speed hence only the roll damping for that speed required to be tuned. Several roll decays were performed, each with a different initial angle. Only slight tuning of the radius of gyration, k_{xx}, was required to have a good resemblance of the measured and the simulated data for all initial roll angles. A typical comparison between the numerical and the measured roll decay is shown in Fig. 38.18. The initial roll angle is close to the equilibrium roll angle of the non-constrained simulations.

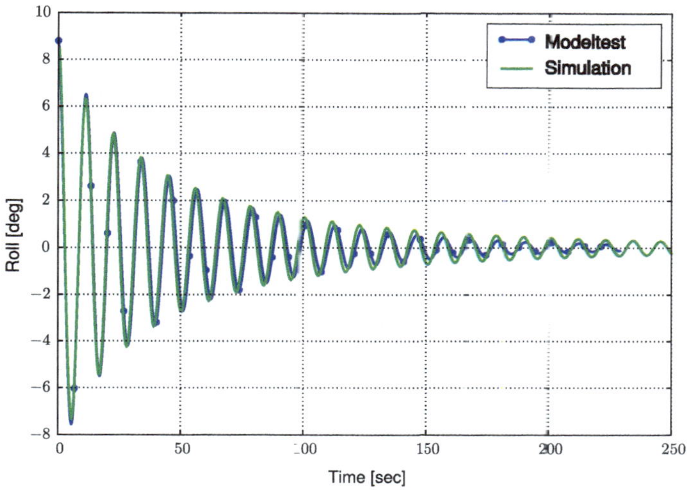

Fig. 38.18 Roll decay for initial angle of 9.0°

38.5.2 Component and Interface Verification

First the "moving mass" module and its interface were tested. An unconstrained FREDYN simulation was done, using the flooding module to calculate the changing mass and its location as a function of time. This output was then used as input to a FREDYN simulation, now without the flooding module but with the moving mass module. The motion output of the second simulation should equal the motion of the first simulation. The roll signal is used as a criterion. Figure 38.19 shows the result of this check and confirms that the influence between the flooding module and vessel simulation program is correct.

The (software) hexapod module was checked by using the motion output of the first simulation and use these as input to drive the hexapod with the flooding module attached. If all is well, the flooding model should output the same floodwater mass and levels as were produced by the first, unconstrained simulation. Results from this are shown in Fig. 38.20.

There are some minor differences between the predicted levels due to minute differences in the heave signal caused by a coordinate transformation. This causes the iterative algorithm in the flooding module to converge in a different way. It is an indication of the sensitivity of the process.

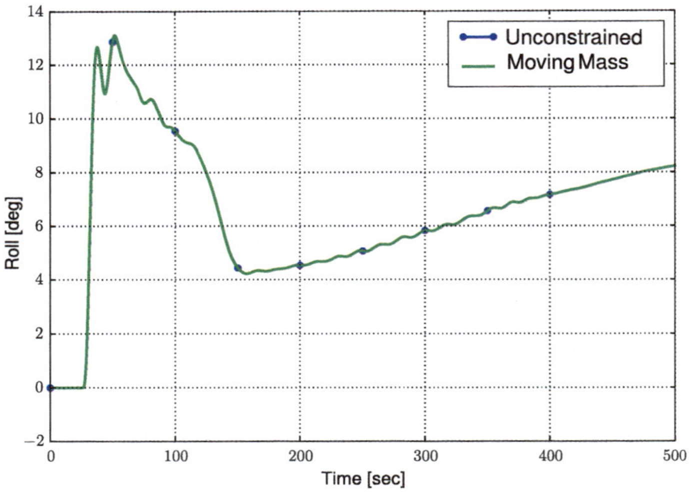

Fig. 38.19 Component and interface verification—moving mass and unconstrained

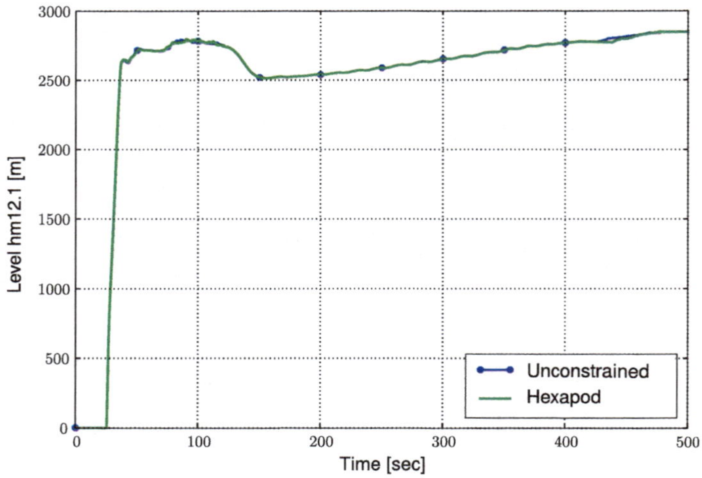

Fig. 38.20 Component and interface verification—level hm12.1

38.5.3 Fully Constrained

During the test the model was fully constrained at zero heel and trim with the draught set at 5.31 m (full scale). This test gives an indication of the performance of the flooding module. The influence of the motions of the vessel on the flooding process is eliminated at this stage.

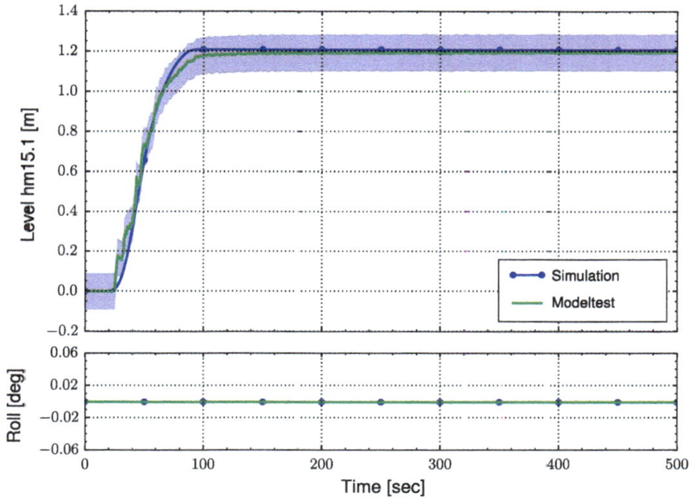

Fig. 38.21 Comparison between predicted and experimental water levels (hm15.1)

Figure 38.21 shows an example of both the simulation and experimental time traces of the water level in a compartment. This compartment is indirectly flooded through a number of openings and other compartments. The air pressure in this compartment remains ambient as the compartment is open to the outside. The measurement uncertainty range is displayed together with the measurement signal itself.

The simulation of the flooding of this compartment shows excellent comparison to the experimental data. The arrival of the water at the position of the probe, the filling rate of the compartment and the final equilibrium level are all predicted very well.

Inspecting the results for compartment S06 (in the double bottom), see Fig. 38.22, shows a large difference between the numerically predicted and the experimental results. This compartment will flood through a exactly horizontal hatch (due to the model being constraint). Video footage of this run showed that a vortex formed during the test-run, allowing the compartment to flood relatively fast. In the simulation a “bubbling” process is simulated resulting in much slower flooding. In other runs however, this bubbling was also seen in the model-test. Distinguishing between these two processes will be extremely difficult in using a Bernoulli based simulation model. It is believed however, that on a moving and flooding ship the vortex process will not have time to develop.

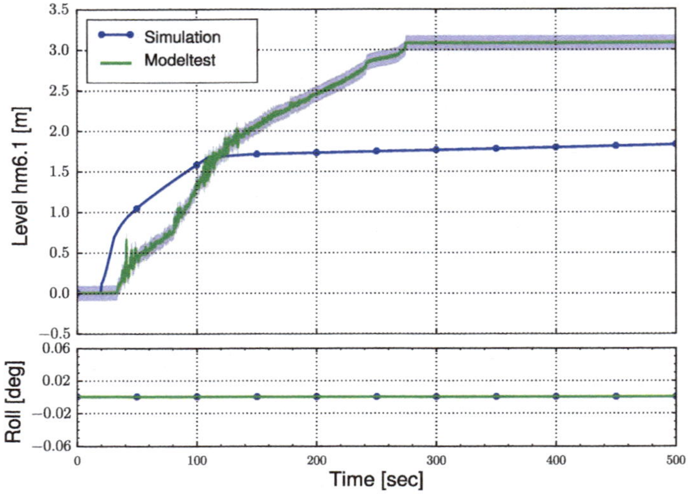

Fig. 38.22 Level comparison of double bottom compartment S06

38.5.4 A Software Hexapod—Forced Motion

For the forced motion analysis the experimental model motions (heave, pitch roll) were used to drive the flooding model. To be able to do this a test application, was developed, which models the functionality of a 6D HexaPod, and which loads a file with motion data and the flooding model. It forces the motion on the flooding model resulting in the regular flooding module output (flooding mass, its centre of gravity and compartment levels as a function of time).

The measured roll, pitch and heave of the selected model test run are shown in Fig. 38.23. All these signals are plotted together with their uncertainty range. However, for the roll and heave the uncertainty values are small relative to the signal itself (95%: 0.1° for pitch and roll, and ~20 mm for the heave) and they are therefore hardly visible in the plot.

Figure 38.24 shows a comparison between the simulation and model tests results for the starboard side, damaged compartment S12 along with the expected uncertainty.

There is a difference between the model test and the simulation slightly outside the limits of uncertainty (light blue range). The compartment S12 which Fig. 38.24 is referring to is directly connected to the sea and is fully ventilated, therefore at equilibrium, the distance between the water level in the compartment and the still water plane should be zero. As can be seen in the plot, at the end of the violent, initial flooding (~50 s) the simulated value is slightly under-predicted and outside the estimated measurement uncertainty. This is probably due to the fact that the simulated floodwater immediately equalizes whereas this takes more time during the model-test. Consequently, a slight pile-up of floodwater in compartment S12 is

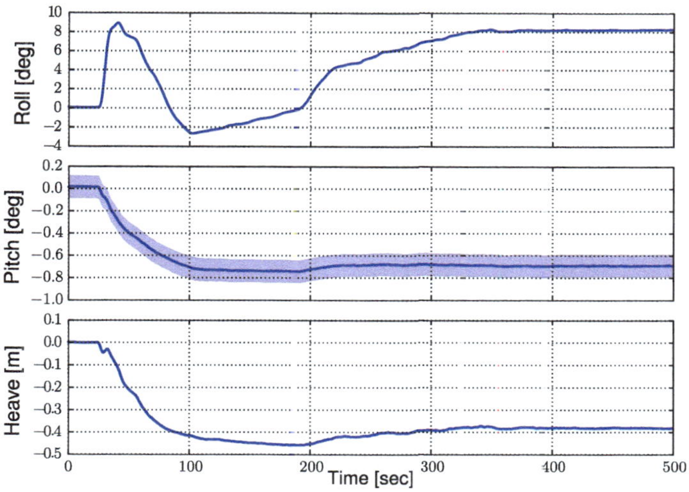

Fig. 38.23 Unconstrained model-test—roll, pitch and heave signal

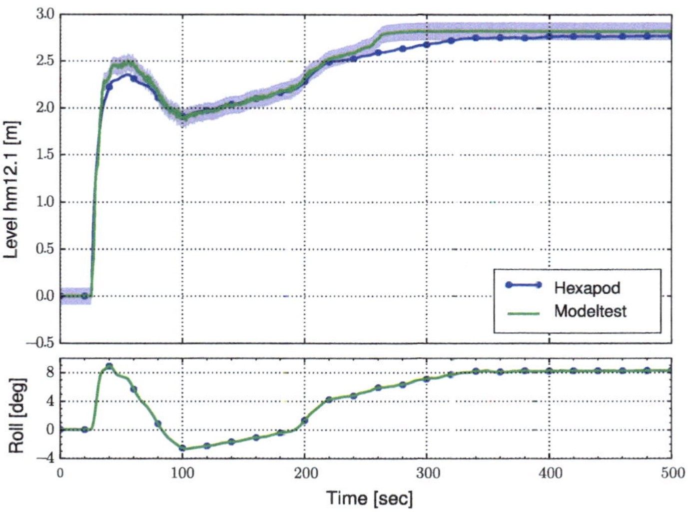

Fig. 38.24 Comparison between hexapod and the experimental water levels (hm12.1)

the result before the water floods to the connected compartments. The differences around 250 s are more difficult to explain. It might be caused by a difference in down-flooding in S06. In contrast to the constrained situation (see Fig. 38.22) the down-flooding process in compartment S06 is now much quicker in the simulation, see Fig. 38.25. An explanation is probably that the vortex that was observed during the constrained run, did not have time to develop in the unconstrained condition.

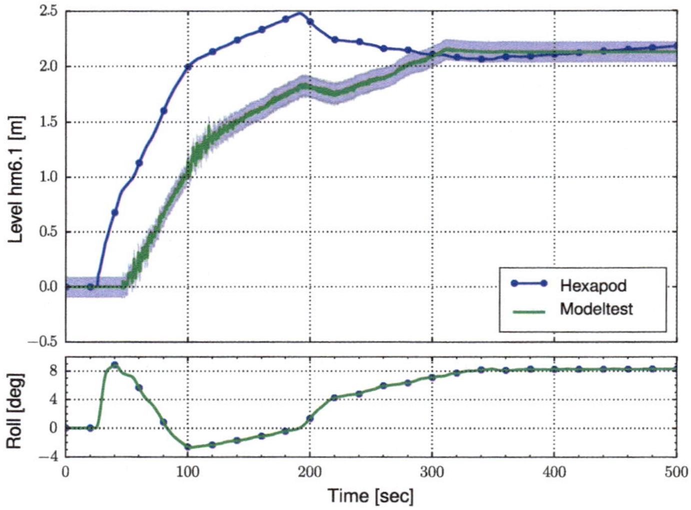

Fig. 38.25 Comparison between hexapod and the experimental water levels (hm6.1)

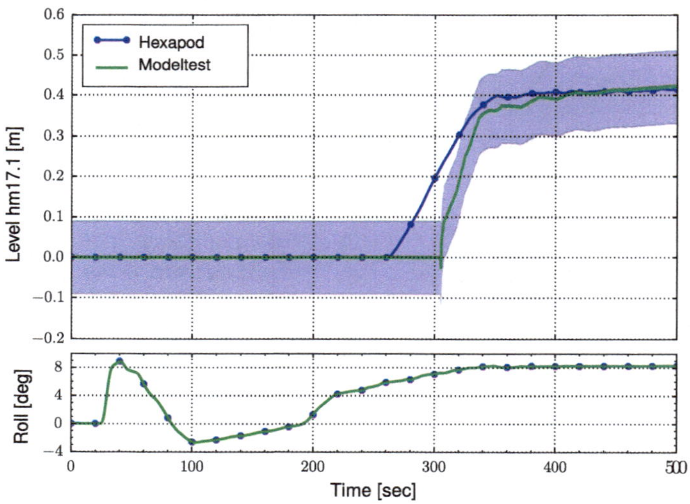

Fig. 38.26 Comparison between hexapod and the experimental water levels (hm17.1)

This will have had an effect upon the floodwater volume, and hence the level, in compartment S12 (connected via S16 to S06). Unfortunately, the level in S16 was not measured, which would have given a further opportunity to check this assumption (Fig. 38.26).

Note also that the down-flooding of S06, see Fig. 38.25, starts sooner in the simulation than in the model-test: the difference is around 25 s (full scale). Again it

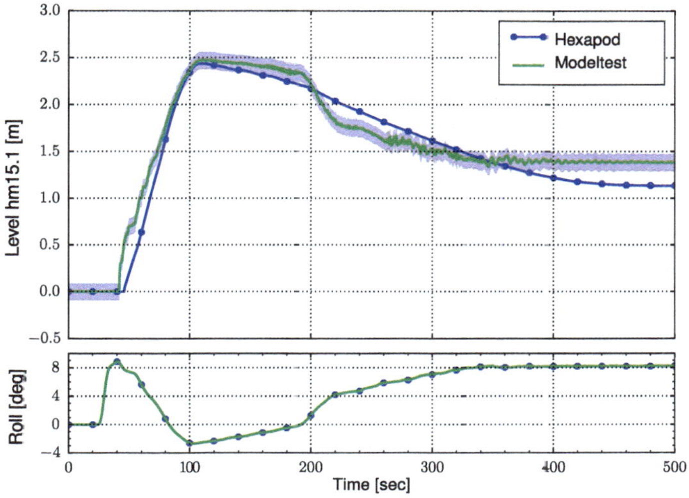

Fig. 38.27 Comparison between hexapod and the experimental water levels (hm15.1)

is likely to be caused by the same lack of fluid dynamics in the flooded compartments that could describe the difference in level hm12.1 between 30 and 60 s, see Fig. 38.24.

The same phenomenon, 'delayed' flooding, is observed when comparing the simulated with the measured level in compartment S17 (one deck above S12):

Again, the flooding of this compartment starts considerably sooner in the simulation than in the model-test. However, because this moment occurs later in the flooding process, it will also have been influenced by earlier events (such as the earlier flooding of S06). Also note that the relative value of the measurement uncertainty is substantially larger in relation to the measured value in this case when compared to for example the situation in hm12.1 (see Fig. 38.24).

The performance is a bit worse in port compartment S15, see Fig. 38.27. This compartment is not directly connected to the starboard damaged compartment S12 but floods via the central compartments S13 and S14. In the constrained situation the levels of the model-test and the simulation were almost exactly the same for this compartment. Now the initial performance is good but after ~100 s the level is decreasing steadily until the equilibrium. This is the moment the vessel starts to roll back from the maximum angle from port to starboard again, see the roll reference plot in Fig. 38.27.

38.5.5 *Moving Mass*

From the output of the Hexapod simulation a time record of the total flooding mass and its centre of mass is obtained. These values are then used to excite the vessel

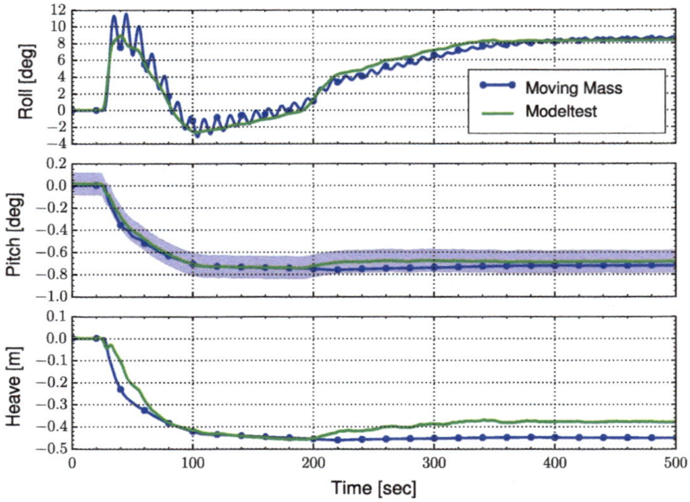

Fig. 38.28 Comparison between the vessel motions for the moving mass and the experimental results

model. The roll, pitch and heave motions of the simulated vessel can then be compared against the model-test values. Figure 38.28 shows the comparison between the experimental and simulated results of the vessel motions. Note that the uncertainty in roll and heave is barely visible. (roll and pitch 95% value is 0.1°, heave ~0.02 m)

The initial numerical roll angle is larger than the measured value. Overall, the trend in the roll angle compares favourably, but starting from the moment the maximum roll angle is reached fluctuations in the roll angle are emerging. They slowly dampen as time progresses. The oscillations have the same period as the natural roll period of the vessel. It could be an indication that there is more damping in model-test. The roll damping was carefully tuned in the intact condition and the simulation roll-decay matches the model-test roll decay almost perfectly (see Fig. 38.18). The origin of the roll oscillations shown in Fig. 38.28 might lie in the internal damping of the floodwater in the internal geometry. The internal damping is present in the model-test but not modelled in the flooding model.

Neglecting the momentum the floodwater, as in the flooding model, in the initial stages of the test, described by Manderbacka and Ruponen (2016), might also play a role. In his paper he describes influence of the impact of floodwater on a longitudinal bulkhead on the motions of a ship.

Figure 38.28 shows a good agreement in the first 200 s between the numerically predicted pitch and the experimentally obtained pitch. Although within the uncertainty range, there is a slight divergence between simulations and model-test after 200 s. The same trend can be observed when the heave is compared.

Even in the decoupled configuration it is difficult to distinguish cause and effect for the divergence between simulation and model-test. Combined, these differences

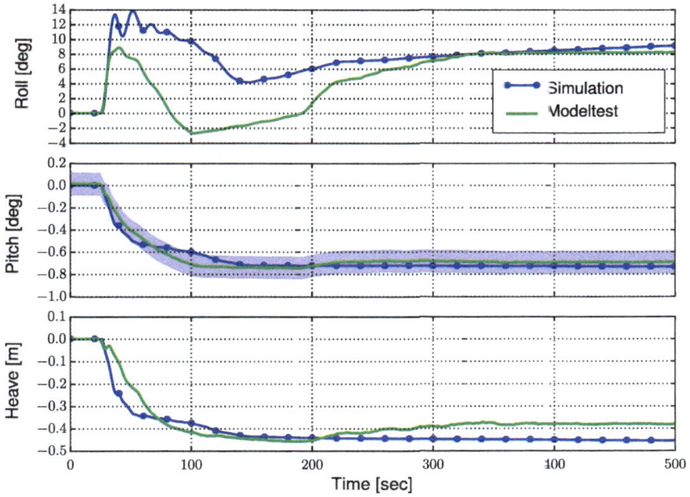

Fig. 38.29 A comparison between the vessel motions for the simulated and the experimental analysis

seem to suggest that there is more floodwater in the simulated model and its centre of gravity is located more forward.

When comparing Fig. 38.26 (the level in S17) with Fig. 38.28, then the difference in results seems to be consistent: S17 is located aft in the compartment geometry. Hence, a larger heave and a more negative pitch value will result in a higher water level in this compartment (between 250 and 400 s). The effect on roll is relatively small since the floodwater volume in S17 remains rather small.

38.5.6 Unconstrained, Fully Coupled Run

The final stage in the validation methodology is the complete coupling of the numerical flooding model with the numerical vessel model. Figure 38.29 shows a comparison between the numerically predicted roll motion and the experimentally obtained values.

It is evident that there is a significant difference between the numerically predicted roll motion and the experimental results. The maximum roll angle, the 'roll back angle' and the equilibrium angle: all are different. In addition, the simulation show roll angle fluctuations that are visible after the initial stages of flooding, that are not seen in the model test results. The oscillations are consistent with the roll behaviour in the last step of the decoupled system, albeit less severe and they seem to be more dampened. Again, a difference of internal damping between the model-test and the simulations might be the cause.

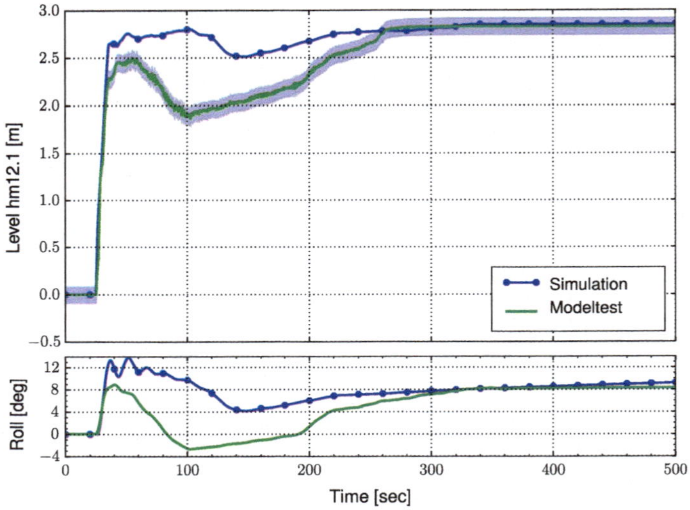

Fig. 38.30 A comparison between the level in S12 for the simulated and the experimental analysis

Finally, the simulation has not reached a stable equilibrium after 500 s, whereas the model-test seems to stabilise around 400 s. This is related to the level still rising in compartment S17 (see Fig. 38.33). But, cause and effect are difficult to distinguish.

Differences are also observed for the pitch and heave data (see Fig. 38.29).

In the pitch signal the difference is less significant and it shows more resemblance with the differences found when looking at the last step of the decoupled system (the moving mass).

The heave signal of the simulation looks very similar to the heave signal of the simulation that was the result of the last step in the decoupled situation (moving mass, see Fig. 38.27).

Overall, and not surprisingly, due to the difference in motions the levels in the compartments also show different behaviour.

In Fig. 38.30, the level in the damaged compartment S12 is shown. The initial level is predicted quite well, until around 50 s, thereafter a deviation occurs which is completely in line with the roll angle differences, see Fig. 38.29. The equilibrium level compares pretty well.

The differences increase for compartments further away from the damage opening. For example S15, see Fig. 38.31.

Surprisingly, compartment S06, filled by down-flooding from S16, shows a slightly improved behaviour, especially when Fig. 38.32 is compared with Fig. 38.25. By all means it is not perfect, but the trend has improved, as has the correct equilibrium level.

The results for compartment S17 show larger differences; due to the large roll angle in the simulation, floodwater enters here in the initial phase of the simulation, see Fig. 38.33.

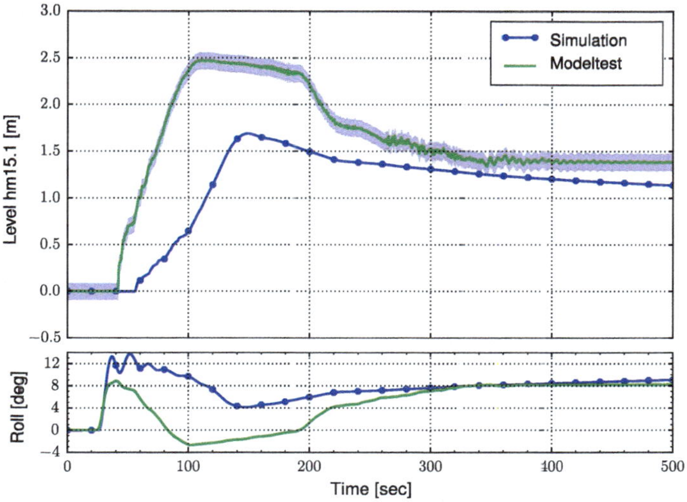

Fig. 38.31 A comparison between the level in S15 for both the simulated and experimental analysis

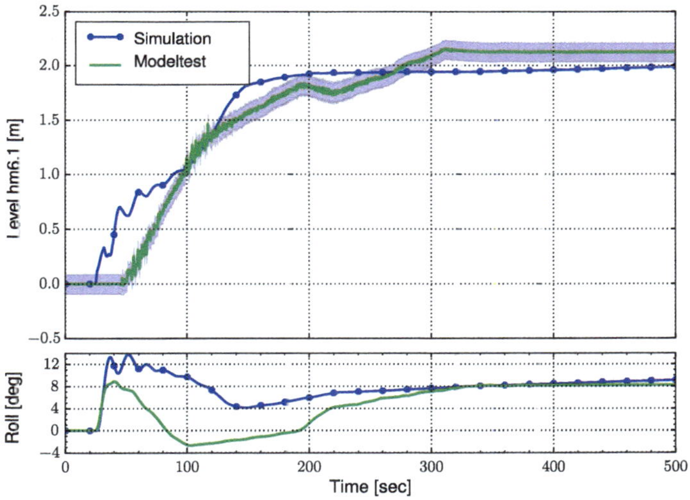

Fig. 38.32 A comparison between the level in S06 for both the simulated and experimental analysis

Thereafter, the simulation results will be different. Comparing Fig. 38.33 with the results for S17 in the decoupled situation (see Fig. 38.26) reveals that aside from the delay of ~50 s between the simulations and the model-test in Fig. 38.26, the characteristics of the signals in these figures reveal other differences: especially noteworthy is the rising level in Fig. 38.33 towards the end of the simulation. The same effect is seen in the roll angle: whereas the model-test seems to have stabilised,

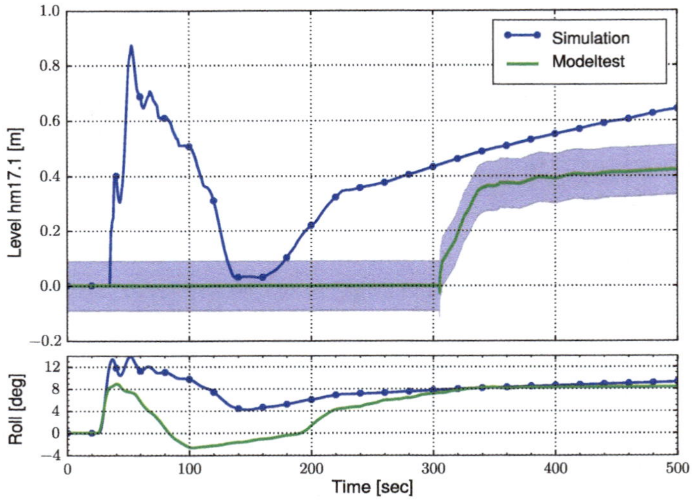

Fig. 38.33 A comparison between the level in S17 for both the simulated and experimental analysis

the roll angle in the simulation still increases. Note that a level of 0.4 m in S17 for these heel and trim values corresponds to a floodwater volume of ~12 tons.

Reviewing all the differences in motions and levels between the simulation and the model test makes it difficult to relate events in the motions with events, or changing characteristics in the levels – and vice versa. In the end that makes it difficult to draw firm conclusions from the comparisons other than that they do not match entirely well.

38.6 Discussion and Conclusions

This paper has provided an overview of the methodology that has been adopted by MARIN and DST to validate the progressive flooding modelling capability within FREDYN.

The flooding of a vessel is a tightly coupled, non-linear, chaotic process. Small changes in the input might result in large differences in output. This applies to both the model-test and the simulation. To minimize the effects implied by this non-linearity it requires a meticulous preparation and execution of the model-test and for the simulation it will require a very detailed knowledge of the "as-built" condition of the model-test model and a meticulous modelling of this model in the simulation.

To effectively deal with these challenges in the validation, a methodology was developed to decouple the vessel model from the flooding model. The methodology included a phased approach where initially both the flooding model and the vessel model were validated separately and then coupled together for the last validation stage. A series of model tests were undertaken in support of this validation process of which at this stage only a part has been analysed.

During the model test program the maximum effort was spent by AMC to reduce the uncertainties, both in the model construction and in the execution of the model-test program. The "as-built" construction was assessed as accurate as possible, and during the execution of the tests a lot of attention was paid to the calibration and the preliminary detection of systematic errors. The resulting dimensions and their associated, estimated uncertainties will be something that cannot be improved upon much further. If a better "as-built" accuracy is required than a larger model is inevitable. Different construction techniques, such as 3D printing might also be worth investigating.

Even the "simple" internal geometry that was used for this validation study was rather complex. Many phenomena, such as vortex ventilation and bubbling ventilation, played a role at the same time during the flooding of the geometry. In combination with the large damage opening (causing the initial, very rapid flooding) and the relatively low transverse stability made it difficult to distinguish between cause and effect. Cause and effect are however crucial when trying to fully understand and explain the behaviour of a tightly coupled system. To get a more detailed insight in the various phenomena during the flooding process, it is better to isolate these using dedicated geometries. For example, use a geometry with one or two down-flooding hatches and a constant water level above the opening to study the down-flooding process, or use a cascaded geometry of a number of compartments connected by various types of opening and opening sizes to study the progression of the floodwater through a geometry.

Accurately and reliably measuring the water level proved to be difficult. A lot of effort was spent on calibration and consistency checks. Unfortunately this was all done for a zero heel and trim such that the influence of non-zero heel and trim on the calibration is unknown. Increasing the level measurement accuracy will be difficult. Either a radically different method will have to be used or a (much) larger model is required, thereby decreasing the relative measurement error. Requirements such as probe size, limited or no accessibility to the probe and the influence of water quality on the measurement also play a role.

The number of measurement probes (6 level probes, 2 air pressure probes) and their location is also important. In hindsight, additional level probes in additional compartments and in the same compartments but at different locations would have helped to understand the differences found between the model-tests and the simulations. In addition, redundant measurements can be used to increase the reliability. Because it is almost impossible to add additional probes after the model is constructed it is worthwhile to study the flooding process in more detail before the model is built. Varying the input parameters in this preliminary study will help to identify the range of possible outcomes and the critical probe locations. Furthermore, this study can determine combinations of heel, trim and heave which results in a certain water level in a number of compartments, but which remain ventilated. The level in all these compartments will be the still water plane. Using this additional information can give a crosscheck on the geometry and the probe location.

The first step in the validation process was to use the measured motions of the model and to force the flooding model to follow these exactly. This step gives an indication of the performance of the flooding model with respect to the prediction of the water levels at the probe locations. The results were quite good for a number of compartments, even when the compartment was located further away from the damaged compartment. For down-flooding the results were not as good. The main factor in the difference is probably related to the fact that during the down flooding, the ventilation of the lower tank switched from bubbling ventilation to vortex ventilation. These two types of ventilation resulted in a large difference in flooding rate: relatively fast flooding in the case a vortex formed (and air can escape more easily) compared to relatively slow ventilation when "bubbling" ventilation was observed. The ventilation in the flooding model is close to bubbling ventilation: vortex ventilation, let alone switching between these two during the flooding simulation process, is not possible.

Another difference between the hexapod simulations and the model-test was that 'delayed' flooding occurred for compartments that were flooded indirectly. This causes differences in timing between the simulations and the model-test which might become a source of deviations, especially between the fully coupled simulations and the model-tests. The source of these delays is the lack of floodwater dynamics in the floodwater model.

In fact, this stage gave the most detailed and most valuable information with regards to the performance of the flooding model and also clearly identified a large array of practical challenges to deal with, such as the difficulties in assessing a sufficiently accurate "as-built" configuration, its influence on the results and the different characteristics of the down-flooding in S06 which was mentioned before.

Using the motions measured during the model-test gave a realistic input to the software Hexapod. An alternative is to also use a hexapod during the model-tests, force a "hardware" geometry (Khaddaj-Mallat et al. 2010) and use the same motions to force the software hexapod with the flooding simulation model. Such a test configuration can provide excellent repeatability and also offer the possibility to vary the motions in a controlled way. The same process can then be repeated in software to study the different reaction on these variations. The definition of the motions input signals will require some thinking: they should, in combination, be representative of a floating, moving vessel subject to (severe) flooding.

Using the outcome of the "Moving Mass" simulation to test the vessel dynamics showed dampened roll oscillations that were more prominent than in all the other simulations. They were not observed in the model test results. It could point at an important difference between the model-test and the simulation and that is the lack of internal damping in the flooding model. Aside from the roll oscillations, the trend in roll, pitch and heave for these simulations compared pretty well.

Finally, a direct comparison between the complete, fully coupled simulation model and the model-test revealed a difference in the roll characteristics. Especially in the intermediate stages, the difference is significant. In addition, the time to reach an equilibrium roll angle is much larger in the simulation than it seemed to be in the model-test. This is probably related to the slow, but continuous, down-flooding of compartment S06 in the simulation. As a result, the roll angle keeps increasing, allowing more floodwater in compartment S17, S20 and S24 again increasing the roll angle (and pitch and heave). Understanding these processes in detail is crucial for the interpretation of the differences between simulation and model-tests.

All the issues discussed here make it clear that the outcome, for example, time to capsize, time to evacuate etc., might be very dependent on small details such as the correct modelling of a down-flooding process in all its details. This is not only important to realize when performing a comparison between flooding simulations and model-test results but also when the simulation model is used to predict the behaviour and safety of a vessel subject to significant flooding. In the last case there will be even more uncertainties (in the input) that need to be dealt with. Aside from uncertainties in the input, there will be in-accuracies in the modelling caused by inevitable simplifications when trying to describe the real world in a simulation model (this also applies when using CFD). In combination with the non-linear, chaotic properties of the combined vessel and flooding model, a large spread in the outcome of the simulations will be possible. The challenge will be to define a process to deal with this. This will require finding a delicate balance between the ability to do fast, and possibly many simulations, a "good enough" description of the important phenomena in the simulation model and a "good enough" description of the flooded geometry.

The ultimate challenge will be to develop a set of tools and procedures that support in the design of ships where the (predicted) spread in outcome after damage is as small as possible and the worst case, as largest roll angle, or shortest time to capsize can be identified reliably and relatively quickly.

Appendix: Internal Geometry

See Figs. 38.34, 38.35, 38.36, 38.37, 38.38 and 38.39.

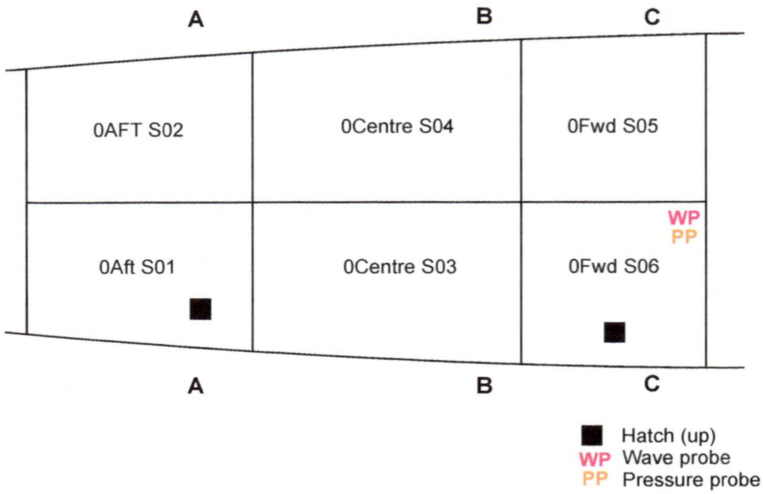

Fig. 38.34 Tank-top—view from top

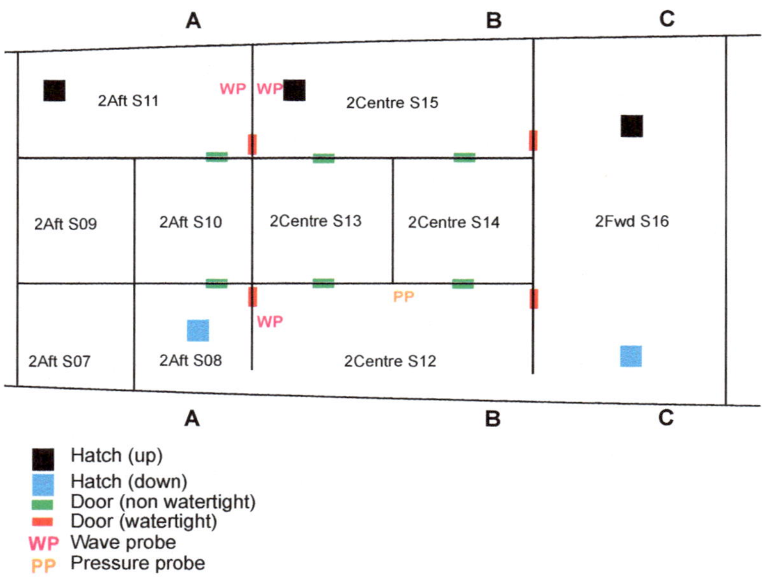

Fig. 38.35 2nd deck—view from top

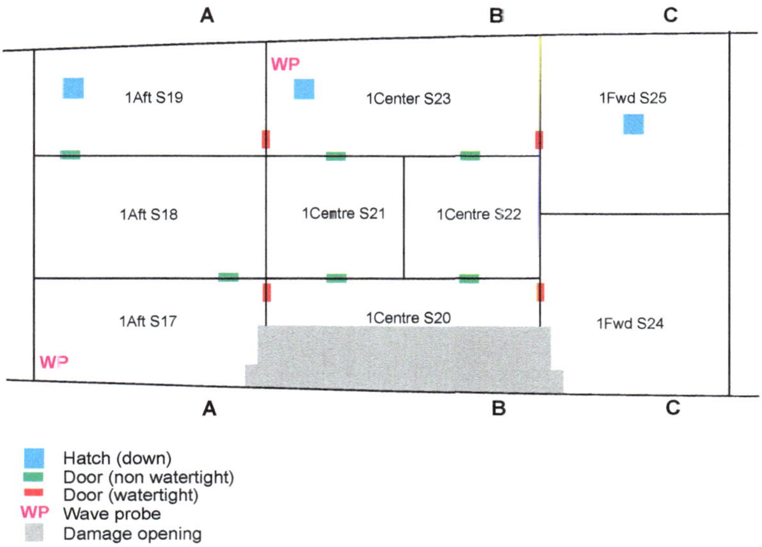

Fig. 38.36 1st deck view from top

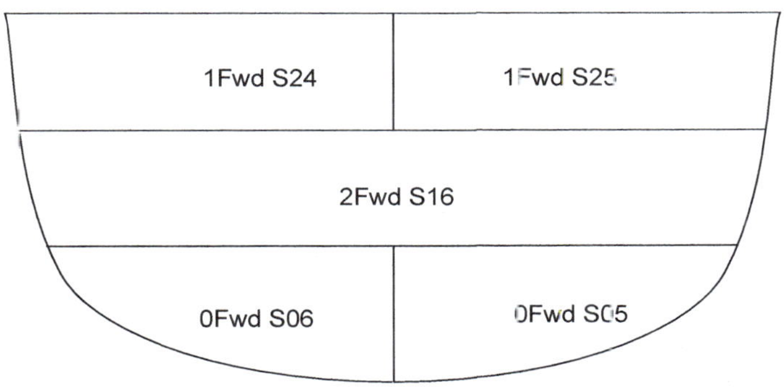

Fig. 38.37 Cross section AA—view from forward to aft

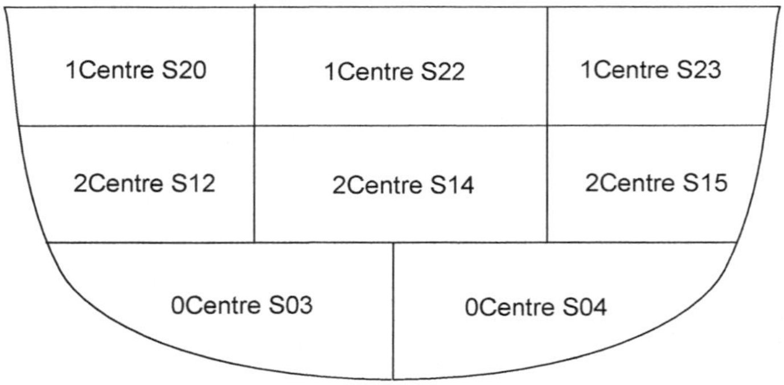

Fig. 38.38 Cross section BB—view from forward to aft

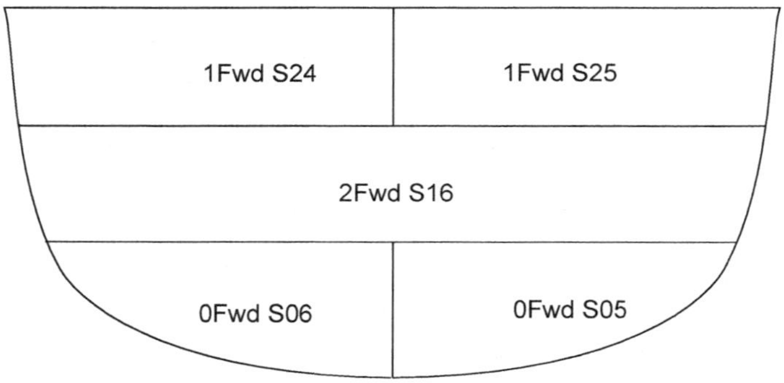

Fig. 38.39 Cross section CC—view from forward to aft

References

de Kat, J. O., Brouwer, R., McTaggart, K.A., and Thomas, W.L. (1994) "Intact ship survivability in extreme waves: New Criteria From a research and naval perspective". *Proc. of 5th International Conference on Stability of Ships and Ocean Vehicles, STAB 94,* Melbourne, Florida, USA.

Khaddaj-Mallat, C, Roussetm J.-M., Alessandrini, B., Delhommeau, B., Ferran, P., (2010) "An Application of the DOE Methodology in Damage Survivability", *Proc. of 10th International Ship Stability Workshop*, Wageningen, The Netherlands, pp. 255–261.

Lebigot, E.O. (2015) Uncertainties: a Python package for calculations with uncertainties, http://pythonhosted.org/uncertainties/.

Manderbacka, T., Ruponen, P. (2016) "The impact of the Inflow Momentum on the Transient Roll Response of a Damaged ship", in *Ocean Engineering*, Vol. 120, pp 346–352.

McTaggart, K.A. (1999) Capsize Risk Assessment Using Fredyn Ship Motion Predictions", (DREA TM 99/149), Defence Research Establishment Atlantic.

MARIN (2014) FREDYN v14.1.1, Getting Started guide, The Netherlands

Ruponen, P. (2007) Progressive Flooding of a Damaged Passenger Ship. Doctoral Dissertation, Helsinki University of Technology

Ruponen, P. (2017) "On the effects of non-watertight doors on progressive flooding in a damaged passenger ship", *in Ocean Engineering,* Vol. 130, pp. 115–125

Ypma, E. (2010) "Model tests in atmospheric and vacuum conditions", FLOODSTAND report D2.5b, Revision 1.01

Bibliography

GRC (2009) PARAMARINE v6.1, User Manual, GRC Ltd, United Kingdom
de Kat, J.O. (1996) "Dynamics of a ship with partially flooded compartment", *The Second Workshop on Stability and Operational Safety of Ships*, Osaka, Japan.
Ruponen, P., Kurvinen, P., Saisto, I., Harras, J (2013) "Air compression in a flooded tank of a damaged ship", in *Ocean Engineering*, Vol. 57, pp. 54–71.
van Walree, F. and Papanikolaou, A. (2007), "Benchmark study of numerical codes for the prediction of time to flood of ships, Phase 1", *Proc. of the 9th International Ship Stability Workshop ISSW*, Hamburg, Germany.
Turner, T., Ypma, E., Macfarlane G. and Renilson, M. R. (2010) "The Development and Application of a Damage Dynamic Stability modelling capability for Naval Vessels", *Proc. International Maritime Conference*, Sydney, Australia
van't Veer, R. and de Kat, J.O., 2000, "Experimental and Numerical Investigation on Progressive Flooding and Sloshing in Complex Compartment Geometries", *Proc. of the 7th International Conference on Stability of Ships and Ocean Vehicles, STAB2000*, Launceston, Australia, Vol. 1, pp. 363–384.

Part XI
Requirements and Operation: Developments in Intact Stability Regulations

Chapter 39
Research Towards Goal Based Standards for Container Shipping

Vladimir Shigunov, Helge Rathje and Ould El Moctar

Abstract Analysis and verification of rule-related technical aspects of safe and efficient container shipping are an important part of R&D activities of classification societies. Casualty statistics show that container loss in heavy weather is an important issue for innovative container ship designs. The paper demonstrates two examples of research activities aiming at the reduction of cargo losses. One example is ship-specific operational guidance, assisting the ship master to avoid excessive motions and accelerations in heavy weather. The design accelerations underlying the operational guidance are part of classification rules, requiring understanding of the physics of dynamic loads on containers and lashing. The status of the ongoing research in this area is shown, in particular, the study of the effects of container flexibility and dynamic load amplification, not addressed explicitly in the present classification rules.

39.1 Introduction

Classification societies are committed to maintain technical aspects of existing and new regulations related to safe and efficient container shipping. When new regulations are developed, they should be relevant (i.e. address real problems), feasible (not too restrictive so that they outweigh the expected benefits), consistent with the safety level

V. Shigunov (✉)
DNV GL Maritime, Brooktorkai 18, Hamburg, Germany
e-mail: vladimir.shigunov@dnvgl.com

H. Rathje
Marine Engineering Consultant, Else-Wex-Ring 25, Bad Oldesloe, Germany
e-mail: helge.rathje@gmail.com

O. El Moctar
Institute of Ship Technology, Ocean Engineering and Transport Systems, University of Duisburg-Essen, Bismarckstr. 69, Duisburg, Germany
e-mail: ould.el-moctar@uni-due.de

V. L. Belenky et al. (eds.), *Contemporary Ideas on Ship Stability*, Fluid Mechanics and Its Applications 119, https://doi.org/10.1007/978-3-030-00516-0_39

provided by other measures and efficient (i.e. aiming at the issues where maximum gains can be achieved by ship owners).

In the EU-funded research project SAFEDOR, FSA study for container vessels has been carried out in order to estimate current risk levels for major risk scenarios, develop generic risk-benefit models for future use and identify cost-effective risk-control options. Historical data LMI (2004) were used to determine the frequency of occurrence for different risk categories, based on the casualty data for modern fully cellular container ships for the period 1993–2004.

The world container fleet is relatively young: 71% of ships by number and 81% by the capacity are built less than 16 years ago. Larger container carriers (post-panamax and panamax) comprise 29.1% by number and 60.6% by capacity, whereas smaller vessels (sub-panamax, handysize and feeder) 70.9% and 39.4%, respectively. The results of the study show that incidents occur for all sizes at similar rate: whereas smaller container vessels are known to suffer substantial losses and damages, larger are suspected to be even more vulnerable because of immature technical standards and the associated lack of experience. Because of high rate of innovation in both design and operation of container ships, designers, operators and regulators alike have limited experience regarding cost-effective safety measures for newly built container ships.

The results show that container carriers are a relatively safe ship type in heavy weather. The societal risk (F-N diagram) for container ship crew fits into the ALARP range, thus justifying exploration of cost-efficient risk-control options. However, this risk is dominated by collision and grounding; heavy weather produces the lowest contribution. The individual risk to crew members is also in the ALARP region, dominated by collision (with the contribution 67.9%), fire and explosion (16.7%) and grounding (13.7%); heavy weather contribution (0.3%) is again insignificant.

Environmental risk (the expected quantity of released dangerous cargo from damaged containers) comprises in total about 1.0 t per ship per year, with the largest contributions from collision (53.3%), grounding (26.6%) and fire and explosion (10.3%); heavy weather contribution is 6.4%.

The consequences of heavy weather accidents are dominated by miscellaneous reasons (78% of all accidents in heavy weather, mostly loss of cargo), hull damage (15%) and machinery damage (6%); only 1% of accidents lead to foundering.

This assessment shows that cargo loss and damage due to ship motions in waves is the most significant intact stability problem for container ships, whereas capsize and hull damage are much less relevant. The situation could be different if container ships would sail not on the damage stability boundary, as it is usually now, but on the intact stability boundary due to different subdivision.

Both the SAFEDOR study and data from insurance companies suggest that containers are lost mostly due to excessive ship motions and accelerations in heavy weather (60% of all lost containers according to SAFEDOR results); however, there is large discrepancy regarding the total number of lost containers. According to SAFEDOR results, 100 containers are lost due to heavy weather per year, while according to insurance clubs, this number is at least one order of magnitude higher, comprising 2000–10,000 containers per year.

This leads to differences in the estimation of the average container loss rate: 0.039 lost containers per ship per year and 1.5×10^{-3} container loss events per ship per year (according to SAFEDOR) compared to 0.4 lost containers per ship per year and 0.1 container loss events per ship per year (according to insurance companies). As a possible explanation of these differences, the authors of SAFEDOR results assume significant underreporting in the used data, because container losses are not safety related. This explanation agrees with the estimation of the number of lost containers per accident: 26.7 according to SAFEDOR data vs. 4 according to insurance companies, which implies that LMI (2004) database contains only the largest accidents, whereas smaller loss events are not always reported; a plausible explanation is that loss claims lead to delays.

Consistently with the identified risk levels due to heavy weather, the corresponding risk-control options were prioritized in the SAFEDOR FSA study as medium (exact weight distribution, constructive roll-damping devices, shipboard routing assistance and enhanced weather routing) to low (modified hull shape); therefore, none of these options were assessed in more detail with respect to their cost-effectiveness.

39.2 Counter-Measures

Container losses in heavy weather may occur due to accidental combination of several factors, including large accelerations, wave impacts and green water, dynamic deformations of containers and lashing, pre-damaged containers, twistlocks and lashing and improper loading (e.g. container overweight or heavy containers on top of a stack). The risk of such accidents may increase due to innovative ship designs (e.g. higher container stacks), tighter operating and loading schedules, as well as crew with insufficient experience on modern vessels.

Experience from the investigations of container damage accidents highlights the need for prompt pro-active measures in regulatory framework, including stricter control of container strength, weight and stowage, ship loading and operational performance standards. Presently, cargo safety is addressed by the following regulations:

- containers are designed and built according to ISO standards, thus their structural strength is pre-defined;
- the Container stowage and lashing plan (subject to class approval) specifies allowable weights of container stack and properties of lashing system;
- twistlocks and fully automatic locks are subject to class-specific standards;
- ship-specific accelerations are maintained by and updated in classification rules.

For example, according to GL rules, either rule-based or calculated design accelerations can be used; the former represent a 'safety envelope' over calculated accelerations for a large number of modern container ships, while the latter follow from hydrodynamic analysis in design wave conditions with an appropriate frequency of occurrence, not covering the most extreme scenarios. The level of safety implied by design accelerations is consistent with the ISO standards for container strength and the class regulations for stowage, lashing and locks. Therefore it would not be efficient to simply increase class-controlled safety level without controlling ISO container standards.

Moreover, the control of the entire system of regulations will not be efficient without the supervision of its implementation. Presently, the implementation of standards regarding container cargo safety is not sufficiently controlled. Although classification societies have competence and infrastructure to do this, authorization of such implementation by flag or port authorities is required.

Rule assumptions for design accelerations as well as other relevant design rules are based on the assumption of prudent seamanship, which may imply increased risks for those modern hull forms where crew experience is insufficient; this issue is also not controlled. Thus, one of important missing parts in the current regulatory framework is the control of operational measures, such as the ship-specific operational guidance. Such operational guidance should be consistent with the other regulations, e.g. with rule-based design accelerations, and is expected to increase the safety level in operation up to the level of other risks. In other words, the operational guidance supports the achievement of 'prudent seamanship' implied by other regulations, up to standard service performance, which is particularly urgent for innovative designs.

In addition, such an operational guidance provides a very flexible measure for prompt support of future innovative designs and innovative operational solutions, and can also be used to address issues not related to cargo safety, e.g. wave loads and crew safety in heavy weather and people comfort onboard. Broadly speaking, ship master should not be left alone in heavy weather: regulators should take care of operations as strictly as it is done in design. Although increasing number of ships are employing onboard weather routing (Rathje and Beiersdorf, 2005) or similar decision-support systems, the quality, consistency and safety standards of such systems should be controlled. Development of the requirements to ship-specific operational measures is presently on the IMO agenda.

39.3 Operational Guidance

Excessive motions and accelerations in waves can occur due to rigid-body motions, particularly heave, pitch and roll, due to slamming impacts and whipping responses, as well as due to green water on deck and wave impacts. The purpose of the operational guidance is to indicate the combinations of operational parameters (ship speed and course) that should be avoided for given loading conditions and seaway characteristics. To do this, operational guidance requires some short-term performance measure (criterion) and the boundary between acceptable and unacceptable values of this criterion (standard).

Because this standard specifies short-term safety, a way is required of relating it to the long-term safety level. Two possibilities were proposed in Shigunov et al. (2010):

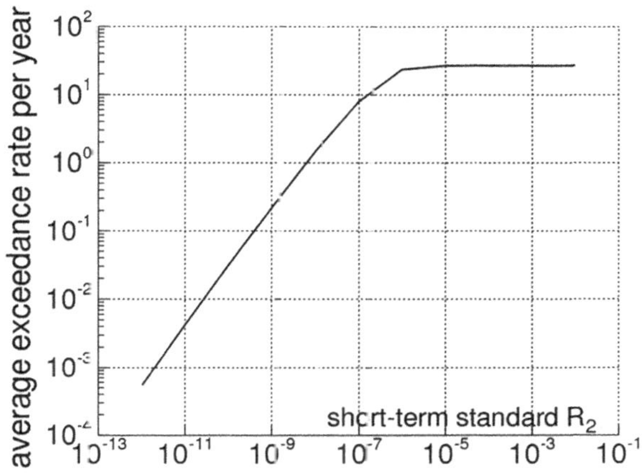

Fig. 39.1 Average annual exceedance rate versus short-term standard

- to determine the value of the short-term standard in such a way that it leads to the required long-term (i.e. average over operational life) safety level
- to set standard value so that it minimizes the difference between additional benefits per time (due to reduced rate of cargo loss) and additional cost per time (due to increased time on route), incurred due to the use of operational guidance.

As an illustration of the first way, the long-term exceedance rate of the maximum (over ship) lateral acceleration g/2 was computed as a function of short-term standard (denoted as R_2) using numerical simulations for an 8400 TEU container ship. The resulting dependency is shown in Fig. 39.1.

Assuming the required long-term safety level as 0.02 container loss events per ship per year, the short-term standard R_2 can be set to 10^{-10} 1/(m s^2). Figure 39.2 shows examples of unacceptable combinations of operational parameters (grey areas) for the load case with GM = 2.3 m in two seaways.

39.4 Simplified Design Assessment Procedure

In order to distinguish between ships requiring and not requiring operational guidance, a simplified design assessment procedure is proposed in SLF 51/INF.2 (2009): numerical simulations are performed for several irregular 'design' sea states. The wave period of these sea states is systematically varied in a broad range, wave height (one per wave period) is specified as a function of the wave period, and the ship forward speed is defined as a function of wave height and wave direction.

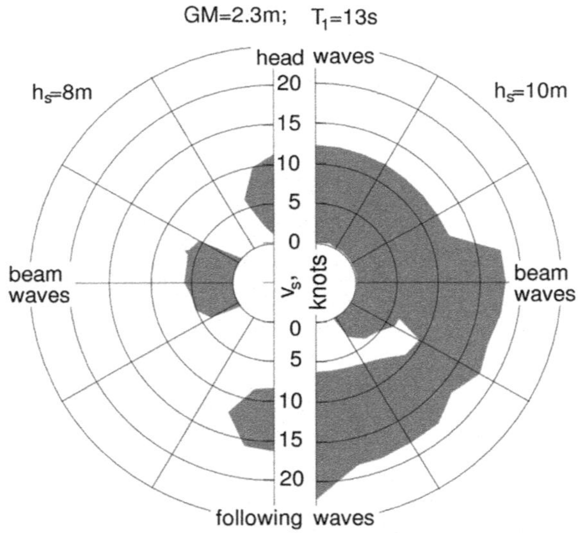

Fig. 39.2 Areas of unacceptable operational parameters for a 8400 TEU container ship with $GM = 2.3$ m in a seaway with the mean wave period 13 s and significant wave height 8.0 m (left) and 10.0 m (right)

39.5 Further Factors

Besides rigid-body motions, further factors are becoming increasingly important for container ships: hull girder flexibility and flexibility of container stacks. An example in Fig. 39.3 shows time history of measured vertical acceleration at the forward perpendicular for a segmented flexible model of an 8400 TEU container ship, indicating significant dynamic amplification of vertical accelerations due to slamming impact and the resulting whipping response, Oberhagemann et al. (2008).

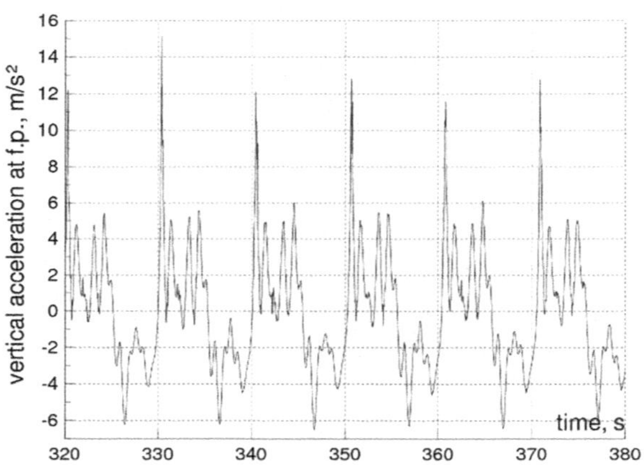

Fig. 39.3 Measured vertical acceleration at the forward perpendicular of an 8400 TEU container ship

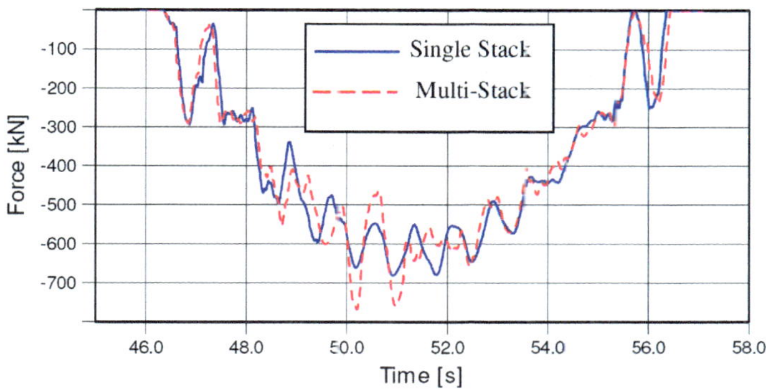

Fig. 39.4 Effect of stack interaction: vertical forces on container corner for a single stack and multiple stacks

Wolf and Rathje (2009) studied the influence of container flexibility on container stack dynamics and loads on containers and showed that the consideration of prevailing dynamic effects due to flexible container stacks on the weather deck is essential for the assessment of stack loading.

The dynamic response of a container stack is highly nonlinear due to clearance in lashing, interaction with adjacent stacks, friction effects etc. Therefore, time-domain simulations were performed using a FE model of container stacks. Containers were modelled as super-elements with interfaces to other elements and with contact and friction effects between stacks; stiffness and mass inertia of the super-elements were condensed from a detailed FE-model of a container. Twistlocks were modelled as spring-damper elements with gap and contact capability; their stiffness was derived from a detailed FE-model. Lashing was not considered and is addressed in the ongoing work. Friction and damping parameters for high-frequency responses were derived from full-scale measurements of the dynamics of stowed containers.

An example study is shown for a container stack carried on the weather deck of a 9200 TEU container ship. Roll motion characteristics are derived from hydrodynamic analysis, leading to design conditions with roll period 18 s and roll amplitude 26°.

Parametric studies were carried out in order to quantify the effects of the cargo distribution over the stack, twistlock stiffness, structural damping and adjacent stack interaction.

The study has revealed that flexibility effects lead to distinctive dynamic amplification of transverse racking forces and, particularly, vertical forces due to successive uplifting and crashing down of the upper containers due to roll motion. Because of this effect, the influence of the vertical cargo distribution is especially significant: container and twistlock loads are higher for stacks with higher centre of gravity.

Stack interaction also has a significant influence: both vertical and transverse loads are amplified due to the interaction of the upper containers in the adjacent stacks, Fig. 39.4.

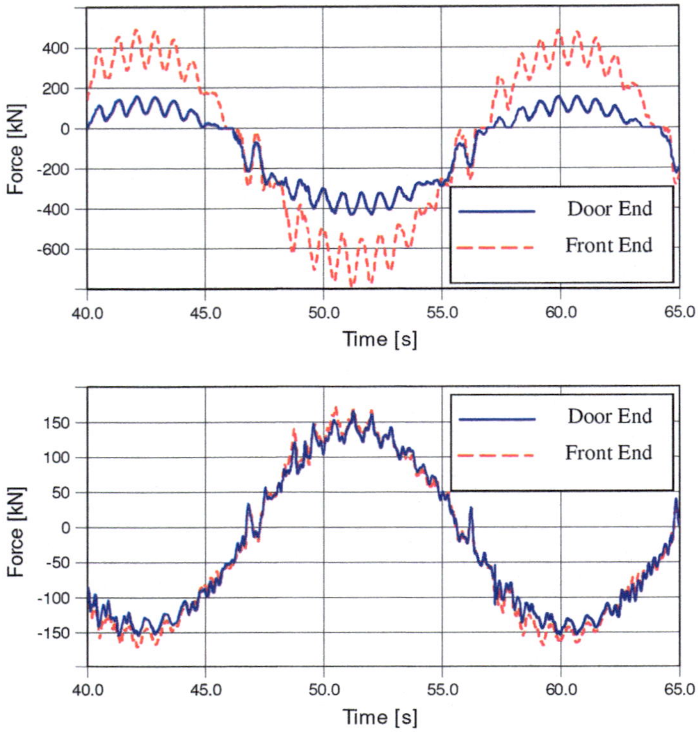

Fig. 39.5 Front-to-rear asymmetry of container loads: vertical (top) and transverse (bottom) forces on the top of the container at the bottom of the stack

The results of the simulations were compared with loads based on classification rules for a single unlashed eight-tier rigid container stack with proper cargo distribution and standard accelerations.

Simulations (Fig. 39.5) show asymmetrical front- to rear-end distribution of container loads: the front end carries higher transverse and, especially, vertical loads because of the higher flexibility of the door end.

This effect is not considered in the present rules for unlashed configurations: for unlashed case, the loads at both ends are assumed identical and, effectively, equal to the average load between the front and rear ends. Therefore, the simulated vertical loads and corner post forces at the door end are lower (respectively, at the front end higher) than those from the rules. On the other hand, the average between the front and rear end lifting force in simulations is about 25% higher than the rule-based value due to dynamic load amplification (container uplifting and bouncing).

39.6 Conclusions

Analysis and verification of rule-related technical aspects of safe and efficient container shipping are important parts of R&D activities of classification societies. The presented results show that cargo loss and damage may be of especial concern for modern container carriers. Mitigation measures are proposed, such as ship-specific operational guidance. Example is shown of a possible approach to operational guidance, reducing the average long-term exceedance rate of a specified threshold of lateral accelerations to the prescribed value. Further factors are identified which may be responsible for cargo losses, particularly flexibility of ship hull girder and container stacks.

Ongoing R&D activities concern further factors responsible for cargo loss and their design limits (e.g. vertical accelerations), cost-benefit analysis over operational life for setting economically sound standards, incorporation of further factors into operational guidance (slamming and whipping, vertical accelerations, dynamic response of container stacks and lashing, crew safety and comfort) and roll-damping devices, to update design rules; however, a big potential for cost-effective improvement of ship safety exists in the control of ship operation in the same way as it is done in design—from the approval of routing assistance and operational guidance to the control of container strength, weight and stowage.

References

MSC 83/INF.8 (2007) FSA – container vessels. Details of the Formal Safety Assessment, submitted by Denmark

LMI (2004) Casualty database, Lloyd Maritime Information, December

Rathje, H. and Beiersdorf, C. (2005) Decision support for container ship operation in heavy seas – Shipboard Routing Assistance, 4-th Conf. Comp. and IT Applications in the Maritime Industries COMPIT, pp. 455–467

Germanischer Lloyd (2007) Rules for classification and construction, I – Ship Technology, Part 1 – Seagoing Ships, Chapter 20 – Stowage and Lashing of Containers. Hamburg

Shigunov, V., El Moctar, O. and Rathje, H. (2010) Operational guidance for prevention of cargo loss and damage on container ships, Ship Technology Research 57(1) 6–23

SLF 51/INF.3 (2008) New generation intact stability criteria, submitted by Germany

Oberhagemann, J., El Moctar, O. and Schellin, T. E. (2008) Fluid-structure coupling to assess whipping effects on global loads of a large containership, Proc. 27-th Symp. on Naval Hydrodynamics, 296–311

Wolf, V. and Rathje, H. (2009) Motion simulation of container stacks on deck, SNAME Annual Meeting and Ship Production Symp. 1, 277–285

Chapter 40
On Regulatory Framework of Direct Stability Assessment

William S. Peters, Vadim L. Belenky and Arthur M. Reed

Abstract Direct assessment of stability, including model tests and numerical simulations, is the ultimate way to evaluate the risk of stability failure for an unconventional vessel. That is why direct assessment is considered to be the highest tier of the second generation of intact-stability criteria, that are being developed by IMO. Direct assessment procedures for stability failure are intended to employ the most advanced state-of-the art technology available, yet be sufficiently practical so as to be uniformly applied, verified, validated, and approved using currently available infrastructure. This paper addresses several principal issues related to the application of numerical simulation in the IMO regulatory framework, including possible requirements for a method that adequately replicates ship motions in waves, validation of such a method, actual assessment procedures and their validation.

Keywords Direct assessment · Intact stability · Numerical simulations

40.1 Introduction

The concept of direct stability assessment has evolved from the idea of performance-based intact-stability criteria during development of second generation intact stability criteria by IMO (Belenky et al. 2008). The principal motivation for developing direct assessment is to reduce the level of empiricism and be prepared for the assessment of novel designs. These new designs may be well outside of the population of ships

---

W. S. Peters (✉)
US Coast Guard, Office of Design and Engineering Standards, Washington DC, USA
e-mail: william.s.peters@uscg.mil

V. L. Belenky · A. M. Reed
David Taylor Model Basin (NSWCCD), West Bethesda MD, USA
e-mail: vadim.belenky@navy.mil

A. M. Reed
e-mail: arthur.reed@navy.mil

V. L. Belenky et al. (eds.), *Contemporary Ideas on Ship Stability*, Fluid Mechanics and Its Applications 119, https://doi.org/10.1007/978-3-030-00516-0_40

used for development of the first generation of the intact-stability criteria codified in the 2008 IS Code.

Accurate computational reproduction of an intact-stability failure in irregular seas is a challenging technical task because the dynamical system describing ship motions in waves is characterized by significant nonlinearities. On the other hand- and fortunately-, intact-stability failures are very rare. This combination of nonlinearity and rarity makes simulation a challenge.

Multi-tiered structure of the second-generation-stability criteria was adopted to make sure that costly numerical assessment procedures are only applied when it is absolutely necessary, i.e., when vulnerability to dynamical stability failures has been established "beyond reasonable doubt" (Peters et al. 2011).

Application of advanced numerical simulations for regulatory purposes and operational advice is relatively new. Several example are: the ABS Parametric Roll Guide (ABS 2004a), Themelis and Spyrou (2007) and Shigunov (2009). As a part of optional class notation ABS (2004a) requires numerical simulations to be carried out to develop an operational guidance system to avoid parametric roll. Themelis and Spyrou (2007) described a numerical simulation of a typical voyage to assess the probability of stability failure using the wave-group method. Shigunov (2009) reported application of advanced numerical simulations to develop operational guidance to reduce loss of containers from container ships.

Mathematical methods and software for numerical simulation of ship motions in waves have been developed during the last three decades (Beck and Reed 2001). However, computational efficiency issues with rigorous physics-based models have been identified by Belknap and Reed (2010). Hybrid codes combining potential flow hydrodynamics with simple models for other forces seems to be the practical way to approach dynamic stability assessment. Examples of this are represented by Shin et al. (2003), Brunswig and Pereira (2006). The most difficult issue, however, is related to the validation of the numerical codes (Reed 2008, 2009; Smith 2012). Another challenge is presented by the rarity of stability failures, a fact that requires application of special extrapolation procedures (see review by Belenky et al. 2012).

One of the first attempts to formulate requirements to direct assessment of stability was presented in Annex 21 of SLF 54/INF.12, followed by the comments and discussion in Annex 22. This paper presents a review and further discussion of this subject.

40.2 Possible Structure of Requirements

Up to now there are only a few cases in which numerical simulations of ship motions were applied to parametric roll and then recognized by classification societies. However, the history of application and recognition of advanced numerical methods in classification society practice is much longer, since finite element analysis is considered as a standard tool today (see, e.g., ABS 2004b). Following this long-term experience, the requirements for direct assessment of intact stability have to be formulated in terms of the method itself and the procedure of its application. These requirements also have to include the conditions for recognition and acceptance by flag administration (or by an organization acting on behalf of the administration) of the method and procedure.

Since direct stability assessment is meant to be a state-of-the art method, it has to be applied to the most realistic model of the wave environment, i.e., irregular waves. Since intact-stability failure is a very rare event, a sample record of sufficient volume cannot be obtained by numerical simulation alone. Therefore, any statistical extrapolation method will be based on a set of assumptions that will also need validation.

Based on the above considerations, the possible structure of the requirements for direct stability assessment can be envisioned as consisting of the followings:

- A method that adequately replicates ship motions in waves.
- A prescribed procedure that identifies the process by which input values are obtained for the assessment, how the output values are processed, and how the results are evaluated.
- Conditions of acceptance and requirements for verification and validation of the method for ship motion replication.
- Conditions of acceptance and requirements for verification and validation of the extrapolation procedure.

40.3 Method for Replication of Ship Motions

40.3.1 Mathematical Model of Waves

Mathematical modeling of irregular waves is more than just a representation of a stochastic process of wave elevations. Calculations of potential hydrodynamic forces (including hydrostatic, Froude-Krylov, diffraction and radiation) require evaluating the pressures around the instantaneous position of the hull relative to the water surface. Therefore, the modeling of irregular waves involves presentation of a large number of dependent stochastic processes. Fourier series have been traditionally used as a method to present these stochastic processes. Wave elevations (as a function of time and space), for example, are presented as

$$\zeta_W(x,t) = \sum_{i=1}^{N} r_i \cos(k_i x - \omega_i t + \varphi_i) \tag{40.1}$$

where r_i is an amplitude of a component, calculated from a wave spectrum; φ_i is a random initial phase (uniformly distributed from 0 to 2π); and k_i and ω_i are wave number and frequency of a component. They are related through the well-known

Dispersion Equation, that expresses the assumption of small waves according to the Airy theory in deep water:

$$k_i = \frac{\omega_i^2}{g}. \tag{40.2}$$

where g is the gravitational acceleration. Inclusion of the dispersion relation into the presentation of the stochastic process makes this wave model hydro-dynamically valid within the Airy theory.

While this model remains the mainstay of numerical simulations of ship motions, it incurs a computational cost when a large sample size is needed. This problem comes from the so-called "self-repeating effect" that is caused by an insufficient number of components for the required record length (Belenky 2011). The presence of the self-repeating effect in a particular frequency set can be revealed by an autocorrelation function calculated from the wave spectrum when using the method of rectangles.

$$R(\tau) = \int_0^\infty s(\omega)\cos(\omega\tau)d\,\omega$$
$$= \sum_{i=1}^{N} s(\omega_i)\cos(\omega_i\tau)\Delta\omega_i \tag{40.3}$$

here $s(\omega_i)$ is the spectral density and $\Delta\omega_i$ is the frequency increment around the i-th component.

Figure 40.1 shows the autocorrelation function (40.3) indicating the self-repeating effect. The repetition of the pattern may be different if the variable- frequency step is used. An increase of the autocorrelation function after the initial decay is a result of numerical error and indicates the limits of the duration of the record length. For example, in Fig. 40.1 the valid duration is about 180 s.

To extend the length of the record, the number of components needs to be increased, which leads to an increase in computational costs. One way to deal with a problem of increased computational costs is increase the number of records instead on number of frequencies.

A promising alternative to the Fourier series representation is the Autoregression Method.

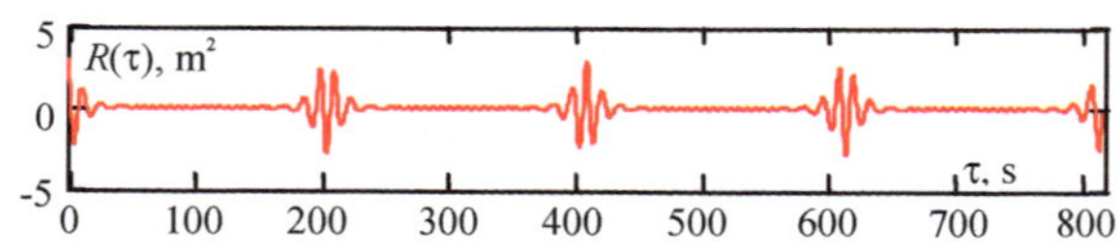

Fig. 40.1 Example of the self-repeating effect

$$\zeta(x,t) = \sum_{i=1}^{N_t}\sum_{j=1}^{N_x} \Phi_{i,j}\zeta(t_i, x_j) + \varepsilon(x,t) \quad (40.4)$$

here Φ is the autoregressive parameter, and ε is white noise. Hydrodynamical validity of the method has been recently addressed by Degtyarev and Reed (2011), while further development of the autoregressive method is focused on predicting the pressures needed to compute forces and moments (Degtyarev and Gankevich 2012), see also Chaps. 2 and 3 of this book.

40.3.2 Mathematical Model of Roll Damping and Hull Forces

Using a hybrid approach for reproducing ship motions is computationally efficient, but the cost for this efficiency is potential inconsistency in the modeling of forces with significant vortex components. Roll damping is the best example of such potential inconsistency. The potential solver of a hybrid code internally reproduces the wave component of roll damping. An empirical evaluation of roll damping that is obtained from a model test or CFD also contains a wave component. As a result the simple addition an empirical to a potential flow model will inevitably lead to "double counting" of the wave component of roll damping.

To deal with this problem, the ABS Guide (2004a) requires that the empirical part of roll damping be "calibrated" such that the resultant damping is close to that measured in a roll decay test. This approach was first successfully applied during a parametric-roll-accident investigation (France et al. 2003). A detailed description is available in Belenky et al. (2011b). The procedure can be expressed in the form of a system of two nonlinear algebraic equations.

$$(k, f) - D(B_1(f), B_2(k)) = 0, \quad (40.5)$$

here, D is a symbolic expression for a run of numerical simulations followed by a standard procedure for roll-decay processing, k is the slope, and f is the intercept of the fitted line; functions $B_1(f)$ and $B_2(k)$ are expressions, respectively, of the linear-roll-damping coefficient which depends on the intercept and the quadratic-roll-damping coefficient witch depends on the slope. This equation can be solved with any appropriate numerical method Practical experience has shown that, on the average, calculations converge after five to six iterations.

A similar approach has been developed and verified for maneuvering hull forces (Yen et al. 2010). The maneuvering coefficients were "calibrated" to allow a hybride code to reproduce a model test done with the Planar Motion Mechanism (PMM) in calm water. It was shown that this procedure offers a reasonable performance when simulating maneuvers in waves.

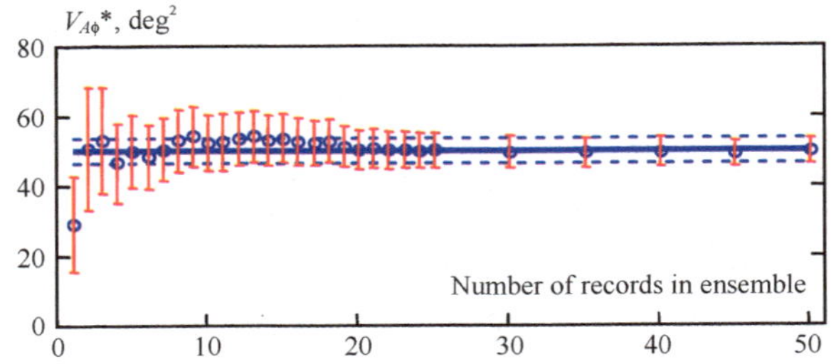

Fig. 40.2 Convergence of ensemble estimate of variance (Belenky and Weems 2011)

40.3.3 Parametric Roll

Since early 2000, when the problem of parametric roll of container ships came into focus, there have been some positive results from numerical simulations of parametric roll. Stability variations in waves are modeled as a part of a body-exact formulation for the Froude-Krylov and hydrostatic forces. As described above, calibration of roll damping is known to work well for modeling parametric roll (France et al. 2003).

Planning numerical simulations in irregular waves requires special care (Reed 2011). The parametrically excited motions in irregular waves "come and go" which leads to practical non-ergodicity (Belenky et al. 2011a). Here, ergodicity refers to the ability to estimate statistics from a single "long-enough" record. While the process of roll motions formally remains ergodic, the required length of a record may be impractically long. However, the use of a collection of independent records mitigates the problem (see Fig. 40.2).

40.3.4 Pure Loss of Stability

The phenomenon of pure loss of stability is caused by the degradation of roll restoring near a wave crest. The body-exact formulation for hydrostatic and Froude-Krylov forces is the best way to predict loss of stability. The main difference with parametric roll is the importance of the time duration of reduced stability around a wave crest. This is one reason why numerical simulations of a ship's motion may need to include surging. As shown by Umeda and Yamakoshi (1993), surging may influence the timing of reduced stability due to specific phase shifts between the processes of surge motions and stability variation.

40.3.5 Surf-Riding and Broaching-to

Contemporary mathematical models describing surf-riding and broaching-to include at least four degrees of freedom: surge, sway, roll and yaw. These models were developed in the 1990s (Spyrou 1995) and were used for developing the modern dynamical theory of broaching-to (Spyrou 1996). Previously, all the predictions for surf-riding and broaching-to were described by ordinary differential equations (ODE). While ODE are still considered acceptable (SLF 53/3/8), work is underway to simulate surf-riding and broaching-to using hybrid codes (Spyrou et al. 2009; Belknap and Reed 2010; Yen et al. 2010) and CFD (Sadat-Hosseini et al. 2011). Because the computational costs of the CFD remains high, its role may only be as a source of coefficients for the ODE and hybrid codes. However, at this point, a model test remains the main source of coefficients.

40.3.6 Dead-Ship Conditions

Within the framework of IMO regulations, capsize in dead-ship conditions was the first situation in which direct stability assessment was considered (MSC.1/Circ.1200 & 1227). The assessment was done using model tests; however, in the future, the use of numerical simulations for predicting capsize in dead-ship conditions cannot be ruled out.

To provide consistency with the existing weather criteria, a code for predicting behavior in dead ship conditions should be able to reproduce the responses of a ship under the action of waves and experiencing a wind gust. Although this hypothetical situation is classic, its numerical simulation may be not simple. There are several issues to discuss: beam seas assumption, hydrodynamic drift force and influence of water on deck.

Beam Seas Assumption

Stability regulations were first formulated in the steamship era (Rahola 1939). The prevailing architecture of those ships included a superstructure located amidships with the topside as well as the hull volume more or less longitudinally symmetric about amidships. In a dead-ship condition, that type of vessel would be turned by the wind into a beam seas orientation. Fleets today are characterized by a much larger variety of architectural types: the superstructure located aft (containerships, bulk carriers and tankers); the superstructure located forward (offshore supply vessels); the superstructure occupying most of ship length (car carriers, RoPax, cruise ships). Thus, today a vessel in dead-ship conditions may not be turned to a beam seas orientation.

Umeda et al. (2007) developed a method to predict the orientation of a vessel in waves and wind under dead-ship conditions, which was based on a system of ODE describing sway, surge, roll and yaw motions to predict the orientation.

Because the heeling moment from the wind is maximum in this orientation, the beam position is still considered the most dangerous in Annex 22 of SLF 54/INF.12. The issue identified here is: when carrying out numerical simulations in dead-ship conditions, should a ship be fixed in beam position even in the presence of significant aerodynamic yawing moment?

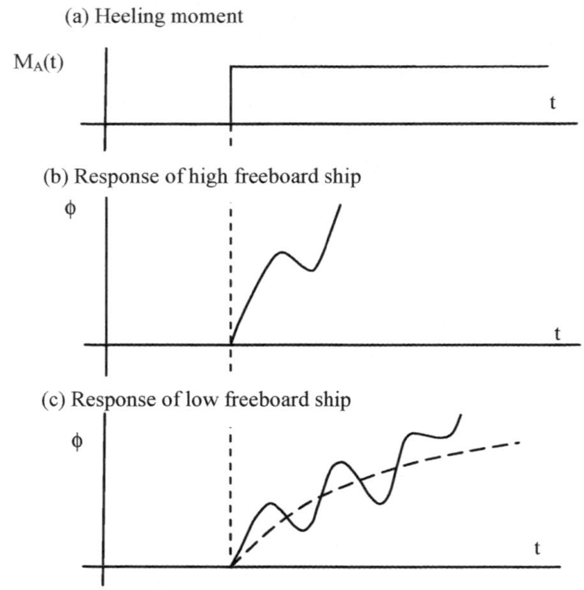

Fig. 40.3 Difference in response to a sudden wind gust (Belenky and Sevastianov 2007)

The answer to this may depend on how quickly aerodynamic forces can turn a ship. The actual orientation may be a more realistic scenario to consider if the turn toward quasi-equilibrium position can be completed quickly, i.e., before the hydrodynamic drift reaction is fully developed. If the turn takes a long time, then it may be reasonable to ignore the aerodynamic turning moment and consider beam seas only.

Hydrodynamic Drift Force

The challenge of modeling this force is in the complexity of flow around the hull with non-stationary vortexes. Thus, model testing remains the main source of data for this force and its moment. Review of some of the experimental data is available from Belenky and Sevastianov (2007). CFD calculations may be a practical way to get the coefficients of force for the direct stability assessment.

Presence of Water on Deck

If a ship has a relatively low free board, water may be shipped on deck after the first semi-period of roll oscillations. This may significantly change the dynamics of further motions. Water on the deck may play the role of additional damping, resulting in the heeling motion growing slower, and causing the ship to capsize after the second or third period (see Fig. 40.3).

The dynamics of a ship with water on the deck can be simulated with a hybrid code (Belenky et al. 2003), which showed that the inertia of water may have significant influence on ship motions. The scenario of water on the deck shown in Fig. 40.3 is

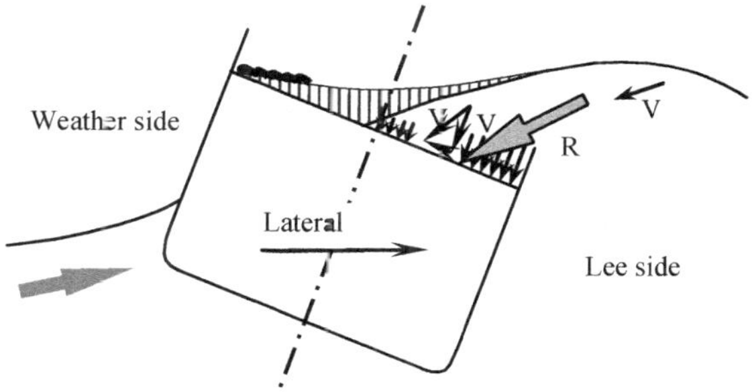

Fig. 40.4 Hydrodynamic reaction on submerged part of the deck (Grochowalski et al. 1998)

only one possibility; another one is shown in Fig. 40.4. It corresponds to the "deck-in-water" situation when the water on the deck is connected with the outside fluid domain. The hydrodynamic force created by the deck-in-water scenario may create a significant additional heeling moment (Grochowalski et al. 1998).

Water on the deck is not only a modeling issue, but raises the question of consistency because water on deck/deck-in-water considerations are not included in the vulnerability criteria development at this time. One possible solution is to consider inclusion of these effects only when they are not dominating factors in the dynamics of heeling and capsizing in dead-ship conditions. The literature on stability in dead ship conditions is quite comprehensive, with the most recent review on this topic available from Ogawa et al. (2008).

40.4 Validation of Numerical Tools

In general, the problem of validation of numerical tools is not new. A standard already exists for numerical fluid dynamics (ASME 2009). However, the field of dynamic stability is much wider, because it includes fluid dynamics and rigid body dynamics, as well as statistical problems.

Possible regulatory use of software for numerical simulation of ship motions makes the validation problem even more difficult. The first question to ask is if the ship motion software reproduces the physical phenomenon "responsible" for a particular stability failure. Thus, the first step is a qualitative validation.

Qualitative validation can be seen as a series of test runs of simulation software for a formal proof that a particular physical phenomenon is actually modeled. For example, from the preceding discussion, it is evident that a tool based on the mathematical model of linear roll is inadmissible for direct stability assessment. Then, how does one demonstrate that roll nonlinearity is actually present in the code? The answer can be found in the application of the methods of nonlinear dynamics. Figure 40.5 shows a response curve (calculated with Large Amplitude Motion Program—LAMP), compared with a simple 1 DOF-roll equation.

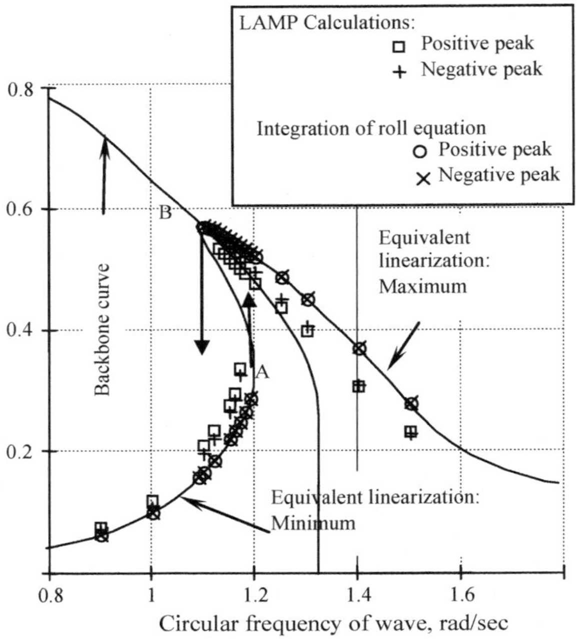

Fig. 40.5 Response curve of roll based on LAMP calculation, numerical integration of nonlinear roll equation and equivalent linearization (Shin et al. 2003)

Qualitative validation requires showing that the numerical code is capable of reproducing known nonlinear effects, i.e., demonstrates the "proper nonlinearity." The example from Shin et al. (2003) includes a calculation of the response curve with LAMP. Calculations are done for a series of regular waves of the same height with systematically changing frequencies. Simulations are run until the steady-state mode is achieved. Then the frequency is slightly changed, while the initial conditions are taken from the steady-state mode of the previous run. There are two series of such runs: from the lower frequencies to the higher and back to the lower. The expected known nonlinear behavior is a hysteresis effect; which results in the existence of a frequency range with two stable solutions. The ability of the code to reproduce this effect is especially important because the same effect was observed experimentally (Francescutto et al. 1994).

The hysteresis effect in roll motions is associated with fold bifurcation. Motion stability analysis can be used for further proof that observed hysteresis is, indeed, fold bifurcation, for which motion stability analysis shows that the eigenvalues of the Jacobean leave the unit circle in a positive direction (See Fig. 40.6a).

$$J = \begin{pmatrix} \frac{\partial\phi(t_0+T)}{\partial\phi(t_0)} & \frac{\partial\phi(t_0+T)}{\partial\dot{\phi}(t_0)} \\ \frac{\partial\dot{\phi}(t_0+T)}{\partial\phi(t_0)} & \frac{\partial\dot{\phi}(t_0+T)}{\partial\dot{\phi}(t_0)} \end{pmatrix} \tag{40.6}$$

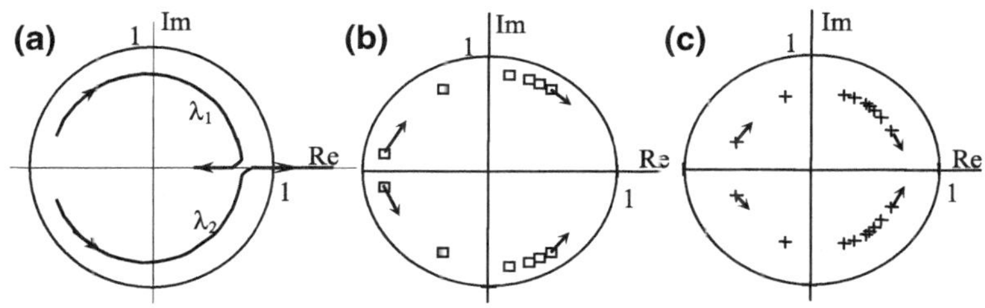

Fig. 40.6 Eigenvalues of Jacobean Matrix for **a** LAMP calculation, **b** Nonlinear roll equation, **c** Theoretical (Shin et al. 2003)

here ϕ is the roll angle, the dot above the symbol means a temporal derivative, T stands for the period of steady-state-roll motions, t_0 is the initial instant in time for determining motion stability.

Computation of the Jacobean consists of a series of short simulations starting at t_0 when the motion, or its derivative, is given a small perturbation (on the order of a quarter or half a degree). Then, the simulation is carried out for one period. The value of the motion displacement is recovered and used to evaluate the derivatives in Eq. (40.6). From this the eigenvalues of the Jacobean can be computed. The above eigenvalues calculations are repeated for each frequency. Comparing the tendency of eigenvalues based on a LAMP solution (Fig. 40.6b) with a theoretical plot (Fig. 40.6a), one can see that they are similar, i.e., do not contradict each other.

The eigenvalues from the numerical simulation cannot reach the axis, as on the theoretical plot because the numerical simulation can only reproduce a stable solution. A similar picture can be seen in Fig. 40.6c where the same method is applied to the ordinary differential equation that describes roll motions.

In general, most of methods of ODE analysis seem to be applicable to the hydrodynamic codes. Spyrou et al. (2009) demonstrated that the continuation method can be used together with LAMP and even find unstable equilibria. Therefore, more interesting results can be expected from "merging" nonlinear dynamics with numerical hydrodynamics.

Available techniques allow a comprehensive qualitative analysis of ship-motion simulation software. Annex 21 of SLF 54/INF.12 contains a table that summarizes these possibilities, and is reproduced here in Table 40.1 for easy reference. The second row in that table contains requirement for the analysis that was shown in Figs. 40.5 and 40.6 as an example. Unfortunately, providing sample analyses for other lines in Table 40.1 is not within the scope of this paper.

The next validation step is a quantitative validation, and the question to be answered is, "How well are the forces modeled?" Belknap et al. (2011) addressed key challenges in this area. One of the challenges is formulating metrics or an acceptance criterion (e.g., how much difference is still acceptable?). A realistic approach to this problem, however, shows that a numerical tool has to be valid only for the particular problem for which it is intended to be used. This issue is considered in

Table 40.1 Possible requirements for qualitative validation

Item	Required for:	Objective:	Acceptance criteria:
Periodic properties of roll oscillator	Software where hydrostatic and Froude-Krylov forces are calculated with body exact formulation	Demonstration of consistency between calculated roll backbone curve (dependence of roll frequency in calm water on initial roll amplitude) and GZ curve in calm water	Shape of calculated backbone curve. The backbone curve must follow the trend of instantaneous GM with increasing heel angle
Response curve of roll oscillator	Software where hydrostatic and Froude-Krylov forces are calculated with body exact formulation	Demonstration of consistency between the calculated roll backbone curve and the calculated roll response curve (dependence of amplitude of excited roll motion on the frequency of excitation)	Shape of the roll response curve. The roll response curve must "fold around" the backbone curve and show hysteresis when magnitude of excitation is increased
Change of stability in waves	Software where hydrostatic and Froude-Krylov forces are calculated with body exact formulation. Additional capability to track the instantaneous GZ curve in waves may be required	Demonstration of capability to reproduce wave pass effect	Decrease of stability when the wave cress is located around midship section (within the quarter of length) and increase when the wave trough is located around midship section (within the quarter of length)
Principal parametric resonance	Software where hydrostatic and Froude-Krylov forces are calculated with a body exact formulation	Demonstration of capability to reproduce principal parametric resonance	Observing increase and stabilization of amplitude of roll oscillation in exact following or head seas when encounter frequency is about twice of natural roll frequency
Surf-riding equilibrium	Software for numerical simulation of surf-riding and broaching	Demonstrate capability to reproduce surf-riding; & turn off yaw motions	Observing sailing with the speed equal to wave celerity when the propeller RPM is set for the speed in calm water is less than the wave celerity. Center of gravity is expected to be located near wave trough

(continued)

Table 40.1 (continued)

Item	Required for:	Objective:	Acceptance criteria:
Heel during turn	Software for numerical simulation of surf-riding and broaching	Demonstrate capability to reproduce heel caused by turn	Observing development of heel angle during the turn
Turn in waves	Software for numerical simulation of surf-riding and broaching	Demonstrate correct modeling of maneuvering forces in waves	Same direction of drift that was observed in a model test
Heel caused by drift and wind	Software for numerical simulation of ship motions in dead ship condition	Demonstrate capability to reproduce heel caused by a moment created by aerodynamic load and drag caused by drift	Observed slowly developed heel angle after applying aerodynamic load

detail by Smith (2012). The concept of "intended use" allows limiting the scope of the validation problem to a more manageable size.

In addition to the force components, a numerical tool should reproduce a response reasonably close to the true values (assuming that a model test can provide such). There are two distinct problems to consider. The first is physical uncertainty (Belknap et al. 2011), the second is statistical uncertainty, which is relevant to irregular waves (Smith 2011, 2012).

The problem of physical uncertainty is addressed by comparisons between the characteristics of roll motions in regular waves. Annex 21 of SLF 54/INF.12 proposes considering the response curves for synchronous and parametric resonance. The acceptance criteria can be based on amplitude for synchronous resonance, and on both amplitude and frequency for parametric resonance. Accounting for significant nonlinearity by relaxing accuracy requirements for large amplitudes makes sense.

Turning and zig-zag maneuvers in calm water may be used as a background for validation of horizontal-plane motions.

Wave conditions (height and length) and ship speed, where broaching is observed, provides another basis for comparison.

Statistical uncertainty is encountered when carrying out validation in irregular waves. In this case two random variables must be compared. Smith (2011, 2012) describes a number of statistical tests that may used to judge the validity of the numerical tools under consideration.

40.5 Procedures

Validated numerical tools are necessary, but not sufficient by themselves to complete a direct- stability assessment. There should be a prescribed procedure of application of the tools, and following such a procedure, multiple applications should reach the same conclusions on the subject vessel. Obviously, the procedure should prescribe how to choose loading and environmental conditions. Beyond the choice of input parameters, the procedure can be seen as instructions on "how to get there from here". The following objectives of the procedure are formulated in Annex 21 of SLF 54/INF.12:

- Confirmation of the vulnerability to a particular mode of stability failure, established with Level 2 vulnerability criteria.
- Evaluation of the safety level and showing the adequacy of the assessment if the likelihood of stability failure for that failure mode is acceptable.
- Development of ship-specific operational guidance, if necessary.

The measure of likelihood of stability failure is one of the main results of direct-stability assessment. The long-term-averaged rate of failure events (λ_a), one of the best candidates for this measure, was proposed by Sevastianov (1963). (For an English version see Sevastianov 1994; or Belenky and Sevastianov 2007). There are several important properties of this value that are briefly reviewed below.

The long-term-averaged rate of events is related with a life-time probability of failure as

$$P(T_L) = \exp(-\lambda_a T_L), \tag{40.7}$$

where T_L is the expected life time of a vessel. In principle this allows comparison of the estimated probability of failure with the achieved safety level. The latter can be evaluated using the same direct stability assessment procedure on an older vessel with a good stability-safety record. The long-term-averaged rate of events is expressed as:

$$\lambda_a = \sum_i W_i \lambda_i. \tag{40.8}$$

here, λ_i is the rate of events (stability failures) in a particular hypothetical situation characterized by a given sea state, and ship speed and heading; and W_i is a statistical weight corresponding to a life-time exposure to such conditions. Determining the sources of data for the statistical weights also is the part of the procedure. Usually data of this type are obtainable from operational statistics of a prototype. An additional advantage of this approach is that the probability (40.7) can be re-evaluated if the mode of operation changes in the future.

In principle, the rate of events, λ_i, can be estimated simply by direct counting of large roll angles or capsizes during the numerical simulations. However, because of the rarity of stability failures and the complexity of using numerical tools, direct counting is not very practical. This problem is already recognized in various IMO documents as the "problem of rarity" (Belenky et al. 2008).

A solution of the problem of rarity is sought by the application of special extrapolation methods to the output of numerical simulations, which produce the rate of events, λ_i. These methods can take into account extreme nonlinearities of the intact-stability problem. A state-of-the-art review of these special extrapolation methods can be found in (Belenky et al. 2012).

Although these methods are based on physical considerations of the mechanisms of stability failures, they are not free from assumptions and therefore must be validated. Validation of the extrapolation methods also presents a challenge because of the same problem of rarity.

How can an extrapolation method be validated in principle? Extrapolation methods require several hours of numerical simulations of ship motions to predict a probability of stability failure with the average time of occurrence much longer than the simulation time. If a large statistical sample (with a sufficient number of stability failures) is available, a portion of the statistical sample can be used to run the extrapolation method, and then the results can be compared with direct counting. However, producing a data sample of such a large size may be neither possible nor practical because of the "problem of rarity" and the computational cost associated with it.

There are two approaches to solve this problem: the first is to use more severe conditions and the second is to use a less complex mode. In the first approach, a validated model is used to produce a large data sample under conditions that are more severe than normal conditions. Thus the stability failures may become more frequent and computational costs will be reduced to a more manageable level. This method is used in reliability engineering, e.g., when the reliability of an electronic component is tested using temperatures much higher than the normal operating temperatures. Then the data which have been obtained are recalculated for normal operating conditions (Meeker and Escobar 1998).

In the second approach, the "problem of rarity" is resolved by using less complex models of ship motions. Such models must possess all of the primary qualitative similarities of the validated numerical tool, but may be significantly less accurate. If the computational costs of running the reduced-complexity model are reasonably low, an extrapolation method can be validated under more or less realistic conditions.

40.6 Conclusions

Direct-stability assessment is intended to become the third tier of the second-generation IMO intact stability criteria that are under development. Although it is to be the newest state-of-the-art stability-assessment method, its use is envisioned only in exceptional cases of novel and unconventional ship designs. Notwithstanding the above, the maritime industry has experience with the application of advanced numerical methods in the regulatory and classification framework, i.e., finite element analysis of ship structures.

Successful application of direct-stability assessment within the regulatory framework requires addressing the following issues: (1) requirements for numerical tools, (2) validation of numerical tools, (3) extrapolation procedures, (4) validation of extrapolation procedures.

Based on the current state-of-the-art of numerical hydrodynamics, it seems that hybrid codes represent the best combination of fidelity and performance. However, numerical tools based on a system of ordinary differential equations are also useful for more complex problems related to stability failures due to broaching-to. In general, the requirement for a tool should be specific to the mode of stability failure. Requirements for validation of numerical tools are also dependent on the mode of stability failure and consist of both qualitative and quantitative validation.

The way in which the procedures are applied is an important part of the direct-stability assessment process because it prescribes the origin of the input, the post-processing methodology and the interpretation of the output. It is also necessary to include extrapolation procedures to handle the "problem or rarity". Extrapolation procedures must also be validated.

Acknowledgments This work was partially funded by ONR under Dr. L. P. Purtell. The authors are grateful to the following colleagues for their fruitful discussions and helpful comments: W. Belknap, B. Campbell, and T. Smith (David Taylor Model Basin, NSWCCD); K. Weems (SAIC); K. Spyrou (National Technical University of Athens); and N. Umeda (Osaka University).
The authors are grateful to Ms. Suzanne Reed for her detailed editing that has greatly improved clarity and readability of the text.

References

ABS (2004a) "Guide for the Assessment of Parametric Roll Resonance in the Design of Container Carriers." American Bureau of Shipping, Houston, TX, 70 p.

ABS (2004b) "Guidance Notes on Safehull Finite Element Analysis of Hull Structures." American Bureau of Shipping, Houston, TX, 52 p.

ASME, (2009) "Standard for Verification and Validation in Computational Fluid Dynamics and Heat Transfer." V V 20 2009, Amer. Soc. Mech. Engin., New York.

Beck, R. F. & A. M. Reed, (2001). Modern computational methods for ships in seaway. Trans. SNAME, 109:1–48.

Belenky, V. L. (2011) "On Self-Repeating Effect in Reconstruction of Irregular Waves" in Contemporary Ideas on Ship Stability, Neves, M. A. S., *et al.* (eds), Springer, pp. 589–598.

Belenky V., J. O. de Kat & N. Umeda (2008) "Towards Performance-Based Criteria for Intact Stability." Marine Tech., 45(2):101–123.

Belenky, V. L., D. Liut, K. M. Weems & Y. S. Shin (2003) Nonlinear roll with water-on-deck: numerical approach, Proc. 8th Int'l Conf. Stability of Ships & Ocean Vehicles (STAB'03), Madrid, Spain.

Belenky, V. L & N. B. Sevastianov (2007) Stability and Safety of Ships: Risk of Capsizing. (2nd ed) SNAME, Jersey City.

Belenky, V. & K. M. Weems (2011), "Probabilistic Properties of Parametric Roll." Chapter 6 of Parametric Resonance in Dynamical Systems, Fossen, T. I. & H. Nijmeijer (Eds.) Springer, NY.

Belenky, V., K. M. Weems, C. C. Bassler, M. J. Dipper, B. Campbell & K. Spyrou (2012) "Approaches to Rare Events in Stochastic Dynamics of Ships." Probabilistic Engineering Mechanics, 28:30–38.

Belenky, V., K. M. Weems, W. M. Lin & J. R. Paulling (2011a) Probabilistic Analysis of Roll Parametric Resonance in Head Seas Contemporary Ideas on Ship Stability, Neves, M. A. S., *et al.* (eds), Springer, pp. 555–572.

Belenky, V., H. Yu & K. M. Weems (2011b) "Numerical Procedures and Practical Experience of Assessment of Parametric Roll of Container Carriers." in Contemporary Ideas on Ship Stability, Neves, M. A. S., *et al.* (eds), Springer, pp. 295–305.

Brunswig, J. & R. Pereira (2006) Validation of Parametric Roll Motion Predictions for a Modern Containership Design. Proc. 9th Int'l Conf. Stability Ships & Ocean Vehicles (STAB '06), Vol. 1, Rio de Janeiro, Brazil, pp. 157–168.

Belknap, W. F. & A. M. Reed (2010) TEMPEST — A New Computationally Efficient Dynamic Stability Prediction Tool. Proc. 11th Int'l Ship Stability Workshop, Wageningen, the Netherlands, pp. 185–197.

Belknap, W. F., T. C Smith & B. Campbell (2011) Addressing Challenges in the Validation of Dynamic Stability Simulation Tools, Proc. 12th Int'l Ship Stability Workshop, Washington, DC. USA, pp. 81–90.

Degtyarev A. B. & I. Gankevich (2012) Evaluation of Hydrodynamic Pressures for Autoregression Model of Irregular Waves, Proc. 11th Int'l Conf. Stability of Ships & Ocean Vehicles (STAB '12), Athens, Greece.

Degtyarev, A. B. & A. M. Reed (2011) Modeling of Incident Waves near the Ship's Hull (Application of Autoregressive Approach in Problems of Simulation of Rough Seas) Proc. 12th Int'l Ship Stability Workshop, Washington, DC, USA, pp 175–188.

France, W. M, M, Levadou, T. W. Treakle, J. R. Paulling, K. Michel & C. Moore (2003). An Investigation of Head-Sea Parametric Rolling and its Influence on Container Lashing Systems, Marine Tech., 40(1):1–19.

Francescutto, A., G. Contento & R. Penna (1994) Experimental Evidence of Strong Nonlinear Effects in the Rolling Motion of the Destroyer in Beam Seas, Proc. 5th Int'l Conf. Stability of Ships and Ocean Vehicles (STAB '94), Vol. 1, Melbourne, Florida, USA.

Grochowalski, S., C. C. Hsiung, Z. J. Huang & L. Z. Cong (1998). Theoretical Modelling of Ship Motions and Capsizing in Large and Steep Waves Trans. SNAME, 106:241–267.

IMO MSC.1/Circ.1227 (2007) Explanatory Notes to the Interim Guidelines for Alternative Assessment of the Weather Criterion, London, UK, 23 p.

IMO SLF 53/3/8 (2010) "Comments on Proposed Criteria for Surf-riding and Broach-ing." submitted by Japan and the United States, London, 2010.

IMO SLF 54/INF.12 (2011) Information Collected by the Correspondence Group on Intact Stability, Submitted by Japan, London, UK, 147 p.

Meeker, W. O. & L. A. Escobar (1998) Statistical Methods for Reliability Data. Wiley, New York, 680 p.

Ogawa, Y, N. Umeda, D. Paroka, H. Taguchi,, S. Ishida, A. Matsuda, H. Hashimoto & G. Bulian (2008) "Prediction Methods for Capsizing under Dead Ship Condition and Obtained Safety Level — Final Report of SCAPE Committee (Part 4)", Proc. Osaka Colloquium on Seakeeping and Stability of Ships, Osaka, Japan, pp. 253–265.

Peters, W., V. Belenky, C. Bassler, K. Spyrou, N. Umeda, G. Bulian & B. Altmayer (2011) "The Second Generation of Intact Stability Criteria An Overview of Development." Trans SNAME. Vol. 119.

Rahola, J. (1939) "The Judging of the Stability of Ships and the Determination of the Minimum Amount of Stability Especially Considering the Vessel Navigating Finnish Waters." PhD Thesis, Technical University of Finland, Helsinki, viii + 232 p.

Reed, A. M. (2008) Discussion of: Belenky, V., J. O. de Kat & N. Umeda (2008) "To-wards Performance-Based Criteria for In-tact Stability." Marine Tech., 45(2):122–123.

Reed, A. M. (2009) A Naval Perspective on Ship stability, Proc. 10th Int'l Conf. Stability of Ships & Ocean Vehicles (STAB '09), St. Petersburg, Russia, pp. 21–43.

Reed, A. M. (2011) 26th ITTC Parametric Roll Benchmark Study, Proc. 12th Int'l Ship Stability Workshop, Washington DC, USA, pp. 195–204.

Sadat-Hosseini, H., P. Carrica, F. Stern, N. Umeda, H. Hashimoto, S. Yamamura & A. Mastuda (2011) CFD, system-based and EFD study of ship dynamic instability events: Surf-riding, periodic motion, and broaching. Ocean Engineering, 38:88–110.

Shigunov, V. (2009) Operational Guidance for Prevention of Container Loss. Proc. 10th Int'l Conf. Stability of Ships & Ocean Vehicles (STAB '09), St. Petersburg, Russia, pp. 473–482.

Shin, Y. S., V. L. Belenky, W. M. Lin, K. M. Weems & A. H. Engle (2003) "Nonlinear time domain simulation technology for sea-keeping and wave-load analysis for modern ship design." Trans SNAME. 111:557–578.

Smith, T. C. (2011) Statistical Data Set Comparison for Continuous, Dependent Data, Proc. 12th Int'l Ship Stability Workshop, Washington, DC. USA pp. 75–80.

Smith, T. C. (2012) Approaches to Ship Motion Simulation Acceptance Criteria, Proc. 11th Int'l Conf. Stability of Ships & Ocean Vehicles (STAB '12), Athens, Greece.

Sevastianov, N. B. (1963). "On probabilistic approach to stability standards." Trans. Kaliningrad Institute of Technology, 18:3–12. (in Russian).

Sevastianov, N. B. (1994). "An algorithm of probabilistic stability assessment and standards" Proc. 5th Int'l Conf. Stability of Ships & Ocean Vehicles (STAB '94), Vol. 5, Melbourne, Florida, USA.

Spyrou, K. (1995). Surf-riding and oscillations of a ship in quartering waves, Journal of Marine Science and Technology, Vol. 1, Issue 1, pp. 24–36.

Spyrou, K. (1996). "Dynamic instability in quartering seas: the behaviour of a ship during broaching." J. Ship Research, 40(1):46–59.

Spyrou, K., K. Weems & V. Belenky (2009) Patterns of Surf-riding and Broaching-to Captured by Advanced Hydrodynamic Modeling Proc. 10th Int'l Conf. Stability of Ships & Ocean Vehicles, St. Petersburg, Russia.

Themelis, N. & K. J. Spyrou (2007) Probabilistic Assessment of Ship Stability. Trans. SNAME, 117:181–206.

Umeda, N., S. Koga, J. Ueda, E. Maeda, I. Tsu-kamoto & D. Paroka (2007) Methodology for Calculating Capsizing Probability for a Ship under Dead Ship Condition, Proc. 9th Int'l Ship Stability Workshop, Hamburg, pp. 1.2.1–1.2.19.

Umeda, N. & Y. Yamakoshi (1993). Probability of Ship Capsizing Due to Pure Loss of Stability in Irregular Quartering Seas, Naval Architecture and Ocean Engineering, Vol. 30.

Yen T. G., S. Zhang, K. Weems & W-M. Lin (2010) Development and Validation of Numerical Simulations for Ship Maneuvering in Calm Water and in Waves, Proc. 28th Symp. Naval Hydro., Pasadena, California.

Chapter 41
A Probabilistic Analysis of Stability Regulations for River-Sea Ships

Igor Bačkalov

Abstract The present contribution focuses on the recent developments in the field of stability regulations for the river-sea ships in Europe. The river-sea ships are primarily inland vessels that operate on the coastal sea routes on a regular basis. Due to their particular service and design, the river-sea ships may be subject to special stability requirements that incorporate seakeeping considerations to an extent. In the present analysis, some of the existing stability rules (Russian River Register) and working proposals of regulations (UNECE Resolution No. 61) intended for the river-sea ships are evaluated from the probabilistic point of view. It is demonstrated that, in some cases, these regulations are insufficient and not strict enough. Critical analysis is followed by some guidelines for possible improvement of regulations and development of new intact stability criteria for river-sea ships.

41.1 Introduction

In the recent years, several initiatives for improvement of safety of river-sea ships were launched on both the national and international level. In some countries, such as Belgium and France, new regulations already came into force. In the same time, the international (European) regulatory framework, i.e. United Nations Economic Commission for Europe (UNECE) Resolution No. 61 (2006a) was to be updated with the provisions for stability of river-sea ships as well. The present investigation provides an overview of the afore-mentioned developments, with special emphasis on the requirements set forth in the Russian River Register (2008). The principal reason for this particular focus is the fact that the criteria (deterministic in nature) originally established by the Russian classification societies (River and Maritime Register) serve as the basis for development of the river-sea ships stability regulations on the European level.

I. Bačkalov (✉)
Department of Naval Architecture, Faculty of Mechanical Engineering, University of Belgrade, Kraljice Marije 16, 11120 Belgrade, Serbia
e-mail: ibackalov@mas.bg.ac.rs

V. L. Belenky et al. (eds.), *Contemporary Ideas on Ship Stability*, Fluid Mechanics and Its Applications 119, https://doi.org/10.1007/978-3-030-00516-0_41

The stability requirements imposed by the Russian River Register (2008) as well as the regulations proposed by UNECE (2010) were analyzed by means of probabilistic tools, previously described in e.g. Hofman and Bačkalov (2005). Probability of a partial stability failure (flooding through unsecured openings) and lateral accelerations in the working areas positioned furthermost from the rolling axis were calculated in a series of numerical experiments performed on the sample ships, whose main particulars are typical for river-sea navigation in Europe.

Numerical tests reveal that the level of safety attained by present stability standards considerably varies depending of ship characteristics. Furthermore, it was found that vessels satisfying examined regulations could not be considered as safe from the probabilistic point of view in a number of cases. Nevertheless, stability requirements of Russian River Register ("acceleration criterion") constitute a solid basis for development of the next generation of intact stability criteria for river-sea ships.

It should be noted that the contribution provides results of an independent analysis, as the author is not affiliated with either UNECE or Russian River Register.

41.2 Stability Regulations for the River-Sea Ships: State-of-the Art and Prospects

The terms "river-sea ship" and "sea-river ship" are sometimes used to describe the same vessel type. On the other hand, either of the terms may also be used to refer to very dissimilar ship types. Here, however, a principal distinction would be made between the two. The term "sea-river ship" would be used for a seagoing vessel that also operates on inland waterways. Unlike that, a "river-sea ship" would represent primarily an inland vessel that, under certain conditions, performs sea voyages on a regular basis.

Clearly, an inland vessel is not designed for the maritime environment; therefore, additional safety measures, including special stability provisions, are required. So far, process for improvement of river-sea ships safety resulted in new regulations applicable on the local level (in Belgium and France), while the European regulations are underway.

41.2.1 Belgium

In 2007, following the research carried out by the Ghent University (see Truijens et al. 2006), Belgian Federal Authorities launched safety regulations intended for the so-called estuary ships. Estuary ships are basically inland navigation vessels which are allowed to operate on the sea routes between Belgian deep-sea ports and the West Scheldt estuary. The regulations, which are published in the Belgisch Staatsblad/Moniteur Belge (2007), impose several probabilistic requirements related to slamming, bow diving, shipping of green water, rolling, habitability and ship strength. Sufficient safety should be demonstrated by direct calculations. Wave loads are obtained from measured directional spectra. Regardless of the use of linear seekeping theory and simplified treatment of wind in analysis of roll motion (which required establishment of some deterministic margins) Belgian regulations introduced an advanced approach to ship stability. Furthermore, for the first time, probabilistic calculations and risk analysis became an integral part of the intact ship stability regulatory framework (see for instance Vantorre et al. 2012). The part of the regulations addressing ship rolling in beam wind and waves was thoroughly analyzed by Bačkalov (2010).

41.2.2 France

"Port 2000" is the new container terminal in the Le Havre seaport, designed to facilitate the largest container ships. Le Havre is situated near the river Seine estuary, enabling waterborne transport connection between the major maritime port and the hinterland, in particular with the Greater Paris area. During 2007, in order to provide access of inland container vessels to the new container terminal, French transport authorities have issued two decrees, published in Journal officiel de la République Française (2007), dealing with safety of inland vessels in maritime environment. Inland container vessels using the North access to the Port 2000 (involving a short sea trip) were required to prove sufficient stability in waves, for a number of loading conditions, speeds and wave headings, by means of model experiments or computer simulations approved by a classification society. At the time, the navigation was limited to wave heights up to 1.2 m and mean wind speeds up to 21 km. However, by the end of 2014, the regulations were updated with a novel set of rules (see Journal officiel de la République Française 2014) very much in line with the safety requirements for Belgian estuary ships. Consequently, the navigation was extended to significant wave height $H_{1/3} = 2$ m.

Belgian experiences with estuary navigation and specific navigation requirements in the port of Le Havre have motivated Permanent International Association of Navigation Congresses (PIANC) to appoint an expert group on "Direct Access of Maritime Ports by Inland Waterway Vessels" in 2011, which clearly indicates a necessity to thoroughly investigate the problem of river-sea ships safety.

41.2.3 Europe

On the European level, consideration of safety requirements for the river-sea ships took a different course. The appropriate regulations are introduced through the UNECE Resolution No. 61 (UNECE 2006a). The Resolution represents a step towards the Europe-wide harmonization of technical regulations for inland vessels. Chapter 20 of the Resolution No. 61, dedicated to "special provisions for sea-going ships", was originally left void. In 2006, upon invitation by UNECE Inland Transport Committee, Russian Federation (which has a considerable number of registered ships for river-sea navigation and, consequently, the appropriate experience) submitted the basic document on technical requirements for the river-sea ships (UNECE 2006b). In this document, the regulations used in Russian Federation, including ship stability rules, were indicated as the starting point for development of the appropriate European rules.

During 2010 and 2011, the Group of Volunteers submitted the draft versions of Chapter 20B: Special provisions applicable to river-sea navigation vessels, to the UNECE Working Party on the Standardization of Technical and Safety Requirements in Inland Navigation (SC.3/WP.3). At the time, it was expected that the draft of Chapter 20B could be finalized in spring 2012. The first draft of the Chapter 20B (UNECE 2010) submitted in the 36th session of SC.3/WP.3, was used as the reference document in the present analysis.

UNECE (2010) defines a river-sea vessel as a "vessel intended for navigation on inland waterways and suitable for restricted navigation at sea". River-sea vessels are divided in classes by "navigation areas" they are allowed to operate on:

- Class RS 6.0, in waves up to $H_{3\%} = 6$ m;
- Class RS 4.5, in waves up to $H_{3\%} = 4.5$ m;
- Class RS 3.5, in waves up to $H_{3\%} = 3.5$ m;
- Class RS 3.0, in waves up to $H_{3\%} = 3$ m;
- Class RS 2.0, in waves up to $H_{3\%} = 2$ m,

where $H_{3\%}$ stands for wave height with 3% probability of exceedance.

Stability should be verified for a number of loading conditions. Stability is deemed as sufficient if the ship satisfies the following: minimal metacentric height, parameters of static stability diagram, a weather criterion (corresponding to wind speed of 18 m/s) as well as supplementary provisions for specific ship types. The latter include "dry bulk cargo ships" whose stability is additionally checked by the "acceleration criterion". Acceleration criterion determines limiting operational conditions of a river-sea ship, i.e. permitted wave heights that would allow safe navigation. Estimated acceleration a_{est} in transverse direction should not be larger than 0.3, in g-fractions.

$$k^* = \frac{0.3}{a_{est}} \geq 1. \tag{41.1}$$

Estimated acceleration is calculated as:

$$a_{est} = 1.1 \times 10^{-3} \cdot B \cdot X_1^2 \cdot \varphi_1, \tag{41.2}$$

Table 41.1 Wave heights limitations based on acceleration criterion, UNECE (2010)

k^*	<0.5	$\geq$0.5 & $\leq$1	$\geq$1 & $\leq$1.5	$\geq$1.5 & $\leq$2	>2
$H_{3\%}$ (m)	2	3	3.5	4.5	6

where B is the beam, X_1 and roll amplitude φ_1 are determined in compliance with the 2008 Intact Stability Code (A/2.3).[1] In case that $k^* < 1$, Administration may permit a vessel to operate with wave height restrictions, "upon well-grounded presentation by the owner". Permitted wave heights, as a function of the k^* value, are given in the Table 41.1.

41.2.4 Russian Federation

As outlined before, the approach based on the acceleration criterion, proposed by the UNECE (2010) is adopted from the Russian stability regulations for the river-sea ships. There are two classification societies in Russian Federation, concerned with the safety of ships for combined navigation: Russian River Register (RRR) and Russian Maritime Register of Shipping (RS). Stability assessment of the river-sea ships in both registers includes the "acceleration criterion", i.e. a condition similar to the formula (41.1). Although both registers refer to the "river-sea ships", classes assigned by RRR and RS as well as associated navigation areas are substantially different and so are the formulas and related coefficients. Using the terminology herein adopted, RS in fact deals with the sea-river ships. The river-sea vessels, as defined by UNECE (2010), are within the scope of the RRR.

Russian Maritime Register (2005) specifies two classes of ships for combined navigation:

- Class IIСП, in waves up to $H_{3\%} = 6$ m;
- Class IIIСП, in waves up to $H_{3\%} = 3.5$ m.

Russian River Register (2008) indicates three classes of ships for combined navigation:

- Class M-СП, in waves up to $H_{3\%} = 3.5$ m;
- Class M-ПР, in waves up to $H_{3\%} = 2.5$ m;
- Class O-ПР, in waves up to $H_{3\%} = 2$ m.

In addition to prescribed minimal metacentric height, static stability diagram parameters and weather criterion, dry bulk cargo river-sea ships of class M-СП are required to satisfy the acceleration criterion $a_{est} \leq 0.3\ g$. Estimated acceleration is calculated as:

$$a_{est} = 1.1 \times 10^{-3} \cdot B \cdot m_1^2 \cdot \theta_m. \quad (41.3)$$

[1]In UNECE (2010), roll amplitude φ_1 is denoted by θ_r.

Table 41.2 Wave heights limitations based on acceleration criterion, RRR (2008)

k^*	<0.5	≥0.5 & ≤1	>1
$H_{3\%}$ (m)	2.5	3	3.5

The coefficient m_1 is derived based on the natural frequency of the ship, while θ_m represents the resonant roll amplitude, calculated according to the rules of the Russian River Register. Again, wave height restrictions apply when $k^* < 1$ (Table 41.2). According to Lesyukov (1974), a_{est} should correspond to the estimated acceleration in the working spaces furthermost positioned from the rolling axis, such as bridge. Nevertheless, estimated accelerations (as defined by RRR and UNECE) do not depend on the bridge height. Instead, it seems that the wheelhouse position (or another appropriate location) typical for standard ships of the specified class is incorporated in coefficients contained in formulas (41.2) and (41.3).

Although formulas (41.2) and (41.3), given by UNECE (2010) and Russian River Register (2008) respectively, are quite alike, the results obtained by their application on the same vessel may be very dissimilar, as it is going to be demonstrated later in this paper.

41.3 Probabilistic Analysis

Probabilistic analysis used in the present investigation is a two-phase approach. In the first phase, the roll motion of the sample vessels in the dead-ship condition, subjected to simultaneous action of wind and waves, is derived. The time history of roll motion is subsequently statistically analyzed, in an aim to estimate the probability of an event considered as critical for the safety of the examined vessel. The method was previously described in detail elsewhere, e.g. in Hofman and Bačkalov (2005), Bačkalov et al. (2010).

41.3.1 Sample Ships

Following the UNECE (2010) definition of the river-sea ships, the selected sample vessels *V* and *ST* have the hull form of the standard European inland ships, whereas their main particulars are typical for combined navigation (Table 41.3). Nevertheless, the ships *V* and *ST* have minimal freeboards in accordance with the UNECE (2010) and Russian River Register (2008) requirements, while in practice, river-sea ships may have higher freeboards than required. The vessels of the same size and hull form, V_{sp} and ST_{sp}, but with some features that are more common in shipbuilding practice will be introduced and analyzed later in the text. An outline of a typical inland vessel featuring high *L/B* ratio and a box-shaped cargo hold is shown in Fig. 41.1;

Table 41.3 Sample ships

Name	Length L (m)	Beam B (m)	Draught d (m)	Depth D (m)
V	136	16.5	3.5	5.7
ST	83.6	12	2.9	3.9

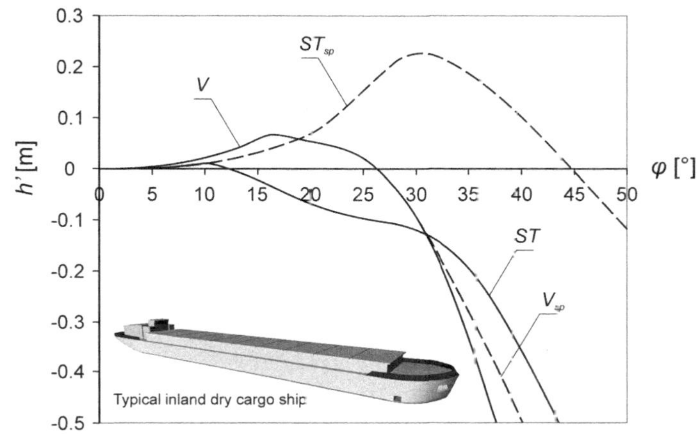

Fig. 41.1 Residuary stability levers h' of the examined vessels: dashed lines (also denoted by subscript sp) correspond to residuary stability levers of V-type and ST-type vessels with some specific design features common in shipbuilding practice

the residuary stability levers [see Eqs. (41.8) and (41.9)] of the analyzed vessels are given in the same Figure.

41.3.2 Numerical Experiments Setup

Numerical experiments should indicate whether sample ships, which fulfil stability requirements prescribed by the regulations, may be considered as safe from the probabilistic point of view, in waves determined based on the acceleration criteria (Tables 41.1 and 41.2) and mean wind specified by class notation.

In the present analysis, roll motion of the sample ships, subjected to stochastic action of beam wind and waves, is modelled using the one degree of freedom (1DOF) nonlinear differential equation of roll motion.

$$\left(J_x + m_\varphi\right) \cdot \ddot{\varphi} + M_{damp}(\dot{\varphi}) + M_{st}(\varphi) = M_{wave}(t) + M_{wind}(t) \tag{41.4}$$

where J_x is the moment of inertia for the x axis, m_φ is the added mass of roll, M_d is the damping moment, M_{st} is the righting moment, and M_{wave} and M_{wind} are the exciting wave and wind moments, respectively.

The wind speed, v_w, is not constant, but consists of the mean wind speed, $\bar{v}_w$, and wind speed fluctuations:

$$v_w(t) = \bar{v}_w + \sum_{n=1}^{N} A_n \cos(\omega_n t + \alpha_n), \tag{41.5}$$

where ω_n is the frequency and α_n is the phase shift of n-th wind component. Gusting wind amplitudes, A_n, are obtained from the semi-empirical Davenport wind spectrum S_D:

$$A_n = \sqrt{2S_D(\omega_n) \cdot d\omega}, \tag{41.6}$$

where:

$$S_D = \frac{4K \cdot \bar{v}_w^2 X_D^2}{\omega_n \left(1 + X_D^2\right)^{\frac{4}{3}}}, \; X_D = \frac{600\omega_n}{\pi \cdot \bar{v}_w},$$

The dimensionless terrain roughness coefficient is taken as $K = 0.003$ (the value appropriate for open sea).

In the present analysis, wave amplitudes, B_n, are acquired from the Bretschneider wave spectrum S_B:

$$B_n = \sqrt{2S_B(\omega_n) \cdot d\omega}. \tag{41.7}$$

Righting moment is the restoring term in Eq. (41.4):

$$M_{st}(\varphi) = g\Delta \cdot GZ(\varphi) = g\Delta \cdot \left[h'(\varphi) + GM \sin\varphi\right], \tag{41.8}$$

where *GZ* is total stability lever (righting lever), Δ is ship displacement and *GM* is metacentric height. Residuary stability lever of the vessel was approximated by a high order odd polynomial:

$$h'(\varphi) \approx \sum_{n=0}^{N} a_{2n+1}\varphi^{2n+1}. \tag{41.9}$$

The roll damping moment is nonlinear, consisting of the quadratic term:

$$M_{damp}(\dot{\varphi}) = \beta \cdot \left(J_x + m_\varphi\right) \cdot \dot{\varphi}|\dot{\varphi}|. \tag{41.10}$$

Since the roll damping significantly affects calculated roll motion and, consequently, ship safety considerations, proper assessment of roll damping coefficients is an important and a delicate matter. This is particularly true for inland vessels, as typical inland forms fall out of scope of applicability of well-established semi-empirical methods for roll damping estimation, while the available experimental data

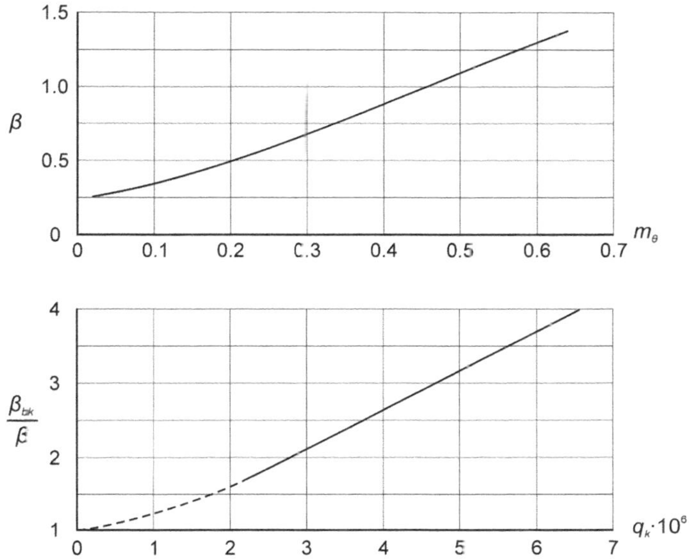

Fig. 41.2 Quadratic coefficient of roll damping β, and increase of quadratic roll damping coefficient due to bilge keels β_{bk}, Basin and Anfimov (1961)

are scarce. This difficulty was highlighted in several papers so far (see for instance Bačkalov et al. 2010). In the present analysis, nonlinear roll damping was assessed using the recommendation given by Basin and Anfimov (1961), Fig. 41.2. Quadratic roll damping coefficient β is expressed as a function of m_θ:

$$m_\theta = \left(\frac{r_s}{\rho_{xx}}\right)^2 \frac{GM}{B}, \tag{41.11}$$

where ρ_{xx} is the roll radius of gyration and r_s is the average distance between the roll axis and the hull surface, depending on the waterplane coefficient and *B*/*d* ratio.

Quadratic coefficient β was derived based on the results of model tests with single- and twin-screw inland vessels and seagoing ships, without the influence of forward speed and bilge keels.

Assuming that the river-sea vessels are normally equipped with bilge keels, the following recommendation (also based on the model experiments, see Basin and Anfimov 1961) for quadratic damping coefficient of the ship with bilge keels, is adopted, whereby:

$$\frac{\beta_{bk}}{\beta} = f(q_K), \; q_K = \frac{s_K}{LB}\left(\frac{d_{bk}}{B}\right)^3 \frac{KG}{d} \times 10^{-3}, \tag{41.12}$$

where s_K is total area of bilge keels, d_{bk} is the normal distance of bilge keels from vessel's rolling axis and KG is vertical position of vessel's centre of gravity.

Equation (41.4) is solved numerically using Runge-Kutta method, in order to obtain time record of roll. In the second phase of analysis, time history of roll is statistically analyzed in order to acquire mean angle of roll $\bar{\varphi}$ and standard deviation of roll σ_φ. Using these statistical values, as it was demonstrated by Bulian and Francescutto (2004) the probability that the vessel would heel to a critical heel angle ϕ in a given period of time t_s, is determined as:

$$P = 1 - \exp\left\{-N_c \exp\left[-\frac{1}{2}\left(\frac{\phi - \bar{\varphi}}{\sigma_\varphi}\right)^2\right]\right\}, \tag{41.13}$$

where N_c is the number of cycles related to number of zero crossings. In the present analysis, critical angle ϕ is the angle of flooding through unsecured openings, defined as the angle at which the top of hatch coaming enters the water. Probabilistic criterion of stability used in the investigation permits probability of flooding $P_{fl} = \boldsymbol{O}(10^{-4})$ in two hours of specified weather conditions. This probabilistic criterion was adopted in Bačkalov (2010) assuming that flooding through unsecured openings may occur once in a lifetime of a coastal ship (which typically makes 300 round trips a year during 20 years of operation, whereby each trip takes couple of hours) and it was preserved in the present investigation for the sake of comparison. Clearly, the choice of an appropriate safety level is a sensitive issue, open to discussion.

Furthermore, it should be noted that the formula (41.13) is used assuming that the ship roll is Gaussian and ergodic process. These approximations enable assessment of important safety-related quantities; hence they are utilized in a number of comprehensive studies dealing with probabilistic approach to ship stability in realistic weather conditions (see e.g. Bulian and Francescutto 2004; Vassalos et al. 2004).

41.3.3 Discussion of Results

Probability of flooding in weather conditions defined by the Russian River Register (2008) (mean wind speed $\bar{v}_w = 24$ m/s, wave height based on the acceleration criterion, see Table 41.2) for examined sample vessels *ST* and *V* is given in Fig. 41.3 (full lines). Breaks in the curves corresponding to the same sample vessel are a consequence of the step change in permitted wave heights (see also Fig. 41.4). The roll standard deviation and the mean roll angle are estimated from 628 s long simulations, with 0.1 s step, whereby the first 28 s are omitted so as to exclude the transient effects.

Although the application of acceleration criterion should secure sufficient safety in given conditions over the range of realistic metacentric heights (provided that the other stability criteria of RRR are fulfilled as well), obviously this is not the case as the probabilities of flooding of *V* and *ST* sample ships (solid lines in Fig. 41.3) attain

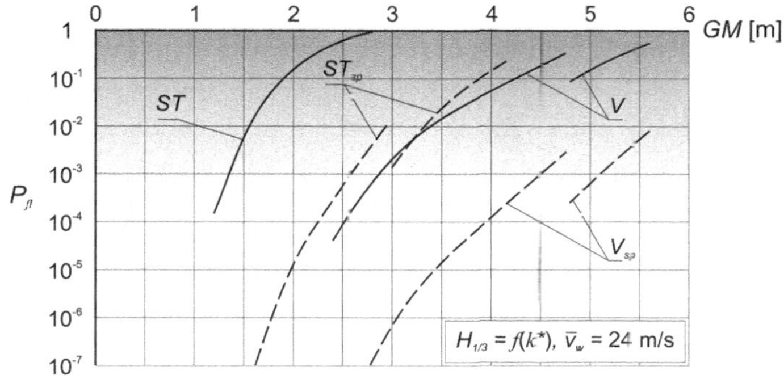

Fig. 41.3 Probability of cargo hold flooding (P_{fl}) of the examined river-sea ships; dashed lines (also denoted by subscript *sp*) correspond to P_{fl} of *V*-type and *ST*-type vessels with some specific design features common in shipbuilding practice

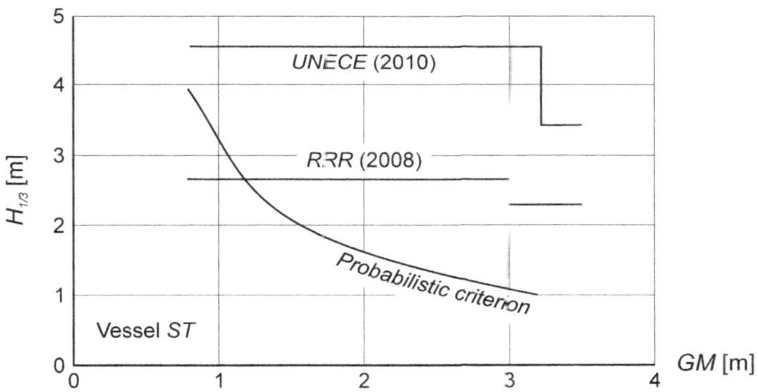

Fig. 41.4 An example of permitted wave heights according to the existing rules and proposals, and the probabilistic criterion based on the risk of flooding

values higher than 10^{-4}, for a large span of *GM*. In other words, the sample vessels *V* and *ST* fail to fulfill the probabilistic criterion, although they meet the terms of the acceleration criterion.

According to available data (see comprehensive study of Egorov 2007) dynamic stability related accidents do not seem to be as frequent as probabilistic analysis indicates. Nevertheless, river-sea ships are highly susceptible to wave climate: in some 70 accidents analyzed by Egorov (2007), almost half occurred when class-prescribed wave height limitations were exceeded (either intentionally or due to forecast errors). In addition, in over 30% of cases, a violation of ICLL (International Convention on Load Lines) conditions of assignment requirements for construction of hatch covers, coamings, air pipes and watertight doors (i.e. possible points of water ingress) was reported. So, is there really a disagreement between numerical experiments and operational experience?

In an attempt to resolve this dilemma (that is, why sample vessels which fulfil the acceleration criterion fail to satisfy the probabilistic criterion for a number of loading cases), a new series of numerical experiments was performed on sample vessels whose characteristics were in accordance with the current shipbuilding practice. For instance, V-type ships may have exceptionally high hatch coaming, up to 3.5 m (instead of 2 m so far used in the analysis); ST-type ships, on the other hand, commonly have freeboards much higher than minimal, i.e. $D = 6$ m instead of 3.9 m (see e.g. Egorov 2007). Probabilistic results obtained for such vessels are also given in Fig. 41.3 (dashed lines V_{sp} and ST_{sp}). Obviously, these vessels may be considered as much safer from the probabilistic point of view. The ships that are known to perform well in terms of stability in class-defined weather conditions, satisfy the probabilistic criterion as well, for a reasonable span of metacentric heights. However, the higher safety is not a consequence of stability regulations. In case of vessel V_{sp}, for instance, high coamings represent strong structural elements which contribute to overall longitudinal strength by increasing hull section modulus (see Egorov and Ilnitskyy 2006).

The goal of the next series of numerical tests was to calculate the wave heights that would yield a probability of flooding of the same level (i.e. $\boldsymbol{O}(10^{-4})$) for any GM in the examined range. The mean wind speed was assumed to be 24 m/s, as required by the RRR. Using the probabilistic analysis, the upper boundary of weather conditions in which a sample vessel of a particular set of characteristics (metacentric height, angle of flooding) could safely operate was defined. Wave height limitations obtained by application of the probabilistic criterion, RRR and UNECE requirements are compared in Fig. 41.4. In general, present regulations would allow sailing in wave heights higher than those permitted by the adopted probabilistic condition.

Following the RRR safety concept, each numerical test included the calculation of time history of lateral acceleration in the helmsman position in the wheelhouse, in the centreline of the ship, divided by gravity acceleration g:

$$a_T = \frac{l \cdot \ddot{\varphi}}{g}, \tag{41.14}$$

where l represents the distance of the bridge from the rolling axis. The probability that the limiting value $a_T = 0.3$ would be attained in the specified point on the ship was also derived for a range of metacentric heights. This probability, denoted as $P(a_T > 0.3)$, should be restricted based on the assessment of impact of high lateral accelerations on ship safety (which, however, falls out of the scope of present research). Other limiting values for lateral accelerations could be considered, see for instance Shigunov et al. (2011) and Ross (2009). For the purposes of present analysis, it will be assumed that $P(a_T > 0.3) = \boldsymbol{O}(10^{-3})$ may be permitted. Thus, in this way an additional safety requirement, "modified acceleration criterion" could be introduced. The safety assessment of a river-sea ship may be carried out in the following manner (Fig. 41.5). The probability of flooding is considered as the basic safety criterion. Consequently, the first step would be to calculate the upper limit of permitted significant wave heights $H_{1/3}$ based on the condition $P_{fl} = \boldsymbol{O}(10^{-4})$. This is carried out for

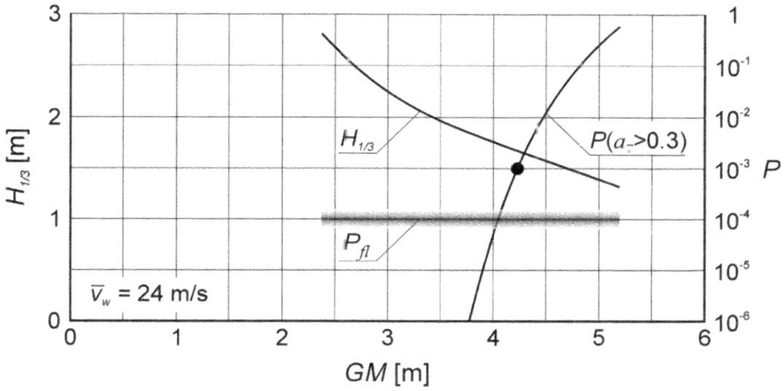

Fig. 41.5 Safety assessment of sample ship V using modified acceleration criterion

a range of metacentric heights (curve $H_{1/3}$). Then, for the obtained wave heights, the probability that the acceleration 0.3 g would be attained at the helmsman's position is estimated, for the same range of GM. In the given example shown in Fig. 41.5, the modified acceleration criterion is exceeded in waves corresponding to an acceptable probability of flooding if $GM \geq 4.2$ m (indicated with a black dot); permitted wave heights, therefore, should be lowered accordingly from that point for greater values of GM.

In Fig. 41.6, the modified acceleration criterion is used to compare two designs of the same length, breadth and draft but different freeboard. Due to a relatively high freeboard, the probabilities of cargo hold flooding of the sample vessel ST_{sp} are much lower than corresponding values for the vessel ST; that is, ST_{sp} could safely operate in higher sea states than ST, in some cases even higher than prescribed by RRR. However, a higher freeboard also implies that the bridge (i.e., wheelhouse) on the vessel ST_{sp} is located farther from the rolling axis within the same range of metacentric heights. As a consequence, considerably higher lateral accelerations are attained in the wheelhouse.

So, although ship ST_{sp} may be regarded as safer from one point of view (as far as the flooding through unsecured openings is concerned), her crew would experience ship motion induced disturbance that could lead to navigation errors and ultimately affect the safety of the vessel. In order to diminish these effects, following the modified acceleration criterion, the ship ST_{sp} would have to sail in wave heights lower than determined by the probabilistic criterion of flooding, in case the GM exceeds ≈ 1.5 m.

So far, the analysis focused on the stability requirements of the Russian River Register (2008). The UNECE (2010) proposal generally prescribes higher wave heights, as demonstrated in Fig. 41.4. Therefore, in the next series of experiments, the P_{fl} in weather conditions defined by UNECE (2010) was estimated for the "safer" sample vessels V_{sp} and ST_{sp} (Fig. 41.7). UNECE (2010) clearly overestimates the safety of vessels in waves. Though this could be the subject of a separate analysis, some possible reasons could be the following. The UNECE (2010) proposal covers the complete wave height range otherwise considered by the Russian River and

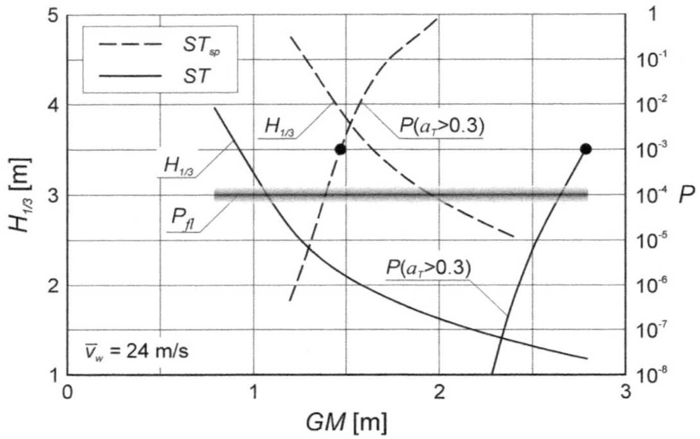

Fig. 41.6 Comparison of two designs using a modified acceleration criterion

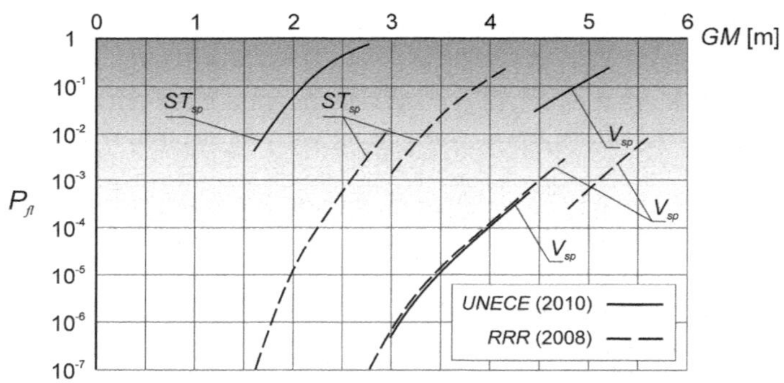

Fig. 41.7 Probability of cargo hold flooding (P_{fl}) of V_{sp} and ST_{sp} ships in weather conditions defined by UNECE (2010) and Russian River Register (2008)

Maritime Registers together. This could represent an effort of the authors of UNECE (2010) to extend the operation of the river-sea ships to the higher sea states, or perhaps to integrate the rules for sea-river and river-sea ships in one document.

However, the Russian classification societies use different calculation procedures to treat these two ship types, taking into account their specific features. A uniform, deterministic approach used in UNECE (2010) apparently fails to encompass both types.

Nevertheless, the revised Resolution 61 (UNECE 2013) did not bring forward the concept of an acceleration criterion. According to the adopted, amended version of Chapter 20B, the stability of river-sea ships is to comply with the weather criterion, as defined in the 2008 Intact Stability Code for seagoing ships, and minimal metacentric height requirements.

41.4 Conclusions

The goal of the contribution was to present an overview and an analysis of the recent developments in the field of stability regulations for river-sea ships. The rules of the classification societies of the Russian Federation are probably the most comprehensive ones for this type of ships, and as such were used as the basis for development of new safety requirements for the river-sea ships on the European level (Chapter 20B of the UNECE Resolution No. 61). Therefore, a probabilistic investigation of the stability requirements laid down by Russian River Register (2008) was carried out. The author was not involved in development of the analyzed regulations; hence the investigation presents an independent, personal view.

It should be noted that Russian classification societies recognized the perils of excessive accelerations and incorporated a seakeeping approach as a remedy. However the present analysis demonstrated that a deterministic acceleration criterion, which is used to define operational limits of river-sea ships, failed to provide sufficient safety, from the probabilistic point of view, in a number of realistic cases. In the performed numerical tests, wave heights permitted by the Russian River Register (2008) and particularly UNECE (2010) generated high probabilities of flooding of the cargo hold for sample vessels which fulfilled stability criteria laid down by regulations. However, vessels with high freeboards and hatch coaming heights, common in shipbuilding practice, performed much better in tests, which may explain the low rate of accidents related to dynamic stability of river-sea ships in waves.

As it was correctly pointed out by Shigunov et al. (2011), physical and mental fatigue, induced by ship motions, may influence the crew performance and lead to increase of errors, which in turn adversely affect safety. Numerical examples given in this investigation, related to accelerations on the bridges of analyzed river-sea ships (Fig. 41.6) demonstrated that safety assessment is a complex and multifaceted task. Although ship ST_{sp}, designed according to current shipbuilding practices, may be safe from heeling to a critical angle in a number of realistic cases, high accelerations on the bridge could trigger an accident. Indeed, navigation errors were noticed in about 10% of river-sea ship accidents examined by Egorov (2007).

Criticism of a particular "chapter" in regulations (such as stability requirements) which constitute a part of a larger set of (classification) rules, may seem as incomplete without taking into account other rules that also affect safety (e.g. structural requirements or guidelines for navigation in a specific area). However, the author believes that sufficient stability in given weather conditions should be targeted and accomplished by stability regulations themselves. Presently, as it was demonstrated in the investigation, it may be an unintentional consequence of application of another (unrelated) rule.

Finally, it should be noted that certain developments in the field of safety regulations for river-sea ships took place since the presented research was completed. On the European level, the stability requirements of the revised Resolution 61 maintained a deterministic character, whereby the idea of limiting the navigation area of river-sea ships, based on estimated lateral accelerations, seems to be abandoned for the time being. Conversely, the French regulations related to the operation of inland vessels in the port of Le Havre, evolved further, by embracing a probabilistic approach. The introduction of probabilistic ship stability analysis in river-sea navigation is perhaps unexpected, but hopefully encouraging for further development of the next generation of intact stability criteria.

Acknowledgements The author would like to acknowledge the contribution of Professor Milan Hofman (Department of Naval Architecture, Faculty of Mechanical Engineering University of Belgrade) whose pioneering work on probabilistic approach to stability of inland vessels led to the research herein presented.

References

Bačkalov I, (2010) Probabilistic safety of estuary vessels based on nonlinear rolling in wind and waves, Transactions RINA, International Journal of Maritime Engineering, Vol. 152, Part A1, pp 1–8.

Bačkalov I, Kalajdžić M, Hofman M (2010) Inland vessel rolling due to severe beam wind: A step towards a realistic model, Probabilistic Engineering Mechanics, Vol. 25, No. 1, pp 18–25.

Basin AM, Anfimov VN (1961) Ship Hydrodynamics (In Russian), Rechnoy Transport, Leningrad.

Belgisch Staatsblad/Moniteur Belge (2007) Royal Decree concerning inland vessels also used for non-international sea voyages (In Flemish and French), 16.03.2007, N. 2007 – 1187, pp. 14699–14711.

Bulian G, Francescutto A (2004) A simplified modular approach for the prediction of the roll motion due to the combined action of wind and waves, Proc. Inst. Mech. Eng. Vol. 218, Part M, Journal of Engineering for the Maritime Environment, pp. 189–212.

Egorov GV (2007) Risk-based design of ships for restricted navigation areas (In Russian), Sudostroenie, Saint-Petersburg.

Egorov GV, Ilnitskyy IA (2006) New generation of the Russian dry-cargo vessels of river-sea navigation 'Volgo-Don MAX type', Proceedings of the 2nd Conference on Coastal Ships & Inland Waterways, London.

Hofman M, Bačkalov I (2005) Weather Criterion for seagoing and inland vessels – some new proposals, Proceedings of International Conference on Marine Research and Transportation, Ischia, pp 53–62.

Journal officiel de la République Française (2007) Arrêté du 10 janvier 2007 relatif à la navigation de bateaux fluviaux en mer pour la desserte de Port 2000, 27.01.2007, Texte 18 sur 122.

Journal officiel de la République Française (2014) Arrêté du 15 décembre 2014 relatif à la navigation de bateaux porte-conteneurs fluviaux en mer pour la desserte de Port 2000 et des quais en Seine à Honfleur, 26.12.2014, Texte 12 sur 302.

Lesyukov VA (1974) Theory and design of inland vessels (In Russian), Transport, Moscow.

Ross JM (2009) Human factors for naval marine vehicle design and operation, Ashgate Publishing Limited.

Russian Maritime Register of Shipping (2005) Rules for classification and construction of sea-going ships, Saint-Petersburg.

Russian River Register (2008) Rules for classification of ships for inland and combined (river-sea) navigation (In Russian), Moscow.

Shigunov V, Rathje H, El Moctar O, Altmayer B (2011) On the consideration of lateral accelerations in ship design rules, Proceedings of the 12th International Ship Stability Workshop, Washington D.C., pp 27–35.

Truijens P, Vantorre M, Van der Werff T (2006) On the design of ships for estuary service, Transactions RINA, International Journal of Maritime Engineering, Vol. 148, Part A2, pp. 25–39.

United Nations Economic Commission for Europe, UNECE (2006a) Resolution No. 61, Recommendations on Europe-Wide Technical Requirements for Inland Navigation Vessels, New York and Geneva.

United Nations Economic Commission for Europe, UNECE (2006b) Drafting of ECE technical requirements for sea-river vessels", Submitted by Russian Federation, ECE/TRANS/SC.3/2006/8, Geneva.

United Nations Economic Commission for Europe, UNECE (2010) Draft proposal on Chapter 20B "Special provisions applicable to river-sea navigation vessels", Submitted by the group of volunteers on Resolution No. 61, ECE/TRANS/SC.3/WP.3/2010/6, ECE/TRANS/SC.3/WP.3/2010/6/Add.1, Geneva.

United Nations Economic Commission for Europe, UNECE (2013) Resolution No. 61, Recommendations on Europe-Wide Technical Requirements for Inland Navigation Vessels, Revision 1, Amendment 1, New York and Geneva.

Vantorre M, Eloot K, Delefortrie G (2012) Probabilistic regulation for inland vessels operating at sea as an alternative hinterland connection for coastal harbours, European Journal of Transport and Infrastructure Research, Vol. 12, Issue 1, pp 111–131.

Vassalos D, Jasionowski A, Cichowicz J (2004) Issues related to the Weather Criterion, International Shipbuilding Progress, 51, no. 2/3, pp 251–271.

Part XII
Requirements and Operation: Developments in Damage Stability Regulations

Chapter 42
Issues Related to Damage Stability

Andrew Scott

42.1 Introduction

The regulatory framework which encompasses the subject of damage stability is subject to almost continuous review and updating to reflect changes in ship size, type and design and to meet demands for increased safety. Also, in recent years there have been huge improvements in the computer hardware and software employed to perform the complex calculations necessary to demonstrate that a new ship design meets whatever damage stability standards are applied by the authorities. Routine arithmetical calculations that throughout most of the twentieth century would have taken an entire design office full of staff armed with slide rules or calculators several weeks to perform and validate can now be accomplished in a matter of seconds, although preparing and checking the input data and verifying the output as well as writing, checking and maintaining the software obviously take much longer.

Increasingly powerful computers with ever-improving software have opened new possibilities for exploring the more complex aspects of damage stability. The aim of trying to simulate on a computer what happens to a damaged ship in a seaway in real time would now seem to be achievable and some would claim that it already has been achieved to some degree. Of course, approximations and simplifications are still required as we cannot yet pretend that we can describe or deal with dynamic flooding events at molecular level. There is, however, some confidence that the gap between computer simulation and reality is already close enough to allow reasonably accurate predictions to be made as to the likely outcome of a given damage scenario, information which, of course, would be invaluable to a ship's Master in helping to decide whether to take the risky step of abandoning ship.

A. Scott (✉)
Maritime and Coast Guard Agency, Spring Place, 105 Commercial Road, Southampton SO15 1EG, England, UK
e-mail: andrew.scott@mcga.gov.uk

V. L. Belenky et al. (eds.), *Contemporary Ideas on Ship Stability*, Fluid Mechanics and Its Applications 119, https://doi.org/10.1007/978-3-030-00516-0_42

Changes to the design regulations affecting damage stability have, of necessity, come about more slowly as not every designer or operator has access to the very latest computer systems and so the principle that the convoy must advance at the speed of the slowest ship must be observed by the regulators. Also, and regrettably, the regulations themselves tend to be amended in response to major disasters, indeed the origins of SOLAS itself can be traced back to the loss of the "Titanic" in 1912.

42.2 Damage Stability Calculations; Historical Perspective and Underlying Assumptions

Until the probabilistic 2009 SOLAS amendments (SOLAS Consolidated Edition 2014) came into force on January 1st, 2009, the damage stability of passenger ships was linked primarily to subdivision and the provision of adequate reserve buoyancy should the hull suffer penetrating collision damage. The subdivision would be designed such that the ship would not submerge the bulkhead deck or plunge (trim excessively) nor would it capsize due to insufficient residual stability following analysis of a complete set of side damage scenarios of prescribed extent along the length of the ship and penetration depth (the vertical extent was assumed to be without limit).

At this point it is worth noting that the damage stability regulations have generally assumed:-

- that ships are always vulnerable to collision damage (i.e. it is not the case that a ship is unlikely to suffer collision damage because it operates in an area remote from other shipping, for example);
- that the momentum of each of the colliding ships is comparable, such that penetrating damage could potentially occur to both ships (i.e. the regulations do not deal with situations in which a much larger ship simply "rolls over" and/or pushes aside a smaller ship, as occurred in the tragic loss of the small passenger ship "Marchioness" following impact with the dredger "Bowbelle" on the River Thames in London on 21st August 1989)[1];
- that side collision damage puts the vessel at more risk than bottom damage and that such collision damage is more dangerous than "allision" (contacting a static object);
- that the bulkhead deck, or the deck up to which the main watertight transverse bulkheads are carried, was itself at least weathertight, if not watertight; one of the main contributing factors to the loss of the "Titanic" was cited to be that the main watertight bulkheads did not fully extend to the bulkhead deck so water

[1] Marchioness- Bowbelle Formal Investigation http://webarchive.nationalarchives.gov.uk/nobanner/20141008142043/http://assets.dft.gov.uk/marchioness-bowbelle.org.uk/index.htm.

progressively over-topped successive transverse bulkheads as the ship increasingly trimmed by the head[2];

- that passenger ships were predominantly subdivided transversely to minimise the effects of asymmetrical damage. In locations where this was not possible, cross-flooding arrangements would be installed to mitigate adverse list and these devices would preferably act automatically rather than by human intervention;
- that the potentially dangerous effects of "transient asymmetrical flooding", whereby the de-stabilization arising from the effect of the sudden inrush of flood water to the penetrated side of the impacted ship prior to the mitigating effects of equalisation past obstructions or through cross-flooding, could be ignored. This phenomenon was recognised relatively recently during the investigation into the loss of the ro-ro passenger ship "European Gateway" in 1982 (Spouge 1986). It is ignored in the regulatory damage stability assessment even now because it is largely a dynamic effect (therefore difficult to compute) and specifying in advance the amount and inflow rate of incoming flood water is problematical, depending largely as it does on the size of the damage opening;
- that penetrated tanks and other spaces were essentially empty upon impact (with standard allowances for permeability) and that flood water quickly rose to the outside water level, albeit with some allowance for "stage flooding" and "progressive flooding" during the equalisation process in case a more dangerous situation arose before final equilibrium was achieved. Nowadays, for liquid bulk carriers such as tankers, the effect of the loss of tank contents followed by eventual replacement with incoming flood water of heavier density to the outside water level cannot be ignored as it can be a dominating factor in the determination of the final equilibrium position and can vary according to initial depth of filling and the density of the cargo. As a result, IMO have now introduced guidelines on how this should be computed ("Guidelines for Verification of Damage Stability Requirements for Tankers") (MSC.1/Circ.1461 2013);
- that the effects of wind, tide, waves and damaged ship motion in a seaway can be ignored; the standard regulatory damage stability calculations are "static" in this respect as quantification of any dynamic effects was deemed to be too "difficult" for regulatory and approval purposes, although such calculations are now increasingly possible with modern computer systems, some of which can operate in the time domain;
- that, although the de-stabilizing dangers of the so-called free liquid surface effect in partially-filled tanks has been well understood for a long time and allowed for in a simplified way in undamaged ships, its cumulative effect in the undamaged, partially-filled tanks of an impacted ship could be ignored. Before the advent of computers, accurate determination of the liquid level in the undamaged tanks of a damaged ship, allowing for list and trim to obtain the correct static balance at a whole range of pre-set heel angles (and trims) would have been prohibitively time-consuming and expensive, so was seldom even attempted. With modern com-

[2]British Wreck Commissioner's Inquiry "Report on the Loss of the "Titanic"". http://www.titanicinquiry.org/BOTInq/BOTReport/botRep01.php.

puters it is technically and economically feasible but has still never been explicitly allowed for in the damage stability regulations. A most dangerous free-surface related phenomenon is the so-called "Water-on-Deck" (WOD) effect in ships, typically ro-ro's, which have a large open deck area for transporting vehicles. Any water accumulation in the garage space above the bulkhead deck from whatever source (for instance through the bow doors as for the "Herald of Free Enterprise" (Sheen 1987) and "Estonia" (1994), or fire-fighting ("Al-Salam Boccaccio 98") (2010) can result in a very rapid loss of stability. For European ro-pax ships this effect is mitigated by the so-called "Stockholm Agreement" which contains additional stability requirements and criteria to allow for the effect of WOD (Directive 2003/25/EC 2003).
- that passenger ships should be designed to ensure that the larger the ship and therefore, in general, the more passengers it carries, the safer it will be.

From the time the SOLAS regulations first appeared after the loss of the "Titanic" until the introduction of the 2009 Amendments the underlying approach to passenger ship subdivision and damage stability was primarily deterministic in nature except for the probabilistic method introduced in 1973 (IMO Res. A265(VIII)) (Resolution A.265(VIII) 1969) which was deemed to be an acceptable alternative to the deterministic regulations in force at the time.

IMO Res. A265(VIII), which retained some deterministic elements such as the presumption that the maximum side damage penetration depth would be B/5, was little utilized by designers in the early stages as it was perceived to be both more complex and more onerous to comply with than the equivalent deterministic regulations. In the mid 1980s and beyond, however, it became more utilized, especially for ro-ro passenger ships with combined longitudinal and transverse subdivision (typically those with a long lower hold protected by B/5 longitudinal bulkheads) and as the deterministic SOLAS damage stability criteria themselves gradually became more onerous, particularly following the amendments introduced in 1990, commonly abbreviated to SOLAS90 (see below for more details) (SOLAS Consolidated Edition 1992).

At around the same time as IMO Res. A.265(VIII) first appeared in 1973, IMO started to introduce measures to try to reduce pollution from oil tankers both in response to accidents, such as the infamous "Torrey Canyon" disaster off the Scilly Isles in 1967 which resulted in the spillage of 120,000 tonnes of crude oil (Petrow 1968), and increasing pollution from routine operations such as tank cleaning. As a result, the MARPOL Convention was adopted on 2 November 1973 at IMO. The Protocol of 1978 was adopted in response to a spate of tanker accidents in 1976–1977. As the 1973 MARPOL Convention had not yet entered into force, the 1978 MARPOL Protocol absorbed the parent Convention. The combined instrument entered into force on 2 October 1983. In 1997, a Protocol was adopted to amend the Convention and a new Annex VI (covering air pollution from exhaust gases) was added which entered into force on 19 May 2005.

From the beginning, MARPOL included provisions for subdivision and damage stability, with deterministic regulations defining the assumed extent of side and bottom damage, the maximum outflow of oil and the residual stability criteria. It has been updated by amendments through the years, most notably in response to, the grounding of the "Exxon Valdez" in Prince William Sound, Alaska in March 1989 (Marine Accident Report 1989). The 1992 amendments to MARPOL Annex I made it mandatory for new oil tankers to have double hulls and introduced a phase-in schedule for existing tankers to have them fitted, provisions which were subsequently revised in 2001 and 2003.

IMO introduced variants of the deterministic damage stability regulations throughout the 1980s for ships carrying, for example, dangerous chemicals in bulk (IBC Code, in 1983), liquefied gases in bulk (IGC Code, mandatory since 1986), offshore supply vessels (OSV Code, in 1981), mobile offshore drilling units (MODU Code in 1989), special purpose ships (SPS Code, in 1984) and others, each with their own characteristics to suit the operations that the ships were involved in. This period happened to coincide with the era of increasing computing power (hardware) and more advanced software, previously mentioned, which enabled smaller teams of designers in shipyards or consultancies to undertake the more specialised damage stability calculations now required. By and large, the deterministic damage stability regulations for tankers etc. have changed little in the intervening years.

More recently was it recognised that some tankers, particularly parcel tankers, could be at risk by virtue of the complexity of the post-damage fluid loss mechanism giving rise to final equilibrium situations not easily predictable other than by the use of on-board (or shore-based) damage stability computer systems, even allowing for some simplifying assumptions. This resulted in IMO introducing, in 2013, the new guidelines mentioned earlier (MSC.1/Circ.1461 2013). There is a retrospective element to the installation and use of such computer systems which, ideally, should be capable of calculating, before the ship sails, the residual stability characteristics of the standard set of damage scenarios appropriate to each Code for the actual departure loading condition thereby verifying in advance either that the ship fully complies with all the required stability criteria or, if not, alerting the loading officer of the need to take corrective action.

Such systems should utilise an accurate computer model of the ship and internal tanks to "balance" the ship by direct calculation so that the final equilibrium position can be computed allowing both for the loss of buoyancy (with potential fluid loss) in the damaged tanks and the cumulative free surface effects in any undamaged partially-filled tanks. The process of upgrading computer systems on vulnerable tankers is currently under way and should be complete by 2021. Existing ships, such as crude oil tankers, which load to a predictable pattern and carry oil only within a narrow range of densities and whose damage stability characteristics can therefore be pre-calculated for inclusion in an approved stability booklet may not necessarily be required to fit a new computer system, although many already employ less advanced versions.

The idea of providing on board damage stability computers or shore-based support for providing information to the Master has more recently been applied to so-called "Safe Return to Port" passenger ships (generally those of around 120 m in length and over, as described in SOLAS2009 amendments, Chapter II-1 regulation 8-1)

(SOLAS Consolidated Edition 2014). At first required only for new passenger ships constructed on or after 1st January 2014, IMO has now introduced guidelines in the form of MSC Circulars to cover all passenger ships included under the "Safe Return to Port" category (ref: MSC.1/Circ.1400, MSC.1/Circ. 1532 and a further circular to be approved at MSC 99 for passenger ships constructed before 1 January 2014).

For dry cargo ships, prior to 1992, there were no damage stability requirements other than those in Regulation 27 of the 1966 Load Line Convention which were designed only to ensure floatability in the event of fore end and engine room damage for larger ships (including tankers). Then in 1992 probabilistic rules for dry cargo ships of over 100 metres in length were introduced by IMO (see SOLAS90 Chapter II-1, Part B-1, Regulations 25-1 to 25-9) and shortly afterwards extended to include those of over 80 m.

A view gradually emerged in the mid to late 1990s that, for various reasons, the probabilistic approach to calculating damage stability was preferable to the deterministic methods used hitherto. It was therefore decided at IMO in the late 1990s to try to harmonize the 1992 dry cargo ship and the 1973 passenger ship damage stability regulations into a single set of new SOLAS Chapter II-1 probabilistic formulations whilst maintaining the safety standards set by the damage stability regulations in force at the time (SOLAS90) (SOLAS Consolidated Edition 1992).

Consideration was given to extending the probabilistic approach to damage stability to other internationally-operating ship types but finally it was decided to restrict it to dry cargo ships and passenger ships. It was thought, for example, that correctly accounting for fluid loss in damaged tankers and other bulk liquid carriers could not be readily accomplished probabilistically given the potential for a multitude of input variables, such as tank filling depths, cargo densities and loading patterns.

It was during the later stages of the harmonization process, which culminated in the 2009 SOLAS amendments, that certain difficulties with some of the formulations and text emerged which eventually necessitated further significant amendments which will enter into force on the first of January 2020, and will be known as the 2020 SOLAS amendments.

42.3 Doubts Arise Over the SOLAS 2009 Amendments, Particularly for Ro-Pax Ships

So profound were the changes being made to Chapter II-1 of SOLAS that even before the 2009 amendments came into force some disquiet, particularly over the safety levels of ro-ro passenger ships, had begun to arise. For one thing, the Europeans had hitherto additionally required ro-ro passenger ships constructed before 1 January 2009 to comply with the Stockholm Agreement which had been introduced after the loss of the "Estonia" to allow for the so-called "water-on-deck" (WOD) effect whereby flooding of the open car deck space could induce extremely rapid capsize (Directive 2003/25/EC 2003). The understanding was that this additional requirement would be phased out, along with the SOLAS90 regulations and IMO Res. A265(VIII) on 1st January 2009. The 2009 amendments, in accordance with probabilistic principles, made no distinction between ro-ro passenger ships and conventional passenger ships (such as cruise liners) even though the former were known to be at increased risk of rapid capsize due to water accumulation on the open car deck. This seemed to be an obvious deficiency and worthy of further investigation.

In the UK the concern was always that the loss mechanism for ro-ro passenger ships is quite different from that of conventional passenger ships. The latter can technically capsize (however that is defined) but stay afloat for a lengthy period thanks to the reserve buoyancy provided by the superstructure (which is often neglected in the damage stability calculations and so acts as an unquantified "safety margin") and the relatively slow speed of progressive flooding, allowing more time for evacuation. A ro-pax ship in contrast can capsize and sink in a matter of seconds from an apparently "stable" situation once sufficient water accumulates on the large open car deck through wave action, leading to potentially much higher casualty rates. Even a survival period of 30 min would scarcely provide sufficient time for an orderly evacuation of the ship, especially in adverse environmental conditions.

The focus of attention between completion of the HARDER project, which undertook the applied research needed to formulate the new harmonized probabilistic regulations, and the entry into force of the SOLAS 2009 amendments was therefore to seek re-assurance that the new regulations were at least as effective in allowing for the dangerous WOD effect in ro-pax ships as SOLAS90 and the Stockholm Agreement, imperfect as the latter combination may have been.

There are two main designs of ro-pax ship—one which is entirely transversely sub-divided below the car deck (usually employed on short crossings with rapid turnaround times) and one which combines longitudinal and transverse subdivision forming a long, lower hold (LLH) which is accessible by an internal ramp or ramps for loading wheeled vehicles. The loss mechanism for a LLH ro-pax ship may be quite different from one which is purely transversely sub-divided below the vehicle deck.

An unpublished UK study (RP 564) carried out in 2007 on behalf of the MCA by Safety at Sea on an existing LLH ro-pax ship compliant with IMO Res. A265(VIII) showed that damaging the LLH, one wing compartment and the vehicle deck resulted in margin line immersion as the LLH slowly filled followed by very sudden loss as water rapidly spread over the car deck. The vessel being studied sank in less than 20 min (real time) in almost calm conditions, the primary cause being the immersion of the margin line quickly followed by complete loss of reserve buoyancy.

This rather alarming result triggered another research project funded jointly by the UK and Netherlands, which commenced late in 2007 and was completed early in 2009. As little published work was available on the design of new ships optimized to meet the forthcoming SOLAS2009 amendments (without the Stockholm Agreement for ro-pax ships), this project (RP592) aimed to remedy this gap for a range of ship types, including cruise liners, ro-pax, small cargo ships, car carriers etc. The design work was undertaken by Safety at Sea in association with Deltamarin and late in 2008 it became clear from these studies that it was possible to design a LLH ropax ship fully compliant with the SOLAS 2009 amendments which had little or no resistance to rapid capsize, even in calm conditions, following plausible damage scenarios involving the long lower hold (which, in the optimized design in question, was now only protected by B/10 longitudinal bulkheads).

This result (which, later in 2009, was independently confirmed by another research project (EMSA 1), funded by the EU) was reported to COSS, the EC committee responsible for EU maritime regulations, at a meeting on 1st November 2008. It led soon afterwards to a decision by the EC to take the precautionary measure of retaining the Stockholm Agreement for new European flag ships designed and constructed to comply with the SOLAS 2009 amendments. This combined regulatory arrangement is still in place for European-flag ro-pax ships and foreign flag ships operating in European waters and will remain so until the amendments which have been made to SOLAS 2009 for entry into force on 1 January 2020 are clearly proven to show that the WOD effect has been fully accounted for without recourse to the use of the Stockholm Agreement. Research work sponsored by the EU is currently investigating this issue and the results are expected to be announced later in 2018 whereupon a decision will be made by the EC as to whether the Stockholm Agreement will continue to apply to new ro-pax ships constructed after the 2020 amendments to SOLAS enter into force.

42.4 Research Justifies Concerns, More Concerns Emerge—Remedies Are Proposed

These uncertainties over the degree of safety provided by the 2009 amendments to SOLAS, particularly for ro-pax ships fully justified further research work, and this was duly undertaken in the period from 2010 until 2015 in the form of two major projects sponsored by EMSA (known as EMSA2 and EMSA3), another project (GOALDS) which was funded under the EC FP7 program and another by the UK (RP625).

Full details of these projects can be found on-line and all of them reported their findings to the IMO SLF Sub-Committee (now SDC) and to the Maritime Safety Committee (MSC). By and large the concerns over the s-factor used for ro-pax ships in the SOLAS 2009 amendments were found to be fully justified and a new formulation was proposed and accepted at SLF 55, early in 2013, based on increasing both the residual maxGZ (from 0.12 to 0.20 m) and the residual range of stability (from 16 to 20°), hence becoming known as the "20/20" solution:

$$s_{final,i} = K \cdot \left[\frac{GZ_{\max}}{TGZ_{\max}} \cdot \frac{Range}{TRange} \right]^{\frac{1}{4}}$$

where

$GZ_{\max}$ is not to be taken as more than $TGZ_{\max}$;
Range is not to be taken as more than *TRange*;

TGZ_{max} 0.20 m, for ro-ro passenger ships each damage case that involves a ro-ro space, $= 0.12$ m, otherwise;

$TRange$ 20°, for ro-ro passenger ships each damage case that involves a ro-ro space, $= 16°$, otherwise;

$$K = 1 \text{ if } \theta_e \leq \theta_{min}$$
$$K = 1 \text{ if } \theta_e \leq \theta_{max}$$
$$K = \sqrt{\frac{\theta\text{max} - \theta_e}{\theta\text{max} - \theta\text{min}}} \quad otherwise,$$

where

θ_{min} is 7° for passenger ships and 25° for cargo ships; and

θ_{max} is 15° for passenger ships and 30° for cargo ships.

These amendments will enter into force for new ro-pax ships constructed on or after January 1st, 2020. Graphically, this change can be seen below in which the new GZmax and range requirement for ro-pax ships is compared with earlier SOLAS damage stability criteria to depict how they have steadily become more onerous over the years (Fig. 42.1).

Alternative proposals based on minimizing water accumulation on the vehicle deck by increasing residual freeboard were rejected by the SLF Sub-Committee who thereby invested full confidence in the ship resisting the threat of rapid capsize from the WOD effect by increase of residual stability.

In the alternative freeboard method (see SLF 55/INF.10), based on some of the parameters in the Stockholm Agreement (such as the WOD effect equating to zero if the residual freeboard in way of the damage opening is greater than 2 m), an increased residual maxGZ of 0.16 m and residual range of 25° would only be a requirement in individual damage cases involving breaching the garage space if the residual freeboard was less than 2 m. Advocates of the method believed that this would encourage designers to try to keep water off the deck altogether (by increasing residual freeboard) rather than allowing water to accumulate and then attempt to counteract the WOD effect by increasing the residual damage stability.

Probabilistically speaking, of course, the ro-pax ship could still comply with the $A \geq R$ criterion by accumulating sufficient attained index A from damage cases other than those involving the garage space with the attendant risk of WOD. This fact emerged quite early in the research projects, the concern being that permitting ro-pax ships to capsize in a whole series of damage scenarios involving the WOD effect would be accepting that there could be heavy loss of life in these circumstances, yet the ship could still be deemed to be fully compliant with the SOLAS

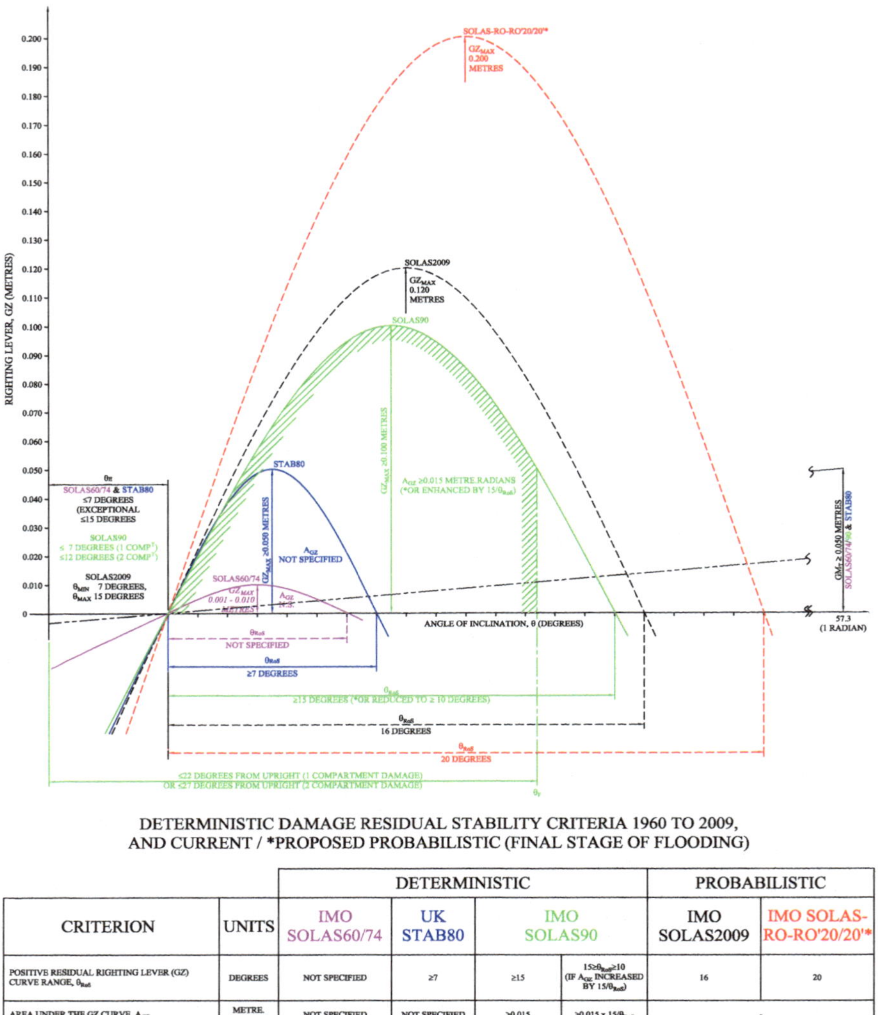

DETERMINISTIC DAMAGE RESIDUAL STABILITY CRITERIA 1960 TO 2009,
AND CURRENT / *PROPOSED PROBABILISTIC (FINAL STAGE OF FLOODING)

		DETERMINISTIC				PROBABILISTIC	
CRITERION	UNITS	IMO SOLAS60/74	UK STAB80	IMO SOLAS90		IMO SOLAS2009	IMO SOLAS-RO-RO'20/20'*
POSITIVE RESIDUAL RIGHTING LEVER (GZ) CURVE RANGE, $θ_{RoS}$	DEGREES	NOT SPECIFIED	≥7	≥15	15≥$θ_{RoS}$≥10 (IF A_{GZ} INCREASED BY 15/$θ_{RoS}$)	16	20
AREA UNDER THE GZ CURVE, A_{GZ}	METRE. RADIANS	NOT SPECIFIED	NOT SPECIFIED	≥0.015	≥0.015 x 15/$θ_{RoS}$	-	
MAXIMUM RESIDUAL RIGHTING LEVER GZ_{MAX}	METRES	0.001 TO 0.010 (UK)	≥0.050	≥0.100		0.120	0.200
ANGLE OF HEEL DUE TO UNSYMMETRICAL FLOODING AFTER EQUALISATION (DAMAGE EQUILIBRIUM ANGLE), $θ_E$	DEGREES	≤7	≤7	≤ 7 (1 COMPARTMENT DAMAGE) ≤12 (2 COMPARTMENT DAMAGE)		MINIMUM 7 MAXIMUM 15	
POSITIVE RESIDUAL METACENTRIC HEIGHT, GM_T	METRES	≥0.050	≥0.050	≥0.050		-	

Fig. 42.1 This figure shows how the damage stability criteria based on the residual GZ curve have gradually become more onerous over the years. The latest curves are shown as dotted lines as they represent probabilistic criteria to ensure that s = 1 and are not absolute requirements as is the case with the earlier deterministic damage stability criteria. The red dotted so-called "20/20" line only applies to damage cases to ro-pax ships in which the vehicle deck space has been breached

2009 amendments. The lower the required index, R, the greater proportion of all the potential damage cases to be examined may be permitted to capsize, potentially rapidly, (with s = 0).

42.5 Some Further Issues with the SOLAS 2009 Amendments

Without going into too much detail over the extensive arguments which occurred as the various research projects reported their results to IMO, it is sufficient to say that it was also generally agreed that the "R" index for passenger ships (including ro-pax) in the SOLAS 2009 amendments was too low. A decision was reached at SDC3 in January 2016 to increase "R" in regulation 6.2, which eventually resulted in the regression formulae shown in Table 42.1 and shown graphically in Fig. 42.2 as finally agreed at MSC 98 in June 2017 to be used in the SOLAS 2020 Amendments.

From the above, it is worth noting that, unlike the s-factor, the amended R index continues to apply to both ro-pax and conventional passenger ships. Also, it now depends solely on N (the number of persons on board) and no longer varies according to the degree of lifeboat provision, which will now be determined directly in SOLAS Chapter II-2. Whilst it is undoubtedly true that lifeboats do provide a tried and tested method of disembarking passengers in emergencies it was no longer felt appropriate to link the survivability of the ship to lifeboat provision. Therefore, R is now based solely on the number of persons on board and continues to increase with N but to a lesser degree than was the case with SOLAS90.

The argument here was that too steep a gradient implies that the value of individual human lives somehow decreases rapidly as the total numbers involved in the incident decreases, which is no longer considered acceptable. For this reason, some argued that the R versus N curve should be almost horizontal, until N decreases to the point where the ship is too small to provide the necessary subdivision. The pronounced discontinuity at N = 1000 in the curve below is an attempt to keep the line more horizontal.

Table 42.1 Agreed new formulations for required index "R" for use in the 2020 SOLAS Amendments

Persons on board	R
$N < 400$	$R = 0.722$
$400 \le N \le 1350$	$R = N / 7580 + 0.66923$
$1350 < N \le 6000$	$R = 0.0369 * \text{Ln}\ (N + 89.048) + 0.579$
$N > 6000$	$R = 1 - (852.5 + 0.03875 * N) / (N + 5000)$

N = total number of persons on board

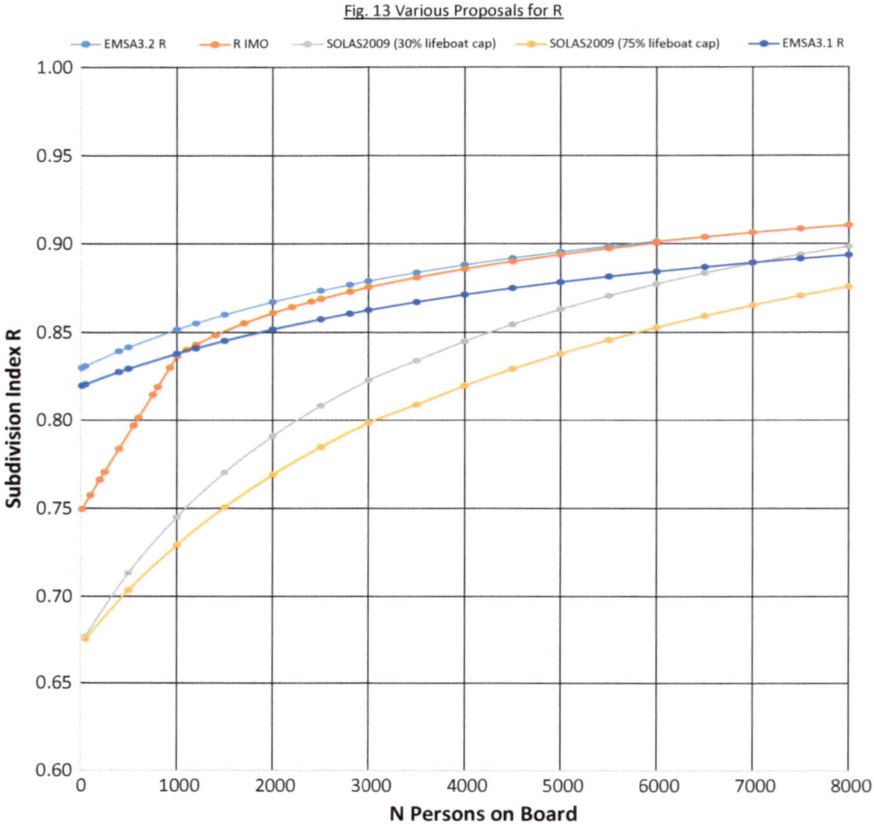

Fig. 42.2 This figure shows the new "R" index (R IMO SOLAS 2020) in comparison with "R" from the SOLAS2009 amendments, which was also based on length and lifeboat provision. The EMSA curves represent proposals made by the EMSA3 project (see SDC 3/3/7 for more details)

Although the "R IMO" curve shown above, which is derived from the formulae in the table, as proposed by SDC3, was acceptable to most delegates at MSC96 in May 2016 there were some objections, particularly as some countries felt that the increase in "R" was too great for smaller ships (N < 1200). So, the issue was debated again at MSC97 in November 2016 before "R IMO" was finally approved by IMO at that session.

Sooner or later, the EC also must decide whether the above-mentioned changes to the s-factor and R will provide sufficient re-assurance that the Stockholm Agreement can be safely revoked for EU-flag ro-pax ships constructed after January 1st, 2020. To this end EMSA initiated a further research project in August 2016 to provide the necessary supporting evidence for a decision to be made. The project will take around 16 months and so should produce results some time in 2018.

My own personal view is that the Stockholm Agreement should be retained for the time being and that there should be an obligation for each new ro-pax design to be model tested. The main reason for this is that the "*s-factor*" is a key element within the entire probabilistic regulatory framework. Whereas the theory used to develop the "*p-factor*" from the IMO accident database which is used to determine the likelihood of a ship being damaged in a given location is fairly clear and verifiable, that used to develop the "*s-factor*" is much less so.

There is no clear definition of "*s*" either in the SOLAS regulations or the associated Explanatory Notes (Resolution MSC.281(85) 2008). The method used to justify the SOLAS2009 equation for "*s*" was shown to be flawed, especially when applied to ro-pax ships. Even the accepted "20-20" revision, although a step in the right direction, has very little theoretical or model-testing evidence to support it. Some recent research work has indicated that the "*20/20 s*" has only a very minor effect on reducing "A" for medium-to-large size ro-pax ships but a startlingly large effect (up to 12%) for small ro-pax ships. To date no explanation for this wide variation has been proposed.

The "*s-factor*" ought to have some predictive capability and provide reasonable correlation with model test results and the real world. For example, a ship model should survive in any random seas with SWH up to 4 m for a damage case with s = 1 in 10 out of 10 test runs. The SOLAS 2009 "*s-factor*" had rather poor predictive capability for model test results in this respect and it is surely vital that the proposed new "20/20" solution should be proven to have a much better predictive capability for WOD damage cases than the previous formulation.

The only way to demonstrate this is to validate the "*s 20/20*" predictive capability against model test results. Once a reasonably good correlation is proven unequivocally only then should consideration be given to removing the Stockholm Agreement as a standard for ro-pax survivability for EU-flag ships.

There is an outstanding issue to be resolved over the assumed watertightness of the bulkhead deck which will be debated at IMO in the SDC Sub-Committee from 2018 onwards over a 3-year period. Any amendments, if found necessary, will not enter into force until the 1st January 2024, according to IMO rules.

If that happens it will be most interesting to see what effect both this and the other forthcoming amendments to "*s*" and "*R*", described above, will have on future passenger ship and ro-pax designs.

Finally, readers may find a summary of much of the above material in a presentation given to the SNAME (Western European Section) on 6th February 2018 entitled "*SOLAS 2020 Damage Stability Regulations*" by the author and his co-presenter Hutchinson (2018).

References

SOLAS Consolidated Edition, 2014, *Consolidated text of the International Convention for Safety of Life at Sea, 1974, and its Protocol of 1988: articles, annexes and certificates. Incorporating all amendments in effect from 1 July 2014,* International Maritime Organization, London, United Kingdom, 2014. ISBN 978-92-801-1594-9.

"The Technical Investigation of the sinking of the ro/ro ferry "European Gateway"" by J.R. Spouge published in the Transactions of the Royal Institution of Naval Architects, 1986.

MSC.1/Circ.1461 *Guidelines for Verification of Damage Stability requirements for Tankers*, International Maritime Organization, London, United Kingdom, 8th July, 2013. (Approved by the Maritime Safety Committee at its ninety-second session 12th to 21st June, 2013).

Flooding and capsize of ro-ro passenger ferry Herald of Free Enterprise with loss of 193 lives; Formal Investigation Report by Hon. Mr Justice Sheen, Wreck Commissioner. Report of Court No. 8074, first published 1987. https://www.gov.uk/maib-reports/flooding-and-subsequent-capsize-of-ro-ro-passenger-ferry-herald-of-free-enterprise-off-the-port-of-zeebrugge-belgium-with-loss-of-193-lives.

Final Report on the capsizing on 28th September 1994 in the Baltic Sea of the Ro-Ro passenger vessel MV Estonia by the Joint Accident Investigation Commission of Estonia, Finland and Sweden; published December 1997. http://www.onnettomuustutkinta.fi/material/attachments/otkes/tutkintaselostukset/en/vesiliikenneonnettomuuksientutkinta/1994/mv_tutkintaselostus/mv_tutkintaselostus.pdf.

Information resources on the "Al Salam Boccaccio 98", IMO Maritime Knowledge Centre, last updated January 2010. http://www.imo.org/en/KnowledgeCentre/InformationResourcesOnCurrentTopics/InformationResourcesOnCurrentTopicsArchives/Documents/AL%20SALAM%20BOCCACCIO%2098%20_January%202010.pdf.

Directive 2003/25/EC of the European Parliament and of the Council of 14 April 2003 on specific stability requirements for ro-ro passenger ships, Official Journal L123, of the European Union, Brussels, Belgium, 17th May, 2003.

Resolution A.265(VIII) *Regulation on subdivision and stability of passenger ships as an equivalent to Part B of Chapter II of the International Convention for the Safety of Life at Sea, 1969*, Assembly, International Maritime Organization, London, United Kingdom, adopted on 20th November 1973.

SOLAS Consolidated Edition, 1992, *Consolidated text of the International Convention for Safety of Life at Sea, 1974, and its Protocol of 1988: articles, annexes and certificates. Incorporating all amendments up to and including the 1990 amendments (1991 amendments as an appendix),* International Maritime Organization, London, United Kingdom, 1992. ISBN 978-92-801-1594-5.

"*The Black Tide: in the wake of the Torrey Canyon*" by Richard Petrow, published 1 January, 1968.

Marine Accident Report: Grounding of the U.S. Tankship "Exxon Valdez on Bligh Reef, Prince William Sound, near Valdez, Alaska March 24, 1989 by National Transportation Board. http://docs.lib.noaa.gov/noaa_documents/NOAA_related_docs/oil_spills/marine_accident_report_1990.pdf.

Resolution MSC.281(85) *Explanatory Notes to the SOLAS Chapter II-1 Subdivision and Damage Stability Regulation*, Maritime Safety Committee, International Maritime Organization, London, United Kingdom, 4th December 2008.

"*SOLAS 2020 Damage Stability Regulations*" by A.L. Scott and K.W. Hutchinson – a power-point presentation given to the Western Europe Section of SNAME at Lloyds Register of Shipping, 71 Fenchurch Street, London on 6th February 2018.

Chapter 43
Damage Stability Making Sense

George Mermiris and Dracos Vassalos

Abstract Although aviation, nuclear, processing, etc. industries have long ago adopted and established preventative frameworks and procedures to safeguard against unwanted outcomes of daily operations, the maritime industry still places the emphasis on the mitigation of consequences following an accident. Despite the widely expressed opinion that prevention is the way forward, curing occupies a central position not only in every day practice but in the underlying regulatory framework as well. Contrary to this approach, the work presented here aims to create the necessary momentum towards rationalisation of the fundamental choices made during the design process, thus attracting attention to areas where prevention strategies can find fertile ground and be fruitful and cost-effective. The methodology addresses the occurrence of a collision event and the crashworthiness capacity of a ship as prerequisites for its survivability assessment, with promising results to encourage further development.

Keywords Accident prevention · Collision · Crashworthiness

43.1 Introduction

Traditionally, the damage stability and survivability performance of a ship are treated under the assumption that the hull is breached following a collision event. This approach has received considerable attention and significant effort has been spent in collating the required information for dimensioning the damage opening (SOLAS Ch. II-1).

Even though the probability of pertinent events that can compromise the watertightness of the hull, like collision and grounding, are consistently accounted for in quantitative risk analyses, the compulsory use of the Attained Index of subdivision,

G. Mermiris · D. Vassalos (✉)
The Ship Stability Research Centre, Department of Naval Architecture, Ocean and Marine Engineering, University of Strathclyde, Glasgow, Scotland, UK
e-mail: d.vassalos@strath.ac.uk

V. L. Belenky et al. (eds.), *Contemporary Ideas on Ship Stability*, Fluid Mechanics and Its Applications 119, https://doi.org/10.1007/978-3-030-00516-0_43

Eq. (43.1), discourages any focus on the associated causal factors and, in the particular case of collisions, on crashworthiness. As a result, accidents still happen, much more frequently than they should, and ships are lost with significant price for human life and the environment.

One key reason for this state of affairs relates to the fact that rule making in our industry focuses on damage limitation (cure) rather than damage prevention. Hence, the industry is pursuing happily a very ineffective means of sorting bad image and reputation. This being the case, the time for diverting attention towards an approach that makes sense of damage stability is long overdue but, fortuitously, ripe. More specifically, the emergence of the design for safety philosophy and the development of risk-based design methodology allows due attention on the risk pertinent to each vessel category in a scientific and all-embracing way, capable of balancing risk reduction and mitigation with other design objectives cost-effectively.

The work presented here demonstrates that in order to integrate safety against collision in the design process, it is necessary to rationalise the survivability assessment as it is presented by Vassalos (2004). This can be achieved by addressing the probability of collision occurrence, the probability of water ingress due to collision, the probability of capsize due to the ensuing water ingress and the consequential loss (Fig. 43.1). Such an integrated approach has been the focus in SSRC over the past 5 years, reaching the stage where potential benefits from trying to make sense of damage stability are demonstrable. This offers new inroads for the integration of safety against collision in the design process by drawing information from and feeding knowledge to the ship operation in an unprecedented way.

43.2 The Regulatory Framework

The assessment of the risk level following a ship collision event is presently performed according to Wendel's (1960), probabilistic approach, which is practically implemented with the Attained Index of subdivision A (IMO 2009):

$$A = \sum_{j=1}^{J} \sum_{i=1}^{I} w_j p_i s_{ij} \tag{43.1}$$

where

- j the counter for loading conditions;
- i the counter for damaged compartments or groups of adjacent compartments;
- J the number of loading conditions;
- I the number of damaged cases (single or groups of adjacent compartments) for each loading condition;
- w_j probability mass function of the loading conditions;
- p_i probability mass function of flooding extent of a compartment or group of compartments for loading condition j$\left(\sum_i P_i = 1\right)$;

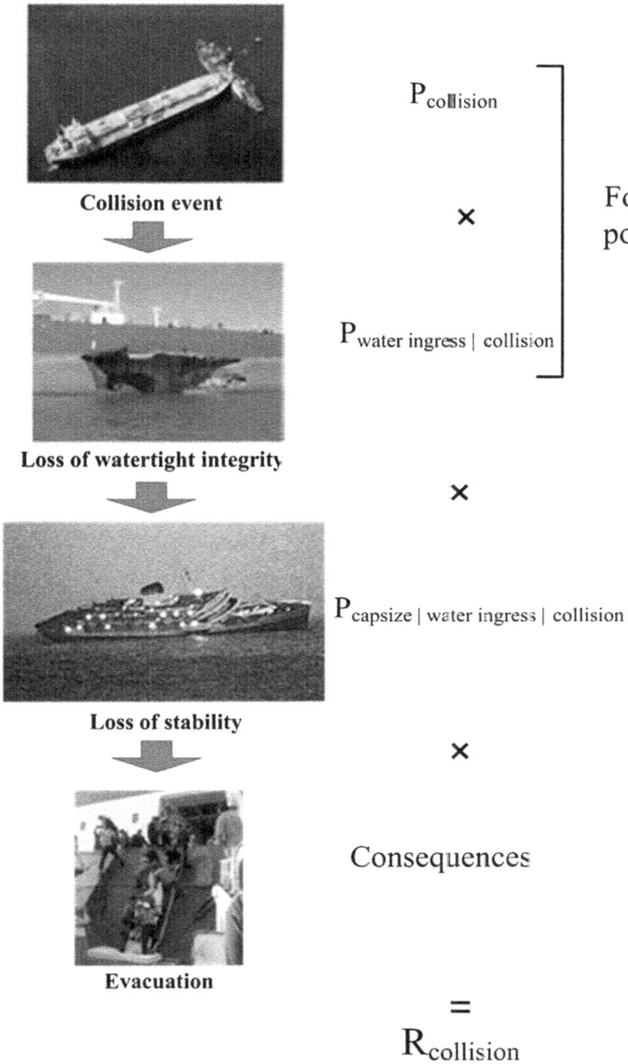

Fig. 43.1 Sequence of events in flooding scenario with the corresponding probability elements for the collision risk assessment

s_{ij} the average probability of surviving the flooding of a compartment (or group of compartments), for loading condition j.

Index A is the weighted average of the probability of survival, i.e. its expected value E(s), of all damage cases for a ship. As long as the value of A is greater than a prescribed threshold value (index R), the safety level of the ship is considered satisfactory, at least from a regulatory point of view.

43.2.1 A Critique on the Current Approach

The philosophy of this regulation is attractive (due to its scientific foundation on probability theory) and special (as few precedent frameworks, if any, have ever adopted a similar approach). However, the framework is based on statistical analysis of past accidents and unavoidably builds on the fact that a collision has occurred and the watertightness is lost (otherwise the accident would not be considered). Instead of using statistical information for rationalising the choices of the damage scenarios and benchmarking the results of structural analyses, the regulation puts emphasis on the identification of all damage cases that would compromise survivability. That is, irrespective of how improbable 5-compartment damage would be, this scenario will still be considered in the assessment. Hence, the process changes into a vulnerability analysis.

A closer look at the provisions of the framework will reveal determinism and inconsistency, as it is explained next:

(i) The calculation of the probability of flooding is conditional on the collision occurrence, i.e. the probability of collision $P_{collision} = 1.0$. However, modern communication and IT developments in combination with improved training of the navigation officers contribute significantly towards the traffic management even in the most congested waters.

(ii) The probability of flooding is also conditional on the probability of water ingress due to collision, i.e. the ship shell is breached and the penetration is of sufficient size to cause large scale flooding of one or more compartments instantly. Therefore, $P_{water\ ingress\ |\ collision} = 1.0$. Yet, the fact that a collision occurring does not mean that the watertightness of the hull is lost. Statistical data and computer simulations clearly indicate that the overall damage can range between denting and breaching of the side shell, with large variation of the damage opening (Fig. 43.2). In any case, instant flooding is expected to be very remote.

(iii) The calculation of the p-factor is solely based on the location of transverse and longitudinal bulkheads. At the same time, the crashworthiness of the side panel of each compartment, i.e. its capacity to absorb impact energy (Vredeveldt 2005), is ignored.

(iv) In the process of the above calculations, the operational profile of the struck ship should be taken into consideration for the following reasons: (i) in the case of $P_{collision}$, information on the traffic density and the geographical restrictions will indicate the level of congestion in a seaway, whereas (ii) in the case of $P_{water\ ingress\ |\ collision}$ it will offer an estimation of the available kinetic energy and bow geometries (as it will be explained in the next section) that can compromise

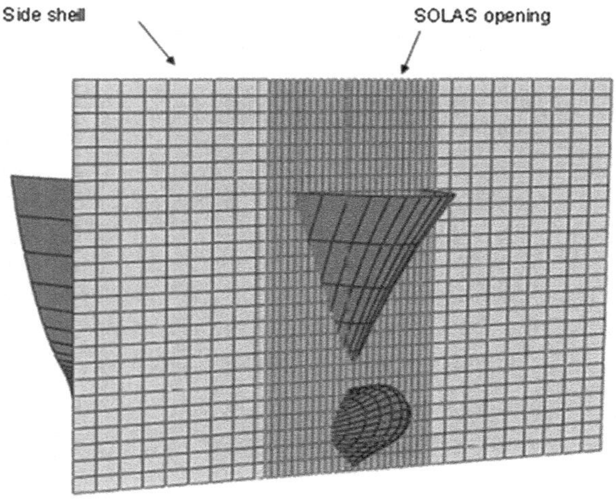

Fig. 43.2 Actual damage as opposed to SOLAS opening

the side shell. This way, a ship, which operates in coastal waters and in open sea, will experience different collision risk levels but because the operational profile is not accounted for in the regulation, the p-factor will remain the same.

As a result, the level of assumptions in the calculation of the p-factor renders the value of A questionable. More importantly though, index R is derived on the basis of a sufficient number of A-values of ships that have survived the elements over their life-cycle and represents an acceptable level of safety standard (HARDER 2003). But if R is based on values of A, the value of which is fraught with uncertainty, then R is also uncertain and the level of safety it represents is questionable.

43.3 The Proposed Model

Conventionally, the environment (in terms of wind, waves, etc.) in which a ship operates largely defines its design characteristics with respect to hydrodynamic and structural performance. In addition to the imposed loading on the hull girder, the operational environment also provides information concerning the accidental loading on the ship (congestion levels, speed and direction of the surrounding traffic, etc.), which until recently was of secondary or no importance during design. With this information readily available, the calculation of the p-factor can be rationalised as it is described in the following two sections.

43.3.1 Probability of Collision

The assessment of the probability of collision is based on the concept of *ship domain*, as it was introduced in the late 70s (Goodwin 1979), and treated in various contexts and studies (Hansen et al. 2004; Filipowicz 2004), etc. It was initially defined as a circular area surrounding a ship and if an object entered this area then a collision was assumed.

In the proposed model, the shape of the domain is retained but its diameter varies as a function of operational and design parameters. When the domain diameter becomes equal to or less than the ship length then a collision occurs.

The elements of the model that define the ship domain are:

1. **The ship length (L)** is indicative of the size of the vessel in a seaway and it is inversely proportional to the diameter of the domain.
2. **The response time (R)** is the necessary time for the vessel to advance at 90° and it defines how fast the ship will respond to a command for an evasive manoeuvre (ignoring any depth effects). R is reciprocal to the size of the domain as well.
3. **The speed of the vessel (V)** is important from an operational point of view. Its value reflects the conditions (traffic density, visibility, time schedule, etc.) under which the vessel steams and its variation depends on the geography of the navigational area.
4. **The traffic density (ρ)**, i.e. the number of ships per unit area, in a seaway can impose further restrictions to the speed range. Evidently, speed and traffic density are inversely proportional to the domain size as well.
5. **The transverse channel width (C)** defines the topological boundaries of the course of the ship in a waterway. It varies proportionally to the domain size and, according to Kriastiansen (2005), it is related to the traffic density:

$$\rho = \frac{N}{V'C} \tag{43.2}$$

where N is the number of ship passages per unit time (e.g. annually), and V' is the speed of the surrounding traffic.

6. Over the years, authors like (Fujii et al. 1974) and accident investigators, e.g. (MAIB 2005), have stressed that collision accidents (i) never occur instantaneously and without the right initial conditions (low visibility, early morning hours, etc.), and (ii) can be attributed to miscalculations, over-confidence, lack of communication, etc. When everything is orchestrated properly, then there is always a critical *point of no return*, which is measured consistently in the range of a few minutes (Cahill 2002)!

The fact that ship collisions always occur for a very specific set of initial conditions suggests that existing methodologies are fragmented (attributing the accident to human factors and adverse weather conditions or bad maintenance of hardware) and inadequate (the irreversibility of the situation is ignored).

Table 43.1 Examples of high and low entropy situations

High entropy	Low entropy	Remarks
Disorder, disorganisation, thorough mix-up	Order, high degree of organisation	Existence of a Vessel Traffic System (VTS) in the area of navigation
Great uncertainty	Near certainty, high reliability	Information about wind gusts, when close quarter manoeuvring is required
Great surprise	Little or no surprise	The familiarity of the navigator with the area of operation and the dominant conditions

In the proposed methodology the "softer" aspects of an accident are accounted for as disorder or uncertainty, i.e. in the form of **entropy of a situation (H)** (Williams 1997), which is expressed as:

$$H = \sum_{j=1}^{M} \sum_{i=1}^{N_j} P_{ij} \log_2 \left(\frac{1}{P_{ij}} \right) \tag{43.3}$$

where:

i counter for the number of states of each event,
j counter for the number of events,
M number of events,
N_j number of states of event j,
P_{ij} probability of occurrence of the state i and the event j, where $\sum_i P_i = 1$.

As the value of entropy increases, the more imminent a collision is. Examples of high and low entropy values are presented in Table 43.1.

Establishment of threshold values for entropy is an ongoing development but this concept allows a broad range of critical information to be consolidated into a single number with widely accepted meaning.

In summary, the domain diameter is expressed as:

$$D = \frac{C}{VLR\rho} 10^{-H} = \frac{V'C^2}{VLRN} 10^{-H} \tag{43.4}$$

The probability of collision per unit time can be obtained with a Monte Carlo sampling of the entailed parameters.

With Eq. (43.4), the point of no return is substantiated (due to its non-linear character) since the contribution to the entropy level of each of the participating events can be determined at successive instances and the escalation of a situation can be quantified, thus providing better decision support to the navigator, the port authorities, etc. An example of this is the comparison between navigation in open and closed

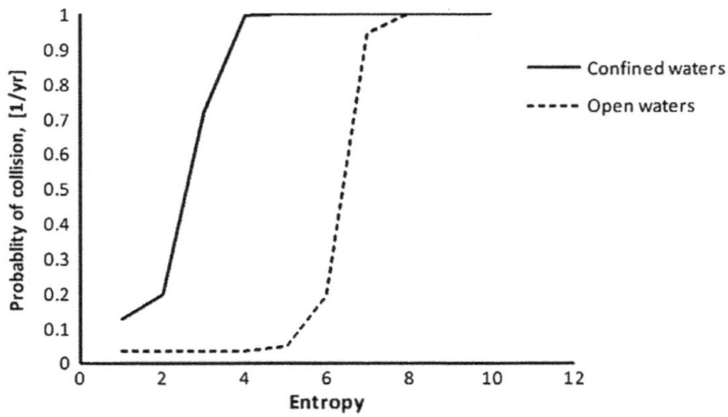

Fig. 43.3 Entropy variation for open (Pedersen and Zhang 1999), and confined waters (Øresund 2006)

waters for a ROPAX ship (Fig. 43.3). In the former case a collision event is guaranteed for values of entropy approximately equal to 4.0, whereas in the latter case the entropy levels will have to be doubled. The fact that space availability allows longer decision-making times is reflected in the proposed model and justifies the choice of entropy as an aggregate measure for quantitative and qualitative information.

It should be stressed that Eq. (43.4) is applicable when the ship is in sailing mode and when collision with other ships is considered; otherwise the element of speed of surrounding traffic (V') becomes meaningless.

43.3.2 Probability of Water Ingress Due to Collision

The extent of the structural damage following a collision event is tightly connected to the crashworthiness of the side shell panels as it was stressed earlier. Although the highly non-linear failure of the structure intuitively calls for sophisticated analysis with the Finite Elements (FE) technique, the very nature of FE is prohibitive for early design application (where most of the main characteristics of a ship are decided) due to long modelling, processing and post-processing times, and because such results cannot be communicated easily to the rest of the design tools. This being the case, the designers can either consider a small number of selected damages (i.e. check the vulnerability of the hull) or ignore such input and resort to using damage openings as prescribed in SOLAS.

The proposed approach is founded on the absorption of the kinetic energy of the striking ship by a restricted portion of the structure of the struck ship. The phenomenon is governed by (i) the magnitude of the kinetic energy, (ii) the structural configuration of the struck panel, and (iii) the geometry of the striking bow (assumed rigid here). The first two aspects can be derived from the operational profile of the ship in terms of the surrounding traffic (i.e. the size and the speed of other vessels), and its structural configuration respectively. The latter complements the expectation of breach occurrence considering that the sharper the contact edge of the striking body is, the easier the panels of the side shell will rupture (i.e. with less expenditure of kinetic energy), as it is confirmed by numerical simulations and experiments.

The remaining factors, which affect the development of a collision event are related to the angle between the two ships (as the angle increases the sharpness of the striking bow is reduced), their inertia, i.e. their virtual (real plus added) mass before (striking ship) and after the contact (struck ship), and the friction during the penetration.

The link of the side structure deformation and the striking body geometry is the *principal radii of curvature* of the latter, which provides a measure of its sharpness at the contact points. The radii of curvature of a three-dimensional surface can be obtained by its parametric definition:

$$x = x(\mathrm{p}, \mathrm{t}, \mathrm{w_O}),\ y = y(\mathrm{p}, \mathrm{t}, \mathrm{w_O}),\ z = z(\mathrm{p}, \mathrm{t}, \mathrm{w_O})\mathrm{p}, \mathrm{t} \in [0, 1] \tag{43.5}$$

where x, y and z are real, continuous and at least twice differentiable functions (with respect to either of the two parameters) in a right-handed coordinate system and $\mathrm{w_O}$ is the indentation of the panel since in the current context interest lies in the necessary deformation to cause rupture. The geometry of the striking body is represented with a Bezier surface, whereas the struck surface deformation is modelled with the *Witch of Agnesi* function, which allows for explicit consideration of the deflection $\mathrm{w_O}$ as a function of radii of curvature of the striking body:

$$\begin{aligned}
\mathrm{u}(\mathrm{x}, \mathrm{y}, \mathrm{w_O}) &= \mathrm{C_x}\frac{\mathrm{w_O}}{\left(1+\left(\frac{\mathrm{x}}{\mathrm{r_1}}\right)^2\right)}, \\
\mathrm{v}(\mathrm{x}, \mathrm{y}, \mathrm{w_O}) &= \mathrm{C_y}\frac{\mathrm{w_O}}{\left(1+\left(\frac{\mathrm{y}}{\mathrm{r_2}}\right)^2\right)} \\
\mathrm{w}(\mathrm{x}, \mathrm{y}, \mathrm{w_O}) &= \mathrm{C_{xy}}\frac{\mathrm{w_C}}{\left(1+\left(\frac{\mathrm{x}}{\mathrm{r_1}}\right)^2+\left(\frac{\mathrm{y}}{\mathrm{r_2}}\right)^2\right)}
\end{aligned} \tag{43.6}$$

where:

u, v and w are the deformation functions along x (longitudinal), y and z (vertical to its plane) directions of the stiffened panel.
$\mathrm{C_x}$, $\mathrm{C_y}$ and $\mathrm{C_{xy}}$ are constants accounting for the stiffening along the x, y and the x-y directions, respectively.
$\mathrm{r_1}$ and $\mathrm{r_2}$ are the radii of curvature of the striking bow at the point of contact.

Because of the substantial deformations experienced by the stiffened panel, the accumulated strain energy is dominated by membrane action and is expressed as:

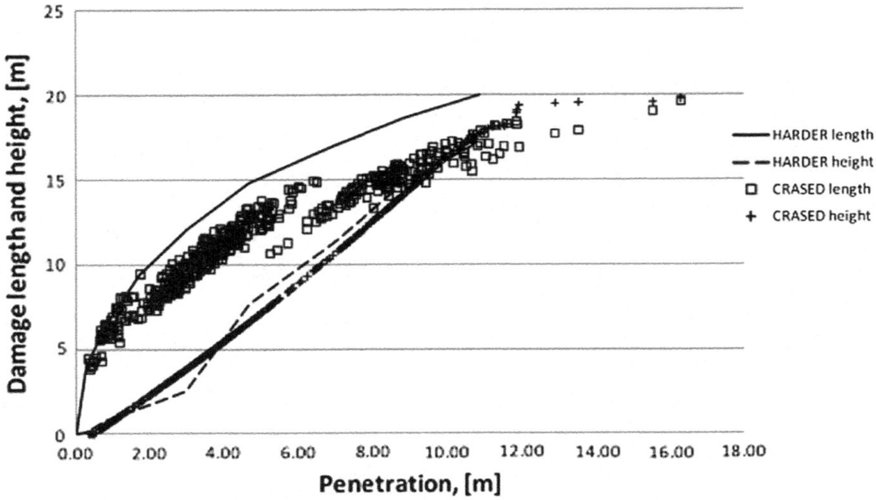

Fig. 43.4 Comparison between CRASED and statistical data

$$V_{mem} = \frac{1}{2}\int_0^L \int_0^B (N_x\varepsilon_x + N_y\varepsilon_y + N_{xy}\gamma_{xy})\,dy\,dx \quad (43.7)$$

where N_x, N_y and N_{xy} are forces per unit length of the plate edge and ε_x, ε_y and γ_{xy} are the corresponding strains for large deflections (Timoshenko and Woinowski-Kreiger 1964).

The necessary energy for rupture initiation is obtained from the experimental work of Jones and Birch (2006), where the diameter of the indenter is taken into account when measuring the responses of plates subjected to low speed (in the range of ships' speeds) collisions.

The above model is implemented in the CRASED (*CR*ashworthiness *AS*sessment for *E*arly *D*esign) program. Its results are compared with the statistical data obtained in HARDER for the case of a ROPAX colliding with a similar ship. The length and breadth of the damage opening is presented in Fig. 43.4 as a function of penetration.

43.3.3 The Integrated Model

Putting the two elements of probability together (for a particular waterway or a set of routes) will provide a concise picture of the flooding probability and its extent due to collision and will highlight potential deficiencies (e.g. in structural arrangement and watertight subdivision) that need to be addressed at design level. This way, the operational profile of a new ship and its physical properties are mutually contributing to the derivation of the ship collision risk levels.

43.4 Conclusions

Although the probabilistic framework for damage stability is moving in the right direction for the quantification of safety levels of ships, its implementation is inconsistent as the weight is placed on the vulnerabilities of a ship. This way, any realistic treatment of the operational risks, and with it any serious attempt to build on prevention, is lost. The methodology proposed here aims to address this issue and, considering that accidents still happen despite the substantial effort spent for analysis and regulation, to create a momentum of thinking for rationalising the ship survivability assessment and the shipping operations in general.

References

Cahill, R. A. (2002). "Collisions and their causes", The Nautical Institute, 3rd Edition

Filipowicz, W. (2004). Vessel traffic control problems", The Journal of Navigation, **57**, pp. 15–24

Fujii, Y., Oshima, R., Yamanouchi, H., Mizuki, N. (1974). "Some factors affecting the frequency of accidents in marine traffic: I-The diameter of evasion for crossing encounters, II-The probability of stranding, III-The effect of darkness on the probability of collision and stranding, IV-Visual range and the degree of risk", The Journal of Navigation, **27**, 234–252

Goodwin, E.M. (1979). "Determination of ship domain size", Published in "Mathematical aspects of marine traffic" based on the Proceedings of the "Conference on Mathematical Aspects of Marine Traffic" held at Chelsea College London in September, 1977, organized by the Institute of Mathematics and its Applications edited by S.H. Hollingdale, London Academic Press

Hansen, P. F., Ravn, E. S., Hartman, J. P. and Sorensen, A. (2004) "FSA of the navigational safety in Baltic west", Proceedings of the 3rd International Conference on Collision and Grounding of Ships (ICCGS), Izu, Japan

HARDER, (2003). "Final Technical Report", Det Norske Veritas, Harmonisation of Rules and Design Rationale (HARDER), GRD1-1999-10721

IMO, (2009). "Adoption of amendments to the international convention of Safety Of Life At Sea, 1974, as amended", Resolution 216(82), Annex 2

Jones, N. and Birch, R. S. (2006), "Low velocity perforation design of metal plates", Structures Under Shock and Impact IX, WIT Transactions on the Built Environment, Vol. 87, pp. 179–186

Kriastiansen, S. (2005). "Maritime Transportation: Safety Management and Risk Analysis", Elsevier Butterworth-Heinmann

MAIB, (2005). "Report of the investigation of the collision between Cepheus J and Ileksa in the Kattegat, 22 November 2004", Marine Accident Investigation Branch, Report Number 12/2005

Øresund, (2006). "Navigational safety in the sound between Denmark and Sweden (Øresund): Risk and Cost-benefit analysis", The Royal Danish Administration of Navigation and Hydrography, The Danish Maritime Authority, The Swedish Maritime Administration

Pedersen, P. T. and Zhang, S. (1999). Collision analysis for MS Dextra. SAFEREURORO Spring Meeting, Paper No.2, Nantes, France

Timoshenko, S. P. & Woinowski-Kreiger, S. (1964). "Theory of plates and shells", McGraw-Hill Publishing Co

Vassalos, D. (2004). "A Risk-based Approach to Probabilistic Damage Stability", Proceedings of the 7th International Ship Stability Workshop, Shanghai, China

Vredeveldt, A. (2005). "Crashworthiness", Training Course on Risk-Based Ship Design, Universities of Glasgow and Strathclyde, Glasgow, UK, 7–8 June, 2005

Wendel, K. (1960), "Die Wahrscheinlichkeit des Uberstehens von Verletzungen", Schiffstechnik, Vol. 7, No. 36, pp. 47–61
Williams, G. P. (1997). "Chaos Theory Tamed", Joseph Henry Press

Chapter 44
Coupling of Progressive Structural Failure and Loss of Stability in the Safe Return to Port Framework

Seungmin Kwon, Qi Chen, George Mermiris and Dracos Vassalos

Abstract This paper addresses the survivability assessment of damaged ships with respect to the coupled effects of structural degradation and damage stability in the context of the Safe Return to Port (SRtP) framework for passenger ship safety. The survivability is evaluated in the time domain with varying wave loads. The proposed methodology is demonstrated through application to two diverse but safety-critical ship types, namely a RoPax with side damage, and an Aframax tanker with asymmetric damage at the bottom.

Keywords Safe return to port · Progressive structural failure · Damage stability

44.1 Introduction

Although substantial effort has been spent in the design stage and operational procedures of ships towards prevention and mitigation of accidental events and their ensuing consequences, the residual risk remaining is sufficient to fuel progressively higher expectations on the acceptable level of safety at sea by the society at large. The recurrence of high profile accidents demonstrates this situation with little margin for dispute and stresses the urgency for more comprehensive understanding of the underlying phenomena of a ship in distress.

S. Kwon · Q. Chen · G. Mermiris · D. Vassalos (✉)
The Ship Stability Research Centre, Department of Naval Architecture Ocean and Marine Engineering, University of Strathclyde, Glasgow, Scotland, UK
e-mail: d.vassalos@strath.ac.uk; d.vassalos@na-me.ac.uk

S. Kwon
e-mail: seung.kwon@strath.ac.uk

Q. Chen
e-mail: q.chen@strath.ac.uk

G. Mermiris
e-mail: g.mermiris@strath.ac.uk

V. L. Belenky et al. (eds.), *Contemporary Ideas on Ship Stability*, Fluid Mechanics and Its Applications 119, https://doi.org/10.1007/978-3-030-00516-0_44

In response to this need, the SRtP framework for passenger ship safety has set the foundation for a series of developments by advocating zero tolerance to loss of life following an accident and allowing 100% survivability for a specified interval of time (3 h recommended) if the casualty threshold criteria are exceeded, or indefinite survivability otherwise. In the latter case, the ship should be able to return to port under its own power or remain upright, afloat and with sufficient functionality to maintain passengers and crew until help arrives (IMO 2004).

In this context the authors have developed a methodology for evaluating the coupled stability deterioration and progressive structural failure of a damaged ship for a succession of sea states until capsize or global structural failure occurs. The structural degradation of the hull girder is captured by parametric models of damage propagation developed on the basis of FE analyses (Kwon et al. 2011). The stability of the ship following initial damage is assessed with PROTEUS3 (Jasionowski 2001).

The applicability of the developed methodology is demonstrated with two case studies, namely an Aframax tanker with asymmetric bottom damage, and a RoPax with side shell damage amidships. The two ship types are selected on the basis of their robustness in stability and longitudinal strength respectively, and set the foundation for a comprehensive treatment of post-damage survivability of safety-critical ships.

44.2 Methodology for Survivability Assessment

The key feature of this approach is the survivability assessment of ships with the coupling effect of damage evolution in the time domain. The progressive structural failure assessment of a damaged ship is achieved by crack propagation analysis under varying wave loading. In addition, when the damaged ship is flooded the water ingress and egress from the damaged compartment (due to the ship motion and the waves) constitutes a source of loading that deteriorates the structure locally and induces further crack propagation. In turn, the damage extension will exacerbate flooding and the cycle repeats itself until either stability is totally lost or residual strength becomes insufficient to sustain the applied loads.

For every time step of the calculation, the crack propagation is analysed and the effect of damage evolution on the survivability of the damaged ship is assessed in terms of criteria pertaining to residual strength and damage stability. In principle, such criteria can be expressed as the bending moment (BM) and the metacentric height (GM) respectively, and they should account for the dynamics of the situation by focusing on the time to break and the time to capsize respectively (Vassalos 2009). A high level description of the proposed methodology is shown in Fig. 44.1.

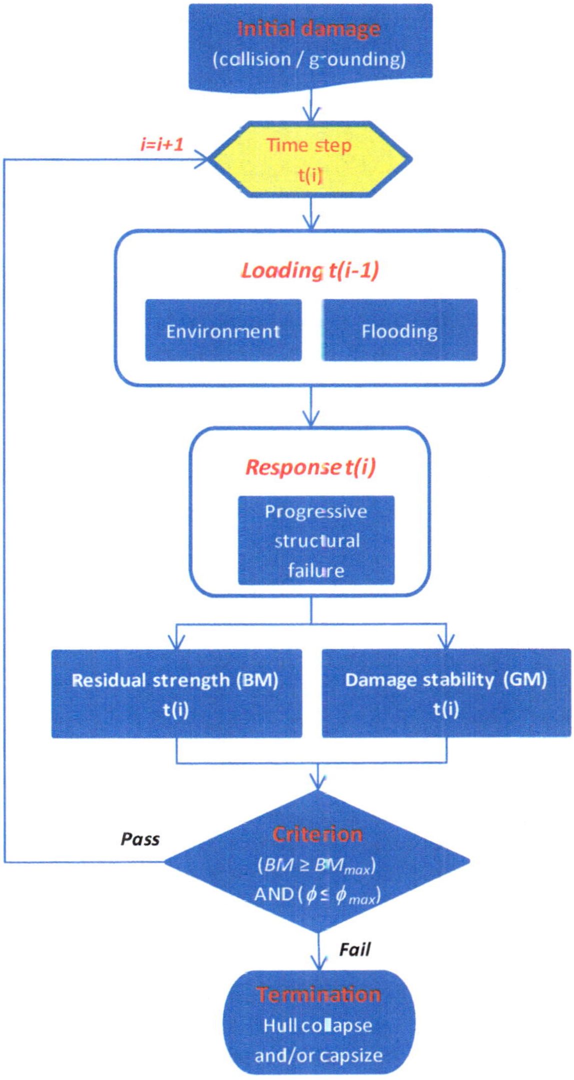

Fig. 44.1 High level process of the proposed methodology

44.2.1 Loading

Environmental loading

The load induced by the waves is the primary source for crack propagation and it is calculated either by an in-house 3D panel code with Green's function implementation (Xie 2011), or class guidelines (DNV 2012). The obtained wave and still water bending moments form direct input to the damage propagation analysis. On the other hand, the total bending moment will be compared with the residual strength of the damaged ship for every time step of the process. Also the calculated ship motions are used in flooding simulation.

Flooding loading

The flooding of the damaged compartment will induce inflow and outflow of water as the ship moves in waves, and the ensuing pressure differential effect on the crack propagation needs to be taken into account in the progressive failure of the structure. Depending on the size and the location of the damage opening, the size of the compartment, and the ship motions, the induced flooding pressure and its extent is defined. In principle, this calculation should be based on CFD simulations and in particular the solution of the Navier-Stokes equations with the VOF method (Gao et al. 2010). However, despite the increased resolution that this model could contribute to the overall methodology, its inherently complicated and computationally-intensive nature does not allow its direct integration in the methodology at the current stage of development and constitutes an item of future work.

44.2.2 Progressive Structural Failure

The crack growth modelling emanating from or near the damaged ship parts is based on the Linear Elastic Fracture Mechanics (LEFM) and Paris Law, in which 'time' is considered implicitly by the number of cycles, N, where 'N = time/period'. This approach has been adopted not only in the shipbuilding industry by Dexter and Pilarski (2000) and Dexter and Mahmoud (2004), but in the aviation industry by Farahmand et al. (2007) as well.

A crack growth model proposed by McEvily and Groeger (1977) has been modified to include the effective range of SIF and the material constants are chosen so as the linear region of the model fits with the original linear Paris Law (Kwon et al. 2011). Also, instead of conducting FE analysis, the Stress Intensity Factor (SIF) has been obtained through knowledge-intensive models, in order to overcome cumbersome numerical FEA and to utilise exploitation of accumulated knowledge (Tada et al. 2000).

A procedure for evaluating the progressive structural failure using the crack growth model is shown in Fig. 44.2. The extended crack size in the time domain would be used to assess the residual hull girder strength, and to update the opening of the flooded compartment and the adjacent ones if its boundaries are breached. In this manner, all the time-dependant information such as loading, damage and crack size is updated for each time step of the simulation.

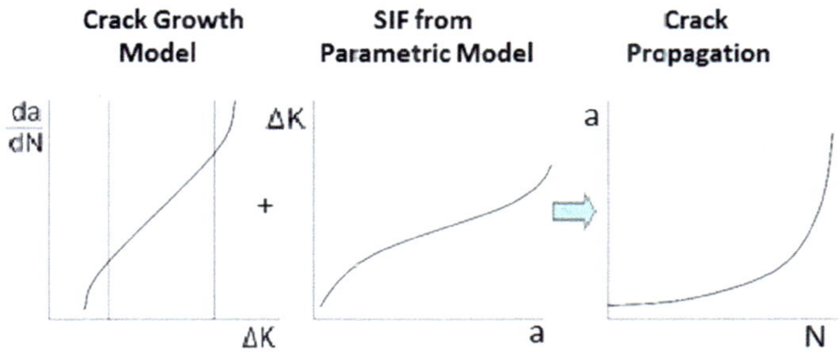

Fig. 44.2 Procedure of progressive structural failure analysis

44.2.3 Residual Strength

The residual strength assessment of a damaged ship is based on the moment-curvature relationship for stiffened panels known as Smith's method (Smith 1977). The relationship is obtained by imposing a curvature from sagging to hogging on the hull girder, which is assumed to consist of several beam-column elements. For each curvature, the average strain of each element is calculated and the stress imposed on it is obtained from a corresponding load-end shortening curve. The moment sustained by the whole section is obtained by summing up the moments of each element induced by axial force and distance of each element from the neutral axis of the section. The ultimate bending strength of the section is the maximum bending moment in hogging and sagging conditions.

The load-end shortening curve of each element is based on IACS rules (2008), in which the effect of plate induced failure, the flexural bucking failure of a column, and the tripping failure of a stiffener are considered as failure modes.

Validation of this approach was made by Kwon et al. (2010), through comparison with FEA results. For each time step, the damage evolution reduces the effective width of plates and the number of effective stiffeners and results to the reduction of the ultimate residual bending strength.

44.2.4 Damage Stability

The time to sink and/or capsize following damage is a critical factor in determining the level of safety of a damaged ship. In this case the dynamic response of the ship and the progressive flooding of its compartments in a random seaway form a highly non-linear dynamic system, the behaviour of which is modelled and simulated in the time domain by PROTEUS3 (Jasionowski 2001). The main features of the software are:

(i) The ship hydrodynamics are derived from properties of the intact hull and they are based either on asymmetrical strip theory formulation with Rankine source distribution or a 3D panel code, both accounting for the non-linearities arising from instantaneous variation of the mean ship position and large amplitude motions.
(ii) The effects of floodwater dynamics described by a full set of non-linear equations are derived from rigid-body theory.
(iii) The water ingress/egress is based on Bernoulli's equation and the floodwater motions are modelled as a Free Mass on Potential Surface (FMPS) de-coupled system in an acceleration field.

The ship geometry is defined in sections both for the hull and the internal compartments. The necessary environmental conditions for the simulations are generated according to JONSWAP or Pierson-Moskowitz spectra. The output of the calculation concerns ship motions, floodwater mass variation and motion, flow of floodwater through openings, environmental forces, etc.

44.2.5 Criteria

The survivability of damaged ships is assessed through comparison of the results from the residual strength analysis and the damage stability analysis with a set of criteria. That is, the damaged ship would be regarded as capsized if the roll angle exceeds 30% instantaneously or if steady (average) heel is greater than 20% for a period longer than three minutes according to EC (2003). On the other hand, the ship is considered to lose its structural integrity if the ultimate residual bending moment becomes lower than the applied one due to wave loading or when major structural elements are severed completely due to unstable crack propagation.

44.3 Case Studies

The verification of the proposed approach is conducted by application of the methodology to an Aframax tanker with bottom damage, and a RoPax with damage on the side shell. In the analysis only the environmental loading (in deep water) is considered.

Table 44.1 Principle dimensions of the tanker under consideration

Type of dimension	Value (m)
Length C. A.	250.17
Length B. P.	239.00
Breadth MLD.	44.00
Depth MLD.	21.00
Draught MLD. (Design)	14.60

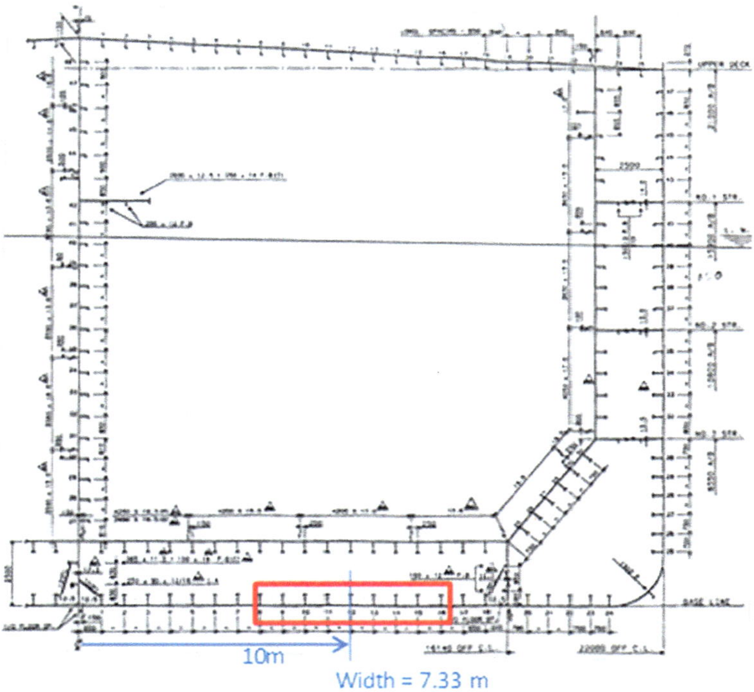

Fig. 44.3 Midship section of the tanker with initial bottom damage

44.3.1 Tanker

The ship used in the analysis is an 112,700 DWT crude oil carrier fitted with double side structure. The ship has six cargo tanks on each side. Its principal dimensions are summarized in Table 44.1 and the midship section is shown in Fig. 44.3.

The initial damage is located at the starboard side of the bottom structure close to mid length. The width of the initial damage is defined as 7.33 m as shown in Fig. 44.3 according to ABS guidelines (1995). A circular damage opening is assumed for convenience in the calculations.

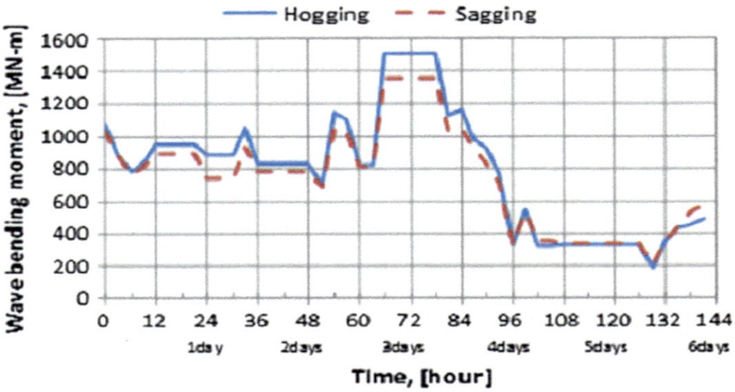

Fig. 44.4 Wave bending moment based on the wave data from the *Prestige* accident report

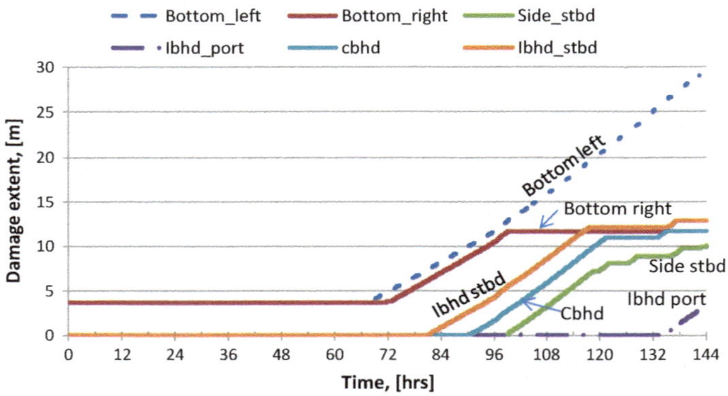

Fig. 44.5 Asymmetric grounding damage propagation on the bottom of a tanker

A fully loaded condition is considered for the wave loading analysis. The wave data is obtained from the BMA report (2004). The resultant wave bending moments for head sea conditions are shown in Fig. 44.4.

The result of the asymmetric damage propagation is obtained from the progressive structural failure analysis and shown in Fig. 44.5. Accelerated propagation (unstable) starts at 68 and 73 h for the port and starboard bottom damage respectively. It is identified that the starboard bottom damage spreads to the starboard side inner bulkhead (bottom girder is included) and side shell in sequence, whilst the port damage propagates to the centre bulkhead and to the port side inner bulkhead (and bottom girder) gradually.

Considering that the crack propagation on the bulkheads exceeds the height of the inner bottom translates to the breach of the corresponding cargo tank. That is, oil outflow is expected to appear at 80 and 90 h after the damage initiation from the starboard and port cargo tanks respectively.

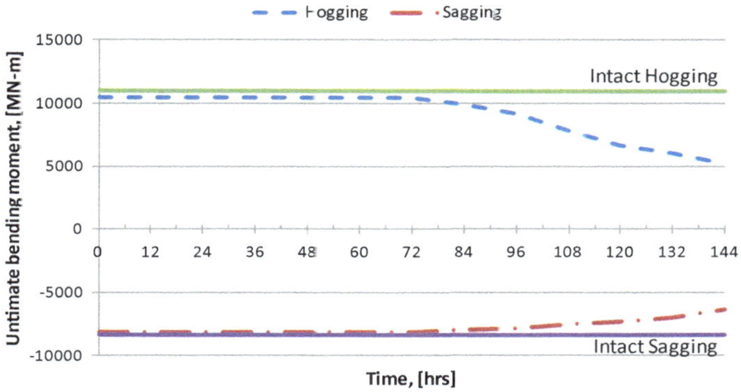

Fig. 44.6 Degradation of the ultimate residual strength of the tanker

As damage propagates, the ultimate residual strength of the hull girder is calculated, as shown in Fig. 44.6. The structural degradation starts at 72 h after the initial damage occurred, and coincides to the unstable damage propagation of the bottom. From this point onwards, the ultimate hull girder strength is decreased by 53 and 23% for hogging and sagging respectively, compared to the bending moment capacity of the intact condition. The vessel is regarded as lost at 144 h, i.e. when the bottom plate is severed.

The stability deterioration of a ship is measured by the time to capsize or sink after damage. For this calculation the permeability of the damaged compartment is assumed 0.95. At 92 h, i.e. when the damage size exceeds the centre bulkhead, the symmetrical compartment to the initially damaged one is considered flooded with identical permeability. Figure 44.7 shows the comparison of GZ curves for the fully loaded condition. The residual GZs for damage conditions prove that a wide safety margin is left following the damage occurrence, especially even after the flooding extends to the port side tank.

With respect to the time-domain numerical simulation, the initial transient flooding has been modelled by setting that the water enters the compartments (after 92 h) very fast. The survivability simulations are based on the same set of wave data as the one used for the wave loading calculations (BMA 2004). Two headings of JONSWAP spectrum (90 and 180%) are simulated to examine the influence of ship heading on damage stability. As expected no capsize occurs during the whole simulation time, regardless of the heading.

Table 44.2 presents the entire simulation process in ten consecutive steps. The output information (e.g. mode of motion, volume of flooded water) at the final stage in one step is accounted as the initial conditions for the successive step. The total simulation time is in line with the analysis for crack propagation, which lasts for 6 days (144 h). Figure 44.8 provides snapshots of flooded compartments in the damage area of the ship. The maximum rolling during the simulation does not exceed 5%.

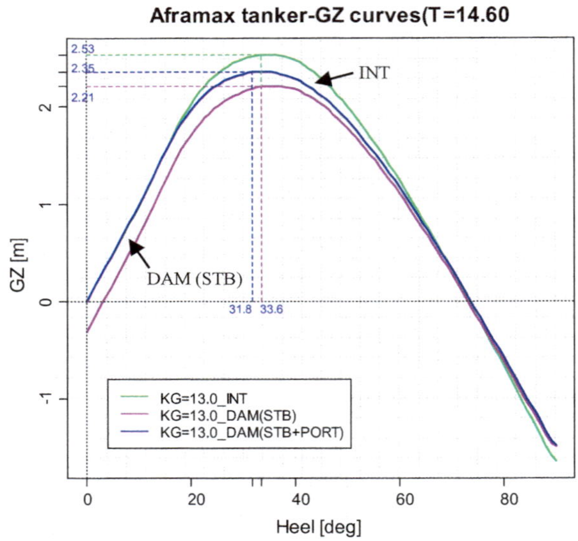

Fig. 44.7 GZ curves calculated by PROTEUS3 for the tanker under consideration

Table 44.2 Summary results for head seas conditions

	H_s (m)	Cumulative simulation time (h)	$\phi_{max,180}$ (deg.)	Reduction of BM	
				BM_{hog} (%)	BM_{sag} (%)
Step 1	5.5	3	3.13	4.8	1.9
Step 2	5.1	6	3.27	4.8	1.9
Step 3	4.5	36	3.23	4.8	1.9
Step 4	4.0	54	3.12	4.8	1.9
Step 5	5.0	66	3.08	4.8	1.9
Step 6	6.5	81	2.93	8.1	3.7
Step 7	4.7	92	3.20	13.5	5.5
Step 8	3.25	102	0.01	24.0	8.0
Step 9	2.5	132	0.03	45.3	15.8
Step 10	3.5	144	−0.09	52.9	23.3

Because the flooding compartments concerned in this damage case are symmetric, capsize is highly unlikely during the long period of simulation time.

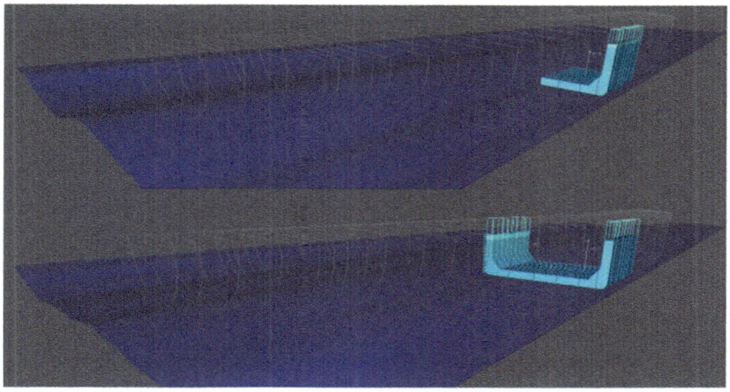

Fig. 44.8 Numerical simulation of the bottom flooding before and after the 92 h stage

Table 44.3 Principle dimensions of the RoPax under investigation

Type of dimension	Value (m)
Length O. A.	194.30
Length B. P.	170.00
Breadth MLD.	27.80
Depth to strength deck	14.85
Depth to vehicle deck	9.00
Draught MLD. (Design)	6.25

44.3.2 RoPax

The next application is made with a middle-sized RoPax vessel that has a vehicle deck at 9.0 m and a strength deck at 14.85 m above the baseline. The principal dimensions of the vessel are summarized in Table 44.3 while the cross section of interest and the general arrangement are illustrated in Figs. 44.9 and 44.10 respectively.

The initial damage is assumed to be located on the starboard side of the side shell near the middle of the ship. The height and depth of the damage are defined as 5.88 and 2.2 m respectively according to ABS guidelines (1995). The damage length is chosen to be the same as the web frame spacing (4.8 m).

Because of the sensitivity of the damage extent and the ensuing flooding conditions of the RoPax ship, two damage cases (of the same size but different location) are investigated for comparison purposes as shown in Fig. 44.9. In this manner, it will be demonstrated that uncertainty management is an inherent ingredient of the proposed methodology.

The results of the progressive structural analysis are shown in Fig. 44.11 and summarized in Table 44.4. The initial damage on the side shell is stable for the first 111 and 26 h in Case 1 and 2 respectively, after which unstable propagation upwards starts causing sequential breach of the joints with the deck plates above the initial

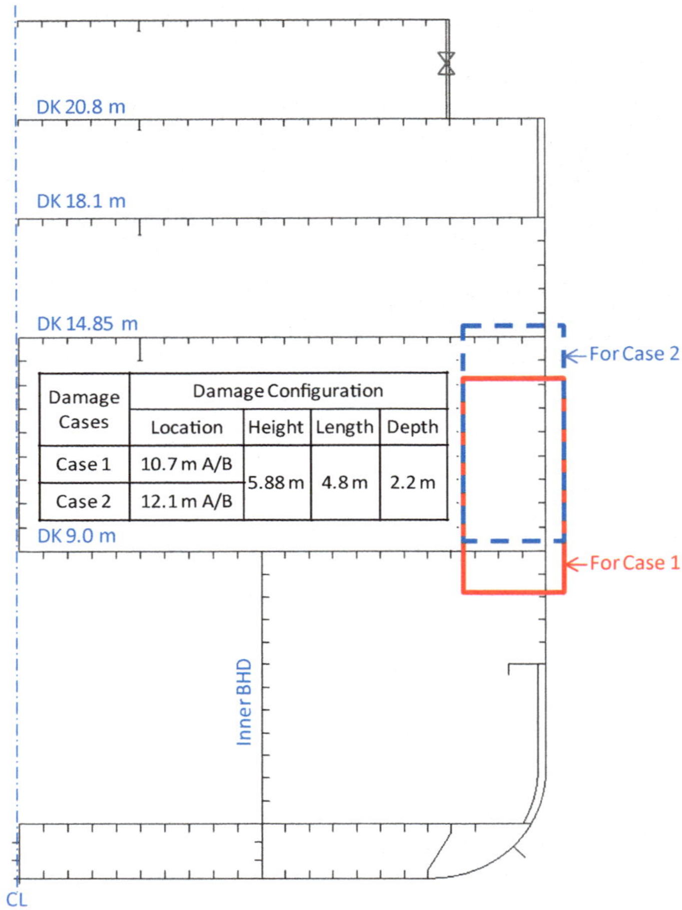

Damage Cases	Damage Configuration			
	Location	Height	Length	Depth
Case 1	10.7 m A/B	5.88 m	4.8 m	2.2 m
Case 2	12.1 m A/B			

Fig. 44.9 The cross section of the RoPax with the initial collision damage cases

damage location. The cracks spreading to decks are found to sever the deck plates as the unstable propagation is fuelled by the high stress concentration levels.

In the Case 1, the cracks on the vehicle deck and the lower part of the side shell propagate a negligible amount during the analysis. This is contributed to the fact that the initial location of the neutral axis of the damaged section is above the crack tips causing compressive bending stress on them during most of the analysis. Although the similar phenomenon happens to Case 2 at the initial stage, the elevated initial damage location causes earlier unstable damage propagation in deck plates, of which loss is sufficient to lower the neutral axis of the damaged section. As a result, the compressive bending stress on the lower side shell crack changes into tension at 45 h.

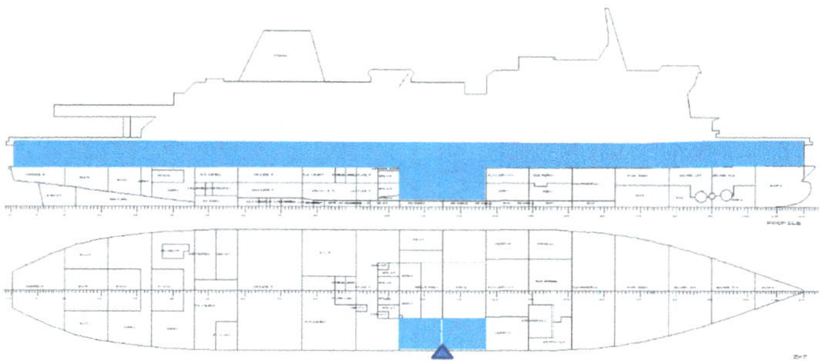

Fig. 44.10 General arrangement of RoPax and the flooding compartments

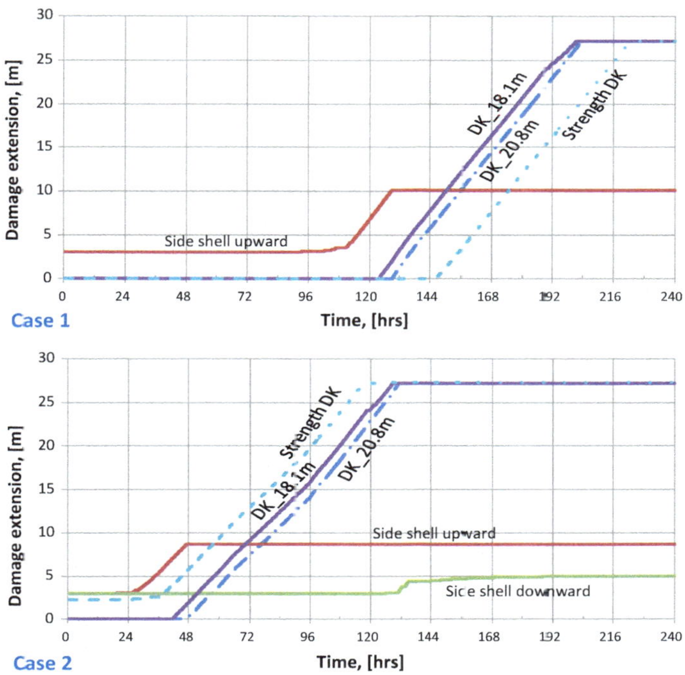

Fig. 44.11 Damage propagation for collision damage on side shell of the RoPax

Consequently, the lower crack of the side shell will propagate below the vehicle deck at 131 h causing extended flooding.

The ultimate residual strength of the vessel at the damaged section is reduced as shown in Fig. 44.12. In general, as the unstable propagation of the upper side shell crack arises, the ultimate residual strength in sagging starts to decrease followed by similar deterioration of the ultimate bending capacity in hogging, which starts the damage propagation on deck plates becomes unstable. The rapid deterioration of

Table 44.4 Summary of the progressive structural failure results

	Case 1	Case 2
Time, side upper unstable (h)	111.4	25.7
Time, Strength deck breach (h)	113.1	Initial
Time, deck 18.1 m breach (h)	121.8	38.3
Time, deck 20.8 m breach (h)	128.9	47.3
Time, vehicle deck breach (h)	Initial	130.7
Final damage size under vehicle deck (m)	1.26	1.89

the longitudinal strength continues until the cracks on the decks above the damage sever the entire deck plates across the hull breadth. In both cases, the damaged cross section loses its residual bending capacity by 39 and 45% for hogging and sagging respectively in comparison to the intact condition. The vessel is regarded as lost at 223 and 117 h when the strength deck is totally severed for Case 1 and 2 respectively. The loss of multiple decks induces further reduction of strength in sagging by lowering the instantaneous neutral axis of the damaged section. It should be noted that the limitation of the damage propagation to the starboard side shell prevents further reduction of the residual strength and flooding extension.

With respect to flooding, the basic difference between Case 1 and 2 is the location of the initial opening. In Case 1 the side tanks below the vehicle deck are considered, whereas in Case 2 the lower vertical boundary of the damage opening ($Z = 9.16$ m) is above the water line ($T = 6.25$ m). The stability curves for the RoPax ship at the intact and both damage conditions are presented in Fig. 44.13.

According to the results of the damage propagation outlined in Table 44.4, it is assumed that the damaged RoPax has a fixed opening in Case 1 because crack propagations on the vehicle deck and the lower part of side shell are negligible. Contrary to this, an additional opening is required for Case 2 due to the breach of the vehicle deck at 131 h after the initial damage.

The simulation results for Case 1 are presented in Table 44.5 (3-h interval, $KG =$ 12.9 m, $Hs = 2.0 \sim 4.0$ m, and 20 simulations per sea state). As expected the worst heading is the beam seas regardless of wave height. The $Hs = 2$ m is identified as the lower capsize boundary, as no capsize occurrence is observed. Figure 44.14 shows an example of a time series of the simulation with $Hs = 4.0$ m that results in capsize after 4.6 min as the rolling angle exceeds 30%.

The simulation results for Case 1 are illustrated in Fig. 44.15, where the vulnerability to flooding against time is shown. In this case structural degradation is not of any concern for the survivability of the ship.

The results for Case 2 are presented in Table 44.6. This time only a small amount of floodwater accumulates on the vehicle deck in 131 h' simulation even at $Hs =$ 4 m in beam seas. The observed maximum list of the ship is close to 2% (Step 1). Subsequently, a further opening is added to reflect that the crack propagation reaches

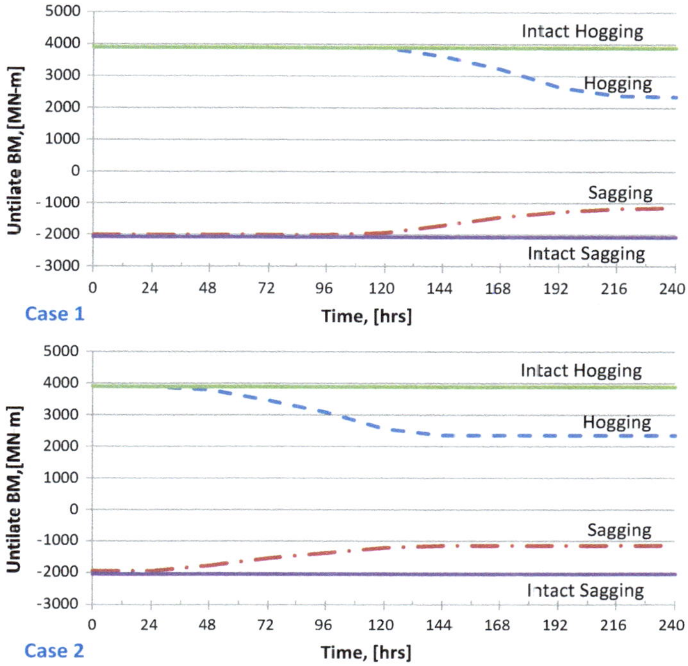

Fig. 44.12 Decrease of the ultimate residual strength of the RoPax at the damaged section due to the progressive structural failure

Fig. 44.13 GZ curves for the RoPax ship

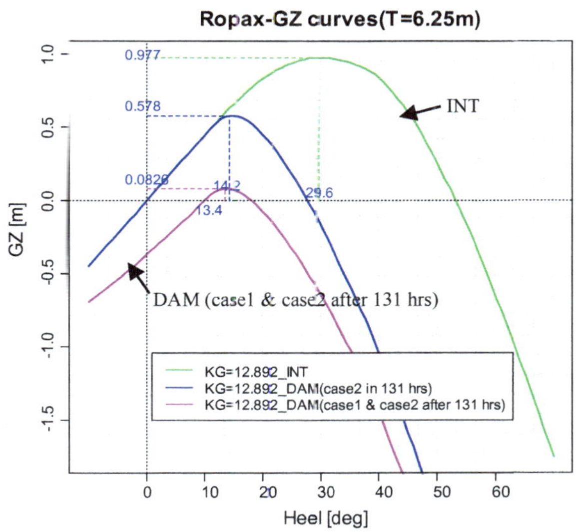

Table 44.5 Numerical simulation matrix (Case 1)

Hs (m)	Heading (deg.)	No. of capsizes	No. of runs	Probability of capsize, P_{cap} (t_{cap} = 3 h | Hs)
2.0	180	0	20	0
	270	0	20	0
2.5	180	0	20	0
	270	12	20	0.6
3.0	180	1	20	0.05
	270	20	20	1
3.5	180	17	20	0.85
	270	20	20	1
4.0	180	20	20	1
	270	20	20	1

Table 44.6 Summary of results for beam seas (Case 2)

	Hs (m)	Simulation time (h)	Probability of capsize P_{cap}(Time | Hs)	Reduction of BM	
				BM_{hog} (%)	BM_{sag} (%)
Step 1	2.0	131	0	38.9	43.3
	4.0				
Step 2	2.0	134	0	38.9	43.8
	2.5		0.4		
	3.0		1		
	3.5				
	4.0				
Step 3	2.0	144	0	39.0	44.1
	2.5		0.9		

almost 1.0 m below the vehicle deck after 134 h (Step 2) and 1.3 m after 144 h (Step 3).

The same range of wave heights as for Case 1 is applied for the survivability simulations. Base on the obtained capsize probability, it appears that the ship's vulnerability in Case 2 is close to Case 1 from Step 2 onwards, where an enlargement of the flooding extent takes place.

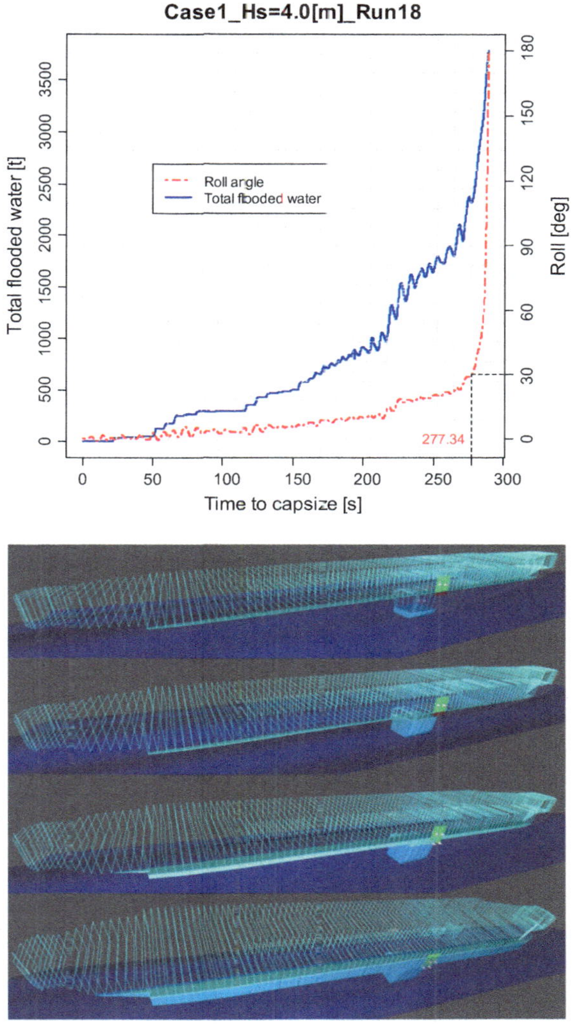

Fig. 44.14 Time series of a typical capsize simulation in beam sea given Hs = 4 m (Case 1)

44.4 Conclusions and Future Work

Based on the results of the two case studies of the developed methodology the following conclusions can be drawn:

– The survivability assessment following accidental damages is carried out in the *time domain* by taking into account the *coupling* of the progressive structural failure with the damage stability as well as the hull girder residual strength.

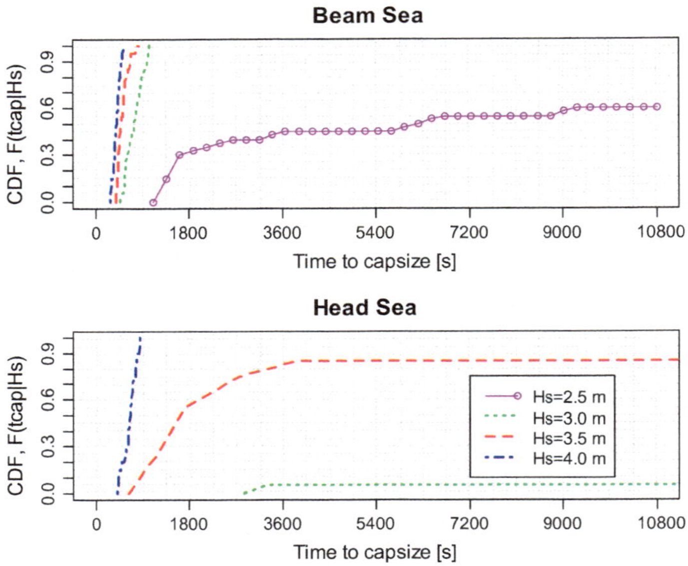

Fig. 44.15 Cumulative probability distribution of time to capsize $F(t_{cap})$ given the specific loading, flooding extent, sea states, within the simulation time of 3 h (Case 1)

- For the tanker case, it is demonstrated that the degradation of the structural integrity becomes a priority over the stability deterioration due to the initial damage.
- On the other hand, the RoPax vessel is more susceptible to loss of stability, as it has been extensively documented in the literature. However, the geometry and location of the initial damage in combination to unfavourable environmental conditions can also threat the structural capacity of the ship.

As discussed at the beginning of this paper, this methodology is still at its infancy, and the material presented here addresses only the main elements of it. Along these lines, the topics that will be addressed in the immediate future are:

- Investigating the effect of the various damage configurations (shapes, multi-openings) on the progressive structural failure analysis;
- Developing a platform that will embrace all the relevant tools (including systems availability) and will enable their consistent implementation for SRtP studies; and
- Coupling of structural FE models with CFD for flooding loading assessment.

References

ABS, 1995, "Guide for Assessing Hull Girder Residual Strength", American Bureau of Shipping

Bahamas Maritime Authority (BMA), 2004, "Report of the Investigation into the Loss of the Bahamian Registered Tanker '*Prestige*' off the Northwest Coast of Spain on 19th November 2002", The Bahamas Maritime Authority

Dexter, R.J., Pilarski, P.J., 2000, "SSC-413: Effect of Welded Stiffeners on Fatigue Crack Growth Rate", Ship Structure Committee

Dexter, R.J. and Mahmoud, H.N., 2004, "SSC-435: Predicting Stable Fatigue Crack Propagation in Stiffened Panels", Ship Structure Committee

DNV, 2012, "Hull Structural Design, Ships with Length 100 metres and above", Rules for Classification of Ships, Part 3 Chapter 1, Det Norske Veritas

EC, 2003, "Directive 2003/25/EC of the European Parliament and of the Council on Specific Stability Requirements for Ro-Ro Passenger ships", European Commission

Farahmand, B., Saff, C., Xie, D. and Abdi, F., 2007, "Estimation of Fatigue and Fracture Allowables for Metallic Materials under Cyclic Loading", Proceedings of the 48th Conference of the American Institute of Aeronautics and Astronautics, AIAA-2007-2381

Gao, Z., Vassalos, D. and Gao, Q., 2010, "Numerical Simulation of Water Flooding into a Damaged Vessel's Compartment by the Volume of Fluid Method", Ocean Engineering, Vol. 37, pp. 1428–1442

IACS, 2008, "Common Structural Rules for Double Hull Oil Tankers", International Association of Classification Societies

IMO, 2004, "Report of the Maritime Safety Committee on its Seventy-eighth Session", International Maritime Organization, MSC 78/26

Jasionowski, A., 2001, "An Integrated Approach to Damage Ship Survivability Assessment", PhD Thesis, University of Strathclyde, UK

Kwon, S., Vassalos, D., and Mermiris, G., 2010, "Understanding Potential Risk from a Coupled Problem of Flooding and Structural Degradation of a Damage Ship", Proceedings of the 4th International Maritime Conference on Design for Safety

Kwon, S., Vassalos, D. and Mermiris, G., 2011, "Progressive Structural Failure and Residual Strength of Damaged Ships", Proceedings of the International Conference on the Damaged Ship, The Royal Institution of Naval Architects, London, UK

McEvily, A.J. and Groeger, J., 1977, "On the Threshold for Fatigue-Crack Growth", Proceedings of the 4th International Conference on Fracture, Vol. 2, pp. 1293–1298

Smith, C.S., 1977, "Influence of Local Compression Failure on Ultimate Longitudinal Strength of a Ship's Hull", Proceedings of the International Symposium on Practical Design in Shipbuilding (PRADS), pp. 73–79

Tada, H., Paris, P.C. and Irwin, G., 2000, "The Stress Analysis of Cracks Handbook", Third Edition, ASME Press

Vassalos, D., 2009, "Risk-Based Ship Design", In: Papanikolaou, A. D. (ed.) Risk-Based Ship Design: Methods, Tools and Applications, Springer

Xie, N., 2011, "A Numerical Prediction of Global Wave Loads for Intact and Damaged Ships", Ship Stability Research Centre, Internal Report

Chapter 45
Impact of Watertight Door Regulations on Ship Survivability

James Person

Abstract When demonstrating compliance with the International Convention for the Safety of Life at Sea (SOLAS), Chapter II-1 subdivision and damage stability regulations, it is assumed that all watertight doors are closed and the related internal watertight subdivision is 100% effective. Unfortunately, casualty history indicates that this is not always the case. Contributing to this situation are provisions in the SOLAS regulations that permit some watertight doors to remain open or be open for extended periods of time during navigation under certain conditions. This paper provides background information on the SOLAS requirements for watertight doors and discusses whether this regulatory treatment is still appropriate for passenger ships of the future. Originally written in June 2011, an update is included to indicate the latest SOLAS regulatory developments as of June 2017.

Keywords Watertight doors · Damage stability · Passenger ships · SOLAS

45.1 Introduction

Inherent in the application of SOLAS Chapter II-1 subdivision and damage stability regulations is an assumption that all watertight doors will be closed in a flooding casualty and that the related internal watertight subdivision will be 100% effective. Unfortunately, casualty history indicates that this assumption is not always valid. Contributing to this situation is SOLAS regulation II-1/22.4, which is a long-standing provision that allows Administrations to permit certain watertight doors to remain open during navigation if considered absolutely necessary for the safe and effective operation of the ship's machinery or to permit passengers normally unrestricted access throughout the passenger area. Considering the substantial evolution of SOLAS passenger ship safety standards over the last two decades and the current SOLAS 2009 probabilistic damage stability regulations that provide new subdivision

J. Person (✉)
U.S. Coast Guard, Washington, DC, USA
e-mail: James.L.Person@uscg.mil

V. L. Belenky et al. (eds.), *Contemporary Ideas on Ship Stability*, Fluid Mechanics and Its Applications 119, https://doi.org/10.1007/978-3-030-00516-0_45

design flexibility, this paper questions whether regulation II-1/22.4 is still appropriate for passenger ships of the future.

45.2 Background of SOLAS Watertight Door Regulations

The 1948 SOLAS Convention requirement for keeping watertight doors closed only applied to passenger ships and was as follows (regulation II/12(k)): "All watertight doors shall be kept closed during navigation except when necessarily opened for the working of the ship, and shall always be ready to be immediately closed." This requirement was not changed in the 1960 or 1974 SOLAS Conventions (regulation II/13(n)), and although the 1981 SOLAS amendments (resolution MSC.1(XLV)) changed the regulation format, the requirement remained the same (regulation II-1/15.14).

Following the *Herald of Free Enterprise* casualty in 1987, several sets of SOLAS amendments were adopted. The 1989 SOLAS amendments (resolution MSC.13(57)) substantially revised the watertight door requirements in regulation II-1/15. Although the primary impact was to significantly upgrade the control system requirements for power-operated sliding watertight doors on new passenger ships (constructed on or after 1 February 1992), changes were also made to the requirements for opening watertight doors during navigation. These changes were an effort to clarify and bound the existing provision "when necessarily opened for the working of the ship". Apparently it was felt at the time that many watertight doors were being allowed to remain open under the vagueness of this provision. The new provisions in regulation II-1/15 were as follows:

> 9.2 A watertight door may be opened during navigation to permit the passage of passengers or crew, or when work in the immediate vicinity of the door necessitates it being opened. The door must be immediately closed when transit through the door is complete or when the task which necessitated it being open is finished.
>
> 9.3 Certain watertight doors may be permitted to remain open during navigation only if considered absolutely necessary; that is, being open is determined essential to the safe and effective operation of the ship's machinery or to permit passengers normally unrestricted access throughout the passenger area. Such determination shall be made by the Administration only after careful consideration of the impact on ship operations and survivability. A watertight door permitted to remain thus open shall be clearly indicated in the ship's stability information and shall always be ready to be immediately closed.

With the exception of the retroactive application of the watertight door control system requirements (to passenger ships constructed before 1 February 1992) following the *Estonia* casualty in 1994, the SOLAS watertight door requirements for passenger ships have remained the same. The SOLAS 2009 amendments separated the watertight door requirements into design (regulation II-1/13) and operational (regulation II-1/22) sections, but the requirements did not change. Therefore, the current watertight door requirements originate from the 1989 SOLAS amendments.

45.3 Guidance on Watertight Doors Permitted to Remain Open

Following the 1989 SOLAS amendments that attempted to clarify and restrict conditions when watertight doors could be opened during navigation, there has been little activity at the International Maritime Organization (IMO) with respect to watertight door requirements until recently. In 2006 the Stability, Load Lines and Fishing Vessel Safety (SLF) Sub-Committee initiated work to establish IMO guidance on the SOLAS 2009 regulation II-1/22.4 provision "Such determination shall be made by the Administration only after careful consideration of the impact on ship operations and survivability." The SLF Sub-Committee developed the survivability related guidance and the Ship Design and Equipment (DE) Sub-Committee developed the ship operations related guidance. That joint effort resulted in the December 2010 MSC.1/Circ.1380 *Guidance for watertight doors on passenger ships which may be opened during navigation.*

The MSC.1/Circ.1380 guidance represents substantial compromise from the initial SLF and DE proposals and essentially requires a risk assessment but only stipulates that the minimal stability criteria need be met when operating in high risk navigation areas. As a result several countries reserved their position on MSC.1/Circ.1380, indicating the guidance is not adequate and represents a permanent degradation of the subdivision of the ship. They proposed that no watertight doors should be allowed to remain open when the ship is operating in high risk areas, and under such conditions watertight doors should only be allowed to be opened for passage and closed immediately afterwards. Then in conditions of low risk, certain watertight doors may be allowed to remain open following satisfaction of the minimal stability criteria.

45.4 Observations Regarding SOLAS Requirements

The SOLAS 2009 passenger ship subdivision and damage stability requirements incorporate a probabilistic methodology that allows new subdivision arrangement flexible. However, the SOLAS 2009 watertight door requirements reflect a standard that was developed in 1989. The passenger ship subdivision and damage stability requirement at that time was a deterministic standard with floodable length, factor of subdivision and prescriptive main transverse watertight bulkhead requirements.

The overall SOLAS passenger ship safety standards have been significantly raised since 1989. The higher safety bar is a result of both improved technology and a reduced societal tolerance for casualties and loss of life.

45.5 Proposed Way Ahead for Watertight Door Requirements (*author's opinion in June 2011*)

Given the evolution of SOLAS passenger ship safety standards and the SOLAS 2009 subdivision design flexibility that was not available when the watertight door requirements in regulations II-1/22.3 and II-1/22.4 were established, it is time to consider:

- deleting regulation II-1/22.4 that allows certain watertight doors to remain open during navigation; and
- revising regulation II-1/22.3 to further restrict when watertight doors may be opened during navigation (consider possible limits on: duration of time open; door location; number of doors open; risk of navigating area; etc.).

In this task it may be necessary to account for passenger ship size differences based on the general premise that larger ships have more arrangement flexibility, which is consistent with the legacy provision now in regulation II-1/4.3. In this context, 400 persons is an established break point and is currently used in regulation II-1/8.

With respect to personnel safety concerns of frequent passage through normally closed watertight doors, it is suggested that this risk can be reduced through future innovative designs that eliminate these arrangements.

Passenger ships of the future will be required to meet high safety standards. Therefore, careful consideration should be given to ensure that watertight door requirements don't impact the flooding survivability by compromising the subdivision and damage stability requirements.

45.6 Update on Watertight Door Requirements (*the latest SOLAS regulatory developments as of June 2017*)

Following the *Costa Concordia* casualty in January 2012, IMO initiated a broad review of passenger ship safety requirements to consider the need for further improvements. As part of that initiative, the Ship Design and Construction (SDC) Sub-Committee (a reorganization/consolidation of the DE and SLF Sub-Committees) was instructed to review the conditions under which watertight doors may be opened during navigation. This included a review of SOLAS regulations II-1/22.3 and II-1/22.4, and the related guidance in MSC.1/Circ.1380. The results of that work were to recommend:

- deletion of regulation II-1/22.4 that allows certain watertight doors to remain open during navigation; and
- revision of MSC.1/Circ.1380 *Guidance for watertight doors on passenger ships which may be opened during navigation* to further restrict when watertight doors may be opened during navigation.

The SDC Sub-Committee finalized the regulation amendment and revised MSC circular in February 2015, and the IMO Maritime Safety Committee (MSC) subsequently approved them in principle in June 2015. The amendment to SOLAS regulation II-1/22 was adopted by MSC in June 2017, with an entry into force date applicable to passenger ships constructed on or after 1 January 2020. In addition to the deletion of regulation II-1/22.4 that allows certain watertight doors to remain open during navigation, the amendment also added a new provision to regulation II-1/22.3 that will require Administrations to authorize watertight doors that may be periodically opened during navigation after considering the impact on ship operations and survivability. MSC also approved the associated revised MSC circular as new MSC.1/Circ.1564 *Revised guidance for watertight doors on passenger ships which may be opened during navigation.*

In addition, the Ship Systems and Equipment (SSE) Sub-Committee is now considering potential requirements for anti-crushing protection devices that could be fitted on watertight doors to improve personnel safety during the daily operation of watertight doors.

45.7 Conclusions

Looking ahead, watertight doors will no longer be allowed to remain open during navigation on new passenger ships constructed on or after 1 January 2020. In addition, the ship's Administration will authorize any watertight doors that may be periodically opened during navigation after considering the impact on ship operations and survivability, and when navigating in potentially hazardous conditions these watertight doors are to be kept closed except when a person is passing through them. These IMO regulatory improvements should significantly help ensure that watertight doors don't inadvertently compromise a passenger ship's survivability during a flooding casualty.

Disclaimer The opinions expressed in this paper are only those of the author and do not represent those of the U.S. Coast Guard.

Chapter 46
Damage Stability of Passenger Ships—Notions and Truths

Dracos Vassalos

Abstract A painstaking evolutionary development on damage stability of ships is giving way to unprecedented scientific and technological developments that has raised understanding on the subject as well as the capability to respond to the most demanding societal expectations on the safety of human life and to do so cost-effectively. Within less than half a century, damage stability calculations catapulted from **one** scenario per newbuilding (QE II, mid-1960s over a few months) **to tens of thousands of scenarios** (modern cruise liners in 2010s in a few weeks). Given the steepness of the learning curve and the pace of developments, it is understandable that certain notions were accepted as truths without due rigor and, as such, continue to shape contemporary thinking and developments. This paper draws attention to various issues that are emerging as knowledge grows and proposes a way forward for establishing a stronger foundation to safety assurance in the maritime sector and for future developments on the subject.

Keywords Damage stability · Vulnerability
Emerging issues with "old" and "new" ships and rules
A verification framework for maritime safety

46.1 Introduction

With artefacts on human endeavours at sea dated as far back as 6500 B.C., it is mind boggling to think that it was not until 250 B.C. when the first recorded steps to establish the foundation of Naval Architecture, *floatability* and *stability*, were made by Archimedes. It is even more astonishing that it took nearly two millennia, after this, before the first attempts to convey the meaning of stability to the designers of the day took place in the 18th century by Hoste and Bouguer. Pertinent regulations,

---

D. Vassalos (✉)
The Ship Stability Research Centre, Department of Naval Architecture, Ocean and Marine Engineering, University of Strathclyde, Glasgow, Scotland, UK
e-mail: d.vassalos@strath.ac.uk

V. L. Belenky et al. (eds.), *Contemporary Ideas on Ship Stability*, Fluid Mechanics and Its Applications 119, https://doi.org/10.1007/978-3-030-00516-0_46

especially after accidents involving water ingress and flooding, were introduced even much later; notably, the first Merchant Shipping Act of 1854 addressing subdivision and leading eventually and after heavy loss of life to the adoption of the first internationally agreed system of subdivision in SOLAS 1929. Indeed, the first specific criterion on residual static stability standards was introduced at the 1960 SOLAS Convention. This "tortoise" pace gave way to the steepest learning curve in the history of Naval Architecture with the introduction of the probabilistic damage stability rules in the late sixties as an alternative to the deterministic requirements. Prompting and motivating the adoption of a more rational approach to damage stability and survivability, probabilistic rules instigated the development of appropriate methods, tools and techniques capable of meaningfully addressing the physical phenomena involved. The ensuing improved technical capability virtually at worldwide has, in turn, been fuelling innovation in the shipping sector to meet the demand for larger, faster, more complex and specialised ships. Within 50 years, this new impetus has climaxed to the "zero tolerance" concept of Safe Return to Port in July 2009 and to an open proclamation by the Secretary General of the International Maritime Organization (IMO), stating that deterministic regulations have no future (International Conference on "The Future of Maritime Safety" IMO, June 2013). However, this is taking place in an industry that is fragmented, undermanned, intensely competitive and, above all, with a traditional deterministic and reactive mindset that sees all new developments as a hard-to-swallow "pill", likely to cause more ills than it can cure.

Deriving from this pace of developments and in the absence of a nurturing environment, the safety regime in the maritime industry is full of conceptual gaps that tend to undermine progress and safety at large. This paper draws attention to a few obvious issues, attempting to lay the foundation for a more concerted effort at international scale to "put things right". The best available means has always been legislation and, fortuitously or otherwise, the right instruments are in place to affect such change most effectively.

46.2 Damage Stability Legislation

46.2.1 Prevailing Regulatory System

The maritime regulatory system consists of internationally agreed standards at IMO, regionally agreed regulations, national standards, International Association of Classification Societies (IACS) Common Structural Rules and Unified Requirements, classification rules of individual Classification Societies and other technical standards. The regulatory system is a result of a continuous amendment process, mostly as a result of major accidents, tending to address the safety deficiencies resulting in the latest accident—in some cases only peripherally. Table 46.1 and Fig. 46.1 tell the story!

Table 46.1 Modern Ferry accidents in the western world (Vassalos 1999)

1953	Princess Victoria capsized and sank when large waves burst open the stern door in rough weather with the ***car deck*** and starboard engine room ***flooded*** (134 dead)
1974	Straitsman capsized and sank whilst approaching berth with the vehicle door partly open and, as a result of squat, ***flooding the vehicle deck*** (2 dead)
1987	Herald of Free Enterprise capsized when the bow wave and bow-trim combined to bring the open bow door underwater, leading to ***flooding of the vehicle deck*** (193 dead)
1987	Santa Margarita Dos capsized in port in Venezuela due to heeling while loading vehicles as a result ***of flooding of the vehicle deck*** (5 dead)
1994	Estonia capsized and sank due to ***flooding of the vehicle deck*** (852 dead)
2006	Al Salam Boccaccio '98 capsized and sank due to ***flooding of the vehicle deck***, following fire (1002 dead)

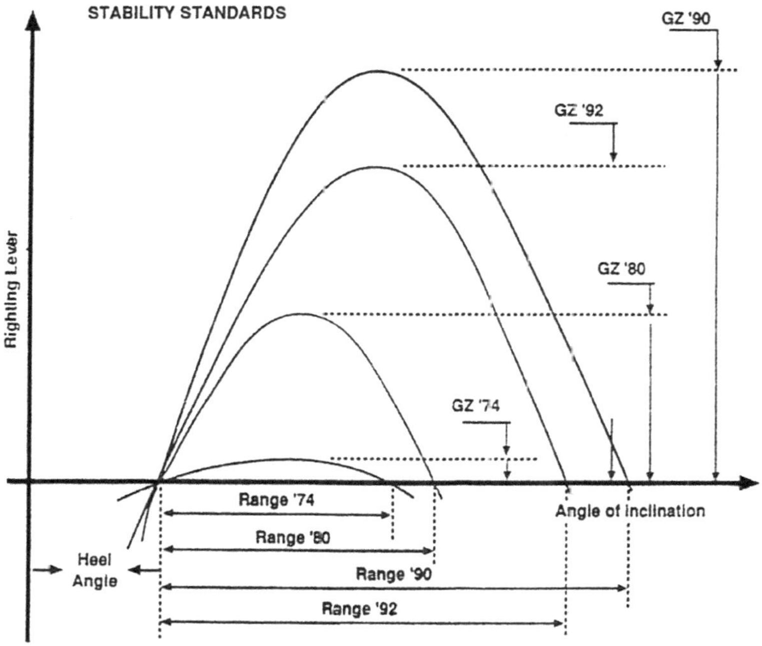

Fig. 46.1 Deterministic damage stability standards for passenger ships

The prevailing regulatory system is often referred to as prescriptive, implying that specific solutions are prescribed with little room for innovation. This may be cost-effective for standard ship designs like bulk carriers and oil tankers, resulting in very short delivery time, effective production and low costs. However, such regulations could inhibit innovation, as even designs that would increase ship safety may be violating existing prescriptive regulations.

46.2.2 Emerging Regulatory System

At IMO level, SOLAS Ch. II-1, Regulation 5 (certain systems—exemptions) allows for equivalent design of "fitting, material, appliances or apparatus" provided "it is at least as effective as that required by the present regulations." This, in principle, facilitated equivalent safety and risk-based approaches but wide use is not documented. Today, the most widely used and known regulation SOLAS-II.2/17 addresses Alternative Design and Arrangements for fire safety. The associated guideline (IMO MSC 2001) advocates use of engineering analysis (i.e., risk analysis), a strategic goal adopted by IMO to "develop a regulatory framework for assessing alternative designs and arrangements so that new concepts and technologies may be permitted, which provide a level of safety at least equivalent to that provided by the prescriptive regulations." Other regulations in place that are, in principle, goal-based include: the new harmonised probabilistic rules for damage stability (this is implied but NOT explicit and has as a result not been utilised by the industry, with the exception of a few isolated cases on an informal basis), Safe Return to Port (Chapter II-2) and Regulation 38, Chapter III on Life Saving Appliances (IMO 2006). Assisting in this direction are regulations such as the Stockholm Agreement (IMO Resolution 14 1995), utilising a performance-based approach by means of numerical and physical model experiments.

However, legislation for "Alternative Design and Arrangements" and equivalent safety are now applicable to damage stability (albeit in principle only), fire safety, life saving appliances, ship systems and crashworthiness (again implicitly), thus offering probabilistic and risk-based approaches to safety for all these issues. As a result, safety equivalence will not be sufficient as an instrument to cater for innovative ships, which incorporate more than one risk-based design element. The safety implications between damage stability and fire safety on one hand and life saving appliances on the other are too complex to be handled in isolation and without the use of holistic risk assessment and optimisation. In this context, optimisation should target not only optimum balance between safety and other design objectives but also between different safety objectives. The term to be used in this case is **risk balance**. As an example, even with very high survivability one would not eliminate the use of life saving appliances. On the other hand, efforts targeting safety enhancement could become much more effective. For example, with risk due to flooding-related accidents dominating the safety of passenger ships (some 90% of the risk), the focus on safety enhancement of passenger ships must clearly address damage stability as a priority. Ultimately, it is expected to see risk-based design being applied at ship/platform level as an holistic optimisation process. This will need to be supported by risk-based regulations (regulations justified by risk analysis based on agreed risk acceptance criteria) and approved Risk Control Options (design solutions or measures for risk prevention/mitigation).

46.2.3 Risk Acceptance Criteria

A common way of presenting risk graphically, in terms of fatalities, is by using the so-called F-N diagram, the plot of cumulative frequency of N or more fatalities, together with related risk acceptance criteria, Fig. 46.2, (MSC 72/16, 14 Feb. 2000). In addition, some form of aggregate information is used, such as the expected number of fatalities E(N), often referred to as the potential loss of life, PLL. This figure, innocent as a notion, hides **three** very important truths, as described next.

A cursory look at this diagram would reveal that tolerability drops by one order for every order of increase in fatalities. It is well-known that the public is much more averse to accidents, which kill many people at once than to others, which kill more people one at a time or in small "bunches". However, this has led to the situation where we have hundreds of people losing their lives in our industry every single day going unnoticed. For example, the total death toll at sea in the Philippines has been estimated to be in the region of 20,000–40,000 per year if the small outriggers, pump boats and motor launches, which operate unscheduled services between islands, are to be included in the statistics (Spouge 1990).

This is not a problem specific to Philippines. The global fishing industry statistics from IMO, the International Labour Organization (ILO), and the Food and Agriculture Organization (FAO) of the United Nations, demonstrate a huge safety problem, as outlined next:

- Each year there is an average of 24,000 fatalities and 24 million non-fatal accidents.
- The fishing fatality rate is estimated at 80 deaths/100,000 individuals per annum, which is 79 times higher than the overall occupational fatality rate.

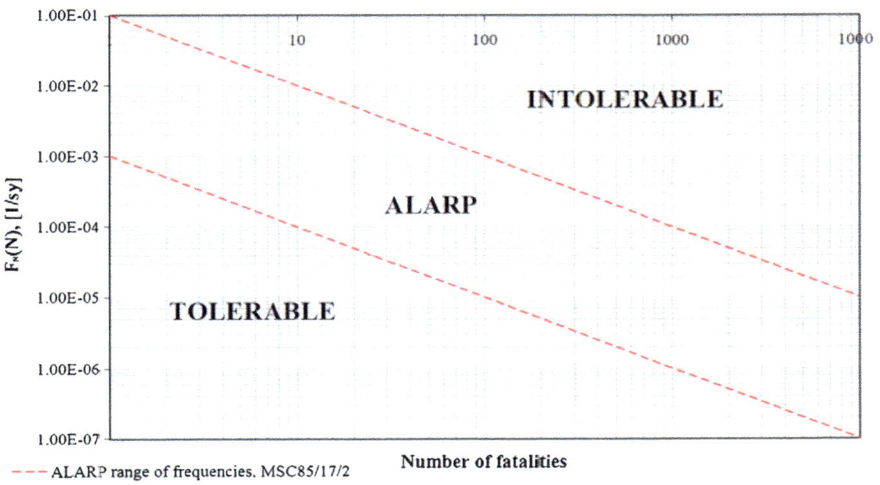

Fig. 46.2 FSA—decision parameters and risk acceptance criteria

- The risk of a fishing accident in EU waters is 2.4 times greater than the average of all EU industry sectors.

The principle that accidents involving large loss of life should be less likely than those involving a lesser loss was introduced by SOLAS 1929 and has not been addressed since. This notion, much as it sounds reasonable, is being severely challenged by a very different truth.

The second truth relates to the "value of life" as a currency for decision making in the risk assessment process. Risk acceptance implies two conditions: (a) the evaluated risk (profile) needs to be within the risk tolerability region, referred to as ALARP and (b) it needs to be **A**s **L**ow **A**s **R**easonably **P**racticable within that region, the latter being decided on the basis of cost-effectiveness criteria, adopted by IMO (Skjong and Ronold 1998). The basic principle on the latter relates to the cost society is prepared to invest for averting one fatality (CAF = $3M adopted), which reflects some average value deriving from developed countries (Skjong 2009), shown here as Fig. 46.3.

Notwithstanding the logic of this argument and the wisdom of this mechanism, there are side effects to it, namely: (a) a discriminatory incentive between developed and developing countries to invest on safety and (b) a discriminatory view on the value of life between developed and developing countries, which has nothing to do with propensity to invest for safety. The facts are clear to see. The world's worst shipping accident concerning loss of human life is that of the Philippine ferry Dona Paz in 1987, following a collision with an oil tanker. In the ensuing fire 4400 people died (**who is** aware of this?). The same year, Herald of Free Enterprise capsized resulting in the loss of 193 lives (**who is not** aware of this?). Sulpicio Lines, the operator,

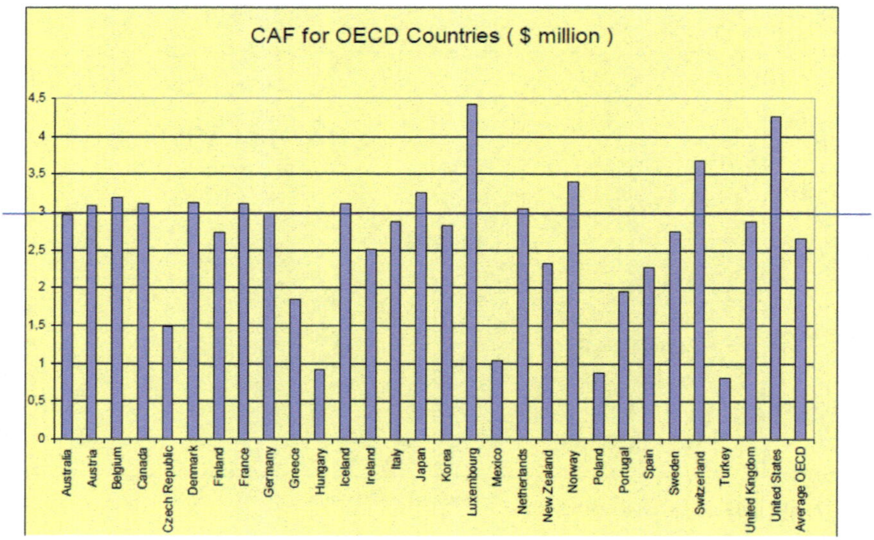

Fig. 46.3 Societal indicators (risk-based regulation & design in the maritime industry)

offered compensation of £580 for passengers killed on the Dona Paz, which was the insurance required by the Philippine maritime authorities (Spouge 1990). The corresponding compensation for the Herald of Free Enterprise was £500,000.

The third truth, still on risk acceptance criteria, relates directly to tolerability. On the basis that there are some 500 RoPax vessels worldwide carrying 1000 persons or more, the proposed criterion of just below 1E-4 in Fig. 46.2 for accidents with 1000 or more fatalities, implies that it is tolerable for such an accident to happen once every 20 years on average, assuming that RoPax population remains at current levels. To which extent this **truth** is understood widely is uncertain but this level, being considered tolerable by the maritime industry, deviates considerably from risk acceptance criteria of other comparable industries (Jasionowski 2012).

46.3 Damage Stability Verification

46.3.1 Validation and Verification

Safety legislation is one "thing" and safety verification an altogether different issue. Before elaborating further, it is important to elucidate the meaning of the two words: validation and verification. Validation ensures that the right model/method/formulation is being used whilst verification focuses on using the result right. This simple observation and understanding has, to date, eluded all of us working in maritime safety. This section will attempt to elucidate why verification is not receiving due attention by our industry.

Prescriptive safety legislation is, in principle, a risk control measure aimed primarily at risk mitigation post-accident for the specific hazard in question. The prevailing notion is that safety is ensured through compliance with pertinent regulations, which as we know only too well it has not worked all that well in the past. Key reason for this remains the fact that prescriptive rules are not risk informed. Prescriptive and performance-based instruments, described next, help illustrate this better:

- SOLAS 74: 1-compartment standard aimed at preventing ships from sinking/capsizing if any **one** compartment is breached in calm water (static GZ curve characteristics). ***Little is known of the implied safety.***
- SOLAS 90: 2-compartment standard aimed at preventing ships from sinking/capsizing if any **two** adjacent compartments are breached in calm water (static GZ curve characteristics). ***Little is known of the implied safety.***
- Stockholm Agreement: as per SOLAS '90 but with a pre-defined level of water on deck depending on freeboard and in operational sea states of up to 4 m significant wave height (IMO Resolution 14, 1995). This instrument represents a step change in damage stability verification in that the opted route for compliance is sought through ***performance-based assessment*** (model experiments). Still, because it is based on SOLAS '90, ***little is known of the implied safety***.

- Harmonised Damage Stability Regulations, SOLAS Chapter II-1 (SOLAS 2009): in principle, assessment of all known damage cases, based on accident statistics, accounting for loading conditions and sea states at the time the accident occurred. Contrary to previous deterministic instruments, this is a goal-based regulation (Attained Index of Subdivision, A > Required Index of Subdivision, R) and involves potentially thousands—rather than tens as in the first three—of damage cases. ***Adherence to the regulation is meant to ensure that the majority of cases are survivable (typically 80%).***

All four instruments were ***validated*** using a specific data set of existing ships (this was aimed to ensure that the chosen GZ characteristics are the ***right*** ones). However, principally the same characteristics are used to date for completely new and very different ship designs. This trend climaxed with SOLAS 2009 when, in an attempt to harmonise the probabilistic rules for damage stability, cargo ships data was adopted as the basis for deriving the harmonised solution applicable to both cargo and passenger ships (i.e., the validated solution—formulae—are ***not being used right***). Hence, verification of safety "went out of the window" …**and it stayed there.** There is more to this point as explained in the following.

46.3.2 Contemporary Developments

Over the recent past, knowledge on damage stability has been considerably enhanced with the development of advanced numerical simulation tools (Jasionowski 2001), offering fast, accurate and reliable tools for performance-based evaluation and verification of damage stability (tens of thousands of damage cases in a few weeks). This capability has revealed two serious problems:

- Damage stability is the Achilles heel of passenger ships.
- The simple regression formulae used to assess damage stability of passenger ships (s-factors and Index-A) fail to capture the increasing complexity of the watertight arrangements of these ships (see Fig. 46.4).

The cumulative probability distributions for time to capsize, for the two ships shown in Fig. 46.4, are derived using two approaches: (a) an analytical expression based on the formulation of SOLAS 2009 (Jasionowski 2006); and (b) numerical time-domain simulations using Monte Carlo sampling of pertinent statistical distributions of damage characteristics and sea states used in these rules (500 damage cases). Both ships comply with SOLAS 2009. Moreover, as explained in (Vassalos 2004), the values of these distributions tend to 1-A within a reasonable time period. Therefore, based on these results, the following observations can be made:

- For the RoPax, close agreement between calculation and performance-based assessment of damage stability is observed. However, approximately one in three collision and flooding events would lead to capsize within approximately half an hour. This is not a good standard.

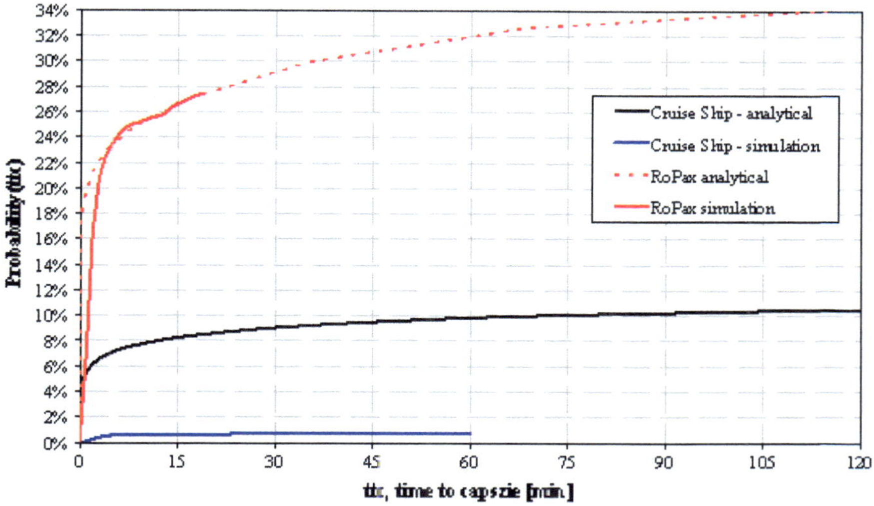

Fig. 46.4 Cumulative probability distributions for time to capsize for a traditional RoPax and a large newbuilding cruise ship (Jasionowski and York 2007)

- Application of SOLAS 2009 indicates that the cruise ship will follow a similar fate in approximately one in every 10 events whilst using first-principles time-domain simulation tools the rate becomes approximately 1:100. This is a big difference, implying primarily that because cruise ships are different (complex internal architecture with survivability governed by local details rather than global parameters) the s-factor in SPLAS 2009 is not describing the survival capability of cruise ships in a meaningful manner (EU Project GOALDS: Goal-Based Damage Stability, 2012, as described in the following).

46.3.3 Project GOALDS

Based on these results and aiming to address the aforementioned "liberty" of using cargo ships characteristics as a basis for assessing damage stability of passenger ships, a large-scale EC Research Project was set up (Goalds 2009–2012) to address these issues and to derive a verifiable goal-based (hence, performance-based) formulation for the damage stability of passenger ships. GOALDS helped produce the following revelations (more to the point, it helped reinforce some hard-learned truths):

New Formulation for Probability of Survival (s-factor)

As detailed in (Cichowicz et al. 2011), the new s-factor formulation is as follows

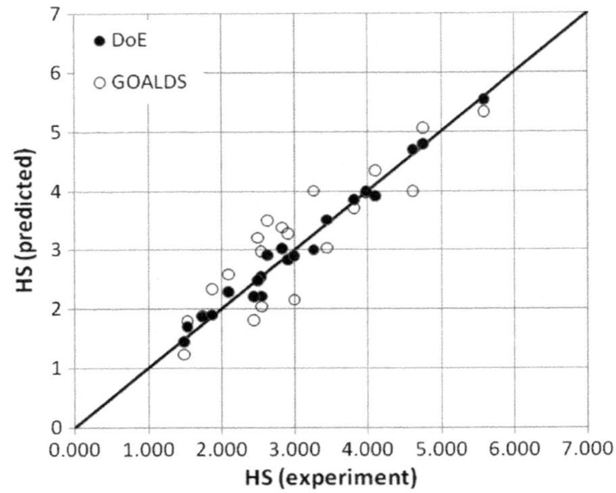

Fig. 46.5 DoE applied to experimental data based on A_{GZ}, GM_f, *Range and* V_R

$$H_{S\,crit} = \frac{A_{GZ\,E}}{\frac{1}{2}GM_f \cdot Range} V_R^{1/3}$$

where,

$H_{S\,crit}$	the critical significant wave height
$A_{GZ\,E}$	an effective area under the GZ curve taken up to the heel angle corresponding to the submergence of the opening in question
GM_f, $Range$, V_R	residual GM, range and volume respectively

The much-awaited re-formulation of the s-factor has now taken place. The new formulation has all the characteristics that one would intuitively expect, in particular:

- The formulation is simple, rational and readily calculable.
- It is much along the lines of the SOLAS 2009 formulation but uses as a basis Hs critical for survivability, consistent with the Safe Return to Port philosophy.
- Water on deck is accounted for albeit not included explicitly in the formulation.
- It accounts for scale as we all suspected it should.
- It encourages use of a watertight envelop above the traditional bulkhead deck.
- Validation studies demonstrate a high degree of correlation with all available experimental data from HARDER, GOALDS and EMSA projects (Fig. 46.5). The outcome expressed as a combination of linear, quadratic and interaction terms in the response surface model used resulted in 0.99 correlation with the experimental data, which can be considered as a satisfactory test concerning completeness of the parameter set.

Countering the aforementioned achievements are the following worrying observations:

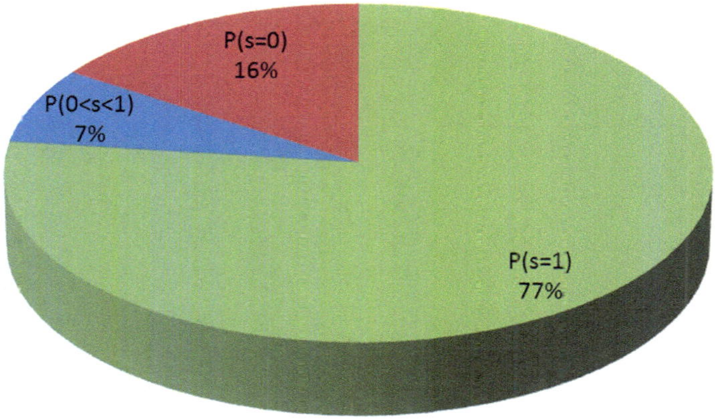

Fig. 46.6 Typical outcome of damage stability calculations for a collision accident of a RoPax vessel

- Comparison between A-Index calculations using SOLAS and GOALDS s-factor formulation revealed little difference. A cursory look of Fig. 46.6 helps clarify this point (at least for RoPax vessels).

Figure 46.6 illustrates that 77% of damages are survivable, 16% non-survivable and the remaining 7% of ambiguous survivability. Thus, irrespective of the calculation error, known to be present in the s-factor for individual damage cases in the SOLAS formulation, a better formulation capable of resolving the inherent ambiguity will only improve the situation marginally. This is what was observed in GOALDS but the reasons were not quite appreciated. What is **worrying**, however, is that a large proportion of the feasible damages a RoPax vessel can suffer results in no stability, whatsoever. This is a design vulnerability in need of attention. (see Chap. 47).

- There are only two points in the data set representing cruise ships (one from HARDER and one from GOALDS). The rest is RoPax data.
- Most RoPax ships are SOLAS '90 designs, hence not the best sample of ships to use as a basis for high survivability designs (see Sect. 46.5).
- SOLAS '90 designs are of simple configuration, hence easy to predict their performance when flooding takes place following collision/grounding. None of the RoPax ships used include subdivided side casings or car decks, for example.

As a result, calculations of Index-A alone, using even the latest derived s-factors in Project GOALDS, does not constitute a sufficient means for damage stability verification of modern cruise ships and RoPax, particularly with complex watertight arrangement. This point is further elaborated in the following.

Comparison of Index-A with Performance-Based Assessment of Survivability (PBS)

PBS in this case entails numerical simulation "tools" using Monte Carlo sampling of pertinent probability distributions (damage characteristics, loading and sea states

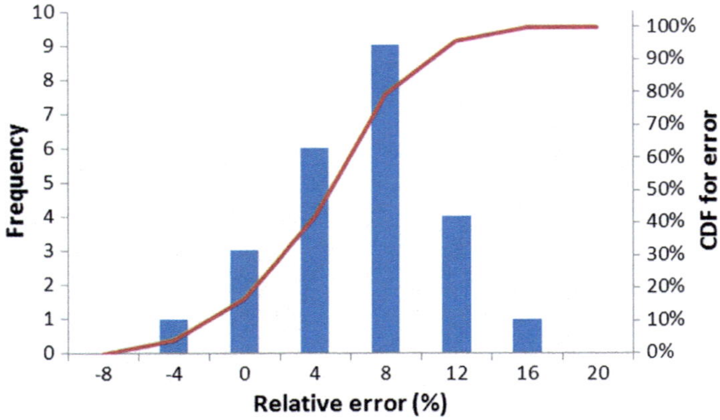

Fig. 46.7 Typical error distribution of Index-A compared with PBS for RoPax

during collision). In the case studies considered, 300 damage cases (uncertainty 10% in CDF relating to number of samples) were generated for the RoPax vessels and 500 for the cruise ships (uncertainty 4%). Each damage case was simulated for 30 min for RoPax and 60 min for cruise vessels. Index-A was calculated for the sample ships using NAPA. Unsurprisingly, the results of this study replicate Fig. 46.4, namely, Index-A calculations show good agreement with PBS for traditional RoPax designs but not so in the case of cruise ships. A typical error distribution for RoPax is shown in Fig. 46.7. The histogram is symmetrical around a mean value of approximately 5.0 with a standard deviation of 5.0. As indicated earlier, this error is expected to grow with complexity of watertight architecture, which is inevitable for the requisite higher survivability being sought.

A similar comparison between Index-A calculations and numerical simulations for cruise vessels is shown here in Fig. 46.8. GOALDS results show a significant improvement over SOLAS (Fig. 46.4) but the error is still large.

Having gained an increased understanding and made these observations, what can we do and what can we not do with the ensuing results/knowledge? It is clear that performance assessment of passenger ship survivability, using Index-A calculation, is far from perfect when compared with PBS, using modern "tools". Some reasons include the use of deterministic elements in the Index-A calculation, interpretation of the various elements involved in both calculations and simulations and so on but, ultimately, inability to deal with difficult physical phenomena, such as flooding of increasingly complex passenger ship watertight arrangements, constitutes the kernel of the problem. Notwithstanding this, the fact that results from Index-A calculations follow consistently the same trends as PBS results, renders ***Index-A a very attractive "tool" for survivability assessment and decision-making in concept design phase, including design optimisation***.

However, before any passenger ship sets sail, damage stability needs to be verified by more rigorous means. The profession must realise that this is the only way forward

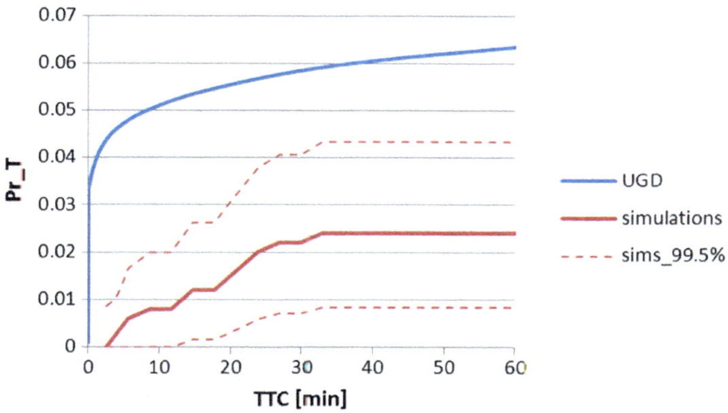

Fig. 46.8 Comparison between Index-A and simulations for a typical cruise vessel (UGD stands for Unified Geometric Distribution)

and take action to develop and implement the right verification process and means to achieve this.

Can we imagine naval architects today using the Froude formula alone for resistance or the Watson formulae for powering calculations, without the use of large scale model tests and extensive CFD (Computational Fluid Dynamics) studies for verification of power requirements and performance for any ship design? Why is shipping safety subjected to this degree of unwarranted and dangerous oversimplification?

Clearly we need to (and we can) do better!

46.4 Passenger Ship Vulnerability

46.4.1 Basic Definition

"Vulnerability" is a word being used extensively in the naval sector but not so in the merchant shipping world. Hence, a definition here is in order. The way this term has been used by the Ship Stability Research Centre relates to "the probability that a ship may capsize within a certain time when subjected to any feasible flooding case." As such, vulnerability contains (and provides) information on every parameter that affects damage ship survivability. A simple example is provided next.

Figure 46.9 indicates that there are 3 possible flooding cases of known (available statistics) frequency and calculable (available "tools") probability of surviving, say 3 h, which is shown in the Figure. With this information at hand, Vulnerability to collision flooding of this simple example is:

$$0.5 * 0.72 + 0.35 * 0.01 + 0.15 * 0.99 = \mathbf{51.2\%}$$

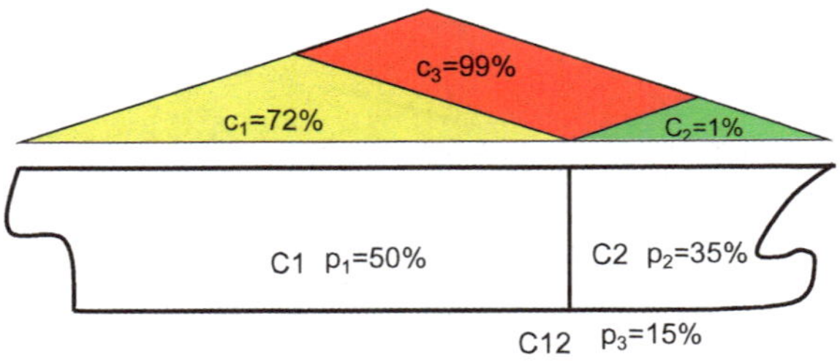

Fig. 46.9 Vulnerability to collision flooding

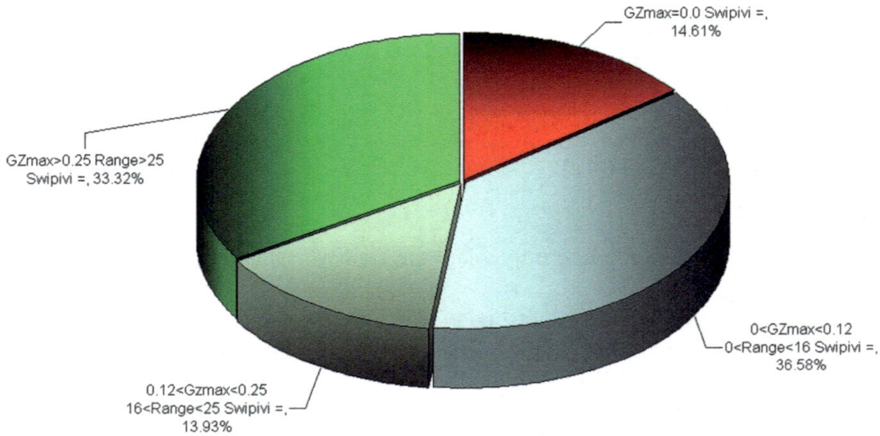

Fig. 46.10 Design vulnerability (watertight integrity—non-survival cases for typical RoPax)

46.4.2 Design Vulnerability

The vulnerability to collision/grounding damage of passenger ships is well documented through a number of accidents claiming many lives. Such vulnerability relates to **Water On Deck** (WOD), leading to progressive flooding and rapid capsize of the ship. Whilst for Ro-Ro passenger vessels this design vulnerability is well understood, for cruise ships it has been brought to light as recently as the early 2000s. The latter case relates to the service corridor on the strength deck, which acts as conduit for floodwater to spread along the ship, leading to down flooding through deck openings/stairwells and ultimately to sinking/capsize of the ship. Figures 46.10 and 46.11, provide typical results demonstrating such vulnerability in the design of RoPax and Cruise vessels, respectively.

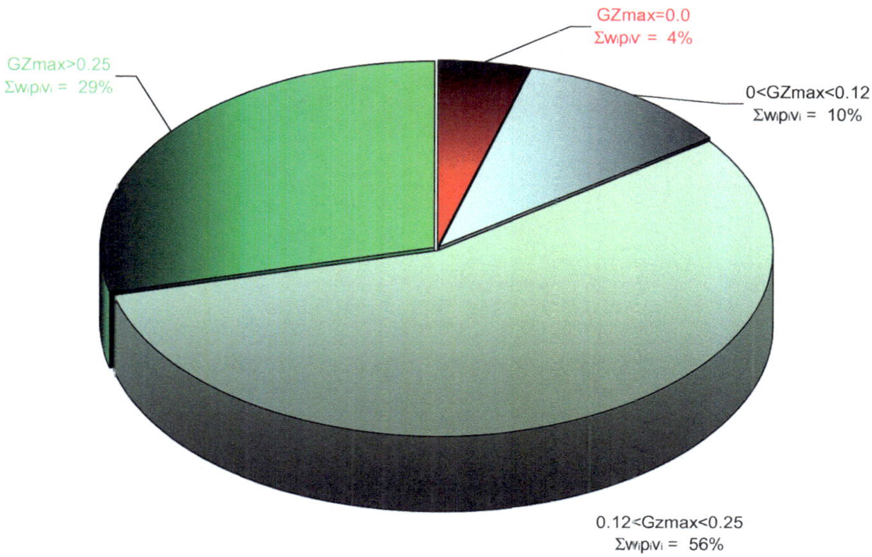

Fig. 46.11 Design vulnerability (watertight integrity—non-survival cases for typical cruise ship)

46.4.3 Vulnerability in Operation

A threat that exacerbates further the design vulnerability of passenger ships to collision/grounding damages, probably at the heart of many catastrophes is vulnerability in operation. This is an issue that has been attracting serious attention at IMO over the past few years and new legislation is now in place. It aims to address the fact that most passenger ships are operated with a number of Watertight (WT) doors open, thus exacerbating considerably the design vulnerability of these ships. Figures 46.12 and 46.13 demonstrate this rather emphatically by considering the well known Estonia case, as designed and at the time of her loss. In the latter case (because of open WT doors) the vulnerability of the vessel was at 68%; 3.5 times higher than her design vulnerability of 19%.

46.4.4 Watertight Doors—Current Legislation

SOLAS Regulation II-1/15.9.3: "Certain watertight doors may be permitted to remain open during navigation only if considered absolutely necessary; that is, being open is determined essential to the safe and effective operation of the ship's machinery or to permit passengers normally unrestricted access throughout the passenger area. Such determination shall be made by the Administration only after **careful consideration** of the impact on ship operations and survivability. A watertight door

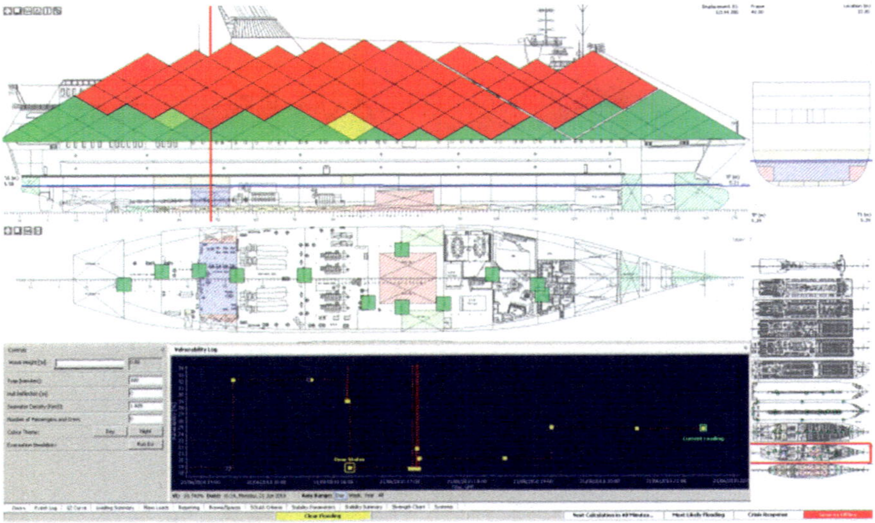

Fig. 46.12 MV Estonia—design vulnerability

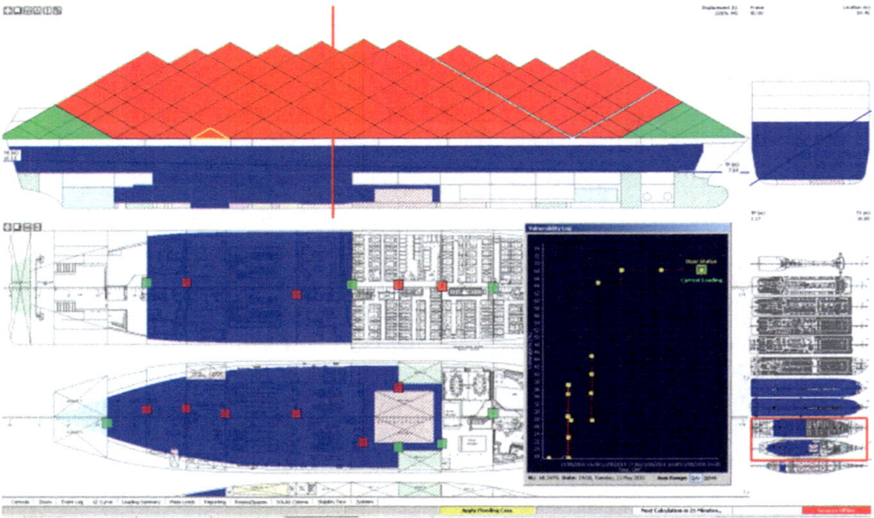

Fig. 46.13 MV Estonia—as operated at the time of her loss

permitted to remain thus open shall be clearly indicated in the ship's stability information and shall always be ready to **be immediately closed**."

However, careful consideration by an Administration would necessitate due ability by the Administration to address the impact on survivability of open watertight doors so as to ensure rational decision-making. The worst watertight door shown

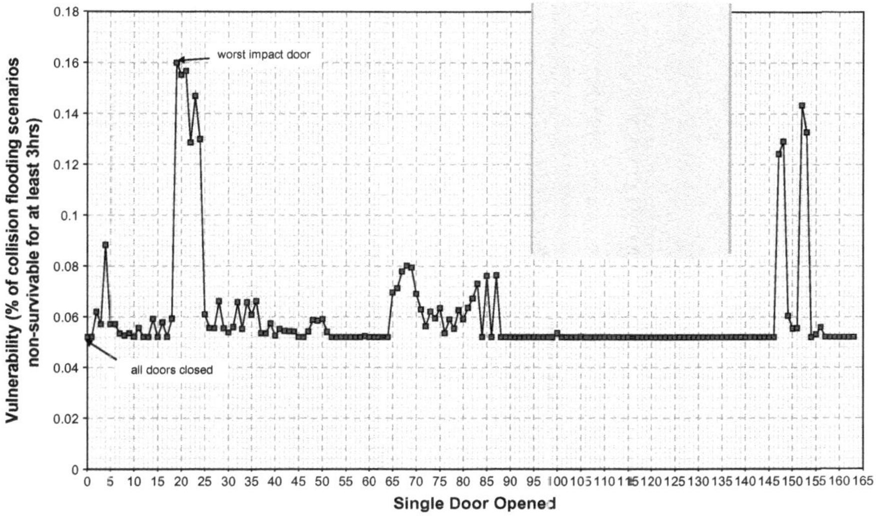

Fig. 46.14 Watertight door exemptions—cruise ship example

in Fig. 46.14 increases ship vulnerability four-fold: yet it is Class-exempt. This is simply the result of the floatability assessment required by IMO. The requirement for a watertight door to be ready to be immediately closed is another myth. Measurements in the field indicated that such doors take on average 2.5 min to close (1 min before "pressing the button" for obvious safety reasons and 1.5 min for the door to close) by which time the damage is done.

46.5 New Ships Versus Existing Ships

46.5.1 Regulatory Gaps

The Compartment "Currency"—Weakness in Old SOLAS

Adoption of probabilistic rules for damage stability and the use of damage statistics led to clearer understanding of the fact that high-energy impact collisions penetrate the ship's hull up to the centre line (in fact at a rate of 45%). Hence, side compartments (B/5, B/10 and so on) offer reduced protection from a survivability perspective. These different compartment "currencies" in old and new SOLAS (for example SOLAS '90 and SOLAS 2009, respectively) cause confusion when used to judge the level of damage survivability, making it imperative to use transparency in communicating compartment-based damage stability standards.

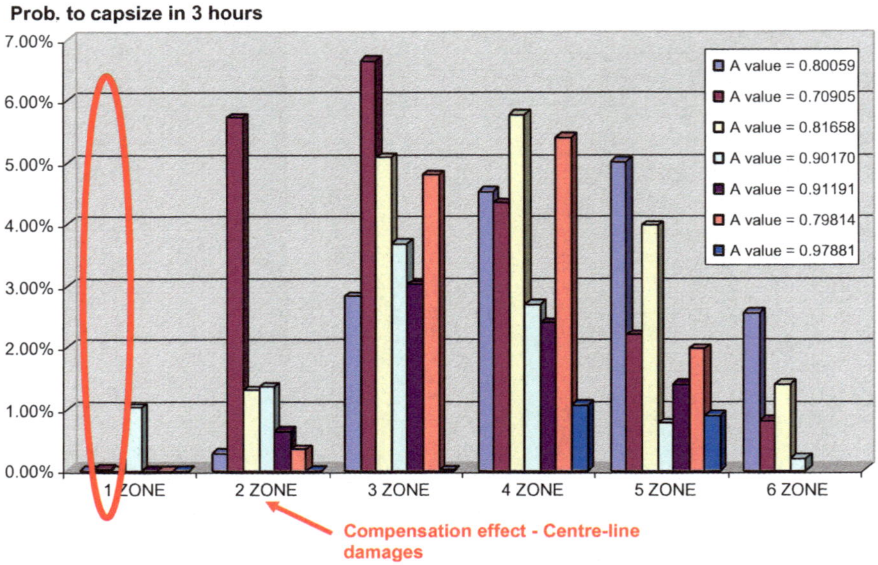

Fig. 46.15 Compensation effect—newbuildings experience (passenger ships)

The "Compensation" Effect—Weakness in New SOLAS

Only survivable damage cases contribute to the value of Index-A (an *aggregate statistic*). As such, it is implied that even if a vessel achieved the required index of subdivision, there may be cases, which are likely and which have a low probability of survival—hence a high risk of sinking/capsize. The term introduced in the literature to describe this (Vassalos 2008) is the "compensation effect" and as Fig. 46.15 indicates, even vessels with very high index of subdivision, may fail to survive what is, in principle, one- and 2-compartment damages.

Safety Equivalence—Weakness in Knowledge

The safety standard in SOLAS 2009 was established by using a sample of SOLAS '90 ships as reference, aiming to ensure that the same damage stability standard is maintained in the new rules. However, evidence from performance-based assessment of damage survivability (recently gained knowledge) of existing and new ships demonstrates that this is hardly the case (Fig. 46.16). The vulnerability of SOLAS '90 ships is considerably higher than ships designed to SOLAS 2009 rules. Yet another reason for paying particular attention on the damage survivability of existing ships.

Notwithstanding the dissimilarity in old and new SOLAS standards, striving for higher standards for passenger ships, one MUST NOT forget that if these were applied only to new ships and not to all existing ships (retrospectively as it was done, for example, with the Stockholm Agreement) we would only scratch the surface of the problem. Moreover, this would mean that we reinforce a **two-tier safety**, allowing

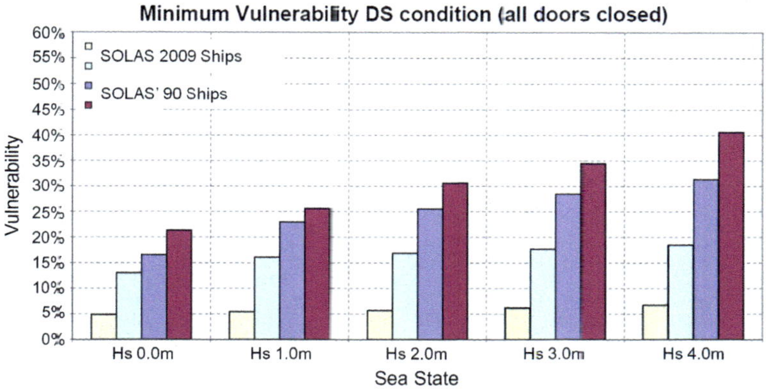

Fig. 46.16 Comparative assessment of vulnerability between SOLAS '90 and SOLAS 2009 passenger ships; clearly the safety level is not the same

the vast majority of ships to continue to operate, despite knowing that their safety standard is not good enough. This is a significant regulatory gap that needs attention! This is the focus of the next chapter.

46.6 A Way Forward

46.6.1 Grandfather Clause[1]

"International regulation would fall apart if the grandfather clauses were removed!!!" This is the response of the former MSC Chairman (Tom Allan) when I made the suggestion to him of retrospective legislation, including the exclamation marks. I am certain, the same argument was made prior to the removal of the margin line, freeboard, factorial system of subdivision and other "grandfather concepts" but the world is still standing, financial problems apart.

What this paper tried to bring forward is many such concepts, which for lack of better knowledge at the time, they were adopted and most of these served our industry well. But the time has come for us to review critically the maritime safety system and lay the foundations for a modern, sustainable system. Key to this, particularly with

[1]Grandfather clause is a legal term used to describe a situation in which an old rule continues to apply to some existing situations, while a new rule will apply to all future situations. The term originated in late-19th-century legislation of U.S. ***Southern*** states, which created new literacy and property restrictions on voting, but exempted those whose ancestors (***grandfathers***) had the right to vote before the ***Civil War***. The intent and effect of such rules was to prevent poor and illiterate African American former slaves and their descendants from voting, but without denying poor and illiterate whites the right to vote. The term *grandfather clause* remains in use but with no connotation regarding the justness of these provisions when applied in other areas.

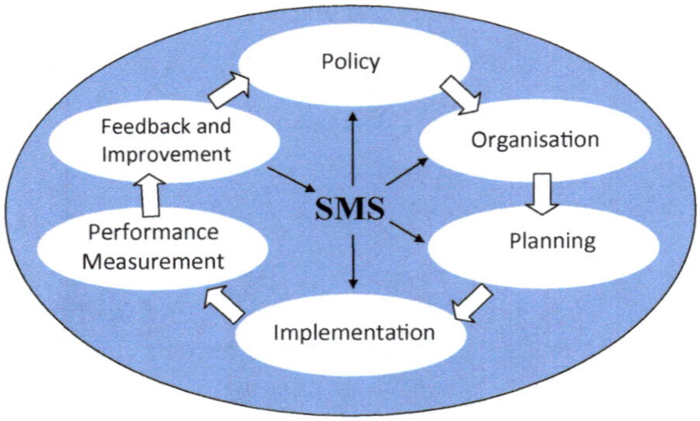

Fig. 46.17 A virtuous life-cycle safety management

reference to damage stability of passenger ships, is retrospectively applied legislation. ***There may be a need for a debate, at IMO level, to ensure that the time is ripe to rip the Grandfather clause apart but from a technical and practical perspective, if it can be done it should be done.*** This Chapter puts forward a workable framework to achieve this, which focuses on life-cycle risk management, as outlined next.

46.6.2 Life-Cycle Safety (Risk) Management

Safety Management is a life-cycle process, starting at the concept design stage and continuing throughout the life of the vessel. In this process, safety must be monitored and reviewed to ensure changes in design/operation are reflected in the way safety is managed. The safety management process must be formal and transparent to allow the operator to nurture a safety culture and to manage safety cost-effectively. More importantly, a formal process facilitates measurement of safety performance, which constitutes the basis for continuous improvement, as indicated by the Virtuous Cycle in Fig. 46.17 (see Footnote 1).

This brings to the fore the role of ship operation in life-cycle safety management, which deserves special attention. This thinking is largely in line with the Safety Case approach (Safety Case 2005) of Health and Safety Executive (HSE) and the IMO Guidelines on FSA and on Alternative Design and Arrangements (MSC\Circ.1002, 1212) with the focus clearly on safety performance verification. The Safety Case approach to safety management is more ship-specific rather than ship type specific as in the case of the FSA. As such, it is a "living instrument", starting with the first concept of design and spanning the whole life cycle of the vessel.

The first 3 steps in Fig. 46.17 point to the fact that safety management is an all- stakeholders affair. The next step (design phase) involves safety performance evaluation and verification activities (e.g., engineering analysis, model tests, etc.) aiming at minimising risk cost-effectively. The last 2 steps (operational phase) ensure that adequate measures are being taken, using safety performance measurement and feedback, to manage the residual risk, i.e., to render the residual risk acceptable.

The focus on dealing with residual risks, naturally leads to the need for a SMS (Safety Management System), outlining the organisation and procedures required to maintain an acceptable level of safety throughout the life of the vessel. This has to be aligned with the ISM (International Safety Management) Code implemented onboard. Pertinent activities include aspects of onboard and shore-based Safety Centres, monitoring systems and emergency Decision-Support for crisis management.

In conclusion, a formal process should address risk at the ship **design** stage (risk reduction/mitigation), in **operation** (managing residual risk) and, ultimately, in **emergencies** (crisis management), ensuring in all cases an acceptable level of risk (safety assurance). This constitutes the basis of the verification framework described in the next section.

46.6.3 A Damage Stability Verification Framework

Existing Ships

Deriving from the foregoing and with emphasis on existing ships, two main phases are envisaged:

1. ***Design Upgrade Phase:*** Recent knowledge, tools and experience has demonstrated that there is a great deal that can be achieved to improve the damage stability of existing passenger ships with minor modifications, addressing curtailment of progressive flooding by focusing on removing certain openings, introducing semi-watertight doors and such like. This will enhance survivability substantially. The outcome should be **verified** through performance-based survivability assessment; first principles tools or model experiments.
2. ***Operational Phase:*** As it is more than likely that design measures alone will not be sufficient to raise the standard of existing ships to the requisite level, the next step should address management of the residual risk, along the lines discussed in the foregoing. Namely, the ship owner/operator will need to demonstrate, in a **verifiable** way, how residual risk is to be managed (rendered acceptable).
3. ***Emergencies:*** Preparedness is the key to handling emergencies with severe time constraints, such as accidents involving flooding, and limitations in human factor performance. This will require a **verifiable** emergency response system for crisis management along the lines described briefly in the previous section.

New Ships

The procedure outlined in Sect. 46.6.2 will apply directly to newbuildings, subject to an appropriate verification framework.

For the design phase, we need to look no further than developing for damage stability the same framework as that already existing at IMO for fire, namely Alternative Design and Arrangements for Fire Safety and associated Guidelines. For the operational phase and for emergencies, a verification process needs to be established involving the instruments already in place as explained in Sect. 46.6.2 (Safety Case, SMS, ISM Code). Possible actions would involve the following:

- Industry study to investigate feasibility of retrospective application of new rules for damage stability through (a) design upgrades, (b) operational measures and (c) focus on emergency response.
- Establish, through the IMO route of Alternative Design and Arrangements, a similar regulation for damage stability (as that for fire) for verification through performance-based assessment of typical cases from the industry study. This would also enable setting in place a suitable framework for approval of damage survivability for newbuildings.
- Set in place through appropriate legislation at IMO and certification through Administration/Class a system for monitoring damage stability onboard (linked to SMS) to provide feedback for verification and management of residual risk, including crisis management.

46.7 Concluding Remarks

- A painstaking evolutionary development on ship stability is giving way to unprecedented scientific and technological developments that has raised understanding on the subject as well as the capability to address demanding societal expectations on human life safety and do so cost-effectively.
- This enhanced capability has brought with it knowledge on the level of safety ships are being designed to and on the way ship operation impacts on this.
- As a result, notions about ship stability and safety are also giving way to some important truths: ships are vulnerable platforms, particularly to flooding following collision/grounding accidents, a fact that is further exacerbated by accepted operational practices.
- This demands immediate action, led by industry, aiming to establish a workable framework and process for raising and upholding damage stability standards and to do so equitably across the whole fleet.
- More specifically, this paper is putting forward a strong case and means to argue that **the time is ripe to rip the Grandfather Clause apart** and to move on.
- In a wider sense this, in turn, necessitates raising awareness to stimulate and nurture a maritime industry safety culture and a safety regime that aims for and supports continuous safety improvement.

References

VASSALOS, D., "Shaping Ship Safety: The Face of the Future", Journal of Marine Technology, Vol. 36, No. 4, 1999, pp. 1–20.

IMO (2001), "Guidelines on alternative designs and arrangements for fire safety", MSC/Circ. 1002.
IMO (2006), "Guideline on Alternative Designs and Arrangements for SOLAS chapter II-1 and III", MSC/Circ.1212.
IMO Resolution 14, "Regional Agreements on Specific Stability Requirements for Ro-Ro Passenger Ships" – (Annex: Stability Requirements Pertaining to the Agreement), adopted on 29 November 1995.
FSA, "Guidelines for Formal Safety Assessment for Use in the Rule-Making Process", MSC.1-Circ.1023 - MEPC.1-Circ.392, 2002.
SPOUGE, J., "Passenger Ferry Safety in the Philippines", Transactions, RINA 1990.
SKJONG, R. and RONOLD, K., "Social Indicators and Risk Acceptance", Offshore Mechanics and Arctic Engineering Conference, OMAE, Lisbon 1998.
SKJONG, R., "Risk Based Regulation, Risk Based Design & SAFEDOR", Guest Lecture NTNU, 2009-10-21.
JASIONOWSKI, A., "Study on the Specific Damage Stability Parameters of RoRo Passenger Vessels according to SOLAS 2009, including Water on Deck Calculation", Project No EMSA / OP /08 / 2009, SSRC, University of Strathclyde, Glasgow, 2012.
JASIONOWSKI, A., "An Integrated Approach to Damage Ship Survivability Assessment", PhD Thesis, University of Strathclyde, 2001, UK.
JASIONOWSKI, A., "Fast an accurate flooding prediction – analytical model", SAFEDOR, D2.1.3, November 2006.
VASSALOS, D., "A risk-based approach to probabilistic damage stability", 7th International Workshop on the Stability and Operational Safety of Ships, November 2004, Shanghai, China.
JASIONOWSKI, A., YORK, A., "Investigation Into The Safety Of Ro-Ro Passenger Ships Fitted With Long Lower Hold", UK MCA RP564, 10 July 2007, Draft Report MCRP04-RE-001-AJ, Safety at Sea Ltd.
GOALDS 2009–2012, "Goal-Based Damage Stability", European Commission Project FP7 Project 233876, 3009-2012.
CICHOWICZ, J., TSAKALAKIS, N., and VASSALOS, D., "Survivability of Passenger Vessels - Re-engineering of the s-Factor", Proceeding of the 12th International Ship Stability Workshop, 12–15, June 2011, Washington D.C., USA.
VASSALOS, D., JASIONOWSKI, A., TSAKALAKIS, N. and YORK, A., "SOLAS '90, Stockholm Agreement, SOLAS 2009 - The False Theory of Oranges and Lemons", 10th International Stability Workshop, Taejon, Korea, 2008.
SAFETY CASE, "The Offshore Installation (Safety Case) Regulations, UKSI 3117", 2005.

Chapter 47
Defining Rational Damage Stability Requirements

Nikolaos Tsakalakis, Dimitris Konovessis and Dracos Vassalos

Abstract The major benefit of switching from the deterministic frameworks for damage stability of the past to the current performance-based state of the art is the ability to have a measurement of the level of survivability of any given design. The required level of survivability is probably the key parameter in any probabilistic framework, in essence answering the question "how safe is safe enough?". To this end, survivability analysis results on representative cruise and Ro-Pax ships can be related to design and operational parameters with a view to define and quantify the relationships between damage survivability characteristics following a collision and time available for evacuation with potential outcomes in terms of people potentially at risk. For this paper, established numerical methods for the measurement of performance-based survivability have been utilised and used as benchmark against available analytical methods in an attempt to define a rational requirement for the level of survivability.

Keywords Damage stability · Survivability assessment · SOLAS 2009
Safety level · Passenger ships · Evacuation · Risk · Index A

47.1 Introduction

A comprehensive performance-based survivability assessment of 4 passenger vessels has taken place within EU funded project GOALDS. The target was to provide an answer regarding the safety level offered by modern, state-of-the-art ships, firstly

N. Tsakalakis
Alpha Marine Consulting, Piraeus, Greece

D. Konovessis
Singapore Institute of Technology, Singapore, Singapore

D. Vassalos (✉)
The Ship Stability Research Centre, Department of Naval Architecture, Ocean and Marine Engineering, University of Strathclyde, Glasgow, Scotland, UK
e-mail: d.vassalos@strath.ac.uk

V. L. Belenky et al. (eds.), *Contemporary Ideas on Ship Stability*, Fluid Mechanics and Its Applications 119, https://doi.org/10.1007/978-3-030-00516-0_47

for the development of an analytical methodology for the estimation of risk from flooding but also for the development of a rational level for the required index of subdivision. Independent as well as in-house developed tools provided us with the capability to perform the survivability assessment in various ways and compare them against each other, thus securing a reliable result.

47.1.1 The R-Factor

The minimum level of survivability a vessel should offer is postulated by the R-factor. This applies either to new ship designs or existing ships and should reflect the societal acceptance of risk from loss of life, in an attempt to place more emphasis on passenger and crew survival. This represents an area of research highlighted for attention and improvement [as reported, for example, in (Vassalos 2007; Spyrou and Roupas 2007)]. A typical F-N diagram contains 3 zones according to severity as can be seen in Fig. 47.1, where an F-N curve for collision night-time scenarios of a typical Ro-Pax is plotted using the Univariable Geometric Distribution (UGD) method (Jasionowski et al. 2007). A common misperception is that if the F-N curve lies within the boundaries of the ALARP region the design is acceptable. Cost-benefit analysis though, could as well prove that the F-N curve should in fact be as noted by the acronym: As Low As Reasonably Practicable, thus not low enough in the example of the figure. The ALARP region boundaries have been obtained by Skjong et al. (2007) and could change.

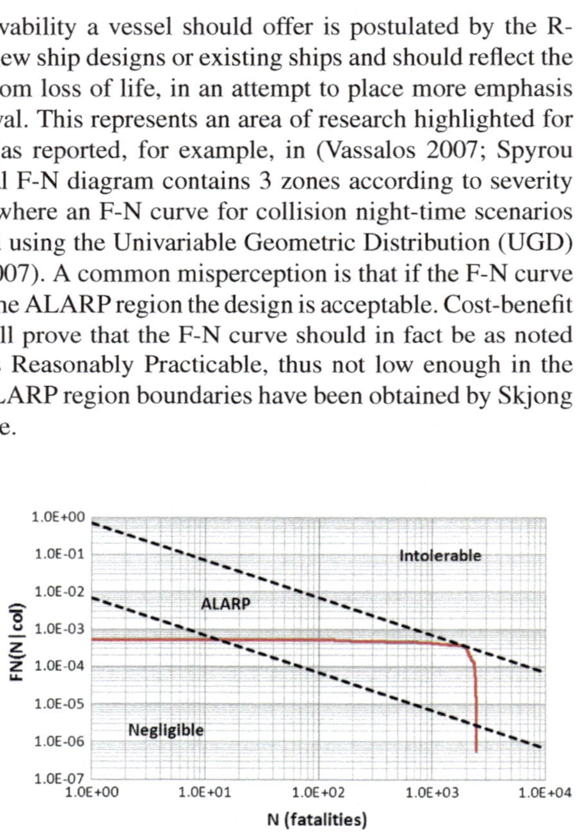

Fig. 47.1 Typical F-N diagram where the distinction between the consequences can be seen

47.2 Method of Approach

47.2.1 Numerical PBS Assessment

PROTEUS3 is a state-of-the-art time-domain numerical simulation tool capable of handling complex geometries such as the ships used in the study. Each ship variant is simulated for 30 min in a number of randomly generated damage scenarios. Scenarios include damage opening size and location as well as environmental conditions. 300 scenarios are automatically generated for each ship, using the Monte Carlo sampling technique (see Fig. 47.2). The probability distributions used for this purpose are the same that were used for the development of the P-factor currently in SOLAS 2009, derived from EU-funded project HARDER (1999–2003).

The result of a series of simulations can be seen in Fig. 47.3. Both the probability density and the cumulative distribution functions for time to capsize are visible. The information acquired from the marginal value of the CDF is the probability of the vessel to capsize in 30 min given the specific loading condition.

The vessel's survivability is given by the compliment of this value and should correspond to the Index A for the specific loading condition. Numerical PBA is the most accurate measure of survivability readily available now.

47.2.2 Analytical PBS Assessment

Due to the complexity and time involved with numerical simulations, an easier method has been devised, namely the Univariable Geometric Distribution (UGD)

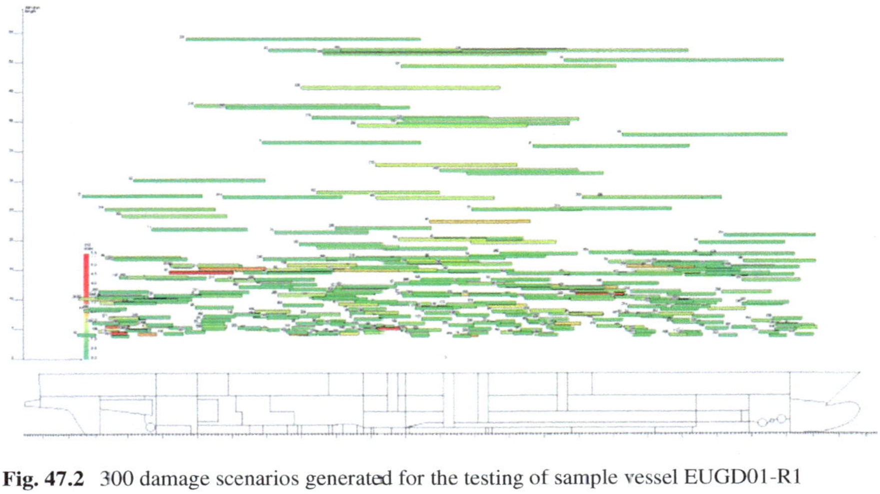

Fig. 47.2 300 damage scenarios generated for the testing of sample vessel EUGD01-R1

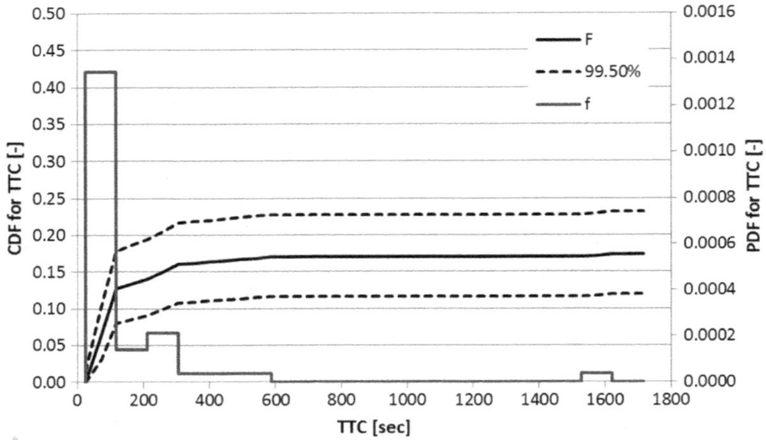

Fig. 47.3 Probability distribution for time to capsize given specific loading condition and 30 min

(Jasionowski et al. 2007). It is based on an analytical model for the estimation of time to capsize and results in the same marginal probability for time to capsize as given from Numerical PBSA, thus the two can be directly compared. The results of the analytical method are not quite as accurate as those of the numerical one but they have through various studies proven to be of comparable value (Jasionowski et al. 2008; Tsakalakis et al. 2011). Given the significant reduction in calculation time, the analytical method can be considered as a reasonable alternative. When combined with information on time needed to evacuate the studied vessel, the analytical software can provide information on potential loss of life as well. Typical results of UGD can be seen in Figs. 47.4 and 47.5 where the comparison with the CDF for time to capsize of numerical PBSA of Fig. 47.3 can be made.

47.2.3 Evacuation and Potential Loss of Life

The concept of potential loss of life is summarized in Eqs. 47.1, 47.2 and 47.3 (Vassalos and Jasionowski 2006; Jasionowski et al. 2007):

$$Risk_{PLL} \equiv E(N) \equiv \sum_{i=1}^{N_{max}} F_N(i) \tag{47.1}$$

where N_{max} is the maximum number of persons onboard and the F-N curve is given as:

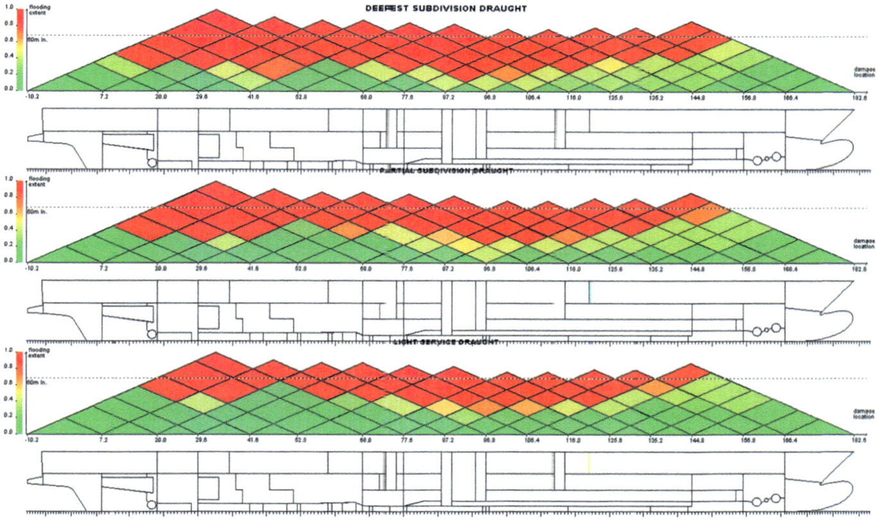

Fig. 47.4 Distribution of conditional probability for time to capsize, given loading conditions and damage extent occurred (color code for the triangles in accordance with attained s-factor: red, s = 0; green, s = 1 and yellow, 0 < s < 1)

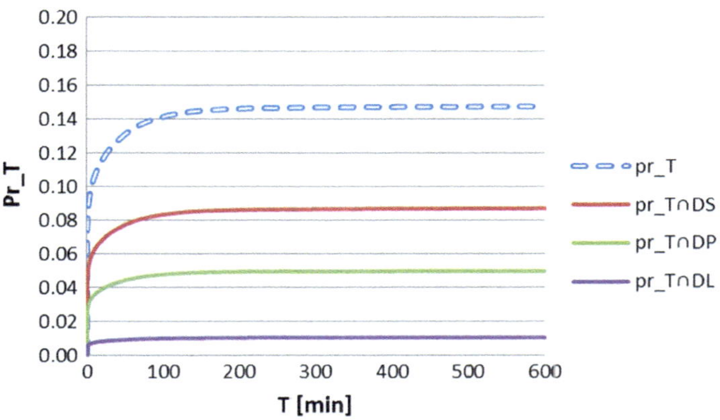

Fig. 47.5 Distributions of probability for time to capsize: unconditional (top) and conditional, given loading condition

$$F_N(N) = \sum_{i=N}^{N_{max}} fr_N(i) \tag{47.2}$$

The frequency $fr_N(N)$ of occurrence of exactly N fatalities per ship year is modelled as:

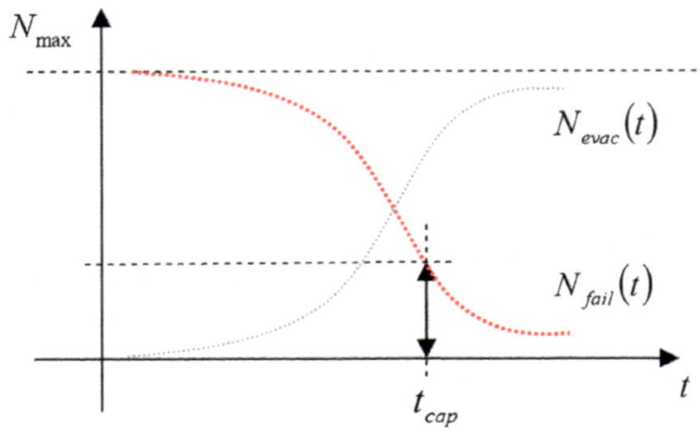

Fig. 47.6 Interplay between time to capsize and evacuation time

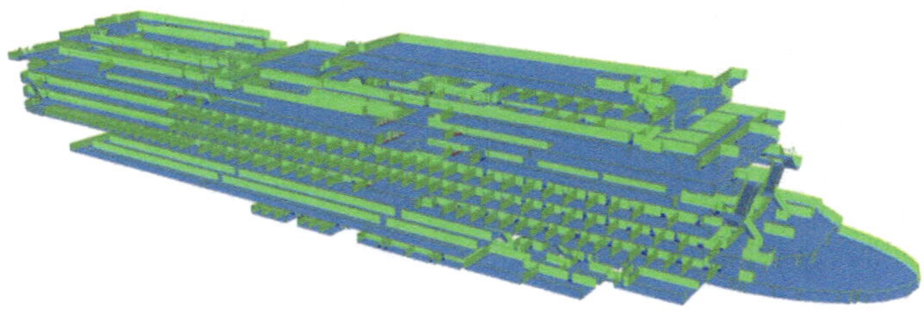

Fig. 47.7 Geometry of a RoPax ship as modeled in EVI®

$$fr_N(N) = \sum_{j=1}^{n_{hz}} fr_{hz}(hz_j) \cdot pr_N(N|hz_j) \tag{47.3}$$

where, n_{hz} is the number of loss scenarios considered, and hz_j represents a loss scenario, identifiable by any of the principle hazards (such as, collision, grounding, or fire). $fr_{hz}(hz_j)$ is the frequency of occurrence of scenario hz_j per ship year, and $pr_N(N|hz_j)$ is the probability of occurrence of exactly N fatalities, given that loss scenario hz_j has occurred.

This can be demonstrated graphically by Fig. 47.6. In short the number of fatalities is the number of people that failed to evacuate when the capsize event took place.

The evacuation analysis has been performed by means of EVI, a software package developed at the Ship Stability Research Centre, which uses as input a number of parameters such as ship geometry, passenger demographics and ship motions due to environment, to determine the time required to evacuate (Vassalos et al. 2001). It has been used for a number of years in a wide variety of studies and has been benchmarked against available real data. A typical model of a RoPax ship as used in EVI® can be seen in Fig. 47.7 while Fig. 47.8 shows an example of an objective completion curve.

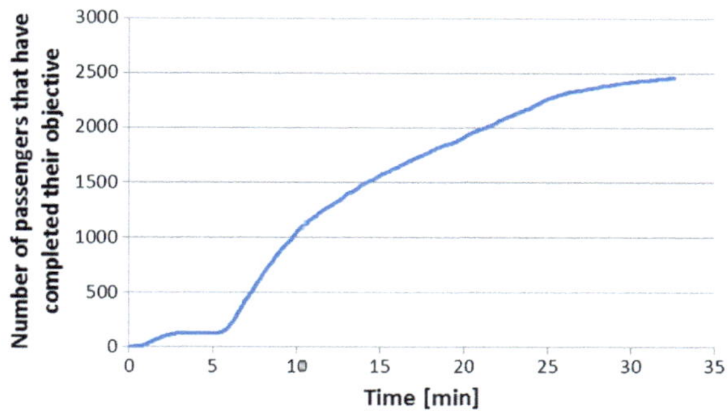

Fig. 47.8 Objective completion curve; note the approximately 5 min lag in the beginning of the simulation due to response time of passengers sleeping in their cabins

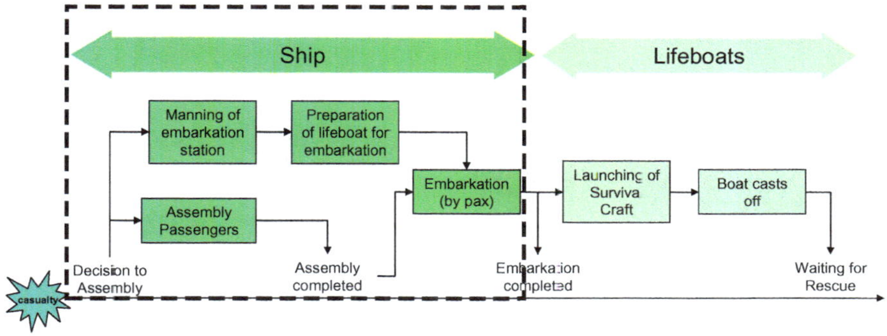

Fig. 47.9 Total evacuation process and mustering part

Objective completion curves have been obtained with EVI® for all vessels in day and night conditions. The objective used in this study is mustering and not abandonment of ship as shown in Fig. 47.9 (SAFEDOR 2006). This would additionally involve embarkation and launching of the lifeboats and clearing the vessel, which are all very crucial elements of ship abandonment that wrongfully tend to be overlooked at the moment since they would require separate measurements and models.

47.3 Sample Vessels

The study vessels consist of two RoPax ships and two cruise ships, the attributes of which can be seen in Table 47.1. The ships have been selected in such a way as to ensure coverage of the design space. This also ensures that should any similarities or differences exist between these ship types they would be pronounced so as to be dealt with accordingly.

Table 47.1 Principal dimensions of study ships

	R1	R2	C1	C2
Number of passengers	1400	800	3840	2500
L_{OA} (m)	194.3	97.9	311	295
L_{BP} (m)	176.0	89.0	274.7	260.6
Breadth moulded (m)	25.0	16.4	38.6	32.2
Deepest subdivision load line (m)	6.55	4.0	8.6	8.0
Depth to bulkhead deck	9.1	6.3	11.7	10.6
Displacement (tn)	16.6	3.4	62.5	45.0
Service speed (kn)	27.5	19.5	22.6	22.0

47.4 Results

The study involved getting the F-N curves and PLL values for various collision frequencies for the study ships that range from 5.0E−04 to 1.0E−02. It appears though that no matter the frequency, the potential loss of life is quite high, not only for the vessels carrying thousands of passengers (Fig. 47.10) but also for smaller vessels as pictured in Fig. 47.11.

One could easily argue that both these cases definitely require risk control options irrespective of cost effectiveness. The resulting PLL for all the ships and frequencies is visible in Table 47.2.

It can be generalised that PLL is increased according to the order of magnitude of passengers carried by the vessel. What is more, it appears that the effect that frequency has on PLL is linear as can be seen in Fig. 47.12 where the average PLL values for night and day cases have been plotted against frequency.

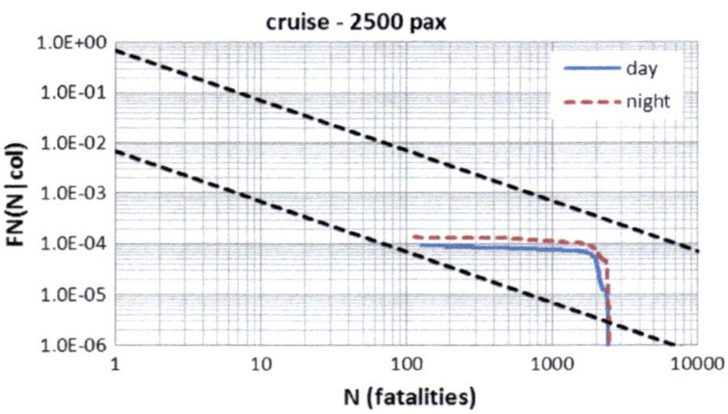

Fig. 47.10 F-N diagram for one of the cruise vessels with frequency of collision equal to 1.0E−03. ALARP region obtained from Skjong et al. (2007)

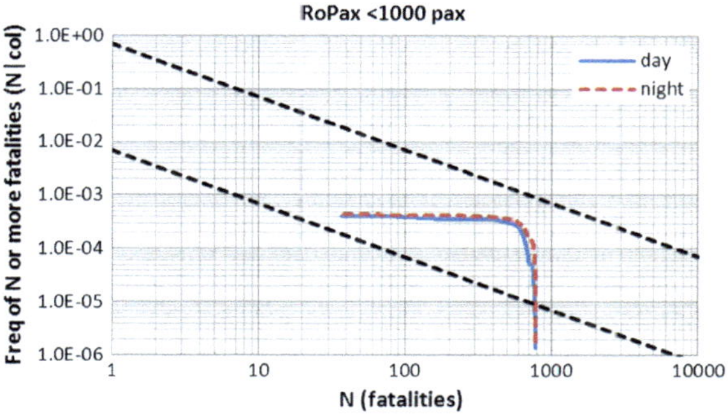

Fig. 47.11 F-N diagram for the smaller one of the RoPax vessels with frequency of collision equal to 5.0E−03. ALARP region obtained from Skjong et al. (2007)

Table 47.2 Resulting PLL for the ships, frequencies and cases tested

f	C1	C2	R1	R2
Day				
1.0E−02	2.22E+00	4.20E+00	1.44E+00	1.57E+00
5.0E−03	1.13E+00	2.16E+00	7.19E−01	7.37E−01
1.0E−03	2.19E−01	4.02E−01	1.43E−01	1.46E−01
5.0E−04	1.08E−01	2.10E−01	7.18E−02	7.62E−02
Night				
1.0E−02	2.84E+00	5.92E+00	2.94E+00	1.37E+00
5.0E−03	1.52E+00	2.81E+00	1.40E+00	7.22E−01
1.0E−03	2.89E−01	6.01E−01	2.91E−01	1.41E−01
5.0E−04	1.49E−01	2.39E−01	1.46E−01	6.92E−02

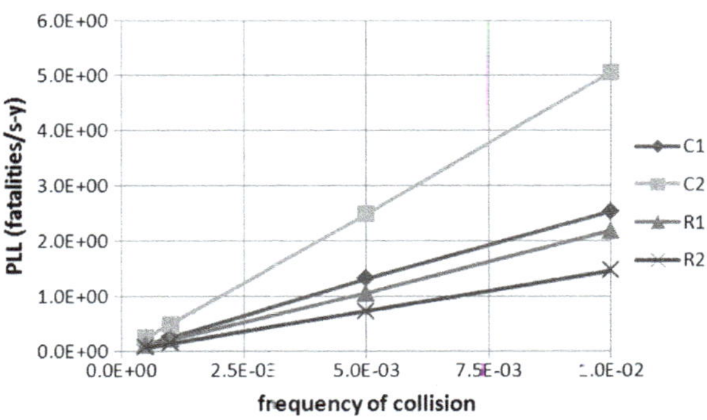

Fig. 47.12 Average PLL for day and night versus frequency of collision

47.5 Conclusions

Although there is still a great distance to be covered with respect to establishing a reasonable required level of survivability, this study has provided with a few insights to the right direction. The most important of those is probably that no matter what the frequency of incident compromising the ship safety, the large number of guests in modern ships suggest that if the potential is there the consequences will be catastrophic. Adding the fact that there are experiments that suggest that modern ships could capsize within a few minutes thus leaving little time for evacuation only strengthens the suggestion that the required index should be revised. Furthermore, this study has shown that irrespective of how safe a ship is, risk will always increase as the number of passengers carried increases.

Acknowledgements The authors would like to express their gratitude to their colleagues within the Ship Stability Research Centre (SSRC) of the University of Strathclyde as well as all those involved in project GOALDS for their valuable contribution.

References

HARDER (1999–2003): "Harmonisation of Rules and Design Rationale". EC funded project, DG XII-BRITE, 2000–2003

Jasionowski et al.: "Ship Vulnerability to Flooding", 3rd International Conference for Maritime Safety, Berkeley, California, 2007

Jasionowski et al.: "Investigation Into The Safety Of Ro-Ro Passenger Ships Fitted With Long Lower Hold Phase 2", UK MCA RP592, Report MCRP05-RE-001-AJPB.16, Safety at Sea Ltd, 2008

SAFEDOR, "Design, Operation and Regulation for Safety". Integrated Project, www.safedor.org, 2006

Skjong et al.: "Risk Evaluation Criteria", SAFEDOR Deliverable 4.5.2, 2007

Spyrou, K.J. & Roupas, I.: "Damaged ship survivability: a step beyond Wendel", International Shipbuilding Progress, 54, 285–303, 2007

Tsakalakis, N., Puisa, R., Chen, Q.: "A critical look At SOALS Ch. II-1 with respect to floodable length of compartments in passenger ships", Work Package 2 report, UK MCA Research Project 625, 2011

Vassalos, D., Kim, H., Christiansen, G. and Majumder, J.: "A mesoscopic model for passenger evacuation in a virtual ship-sea environment and performance-based evaluation", Proceedings of the International Conference on Pedestrian and Evacuation Dynamics, 2001

Vassalos, D., Jasionowski, A.: "Conceptualising Risk", Proceedings of the 9th International Conference on Stability of Ships and Ocean Vehicles, 2006

Vassalos, D.: "Safe Return to Port—A Framework for Passenger Ship Safety", 10th International Symposium on Practical Design of Ships and Other Floating Structures, Houston, USA, 2007

Part XIII
Requirements and Operation: Design, Planning and Onboard Guidance

Chapter 48
Design Requirements for Stability and Minimal Motions in a Storm

Vasily N. Khramushin

Abstract The hydrodynamics of a ship in a storm is not limited to the hull below the calm-water waterline. In a storm, the operating waterline varies between the bilge and the deck, causing unpredictable wave forces on the hull as well as the possibility of slamming on flat surfaces and the flared sections of the vessel. The present work describes the early stage of a design process for a hull form that accounts for the range of changing of the waterline in order to insure stability under severe heave. With this approach, it is possible to reduce the metacentric height, which minimizes roll resonance. The concept is part of a consistent ship design process; conventional naval architectural approaches will still be needed for successful solutions for reducing the pitching and yawing of the vessel and as a necessary condition for using active stabilizers and other seaworthiness improvements.

Keywords Design for seaworthiness · Stability · Rolling · Pitching
Efficiency of operation

48.1 Introduction

From a design standpoint, there are contradictions regarding the dynamical properties of a ship; improving one leads to the deterioration of another.

An attempt to increase the initial static stability may result in an increased rolling of the ship in severe seas, which may lead to potentially catastrophic problems with the strength of the hull structure. It is a paradox—attempting to improve static stability may lead to an increased likelihood of capsizing as a result of strong resonant rolling or broaching. Here, a different approach is discussed, the main idea of which is to design the lines in such a way that the interaction with waves is dramatically decreased; creating a ship that will not resist waves and wind (Khramushin 2014).

V. N. Khramushin (✉)
Seaworthiness of Vessels in Storm Conditions Subsection of the Alexey Krylov Russian Science-Engineering Society of Shipbuilders, Yuzhno-Sakhalinsk, Russia
e-mail: v.khram@gmail.com

V. L. Belenky et al. (eds.), *Contemporary Ideas on Ship Stability*, Fluid Mechanics and Its Applications 119, https://doi.org/10.1007/978-3-030-00516-0_48

Significant difficulties in the operation of a vessel may occur when the freeboard and reserve buoyancy are too large for the vessel, particularly when the requirements for prevention of the shipping of green water are implemented without consulting the operators. Another example is installation of unwieldy life-saving equipment that can pose a danger for the crew.

In the present paper, some elements of hull shape optimization and general ship geometry are considered from the standpoint of achieving the best seaworthiness of the vessel. However, as optimized lines may significantly change the dynamics of a ship in waves, the development of appropriate guidance for the crew has to be considered.

48.2 Marine Design Experience

As a ship designed for seaworthiness, consider a hull form with a sharp cruiser stern and no increased freeboard in the bow section. Such a vessel has no flare and no bulbous bow, as too much volume forward makes a ship pitch. Such a hull form is capable of maintaining operability in all-weather conditions. Such a design has almost constant waterplane area as a function of draft and low damping in heave. As a result, this vessel's stability in waves is essentially constant; the minimum stability therefore occurs at the design condition. To remedy this situation, the static stability with varying draft must be changed.

Many modern vessels have a wide transom stern and with flare and increased freeboard at the bow, and experience intensive pitching in longitudinal waves. In this mode of sailing, the hydrodynamic effect of excessive volumes in the extremities is essential when the hull of a vessel moving through the wave crest and over the wave through. However, the hydrostatic approach to this problem is not contrary to the implementation of the proposed scheme of the hull design. The objective of this work is to find a compromise between the proposed design scheme and modern hull forms.

48.2.1 Hydrostatic Peculiarities of a Highly Seaworthy Vessel

The hydromechanics of a ship in extreme seas includes some design paradoxes, which significantly affect the seaworthiness of non-optimized vessels, and, in fact, presents the possibility of optimal all-weather performance or special modes of vessel operation in accordance with the vessel's design purpose.

Some elements in hull form design are well established in maritime history, though now their use is somewhat weakened in favor of efficiency of cargo handling, increased volumetric capacity, or sometimes just for aesthetic reasons.

Identification of and optimal resolution of the hydrometrical paradoxes is defined by accounting for the adverse effects of weather through specifically designed lines. Let us consider some specific elements of the hull form:

The metacentric radius *BM* has cubic dependence on the breadth *B*.

$$BM = \frac{J_X}{V} = \frac{1}{V}\frac{2}{3}\int_{x_K}^{x_H} y^3 dx \bigg|_{WL} \tag{48.1}$$

where V is the displacement; J_X is the transverse moment of inertia of the waterplane area relative to the longitudinal x axis; and y is the ordinate of the waterplane, the width of the body at a particular x value of the current waterline *WL*.

In a storm conditions, external forces and moments are created by the curved sea surface and this affects the stability of a ship:

$$m_\alpha = \gamma \cdot BM \cdot V \cdot \sin(\alpha) = \gamma \cdot J_X \cdot \sin(\alpha) \tag{48.2}$$

where m_α is the exciting moment; $\gamma = 1025$ kg/m^3 is the density of water; and α is the angle of wave slope.

Note that this moment is not balanced by the weight of the vessel and reaches its maximum value for vessels with a wide transom sterns. This moment may be dangerous for multihull vessels, because it is directly dependent on the area of waterplane, since the metacentric radius (*BM*) changes as the cube of the offsets.

The formula for the metacentric radius is consistent with the subsequent results of hydrostatic calculations at large roll angles. To be able to assess the stability of the vessel as a rigid body in a storm, fix the center of gravity at the z-coordinate of the metacenter at a given draft.

To achieve smoothness of roll in a storm, any captain always tries to reduce the metacentric height by raising the *KG* of the vessel towards the *KB*-value (Fig. 48.1). This decreases roll motions by suppressing the external forces by the inertia of the ship. This process is described by the "captain's" formula for initial stability:

$$T_\phi = \frac{C \cdot B}{\sqrt{GM}} = \frac{C \cdot B}{\sqrt{BM + KB - KG}} \tag{48.3}$$

where: $C \approx 0.8$ [cm/m$^{1/2}$] is a dimensional empirical coefficient with a characteristic value for a conventional mono-hull cargo vessels; *KB* is the distance from the base plane to the center of buoyancy; and *KG* is the distance from the base plane to the center of gravity.

If the center of gravity G is significantly above the waterline, the ship becomes susceptible to wave action as the point of application of wave forces is far away from G (Fig. 48.1). The contours of the stations in the midship section of the optimized vessel (Fig. 48.2) are close to a circle, and it has a beam-to-draft ratio (*B*/*d*) of about two (Khramushin 2010). Accordingly, the metacenter is located at the waterline, very

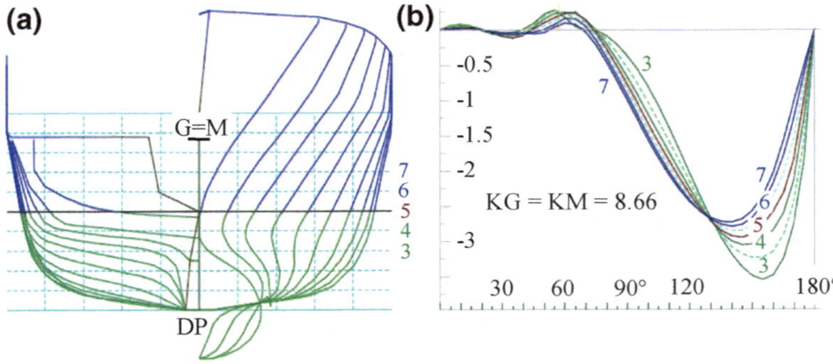

Fig. 48.1 **a** Body plan for conventional hull form; **b** GZ curves of the ship for five waterlines, the center of gravity is placed at the position of the metacenter at the design draft. Static stability is insufficient and the most probable stable position of the ship hull is "keel up"

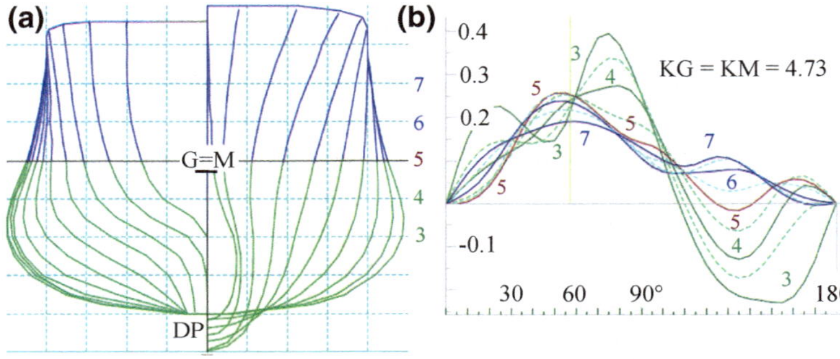

Fig. 48.2 **a** Body plan for optimized hull form; **b** GZ curves of the ship for five waterlines, the center of gravity is located at the position of the metacenter at the design draft. The *GM* value of the optimized ship can be safely brought to zero at the design draft. *GM* will be positive for any other draft and there is a large reserve of stability

close to the center of gravity. As a result, the resultant moment of the external forces becomes relatively small and does not lead to significant roll motions.

The hydrostatic and Froude-Krylov moment in pitch of a long vessel cannot be reduced by increasing the *KG*, because the longitudinal metacenter is too high. Thus the lines can only be optimized to decrease external forcing, allowing the ship to smoothly pierce through the waves.

Once the hydrostatic and Froude-Krylov forces are minimized, a further decrease of pitch and roll motions can be achieved by considering the hydrodynamics of the hull in trochoid waves. This is a separate problem considered in Khramushin (2009).

Here we focus on a simple optimization of lines by considering Froude-Krylov and hydrostatic forces.

48.3 Storm Operational Stability

The expected technical result is a displacement ship with a relatively deep draft and pointed extremities (cruise stern and "reclined stem"—stem inclined back). The reserve buoyancy is concentrated in the midship section. The *GM* is minimal at the design draft, increasing with any variation of the draft. This is achieved by special design of the lines (Khramushin 2012). The next subsection is focused on the analysis of the hydrostatics of the midship section only.

48.3.1 Analysis of Metacentric Height at Design Draft

The stability of a wall sided ship is characterized by the value of *GM*. The variation of *GM* during heave motions is caused by changes of the vertical position of the center of buoyancy *KB* and volumetric displacement. The *GM* increases with an increase of *KB* and decreases with an increase of the displaced volume. We also fix the center of gravity. The transverse moment of inertia of the waterplane J_x is constant since wall-sidedness is assumed,

$$GM = KM - KG = BM + KB - KG = \frac{J_X}{V} + KB - KG \tag{48.4}$$

All of the quantities in Eq. (48.4), except *KG*, are computed as integrals of ship offsets and are functionally related to the draft. They can be seen in the hydrostatic curves in Figs. 48.3b and 48.4b

In the case of a wall-sided midship section, the minimum of the vertical position of the metacenter *KM* occurs for $B/d = 3$. Accordingly, for a ship with $B/d > 3$, the minimum *KM* is achieved for a flared configuration due to the rapid increase in J_X

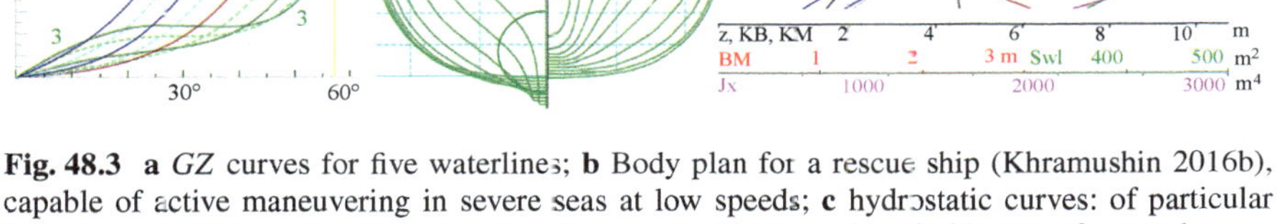

Fig. 48.3 **a** *GZ* curves for five waterlines; **b** Body plan for a rescue ship (Khramushin 2016b), capable of active maneuvering in severe seas at low speeds; **c** hydrostatic curves: of particular importance is *KM* as a function of draft; *A* is area of midship section, *Swl* is area of waterplane

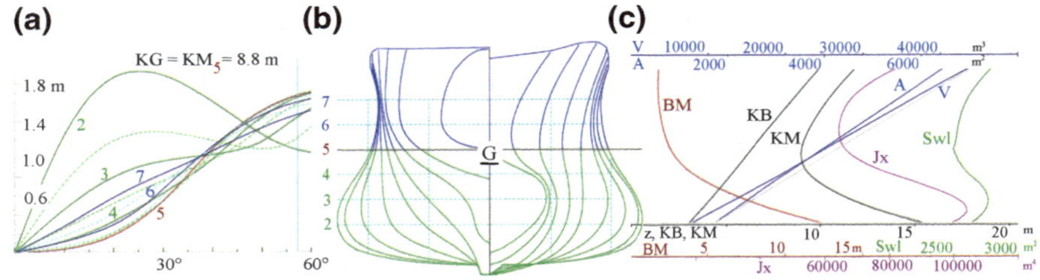

Fig. 48.4 **a** *GZ* curves for five waterlines; **b** Body plan for a universal cargo ship (Khramushin 2016a), capable of operation in a storm with minimum motions at zero *GM* values; **c** hydrostatic curves: of particular importance is the *KM* as a function of draft, *A* is area of midship section, *Swl* is area of waterplane

on the hydrostatic curves (see Figs. 48.3c and 48.4c). The behavior of such ship in a storm includes significant pitch and roll, though she is capable of contouring waves.

A vessel with good seaworthiness should have $B/d = 2$, for which a minimum *KM* can be achieved at the design draft. Such a ship must have a tumblehome shape at the waterline.

Ship lines with a round midship section does not have a clear minimum of *KM*. It therefore becomes important to maintain positive metacentric height, because the reserve stability is small and roll motions may reach large amplitudes. An increase in static stability leads to large-amplitude roll motions in stormy weather.

The minimum of the *KM* is shown most clearly for the configuration in Fig. 48.4. The stations have concave shape at the design waterline—the *KM* as a function of draft has a minimum around the design waterline *WL*.

48.3.2 Brief Description of the Drawings

Stability in storm conditions should be provided by the shape of the entire hull, as her draft may change very significantly. This effect can be easily achieved for a ship design with concave stations at the waterline (Fig. 48.4) and, with the some difficulties, for a ship with round stations (Fig. 48.3).

Hydrostatic curves in Figs. 48.3c and 48.4c provide a quantitative description of the forms. The stability is associated with the condition of mutual compensation of the changes of moment *KB V* relative to base plane, through the moment of inertia of the waterplane J_X. Geometrically, it is the choice between wall-sided, flared or tumblehome configuration of the midship section at the waterline. In fact, this is the optimization of the angle of side slope at the design draft. The algorithm for computing the optimal characteristics of the ship's hull is implemented in the computer program "Hull" (Khramushin 2010a).

The *GZ* curves (Figs. 48.3a and 48.4a) are computed for a variety of drafts at a fixed value of *KG*: the curve with zero metacentric height (labeled "5") corresponds to the design draft and other curves (3, 4, 6 and 7) show positive initial stability as the draft varies.

All calculations were performed for a fixed position of the center of gravity *KG*, placed exactly that the metacenter at the design draft, where the initial stability is equal to zero ($GM = 0$).

48.4 Lines at Bow and Stern

The hull form is optimized using a series of calculations of Michell's integral (Michell 1898). The physical and geometrical interpretation of the latter reveals the relationship between wave-making process and the wave forces acting on a ship (Khramushin 2010a).

$$a(\lambda) = \frac{M}{\lambda} \cdot \left| \int_{\Omega} q(x_0, y_0) \cdot e^{k \cdot (-z_0 + i\omega_0)} \delta\Omega_0 \right|$$

$$R_x = -\frac{\pi}{2} \cdot \rho \cdot V_0^2 \int_{\Lambda}^{0} a^2(\lambda) \cdot \frac{\delta\lambda}{\lambda \cdot M^3 \sqrt{M^2 - 1}} \qquad (48.5)$$

$$M = \sqrt{\Lambda/\lambda}; k = 2 \cdot \pi/\lambda$$

where R_x is the wave-making resistance force at a speed V_0; a is wave amplitude, λ is incident wave length; Λ is the maximum length of the transversal wave in Kelvin pattern; k is the incident wave number; x_0, y_0 are the coordinates of a point at a hull surface; ω_0 is the magnitude of the phase angle for a particular ship's wave radiated by the hull.

Figure 48.5 shows a physical-geometrical interpretation of the hydrodynamic phenomena described by Mitchell integral (5). Wave-making can be seen as a reflection of the incident wave from the hull (Khramushin 2010a, 2012). The reflected wave propagates away from the hull with the angle ϑ. Indeed, the group velocity V_{fr} is twice less than the phase velocity (celerity) V_w. The length of the reflected waves varies from 0 to Λ depending on forward speed.

Several areas are identified in Fig. 48.5. A ship-generated wave cannot detach from the hull in the areas *Ae* and *Be*, where the wave is a result of a summation of short waves generated by a jet flow adjacent to the hull. To damp the diverging waves in the areas *Aw* and *Bw*, the bilge has a smoothly twisted surface that stabilizes the flow at the parallel mid-body area Λ and compensates for the energy of the generated transversal wave.

Historically, this phenomenon was interpreted in terms of wave-making and wave-exposed ship length Λ. Outside of Λ (but inside *LWL*), inverse trimming moments

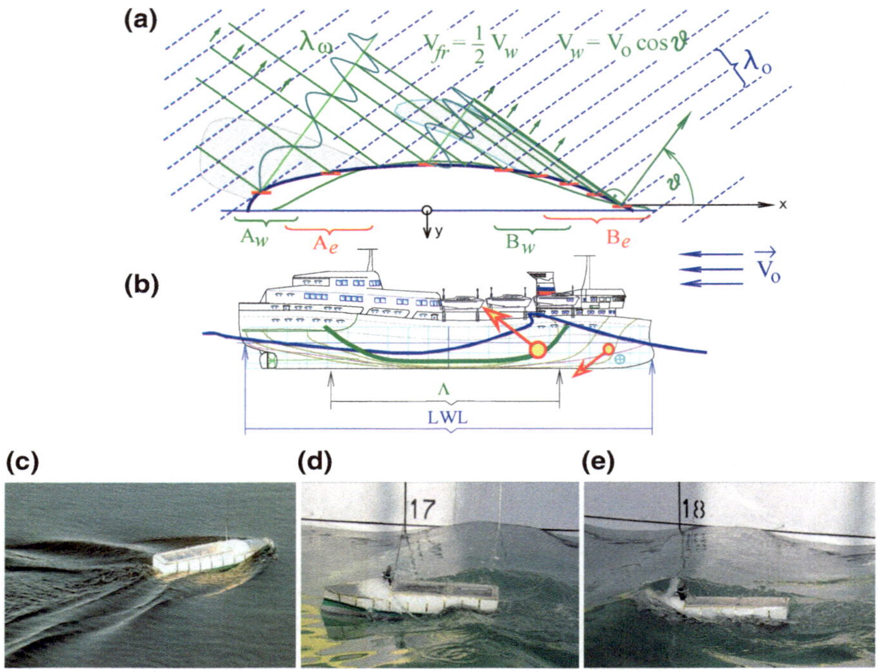

Fig. 48.5 **a** wave-making as a reflection of incidence waves; **b** compensation of trimming moments; **c** calm-water trim compensation by pulling the ship wave under the bottom; **d** even-keel heaving upwards after the wave crest passes; **e** diving under the wave crest

are generated. These inverse trimming moments prevent forward speed trim in calm water at relatively high speed. They also compensate trimming moments and pitch while sailing in heavy weather.

The Decrease of the pitch motions improves seakeeping of an optimized ship and allows:

- Keeping speed and heading while maintaining safe and comfortable conditions for crew and passengers;
- Stabilizing of green water shipping of upper decks to an acceptable level, with design solutions for prevention of surf-riding and broaching-to;
- Decreasing accelerations at bow and stern and improve the reliability of securing of cargo.

48.4.1 Stem and Bilge

Stem and bilge of a fast ship (Fig. 48.6a) or of a relatively slow cargo vessel (Fig. 48.6b) are designed to provide the best forward speed and seakeeping, prevent ice accretion and enable autonomous operations in heavy ice conditions.

An extensive series of model tests and numerical simulations (Khramushin 2003, 2011; Khramushin et al. 2004) have revealed elements of geometry and structural features of a stem and bow lines (Khramushin 2017a).

- Wave-piercing bow with inverted stem (Fig. 48.6a, line 3) or plumb and sharp stem (Fig. 48.6b, line 3) within the interval of variable drafts for sailing in moderate waves (approximately one-third of the design draft); Such a bow configuration is helpful for no-impact interaction with waves and keeping ice field edges from getting under the bilge or the bottom.
- "Storm undercut" at the lower part of the stem (Fig. 48.6, lines 6) with average angle 20°–30° relative to the keel line to free yawing in the case of heavy roll and pitch while sailing in heavy seas.
- Special strengthening of the most forward part of the stem (Fig. 48.6, lines 4) for taking impact loads from ice breaking at speeds below 6 knots. This feature helps to safely reduce the speed down to 3–4 knots when interacting with ice. Functionality also includes the prevention of pulling pieces of broken ice under the bottom. It is achieved by creating a maximum upward force at the low edge of a large ice piece where it is heated up to the temperature of surrounding water.
- Upper parts of bow frames may have a little flare to prevent bow-diving at high speeds (Fig. 48.6a). For the low-speed vessel (Fig. 48.6b), those frames have

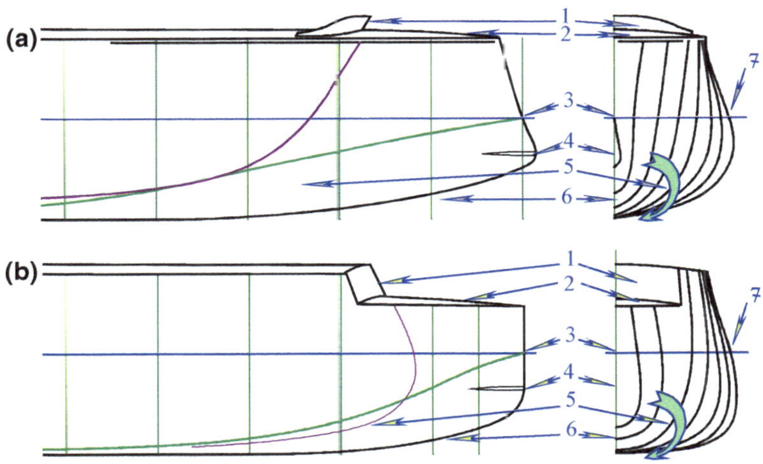

Fig. 48.6 Bow of a fast ship (**a**) and slow cargo vessel (**b**) adapted for active maneuvering or course-keeping in heavy weather and moderate ice conditions

tumblehome geometry at their upper part and decreased freeboard height in the bow section. This configuration helps to compensate pitch while shipping green water.

- Wave breaking structure for a fast ship (Fig. 48.6a, line 1) and fore bulkhead of the shelter deck superstructure of a cargo vessel (Fig. 48.6b, line 1) protect the crew on the deck from direct action of waves in heavy weather.
- Bulbous bow forms shell plating as a helicoid in varying draft area (Fig. 48.6, line 3). The helicoid starts from the stem and terminates at the area where diverging ship waves separates from the hull. The stretch of the helicoid is meant to be comparable with the wavelength in heavy seas, so the crests of the incident waves can be sucked under the bow bilge and bottom.
- Immediately behind the area of diverging wave separation, the bulging frames form a blister in the middle body (Fig. 48.6, line 7). This blister reduces roll motions at any wave heading. It also helps pushing large ice pieces under surrounding ice fields and prevents access of those ice pieces to propeller and rudder.

48.4.2 Sternpost and Stern Overhung

The sternpost, stern bilge and stern overhang can be optimized for seakeeping in heavy weather conditions (Khramushin 2017). The optimized stern lines for a two-screw configuration are shown in Fig. 48.7. The same approach can be used for single-screw ships as well. The installation of active pitch and roll stabilizing fins is possible immediately behind the propellers (considered in separate subsection). The lines were optimized to minimize the trim caused by forward speed in calm water and heavy seas. The following features help achieve those goals:

- A helicoid is formed in the area of varying drafts stretching from the blisters (Fig. 48.7, line 3) to the stern overhang (Fig. 48.7, line 1). It twists the flow and pulls it out of the bottom in the vicinity of the side shell plating. It compensates for the wake depression and prevents separation of high-frequency components of ship diverging wave. As a result, ship wave energy concentrates in low-frequency

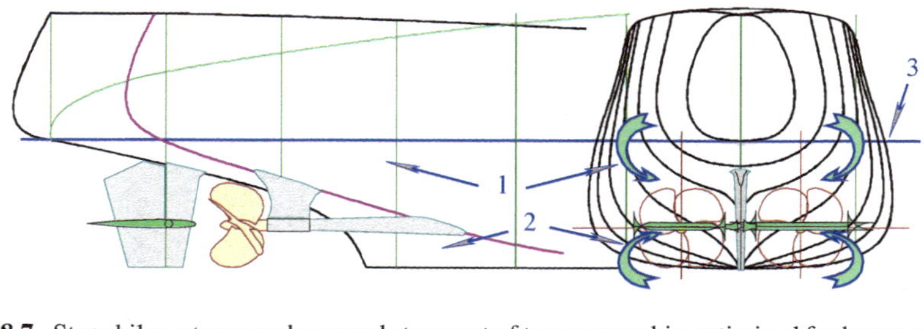

Fig. 48.7 Stern bilge, stern overhung and sternpost of two-screw ship optimized for heavy weather seakeeping: (1) upper helicoid surface (2) lower helicoid surface (3) design water line

part of the spectrum. It also leads to a phase shift. The combination of this phase shift and the low-frequency energy concentration causes the wave to interfere with the transverse wave and damp it, resulting in overall decrease of wave-making.

- The helicoid on the level of varying drafts leads to decreases in the volume above the water and naturally makes the stern lines sharper. The sharp stern decreases the wave forces, pitch, heave and wave impacts. As the surging wave forces are also decreased, the likelihood of surf-riding and broaching-to is lessened, including dangerous wave heaving and accidental power losses.
- Sharper cruiser stern brings fewer disturbances to the flow field below the stern overhang. In the case of the loss of main engine, active motions stabilizers can be used as emergency propulsors (see subsection on motion stabilizers). To facilitate the use of the stabilizers in an emergency situation, the lines are optimized for controllability at minimum thrust.
- The second helicoid stretches from the bottom to the sternpost, Fig. 48.7-line 2. It twists the flow on a deeper (supporting) depth. It is required for mutual compensation of the upper and lower twisted flows under the stern overhang, in the location of the rudder and motion stabilizers. It prevents access of broken ice to the propeller/rudder/motion stabilizers location and helps to create a clean canal in the ice.
- Tumblehome side configuration on the waterline level at the mid-ship section helps with hydrodynamic compensation of roll motions. It intercepts large pieces of broken ice lifted by ship wave crests, prevents their sinking and does not let them to be sucked under the stern overhung.

48.5 Architecture and Propulsion

One of the aspects of seakeeping is the usability of deck equipment in conditions of large-amplitude ship motions and cold winds, prevalent in subpolar regions (Khramushin 2015). Consider two designs: a ship with relatively high freeboard and a trawler with a very low working deck (Khramushin 2016c).

48.5.1 The Freeboard and the Upper Deck of the Ship

The idea is to control the water on deck rather than trying to avoid green water shipping. The objective is to enable a ship to sail any heading in heavy weather and maintain safe conditions for the work on the deck.

The freeboard height is set to achieve approximately uniform green water shipping along the ship length. The green water is trapped at the deck at the weather side by uninterrupted deckhouses and bulky deck equipment that prevents the water to flow to the other side. The bulwark and freeing port are designed to keep most of water for a half of roll period.

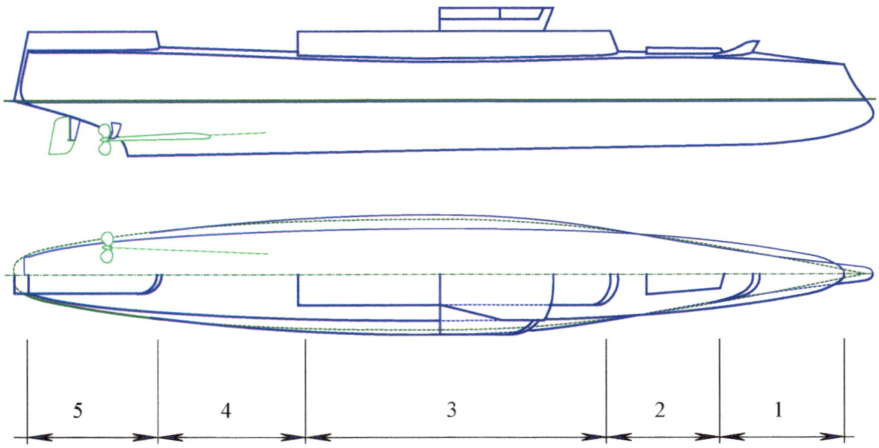

Fig. 48.8 A patrol boat: (1) forecastle (2) bow section (3) midship section (4) stern section (5) poop deck

Accumulation of water, however, should be avoided on forecastle and poop decks. A high forecastle breakwater helps to throw overboard a significant portion of the green water brought by an encounter with a large wave crest. The area of the poop deck, vulnerable for green water accumulation, is decreased by using a narrow cruiser stern, side shelves or stern deckhouses and large deck equipment. A cruiser stern configuration also decreases surging wave forces (compared to transom stern configuration) thus lessen chances of surf-riding and broaching.

These ideas are implemented differently for different classes of ships. Consider the architecture of a patrol boat (Fig. 48.8) and a fishing trawler (Fig. 48.9) as examples. The patrol boat in Fig. 48.8 has a relatively high freeboard, tumblehome top configuration and small area upper deck. The inverted stem provides good wave piercing capability in heavy weather. The long deckhouse prevents the flow of the green water towards the lee side.

The fishing trawler in Fig. 48.9 has a low freeboard and maximum width working deck. The low freeboard provides safe working conditions in heavy seas and high winds (Khramushin 2014a). The necessary buoyancy reserve is provided by a watertight superstructure and deckhouses.

The following functional areas of the upper deck are shown in Figs. 48.9 and 48.10:

1. Increased vulnerability for green water shipping limits access of the crew for the deck equipment; remote control is needed for operating equipment.
2. It is the location for weapon system and hold hutches for the patrol boat (Fig. 48.8); watertight superstructure is fitted for the trawler (Fig. 48.9) that is included in total buoyancy reserve.

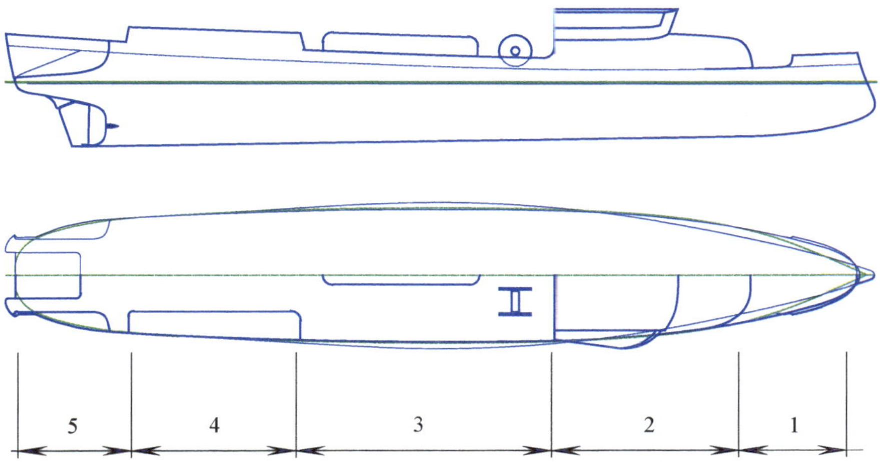

Fig. 48.9 Fishing trawler: (1) forecastle (2) bow section (3) midship section (4) stern section (5) poop deck

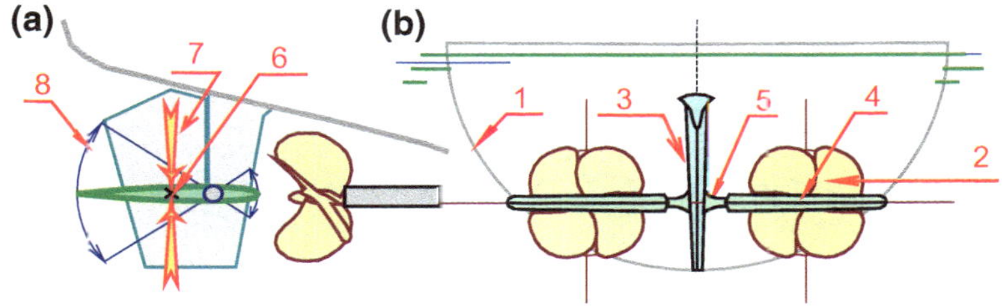

Fig. 48.10 Propulsion and control: **a** profile **b** body plan: (1) Optimized ship hull; (2) Main propulsor; (3) Rudder; (4) Active roll and pitch motions stabilizer, may be connected with a nuzzle in case of ducted propeller; (5) Axis of motion stabilizer is located at the center of propeller flow and allows ±30° rotation; (6) Center of the area of the motion stabilizer; (7) Lifting force on the motion stabilizer; (8) Angle of rotation of the motion stabilizer

3. A long deckhouse is located at the midship section of the patrol boat (Fig. 48.8) and prevents green water flow to the leeward side. The trawler (Fig. 48.9) has a wide deck with bulwark protecting trawl and cable winches separated by a deckhouse in the diametric plane. Total area of freeing ports is sufficient for outflow of the green water during roll semi-period, while inflow of the green water occurs practically instantly during the next semi-period of roll upon encounter of the next wave crest.
4. This stern section is the least vulnerable for heave at forward speed, so it is a good place of helicopter pad of the patrol boat at Fig. 48.8. It is the best place for the critical fishing operation for the trawler in Fig. 48.9. This location of the trawl deck is protected by deckhouses and is well suited for unloading the catch from the trawl, cable switching, and parallel preparation for the new trawl cycle.

5. The poop deck is the most comfortable location of the upper deck while a ship has forward speed. However, when stopped or at slow speed (for example, the trawl is lifted), a wide stern may be vulnerable to the wave impact. For the patrol boat in Fig. 48.8, it is the place for a hangar and a ramp for launch and recovery of autonomous vehicles, as those operations are complex and dangerous. For the trawler in Fig. 48.9, it is the place for trawl slip.

48.5.2 Active Stabilizer of Heave and Pitch Motion/Emergency Propulsion for Heavy Weather

Minimization of the external forces by simultaneously decreasing the excitation and restoring forces and moments may lead to some negative consequences of another nature that did not initially seem important (Khramushin 2010b). The stabilization of motions by decreasing the moment of inertia of the waterplane area may lead to excessive heel caused by wind or a turn, as well as excessive trim at acceleration or reversing. Active pitch and roll motion stabilizers can be used to compensate for those usually residual effects; but the active stabilizers should have sufficient energy to overcome these conditionally "insignificant" external forces with sizable reactions.

When a ship is steaming full speed ahead in heavy weather, a directionally stabilized fluid flow near ship hull can be found only below the stern overhang immediately aft of the propellers (Fig. 48.10). Thus, it seems to be the best place to generate forces for active stabilization of pitch and roll for both heavy weather and sharp turns in calm water.

In the case of the loss of the main machinery, the stern is subjected to significant wave-induced motions. Pitch becomes worse without forward speed because of the relatively full waterplane and large volume above the stern overhung. Large-amplitude oscillations create hydrodynamic conditions for the motion stabilizer to work as a flapping fin propulsor, so it can be activated in the dangerous situation of heavy weather and failed main engines. This emergency propulsor is passive in a sense that it does not require and additional power or control; free turn of the stabilizer up to $\pm 30^\circ$ decreases likelihood of wave impacts, which is also important for active stabilization mode.

48.6 Conclusions

The paper presents ideas how to optimize the initial elements of a ship's lines based on hydrostatic and hydrodynamic considerations. The idea is to decrease wave forces, thus minimize the motions in waves.

Roll motions are decreased by decreasing initial stability, but not let the GM become negative. Pitch motions are descries by special design of lines bow and stern, including application of wave piercing bow. The adverse effects of green water shipping are mitigated by deckhouses, preventing of the water flow to the other side and keeping deck working spaces green water free. Moreover the green water is used to create an anti-rolling tank effect. Design the lines around of the stern overhang is proposed with action of large amplitude wave in mind rather than calm water powering optimization. The adverse effect of very low GM is compensated by a fin stabilizer.

These solutions may be applicable to are to all classes of small- and medium-displacement ships and appear to be the most effective when applied simultaneously with optimization of their geometric dimensions and inertial characteristics for geographically-specific navigation conditions.

All of these design solutions and seaworthiness estimations must be explained by operational guidance onboard of ship's bridge and on shore services. This knowledge can enhance the operability of the ship in complex storm and ice conditions. These hydrostatic assessments make up the foundation for further, deeper design optimization based on the fluid mechanics interaction between the hydro- and aero-mechanics of the ship in storm sea and hurricane winds.

Acknowledgements The author would like to express his sincere gratitude to his teachers in the Kaliningrad Marine College Alexandre Kamyshev, Georgy Malenko, Dalen Bronshtein, as well as the professors of Naval Architecture Drs. Walter Amfilohiev, Alexei Kholodilin and Sergei Antonenko.
The author is grateful to Dr. Art Reed for his detailed editing that has greatly improved clarity and readability of the text.

References

Khramushin, V. N. (2003) Research explorations of the ship's seaworthiness. Far Eastern Branch of Russian Academy of Sciences publishing, Vladivostok 172p. in Russian.

Khramushin, V. N., Antonenko, S.V., Komaritsyn, A.A., Brovko, P.F., Nedorez, Yu.I., Soldatenkov, A.E., Surov, O.E., Fain, A.V., Shustin, V.A. (2004) History of evolution of seakeeping of a ship in storm conditions: from antiquity to the present. Sakhalin book publishing, Yuzhno-Sakhalinsk, 280 p. in Russian.

Khramushin, V. N. (2009) "Key design solutions and specifics of operation in heavy weather (Fluid Mechanics Approach to Stabilization of Ship in Heavy Seas)" Proc. 10th International Conf. on Stability of Ships and Ocean Vehicles, S-Petersburg, Russia. pp. 473–482.

Khramushin, V. N. (2010) Hydrographic and Patrol Service Ship. Patent RU-2384456 from 2010.03.20.

Khramushin, V. N. (2010a) Hull – creation of the analytical shape of the hull of the ship, the wave resistance calculation, the hydrostatic curves and diagrams of stability of sea vessels. Patent owner: Sakhalin State University. RosPatent 2010615849 from 2010.09.08.

Khramushin, V. N. (2010b) Active Stabilizer of Pitching and Rolling Motion of ship – Stormy Emergency Propeller. Patent RU-2384457 from 2010.03.20.

Khramushin, V. N. (2011). Exploratory study of ship stormy seaworthiness (The history of evolutionary development and engineering solutions of hull contours and ship architecture as the joint product of marine science and good seamanship). LAMBERT Academic Publishing, 288 p. in Russian.

Khramushin, V. N. (2012) Vessel of increased stability in heavy gale. Patent owner: Sakhalin State University Patent RU-2487043 from 2012-11-01.

Khramushin, V.N. (2014) "Features architecture of mean ship to navigation in heavy, stormy and ice conditions on the northern seas", Proc. of the 14th Intl Ship Stability Workshop, Kuala Lumpur, Malaysia. pp. 47–57.

Khramushin, V.N. (2014a) Fishing Boat for Northern Seas. Patent owner: Sakhalin State University. Patent RU-2535382 from 2014-04-27.

Khramushin, V.N. (2015) "Target ship design and features of navigation for motion stabilization and high propulsion in strong storms and icing", Proc. of the 12th Intl Conf. on the Stability of Ship and Ocean Vehicles, Glasgow, UK, pp. 1017–1025.

Khramushin, V.N. (2016a) Multi-purpose transport ship. Patent owner: Sakhalin State University. Patent RU-2603709 from 2016-11-27.

Khramushin, V.N. (2016b) Marine rescue research ship. Patent owner: Sakhalin State University. Patent RU-2603818 from 2016.11.27.

Khramushin, V.N. (2016c) Freeboard and upper deck of the ship. Application for the invention: No. RU-2016150045 from 2016-12-19.

Khramushin, V.N. (2017) Stern and aft of ship. Aft end of ship stabilized for storm sailing. Patent RU 2607135 from 2017.01.10.

Khramushin, V.N. (2017a) Stem and bow bilge of ship. Fore end of high-speed surface ship or relatively low-speed civil ship of high thunderstorm navigability and ice flotation in autonomous navigation. Patent RU 2607136 from 2017.01.10.

Michell, J. H. (1898) "The wave resistance of a ship" Philosophical Magazine, ser. 5. London, Vol. 45, Issue 272, pp. 106–123.

Chapter 49
Further Perspectives on Operator Guidance and Training for Heavy Weather Ship Handling

Laurie J. Van Buskirk, Philip R. Alman and James J. McTigue

Abstract Historically, mariners have received minimal formal training in heavy weather shiphandling, relying on mentoring and hands on experience to develop shiphandling skills for dangerous environmental conditions. Maritime organizations are increasingly turning to technology to reduce the inherent risks of heavy weather, including operator guidance systems and simulation to train watch personnel. Shiphandling simulators are on the cusp of extending training capabilities from simple maneuvering situations to highly realistic heavy weather scenarios, resulting in vastly improved training effectiveness. This is especially critical as actual time spent afloat may represent proportionately less of a mariner's total career.

Keywords Shiphandling simulation · Heavy weather training
Operator guidance

49.1 Introduction

Heavy weather presents mariners with significant risk of structural damage, loss of cargo, crew injury, and the potential for environmental damage (e.g., oil spills). Damage from heavy weather incurs significant costs to the maritime industry, both in property and environmental damage. In most cases, ships try to avoid heavy weather if possible, but some storms cannot be avoided, or prove to be worse than originally forecasted, leaving shiphandlers to deal with seas and winds for which they may have received little formal training.

All ships can be at risk of capsizing in extreme seas, and that risk can be exacerbated by poor shiphandling decisions. Current heavy weather training follows

The opinions expressed in this paper are those of the authors and not necessarily those of the Naval Sea Systems Command or the United States Navy.

L. J. Van Buskirk · P. R. Alman (✉) · J. J. McTigue
Naval Sea Systems Command (NAVSEA), Washington, DC, USA
e-mail: jackstay5@aol.com

V. L. Belenky et al. (eds.), *Contemporary Ideas on Ship Stability*, Fluid Mechanics and Its Applications 119, https://doi.org/10.1007/978-3-030-00516-0_49

two basic precepts: avoid extreme weather, and if the weather cannot be avoided, adhere to "rule of thumb" procedures and techniques to assist in safely riding out the storm. Advances in meteorological technology have significantly enhanced the ability to avoid severe weather by providing concise, real time understanding of the current and predicted weather environments, as well as storm mapping. However, on occasion, seamen must face the elements. It is at this point that correct and time-sensitive shiphandling decisions must be made, often in a high-stress environment that increases the potential for human error, and heavy weather training becomes critical.

49.2 Shiphandling Training

Historically, shiphandling training has focused on building skill sets for normal seaway and restricted maneuvering situations, such as entering and exiting ports and special evolutions at sea. The focus has been on understanding basic shiphandling characteristics and techniques as bounded by a ship's size, propulsion, ship control, and steering capabilities. Mariners have received minimal formal training in heavy weather shiphandling, relying instead on personal mentoring and hands on experience in specific ship types or classes with known handling characteristics to impart the ability to cope with difficult and dangerous environmental conditions. The reality of shiphandling in heavy weather is that normally only the most experienced shiphandlers are engaged in ship control in severe weather, so junior officers get little actual hands-on experience. Because heavy weather is normally avoided, even the most seasoned mariners may have only limited experience in higher sea states. This training gap in appropriate shiphandling procedures in heavy seas contributes to a higher risk of damage and loss when heavy weather is encountered.

49.2.1 Heavy Weather Shiphandling Training Objectives

In addition to the paucity of actual heavy weather shiphandling experience facing many of today's mariners, the advent of a variety of new hull forms makes it increasingly important to educate shiphandlers on the unique handling characteristics of these designs, particularly in higher sea states. In some cases, traditional shiphandling methods may not be appropriate for some of the more innovative designs, so relying on traditional responses in emergency situations may in fact exacerbate the danger. This is an important consideration in the training solution, as care must be taken to mitigate the possibility of negative transfer between traditional shiphandling techniques and those required for non-traditional hull forms. Shiphandling training, and in particular, heavy weather shiphandling training, should focus on providing tools that complement existing training and focus on ensuring the safety of ship and crew.

Regardless of the hull form, mariners must have a practical knowledge of sea characteristics and the ability to "read" and predict conditions based on clues ascertained from the surrounding environment. This type of information can be covered through classroom training, is currently included in most shiphandling training programs, and provides the foundation for heavy weather operations. Higher sea states, however, require separate skills outside of the scope of shiphandling in calm seas. Certain standard operating procedures can improve the likelihood that at-sea maneuvering events do not result in catastrophic loss or damage. To effectively train for heavy weather, the shiphandler must learn to correctly interpret several basic elements of dynamic information (presented either by display or visual/physical recognition):

- Own ship stability data
- Wave direction, length, height, and periodicity
- Wind speed and direction
- Ship motions (roll, pitch, yaw, surge, sway, and heave)
- The combined dangers/effects of all of the above (slamming, pounding, pooping, surf-riding, broaching, and ultimately, capsizing)
- Appropriate mitigation techniques
- Casualty situations (structural damage, flooding, loss of power/steering, etc.).

Each hull form has its own unique stability characteristics. Factors such as list, trim, displacement, ballasting, KG, and GZ are all important for the shiphandler to know and understand in order to successfully maneuver in heavy weather. Paramount for the shiphandler is the ability to understand the combined effects of environmental conditions and how they impact the unique shiphandling characteristics of the ship. A dynamic capsize can seem to be the result of unrelated events, but in reality, it is comprised of a cascading series of events and conditions that must be understood in order to properly interrupt the sequence and avoid catastrophic consequences.

There are basic tenets of good shiphandling that hold true in any situation, such as maintaining power, buoyancy, and stability; avoiding beam seas; and adjusting course and/or speed to minimize pitch and slam. However, once in heavy weather, understanding the combined effects of wind and waves on the specific hull form is critical (Alman et al. 1999). Simple classroom training can provide a basic understanding of these effects, but the best form of instruction is simulation, through which the trainee can practice decision-making skills and experience the results of both correct and incorrect actions. These aerodynamic and hydrodynamic effects were heretofore difficult to simulate but modern advances in physics-based ship motion software coding are now providing critical enhanced capability. This opens up the potential for rigorous hands-on training in a simulated environment, allowing routine training in the most dangerous of sea conditions, without jeopardizing personnel or ships.

Heavy weather training curricula should also include instruction on the use of basic calculations for estimating wave encounter period. This can be a useful tool during heavy sea states when technology is unreliable and/or unavailable. When simulation opportunities are added to this type of training, they allow the operator to effectively test his/her understanding of the principles, and to practice mitigation strategies appropriate for the ship type. This type of training helps solidify appropriate reactions when faced with time-sensitive decisions in actual heavy weather situations. There are basic mitigation strategies, or "rules of thumb," to assist the operator in maintaining a stable ship environment, such as the International Maritime Organization's "Revised Guidance to the Master for Avoiding Dangerous Conditions in Heavy Seas".[1] Guidance of this nature is useful, but should directly relate to the specific characteristics of the ship in question. For instance, some of the newer container ships appear to be susceptible to head sea parametric rolling, something not historically noted as a shiphandling concern. The magnitude of risk of a stability failure or capsize event can vary significantly between ship designs, as can the mode of failure. Consequently, the criticality of being able to recognize potentially severe conditions and make the correct judgment call with respect to the shiphandling decisions needed to mitigate risks assumes a degree of importance that cannot be underestimated.

The shiphandler should be trained to recognize 'cues' or precursors leading to an imminent dynamic stability event such as wave capture, bow plunging, or broaching to name a few, and understand the correct action necessary to get the ship out of danger in these situations. Ship motions are multi-dimensional, and shiphandlers need to thoroughly understand the implications of their ship's response to heavy weather (i.e., its motions), the dangers certain combinations present, and how to correct for them. For many ships, the arrangement of hull and superstructure create significant windage and there may be large differences across various load conditions. The lateral distribution of windage can create lee/weather helm characteristics similar to that of a vessel under sail. A ship trying to ride out a storm in head seas may need sufficient headway to maintain controllability, but at the same time, may suffer significant or dangerous structural damage as a result of wave impact, making it necessary to come about into stern seas. A ship with insufficient power may get caught "in irons" if trying to steer thorough head seas and come around to a new course. Multiple factors are in play at any given time, and maneuvering decisions need to be balanced against handling capabilities accordingly. The shiphandler must weigh the amount and rate of turn to minimize slamming or pounding when turning into the wind, and rolling when turning away from the wind. Each ship motion imparts key information to the shiphandler. For instance, a long-hanging roll implies a loss of stability in following seas, but might be interpreted by an inexperienced shiphandler as an improvement on how the ship is riding. Avoiding a roll event may be as simple as altering course to ensure the period of encounter is as different as possible from the ship's natural roll period, while in the same situation, changing speed alone will not correct for roll occurrence (see Footnote 1). Here again, the opportunity to test these skills in a simulator allows the shiphandler to hone his "seaman's eye" and get an accurate assessment of what can and cannot be done safely, so that when faced with an actual emergency, appropriate decisions can be made.

Over the past several years, the authors have worked with the Operator Guidance and Training Working Group (OGTWG), part of Cooperative Research Navies

---

[1] *Revised Guidance to the Master for Avoiding Dangerous Situations in Adverse Weather and Sea Conditions*. Ref. T1/2.04 (11 January 2007). MSC.1/Circ. 1228.

(CRNAV), to help define heavy weather shiphandling training objectives for the Naval Watch Officer. In addition to basic shiphandling objectives already routinely contained in shiphandling curricula, the following recommended additions have been identified: better meteorological training; training on available decision aids; enhanced static, dynamic, and damaged stability training (including how to avoid/escape from hazardous situations, recognizing and understanding non-survivable conditions, and consequences of damage or system failures in heavy weather); and discussions/assignments on heavy weather stability. Several workshops have been held over the years using full mission bridge simulators with heavy weather simulation capability. During these workshops, a number of simulator scenarios were tested to help develop these recommendations. Additional benefit can also be gained by using a full-mission shiphandling simulator with enhanced heavy weather rendering and ship capsize modeling, and (if possible) by incorporating a classroom physics-based model simulator with an interface that can support changing factors such as course, speed, KG, wave height, etc.

One key advantage of adding simulator training is that it allows a scenario to be replayed (multiple times if desired) and the operator to practice different mitigation techniques as environmental conditions change. If a "bad" decision is made, the consequences should be clearly apparent, and the operator can try again and experience the results from a different set of shiphandling maneuvers. Repetition can progressively enhance the degree of training transfer, while the risk of transfer failure is reduced (Foxon 1993). The trainee can also dissect the actions taken to better understand when naval architectural limits are reached and resulting damage can be anticipated. This type of training experience can provide lasting impressions on trainees, and can also facilitate development of a shiphandling "fault tree" specific to each ship type.

49.2.2 Training Proficiency

One of the main issues with any type of proficiency is the rate at which it decays when it is not used. Higher order cognitive skills and team behaviors (such as shiphandling in heavy weather) are extremely perishable (Cannon-Bowers et al. 1998). The infrequency with which most shiphandlers have to face severe weather puts them at risk of having a much lower proficiency level than would be desired when confronted by those conditions. Today, maritime organizations (including navies) are increasingly turning to simulation tools as a means of providing required training to watch officer and bridge personnel in order to meet qualification requirements.

The effectiveness of training transfer is directly linked to how well training devices duplicate the actual environment (e.g., simulation fidelity).[2] Simulators have long been used in the aviation world as a principal (and economical) form of training. Airlines have been able to amortize the cost of a simulator in less than two years. For

[2]Allen (1986), Alessi (1988), Hays and Singer (1989), Gross et al. (1999).

instance, Boeing 767 aircraft full flight simulator training costs approximately $400 per hour, while actual aircraft training time costs between $7000 and $8000 per hour (Thompson et al. 2009). Simulator use has also increased significantly over the past 20+ years for shiphandling, though primarily for such tasks as open water and harbor maneuvering, man overboard practice, and for naval vessels, steaming in formation, and special evolutions. Shiphandling simulation also has to incorporate the element of motion in a seaway, which is difficult to accurately model in higher sea state conditions. Recent improvements in software coding capabilities are redefining the limits of shiphandling training possibilities. Shiphandling simulators are beginning to have the technical capacity to extend their training capabilities from providing traditional calm water/low sea state and restricted waters maneuvering to presenting highly realistic heavy weather scenarios, resulting in improved knowledge and effectiveness under the most severe circumstances. This is especially critical as, in many contemporary instances, actual time afloat may represent proportionately less of a mariner's total career. Consequently, the integration of a heavy weather shiphandling training capability into an overall maritime training program should be approached carefully, with a structured set of goals.

49.2.3 Simulator Fidelity

Simulation quality and human capabilities are critical factors in training effectiveness and efficiency. Simulator fidelity is potentially the most important aspect of simulator quality, and is also a critical factor in the cost effectiveness of simulation device design. It is normally understood to mean the degree to which the simulation replicates the actual environment, and there is a strong link between it and transfer of training (Liu et al. 2009). There are two principal aspects of simulator fidelity—physical fidelity (the replication of sights, sounds, and the "feel" of the actual environment), and psychological or cognitive fidelity (the replication of such things as communication, situational awareness, etc.), and these aspects have subsets which are not mutually exclusive, but rather, have a large degree of overlap. These include: visual and auditory fidelity (how well the simulation replicates known visual and auditory stimuli of the actual environment); equipment fidelity (how well the simulator replicates the actual equipment/systems the operator is expected to use); motion fidelity (how well motion cues experienced in the actual environment are replicated); task fidelity (the tasks and maneuvers executed by the user); and functional fidelity (how the device functions and provides realistic stimuli in the simulated environment). All of these must be considered in the overall simulation solution equation.

Shiphandling simulators have become quite good at representing most of these aspects of simulation. Technology has significantly enhanced visual fidelity in recent years. For instance, harbors now used in simulators are extremely realistic, with recognizable structures, piers, buoys, lights, navigational aids, etc. Environmental factors such as fog, low light levels, rain, lightning, thunder, and other characteristics can be added into the simulation, as can other vessels, numerous types of aircraft, small boats, and even birds and people in the water. Ship sounds, such as whistles and alarms, and communications equipment have been accurately replicated. Equipment fidelity, the extent to which a simulator can emulate or replicate the equipment being used, which includes all the software and hardware components of the system (Zhang 1993), can prove to be more of an issue for some ship classes that have unique bridge or engineering equipments, but most bridge equipments are of sufficient similarity to provide adequate training transfer for routine evolutions. However, portraying realistic sea conditions in higher sea states has proven to be a challenge.

49.2.4 Heavy Weather Simulator Models

To provide accurate seaway representations, a heavy weather shiphandling simulator must be driven by a physics-based hydrodynamics model (such as FREDYN) which is capable of providing non-linear, six degree of freedom motion in the large amplitude motions resulting from exposure in a severe seaway. A principal requirement for the hydrodynamics model is that it should be executable in time domain at a time scale that is at least as fast as real time and validated for use in training. Development of numeric codes providing this capability is an evolving science. The non-linearities associated with seakeeping computations are associated with viscosity, pressure, free surface, and body geometry. Currently, fully non-linear codes are not suitable for integration into the simulator environment because excessively long execution times are in excess of real time. Some codes have adopted short cuts by blending linear and non-linear theories. These blended codes are significantly faster and are capable of engineering accuracy (Beck and Reed 2001). The code used also must be capable of fidelity that can replicate behavior characteristics for specific ship classes in the heavy weather environment. These general characteristics are a functional requirement of the previously identified training objectives.

Numerous commercial shiphandling simulation tools are available. Determination of the appropriateness of any simulator should include the verification, validation, and accreditation (VV&A) of the model used to run the simulation. The VV&A of the model is a necessity, and should include conceptual validation, design verification, code verification, results validation and accreditation, which must be specific for the application. Specific intended uses of the tool should be clearly defined as part of this process. This will help ensure that desired training transfer can actually be achieved by the simulator.

49.3 Operator Guidance

There are several commercially available systems designed to provide operator guidance on ship motions and limitations, give warnings of impending difficulties, and serve as decision aids in situations such as extreme roll motions/parametric rolling, bow impact, green seas on deck, and broaching. These are real-time systems that display the ship's position in relation to pre-calculated sea-keeping operational risk limits. Some can also be interfaced with weather routing systems to predict ship motions based on forecasted weather under different motion parameters, and define the operational limits for route planning, as well as recommend tracks that avoid areas with forecasted excessive motion.

The emergence of these new operator guidance systems also supports the inclusion of heavy weather shiphandling into training curricula. These capabilities offer shiphandlers a tool that can automatically calculate safe operating environments and provide course and speed options to minimize hazards based on real-time wind and sea data. This can improve operational safety and provide an enhanced capability to continue a ship's mission in certain situations. More importantly, these operator guidance tools can incorporate actual hull form data for unique ship types and help prevent catastrophic consequences for an operator who does not have a significant experiential base in that platform. When coupled with physics-based ship motion simulator training opportunities, this decision aid can significantly enhance the overall training experience, allowing the operator to test the limits of the ship and "experience" the consequences of erroneous shiphandling decisions, even taking the ship to the point of capsizing to better understand the dynamics of each shiphandling decision.

49.4 Conclusion

Current technology advances are beginning to offer the ability to integrate multiple simulators, which create even greater "virtual reality" potential for heavy weather training. Simulation of various casualties can provide shiphandlers with training opportunities to better prepare them for decision-making under duress. Decision aids in the form of operator guidance capabilities are becoming more refined, and combining these capabilities with heavy weather shiphandling training could significantly reduce the incidence of mishaps in heavy weather.

As we look to the future, the potential for heavy weather simulator training is extremely encouraging, and this valuable resource should be a standard part of all shiphandling training. Simulators are on the cusp of providing highly realistic heavy weather scenarios, resulting in vastly improved knowledge and effectiveness under the most severe circumstances.

References

Alessi, S. M. (1988). "Fidelity in the Design of Instructional Simulations." Journal of Computer-Based Instruction, 15(2), 40–47.

Allen, J. A. (1986). "Maintenance Training Simulator Fidelity and Individual Difference in Transfer of Training." Human Factors, 28(5), 497–509.
Alman, P. R., Minnick, P. V., Sheinberg, R., Thomas, W. L. III (1999). "Dynamic Capsize Vulnerability: Reducing the Hidden Operational Risk", SNAME Transactions, Society of Naval Architects and Marine Engineers, Vol. 107, New York.
Beck, R. F., and Reed, A. M. (2001). "Modern Computational Methods for Ships in a Seaway," SNAME Transactions, Society of Naval Architects and Marine Engineers, Vol. 109: 1–51, Jersey City, NJ.
Cannon-Bowers, J. A., Burns, J. J., Salas, E., and Pruitt, J. S. (1998). "Advanced Technology in Scenario-Based Training". In Cannon-Bowers, J. A., and Salas. E. (Eds.) Making Decisions Under Stress (pp. 365–374), Washington, D.C.: American Psychological Association.
Foxon, M. (1993). "A Process Approach to the Transfer of Training." Australian Journal of Educational Technology, 9(2), 130–143.
Gross, D. C.; Pace, D., Harmoon, S.; and Tucker, W. (1999). "Why Fidelity?" In the Proceedings of the Spring 1999 Simulation Interoperability Workshop.
Hays, R.T.; and Singer, M.J. (1989). Simulation Fidelity in Training System Design. New York: Springer-Verlag.
Liu, D., Macchiarella, N. D., and Vincenzi, D. A. (2009). "Simulation Fidelity" in D. A. Vincenzi, J. A. Wise, M. Mouloua, and P. A. Hancock (Eds.) Human Factors in Simulation and Training. Boca Raton, FL: CRC Press.
Revised Guidance to the Master for Avoiding Dangerous Situations in Adverse Weather and Sea Conditions. Ref. T1/2.04 (11 January 2007). MSC.1/Circ. 1228.
Thompson, T. N., Carroll, M. B., and Deaton, J. E. (2009). "Justification for Use of Simulation" in D. A. Vincenzi, J. A. Wise, M. Mouloua and P. A. Hancock (Eds.) *Human Factors in Simulation and Training*. Boca Raton, FL: CRC Press.
Zhang, B. (1993). "How to Consider Simulation Fidelity and Validity for an Engineering Simulator." American Institute of Aeronautics and Astronautics, 298–305.

Chapter 50
Onboard Analysis of Ship Stability Based on Time-Varying Autoregressive Modeling Procedure

Daisuke Terada and Akihiko Matsuda

Abstract In this study, it is clarified that a dynamical system of rolling motion can be approximated by a time-varying autoregressive model, which is a kind of statistical model. As the result, when it is possible to measure the time series of rolling motion, a ship's stability can be judged based on time series analysis by using a time-varying autoregressive modeling procedure. As for the verification of this method, the results of the model experiment for parametric roll resonance were used. It was confirmed that an evaluation of ship safety is possible based on the proposed procedure.

Keywords Time-varying autoregressive model · Kalman filter
Time-varying characteristic root

50.1 Introduction

It is important for ship operators to understand the stability of an operating ship. Fundamentally, before a ship reaches the construction stage, its stability characteristics are regulated by various rules of the International Maritime Organization (Umeda 2007) . However, ship stability varies in an actual voyage because of the varying of the loading conditions, climatic conditions navigation routine and so on. Therefore, from the viewpoint of safe operation, we should consider ship stability as a property of a stochastic dynamical system and should directly evaluate onboard measured ship motion.

Ozaki (1986) showed that a time series model, a stochastic dynamical system, and a nonlinear physical process are indirectly related. According to this study, coefficients of the time series model contain information on damping, natural frequency, and parameters of the stochastic dynamical system, which is determined by the phys-

D. Terada (✉)
Mechanical Systems Engineering, National Defense Academy, Yokosuka, Japan
e-mail: dterada@nda.ac.jp

A. Matsuda
National Research Institute of Fisheries Engineering, Hasaki, Japan

V. L. Belenky et al. (eds.), *Contemporary Ideas on Ship Stability*, Fluid Mechanics and Its Applications 119, https://doi.org/10.1007/978-3-030-00516-0_50

ical process. Therefore, if we measure actual ship motion, we can evaluate a ship's stability onboard using an analysis of the time series based on the appropriate modeling procedure (Kitagawa 2010). A class of an autoregressive model is especially useful because only the time series data of the rolling motion is used for the analysis. It should also be noted that a linear stochastic dynamical system can be approximated by an autoregressive moving average model (Bartlett 1946). However, in this approach an external excitation is treated as a white noise sequence. This is not relevant in ship stability evaluation because the main factor in the external excitation is ocean waves that cannot be represented with a white noise. Ozaki's method also uses the white noise assumption. To resolve this issue, an appropriate time series model should be used.

In this study, we first consider stability as a property of a stochastic dynamical system. Then, we show that the stochastic dynamical system can be represented with a time-varying autoregressive (TVAR) model, through noise whitening. Finally, based on the results of model experiments with respect to a parametric roll resonance (Hashimoto et al. 2005) , the validity of the proposed method is demonstrated.

50.2 Time-Varying Autoregressive Modeling Procedure

50.2.1 Discretization of the Second-Order Nonlinear Dynamical Model

In this study, we focus on the nonlinear rolling response of a second-order nonlinear stochastic dynamical (SNSD) system. Assuming external excitation is a random process, the following parametric roll resonance equation (Umeda and Taguchi 2003) is considered as a SNSD model, in the Stratonovich sense (Ozaki 1986):

$$\ddot{x}(t) + a_1\dot{x}(t) + a_2|\dot{x}(t)|\dot{x}(t) + a_3(1 + a_4 \cos a_5 t)x(t) = u(t) \qquad (50.1)$$

Here, $x(t)$ is a roll angle; the notations ($\ddot{}$) and ($\dot{}$) above $x(t)$ represent a differential operator with respect to time; a_1 is a coefficient of roll damping; a_2 is a coefficient of nonlinear roll damping; a_3 is a square of the natural roll frequency; a_4 is a ratio of the amplitude of the parametric roll resonance to the transverse metacentric height (GM); a_5 is an encounter frequency with respect to the waves; and $u(t)$ is a random external excitation. The process $u(t)$ is assumed to have a variance σ^2. However, the assumption of white noise is not required. Equation 50.1 can be also written as the following vector representation:

$$\dot{\boldsymbol{x}}_t = f(\boldsymbol{x}_t) + \boldsymbol{u}_t \qquad (50.2)$$

Here,

$$\boldsymbol{x}_t = (x_1, x_2)^{\mathrm{T}} = (\dot{x}(t), x(t))^{\mathrm{T}},$$

$$f(\boldsymbol{x}_t) = (f_1(\boldsymbol{x}_t), f_2(\boldsymbol{x}_t))^{\mathrm{T}},$$

$$f_1(\boldsymbol{x}_t) = -a_1\dot{x}(t) - a_2|\dot{x}(t)|\dot{x}(t) - a_3(1 + a_4 \cos a_5 t)x(t),$$

$$f_2(\boldsymbol{x}_t) = \dot{x}(t),$$

$$\boldsymbol{u}_t = (u(t), 0)^{\mathrm{T}},$$

and the notation T is the transpose of the vector or matrix. According to Ozaki (1986), by using a sampling interval Δt Eq. 50.2 can be discretized as

$$\mathbf{x}_{n+1} = \mathrm{EXP}(\mathbf{K}_n \Delta t)\mathbf{x}_n + \mathbf{B}_n \mathbf{u}_{n+1} \tag{50.3}$$

where

$$\mathbf{x}_n = (x_1, x_2)^{\mathrm{T}} = (\dot{x}_n, x_n)^{\mathrm{T}},$$

$$\mathbf{K}_n = \frac{1}{\Delta t}\mathrm{LOG}(\mathbf{A}_n),$$

$$\mathbf{A}_n = \mathbf{I} + \mathbf{J}_n^{-1}\{\mathrm{EXP}(\mathbf{J}_n \Delta t) - \mathbf{I}\}\mathbf{F}_n,$$

$$\mathrm{LOG}(\mathbf{A}_n) = \sum_{k=1}^{\infty} \frac{(-1)^k}{k}(\mathbf{A}_n - \mathbf{I})^k,$$

$$\mathrm{EXP}(\mathbf{J}_n \Delta t) = \frac{1}{\mu_1 - \mu_2}\begin{pmatrix} \mu_1 \exp(\mu_1 \Delta t) - \mu_2 \exp(\mu_2 \Delta t) & \mu_1\mu_2\left[\exp(\mu_2 \Delta t) - \exp(\mu_1 \Delta t)\right] \\ \exp(\mu_1 \Delta t) - \exp(\mu_2 \Delta t) & \mu_1 \exp(\mu_2 \Delta t) - \mu_2 \exp(\mu_1 \Delta t) \end{pmatrix},$$

$$\mathbf{J}_n = \begin{pmatrix} -a_1 - a_2|\dot{x}_n| - a_2\frac{\dot{x}_n}{|\dot{x}_n|}\dot{x}_n & -a_3(1 + a_4 \cos a_5 n \Delta t) \\ 1 & 0 \end{pmatrix},$$

$$\mathbf{F}_n = \begin{pmatrix} -c_1 - c_2|\dot{x}_n| & -c_3(1 + c_4 \cos c_5 n\Delta t) \\ 1 & 0 \end{pmatrix},$$

and

$$\mathbf{B}_n = \mathbf{U}\begin{pmatrix} \sqrt{\lambda_1} & 0 \\ 0 & \sqrt{\lambda_2} \end{pmatrix}$$

Here, $\mathbf{u}_n$ is bivariate colored noise with a zero-mean and a finite covariance $\mathbf{I}\sigma^2$, whereas μ_1 and μ_2 are eigenvalues of the matrix $\mathbf{J}_n$, and $\mathbf{U}$ is a unitary matrix that consists of the following elements:

$$u_{11} = \sigma_{12}/\sqrt{\{\sigma_{12}^2 + (\lambda_1 - \sigma_{11})^2\}},$$

$$u_{22} = \sigma_{12}/\sqrt{\{\sigma_{12}^2 + (\lambda_2 - \sigma_{22})^2\}},$$

$$u_{12} = (\lambda_1 - \sigma_{11})/\sqrt{\{\sigma_{12}^2 + (\lambda_1 - \sigma_{11})^2\}},$$

$$u_{21} = -u_{12}.$$

Here, λ_1 and λ_2 are calculated by solving the equation

$$\lambda^2 - (\sigma_{11} + \sigma_{22})\lambda + \sigma_{11}\sigma_{22} - \sigma_{12}^2 \tag{50.4}$$

where

$$\sigma_{11} = \frac{1}{(\alpha_1 - \alpha_2)^2}\Big[\frac{\alpha_1}{2}\{\exp(2\alpha_1\Delta t) - 1\} + \frac{\alpha_2}{2}\{\exp(2\alpha_2\Delta t) - 1\} - \frac{2\alpha_1\alpha_2}{(\alpha_1 + \alpha_2)}\{\exp((\alpha_1 + \alpha_2)\Delta t) - 1\}\Big],$$

$$\sigma_{22} = \frac{1}{(\alpha_1 - \alpha_2)^2}\Big[\frac{1}{2\alpha_1}\{\exp(2\alpha_1\Delta t) - 1\} + \frac{1}{2\alpha_2}\{\exp(2\alpha_2\Delta t) - 1\} - \frac{2}{(\alpha_1 + \alpha_2)}\{\exp((\alpha_1 + \alpha_2)\Delta t) - 1\}\Big],$$

$$\sigma_{12} = \frac{1}{(\alpha_1 - \alpha_2)^2}\Big[\frac{1}{2}\{\exp(2\alpha_1\Delta t) - 1\} + \frac{1}{2}\{\exp(2\alpha_2\Delta t) - 1\} - \{\exp((\alpha_1 + \alpha_2)\Delta t) - 1\}\Big],$$

$$\sigma_{12} = \sigma_{21},$$

and α_1 and α_2 are eigenvalues of the matrix $\mathbf{K}_n$. Note that the above discussion is satisfied by nonlinear models, although $\mathbf{F}_n$ and $\mathbf{J}_n$ change in the considered model.

50.2.2 Bivariate Time-Varying Autoregressive Model

In the discrete model introduced in the previous section, the external excitation is treated as a colored noise. It has to be obtained from a white noise. Following Yamanouchi (1956), the colored noise was approximated by a discrete autoregressive (DAR) process in order to implement the whitening. Therefore, $\boldsymbol{\varepsilon}_n \equiv \mathbf{B}_{n-1}\mathbf{u}_n$, in Eq. 50.3; this term can be approximated by an m-th order DAR process. Then, the colored noise is represented as

$$\boldsymbol{\varepsilon}_n = \sum_{i=1}^{m} \mathbf{D}_i \boldsymbol{\varepsilon}_{n-i} + \mathbf{w}_n,\ \boldsymbol{\varepsilon}_n = \mathbf{w}_n \quad for\ i = 0 \tag{50.5}$$

where $\mathbf{w}_n$ is the Gaussian white noise with a zero-mean and a covariance matrix $\sigma^2\mathbf{I}$, and $\mathbf{I}$ is the 2×2 identity matrix. It should be noted that in this subsection σ^2 is different from the previous subsection. On the other hand, because the relation

$$\begin{aligned}
\boldsymbol{\varepsilon}_n &= \mathbf{x}_n - \mathbf{A}_{n-1}\mathbf{x}_{n-1} \\
\boldsymbol{\varepsilon}_{n-1} &= \mathbf{x}_{n-1} - \mathbf{A}_{n-2}\mathbf{x}_{n-2} \\
&\ \vdots \\
\boldsymbol{\varepsilon}_{n-m} &= \mathbf{x}_{n-m} - \mathbf{A}_{n-m-1}\mathbf{x}_{n-m-1}
\end{aligned} \tag{50.6}$$

we can obtain the following bivariate TVAR (BTVAR) model from Eqs. 50.5 and 50.6:

$$\mathbf{x}_n = \sum_{i=1}^{m+1} \mathbf{C}_i \mathbf{x}_{n-i} + \mathbf{w}_n \tag{50.7}$$

Here, $\mathbf{C}_i$ ($i = 1, \ldots, m+1$) are the TVAR coefficients, which are represented as follows:

$$\mathbf{C}_1 = \mathbf{D}_1 + \mathbf{A}_{n-1},$$

$$\mathbf{C}_2 = \mathbf{D}_2 - \mathbf{D}_1\mathbf{A}_{n-2}, \ldots,$$

$$\mathbf{C}_m = \mathbf{D}_m - \mathbf{D}_{m-1}\mathbf{A}_{n-m},$$

$$\mathbf{C}_{m+1} = -\mathbf{D}_m\mathbf{A}_{n-m-1}.$$

As such, the SNSD system can be approximated by the bivariate TVAR (BTVAR) model. Therefore, when we apply the idea of a stationary autoregressive modeling procedure (Kitagawa and Gersch 1996) to this problem, we can judge the stability of the system at time step n by using the characteristic roots of the following characteristic equation associated with the TVAR operator:

$$\mathbf{C}_n(B) = \mathbf{I} - \sum_{l=1}^{m} \mathbf{C}_l(n) \cdot B^l = 0 \tag{50.8}$$

Here, B is the "time shift operator," which is defined by

$$B^l \cdot x(n) = x(n-l).$$

Note that the order of the TVAR model has changed from $m+1$ to m., If the characteristic roots at time step n, calculated from Eq. 50.8, lie inside the unit circle, then the system is stable. This is an important point, and we can evaluate the stability of the system using the information for the location of the characteristic roots.

It should also be noted that in Eq. 50.7, each element of the TVAR coefficients matrices $\mathbf{C}_i$ $(i=1, \ldots, m+1)$ contains information such as the damping coefficient, natural frequency, and so on. Therefore, it is also possible to estimate them from each element of the TVAR coefficient matrices $\mathbf{C}_i$ $(i=1, \ldots, m+1)$. In the case of the linear stochastic system, it was shown that the damping coefficient and natural frequency can be estimated by this approach (Terada et al. 2016) .

50.2.3 Scalar Time-Varying Autoregressive Model

In the previous subsection, we introduced the relationship between the dynamics of nonlinear rolling motion and the BTVAR model. When it is possible to measure the roll angle and rate, the BTVAR modeling procedure is most effective for judging the stability of a rolling motion. However, it could be difficult to measure these two processes at the same time depending on the measurement device. In this case, the model order need to be adjusted t. For simplicity we consider the following scalar time-varying autoregressive (TVAR) model for the roll angle or rate.

$$y(n) = \sum_{j=1}^{m} c_j(n) \cdot y(n-j) + w(n) \tag{50.9}$$

With respect to Eq. 50.9, the characteristic equation is expressed as follows:

$$c_n(B) = 1 - \sum_{l=1}^{m} c_l(n) \cdot B^l = 0 \tag{50.10}$$

To evaluate the stability of the system, we use the maximum absolute value of the characteristic roots calculated in Eq. 50.10. If the maximum absolute value of the characteristic roots is less than 1, we can regard the dynamical system as stable.

50.2.4 Estimation of TVAR Coefficients

Since these are $m \times N$ autoregressive coefficients in Eq. 50.9, an attempt to fit the parameters by the least squares method (or any other ordinary means) to N observations, $y(1)$, …, $y(N)$, will yield poor estimates. To address this difficulty, it was assumed that the unknown TVAR coefficients were random variables with a Gaussian distribution. In other words, the following stochastically perturbed difference equation constraint model was introduced:

$$\Delta^k c_j(n) = v_j(n) \tag{50.11}$$

where $v_j(n) = [v_1(n), \ldots, v_m(n)]^{\mathrm{T}}$ denotes the m-th order Gaussian white noise process with a zero-mean of covariance matrix $\mathbf{Q}$, which is assumed to be a diagonal matrix with values τ^2, and Δ is the difference operator at time step n, which is defined by

$$\begin{aligned} \Delta c_j(n) &= c_j(n) - c_j(n-1) \\ \Delta^k c_j(n) &= \Delta^{k-1}(\Delta c_j(n)) \end{aligned} \tag{50.12}$$

and k is the order of difference operator, which equals 1 or 2 in this study. In order to efficiently estimate the unknown TVAR coefficients, the following state space representation is introduced regarding Eqs. 50.9 and 50.12:

$$\mathbf{x}(n) = \mathbf{F}\mathbf{x}(n-1) + \mathbf{G}\mathbf{v}(n) \tag{50.13}$$

$$y(n) = \mathbf{H}(n)\mathbf{x}(n) + w(n) \tag{50.14}$$

In Eqs. 50.13 and 50.14, $\mathbf{F}$ and $\mathbf{G}$ are, respectively, the $km \times km$ and the $km \times m$ matrices, $\mathbf{H}(n)$ and $\mathbf{x}(n)$ are the km vectors. These figures can be expressed as

$$\mathbf{x}(n) = \begin{cases} [c_1(n), \cdots, c_m(n)]^{\mathrm{T}} : k = 1 \\ [c_1(n), c_1(n-1), \cdots, c_m(n), c_m(n-1)]^{\mathrm{T}} : k = 2, \end{cases}$$

$$\mathbf{F} = \begin{bmatrix} \mathbf{F}^{(k)} & & \\ & \ddots & \\ & & \mathbf{F}^{(k)} \end{bmatrix} = \mathbf{I}_m \otimes \mathbf{F}^{(k)}, \mathbf{F}^{(1)} = 1, \quad \mathbf{F}^{(2)} = \begin{bmatrix} 2 & -1 \\ 1 & 0 \end{bmatrix}$$

$$\mathbf{G} = \begin{bmatrix} \mathbf{G}^{(k)} & & \\ & \ddots & \\ & & \mathbf{G}^{(k)} \end{bmatrix} = \mathbf{I}_m \otimes \mathbf{G}^{(k)}, \mathbf{G}^{(1)} = 1, \quad \mathbf{G}^{(2)} = \begin{bmatrix} 1 \\ 0 \end{bmatrix}$$

$$\mathbf{H}(n) = (y(n-1), \cdots, y(n-m)) \otimes \mathbf{H}^{(k)}(n),$$

$$\mathbf{H}^{(1)}(n) = 1\,, \; \mathbf{H}^{(2)}(n) = \begin{bmatrix} 1 & 0 \end{bmatrix},$$

$$\mathbf{Q} = \begin{bmatrix} \tau^2 & & \\ & \ddots & \\ & & \tau^2 \end{bmatrix}, R = \sigma^2.$$

where $\mathbf{I}_m$ is the $m \times m$ identity matrix, and $\otimes$ denotes the Kronecker product. It is known that a Kalman filter is effective on a linear state space representation expressed by Eqs. 50.13 and 50.14. When the initial conditions are set as $\mathbf{x}_{0|0}$ and $\mathbf{V}_{0|0}$, the state estimation can be implemented using the following formulas:

Time update:

$$\begin{aligned} \mathbf{x}_{n|n-1} &= \mathbf{F}_n \mathbf{x}_{n-1|n-1} \\ \mathbf{V}_{n|n-1} &= \mathbf{F}_n \mathbf{V}_{n-1|n-1} \mathbf{F}_n^{\mathrm{T}} + \mathbf{G}_n \mathbf{Q}_{n-1|n-1} \mathbf{G}_n^{\mathrm{T}} \end{aligned} \qquad (50.15)$$

Measurement update:

$$\begin{aligned} \mathbf{K}_n &= \mathbf{V}_{n|n-1} \mathbf{H}_n^{\mathrm{T}} \left(\mathbf{H}_n \mathbf{V}_{n|n-1} \mathbf{H}_n^{\mathrm{T}} + R_n \right)^{-1} \\ \mathbf{x}_{n|n} &= \mathbf{x}_{n|n-1} + \mathbf{K}_n \left(y(n) - \mathbf{H}_n \mathbf{x}_{n|n-1} \right) \\ \mathbf{V}_{n|n-1} &= (\mathbf{I} - \mathbf{K}_n \mathbf{H}_n) \mathbf{V}_{n|n-1} \end{aligned} \qquad (50.16)$$

Smoothing:

$$\begin{aligned} \mathbf{A}_n &= \mathbf{V}_{n|n} \mathbf{F}_n^{\mathrm{T}} \mathbf{V}_{n+1|n}^{-1} \\ \mathbf{x}_{n|N} &= \mathbf{x}_{n|n} + \mathbf{A}_n \left(\mathbf{x}_{n+1|N} - \mathbf{x}_{n+1|n} \right) \\ \mathbf{V}_{n|N} &= \mathbf{V}_{n|n} + \mathbf{A}_n \left(\mathbf{V}_{n+1|N} - \mathbf{V}_{n+1|n} \right) \mathbf{A}_n^{\mathrm{T}} \end{aligned} \qquad (50.17)$$

The variance $\sigma^2(n)$ of the observed noise $w(n)$ is assumed constant, then a reduction in the dimensions of parameters can be made, and the state estimation can be efficiently implemented. Furthermore, the optimum value of the model order m can be obtained by minimizing the following Akaike information criterion (Kitagawa and Gersch 1996):

$$\mathrm{AIC}(m) = N\left(\log 2\pi\hat{\sigma}_m^2 + 1\right) + \sum_{n=1}^{N} \log \tilde{d}_{n|n-1} + 2 \tag{50.18}$$

where $\tilde{d}_{n|n-1}$ is the covariance matrix of the conditional distribution of $y(n)$ given the distribution at the time stamp $n-1$.

In this study, to calculate the characteristic roots shown in Eq. 50.10, we used the estimated values from the smoothing process shown in Eq. 50.17.

50.3 Results and Discussion

In this section, we test the proposed procedure with data from the free-running model experiments in irregular waves (Hashimoto et al. 2005) . Two data sets were chosen for analysis: with parametric rolling and without parametric rolling (Figs. 50.1 and 50.2, respectively).

These figures show the measured time series (Figs. 50.1a and 50.2a) and maximum absolute values of the characteristic roots computed by the TVAR procedure (Figs. 50.1b and 50.2b) These maximum absolute values of the characteristic roots are shown with a dotted lines in Figs. 50.1b and 50.2b, while the solid line indicates the radius of the unit circle.

The measured time series shown in Fig. 50.1a, has the amplitude and the phase change with time unlike the time series of regular waves. The maximum absolute values of characteristic roots in Fig. 50.1b are outside of the unit circle from around 50 to 80 s and from 100 to 160 s, at which point the variance of the measured time series becomes large. The stable and unstable states of the system are intermingled in time. Therefore, the dynamic stability of the ship can be regarded as insufficient.

Figure 50.2a shows the maximum roll angle is over 10° so the rolling motion is not small. However, Fig. 50.2b shows that the maximum absolute values of characteristic roots are constant and remain within the unit circle, except for an effect of the initial estimate. Therefore, it can be seen that the system is stable and the dynamic stability of ship is sufficient (Table 50.1).

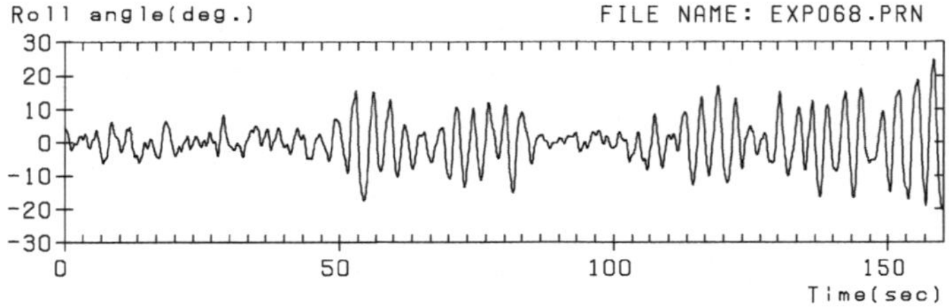

(a) Measured time series

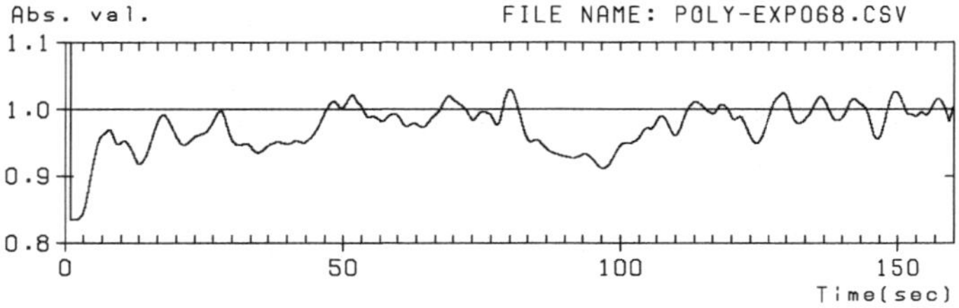

(b) Time series of the maximum absolute values of the characteristic roots

Fig. 50.1 Results in parametric roll resonance

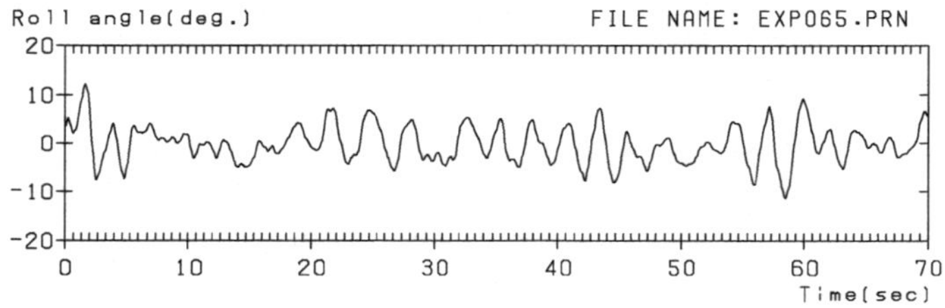

(a) Measured time series

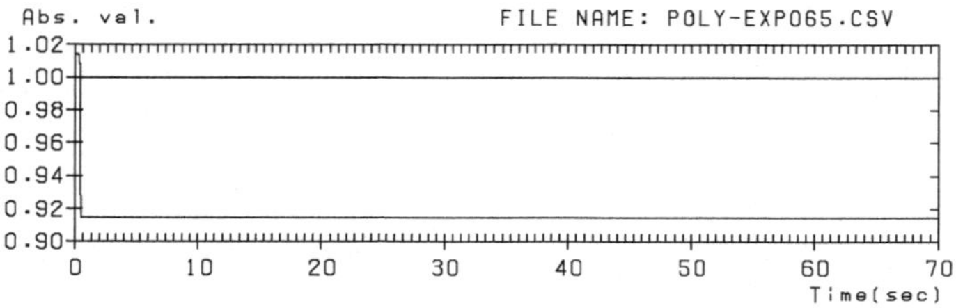

(b) Time series of the maximum absolute values of the characteristic roots

Fig. 50.2 Results without parametric roll resonance

Table 50.1 Principal perpendiculars of a post-Panamax container ship

Items	Ship	Model
Lpp: L	238.8 m	2.838 m
Breadth: B	42.8 m	0.428 m
Depth: D	24.0 m	0.24 m
Mean draft: T	14.0 m	0.14 m
Block coefficient: Cb	0.630	0.630
Metacentric height: GM	1.08 m	0.0106 m
Natural roll period: $T\varphi$	30.3 s	3.20 s

50.4 Conclusions

This study was focused on computing the characteristic roots of the TVAR process and using them to evaluate ship safety in case of large-amplitude rolling motion. The procedure is meant to utilize the measured ship motion data to compute time-varying estimates of the characteristic roots. The effectiveness of the proposed procedure was demonstrated using the data of free-running model experiments. The results can be summarized as follows:

(a) The calculated characteristic roots clearly show different tendencies for the model test cases with and without parametric roll resonance in irregular waves
(b) The proposed TVAR modeling procedure is effective for evaluating stability with a large-amplitude rolling motion.

Therefore, we conclude that the proposed procedure is a powerful tool for the determination of ship stability.

This chapter is a revised and edited version of the paper, originally presented at 12th International Ship Stability Workshop in June 2011.

Acknowledgments This work was supported by a Grant-in-Aid for Young Scientists (B) from the MEXT (No. 21760672). Authors would like to thank Dr. Hirotada HASHIMOTO (Kobe University), Prof. Naoya UMEDA (Osaka University), Prof. Genshiro KITAGAWA (University of Tokyo) and Prof. Emeritus Kohei OHTSU (Tokyo University of Marine Science and Technology) for helpful suggestions. The authors would like to thank Enago (www.enago.jp) for the English language review.

References

Bartlett, M. S. (1946), "On the Theoretical Specification of Sampling Properties of Autocorrelated Time Series," Journal of the Royal Statistical Society, Series B8, pp. 27–41.

Hashimoto, H., Matsuda, A. and Umeda, N. (2005), "Model Experiment on Parametric Roll of a Post-Panamax Container Ship in Short-Crested Irregular Seas (in Japanese)," Conference Proceedings of the Japan Society of Naval Architects and Ocean Engineering, Vol. 1, pp. 71–74.

Kitagawa, G. and Gersch, W. (1996) Smoothness Priors Analysis of Time Series, Springer-Verlag, New York.

Kitagawa, G. (2010) Introduction to Time Series Modeling, Chapman & Hall.

Ozaki, T. (1986), "Local Gaussian Modelling of Stochastic Dynamical Systems in the Analysis of Nonlinear Random Vibrations," in Essays in Time Series and Allied Processes, Festshrift in honour of Prof. E.J. Hannan, Probability Trust.

Terada, D., Hashimoto, H. and Matsuda, A. (2016), "Estimation of Parameters in the Linear Stochastic Dynamical System Driven by Colored Noise Sequence," Proceedings of the ISCIE International Symposium on Stochastic Systems Theory and its Applications, Volume 2016, pp. 125–131.

Umeda, N. and Taguchi, H. (2003), "Parametric Roll Resonance (in Japanese)," Symposium Proceedings of JTTC, pp. 217–235.

Umeda, N. (2007), "Strategy towards Development of Performance-Based Intact Stability Criteria at IMO(in Japanese)," Conference Proceedings of the Japan Society of Naval Architects and Ocean Engineering, Vol. 5E, pp. 19–22.

Yamanouchi, Y. (1956), "On the Analysis of ship's Oscillations as a Time Series (in Japanese)," Journal of Marine Science and Technology, pp. 47–64.

Chapter 51
FLO/FLO Heavy Lift Critical Stability Phases

Paul Handler, Vincent Jarecki and Hendrik Bruhns

Abstract The US Navy has used FLO/FLO heavy lift transport as an alternative to towing for the transport of damaged vessels as well as transport of smaller vessels not suited for ocean transit. There are critical stability considerations that have to be assessed prior to conducting a heavy lift operation, specifically "Draft at Instability" and "Minimum Stability". During de-ballasting of the heavy lift ship with the lifted vessel on the docking blocks, the reaction of the docking blocks on the lifted vessel is effectively the same as removing weight from the lifted vessel's keel (similar to what occurs when a floating dry-dock is de-ballasted). This raises the lifted vessel's centre of gravity thus reducing the lifted vessel's metacentric height (GM) until the lifted vessel's GM is zero. During a critical part of the FLO/FLO heavy lift operation the cargo deck of the FLO/FLO heavy lift ship will be completely submerged. During this phase, only the water-plane of the hull structure which extends above the cargo deck will provide stability for the heavy lift ship. The heavy lift ship will pass through a phase of "Minimum Stability", which should not occur at the same time that the lifted vessel assumes its "Draft at Instability". The lifted vessel and the heavy lift ship may roll out of phase, causing landing problems, or causing the lifted vessel and/or the FLO/FLO heavy lift ship to become unstable, assume a large list or even capsize. Computer naval architectural software programs HECSALV and POSSE are effective tools in modelling critical stability phases for FLO/FLO heavy lift operations. This paper discusses the critical stability phases of a FLO/FLO heavy lift operation, and the methods and practices to plan for and mitigate the effects of reduced stability at these phases.

P. Handler (✉)
Military Sealift Command, Norfolk, VA, USA
e-mail: paul.handler@yahoo.com

V. Jarecki
Naval Sea Systems Command (NAVSEA), Washington, DC, USA
e-mail: vincent.jarecki@navy.mil

H. Bruhns
Herbert-ABS Software Solutions LLC, Alameda, CA, USA
e-mail: hbruhns@herbertsoftware.com

V. L. Belenky et al. (eds.), *Contemporary Ideas on Ship Stability*, Fluid Mechanics and Its Applications 119, https://doi.org/10.1007/978-3-030-00516-0_51

Keywords Heavy lift · Stability phases · Draft at instability · FLO/FLO Minimum stability · Semi-submersible

51.1 Backround

In recent years the US Navy has been relying increasingly on heavy lift transport as an alternative to towing for the transport of damaged vessels as well as transport of smaller vessels not suited for ocean transit. Although there are different types of heavy lift vessels, this paper focuses on FLO/FLO heavy lift transport utilizing semi-submersible ships.

A semi-submersible heavy lift ship is an ocean-going vessel capable of submerging its open deck below the water's surface in order to allow another vessel to be floated over it and, subsequently, landed on docking blocks (which are pre-set on the deck of the heavy lift ship). When the heavy lift ship is subsequently de-ballasted it lifts the other vessel out of the water in a process very similar to the operation of a floating dry-dock. The lifted vessel can then be transported to its destination on the deck of the heavy lift ship. At the destination, the lifted vessel is discharged by reversing the ballasting sequencing process. In other words, it is a "float-on, float-off" or FLO/FLO operation.

In 2000, the *USS Cole* was transported from Yemen aboard *M/V Blue Marlin*, a FLO/FLO heavy lift transport vessel owned and operated by the Dutch firm, Dockwise Shipping BV. Similar operations were done to transport the *USS Samuel B. Roberts* (FFG 58) to the United States following mine damage in 1988, and in 1991/1992 during Operations Desert Shield and Desert Storm. The *Samuel B. Roberts* was transported to the United States on the Dutch-flag FLO/FLO heavy-lift ship, *M/V Mighty Servant 2*. The first US flag FLO/FLO heavy-lift ship was the *M/V American Cormorant*, a vessel reflagged in 1985 from Norwegian flag, and employed with the U.S. Military Sealift Command (MSC) to pre-positioning status with port-opening lighterage and small watercraft equipment that was deployed during Operation Desert Shield. This FLO/FLO heavy-lift ship was converted in 1982 from the oil tanker *M/T Kollbris* to the *M/S Ferncarrier*.

Since the US Navy does not own any FLO/FLO heavy lift ships, it relies on MSC to charter commercial FLO/FLO heavy lift vessels to perform heavy lift operations. In the last 10 years, MSC contracted FLO/FLO heavy-lift transport vessels to move forward-deployed MHCs (coastal mine hunters) and MCMs (mine countermeasure ships) from Ingleside, Texas, to the Persian Gulf and the Far East, as well as to move US Army watercraft to Kuwait.

FLO/FLO heavy lift transport is particularly suitable for transport of mine hunters and mine countermeasure ships. These ships have specially constructed hulls and engines that are not built for ocean service. During the transit from Texas to the Persian Gulf or the Far East, the engines and hulls these ship types would incur such wear and tear as to possibly require extensive dry-docking and repairs upon arrival, and possibly not making their service possible at all. Apart from protecting these ships, carrying them aboard a FLO/FLO heavy lift ship saves a significant amount of time. Whereas it would take about 60 days for a mine hunter to travel to the Persian Gulf on its own power, it takes about 40 days for this transport aboard a heavy lift ship.

Prior to and during a FLO/FLO heavy lift operation, there are critical stability considerations that must to be planned and addressed, specifically "Draft at Instability" and "Minimum Stability".

51.2 Discussion

51.2.1 Draft at Instability

During ballasting/de-ballasting of the FLO/FLO heavy lift ship with the lifted vessel on the docking blocks, the reaction on the docking blocks is equal to the difference in displacement of the lifted vessel at the floating draft (i.e., the lifted vessel's weight) and the displacement at the waterline under consideration in the landed condition. The effect of keel block reaction is a virtual rise in the centre of gravity of the lifted vessel and subsequent reduction in the vessel's GM (metacentric height). The keel blocks are assumed to be knife edge supports and offer no resistance to overturning—no side blocks are assumed.

As the FLO/FLO heavy lift ship is being pumped out (i.e., de-ballasted) or the vertical lift is being raised, the keel block reaction increases. The vertical centre of gravity will eventually rise to a position where the virtual metacentric height (GvM1) at the new waterline (WL1) and the righting arm for small angles of heel will become zero. The ship's draft at this condition of zero GvM1 is called the draft at instability (see Fig. 51.1).

If the draft at instability is reached before the ship has fully landed fore and aft on the keel blocks, then the ship will become unstable. On the other hand, if the draft at instability is not reached until after the ship has fully landed fore and aft on the keel blocks, then the ship remains stable throughout the landing operation.

Generally, if the difference between the draft at full landing and the draft at instability is 1 ft or greater, then the ship has acceptable stability for landing on the keel blocks. If the difference between the draft at landing and the draft at instability is less than one foot but equal to or greater than 6 in., then pre-positioned side blocks will be required. If the difference between the draft at landing and the draft at instability is less than 6 in., then the ship does not have acceptable stability for landing.

If calculations show that the ship will become unstable before landing fore-and-aft, action must be taken to remedy the situation. The first and most obvious remedy would be to remove the trim of a ship to conform to the profile of the docking blocks. A ship in this condition would land fore-and-aft at approximately the same time. If the trim cannot be altered to conform to the profile of the docking blocks, the

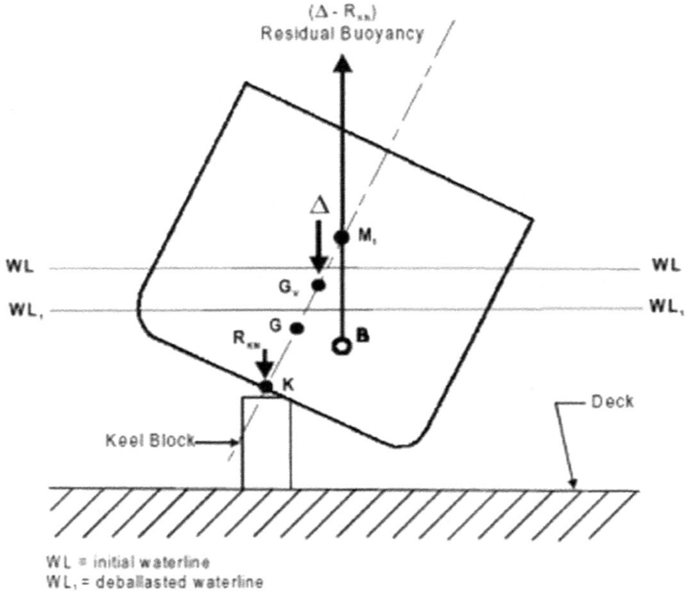

Fig. 51.1 Draft at instability

vertical centre of gravity of the ship must be lowered to improve GM. This may be accomplished by the addition, removal, or relocation of weights, or reduction of free surface on board the ship.

51.2.2 Minimum Stability

During a FLO/FLO heavy lift operation, stability of the FLO/FLO heavy lift ship/lifted vessel system are considered in 5 phases (see Fig. 51.2):

1. FLO/FLO Heavy lift ship at full submergence with the lifted vessel afloat;
2. The lifted vessel makes contact with the keel blocks;
3. The lifted vessel is hard and fully landed on the keel blocks, with the FLO/FLO heavy lift ship trimmed to bring the cargo deck out of the water;
4. With the forward end of the FLO/FLO heavy lift ship cargo deck out of the water, trim is removed to bring the rest of the cargo deck out of the water;
5. FLO/FLO heavy lift ship fully de-ballasted with the lifted vessel secured on the cargo deck.

As the cargo deck of the FLO/FLO heavy lift ship goes into or out of the water, the FLO/FLO ship's stability changes rapidly and substantially. When the cargo deck of the heavy lift ship is completely submerged (i.e., Phases 1, 2, and 3 of Fig. 51.3),

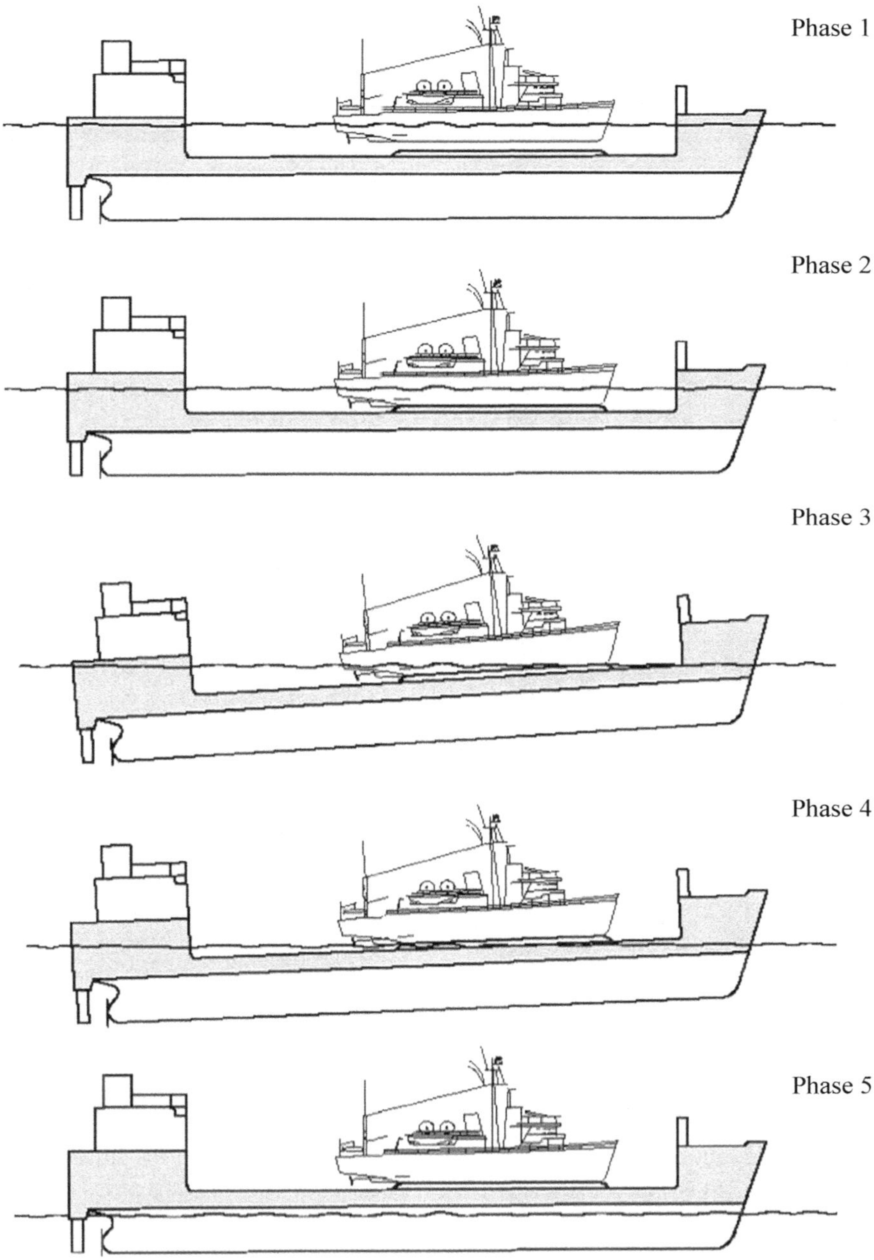

Fig. 51.2 Phases of stability during a FLO/FLO operation

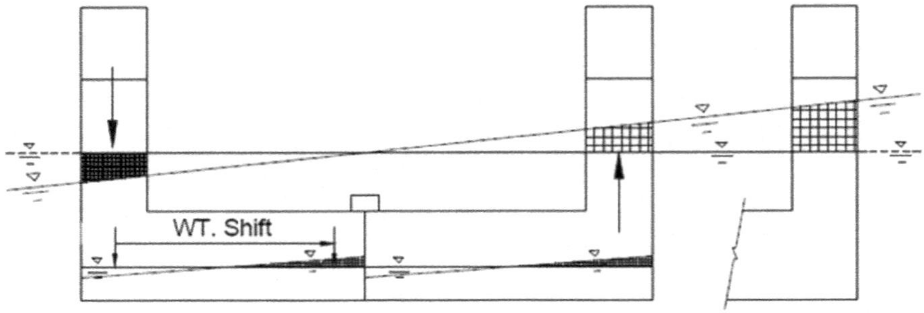

Fig. 51.3 Minimum stability

stability is a function of the water-plane cut by the hull structure which extends above the cargo deck. For a FLO/FLO ship design with wing walls, this can be illustrated as shown in Fig. 51.3.

The wing walls cut the water-plane and provide the stabilizing, buoyant force. As the ship takes on a list, the wing wall on the low side becomes more deeply immersed and a stabilizing buoyant force develops which tries to right the ship. The wing wall on the high side conversely loses buoyancy which also has a stabilizing effect. However, the lower hull and cargo deck stay below the waterline and do not contribute to the righting moment. During this time, the FLO/FLO heavy lift ship will pass through a phase of minimum stability (minimum GM).

Additionally, during this phase, the water levels in the ballast tanks change, which result in a free surface effect to further reduce stability. As the ship lists, the water will level out. A wedge of water will shift from the high side to the low side. This will shift the centre of gravity towards the low side and will tend to increase the list.

To control the effects of reduced stability at the minimum stability phase, it is usually necessary to apply trim and/or list to the heavy lift ship to increase the water-plane area. This practice relies heavily on the master's experience and sometimes trial and error. If too much trim is applied to the FLO/FLO heavy lift ship the lifted vessel can slide off the blocks; not enough trim and the FLO/FLO heavy lift ship can heel over due to insufficient GM.

The phase of minimum stability must be compared to the minimum stability conditions for the lifted vessel. The phase of minimum stability of the FLO/FLO heavy lift ship should not occur at the same time that the lifted vessel assumes its draft at instability. If this happens, the lifted vessel and the FLO/FLO heavy lift ship may roll out of phase, causing landing problems, or causing the lifted vessel and/or the FLO/FLO heavy lift ship to become unstable, assume a large list or even to capsize.

The US Navy requires that during operations involving lifting of U.S. Naval vessels, the FLO/FLO heavy lift ship must maintain a GM (including free surface correction) of no less than 3.28 ft (1 m). The trim of the FLO/FLO heavy lift ship of up to 3° may be accepted to meet the minimum GM.

The 1 m minimum GM may be waived if the lifted vessel is hard and fully landed on the blocks before the phase of minimum stability. However, in no case may the minimum GM (not accounting for the list) be less than 0.5 ft (0.15 m) in all phases of the operation, including the free surface effect.

51.2.3 Modelling with HECSALV or POSSE

The US Navy uses the Program of Ship Salvage Engineering (POSSE), developed by Herbert-ABS Software Solutions LLC. The commercial version is known as HECSALV, and is utilized to model stability, motions, structural loading, and salvage of stranded and/or damaged vessels.

Recently, the US Navy, the Royal Navy, and Herbert-ABS cooperated to expand and refine POSSE/HECSALV to model the entire FLO/FLO operation. This includes stability, structural loading, and motions of the transport vessel and warship, both separately prior to the lift and as a combined system during the lifting and transport operation.

To analyse a FLO/FLO heavy lift operation in POSSE, a full model must be developed for both the heavy lift ship and the lifted ship (see Figs. 51.4 and 51.5). Each model includes all the data necessary to define and analyse the loading of both vessels and includes the full geometry of the hull and tanks defined using offsets, tankage details, strength limits, etc. Once modelled there are two fundamentally different calculation methods available in HECALV/POSSE to analyse a FLO/FLO heavy lift operation. These two approaches include a rigid ship analysis and a flexible ship analysis. For both approaches, tools are provided to define the docking block plan either from an existing docking plan, or develop a new docking plan for special situations such as a damaged vessel.

The rigid ship analysis treats both the lifted vessel and the FLO/FLO ship as rigid bodies and seeks a solution in the form of a classic static equilibrium, in which the weight of the lifted vessel is applied to the FLO/FLO ship through the blocks, which are internally represented as miscellaneous weights The buoyancy of the FLO/FLO ship is instead applied to the lifted vessel through the blocks, represented internally as pliable grounding pinnacles. Using the available time sequence tools in the program, the rigid ship approach can be used to evaluate the detailed stability and strength status of the heavy lift ship and the lifted ship at any step in the sequence so that both the "Draft at Instability" of the lifted vessel and the "Minimum Stability" of the combined lifted vessel/heavy lift ship can be determined. The rigid ship approach adopts an iterative methodology to find the equilibrium solution, where it first analyses the lifted ship assuming a set of dock block locations to get a ground reaction, then applies those loads to the FLO/FLO ship, in which the dock block locations are adjusted to reflect the new FLO/FLO ship drafts and heel. This, in turn, results in the computation of a new set of reactions, then applying those reactions to the FLO/FLO ship, and so on. This process converges when the reaction doesn't change very much. It handles only pure vertical forces

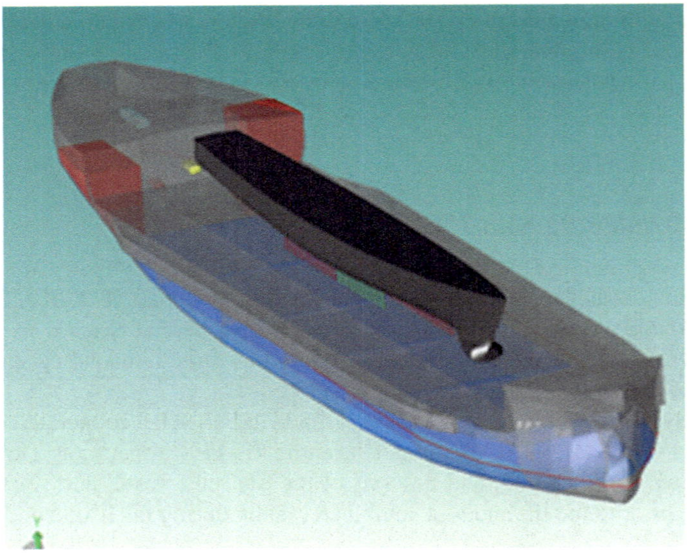

Fig. 51.4 HECSALV 3D graphic

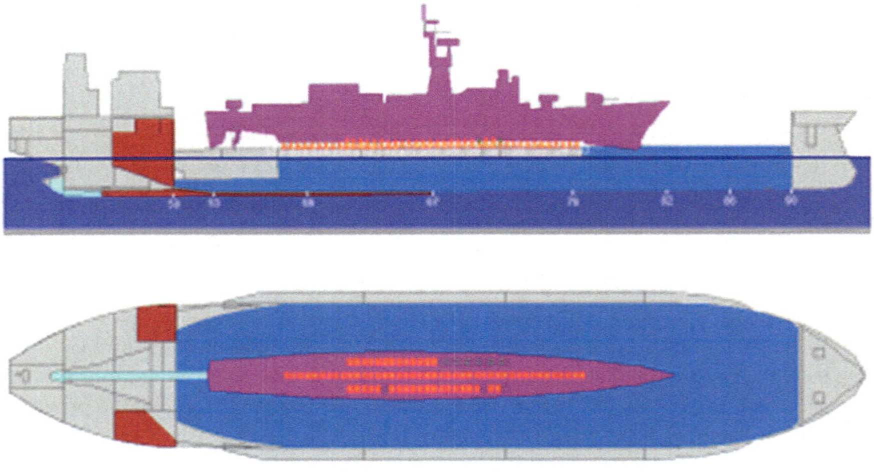

Fig. 51.5 Heavy-lift model

The other method available in HECSALV/POSSE to analyse heavy lift operations is the flexible ship analysis. This analysis uses two beam finite element (FE) models connected by a set of rigid links and a set of unidirectional springs.

The beams are located at the base of the FLO/FLO ship and the height of the centre of gravity of the lifted vessel. This FE model can be viewed for any FLO/FLO-Lifted vessel configuration when the flexible ship analysis is run. This analysis approach can be used for a more detailed analysis of block reactions and strength.

51.3 Conclusions

Prior to and during a FLO/FLO heavy lift operation, there are stability considerations that need to be planned for and addressed. The two critical stability considerations are the draft at instability and minimum stability. The Program of Ship Salvage Engineering (POSSE) developed by Herbert-ABS Software Solutions LLC has recently been expanded to include the capability to model an entire FLO/FLO operation. This expansion includes two fundamentally different calculation methods which are effective tools in modelling the critical stability phases for heavy lift operations, a rigid ship analysis and a flexible ship analysis. Further evaluation and experience will better refine these tools.

Part XIV
Requirements and Operation: Stability of Naval Vessels

Chapter 52
Developing a Shared Vision for Naval Stability Assessment

Douglas Perrault and Steve Marshall

Abstract The design and operating philosophy for naval vessels is driven by necessity of strategic capability, including the capability to deploy at any time to any location. A key enabler for this capability is the hydrodynamic stability of the vessel. The Naval Stability Standards Working Group, in conjunction with the Cooperative Research Navies Dynamic Stability Project, have been working to develop a clear understanding of the limitations and range of applicability of stability criteria and methods of assessing dynamic stability. They have investigated the relationship between GZ and form parameters and the probability of capsize, and have performed studies into the nature of the control variables used and the collinearity of the ship specific assessment parameters. This work will lead to a set of rational criteria for the stability of naval frigates, with a future expansion of the methods to other types of naval platforms.

52.1 Introduction

The design and operating philosophy for naval vessels is driven by the necessity of strategic capability; capabilities that may be called upon at any time and that may require operations in any area of the globe. It is, therefore, not always possible to avoid unfavourable and sometimes even severe environmental conditions. It is incumbent on the ship and crew to be able to contend with the conditions. To this end they must be provided with guidance and training to develop an instinctual

D. Perrault (✉)
Defence Research & Development Canada Atlantic Research Centre, Dartmouth, NS, Canada
e-mail: Doug.Perrault@DRDC-RDDC.GC.Ca

S. Marshall
Naval Authority Group, Bristol, UK
e-mail: Steve.Marshall573@MoD.Gov.UK

V. L. Belenky et al. (eds.), *Contemporary Ideas on Ship Stability*, Fluid Mechanics and Its Applications 119, https://doi.org/10.1007/978-3-030-00516-0_52

understanding of the risks as well as any and all means of minimizing those risks. A clear and complete comprehension of the capability of the platform in its current state is fundamentally imperative for the officers and crew to implement appropriate procedures for minimizing the risk. One of the cornerstones of defining the state of the ship is achieved through design and life-cycle management based in accordance with appropriate standards, including naval stability standards. While stability standards include criteria for both intact and damaged vessels, the discussion herein is limited to intact surface ships.

In examining the stability required of naval vessels, the first step is to make a critical assessment of the current standards, and then address the issues that arise. The critical review is given in the next section, with the following section exploring a methodology to develop rational criteria for naval standards. A summary of parameters that hold the potential of being indicators of stability is given in Sect. 52.4. Some of the results from the ongoing work are presented in Sect. 52.5, followed by a short conclusion and a look at future directions the work is moving into.

52.2 A Critical Review of (Intact) Naval Stability Standards

Currently the stability of naval vessels is assessed using hydrostatic methods, based on concepts that date back over two centuries. The criteria against which such assessments are made have also evolved over a significant period.

One of the most influential developments in naval stability assessment was a study undertaken by Sarchin and Goldberg published in 1962 (Sarchin and Goldberg 1962). This study proposed a standard set of empirically-defined stability criteria for the United States Navy, based upon USN war-time experience that included the capsize of three destroyers in a Pacific typhoon in 1944. This set of criteria has formed the basis for many of the current naval stability standards. The resultant standard, and its derivatives, have served the navies admirably with no known intact losses over the last fifty years. However, Sarchin and Goldberg did anticipate modifications to their criteria with the passage of time and experience, and advocated calculation methods which would, '... embrace the parameters of ship motion, stability on a wave slope ...' (Sarchin and Goldberg 1962). Indeed some navies have subsequently supplemented their standards with additional criteria from other sources. For example, the Royal Navy standard (RN 2011) added enhanced transverse righting arm (GZ) curve shape criteria for angles 0°–30°, 0°–40° and 30°–40°. The ranges of angles are based on the criteria proposed by the Intergovernmental Maritime Consultative Organisation (IMCO, as of 1982 the International Maritime Organization—IMO) in Resolution A.167 (IMCO 1968) and originating from the investigations of Rahola (1939).

Current hydrostatics-based standards attempt to incorporate some consideration of dynamic issues through the application of gust factors as applied to wind heeling levers, the use of roll back angles and area ratios, and in some cases the consideration of the reduction of the righting arm when the vessel is balanced on a wave (Arndt et al. 1982; van Harpen 1970). However, despite such augmentations, the standards remain essentially hydrostatic in nature since they do not directly consider the dynamic behaviour of a ship with forward speed in waves. Research (Ypma and Harmsen 2012) has shown that the classic quasi-static criteria included in today's naval standards do not capture all of the parameters that are important in predicting dynamic capsize, particularly when many modes are the result of non-linear events such as broaching, surf-riding, and dynamic loss of stability. However, the onset of such events is thought to be at least to some extent influenced by characteristics embodied in the transverse righting arm (GZ), and therefore the provision of sufficient hydrostatic transverse stability is an important factor in providing, to some unknown degree, a margin of safety against dynamic capsize.

52.2.1 Loss of Engineering Integrity

It is often assumed that a margin of safety against dynamic capsize is encapsulated, or embodied, within the criteria of the various naval standards, based on the further assumption that the originators of the criteria employed margins (factors of safety) to account for the uncertainty associated with the analysis of their underlying data. Since there have been no known intact losses of warships designed to the Sarchin and Goldberg (1962) criteria and/or its derivatives, it is commonly held that an acceptable level of safety against capsize has been attained through their application. However, the exact rationale behind the determination of the margins of safety is not clear; i.e., there is no evidence to substantiate the assumptions. This is also the case with respect to the detailed information about the fundamental vessel data sets upon which the criteria were based, possibly more so with Rahola than Sarchin and Goldberg. Although the data sets are no longer available, it is known that both of sets of criteria were empirically derived from vessel data, and therefore designs, that pre-dated World War 2 (Rahola 1939; Sarchin and Goldberg 1962).

This loss of traceability to base evidence and lack of alternative evidence leads to dependence on the criteria out of necessity. The perceived low level of capsize risk give some measure of confidence, even if the confidence is not well defined and may even be misplaced in some cases. Additionally, until more recent years the limitations in computational power has precluded more detailed studies into the phenomena. All these factors, taken together, have resulted in a hesitancy to question either the validity or applicability of the criteria; i.e., the existing naval criteria have never been deliberately and methodically validated.

52.2.2 Issues of Application

This situation has resulted in two almost contrary practices with respect to the application of naval standards.

1. Rigid adherence; whereby full compliance to all criteria within a standard is demanded and even extremely small shortfalls against any single criteria cannot be tolerated.
2. Broad application; with the exception of the applicable nominal wind speed used, it is not unusual to see the same standard, and therefore criteria, being applied to a surprisingly broad range of vessels (from harbour tugs to aircraft carriers).

It can only be assumed that these practices are the result of the lack of alternatives and the perception of good service rendered.

52.2.3 Issues of Applicability

The Sarchin and Goldberg (1962) intact criteria were developed based largely on two destroyer designs, the Farragut and Fletcher classes, which pre-date World War 2. It is questionable whether these hull forms exemplify their modern equivalents in the context of stability and dynamics.

Quite radical departures from conventional displacement designs are becoming increasingly common, ranging from the application of tumblehome, deep V and wave piercing bow forms to the inclusion of more hull-integrated watertight superstructure. When more radical hulls are considered, there are concerns that although they meet the current quasi-static stability criteria, they may in reality exhibit poor, if not catastrophic, dynamic stability characteristics when model tests are conducted. The hull form stability issues may be resolved by tapering the stern and increasing the draft of shallow stern sections, thus reducing the height of the center of gravity significantly, and increasing the aft body flare. Such cases highlight the risk and potential limitations of employing traditional criteria to non-traditional forms and the inability of the criteria to reflect these dynamic issues. Therefore, the level of safety attained may not necessarily be the same as that which was assumed delivered through the application of the criteria. In fact, the level of safety provided by the existing criteria is not known.

If we are to continue to apply these traditional static stability measures based on Sarchin and Goldberg (1962) to modern designs, or develop alternatives, we need to address the following questions:

- What probability of capsize have we accepted through the application of current criteria to traditional hull forms?
- What are the limitations of applicability of current criteria with respect to hull form?
- Are there additional characteristics embodied in the transverse righting arm (GZ) curve that may be used?
- Does the application of such additional criteria to the modern form result in a significantly different probability of capsize?
- How should the stability of such vessels be assessed if the current criteria and quasi-static approach are no longer considered applicable?

52.2.4 Through-Life Stability Management

Since the consequences, with respect to capsize risk, of even marginal non-compliance with any criterion are unknown, it is currently the practice to demand total compliance with respect to all applicable stability criteria. This is easily achieved at the start of a warship's life; however, it becomes increasingly difficult later in life due to increasing KG and displacement. Eventually this growth results in failure to meet one or a number of criteria, and, therefore, the placement of restrictions upon the vessel or class. Often restrictions are imposed that limit the use of consumable fluids, in order to maintain the minimum centre of gravity needed for compliance. These restrictions in turn affect operational range and endurance. In more severe cases, limits may be placed on the wind speed and sea state in which the vessel is allowed to operate. Such restrictions are highly undesirable for naval combatants with a responsibility of rapid deployment to any area. Occasionally, highly expensive life-extension programs are required, which at best involve the costly fitting of solid ballast. It should be noted that this practice, as expounded by Sarchin and Goldberg (1962), is often, though not exclusively, brought about by failure to meet damage stability requirements.

52.2.5 Changing Procurement

The safety philosophy of naval shipping differs somewhat from that of their mercantile counterparts. With regard to stability and preservation of life, IMO conventions [e.g., SOLAS (IMO 2009)] are "Being desirous of promoting safety of life at sea … to ensure that, from the point of view of safety of life, a ship is fit for the service for which it is intended"; whereas, in addition safety of life, the safety of the naval ship itself is included in order to safeguard and deliver military capability.

Increasingly, commercial standards are being adopted in favour of defence standards, the rationale being that they offer better value for money. This may be true in many instances, provided that the role and fitness for purpose of such standards are fully compatible with the required naval capability. In some cases, however, commercial standards can be inferior to military standards in the context of achieving naval operations. Adopting a standard that has been developed to reflect a non-military role in a specific global environment and then operating the vessel outside those boundaries may result in a significant difference between the perceived and actual levels of safety.

A more clear understanding of the sensitivity of the various measures as indicators of the risk of capsize, in combination with the determination of the level of safety assured by compliance with stability criteria, will allow the implications of adopting non-military standards in a military context to be fully exposed and comprehended. This should be true whether the criteria originate from naval standards or statutory legislation. Once the implications are comprehended, it will enable a logically based and cost effective rationalisation of the standards needed in modern military procurement.

The duty of care is considered discharged through regulation by certification against standards which must be fit for purpose and fully expose the associated risk(s). Given that the associated level of safety rendered by broad application of current naval standards is unknown and that the standards suffer significantly from a loss of engineering integrity, to assume the key hazard of intact capsize has been adequately addressed based purely on the historical perception of good service can only be considered unconfirmed.

52.2.6 Answering the Challenge

The Co-operative Research Navies (CRNav) Dynamic Stability group was established in 1989 to undertake research into the underlying physical phenomena and characteristics of dynamic stability. The work has led to the development and application of suitable dynamic stability simulation tools in pursuit of this objective. In light of the significant advances made by the group, it was considered that the concerns with current stability standards could now be investigated in more detail.

Since the responsibility for naval stability standards rests with the naval organisations, it was considered appropriate to form a new group lead by the navies. The Naval Stability Standards Working Group (NSSWG) was formed in 1999 from the naval members of the CRNav group, as follows:

- Department of Defence (Australia).
- Department of National Defence (Canada).
- Ministère de la Défense (France).
- Ministerie van Defensie (Netherlands).
- Ministry of Defence, DPA (UK).
- United States Coast Guard Naval Engineering (USA).
- Naval Sea Systems Command (USA).

The naval members are supported by the following guest members from their associated research organisations:

- Australian Maritime College.
- Defence Research and Development Canada.
- Bassin d'Essais des Carènes (France).
- Maritime Research Institute (Netherlands).
- QinetiQ (UK).
- Naval Surface Warfare Center (USA).

The objective of the group is:

'To develop a shared view on the future of naval stability assessment and develop a draft set of stability guidelines which can be utilised by the participating navies at their discretion.'

At a practical level, this involves identification of methods of relating stability criteria to risk. In the short-term, this means identification of the level of safety extant in the current standards, by focusing on the strengths and weaknesses of existing standards, and using a standard set of environmental conditions. In the long term, it means developing methodologies for assessing stability characteristics and practical limits for both design and life-cycle management.

The intent of the work undertaken is to provide advice to designers, life-cycle managers, and operators in the form of criteria, methods, and data that consistently and rationally indicate the risk to the ship. In early design this should be in the form of criteria that are robust and have the limits of applicability clearly defined. In later stages of design, more involved yet still cost-effective methods will establish the inherent stability of the hull. This more detailed stability analysis should also provide the basis for guidance to the operator in the stability booklet. It could also be used to populate a real-time operator guidance system that could be used as a decision aid for the bridge watchkeeper. In conjunction with modern sensing systems, the stability database would identify optimal paths for minimizing the risk to ship and crew. The database and methodology would also be updated throughout the life of the ship, especially for refits and major equipment upgrades/changes.

To answer the challenges, the investigation into stability of intact ships was broken down into three consecutive stages. The first stage is an investigation into the relative merits of static, quasi-static and alternative measures for indicating the risk of extreme dynamic behaviour, and therefore the probability of exceeding a critical roll angle (PECRA) associated with a capsize.[1] The second stage is to assess the level of safety currently accepted through the application of the existing naval standards (RAN 2003; RCN 2012; DM 2001; RN 2011; USN 1975; MN 1999). Originally the third stage was to integrate the effects of the operator (e.g., heading selection and voluntary speed reduction). The NSSWG was requested to provide its knowledge and experience to aid the development of the Naval Ship Code [NSC, also known as ANEP 77 (2014)] developed by NATO and maintained by the International Naval Safety Association (INSA). The third stage was refocused and entered into quickly to establish an engineering solution appropriate for the NSC. Operator influence will be addressed at a later date.

52.3 The Probabilistic Methodology

The work towards rational standards is founded on a statistical analysis of the simulated behaviours of ships. The roots of the work began with the development of a capability to simulate the motion of a frigate in high sea states. FREDYN is a

[1]The probability of capsize is characterized by the probability of exceeding a critical roll angle (PECRA), and although in the present case it is related to capsize, the critical roll angle may also take on a number of other important connotations, such as machinery or weapon limits. Therefore, the methodology can be applied to problems other than capsize.

non-linear, semi-empirical, time-domain software for simulating ship motions in environmental conditions from calm water to severe wind and waves. It allows for studies in stability, seakeeping and manoeuvring. FREDYN is appropriate for any type of a relatively slender mono-hull with a Froude number less than 0.5. Specific to the current study, FREDYN is capable of predicting a range of capsize modes in regular and irregular waves. Only time-domain simulations can capture the transient and non-linear behaviours expected. It is also expected that pursuing this approach will shed light on the physical phenomena involved.

52.3.1 Determination of Probabilities

PECRA is determined by running multiple, time-domain simulations in FREDYN (Ypma and Harmsen 2012) of a ship in a specific loading condition at a mean speed and heading (the operating point of the vessel) in waves of a given significant height and modal period (the environmental condition). The time series of roll responses are used to determine the PECRA. The probability outcomes are later used as the regressands (response variables) in regression analysis investigating relationships with parameters associated with ship stability. PECRA in a given duration is given by:

$$P(\emptyset > \emptyset_{CRITICAL}) = \sum\sum\sum\sum P(V)P(\beta)P(H_S, \tau_P)P(\emptyset > \emptyset_{CRITICAL}|V, \beta, H_S, \tau_P) \tag{52.1}$$

where V is the vessel's speed, β is the vessel's heading relative to the waves, H_S is the significant wave height, τ_P is the peak wave period, and their joint probability density is $P(H_S, \tau_P)$. Each independent variable (X) is discretized into N_X different values. The final probability term, $P(\emptyset > \emptyset_{CRITICAL}|V, \beta, H_S, \tau_P)$, is the conditional probability of exceeding the critical roll angle given a specific combination of speed, heading, and seaway conditions. This term is determined by finding the probability of exceedance for maximum roll angles from multiple FREDYN numerical simulations using a fitted distribution or a distribution-free probability method.

Since 1999, the objectives of the NSSWG have been pursued through the three phases of study for intact ships. Phase 1 (FREDYN version 8.2) used a strip theory approach to look at relationships between the risk of capsize and various stability-related and ship-form parameters. Details of the Phase 1 work can be found in Perrault et al. (2010). Phase 2 (FREDYN version 9.9) used panel methods and the emphasis of the study shifted to looking for the level of safety inherent in the current naval stability standards. In addition to using a panel method for the Froude-Krylov forces, the Frank Close Fit Method (Frank and Salvesen 1970) was implemented to replace a conformal mapping method, the roll damping method was improved, and the ship motion algorithm was upgraded. Phase 3 (FREDYN version 10.2) was conducted after a complete rewrite of the software to modularize the code. The Phase 3 study still used panel methods, but included a more accurate modeling of the effects of deck-edge immersion, as well as an automatic determination of the retardation function time interval and time step. The focus in Phase 3 was narrowed to finding criteria that would be suitable for stability standards, in particular the Naval Ship Code (ANEP 77 2014).

What is common to all three phases is the general approach to determining the probability of exceeding the critical roll angle (PECRA). Simulations were run for each ship in the specific load conditions, at standard operating points and environmental conditions.

52.3.2 Selection of Ships

Frigates are the most common combatant employed by the navies. They are used for a wide range of roles and are generally the smallest warships designed with a world-wide area of operations. To achieve the capability for worldwide deployment they must be able to contend with severe environmental conditions. Their size means they are generally more vulnerable to the effects of severe environments than are larger combatants and auxiliaries which may operate in the same environments. They are also exposed to more weather-related hazards than smaller ships that are operating under weather restrictions which preclude operations beyond specific regions. Because of the number of frigates in use and the number of different roles they are called upon to perform, it is more probable for them to encounter severe weather and risk of extreme motion. Thus frigates were selected as the best candidates for examining their behaviour in extreme conditions.

The core set of ships investigated was largely the same for all three phases, and included slender hulls with twin propellers and one or two rudders. Several different load conditions were explored for each ship, with each load condition delineated by draft (T) and vertical position of the center of gravity (KG). The radii of gyration were held constant for a given ship for all load conditions (TKG). Some of the load conditions were common in two or more phases, but most were not, because the load conditions were chosen to meet the specific objectives of the phase.

Each vessel is of a class that is either currently in service or has seen significant periods of service. The designs can be considered to span at least the last 40 years. Some of the designs pre-date the inception of the Sarchin and Goldberg criteria, but were required to meet the criteria later in life. The majority were designed from the outset to meet either Sarchin and Goldberg or derivatives of that standard.

The hulls were "simplified" in that they are watertight up to the weather deck with no down-flooding points. This is to provide a common baseline condition to make comparison between ships easier. Additionally, the superstructures were not considered to contribute to buoyancy or stability, but were included in calculation of wind heeling moment arms.

52.3.3 Selection of Operating Points

The speeds used were based on agreed typical distribution for frigates. Three speeds were chosen and used for the first 2 phases, while the third phase used only the most common speed. Subsequent work (Perrault 2015) showed that a single speed is inadequate to satisfactorily capture the inherent capability of the ship and the PECRA.

The headings of the ship relative to the uni-directional wave train were 7 headings covering 0°–180°, following the standard assumption that the symmetry of the ship will make the results from 180° to 360° a mirror image of those from 0° to 180°. All headings were considered to be equally likely. The probabilities at 0° and 180° are half those at the other headings, the other half being associated with the 180° to 360° headings. Additionally, the 0° and 180° headings were changed to 1° and 179° to mimic the asymmetry of real vessels (the 0° and 180° headings would not show roll as the numerical model is symmetric).

It is important to note that these operating distributions are independent of any operator action; there are no voluntary heading related speed reductions or heading changes. Therefore the calculated probability of exceeding the critical roll angle should be considered a baseline value reflecting the vessel design, including the influence of the stability standards on the design, and not the added influence of the good seamanship of the operator.

52.3.4 Selection of Environmental Conditions

Intact capsize risk usually involves encountering a critical environment in a manner such that one or more capsize mechanisms are invoked (de Kat et al. 1994; Alman et al. 1999), so the probability of capsize is clearly related to the probability of occurrence of that environment. As discussed in McTaggart and de Kat (2000) a slightly modified version of the Bales North Atlantic Scattergram (Bales et al. 1981) was selected to define the joint probability distribution of the significant wave height (H_S) and peak wave period (τ_P) that characterizes the simulated environment. The scattergram gives the counts of the number of "observations" of each wave height and period over a 1 year period. The data is generated by the Spectral Ocean Wave Model (SOWM, see Bales et al. 1981), which derives the wave characteristics based on wind conditions. Buckley and William (1988) has compared the generated data with data from buoys along the east and west coasts of North America and has noted some discrepancies, but acknowledges the general utility of the SOWM results. This is the same data used in STANAG 4194, Standardized Wave and Wind Environments and Shipboard Reporting of Sea Conditions, in particular to develop the NATO sea state table in appendix D of the STANAG (STANAG 4194 1994).

Figure 52.1 shows the contour plot of (the order of magnitude of) the probability of observing the wave height-period combination according to Bales, as modified by

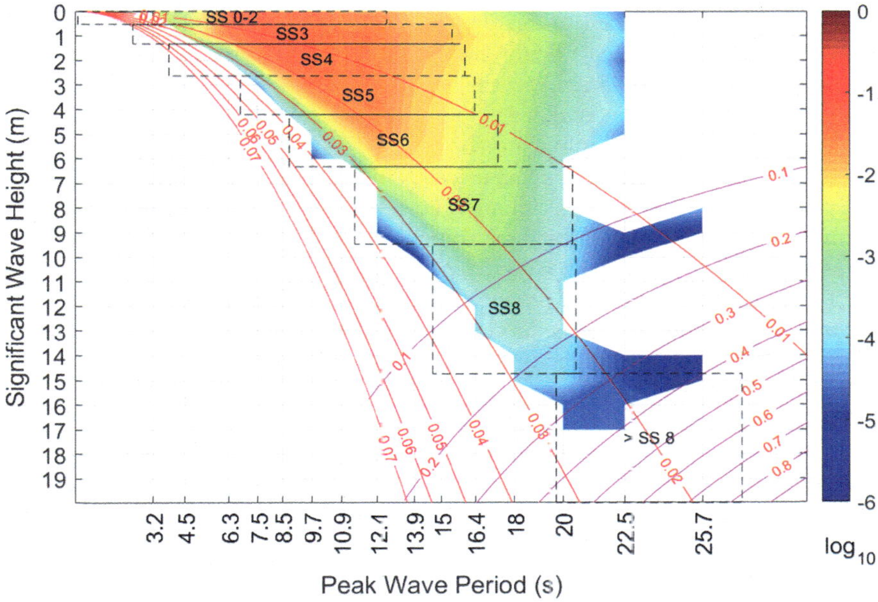

Fig. 52.1 Order of magnitude of probabilities of occurrence in Bales (modified) North Atlantic wave table

McTaggart and de Kat (2000). Wave steepness (significant wave height divided by wave length) is taken into account so that waves that are too steep to exist are not included. This results in the lower left corner being empty; other empty areas are the result of not having data for the specific height-period combination. Overlaid on the plot are wave steepness contours (lines sweeping down from the top left corner). Buckley and William (1988) gives an observed limit equivalent to about 0.05, based on significant wave height and peak wave period from buoy spectral data along the east and west coasts of North America. Note that the Bales data only has valid elements below the limit (Fig. 52.1).

The lines crossing the steepness contours are contours of constant (normalized) energy due to the incident wave. The average energy per unit meter along the wave (perpendicular to the direction of wave travel) is given by:

$$E = \frac{1}{16\pi}\rho g^2 H^2 \tau^2 \tag{52.2}$$

This is clearly a function of the wave height and period only (for a given density of water). The energy is normalized by the highest value, which would be at the largest values of height and period; hence the contours show an increase towards the lower right corner.

The dashed boxes added to this particular figure indicate the NATO STANAG 4194 (STANAG 4194 1994) Sea State definitions for reference. The ends of the boxes are at the 5th and 95th percentile wave periods.

Each wave condition is realized using a Bretschneider (long-crested) wave system, characterized by the significant wave height (H_S) and modal period (τ_P). The spectrum is built by summing regular waves of different amplitudes, wavelengths and phase angles. There are an infinite number of ways to realize the spectrum in a numerical simulation, with each realization accomplished by a different choice of the pseudo-random seed number used to generate the component wave phase angles. Each realization is capable of producing a unique time series of wave conditions, and thus ship responses. This is the key to generating probabilistic results: under the assumption that any one of the unique realizations is equally likely to occur, performing multiple simulation runs (where each run is a unique realization) generates a statistical sample leading to the PECRA.

The use of a long-crest sea without wave spreading reduces the complexity of the problem and is assumed to be conservative as all the wave energy is directed in a single direction. This is expected to result in a more pessimistic estimate of risk (i.e., greater risk).

Since wave conditions are related to the prevailing wind conditions, a simplification was employed that assumed that winds were not only collinear with waves,[2] but could be related to the significant wave height via a linear relationship based on observed conditions at sea:

$$U_W = 1.657H_S + 3.14 \tag{52.3}$$

where U_W is the mean wind speed (m/s) at 10 m above sea level and H_S is the significant wave height. This is a standard method in FREDYN.

No currents were included in the simulations.

The same set of operating points and environmental conditions was used in Phase 2 as in Phase 1, but in Phase 3 there were fewer wave heights and periods and only one ship speed. The same operating points and environmental conditions were used in all cases within a given phase, but the number of seaway realizations was not necessarily the same for each ship, or even for each load condition for the same ship. The number of realizations depended on the quality of the probability result; (small) batches of simulations were added when the uncertainty in the probability result was higher than acceptable. In the first two phases, the uncertainty was calculated based on the particular distribution fit to the results, while in the third phase the uncertainty was calculated as results were added, based on comparing PECRA before and after the added results.

---

[2]Winds generate waves either locally ("seas") or at a distant location ("swell"). Winds lead waves, so it is often true that wind and waves are not collinear. Making them collinear is assumed to be conservative since the influence of each is concentrated in a single direction, likely giving a worst-case scenario.

52.3.5 *Post-simulation Probability Analysis*

For Phases 1 and 2 a block maxima method, called PCAPSIZE (McTaggart and de Kat 2000), was used to determine the probability of exceeding the critical roll angle (which for Phases 1 and 2 was 90°) within one hour.

For Phase 3 an envelope-peaks-over-threshold (EPOT) method called LORELEI (Ypma and Harmsen 2012) was developed to obtain the probability of exceeding the critical roll angle (which for Phase 3 was 70°) within an hour. This method makes fuller use of the time-series data and thus theoretically provides a more accurate value.

The probabilities from either method become the regressands (observed values) in subsequent regression analysis. These observations are keyed to the ship loading conditions.

52.4 Stability Parameters

The other aspect of the problem is to find the regressors for regression analysis: those parameters associated with the ship that are candidates for being key indicators of PECRA. The logical starting point in the investigation is the stability standards and the hydrostatic methods traditionally used to assess stability capability. The seminal paper by Sarchin and Goldberg (1962) formed the basis of, or greatly influenced, the standards of the US and its allies, while the foundational work of Wendel (Arndt et al. 1982) provided the basis for the German and Dutch naval standards (as well as other nations). The former work was based on US experience during World War 2, including the tragic (intact) loss of several vessels during a typhoon in 1944. It works with the Calm-Water (Still-Water) GZ Curve and heeling levers corresponding to winds of up to 100 knots. The latter work additionally applied the concept of balancing the ship on a wave.

A set of parameters was selected to represent the majority of those used to evaluate stability performance in the various naval standards.

52.4.1 *Basic Parameters*

Some of these parameters significantly pre-date Sarchin and Goldberg (1962). As such they have been applied by some naval organizations for a very significant period of time and are the framework upon which such standards as NES109 (RN 2011) were built (see Fig. 52.2; Table 52.1).

Table 52.1 Stability assessment parameters from GZ Curve—angles and lever arms

Parameter	Description	Source
GM	The metacentric height (fluid) for the ship at the given loading condition. Assessed for GZ curves without wind heeling levers only	Bouguer c. 1740
ϕ_{SE}(*phiSE*)	The angle of Static Equilibrium for the ship at the given loading condition, in a particular *balance state*. This angle is typically, but not necessarily, 0° for a ship with no heeling lever (e.g., wind)	
	When a beam wind is applied, it is the angle at which the wind heeling lever arm curve first intersects the *balance state* GZ curve	RN c. 1900 S & G (1962)
ϕ_{VS} (*phiVS*)	The angle of Vanishing Stability for the ship at the given loading condition, in a particular *balance state*	
	When a beam wind is applied, it is still the angle of vanishing stability, but it may occur at the angle where the wind heeling lever arm curve intersects the *balance state* GZ curve a second time, if the intersection is above the GZ = 0 axis	
RPS	Range of positive stability for the ship at the given loading condition, in a particular *balance state*. If there is no down-flooding or other influences, this will be $\phi_{VS} - \phi_{SE}$	RN c. 1900 BV (DM 2001), vH (1970)
RRPS	The residual range of positive stability for the ship at the given loading condition, in a particular *balance state*, with a beam wind applied. (See also ϕ_{VS})	
$\phi_{GZ_{max}}$ (*phiGZmax*)	The angle at which the maximum righting lever arm occurs for the ship at the given loading condition, in a particular *balance state*	RN c. 1900
	The angle at which the maximum residual righting lever arm occurs for the ship at the given loading condition, in a particular *balance state*, with a beam wind applied. The residual righting lever is the righting lever remaining above the wind lever curve	

(continued)

Table 52.1 (continued)

Parameter	Description	Source
GZ_{max} (*GZmax*)	The maximum righting lever arm of the ship at the given loading condition, in a particular *balance state*	RN c. 1900
	The maximum residual righting lever arm of the ship at the given loading condition, in a particular *balance state*, with a beam wind applied	vH (1970)
ϕ_{REF} (*phiREF*)	The reference angle for the ship at the given loading condition, in a particular *balance state*, with a beam: $\phi_{REF} = \begin{cases} 35^\circ & if\ \phi_{SE} \leq 15^\circ \\ 5^\circ + 2 \times \phi_{SE} & otherwise \end{cases}$	BV (DM 2001), vH (1970)
GZ'_{REF} (GZphiREF)	The residual righting lever arm at ϕ_{REF} for the ship at the given loading condition, in a particular *balance state*, with a beam wind	BV (DM 2001)
A_{ratio}	The ratio of areas A_1/A_2 for the ship at the given loading condition, in a particular *balance state*, with a beam wind	S & G (1962)
	A_1: The area under the *balance state GZ* curve, above the *GZ=0* axis and the wind heeling lever arm curve, between ϕ_{SE} and ϕ_{VS} ($A_{\phi_{SE}-\phi_{VS}}$ assuming no down-flooding)	
	A_2: The area above the *balance state GZ* curve, and under the wind heeling lever arm curve, between ϕ_{SE} and the *roll-back* angle, ϕ_{RB}, where the difference, $\phi_{SE} - \phi_{RB}$, is typically 25°	

52.4.2 Sarchin and Goldberg

Other measures were derived from an energy balance approach. These assess the relationship between the shape and area characteristics of the calm water righting curve against an assumed environmentally induced heeling curve. The energy balance assessment parameters selected are given in Fig. 52.3 and Table 52.1. These measures were proposed by Sarchin and Goldberg (1962) and form the core of many of the current naval stability standards [e.g., USN (1975), RN (2011), RAN (2003), RCN (2012), and MN (1999)].

In the original Sarchin and Goldberg (1962) criteria and therefore the US Navy standard (USN 1975), these parameters are related to the application of a beam wind heeling arm as detailed in Table 52.2.

52.4.3 Wendel

A different approach is achieved by employing righting curves that have been determined with the vessel being balanced on a crest or in a trough of a wave of an assumed proportion to the vessel. Figure 52.4; Table 52.1 illustrate the wave adjusted *GZ* assessment parameters selected from those embodied in van Harpen (1970) which form the basis of the Royal Netherlands Navy criteria and are based on BV1030-1 (DM 2001), the German Federal Navy standard, which originates in the work of Wendel (Arndt et al. 1982).

These measures take the effect of waves on the transverse stability into account by calculating the righting arms with the vessel balanced on a sinusoidal wave of a height *H* (m) which is determined according to:

$$H = \frac{\lambda}{10 + 0.05\lambda} \tag{52.4}$$

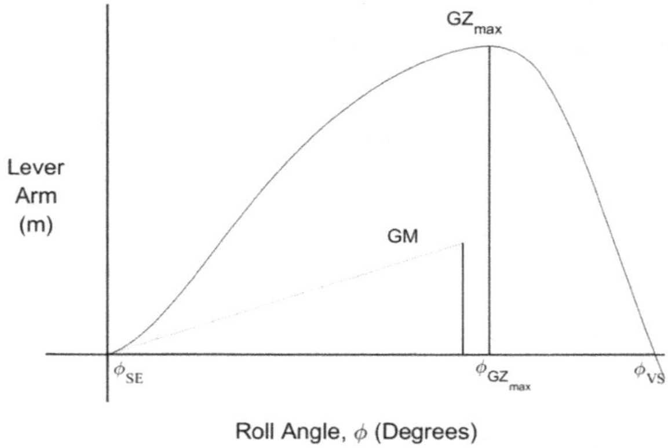

Fig. 52.2 Basic righting arm parameters—fully static angles and lever arms

Fig. 52.3 Illustration of the Sarchin and Goldberg (1962) criteria

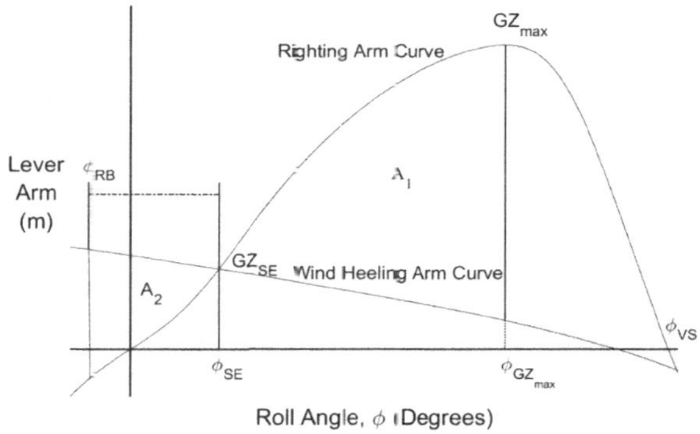

Table 52.2 Heeling terms for energy balance and wave adjusted analysis

Parameter	Definition	Origin	Naval standard
l_w	The wind heeling arm $l_w = \frac{0.0195\, V^2 A_w h \cos^2 \phi}{\Delta \times 1000}$ V = nominal wind speed (kts) A_w = lateral sail area (m^2) h = height of center of area above half draft (m) Δ = displacement (tonnes)	Sarchin and Goldberg (1962)	DDS079 (USN 1975) CFTO (RCN 2012) *DEF (AUST)* (RAN 2003) NES 109 (RN 2011)
K_w	The wind heeling arm $K_w = \frac{p_w A_w h}{\Delta} \times (0.25 + 0.75 \cos^3 \phi)$ A_w = lateral sail area (m^2) h = height of center of area above half draft (m) Δ = displacement (tonnes) $p_w = C_w \frac{\rho_a}{2} V_a^2$ C_w = lateral windage coefficient ($s^2\ m^{-1}$) ρ_a = air density (tonnes m^{-3}) V_a = wind speed (m s^{-1})	Wendel (see Arndt et al. 1982)	vH (1970)
K_v	The free surface heeling arm $K_v = \frac{\sum_{j=1}^{n} \rho_j i_j}{\Delta} \sin \phi$ ρ_j = density of contents of each slack tank (tonnes m^{-3}) i_j = moment of inertia of each free surface (m^4) Δ = displacement (tonnes)	Wendel (see Arndt et al. 1982)	vH (1970)

where the wavelength, λ is set equivalent to the design waterline length of the vessel.

The wave-balanced *GZ* curves are determined for the cases where the vessel is balanced with the crest at mid-length and with the trough at mid-length and also for

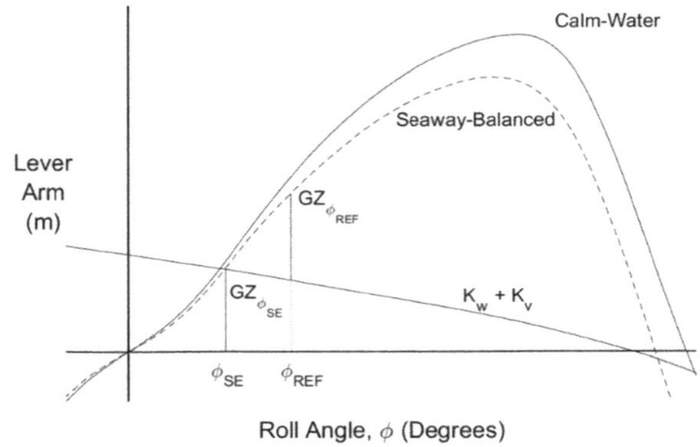

Fig. 52.4 Illustration of the van Harpen (1970) criteria. See also Arndt et al. (1982)

what is termed the seaway-balanced righting arm which is the mean of the former curves:

$$GZ_{seaway} = \frac{GZ_{trough} + GZ_{crest}}{2} \tag{52.5}$$

As part of the van Harpen (1970) criteria, an additional GZ parameter, the residual righting arm, GZ'_{REF}, is determined at a reference angle, ϕ_{REF} (see Fig. 52.4 and Table 52.1).

As applied in van Harpen (1970) and BV1030-1 (DM 2001), these measures are related to the application of a heeling arm that is a combination of the beam wind heeling and a free surface heeling arms, $K_w + K_v$, as detailed in Table 52.2. Note that the beam wind heeling arm, K_w, differs from that used for the Sarchin and Goldberg criteria, in that the former employs a $\cos^3(\cdot)$ relationship and the latter a $\cos^2(\cdot)$. Because the question of how to model the wind is not settled, for the sake of simplicity only the Sarchin and Goldberg beam heeling arm is considered in this investigation.

All standards suggest the use of various wind speeds for different vessels and operational environments. The full set of wind speeds examined herein is: 50, 60, 70, 80, 90, and 100 knots.

52.4.4 Form Parameters

In order to aid the subsequent analysis and allow some degree of discrimination between traditional and more modern hull forms a number of form parameters have also been selected for analysis. These are listed in Table 52.3.

Table 52.3 Form assessment parameters

Parameter	Description
L	Length on waterline (m)
L_{aft}	Length on waterline aft of midship (m)
L_{fwd}	Length on waterline forward of midship (m)
B	Breadth on waterline (m)
T_{Mean}	Mean draft (m)
F_{Mean}	Mean freeboard (m)
A_{MS}	Midship area (m^2)
A_{WP}	Waterplane area (m^2)
A_{WPaft}	Waterplane area aft of midship (m^2)
A_{WPfwd}	Waterplane area forward of midship (m^2)
∇	Volume of displacement in loading condition (m^3)
∇_{aft}	Volume of displacement aft of midship (m^3)
∇_{fwd}	Volume of displacement forward of midship (m^3)
RoB	Reserve of Buoyancy (m^3)
VCB	Vertical center of Buoyancy (m)
LCG	Longitudinal center of gravity (m)
KG	Vertical centre of gravity (fluid) (m)
A_{RR}	Relative rudder area (%)

52.4.5 *Expansion of Parameter Set*

The parameters that are normally used only with a particular GZ curve and wind lever curve were extended for use with all four wave balance curves and all wind conditions, except for GM which was only evaluated for the curves without wind heeling levers applied. Areas between major angles (see Table 52.4) were included in the parameter set. Note that the areas at higher angles do not account for downflooding as this would make comparing results between ships more difficult. The 1st moment of area of the righting arms, with and without the application of the various heeling arms, is included.

Each parameter used from Tables 52.1, 52.2, 52.3 and 52.4 is prefixed by a code (*bwww*) which defines the wave balance and the wind speed used. The first letter designates the wave balance condition and the following three digits define the wind speed applied:

$b \in \{n, c, t, s\}$ corresponding to the balance state $\in$ {*'calm-water' (no wave)*, *'crest-balanced'*, *'trough-balanced'*, *'seaway-balanced'*}.

Table 52.4 Stability assessment parameters from GZ curve—areas under the GZ curve

A_phi1tophi2		The area under the *balance state GZ* curve between two specific roll angles	
		The residual area under the *balance state GZ* curve between two specific roll angles, above the *GZ* = 0 axis and the wind heeling lever arm curve	
M1xA_phi1tophi2		The 1st moment (about the *GZ* = 0 axis) of the area under the *balance state GZ* curve between two specific roll angles	
		The 1st moment (about the GZ = 0 axis) of the residual area under the *balance state GZ* curve between two specific roll angles, above the *GZ*=0 axis and the wind heeling lever arm curve	
M1yA_phi1tophi2		The 1st moment (about the $\phi = 0$ axis) of the area under the *balance state GZ* curve between two specific roll angles	
		The 1st moment (about the $\phi = 0$ axis) of the residual area under the *balance state GZ* curve between two specific roll angles, above the *GZ =0* axis and the wind heeling lever arm curve	
Case 1:	*phi1 =phiSE*	*phi2 =phiVS*	CRN (*calm water*
Case 2:	*phi1 =phiSE*	*phi2 =phiGZmax*	*areas)*
Case 3:	*phi1 =phiGZmax*	*phi2 =phiVS*	BV1030-1 (wave
Case 4:	*phi1=phiSE*	*phi2 =phiREF*	*balance areas*)
Case 5:	*phi1 =phiREF*	*phi2 =phiVS*	

www ∈ {*050, 060, 070, 080, 090, 100*} corresponding to the wind speed ∈ {*50, 60, 70, 80, 90, 100*} knots.

Several functions were developed and used to investigate the calm water *GZ* curve and the wave adjusted curves with and without a wind lever applied. This results in 28 cases altogether for each loading condition of each ship (see Fig. 52.5).

52.5 Results

One study (Perrault 2015) showed that the assessed parameters associated with current stability standards show mixed results as indicators of PECRA. The results of the study indicate reasonable relationships, in many instances, between risk of exceeding the critical angle and those GZ parameters that are employed in current naval standards. This tends to validate the use of these parameters. However, the variation in relative ranking of the parameters for each ship would indicate that few if any of the parameters can be used across all ships.

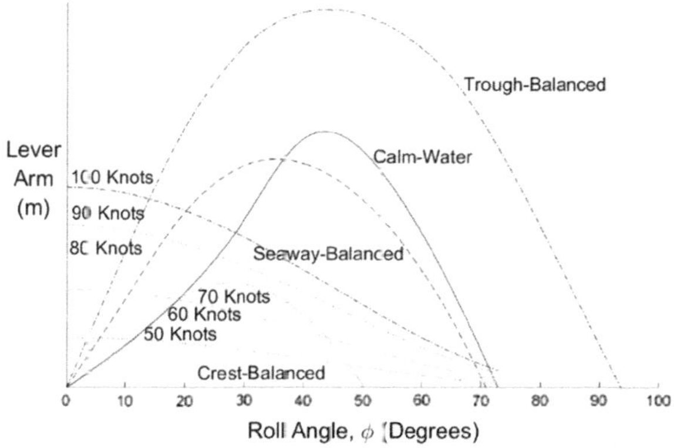

Fig. 52.5 Range of righting arm and wind heeling arm curves

In general, the van Harpen criteria (wave balanced GZ curves) provided stronger correlation than the nominal (no wave balancing) GZ curve parameters.

It should also be noted that the form parameters are less useful than GZ parameters for indicating the risk of extreme motion. This may be because risk of capsize is related to geometry and inertial properties of the ship, and the latter are not reflected in the form parameters.

On an individual parameter basis, many naval standards employ criteria, or measures, that are redundant due to collinearity or superfluous. Additionally, although many standard parameters show high linear correlation with probability of extreme motions, there are parameters not currently used in the standards that have higher correlation.

When the ships are considered as a group, none of the standard parameters has a strong correlation with the probability of exceeding the critical roll angle.

Perrault (2015) has also looked at probability data generated to investigate relationships between the PECRA and ship form and stability parameters. The probabilities were produced for several ships at a number of loading conditions, and for a standard set of operating points (speeds and headings) and environmental conditions (wave heights and periods).

Similar simulations were run for each of three phases which each had a different goal. There are notable differences in the results between the phases, but because of the number of changes in modelling capabilities and choices between the phases, it is not possible to attribute the changes to specific choices.

Within each phase, a careful examination of the probabilities for each ship did not provide any clear patterns related to the typical appendages, or due to geometric parameters, whether expressed in dimensional or non-dimensional form. However, the set of ships used represents a relatively small sample of closely related hull forms with similar features, and it is possible that a larger sample, using more divergent ship types may identify relationships between PECRA and geometry.

When differences between loading conditions for each ship were studied, there was clear evidence of the expected variation due to draft and, more strongly, vertical center of gravity, but these expected variations were not observed in all cases. This suggests greater complexity, and perhaps the influence of other factors. More investigation is warranted.

The study did not investigate the data at the level of each combination of control input variables, because the number of combinations is essentially too large to manually observe. Instead, marginal sums and maxima over operating points and/or environmental conditions provided the basis of analysis. It is possible that there may be some method to examine the large data set, but it is thought that such an investigation would be more suitable when a specific behaviour or anomaly is in view. It was noted that the contours of the order of magnitude of PECRA tended to align with wave steepness, indicating that future work with wave steepness and energy is needed.

The question of how to efficiently and accurately characterize PECRA was addressed by looking at the range and resolution of the input control variables. It was found that the environmental conditions might be reduced in range, but probably need to be increased in resolution. It was also found that the both the range and resolution of the operating points may need to be increased, particularly in terms of the range of speeds. Further investigation is in progress.

A study (Perrault 2016) on collinearity of the GZ and form parameters indicated that GZmax and many other GZ parameters have strong correlations over the set of ships. Parameters associated with the reference angle from the German and Dutch standards showed mixed correlation results (i.e., not robust over the ship set for all wind-wave cases). They were, however, not always available for all wind-wave cases.

The following groups of parameters are suggested as regressors:

- Independent of wave balance or wind speed:
  - Mean freeboard—representative of the group including relative rudder area and reserve of buoyancy.
  - Mean draft—representing the group containing VCB, A_{MS}, A_{WP}, A_{WPaft}, A_{WPfwd}, ∇, ∇_{aft}, and ∇_{fwd}.
  - KG.
- Wind and wave influenced:
  - GZmax—representing most of the other GZ parameters.
- Independent regressors:
  - Parameters associated with the REF angle from the German and Dutch standards. With these it is clear that the wave balance and wind speeds influence the data.

This collinearity study does not show which parameters are useful as indicators of capsize, only that some parameters are redundant in some or all cases of wind speed and wave balance.

From investigations into parameter fitting it was found that the area under the GZ curve to an angle of 70° provided a better single parameter fit than other parameters previously employed in either naval or mercantile standards: the correlation with the PECRA was superior. Following analysis of the Farragut and Fletcher Class in different loading conditions that survived and capsized in the 1944 typhoon, and regression of modern frigate performance, a new GZ area criterion of 0.38 rad m was proposed.

52.6 Conclusions

Since 1999, the members of the Naval Stability Standards Working Group, with full support from the Co-operative Research Navies Dynamic Stability Project, have been working towards a set of rational criteria for the stability of naval frigates. The objective has been to develop a clear understanding of the limitations and range of applicability of the criteria and methods of assessment in order to expand the methodology to other types of naval platforms. They have investigated the relationship between GZ and form parameters and the probability of capsize. They have performed studies into the nature of the control variables used and the collinearity of the ship specific assessment parameters. They have also been successful in providing input to the Naval Ship Code.

52.7 Future Directions

Although the group has made significant progress in advancing a shared vision for naval stability assessment, there is more work to follow. In particular, work is underway to assess the robustness of the stability criterion developed, and to investigate the applicability to other ship types, beginning with smaller ships (<110 m). There is also work ongoing to define the simulation requirements in terms of minimum number of simulations and resolution of operating points and environmental conditions to adequately characterize PECRA.

Work has also begun on applying a similar methodology to stability of damaged ships. FREDYN does have a state-of-the-art flooding module and there have been experiments done to validate the functionality. This validation is currently in progress. The NSSWG is also in the process of applying the lessons learned from the work on intact frigates to the planned investigations into damage stability, with its major multiplication in complexity.

Acknowledgements The work on modernizing and rationalizing naval stability standards has progressed through the dedicated efforts of many naval officers, naval stability authorities, scientists, and technical experts who have participated in or contributed to the work of the Co-operative Research Navies Dynamic Stability Project and the Naval Stability Standards Working Group. These men and women have often performed the work on their own time while keeping up a demanding regular workload. Though they are too many to name, the authors gratefully acknowledge the influence of their work and energy on this chapter.

References

Alman, P.R., Minnick, P.V., Shienburg, R., Thomas, W.L. III, 1999: *Dynamic Capsize Vulnerability: Reducing the Hidden Operational Risk*, SNAME Transactions, Vol. 107.

ANEP-77, 2014: *Naval Ship Code*, North Atlantic Treaty Organization (NATO), Ed. F. Maintained by International Naval Safety Association (INSA).

Arndt, B., Brandl, H., Vogt, K., 1982: *20 Years of Experience: Stability Regulations of the West-German Navy*, Second International Conference on Stability of Ships and Ocean Vehicles, STAB '82, Tokyo, Japan.

Bales, S.L., Lee, W.T., and Voelker, J.M., 1981: *Standardized Wave and Wind Environments for NATO Operational Areas*, David Taylor Naval Ships Research and Development Center, (Report DTNSRDC/SPD-0919-01).

Buckley, William H., 1988: *Extreme and Climatic Wave Spectra for Use in Structural Design of Ships*, Naval Engineers Journal, Vol 100, No 5.

de Kat, J. O., Brouwer, R., McTaggart, K.A., and Thomas, W.L., 1994: *Intact Ship Survivability in Extreme Waves: New Criteria From a Research and Naval Perspective*, Fifth International Conference on Stability of Ships and Ocean Vehicles, STAB '94 Conference, Melbourne, Florida, USA.

Deutsche Marine (DM), 2001: *BV1030-1, Construction Regulations for German Naval Vessels: Stability of Surface Combatants*, Federal Office of Defence Technology and Procurement, Koblenz.

Frank, W., and Salvesen, N., 1970: *The Frank Close-Fit Ship-Motion Computer Program*, Report 3289, Naval Ship Research and Development Center.

IMCO 1968: *Resolution A.167(ES.IV), Recommendation on Intact Stability for Passenger and Cargo Ships Under 100 Metres in Length*, Inter-governmental Maritime Consultative Organization. Superseded by A.749(18), Amended by A.206(VII).

IMO 2009: *International Convention for the Safety of Life at Sea (SOLAS) Consolidated Edition 2009*, International Maritime Organization.

Marine National (MN), 1999: *Instruction Generale N° 6018 A, Stabilité Des Batiments de Surface de la Marine Nationale*.

McTaggart, K.A., de Kat, J. O., 2000: *Capsize Risk of Intact Frigates in Irregular Seas*, SNAME Transactions, Vol 108.

Perrault, D. E., Hughes, T., Marshall, S., 2010: *Naval Ship Stability Guidelines: Developing a Shared Vision for Naval Stability Assessment*, International Journal of Maritime Engineering, Trans RINA, Vol 152, Part A3.

Perrault, D. E., 2015: *Exploration of the Probabilities of Extreme Roll of Naval Vessels*, Twelfth International Conference on Stability of Ships and Ocean Vehicles, STAB '15, Glasgow, Scotland, UK.

Perrault, D. E., 2016: *Correlations of GZ Curve Parameters*, 15th International Ship Stability Workshop, ISSW 2016, Stockholm, Sweden.

Rahola, J., 1939: *The Judging of the Stability of Ships and the Determination of the Minimum Amount of Stability Especially Considering the Vessel Navigating Finish Waters*, PhD Thesis, Technical University of Finland, Helsinki.

Royal Australian Navy (RAN), 2003: *DEF (AUST) 5000, ADF Maritime Material Requirements Set, Volume 3: Hull System Requirements, Part 2: Stability of Surface Ships and Boats*.

Royal Canadian Navy (RCN), 2012: *C-03-001-024/MS-002, Canadian Forces Technical Order: Stability and Buoyancy Requirements Surface Ships*.

Royal Navy (RN), 2011: *DEFSTAN 02-109 Part 1, Issue 2, Stability Criteria for Surface Ships, Part 1 Conventional Ships*, U.K. Ministry of Defence.

Sarchin, T. H., and Goldberg, L. L., 1962: *Stability and Buoyancy Criteria for U.S. Naval Surface Ships*, SNAME Transactions, New York Vol. 70.

STANAG 4194, 1994: *Standardized Wave and Wind Environments and Shipboard Reporting of Sea Conditions, Edition 2*, North Atlantic Treaty Organization (NATO).

United States Navy (USN), 1975: *Naval Ship Engineering Center, Design Data Sheet - Stability and Buoyancy of U.S. Naval Surface Ships*, DDS 079-1, U.S. Navy, Naval Sea Systems Command, Washington DC.

van Harpen, N.T., 1970: *Eisen te stellen aan de stabiliteit en het reserve drijfvermogen van bovenwaterschepen der Koninklijke Marine en het Loodswezen*, Ministrie van Defensie, Marin, Hoofdafdeling Materieel, Bureau Scheepsbouw, Den Haag, Nederland, Rapport 21183/21021/SB.

Ypma, E., Harmsen, E., 2012: *Development of a New Methodology to Predict the Capsize Risk of Ships*, 11th International Conference on the Stability of Ships and Ocean Vehicles, Athens, Greece.

Chapter 53
Approaches for Evaluating Dynamic Stability in Design

Philip R. Alman

Abstract There are many ways of treating dynamic stability. No single approach is always best, but must be defined relative to each design and each yields a fidelity proportionate to resources and technological maturity. During the ship design process choices must be made that balance the approach within a wide trade space encompassing ship design characteristics, operational doctrine, technical risk management, operational safety, cost and schedule. Existing static approaches do not directly account for ship dynamics. There is a clear need to develop a framework for integration of technical approaches into the ship design/acquisition process. The objective of this paper is to define a basis for outlining the range of intact dynamic stability methodologies that can be employed to naval ship design that address dynamic stability in such a way as to minimize technical and safety risks in an economical manner. The paper summarizes ongoing work by the Naval Stability Standards Working Group (NSSWG), and outlines relevant technical approaches suitable for employment on naval ship designs from preliminary/concept design stages through to operator guidance.

Keywords Dynamic stability · Risk management
Naval stability standards working group · Static stability · Probabilistic
Empirical · Criteria

The opinions expressed in this paper are those of the author and not necessarily those of the Naval Sea Systems Command or the United States Navy.

P. R. Alman (✉)
The Naval Stability Standards Working Group, Naval Sea Systems Command (NAVSEA), Washington DC, USA
e-mail: jackstay5@aol.com

V. L. Belenky et al. (eds.), *Contemporary Ideas on Ship Stability*, Fluid Mechanics and Its Applications 119, https://doi.org/10.1007/978-3-030-00516-0_53

53.1 Background

There is no single approach that is best for addressing dynamic stability as part of a ship design effort. Many factors encompassing design characteristics, technical maturity, methodology, resources, cost, and safety must be balanced to find the most appropriate treatment. Risk management techniques are well suited to defining the most cost-effective approach for treating dynamic stability in the design process.

The Naval Stability Standards Working Group (NSSWG) has worked to define these issues over a number of years. The NSSWG has representatives from Canada, Great Britain, Australia, France, United States, and the Netherlands. The development of specific methodologies addressing dynamic stability has been in the work plan for that group since its inception. As efforts have progressed, it has become increasingly clear that a wide range of approaches would have to be defined to meet all the requirements of every Navy.

Historically, dynamic stability has been represented by static measures including GZ area margins, and variation of GZ on prescribed waves, and other empirical rules. This approach is relatively simple and the least onerous for cost and schedule. Treatment of dynamic stability based on vessel dynamic response is still in the research and development stages. Even so, there are many approaches that can yield useful information, but no means to knit them into a coherent process. Thus there is a clear need to develop a framework for integration of intact dynamic stability assessment into the ship design/acquisition process.

53.2 NSSWG Definitions for Intact Dynamic Stability

There are three principle factors affecting dynamic stability:

1. The static restoring moment
2. The dynamic response (including damping and added moment of inertia)
3. The hydrodynamic forces on the vessel from waves/wind.

Estimating and understanding these three factors and their relationship to stability failure modes, and developing appropriate safety margins governing allowable KG and Displacement for the ship design forms the basis for risk control in the acquisition process.

The Naval Stability Standards Working Group (NSSWG) uses the categories below as the basis for stability discussion.

Static Capsize—A static capsize may occur suddenly when a disturbance is encountered that is sufficient to overcome the ship's inherent ability to remain in an equilibrium state at or near upright. The event has traditionally been characterized by parameters which relate to a reduction in the righting arm lever (or GZ curve) which represents the static stability of a vessel independent of forward speed and time. Conditions that could lead to static capsize include improper loading, lifting or topside

icing (increasing VCG); towing, wind, or load shift, (increasing heel angle); trapped fluids on deck (increasing free surface effects); and loss of watertight integrity (loss of buoyancy/water plane area).
Dynamic Capsize—A Dynamic Capsize is defined as a very large amplitude roll caused principally by seaway and wind excitation on a moving vessel or as a function of time. This wind and wave action may lead to equipment damage, personnel injury, loss of system functionality and/or weather-tight/watertight integrity from which the ship is unable to maintain its intact upright state. A dynamic capsize is characterized as a time-dependent event occurring in unrestrained 6 degrees of freedom motion. The loss of dynamic stability may occur under a variety of conditions (intact or damaged) once the forcing function exceeds the available restoring force.

Large Amplitude Motions—Large amplitude motions are a part of dynamic stability considerations and include large roll angles, "knock downs," yaw, lateral accelerations, pitch, etc. These motions are caused by the dynamics of the vessel as it is excited by wind and seaway. Large amplitude motions in the non-linear range tend to be in the range of roll angles where the GZ curve is softening but still able to provide sufficient restoring force to resist capsize. Dynamic capsize occurs once roll has reached an extreme point on the GZ Curve, and restoring force can no longer bring the ship back to an upright position.

Static Stability Standards and Practice

Navies assess stability using static methodologies. Existing stability criteria are a composite based around compliance with specific safety elements. In the case of the Design Data Sheet (DDS)-079-1 these are the following:

Principal Safety Elements in DDS-079 Criteria

Intact Ship

- Beam Winds Combined With Rolling
- Lifting of Heavy Weights
- Crowding of Passengers to One Side
- High Speed Turning
- Topside Icing.

Damaged Ship

- Stranding Involving Moderate Flooding
- Bow Collision
- Battle Damage Involving Extensive Flooding.

Flooded Ship

- Beam Winds Combined With Rolling
- Progressive Flooding.

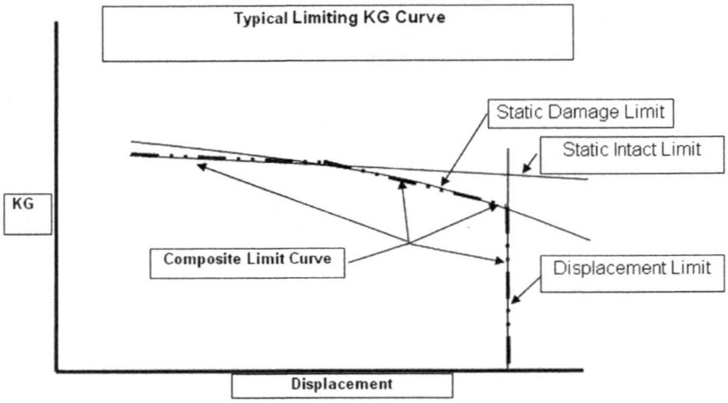

Fig. 53.1 Typical limiting KG curve and components

Each of the safety elements listed above is defined through various criteria. Naval ships must comply with the most restrictive limit resulting from the application of several criteria such as beam wind, passenger crowding, icing, high speed turning, and damage stability (Alman et al. 1999).

In general a range of loading conditions is bounded by the envelope established by the governing limits. This limit becomes a composite curve as shown in the Fig. 53.1. An acceptable loading condition is one which the KG is below the limiting curve.

Historically static stability criteria do not directly address dynamic stability and large amplitude motion; although it is generally acknowledged that the margin of safety for seaway motions is included as the so called A1/A2 area ratio and roll back angle. The historical record supports the adequacy of this approach. However, the adequacy of such factors of safety using static methods may not be adequate when applied to hull forms with novel features. Consequently there is a need to integrate dynamic stability methodologies into the criteria stability criteria.

Intact Dynamic Stability Assessment Methodologies

There are many ways to categorize dynamic stability assessment methodologies, the definitions of which are still under discussion. Discussion of these methodologies is best handled in the context of a risk management process.

The starting point is to form a lexicon by which everybody involved in the risk management process can talk from the same common understanding.

One example is provided in Belenky et al. (2008). Four basic approaches were described which can be summarized as: probabilistic performance-based criterion, deterministic performance-based criterion, probabilistic parametric criterion, and deterministic parametric criterion.

Within the NSSWG, ongoing efforts have been based around a categorization of dynamic stability methodologies as "Empirically Based Rules," "Rules Based on Probabilistic Dynamic Approaches," "Direct Probabilistic Assessment" and "Relative Probabilistic Based Assessment". Although these don't agree exactly with those of Belenky et al. (2008), they are complimentary and generally convey the similar concepts based on naval ship stability practices. The NSSWG categorizations are defined as follows:

1. *Empirically Based Rules*—Development of criteria based on a set of "rules" established from a study of hull form characteristics using engineering principles based on evaluation of design characteristics such as the GZ curve. A suitable body of ships is assessed to form the basis for establishing criteria. The resultant criteria are typically binary and expressed as "pass/fail" and will have factors of safety to account for physical properties which can not be fully modeled. Typically static stability criteria fall into this group. This rules-based methodology is largely based on heuristics—experience with previous designs. It may not be readily applicable to evolutionary or novel designs.
2. *Rules Based on Probabilistic Dynamic Approaches*—A probabilistic study for a series of ship types is used as the basis to determine suitable design characteristics to be used as part of dynamic stability criteria. Design characteristics are identified as being the most closely correlated to capsize probability for the type of ship assessed. A suitable criterion is then derived for the design characteristics identified which provides a reasonable mitigation of capsize risk. The NSSWG has been actively developing this approach as reported in Perrault et al. (2010), see also Chap. 52.
3. *Direct Probabilistic Assessment*—Direct determination of a capsize probability for seaway environments using a validated simulation tool and/or a series of model tests. The resultant capsize probability is assessed as acceptable or unacceptable based on some risk level established for specific seaway operations or for lifetime risk. Some risk comparison can be made using tools such as Farmer's curves (Ayyub 2003) to establish acceptable risk levels in comparison to other occupation or modes of transportation. In Peters (2010) a discussion is provided on approaches to establish acceptable risk levels for naval frigates see also Chap. 54. The authors conclude that an acceptable risk of capsize for a naval frigate on an annual basis could be approximately 1/10,000.
4. *Relative Probabilistic Assessment*—A probability index is established based on comparisons of the design ship capsize probability to a known baseline ship operating in identical conditions. The resultant probability index is assessed as acceptable or unacceptable based on a relative measure against the baseline. A probability index must be developed for the baseline ship as part of the comparison. The assessment is done for the baseline ship when in compliance with an existing static criteria. The index must not only have the baseline determined by the existing ship, but must have a rationally derived scale in order to provide meaningful comparisons between the existing ship and the design ship. Note that the baseline ship will have been assessed by one of the above methods by necessity.

The Intact Dynamic Stability "Tool Kit"

The categories defined above provide the building blocks from which integration of dynamic stability into the design process can begin. The integration is centered on developing a measure of the risks associated with the proposed hull form, definition of the tools available, their fidelity and the resources necessary to use them. Thus a "Tool Kit" of technical approaches is developed. Each tool in the kit has a fidelity and cost associated with its application.

The dynamic stability risk characterization of the hull form should be made through a set of measures. The characterization can be made qualitatively at initial stages but should move into development of quantitative (e.g., probabilistic) measures as the design develops. These risk measures can be broadly characterized as follows:

- Heuristic/Historical Experience (Qualitative)
- Early design assessment/rules of thumb developed from simple design parameters (Qualitative/Quantitative)
- Simulation and/or Test Data (Quantitative).

Determination of the appropriate approach might be accomplished in the context of a risk assessment. The "tool kit" represents the means by which hazards and consequences can be quantified and managed.

For example, the use of vulnerability criteria as proposed by Bassler et al. (2011) very good starting point establishing both the early stage risk and mitigation through the Level 1 and Level 2 vulnerability criteria.

The risk characterization should be revisited several times as the design matures.

Measures for risk mitigation must also be considered along with the risk. In a formal sense risk may be thought of as fitting into the following, Ayyub (2003).

- Risk Reduction or Elimination
- Risk Transfer
- Risk Avoidance
- Risk Absorbance.

For dynamic stability, some of the most prominent mitigation measures can be generally though of as follows:

- Criteria (Risk Reduction)
- Operational Restrictions (Risk Avoidance)
- Operator Guidance (Risk Avoidance)
- Training (Risk Avoidance).

The addition of training and operator guidance specifically to reduce or avoid a dynamic stability risk is an attractive option. In general operator guidance can be as follows:

- Simple rules of thumb compiled from historical experience/data

- Operator guidance based on dynamic stability assessment to produce either polar plots and/or rules of thumb based on specific loading conditions, speeds, headings and environmental conditions, and vulnerabilities.
- Training involving real time simulation and classroom lectures.

Each has an associated cost, fidelity and effectiveness.

53.3 Process for Dynamic Stability Risk Characterization

In broad terms there are several types of risk. Also interrelated are the risks associated with technological maturity and programmatic costs.

Early in a design it may not be possible to develop a quantitative risk assessment for dynamic stability due to a lack of available data. Decisions may have to be made based on judgment, past experience and historical evidence. For some designs this may be sufficient and the process can end there with the application of static criteria. More radical hull form designs may have to be approached with the object of developing a quantitative risk assessment.

The quantitative risk assessment should consider several factors some of which are outlined below.

1. Dynamic Stability Risk Inherent in the Hull Form
    a. *Quantification of Risk*
        i. *Data*
        ii. *Availability and Reliability of Data*
        iii. *Historical Experience*
    b. *Maturity of Technology*
        i. *'Measures' of Risk; i.e., Criteria*
        ii. *Fidelity of Risk Assessment*
    c. *Resource Requirement*
        i. *Cost of R&D*
        ii. *Cost of Implementing*
2. Measures for Risk Mitigation
    a. *Criteria*
    b. *Operator Guidance*
    c. *Operational Limits*
    d. *Design Changes.*

A technical risk assessment team should be established. This team would be composed of a group of subject matter experts covering stability, seakeeping, analytical tools, model testing and ship handling.

The team starts by attempting to quantify the technical risk associated with the hull form. The risk is estimated based on availability of data; historical, analytical or model testing. Lack of available data ("Unknown") for an informed judgment could

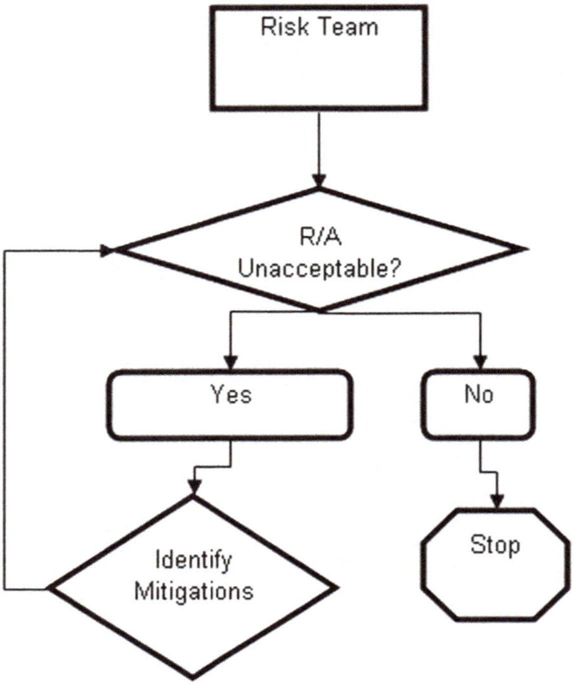

Fig. 53.2 Dynamic stability risk assessment process

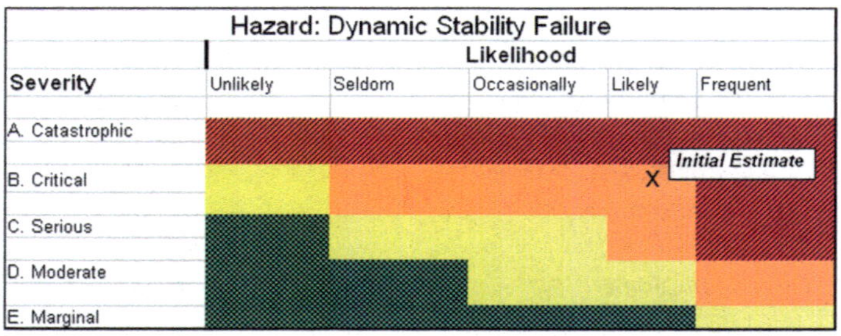

Fig. 53.3 Initial hull form risk assessment

make the risk high. Other sources of data and their fidelity are evaluated accordingly. Mitigations are also identified. The process is iterated until the risk is considered to be in an acceptable range (Fig. 53.2).

The results of an initial risk estimate for a hull form "A" might look like Fig. 53.3. In this case a review of available data suggests that there is a "likelihood" of a "critical" dynamic stability failure in a specified operational time frame.

Table 53.1 Notional tool fidelity ranking

Hull Form A Notional tool fidelity for risk estimation		
Method	Effectiveness cost	
1	Limited	Less
2	Medium	Moderate-high
3	High	High
4	Very high	Very high

1—Heuristics/Historical Studies
2—Simulation Based Methods
3—Systematic Regular Wave Tests
4—Extensive Random Wave Testing

It is also important to consider available technology and its fidelity or 'maturity' as part of this process. The available "tools" may be categorized as follows:

- Heuristics/historical studies
- Simulation-based methods
- Systematic Model testing in regular, unidirectional waves to develop an index
- Direct results of (extensive) model testing in irregular, multidirectional waves.

The team must answer the question "how much do I believe the data and what is the cost impact"? Table 53.1 illustrates how a series of methodologies or "tools" might be ranked for fidelity and cost in developing the risk of dynamic stability failure. Actual metrics would have to be developed for a ranking process.

The process of developing the data required to assess the dynamic stability risk may require a considerable outlay of resources involving personnel and lead time and funding. This should be assessed early on in the design when it is still possible to make hull form changes. The cost of developing the required methodology to refine the risk estimate needs to be addressed and balanced against the benefit of the hull form.

Mitigations should be defined and addressed immediately. The mitigations are defined such that the severity and probability of the risk may be controlled or reduced. The mitigations are also developed based on an understanding of the nature and the magnitude of the assessed risk for the hull form.

In many cases the outcome should simply be a validation of existing practice. For instance an assessment of a conventional hull form 'should' confirm the adequacy of existing stability techniques in managing the risk. In other cases, the risk assessment should serve as a warning flag of potential dynamic stability problems and provide a basis from which to develop an outline of the technical and programmatic challenges associated with addressing dynamic stability for the proposed design. Cost benefit analysis should be developed for the decision process.

Specific risk management techniques for ranking dynamic stability methods and mitigations should be developed according to the needs of the Navy or organization conducting the assessment. There are many references covering application of specific risk management 'tools'. A good example of the application of risk management to submarine weight engineering is provided by Tellet et al. (2007). Similar approaches could be adapted to dynamic stability risk management.

53.4 Example Approaches for Defining Dynamic Stability Risk Mitigation

1. *Early design assessment/rules of thumb developed from simple design parameters*—This approach uses simple design parameters resulting from studies of static stability characteristics on waves, or model test/simulation data using one of the criteria-based approaches. Results may include rules of thumb for distribution of waterplane area, vertical prismatic coefficient, specifications for righting energy and minimum positive GZ. The results are used for guidance during design but not as specific criteria to set the displacement/KG curve. The displacement/KG curve is developed based solely on compliance with unmodified intact static/damage static criteria in the traditional manner. This approach is fairly easy to implement providing sufficient studies have been conducted to provide a basis for the rules of thumb. While it can provide design guidance, these approaches are most useful in highlighting design characteristics which may be problematic from a dynamic stability perspective and will require more rigorous investigation. An example of the structure of such an approach can be found in Belenky et al. (2009).
2. *Integrate dynamic stability into existing stability criteria to produce a unique dynamic stability limit or modified static stability limit*—In this approach dynamic stability becomes one of the safety elements in the existing criteria. This results in a more formalized process. Consequently some strategy to augment existing criteria must be found by identifying the safety element associated most closely associated with dynamic stability. That safety element can be modified by one of the four methodologies defined above to address dynamic stability. This then produces a new dynamic stability limit as a function of mass properties and KG. This new limit is used in combination with the intact, damage and other limits to set the displacement/KG limit for the operation of the ship.
   It is interesting to note that the watertight/weather tight boundaries used for static stability assessments may not directly coincide with the weather deck of the ship. This can make integration of dynamic stability/static stability limits problematic as the buoyant volume and restoring force and wave forcing used in large amplitude motions may not match that of the static criteria limits.
   In the modified criteria, mass properties are maintained within the resulting envelope throughout service life as shown in Fig. 53.4.
   The complexity of the criteria in both definition and implementation is directly related to the methodology. Criteria-based approaches using design parameters,

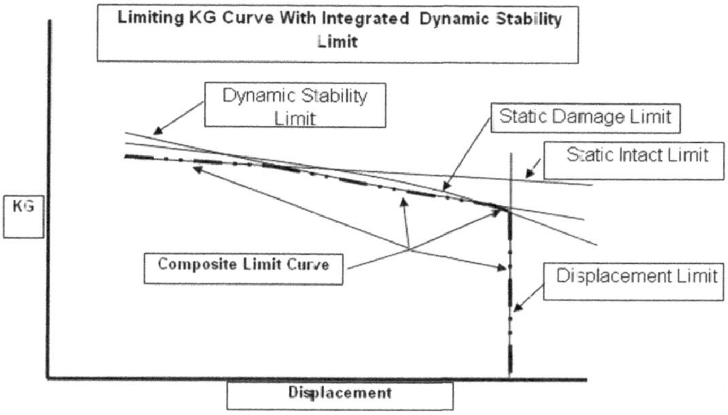

Fig. 53.4 Typical limiting KG curve with integrated dynamic stability limit

and GZ curve assessment techniques are more readily implemented and socialized throughout the design community, although they may not provide sufficient flexibility to address designs outside of the data base from which they were developed.

Novel hull forms will rely more heavily on relative probabilistic and direct probabilistic approaches as they are likely outside of any data base used for development of criteria (Ayyub et al. 2006). There may also be methodologies based on a "simplified deterministic waves approach" (Bassler et al. 2011). These approaches provide for the greatest flexibility but are the most challenging to implement as criteria and enforce through out the acquisition process. The cost associated with these approaches can be daunting as extensive engineering and risk studies are necessary to demonstrate compliance.

The complexity of the approach chosen bears a direct relationship to the perceived risk and/or the factors of safety assigned. Table 53.2 illustrates a notional ranking for effectiveness of criteria in mitigating dynamic stability risk on a design for a notional hull form "A."

3. *Operator guidance based on dynamic stability assessment to produce either polar plots and/or rules of thumb based on specific loading conditions, speeds, headings and environmental conditions and/or Operability Envelopes*—Another complimentary approach is to provide operator guidance as a means of risk mitigation for dynamic stability. Dynamic Stability operator guidance may be as simple as rules of thumb or it may involve a direct probabilistic assessment of dynamic capsize risk or large amplitude motions risk. Key motion parameters are identified and assessed for specific seaway environments, and limits are imposed based on application of risk methodologies. These limits are displayed as polar plots and form the basis for operational guidance to the ship handler (Fig. 53.5). In some cases when operator guidance is provided, it may be considered sufficient to minimize dynamic stability risk without new dynamic stability criteria. Sim-

Table 53.2 Notional criteria ranking

Hull Form A Notional criteria* fidelity for risk mitigation		
Method	Effectiveness cost	
1	Limited	Less
2	Medium	Moderate-high
3	High	High
4	Medium-high	High

1—Empirically Based Rules
2—Rules Based on Probabilistic Dynamic Approaches
3—Direct Probabilistic Assessment
4—Relative Probabilistic Assessment
*Criteria using mitigation method 2

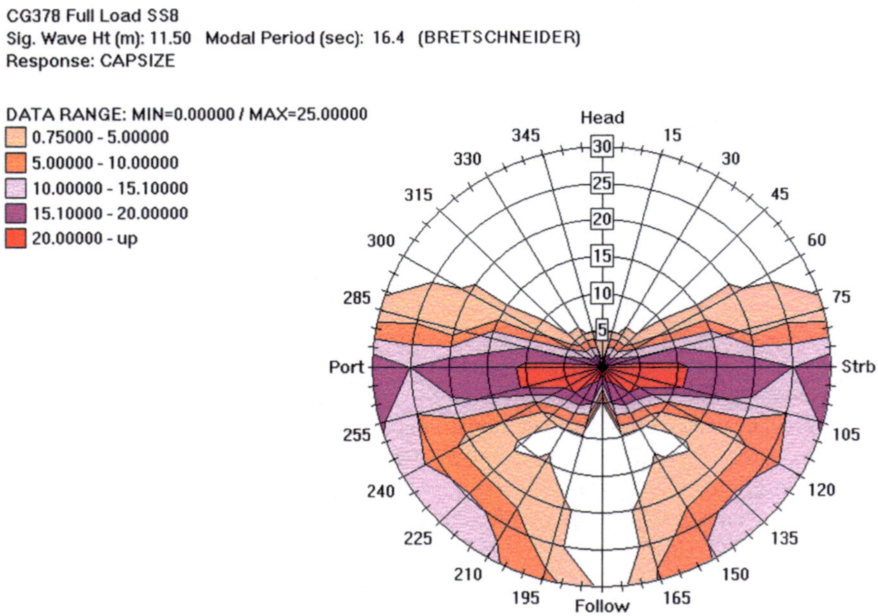

Fig. 53.5 Example capsize risk polar plot

ulation or model testing maybe required developing the appropriate polar plots. Some training and socialization is required to implement the operator guidance. There appears to be an unquantified margin between safe operability and acceptable intact stability implied by current standards. In many cases, safe operability is determined by practice of good seamanship. In spite of the margin being unquantified, it is relatively easy to determine operability envelopes and specify them as part of an acquisition. Dynamic stability events occurring inside the

Table 53.3 Operator guidance SOE ranking

Hull Form A		
Notional OG/SOE fidelity for risk mitigation		
Method	Effectiveness cost	
1	Limited	Less
2	Medium	Moderate-high
3	Medium-high	High
4	High	High

1—Empirically Based Rules of Thumb
2—Polar Plots/Rules of Thumb
3—Polar Plots/Training
4—SOE Restrictions

operability envelope would be expected to have a very low probability of occurrence and this may be checked by simulation and/or model testing as required and supplemented by existing operability criteria (e.g., IMO/SLF 49). The operability approach doesn't rely on an annual or lifetime risk which is likely to be non-discriminate (i.e. in all headings, sea states, etc.) without the influence of the operator or operability factors, and therefore very high.
In development of the operability envelope approach three questions should be addressed:

- What is tolerable from a corporate and societal viewpoint?
- What inherent level of risk is associated with current standards?
- What level of risk is inherent in good ship-handling (reaction to cues)?

Training for the crew should be developed which addresses the use of the operator guidance system, identification of cues, and how to identify and manage risk when in heavy weather. Shaw (2001) The Operator Guidance and Training Working Group (OGTWG) is a group of naval operators convened by invitation of the NSSWG to provide input and insight into the issues involved with operating ships in high seas. Work done to date by the OGTWG has identified appropriate class room and simulator curricula associated with specific bridge team positions. Table 53.3 lists a notional ranking of operator guidance/operational limits that may be identified for the risk assessment.
4. *Changes to Hull Form*—If approached early in the design the most effective mitigation may be the identification of specific design changes that reduce the dynamic stability risk. However it may not be possible to make sufficient geometry changes or mass property changes and still meet requirements for the overall design. In that case some combination of approaches to dynamic stability risk mitigation should be identified that includes hull form changes to the extent possible, coupled with operator guidance, operational limits and criteria.

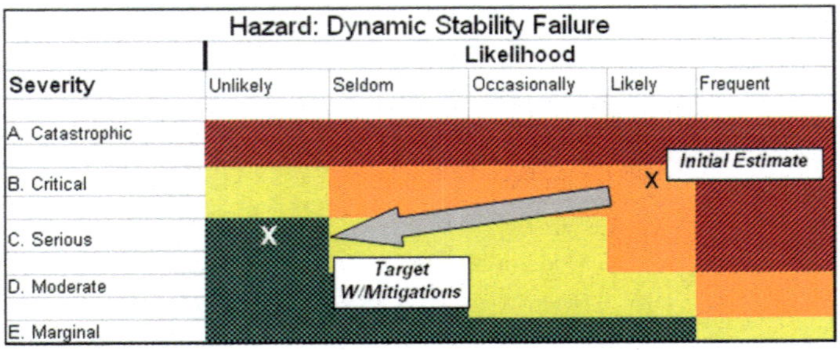

Fig. 53.6 Estimated reduction in hull form risk after mitigation

53.5 Final Hull Form Risk Ranking

Finally, a combination of options assembled from the tables could be assessed for mitigation effectiveness and cost. The best combination will be the one that provides the most effective risk reductions and least cost, taking into account the limitations on both these measures.

Risk reduction/cost plots can be used as a tool to select the best combination of options. For notional "Hull Form A," it could be determined that the best options are achieved using a combination of the following

- Rules Based Probabilistic Dynamic Approaches
- Polar Plots/Rules of Thumb.

It may take several iterations to finally get to an acceptable risk for the hull form as shown in Fig. 53.6.

53.6 Conclusion

The process of developing rational approaches for consideration of dynamic stability is in its infancy. Through intelligent use of analytical tools, test data, and historical evidence it is possible to establish a rational process to manage and reduce the risk of a dynamic stability event occurring at sea. The tools employed to accomplish this should be used carefully and with an eye to economy without sacrificing safety. Risk management techniques provide a rational framework to accomplish this goal. Although not addressed in this paper, similar processes can be tailored to damage dynamic stability.

Acknowledgements The Naval Stability Standards Working Group, David Tellet for his inspiration and advice, Martin Rennilson, Doug Perrault, Steve Marshall, Karl Stambaugh, Brad Campbell for their thoughts and discussions.

References

Alman, P.R., Minnick, P.V., Sheinberg, R., Thomas, W.L., "Dynamic Capsize Vulnerability: Reducing the Hidden Operational Risk," SNAME Transactions, Vol. 107, New York, 1999

Ayyub, B.M. Risk Analysis in Engineering and Economics, Chapman and Hall, 2003. pp 95–109

Ayyub, B., Kaminsky, M., Alman, P.R., Engle, A.H. Campbell. B.L., Thomas, W.L. "Assessing the Probability of the Dynamic Capsizing of Vessels," Journal of Ship Research, SNAME, December 2006, 50(4), 289–310

Bassler, C.C., Belenky, V., Bulian, G., Francescutto, A., Spyrou, K., Umeda, N. "A Review of Available Methods for Application to Second Level Vulnerability Criteria", Chapter 1 of Contemporary Ideas on Ship Stability, Neves, M.A.S et al., eds., Springer, 2011, ISBN 978-94-007-1481-6, pp. 3–24

Belenky, V., de Kat, J.O., Umeda, N., "Towards Performance-Based Criteria for Intact Stability" Marine Technology, 2008, 45(2), 101–123

Belenky, V. Bassler, C.C., Spyrou, K., "Dynamic Stability Assessment in Early-Stage Ship Design", Proc. 10th Int. Conf. on Stability of Ships and Ocean Vehicles (STAB), St. Petersburg, Russia, 2009, pp. 141–154

Perrault,D., Hughes, T. Marshall, S., "Developing a Shared Vision for Naval Stability Assessment", Proc. of 11th Intl. Ship Stability Workshop (ISSW), Wageningen, The Netherlands, 2010, pp. 115–120

Peters, A., "Tolerable Risk of a Naval Vessel", Proc. of 11th Intl. Ship Stability Workshop (ISSW) Wageningen, The Netherlands, 2010, pp. 93-107

Shaw E. J., "Practical experience and Operational Requirements for On-Board Risk Management Under Marginal Stability Conditions", Proc. of 5th Intl. Ship Stability Workshop (ISSW), Trieste, Italy 2001

Tellet, D., Cimino, D., "Marine Vehicle Weight Engineering," pp 231–245, 2007, SAWE

Chapter 54
Tolerable Capsize Risk of a Naval Vessel

Andrew J. Peters

Abstract Many of the operations and duties conducted by naval ships involve a degree of risk. This risk is somewhat unavoidable due to the nature of operating a warship at sea, where operational requirements can put the vessel and crew in harms way. One of the hazards that the crew is subjected to while on operations is that of the weather. The objective of this chapter is to discuss the tolerable risk associated with the loss of a naval vessel due to the weather conditions. A review of tolerable risk and potential methodologies of calculating an annual probability of loss of the vessel which uses time domain simulations and statistics of observed weather conditions aboard naval ships are presented.

54.1 Introduction

Many of the operations and duties conducted by navies involve a certain degree of risk. This risk is somewhat unavoidable due to the nature of operating a naval vessel at sea, where operational requirements can put the vessel and crew in harms way. One of the continual hazards that the crew is subjected to whilst on operations is that of the weather.

Many navies, such as the UK's Royal Navy, now have a duty of care to ensure the level of risk they expose the ship's company to is commensurate with the benefits gained. It is this basis which is the principle of tolerable risk as described in the tolerability principles internally published by the Mininstry of Defence (MoD). Navy ships are exposed to many hazards when at sea, like most commercially operated ships, but unlike commercial ships they may not be able to avoid heavy weather conditions due to operational requirements.

By using the theory and application of risk tolerability principles, as used by the UK's Health and Safety Executive (HSE) and adopted in most industries, an assessment of tolerable risk can be made (HSE 2001). This methodology is available for

A. J. Peters (✉)
QinetiQ, Haslar Marine Technology Park, Gosport, UK
e-mail: AJPeters@QinetiQ.com

V. L. Belenky et al. (eds.), *Contemporary Ideas on Ship Stability*, Fluid Mechanics and Its Applications 119, https://doi.org/10.1007/978-3-030-00516-0_54

any business that deals with risk to the workforce or to the general public, including the UK MoD. The UK MoD assess the tolerability of risks associated with all areas of military equipment and operations. These tolerability principles could be applied to provide a suitable tolerable risk for the annual capsize risk of a naval vessel.

In 1990, the Cooperative Research Navies (CRNAV) Dynamic Stability group was established with the aim of deriving dynamic stability criteria for naval vessels. To derive such criteria, the group needed to evaluate in-service and new ship designs in moderate to extreme seas, in terms of their relative safety and probability of capsize. This would ensure that new vessels continued to be safe, whilst avoiding high build and life-cycle costs associated with over-engineering.

To achieve these objectives, the numerical simulation program FREDYN was developed and continues to be applied extensively both to intact and damaged ships. This time-domain program is able to take account of nonlinearities associated with drag forces, wave excitation forces, large-angle rigid-body dynamics and motion control devices. The current CRNAV group comprises of representatives from UK MoD, the Australian, Canadian, French and the Netherlands navies, as well as the U.S. Coast Guard, Defence Research & Development Canada, (DRDC), Maritime Research Institute in the Netherlands (MARIN) and QinetiQ.

This Chapter discusses the concept of tolerable risk, which is the willingness to accept a risk so as to secure greater benefits. Using an accepted framework known as the Tolerability of Risk (TOR), decisions as to whether risks from an activity are unacceptable, tolerable or broadly acceptable can be made. These principles of tolerable risk are examined in association with the loss of a naval vessel due to the weather conditions.

54.2 Broad Principles of Risk Assessment

Some may argue that any risk is unacceptable, but in reality, the risk of suffering harm is an unavoidable part of living in the modern world. However, some risks can indeed be deemed acceptable for the following reasons (Rowe 1983).

- Threshold condition: A risk is perceived to be so small that it can be ignored.
- Status quo condition: A risk is uncontrollable or unavoidable without major disruption in lifestyle.
- Regulatory condition: A credible organisation with responsibility for health and safety has established an acceptable level of risk.
- De Facto condition: An historic level of risk continues to be acceptable.
- Voluntary balance condition: The benefits are deemed worth the risk by the risk taker.

In recent times there is an expectation for a society free from involuntary risk. The concept of risk is often used in everyday discussions where people often describe taking a risk in relation to taking a chance of adverse consequences to gain some benefit.

Risk, however, is defined as 'the combination of the likelihood and consequence of an unplanned event leading to loss or harm' (HSE 2001). The way in which society treats risk depends upon the individual perception of how the risk relates to them. There are many factors involved and it is down to how well the process giving rise to the risk is understood, how equally the danger is distributed and how individuals can control their exposure (MoD). Studies have shown that hazards give rise to concerns which can be put into two categories:

Individual Concerns:

This is associated with how the hazard affects an individual and all things that they value personally. Individuals are more likely to happily accept higher risks of hazards that they choose to accept rather than any hazards imposed upon them, unless they are considered negligible. If the risks provide benefits they will want the risks to be kept low and be controlled (HSE 2001).

Societal Concerns:

This is the impact of hazards on society and if they were to happen would result in a socio-political response with repercussions for those responsible for controlling the hazard. These concerns are often associated with hazards that if they were to occur would cause significant damage and multiple fatalities. Examples would include Nuclear Power stations, rail travel and genetic engineering. Concerns due to multiple fatalities from a single event/effect are known as societal risk (HSE 2001).

54.3 Characterising the Issues in Terms of Risk

To examine the risk associated with the loss of a naval vessel the first stage must involve framing the issues relating to the risk. This will result in characterising the risk both quantitatively and qualitatively to look at how it may occur and what effect it will have on those involved and society at large.

A risk assessment is normally conducted when characterising the issues affecting the risk, which includes identifying the hazards which would lead to harm, what the likelihood of it occurring would be and what harm and consequences would be experienced if it was to happen.

This stage of the assessment often assesses the individual risk and then moves to look at the effect on societal concerns to first identify if the hazards should be considered at all or could be regulated sufficiently.

The analysis of this for the loss of a naval vessel in heavy weather can be, in some cases, simplified in certain aspects. The outcome of a vessel capsizing in bad weather will inevitably result in the fatalities or extreme harm to the majority of the crew onboard and would result in the material loss of the platform. An event of this type leads towards examining the societal risk aspects due to the outcome resulting in multiple deaths and loss of a naval asset. The additional repercussions that the navy and government would have to deal with are also associated with societal risk.

54.4 Tolerability Principles

Once a risk has been assessed it must be examined to identify if the level of the risk is broadly acceptable, tolerable or unacceptable and whether the hazard should be even considered. It is therefore not surprising that a lot of work in determining criteria for these acceptability levels has been conducted (HSE 2001).

Criteria used by regulators in the health and safety field have shown that they can fall into three 'pure' criteria (HSE 2001):

Equity based criteria

These have the premise that individuals have the unconditional right to a certain level of protection, i.e. which is usually acceptable in normal life. This often results in a level of risk that cannot be exceeded. If the risk level after analysis is above this level and suitable control measures cannot be introduced to lower the risk, the risk is deemed unacceptable. For naval vessels these criteria will be relevant.

Utility based criteria

These criteria apply to the comparison between incremental benefits of the measures to reduce the risk, the risk of injury and the costs of the benefit. These criteria therefore look at comparing, in monetary terms, the cost of the benefits (statistical lives saved) of the preventative measure compared to the cost of implementing it.

Technology based criteria

These criteria essentially reflect that a satisfactory level of risk is obtained when 'state of the art' measures are employed to control the risks. For a naval vessel this could include advanced heavy weather training or onboard operator guidance systems.

54.5 Tolerability of Risk

These criteria described above can be used on their own although a combination is often a better approach. The HSE have incorporated them in a framework known as the Tolerability Of Risk. This methodology breaks the level of risk down into three regions. These are described in Fig. 54.1 with the 'ALARP triangle' and are described in detail as follows:

Broadly acceptable risk region

Risks that fall into the broadly acceptable risk region are deemed insignificant. Regulators would not require any additional measures to reduce the risks further than they already are. Further actions would only be required if lowering the risk was practical or where there is a legal requirement to lower it further. Regulators are required to regularly monitor the risk to ensure that it remains in this region. The level of risk at

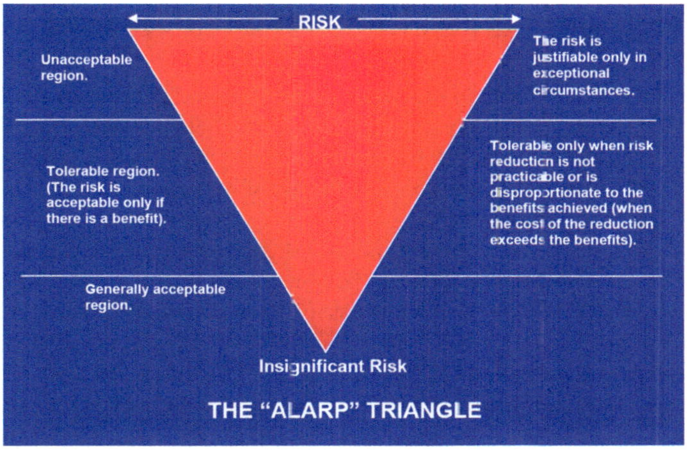

Fig. 54.1 ALARP triangle

this level is comparable to what people regard as insignificant or trivial in their day to day lives (HSE 2001).

Tolerable risk region (As Low As Reasonably Possible—ALARP)

This region lies between the broadly acceptable and intolerable regions. Risks in this region relate to those risks that people are willing to tolerate in order to gain from the benefits. This means that the risk is deemed tolerable where society desires the benefits of the activity and only if further risk reduction is impracticable or the penalties are grossly disproportionate to the improvement gained. The levels of the risks must be assessed and the results used correctly to determine control measures. The assessment method must use the best available scientific knowledge (HSE 2001).

Intolerable risk region

The risk in this region cannot be justified except in extraordinary circumstances. Control measures are required to drive the risk downwards into one of the lower risk regions.

The aim for any activity would be to have the risks fall into the broadly acceptable region. However, the practicability of achieving this, for example with a naval vessel operating in open ocean conditions, may be difficult to achieve without unacceptable restrictions on the ship and operation. Therefore as the intolerable region by its nature cannot be acceptable in anything but extraordinary circumstances, the As Low As Reasonably Possible (ALARP) region is realistic for naval vessels, with measures such as training and heavy weather guidance to assist in controlling the risk of capsize (HSE 2001).

54.6 Tolerability Limits—Individual Risk Boundaries

The term 'Individual risk' is used to describe the level of risk of fatality of an individual that is exposed to a particular activity. UK HSE guidelines state that an annual 1 in a million probability of fatality is a very low level of risk and should be used to define the boundary between the broadly acceptable and the Tolerable regions of risk (HSE 2001).

The UK HSE guidelines for a hypothetical person exposed to hazards in the workplace have defined the maximum tolerable risk of fatality as 1 in 1000 per year (10^{-3}) and 1 in 10,000 (10^{-4}) for the risk of fatality to a member of the general public. This is referred to as the basic safety limit and is the cumulative value of risk an individual is exposed to. This measure is applied to investigate the risk to a hypothetical worker working in a particular industry, such as offshore for example, and used to compare to levels in other industries. It provides a base line for comparison and assessment of changes to the level of risk.

Individual risk however cannot be used on its own for larger events which, if they occur, will result in higher numbers of fatalities. Group risk or societal risk as it is commonly known is used to describe the relationship between the probability of an unplanned event and the number of people affected by the event. It applies to those activities which present major implications for society such as a high number of fatalities, the loss of a major asset, environmental and political damage. Societal risk is not just calculated by taking the individual risk and multiplying it by the total number of fatalities from a single event, but is often complex and has many influences on its level (HSE 2001).

54.7 Tolerability Limits—Societal Risk Boundaries

For large events which impact on society as a whole, the societal risk will be the dominating factor rather than individual risks. Events which involve multiple fatalities will attract wide social interest and the societal risk encompasses both societal risk and society's reaction to an event.

When considering what society considers tolerable, there are several aspects which influence the response of society to the event and hence certain events are considered more tolerable than others. For example:

1. Acts by God or nature are considered more tolerable than those of human error.
2. Risks are more tolerable if we have control or have had participation in the decision leading to the risk e.g. car accidents are deemed more tolerable than aircraft accidents.
3. Risks are not tolerable if we cannot see the benefit for ourselves.
4. Familiarity makes a risk more tolerable. For example, a car accident is more tolerable than a nuclear radiation accident.

5. A large number of accidents spread over a fairly long period of time is more tolerable than a large number of incidents in close succession.
6. Less tolerable with risk towards the innocent and vulnerable.
7. Personal experience.

These and many other factors come into the society's response to an incident; particularly the knowledge of the hazard, whether the hazard was man made or natural and whether the potential victims are particularly vulnerable, e.g. children and the elderly (HSE 2001).

Media coverage can significantly influence society's level of tolerability to a risk. For example, there are few car crashes reported in the press. However, aeroplane crashes or passenger ship accidents always are, when there are far fewer of these incidents. This makes society much more wary of ships and aeroplanes than driving a car.

The loss of a naval vessel due to capsizing in heavy weather would be classed as a significant event, due to the loss of a high proportion of the crew, the naval asset and the political damage associated with it. However, the hazard in this case is from nature and it is understood by society that naval personnel are exposed to greater risks whilst on operations, such as a search and rescue mission in heavy weather, and may accept a higher risk as being tolerable in that case.

The complexity of developing tolerable limits for events that would raise societal concerns is complex, so a way of conveying this information has been accepted. It uses the concept of FN curves, where the F denotes frequency and the N denotes the number of fatalities. These diagrams provide relationship data on the frequency of the fatal accident (plotted on the y axis) and the number of fatalities resulting from it (plotted on the x-axis). These curves can be used to graphically describe limits of risk acceptance. The curves can be generated by defining different combinations of consequence (i.e. fatalities) and the related frequency that gives negligible, acceptable and unacceptable risk respectively (HSE 2001).

The UK HSE has realised the complexity involved in analysing societal risk and has produced guidelines to define the acceptable borders between the tolerable and intolerable regions. This guidance is based on a FN criteria point for a single accident which occurs with a frequency of 2×10^{-4} events per year (1 in 5000) which results in 50 fatalities. This result is then extended on the FN diagram by applying a line with a slope of -1, using logarithmic scales on both axes, which is then defined as the risk neutral line i.e. a linear relationship between frequency and consequence. The broadly accepted region is taken as 2 orders of magnitude below this criteria (<1 in 500,000). These zones of tolerability are shown pictorially in Fig. 54.2. This FN diagram provides a framework in which to assess the risk tolerability of society of a particular event.

In assessing an event such as the capsizing of a naval vessel, both the individual and societal risks need to be evaluated as they incorporate different concepts.

Excluding the other hazards that the crew on board Royal Navy warships are exposed to in this study, the HSE guidelines can be used to assess what could be used as the tolerable risk of loss of a naval vessel.

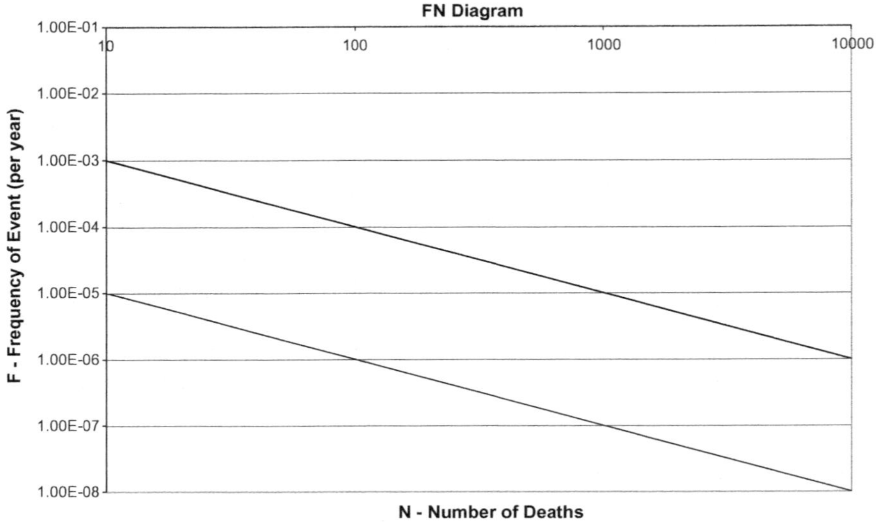

Fig. 54.2 FN diagram

54.8 Risk Associated with Transportation

Examining various forms of transport, identifying how these industries deal with risk and what society deems acceptable allows direct comparison for what could be deemed acceptable to society for the maritime industry and naval vessels.

The risks involved in the air transportation industry are those which most people are aware of and accept when they fly. An accident survey of 1843 aircraft accidents from 1950 through 2006 (Winter-travel.org) determined the causes of the accidents to be as follows (The survey excluded military, private, and charter aircraft):

- 53%: Pilot error
- 21%: Mechanical failure
- 11%: Weather
- 8%: Other human error (air traffic controller error, improper loading of aircraft, improper maintenance, fuel contamination, language miscommunication etc.)
- 6%: Sabotage (bombs, hijackings, shoot-downs)
- 1%: Other cause.

However, the risk of being involved in a crash on a single flight is, on average, 1 in 6 million (Livescience.com), depending upon airline, in comparison to the likelihood of dying in a car journey of 1 in 5000. This means that for anyone flying, the individual is much more likely to die on the journey to the airport rather than during the flight itself. Fear of flying is common, mainly due to lack of personal control, understanding and the general concept of being at high altitude. People are perfectly happy to drive

Table 54.1 Individual risk of fatality for transport modes

Travel mode	Fatalities per 10^8 passenger (km)	Fatalities per 10^8 passenger hour (FAR)
Motorcycle	9.7	300
Bicycle	4.3	60
Foot	5.3	20
Car	0.4	15
Van	0.2	6.6
Bus/coach	0.04	0.1
Rail	0.1	4.8
Water	0.6	12
Air	0.03	15

cars frequently, as they are in control and are happy to disregard the fact that there are 50,000 fatalities on highways every year. To put this into perspective, statistically a person would have to fly once a day every day for over 15,000 years in order to be involved in an aircraft accident.

When discussing modes of transport, there are a number of ways in which to define a fatality risk measure. The potential loss of life (PLL) measure is a basic measure of risk of fatality per year that is often used to define accident rates (Kristiansen 2005). However, this criterion has the shortcoming of not incorporating any exposure time into the measure. It is also important to make the distinction between individual and societal risk. The most common risk measures for individual risk are the Average Individual Risk (AIR) and Fatal Accident Rate (FAR). The AIR measure is calculated by dividing the PLL measure by the number of people exposed e.g. the number of crew on a naval ship. The FAR measure is calculated by dividing the PLL value by the total number of man hours of exposure and multiplying by a 10^8 scaling factor. This gives the number of fatalities per 10^8 h of exposure to the hazard.

These measures provide a good means of comparing risks from travelling by various modes of transport, as shown in Table 54.1 (HSE 2009).

As can be seen from Table 54.1, travelling by sea is one of the least risky modes of transport. The FAR value can be calculated for travelling on UK ferries and is 8.8 fatalities per 10^8 h of exposure (Kristiansen 2005). Compared to the other modes of transport, UK ferries can be seen to be one of the safest forms of transport.

Regarding the risk of capsize of a navy vessel, consideration should be made to the exposure time and particularly the exposure to the heavy weather conditions where capsize is more likely to occur.

Other areas of the marine industry and other wider industries can be used to provide further comparison of the level of risk a person working in that industry is exposed to during their working life. These results for wider industry provide an indication to what society generally regards as acceptable.

The UK HSE (HSE 2009) provides statistics comparing the risk of fatalities in various UK industries, Table 54.2.

Table 54.2 Individual risk of fatality in UK Industries

Industry	Annual individual risk of fatality
Agriculture	8.10×10^{-5}
Construction	3.70×10^{-5}
Offshore	4.00×10^{-5}
Services	0.35×10^{-5}

Table 54.3 Individual risk of fatality in industries worldwide

Industrial activity	Fatalities per 1000 worker-years
Mining	0.9–1.4
Construction	0.3
Industry	0.15
Shipping	1.9–2.1
Fishing on the Continental self	2.3
Fishing	1.5

Table 54.4 Commercial vessel annual risk of vessel loss

Vessel type	Total loss rate (per 1000 ship years)	Annual probability of ship loss
General cargo	5.4	5.4×10^{-3}
Bulk dry	3.3	3.3×10^{-3}
Oil tanker	1.5	1.5×10^{-3}

These UK statistics are lower when compared with statistics from other parts of the world, Table 54.3 (Kristiansen 2005).

These statistics illustrate that the highest individual risks in UK industry are generally around 10^{-5}–10^{-4} fatalities per year, compared to the 10^{-3}–10^{-4} level for industries worldwide which are at the tolerable limit defined by the UK HSE.

Over the last few decades, extensive resources have been used to reduce the risks involved with the shipping industry. The long term trend of loss frequency has been studied (Lacaster 1996) and it was concluded that the annual loss rate had been reduced by a factor of 10 in the twentieth century, from more than 3% in 1900 down to 0.3% in 1990. However, the greatest level of reduction was early in the century and the level of reduction has levelled off in recent years.

Investigations into the risk of loss of merchant vessels using Lloyd's world casualty statistics has been conducted (MSC 2003). In that study, the total loss rate for different types of merchant ships are analysed, Table 54.4.

On examination of fatalities from the loss of these different vessel types, it was found that there were 170 fatalities per year on general cargo ships that were lost. This relates to 1.8 deaths for every complete vessel loss. Taking the typical number of crew on this type of vessel, the individual risk of death for a general cargo ship

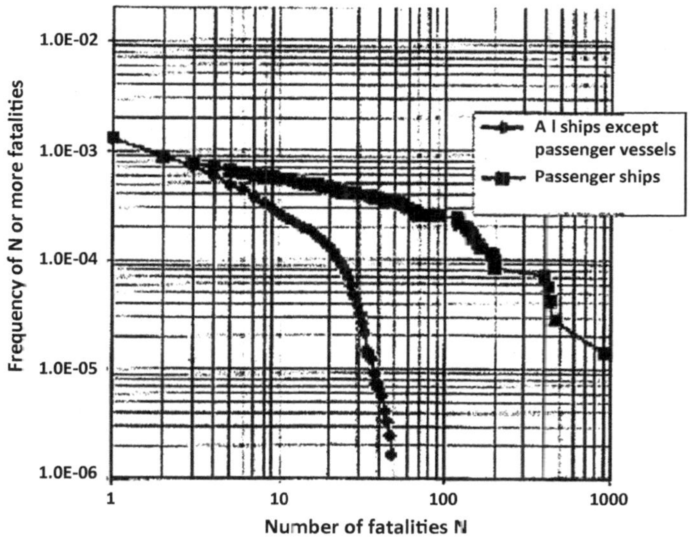

Fig. 54.3 Frequency of accidents involving N or more fatalities DNV (1998)

crew member is calculated as 3.7×10^{-4} (MSC 2003). This is the highest of the vessel types, with many of the other vessel types having a probability of individual risk of death close to the 1×10^{-4} level. RoRo passenger vessels were found to have an individual risk of death of 7×10^{-5}. The relatively large public focus on marine accidents reflects society's considerable awareness of these fatalities.

As described above, regarding multiple fatalities and societal risk, an FN diagram is often used to convey acceptable risk levels for events with multiple fatalities. However, the FN diagram can be used to describe both required and the prescribed risk levels.

Figure 54.3 is based on data from DNV in 1998, which shows the observed FN values for passenger ship accidents (upper curve) and cargo ships accidents (lower curve). For the passenger ships, it can be seen that small single fatality incidents occur with a frequency of approximately 10^{-3} per year, whereas an extreme casualty event (approximately 1000 casualties) occur with a frequency of 10^{-5} per year.

54.9 Tolerable Capsize Risk of a Naval Vessel

It cannot be assumed that the loss of a frigate from capsize would result in an approximate 10% fatality rate among the crew on board, as found in the commercial vessels statistics. As the duty stations of the crew on naval ships are distributed throughout the vessel, many are below the weather deck which is quite different to commercial vessels where the majority of the crew will be in the vessel's superstructures. It therefore can be assumed that the loss of at least 50% of the crew would be a more realistic value, as the crew stationed below 2 deck on a frigate would be unlikely to escape if the vessel capsized.

From the commercial vessel statistics, the probability of loss of the vessel is in the order of 10^{-3}, Table 54.1. This is just within the tolerable region, with up to 10 fatalities from any incident. From the loss of the general cargo vessels, the average fatality rate has been found to be 1.8 deaths per vessel loss. For the capsizing of a naval vessel, the number of fatalities would be significantly higher.

The FN diagram statistics for passenger ships, Fig. 54.3, show a probability of 8.5×10^{-3} for 100 fatalities and 1×10^{-4} for 200 fatalities per year. A passenger ship could be considered to be similar to a naval vessel, as there are a high number of personnel on board compared to a cargo ship. As this is the observed level of probability, it could be taken that this is acceptable to society, as it is a historically accepted level of risk.

From the risk analysis, it is clear that the capsizing of a naval vessel will result in a significant number of fatalities, as a medium sized vessel (such as a frigate) could have approximately 200 crew members. Based on the HSE tolerability limits, this would require the probability of the loss of the ship to be 6×10^{-5} per year to be within the tolerable region. This is slightly higher than the credible failure risk assumed for submarines, which is taken as a minimum of 10^{-6} for a 90 day patrol, where a failure event will also likely result in the fatalities of the entire crew (MAP 73 - MoD defence standard). This 10^{-6} value is on the tolerable and generally acceptable regions boundary of the UK HSE FN diagram.

Having around 100 fatalities (50% of a frigate crew) in the tolerable risk region would require an annual probability vessel loss of less than 1×10^{-4}. The generally acceptable region would require annual probability of loss of less than 1×10^{-6}. This would also result in an individual risk to the crew members at a similar magnitude as other areas of the marine and wider UK industry.

Considering all these points, it is suggested that a tolerable region boundary of 1×10^{-4} would be a suitable level for the annual risk of loss of a navy vessel in heavy weather and would be comparable to other areas of the marine industry and other major events. A value of 1×10^{-4} annual capsize risk was therefore found to be a suitable level for the tolerable risk boundary for the loss of a naval frigate at sea.

However, the manner in which extremely rare independent events are combined adds a final additional complexity to the problem, as probability theory has the combined probability of different independent events defined as the sum of the independent risks. This suggests that the other potential risks of loss of the ship and crew at sea should therefore be considered and subtracted from the 1×10^{-4} risk level to produce the tolerable limit of annual loss of the frigate and crew due to capsize. If these other potential risks have a probability of occurrence that is several orders of magnitude lower than the vessel capsizing, then tolerable risk value presented could still be closely related to that of the vessel capsizing. In a similar way to capsizing, naval vessels have almost never been known to be totally lost to fire, for example, while at sea (in peace time in recent years). Further investigation is required to identify the other potential risks of loss for a warship while at sea to identify how these risks realistically combine to produce an overall capsize risk that compares with other areas of industry.

54.10 Assessing the Risk of Capsize of a Naval Vessel

Assessment of the probability of a vessel capsizing is a significant aspect of assessing the risk. Calculating the probability of the vessel capsizing can be conducted with modern computational tools, such as FREDYN, which can model a vessel in extreme wind and waves. However, there are many areas of uncertainty that are inherent in the calculations that require careful consideration.

In order to accurately calculate the capsize probability of a naval vessel, a simulation tool is required to examine all possibilities of sea state and operational loading conditions to provide assessment of all realistic operational scenarios. The numerical simulation program FREDYN was developed by the Maritime Research Institute Netherlands (MARIN) for the Cooperative Research Navies working group and continues to be applied extensively to both intact and damaged ships. This time-domain program is able to take account of nonlinearities associated with drag forces, wave excitation forces, large-angle rigid-body dynamics and motion control devices. The FREDYN program permits investigations into the dynamics of intact and damaged vessels operating in realistic environments.

54.11 Calculation of Annual Capsize Risk Probability

FREDYN simulations can be used to evaluate the critical roll (capsize) behaviour of a vessel in a range of realistic operating load conditions. This procedure was developed by McTaggert in 2000 and is described further in his paper (McTaggert and DeKat 2000). The method, adopted by the CRN working group, is largely based upon the method described fully in his report (McTaggert 2000) and is used for evaluating capsize risk of intact ships in random seas. This approach for predicting ship capsize risk combines the time domain simulation program FREDYN with probabilistic input data for wave conditions and ship operations (speed and heading). For a ship in a seaway of duration D (e.g. 1 h) the probability of capsize $P(C_D)$ is:

$$P(C_D) = \sum_{i=1}^{N_{V_s}} \sum_{j=1}^{N_\beta} \sum_{k=1}^{N_{H_s}} \sum_{l=1}^{N_{T_p}} P_{v_s}(v_{s,i}) P_\beta(\beta_j) \times P_{H_s,T_p}(H_{s,k} T_{p,l}) P(C_D | V_s, \beta, H_s, T_p) \tag{54.1}$$

where: V_s = ship speed, β = wave heading relative to ship, H_s = wave significant height, T_p = wave modal period.

The last term is a conditional probability of capsize in a given wave condition and ship heading relative to the waves.

Limited Gumbel distributions are used to fit to the maximum roll angles recorded in each of the seaway conditions, in order to calculate the capsize probabilities. A second, distribution free method, is also possible and was investigated with a new set of data calculated in a recent study. However, the limited Gumbel distributions have been shown by members of the CRN group to provide the best data fit and better predictions at the higher roll angles, which is the area of most interest for capsize prediction (McTaggert 2000). The Gumbel fit uses the upper 30° range of the simulation and fits to a minimum of 10 data points. This work was validated on large numbers of simulations (400+) by McTaggert. However this number of runs was not feasible for any routine calculations, as the time to compute would be very lengthy. Realistically, the number of simulations has to be between 10 and 50. The sensitivity of using this number of runs was also investigated by McTaggert and was shown to give very good results (McTaggert 2000). Recent studies with the CRN group have shown that for other frigate types there may be a need for a greater number of simulations to produce statistically reliable results. Current investigations by CRN members are onward to identify if using the peaks over threshold methodology produces better fidelity of results, as the roll motion peaks during the whole simulation are used in the calculation of the capsize probability rather than just the maximum roll angle in each simulation.

The probability of capsize is calculated based on a time period of 1 h and can be computed using Eq. 54.1. The associated annual probability of capsize can be calculated from the following equation, using the 1 h capsize risk (McTaggert 2000):

$$P(C_{Annual}) = 1 - (1 - P(C_D))^{\alpha \times 1\ year/D} \tag{54.2}$$

where α is the fraction of time spent at sea and D is duration (hours).

54.12 Uncertainty in Risk Calculation

In the assessment of uncertainty and the application of safety factors to areas of uncertainty, the HSE recommends making use of sensitivity analysis and comparative risk assessments for novel hazards that have a similarity to the case under investigation (HSE 2001). In the engineering world, safety factors are calculated to take into account the uncertainties in materials, calculation methods, etc. This principle is particularly exploited in the world of ship structures. In general engineering, safety factors between 1.25 and 5 are often used, dependant on the level of knowledge and uncertainty of the material and the environment, stress and load a structure is to be subjected to. The aerospace and automotive industry use factors in the region of 1.15 and 1.25, due to the costs associated with structural weight. The testing and quality control is also higher in these industries, with significant modelling (computationally and physically) of the material stresses involved.

The submarine world uses safety factors of a similar magnitude to the aerospace world, with significant physical and computational models used to ensure accurate understanding of the influences.

When assessing risks, it is usually required for uncertainty in the calculations to be taken into account when there is lack of, or incomplete data (HSE 2001).

When examining the risk of loss of a naval vessel, the uncertainty in the outcome of the event i.e. what would happen if the vessel was to capsize, is actually very low due to the fact that it would result in the inevitable total loss of the vessel and a large number of the crew onboard. However, the uncertainty associated with the calculation of the probability of the event occurring is greater and must be adequately handled in order to calculate realistic values of risk for the vessel.

Knowledge uncertainty is one of the areas that must be dealt with (HSE 2001). This occurs when there are sparse statistics or random errors; for example, in experiment data used to define the probability of the event occurring (HSE 2001). Although many commercial vessels are lost each year and the statistics are available, in the case of the loss of a naval vessel in heavy weather, the statistics are very sparse and mainly representative of outdated designs of hullforms.

Modelling uncertainty is the term given to the uncertainties in the mathematical terms used in a numerical model used to assess risks. This is also closely linked with limited predictability associated with an outcome that is sensitive to the assumed initial conditions of the system under investigation and affects the final state i.e. the initial conditions of the ship affecting whether it capsizes in a certain wave condition or not.

It is clear that there are potential levels of uncertainty in the modelling of the risk of loss of a naval vessel using simulation tools such as FREDYN. Some of the main areas of uncertainty are related to the following:

- The probability of the vessel being in the waves and level of exposure.
- The probability of the speed and heading combinations in heavy weather.
- The simulation time i.e. the length of time the ship is in the waves.
- The number of simulations used in the prediction of the capsize event.
- The vessel loading condition.
- The angle used to define the capsize event.
- The autopilot in the simulations.
- Roll damping characteristics.

Techniques have been developed under what is defined as the 'precautionary principle' to handle uncertainty when dealing with calculating risk (HSE 2001). Uncertainty can be overcome by constructing the most credible scenarios of how the hazards might be realised.

Sensitivity of the annual capsize risk calculation

The variables listed above, which are input parameters into the FREDYN capsize simulations, can be investigated using standard sensitivity type approach to assess the sensitivity of the inputs on the output probability of the capsize event. This would allow scenarios from the most likely to the worst case to be established and allow suitable safety factors to be derived and accounted for in the assessments.

The probability of the vessel being in the waves can cause unrealistically high probabilities of capsize by using the Eq. 54.2. A Bales wave climate statistics table (Bales et al. 1981) for the North Atlantic is often used to provide the probability of the waves occurring during the year, which is multiplied by the probability of the capsize event in those conditions. This can cause an unrealistically high annual probability of loss of the vessel, as the probability of the largest waves occurring with a high probability of loss of the vessel have a large influence on the overall annual capsize risk.

The capsize risk associated with the current calculations suggests that the probability of the vessel actually encountering the worst sea conditions is over estimated in the scenarios. A more realistic probability of the vessel encountering the waves is required.

A study was made for the UK MoD (BMT 2006) which analysed the wave condition records made by the RN bridge teams in the 6 hourly records, which are kept by all Royal Navy ships whilst at sea. This data was collected for 78 Royal Navy vessels from 1968 to the present day. The data was also analysed from 1985 to the present day, to reflect the change in conditions encountered following the end of the cold war. This equates to over 168 years of Royal Naval ships at sea, which provides a substantial data set of more realistic wave statistics for the calculation of an annual capsize risk.

Using this wave height data and the Bales wave scatter table to provide the distribution of wave periods at each wave height condition resulted in a factored wave scatter table with a more realistic probability distribution for the vessel encountering the waves in a year. The change in probability distribution of wave height from the new data compared to the standard Bales scatter table is shown in Fig. 54.4.

It is clear from Fig. 54.4 that there is a distinct difference in the distribution of the wave height data that vessels have historically encountered compared to the annual probability of the waves occurring. The main significant factor is that the Royal Navy ships do not historically experience the larger waves as the standard annual wave statistics would suggest. This could be partly due to avoiding storms in certain cases, but not completely.

In Eq. 54.2 above for the calculation of the annual capsize risk, the hourly capsize risk that is generated from the simulations is effectively extrapolated up for each hour the vessel spends at sea. In the moderate wave heights, the maximum roll angles that are recorded are used to predict the probability of exceeding the 70° capsize angle. The wave height conditions recorded on the navy ships are made every 6 h, which is also a realistic time frame for a large storm sea to remain relatively constant. Calculating results for the probability of capsize over 6 h simulations may provide better results, which would equal the time between measurements made onboard. With a naval ship at sea approximately 30% of the year this equates to 440 6 h time periods.

To evaluate the effect of the simulation time, a number of calculations have been performed with different simulation run lengths, from 30 min to 6 h, as well as

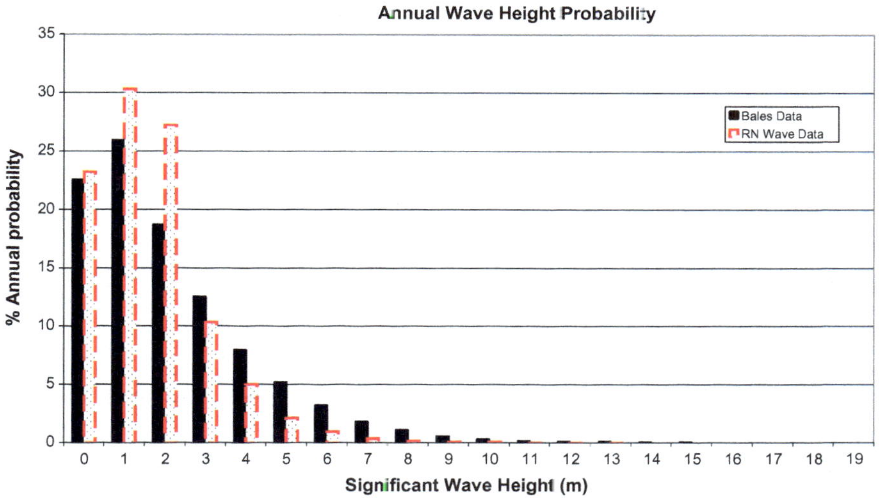

Fig. 54.4 Wave height probability of encounter

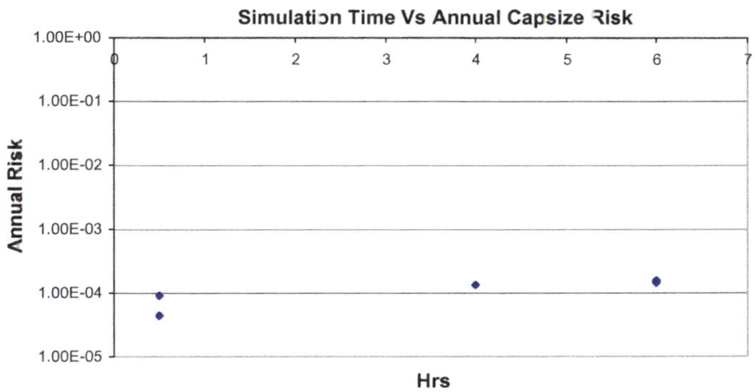

Fig. 54.5 Effect of simulation time on annual capsize risk

different numbers of realisations between 10 and 50. The results show that the effects on the annual capsize risk are very small after 2 h of simulation and increasing the number of simulations makes little difference to the annual capsize risk at this run length, Fig. 54.5. This shows that this has little effect on the probability of the capsize event for this vessel. A wider study is required to identify if this is the same for other vessels and load conditions.

The selection of the ship speeds can have a large effect on the capsize risk and unrealistic speeds should be avoided in the simulations. To achieve the most realistic annual capsize probability, the actual operation of the ship in heavy weather is required to be accounted for in the calculations. Standard heavy weather seamanship training instructs operators to not go faster than 60% wave speed in heavy weather.

This means that the vessel speed selection should be made as realistic as possible. Selecting speeds above 90% wave speed (30% safety factor for the operator) is unrealistic and will result in unrealistic capsize probabilities.

An even probability of heading is also usually assumed for the simulations. This could be considered to be precautionary, as in the very worst conditions the operator would avoid stern sea condition based on their experience, which is difficult to account for. Variation in the risk should be reviewed by removing certain headings, such as stern seas in the worst wave conditions. Selecting the accurate point this decision is made will require further discussion with operator training schools.

The roll damping characteristics of the model in the simulation will require investigation as to how it effects the risk calculation. The damping characteristics will be required to be set up as close as possible to the real vessel by comparing roll decay and roll period information. A systematic variation of the damping input parameters would provide the influence of the results to a specified variation of the damping characteristics. A suitable safety factor could then be derived to account for the variation on the modelling of the roll damping.

The autopilot control in the simulation will have an effect on the survival of the vessel. Using a systematic variation of the autopilot control parameters, the variation in risk could be derived based on those changes to the autopilot. A factor could then be derived based on the variation of the autopilot parameters.

The load condition of the vessel also needs to be considered in the annual capsize probability, as a vessel in a deep loading condition will often be inherently safer than in a light seagoing condition. It is therefore important to calculate at least two load conditions and use the typical operational profile to define the time the vessel would spend at each loading condition. This can then be realistically accounted for in the annual capsize probability. Operational procedures to ballast down with the forecast of heavy weather should also be accounted for in the calculations.

54.13 Results and Conclusion

In reviewing current Health and Safety guidelines, along with comparison with other modes of transport, other industries and the commercial marine industry guidelines for individual and societal risk have been described and can be used to examine the acceptable level of risk for capsize in heavy weather for the loss of a naval vessel. A value of 1×10^{-4} annual capsize risk was found to be a suitable level for the tolerable risk boundary for the loss of a naval frigate at sea. The magnitudes and method of combining other very low risks of loss of the ship and crew at sea, needs to be further investigated and considered in defining the tolerable limit of annual capsize loss. If the other potential risks of vessel loss are found to be several orders of magnitude lower probability of occurring than the vessel capsizing, then the tolerable risk value presented will still relate predominantly to that of the vessel capsizing. Therefore, this could provide an overall capsize risk that can be compared with other areas of industry.

In order to calculate suitable levels of capsize risk, sensitivity analysis is required to assess the input parameters to identify the most realistic scenarios and the potential variation in the capsize risk due to realistic variation of the input parameters. By undertaking this analysis, realistic risk levels and safety factors can be calculated to evaluate the annual capsize risk of a naval vessel for comparison with the tolerable risk level deduced.

Acknowledgements The authors would like to gratefully acknowledge the permission granted by the UK MoD for publishing the results of the investigation.

References

Bales S.L, Lee, W.T. and Voelker J.M. 1981 Standardised Wave and Wind Environments for NATO Operational Areas
DNV 1998, FSA of passenger vessels Report no. 97-2503 MAP 01-73 UKMoD
HSE Reducing Risks, Protecting People, Crown Copyright 2001
HSE Website Statistics 2009
Kristiansen, S, 2005 Maritime Transportation, Safety Managements and Risk Analysis
Lacaster J, Engineering Catastrophies: Cause and Effect of Major Accidents, 1996
McTaggert K.A, 2000 'Ship Capsize Risk in a Seaway using fitted Distributions to Roll Maxima, Transactions of ASME, Journal of Offshore Mechanics and Artic Engineering
McTaggert K.A, DeKat J.O 2000 'Capsize Risk of Intact Frigates in Irregular Seas', SNAME Transactions
MSC 77/25/4 Comparison of general cargo ship losses and fatalities- 2003 - RINA
Rowe W.D 1983. Acceptable levels of Risk for Undertakings. Colloquium Ship Collisions with Bridges and Offshore Stuctures. Copenhagen
Tolerability Principles—Warships in Harbour, MoD internal publication
http://www.livescience.com/environment/050106_odds_of_dying.html
http://winter-travel.org/travel-news/risk-of-dying-in-plane-crash-compared-to-other-activites/
40 years of wave conditions, BMT report 2006

Chapter 55
Thoughts on Integrating Stability into Risk Based Methods for Naval Ship Design

Philip R. Alman

Abstract Design for Safety (DFS), Goal Based Standards (GBS) and Formal Safety Assessment (FSA) are powerful tools which establish a framework for integrating stability into a risk based design process. They provide a foundation for the development of novel designs which can provide insight that is not attainable through any other means. Naval ships are complex systems, sometimes operating in an environment defined by risk acceptance and risk taking beyond those of their commercial counterparts. The hazards seen by a naval ship in its service life may not be foreseen during design. The development of a design for safety process for naval ships should be capable of reflecting the nature of the military mission. Concurrently, there is certain fidelity inherent in the process that should be carefully defined. Three cases to categorize the risk assessment 'fidelity' are defined and discussed. These highlight the dangers of overstating and understating risk. Lastly the challenges of defining intact and damage stability risk in light of the sensitivity to the state of knowledge for naval ships are discussed.

Keywords Dynamic stability · Risk management · Formal safety assessment
Design for safety · Goal based standards · Dynamic stability · Static stability
Probabilistic methodology

55.1 Introduction

The maritime world and especially naval engineering has always dealt in the currency of risk management. Even the origins of the word Risk—*Riscum* (Greek/Latin) has a maritime origin, referring to a difficulty or reef to be avoided at sea (Ayyub 2003).

The opinions expressed in this paper are those of the author and not necessarily those of the Naval Sea Systems Command or the United States Navy.

P. R. Alman (✉)
Naval Sea Systems Command (NAVSEA), Washington, DC, USA
e-mail: jackstay5@aol.com

V. L. Belenky et al. (eds.), *Contemporary Ideas on Ship Stability*, Fluid Mechanics and Its Applications 119, https://doi.org/10.1007/978-3-030-00516-0_55

The defining difference between the commercial and naval world is in the acceptance and management of risk. In the commercial maritime world risks are to be managed and balanced against economic goals and increased risk is to be avoided if it results in reduced safety. While the approach to risk management in the naval maritime world is the same as in the commercial world, there are times when increase risk acceptance is doctrine brought on by necessity, and the balance is weighed not against economic benefit and measures of safety but against policy and strategic objectives. This is the challenge for naval ship stability where the key is to provide a ship with a robust stability capability sufficient to survive in an unpredictable environment.

55.2 Definitions

Hazard	A phenomenon or 'event' of potential harm
Reliability	Ability of system or component to fulfill a design function
Consequence	Degree of damage or loss due to some 'failure' or hazard
Risk	Potential of loss associated with exposure to an event

$$\boldsymbol{Risk} = \boldsymbol{Probability} \times \boldsymbol{Consequence}$$

Risk is defined in terms of a probability of an *'event'* or *'hazard'* and a consequence. The risk assessment is the tool by which all of the anticipated hazards are assembled and ranked according to their relative risk. The fidelity of the risk assessment is based on the accurate definition of the hazards, rigorous calculation of the probabilities and correct determination of the consequences at any given point in time. As the environment changes, so do the hazards and the risks (Papanikolaou 2009; Ayyub 2003).

If done properly the use of risk assessment as a tool can provide a great deal of insight into inherent weakness associated with a particular design or its operation (Shaw 2001). But if not done properly, or done in a haphazard manner, the results can be misleading and actually overlook hazards that may prove fatal. The term "unintended consequences" has become almost cliché but is really evocative of an outcome of a process, policy, or system where the risks were not properly or thoroughly defined.

55.3 Risk Management Processes Applicable to Naval Ship Safety

Risk management processes can be applied to any aspect of acquisition, design development or operation. Within the marine industry, several approaches have been approaches have been developed to address ship safety. These include:

1. Design for Safety (DFS)
2. Formal Safety Assessment (FSA)
3. Safety Case
4. Goals Based Design

The concept of design for safety is an overarching risk based process which includes safety as an objective of the design. In broad terms this is characterized as Risk Based Design and Approval:

Design

- Safety Performance/Mission Capability
- Prediction Tools
- Risk Models
- Optimization

Approval

- Approval Procedures
- Risk Evaluation and Acceptance Criteria
- Framework for Risk Approval

$$\mathbf{Risk_{design} \leq Risk_{Acceptable}}$$

In principle a formal safety assessment is a decision tool providing a systematic process for assessing the effectiveness of regulations at the International Maritime Organization (IMO). The FSA approach is based on five steps:

1. Hazard Identification
2. Assessment of Risk
3. Mitigation Options
4. Cost/Benefit
5. Recommendations (Papanikolaou 2009)

The concept of the formal risk assessment or design for safety approach has been embraced by IMO in MSC/Circ. 1023 as “Guidelines for Formal Safety Assessment (FSA) for Use in the Rulemaking Process” (IMO 2002). NATO has followed a similar course in adopting Goal Based Standards (GBS) as a basis for the “Naval Ship Code” ANEP-77 (NATO 2010). Goal based standards outlines a process a goal or ‘safety objective’ is defined through a series of tiers or a framework for verification through design construction and operation.

In ANEP-77, the goal based standards approach is anchored on five ‘tier’s which are outlined to be:

- Tier 0—Aim (Philosophies and Principles)
- Tier 1—Goal
- Tier 2—Functional Areas
- Tier 3—Performance Requirements

- Tier 4—Verification Methods
- Tier 5—Justification

The "Goal Based" approach provides a systematic framework for certification of a ship to meet the aims of ANEP-77. Performance requirements are defined based on the Concept of Operations "CONOPS" and verified using appropriate criteria. Although the Goal Based approach contained in ANEP-77 provides for implementation of FSA approaches, it is not the same as an overarching "Design for Safety" approach in that ANEP-77 allows a Naval Administration to apply describes FSA in the Tier 4 verification of specific performance requirements associated with defined functional areas. Prescriptive standards can still be used to verify the 'goals' (NATO 2010).

55.4 Integrating Stability into Risk Management Process

The integration of Naval Ship stability into a risk based process requires the ability to articulate a risk based on a probability and a hazard. This can include a wide range of tools which can encompass many levels of technical maturity. For stability these may include heuristics, static design criteria, and model test simulation of a failure event probability (Alman/NSSWG 2010), see also Chap. 53. Each provides a level of information as to what the risk might be for a stability failure for a particular ship design.

The goal based approach in ANEP-77 can be satisfied using existing design criteria, but Tier 4 does provide for inclusion of a FSA process. The FSA process outlined in ANEP-77 does not provide details as to how the hazards should be formulated, or the probabilities of failure and risk determined. It is left to the Naval Administration to establish the acceptable level. Implicit is how an acceptable risk is defined. Within the Goal Based (GB) approach, it is possible that risk levels will not get truly balanced across the design, depending on how the FSA is addressed over the functional areas. Consequently, with increasing use of the risk management processes, it will become increasingly necessary to develop probabilistic metrics for intact and damage stability.

If an overarching "design for safety" process is to be conducted for a Naval Ship, it must be applied form the top down, and Across the Tier 3 functional areas of which stability is one. Hazards must be identified accurately and completely.

A set of safety goals driven by design considerations for a commercial ship has been outlined as follows:

- No Accidents leading to total ship loss
- No loss of human life related to accidents
- Minimal environmental impact from Operation
- Minimal environmental impact from accident
- Vessel to remain afloat and upright in all loading conditions
- Vessel to remain afloat in case of water ingress and flooding

- Ship structure to withstand all foreseeable loads
- High passenger comfort (motions) (Papanikolaou 2009)

Similar safety objectives could be developed for major naval combatant but with considerations recognizing the nature of the hazards.

- Operations
  - No Accidents leading to total ship loss
  - No loss of human life due to ship related accidents
  - Minimal environmental impact in operation
  - Minimal environmental impact from accident
  - Vessel to remain afloat and upright within NA limits in all operational conditions including extreme seaways environments
  - Vessel to remain afloat and upright in case of flooding due to accident
  - Maintain structural integrity over lifetime
- Battle Damage
  - Vessel to remain afloat and upright in case of battle damage
  - Residual Structural Strength after battle damage
  - Other considerations.

55.5 The Challenge of the Naval Ship Environment

Past experience is a powerful tool to facilitate the process of hazard identification, but over reliance on past experience and historical data may be misleading if the knowledge base is derived from experience gained from a class of ships or operating environment that is not reflective of the new design. This paradigm is captured in the cliché of "designing for the last war" in which the design of a weapons system is based on missions and hazards and doctrine from the last war (comfort zone of the experts) the result being that the weapons system may be unsuitable for the required missions of the next war in which it is actually fielded. Generally engineers like to focus on known data and feel uncomfortable when speculating on the unknowns (Ayyub 2003).

Dealing with ignorance of the future is a critical aspect of the naval risk management process. Although risk is always minimized as much as possible there are times in the Naval Environment when risk avoidance is not a 'constant' and in fact at times risk 'acceptance' may be doctrine. In other words in the naval environment safety may not always be paramount and may sometimes be supplanted by overarching tactical or strategic goals.

The decision by ADM Halsey to sail through the 1944 Pacific Typhoon rather than accept delay caused by weather routing is an example of the perceived strategic necessity overtaking strict adherence to safety (Alman et al. 1999). Rather than viewing heavy weather as a hazard to be avoided, in the Pacific theater of operations during World War II, storm fronts were sometimes deliberately used as coverage to obscure fleet maneuvers in advance of attack or as a means to provide a haven during a withdrawal (Orville CAPT USN 1945).

The merchant marine was also impacted. On August 9, 1941, President Roosevelt unilaterally suspended the International Load Line Convention, due to "changing circumstances created by the war in Europe" (Whiteman 1970). While it is not known what impact suspension of the Convention might have had on ship safety, in theory it would be possible to load to deeper drafts if needed without violating international law. Transportation of supplies and equipment across the Atlantic to support the allies in Europe was absolutely vital. The greatest risk to safety faced by the allied merchant marine at that time came from enemy action. During the Battle of the Atlantic 1939–1945, approximately 2177 ships were lost along with 30,132 personnel (Horodysky 2007).

Nothing embodies risk acceptance better than two quotes from John Paul Jones:

> *"I wish to have no connection with any ship that does not sail fast; for I intend to go in harm's way."*
>
> *"It seems to be a law of nature, inflexible and inexorable, that those who will not risk cannot win." (Morison* 1959*)*

If we deconstruct his statement in terms of a goal based approach, John Paul Jones doctrine might be expressed as follows:

- Tier 0 Aim—"Win"
- Tier 1 Goals—A ship to "Get into Harm's Way"
- Tier 2 Functional Requirements -

  Speed
  - Tier 3 Performance Requirement – Faster than the enemy

  Armament
  - Tier 3 Performance Requirement – Superior fighting capability

All other functional requirements as necessary to support Tiers 0 and 1.

Note; he did not say the following:

> *I want to have no connection with any ship that is not safe for I intend to keep out of harm's way.*

Although we live in a complex modern world, is it unreasonable to think that his doctrine will still be paramount during wartime? If so, finding the right balance of safety and capability is the challenge faced by the naval ship community.

In this context risk management or Risk Based Design can eventually help to provide the tools to follow an equitable process but must acknowledge the complexity of conflicting requirements and should be based on a very clear and realistic understanding of the nature of risk acceptance in the naval community. There are times when mission requirements may force increased risk acceptance and consideration of hazards that were not planned for. In this context lies the challenge of defining the hazard environment for a naval ship. The hazards are changing with time over

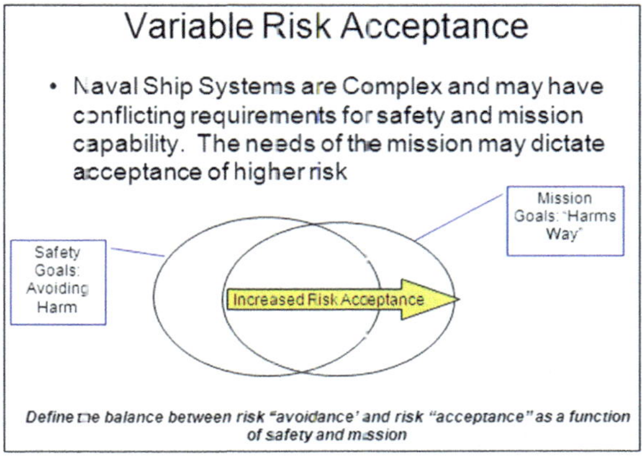

Fig. 55.1 Variable risk acceptance

the life of a ship and are frequently unpredictable. This is especially true when ships have projected service lives of perhaps 40–50 years. The ship designer may be completely ignorant of how missions and threats will evolve in the future. (Cavas 2011) This can perhaps be illustrated as a Venn diagram depicting Commercial shipping and Naval Shipping Hazards. Generally both have an overlap or set intersection of common hazards, but there may be times when there is a necessity for increased risk acceptance brought on by new hazards and an increased willingness to acknowledge the possibility of loss so that the hazard space for a naval ship increases as illustrated in Fig. 55.1.

55.6 Naval Ship Hazard Space

If we allow that naval ships have a unique set of changing hazards resulting from doctrine, mission or wartime employment; it is also true that there is a subset that is the same for all ships whether naval or commercial. The total naval ship hazard space can be thought of as those hazards which are common to all ships plus those hazards which reflect the unique military nature of the ship mission as shown in Fig. 55.2. Some common hazards are outlined in Fig. 55.3.

The hazard space can be further refined by looking at an existing ship class with similar mission capability. But while this can help to further refinement of the hazards, it may not be sufficient to adequately address the entire hazard space. At this point it may be necessary to do scenario development and some amount of crystal ball gazing. All this though ultimately means that there is an "ideal" set of hazards that can be imagined with a degree of realism or fidelity, and then there is the actual set of hazards, which include those things that cannot be envisioned. The goal is to minimize the later.

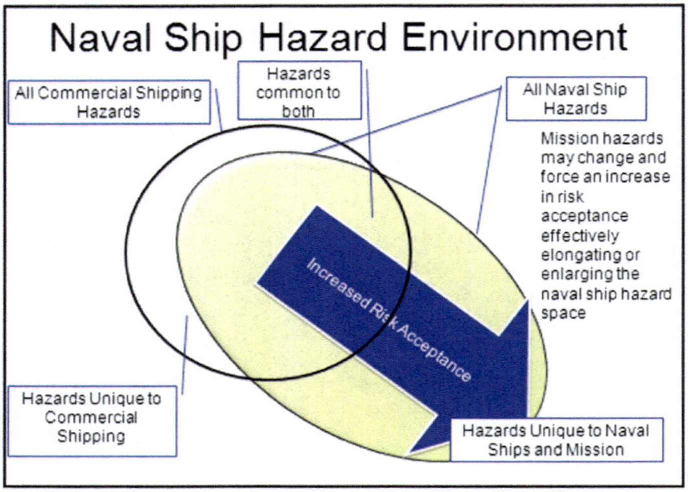

Fig. 55.2 Effect of increased risk acceptance on hazard space

Fig. 55.3 Example common hazards for commercial and naval ships

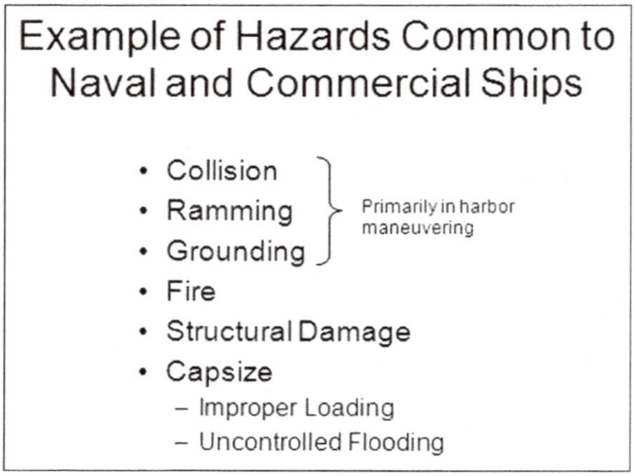

The principle safety elements outlined in the Design Data Sheet (DDS)-079 are an articulation of basic naval ship stability hazards.

Principal Safety Elements in DDS-079 Criteria

Intact Ship

- Beam Winds Combined With Rolling
- Lifting of Heavy Weights
- Crowding of Passengers to One Side
- High Speed Turning
- Topside Icing

Damaged Ship

- Stranding Involving Moderate Flooding
- Bow Collision
- Battle Damage Involving Extensive Flooding

Flooded Ship

- Beam Winds Combined With Rolling
- Progressive Flooding

55.7 The "Unknown" Risk: Assessment Fidelity

"Risk is a characteristic of an uncertain future and is neither characteristic of the present nor the past" (Ayyub 2003). The risk assessment can be thought of as an approximation or simulation of the true ship risk that exists in the real world. This concept has been characterized as the state Knowledge and Ignorance. At any snapshot in time there is a state of knowledge about a complex system that is a measure of truth and fallacy or ignorance. Some hazards, characteristics of performance, probabilities, and consequences we may know as truth from experience. Others we may mistakenly believe as significant or not; still others we may be completely ignorant and unaware because we have limited or no experience (Ayyub 2003). The degree to which the risk assessment approaches truth can be thought of as 'fidelity.' The fidelity of a risk assessment might be thought of as a characterization of how thoroughly the hazards are identified, the probabilities are calculated and the consequences understood. *It is not be possible to ever achieve 100% fidelity.* This would mean that every hazard has been identified, every probability exactly calculated and every consequence completely defined. There are going to be unknowns even in the most thorough process.

Hazards
How well is the system understood? How complex are the systems? Is this a conventional ship? Novel ships may not have much experience base. Areas of operation and potential environment. Is there an experience base to draw on?
Probabilities
All sorts of calculations or estimates can be made to develop a probability. What is the fidelity of the prediction or simulation tool? Is the event being modeled properly and with sufficient accuracy? *"They couldn't hit an elephant at this distance…"*[1] (McMahan 1884).

[1]Last words of General Sedgwick, USA,—Spoken after rebuking his subordinates for taking cover, shortly before being killed by Confederate sharpshooter at Spotsylvania Courthouse. One could say he incorrectly assessed his probability of being hit as nil; perhaps the result of ignorance of the sharpshooter's true likelihood of hitting the target. Other "experts" around had disagreed and took steps to mitigate the risk—which the General adamantly felt were unnecessary. Unfortunately he couldn't revise his estimate later.

Consequences
Is the severity correctly understood? Have undesirable events been adequately developed and by the right people?

For a formal safety assessment for a naval ship, each of these questions must be addressed across systems so that the final risks can be properly ranked. If not properly ranked, the resources may not get applied to the highest risk. In this context, a great challenge for stability is the development of a correct representation of hazards and associated probabilities as well as a realistic portrayal of the cost/benefit of mitigations. A great deal of work has been done to look at probability of damage from collision, ramming and groundings and the characterization of extent and penetration for flooding. However, if the impact of seaway dynamics is to be considered in damage stability the stochastic nature of the environment and motions must be appropriately modeled to attain a realistic fidelity. The stability community is just starting to address flooding dynamics and intact dynamic stability.

In MSC/Circ. 1023 "Guidelines for Formal Safety Assessment," The International Maritime Organization advocates a screening approach based on the progressive employment of more refined tools and increased detail as a means to scope the requirements for the risk assessment. This is an important step as is similar to the approach advocated by the Alman/NSSWG (2010), see also Chap. 53 for assessing dynamic stability tool fidelity in an iterative process evolving through a team collaborative process as shown in Fig. 55.4.

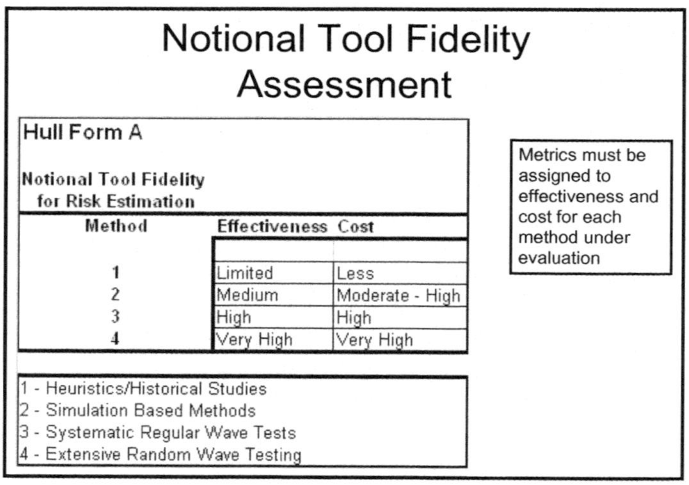

Fig. 55.4 Notional tool fidelity

55.8 Risk Fidelity: Case I, Case II, Case III

One way to think of the maturity of the process is as a 'fidelity' meaning how closely the risk model represents the real life risks of the ship. This is a challenging prospect in the application of risk assessment to naval ship design due to the vast complexity of the systems. Of course in development of a FSA process one always assumes that the process yields results of realistic fidelity. When critical decisions weighing the balance between capability, performance and safety are being made, how do we know this is true? Certainly this applies to the statistical methodology used to estimate probabilities, but it also affects the definition of hazards as well as the consequences. The effect of risk fidelity is shown in Fig. 55.5. In this case there is a region in which the risk model and actual risks coincide.

It is possible to define the potential outcomes of a risk assessment in terms of cases. In this context there are three possible cases for defining the fidelity of the model:

- **Case I:** The risk model predicts the real risks accurately
  - *Hazards are correctly envisioned and*
  - *Consequences are correctly understood and*
  - *Probabilities are correctly modeled*
- **Case II:** The risk model over predicts actual risks
  - *Hazards are mistakenly defined and/or*
  - *Consequences are over estimated and/or*
  - *Probabilities modeled are too high for the hazard*

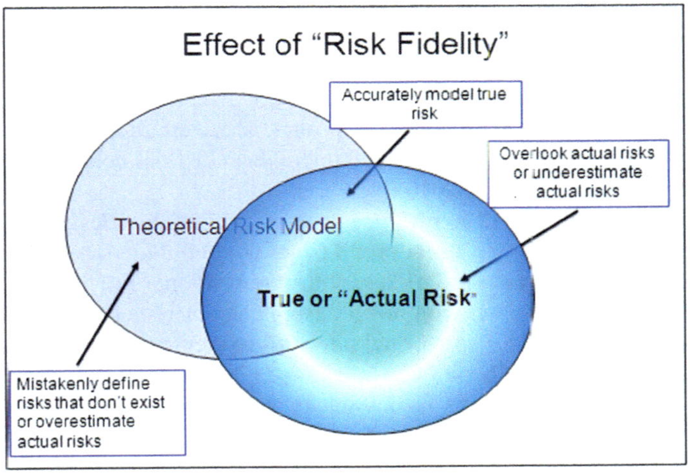

Fig. 55.5 Risk assessment fidelity

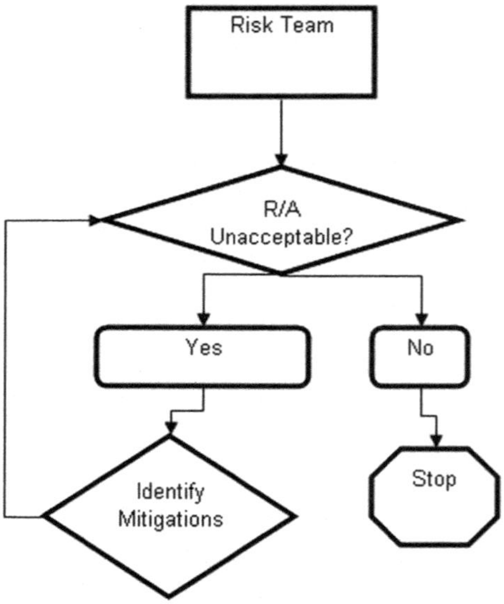

Fig. 55.6 Risk assessment team process

- **Case III:** Risk model under predicts actual risks
  - *Hazards are not identified or are minimized and/or*
  - *Consequences are not identified or are minimized and/or*
  - *Probabilities modeled are under estimated*

We would like to think that the process is confined to Case I, but clearly there are always varying degrees of all three cases present. Additionally, personalities and institutional pressure may weigh into leaning to Case II or Case III results. Clearly the challenge is to develop Case I, minimize Case II and avoid Case III altogether.

For example, for intact dynamic stability the risk estimate can be greatly influenced by the probabilistic model used, the technical tool used to develop events, and the environmental model which defines the exposure. These are subject to technological maturity, and expert opinion. A risk assessment team process could be as shown in the Fig. 55.6 (Alman/NSSWG 2010).

If there is a great deal of experience with a specific ship design or class, the associated hazards for the most part might be pretty well understood by those with experience. Consequently it is more likely with a conventional hull form fall into Case I. However, for a naval vessel with novel design characteristics, new technologies, unique missions, the need to use a formal safety assessment as a tool becomes greater. At the same time, the likelihood of a Case II and Case III fidelity occurring during the process also increases. It may be possible to employ strategies to attempt to head off Case II and Case III from dominating results.

These could include:

- Identification of Scenarios to explore possible failure events

Fig. 55.7 Technological maturity effectiveness and cost

Hull Form A

Notional Tool Fidelity for Risk Estimation

Method	Effectiveness	Cost
1	Limited	Less
2	Medium	Moderate - High
3	High	High
4	Very High	Very High

1 - Heuristics/Historical Studies
2 - Simulation Based Methods
3 - Systematic Regular Wave Tests
4 - Extensive Random Wave Testing

- Evaluation against a baseline design
- Examining Factors of Safety in Criteria

Figure 55.7 provides an example case showing how a team might assess technological maturity to address Case II and Case III.

55.9 Leveraging Experience

The application of a formal safety assessment for a naval ship design could follow an iterative or screening process as a means to leverage experience. The use of a screening process can allow informed judgment to be the basis for deciding where in the assessment it is necessary to increase risk fidelity in critical areas (IMO 2002). Three basic design categories can be identified. These are equally applicable to naval ship design (Papanikolaou 2009):

1. Partial risk-based design using safety equivalence of one function
2. Partial risk based design using safety equivalence and safety balance addressing several functions
3. Complete risk based design

In development towards a complete risk based design it may be possible to develop similar approaches as a stepping stone or tuning process for a novel or alternative design using techniques such as benchmarking or design enhancement.

Benchmarking—Conduct a risk assessment on a benchmark design and the new design as an overarching total system process. Establish a relative risk of system areas based on prior knowledge of the performance of a similar system with a safe history. This would require an assessment of risk for a benchmark ship. The new design then follows an equivalent Risk Assessment process and uses the benchmark

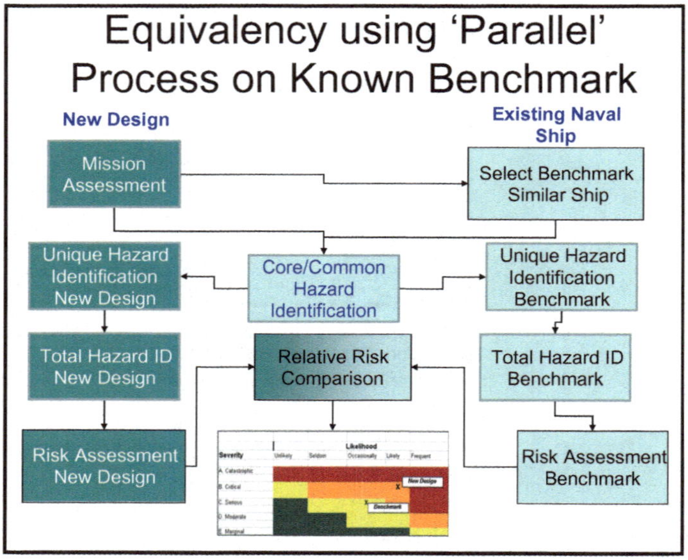

Fig. 55.8 Risk assessment benchmark process

Fig. 55.9 Design enhancement for safety

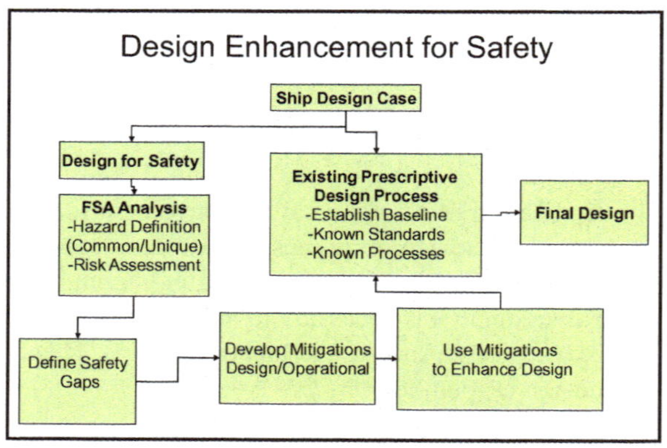

risk level as a target. Figure 55.8 process an example flow diagram to illustrate how this process might look.

Safety Design Enhancement—Conduct a risk assessment a new design which complies with existing criteria. Use the results to identify specific system design weaknesses that should be enhanced. A flow diagram for this might look as in Fig. 55.9.

55.10 Stability—Integrating Dynamic Stability into Acquisition Discussion

Successful integration of stability into an FSA process requires the ability to characterize a risk. It may be a risk associated with maintaining naval architecture limits, and managing mass properties during design and service life (Tellet 2011). The risk can be qualitative as in a probability of a stability failure, or it can start as a subjective process based on heuristics associated with a particular hull form. The process might start with assembling a 'tool' kit defining available resources for developing measures of risk (Alman/NSSWG 2010), see also Chap. 53.

An iterative process to allow 'screening' should be used to identify areas of concern early on and allow employment of focused resources.

A multi tiered approach to identify dynamic stability hazards in heavy weather might follow an approach as shown in Figs. 55.10 and 55.11 (Belenky et al. 2008; Alman/NSSWG 2010), see also Chap. 53.

Since a stability failure could be a catastrophic event resulting in loss of ship and/or life, the goal within the risk assessment process should be focused on attaining Case I fidelity. Also important to stability is the ability to address what might be unknown about the performance or operation of the design. If we consider that Case III (failure to recognize or under predict risk) can in part be attributed to lack of knowledge, then this is an important task in the process.

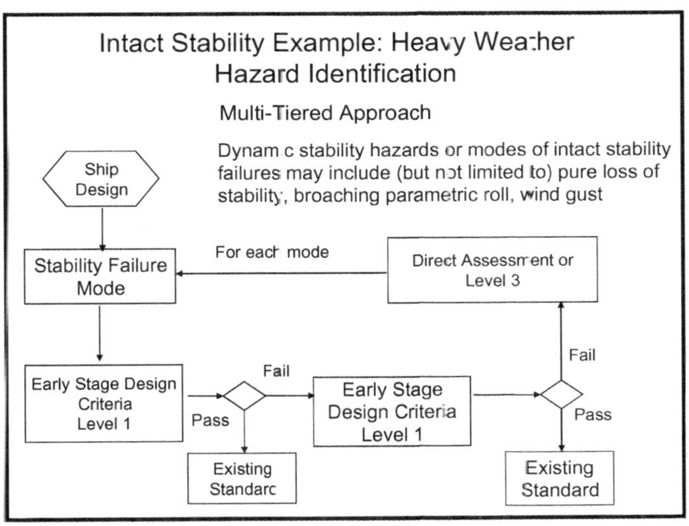

Fig. 55.10 Heavy weather hazard identification

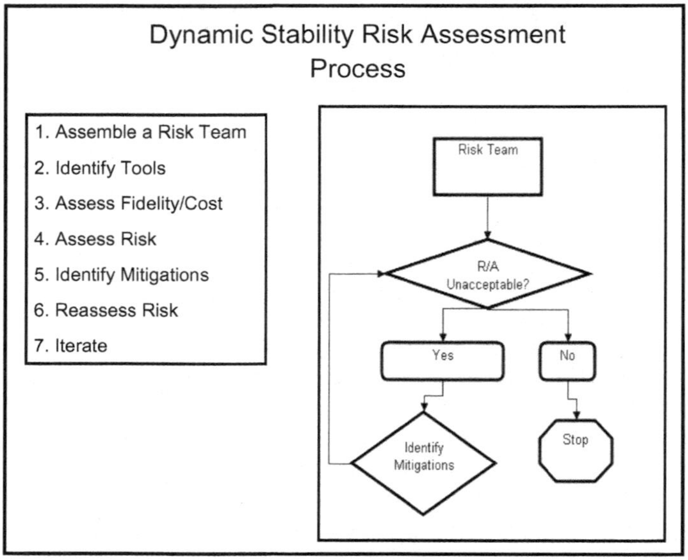

Fig. 55.11 Dynamic stability risk assessment process

55.11 Dealing with the "Unknown" in Stability

As discussed earlier, one of the chief difficulties faced in applying risk assessment processes to naval ships is addressing the unknown. It may be a prudent exercise as part of a stability risk assessment to include an assessment of potential 'unknowns' about vessel performance, operational environment, and mission requirements that might significantly change the risks or hazards. It may be a useful exercise with any new design to try to characterize the state of knowledge about the design, hull form characteristics, CONOPs, environment, special geographic areas of operation.

Categorization of the "knowns" and "unknowns" for the stability system and larger ship system as a whole may provide valuable information. A series of 'what if' exercises using a team of experts to define potential scenarios could be drawn together, collated and evaluated by the risk assessment team (Alman et al. 1999) . Methodologies such as development of Bayesian networks can also be used as a tool to help define the probabilistic relationship between uncertain or unknown actions or decisions (Ayyub 2003). A detailed discussion of the use of such a process is presented in Papanikolaou (2009). Known "facts" about the hazards and risks for a specific design such as operational doctrine, mission, environment etc could be categorized and ranked against the "unknowns." This process could include an evaluation of "Tool" fidelity, criteria etc. A notional ranking is shown in Fig. 55.12.

Information such as this can be used subjectively to "adjust" the risk during the screening process to aid in defining critical areas for further risk development. The

Hull Form A		
Notional Unknown Evaluation for Risk Estimation		
Ranking	Known	Unknown
1	Limited	High
2	Medium	Moderate - High
3	High	Moderate- Low
4	Very High	Low

Fig. 55.12 Tabulation of "Knowns" and "Unknowns"

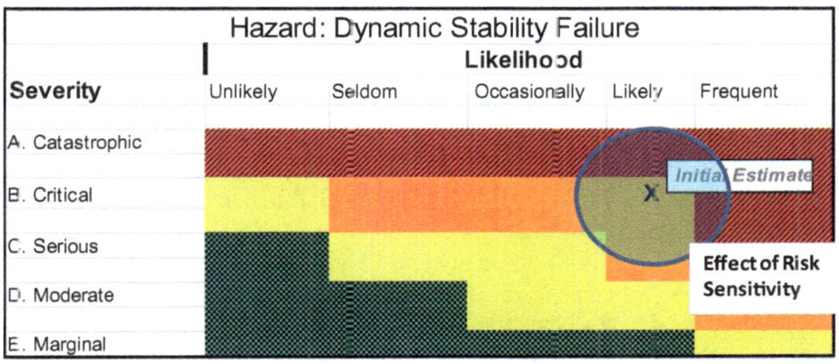

Fig. 55.13 Notional dynamic stability risk sensitivity

goal is to progressively increase the "Known" domain while continuously decreasing the "Unknown" domain.

Risk sensitivity could be developed based on changing requirements, 'what if' exercises, scenarios as an aid to judge the robustness of a new design. This may be useful in highlighting potential design weaknesses arising from unique stability failure modes. In Fig. 55.13, the initial risk estimate is displayed as an 'X' indicating in this case a dynamic stability failure risk. The effect of a sensitivity study may indicate that there is a region which actually may encompass some possible frequent and catastrophic events. The conditions, assumptions, operational environment etc, under which those events occurred should be examined in greater detail to determine if they are really indicative of potential high risk for the design.

55.12 Conclusions

Risk assessment processes can yield valuable information on the safety and capabilities of a design. The maritime community is developing techniques to incorporate risk assessment to develop informed safety decisions regarding design, regulations and criteria. All have benefits, but must be applied consistently across systems in developing risks. A risk assessment process to assess safety for a naval ship should recognize from the onset that there are significant unknowns associated with risk acceptance that affect potential hazards and acceptable probability levels. This impacts the "fidelity" of the risk assessment. Even so, by acknowledging this, it should be possible to develop processes that recognize and isolate those unknown factors and improve or understand the "fidelity" of the risks identified.

Acknowledgements Several people were instrumental in helping guide the direction of this paper; Prof. D. Vassalos for his guidance in discussing risk based ship design, Mr. D. Tellet, Dr. V. Belenky, Prof. B. Ayyub and Captain J. McTigue.

References

Alman, P.R., Minnick, P.V., Sheinberg, R., Thomas, W. L. III, (1999). "Dynamic Capsize Vulnerability: Reducing the Hidden Operational Risk," SNAME Transactions, Vol. 107.
Alman, P. (2010). Approaches for Evaluating Dynamic Stability in Design. Proc. 11th Intl. Ship Stability Workshop. Wageningen, The Netherlands.
Ayyub B. (2003). Risk Analysis in Engineering and Economics. Chapmann and Hall.
Belenky V., DeKat J, and N. Umeda. (2008). Toward Performance-Based Criteria for Intact Stability. Marine Technology, Vol 45, No 2 pp. 101–123.
Orville, H. T. (1945). Weather is a Weapon. All Hands, March, pp. 5–9.
Cavas. (2011). Past Imperfect - Like Fast Carriers, Littoral Combat Ship enters age of experimentation. Armed Forces J. April.
Shaw, E. (2001). Practical Experience and Operational Requirements for Onboard Risk Management Under Marginal Stability Conditions. Proc. of 5th Intl Ship Stability Workshop. Trieste, Italy.
Horodysky, T. (2007). American Merchant Marine in WWII. Retrieved April 20, 2011, American Merchant Marine at War: January 31 http://www.usmm.org/ww2.html.
IMO. (2002). MSC/Circ.1023 Guidelines for Formal Safety Assessment (FSA). London.
McMahan, (1884). The Death of General John Sedgewick. In Johnson, Battles and Leaders of the Civil War Vol 4 (p. 175). Century War Books.
Morison. (1959). John Paul Jones: A Sailor's Biography. Boston: Little, Brown, and Company.
NATO. (2010). Naval Ship Code.
Papanikolaou. (2009). Risk Based Ship Design, Springer-Verlag. Berlin.
Tellet, D., (2011). Incorporating Risk into Naval Ship Weight Control. Proc 12th Intl. Ship Stability Workshop. Washington DC, USA
Whiteman, M. (1970). Digest of International Law, p. 485.

Author Index

V. L. Belenky et al. (eds.), *Contemporary Ideas on Ship Stability*, Fluid Mechanics and Its Applications 119, https://doi.org/10.1007/978-3-030-00516-0

Subject Index

V. L. Belenky et al. (eds.), *Contemporary Ideas on Ship Stability*, Fluid Mechanics and Its Applications119, https://doi.org/10.1007/978-3-030-00516-0

V

W